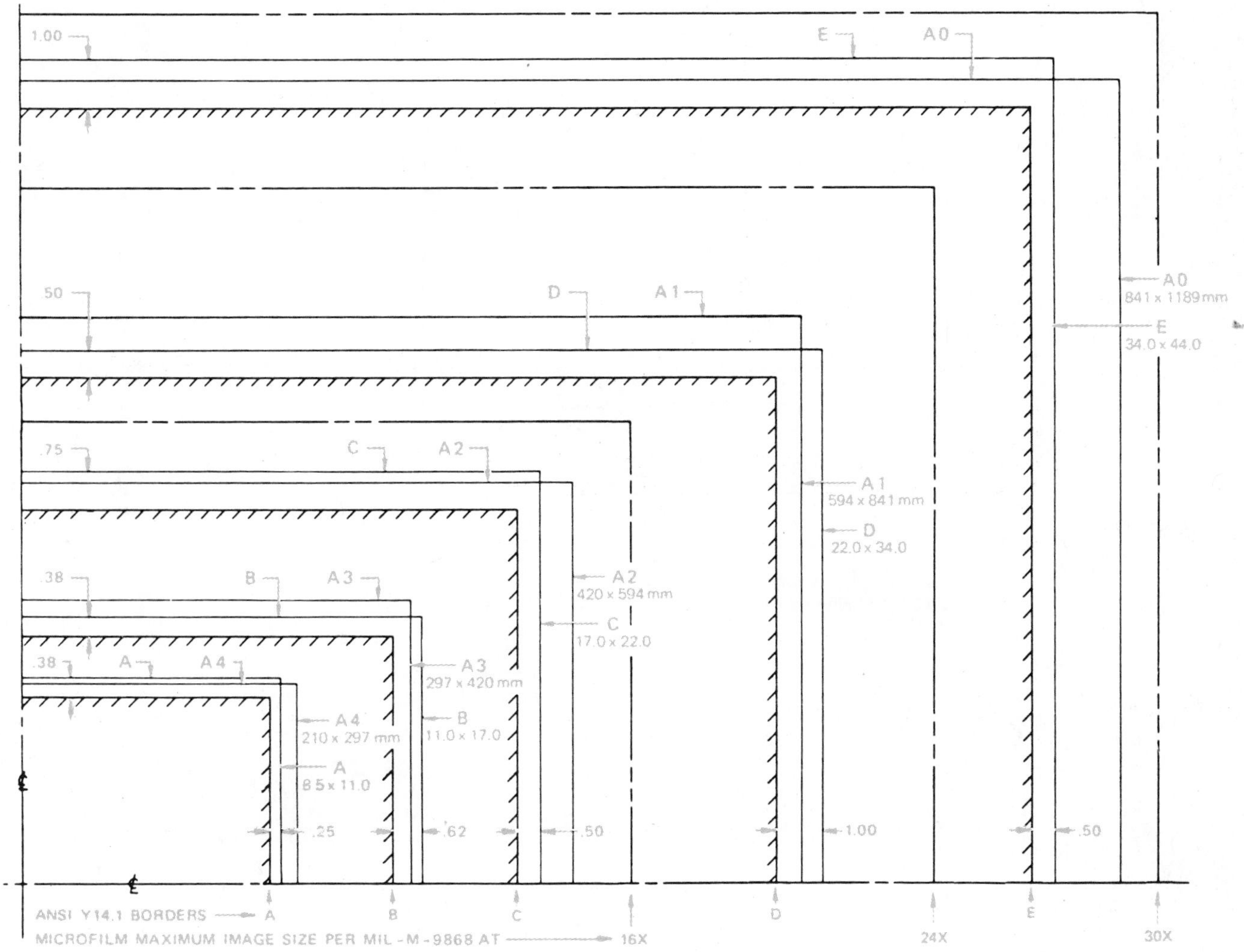

Comparison of ANSI Y14.1 Drawing Format Sizes with ISO Paper Sizes

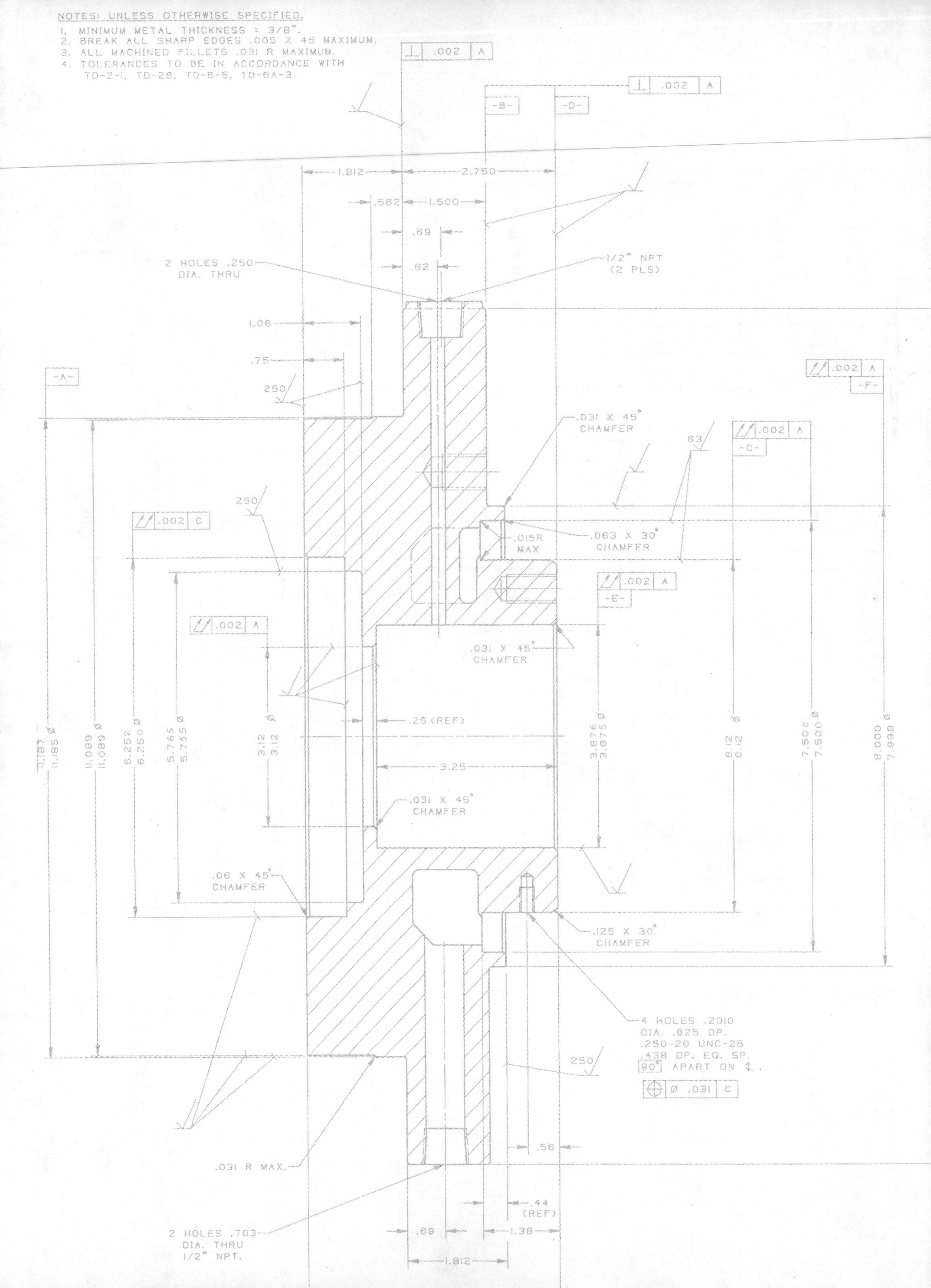
NOTES: UNLESS OTHERWISE SPECIFIED.
1. MINIMUM METAL THICKNESS = 3/8".
2. BREAK ALL SHARP EDGES .005 X 45 MAXIMUM.
3. ALL MACHINED FILLETS .031 R MAXIMUM.
4. TOLERANCES TO BE IN ACCORDANCE WITH TD-2-1, TD-28, TD-6-5, TD-6A-3.
.002 A
-B-
-D-
1.812
2.750
.562
1.500
.69
.62
2 HOLES .250 DIA. THRU
1/2" NPT (2 PLS)
1.06
.75
-A-
250
.031 X 45° CHAMFER
63
.002 A
-C-
-F-
.015R MAX
.063 X 30° CHAMFER
.002 C
.002 A
-E-
.031 X 45° CHAMFER
.25 (REF)
3.25
11.187 Ø 11.185
11.089 Ø 11.089
6.252 Ø 6.250
5.765 Ø 5.755
3.12 Ø 3.12
3.876 Ø 3.875
8.12 Ø 8.12
7.502 Ø 7.500
8.000 Ø 7.999
.06 X 45° CHAMFER
.125 X 30° CHAMFER
4 HOLES .2010 DIA. .625 DP. .250-20 UNC-2B .438 DP. EQ. SP. 90° APART ON ℄.
Ø .031 C
.031 R MAX.
.56
.44 (REF)
2 HOLES .703 DIA. THRU 1/2" NPT.
.69
1.38
1.812
4.56

Principles of Engineering Drawing

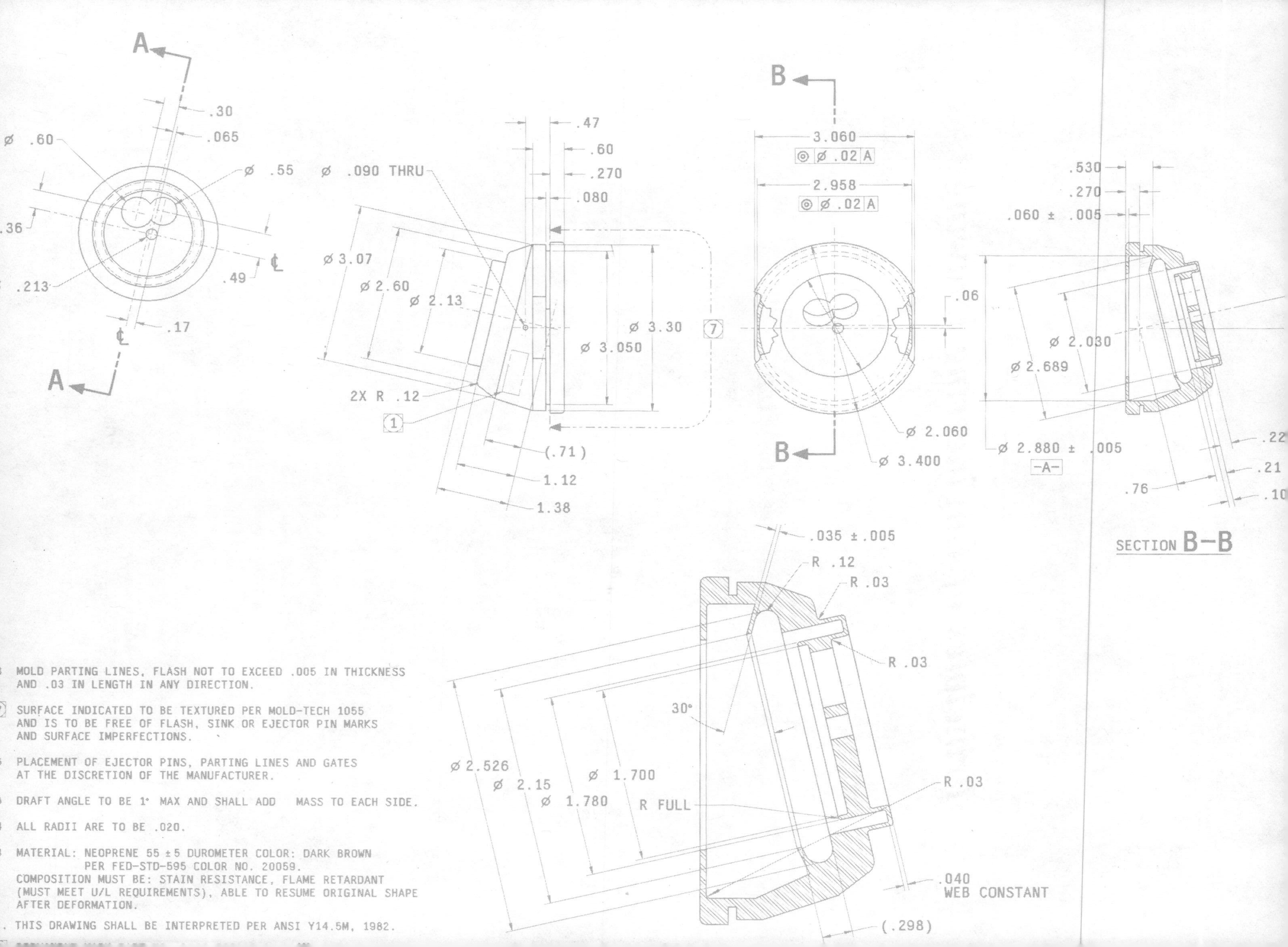

A
A
Ø .60
.30
.065
Ø .55
.36
Ø .213
.49
.17
℄
℄
Ø .090 THRU
.47
.60
.270
.080
Ø 3.07
Ø 2.60
Ø 2.13
Ø 3.30
Ø 3.050
7
2X R .12
1
(.71)
1.12
1.38
B
B
3.060
◎ Ø .02 A
2.958
◎ Ø .02 A
.06
Ø 2.060
Ø 3.400
.530
.270
.060 ± .005
Ø 2.030
Ø 2.689
Ø 2.880 ± .005
-A-
.76
.22
.21
.10
SECTION B-B
.035 ± .005
R .12
R .03
R .03
30°
Ø 2.526
Ø 2.15
Ø 1.700
Ø 1.780
R FULL
R .03
.040
WEB CONSTANT
(.298)
8 MOLD PARTING LINES, FLASH NOT TO EXCEED .005 IN THICKNESS
AND .03 IN LENGTH IN ANY DIRECTION.
7 SURFACE INDICATED TO BE TEXTURED PER MOLD-TECH 1055
AND IS TO BE FREE OF FLASH, SINK OR EJECTOR PIN MARKS
AND SURFACE IMPERFECTIONS.
6 PLACEMENT OF EJECTOR PINS, PARTING LINES AND GATES
AT THE DISCRETION OF THE MANUFACTURER.
6 DRAFT ANGLE TO BE 1° MAX AND SHALL ADD MASS TO EACH SIDE.
4 ALL RADII ARE TO BE .020.
3 MATERIAL: NEOPRENE 55 ± 5 DUROMETER COLOR: DARK BROWN
PER FED-STD-595 COLOR NO. 20059.
COMPOSITION MUST BE: STAIN RESISTANCE, FLAME RETARDANT
(MUST MEET U/L REQUIREMENTS), ABLE TO RESUME ORIGINAL SHAPE
AFTER DEFORMATION.
2. THIS DRAWING SHALL BE INTERPRETED PER ANSI Y14.5M, 1982.

Principles of Engineering Drawing

Louis Gary Lamit
DeAnza College

Kathleen L. Kitto
Western Washington University

Manual Illustrations–John J. Higgins
CAD Illustrations–John I. Shull
Assistant CAD Illustrator–Victor E. Valenzuela
Technical Assistance–Vernon Paige, Dennis Wahler

WEST PUBLISHING COMPANY
MINNEAPOLIS/ST. PAUL NEW YORK LOS ANGELES SAN FRANCISCO

DEDICATION

I wish to dedicate this book to Monica for my daughter Angela, and to Margie for my son Jamie and my daughter Corina.
Louis Gary Lamit

In memory of Elizabeth, whose guiding hand, devotion and love, made everything possible.
and to
Sue, Debbie, Rosalie, Gerry, and Michael.
Kathleen L. Kitto

Copyediting: Lorretta Palagi
Text Design: Roslyn Stendahl/Dapper Design
Layout: Geri Davis, Quadrata
Composition: Carlisle Communications, Inc.
Illustration and photo credits follow the index.
Production, Printing and Binding by West Publishing Company.

WEST'S COMMITMENT TO THE ENVIRONMENT

In 1906, West Publishing Company began recycling materials left over from the production of books. This began a tradition of efficient and responsible use of resources. Today, up to 95 percent of our legal books and 70 percent of our college and school texts are printed on recycled, acid-free stock. West also recycles nearly 22 million pounds of scrap paper annually—the equivalent of 181,717 trees. Since the 1960s, West has devised ways to capture and recycle waste inks, solvents, oils, and vapors created in the printing process. We also recycle plastics of all kinds, wood, glass, corrugated cardboard, and batteries, and have eliminated the use of styrofoam book packaging. We at West are proud of the longevity and the scope of our commitment to the environment.

TEXT IS PRINTED ON 10% POST
CONSUMER RECYCLED PAPER

PRINTED WITH
SOY INK

By WEST PUBLISHING COMPANY
610 Opperman Drive
P.O. Box 64526
St. Paul, MN 55164–0526

Printed in the United States of America
01 00 99 98 97 96 95 94 8 7 6 5 4 3 2 1 0
Library of Congress Cataloging in Publication Data
Lamit, Louis Gary, 1949–
Principles of engineering drawing / Louis Gary Lamit, Kathleen L. Kitto.
p. cm.
Includes index.
ISBN 0-314-02805-6
1. Mechanical drawing. I. Kitto, Kathleen L. II. Title.
T353.L264 1994
604.2--dc20 93-33033
CIP

Contents in Brief

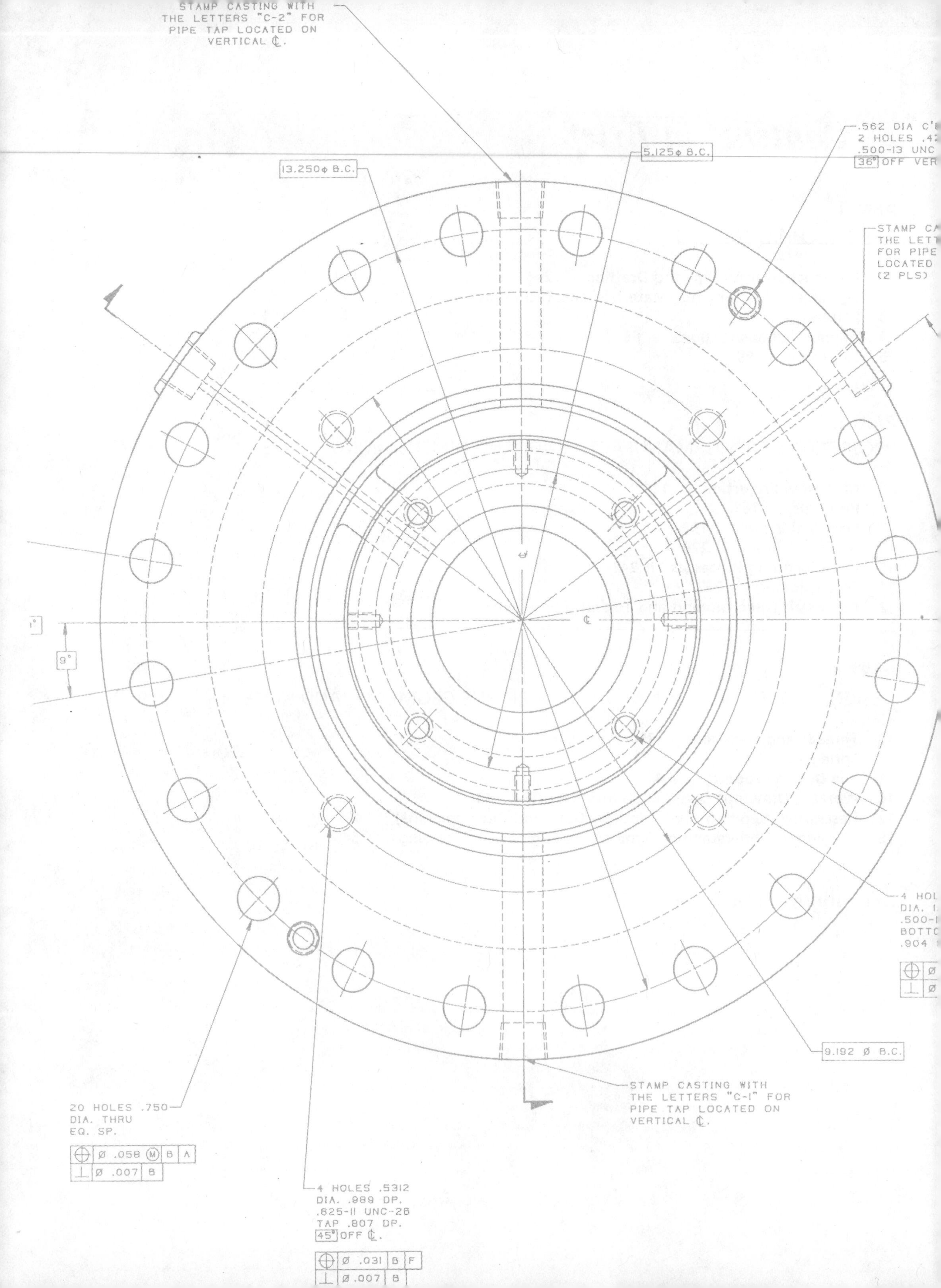

STAMP CASTING WITH THE LETTERS "C-2" FOR PIPE TAP LOCATED ON VERTICAL ℄.
.562 DIA C'
2 HOLES .4
.500-13 UNC
36° OFF VER
13.250ϕ B.C.
5.125ϕ B.C.
STAMP CA
THE LETT
FOR PIPE
LOCATED
(2 PLS)
9°
℄
4 HOL
DIA. I
.500-I
BOTTO
.904
9.192 Ø B.C.
STAMP CASTING WITH THE LETTERS "C-1" FOR PIPE TAP LOCATED ON VERTICAL ℄.
20 HOLES .750 DIA. THRU EQ. SP.
Ø .058 Ⓜ B A
Ø .007 B
4 HOLES .5312 DIA. .989 DP. .625-11 UNC-2B TAP .807 DP. 45° OFF ℄.
Ø .031 B F
Ø .007 B

Contents

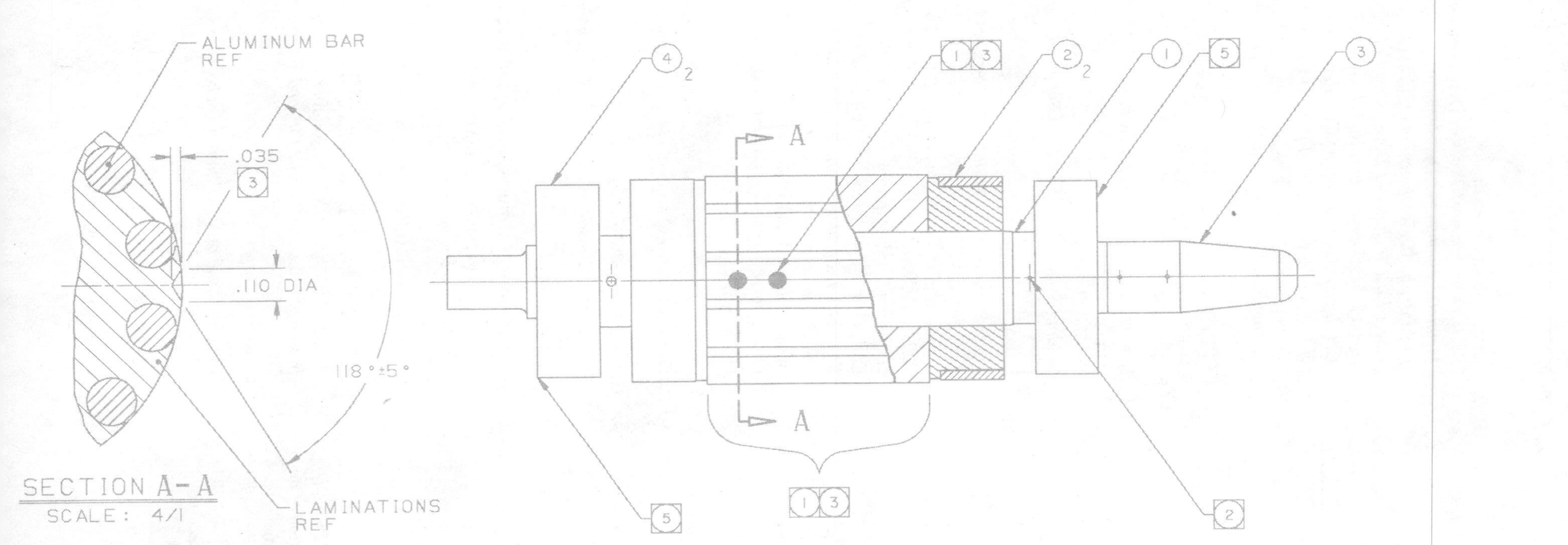

NOTES: (UNLESS OTHERWISE SPECIFIED)

1. BALANCE TO WITHIN 3 MILLIGRAM-INCHES BY DRILLING BETWEEN ALUMINUM BARS AS SHOWN (BOTH ENDS). DEBURR HOLES AS REQUIRED.
2. .020 DIA (REF) IN ITEMS 1 AND .040 DIA (REF) IN IN ITEM (3) TO BE ASSEMBLED ALIGNED.
3. MAX DEPTH OF MATERIAL REMOVED FOR BALANCING TO BE .035.
4. BEARING PRESS FORCE TO BE BETWEEN 100 AND 300 LBS.
5. LETTERING SIDE OF BEARING TO FACE OUTWARD.
6. OPTION: BEARING 356122 (K5) MAY BE USED ON TOP AND/OR OPTIONAL BEARING 356124 (K1419) MAY BE USED ON BOTTOM AS REQUIRED.

ITEM	QTY	PART NO.	DESCRIPTION
4	2	141-356123	BEARING (K6)
3	1	171-349020	OIL PICKUP
2	2	130-349008	END RING
1	1	223-349005	ARMATURE SHAFT ASSY (GRINDING)

MATERIAL

FINISH

UNLESS OTHERWISE SPECIFIED

DIMENSIONS ARE IN INCHES
TOLERANCES
.X ±.050 ANGULAR ±0° 30'
.XX ±.020
.XXX ±.005 MACH. SURF. √

INTERNAL THD HEIGHT 55% MIN
THREADS: CLASS 2A OR 2B
REMOVE BURRS & SHARP EDGES .020 MAX.
MACH. FILLET RADIUS .020 MAX.
MACH. SURF. FLAT WITHIN .001 IN./IN.
OTHER SURF. FLAT WITHIN .005 IN./IN.
CONCENTRICITY MACH. SURF.
T.I.R. WITHIN 1/2 SUM OF DIAS. TOLS., .001 MIN.

DO NOT SCALE DRAWING

	BY	DATE
DR	D.M.DUARTE	5/8/89
CHK		
DSGN		
ENGR		

MOD L10

TITLE ARMATURE AND END RING ASSEMBLY (BALANCING)

SIZE	CODE IDENT NO.	DWG NO.
C	07978	223-349006

SCALE 2X | 1ST USE 349001 | SHEET 1 OF

Preface

Developing this Principles of Engineering Drawing text was challenging and rewarding. The text is designed for a one or two semester course in manual drafting and design. In addition to the traditional topics, it contains a comprehensive chapter on geometric dimensioning and tolerancing, a chapter devoted specifically to the design process and design for manufacturability, a chapter covering the basics of descriptive geometry including intersections and developments and an entire chapter on understanding the symbols used on engineering drawings in welding, piping, electronics, and the fluid power industry. Obviously, the more difficult question in developing this text was what to exclude. The text is written to be concise and easy to read and, therefore, easy to understand.

Current industry drawings are used extensively in this text because the authors feel that *drafting is a process, a tool, used to communicate manufacturing and construction information.* The purpose of a drawing is to give manufacturing the technical data required to make a part or produce a product. The product of a technical drawing is the efficiently manufactured, economically produced, and profitable item that is created for a consumer or for the government. The part itself is the goal, not the drawing of the part. Without proper part definition (graphics and dimensions), the drawing is useless and the part cannot be manufactured.

ILLUSTRATIONS IN THE TEXT

Unlike many of the traditional texts that have only one style of drawing, this text is dedicated to letting the student see what the real world of drafting, design, and engineering is like. No drawing in this text is over 15 years old. Most projects are from industry and have been manufactured within the last 5 years. Drawings in the text include a range of manually drawn illustrations and engineering details, CAD-plotted projects using pen, electrostatic, and laser plotters, and professional technical illustrations. The variety of drawing types and styles found in the text include the following:

- *Instructional drawings* were created by professional illustrators to introduce a concept or guide a student through a series of steps to accomplish a specific construction or lay out a part. These are done with CAD and manual technical illustration techniques.
- *Industry drawings* are directly from a company source, no touch or changes have been made. Some are CAD and some are manual drawings.
- *Example art* illustrates a concept and is normally redrawn from an industry source.
- *Exercises* are placed at the end of each chapter. Each exercise is on an $8\frac{1}{2} \times 11$ in. "A" size sheet with a .25 in. grid.
- *Problems* are found at the end of most chapters and include a variety of CAD, sketch quality projects, and manually drawn instrument drawings. Problems range from simple "A" size one-view drawings to multiple-view "D" size projects. Problems can be completed as 2D or 3D CAD, instrument, or sketch assignments.

TEXT ORGANIZATION

The text is organized to introduce the student to graphic communication in the traditional sequence, with the addition of new innovations in the design process. Chapter 1 (section 1.14) covers the text's specific organization. This text provides the basic material for a one or two semester manual drafting course. In addition, it contains a chapter covering the basics of descriptive geometry including intersections and developments. Chapter 18 is a comprehensive chapter on symbols used in welding, piping, electronics and fluid power.

TEXT FEATURES

A variety of traditional and unique features are incorporated in the text including:

- Performance-based learning objectives begin each chapter.
- Industry-based drawings and examples of *recent* designs are shown.
- *Sketching* has been integrated into most of the chapters. Chapter 5 is devoted entirely to sketching. Other chapters in the text have many sketching exmples, problems or exercises. In fact, all the exercises could be assigned as sketching projects.
- A number of chapters are unique for a Principles of Engineering Drawing text:

1. Chapter 12 provides a comprehensive coverage of geometric dimensioning and tolerancing.
2. Chapter 15 covers the design process and design for manufacturability in depth.
3. Chapter 16 covers assemblies and detail drawings using industry examples.
4. Chapter 17 covers the basics of descriptive geometry including intersections and developments.
5. Chapter 18 contains a comprehensive guide to symbols used on engineering drawings in the welding, piping, electronics, and fluid power industry.

- Industrial-based *interest boxes* (by Pat Courington) focus on career opportunities, historical information, and specific engineering concepts as they relate to the material in each chapter. Interest boxes contain photographs or illustrations to visually present the concept or explain the historical significance of the material.

- *Chapter exercises* are provided at the end of each chapter. The exercises are designed to be completed at specific intervals in the chapter. The exercises are suggested at logical stopping points within each chapter so the student can understand the material incrementally.
- *Chapter review questions* include eight true and false, eight fill-in-the-blank and eight short answer questions. Answers to all quiz questions are given in the solutions manual.
- *End-of-chapter problems* close every appropriate chapter. Problems are numerous and of sufficient level of difficulty to provide real world examples of industrial projects or to teach instructional concepts. Simple to advanced projects are provided in most chapters.
- The text ends with a comprehensive *appendix*. The appendix was designed to be used in industry as a reference on the job.

SUPPLEMENTS

In addition to the text, a series of ancillaries is available to make the transition to a new text as comfortable as possible:

- *Transparency acetates* (about 200) and *transparency masters* (about 200) provide the instructor with a comprehensive set of lecture materials that can be used to display the chapter concepts on an overhead projector.
- The *solutions manual* (by James Wilson) offers answers to all text quiz questions and solutions to many of the exercises and problems found at the end of each chapter. The solutions manual is also available as an AutoCAD disk.
- A selection of *workbooks* offer additional drawing exercises with worksheets for student. Both manual workbooks and an AutoCAD-based workbook should be available. *Worksheets* prepared by L. Gary Lamit are full-sized versions of the 1/4 page grids shown for each chapter exercise and are available packaged with the text for almost no charge.
- An *instructor's resource manual* by the authors offers teaching suggestions for a variety of different programs.

Acknowledgments

No project of this magnitude could be completed without the help and devotion of many individuals. Without their contributions this project would not have been completed. Thank you to Debra Pratt, Rachael Svit, Elena Verne, Irene Guerrero, Pat Scheetz, Valarie Prouty, Jaime Guerrero, Dennis Wahler, Vernon Paige, and Sam Levy for their contributions to specific chapters in the text; to Pat Courington for the chapter objectives and interest boxes; to Lou Moegenburg for his thoughtful contributions to the text and for his tireless review of the art manuscript; to James Wilson for the solutions manual; and to Kenneth Stibolt for a manual workbook.

John Higgins completed all the manual illustrations and John Shull was the CAD illustrator. Several other individuals from De Anza College made significant contributions to the text: John Allan, Lorn Beall, Mike Engle and Vernon Paige. The CAD illustrators at Western Washington University were John Ramalho, Eric Schueler, and Ernie Antin. We would also like to thank all the administrators and staff members at De Anza College and Western Washington University who encouraged us and supported us.

Houston Instruments donated the use of a plotter and digitizer and Kurta Corporation donated the use of a digitizer tablet. T & W Systems donated CADAPPLE and VersaCAD software. Autodesk donated copies of AutoCAD. Many of the illustrations were completed on AutoCAD and a Computervision system (and Personal Designer) that was donated to DeAnza College.

The reviewers of this text provided meaningful and insightful suggestions that added greatly to the completed book:

Brian Bennett—Morrison Institute of Technology (IL)
James R. Brock—Mesa State College (CO)
Richard C. Ciocci—Harrisburg Area Community College (PA)
Pat Courington—Valencia Community College, East (FL)
Dr. Debra Edwards—Appalachian State University (NC)
David D. Gloyeski—Dickson Tennessee Area Vo Tech
John Frostad—Green River Community College (WA)
Clare Lomheim—Southern Alberta Institute of Technology, Calgary
Mark H. Miller—New England Institute of Technology (RI)
Louis A. Moegenburg—University of Wisconsin-Stout
William E. Moore—Tidewater Community College (VA)
Dr. Carlton Salvagin—State University College-Oswego (NY)
David Sinclair—American River College (CA)
J. Pat Spicer—Western Illinois University
Lyle Wilson—Tarrant County Junior College, Northeast Campus (TX)

A special word of thanks is offered to the many instructors who provided comments in reviews and a massive survey for another text, *Technical Drawing and Design,* from which much of this material is adapted; we wish we could list you all here.

At West, we wish to thank the following individuals for their devotion and dedication to making this project a success: Holly Henjum, Jayne Lindesmith, Erin Ryan, Christine Hurney, Cliff Kallemeyn, and all the others. We would also like to thank Liz Riedel for always being there, for her enormous patience and for the million things she did for us during the project. Finally, a special thank you to Chris Conty for believing in the project and for giving us the chance to see it become a reality.

L. Gary Lamit
Cupertino, California
Kathleen L. Kitto
Bellingham, Washington

About the Authors

Louis Gary Lamit is the former department head of drafting and CAD facility manager and is currently an instructor at De Anza College in Cupertino, California, where he teaches computer-aided drafting and design as well as basic drafting.

Mr. Lamit has worked as a drafter, designer, numerical control (NC) programmer, and engineer in the automotive, aircraft, and piping industries. A majority of his work experience is in the area of mechanical and piping design. Since leaving industry, Mr. Lamit has taught at all levels.

Mr. Lamit has written a number of textbooks including *Industrial Model Building* (1981), *Piping Drafting and Design* (1981), *Descriptive Geometry* (1983), and *Pipe Fitting and Piping Handbook* (1984) for Prentice-Hall; *Electronic Drafting and Design* (1985; 2e 1993), and *CADD* (1987) were published by Charles Merrill (Macmillan).

Mr. Lamit received a BS degree from Western Michigan University in 1970 and did masters work at Michigan State University. He has done graduate work at Wayne State University in Michigan and University of California at Berkeley and holds an NC programming certificate from Boeing Aircraft.

Kathleen L, Kitto is an associate professor in the manufacturing engineering technology program at Western Washington University in Bellingham, Washington. Her teaching assignments include engineering graphics, computer integrated manufacturing, robotics, statics, strength of materials and machine design.

Ms. Kitto's first assignment after graduation was the design of high temperature test frames to study the high temperature creep behavior of refractories and ceramics used in magnetohydrodynamic (MHD) air preheaters. Later on in industry, she worked on the flow loops for large scale, one-of-a-kind commercial lasers.

Ms. Kitto has taught at Montana College of Mineral Science in Butte, Montana, at Bellevue Community College in Bellevue, Washington and at Western Washington University in Bellingham, Washington. One of her research interests at Western Washington University has been to develop devices to assist the "differently-abled".

Ms. Kitto received MS and BS degrees (with high honors) in Metallurgical Engineering from Montana College of Mineral Science and Technology in Butte, Montana.

PART I

Basic Techniques

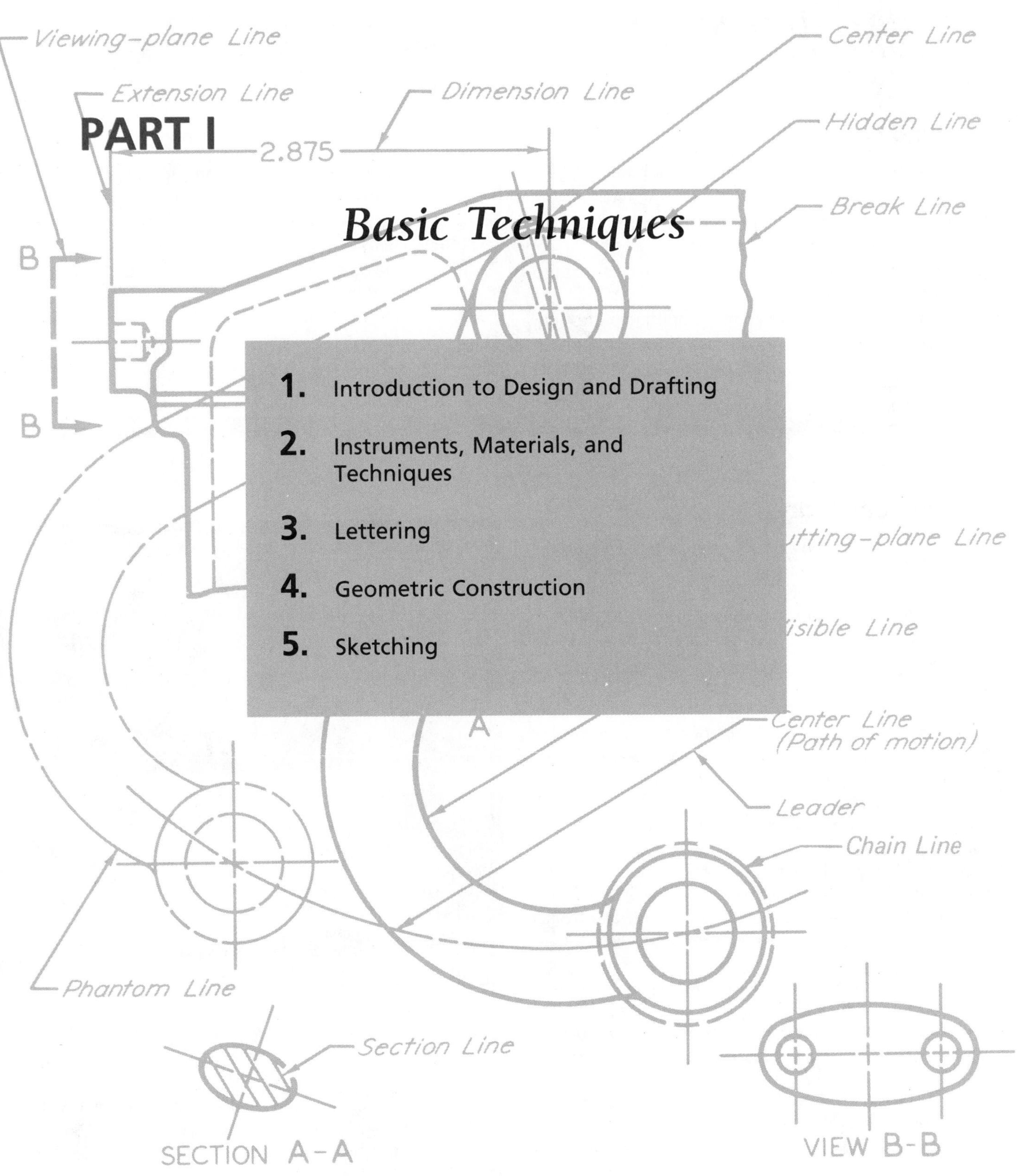

CHAPTER 1

Introduction to Design and Drafting

Learning Objectives

Upon completion of this chapter you will be able to accomplish the following:

1. Recognize design and drafting as tools that allow graphical representation of ideas.
2. Compare possible career fields and sequences that use technical drawing.
3. Define common terms used in the drafting profession.
4. Understand transitions in technical drawing that have taken place from ancient Roman construction projects to modern computer-aided design and drafting.
5. Develop familiarity with and identify technical drawing types and stages in the design process.
6. Define the role of descriptive geometry in solving three-dimensional problems.
7. Identify the various standards of practice used in drafting and design.

1.1 Introduction

Engineering design *(drafting)* uses graphic language to communicate ideas. This language, developed and used by engineers, designers, and drafters, serves as an essential tool from the beginning of a product's development to its production. How do we communicate ideas graphically? What are the components of this graphic language? What is a good drawing? This text answers these questions by covering engineering drawing basics and design.

Engineering drawing is a *language* or a *tool,* not a specialized field. Mechanical drawings play an essential role in design, manufacturing, processing, and production. Every industrial nation employs a large number of drafters and designers. There are more than 400,000 drafters in the United States and Canada.

Drawings are geometric representations of an idea or product that must be processed, manufactured, or constructed. The drawing and design process is used to define, establish, and create. The engineer, designer, and drafter use drawings to communicate technical information. All machines, devices, and products must be graphically designed before they can be manufactured. The cost, the intricacy, and the manufacturability of the item are considered during the beginning of the design stage. After the design has been refined, technical graphics are used to communicate the design data.

You should not look on drafting and design as an end in itself or an island of information. Design drawings are only the first step in the long and complicated process of product development, production, and manufacture.

Drawings are prepared on drafting boards using traditional drafting tools and instruments and with computers. It is not uncommon to see computer-aided design and drafting (CAD) systems interspersed among drafting tables (Fig. 1.1). Some small companies may use traditional drafting exclusively. Many large companies such as IBM, General Motors, Hewlett Packard, (Fig. 1.2) and Ford have converted entirely to CAD for engineering, design, and drafting.

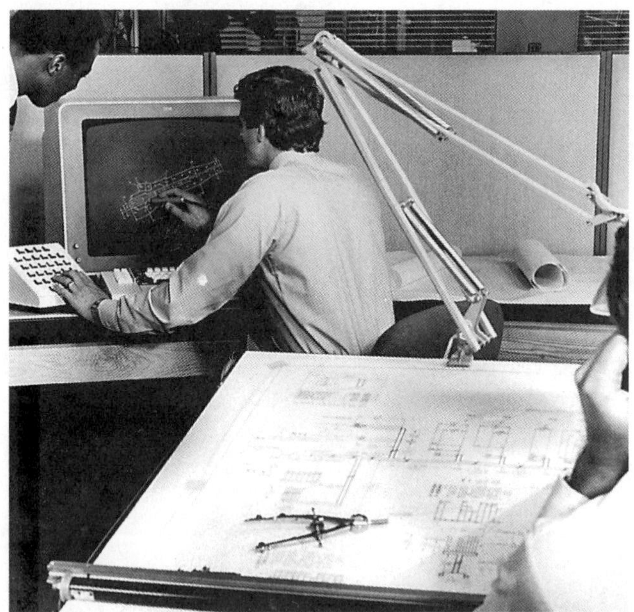

FIGURE 1.1 Drafting and Design in Industry

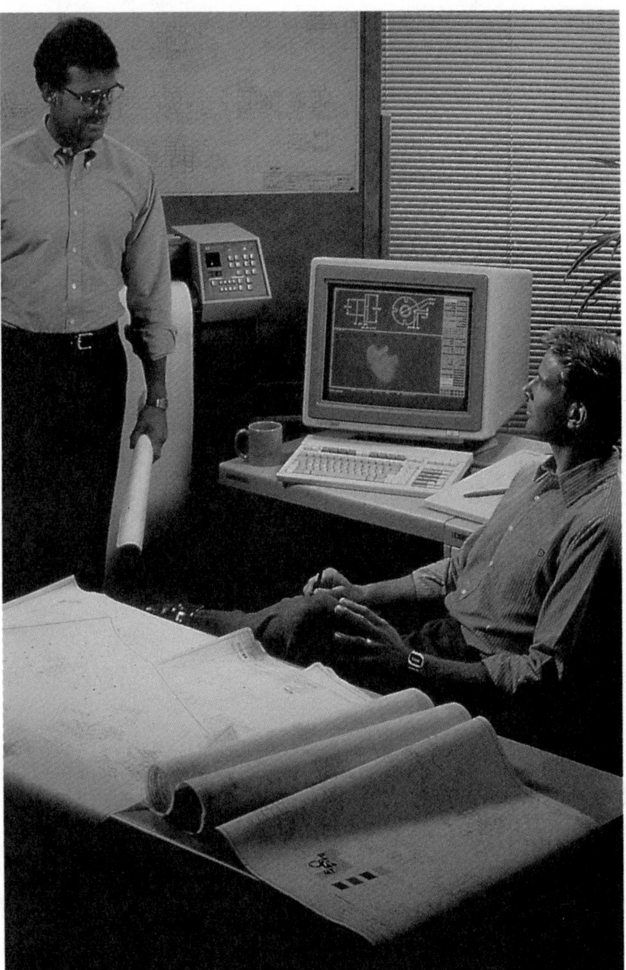

FIGURE 1.2 CAD Systems

1.2 Drafting Careers

There are many sequences you can follow in a career that uses technical drawing. Figure 1.3 shows traditional job categories and the path from drafting trainee to design supervisor. The following list shows job categories and responsibilities:

Job Category	Responsibility
Chief engineer	Management
Engineer	Conceptual design Ideas Calculations
Designer	Design ideas Physical layout
Layout designer	Assemblies Final designs
Detailer	Basic drawings Details Dimensioning
Checker	Drawing and design checks
Technical illustrator	Presentation drawings Manuals Publication quality art

The traditional starting point for a career in drafting is the *drafting trainee* (Fig. 1.3). The drafting trainee normally has had high school or beginning level college courses in drafting, math, and related technical subjects. Some drafting trainees start at the apprentice level, with no drafting experience.

The typical path for starting a career in drafting is to obtain a certificate at a technical school or a one- to two-year associate degree at a community or technical college that offers a drafting and design degree. With this education you enter the job market as a *drafter/detailer* or a *junior drafter.* The entry level depends on the quality of the degree program and the graduate's experience. The junior drafter is required to know considerably more than the drafting trainee. Mastery of the

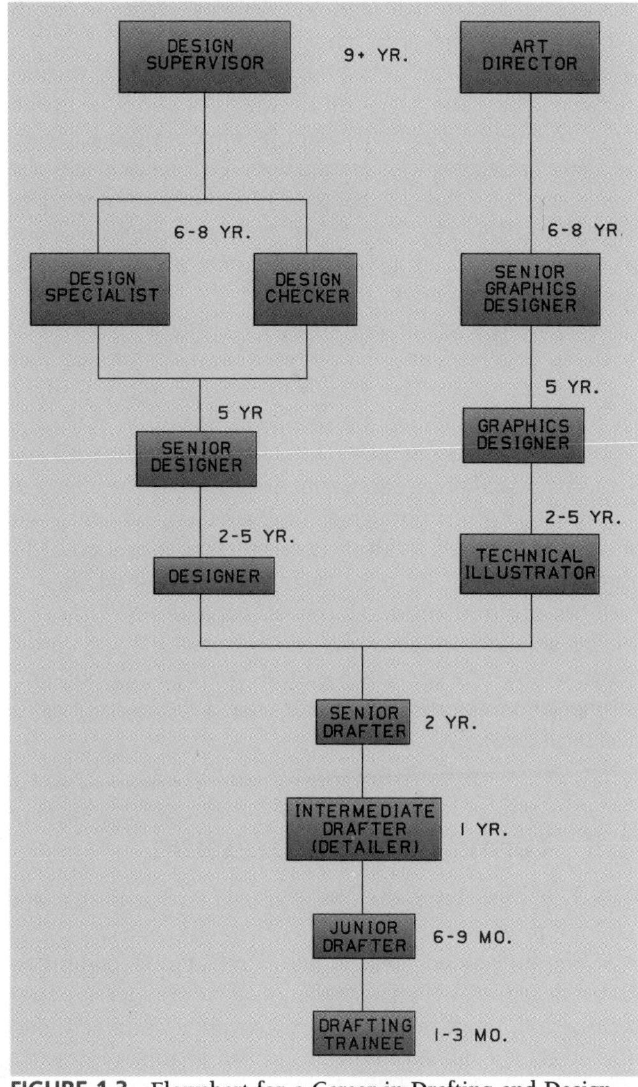

FIGURE 1.3 Flowchart for a Career in Drafting and Design

use of instruments, materials, and drafting techniques including lettering, geometric construction, freehand sketching, projection techniques, sectioning, dimensioning, and tolerancing is essential. The primary responsibility of the junior drafter is to prepare detail drawings.

The *senior drafter* (or *layout designer*) position requires a minimum of two to five years of experience in a particular engineering discipline. Layout designers refine the engineer's and designer's sketches, including investigating alternative design possibilities. Layout designers are required to understand drafting conventions and standards, know how to determine clearances and fits, and make the necessary calculations for an accurate design. Knowledge and understanding of shop practices, procedures, manufacturing techniques, and basic production methods are important. After two to seven years of experience, you may qualify as a junior designer or a designer. A designer refines the designs established by engineers.

Senior designers are in charge of a design group. The senior designer has between six and twenty years of experience as a designer in a particular field. The senior designer works directly with engineers and checkers.

The *checker* is responsible for the accuracy of the finished drawings. Checkers review the drawings for clarity, completeness, production feasibility, and cost effectiveness. Checkers review all mathematical computations. A checker is schooled in all standards and conventions for a particular engineering discipline. The checker makes sure the original design sketches, drawing layouts, and detail drawings of the project are consistent, accurate, and complete.

The ultimate legal responsibility for a project rests with the *engineer.* Engineers must be registered in their state to stamp or certify a project. The *design supervisor* coordinates, supervises, and schedules work assignments.

Whether you work on a drafting board or on a CAD system, the basic knowledge required for a particular engineering field is learned through a combination of schooling and on-the-job training. With the increased sophistication of today's technology, the requirements for entry-level drafting and design positions have increased dramatically.

You should attempt to gain exposure and training on different CAD software and hardware packages. In addition, strong communication skills are essential for a successful technical career.

1.3 Terms of the Profession

This text uses terms that are common in the design and drafting profession.

Computer-aided design and drafting or **computer-aided design** (CAD) refers to the use of a computer to design a part and produce technical drawings. Two-dimensional (2D) CAD is confined to the layout and graphic representation of parts using traditional standard industry conventions. Drawings are plotted on paper. While 2D CAD is limited to detailing and drafting, three-dimensional (3D) CAD is usually the starting point for design (Fig. 1.4).

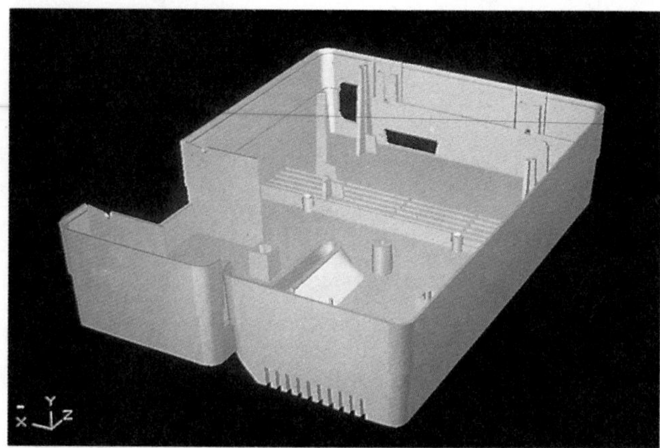

FIGURE 1.4 3D Mold Design

Engineering graphics is a term used to describe the use of graphical communication in the design process. Drawings represent design ideas, configurations, specifications, and analysis for an engineering project.

Manual drafting (or **instrument drawing**) is done on a drafting board using paper, pencils, and drawing instruments.

FIGURE 1.5 Models of Cranes

FIGURE 1.6 Power Plant Model

The term **modeling** is used throughout the text to describe the design stage of constructing a 3D physical model or an electronic 3D model of the part. A model can be created by physical modeling (Figs. 1.5 and 1.6), which is used to create a lifelike scale model of the part.

Engineering drawing encompasses all forms of graphic communication: manual, mechanical, freehand, instrument, and computer-generated drawings used by the engineer, designer, or drafter to express and develop technical designs for manufacturing, production, or construction.

Technical illustrations use artistic methods and pictorial techniques to represent a part or system for use by nontechnical personnel. Technical illustrations are widely used in service, parts, owners, and other types of manuals. Sales and advertising also use technical illustrations.

Technical sketching is the use of freehand graphics to create drawings and pictorial representations of ideas. It is one of the most important tools available to the engineer, designer, and drafter to express creative ideas and preliminary design solutions.

1.4 The History of Engineering Drawing

Some of the earliest evidence of the use of drawings is from the construction of the ancient pyramids and temples. There is evidence of the use of technical drawings as far back as 1400 B.C. Drawings were used in ancient Rome to display bridge designs and other construction projects. Leonardo da Vinci used pictorial sketches to develop and explore different inventions and designs.

The beginning of modern technical drawing dates back to the early 1800s. Until this time, graphic communication was more artistic in nature and used pens, ink, and color washes to display pictorial graphic images of a product or construction projects. By the 1900s, drawings were used for the production and manufacture of a wide variety of industrial products.

A series of standards and conventions were established to aid the transfer of information between the engineering/design department and manufacturing/production or construction. Communication between companies, industries, and countries was also made easier by standardization. Today, we have a very strict, standardized method of displaying graphic information.

Before the mid-1800s, instruments for graphical representation were limited to measuring scales, the compass, dividers, paper, and ink. Ink was replaced by the pencil. The T-square evolved into the parallel bar and then into the drafting machine. The newest *tool* in design and drafting is the CAD system.

1.5 Types of Drawings: Artistic and Technical

The two types of drawings are *artistic* and *technical* drawings. Technical illustrations (Fig. 1.7) use artistic techniques. An artistic drawing makes use of many techniques and expressions that are not used in technical drawings. First of all, a technical drawing must communicate the same message to every user or reader of the drawing, whereas an artistic drawing is usually interpreted differently by everyone who sees it. To limit the interpretation to only one possible conclusion, the technical drawing is controlled by accepted standards, drawing *conventions*, and projection techniques.

Engineering drawings are used to transfer technical information. The drawing must contain all of the information required to bring the concept, product, or idea into reality. Dimensions, notes, views, and specifications are required for a complete drawing.

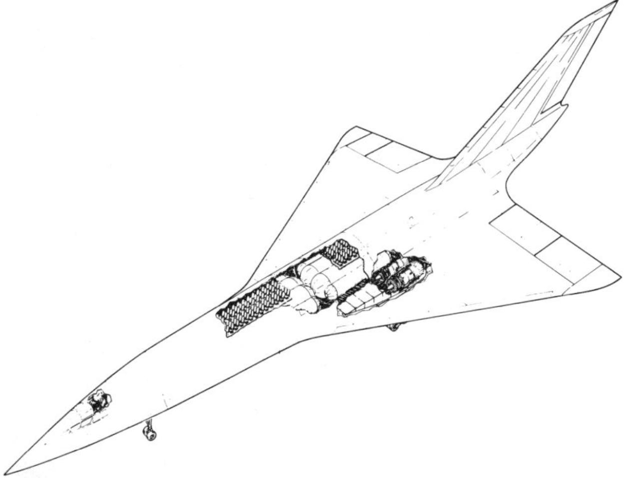

FIGURE 1.7 Technical Illustration of an Airplane

1.6 Types of Engineering Drawings

Various types of drawings are associated with mechanical design and drafting. **Design sketches** are initial design ideas, requirements, calculations, and concepts. They are used to convey the design parameters to the layout designer. **Layout drawings** develop the initial design. A layout drawing must show all the information necessary to make a detail or an assembly drawing.

Assembly drawings show a number of detail parts or subassemblies that are joined together to perform a specific function, while a **detail drawing** shows all information necessary to determine the final form of a part. The detail drawing must show a complete and exact description of the part including shapes, dimensions, tolerances, surface finish, and heat treatment, either specified or implied.

Casting drawings are usually not required. Normal practice is to show the necessary casting dimension along with the machining dimensions on the detail drawing. When a separate casting drawing is used, it contains only information needed for casting, so dimensions for machining and finishing are not included. **Fabrication drawings** are made for parts with permanently fixed pieces. The method of fastening is called out on the drawing with symbols or other standard methods. Welded and riveted parts require fabrication drawings.

1.7 The Design Process

The design process (Fig. 1.8) starts with a concept or an idea. The *first stage* of a project begins with the identification of a particular need for a product. Many times, the product is identified by a need in industry, government, the military, or from the private sector.

The *second stage* involves the creation of a variety of options or design ideas. These ideas might be in the form of sketches and include mathematical computations. The *third stage* is the refinement of the preliminary designs. Possible solutions to the problem are identified.

The *fourth stage* involves refinement and selection of a particular design. Here the project moves into a more formal, finalized state using assembly drawings and models. This stage requires close attention to how the part is to be manufactured and produced and is commonly referred to as the process of *designing for manufacturability* (DFM). This stage requires close attention to how the part will be manufactured and produced. Design for manufacturability (DFM) is where the manufacturability of the part is considered during the design phase. In the past, manufacturability was not considered until the design was released to manufacturing, resulting in costly design and/or tooling modifications.

In the *fifth stage,* detail drawings are prepared. The result is a complete set of working drawings. The *sixth stage* in the design process is the manufacturing and production of a product, or the construction of a system. In manufacturing, design and layout time is allocated for producing dies, tools, jigs, and fixtures.

During the design process, the designer and drafter encounter many situations in which traditional visualization techniques and a mastery of the principles of projection are used in the solution of complex engineering and technical problems.

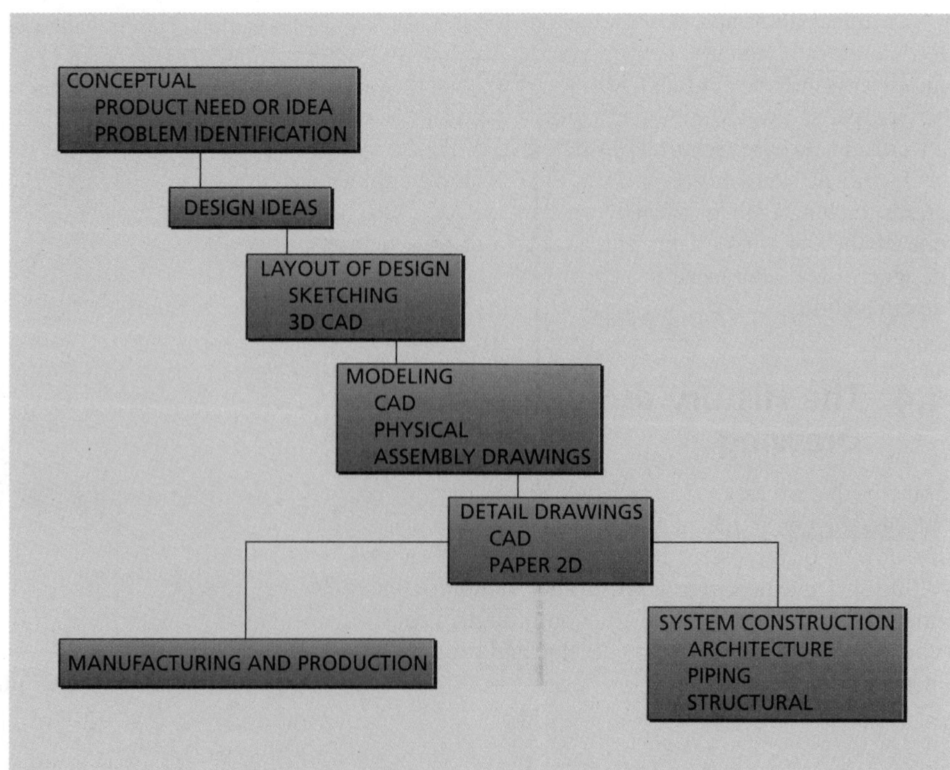

FIGURE 1.8 The Design Process

1.8 Descriptive Geometry

Descriptive geometry uses orthographic projection to solve 3D problems with a 2D graphics procedure. Descriptive geometry applications establish the proper representation and relationships of geometric features. The relationship of elements, such as the true distance between a line and a point or the angle between two planes, is typical of the problems found in descriptive geometry. Gaspard Monge developed the principles of descriptive geometry as a set of projection methods and techniques that are the basis for technical drawing. Figure 1.9 shows a descriptive geometry solution to the angle formed by two intersecting planes. Descriptive geometry also includes *intersections* and *developments* (Figs. 1.10 and 1.11).

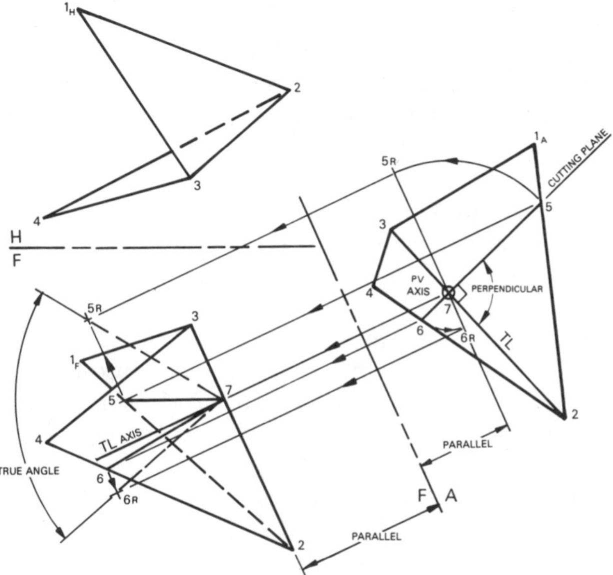

FIGURE 1.9 Descriptive Geometry Problem

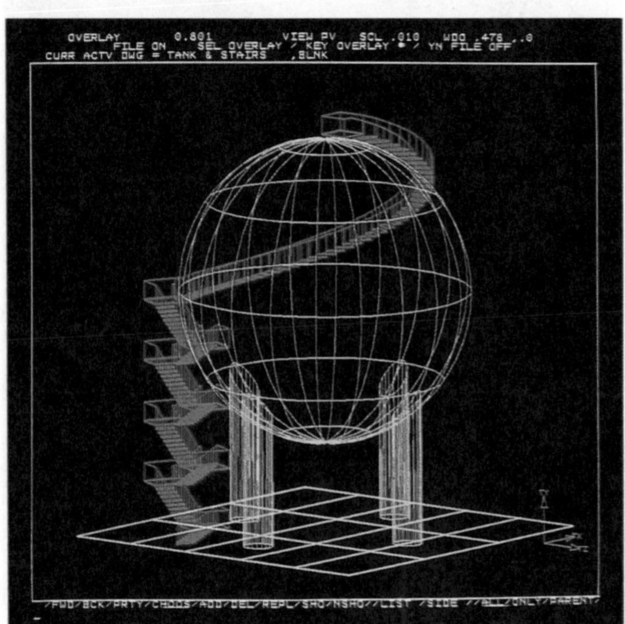

FIGURE 1.10 3D Design of a Holding Tank

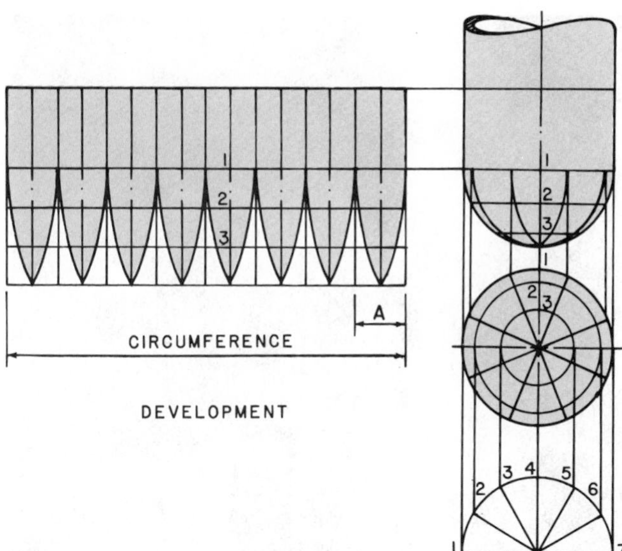

FIGURE 1.11 Development Problem

1.9 Career Fields in Industry

Engineers, designers, and drafters are employed in a variety of fields: civil, electronic, chemical, ceramic, manufacturing, mechanical, nuclear, solar, petrochemical, mining, and metallurgical engineering. All engineering fields employ designers and drafters to refine ideas and take the design into completion. The following list provides an overview of the possible fields of employment for engineers, designers, and drafters:

Mechanical

- Product design
- Manufacturing design: jigs and fixtures, dies, assemblies, and details

Electronic-Electrical

- Circuits, printed circuit boards
- Integrated circuits
- Electrical, electromechanical, computers

Applications for electronic and mechanical design:

- Marine
- Aerospace
- Transportation
- Mining

Architectural, Engineering, and Construction (AE&C)

- Civil: facilities, dams, airports, roads, mapping
- Structural: buildings, plants, power generation
- Piping: solar, nuclear, chemical, process, power, hydroelectric
- Architecture: Commercial, residential, landscape

Technical Illustration

- Product literature: advertising, sales, presentation, service manuals, display

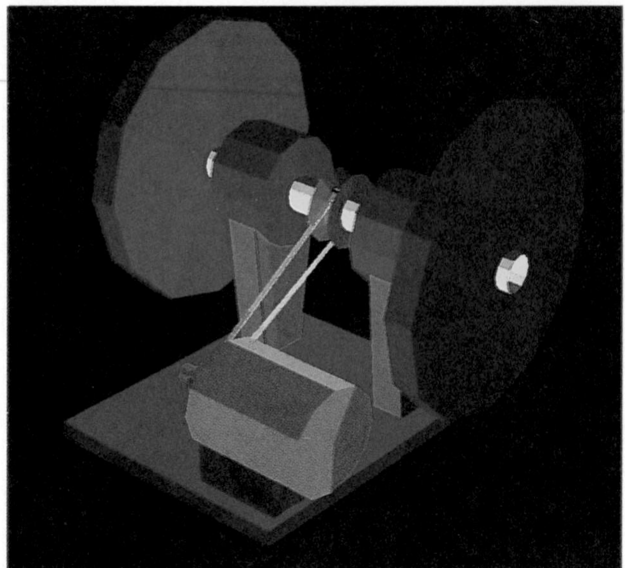

FIGURE 1.12 3D Mechanical Design

FIGURE 1.13 Earthmover Tractor

In mechanical engineering, designers and drafters make assembly drawings of jigs, fixtures, dies, and other types of manufacturing aids to create and produce machine parts and new machinery (Fig. 1.12). This is one of the largest areas for employment for a designer or drafter. The mechanical engineer is concerned with the conceptual development and the calculations involved in creating and developing mechanical devices including items to be used in machinery, automobiles, mechanical equipment (Fig. 1.13), and aerospace products such as airplanes (Fig. 1.14) and helicopters.

Architectural, engineering, and construction is comprised primarily of civil engineering, structural design, piping design, and architecture. The civil engineering and mapping fields (Fig. 1.15) employ drafters and designers to develop highways, roads, railways, and airports. Piping design includes such diverse fields as fossil fuel power plant design (Fig. 1.16), nuclear power plants (Fig. 1.17), solar power, and a wide range of other areas that require industrial piping systems used in the production of chemicals, petrochemical products, food, and beverages.

Architecture (Fig. 1.18) is the design and construction of residential or commercial buildings (larger structures can be included). Structural engineering includes the design and construction of buildings (Fig. 1.19), manufacturing facilities, airport terminals, and power plants.

FIGURE 1.14 3D Wireframe and Surface Model of an Experimental Airplane

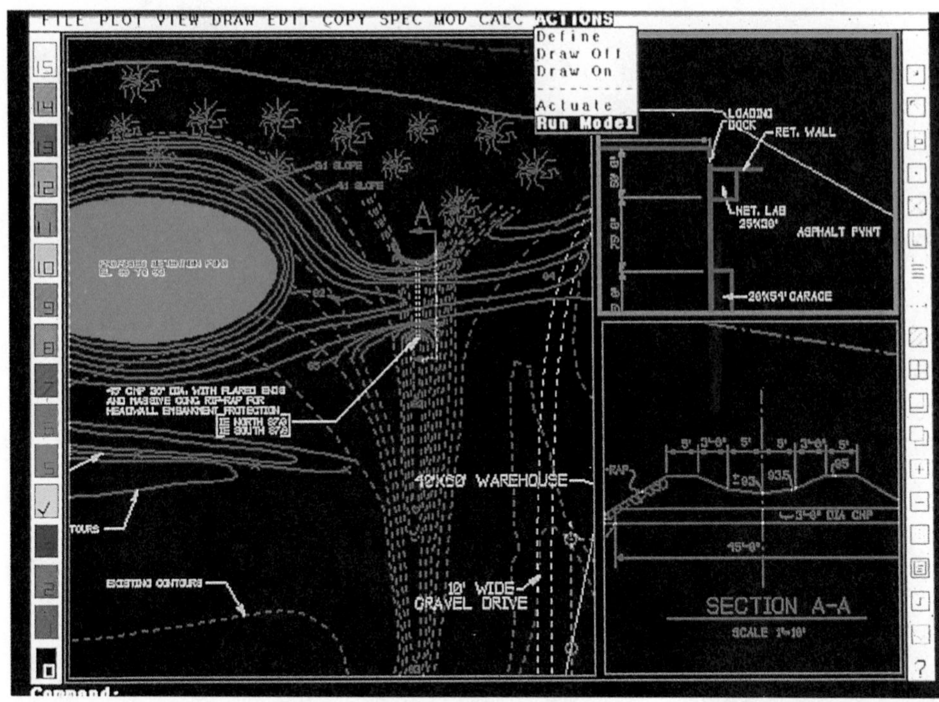

FIGURE 1.15 Civil Engineering Application

FIGURE 1.17 Diablo Canyon Nuclear Power Plant

FIGURE 1.16 Petrochemical Facility

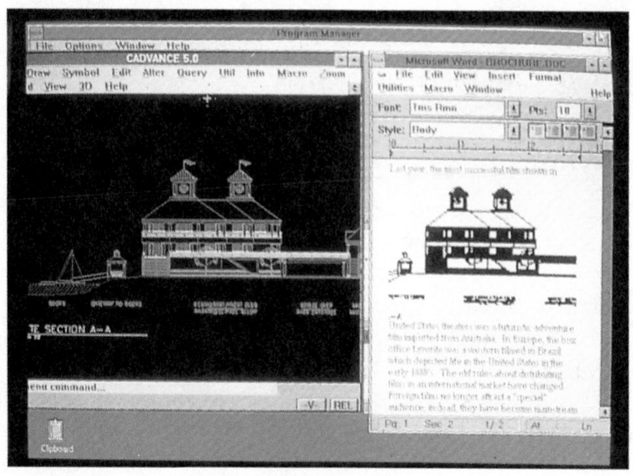

FIGURE 1.18 Architectural Design

FIGURE 1.19 Construction of a Corporate Office Facility

FIGURE 1.21 Integrated Circuit Design

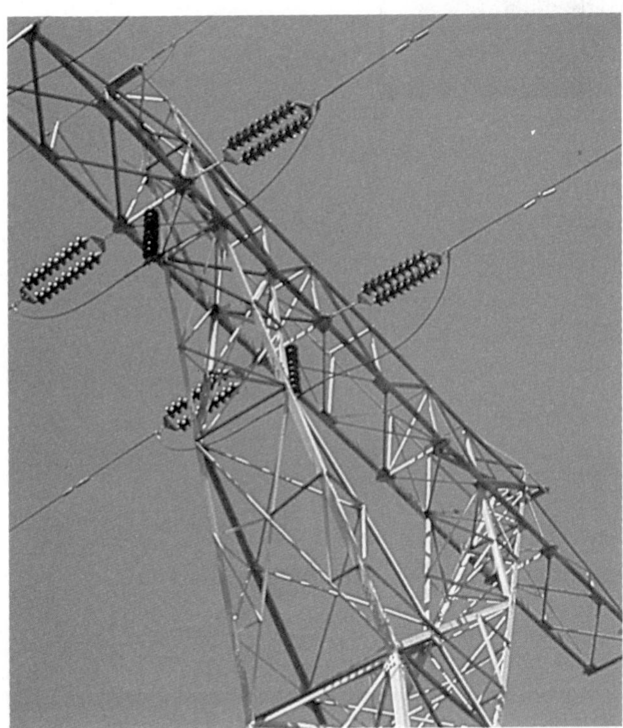

FIGURE 1.20 Power Transmission

FIGURE 1.22 Technical Illustration Using AutoCAD

FIGURE 1.23 Surface Effect Ship Model

Electronic and electrical engineering drafting includes the layout of power systems for the generation, transmission (Fig. 1.20), and utilization of electrical energy, circuits and the design of printed circuit boards, integrated circuits (Fig. 1.21), and computer products.

Mining engineering, aerospace engineering, and transportation engineering all use combinations of mechanical, electronic, and electrical designs.

Technical illustration (Figs. 1.7 and 1.22) is an area where the artistic and mechanical aspects of drafting and design merge. Technical illustrations are pictorial drawings of products, buildings, or other items needed for manuals.

This text is primarily concerned with mechanical design. Mechanical design and drafting are important because they involve the production of devices and designs for a variety of applications. Marine engineering includes the design and manufacture of marine vessels (Fig. 1.23). Aerospace engineering includes the design of engines and other mechanical devices. Transportation engineering includes the design of automobiles, trucks, buses, and trains and their individual components and requires extensive mechanical design (Fig. 1.24).

(a)

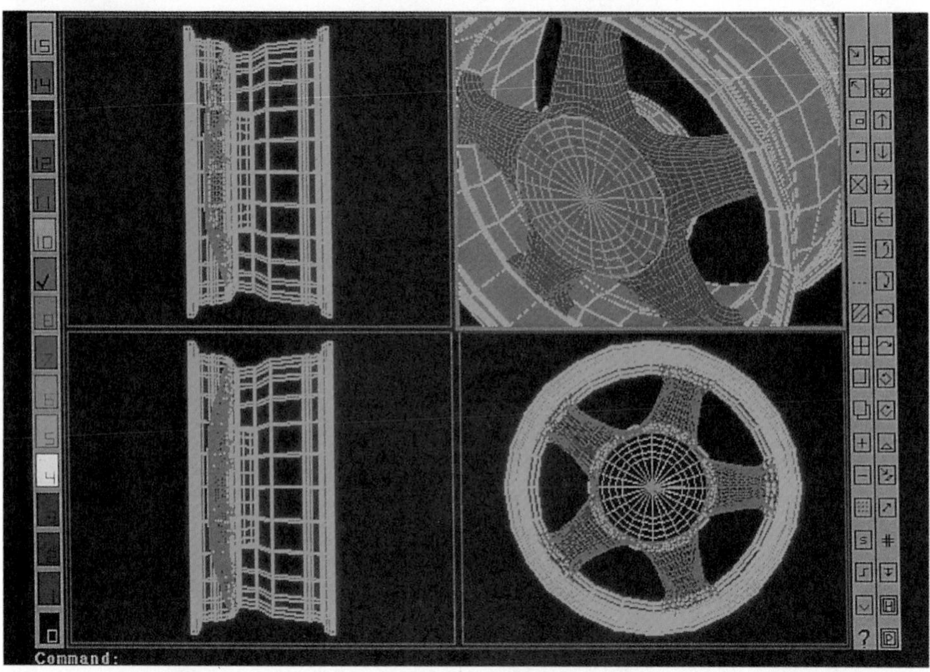

(b)

FIGURE 1.24 CAD 3D Design of a Wheel

1.10 Computers and Engineering Drawing

Computer integrated manufacturing (CIM) is the integration of all phases of production, from design to manufacturing using the computer. **Computer-aided engineering** (CAE), **computer-aided manufacturing** (CAM), and CAD are collectively called CIM. The term CAD/CAM refers to the use of computers to integrate the design and production process to improve productivity. CAM includes **numerical control** (NC), **computer numerical control** (CNC) machining (Fig. 1.25), and the use of robotics in manufacturing.

1.11 Computer-Aided Design

CAD involves any type of design activity that uses the computer to develop, analyze, modify, or enhance an engineering design. CAD systems are based on interactive computer graphics. The drafter creates an image on the CRT screen by entering commands on the computer (Fig. 1.26).

CAD *design* refers to the establishment and definition of the 3D database; *drafting* primarily involves defining, refining, and manipulating the *same* database to provide certain kinds of information.

As a designer or drafter using a CAD system, you must be able to understand the system's *hardware* configuration and its *software* capabilities. Programming ability is not required for operation of CAD/CAM systems. It must be stressed that CAD is a drafting and design *tool*. The method of creating engineering graphics has changed, not the content. Regardless of the type of system, the most common form of output remains the "drawing."

ITEMS OF INTEREST

The Design Process

Even as you read this text, new ideas to give you new sources of pleasure or new sources of frustration are being conceived. Engineers create systems, devices, and processes useful to and sought after by our society. The process by which these goals are achieved in engineering design is a planned sequence of events.

It has been said that "necessity is the mother of invention." Need is the motivating factor in most designs. When Levi Strauss first made what became known as blue jeans, they didn't have the rivets at the pockets. In 1872, Levi was contacted by a tailor from Reno, Nevada who had started riveting the pants he made for his customers. The two men decided to patent this new innovation and in 1873 were awarded the first patent for pocket rivets.

UNITED STATES PATENT OFFICE.

JACOB W. DAVIS, OF RENO, NEVADA, ASSIGNOR TO HIMSELF AND LEVI STRAUSS & COMPANY, OF SAN FRANCISCO, CALIFORNIA.

IMPROVEMENT IN FASTENING POCKET-OPENINGS.

J. W. DAVIS.
Fastening Pocket-Openings.
No. 139,121 Patented May 20, 1873.

Fig 1

Witnesses Inventor

The Patent Certificate for Rivets on Jeans

Some design is an accident. In 1878, a Procter & Gamble worker forgot to turn off the machine that stirred the soap. The soap that resulted had a lot of air bubbles and was so light it could float. He had just invented Ivory soap, by accident!

Curiosity sometimes drives design. The design of the microwave oven came about because Percy Spencer was curious about the amount of heat that was generated from magnetrons, the tubes used in radar during World War II. He could warm his hands by holding them close to the magnetrons. It was not until he found candy melted in his coat pocket that the idea of using the microwave to cook entered his mind. Many experiments later, the *high-frequency dielectric heating apparatus*—a microwave oven!—was invented. Spencer obtained a patent for it in 1953. Today, microwave ovens are an integral part of home, work, and school—all because Spencer was curious.

For years, we have dreamed about "smart homes." Imagine all the electric appliances in your home connected so they electronically communicate with each other. As you return home, your house "senses" your arrival, opens the garage door, unlocks the house, turns on the lights, and turns on the television to your favorite program. As we approach the age when this is indeed possible, it is also easy to imagine the amount of information and technology that is needed to produce such a system.

If a design is to be a success and not a frustration, it must be simple and easy to operate, no matter how much information or technology is used. The designer must be able to transmit precise, clear instructions to the user. Much of our technology today makes devices simpler to use, but requires reams of documentation in the development stage. Information management and our ability to communicate will determine whether our future designs are a joy or a frustrating mess of words and wires.

The Microwave Was Born Out of Curiosity

1.12 Standards

Many agencies control the standards used in drafting and design. The American National Standards Institute (ANSI), the Department of Defense (DOD), and the military standards (MIL) are the three most used standards in the United States. The International Standards Organization (ISO) standards and Japanese standards (JIS) are also used.

ANSI standards are available to drafters and designers at their place of employment. It is important to become familiar with these standards. ANSI-Y14 contains information on drafting practices, dimensioning, projection, descriptive geometry, geometric tolerancing, and a wide variety of other areas associated with drafting and design.

Some companies have not adopted ANSI standards. This text uses ANSI standards as a basis for its drawings. All projects completed from the book are to be drawn using the latest revisions of ANSI standards, conventions, and drawing practices.

1.13 Standards of Measurement

The United States is the only major industrial country in the world still using feet, inches, and decimal equivalents. However, many large companies such as Ford, IBM, John Deere, General Motors, Honeywell, and most electronic, medical instrument, and computer manufacturers have completely converted to the metric system that is called the *Système International* (SI). The English system is now called the *U.S. customary unit.*

Because you may encounter both measurement systems on the job, this text uses a balanced approach and applies both

FIGURE 1.25 CNC Machining Using a CAD Database

FIGURE 1.26 Personal Computer CAD System

systems. Piping, architecture, and structural engineering use units of feet, inches, and fractions, in most cases. The standard of measurement for metric drawings is the millimeter. The U.S. decimal-inch unit is used on many of the illustrations and on many of the exercises and problems at the end of the chapters. In some cases, your instructor may wish you to convert the units of measurement from one system to another.

1.14 Organization of Text

This text is organized into three sections. The first section introduces the subject of design and drafting, describes the instruments and techniques used in drafting, describes geometric construction techniques, and covers basic sketching skills. The second section covers multiview projection, pictorials, section views, auxiliary views, manufacturing processes, dimensioning, and geometric dimensioning and tolerancing. The final section covers threads and fasteners, springs, the design process, working drawings, descriptive geometry and gives a guide to understanding symbols on engineering drawings.

The last section of the text contains the Appendixes. Here, you will find glossaries, abbreviations, standards, conversion charts, and catalog items such as nuts, bolts, washers, bearings, keys, pins, valves, and fittings. You will also find welding, electronic, and piping symbols used on drawings. Consult the Appendixes when working on projects from the text.

Each chapter in the text follows the same sequence. Chapters start with an introduction and continue with an explanation of the material to be covered. Exercises are found at the end of the chapter, but are designed to be completed at specific intervals. You will be prompted at intervals within each chapter to complete exercises designed to test your knowledge of the material just covered. Exercises are on a grid format using one-quarter inch units, and can be transferred directly without the use of dimensions to an $8\frac{1}{2}$ x 11 in. "A" size grid-lined sheet of paper. If metrics are preferred, use metric grid paper with appropriate divisions.

At the end of each chapter (except Chapter 1), there is a quiz composed of true and false, fill in the blank, and answer the following questions. Following the quiz, problems are provided for you to complete. These problems can be assigned in many different ways—either as sketches, ink drawings, manual drafting, or CAD projects. Unlike the exercises, which are confined to an $8\frac{1}{2}$ x 11 in. "A" size format, the size of paper is dependent on the project requirements.

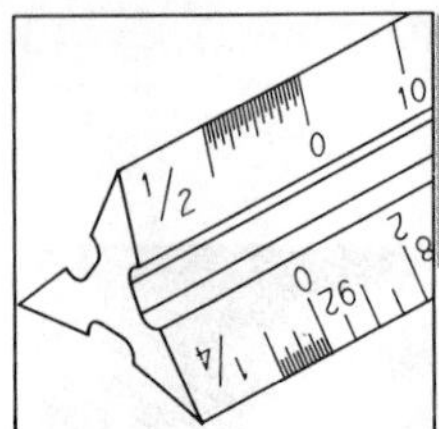

CHAPTER 2

Equipment, Instruments, Materials, and Techniques

Learning Objectives

Upon completion of this chapter you will be able to accomplish the following:

1. Identify equipment and general drafting tools used in technical drawing.
2. Produce drawings using various drafting instruments and appropriate scales.
3. Exhibit knowledge of drafting media and drawing formats.
4. Develop the ability to produce ANSI standard line types while recognizing preferred line precedence.
5. Master techniques for drawing construction and printable lines and curves.

2.1 Introduction

Drafting tools (Fig. 2.1) are used in all engineering and design work. Although computer-aided design (CAD) systems are increasingly found in industry, traditional drafting techniques and tools are still used and will continue to be used in the foreseeable future. The simple lead holder and the complex electronic pen are both important and share the same purpose.

FIGURE 2.1 Drafting Tools

This chapter describes drafting equipment, instruments and materials and introduces you to the techniques for their use. **Equipment** includes drafting boards, drafting machines, print machines, T-squares, triangles, and templates. **Instruments** are precision-manufactured drawing tools, such as the compass and dividers. Drafting **materials** comprise drawing media (vellum and drafting film) and related support items, such as grid underlays, preprinted title blocks, transfer drafting aids, and print paper.

Techniques are methods used to complete a drawing. In all fields of technical drawing and design, symbols, linework, projection procedures, and notation must be in accordance with standard *drafting conventions.*

2.2 Equipment

The most important and conspicuous piece of equipment found in any drafting room is the **drafting table.** Today, board sizes range from hand-carried versions to large-format, stand-alone tables. The table in Figure 2.2 is vertically adjustable and can be tilted to any comfortable angle. Modern tables may be power operated. The table surface must have a pliable surface such as Borco vinyl (or linoleum) or some other covering that permits you to draft without destroying the table surface or marring the drawing medium (vellum, drawing film, etc.).

Light tables are used to prepare printed circuit artwork, to draw pictorial illustrations, and to trace. The drawing surface is a translucent glass or plastic sheet that scatters the rays from the light source. Figure 2.3 shows a light table and reference desk.

FIGURE 2.2 Metal Drafting Table

FIGURE 2.4 Aperture Card Viewing System

FIGURE 2.3 Light Table and Reference Desk

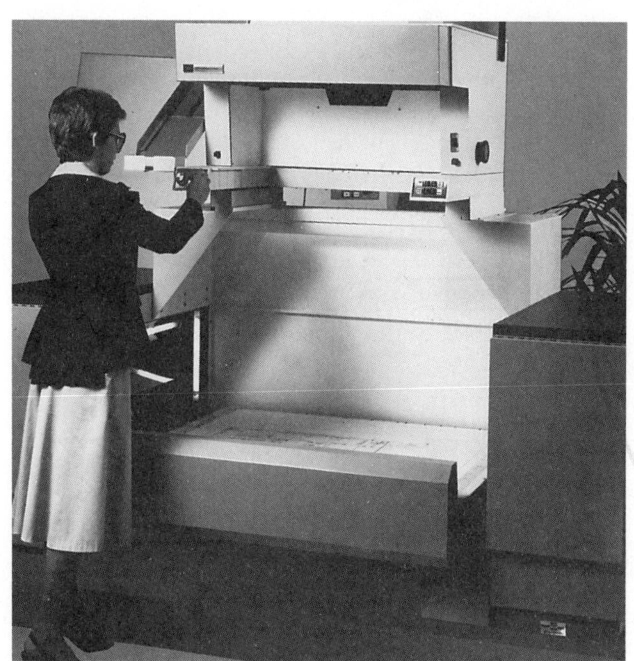

FIGURE 2.5 Microfilm Processor Camera

2.2.1 Storage and Reproduction Equipment

Drawings are *stored* and *reproduced.* Frequently, drawings are stored as originals or as paper prints in multiple-drawer cabinets and in tube storage systems. Because drawings must be cataloged and available to several departments, this method of storage is still used even though it is time consuming and requires considerable office space.

Drawings are also stored on **microfilm** and **microfiche.** Computer graphics systems enable the user to reproduce design data stored on disk or tape. Another form of reprographics uses 35 mm micrographic **aperture cards** or **design data cards** (Fig. 2.4). These card systems allow access to more than 1,000 design drawings in less than $7\frac{1}{2}$ inches of space. When a new or revised drawing is checked and ready for release, it is taken to a processor camera (Fig. 2.5). In seconds, a master data card (an accurately reduced version of the original drawing) is produced. This card is correct in every detail, but it is smaller and easier to use and reproduce than an original drawing. Multiple copies of the data card are then made from the original for distribution. You review the drawing with a display device (Fig. 2.4).

Traditionally, the **blueprint machine** was used to make multiple prints of drawings. However, the term blueprint is no longer accurate because the prints are actually white, or what are sometimes called blueline prints. **Whiteprint machine** would be a more accurate term because most, if not all, reproduction with this method involves developing a print with blue lines and a light background.

When a drawing is completed on a CAD system, the user must either reproduce the drawing from a hard-copy device (photocopier) or plot the drawing with a pen plotter. The pen plotting method produces an accurate original. Multiple copies can then be made from a white printer, an engineering copier, a laser printer, or from input to a data card system.

2.2.2 Straightedges

Originally, the horizontal straightedge device used in drafting was the **T-square** (Fig. 2.6). They are still found in a few drafting classes and are often used for personal drafting. Because the T-square is the most difficult to manage of all straightedge drawing devices, it is said that "if you can draw with a T-square you can draw with anything." Using a T-square is difficult because it is the easiest to misalign of all straightedge devices. The blade portion of the T-square is placed along the edge of a drafting board. Parallel horizontal lines are drawn with the length of the T-square, and parallel vertical lines are drawn with a triangle placed against the top edge of the horizontal length.

Since it is an excellent tool for drawing long horizontal lines, the **parallel straightedge** (Fig. 2.7) is used in industry, especially for creating the large drawings required in the construction trades—architecture, piping design, and civil engineering. The parallel straightedge is attached to the table by cables and pulleys. It remains parallel or at a preset angle to the drafting table as it is moved up or down on the table surface.

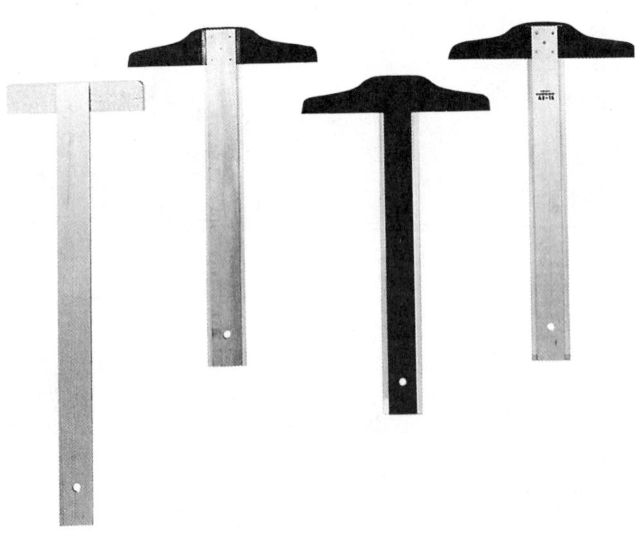

FIGURE 2.6 T-Squares

FIGURE 2.7 Parallel Straightedge

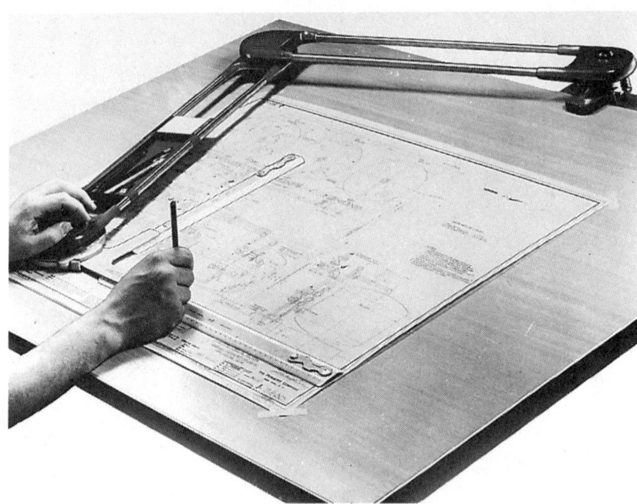

FIGURE 2.8 Arm Drafting Machine

The two standard versions of the **drafting machine** are the drafting arm type (Fig. 2.8) and the track type (Fig. 2.9). The track type is more accurate. The control head on the drafting machine can be rotated to any angle and set by pushing a button to lock in increments of 15°. They are hand-locked for intermediate angles. Drafting machines replace triangles, protractors, and scales.

Regardless of the type of drafting table and straightedge, proper lighting is essential for relaxed, unstrained work. Since the CRT screen is easier to read if it is shaded from external light sources, lighting requirements are different for CAD systems.

2.3 General Drafting Tools

A variety of small tools and equipment are required for drafting (Fig. 2.1), such as special templates, triangles, pencils, lead holders, and technical inking pens. The quality of draft

FIGURE 2.9 Track Drafting Machine

ing is directly influenced by the range and quality of the tools and equipment used. Good quality tools are beneficial for fast, efficient, and precise linework and projection.

Drafting kits are available from a variety of companies and are sufficient for most classes in drafting. Precision, high-quality tools and instruments can be purchased individually, either at a drafting supply store or through a drafting equipment catalog. Table 2.1 lists standard drafting tools. Essential items are distinguished from optional items.

TABLE 2.1 Equipment

Essential Items	Optional Items
Pencils (grades 4H, 3H, 2H, H, HB)	Lead holders
Sandpaper block	Thin-line pencil
Erasers	Electric eraser
Dusting pad or powder	Adjustable triangle
Erasing shield	Symbol templates
Drafting tape	Lettering guide
Drafting brush	Lettering template
Scales (metric, architect, mechanical, civil)	Drop compass
	Beam compass
Protractor	Compass inking attachment
30/60° triangle	Technical inking pens
45° triangle	Ink
Irregular curves	Ink eraser
Templates (circle and ellipse)	Lettering set
Bow compass	Grid paper
Dividers	Drafting table
Drafting board	Flexible curve
Straightedge (T-square, parallel straightedge)	Drafting machine
Calculator	
Paper (vellum, drawing film)	

2.3.1 Pencil Leads and Pencils

Drafting pencils are graded by hardness of lead. The hardness of the lead determines the kind of line that can be drawn. A hard lead makes very sharp and thin lines, but it will lack the darkness and density needed to make a good reproduction. A soft lead will make dark lines, but they are very difficult to keep sharp. "H" lead grades are used for drafting on vellum. The "H" grades are, from hardest to softest: 9H, 8H, 7H, 6H, 5H, 4H, 3H, 2H, H, F, and HB (Fig. 2.10). The recommended hardnesses for lead used on vinyl-topped boards with a good grade of paper are:

1. 6H–3H for layout and construction lines
2. H–HB for reproducible (printable) lines
3. H for lettering

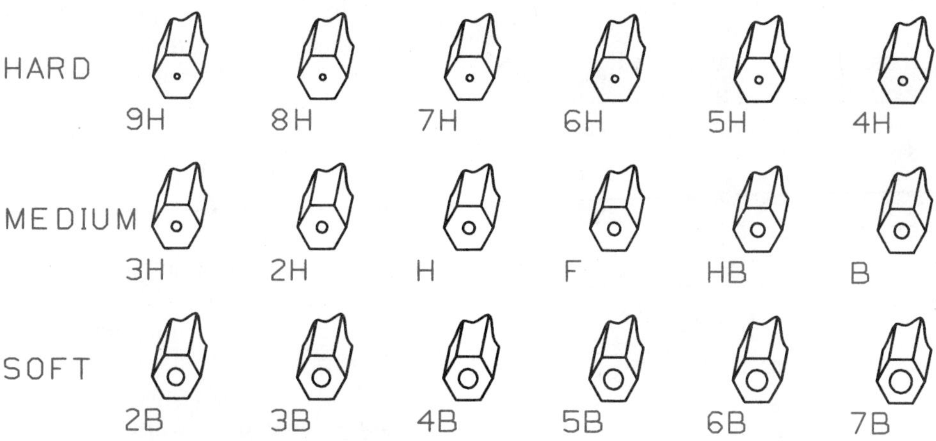

FIGURE 2.10 Pencil Types

The appropriate hardnesses of lead, combined with proper drafting techniques, will produce good reproducible drawings with sharp, dense lines that make good prints. The skills required to use drafting tools and equipment come through practice.

The three types of drafting pencils are the familiar wood pencil, the mechanical lead holder (Fig. 2.11), which uses drafting leads, and the fine-line mechanical pencil. The wood pencil is the least expensive, but is not as convenient to use as mechanical pencil. The wood is cut away in a wooden pencil before it is sharpened with a drafting pencil sharpener or a knife (Fig. 2.12).

The mechanical lead holder (Fig. 2.11) has replaced the wood pencil. It holds a single piece of lead. Having more than one lead holder available with different leads makes it easy to change line weights. This drawing tool is easily sharpened and increases the speed, consistency, and ease of drafting. The length of exposed lead can be adjusted both for sharpening and for drawing.

The automatic drawing pencil (**fine-line pencil**) is an excellent tool for drawing lines and letters of consistent width and never requires sharpening. Fine-line pencils are available in metric sizes: 0.3, 0.5, 0.7, and 0.9 mm (Fig. 2.13). These sizes are used to draw line weights, from 0.25 to 0.35 mm (centerlines, dimension lines, construction lines), 0.5 to 0.7 mm (object lines, diagram lines, hidden lines), and 0.7 mm to 0.9 mm (cutting plane lines, border lines). Unlike the mechanical drafting pencil, the fine-line pencil holds only one type of lead, so you must purchase a number of them. Of course, if you are using a small width fine-line pencil, you can make the line any width desired by thickening the line. The thinner the lead, however, the more frequently it breaks. Thin-line pencils are sometimes difficult to use with lettering guides and templates.

FIGURE 2.11 Mechanical Lead Holders

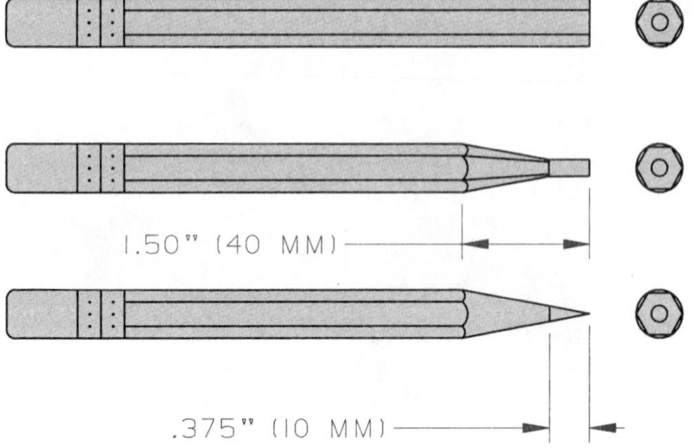

FIGURE 2.12 Sharpening Pencils

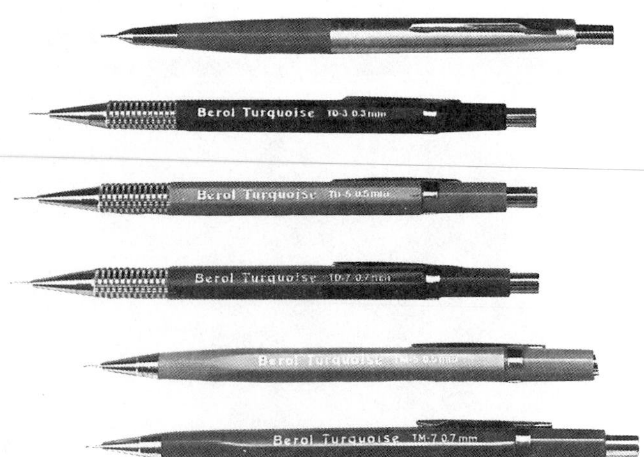

FIGURE 2.13 Fine-Line Mechanical Pencils

Regardless of your pencil type, purchase a variety of leads. Standard leads used on vellum range from soft and dark (6B, 5B, 4B, 2B, B, HB, and F) to the medium hard and dark (H, 2H, 3H). The hardest and lightest types are 4H, 5H, 6H, 7H, 8H, and 9H. Medium leads, 2H and 3H, are used for construction lines and blocking in a drawing, H is used for darkened finished lines. Some drafters prefer HB for finished lines. To be reproduced with good quality, all lines must block light if a whiteprint machine is used, and they must be dark, crisp, and thick enough to be recorded by a camera if a photocopier or micrographics aperture card machine is used.

Plastic leads are used on drawing film. Plastic leads come in three grades (E, K, and CF).

2.3.2 Pencil Sharpeners/Pointers

Lead holders and wood pencils require frequent sharpening or pointing. A sharp conical point is essential to make the thin erasable lines that are required for construction lines. For dark, finished linework, the pencil point is slightly dulled on scrap paper to avoid breaking and to draw wide lines. *To maintain the line thickness. the pencil or lead holder should be rotated as the line is drawn.* A sharpened, then slightly dulled, lead point is required for lettering.

The **pencil pointer** sharpens by cutting only the lead and produces a uniform conical shape with a rather long taper. The taper is three to four times the diameter of the lead. The pencil or lead holder (Fig. 2.14) is put into a hole in the cover of the cutter and rotated around the cutter. Figure 2.15 shows an inexpensive hand-held pointer.

Another way of sharpening wood pencils and lead holders is with the **sandpaper pad/block.** The pencil or lead holder is rotated as the point is sanded (Fig. 2.16). A sandpaper pad can also sharpen lead points on compasses. Compass leads

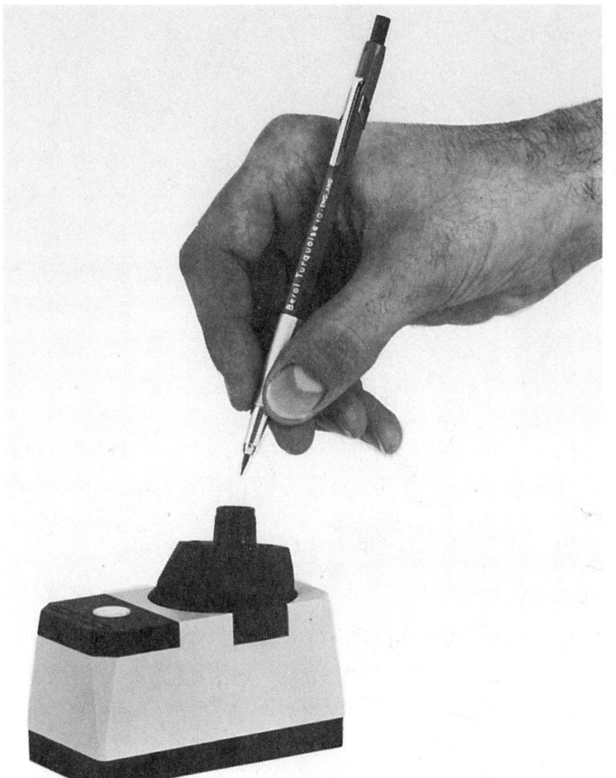

FIGURE 2.14 Mechanical Pencil Sharpener

FIGURE 2.15 Sharpener

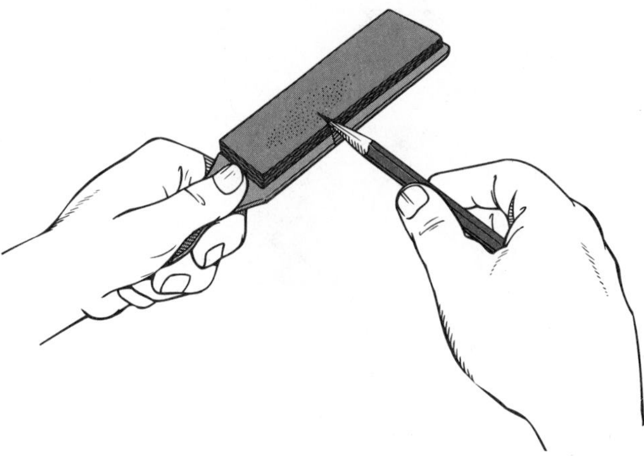

FIGURE 2.16 Sharpening a Pencil with a Sandpaper Block

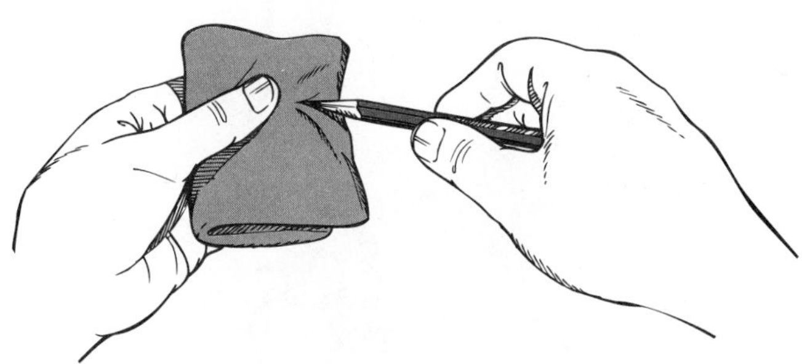

FIGURE 2.17 Cleaning Pencil Tip

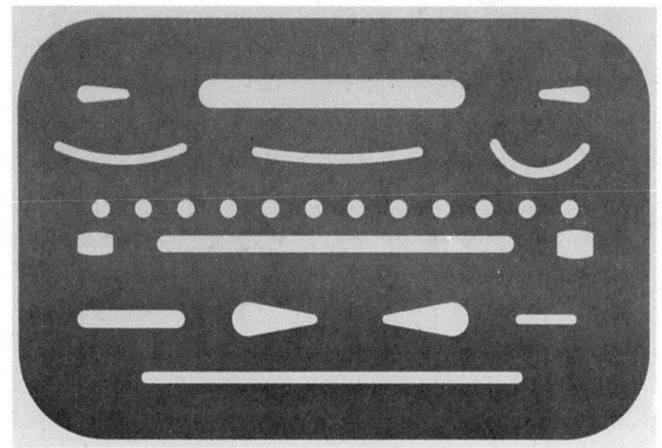

FIGURE 2.18 Erasing Shield

are sharpened as wedge shapes instead of conical points in order to keep the edge sharp longer. After sanding the lead, wipe it clean with a soft cloth or tissue (Fig. 2.17).

2.3.3 Erasers and Erasing Shields

Erasing is a necessary part of drafting and, when done properly, enables you to correct drawings easily. The **eraser** should have good "pick-up" power without smudging. Selection of an eraser is based on the drafting media you are using.

To protect adjacent areas of the drawing that are to remain, most erasing is done through the perforations of a stainless steel **erasing shield** (Fig. 2.18). The erasing shield is firmly held in place on the drawing with one hand while the other hand erases through a selected opening (Fig. 2.19). Eraser crumbs are immediately swept from the drawing with a **drafting brush** (Fig. 2.20). Never rub the crumbs with your hand—each graphite-laden crumb will act as a dull pencil and make smudges.

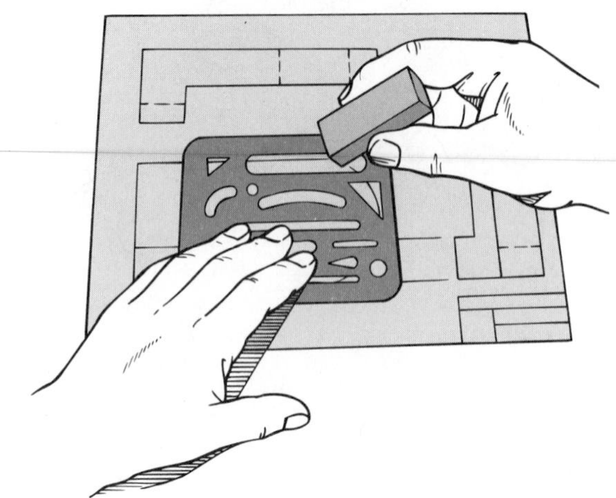

FIGURE 2.19 Using an Erasing Shield

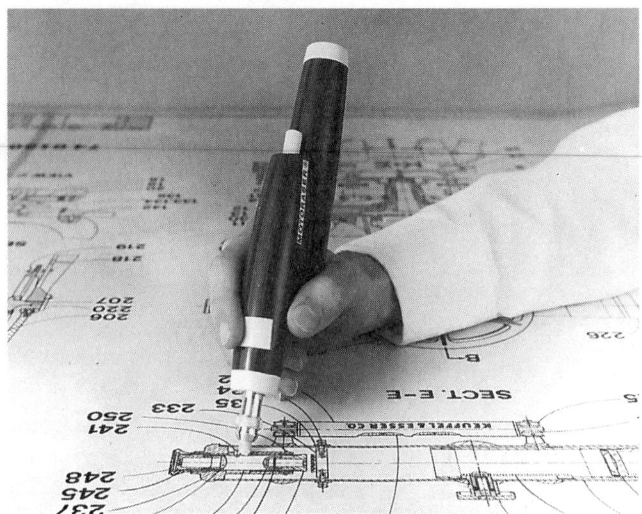

FIGURE 2.21 Battery-Operated Erasing Machine

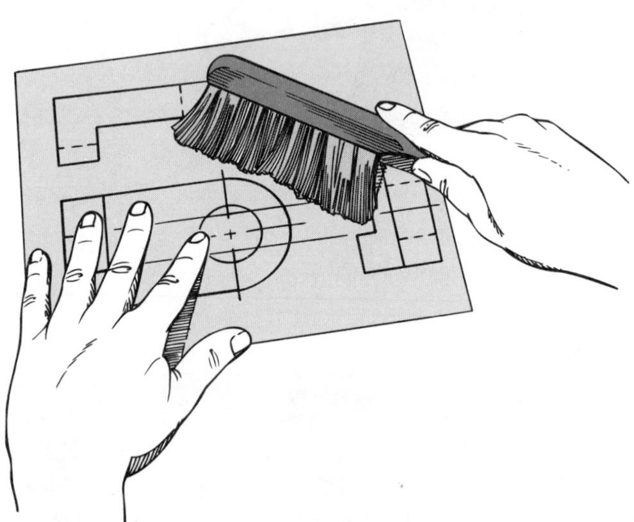

FIGURE 2.20 Brushing the Drawing

Erasers can be purchased in many shapes and sizes, from hand-held to electric, and in many grades. Pink Pearl, white composite, and Art Gum erasers are used for both paper (vellum) and drafting film. Special vinyl erasers are available for erasing inked drawings on drafting film. An **electric eraser** (Fig. 2.21) is essential for erasing ink, but great care is required to avoid rubbing holes in the paper. It also tends to destroy the tooth, or surface, of the drafting film. To erase ink drawings completed on film, the best technique is to use a small amount of moisture applied to a vinyl eraser and carefully rub the area to be erased.

A **dry cleaning pad** (an erasing dust pad) is another useful item for keeping drawings clean and unsmudged. They contain finely ground eraser pieces and powder and can be used to remove dirt and leftover crumbled graphite deposited from drawing and lettering. Do not drag it across the drawing. Instead, after a small portion of the drawing is complete, lightly pat the linework and lettering. Then use a drafting brush to sweep the drawing clean of powder and dirt. Frequent dusting ensures a higher quality drawing. Never use a dry cleaning pad when inking with technical pens. Some drafters prefer to cover the entire drawing with a very light layer of erasing powder while lettering and drawing; others find this method messy and uncomfortable.

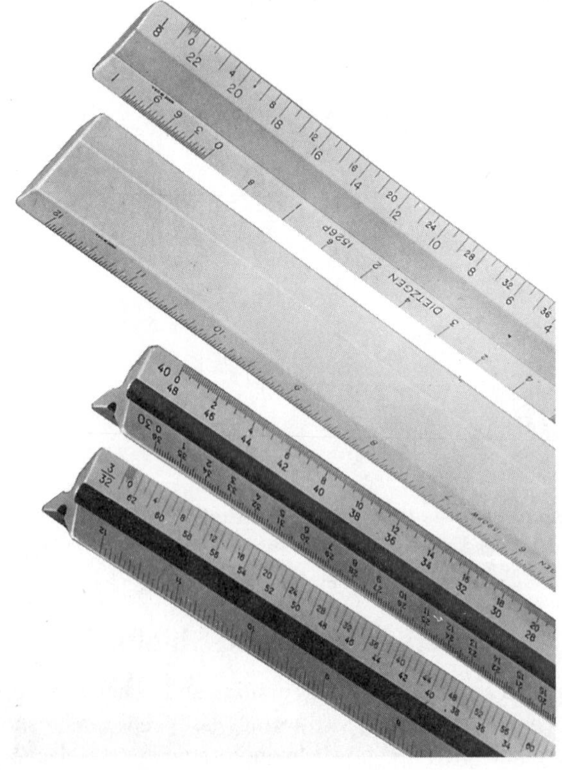

FIGURE 2.22 Scales

2.4 Drawing Scales

All instrument drawings are drawn accurately to a particular scale. Such a drawing is said to be drawn "to scale." Drawings can be drawn full size, or to an enlarged or reduced size. CAD drawings are created full size (1:1) on the computer, but they may be plotted to any size.

Since construction projects—piping, structural, architecture, and civil—are large and the paper size is small, all of these drawings are done at a reduced scale. The instrument used to measure these reduced-size drawings is called a **scale.**

Certain scales are used in construction work: the civil engineer's scale, the architect's scale, and the metric scale for SI projects. The engineer's scale is used to make drawings of very large objects, for example, earthworks, roads, and surveys of property. The architect's scale is used to make drawings of buildings and structures. The basic shape of each of these scales is either two-sided and flat or triangular (Fig. 2.22); each is about 12 inches long. The triangular shape has six surfaces for different sized scales.

The markings on scales are arranged in two ways: *fully divided* and *open divided.* Fully divided scales have each main unit of measurement throughout the length of the scale. The engineer's and the metric scale are fully divided. Open divided scales have each main unit of the scale undivided, except for a fully divided extra main unit at the 0 end of the scale. The architect's scale is open divided.

The scale is a precision instrument and, with proper use, will produce consistent drawings. Do not use the scale (unless it is one of the two scales of the drafting machine) as a straightedge because they do not have edges designed for drawing.

2.4.1 The Civil Engineer's Scale

The **engineer's scale** (triangular) has six scales that are fully divided. Three-sided civil engineering scales are divided into 10, 20, 30, 40, 50, and 60 divisions per inch and are numbered at each tenth division along the length of each scale. The number of divisions per inch is marked at the 0 end of each scale. Usually each division equals 1 foot, but you can assign any unit to the scale divisions. This is designated on a drawing as 1″ = 20′ (1 inch on the scale represents 20 feet). This scale is also used as a decimal scale, where 1″ = 2′ (each division represents one-tenth of a foot on the 20 scale) or 1″ = 200′ (each division represents 10 feet on the 20 scale). Figure 2.23 shows measurements taken along the civil engineer's scale. Here, .50, 3.60, and 4.90 inches are shown measured on the full-size inch scale, which has increments of $\frac{1}{10}$. These measurements could also be in feet if the desired scale was 1″ = 1′.

2.4.2 The Architect's Scale

The **architect's scale** (Fig. 2.24) has a full-size scale on one surface and ten different reduced-size open divided scales. The open divided scale uses only 1 foot units, reading in one direction from the 0 end, with a fully divided 1 foot unit reading in the opposite direction. The number of feet is read along the length of the scale and the number of inches is read in the fully divided unit at the 0 end of that same scale. Both numbers become larger as the distance from the 0 end becomes greater.

Each scale is identified by a number or a fraction at its 0 end. This number does not represent a proportion of size but is an abbreviation for the unit of length in inches that represents 1 foot of real size. Figure 2.24 shows measurements taken along the architect's scale. The following are some architect's scale abbreviations:

Abbreviation	Meaning	Proportion
3	3″ = 1′0″	$\frac{1}{4}$ size
$1\frac{1}{2}$	$1\frac{1}{2}$″ = 1′0″	$\frac{1}{8}$ size
1	1″ = 1′0″	$\frac{1}{12}$ size
$\frac{3}{4}$	$\frac{3}{4}$″ = 1′0″	$\frac{1}{16}$ size
$\frac{1}{2}$	$\frac{1}{2}$″ = 1′0″	$\frac{1}{24}$ size
$\frac{3}{8}$	$\frac{3}{8}$″ = 1′0″	$\frac{1}{32}$ size
$\frac{1}{4}$	$\frac{1}{4}$″ = 1′0″	$\frac{1}{48}$ size
$\frac{3}{16}$	$\frac{3}{16}$″ = 1′0″	$\frac{1}{64}$ size
$\frac{1}{8}$	$\frac{1}{8}$″ = 1′0″	$\frac{1}{96}$ size
$\frac{3}{32}$	$\frac{3}{32}$″ = 1′0″	$\frac{1}{128}$ size

Each of the five open divided scale surfaces has two scales printed on it, reading in opposite directions from each end. Each pair of scales has a 1:2 size ratio and is $3-1\frac{1}{2}$, $1-\frac{1}{2}$, $\frac{3}{4}-\frac{3}{8}$, $\frac{1}{4}-\frac{1}{8}$ and $\frac{3}{16}-\frac{3}{32}$. In the smaller of the two scales, the "foot" numbers are nearer the working edge of the scale. Both scales have alternative foot markers numbered on the larger scale, whereas the other foot markers are for only the smaller size. On the $\frac{1}{4}$ and the $\frac{3}{16}$ scales, only the even numbered and, on the $\frac{1}{8}$ and $\frac{3}{32}$ scales, only the markers divisible by four are numbered. Use caution in making correct readings from all of these foot markers.

The fully divided 1-foot unit at the 0 end of each scale is divided into inches and fractions of an inch and is read from the 0 end. The number of divisions varies with the unit of length that represents 1 foot. The value of the smallest unit varies from $\frac{1}{8}$ in. on the 3 scale to 2 in. on the $\frac{1}{8}$ and $\frac{3}{32}$ scales. Study the various scales to become familiar with the smallest units used on each one. The lengths of the dividing lines vary to make reading the scales easier. The 3, 6, and 9 in. marks are numbered on the 1 and $1\frac{1}{2}$ scales; each inch is marked on the 3 scale.

In Figure 2.24, $3\frac{9}{16}$ in. and $4\frac{1}{2}$ in. have been set off on the full-size inch scale (16), which is divided into increments of $\frac{1}{16}$th inch.

2.4.3 The Mechanical Engineer's Scale

The **mechanical engineer's scale** (Fig. 2.25) is two-sided and flat. One side, the full inch scale, is divided into either decimal units of 0.10 inches or as many as fifty divisions (every

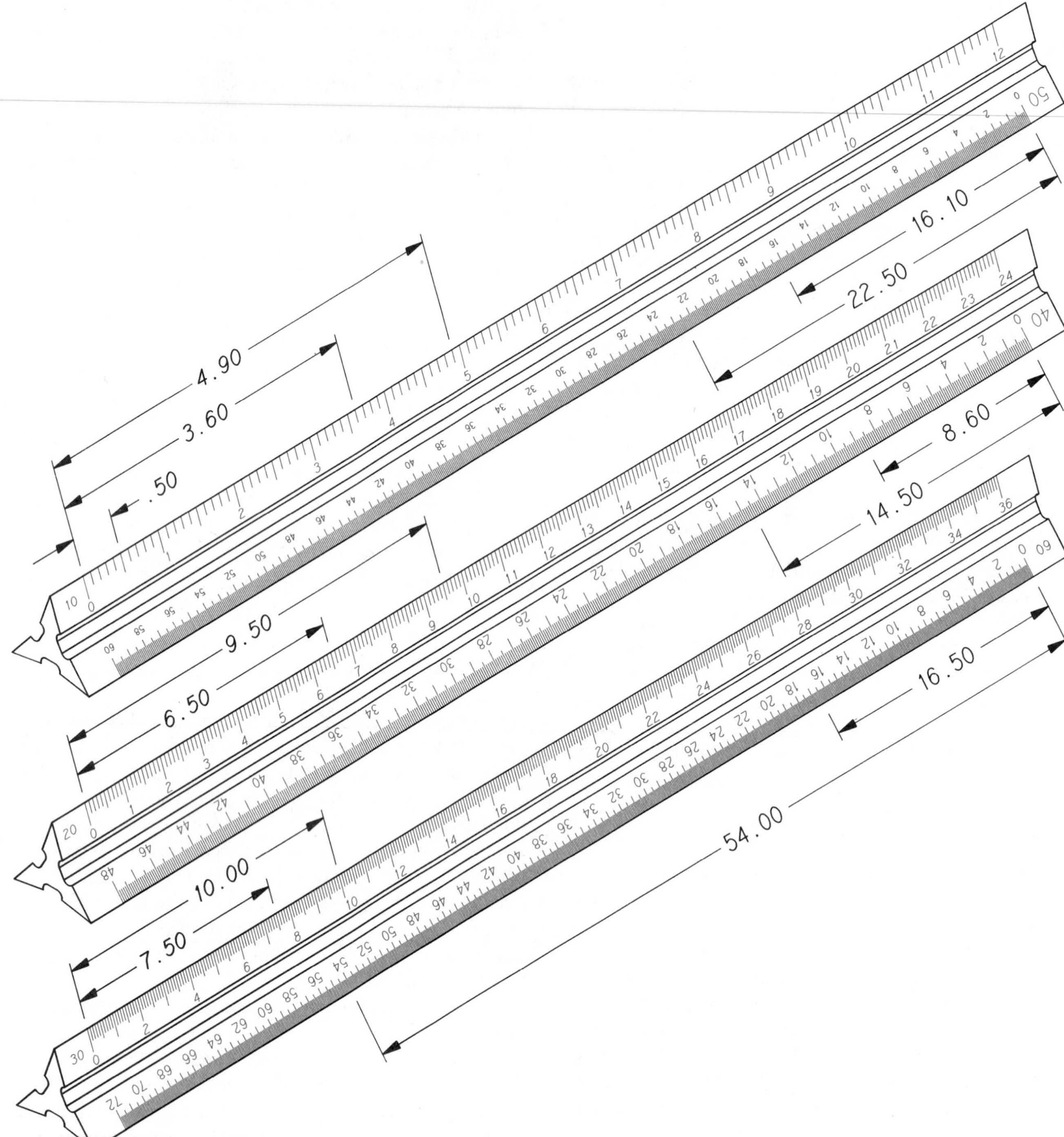

FIGURE 2.23 Civil Engineer's Scale

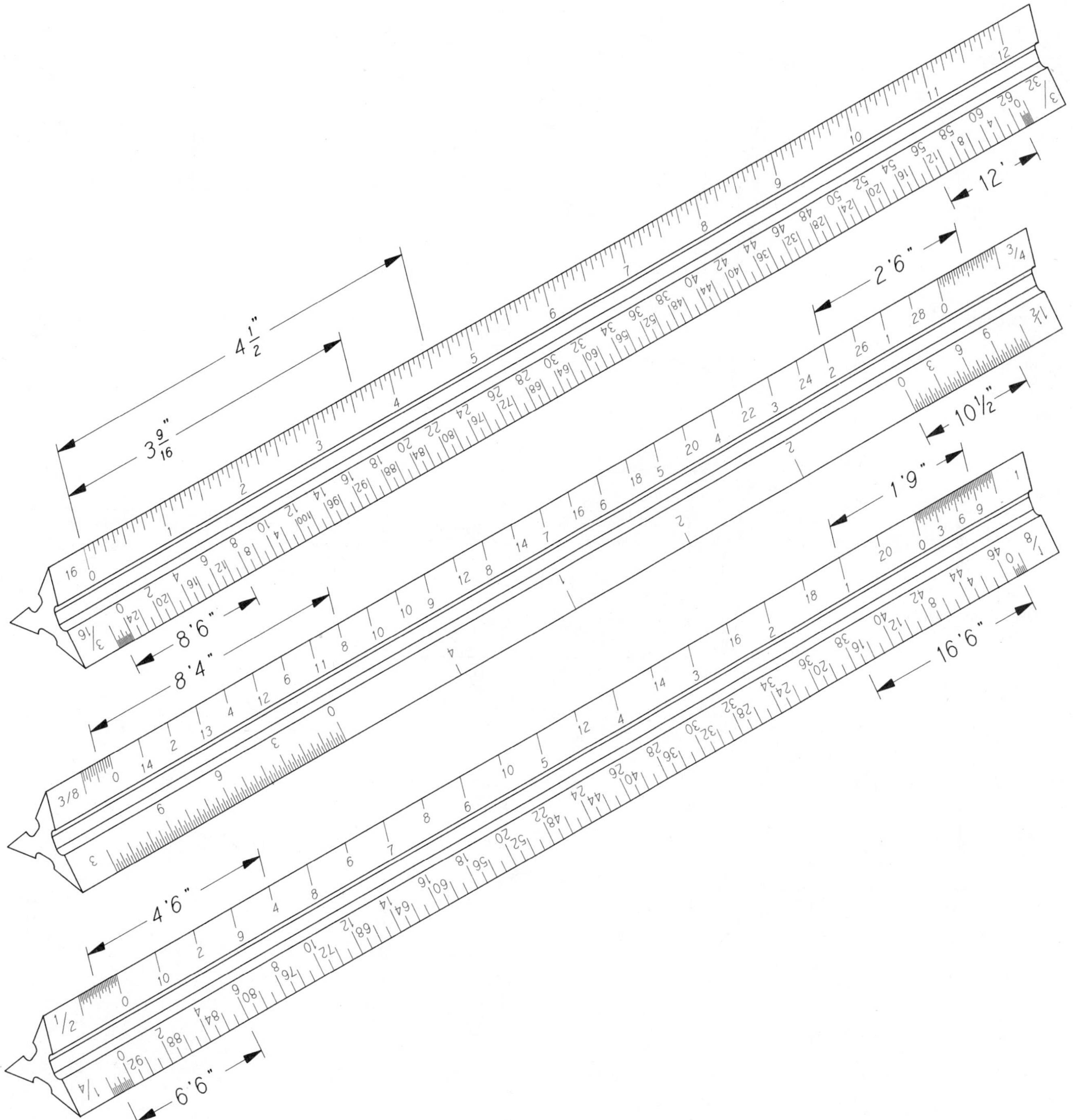

FIGURE 2.24 Architect's Scale

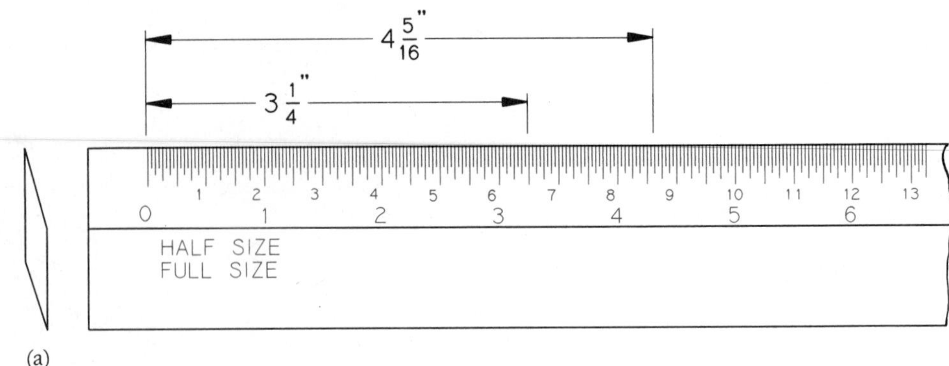

(a)

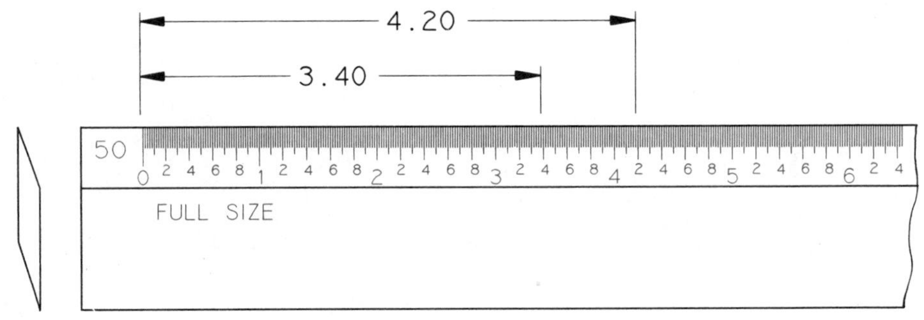

(b)

FIGURE 2.25 Mechanical Engineer's Scale
(a) Mechanical Engineer's Scale
(b) Decimal Scale

.02 inches). The opposite side is half scale (1:2). This scale also can be purchased in a triangular version.

Figure 2.25 also shows a sixteenth scale used by mechanical engineers. This scale is flat and has a full scale fractionally divided on one side and a $\frac{1}{2}$ scale fractionally divided on the other side. Figure 2.25 shows full-size inch measurements of $4\frac{5}{16}$ in. and $3\frac{1}{4}$ in. [Fig 2.25 (a)] and 4.20 and 3.40 full size decimal-inch measurements [Fig 2.25(b)].

2.4.4 The Metric Scale

Metric units, also called SI units (Systeme Internationale d'Unites), are measured with a **metric scale** (Fig. 2.26). To convert customary unit (inch decimal and inch fraction) drawings, multiply the inch value times 25.4 to get the metric equivalent (1 in. = 25.4 mm). As an example, 2.50 in. × 25.4 mm/in. = 63.5 mm. To change a metric value into decimal inches, divide by 25.4. As an example, 50 mm/25.4 mm/in. = 1.96 in.

TABLE 2.2 Scales

Architect's Scale	Mechanical Engineer's Scale	Civil Engineer's Scale	Metric Scale
$\frac{3}{32}$	1 in. = 1 in. (full size)	10 divisions/unit	1:10
$\frac{1}{8}$	$\frac{1}{2}$ in. = 1 in. ($\frac{1}{2}$ size)	20 divisions/unit	1:20
$\frac{3}{16}$	$\frac{1}{4}$ in. = 1 in. ($\frac{1}{4}$ size)	30 divisions/unit	1:25
$\frac{1}{4}$	$\frac{1}{8}$ in. = 1 in. ($\frac{1}{8}$ size)	40 divisions/unit	1:333
$\frac{1}{2}$		50 divisions/unit	1:50
1		60 divisions/unit	1:75
$1\frac{1}{2}$		80 divisions/unit	1:100
3			1:150

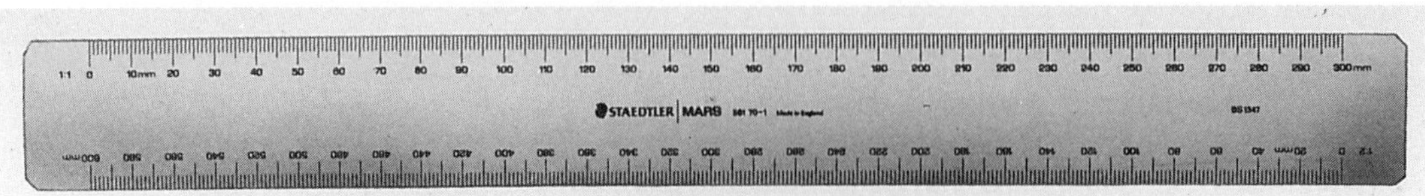

FIGURE 2.26 Flat Metric Scale

The full-size metric scale is divided into major units of centimeters and smaller units of millimeters. There are 10 millimeters in each centimeter. Metric units are being used more often in all forms of design work. To many, the metric scale is much easier to master and use than decimal-inch or fraction-inch scales. Figure 2.27 shows a triangular metric scale with measurements. To set off 80 mm full size (1:1) on the metric scale, start at the 0 end of the 1:1 scale as shown in Figure 2.27 and count to the right until you get to the 8 (8 centimeters = 80 mm). On the 1:5 scale ratio, the 4.5 mark on the scale gives you 4500 mm or 4.5 m.

Table 2.2 compares the four basic scales.

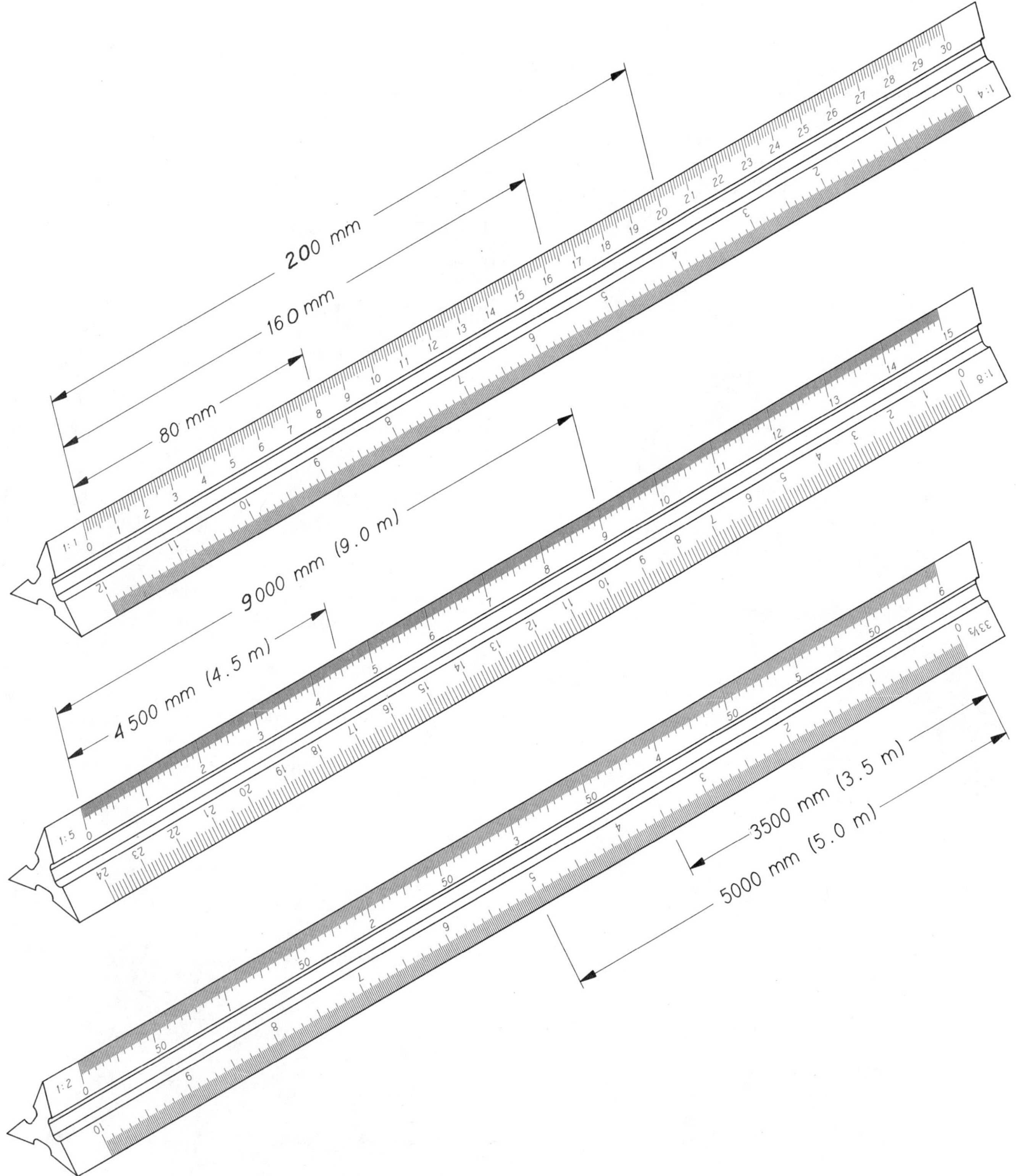

FIGURE 2.27 Triangular Metric Scale

2.5 Drawing Tools

Drawing tools include a variety of items to create geometric figures, measure and layout constructions, and establish features. These items include protractors, triangles, and templates.

2.5.1 Protractors

The **protractor** measures angles and is used to lay out lines at an angle. A 360° protractor (circular) is the easiest to use. Figure 2.28 shows a 180° protractor along with several irregular curves. The center of the protractor is aligned with the intersecting point of the lines to be drawn or measured.

Since it is easy to misread angle measurements, a protractor should be used to check all constructions. Features of a part drawn at angles should always be checked with a protractor.

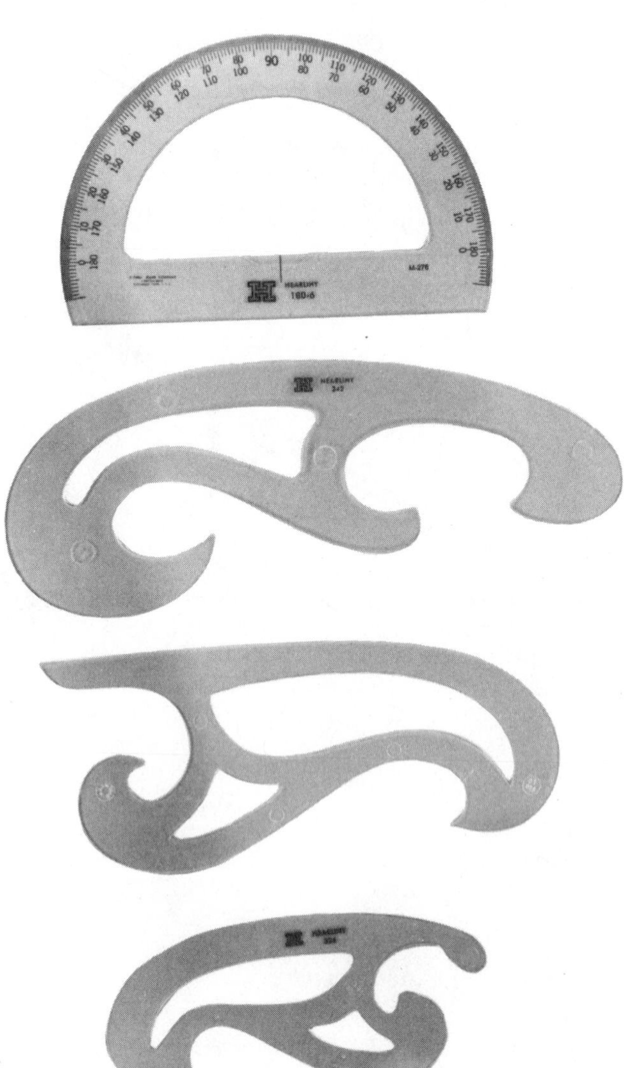

FIGURE 2.28 Protractor and Irregular Curves

FIGURE 2.29 45° Triangle

2.5.2 Triangles

The standard **triangles** are the 45° triangle, the 30/60° triangle, and the adjustable triangle. Triangles are used with a straightedge to draw a vertical lines. Since it is difficult to keep the vertical scale of a drafting machine 90° (perpendicular) to the horizontal scale, triangles are used with drafting machines to eliminate this problem.

The **45° triangle** (Fig. 2.29) is used to draw lines at an angle of 45° with the baseline. The **30/60° triangle** (Fig. 2.30) is used to create lines at 30° or 60° with the baseline. Together the two triangles create angles of 15° and 75°.

The **adjustable triangle** (Fig. 2.31) is the same as a 45° triangle when it is closed, but it can be opened to form two parallel edges. The amount of opening of the triangle is measured by a protractor scale that reads from 0 to 45° and then doubles back from 45 to 90°. Angles to 45° angles are formed by the two edges at the "open" corner of the triangle. Angles 45 to 90° are formed by the sides at the "hinge" corner. By rotating the adjustable triangle into position, you can draw all angles.

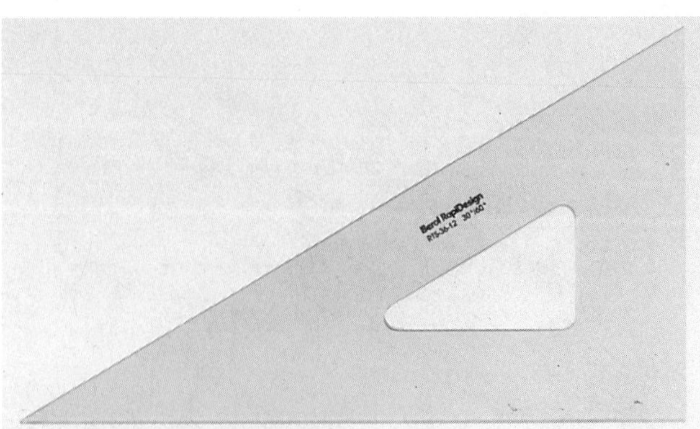

FIGURE 2.30 30/60° Triangle

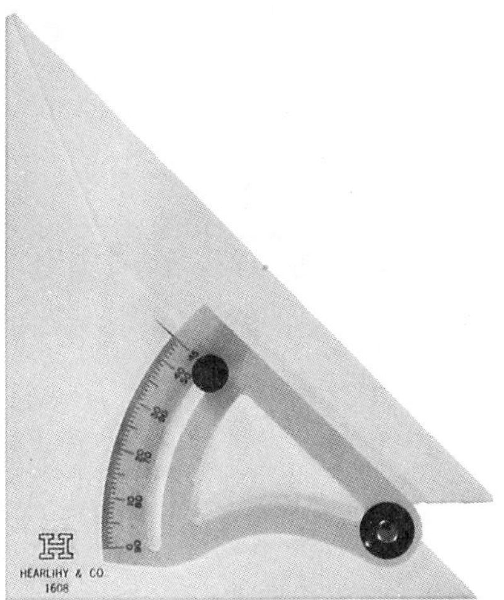

FIGURE 2.31 Adjustable Triangle

2.5.3 Templates

A **template** is a time-saving tool for drafting shapes of all sizes. Standard templates (Fig. 2.32) are essential for the quick, easy construction of circular, square, rectangular, triangular, elliptical, and symbolic shapes. Templates are better than a compass for small-diameter circles. Circle templates are available in all standard sizes for U.S. and metric units.

2.6 Instruments

Instruments include all forms of *compasses* and *dividers*. The drafting instrument set (Fig. 2.33) consists of one or two sizes of compasses, a divider, and accessories. The compass is used, to draw circles and circular arcs. Although various drafting sets are available, they normally contain such obsolete items as the ruling pen (replaced by the technical pen). You need only purchase a medium-sized, high-quality bow compass and a medium-sized dividers.

2.6.1 Bow Compass and Dividers

A good **bow compass** and **dividers** are essential to make accurate constructions in drafting. A bow compass (Fig. 2.34) has a center thumb wheel that is used to set and hold the spacing between the center point and the lead. Dividers (Fig. 2.35) do not have a center wheel and are used to transfer measurements quickly from one view to another. They are extremely useful in construction of mechanical drawings and for descriptive geometry.

The center point for the compass is either a tapered point or a short needle point projecting from a wider shaft that creates a "shoulder." The shoulder acts as a limit to the point's penetration into the paper and board. To restrict the compass point from penetrating the drawing paper and to provide a stable secure centering point from which to swing an arc or a circle, place a small piece of drafting tape (or dot) on the

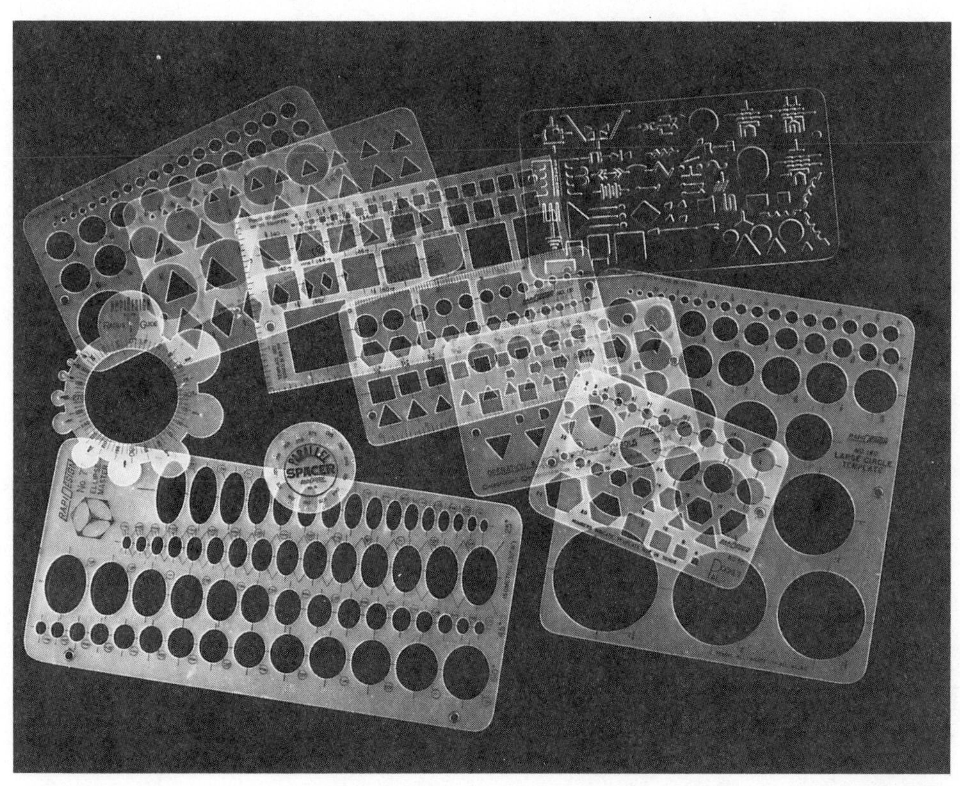

FIGURE 2.32 Templates

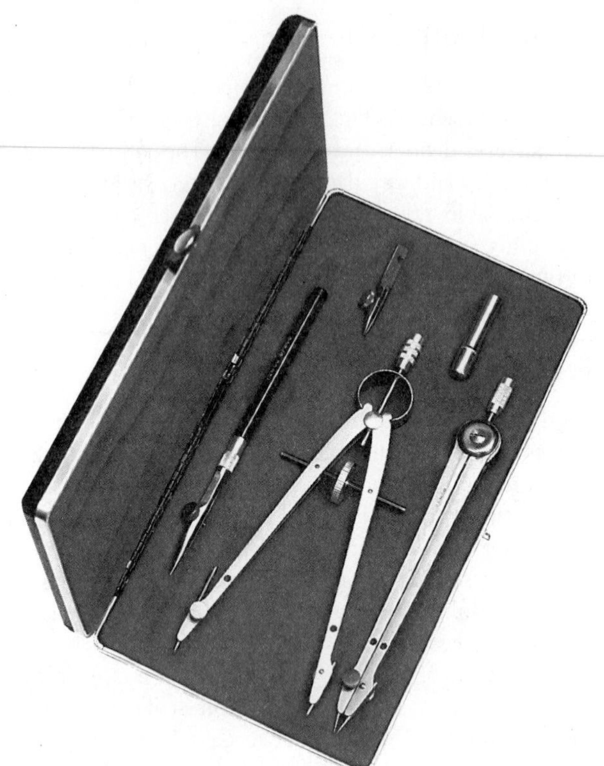

FIGURE 2.33 Drawing Instruments

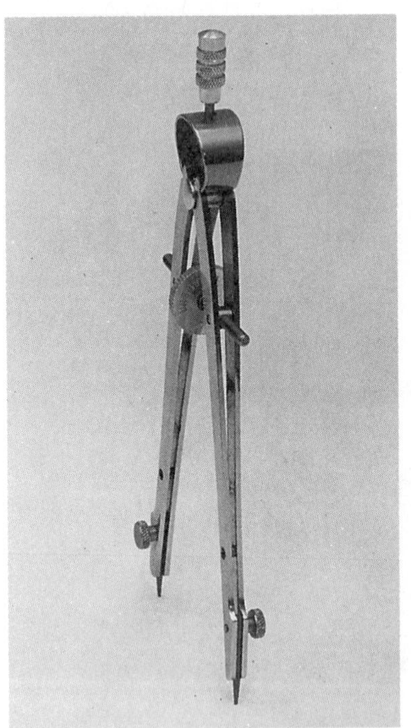

FIGURE 2.34 Bow Compass

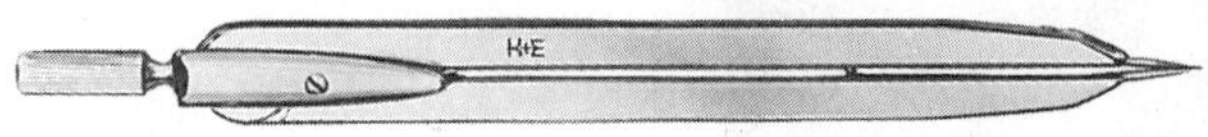

FIGURE 2.35 Dividers

drawing at the center of the arc or circle to be drawn. Circles smaller than 0.50 in. (12 mm) are much easier to draw with a template or a drop bow compass. A compass is best for odd-sized and large circles and for constructions.

The compass lead should be a piece of the drafting pencil lead (same grade lead or softer). Thus, both straight and curved lines will be drawn with the same lead hardness and it will be easier to maintain uniformity. The lead is secured in the compass with about $\frac{3}{8}$ in. (9 mm) exposed and is sharpened with a sandpaper block (Fig. 2.36). Make a flat cut that leaves an oval surface (bevel). The bevel should be about three times as long as the diameter of the lead. The resulting point is chisel-shaped and should have about the same taper, when viewed from the side, as the drafting pencil's. Do not adjust the lead in the compass after it is sharpened because it is almost impossible to reposition the chisel shape properly. Adjust the centering point so that the midpoint of the needle point is even with the end of the lead. The beveled end can be on either side, but most drafters place it on the outside (Fig. 2.37). To create a thin, dark curve, both sides of the lead may be beveled.

Dividers have two identical tapered metal points. Some drafters prefer to replace one metal point with a piece of 4H lead. The lead point can be used to set off dimensions instead of using the two metal points, which tend to mar the drafting medium.

2.6.2 Inking Instruments

Most compasses have ink-pen or technical-pen attachments. When you purchase a compass set, find one equipped with an attachment for holding a **technical pen** (Fig. 2.38).

Technical pens are available in a wide range of pen widths (diameters) (Fig. 2.39), each corresponding to a metric thickness. Inked projects include diagrams and pictorials for technical manuals, sales brochures, and graph and chart presentations.

2.7 Drafting Materials

Drafting materials include drawing media (vellum, film, grid sheets) and preprinted transfer items (title blocks, lettering, symbols).

2.7.1 Media

Traditional drafting media transparent enough to be white printed include vellum and drafting film. Drawing **vellum** is a high-quality, translucent paper. Paper used for the diazo reproduction process must allow light to shine through (be translucent). In addition, pencil on vellum is easy to erase; vellum also takes ink well. Vellum and drafting film are available plain and with fine, blue nonreproductive grids. **Drafting film** is a durable, high-quality, polyester. This drafting medium is excellent for ink and for plastic and combination leads.

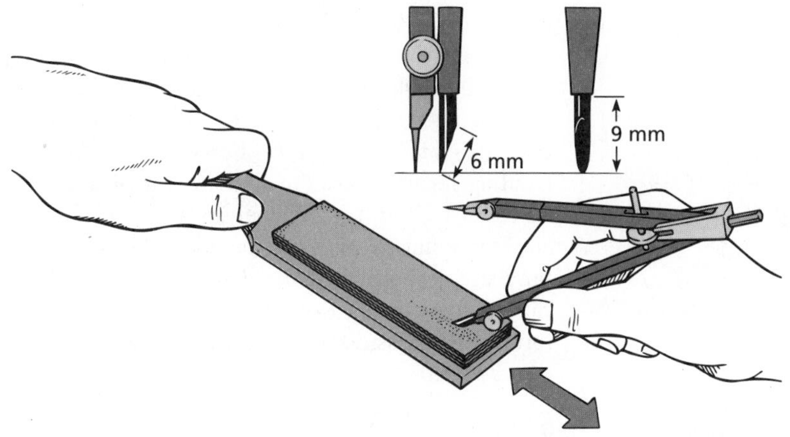

FIGURE 2.36 Sharpening the Lead of a Compass

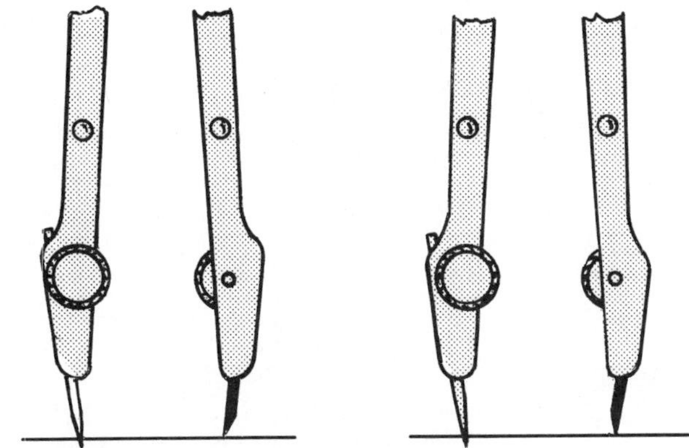

FIGURE 2.37 Positioning Compass Leads

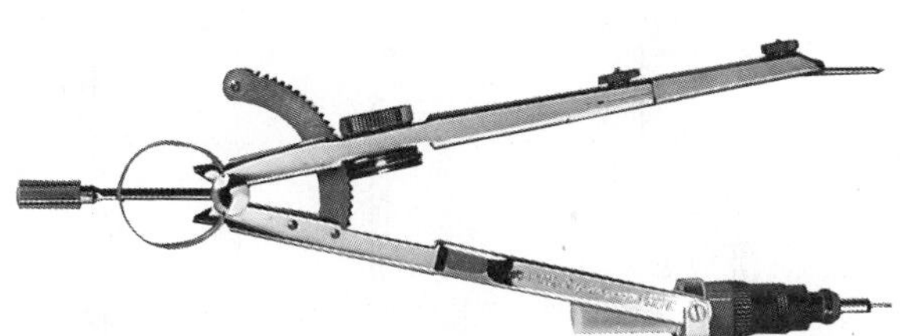

FIGURE 2.38 Bow Compass with Technical Pen Attachment

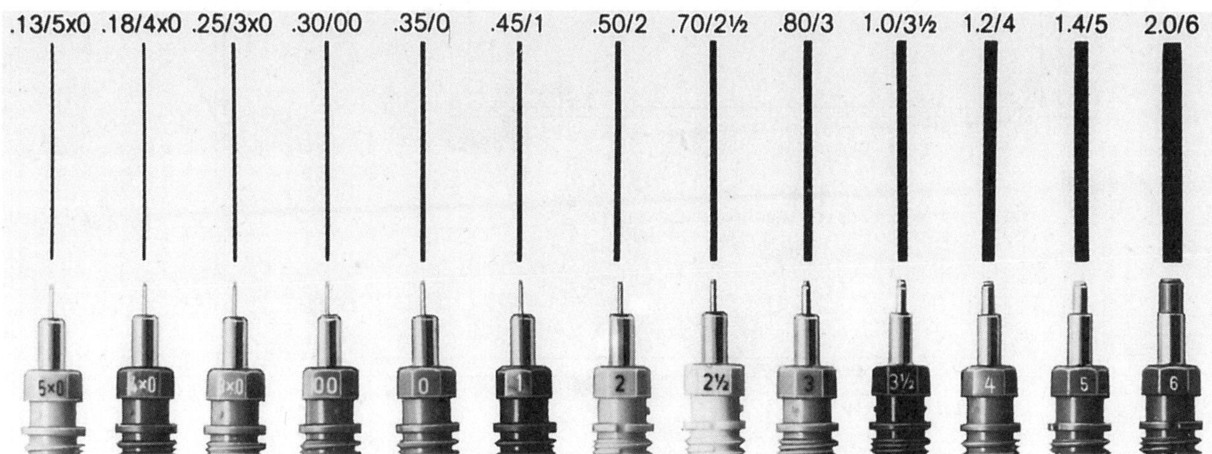

FIGURE 2.39 Pen Sizes

TABLE 2.3 Drawing Sheet Sizes

American National Standard Y14.1, in.	International Standard, mm
A–$8\frac{1}{2}$ × 11	A4–210 × 297
B–11 × 17	A3–297 × 420
C–17 × 22	A2–470 × 594
D–22 × 34	A1–594 × 841
E–34 × 44	A0–841 × 1189
F–28 × 40	

Drafting media are secured to the drafting table with drafting tape. **Drafting tape** (or drafting "dots") is a high-quality version of masking tape that is designed not to pull the finish from the paper surface.

2.7.2 Drawing Sheet Size and Format

Drafting media come in standard sheet sizes and rolls. Table 2.3 compares International Standards Organization (ISO) and ANSI drawing sizes. Rolls of drafting paper and film areavailable in widths of 30, 36, 42, and 54 inches and in lengths of 25 feet or more. International standards establish a series of paper sizes based on width to length proportions. Figure 2.40 illustrates the various ANSI sheet sizes. The margins shown in Figure 2.40 produce net drawing areas that are well within the sheet sizes of both standards. Drafting formats made to this standard can be reproduced on either U.S. or International sheet sizes by contact printing and microfilm projection methods. Most U.S. companies purchase preprinted standard sheets in ISO or ANSI specification. Figure 2.41 compares the ISO and ANSI drawing formats. Sheet size and format are covered in ANSI Standard Y14.1–1980.

2.7.3 Drawing Formats

The size and style of lettering on **drawing formats** is to be in accordance with ANSI Y14.2M. To provide contrasting divisions between major elements, follow this guide:

Thick lines— 0.7 to 0.9 mm (approximately .03 in.):

1. border line
2. outline of principal blocks
3. main division of blocks

Medium lines— 0.45 to 0.5 mm (approximately .02 in.):

1. minor divisions of the title block

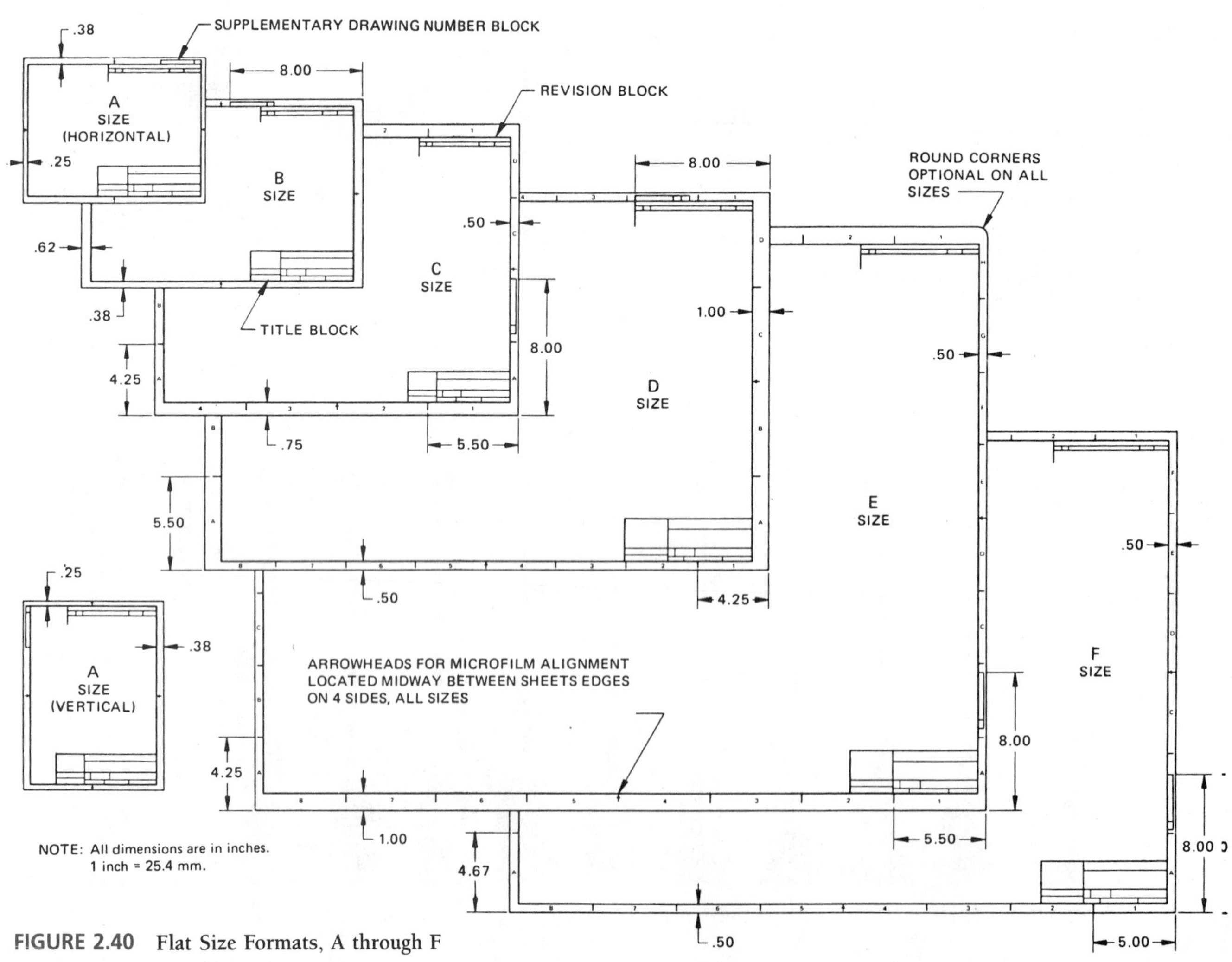

FIGURE 2.40 Flat Size Formats, A through F

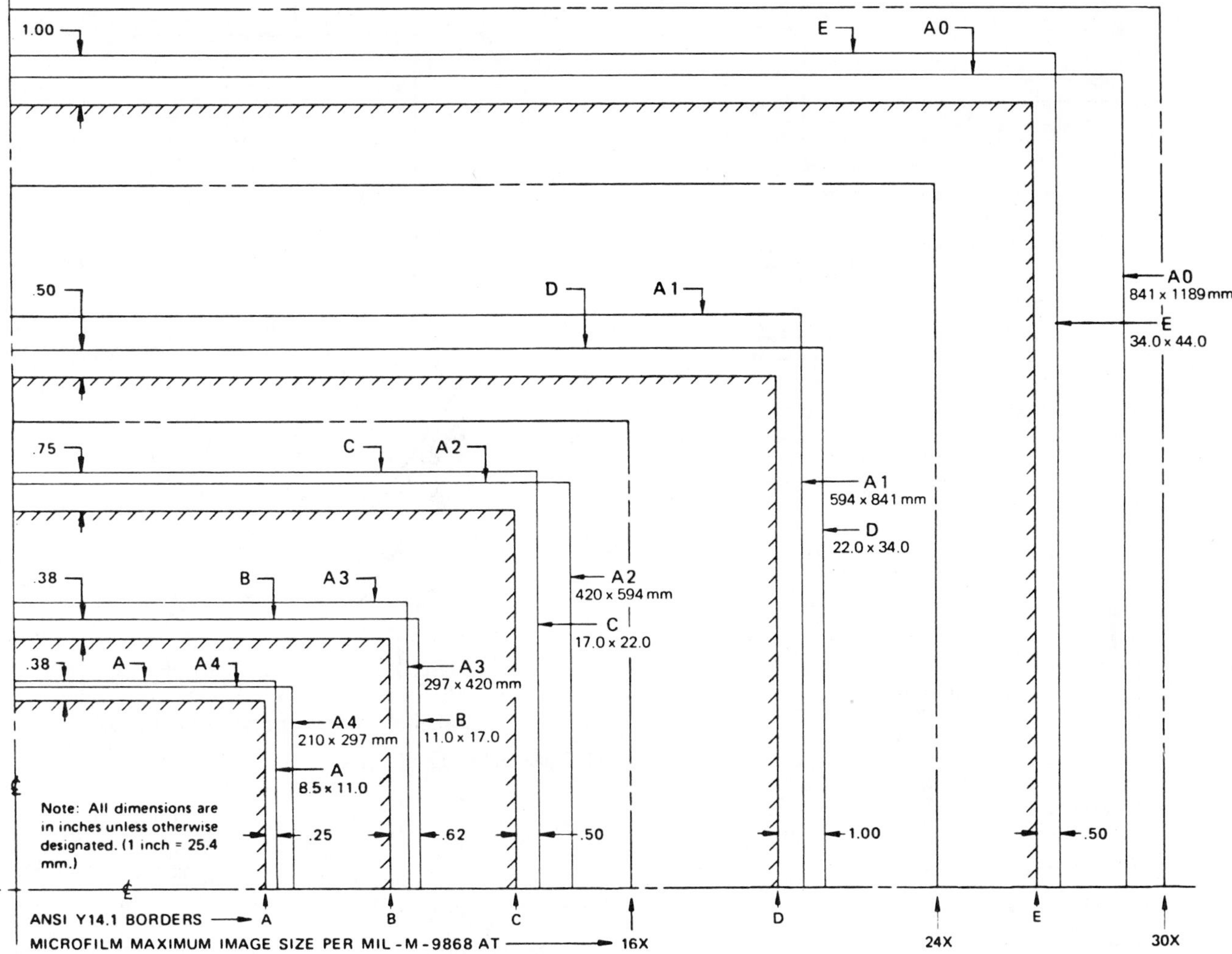

FIGURE 2.41 Comparison of ANSI-Y14.1 Drawing Format Sizes with ISO Paper Sizes

2.7.4 Title Blocks

The **title block** is a very important part of the drawing. It is located in the lower right corner of the format. The title block includes spaces for the following information:

- Company/school name
- Project title/part name
- Scale
- Drawn by
- Material specification
- Date
- Checked by
- Sheet number
- Drawing number
- Standard company tolerances (sheet tolerance)
- Approved by

Figure 2.42 shows the ANSI standard title block layouts for an A–K sheets.

2.8 Basic Drafting Techniques

Drawings are reproduced by processes such as whiteprinting, photocopying, or microfilming, so lettering and linework must be dark and of high quality. This section covers the basics of *linework*. It presents procedures and techniques to help you develop high-quality drawing skills.

2.8.1 Lines

The characteristics of all **lines** on a drawing are that they are black, clean-cut, and precise, with sufficient contrast in thickness. You must understand the process, the intent and the content of a drawing. An understanding of how lines function and what they mean is particularly important.

A straight line is the shortest distance between two points (Fig. 2.43) and is the type of line implied by the word "line." A line that bends is a *curve*. *Parallel lines* are equally spaced along their entire length. The symbol for parallel is //

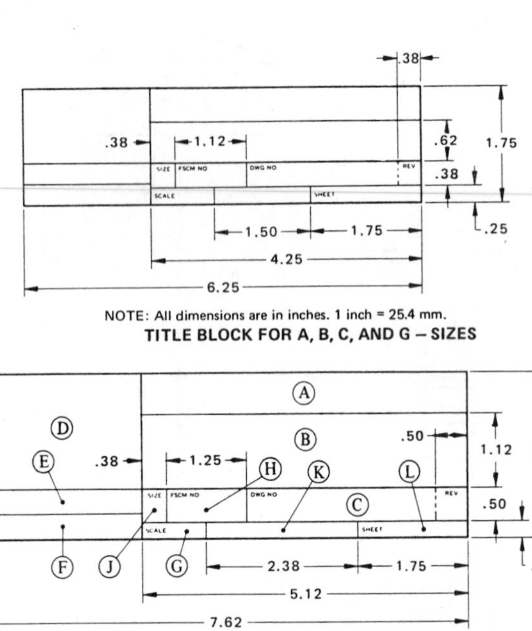

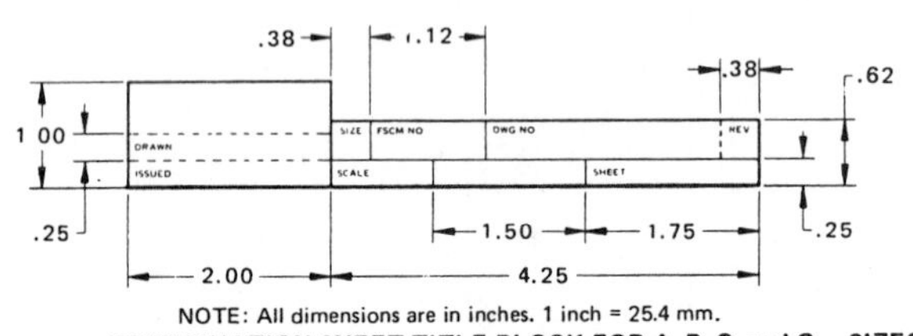

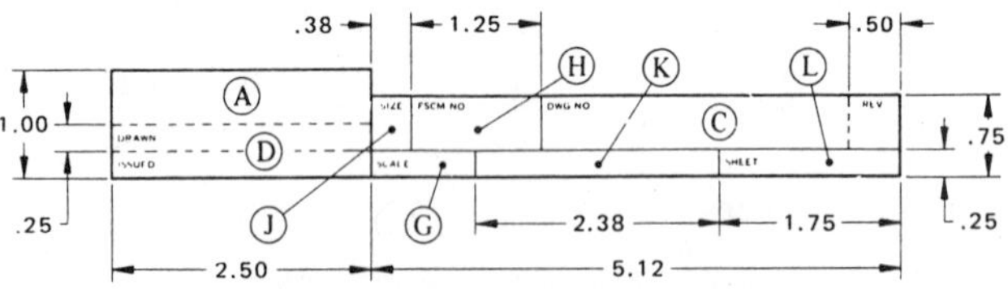

FIGURE 2.42 ANSI Title Blocks Dimensions

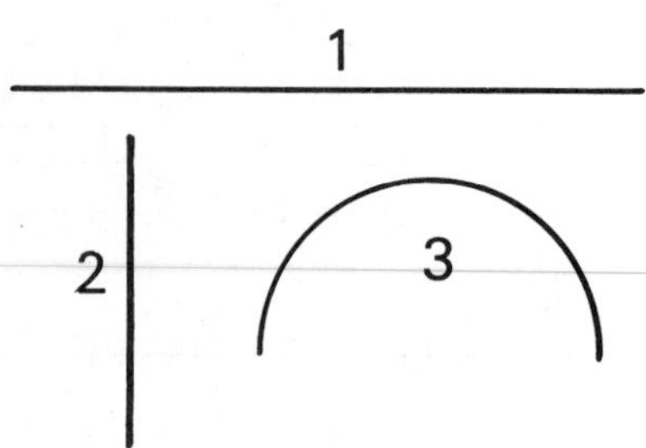

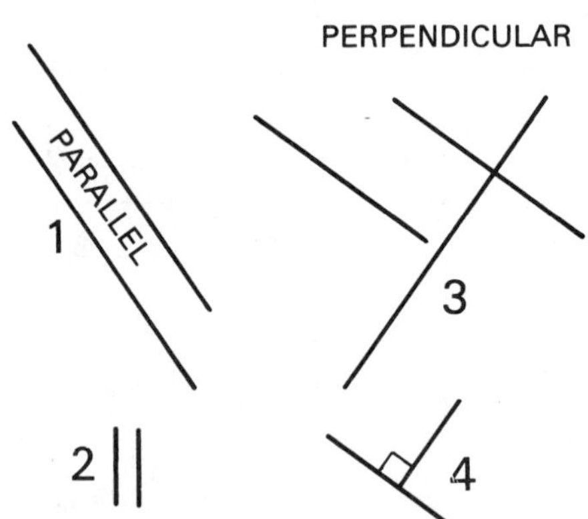

FIGURE 2.43 Line Types

[Fig. 2.43]. *Perpendicular lines* are at an angle of 90° to each other and can be intersecting or nonintersecting (Fig. 2.43). The symbol for perpendicular is ⊥.

Lines in engineering drawings are drawn with different widths to provide specific information. Each line type is actually a symbol that represents a function or idea, or communicates a special situation. The thickness of any line is determined by what it represents and the smallest size to which it will be reduced. To avoid confusion, lines representing the same function must be the same thickness throughout a single drawing. The minimum spacing between parallel lines is determined by how much the drawing will be reduced. Two parallel lines that are placed too close will merge when the drawing is reduced (*fill-in*). Usually 0.06 in. (1.5 mm) minimum parallel spacing meets reduction requirements.

The following list describes the traditional line thicknesses used on engineering drawings:

1. *Fine lines:* Thin, black lines used to provide information about the drawing or to construct the drawing. These include dimension lines, leader lines, extension lines, centerlines, and construction lines.
2. *Medium lines:* Medium-width, solid, black lines used to outline planes, lines, surfaces, and solid shapes. Medium lines are also used for hidden (dashed) lines.
3. *Heavy lines:* Solid, thick, black lines used for the border, cutting plane lines, and break lines. *The newest ANSI standard, however, suggests only two line thicknesses: thin and thick.* All lines listed under 1 and 2 from the preceding list are now drawn with thin lines and all lines listed under 3 are drawn with thick lines. Many companies still use three line thicknesses.

2.8.2 Precedence of Lines

Whenever lines coincide in a view, certain ones take precedence. Since the visible features of a part (object lines) are represented by thick solid lines, they take precedence over all other lines. If a centerline and cutting plane coincide, the more important one should take precedence (usually the cutting plane line). The following list gives the preferred *precedence of lines:*

1. Visible (object) lines
2. Hidden (dashed) lines
3. Cutting plane lines
4. Centerlines
5. Break lines
6. Dimension and extension lines
7. Section lines.

2.8.3 Line Types

Line types and conventions for mechanical drawings are covered in ANSI Standard Y14.2M (Fig. 2.44). Figure 2.45 shows an example drawing that contains each type of line. Every line on your drawing has a meaning.

Besides lines having different configurations such as dashes and spaces, each line type has a weight-thickness (Figs. 2.44 and 2.45). The following list gives the suggested weight (thickness) for lines:

Visible line 0.60 to 0.70 mm
Hidden line 0.45 to 0.50 mm
Section line 0.25 to 0.30 mm
Centerline 0.30 to 0.35 mm
Dimension line 0.30 to 0.35 mm
Cutting plane line 0.70 to 0.90 mm
Phantom line 0.45 to 0.50 mm
Border line 0.70 to 0.90 mm
Break line 0.45 to 0.70 mm

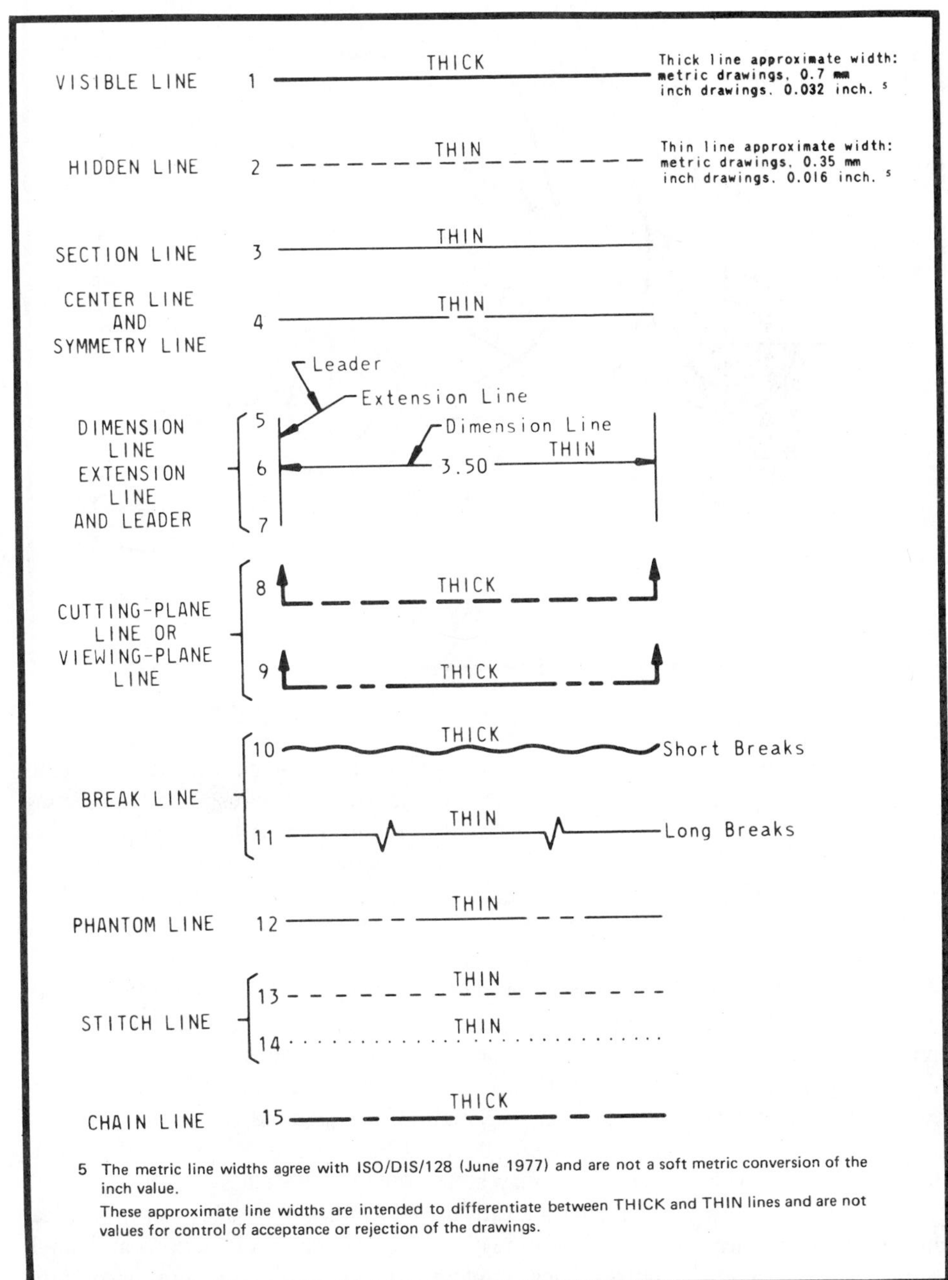

FIGURE 2.44 Standard Line Types and Thicknesses

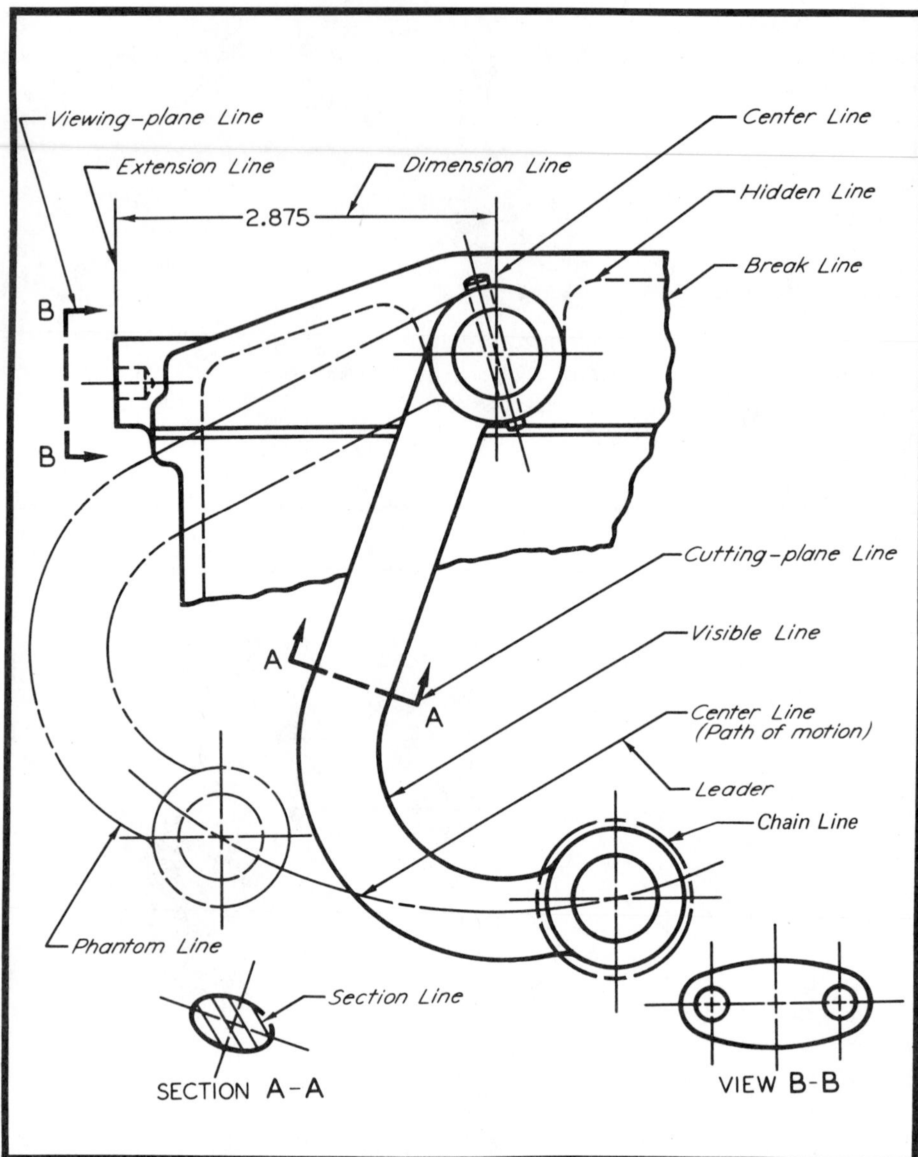

FIGURE 2.45 Application of Line Types

Visible Object Lines Visible object lines (Figs. 2.44 and 2.45) are thick lines that represent the visible edges and contours. They are the most important lines and must stand out from all other secondary lines on the drawing. In mechanical drawing, visible lines are about .032 in. thick (between 0.6 and 0.7 mm).

Hidden Lines Hidden lines (Figs. 2.44 and 2.45) are short, thin dashes, approximately .12 in. (3.0 mm) long, spaced about .03 to .06 in. (0.7 to 1.5 mm) apart. They show the hidden features of a part. Hidden lines always begin and end with a dash, except when a dash would form a continuation of a visible line.

Dashes always meet at corners: a hidden arc starts with dashes at the tangent points. When the arc is small, the length of the dash may be modified to maintain a uniform and neat appearance. Since many hidden lines are difficult to follow, only lines or features that add to the clearness and the conciseness of the drawing are shown. Confusing and conflicting hidden lines should be eliminated. If hidden lines do not adequately define a configuration, use a section view. Hidden lines are approximately .017 in. (0.45 to 0.50 mm) thick.

Centerlines Centerlines (Figs. 2.44 and 2.45) are thin, long and short dashes, alternately and evenly spaced, with long dashes placed at each end of the line. The long dash is dependent on the size of the drawing and normally varies in length from .75 to 2 in. (20 to 50 mm). Short dashes, depending on the length of the required centerline should be approximately .06 to .12 in. (1.5 to 3.0 mm). Very short centerlines may be unbroken with dashes at both ends.

Centerlines are used to indicate the axes of symmetrical parts of features, bolt circles, paths of motion, and pitch circles. They should extend about .12 in. (3 mm) beyond the outline of symmetry, unless they are used as extension lines for dimensioning. Every circle, and some arcs, should have two centerlines that intersect at their center of the short dashes. Centerlines are usually drawn about .012 in. (0.3 mm) thick.

Dimension Lines Dimension lines (Figs. 2.44 and 2.45) are thin lines that show the extent and the direction of dimensions. Space for a single line of numerals is provided by a break in the dimension line.

If possible, dimension lines are aligned and grouped for uniform appearance. Parallel dimension lines should be spaced not less than .25 in. (6 mm) apart, and no dimension line should be closer than .38 in. (10 mm) to the outline of a part feature [.50 in. (12 mm) is the preferred distance].

All dimension lines terminate with an arrowhead (mechanical), a slash, or a dot (architecture). Arrowheads are drawn with a ratio of 1:3. Dimension lines are drawn the same thickness as centerlines, .012 in. (0.3 mm).

Extension Lines Extension lines (Figs. 2.44 and 2.45) indicate the termination of a dimension. An extension line must not touch the feature from which it extends, but should start approximately .04 to .06 in. (2 mm) from the feature being dimensioned and extended the same amount beyond the last dimension line. When extension lines cross other extension lines, dimension lines, leader lines, or object lines, they are usually not broken. When extension lines cross dimension lines close to an arrowhead, breaking the extension line is recommended. Extension lines are drawn the same thickness as dimension lines and centerlines, .012 in. (0.3 mm).

A **leader line** (Figs. 2.44 and 2.45) is a continuous straight line that extends at an angle from a note, a dimension, or other reference to a feature. An arrowhead touches the feature at that end of the leader. At the end of the note, a horizontal bar .25 in. (6 mm) long terminates the leader approximately .12 in. (3 mm) away from midheight of the lettering. Leaders should not be bent in any way except to form the horizontal terminating bar at the note end of the leader. The angle of the leader should not be near vertical or horizontal.

Leaders or extension lines may cross an outline of a part or extension line if necessary, but they usually remain continuous and unbroken at the point of intersection. When a leader is directed to a circle or a circular arc, its direction should be radial. Leader lines are drawn the same thickness as centerlines, dimension lines, and extension lines, .12 in. (0.3 mm).

Section Lines Section lines (Figs. 2.44 and 2.45) are thin, uniformly spaced lines that indicate the exposed cut surfaces of a part in a section view. Spacing is approximately .10 in. (3 mm) and at an angle of 45°. Section lines are drawn slightly thinner than centerlines and dimension lines, .01 in. (0.25 mm).

Phantom Lines Phantom lines (Figs. 2.44 and 2.45) consist of medium-thin, long and short dashes and are used to indicate alternate positions of moving parts, adjacent positions of related parts, and repeated details. They also show the cast, or the rough shape, of a part before machining. The line starts and ends with the long dash .60 in. (15 mm) with about .06 in. (1.5 mm) space between the long and short dashes. A phantom line is drawn approximately as thick as a hidden line, .016 in. (0.45 mm). Phantom lines are similar to centerlines except they have two short dashes between each long dash. The short dashes are drawn approximately .12 in. (3 mm).

Cutting Plane Lines and Viewing Plane Lines Cutting plane lines and viewing plane lines (Figs. 2.44 and 2.45) consist of thick, long and short dashes and indicate the location of cutting planes for sectional views and the viewing positions for removed partial views. These lines start and stop with long dashes [.60 in. (15 mm) or longer]. The short dashes are approximately .25 in. (6 mm) long, with about .12 in. (3 mm) space between them. An alternative method uses medium-length [.38 in. (9 mm)] dashed lines for the total cutting plane. Both methods are acceptable. Cutting-plane lines are normally drawn with a thickness of about .032 in. (0.70 mm) and are the thickest lines on a drawing.

Break Lines Break lines (Figs. 2.44 and 2.45) are thick, freehand, continuous, ragged lines used to limit a broken view, a partial view, or a broken section. For long breaks, where space is limited, a neat break may be made with long, medium thickness, ruled dashes joined by freehand zigzags. For short breaks, the lines are drawn thicker, the same as cutting plane lines, .03 in. (0.7 mm). Long break lines are about as thick as hidden lines, .017 in. (0.45 to 0.50 mm).

Construction Lines Construction lines are used to lay out features and to locate dimensions. They are very thin, light gray lines, and a 6H–3H grade lead is used for construction lines. Construction lines that are drawn lightly enough do not need to be erased. When construction lines are drawn with blue nonreproducible lead, they are not erased.

2.8.4 Placing the Paper or Drafting Film on the Board

Drafting paper is placed on the board in a position that will allow you to properly use drawing tools and to be comfortable while drawing (approximately halfway up on the board and near the working edge of the board or centered).

If you are using a T-square, the working edge of the board is the side against which the head of the T-square rests. With the paper positioned properly, the head of the T-square will make full contact with the working edge of the board. The blade of the T-square is lightly flexible and "gives" as pressure is applied when drawing. Placing the paper near the working edge of the board gives minimum bending of the blade.

When using a parallel straightedge or a drafting machine, the bottom of the paper is aligned first and the corners are then taped.

Unless standard format preprinted sheets are available, the piece of drawing paper is always about 1 inch larger in width and height than the size of the final sheet. This excess paper is trimmed off when the sheet is complete. (A completed sheet is called a drawing.)

The paper is square with the board and taped down with small strips of drafting tape (.5 in. by 1 in. or use drafting dots). One-half of each strip of tape is attached to the paper first. Then, after the paper is pulled snug (but not stretched), the tape is pressed onto the drawing board. Hold the paper in position by laying the straightedge across it and holding the straightedge down with the left arm while taping with the right hand. After the two top corners are taped, release the straightedge. Then tape the bottom corners. The paper should cling tightly to the board without wrinkles, loose edges, or signs of stretching.

2.9 Instrument Drawings

Drawings with straight lines that have been drawn with the aid of a T-square, a straightedge, or a drafting machine and triangle are **instrument drawings.** Lines of each type are uniform in width and density. The lines begin and end so as to form square corners and intersections and are accurately constructed.

First find the center of the sheet by drawing diagonals to center the work on a layout drawing. Drafting a line in an instrument drawing is a two-step process that requires drawing two different kinds of lines. First, the position and the length of the line are determined, and then the line is drawn with correct width and density. The first line, for positioning, is drawn thin and light gray and is called a **construction line.** A second line is drawn exactly over the first line. This second line is dense and uniform and is called a **printable line.** The part or project being drawn is completely blocked-in (laid out) before darkening any lines.

Construction lines may extend beyond corners and intersections of the part. Construction lines are kept thin and gray. Printable lines are drawn using a 2H, H, or HB lead and with different widths.

Drafters use triangles and a straightedge or a drafting machine. Vertical lines are constructed with a straightedge and triangle or a drafting machine. Horizontal lines are drawn with a straightedge that will give consistent parallel lines. Uniform curved lines are drawn with a compass or a template. Only lettering is drawn freehand (with guidelines).

2.9.1 Techniques for Drawing Lines

A properly drawn line is uniform over its entire length. Using a wood pencil, a lead holder, or a fine-line pencil, you can make a line consistent:

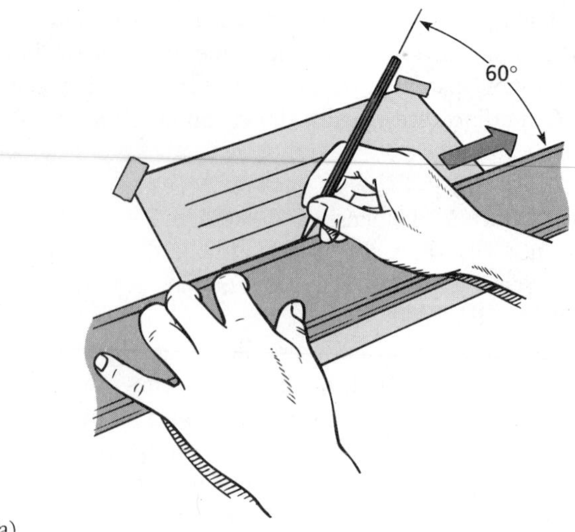

(a)

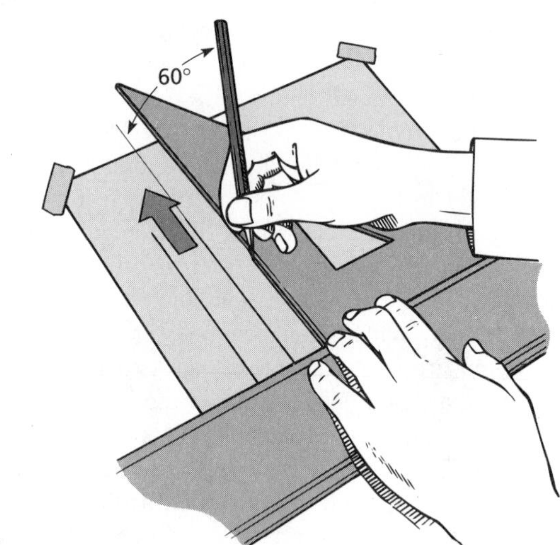

(b)

FIGURE 2.46 Angling the Pencil While Drawing

1. Incline the pencil or the lead holder so that it makes an angle of about 60° with the surface of the paper and then pull it in the direction in which it is leaning [Fig. 2.46(a) and (b)]. Keep the pencil at a consistent angle as you draw the line.
2. Rotate the pencil or the lead holder slowly as the line is drawn to maintain a semisharp conical point. This controls the thickness and quality of the line.

Fine-line pencils are held straight (vertical to the board) instead of at an angle and are not rotated.

How well lines print is determined by their density (ability to block light). Density is controlled by the hardness of the lead and by pressure. The width and the sharpness of the line are determined by the size of the point touching the paper. A sharpened pencil point should be smoothed and rounded on scratch paper after being repointed. Uniform lines require uniform point preparation.

Fine-line lead holders are available in different lead thicknesses. A 0.5- and a 0.7-mm lead holder with H or 2H leads are good for lettering and linework. Construction lines are drawn with 0.3 or 0.4 mm fine-line pencils with 3H or 6H leads or nonreproducible blue lead. They require no sharpening and maintain a high-quality, consistently uniform line.

Construction lines must be drawn with the greatest accuracy possible. Place the pencil point on the paper where the line is to be drawn. Then carefully move the straightedge or triangle up to it so as to just touch the pencil point. Draw a construction line with the pencil point riding along the top edge of the straightedge. Tilt the pencil slightly away from the straightedge. Always pull the pencil; do not push it, except when using a fine-line pencil.

2.9.2 Pencil Position for Printable Lines

Once the lines, corners, and intersections have been positioned with construction lines, the figure must be redrawn with printable lines. These lines are drawn exactly over the construction lines even though they will not extend the full length of the construction lines.

Let the pencil lead ride along the top edge of the straightedge by tilting the pencil slightly toward the straightedge. This will move the point slightly away from the straightedge so that both edges of the line are visible as it is being drawn. Since the construction line is completely visible ahead of the point, it is easy to see that it has been completely covered by the printable line. It is usually necessary to go over a printable line a couple of times in order to build up enough density to make sharp clear prints. Consistent line width is maintained by touching up the lead point as often as necessary.

2.9.3 Drawing Horizontal Lines

Horizontal lines are drawn with the T-square, the parallel bar, or the drafting machine. Place the pencil point at the desired position of the horizontal line and move the straightedge up to the point, just touching it. When the straightedge is positioned, hold it with the left hand and forearm. This will minimize the deflection of the blade (when using the T-square or a drafting machine) as the line is drawn. Draw the line from left to right (Fig. 2.47). Horizontal lines are always drawn along the top edge of the straightedge.

As any line is drawn, some graphite "chalks" off the point and lies as dust on the drawing. To avoid smearing this graphite dust, frequently brush the dust from the drawing. Graphite dust is the source of almost all "dirt" on drawings. Lift all drawing equipment from the board. If you drag the equipment or instruments across the drawing you will smear your linework. Keep the board, your hands, equipment, and instruments clean.

2.9.4 Drawing Vertical Lines

Vertical lines are drawn with the vertical edge of any triangle (or the vertical scale of the drafting machine). Position the straightedge and the triangle at the desired spot with the vertical edge of the triangle to the left (Fig. 2.48). Place the pencil point at the desired position of the vertical line and move the triangle up to the point, just touching it. Hold the straightedge with the left hand and forearm and position the triangle with the fingers of the left hand. Draw the line from bottom to top using the construction-line or printable-line technique.

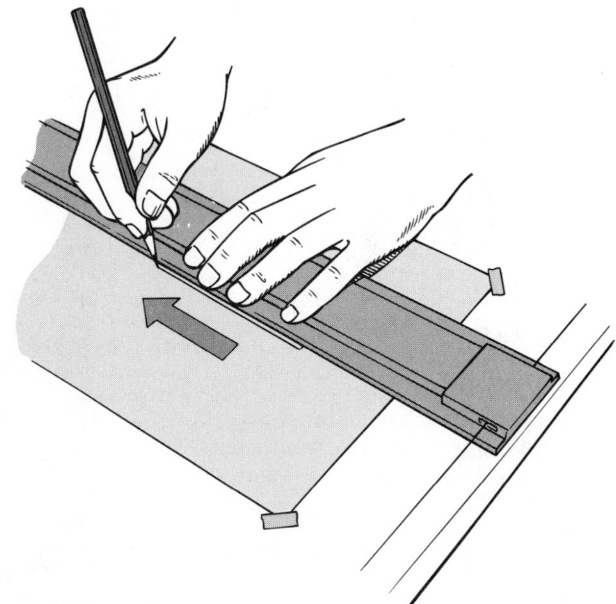

FIGURE 2.47 Drawing Horizontal Lines

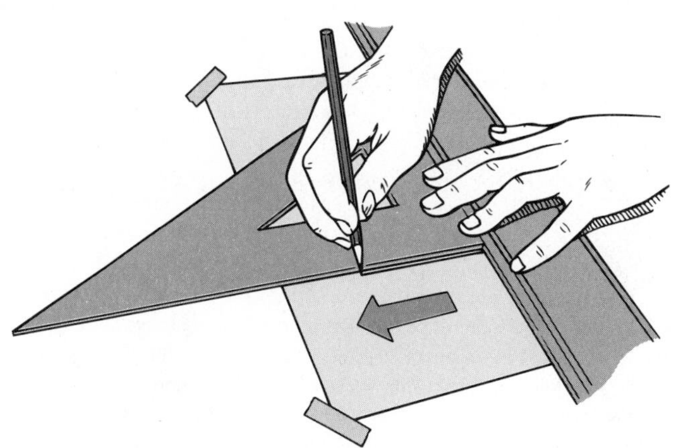

FIGURE 2.48 Drawing Vertical Lines

Vertical lines are usually drawn with an upward motion along the left edge of the triangle. This places the triangle between the hand and the paper and helps keep the drawing clean. When drawing a construction line change only the angle of the pencil. Your goal is a clean, accurate, and quickly constructed drawing.

2.9.5 Drawing Sloping Lines

Sloping lines are drawn much like vertical lines except that the sloping edge of the triangle, adjustable triangle, or drafting machine is used. Lines that slope toward the upper right

ITEMS OF INTEREST — *Reproduction Equipment*

You have probably heard the term *blueprint* to describe an engineering drawing. Blueprinting is a reproduction technique that was, for many years, the only way to duplicate engineering drawings. It is a photographic process in which the original drawing is the negative. The paper for the duplicate print is treated with chemicals that are sensitive to light. After the paper is exposed, it passes through a developing or fixing bath and is then rinsed and dried. The end product is a print (the same size) with blue background paper and white lines.

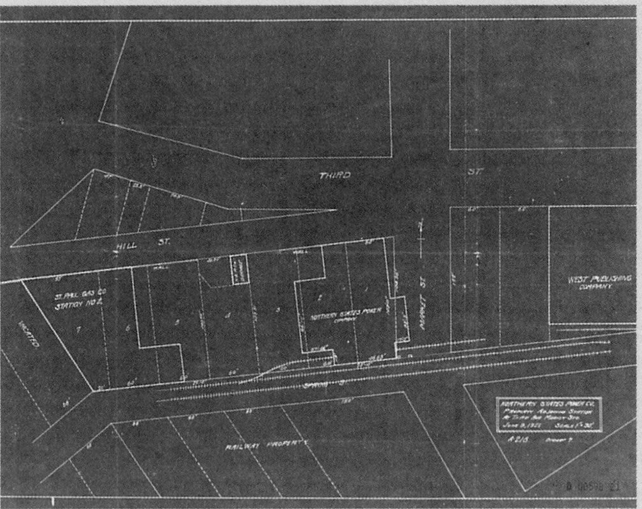

Early Reproduction of an Engineering Drawing (Blueprint)

As more and more engineering drawings were created, it became evident that a new reproduction process that was fast, exact, cost effective, and simple was needed. Considering the number of hours invested in engineering drawings, it seems reasonable that people would also invest many hours trying to create the best reproduction process for those valuable drawings.

The *diazo process* was the answer. This process produces a positive print with dark lines on a white background. Light is transmitted through the original onto chemically treated paper. Developing is completed by one of three processes: dry (utilizing an ammonia vapor), moist, (transferring an ammonia solution to the print), or pressure (a thin film of activator is deposited on the exposed paper). You can easily read marks and notations made directly on the print with this method. Unfortunately, the prints soil easily and the life of a print is relatively short.

The next evolution in reproduction equipment was developed from an idea that originated in 1937. In 1937, a young law student named Chester Carlson developed a method called *xerography* to make copies. (The word *xerography* comes from the Greek words for "dry" and "writing.") A copy made by this method became known as a Xerox. No doubt you have also heard of Xerox, the company that developed and marketed this process throughout the world. Of course, it was only a matter of time until reproduction equipment was made large enough for copying drawings. This process can produce a copy not only from an original, but also from a copy. Enlarged or reduced size copies are also possible.

Today, a copy of a computer-generated drawing can be produced with a laser printer/plotter. A laser plotter uses a laser to form areas of static charge to attract metallic powder to the paper. The process produces sharp, clear prints, is extremely fast, and is inexpensive enough to be used in small engineering offices. Laser printers have become an integral part of the engineering workplace.

When you walk into an engineering firm today, you could see a diazo print, a Xerox print, or a print produced with a laser plotter. Regardless of the method used to produce the print, many engineers, designers, and drafters ask for a "blueprint" of the latest product or assembly even thought blue paper with white lines hasn't been around for many years.

Whatever the next evolution is in reproduction equipment, it will probably be faster, more accurate, easier to use, and more economical—just like all the other versions—yet whatever process we use, we will probably call the print a "blueprint."

corner of the board are easy to draw. Sloping lines must first be accurately measured and laid out with construction lines. Check the angle with a protractor before darkening.

2.9.6 Drawing Curved Lines

Using a compass and an irregular curve to create dark, consistent linework requires practice. Circle, ellipse, and other curved templates are also available in standard sizes.

Arcs, circles, and other **curved lines** require special linework techniques. The compass lead is fixed in the compass and cannot be rotated (requires frequent repointing). Noncircular curves are drawn with a *French curve,* an *irregular curve* (Fig. 2.28), or a *flexible curve.* Curves must be drawn equal in width to the straight lines to produce a uniform drawing.

2.9.7 Using the Bow Compass

The compass lead in a bow compass should be a short piece of the *same lead* used in the drafting pencil. On a construction line drawn on scratch paper, measure a distance equal to the

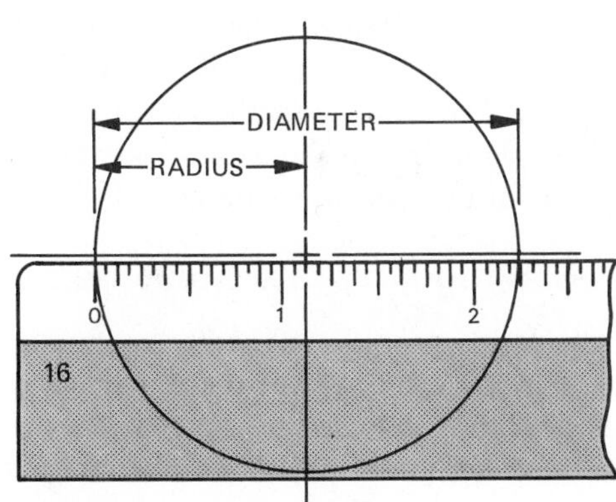

FIGURE 2.49 Measuring a Circle

radius of the circle or arc to be drawn. Set the compass to this distance and draw a construction circle. For an accurate diameter reading, take the measurement along a line that passes through the center point of the circle (Fig. 2.49). In Figure 2.50, the compass is being set by using the scale. This is not the easiest or most accurate method, but it is faster.

The width of the line drawn is determined by the thickness of the lead at the bevel. As a circle is drawn, the point shortens and the line widens. Therefore, a circle is started with a line somewhat narrower than desired. The line is redrawn until it is the correct width and density. A longer taper will hold a line width longer than a short stubby taper.

Figure 2.51 shows the proper method of constructing a circle with a **bow compass.** First, draw the circle with a thin dark line. Then thicken it by resetting the radius slightly and drawing another curve touching the first one. This method ensures crisp black lines with the appropriate thickness. Of course, this method is of no use for drawing a hidden (dashed) line. Dashed curves are drawn with a slightly dulled compass lead point and only one pass to complete the circle.

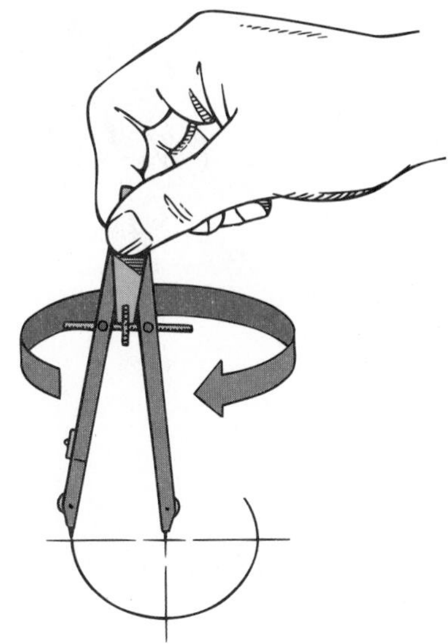

FIGURE 2.51 Using the Bow Compass

2.9.8 Using Dividers

Dividers are used to transfer dimensions and measurements. Dividers are not used to set off distances when accumulation errors could result. The scale is used to measure divisions in these cases.

Figure 2.52 provides an example of how to use dividers. Dividers are held, adjusted, and manipulated with one hand.

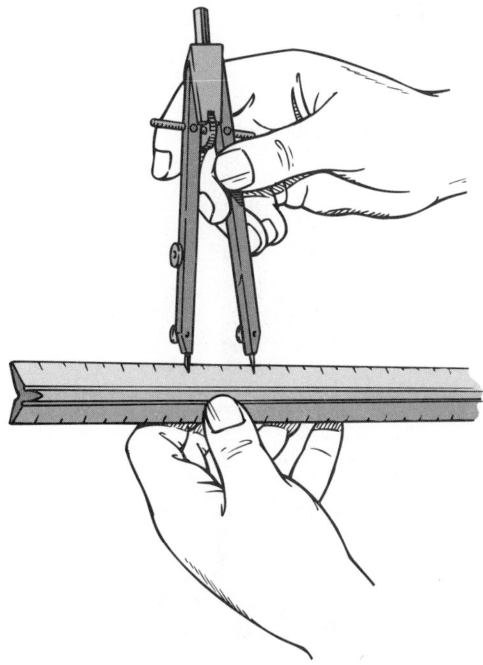

FIGURE 2.50 Setting the Compass Radius

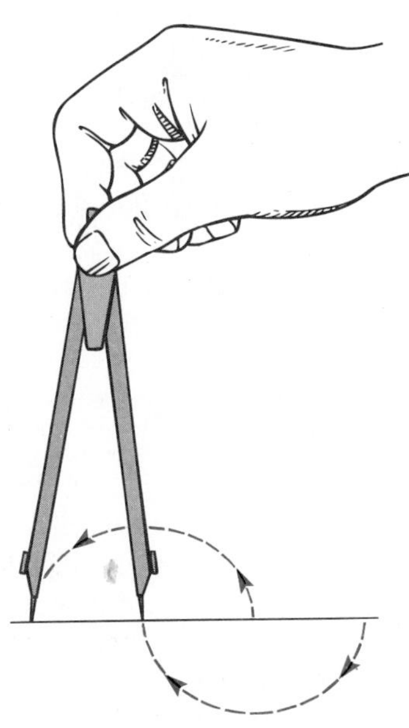

FIGURE 2.52 Using the Dividers

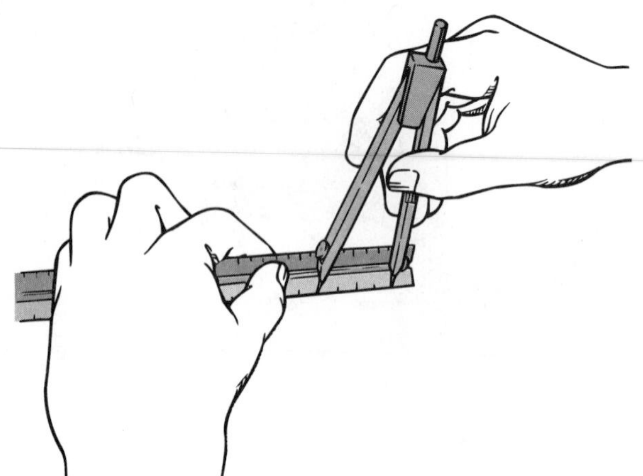

FIGURE 2.53 **Setting the Dividers with a Scale**

Measurements are taken from an existing view or from a scale (Fig. 2.53). Dividers help you construct quick and accurate drawings and are essential for solving descriptive geometry problems.

2.9.9 Using the Irregular Curve (French Curve)

Irregular curves (Fig. 2.28) are used to construct noncircular lines with smooth printable lines. Examples of such curves are the ellipse, the helix, and spirals. Irregular curves are manufactured in many shapes and size.

Curves that are drawn using the irregular curve are completed by plotting a series of points that lie on the curve. Then, a curve is drawn that includes all of these points. Figure 2.54 illustrates the use of the irregular curve. Good results can be obtained by following these steps:

1. Lightly sketch a smooth freehand line to include the plotted points. It is easier to set the irregular curve to a line than it is to match a series of points.
2. Position the irregular curve so that it matches a part of the line.
3. Draw the line that fits the curve, but stop a little before the end of the fit.
4. Reposition the irregular curve to fit the next part of the curve and draw the next portion of the line. Again, the last portion of fit is not drawn.
5. Repeat this process until the curve is complete.

If the sketched curve and the first series of matching the irregular curve to the sketched line are all done on a tracing paper overlay, then the result will be much neater. The ends of each segment of the line are marked as the line matches the irregular curve. Then, the same fits can be used in the next step. When all fits are made, the tracing paper overlay is placed under the drawing and carefully aligned with the curve under the plotted points. The curve is traced onto the drawing with the irregular curve marked on the overlay.

If a smooth curve is desired, plotting the points of an irregular curve is particularly important. A small error in the position of a point can easily cause irregularities in the curve. The spacing of plotted points should be close where the curve is sharpest and further apart where the curve is the straightest.

2.9.10 Making Accurate Measurements

Accurate drafting is possible only with accurate use of the scale to mark measurements. The thickness of the edge of the scale and the distance from the mark on the scale to the surface of the paper is a physical limitation of your scale's accuracy. The most accurate measurements are made by sighting along a line that is perpendicular to the paper.

All scale readings should be marked on the paper with a short, thin dash (Fig. 2.55). The straightedge is then positioned to use the measurement. Measurements put down as dots are often lost; incorrect lines have also been drawn from specks of dust.

If a number of measurements are placed end to end, all of them should be measured from the same points. If the measurements are placed by moving the scale for each measurement to the end of the previous one, cumulative errors result.

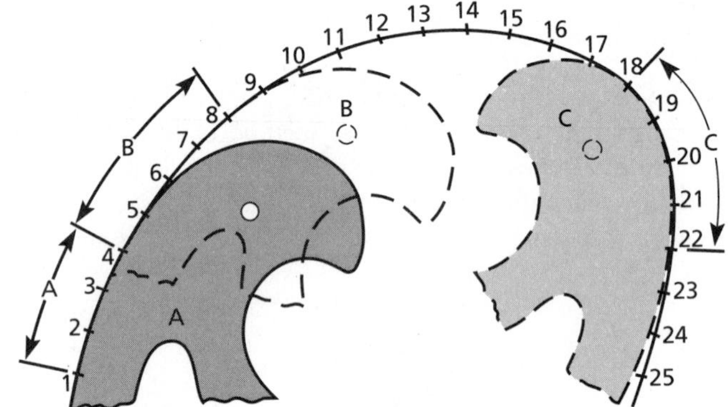

FIGURE 2.54 **Using the Irregular Curve**

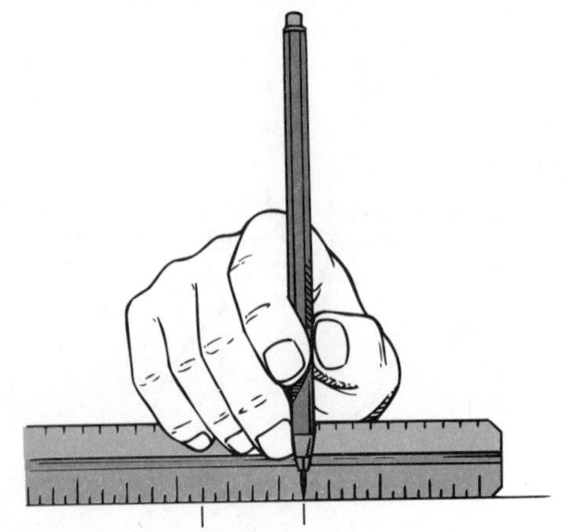

FIGURE 2.55 **Measurements with a Scale**

2.9.11 Keeping Your Drawings Clean

All drawings attract dirt. Cleanliness does not just happen; it is the result of developing correct habits, procedures and techniques:

1. *Clean hands:* Periodically wash your hands to remove accumulations of graphite, perspiration, body oils, and dirt.
2. *Equipment:* Periodically wash with soap and water all tools that touch the paper. Tools that contain wood or metal should be cleaned with a damp sponge. When they become soiled, they must also be scrubbed. The drawing surface should be cleaned regularly.
3. *Graphite:* Most dirt on a drawing is actually graphite. Repeatedly and consistently use the drafting brush and the dust pad to remove graphite dust, before other tools smear it around.
4. *Pencil pointer:* The pencil pointer leaves dust clinging to the lead. Some pointers also push shavings up into the jaws of the pencil. If the dust and the shavings are not removed before drawing starts, they will drop onto the drawing. Thus, after each sharpening, lightly tap the pencil on the side of the desk to dislodge any shavings and then wipe the lead on a piece of tissue. Poking the lead point into a piece of Styrofoam also works well.
5. *Equipment use:* Proper use of the straightedge and triangle always places these instruments between your hands and the paper. Even clean hands put body oils onto the paper; this has a magnetic effect on dirt. When lettering, place a sheet of clean paper under your hands to keep the drawing clean.

2.9.12 Inking Drawings

Ink is frequently used on drafting film or vellum. Drawings used in product literature, technical manuals, and pictorial illustrations are inked to obtain good photographic quality.

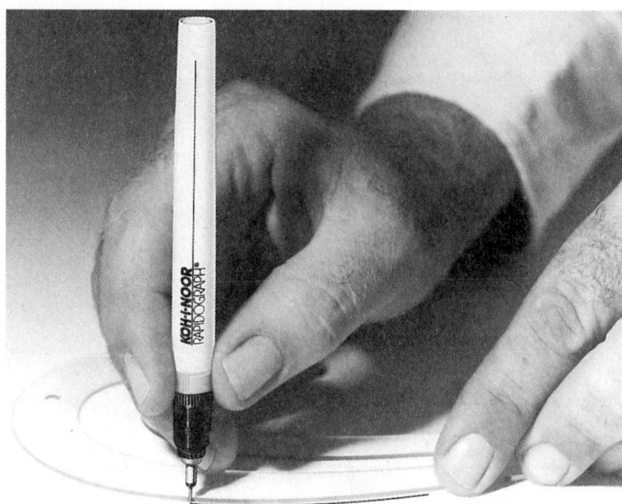

FIGURE 2.56 Inking with a Technical Pen and an Irregular Curve

Ink drawings are first laid out with construction lines and then inked. Light tables are excellent for inking and tracing drawings. Because ink tends to flow between surfaces and to smear, triangles and templates must be raised from the drawing during inking. Specially designed equipment with a ledge or with inking risers prevents the equipment from being flush with the paper.

Ink drawings are prepared with technical pens. Keeping the technical pen almost vertical helps prevent uneven and ragged linework (Fig. 2.56). No more than one pass should be made for thin and medium lines. Extremely thick lines are drawn with an appropriate pen. The ink should be completely dry before you start another portion of the drawing.

You May Complete Exercises 2.1 Through 2.12 at This Time

QUIZ

True or False

1. Plastic leads are used on vellum.
2. 5H and 6H leads are used to darken the final drawing.
3. Construction lines drawn with blue nonreproducible lead do not require erasing before the drawing is darkened.
4. The title block is always placed in the lower left-hand corner of the sheet.
5. Hidden lines always take precedence over centerlines.
6. Object lines are thin, black, and approximately 0.35 mm long.
7. Break lines are normally drawn freehand.
8. A dry cleaning pad is used to remove graphite from a newly sharpened pencil.

Fill in the Blanks

9. A sandpaper pad is used to ________ .
10. Dry cleaning pads are used to ________ and ________ a drawing.
11. An architect's scale is ________ divided.
12. ________ and ________ curves are used to draw odd-sized circular curves and arcs.
13. Technical pens should be held ________ .
14. A mechanical engineer's scale is ________ divided.
15. Always draw on the ________ side of the straightedge.
16. Incline lead holders at ________ degrees to the drafting board when drawing.

Answer the Following

17. Describe the process of drawing a vertical instrument line.
18. Describe three ways to keep your drawing clean.
19. Explain how to sharpen and prepare a wooden pencil for drawing an instrument line.
20. Describe the process of drawing with an irregular curve.
21. Describe the two primary types of drawing media used in drafting.
22. What does *precedence of lines* mean?
23. What line widths are used on a drafting format and title block?
24. Name five types of information included in a title block.

EXERCISES

Transfer the given information to an "A" size sheet of .25 in. grid paper. Complete all views and solve for proper visibility, including centerlines, object lines, and hidden lines. Exercises that are not assigned by the instructor can be sketched in the text to provide practice for the preceding instructional material.

After Completing the Chapter You May Draw the Assigned Exercises

Exercise 2.1 Draw the given design as shown.

Exercise 2.2 Draw the cover plate as shown.

Exercise 2.3 Draw the gage plate as shown.

Exercise 2.4 Draw the design as shown.

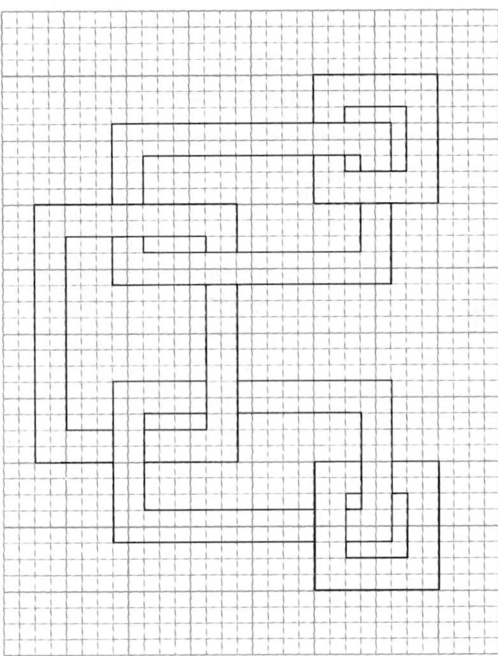

Exercise 2.1

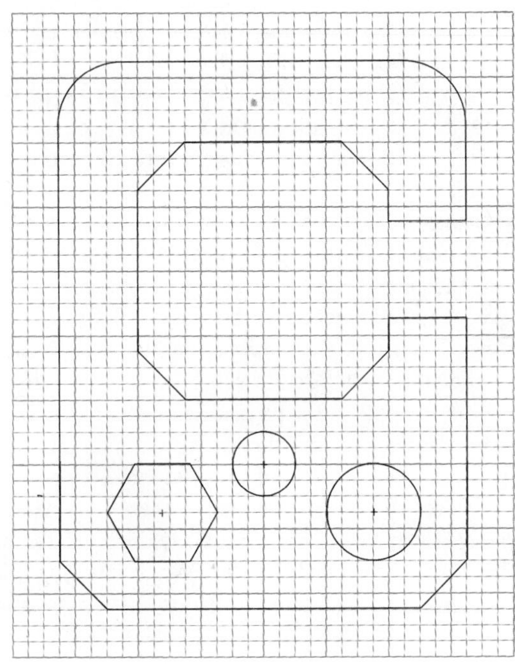

Exercise 2.3

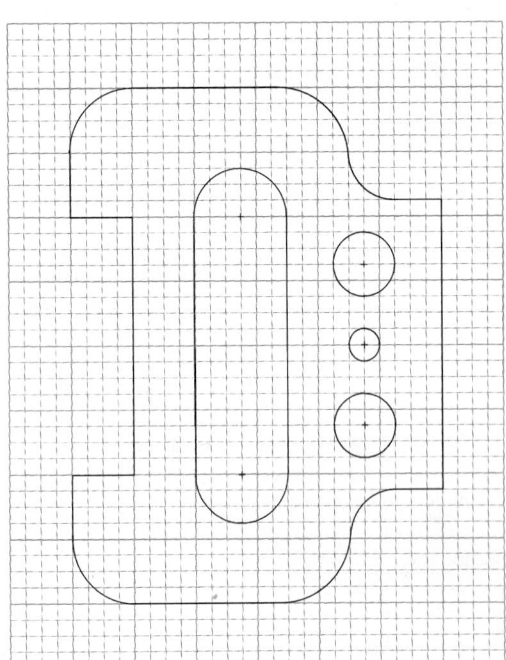

Exercise 2.2

Exercise 2.4

Exercise 2.5 Draw the two gaskets as shown.

Exercise 2.6 Draw the two cover plates as shown.

Exercise 2.7 Draw the complete cone check and guide.

Exercise 2.8 Draw the two guides.

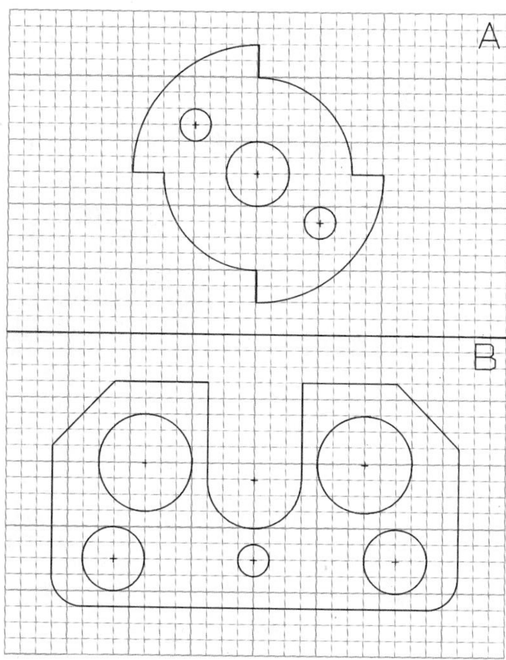

Exercise 2.5

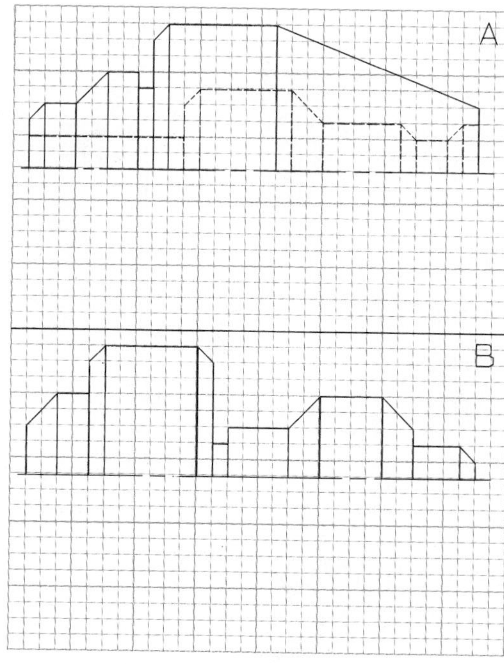

Exercise 2.7

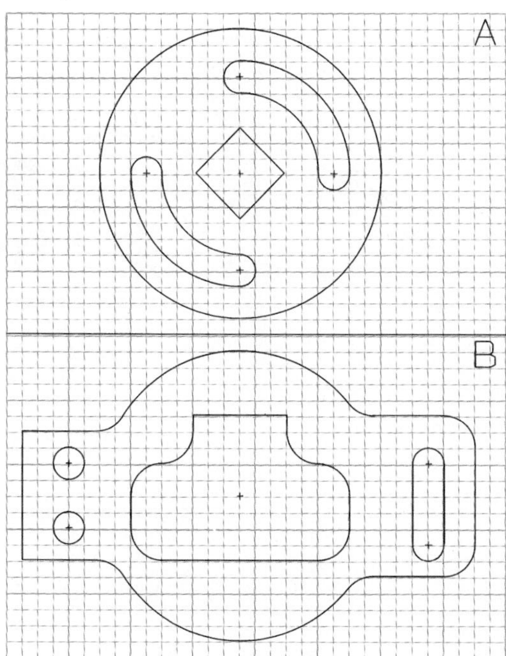

Exercise 2.6

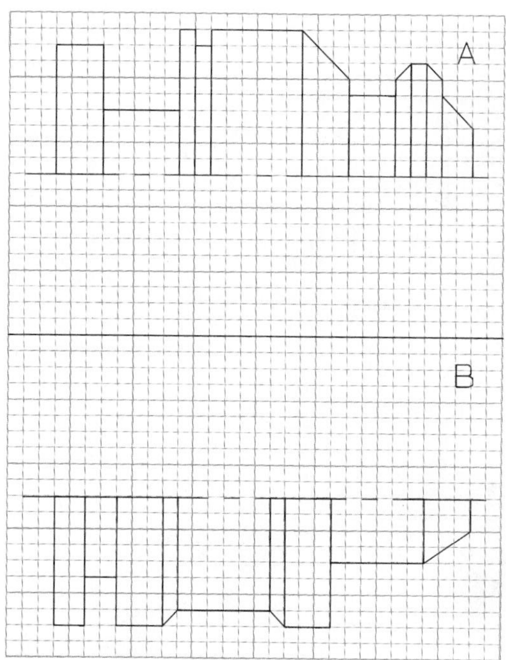

Exercise 2.8

Exercise 2.9 Draw the control plate as shown.

Exercise 2.10 Draw the disk guide as shown.

Exercise 2.11 Draw the mount surface as shown.

Exercise 2.12 Draw the tube gasket as shown.

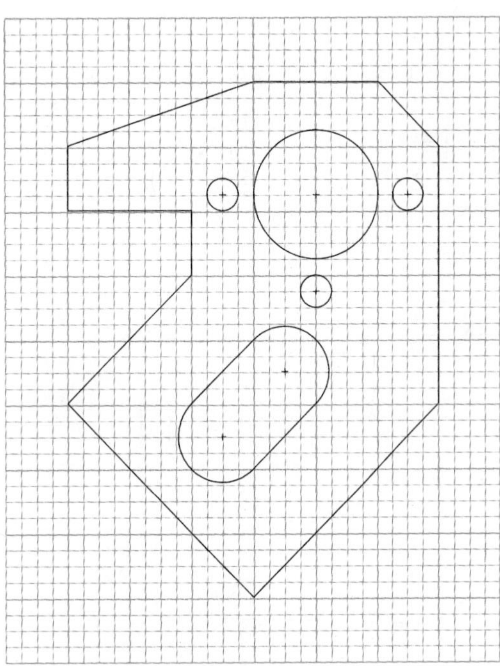

Exercise 2.9

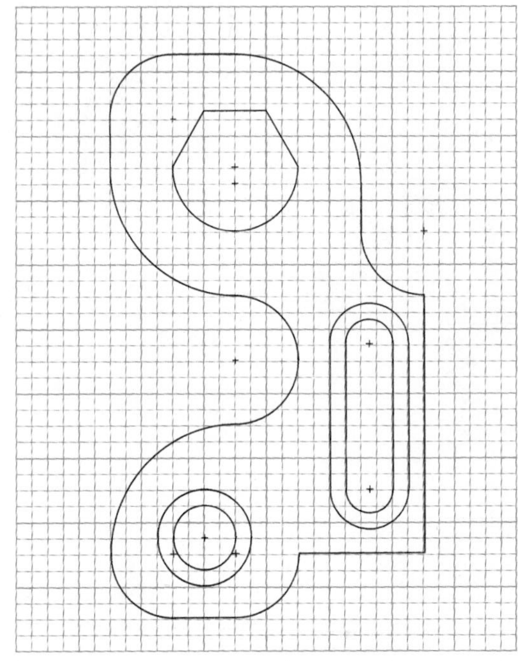

Exercise 2.11

Exercise 2.10

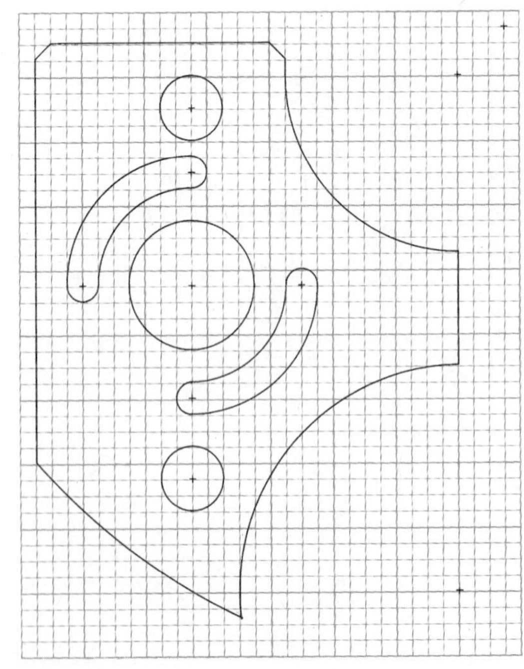

Exercise 2.12

PROBLEMS

Problems 2.1(A) through (K) Draw each problem assigned by the instructor on an "A" size sheet, one drawing per sheet. Establish measurements by using one of the three scales provided. Your instructor may request enlarged or reduced drawings as needed.

Problem 2.2 An "A" size drawing format is called for in this project. Redraw the object shown as Problem 2.2. Do not dimension.

Problem 2.3 Using a "B" size sheet, redraw the part shown as Problem 2.3. This is an ISO standard drawing using metric dimensions (millimeters). Do not dimension.

Problem 2.4 Using a "C" size sheet, draw the part shown as Problem 2.4. Dimension only if assigned by the instructor.

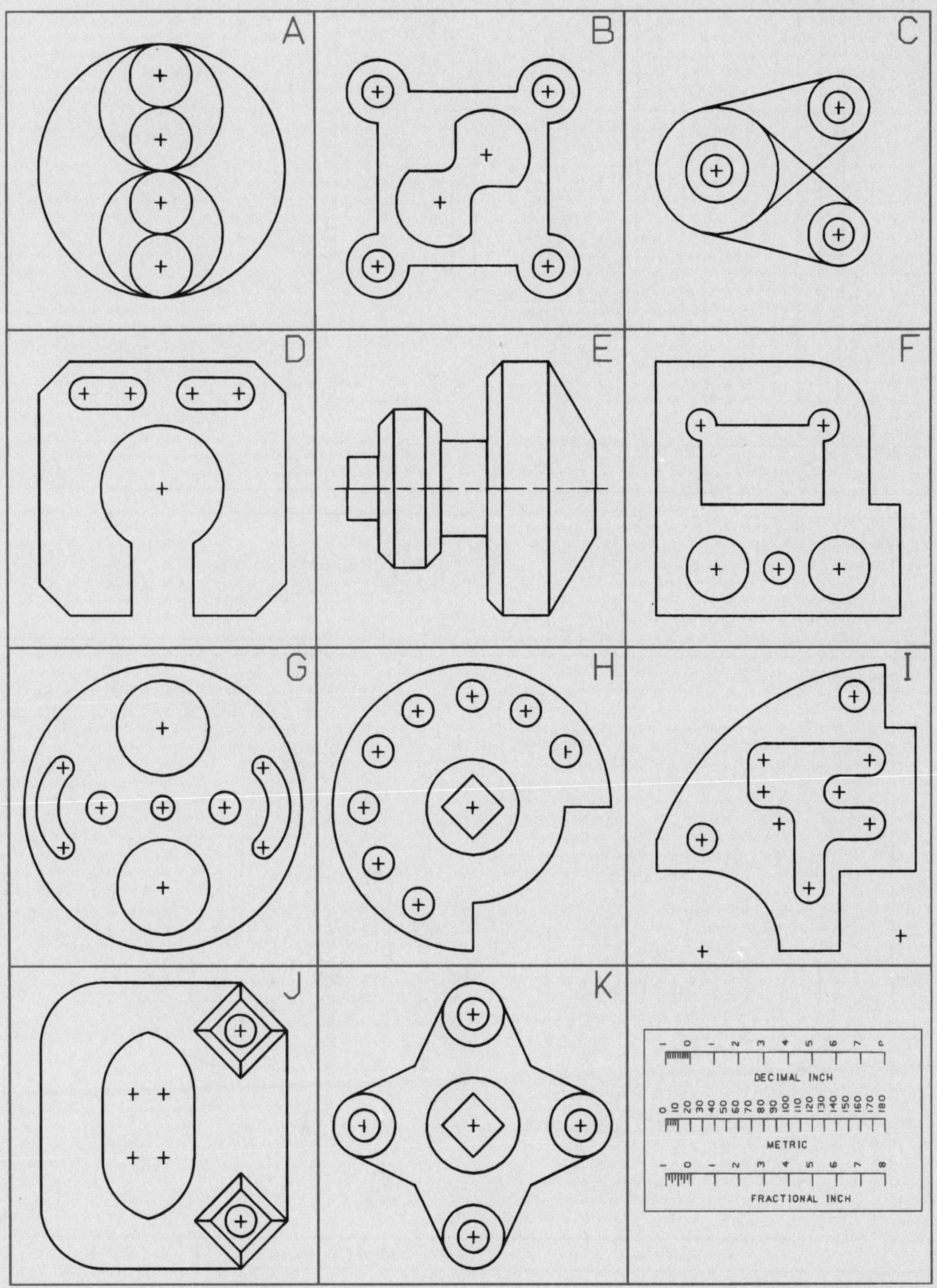

Problem 2.1

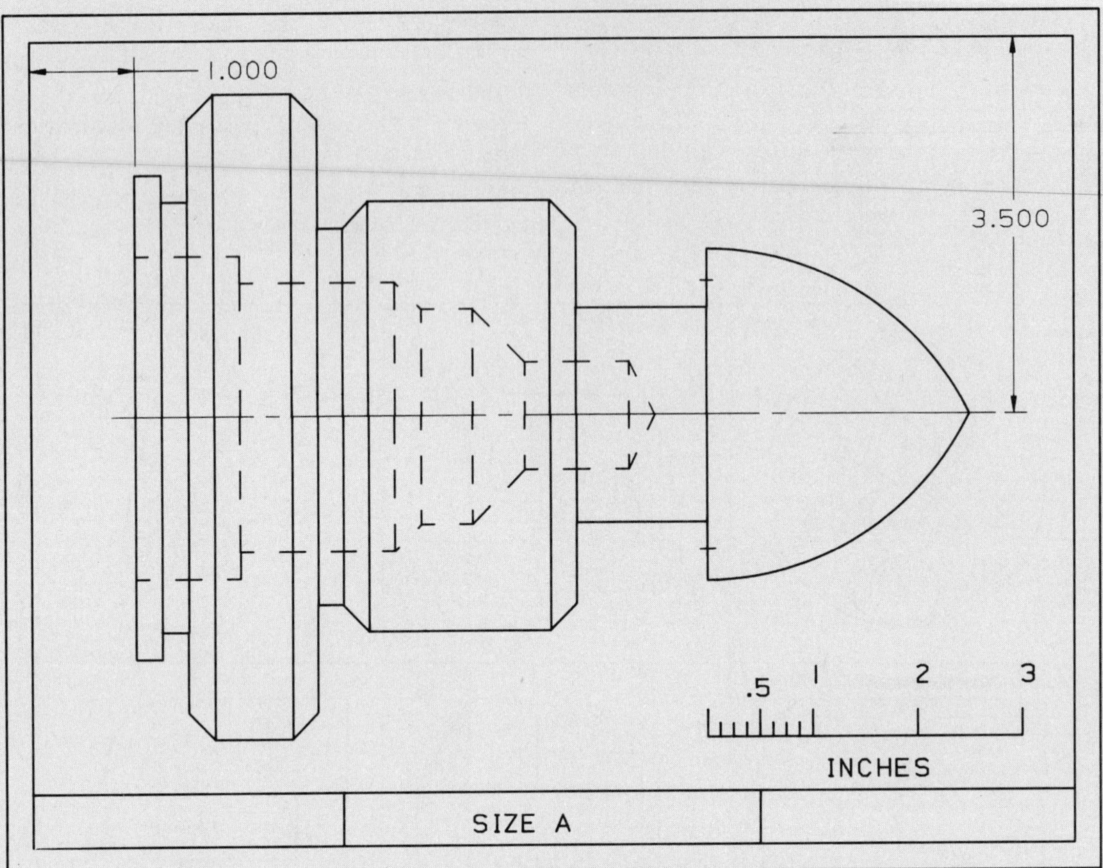

Problem 2.2 Cone Check

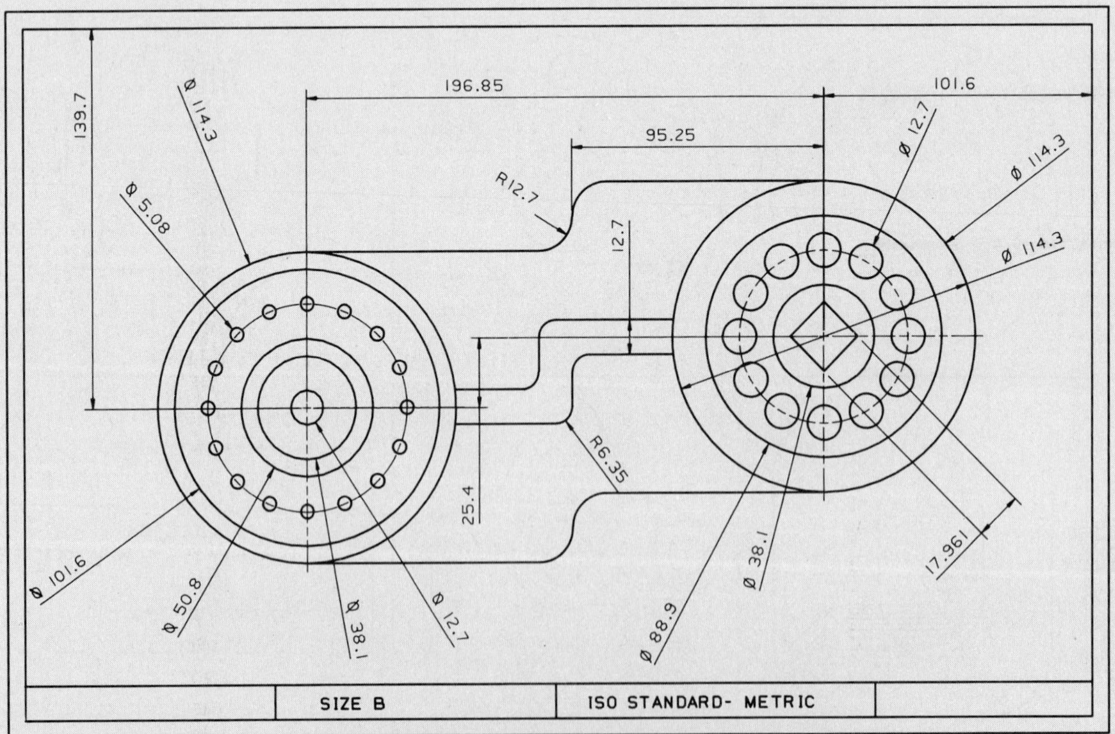

Problem 2.3 Assembly Plate

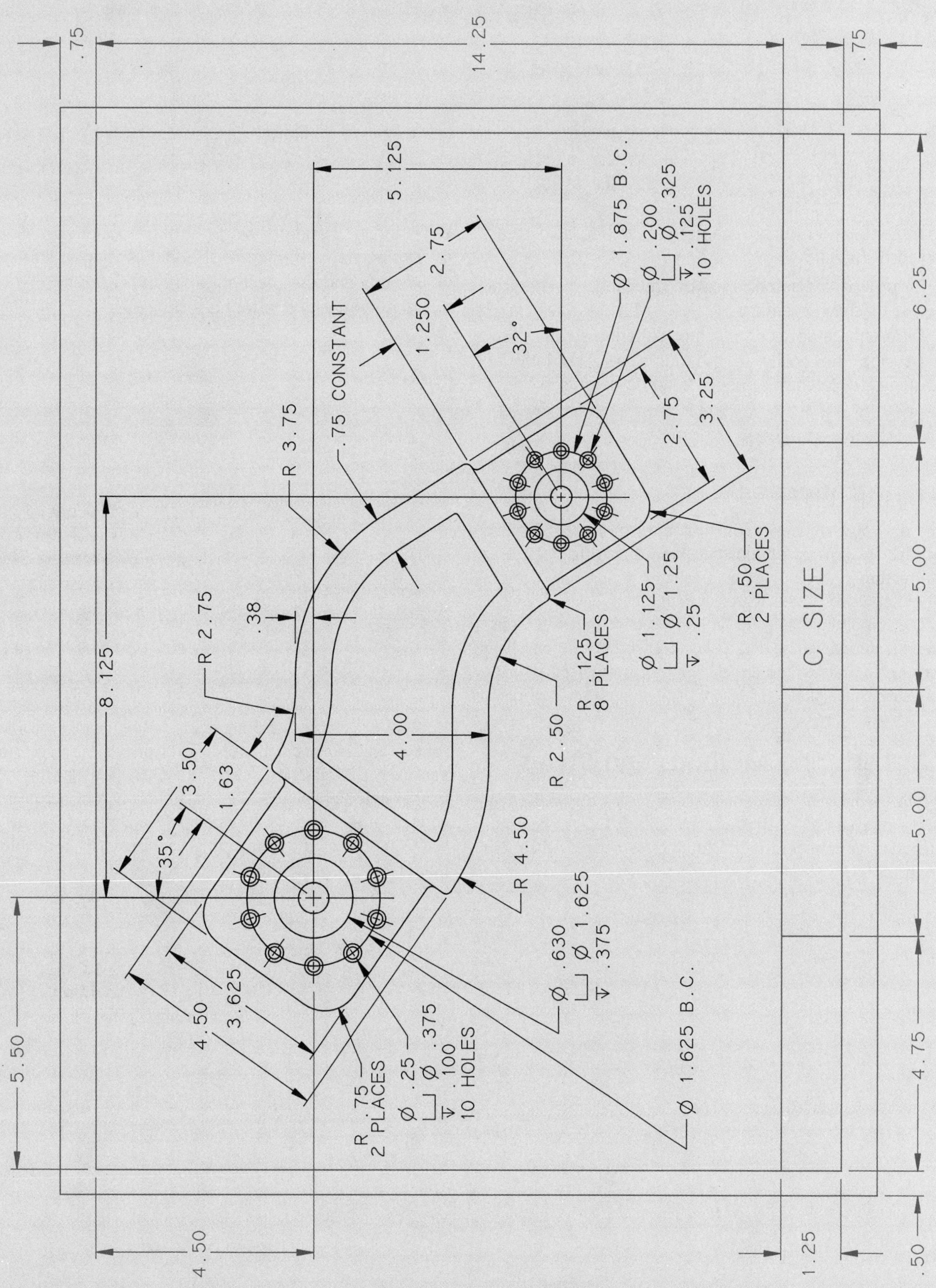

Problem 2.4 Arm

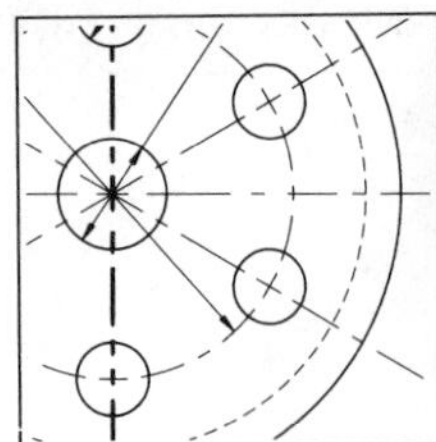

CHAPTER 3

Lettering

Learning Objectives

Upon completion of this chapter you will be able to accomplish the following:

1. Recognize the importance of freehand, mechanical, and machine lettering.
2. Differentiate between common lettering styles.
3. Develop the ability to use guidelines and lettering guides to determine proper lettering heights.
4. Produce standard single-stroke, uppercase Gothic characters with uniform size and spacing.
5. Identify and use mechanical lettering aids.
6. Identify machine lettering techniques.

3.1 Introduction

Line drawings are never complete until they are explained with labels, dimensions, notations and titles (Fig. 3.1). This information is either lettered freehand or inserted using a computer-aided design (CAD) system.

The importance of good lettering cannot be overemphasized. Lettering will *make or break* an otherwise excellent drawing. In drafting and design, sloppy or misplaced lettering causes misconceptions and inaccurate communication of data.

3.2 Lettering Methods and Styles

The three different methods of lettering are manual (freehand), mechanical, and machine. Table 3.1 lists the three categories, equipment, and techniques associated with each group.

TABLE 3.1 Lettering Methods

Types of Letters	Manual	Mechanical	Machine
Vertical	Freehand	Template	Typewriter
Inclined	Lettering aid (slot guide)	WRICO	Printer
Uppercase		Leroy	Dry transfer
Lowercase		Letterguide	Phototypesetter
		Varigraph	CAD

A wide variety of lettering styles, or fonts, are available. A **font** is an assortment of type all of one style. *Single-stroke uppercase Gothic style* is used in mechanical drafting. The Gothic style alphabet does not have short bars or **serifs** at the ends of strokes as does the Roman style alphabet. Figure 3.2 shows a few of the many fonts commercially available in phototypesetting, printing processes, and dry transfer letters.

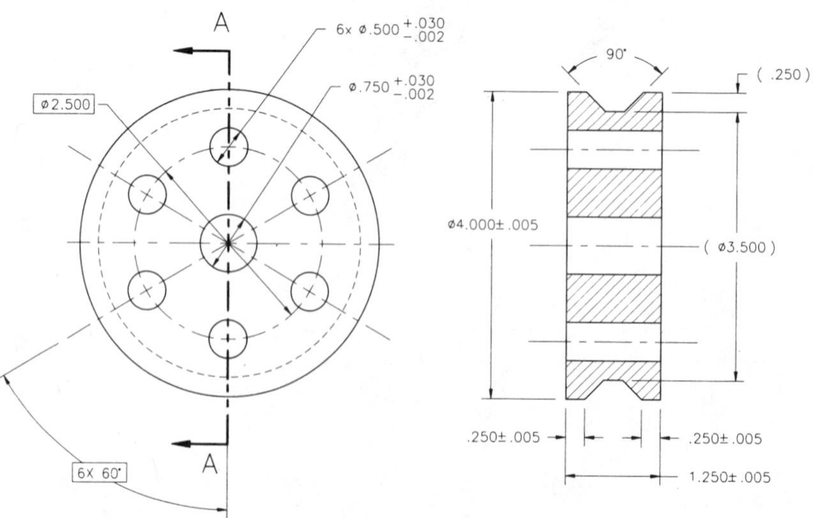

FIGURE 3.1 An Engineering Drawing

CooperBl.Ital.
CooperBlk.Out.
COPPERPLATE
DAVIDA
Dom Casual
Dom Diagonal
DRAFTING STANDARD
Eckmann
Egyptian Exp.
EMBOSS LIGHT
EMBOSS BOLD
Embrionic
Engravers Old Eng.
Eurostile Med.Ex.
Eurostile Bold Ext.
Fanfare
Folio Light
Folio Medium
Folio Bold
Folio Ex. Bold
Folio Bold Cond.

FIGURE 3.2 Examples of Typefaces

ABCDEFGHIJKLMNOP
QRSTUVWXYZ&
1234567890

ANSI Standard Vertical Upper and Lowercase Lettering

ABCDEFGHIJKLMNOP
QRSTUVWXYZ&
1234567890

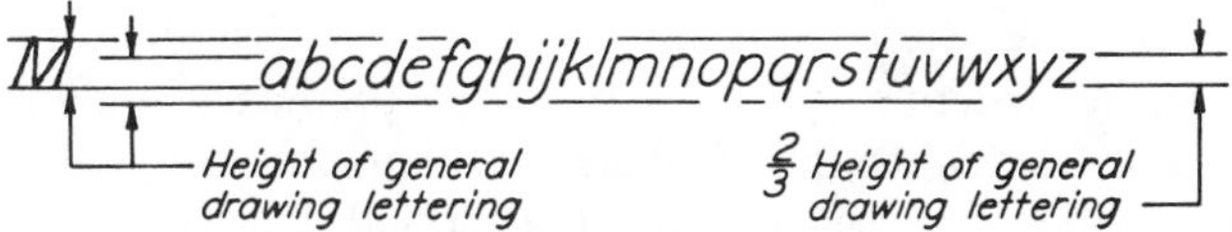

FIGURE 3.4 ANSI Standard Inclined Upper and Lowercase Lettering

The quality of your lettering after reduction will depend on the legibility of the lettering and its height. The recommended minimum freehand and mechanical lettering heights for various size drawings are given in Table 3.2.

3.3 Manual Lettering

Many drawings are still made and revised with freehand lettering techniques. Both **vertical** (Fig. 3.3) and **inclined** (Fig. 3.4) lettering are used in industry. Vertical lettering is preferred because it reduces and microfilms better than inclined lettering. Inclined lettering, however, is easier for some to master and is usually faster to letter.

Uppercase letters are used for all lettering on drawings unless lowercase letters are required for other established standards, equipment nomenclature, or marking. Only uppercase Gothic lettering is used on mechanical and electronic drawings. Piping, architectural, structural, and civil drawings sometimes employ lowercase lettering.

TABLE 3.2 Blended Lettering Heights for Manual and Mechanical Lettering (Uppercase Letters)

Project	Size of Drawing	Height of Manual Letters, U.S. (Metric) Units	Height of Mechanical Letters, U.S. (Metric) Units
Numbers in a title block	A–C*	.250 in., $\frac{1}{4}$ in. (7 mm)	.240 in. (7 mm)
	D and above*	.312 in., $\frac{5}{16}$ in. (7 mm)	.290 in. (7 mm)
Title, section lettering	A–F	.250 in., $\frac{1}{4}$ in. (7 mm)	.240 in. (7 mm)
Zone letters and numerals in borders	A–F	.188 in., $\frac{3}{16}$ in. (5 mm)	.175 in. (5 mm)
Lettering in dimensions, tolerances, notes,	A–C	.125 in., $\frac{1}{8}$ in. (3.5 mm)	.120 in. (3.5 mm)
tables, limits	D and above	.156 in., $\frac{5}{32}$ in. (5 mm)	.140 in. (5 mm)

*Drawing sizes: A – $8\frac{1}{2} \times 11$ in., B – 11×17 in., C – 17×22 in., D – 22×34 in.

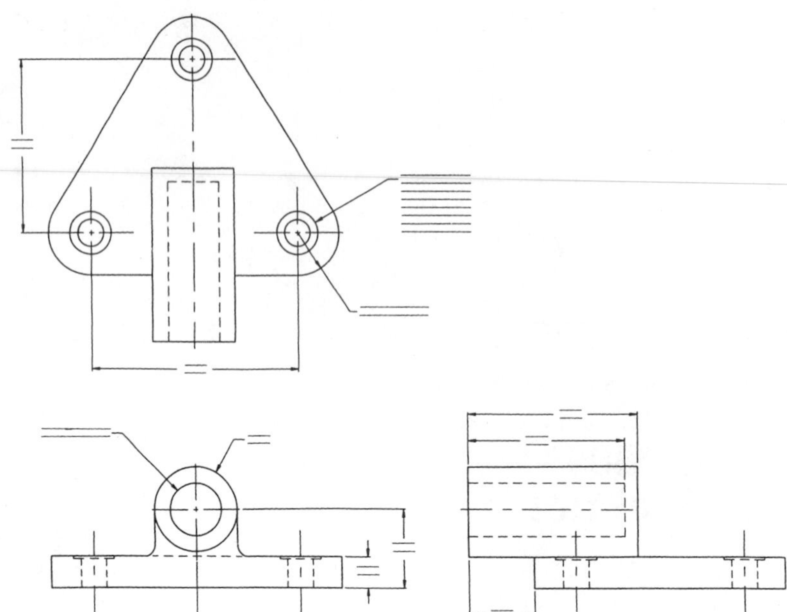

FIGURE 3.5 A Mechanical Drawing Example Using Guidelines

3.3.1 Guidelines and Lettering Heights

Guidelines are used in freehand lettering to determine the height of lettering and to assist you in producing uniform letters. **Guidelines** (Fig. 3.5) are very thin, sharp, light gray and drawn with 6H–3H grade lead or with blue nonreproducible lead. Guidelines are used for all lettering.

Since most lettering is done using capital letters and whole numbers, only two guidelines are necessary. Guidelines can be drawn with a straightedge or with the aid of a line spacing guide, such as the AMES lettering guide (Fig. 3.6) or the Braddock-Rowe triangle. For dimensions, notes, and labels, most lettering is between $\frac{1}{8}$ in. (.125 in.) and $\frac{5}{32}$ in. (.156 in.) high in U.S. units or 3.5 mm and 5 mm high in SI units. Lettering height is determined by the drawing format size. For all problems in the text you should use the standards for lettering heights and guideline spacing shown in Table 3.2. This table corresponds to *Conventions and Lettering* from the American National Standards Institute (ANSI).

The distance between lines of lettering on manually drawn projects, for notes and labels, is equal to the full height of the letter being used. This spacing is best for reproducible, legible letters for reduction and enlargement (when a microfilmed drawing is returned to its original size). When upper- and lowercase lettering is used on D size sheets and larger, the minimum uppercase height is a minimum of 5 mm for metric drawings and $\frac{5}{32}$ in. (.156 in.) for U.S. drawings.

The **freehand lettering guide** (Fig. 3.7) may be used as an aid in lettering. This guide eliminates the need for guidelines since it limits the height of the lettering to the space within the slots. However, it tends to flatten the upper and lower portions of some letters. Guidelines are unnecessary when you use vellum or drafting film with nonreproducible grid lines or when you use a lettering template.

To complete or revise an existing drawing, try to match the existing lettering style. *Vertical and inclined lettering should never be mixed on one drawing* (Fig. 3.8).

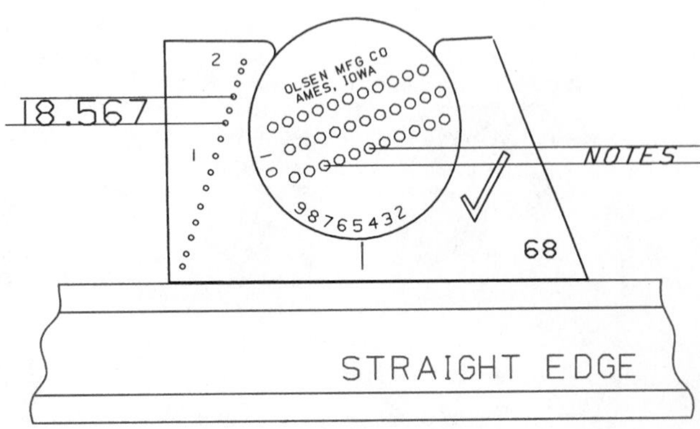

FIGURE 3.6 AMES Lettering Guide

FIGURE 3.7 Freehand Lettering Aid

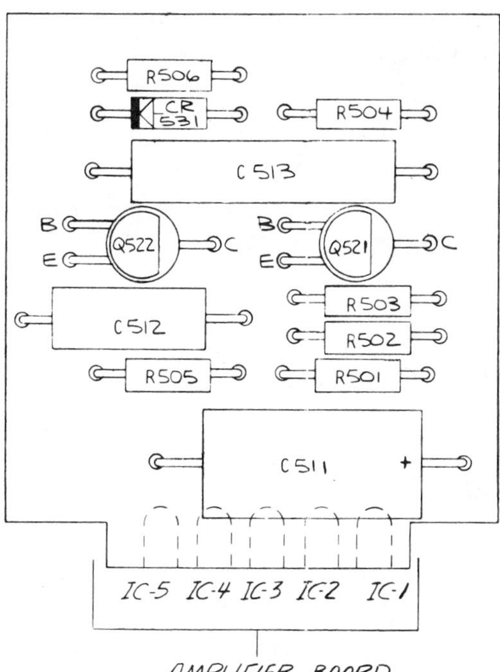

FIGURE 3.8 Mixed Vertical and Inclined Lettering on the Same Drawing. Guidelines were not used.

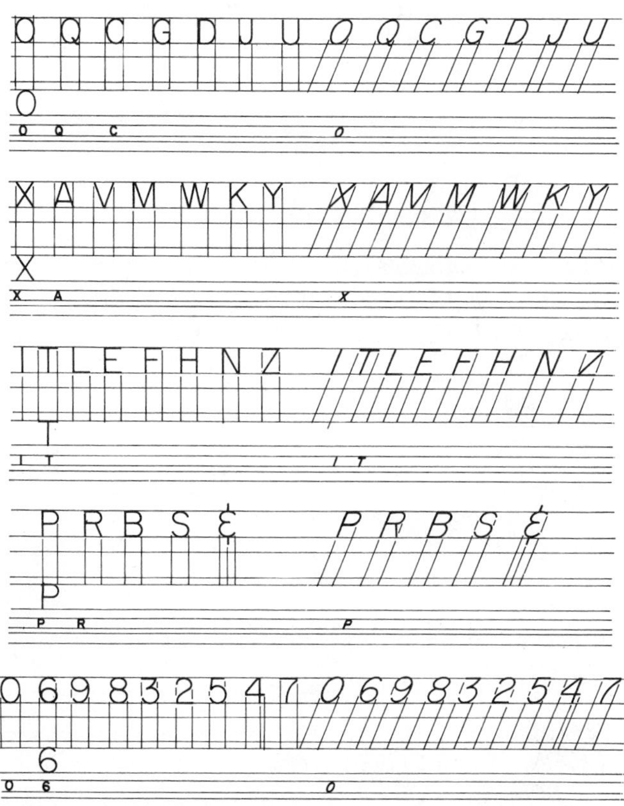

FIGURE 3.9 Using Horizontal and Vertical (or Inclined) Guidelines for Lettering Practice

FIGURE 3.10 Hand Position When Lettering

The lettering page provided in Figure 3.9 shows both horizontal and vertical (or inclined) guidelines. Vertical or inclined guidelines are rarely used on drawings.

Except when special emphasis is required, lettering should not be underlined. If used, the underlining should not be less than .06 in. (1.5 mm) below the lettering.

3.3.2 Pencil Technique

Instrument lines are made dense when the construction line is traced. It is usually impossible to trace freehand lines consistently, so lettering must be drawn with the proper density in only one stroke. To help get the proper density, use a soft lead. Depending on your preference, the H, HB, or F lead may be used for all lettering with good results.

However, soft lead contributes to the dirt on the drawing because it "chalks" easily. Use your drafting brush often and do the lettering last. Always place a sheet of clean paper between your hand and the drawing medium. This helps keep the drawing free of body oils and dirt (Fig. 3.10).

The fine-line pencil is an excellent lettering device. Use a 0.5 or 0.7 mm fine-line pencil. Rotate your pencil or lead holder to maintain consistency of character width.

The suggested hand orientation (Fig. 3.10), stroke sequence, and stroke direction (Fig. 3.11), are provided to help guide you, but are not meant to be interpreted as the only method of lettering. The most important thing is that the end result must conform to the ANSI standard style and quality.

3.3.3 Lettering Strokes, Uniformity, and Form

The strokes of your letters must be consistent in both width and density and should not vary in density from the linework. You should strive for consistent, uniform, well-spaced letters. The stroke sequence is the same for inclined and vertical lettering. Figure 3.11 shows suggested stroke sequences. Lettering examples are also provided below each comment.

The grid pattern shown in Figure 3.11 gives the ideal width and height relationship for single-stroke Gothic lettering. All characters are six units in height and vary in width from the l and I to the W and M. Figure 3.12 shows the six basic stroke sequences for freehand lettering.

CHARACTER	COMMENTS AND EXAMPLES	INCORRECT	POSSIBLE MISTAKES
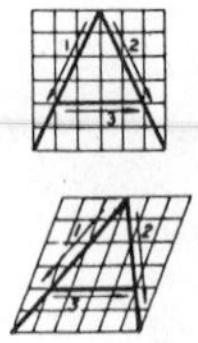 A	MAKE UPPER PART LARGER THAN BOTTOM PART. *ADAPTER, CONNECTOR (CP)*	A	4
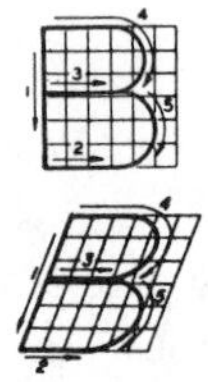 B	*LOWER PART SLIGHTLY LARGER THAN UPPER PART.* BARRIER PHOTOCELL (V) BLOCK, CONNECTING (TB)	B	8
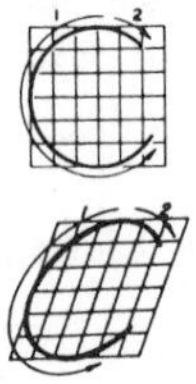 C	FULL OPEN AREA, ELLIPTICAL LETTER BODY. *COUPLER, DIRECTIONAL (DC) CUTOUT, FUSE (F)*	C	O
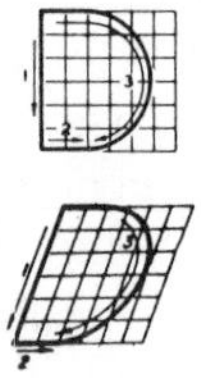 D	*HORIZONTAL BARS AND STRAIGHT LINE BACK.* DIODE, SEMICONDUCTOR (CR) DELAY FUNCTION (DL)	D	O
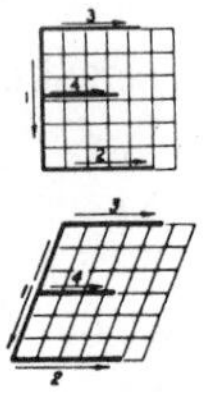 E	SHORT BAR SLIGHTLY ABOVE CENTERLINE. *ELECTRONIC MULTIPLIER (A) EQUALIZER, NETWORK, EQUALIZING (EQ)*	E E	L
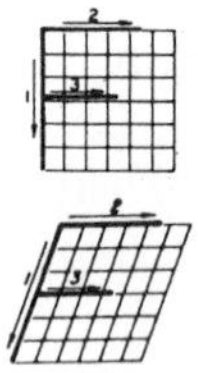 F	*SHORT BAR SLIGHTLY ABOVE CENTERLINE.* FIELD EFFECT TRANSISTOR (Q) FUSE HOLDER (X)	F F	T E
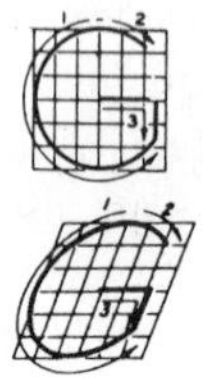 G	BASED ON TRUE ELLIPSE, SHORT HORIZONTAL LINE ABOVE CENTERLINE. *GENERATOR (G) GAP (HORN, PROJECTIVE, OR SPHERE) (E)*	G G G	C O 6

FIGURE 3.11 Stroke Sequence, Comments, Examples, and Possible Errors in Lettering
Continues

CHARACTER	COMMENTS AND EXAMPLES	INCORRECT	POSSIBLE MISTAKES
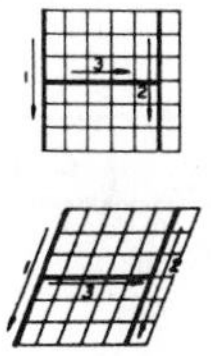	BAR SLIGHTLY ABOVE CENTERLINE. HARDWARE (COMMON FASTENERS, ETC.) (H) HEADSET, ELECTRICAL (HT)	H H	
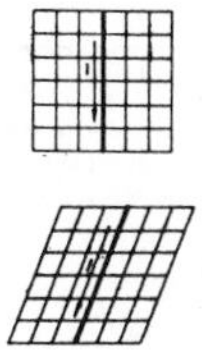	NO SERIFS, EXCEPT WHEN NEXT TO NUMBER ONE (I). INDUCTOR (L) INDICATOR (EXCEPT METER OR THERMOMETER) (DS)	I I	
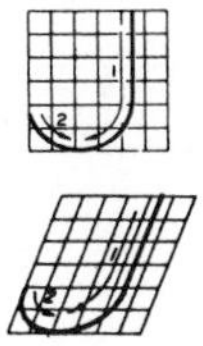	WIDE FULL HOOK WITH NO SERIFS. JUNCTION (COAXIAL OR	J J	
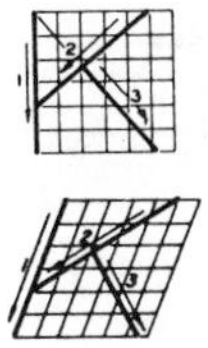	EXTEND LOWER BRANCH FROM UPPER BRANCH. WAVE GUIDE) (CP) JACK (J)	K K	R
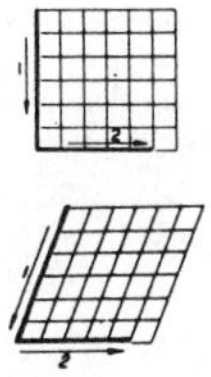	MAKE BOTH LINES STRAIGHT. LOOP ANTENNA (E)	L	
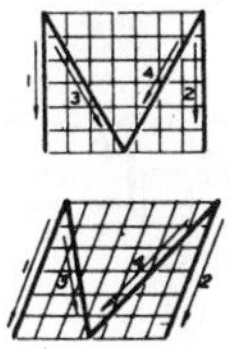	NOT AS WIDE AS W: CENTER PART EXTENDS TO BOTTOM OF LETTER. MICROCIRCUIT (U) MULTIPLIER, ELECTRONIC (A)	M M	
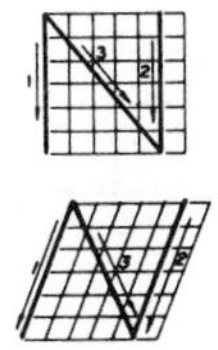	DO NOT CRAM LINES TOGETHER. NETWORK, EQUALIZING (HY) DIODE, TUNNEL (CR)	N N	V U

Continues

CHARACTER	COMMENTS AND EXAMPLES	INCORRECT	POSSIBLE MISTAKES
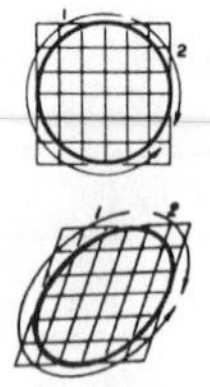	FULL TRUE ELLIPSE. OSCILLOGRAPH (M) OSCILLOSCOPE (M)	0 O	C Q 6
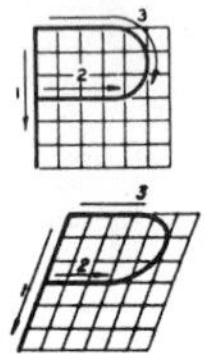	MIDDLE BAR INTERSECTS AT LETTER'S MIDDLE. PHOTODIODE (CR) POTENTIOMETER (R)	P P	K T D
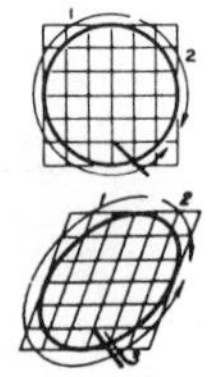	BASED ON TRUE WIDE ELLIPSE. NETWORK, EQUALIZING (HY) SWITCH, SEMICONDUCTOR CONTROLLED (Q)	Q	O
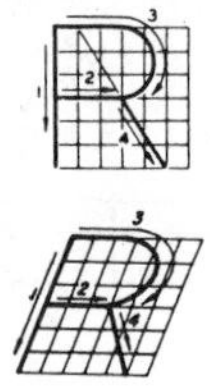	MAKE UPPER PORTION LARGE. REGULATOR, VOLTAGE (V) RESISTOR, THERMAL (RT)	R R	K
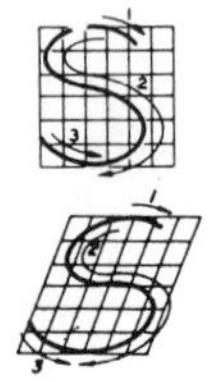	BASED ON NUMBER 8; KEEP ENDS OPEN. SOLENOID, ELECTRICAL (L) SWITCH, INTERLOCK (S)	S	8
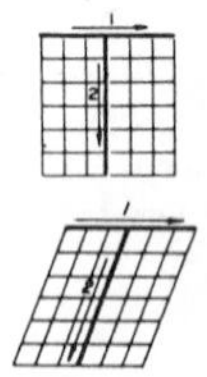	DRAW FULL WIDTH OF LETTER E. THERMOCOUPLE (TC) TRIODE, THYRISTOR (Q)	T T	7
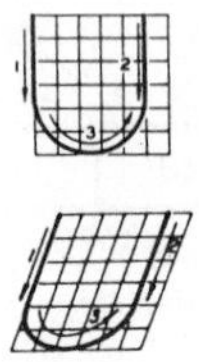	LOWER PORTION ELLIPTICAL, VERTICAL BARS PARALLEL. COMPUTER (A) WAVE GUIDE (W)	U	V

FIGURE 3.11 Stroke Sequence, Comments, Examples, and Possible Errors in Lettering *Continued*

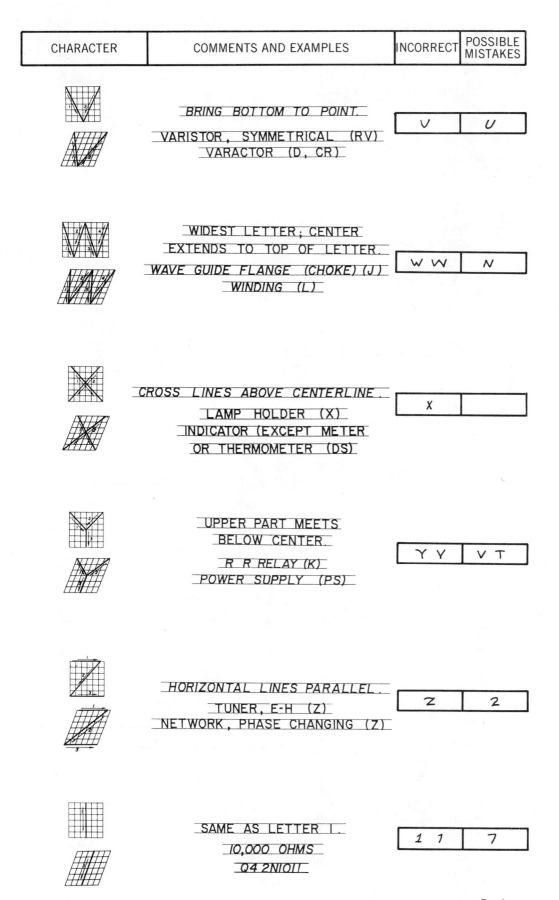

CHARACTER	COMMENTS AND EXAMPLES	INCORRECT	POSSIBLE MISTAKES
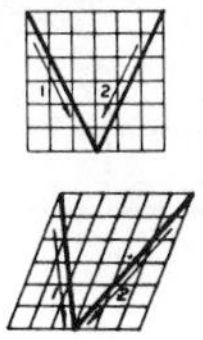 V	BRING BOTTOM TO POINT. VARISTOR, SYMMETRICAL (RV) VARACTOR (D, CR)	V	U
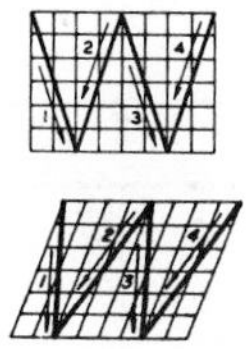 W	WIDEST LETTER; CENTER EXTENDS TO TOP OF LETTER. WAVE GUIDE FLANGE (CHOKE) (J) WINDING (L)	W W	N
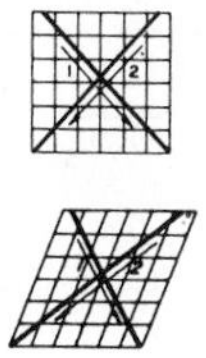 X	CROSS LINES ABOVE CENTERLINE. LAMP HOLDER (X) INDICATOR (EXCEPT METER OR THERMOMETER (DS)	X	
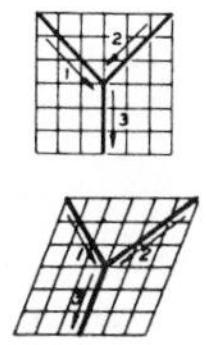 Y	UPPER PART MEETS BELOW CENTER. R R RELAY (K) POWER SUPPLY (PS)	Y Y	V T
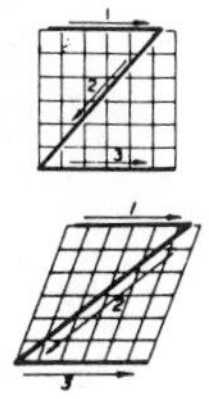 Z	HORIZONTAL LINES PARALLEL. TUNER, E-H (Z) NETWORK, PHASE CHANGING (Z)	Z	2
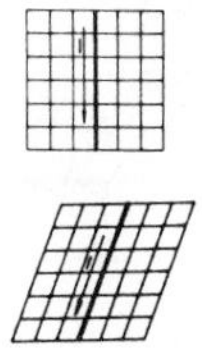 1	SAME AS LETTER I. 10,000 OHMS Q4 2NIOII	1 1	7

Continues

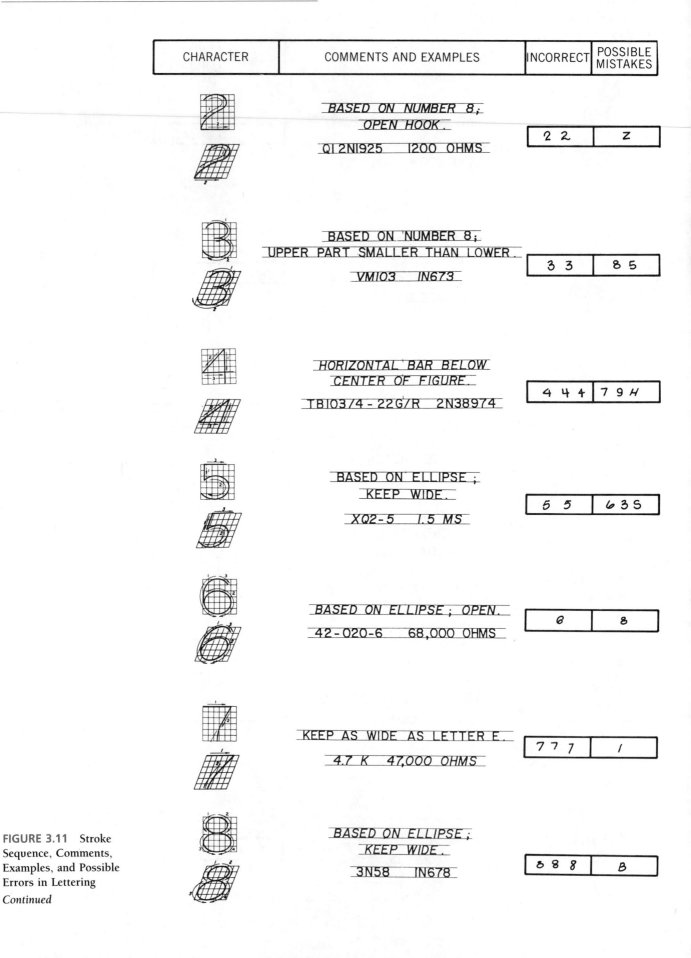

CHARACTER	COMMENTS AND EXAMPLES	INCORRECT	POSSIBLE MISTAKES
2	BASED ON NUMBER 8; OPEN HOOK. QI 2NI925 I200 OHMS	2 2	Z
3	BASED ON NUMBER 8; UPPER PART SMALLER THAN LOWER. VMI03 IN673	3 3	8 5
4	HORIZONTAL BAR BELOW CENTER OF FIGURE. TBI03/4-22G/R 2N38974	4 4 4	7 9 H
5	BASED ON ELLIPSE; KEEP WIDE. XQ2-5 I.5 MS	5 5	6 3 S
6	BASED ON ELLIPSE; OPEN. 42-020-6 68,000 OHMS	6	8
7	KEEP AS WIDE AS LETTER E. 4.7 K 47,000 OHMS	7 7 7	I
8	BASED ON ELLIPSE; KEEP WIDE. 3N58 IN678	8 8 8	B

FIGURE 3.11 Stroke Sequence, Comments, Examples, and Possible Errors in Lettering *Continued*

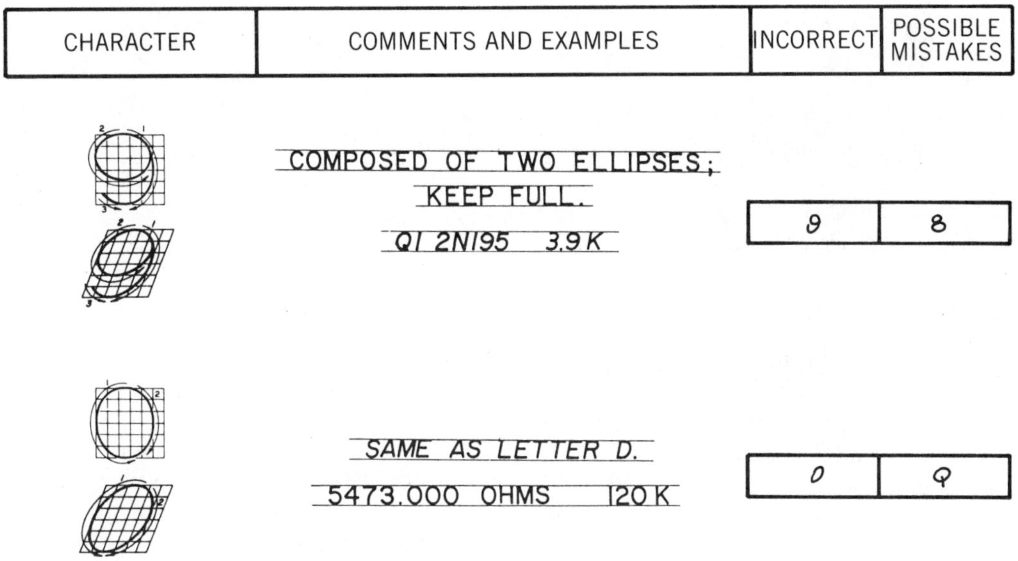

CHARACTER	COMMENTS AND EXAMPLES	INCORRECT	POSSIBLE MISTAKES
9	COMPOSED OF TWO ELLIPSES; KEEP FULL. QI 2NI95 3.9K	9	8
0	SAME AS LETTER D. 5473.000 OHMS I20K	0	Q

FIGURE 3.11 Stroke Sequence, Comments, Examples, and Possible Errors in Lettering *Continued*

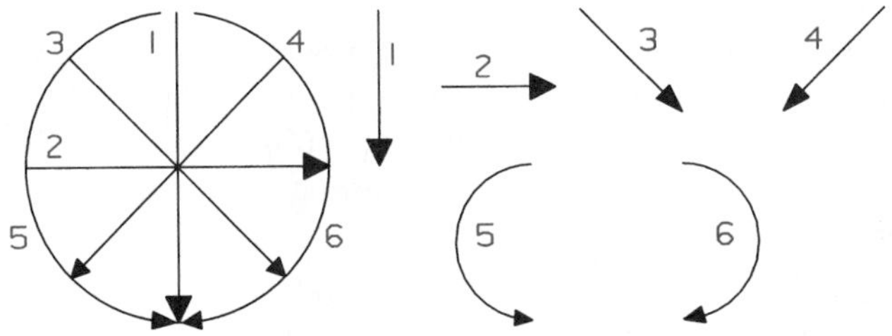

FIGURE 3.12 Basic Lettering Strokes

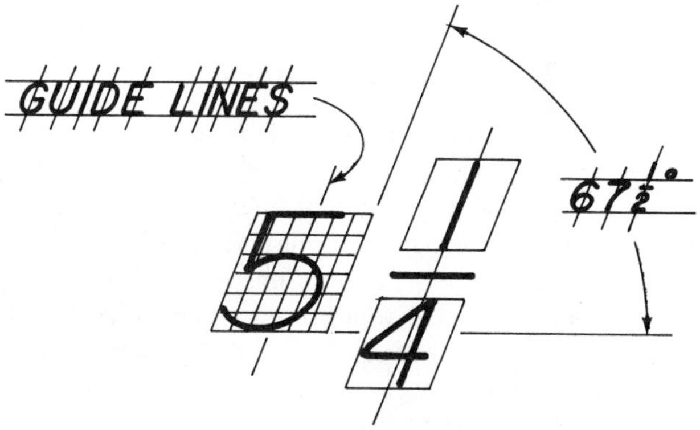

FIGURE 3.13 Using Guidelines for Inclined Lettering

You should strive to develop a lettering style that is comfortable and that communicates the necessary engineering data without confusion and mistakes.

Figure 3.13 shows the typical slant angle for inclined lettering. Any angle between 90° (vertical) and 65° is acceptable unless an individual company has a preferred practice.

3.3.4 Spacing

Spacing is done by eye to create a pleasing and orderly set of words and numbers. The spacing between letters within a word is as important as the spacing between words. The background area between characters should *appear* equal. The spacing between words should be a minimum of six units wide (a letter such as a W or M). The spacing between letters varies because the shape of the adjacent letters varies. Spacing for letters and words should correspond to the following specifications (Fig. 3.14):

- Background areas between letters in words are separated by approximately equal areas.
- Spacing for numerals separated by a decimal point (5.375, 2.54 mm, etc.) are a minimum of $\frac{2}{3}$ of the character height.

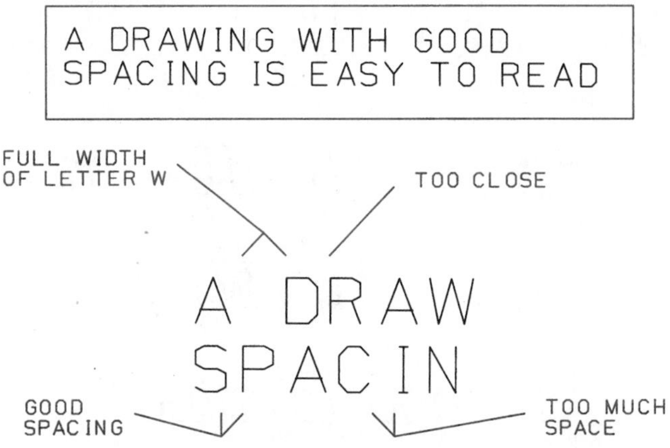

A DRAWING WITH GOOD
SPACING IS EASY TO READ

FIGURE 3.14 Spacing of Letters and Words

- Spaces between words are approximately equal and a minimum of .06 in. (1.5 mm). A full character height for horizontal word spacing is suggested.
- The horizontal space between lines of lettering is at least $\frac{1}{2}$ the height of the characters, but preferably one full character of space is left between lines.
- Sentences are separated by at least one full character height and preferably two character heights if space permits.

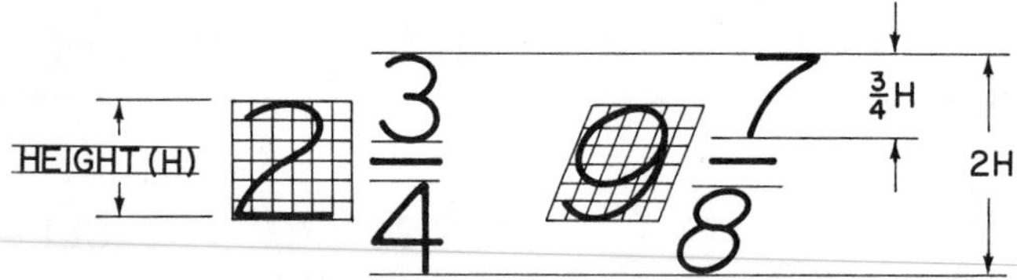

FIGURE 3.16 Fractions

3.3.5 Lowercase Lettering

Lowercase lettering (Fig. 3.15) is seldom used on engineering drawings except in construction drawings. Guidelines for the *waistline* (top of the main body of the letter), *baseline* (bottom of the main body of the letter), *ascender,* and *descender lines* are required for good lettering. Ascender lines or *cap lines* designate the tops of strokes for letters that extend above the waistline, such as b, d, f, h, k, and l. Descender lines or *drop lines* designate the bottom of strokes for letters below the baseline, such as g, j, p, and y.

3.3.6 Fractions on Drawings

Most drawings are dimensioned with decimal-inch or metric units (they are quick to draw since all numbers are placed between two equally spaced guidelines). The tolerance and accuracy required for manufacturing and construction may be loose enough to permit fraction dimensioning (sheet metal work). Fractions are also widely used in piping, civil, architectural, and structural design. Figure 3.16 shows the height ratio of a fraction number to the whole number. *The ANSI standard on lettering states that the height of the fraction number should be the same as that for the whole number.* Most drafting books, and many companies suggest the relationship shown in Figure 3.16.

The division line of the common fraction is drawn parallel to the direction in which the dimension reads and is separated from the numerals by a minimum of .06 in. (1.5 mm) spacing. The numbers must not touch the fraction division bar. The division bar is drawn horizontally between the numbers, except in notes, when the angled division bar is acceptable. Some company standards require the angled division bar, but it is not an ANSI standard.

FIGURE 3.15 Lowercase Lettering

3.3.7 Lettering Composition

By practicing letters and numerals in groups, going from simple to complex, you will learn an easy way to practice on the forms that you need to improve without having to letter the whole alphabet. Consider the following groupings during practice:

- Straight lines only
  A, E, F, H, I, K, L, M, N, T, V, W, X, Y, Z, 1, 4
- Straight and curved lines
  B, D, J, P, U, 2, 5, 7
- Curved lines only
  C, G, O, Q, S, 3, 6, 8, 9, 0

Stability of lettering construction is important to lettering composition. How does the lettering or number look on the paper? The construction of each letter and number is extremely important: proportion, stability, uniformity, balance, consistency, thickness, and density. When you combine these factors to make notes, it is called *composition.*

3.3.8 Lists and Notes

Traditionally, notes have been placed above the title block area on the drawing on the far right. The newest ANSI standards have reversed the placement of notes. *Notes are now to be placed on the lower left or the upper left of the drawing.* Many companies still follow the older practice of placing notes above the title block on the right side of the drawing. Hand-lettered lists (Fig. 3.17) and notes are time consuming. The notes in Figure 3.18 are standard preprinted company notes; notes 7 and 8, however, were added using freehand lettering.

You May Complete Exercises 3.1 Through 3.4 at This Time

ITEMS OF INTEREST

The Alphabet

The alphabet developed as a result of man's need to record events. In fact, our modern alphabet had its origin in Egyptian hieroglyphics. The word *hieroglyphics* means "picture writing" and is the oldest and most primitive of all writing. Some of the letters of the Roman alphabet in use today can be traced back to these crude pictures.

The Greeks adopted symbology from the Phoenicians who had developed a 22-letter alphabet in about 1500 B.C. The adopted system evolved into two distinct alphabets in two parts of Greece. The Western type became the Latin alphabet (about 700 B.C.) and was used throughout the Old World. (The modern English word for *alphabet* comes from the first two letters of the Greek alphabet, *alpha* and *beta*.) The original Roman alphabet of twenty-three characters has remained unchanged except that characters have been added.

People began communicating with each other through all forms of written communication once the alphabet was accepted. Unfortunately, books, even from the earliest times, were prepared by the laborious method of hand copying onto papyrus, parchment, or vellum. The scribes cut quills and made ink from gum and lampblack. Before the fall of the Roman Empire, the copying of books was a thriving and important industry. When Rome fell, the rich patrons of literature were scattered and their libraries were left to be burned. Monks, fearful that all literature would be lost, took on as part of their religious duties, the task of copying classical and religious books.

A Gutenberg Printing Press

About fifty years before Columbus discovered America, Johannes Gutenberg revolutionized graphic communications. Gutenberg, in Mainz, Germany, perfected a way to cast individual letters. As a young man, he had studied the arduous task of scribes and wanted to invent a mechanical printing process to make the scribes' work easier and make books more accessible. Until his time, all lettering was done by hand and it was left to the individual as to how each letter was made or decorated.

Three years after he started, Gutenberg had printed 200 copies of the Bible. Thirty of these were printed on a paper made from animal skins. These thirty copies used the hides of about 10,000 calves!

During the Industrial Revolution, the printing press needed for production was invented. Now with the printing process, more books could be printed and more people could afford to own them. Gutenberg's dream was realized at last!

It seems inconceivable to not know an alphabet or to have printed books. Even though we don't think much about our alphabet, most of what we do to communicate with each other is based on standard alphabets and printed material.

The computer is now the basis for another revolution in communications. The operators of modern CAD systems have a variety of *fonts* to choose from when inserting text into their drawings. A font is a series of patterns created by the CAD program to represent specific letters in certain styles (roman, italic, or script, for example). The font of the text can be changed at will. Computers can virtually link offices together across the country and the world. Regardless of how fast or sophisticated the method or the style of the text, our modern alphabet remains the basis for the way we communicate in written form.

Hieroglyphics

A Modern Printing Shop

REF NO	COMPONENT	PART NO
R-401	33K	216480
R-402	24K	216477
R-403	9.1K	216467
R-404	33	549978
R-405	100K	216491
R-406	430K	216731
R-407	7.5K	216465
R-408	100	595359
R-409	1K	216445
R-410	5.1K	216461
R-411	15K	216472
R-412	47K	216484
R-413	100K	216491
R-414	680	216442
J	JUMPER	1207833
C-421	.15/35 MFD	491255
C-422	150 PF DISC	1207587
C-423	3.3/35 MFD	1207585
C-424	.47/35 MFD	1208599
C-425	.33/35 MFD	1208591
C-426	2.2/35 MFD	1208601
C-427	.0068/100 MFD	492500
C-428	.0027/100 MFD	491309
C-429	150 PF DISC	1207587
Q-441	GREEN	1207577
Q-442	GREEN	1207577
Q-443	BLACK	1207601

FIGURE 3.17 Hand-Lettered Parts List

NOTES:

△1 MARK PER MIL-STD-130 APPROXIMATELY WHERE SHOWN, .093 HIGH CHARACTERS USING ITEM 48

2 SOLDER IN ACCORDANCE WITH NHB5300.4 (3A-1)

3 PARTIAL REFERENCE DESIGNATIONS ARE SHOWN FOR COMPLETE DESIGNATIONS PREFIX WITH UNIT NUMBER AND SUBASSEMBLY DESIGNATIONS

△4 ELECTROSTATIC DEVICE, HANDLE PER DOD-STD-1686

△5 TORQUE 2-2.5 INCH LBS

△6 FINISH: CONFORMAL COAT PER GEN-PS5205 EXCEPT CONNECTOR AND DESIGNATED AREAS SHOWN

△7 BOND ITEM 67 TO ITEM 1 PRIOR TO POPULATION OF CARD ASSEMBLY PER GEN-PS5402 CLASS 7

△8 APPLY FILLET TO COMPONENTS INDICATED AFTER CONFORMAL COATING PER GEN-PS5402 CL II

FIGURE 3.18 Hand-Lettered Notes Added to Preprinted Company Notes

3.4 Mechanical Lettering Aids

The most common lettering device found in a drafting room is the **template.** Although freehand lettering is the rule rather than the exception on manual drawings, templates are used extensively. Templates are available for almost any size and style of lettering and can be adapted for inking (Fig. 3.19). You can produce repeatable, uniform letters and numerals with them. Template lettering, however, takes considerably more time than freehand lettering.

With a template, guidelines are unnecessary, but rest the template against a straightedge so that all the letters are aligned properly. In general, template lettering is used on drawings that are inked, in title blocks, and for section letter identification. The KOH-I-NOOR Rapidometric Guide template in Figure 3.19 is an example of a template designed to be used for inking. Note that the template shelf does not come in contact with the drawing surface.

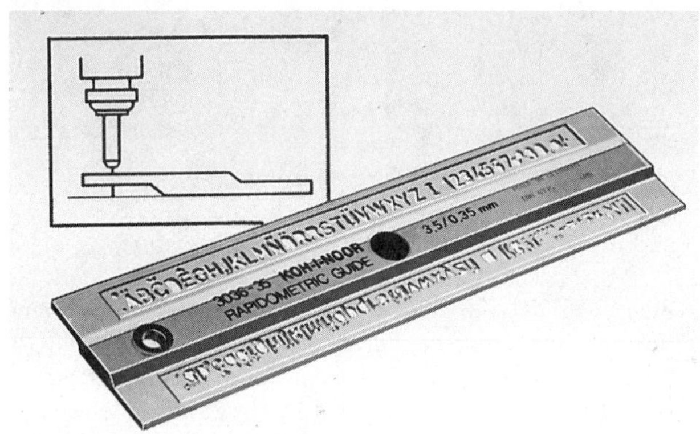

FIGURE 3.19 Lettering Guide Template

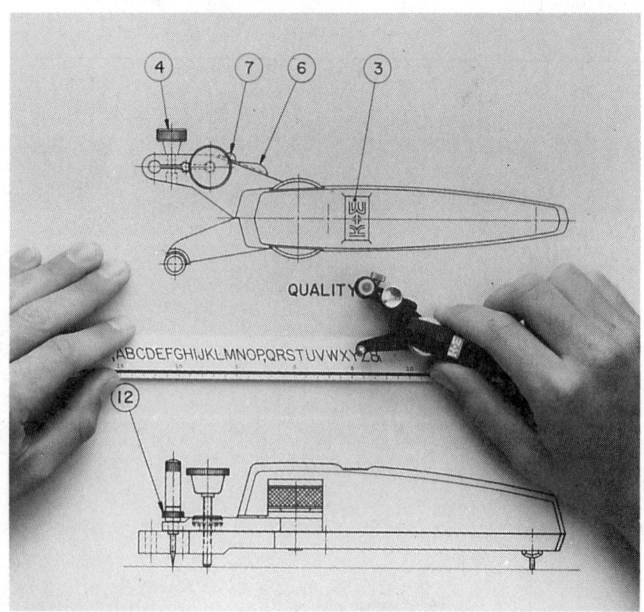

FIGURE 3.20 Lettering with a Leroy Set

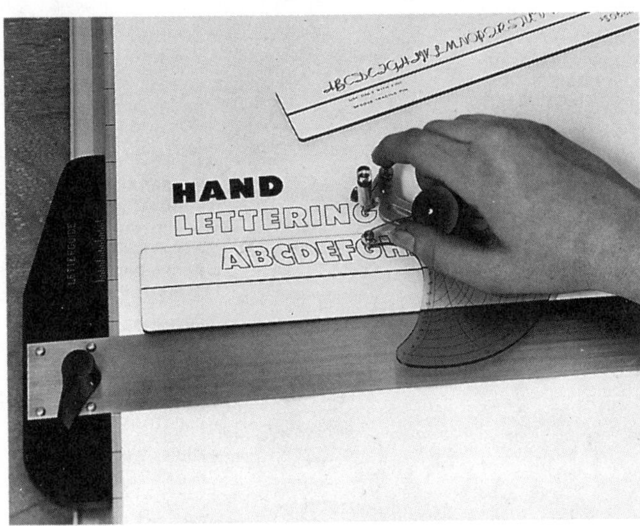

FIGURE 3.21 Letterguide

FIGURE 3.22 Varigraph Lettering Device

The **Leroy lettering set** uses a scriber and a template with a slot guide to produce close-to-perfect letters (Fig. 3.20). The scriber can be adjusted to alter the slant of the lettering.

The **Letterguide** system shown in Figure 3.21 uses an adjustable scriber that can alter the slant of the characters and change the character size from 60% to 140% of the template size. The **Varigraph** lettering headline machine (Fig. 3.22) uses the template system combined with a special holder and scriber mechanism. With it, you can adjust the vertical and horizontal size and the inclination of the letters. Varigraphs are used for headings and graphic art productions.

All hand-operated lettering systems are expensive and take more time than traditional freehand lettering. Mechanical lettering devices and the inking of drawings are limited to drawings for publication. Manuals, catalogs, and sales literature require more precise lettering and linework than design, detail, and assembly drawings.

Mechanical lettering devices enable you to make slightly smaller letters than manual techniques (Table 3.2). The variation in recommended minimum standard letter heights between freehand and mechanical devices is needed because freehand lettering does not reduce and enlarge as accurately as mechanically drawn characters.

3.5 Machine Lettering Devices

In the past, typewriters with specially designed carriages and Gothic typefaces were sometimes used on "A", "B", and "C" size sheets. Figure 3.23 shows a panel drawing where the labels have been typed on the drawing. A special ink ribbon must be used so the characters do not smear.

Dry transfer lettering and **appliques** are confined to artwork or headings (Fig. 3.24). It is time consuming to apply each letter or number separately. With the Kroy or Merlin lettering systems, you dial a sequence of letters or numbers (Fig. 3.25) and obtain a dry adhesive-backed strip to attach to the drawing. Notes, headings, and titles are easy to apply with this system (Fig. 3.26).

Phototypesetting and printing are used for publication-level artwork and drawings when quality is extremely important. Figure 3.27 is an example of phototypeset lettering on an illustration of a pressure vessel module.

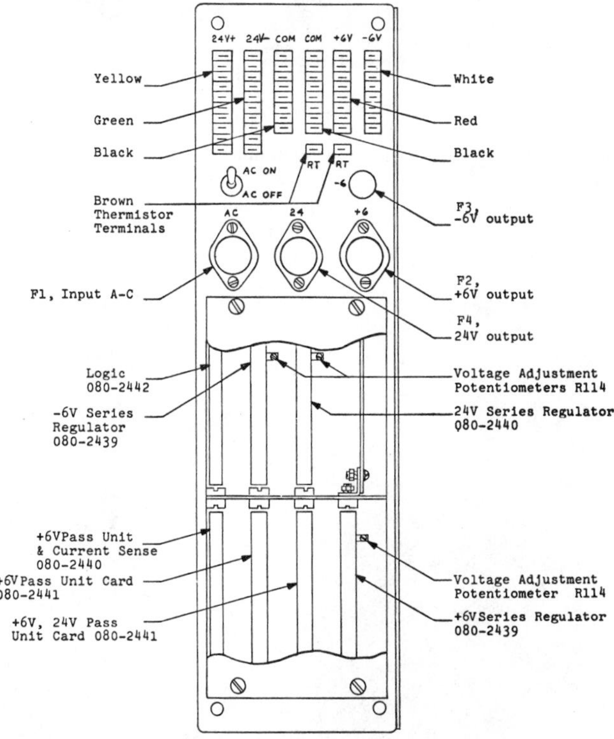

FIGURE 3.23 Typed Lettering on Panel Drawing

FIGURE 3.24 Transfer Lettering

FIGURE 3.25 Kroy Lettering Systems

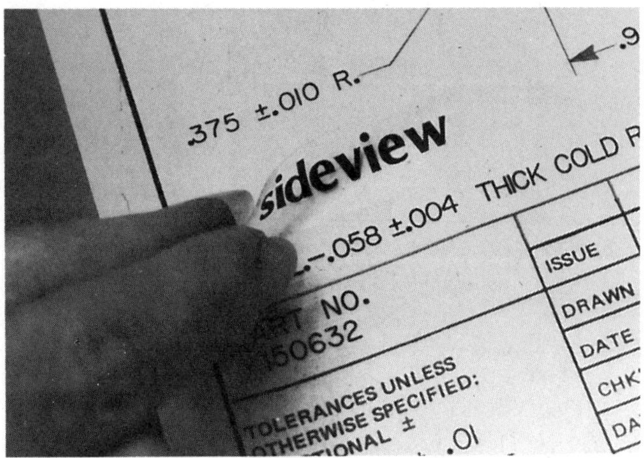

FIGURE 3.26 Kroy Lettering Being Applied to a Drawing

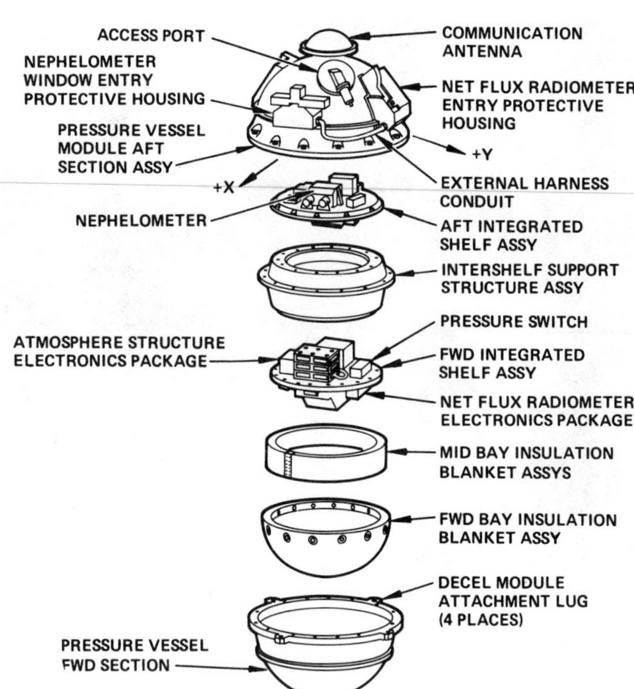

FIGURE 3.27 Typeset Lettering Used on Pressure Vessel Illustration

QUIZ

True or False

1. Inclined lettering .25 in. high is preferred on mechanical drawings.
2. When hand lettering a drawing, the distance between lines is equal to the character height used.
3. The distance between words should be a full four units or equal to the letter J.
4. Vertical lettering is preferred over inclined lettering because it reduces better.
5. Vertical and inclined lettering should not be mixed on one drawing.
6. CAD systems eliminate the need to master freehand lettering.
7. There are eight basic strokes for forming letters and numbers.
8. Guidelines need only be used when learning how to letter.

Fill in the Blanks

9. Lettering is divided into three separate methods: __________, __________ and __________.
10. Guidelines must be used when hand lettering except when using __________.
11. The ANSI standard on lettering states that the height of a fraction number should be __________ to the whole number.
12. Notes and dimensions should be at least __________ in height on a "D" size drawing.
13. Inclined lettering should be approximately __________ degrees.

14. ANSI lists and notes are placed in the ___________ ___________ or ___________ ___________ side of the drawing.
15. Templates and lettering guides should always be placed against a ___________ ___________ when lettering.
16. Guidelines can be drawn with a ___________ ___________.

Answer the Following

17. What ANSI standard covers lettering on engineering drawings?
18. When is lowercase lettering used and on what type of drawings?
19. When are machine lettering devices normally used for a drawing?
20. Explain the difference between manual, mechanical, and machine lettering.
21. Explain the reasons for mastering manual lettering.
22. What is a font?
23. Explain how to determine the spacing between words in a sentence.
24. Describe how guidelines are used to help produce uniform lettering.

EXERCISES

Exercises may be assigned as freehand, template, or machine lettering projects. Transfer the given information to an "A" size sheet of .25 in. grid paper.

After Reading the Chapter Through Section 3.3.8, You May Complete the Following Exercises

Exercise 3.1 Practice lettering using the standard stroke sequence. You may add vertical or inclined guidelines for this exercise.

Exercise 3.2 Letter the sentence and the notes three times each.

Exercise 3.3 Letter the page as shown.

Exercise 3.4 Letter the drawing notes as shown.

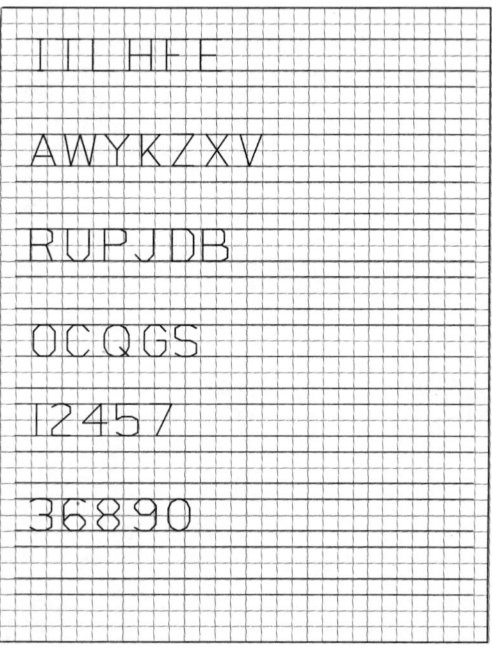

Exercise 3.1

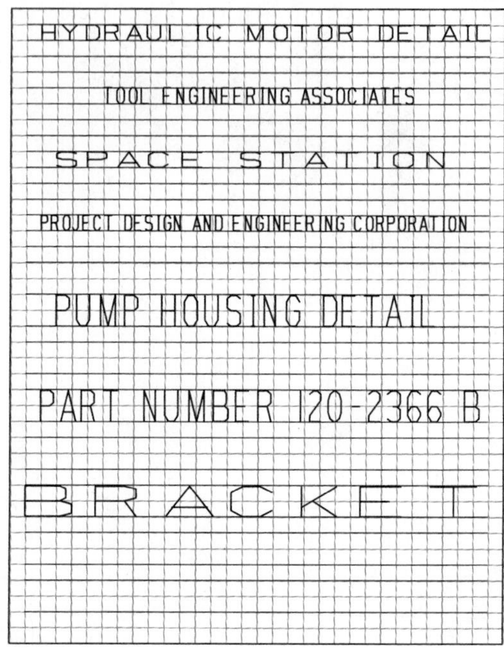

Exercise 3.3

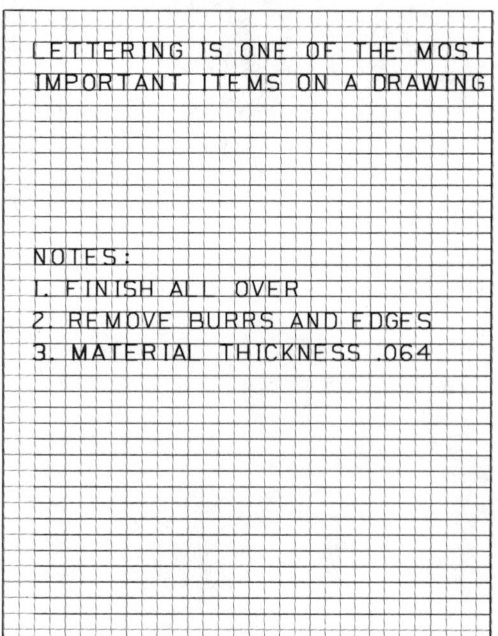

Exercise 3.2

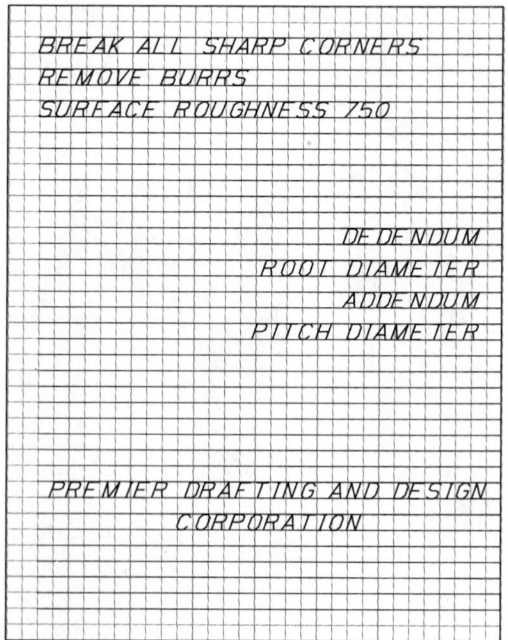

Exercise 3.4

PROBLEMS

Problem 3.1 Using the layout sheet provided in Problem 3.1 complete the lettering assignment as shown. Use this example to layout the drawing format for the following problems.

Problem 3.2 Letter each of the following twice, uppercase lettering. See Problem 3.1 for the page layout.

ANGLE BRACKET ASSEMBLY
PUMP HOUSING DETAIL
DESIGN ENGINEERING AND DRAFTING, INC.

Problem 3.3 Letter the following note three times at 4 mm height in vertical upper case lettering.

HEAT TREATMENT:
MC QUAID-EHN GRAIN SIZE 5–8 HEAT TO
1550 DEGREES F AND QUENCH IN OIL.
DRAW TO BRINELL HARDNESS 241–285.
100% BRINELL REQUIRED

Problem 3.4 Letter the following note three times using .25 inch height inclined letters. The instructor may assign project to be inked using a lettering template or a Leroy set.

NOTE:
1. LOCATING POINTS TO BE CAST FLAT AND SMOOTH
2. CAST FEATURES ARE DETERMINED BY BASIC DIMENSIONS IN RELATION TO LOCATING SURFACES
3. DIMENSIONS IN RELATION TO LOCATING SURFACES.

Problem 3.5 Letter the following specifications using vertical mixed uppercase and lowercase characters $\frac{5}{32}$ in height or 4 mm in height.

1. Casting to be pressure tight when tested at 100 P.S.I.
2. Finish all over 125.
3. Do not apply piece mark.
4. Material thickness .125 in.

Problem 3.6 Reletter the parts list in Figure 3.17 using vertical uppercase lettering.

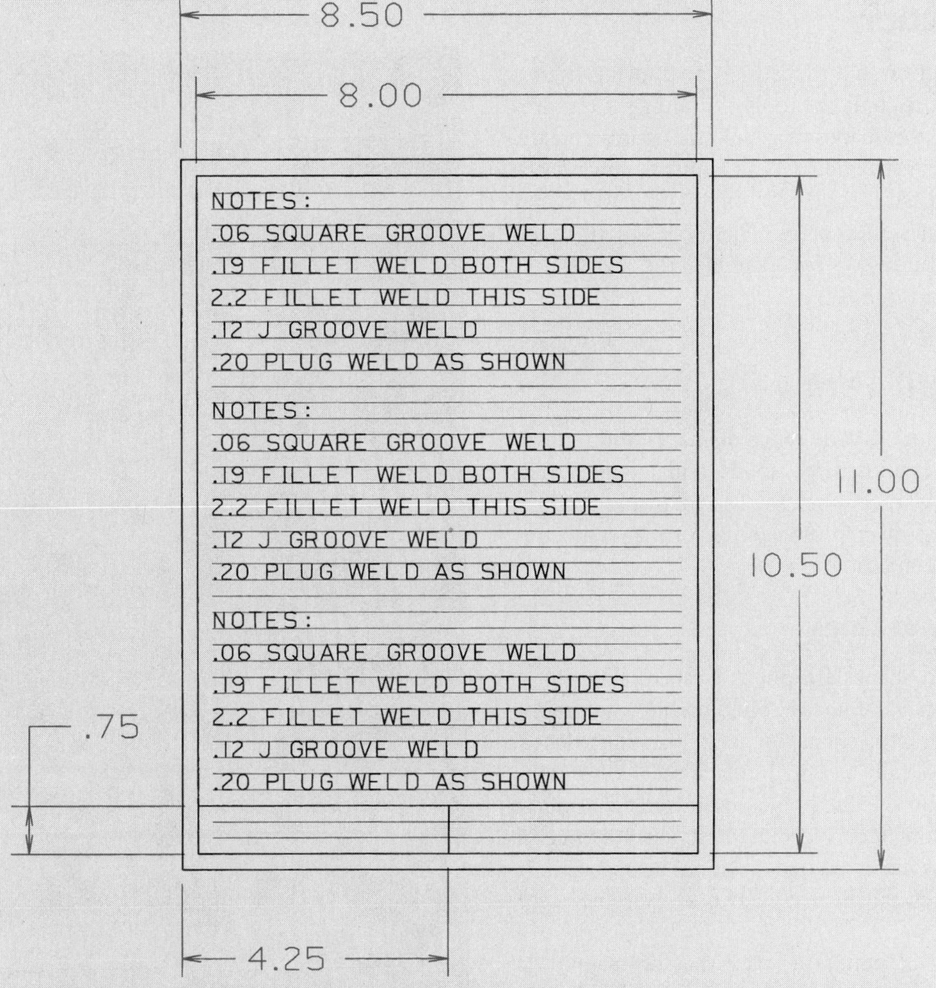

CHAPTER 4

Geometric Constructions

Learning Objectives

Upon completion of this chapter you will be able to accomplish the following:

1. Develop the ability to interpret graphic solutions to common geometrical problems.
2. Define and construct plane geometric shapes: points, lines, curves, polygons, angles, and circles.
3. Define solid geometric shapes: polyhedra, curved surfaces, and warped surfaces.
4. Apply basic construction line drawing techniques.
5. Produce uniformly drawn and scaled examples of commonly used geometric forms and entities.
6. Use geometric construction methods to facilitate feature locations.

4.1 Introduction

Geometric construction is a procedure for drawing figures and shapes that requires only the tools of drafting. It requires an understanding of the shapes of geometric figures and the mechanics for their construction and an ability to solve problems visually. It emphasizes scale, uniformity of linework, and smooth joining of lines and curves. These constructions are used extensively in industry. For example, the stairway in Figure 4.1 is a *cylindrical helix*.

FIGURE 4.1 Helical Stairway

4.2 Geometric Forms

Geometric forms include a wide range of shapes and figures, squares, triangles, arcs and circles, solids, and single-curved, double-curved, and warped surfaces. The following sections provide you with step-by-step procedures for manually constructing common geometric forms.

4.2.1 Points and Lines

Geometric forms and shapes are points connected by lines. The **point** is the primary geometric building block in graphical construction. All projections of lines, planes, surfaces, and solids can be physically located and manipulated by identifying a series of points. These points locate ends of straight lines or are placed along a curved line to establish the line in space. Since a point exists at one position in space, it is located in space by establishing it in two or more adjacent views.

A **line** is a series of points in space, having magnitude (length) but not width. Although a line may be located by establishing any two points and although it may have a specified length, all lines can be extended.

Lines are used to draw edges of plane surfaces and solid shapes. A **straight line** is the shortest distance between two points. The word *line* usually refers to a straight line. A line that bends is a *curve*. When two lines are in the same plane, they are parallel or they will intersect. **Parallel lines** symbolized by //, are the same distance apart along their entire length. Lines that intersect at an angle of 90° are **perpendicular lines,** symbolized by ⊥ . Figure 4.2 shows various types of lines. Geometric constructions require you to draw arcs, circles, and other *curved lines* that use specific linework techniques.

4.2.2 Polygons

A **polygon** is a plane closed figure that has three or more straight sides. A **regular polygon** has all sides of equal length and all angles of equal size. A regular polygon can be *inscribed within a circle* with corners touching the circle, or it can be *circumscribed about a circle* with sides touching the circle.

A **triangle** is a three-sided polygon. The sum of its interior angles always equals 180°. In Figure 4.2, the **equilateral**

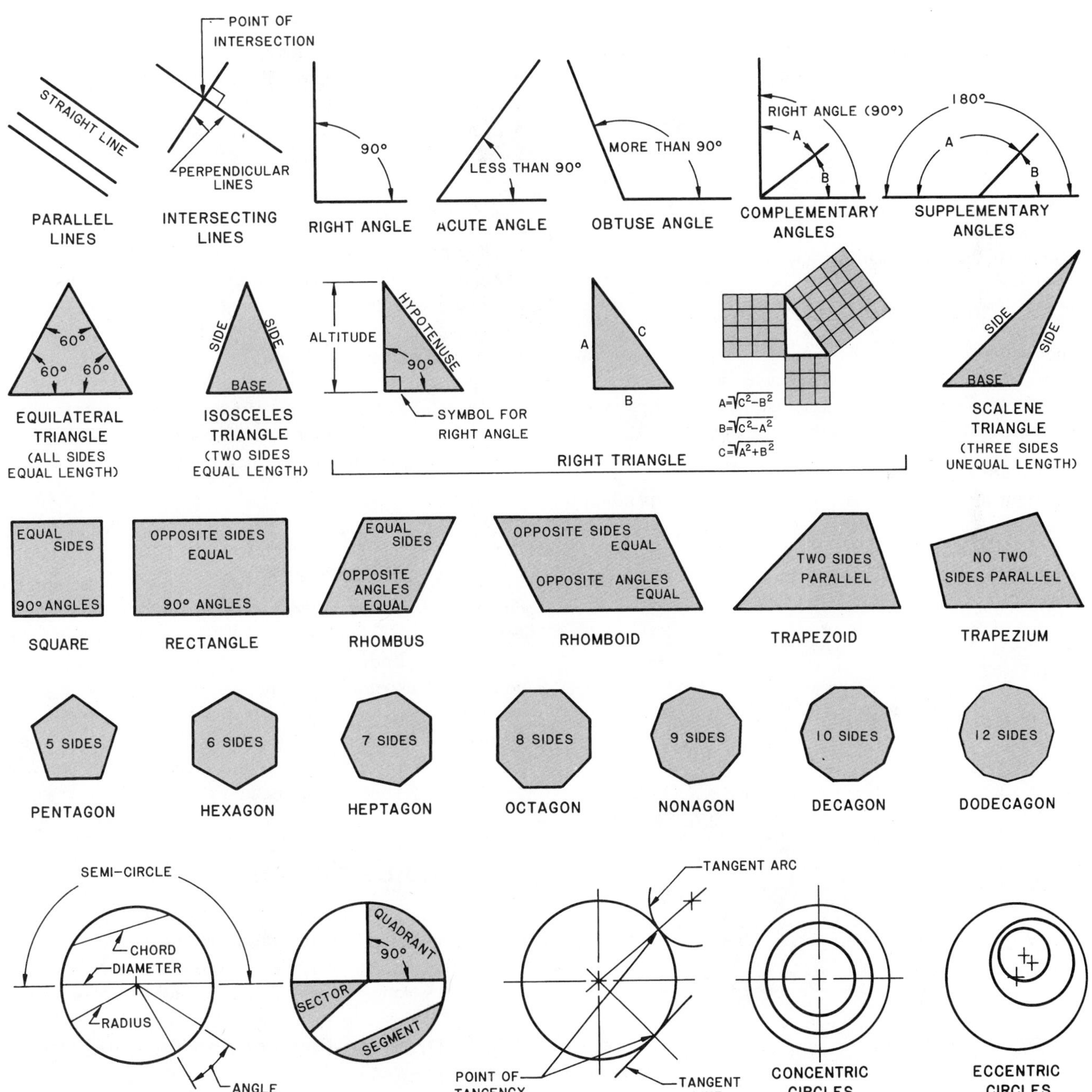

FIGURE 4.2 Geometric Shapes and Items

triangle has equal sides and equal angles and is a regular polygon. The second type of triangle is an **isosceles triangle.** It has two equal sides and two equal angles; the unequal side is the base, and the corner opposite the unequal side is the **apex,** or **vertex.** A line drawn through the apex to the base divides an isosceles triangle into two equal triangles. **Scalene triangles** do not have equal angles or sides.

A **quadrilateral** is a four-sided polygon. The sum of its interior angles is 360°. Figure 4.2 shows the six types of quadrilaterals. The first four quadrilaterals have opposite sides that are equal in length and are called **parallelograms.** The first parallelogram is a **square** because all sides and angles are equal. The second parallelogram is a **rectangle** because its opposite sides are equal and its angles are all the same. The third parallelogram is a **rhombus.** It has four equal sides and its opposite angles are equal. The fourth parallelogram is a **rhomboid** and has opposite sides parallel and opposite angles equal. A **trapezoid** has two sides parallel. When a quadrilateral has no equal sides it is called a **trapezium.**

Figure 4.2 also includes seven other regular polygons: **pentagon** (five sides), **hexagon** (six sides), **heptagon** (seven sides), **octagon** (eight sides), **nonagon** (nine sides), **decagon** (ten sides), and **dodecagon** (twelve sides).

4.2.3 Angles and Circles

Angles, represented by the symbol <, are formed by two intersecting lines (Fig. 4.3). The angle measurement of the distance between lines is typically expressed in degrees and sometimes in radians. Various types of angles are:

- An **acute angle** is less than 90°.
- A **right angle** is 90° and is formed by two perpendicular lines.
- An **obtuse angle** is more than 90° but less than 180°.
- An angle of 180° is a **straight line.**

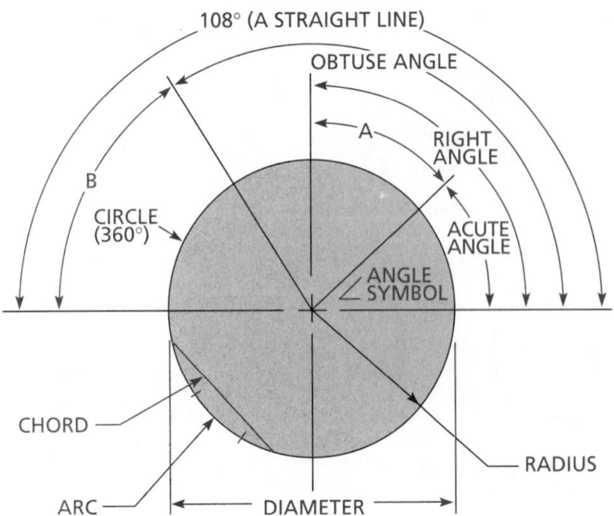

FIGURE 4.3 Angles and Circles

- **Complementary angles** are two angles whose sum equals 90°.
- **Supplementary angles** are two angles whose sum is 180°.

Circles represent holes and solid round shapes on drawings. A full circle is 360°. The parts of a circle are (Fig. 4.3):

- The **circumference** is the distance around a circle.
- The **diameter** is the distance measured from edge to edge and through the center of the circle.
- The **radius** is one-half the diameter measured from the center of the circle to the circumference.
- A **chord** is a straight line that connects two points on the circle's circumference.
- An **arc** is a continuous portion of the circumference from one fixed point to another.
- **Concentric circles** have different radii but have the same center point.
- **Eccentric circles** have different center points and different radii.

4.2.4 Polyhedra

Polyhedra (Fig. 4.4) are solids formed by plane surfaces. Every surface (face) of each form is a polygon. **Prisms** are polyhedra that have two parallel polygon-shaped ends and sides that are parallelograms. The **cube** is a polyhedron that has six equal sides.

A **pyramid** is a polyhedron that has a polygon for a base and triangles with a common vertex for faces. A **tetrahedron** is a pyramid that has four equal sides. Figure 4.4 also illustrates a **right pyramid**, a **truncated pyramid**, and an **oblique pyramid.**

4.2.5 Curved Surfaces

Curved surfaces are divided into two categories: **single-curved** (also called **ruled surfaces**) and **double-curved.** Forms that are bounded by single-curved surfaces include **cones** and **cylinders.** Variations of cones include the **right cone**, the **frustrum of a cone**, the **oblique cone**, and the **truncated cone.** Figure 4.4 also shows a **right cylinder** and an **oblique cylinder.** Double-curved surfaces (Fig. 4.4) are generated by moving a curved line about a straight line axis and include a sphere, a **torus**, and an **ellipsoid.**

4.3 Geometric Constructions

The following section presents step-by-step instructions for drawing geometric constructions including parallel lines, perpendicular lines, angles, circles, polygons, tangencies, tangent arcs, curves, conics, involutes, spirals, and helices.

4.3.1 Drawing Parallel and Perpendicular Lines

Parallel and perpendicular lines are easily constructed using a straightedge and a triangle. In Figure 4.5 an adjustable trian-

REGULAR SOLIDS

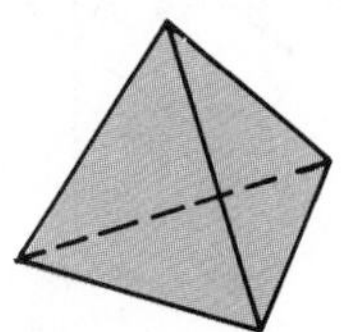

TETRAHEDRON (4 TRIANGLES)

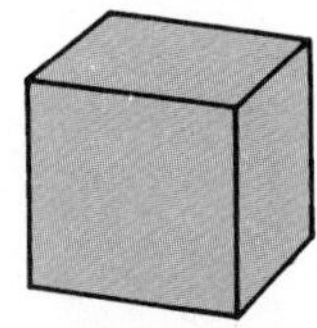

HEXAHEDRON (CUBE)

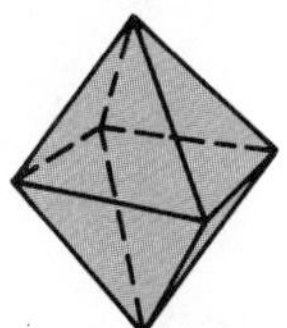

OCTAHEDRON (8 TRIANGLES)

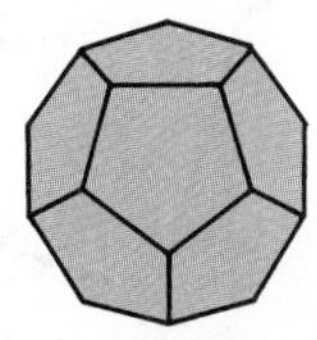

DODECAHEDRON (12 PENTAGONS)

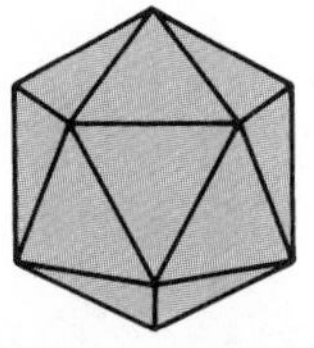

ICOSAHEDRON (20 TRIANGLES)

PRISMS

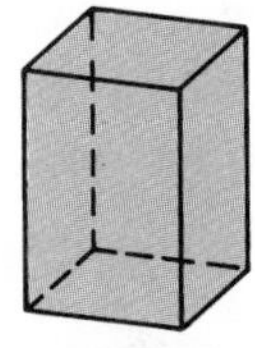

RIGHT SQUARE

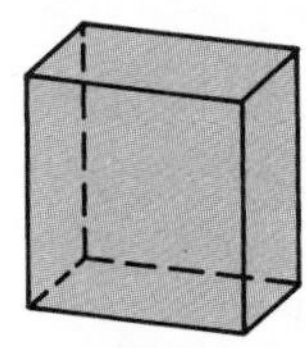

RIGHT RECTANGULAR

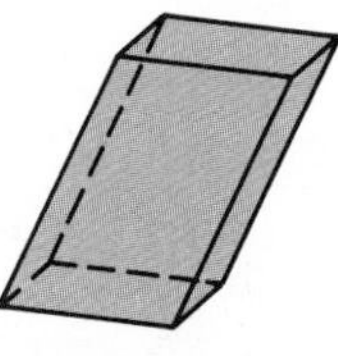

OBLIQUE RECTANGULAR

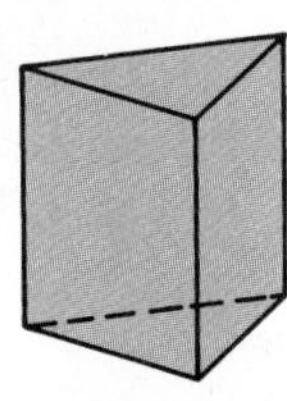

RIGHT TRIANGULAR

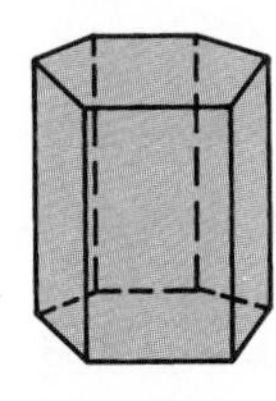

RIGHT HEXAGONAL

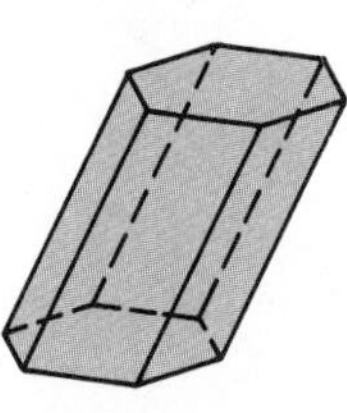

OBLIQUE HEXAGONAL

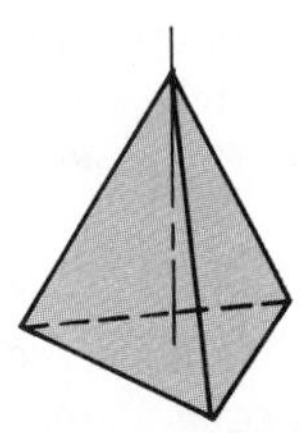

RIGHT TRIANGULAR

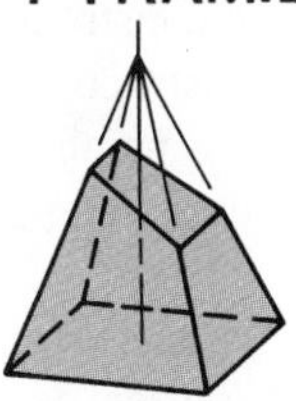

RIGHT SQUARE (TRUNCATED)

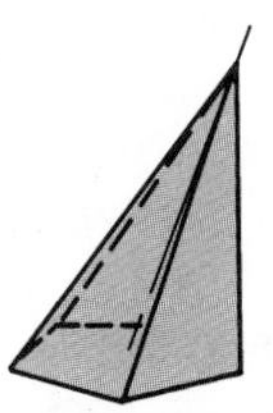

OBLIQUE PENTAGONAL

CONES

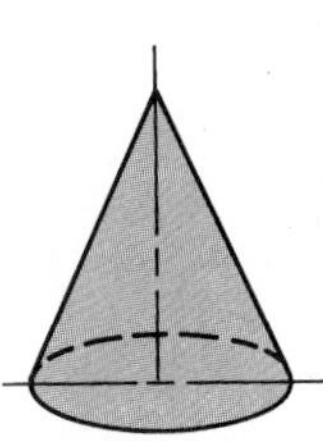

RIGHT CIRCULAR

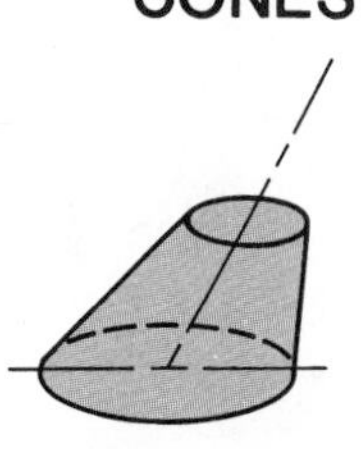

OBLIQUE CIRCULAR (FRUSTRUM)

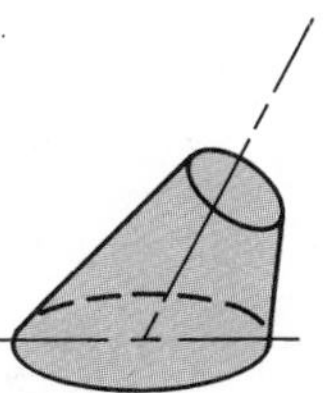

OBLIQUE CIRCULAR (TRUNCATED)

CYLINDERS

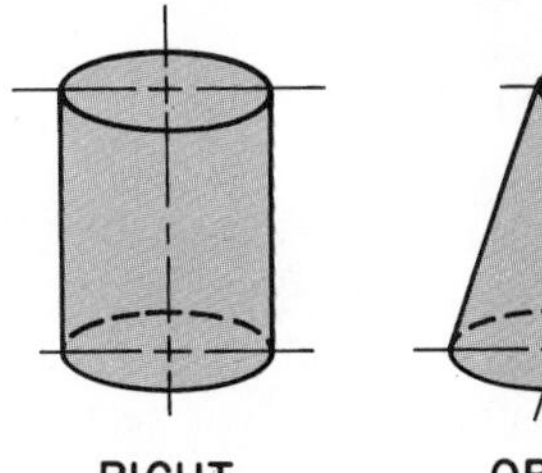

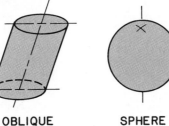

RIGHT CIRCULAR

OBLIQUE CIRCULAR

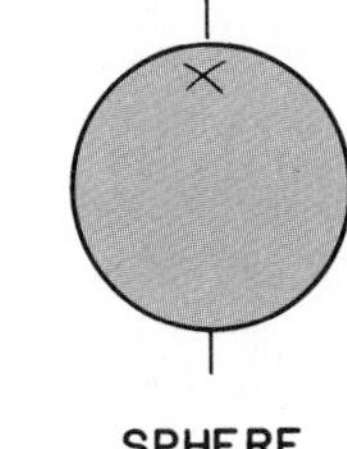

SPHERE

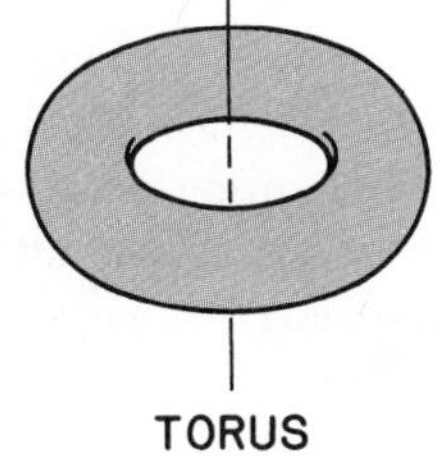

TORUS

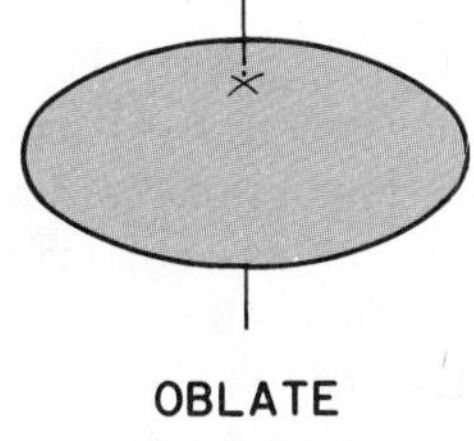

OBLATE ELLIPSOID

FIGURE 4.4 Solids

gle is used in the construction, but any triangle could be used. *Position 1* is the first line drawn. Use the following steps:

1. Move the triangle along the straightedge to *position 2* and draw a parallel line.
2. Rotate the triangle to *position 3*. A line perpendicular to the first two lines is then drawn. Draw on the same edge of the triangle that you used before it was rotated.

4.3.2 Dividing a Line into Equal or Proportional Parts

One way to divide a line is to calculate its length, divide the length by the number of required parts, and then use the result to mark the divisions. However, this method produces an accumulated error. A better way to divide a line equally or proportionally is to use one of the following methods.

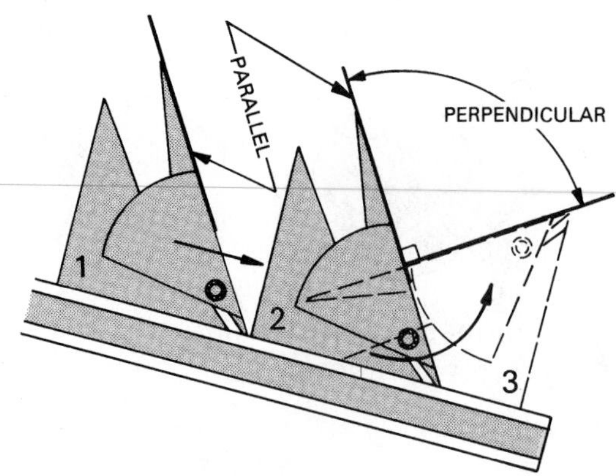

FIGURE 4.5 Drawing Parallel and Perpendicular Lines

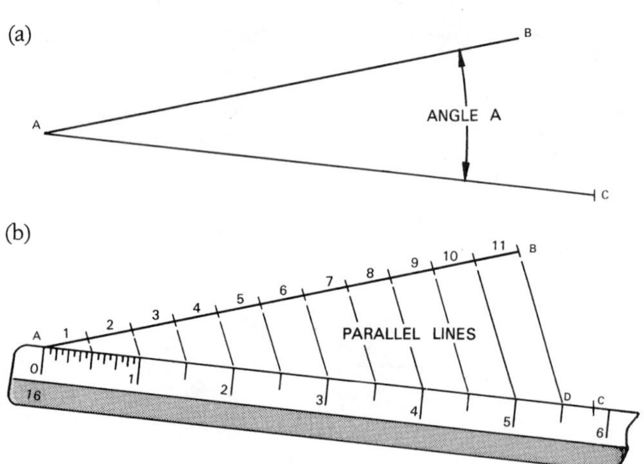

FIGURE 4.6 Dividing a Line into Equal Parts

Figure 4.6 illustrates the **parallel line method** to divide a line equally. Any type of scale and unit of measurement can be used. To simplify the construction and measuring process, choose the unit type and scale that is the most convenient. In Figure 4.6, line AB is to be divided into eleven equal segments. Use the following steps:

1. Draw a construction line AC that starts at either end of line AB. This line is any convenient length (slightly longer works well). Angle A should not be less than 20° or more than 45° or it will be hard to project the divisions from the construction line AC to the original line AB.
2. Find a scale that will approximately divide line AB into the number of parts needed and mark these divisions on line AC. Here, the full-size inch scale was used, with $\frac{1}{2}$ in. marking each division. There are now eleven equal divisions from A to D that lie on line AC.

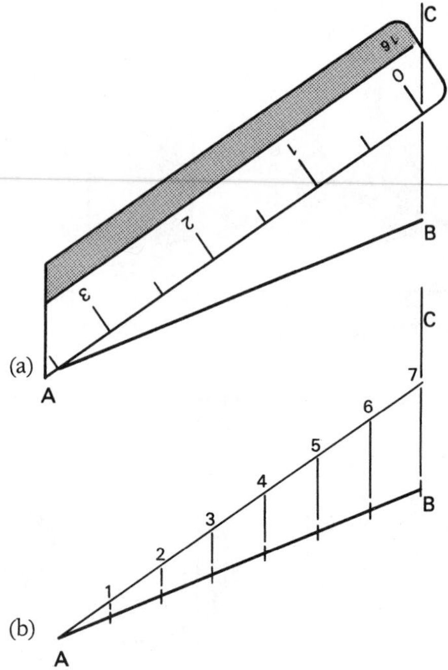

FIGURE 4.7 Dividing a Line into Equal Parts Using the Vertical Method

3. Set the triangle to draw a construction line from point D to point B. Then draw construction lines through each of the remaining ten divisions parallel to line BD by moving the triangle along the straightedge.

It is also possible to use dividers for step 2 to divide the construction lines into the required number of equal parts.

In the **vertical line method** (Fig. 4.7), all of the projection lines are vertical lines and are drawn using a straightedge and any triangle. Any type of scale and unit of measurement can be used. Line AB is to be divided into seven equal parts. Use the following procedure:

1. Draw a vertical construction line BC through point B of line AB.
2. Using point A as the pivot point, position a scale that gives the required number of divisions and equally divides the distance from point A to some point on line BC. Here full-scale U.S. customary units were used—a $\frac{1}{2}$ in. unit of the scale corresponds to each division to mark points 1 through 7. It is necessary to use a scale that gives an overall length of seven units that is longer than the line AB.
3. Using the vertical side of a triangle, draw construction lines from points 1 through 7 to line AB. This establishes seven equally spaced segments along line AB.

In the vertical line and parallel line methods, there is no need to make measurements that are less than one easily measured unit, regardless of the mathematical value of the resulting divisions. The scale is used only to measure equal units. Any scale that will measure equal units may be used.

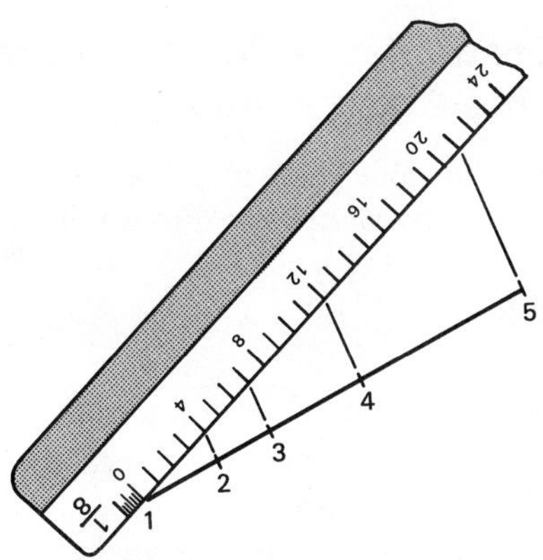

FIGURE 4.8 **Proportional Division of a Line**

4.3.3 Proportional Division of a Line

A line must be divided so the first part is three times as long as the second part. This ratio is written as 3:1 and is read as "three parts to one part" or "three to one." To divide a line proportionally, you use a method similar to that used to divide a line into equal parts.

Figure 4.8 illustrates dividing a line into four parts that have proportions of 4:3:5:9. The following steps are used:

1. Add the proportions of the parts: 4 + 3 + 5 + 9 = 21. This is the number of equal parts that are to be measured on the scale.
2. Draw a construction line at an angle to, and longer than, the given line. Set the scale to make 21 equal divisions. Make the first mark at 4 units, add 3 units and make the second mark at 7 units, add 5 more units and make the third mark at 12 units. Adding 9 units brings the total to 21 units.
3. Project these marks to position points 2, 3, 4, and 5 on the given line using the parallel line method. This creates line segments in proportions of 4:3:5:9.

4.3.4 Bisectors for Lines and Angles

A **perpendicular bisector** of a line *divides that line into two equal parts.* A perpendicular bisector can be constructed using only compass and straightedge (Fig. 4.9) by using the following steps:

1. Set the compass at radius (R) equal to a distance greater than one-half of AB.
2. Using points A and B as centers, draw intersecting arcs to establish intersection points 1 and 2.
3. Draw construction line 1-2 by connecting the two new points. Line 1-2 intersects line AB at its midpoint and is perpendicular to it.

Figure 4.10 shows how to *divide an angle into two equal parts.* Lines AB and BC intersect and form angle ABC. Use the following steps:

1. Set the compass to any convenient radius. For small angles and short lines, extend the lines that form the

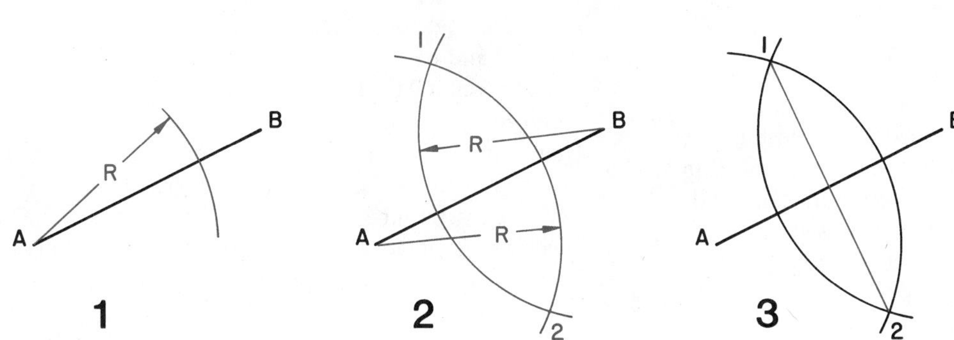

FIGURE 4.9 **Bisecting a Line**

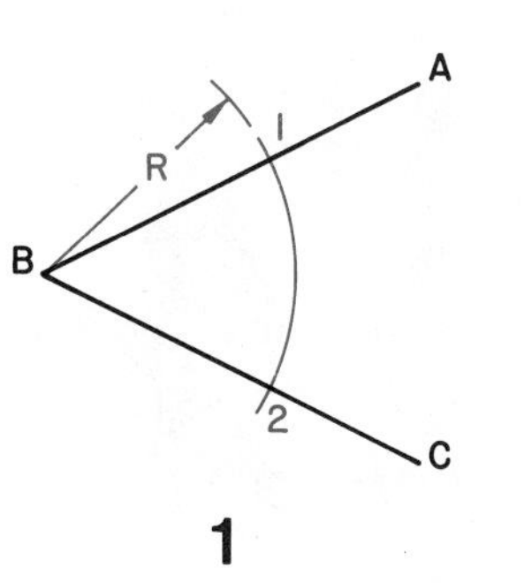

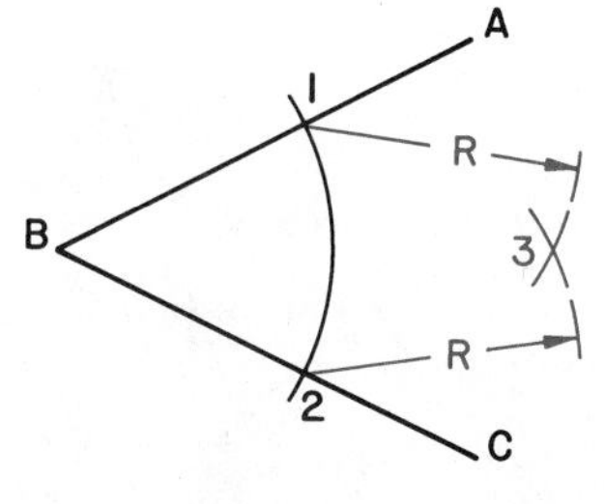

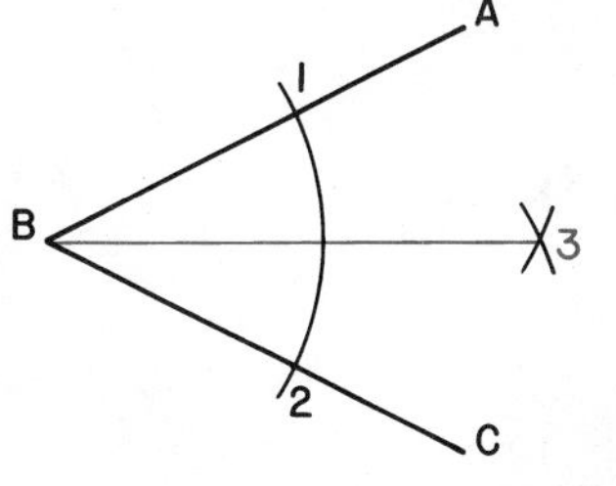

FIGURE 4.10 **Bisecting an Angle**

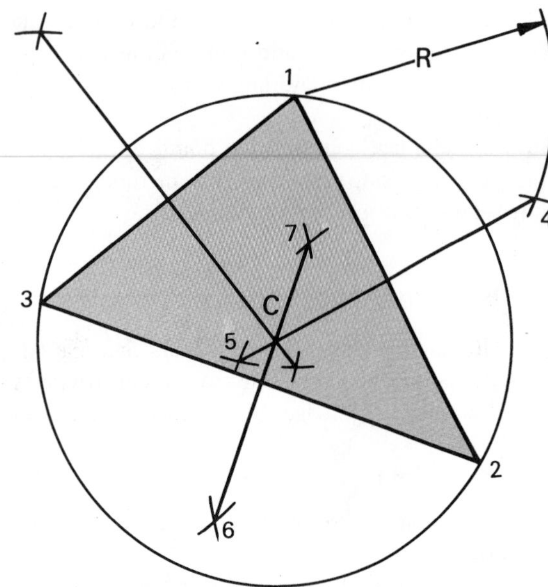

FIGURE 4.11 Finding the Center of a Circle

angle. With point B (vertex) as the center, draw an arc (radius R) that locates points 1 and 2. The length of B1 is equal to B2.
2. Using the radius R, draw arcs from points 1 and 2. Point 3 is the intersection of these two arcs.
3. Draw line 3B. This is the bisector of the angle.

4.3.5 Locating the Center of a Known Circle

The perpendicular bisector of a chord of a circle passes through the center of the circle. If a significant portion of a circle is known, its center can be located by establishing the perpendicular bisectors of any two chords of the circle. In Figure 4.11, chords 1-2, 2-3, and 3-1 form a triangle inside the circle. Bisectors of two of these chords cross at the center of the circle (point C). The third perpendicular bisector, though not necessary, serves as a check.

4.3.6 Construction of a Circle Through Three Given Points

Using the procedure for constructing a perpendicular bisector of a line, you can construct a circle through three given points in space. Figure 4.12 shows points 1, 2, and 3. The following steps were used:

1. Connect the three points with lines and then construct perpendicular bisectors for any two chords of the circle (lines 1-2 and 1-3 here). The perpendicular bisectors intersect at the center of the required circle (C).
2. Draw the circle using the distance from C to any of the three points as the radius (C-1, C-2, or C-3).
3. Check the solution by drawing a perpendicular bisector through chord 2-3.

4.3.7 Inscribed Circle of a Triangle

An **inscribed circle of a triangle** is a circle that is tangent to (touches) each side of the triangle. Figure 4.13 illustrates the procedure for constructing the inscribed circle of a triangle. The given triangle is represented by points 1-2-3. The following steps were used:

1. First, the center of the circle is found by bisecting a minimum of two of the triangle's angles. The angle at point 1 is bisected by drawing arc RA to establish points 4 and 5 (RA is any convenient length).
2. From points 4 and 5 draw equal arcs (RB). Point 6 is the intersection of the two arcs.
3. Draw line 1-6 to establish the bisector of the angle and extend this line beyond point 6.
4. Bisect the angle at point 2 by drawing arc RC to locate points 7 and 8.
5. Establish point 9 by drawing equal arcs (RD) from points 7 and 8.
6. Draw a line from point 2 through point 9 and extend it to intersect the first bisector. The intersection of these two lines determines the center of the circle (C). To check the accuracy of this point, construct a third bisector that will also meet at point C.

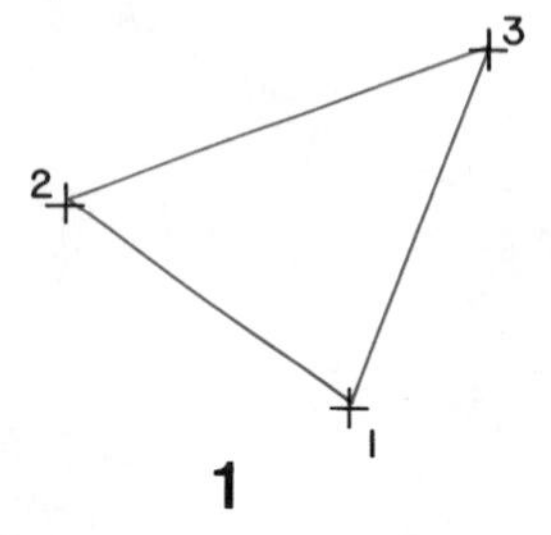

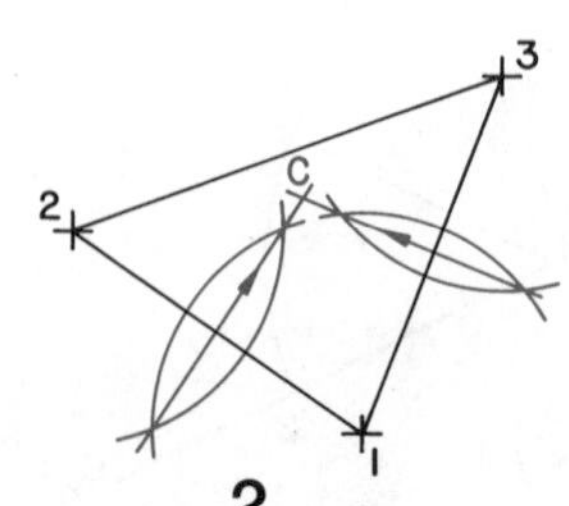

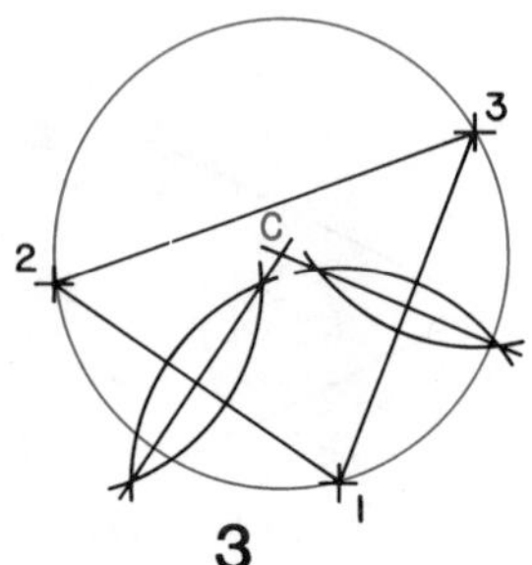

FIGURE 4.12 Drawing a Circle Through Three Given Points

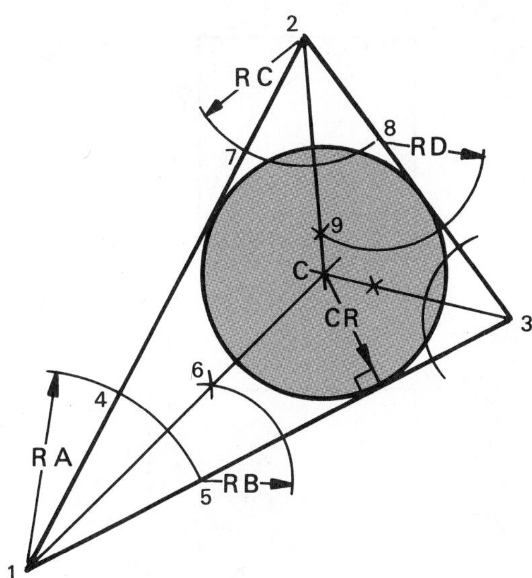

FIGURE 4.13 The Inscribed Circle of a Triangle

7. Draw a line from point C perpendicular to one of the triangle's sides (side 1-3 here) to determine radius CR.
8. To complete the solution, draw the inscribed circle using C as the center and distance CR for the radius.

4.3.8 Circumscribed Circle of a Triangle

A **circumscribed circle of a triangle** touches the three vertex points of a triangle. Constructing a circumscribed circle of a triangle uses the method for constructing a circle through three given points. In Figure 4.11, the perpendicular bisectors of sides 1-2, 2-3, and 1-3 have been drawn. The intersection of the perpendicular bisectors 4-5 and 6-7 establish the center of the circle (C). A third perpendicular bisector (of line 1-3) can be drawn to check for accuracy. The radius of the circle is the distance from C to any of the three points on the triangle (C-1, C-2, or C-3).

4.3.9 Drawing a Triangle with Sides Given

Drawing a triangle, given the sides, is called **triangulation.** Use lines A, B, and C in Figure 4.14 to construct a triangle using the following steps:

1. A, B, and C are given.
2. Draw the baseline (C) and swing an arc as shown.
3. Swing arc B from the end of line C. The intersection of the arcs A and B determines the vertex of the triangle.
4. Draw lines A and B.

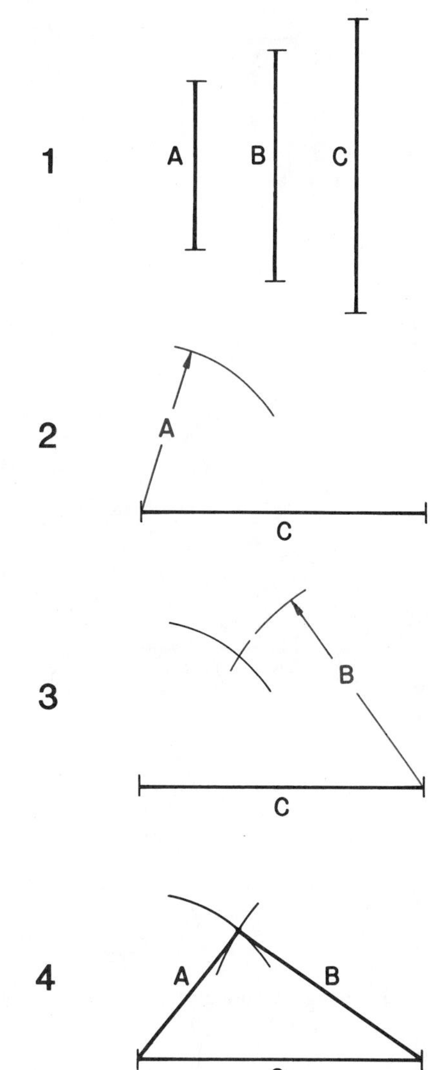

FIGURE 4.14 Drawing a Triangle with Sides Given

4.4 Regular Polygon Construction

Polygons are closed figures having three or more sides. **Regular polygons** have sides of equal length and equal angles. All regular polygons can be *inscribed* within a circle and can also be *circumscribed* about the outside of a circle. (In this case, the circle drawn tangent to the polygon's sides is inscribed in the polygon.)

To draw a particular regular polygon, you must have at least one dimension. The two dimensions that are used for even-number-sided figures (square, hexagon, octagon, etc.) are *across corners* and *across flats.* Across corners is the maximum measurable straight-line distance across the figure and is equal to the diameter of its circumscribing circle. Across flats is the minimum measurable straight-line distance across the figure and is equal to the diameter of its inscribed circle.

4.4.1 Construction of an Equilateral Triangle

The simplest type of regular polygon is the **equilateral triangle**, which has three equal sides and three equal angles

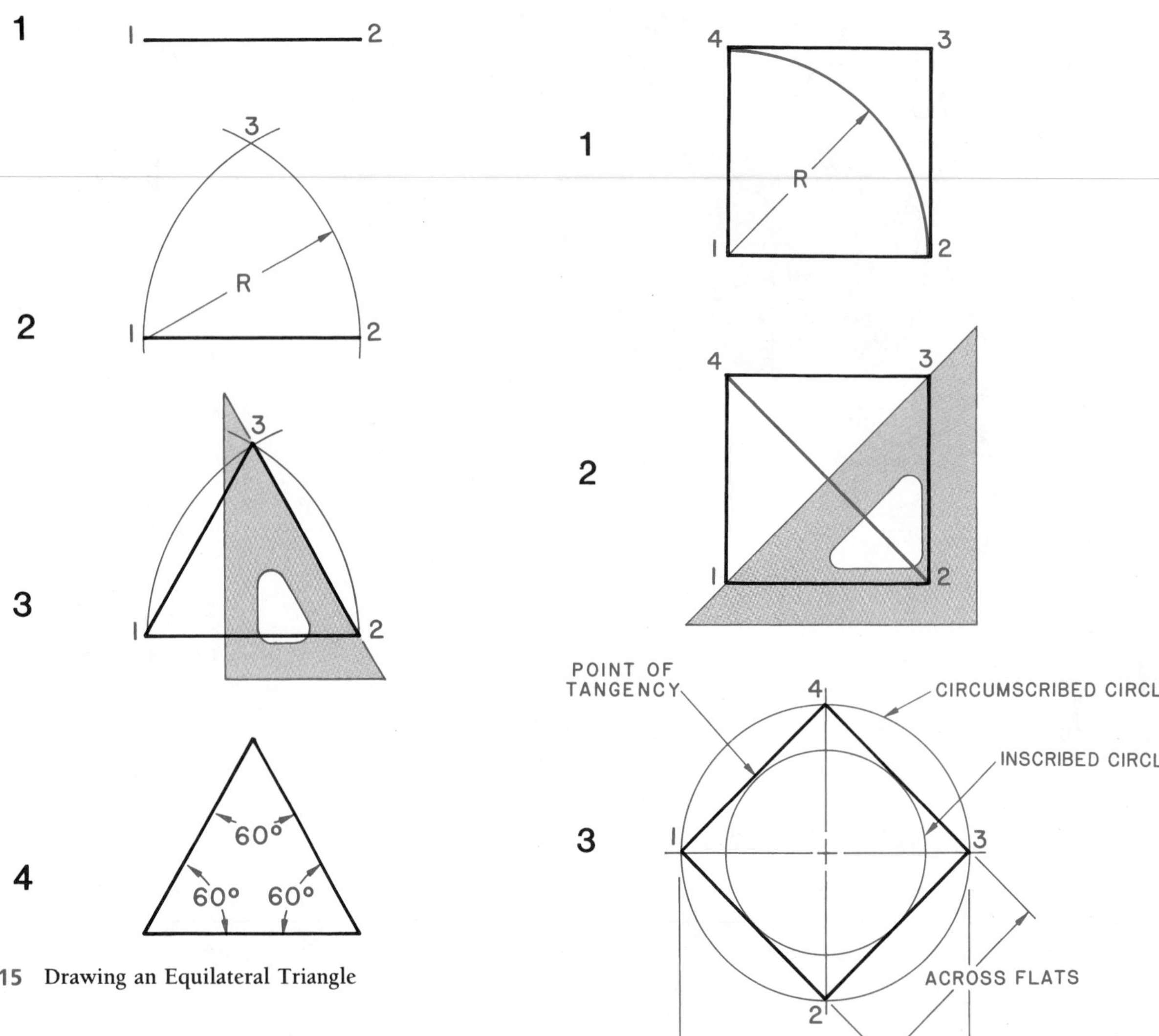

FIGURE 4.15 **Drawing an Equilateral Triangle**

FIGURE 4.16 **Drawing a Square**

(60°). Figure 4.15 shows the steps used to construct an equilateral triangle:

1. Given the length of one side of the triangle (line 1-2), draw the baseline.
2. From endpoints 1 and 2, draw arcs using the side length as the radius. The intersection of the arcs establishes the vertex of the triangle.

Another way to construct an equilateral triangle is to lay out the baseline and then use a 60° triangle to draw each side [Fig. 4.15(3)]. The intersection of the sides establishes the vertex.

4.4.2 Construction of a Square

A **square** has four equal sides and four equal angles. Figure 4.16 shows a square inscribed with a circle. The circle has a diameter equal to the distance across its corners. A circle that is inscribed within a square has a diameter equal to the side of the square. Three methods can be used to construct a square.

In step 1 of Figure 4.16, the base is drawn using the side length. Then an arc R is drawn using point 1 as the center and line 1-2 as the length. The intersection of the arc with a vertical line extended from point 1 establishes the height of the square. The square is then completed.

In step 2 of Figure 4.16, the baseline is drawn and a 45° triangle used to draw lines diagonally through points 1 and 2 to establish points 3 and 4. Points 3 and 4 are at the intersection of the diagonals and lines drawn vertically through points 1 and 2.

Step 3 of Figure 4.16 uses a circle template or a compass to draw inscribed and circumscribed circles to construct a square. The point of tangency is the position where the square's sides touch the circumference.

4.4.3 Construction of a Pentagon

A **regular pentagon** has five equal sides and five equal angles. Figure 4.17 illustrates how to draw a pentagon (the diameter of the circumscribing circle is given):

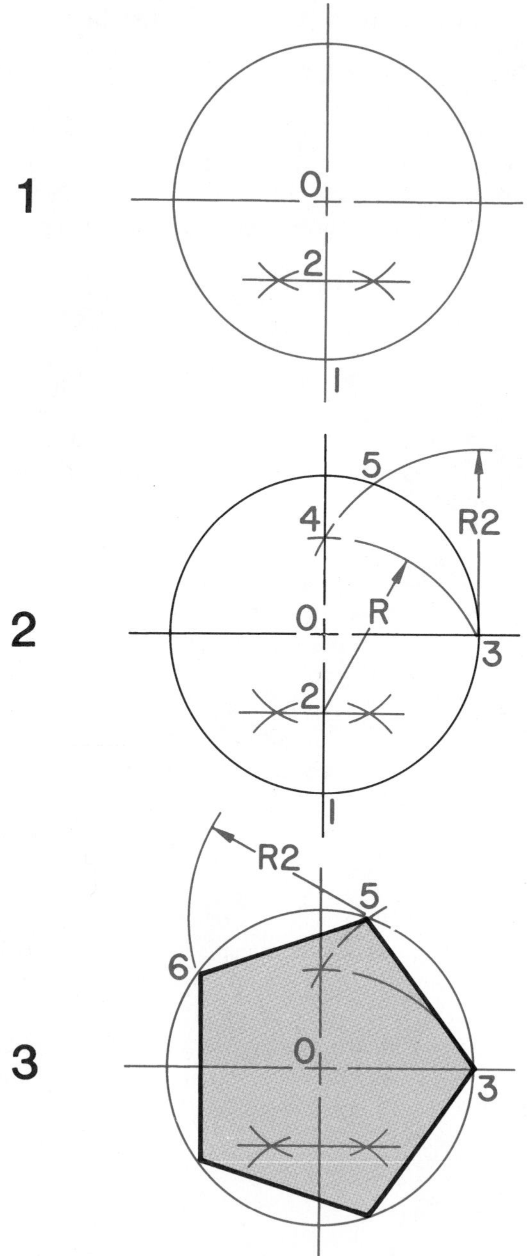

FIGURE 4.17 Drawing a Pentagon

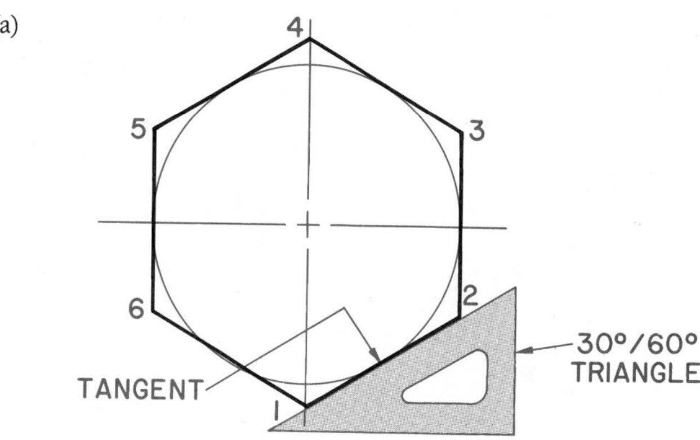

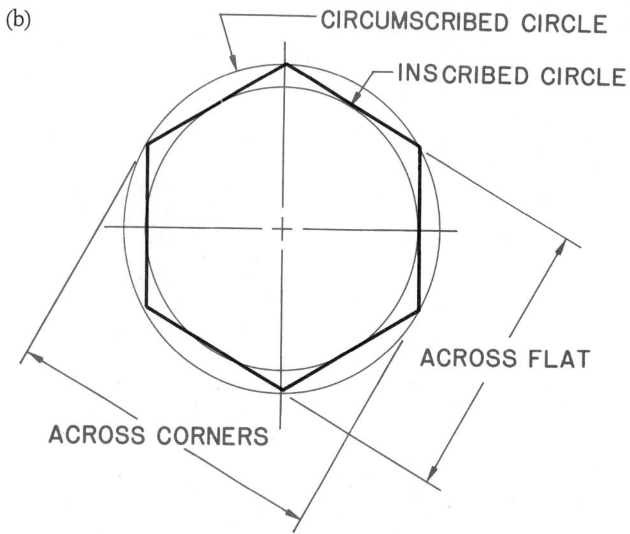

FIGURE 4.18 Drawing a Hexagon Using Inscribed and Circumscribed Circles

1. Draw the centerlines of the figure first and then draw the circle. The center of the circle is point 0.
2. Find point 2 by bisecting line 0-1. Radius R (2-3) is used to establish point 4 on the vertical centerline. The distance from point 3 to 4 (radius R2) is then used to locate point 5 on the circumference of the circle.
3. Draw side 3-5. Use radius R2 from point 5 to establish point 6. Then R2 is used to establish the remaining sides of the pentagon.

4.4.4 Construction of a Hexagon

A **regular hexagon** has six equal sides and six equal angles. Figure 4.18 shows how to construct a hexagon. In this figure, the distance across the flats was known. The distance across the flats is equal to the diameter of the inscribing circle. The following steps were used:

1. First, locate the center of the hexagon.
2. Draw a circle equal to the distance across the flats, then use a 30° angle to construct tangents to the circle.

If you know the distance across the corners, then draw the circumscribed circle first and mark off each side length along the circumference, using a distance equal to the radius of the circle (use dividers) [Fig. 4.19(a)].

Figure 4.19(b) shows an alternative method that can be used to produce a hexagon.

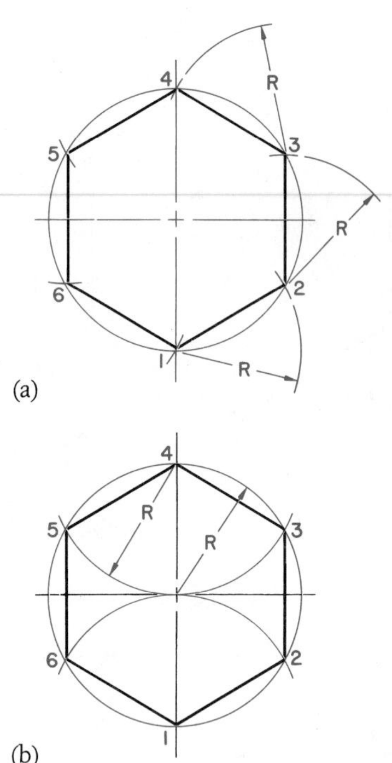

FIGURE 4.19 Drawing a Hexagon

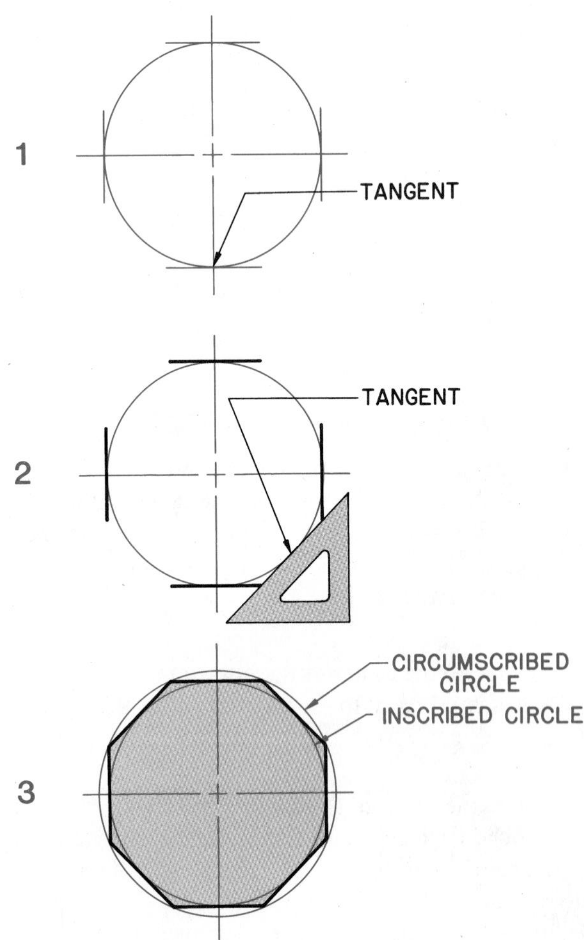

FIGURE 4.20 Drawing an Octagon

4.4.5 Construction of an Octagon

A **regular octagon** has eight equal sides and eight equal angles. To draw an octagon with the distance across the corners known, draw the circumscribed circle first and then mark off the side lengths around the circumference. If you know the distance across the flats, then use a 45° triangle to draw tangent lines to establish the eight sides (Fig. 4.20).

4.4.6 Construction of a Regular Polygon with a Specific Number of Sides

To construct a **regular polygon** with a specific number of sides, divide the given diameter of the circumscribing circle using the parallel line method described earlier (Fig. 4.21). A polygon with seven sides is used as an example. The following steps were used:

1. First, construct an equilateral triangle (0-7-8) with the diameter (0-7) as one of its sides.
2. Draw a line from the apex (point 8) through the second point on the line (point 2).
3. Extend line 8-2 until it intersects the circle at point 9. Radius 0-9 will be the size of each side of the figure.
4. Using radius 0-9, mark off the corners of the polygon and connect the points.

4.5 Tangencies

An arc that touches a line at only one point is tangent to that line and the line is *tangent* to the arc. Two curves can also be tangent. The line and the arc touch at only one place even if they are extended. If a line and an arc are tangent, (1) the tangent line is perpendicular to the radius of the arc at the point of tangency, and (2) the center of the arc is on a line that is perpendicular to the tangent line and extends from the point of tangency.

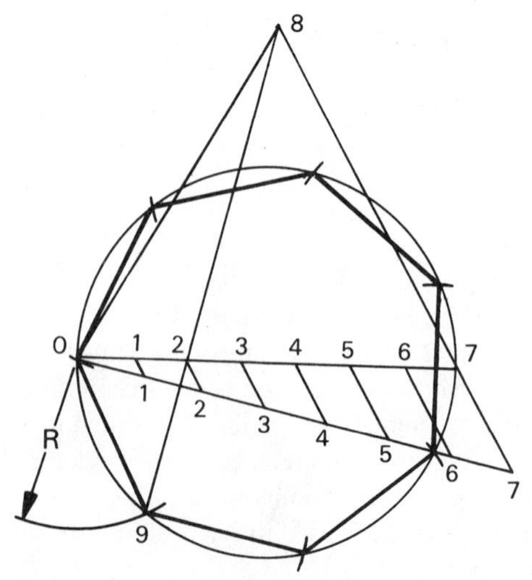

FIGURE 4.21 Drawing a Regular Polygon

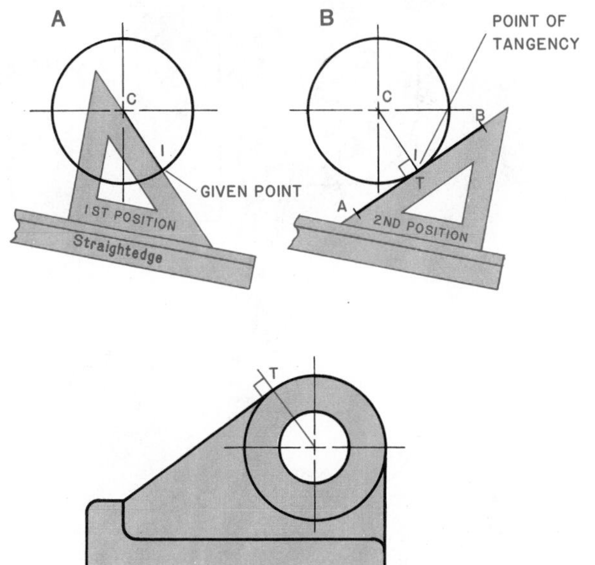

FIGURE 4.22 Drawing a Line Tangent to a Circle

Figure 4.22 illustrates principle 1. To draw a line tangent to the circle at point 1, draw radius C1. Construct line AB perpendicular to the radius line C1 passing through point 1. Line AB is tangent to the circle at point 1.

Figure 4.23 illustrates principle 2. To draw a circle tangent to a given line, first project a line perpendicular to the given line AB from point T (tangent point). The center of the circle will be on this line. Locate the center point by marking off an arc from point T, using the radius of the circle. Using the same radius (line CT), draw the tangent circle.

4.5.1 Line Tangent to Two Circles

Figure 4.24 illustrates the procedure for finding the points of tangency between a line and two circles. A line can be tangent to two circles as shown in Figures 4.24(a) and (b). Four tangency positions are possible. In Figure 4.24(a), the lines are tangent to the outside of the circles. This is called an **open belt tangent.** In Figure 4.24(b), the lines form a **closed belt tangent.** In both examples, the circles are given. The construction is the same as that used for Figure 4.22.

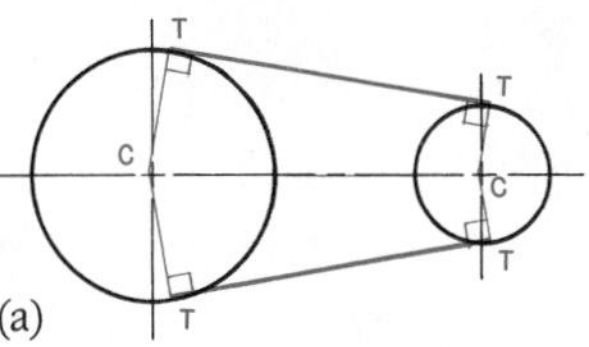

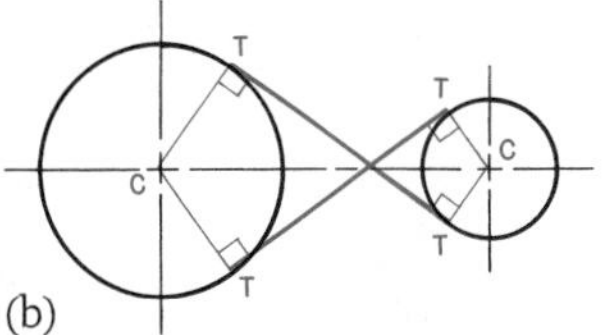

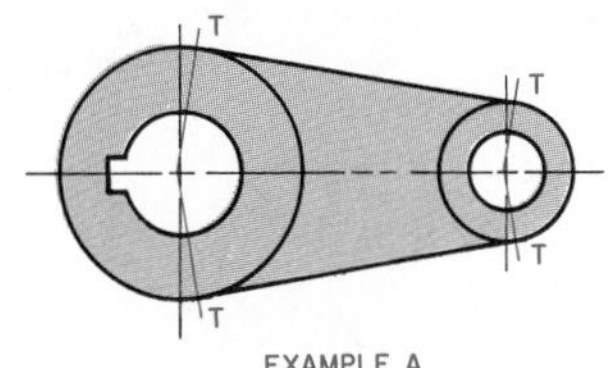

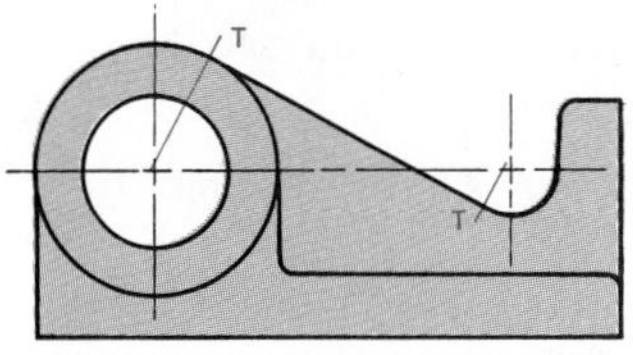

FIGURE 4.24 Tangencies of a Line and Two Circles

4.5.2 Tangent Arcs

There are two methods used to draw an **arc between two perpendicular lines.** Figure 4.25(a) illustrates the construction of the arc 2-3 using only a compass:

1. Extend the two given perpendicular lines to meet at point 1.
2. From point 1 strike a radius equal to the required radius of the tangent arc. The intersection of this radius and the given lines establishes tangent points 2 and 3.

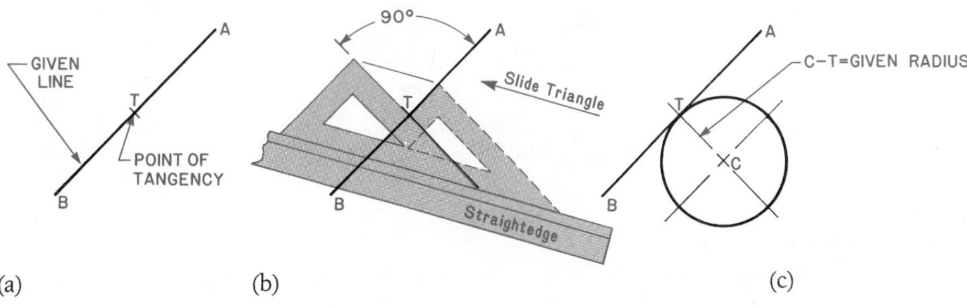

FIGURE 4.23 Drawing a Circle Tangent to a Line

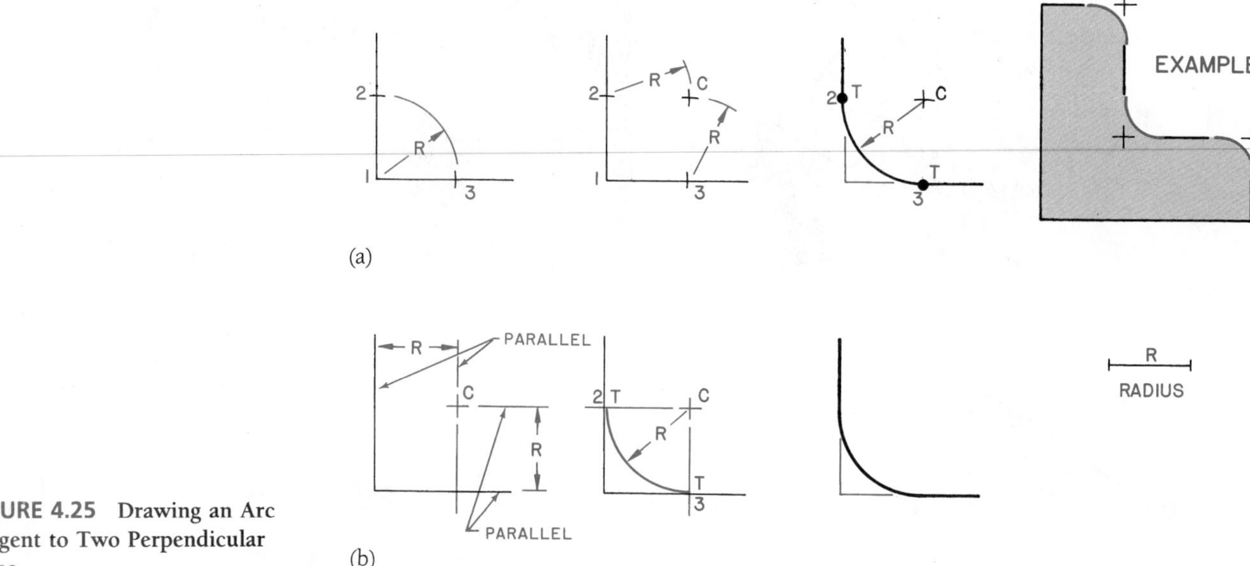

FIGURE 4.25 Drawing an Arc Tangent to Two Perpendicular Lines

3. Using the same radius, strike construction arcs from points 2 and 3. The intersection of these two arcs, at point C, establishes the center of the tangent arc.
4. From point C, draw arc 2-3 tangent to both perpendicular lines.
5. Locate points of tangency on both lines.

Figure 4.25(b) illustrates a second method:

1. Extend the given perpendicular lines so that they meet at point 1.
2. Draw a parallel line distance R from each of the given lines using the required tangent arc radius for dimension R. Point C is at the intersection of these two lines.
3. Locate tangent points 2 and 3 by extending construction lines from C perpendicular to the given lines.
4. From center point C, draw the required tangent arc from point 2 to 3.

To draw arcs that are tangent to nonperpendicular lines, use the same procedure as in Figure 4.25(b). This method can be used for lines at acute (Fig. 4.26) or obtuse (Fig. 4.27) angles. In Figures 4.26 and 4.27, the given lines have been extended to meet at point 1. Use the following steps:

1. Draw construction lines parallel to and at distance R from the given lines using the required tangent arc radius.
2. Where these two lines intersect (point C), draw construction lines perpendicular to the given lines to establish points 2 and 3 as the points of tangency.
3. Draw radius R from point 2 to 3 to establish an arc tangent to both given lines.
4. Darken the lines and the arc to form a continuous smooth figure.

4.5.3 Drawing an Arc Tangent to a Line and an Arc

To construct an arc tangent to a line on one side and an arc on the other (Fig. 4.28) use the following procedure. The line

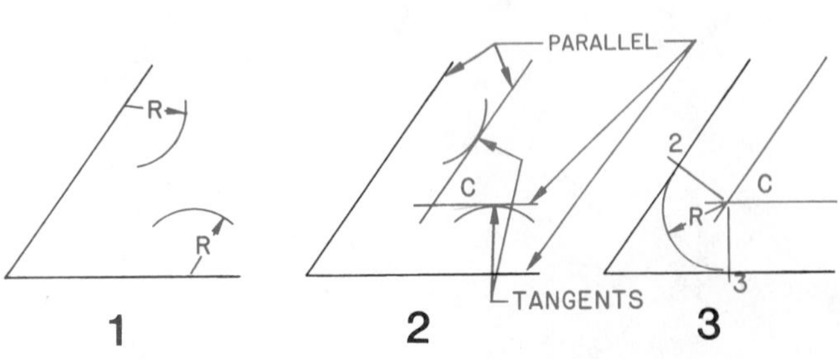

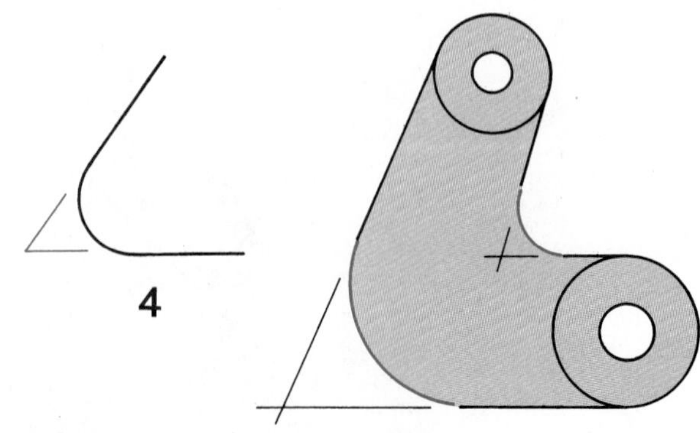

FIGURE 4.26 Drawing an Arc Tangent to Two Lines Forming an Acute Angle

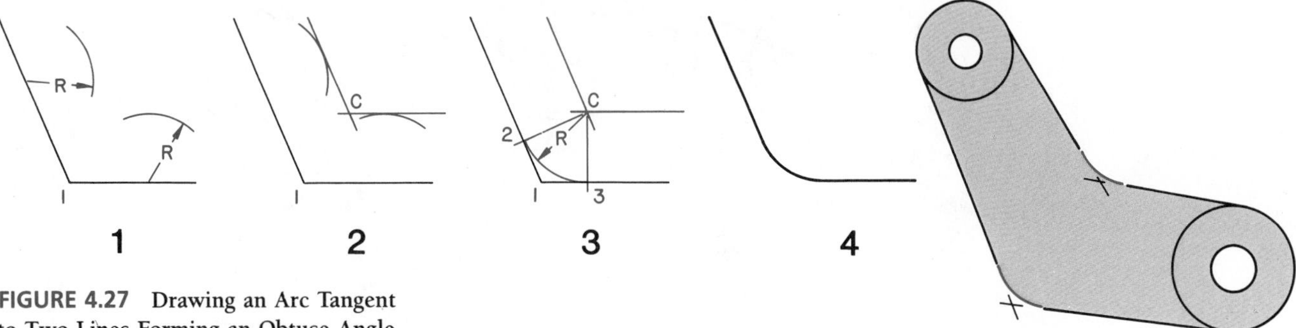

FIGURE 4.27 **Drawing an Arc Tangent to Two Lines Forming an Obtuse Angle**

and the arc are given along with the required radius R for the tangent arc.

1. Draw the given line and arc.
2. Draw a construction line parallel to and at distance R from the given line. Add R and R1 to establish R2. Use R2 to swing a construction arc until it intersects the construction line at point C.
3. Using R, swing an arc tangent to the line and the given arc. Use C as the center point. Draw construction lines from the center of the given arc to C and from C perpendicular to the given line. These construction lines locate the points of tangency (T).
4. Darken the line and the arcs, forming a smooth, consistent line.

4.5.4 Drawing an Arc Tangent to Two Arcs

To construct an arc tangent to two arcs or circles, lay out the given arcs as shown in Figure 4.29. Here R1 and R2 are given along with the distance between their centers. The radius length (R) of the tangent arc is also provided.

1. Add the radius length R to R1. Use this length to draw a construction arc.
2. Add R to R2 and draw another construction arc. The two construction arcs intersect at C.
3. Using the given radius length (R), draw the tangent arc with C as the center.
4. The point of tangency (T) is located by drawing a line

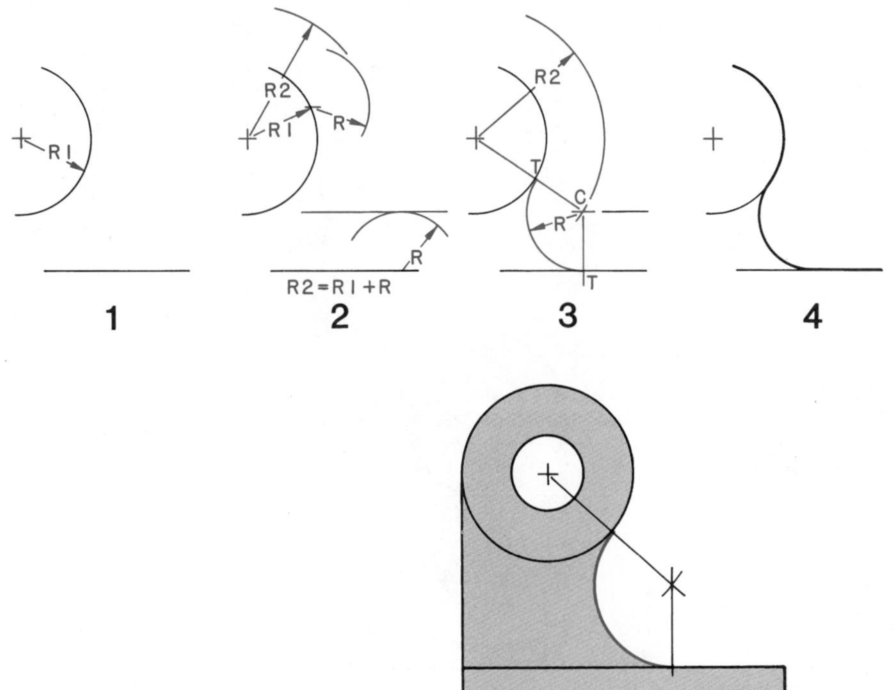

FIGURE 4.28 **Drawing an Arc Tangent to a Line and an Arc**

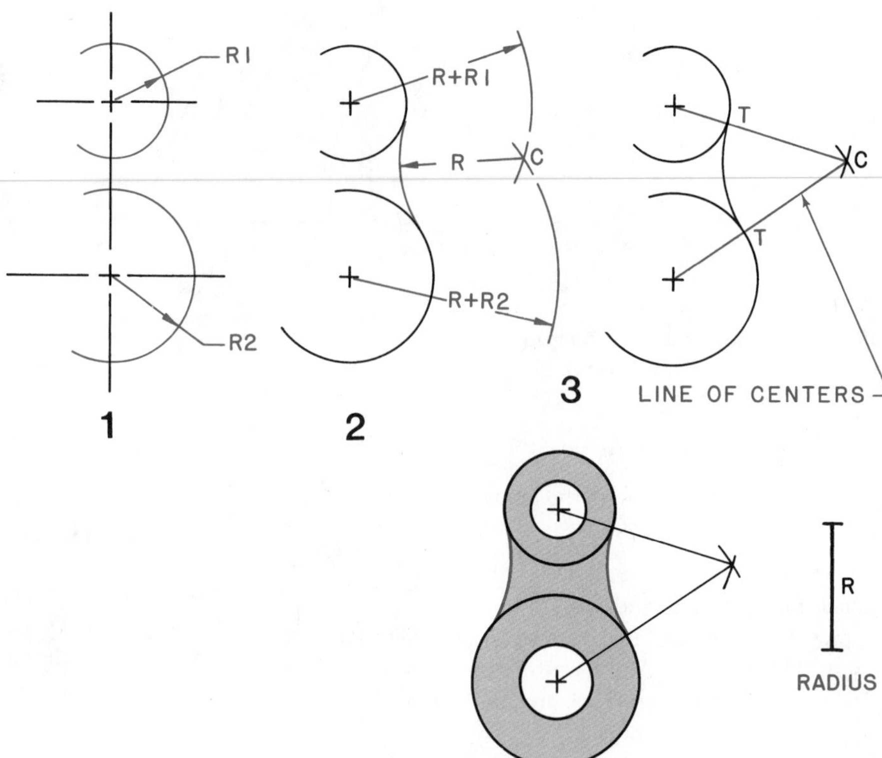

FIGURE 4.29 Drawing an Arc Tangent to Two Arcs

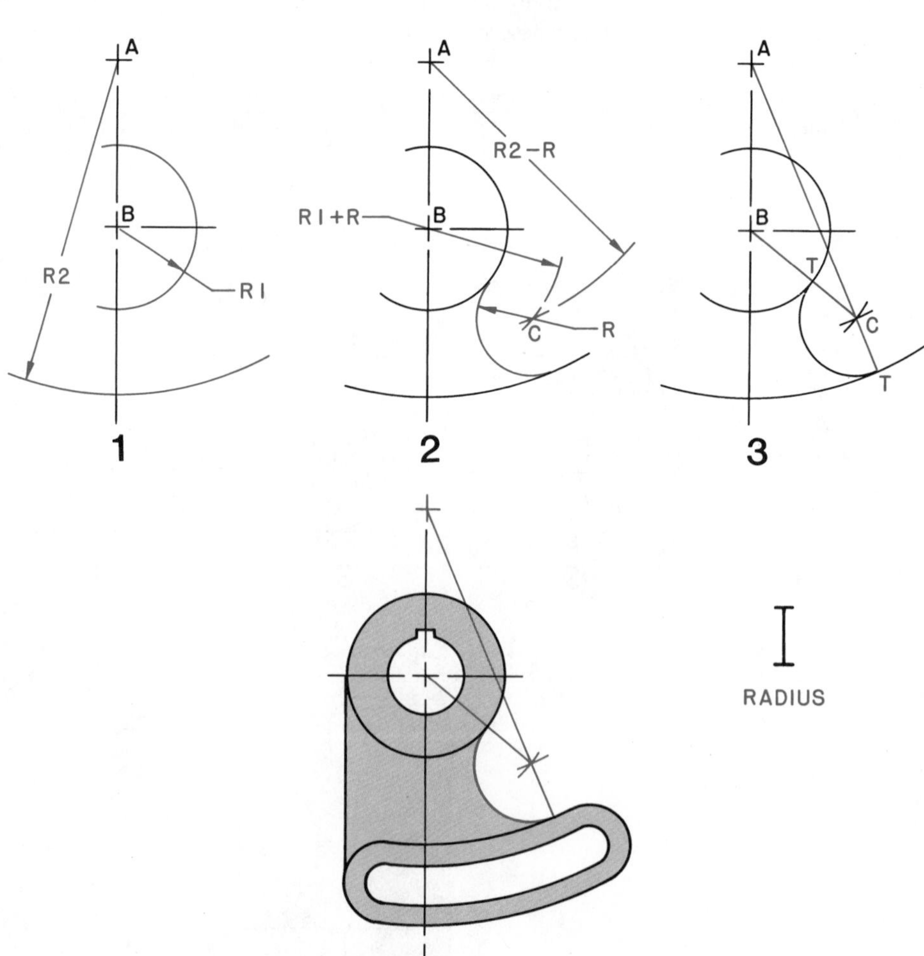

FIGURE 4.30 Constructing a Tangent Arc Between Two Arcs

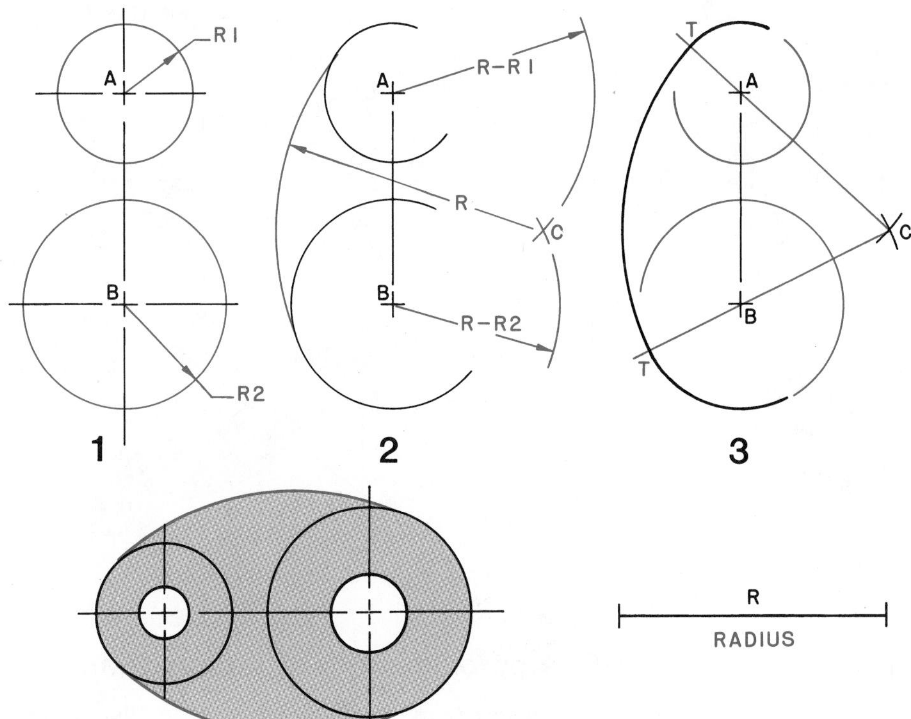

FIGURE 4.31 Drawing an Outside (Enclosing) Arc Tangent to Two Arcs

from C to each center of the given arcs. This line is called the *line of centers.*

4.5.5 Drawing an Arc Tangent to Two Arcs with One Arc Enclosed

In Figure 4.30, a tangent arc joins two arcs. In this example, the tangent arc becomes tangent to the inside of one arc and tangent to the outside of the other. The arcs are given along with their centers, A and B. The radius length (R) of the tangent arc is also provided. The following method is used:

1. Locate centers A and B and construct the two given arcs, Rl and R2.
2. *Add* R and Rl and use this length to draw a construction arc from B. *Subtract* R from R2, and use this length to draw a construction arc from A. The intersection of these two construction arcs locates C. Using the given tangent arc radius (R), draw an arc from C tangent to both given arcs.
3. The point of tangency is located by drawing a line (line of centers) from A through C until it intersects the large arc at T. The other point of tangency (T) can be located by drawing a line from B to C.

4.5.6 Drawing an Arc Tangent to Two Arcs and Enclosing Both

In Figure 4.31, the tangent arc encloses both given arcs. Rl, R2, A, and B are given along with the tangent arc radius R. The tangent arc is drawn by using the following steps:

1. Lay out the two given arcs as shown in the figure (R1 and R2).
2. C is found by first subtracting the radius length R1 from R. Use this length to draw a construction arc. Subtract R2 from R and draw another construction arc. The intersection of these two arcs locates C. Using R, draw an arc with C as its center and its ends tangent to the two given arcs.
3. The exact point of tangency is determined by drawing construction lines from C to A and from C to B, extending both until they intersect the arc as shown. The intersection of these lines and the two given arcs locates the two tangent points (T).

4.5.7 Drawing Ogee Curves

An **ogee curve** is used to connect two parallel lines with tangent arcs. In Figure 4.32, lines l-2 and 3-4, their parallel distance, and their location in space are given. The curve is constructed by the following process:

1. Draw lines 1-2 and 3-4. Connect points 2 and 3 and bisect this new line (2-3) to locate point A.
2. Bisect lines 2-A and 3-A. Extend these bisectors until they intersect perpendiculars drawn from points 2 and 3. This will locate points B and C.

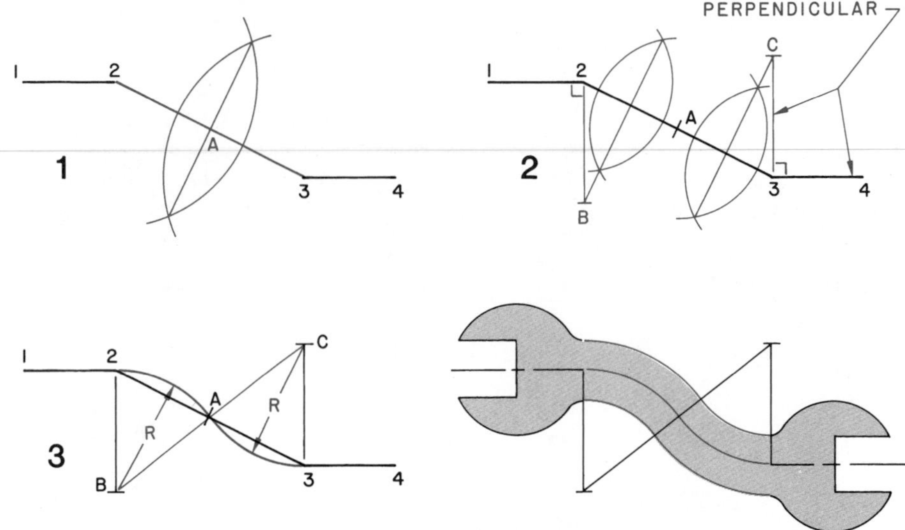

FIGURE 4.32 Drawing an Ogee Curve

3. Draw arcs (R) using the distance from B to 2 (or C to 3). The points of tangency are 2 and 3 for the arcs and the lines, and A for the two arcs.

You May Complete Exercises 4.1 Through 4.4 at This Time

4.5.8 Rectifying Circles, Arcs, and Curves

Circles and arcs can be laid out (*rectified*) along a straight line. Their true length (circumference or arc length) is layed off along a straight line. All rectification is approximate but is still graphically acceptable within limits.

To **rectify** the circumference of a circle means to find the circumference graphically. In Figure 4.33, the circumference of the circle has been established by rectification:

1. Draw line 2-5 tangent to the bottom of the circle and exactly three times its diameter.
2. Draw line C-3 at an angle of 30°.
3. Draw line 3-4 perpendicular to the vertical centerline (line 1-2) of the circle.
4. Connect point 4 to point 5. Line 4-5 will be approximately equal to the circumference of the given circle.

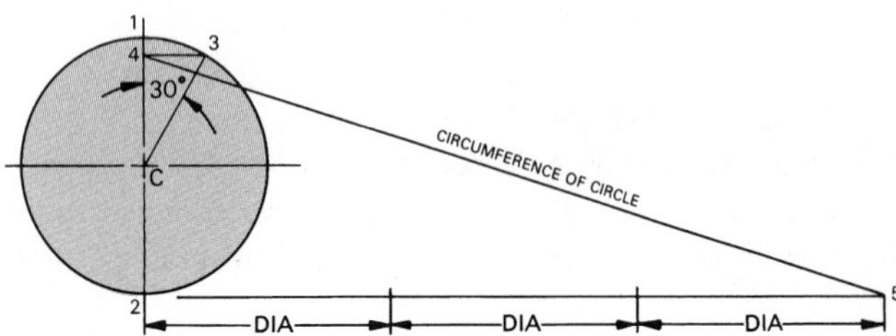

FIGURE 4.33 Rectifying a Circle

4.5.9 Approximate Rectification of an Arc

To rectify an arc or curved line, start by drawing a line tangent to one end. In Figure 4.34, the line was drawn tangent to the curved line at point 1. (Note that it is not necessary to have point A, although it does help to establish the exact tangent points.) The following steps are used:

1. Use dividers to mark off very small equal distances along the curve. The smaller the distance, the more accurate the approximation because each distance will be the

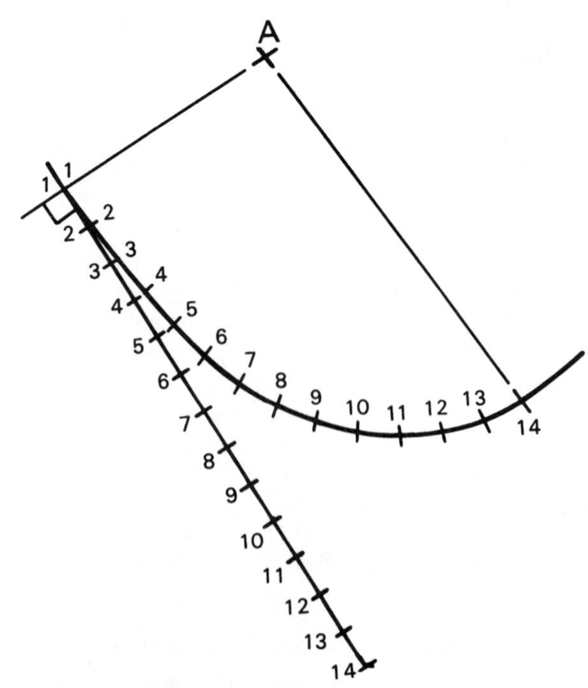

FIGURE 4.34 Rectification of an Arc

chord measurement of its corresponding arc segment and, therefore, will be somewhat shorter than the arc's true length.

2. Starting at the opposite end of the arc, away from the side with the tangent line, mark off equal chords, point 14 to point 13, 13 to 12, 12 to 11, 11 to 10, and so on. Continue marking off each division until less than one full space remains, which is at point 2 in the given example.
3. Without lifting the dividers, start dividing the tangent line into the same number of segments, 2 to 3, 3 to 4, 4 to 5, and so on. The tangent line 1-14 will approximately equal the length of the given arc.

4.6 Conic Sections

A **right circular cone** is one in which the altitude and the axis coincide (the axis in perpendicular to the base). The intersection of a plane and a right circular cone is called a conic section. Five possible sections can result from this intersection (Fig. 4.35). The shapes formed by the sections are:

1. **Parabola** A plane (EV 1) passes parallel to a true length element (edge) of the cone, forming the same base angle (angle between the base and the edge) and resulting in a **parabola.**

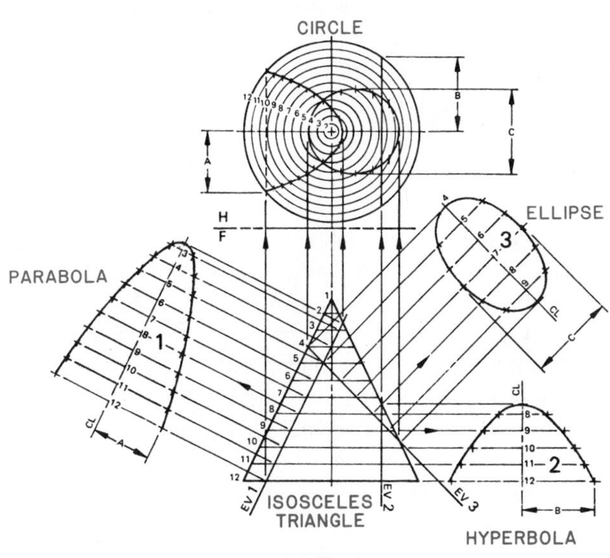

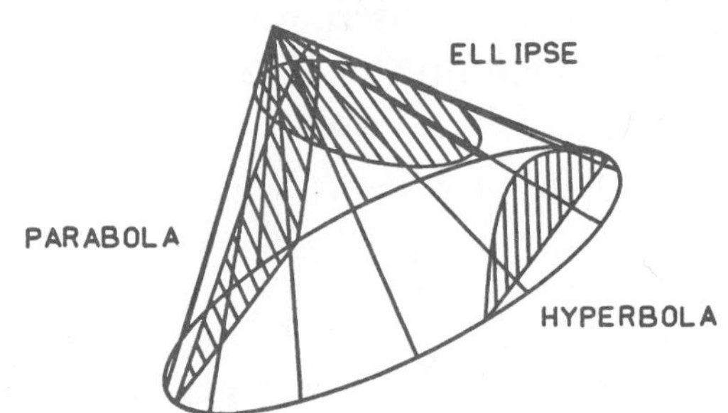

FIGURE 4.35 Conic Sections

2. **Hyperbola** A plane (EV 2) passing through a cone, parallel to the altitude and perpendicular to the base, results in a **hyperbola.**
3. **Ellipse** A plane (EV 3) that cuts all the elements of the cone, but is not perpendicular to the axis, forms a true **ellipse.**
4. **Triangle** A plane that passes through the vertex and is parallel to the axis cuts an isosceles (or equilateral) **triangle** (front view).
5. **Circle** A plane that passes perpendicular to the axis forms a circular intersection. In Figure 4.35, a series of horizontal cutting planes have been introduced in the frontal (front) view, which project as **circles** in the horizontal (top) view.

4.6.1 Intersection of a Cone and a Plane

The **intersection of a cone and a plane** is established by passing a series of horizontal cutting planes through the cone (perpendicular to its axis). In Figure 4.35, the front and top views of the cone are shown along with the edge view of three planes that intersect it. To find the top view and the true shape of each intersection, use the following steps:

1. In the front view, pass a series of evenly spaced horizontal cutting planes through the cone, CPl through CP12.
2. Each cutting plane projects as a circle in the horizontal view.
3. EV 1 intersects cutting planes 3 through 12 in the frontal view. Project intersection points to the top view. The intersection of EV 1 and the cone forms a parabola (1).
4. The true shape of the parabola is seen in a view projected parallel to EV 1. Draw the centerline of the parabola parallel to EV 1, and project the intersection points of the plane (EV 1) and each cutting plane from the front view. Distances are transferred from the horizontal view, as in dimension A.
5. Repeat steps 3 and 4 to establish the intersection of EV 2 and EV 3 with the cone. EV 2 projects as a line in the top view and as hyperbola in a true shape view (2). EV 3 forms an ellipse in the top view and projects as a true size ellipse in view (3).

4.6.2 Ellipse Construction

Two methods of constructing an **ellipse** are covered in this section: the **concentric circle method** and the **four-center method.** The two methods are useful for constructing oddly sized or large ellipses. Both methods *approximate* the shape of a true ellipse. Figure 4.36 illustrates the concentric circle method of constructing an ellipse:

1. Given the major axis A-B and the minor axis C-D, draw concentric circles (circles of a different size with the same center point) using the axes as diameters.

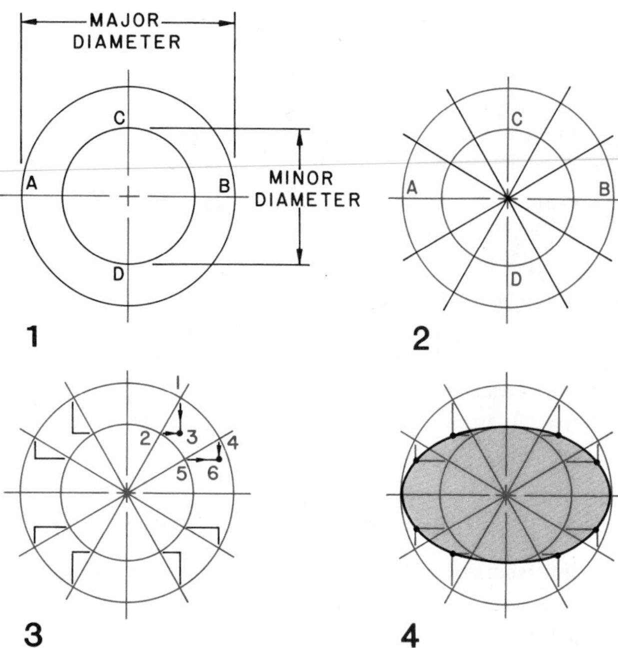

FIGURE 4.36 Drawing an Ellipse Using the Concentric Circle Method

2. Divide the circles into an equal number of sections. Figure 4.36 uses 12 equal divisions.
3. Where each line crosses the inner circle (point 2 or 5), draw a line parallel to the major axis; where the same line crosses the outer circle (point 1 or 4), draw a line parallel to the minor axis. The point of intersection of these two lines (point 3 or 6) will be on the ellipse.
4. Repeat this process for each division of the circles. Use an irregular curve to connect the points smoothly. It is accurate in direct proportion to the number of divisions used and points located.

Figure 4.37 uses the approximate method, also called the four-center method, to construct the ellipse:

1. With the major axis (A-B) and minor axis (C-D) given, connect points B and C.
2. Using the distance from the center of the ellipse to point B as the radius, strike arc R1. Point 1 is the intersection of R1 and the extended minor axis.
3. The distance from point C to point 1 is used to establish R2. Draw arc R2 so that it intersects line B-C at point 2.
4. Bisect line B-2 and extend the bisector 5-6 so that it crosses the minor axis at point 3. Point 3 is the center point for radius R4.
5. Where bisector 5-6 crosses the major axis (point 4), draw radius R5 to establish the sides of the ellipse at point A and point B. R4 is the radius for the upper and lower arc at point C and D of the ellipse.

These two methods work best when *the minor axis is at least 75% of the major axis.* When the minor axis is too small in comparison to the major axis, the top and bottom of the ellipse are flattened. The closer the major axis and the minor axis are in length, the more accurate the ellipse.

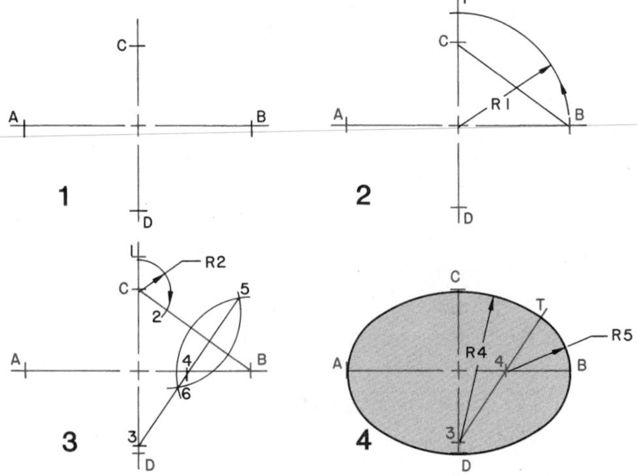

FIGURE 4.37 Drawing an Ellipse Using the Approximate Method (Four-Center Method)

4.6.3 Parabola Construction Using a Rectangle or Parallelogram

A **parabola** is the result of an intersection between a cone and a plane passed parallel to one of its elements. It is a plane curve, generated by a point moving so that its distance from a fixed point, known as the *focus,* is always equal to its distance from a fixed line. Parabolas are used in the design of surfaces that need to reflect sound or light in a specific manner. The construction of a parabola is demonstrated using a rectangle in Figure 4.38(a) and a parallelogram in Figure 4.38(b):

1. Divide side BC into an even number of equal parts and side AB into half as many equal parts.
2. Connect the points along AB and CD to E.
3. Draw parallel lines from the points along BC to where they intersect the lines drawn in step 2. The intersection points are points along the parabola's curve.
4. Connect the intersection points using an irregular curve. The greater the number of divisions, the greater the accuracy of the curve.

4.6.4 Parabola Construction by Establishing the Intersection of a Plane and a Cone

Figure 4.39 illustrates the step-by-step procedure for constructing a parabola by establishing the intersection of a plane and a cone:

1. Draw the given cone and the intersecting plane in the front and top views. A parabola is formed by an

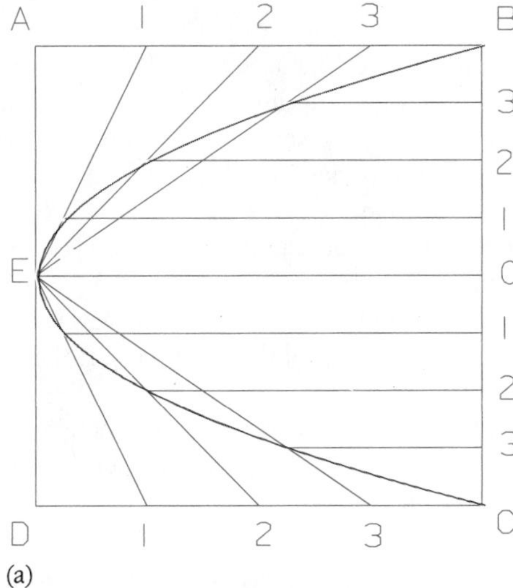

(a)

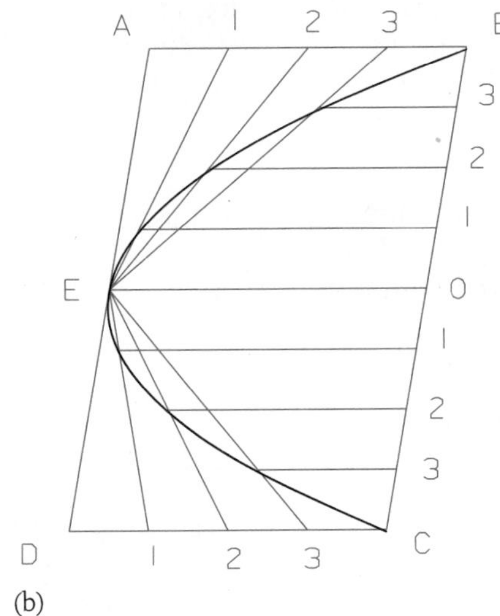

(b)

FIGURE 4.38 Drawing a Parabola

intersecting plane that is parallel to one of the cone's elements-edge lines (front view).

2. Draw any number of concentric circles in the top view. The greater the number of circles, the greater the accuracy of the parabola. Figure 4.39(b) uses only two circles in order to provide a clear picture of the process.
3. Project the circle to the front view. The intersection of the circle (seen as an edge) and the plane establish points X and Y in the front view.
4. Project points X and Y to the top view and complete the top view of the parabola.
5. Using point 2 as the center, draw arcs using lengths 2-Y, 2-X, and 2-1 until they intersect the base plane in the front view. Project these points to the top view.
6. Draw horizontal lines from each intersecting point in the top view until they intersect with corresponding points projected from the front view.
7. Connect these points to form the true view of the parabola.

4.6.5 Connecting Two Points with a Parabolic Curve

Figure 4.40 shows three parabolic curves. In each case, points X, Y, and 0 are given. The following steps are used for the construction:

1. Draw lines X-0 and Y-0.
2. Divide each line into the same number of equal parts and number the divisions.
3. Connect the corresponding points with construction lines.
4. Sketch a smooth curve that is tangent to each of the elements as shown.
5. Use an irregular curve to draw the curve.

(a)

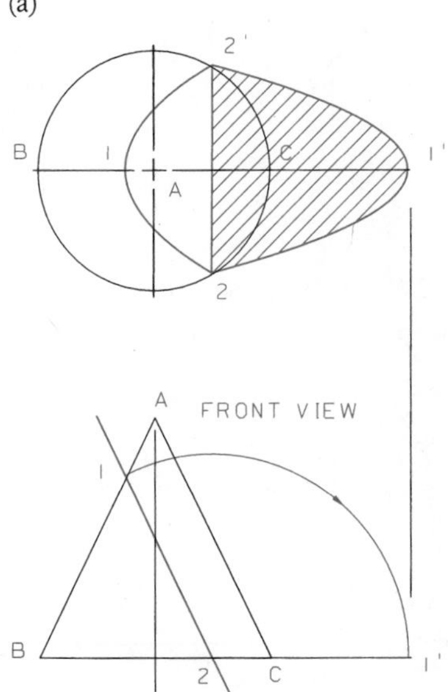

(b)

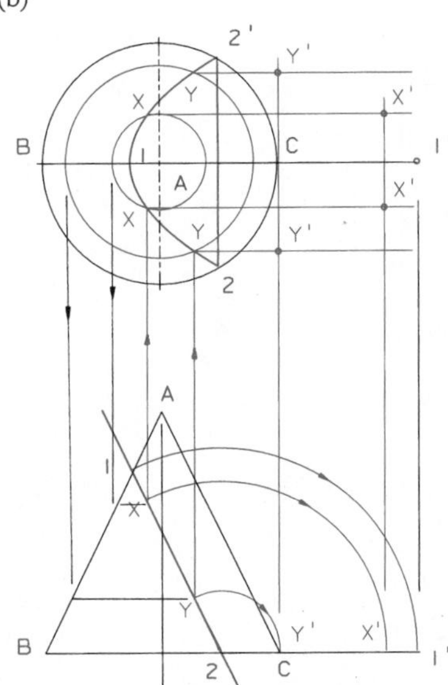

(c)

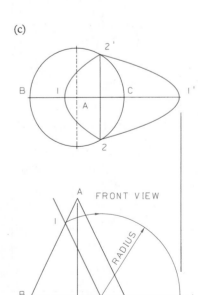

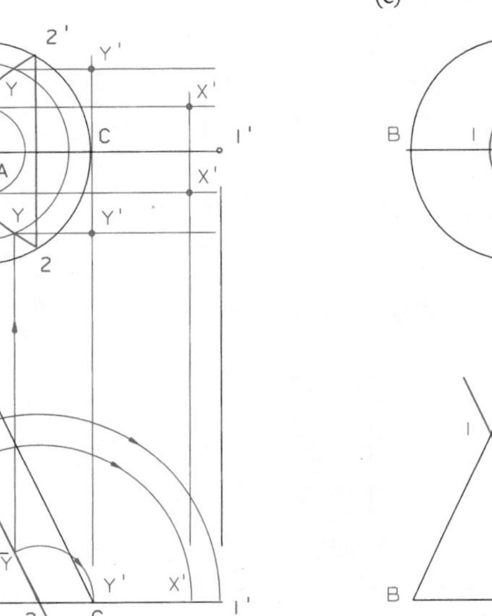

FIGURE 4.39 Construction of a Parabola

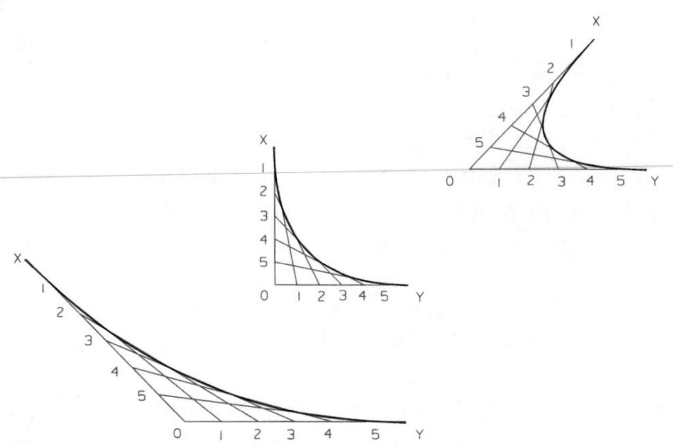

FIGURE 4.40 Drawing Parabolic Curves

4.6.6 Hyperbola Construction

A **hyperbola** is a plane surface (curve) that is formed by the intersection of a right circular cone and a vertical plane. Figure 4.41 illustrates the procedure for drawing a hyperbola:

1. Draw the cone and plane in the top and front view [Fig. 4.41(a)].
2. Construct a number of planes parallel to the base [Fig. 4.41(b)]. These planes form concentric circles when they intersect the cone (top view). The greater the number of planes, the greater the accuracy of the hyperbola.
3. In the front view, the intersection of each edge of the planes intersects the vertical plane and establishes four points in the top view: X, X′, Y, and Y′.
4. The vertical plane appears as an edge in the top view. Draw horizontal construction lines from each intersecting point as shown in Figure 4.41(b).
5. Draw the required arcs in the front view until they intersect the base plane of the cone. Project these points to the top view.
6. The intersection of corresponding points establishes points on the hyperbola in the top view.

4.6.7 Drawing a Spiral of Archimedes

A **spiral of Archimedes** is a plane curve generated by a point moving away from or toward a fixed point at a constant rate while a radial line from the fixed point rotates at a constant rate. Figure 4.42 shows a spiral of Archimedes. To draw one, use the following steps:

1. Draw centerlines with a center point at 0 as shown.
2. Establish an equal number of angles; 12 angles of 30° each were used in the example.

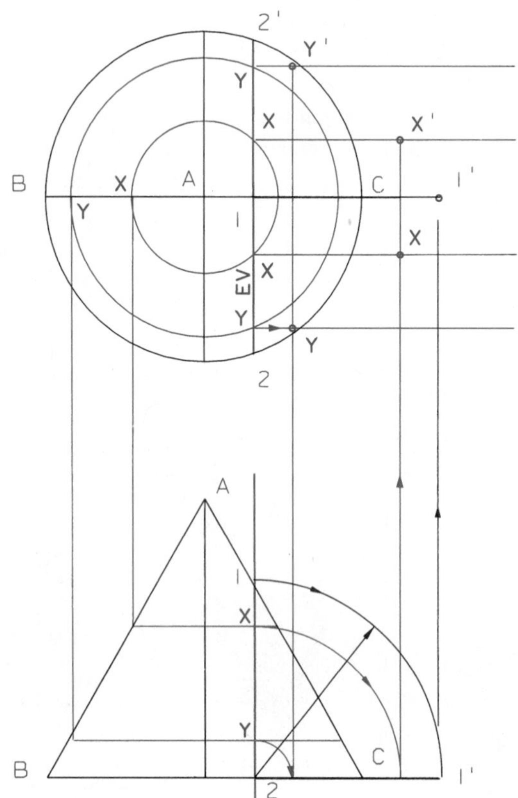

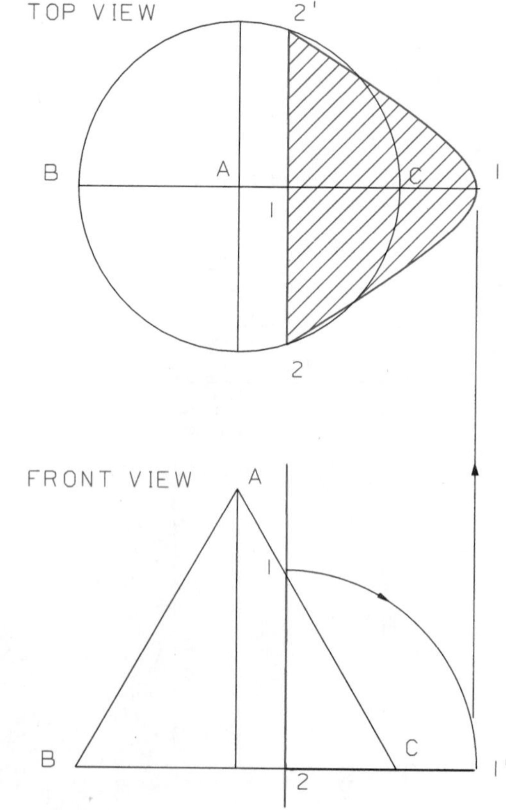

FIGURE 4.41 Construction of a Hyperbola

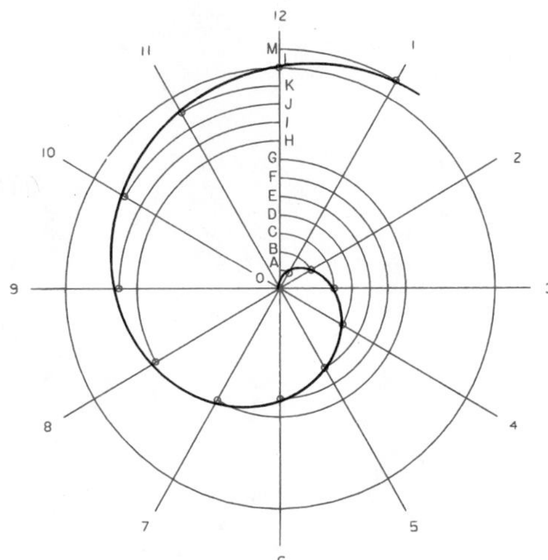

FIGURE 4.42 **Drawing a Spiral of Archimedes**

3. Divide any line into the same number of equal divisions; in the example, line 0-12 is used (divisions A through L).
4. Draw the construction arcs from each point to the corresponding angle.
5. Each intersection of an arc and an angle establishes one point of the spiral.
6. Use an irregular curve to connect the points.

4.6.8 Construction of an Involute of a Line, a Triangle, a Square, or a Circle

An **involute** is a plane curve traced by a point on a thread kept taut as it is unwound from another curve. Figure 4.43 shows four kinds of involutes.

In Figure 4.43(a), the **involute of a line** is constructed by first drawing the given line 1-2. Point 2 is used as the center of the first arc (radius 1-2). Point 1 is used as the center of the second arc (radius 1-3). Point 3 is used as the center of the third arc (radius 3-4).

The **involute of a triangle** is constructed by drawing the triangle [Fig. 4.43(b)]. Use point 3 as the center of the first arc (radius 3-1). Extend lines 2-3, 1-2, and 3-1. Use point 2 as the center of the second arc (radius 2-4). Use point 1 as the center of the next arc (radius 1-5).

The **involute of a square** is constructed by drawing the square and extending each side line [Fig. 4.43(c)]. Use point 4 as the center of the first arc (radius 4-1). Use point 3 as the center for the second arc (radius 3-5). Use point 2 as the center for the third arc (radius 2-6). Use point 1 as the center for the fourth arc (radius 1-7).

The **involute of a circle** is constructed by drawing the circle and dividing it into equal angles [Fig. 4.43(d)]. Draw tangent construction lines from the end of each angle. Mark off along each tangent the length of each circular arc. Use an irregular curve to connect the end points of the arc and tangent lines with a smooth curve.

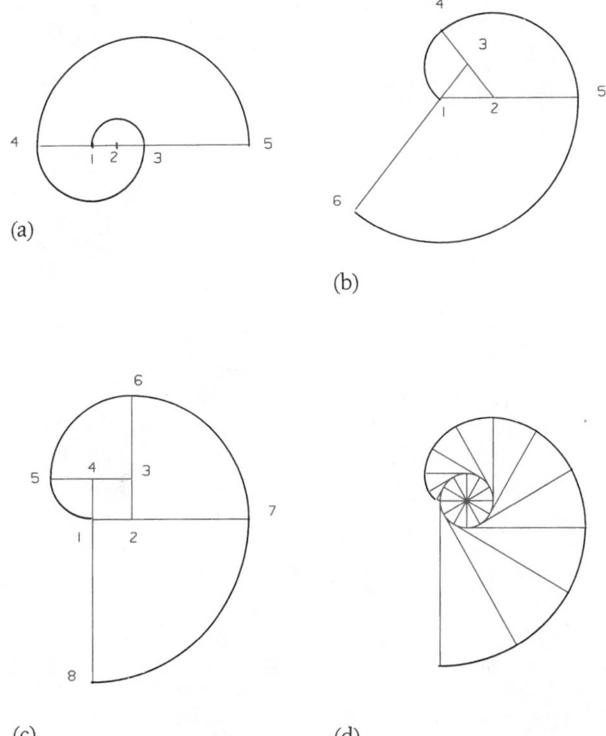

FIGURE 4.43 **Drawing Involutes** (a) Involute of line (b) Involute of triangle (c) Involute of square (d) Involute of circle

4.7 Helices

A **cylindrical helix** is a double-curved line drawn by tracing the movement of a point as it revolves about the axis of a cylinder. Figure 4.44 shows two revolved positions of a cylindrical helix modeled in 3D. The resulting curve is traced on the cylinder by the revolution of a point crossing its right sections at a constant oblique angle. The point must travel about the cylinder at a uniform linear and angular rate. The linear distance (parallel to the axis) traveled in one complete turn is called the **lead.** This type of helix is called a cylindrical helix. A variety of industrial products are based on the

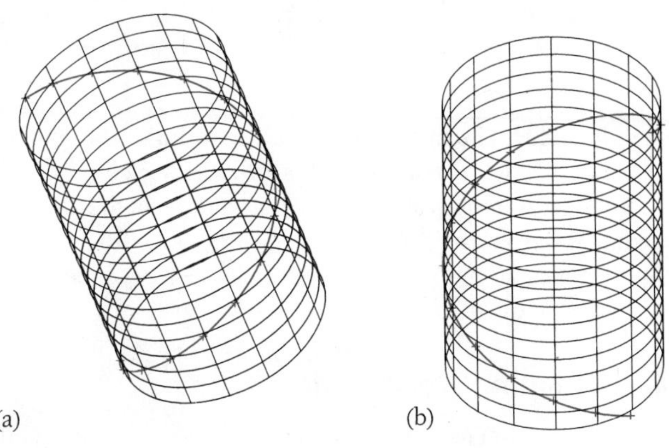

FIGURE 4.44 **3D Model of a Cylindrical Helix**

ITEMS OF INTEREST

Geometry

If you want to study the properties of figures and the relationships between points, lines, angles, surfaces, and solids, you study *geometry.* Geometry actually means "earth measurement." The name can be traced back to the way in which man first used those concepts. Practical geometry grew out of the needs of the Egyptians to survey their land to reestablish land boundaries after periodic flooding of the Nile. The flooding itself left a rich and sought-after soil. The men who made these measurements became known as "rope stretchers" because they used ropes to do their measuring. The Egyptians also used geometry to help build their temples and pyramids.

Around 600 B.C., the Greeks returned from their travels through Egypt and brought with them their first knowledge of geometry. Thales, the most famous of those men returning from Egypt, was the first to show the truth of a geometric relationship by showing that it followed in a logical and orderly fashion from a set of universally accepted statements (axioms or postulates). You may remember axioms and proofs from your geometry class.

Thales's student, Pythagoras, established a society in Italy that was devoted to the study of geometry and arithmetic. His most famous work was the theorem that bears his name. His influence was felt for centuries. Every student of geometry and trigonometry knows the Pythagorean theorem, which relates the lengths of tthe three sides of a right triangle $(a^2+b=c^2)$. The side opposite the right angle is the longest side or the hypotenuse (*c*). The other two sides, *a* and *b*, are the sides opposite the other two angles in the triangle.

While the early Greeks and others were able to make great contributions, it was not until 1796 that a great advancement was made in geometry. A 19-year-old German, Carl Friedrich Gauss, proved it was possible to construct a regular 17-sided polygon using a compass and a rule. The Greeks had only been able to construct regular polygons of 3, 4, 5, 6, 8, 10, and 15 sides. The 17-sided polygon was a major breakthrough. In 1799, Gauss was awarded a Ph.D. for developing the first proof of the fundamental theorem of algebra.

It may be difficult to appreciate how each contribution to geometry made engineering graphics possible. Even the most sophisticated CAD system makes use of fundamental geometric principles that were developed long ago. Operators of modern CAD systems can use the **POLYGON** command to create regular polygons of any number of sides (17 sides or 1000 sides, for example).

What started out as a way to measure the earth developed into a discipline that is the key to solving most engineering problems. All the geometric constructions used today in drafting were developed by individuals building on previous developments of others. No doubt, mankind will continue building on the past for the future.

D1

```
Command: POLYGON
Number of sides: 8
Edge of/<Center of Polygon>: D1
Inscribed in circle/Circumscribed about circle: C
Radius of circle: .75
```

The **POLYGON** Command Using an AutoCAD System

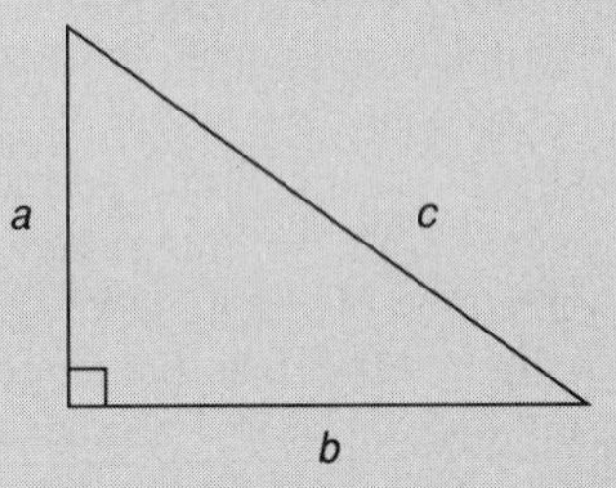

In a right triangle with sides *a* and *b*, and hypotenuse *c*, $a^2 + b^2 = c^2$
Pythagorean Triangle

cylindrical helix including fasteners and springs. The stairway in Figure 4.1 was designed with a cylindrical helix.

If the point moves about a line that intersects the axis, it is a **conical helix.** The generating point's distance from the axis line changes at a uniform rate. A helix can be either *right-handed* or *left-handed.*

4.7.1 Helix Construction

The construction techniques for a cylindrical helix and a conical helix use the same steps. Start the construction by radially dividing the end view (curve) into an equal number of parts (Fig. 4.45). The lead is divided into the same number of parts. Use the following steps:

1. Draw the right-handed cylindrical helix by first dividing the circular end view into equal divisions. Also divide the lead into equal parts (16 was used in the example).
2. Label the points on both views.
3. Project the end view divisions to the front view as vertical elements on the surface of the cylinder. In the front

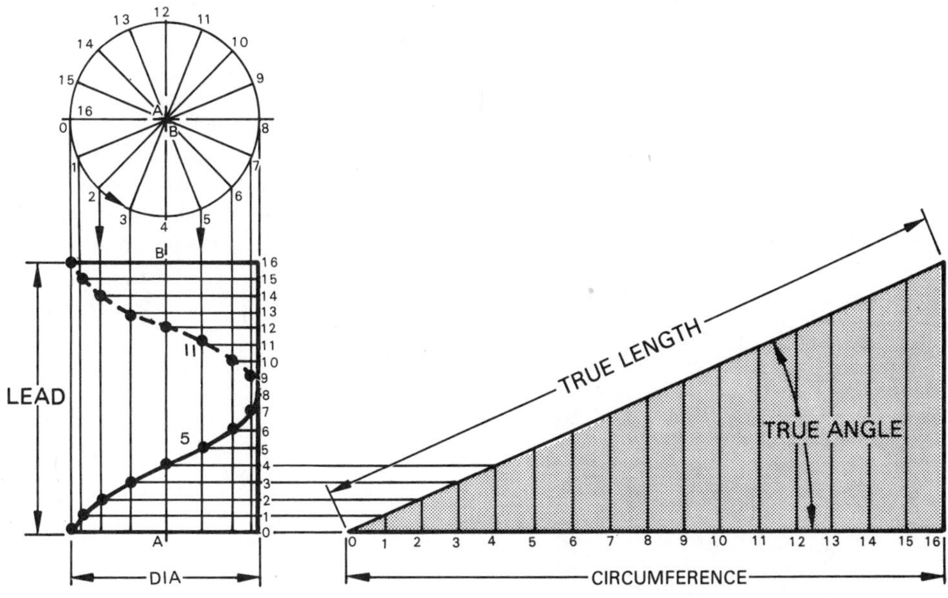

FIGURE 4.45 **Drawing a Cylindrical Helix**

view, establish a series of points on the surface of the cylinder. Each point represents a position of the generating point as it rotates about the axis.

4. You can develop the cylindrical helix by unrolling the cylinder's surface. The helix line is a straight line on the development. The angle the helix line makes with the baseline is called the **helix angle** (true angle).

To construct a conical helix, you must know its taper angle (angle between the cone's axis and an element on the cone surface) and lead. In Figure 4.46, the lead and the circle divisions are established as for a cylindrical helix. Elements determined in the top (end) view appear in the front view as straight lines, intersecting the vertex of the cone. Lead elements are drawn as horizontal lines. Points on the surface of the cone are located at the intersection of related elements.

You May Complete Exercises 4.5 Through 4.8 at This Time

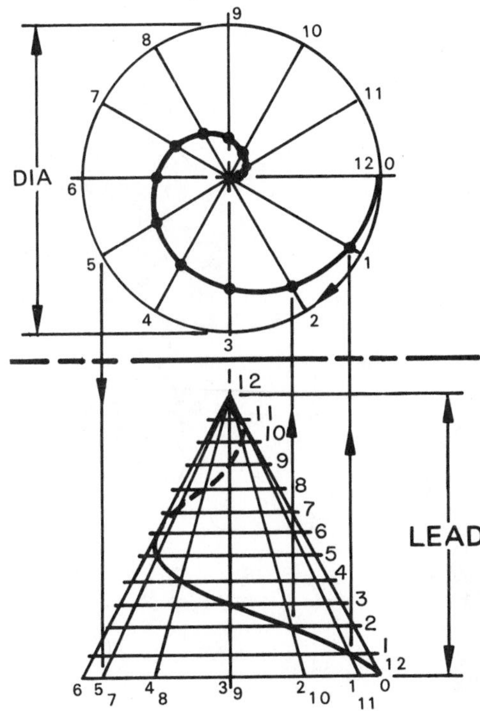

FIGURE 4.46 **Drawing a Conical Helix**

QUIZ

True or False

1. Both hyperbolas and cones are generated from conic sections.
2. In helix construction, the distance traveled by one point for one revolution measured parallel to the axis is called the lead.
3. The concentric method of ellipse construction is more accurate than using a template.
4. Tangent arcs are basically the same thing as fillets.
5. Squares, hexagons, pentagons, and ellipses are regular polygons.
6. A circle can be used to construct all forms of regular polygons.
7. A bisector of a line or an angle divides the line or angle into an equal number of parts.
8. A regular polygon has equal angles.

Fill in the Blanks

9. All geometric forms are composed of ________ and their ________.
10. ________ are used to divide lines into ________ ________ parts.
11. An ________ circle of a triangle will touch all ________ sides.
12. The distance across the ________ of an octagon will be ________ to the diameter of the ________ circle.
13. To ________ a circle means to ________ out its circumference along a straight line.
14. A ________ arc is a curve connecting two entities and is also known as a ________.

15. ________ ________ are equally spaced along their entire length.
16. Geometric forms include: ________, ________, ________, ________, and ________ .

Answer the Following

17. When is it appropriate to use the four-center method of ellipse construction?
18. Give a simple definition of a line and a curve.
19. What are the five types of figures that result from the intersection of a cone and a plane?
20. Define a cylindrical and a conical helix.
21. Why would the graphical method of dividing a line be more accurate than the mathematical method?
22. Name four solid shapes commonly used in industry.
23. Why is it important to learn the manual method of constructing geometric forms such as ellipses instead of just using templates?
24. Describe an industrial application of geometric construction.

EXERCISES

Exercises may be assigned as sketching, instrument or CAD projects. Transfer the given information to an "A" size sheet of .25 in. grid paper. Complete all views and solve for proper visibility, including centerlines, object lines, and hidden lines. Exercises that are not assigned by the instructor can be sketched in the text to provide practice and understanding for the preceding instructional material.

After Reading the Chapter Through Section 4.5.7, You May Complete the Following Exercises

Exercise 4.1(A) Bisect the line, the angle, and the arc.

Exercise 4.1(B) Divide line 1 into eleven equal parts using the graphical method. Divide line 2 into seven equal parts and line 3 into proportional parts having ratios of 3:2:5.

Exercise 4.1(C) Construct a hexagon inside and an octagon around the outside of the given circle.

Exercise 4.1 (D) Draw every possible tangency for the three circles.

Exercise 4.2(A) Draw a 40 mm radius arc (fillet) between the connected lines. Connect the two lines with a tangent arc (fillet) using a 25 mm radius.

Exercise 4.2(B) Draw a 2 in. or a 50 mm radius arc (fillet) between the circle and the line.

Exercise 4.2(C) Construct a 3 in. or a 70 mm inside arc (fillet) on the top right side on the two circles. Draw a 5 in. or a 120 mm outside (enclosing) arc connecting the two circles on the bottom left.

Exercise 4.2(D) Draw an ogee curve using the given lines and points.

Exercise 4.3(A) Given two circles of 2.25 and 3.75 in., draw an ellipse using the concentric circle method. For a metric problem, use diameters of 70 and 90 mm.

Exercise 4.3(B) Given a major diameter of 4.74 in. and a minor diameter of 2.45 in., draw an ellipse using the four-center method. Draw the ellipse so that the major diameter is vertical. Use 120 and 60 mm for a metric problem.

Exercise 4.3(C) Given circles of 2.50 and 3.5 in. in diameter, draw an ellipse using the approximate method. Use 60 and 80 mm for a metric problem.

Exercise 4.3(D) Draw two identical ellipses. Use the concentric circle method for one and the four-center method for the other. Use 4.5 in. for the vertical diameter and 2.75 in. for the minor (horizontal in this case) diameter. If metrics are selected as the unit of measurement, use 110 and 70 mm. Compare the two methods for quality and accuracy.

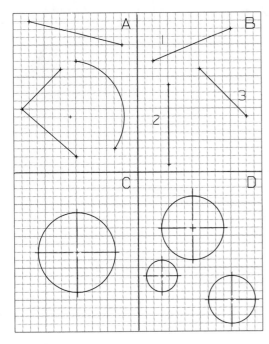

Exercise 4.1

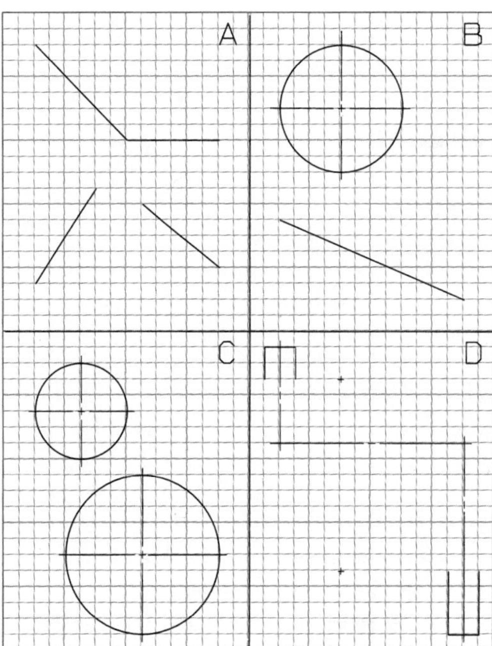

Exercise 4.2

Exercise 4.4(A) and (B) Draw the two figures using geometric construction techniques covered in the chapter.

After Reading the Chapter Through Section 4.7.1, You May Complete the Following Exercises

Exercise 4.5(A) and (B) Given the rectangle, the rise, and the axis, draw a parabola for each of the problems.

Exercise 4.6(A) or (B) Draw a hyperbola using a 4 in. (or 100 mm) diameter for the cone base and a height of 4 in. (or 100 mm). Pass a cutting plane vertically through the cone 1 in. (or 25 mm) to the right of the cone's vertical axis. For (B) use 5 in. (or 120 mm) as the base diameter and 4.5 in. (or 110 mm) as the cone's height. Draw the cutting plane vertically through the cone at 1.35 in. (or 35 mm) to the left of the cone's axis. Only one of these two problems can be done on the exercise page since the opposite space will be needed for construction.

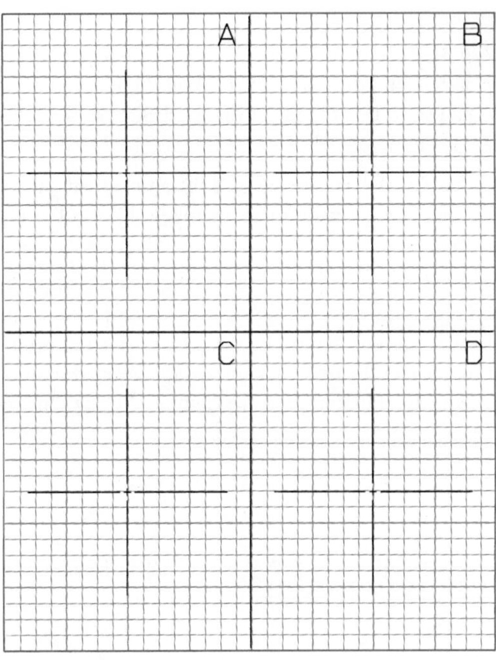

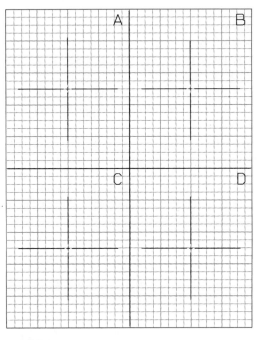

Exercise 4.3

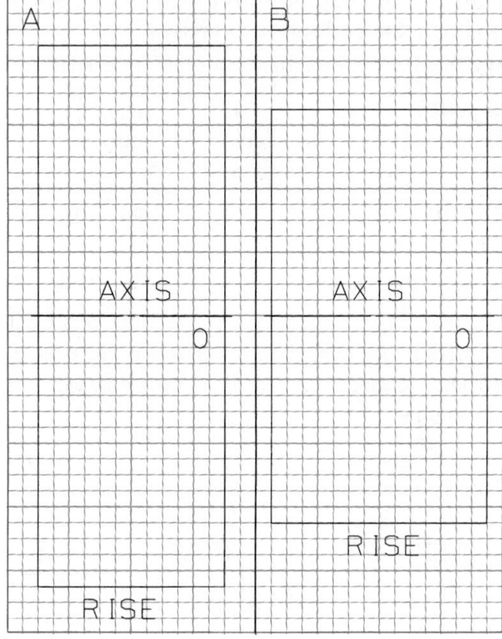

Exercise 4.5

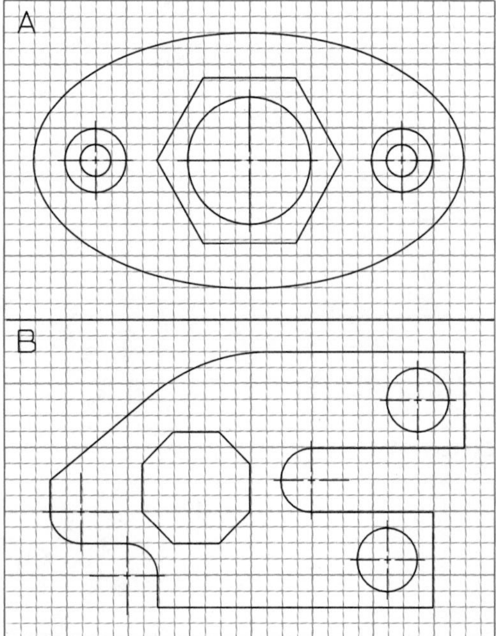

Exercise 4.4

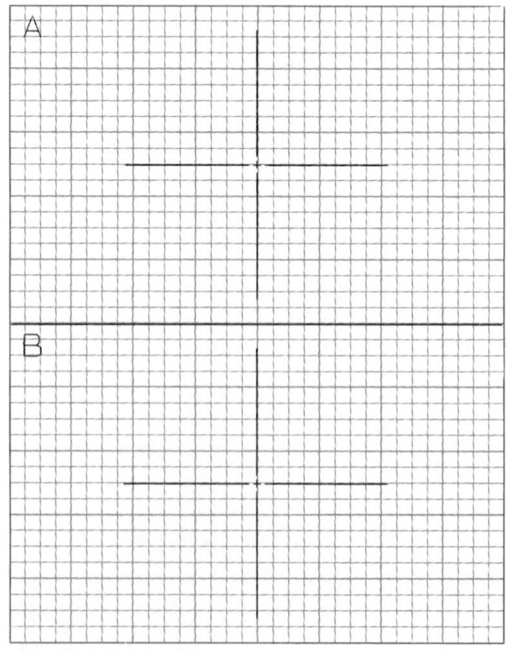

Exercise 4.6

Exercise 4.7(A) and (B) Use the given circle, divisions, and equal angles to construct a spiral of Archimedes. Start at the center and use the division line [vertical in (A) and horizontal in (B)] as the beginning line to draw the arcs needed for construction.

Exercise 4.8(A) Using the given cylinder for the diameter and the height, draw a right-handed helix with a lead of 3 in. Start the helix at the middle of the cylinder at the base where the axis line crosses the baseline.

Exercise 4.8(B) Use the given cone diameter and height to construct a left-handed conical helix. Start the helix on the lower left of the base line. Use a lead of 2.5 in.

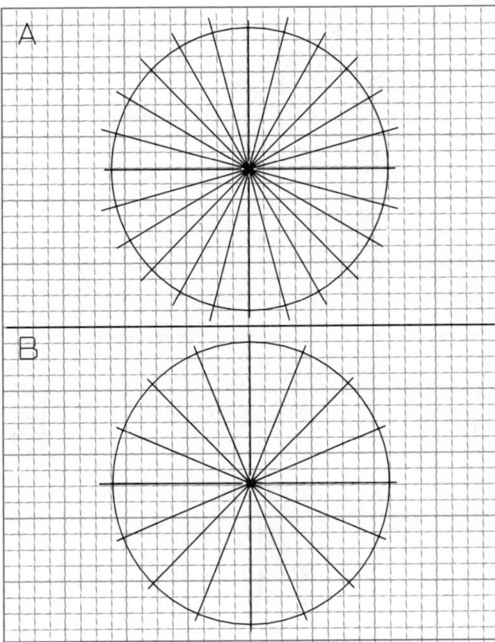

Exercise 4.7

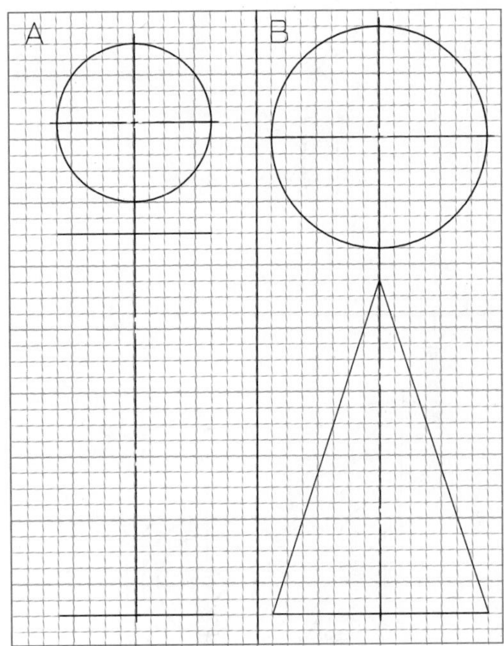

Exercise 4.8

PROBLEMS

Dimensions are provided for construction of each figure. They are not to be considered correct as per ANSI standards. Do not dimension without the instructor's approval.

Problems 4.1 (A) through (G) Transfer the problems to an appropriate size drawing format (one per drawing). Use one of the three scales provided.

Problems 4.2 through 4.13 Draw each of the assigned problems on separate sheets.

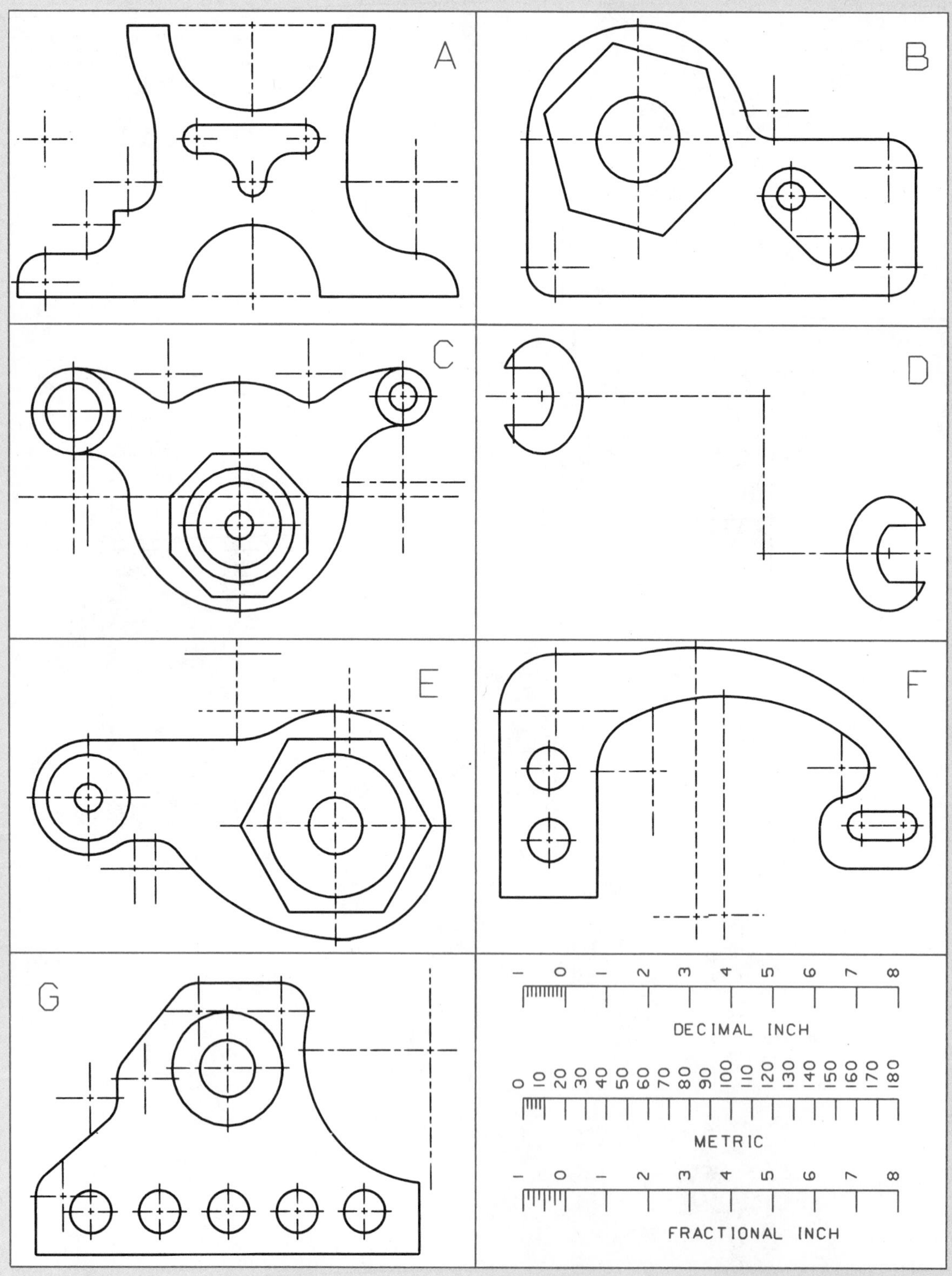

Problem 4.1

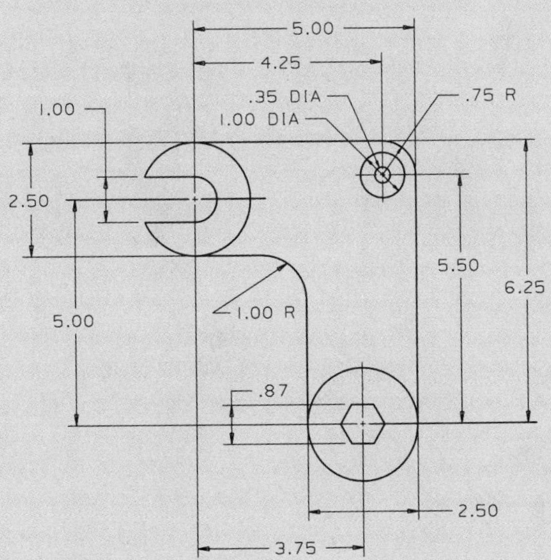

Problem 4.2

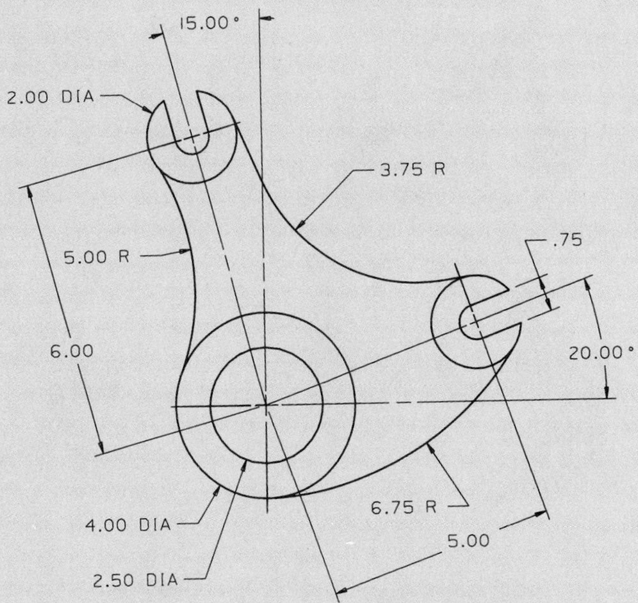

Problem 4.5

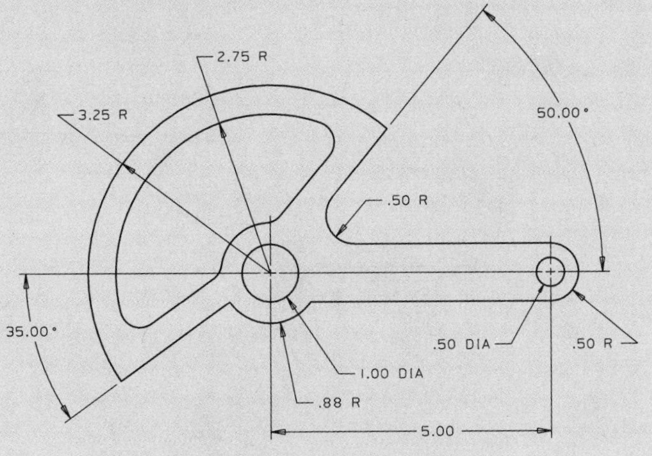

Problem 4.3

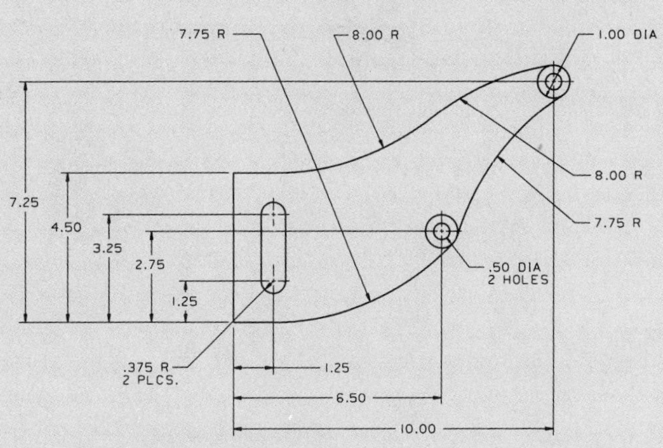

Problem 4.6

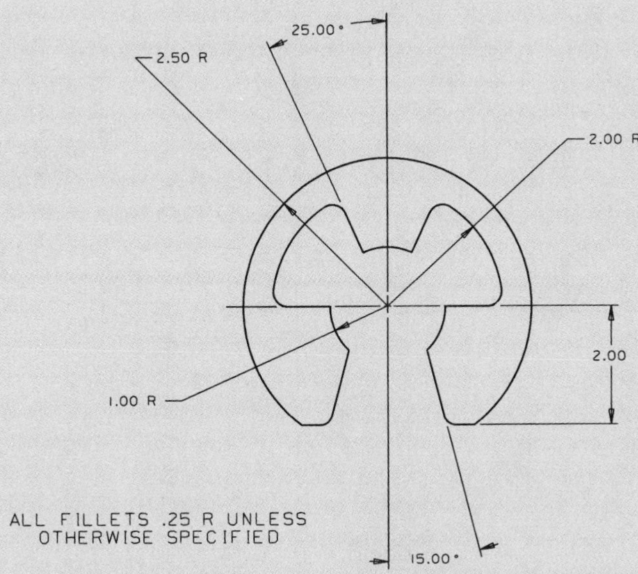

Problem 4.4

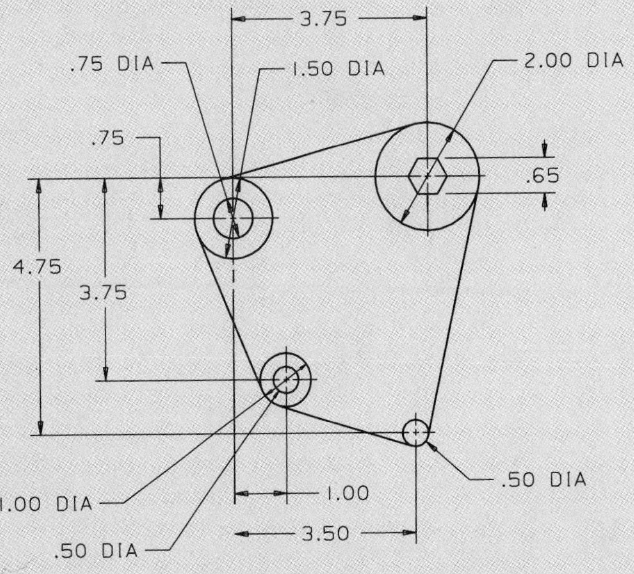

Problem 4.7

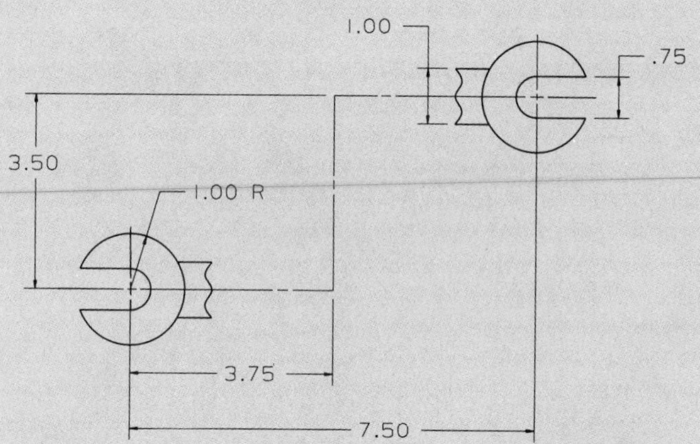

Problem 4.8

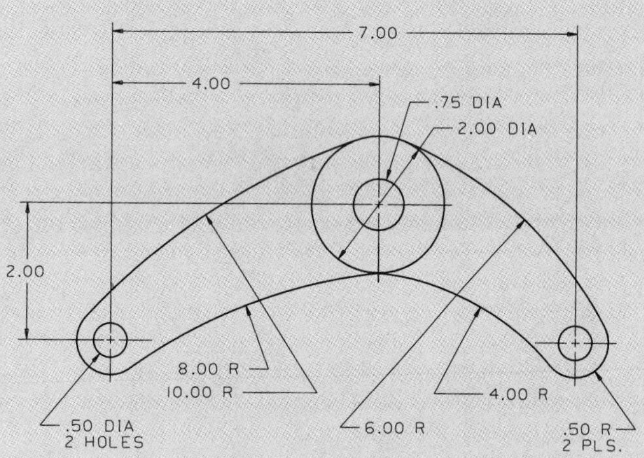

Problem 4.9

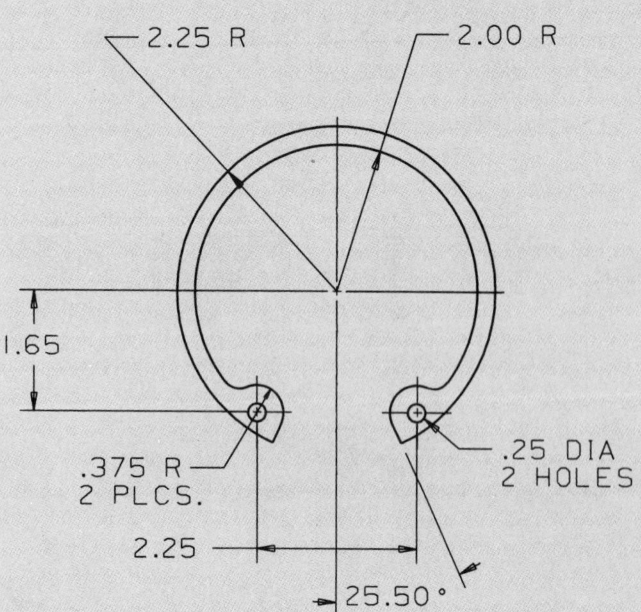

Problem 4.10

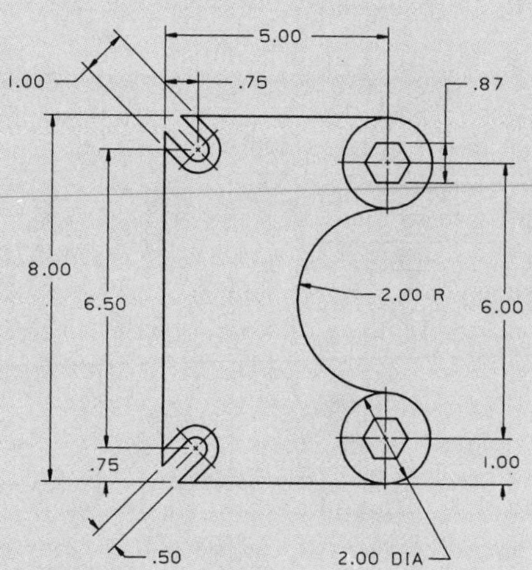

Problem 4.11

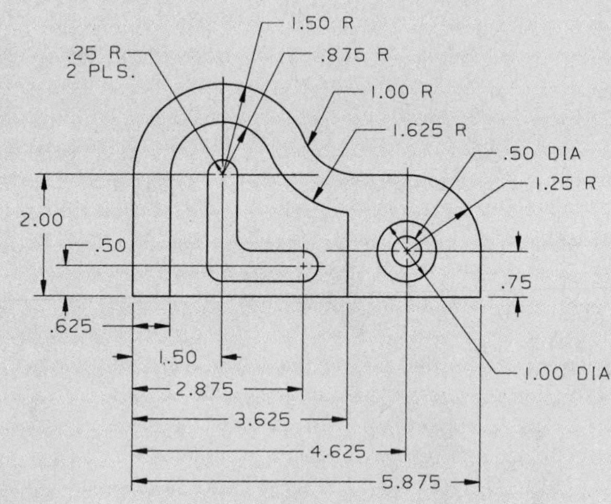

Problem 4.12

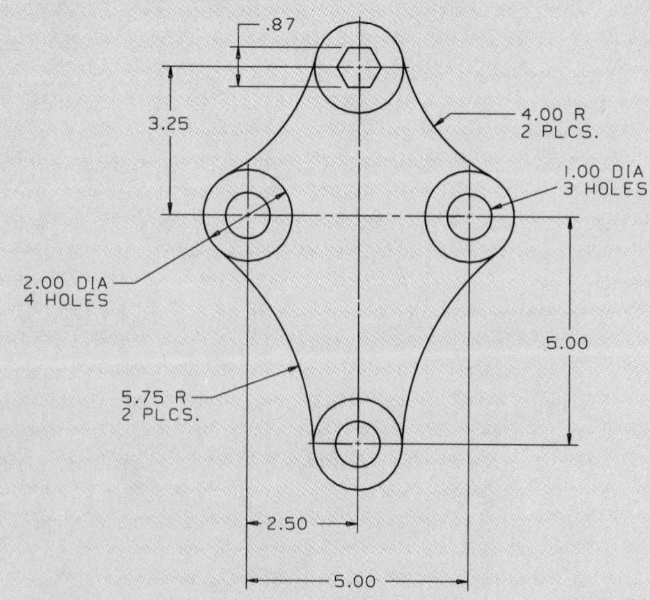

Problem 4.13

Problems 4.14 through 4.24 Draw each of the projects on appropriate size drawing format. These problems use metric measurements. Draw only the front view for problems 4.23 and 4.24.

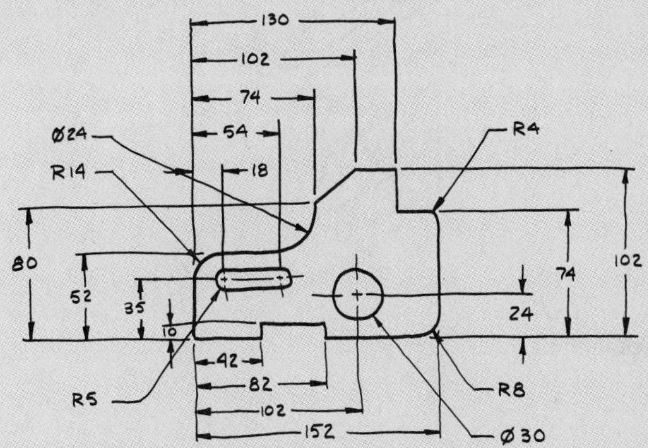

Problem 4.14

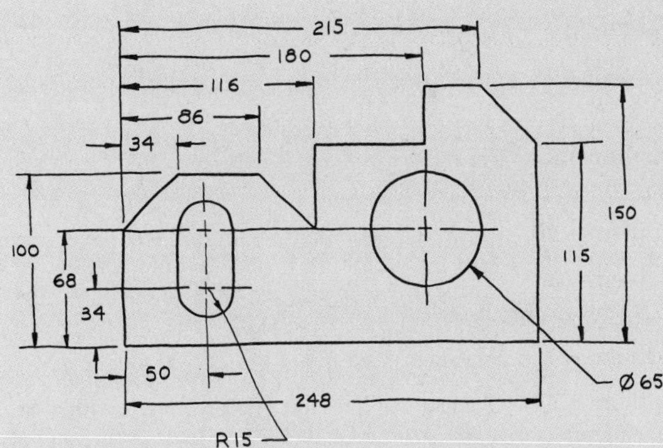

Problem 4.15

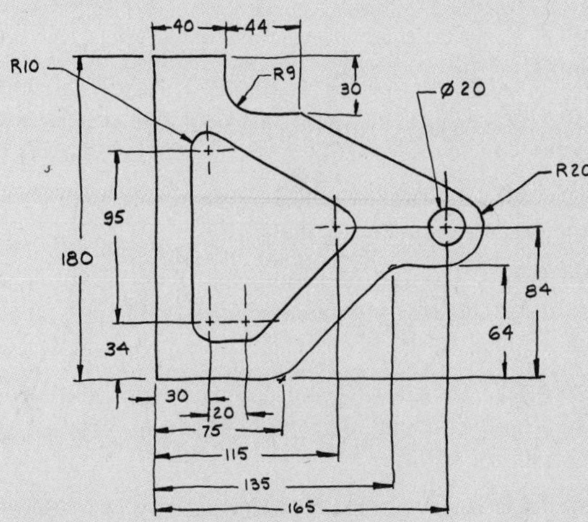

Problem 4.16

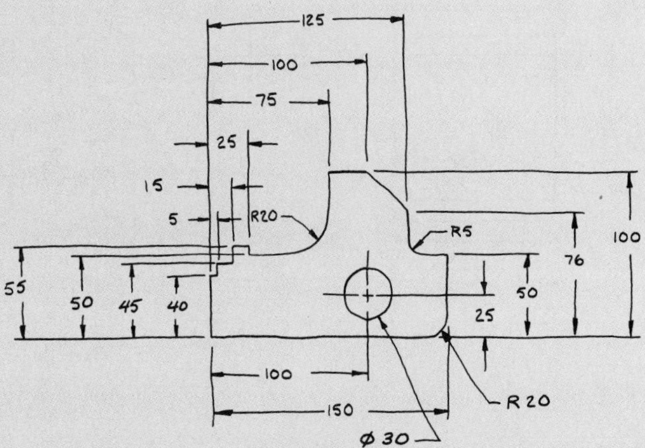

Problem 4.17

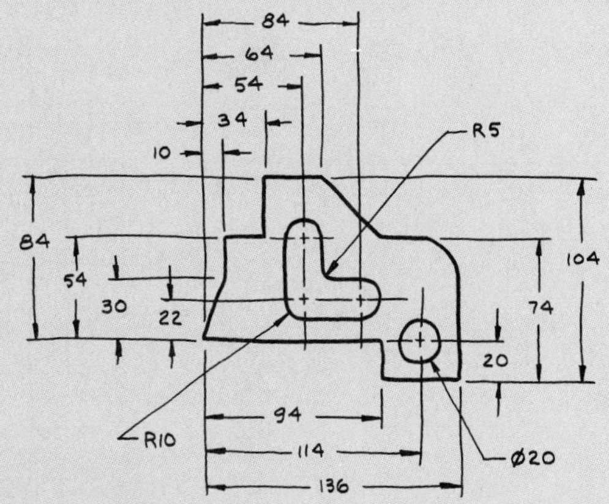

Problem 4.18

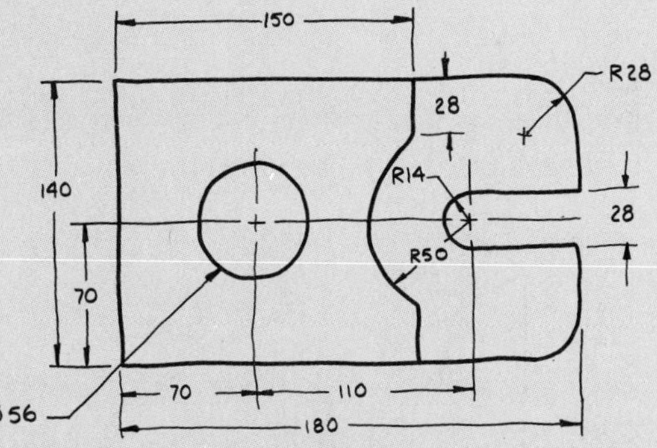

Problem 4.19

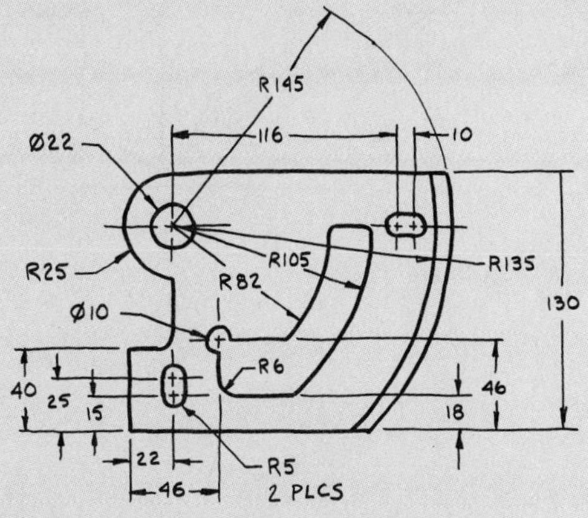

Problem 4.20

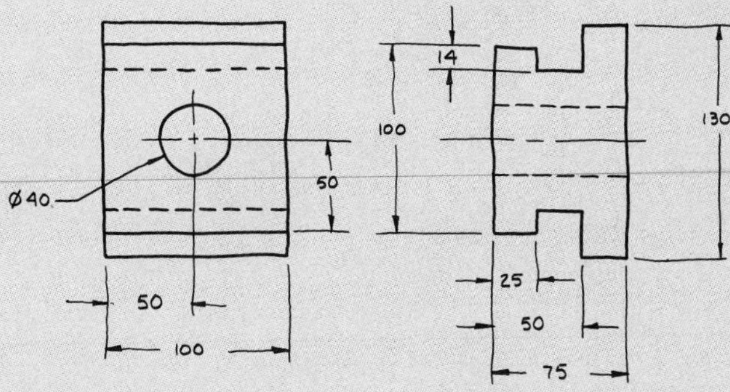

Problem 4.21

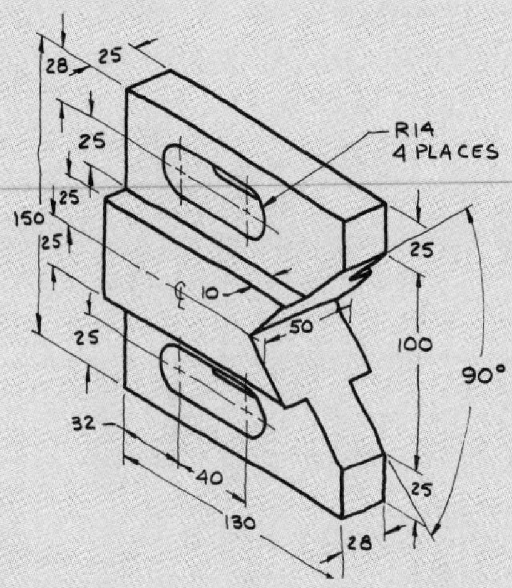

Problem 4.24

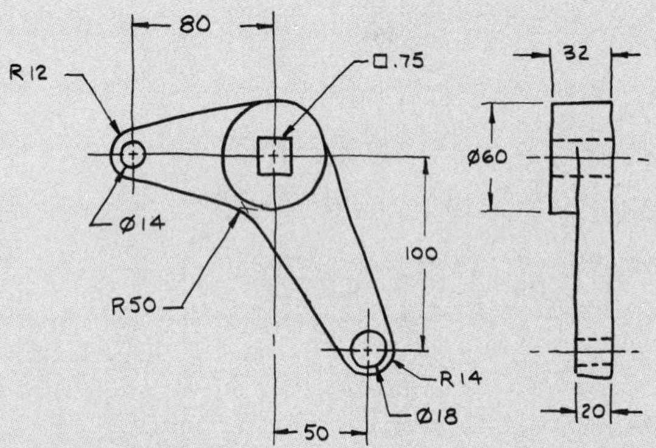

Problem 4.22

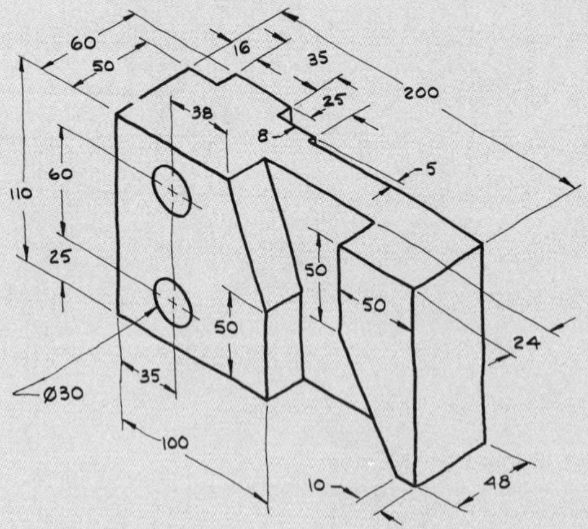

Problem 4.23

Problem 4.25 Divide a 5 in. (120 mm) line into five equal parts.

Problem 4.26 Construct bisectors of the angles of a triangle having a ratio of 3:5:6 units for the sides. Use centimeters or inches as the units.

Problem 4.27 Bisect a 55° angle.

Problem 4.28 Draw a triangle having sides with the proportions 3:4:5.

Problem 4.29 Draw a hexagon that is 3 in. (or 70 mm) across the flats.

Problem 4.30 Draw a hexagon that is 75 mm across the corners.

Problem 4.31 Construct a seven-sided regular polygon in a 5 in. (120 mm) diameter circle.

Problem 4.32 Find the center of a 4 in. (or 100 mm) circle.

Problem 4.33 Connect two lines forming a 35° angle with a 1 in. (or 25 mm) radius arc.

Problem 4.34 Connect two perpendicular lines with a $1\frac{1}{4}$ in. (or 30 mm) radius arc.

Problem 4.35 Draw an ellipse having a major axis of 70 mm and a minor axis of 50 mm.

Problem 4.36 Construct the inscribed circle of a 2 by 3.5 by 4 unit triangle.

Problem 4.37 Draw a circumscribed circle triangle. Use centimeters as units.

Problem 4.38 Find the center of a 4 in. (or 100 mm) circle by perpendicular bisectors. Rectify the circle.

Problem 4.39 Draw a cylindrical helix having an 80 mm diameter base and a height of 140 mm and a lead of 50 mm. Draw as a right-handed helix.

Problem 4.40 Draw a left-handed conical helix with a base diameter of 3 in. (or 70 mm), a height of 4.5 in. (110 mm), and a lead of 2.5 in. (or 60 mm).

Problem 4.41 Draw a spiral of Archimedes using a 5 in. (or 120 mm) diameter with angles of 30° and .125 in. (or 10 mm) divisions.

CHAPTER 5

Sketching

Learning Objectives

Upon completion of this chapter you will be able to accomplish the following:

1. Realize how sketches can be used to transform design concepts into visual communication.
2. Differentiate between pictorial, multiview, and diagrammatic sketches.
3. Recognize the importance of proper proportioning and thorough dimensioning.
4. Apply ANSI standard line weights and symbols while developing sketching techniques.
5. Demonstrate and understanding of multiview projection and selection of views.
6. Produce isometric, oblique, and multiview sketches.

5.1 Introduction

Because **sketching** is one of the primary methods used to communicate graphic ideas in the engineering community, the ability to sketch is an essential and useful skill for all drafters, designers, engineers, and technicians. *A sketch is often used to convey original design ideas.* For example, it is not unusual to see members of any design team making freehand sketches to clarify the design of three-dimensional parts or to explore alternative configurations for an assembly. There are three primary types of sketches: pictorial, multiview, and diagrammatic.

The design team uses the freehand sketch to explore alternative configurations for the part or assembly. Engineers and designers sketch preliminary ideas. Drafters or layout designers then refine those original ideas and requirements and produce drawings.

Sketching is usually done at any location that has a flat surface. The tools used most often for sketching are paper (or grid paper), soft lead pencils, and an eraser. The grid on grid paper speeds the construction of any sketch. Sometimes it is necessary to evolve a layout through a series of sketches. Figure 5.1 shows the evolution of such a layout through the following stages; rough sketch [Fig. 5.1(a)], refined sketch [Fig. 5.1(b)], and the final CAD drawing [Fig. 5.1(c)].

5.2 Materials and Equipment Used in Sketching

Only a few basic items and materials are required for sketching. Any sketch pad should be grid lined or have a cross-hatched underlay grid sheet that can be placed beneath the transparent sketch paper. Nonreproducing grid sheets are also available. Isometric grid-lined paper is available for pictorial sketching. Posterboard grid formats include isometric, oblique, orthographic, and a variety of perspective formats. Figure 5.2 is an illustration of a simple part sketched on isometric grid paper. This grid paper has, in addition to the 30° receding lines in both directions, vertical and horizontal lines for multiview projection. Grid squares are very useful in maintaining the proper proportion for a part because it is easy to count grid squares while sketching.

5.2.1 Drawing Size When Sketching

While the size of a particular sketch is usually unimportant, conveying the proper proportions of the part in a sketch is *essential*. Using grid paper helps to ensure that the sketch is in the proper proportion. Sketches are seldom drawn full size. However, sketches, like all other drawings, must be dimensioned. Drawings of every type should not be measured, but "read". "Reading" a drawing means that you should be able to find every dimension for every part and use that information in later stages of the project. The one exception to the dimensioning rule is when the drawing is a diagram (Fig. 5.1). Diagrams tell a story and do not represent parts or objects. Diagrams can be measured and digitized.

5.2.2 Line Types Used in Sketching

The line types and widths used in freehand sketches are the same as those used in instrument drawing (ANSI standard line weights, types, and symbols). The line quality in a sketch is not perfect. Figure 5.3 shows the typical range of line types that may be encountered in sketches. *Cutting plane lines are the widest; object and hidden lines are medium thickness; extension, dimension, centerline, phantom, and section lines are thin lines.* Lines should be equally black as in instrument drawing. Construction lines are usually not removed. Lettering must be clear and easy to read.

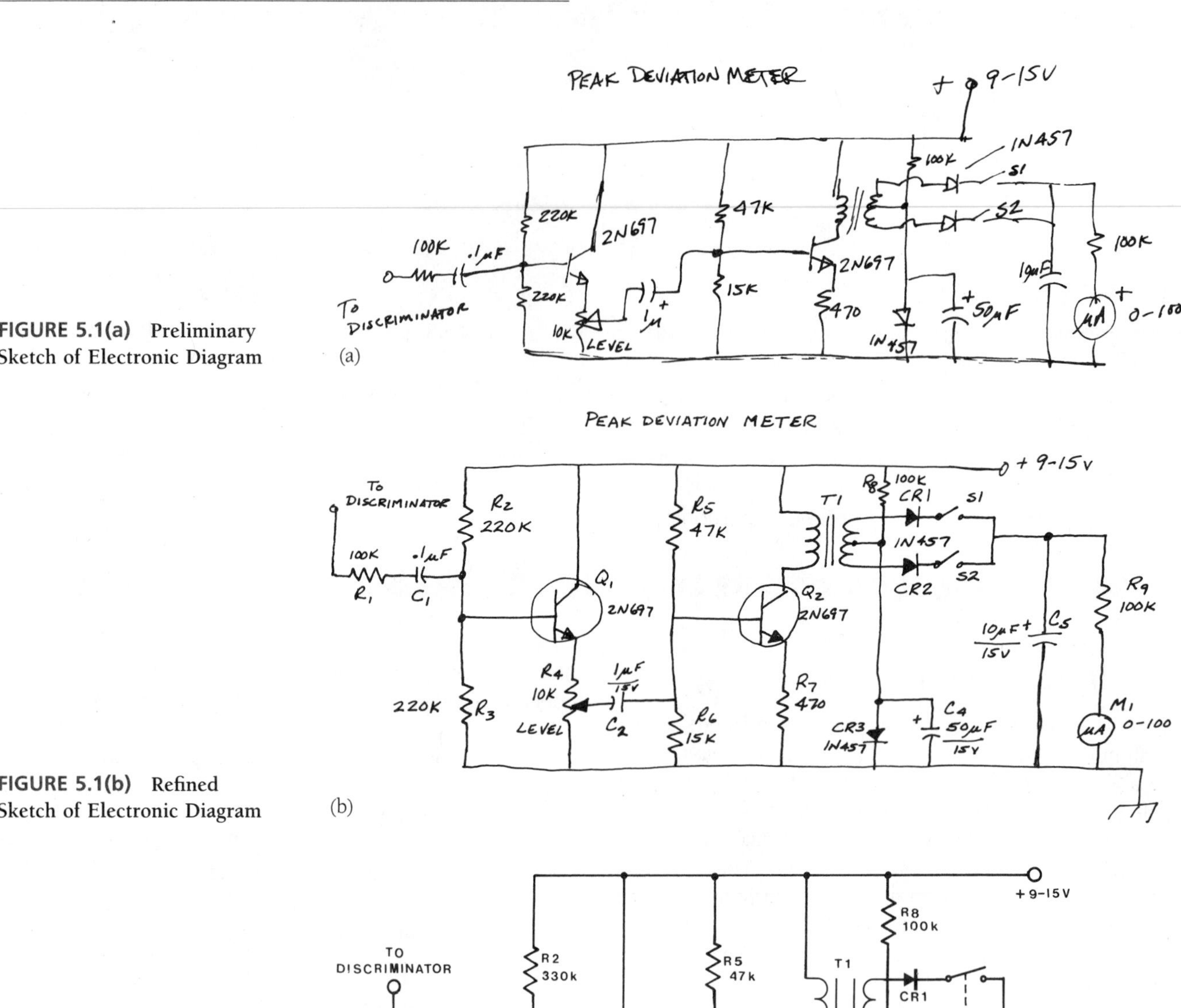

FIGURE 5.1(a) Preliminary Sketch of Electronic Diagram (a)

FIGURE 5.1(b) Refined Sketch of Electronic Diagram (b)

FIGURE 5.1(c) Finalized Drawing of Electronic Diagram (c)

5.3 Sketching Techniques

Sketching skills are developed over a period of time by practice and effort and should be cultivated during your career. Speed in sketching is not important while you are learning the basic techniques. However, later on, sketching with ease and speed will enhance your ability to communicate graphic ideas efficiently.

When sketching, *your pencil should be held at an angle to the paper.* As shown in Figure 5.4, 50° to 60° is recommended for straight lines and 30° to 45° is recommended for circles and arcs. *Rotate your pencil while sketching* because it will help maintain a conical point and reduce the time required for sharpening. In fact, you may prefer to use fine-line (0.7 to 0.9 mm, with H or HB lead) mechanical pencils for sketching.

Hold the pencil 1.5 to 2 in. (30 to 50 mm) from the tip as shown in Figure 5.5. Some drafters prefer to hold the pencil in the flat position as demonstrated in Figure 5.6. Remember, it is not the intent of the text to change the way you hold your

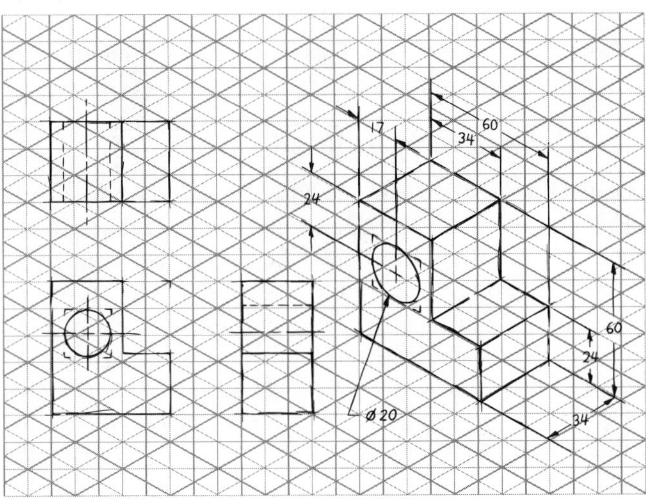

FIGURE 5.2 Grid Paper and Sketching

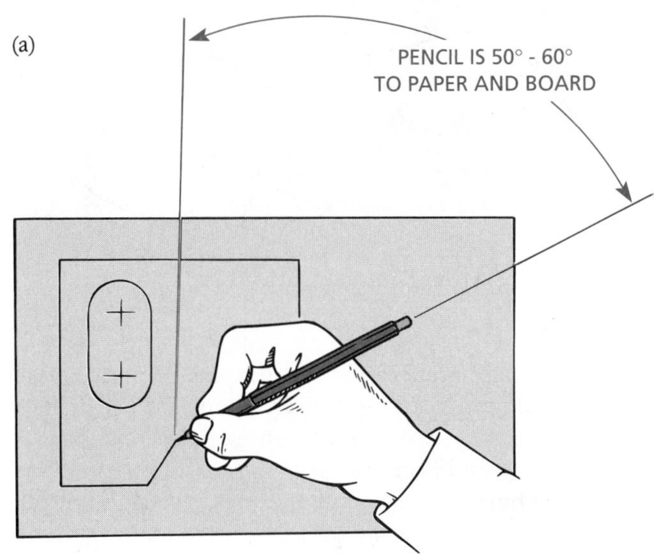

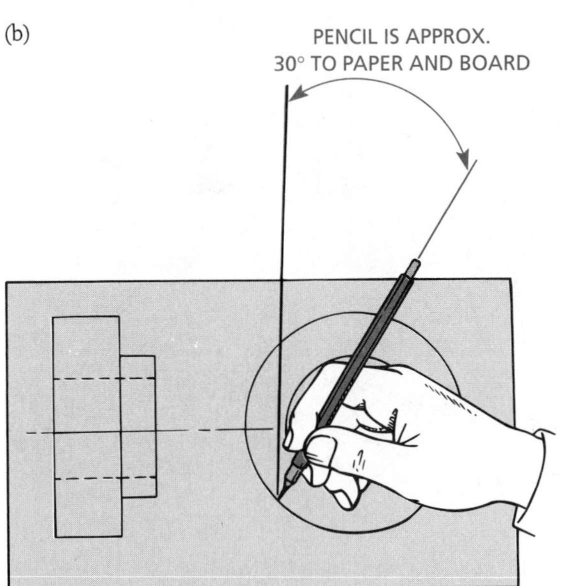

FIGURE 5.4 Angle of Pencil when Sketching

pencil. The information in this section is provided to help develop sketching skills. Left handers may hold their pencils at different angles and orientations.

5.3.1 Sketching Horizontal Lines and Vertical Lines

Horizontal lines are drawn by locating their end points and connecting them with a line. Draw lines using construction lines first and, later, after the design is close to completion, go back and darken them. The pencil is moved from the left to the right (Fig. 5.7). Use short strokes, but try to avoid "feathering" the lines. Pull a wood pencil or lead holder to avoid

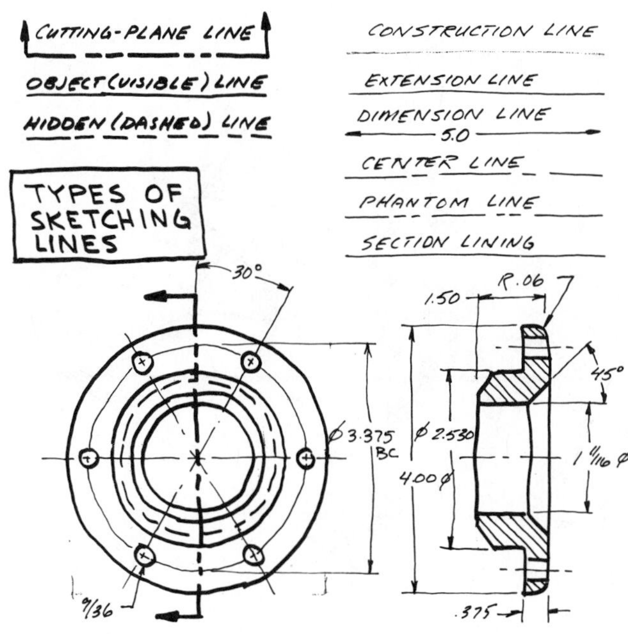

FIGURE 5.3 Line Types for Sketches

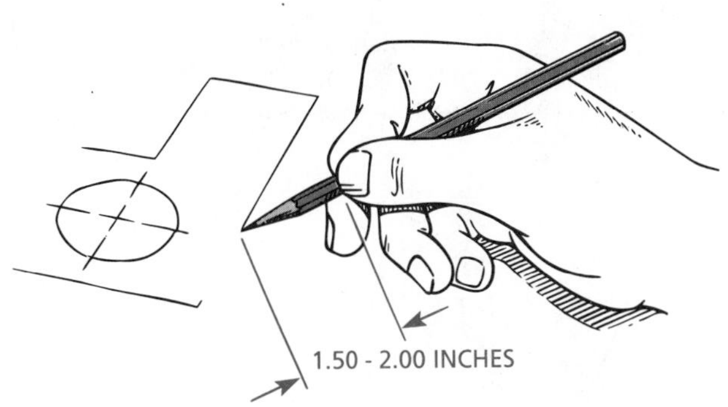

FIGURE 5.5 Holding a Pencil for Sketching

FIGURE 5.6 Sketching Vertical Lines

ripping the paper surface. The lead in fine-line mechanical pencils breaks easily, so you should push them. Some drafters leave a small space between each line segment (the space is unnecessary if grid paper is used).

Vertical lines are drawn with the same general technique. For vertical lines, move the pencil from the top toward the bottom of the paper (Fig. 5.8). Again, grid paper helps ensure that the lines will be drawn vertical. Turn the paper to any convenient position to help speed the process. Some drafters prefer to move the pencil away from the body, from bottom to top or left to right. Try different methods to find the one that works best for you.

5.3.2 Sketching Inclined Lines

Angled lines are drawn by establishing the end points, lightly sketching the line, and, finally, darkening the line. Sketch inclined lines away from you if they are angled to the right (Fig. 5.9) or toward you (or turn the paper) if they are angled to the left. Use the opposite technique if you are left handed.

Since horizontal lines are the easiest to draw, turn the paper so that the line you are sketching is close to horizontal (Fig. 5.10). However, large sketches are often taped to the table so you should also learn to sketch without turning the paper.

Estimating angles is done by drawing two lines perpendicular (90°) to one another. Bisecting this angle gives a 45° measurement. Similarly, dividing the 45° angle into 3 provides a 15° angle and a 30° angle (Fig. 5.11). Always locate the end points by dimensions or give an angle dimension.

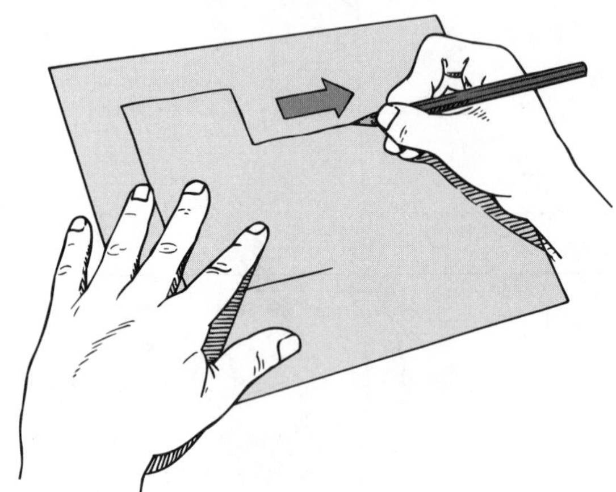

FIGURE 5.7 Sketching Horizontal Lines

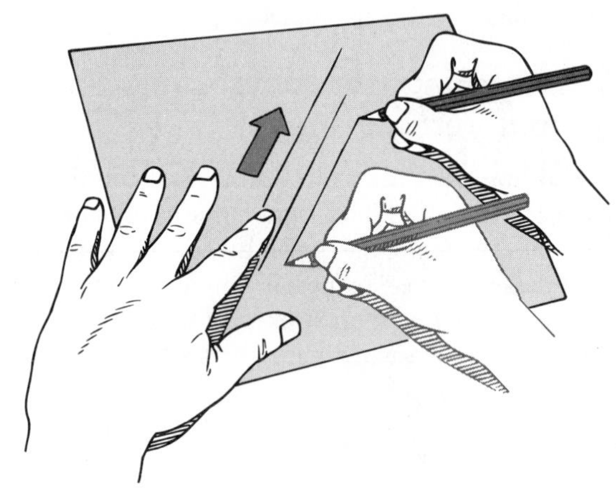

FIGURE 5.9 Sketching Inclined Lines

FIGURE 5.8 Sketching Vertical Lines

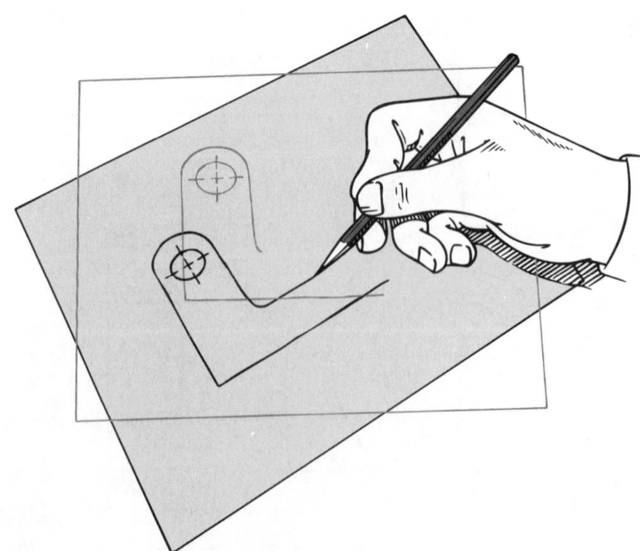

FIGURE 5.10 Turning Paper May Make Sketching Easier

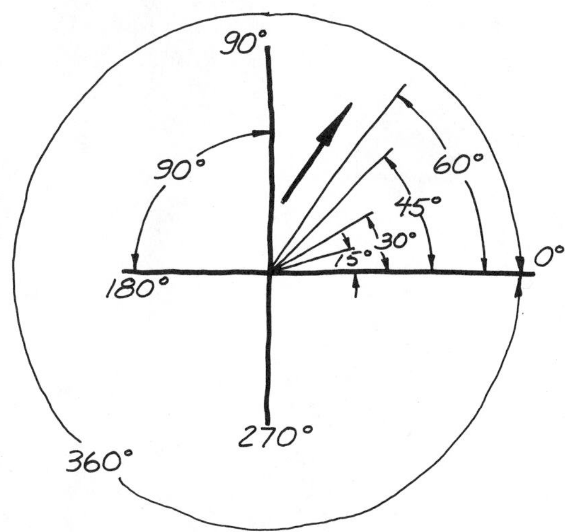

FIGURE 5.11 Typical Angles Used in Sketching

5.3.3 Sketching Arcs and Circles

Learning to sketch arcs and circles can be frustrating. Always start by locating the center point of the circle or arc, then draw the centerlines of the circle. Measure or estimate the size of the circle and lay out the diameter along the centerlines as shown in Figure 5.12. Block out the circle by drawing a square that encompasses it [Fig. 5.12(a)]. Next, draw diagonals and lay out the diameter on the diagonals [Fig. 5.12(b)]. If the circle is large, divide the circle into smaller segments and measure the diameter [Fig. 5.12(c)]. Connect the points by sketching short arcs to complete the circle [Fig. 5.12(d)]. If the sketch is small, rotating the paper helps keeps the circle round.

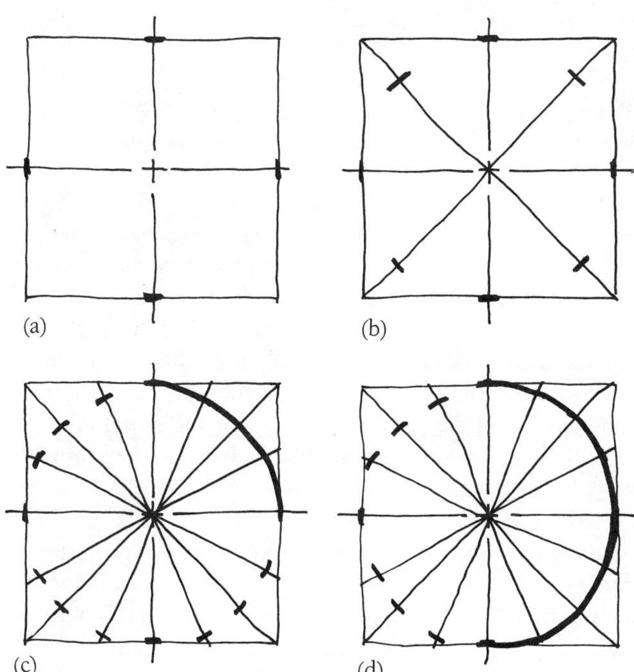

FIGURE 5.12 Sketching a Circle (a) Sketch square (b) Divide into segments (c) Sketch short arcs (d) Connect arcs (points)

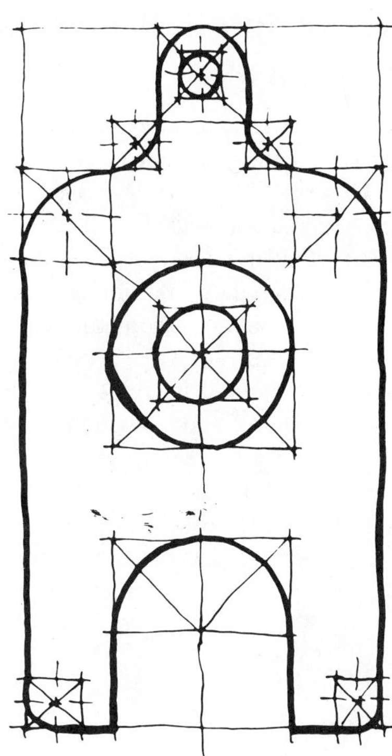

FIGURE 5.13 Blocking Out Circular Shapes when Sketching

Use the same general technique to sketch arcs. In Figure 5.13 several arcs and circles were required. Centerlines were used for every arc and circle, and both circles and arcs were blocked before they were drawn.

5.3.4 Sketching Irregular Curves

Freehand sketching of irregular curves involves establishing an adequate number of points along the curve and then connecting the points with a smooth curve. A lightly sketched construction curve is drawn first; then the irregular curve is darkened. Grid paper makes it easier to establish the controlling points (Fig. 5.14).

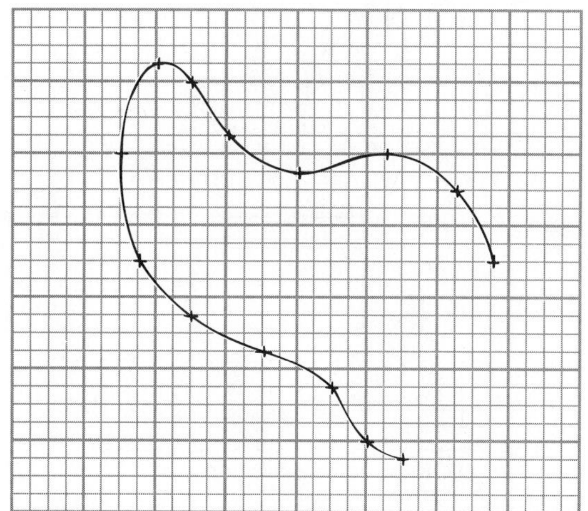

FIGURE 5.14 Sketching Irregular Curves

5.4 Introduction to Projection Techniques

Technical operations usually require two-dimensional (paper) representations to communicate ideas and give physical descriptions of 3D shapes. These projections are divided into two categories, *pictorial* and *multiview.* Pictorials simulate 3D views of the part, while multiviews are two-dimensional projections of the part. This simple division separates single-view drawings (pictorials—oblique, isometric and perspective) from multiview drawings.

Often engineering working drawings are multiviews, while pictorials are used for technical illustrations. In sketching, however,both types may be used to refine design concepts. Figure 5.15 shows each of the four projection types for an angle block. Pictorial projections are single-view drawings that may be used as rough sketches of preliminary ideas, but do not lend themselves to communicating exact technical details. *Perspective* projections are constructed with projecting lines that converge at a point. Although this method provides the most lifelike appearance of the part, it does not show true dimensions. *Oblique* pictorials distort the depth of the part. Since the *isometric* method uses full scale dimensions for all lines that are vertical or parallel to the axes, it is the most common and useful method for engineering sketching.

Multiview drawings are not lifelike because they show the parts in more than one view and are projections. Multiview projection presents the object's top, front, and side in related adjacent views. The theory behind orthographic projection is that the object is rotated by turning it to the appropriate view. For example, rotating it 90° sideways provides a side view. In Figure 5.16, the part was rotated to the right, so the resulting view is a right side view. The three-view drawing (bottom) shows the part aligned between views.

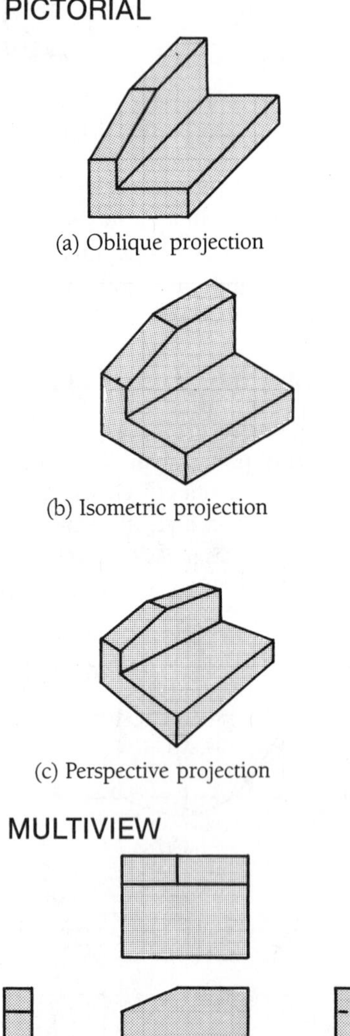

FIGURE 5.15 Projection Methods

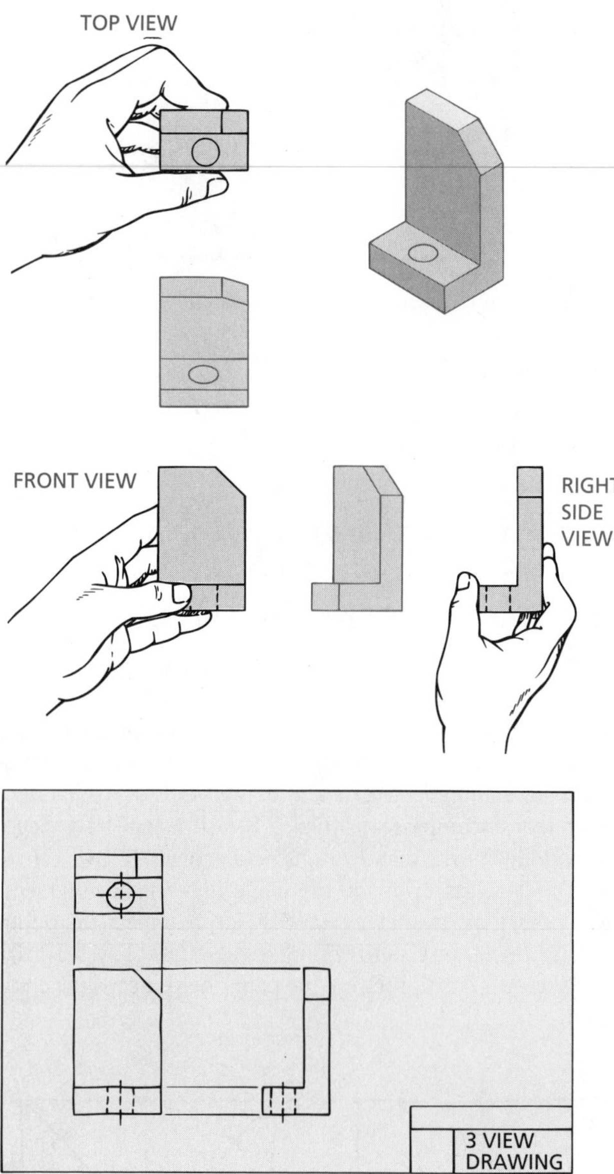

FIGURE 5.16 Three-View Orthographic Projection

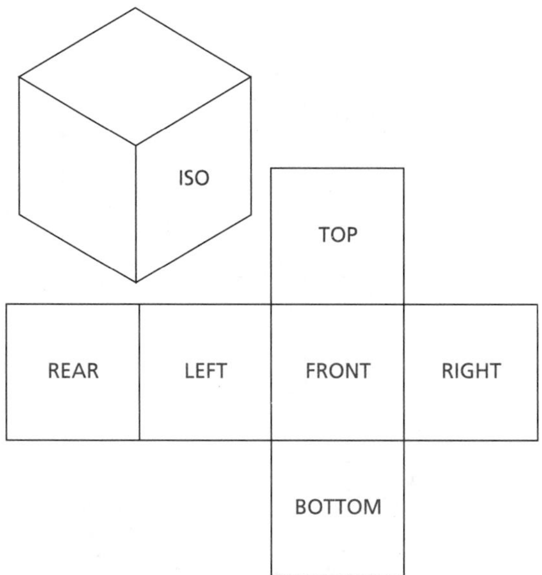

FIGURE 5.17 The Six Standard Views of an Object

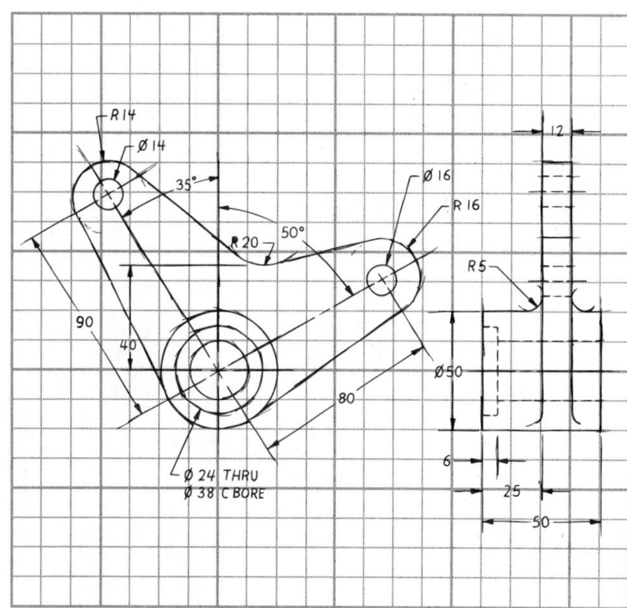

FIGURE 5.18 Two-View Drawing

All dimensions for multiviews are drawn to scale. The three principal views (top, front, and side) can be used to project any number of needed views to provide engineering data. An *auxiliary* view is any projection other than one of the six principal views; top, front, right side, left side, back, and bottom (Fig. 5.17).

5.5 Multiview Projection

Multiview projection describes the features of a part and dimensions in one or more views that are projected at 90° angles to each other. This form of projection is the primary method used in engineering work. Figure 5.18 shows a multiview sketch that communicates ideas, dimensions, and shapes for the manufacture of a rocker arm.

Multiview drawing uses orthographic projection to establish the spatial relationship of points, lines, planes, or solid shapes. Two methods are used to make multiview orthographic projections: the *normal method* and the *glass box method.* In the normal (natural) method, the object is viewed perpendicular to each of its three primary surfaces. In the glass box method, you imagine that the part is enclosed in a transparent box. A view of the part is established on its corresponding glass box surface (plane) by perpendicular projectors originating at each point on the object and extending to the box surface (Fig. 5.19). The glass box is hinged so it can be unfolded onto one flat plane (the paper). Each projection shares a dimension with its adjacent view. For example, the top and front view share the width dimension. In this method, all six sides are revolved outward so that they are in the plane of the paper. All are hinged to the front plane, except the back plane. The back plane is not used very often, but is normally revolved from the left side view when used. Each plane is parallel to the plane opposite from it before it is revolved around its hinge line.

A *hinge line* is the line of intersection between any two adjacent image planes. The left side, front, right side, and back are all elevation views and show the height dimension. The top and bottom surfaces are in the horizontal plane. The depth dimension, width dimension, front, and back are established there.

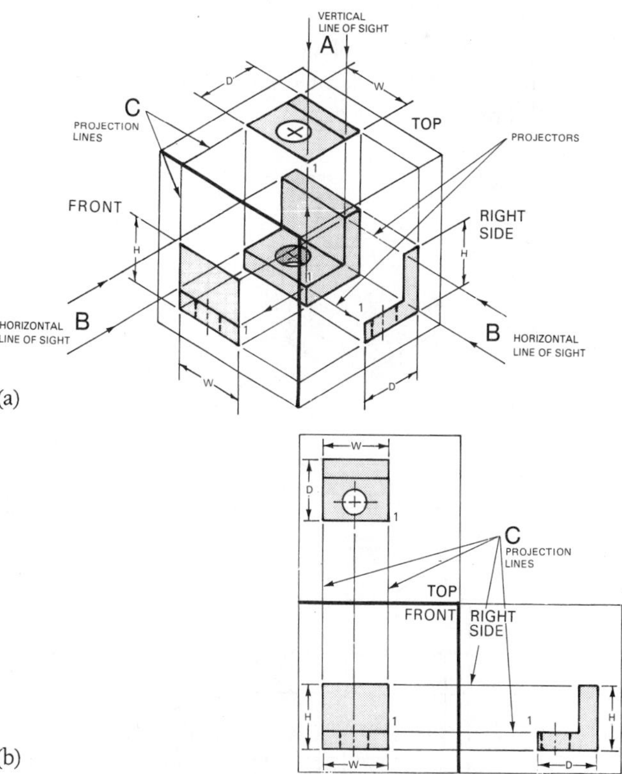

FIGURE 5.19 The Glass Box Method of View Projection

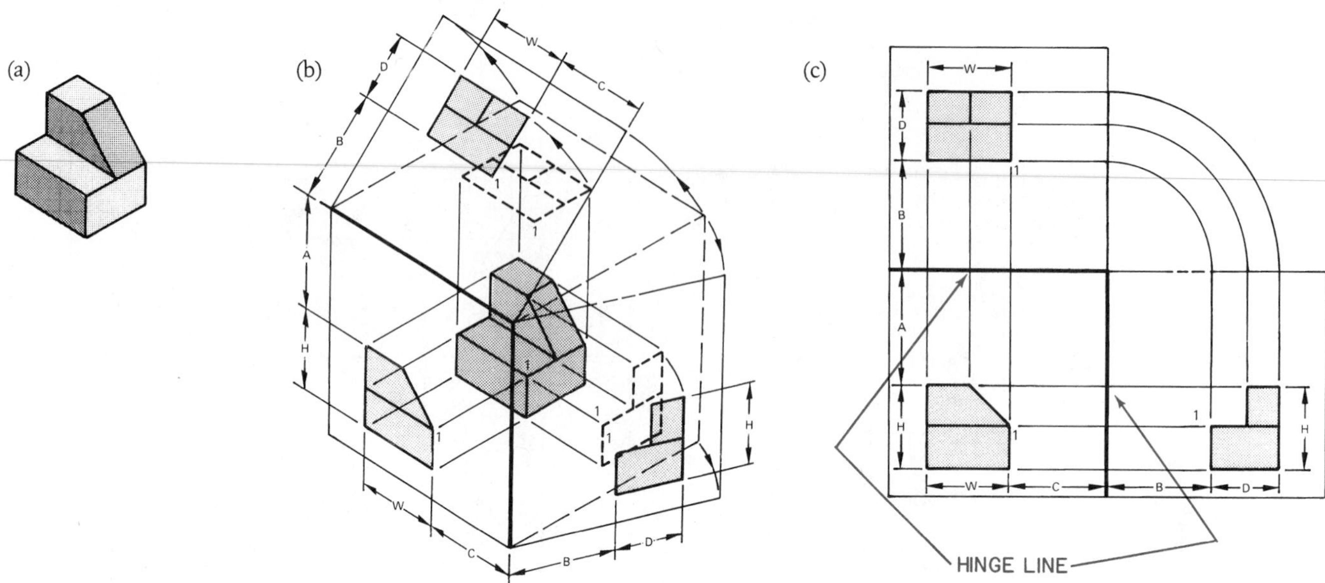

FIGURE 5.20 Unfolding the Glass Box

In the United States and Canada, the six principal views of a part are drawn using *third-angle* projection. In third-angle projection, the *line of sight* goes through the image plane to the object (Figs. 5.19 and 5.20). Assume that the object is projected back along the lines of sight to the image plane. The line of sight is at a right angle to the projection plane and is assumed to originate at infinity. To visualize this, place the plane between you and the object. Your position changes with every view so that your line of sight is always at a right angle to each image plane. A point is projected on the image plane where its projector (line of sight) pierces that image plane. Point 1 in Figure 5.21 is located on the part and is projected onto the three primary image planes.

5.5.1 The Glass Box and Hinge Lines

Hinge lines are the intersection of two perpendicular image planes (Fig. 5.21). Each image plane (surface of the glass box) is connected at right angles to an adjacent view. For example, the top view is hinged to the front view, as is the right side view. Hinge lines are not shown on technical drawings.

5.5.2 Selection of Views

Selecting the proper views, and orientation of those views, requires consideration of the actual part and its natural or assembled position. The front view customarily shows the primary features of the part in elevation. Selection of the top view is

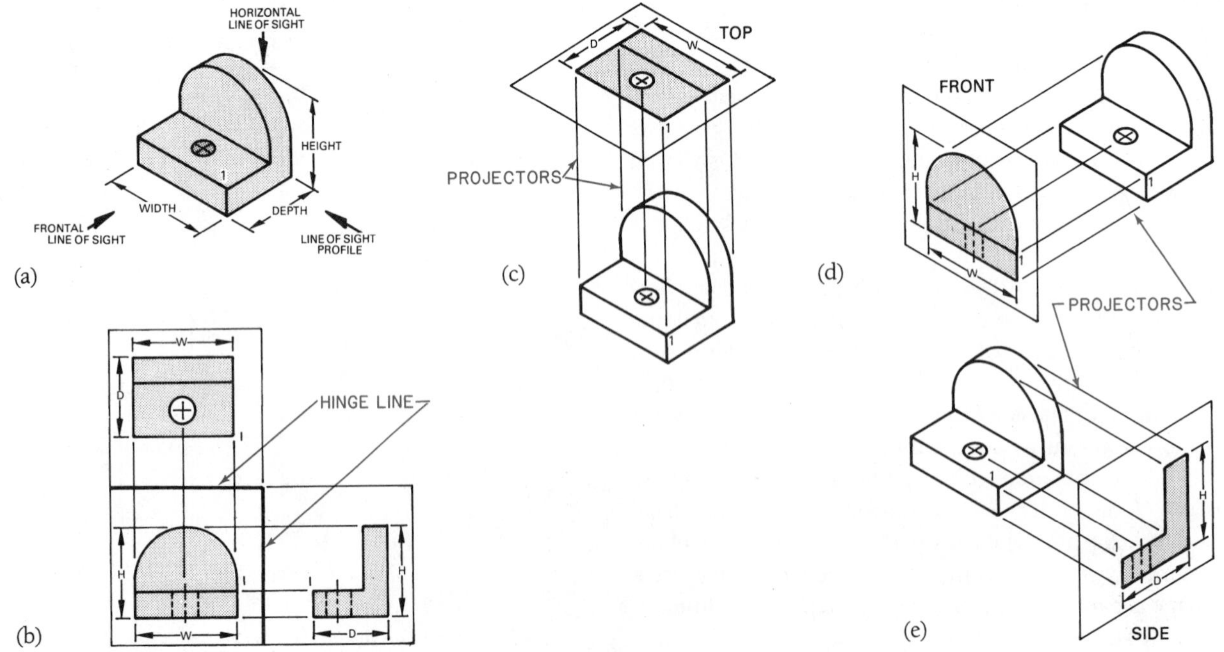

FIGURE 5.21 Line of Sight for Views

ITEMS OF INTEREST

Leonardo, "The Sketcher"

Sketches, illustrations, and technical drawings visually represent the designer's ideas so they may be understood by others. The thought required to sketch an idea and the discussion of ideas with others are good ways to refine proposed solutions to engineering problems.

Prehistoric people recorded their experiences by drawing on cave walls. These cave drawings showed hunting scenes and included people, animals, and tools such as spears and arrows. Who knows, they may have even believed these drawings had the power to make events come true.

A freehand sketch has always been a fast and easy way to put on paper ideas formulated in the mind. Leonardo da Vinci sketched hundreds of plans for his inventions. Today, manufactured parts often begin with a freehand sketch.

When you think of Leonardo da Vinci, you probably think about him as one of the greatest painters of the Italian Renaissance. It is true that he was trained to be a painter and he did produce some of the world's greatest paintings, including the *Mona Lisa.* He also designed machines there were far ahead of his time, such as a flying machine and a parachute. He became one of the most versatile geniuses in history because of his achievements, including scientific inventions.

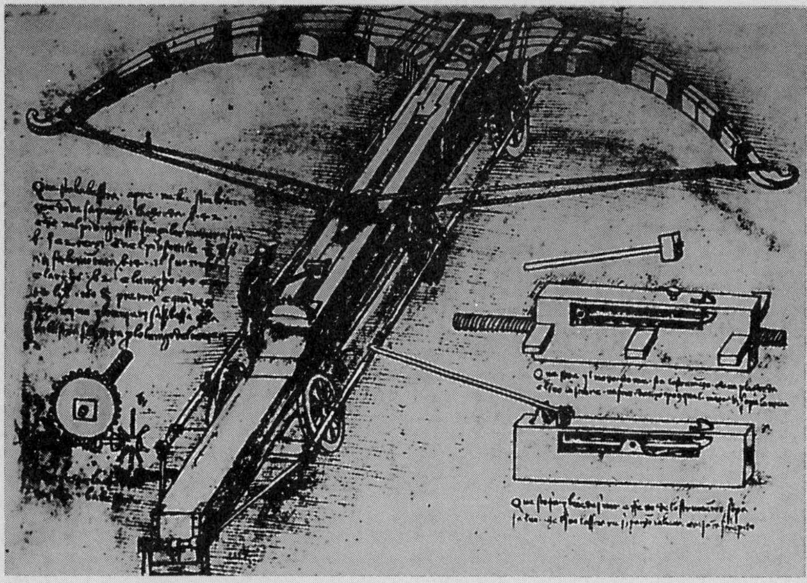

Backward Notes

In approximately 1482, Leonardo went to Milan to be the court artist to the Duke of Milan. One of his duties there was that of a military engineer. He designed artillery and the diversion of rivers. He also designed sets for court pageants. When he was forced to leave that post because of the French invasion, he returned to Florence to serve as a military engineer to that court. During this time, he traveled throughout central Italy preparing sketches for maps that would become important to the history of cartography. Although he never did construct a building, he was held in the highest esteem as an architect. He drew plans ranging from the dome of the Milan cathedral to an enormous bridge over the Bosporus.

During his later years, Leonardo did little painting; instead he produced many sketches of experimental machines and other inventions. These rank among his greatest masterpieces because of their sense of motion and his use of shade and shadows.

Leonardo recorded his ideas in several notebooks, many of which include sketches and drawings that reveal his skill as a drafter and designer. About 4200 pages of his notebooks are still in existence. However, should you decide to read them, be sure to bring a mirror. Leonardo wrote his notes backward!

If all engineers, designers, and drafters recorded their ideas in a similar diligent and elegant fashion, we too might be well known for our graphic communications skills. Leonardo showed us all the value of a sketch or two. His, of course, were also masterpieces.

Leonardo's Mechanical Sketch

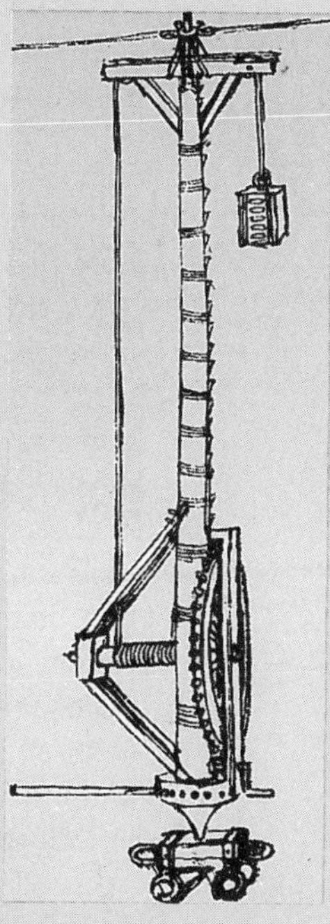

One of Leonardo's Sketches

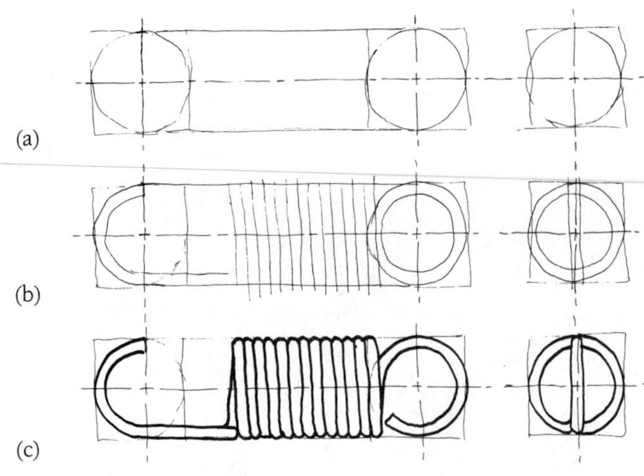

FIGURE 5.22 Blocking-out a Part

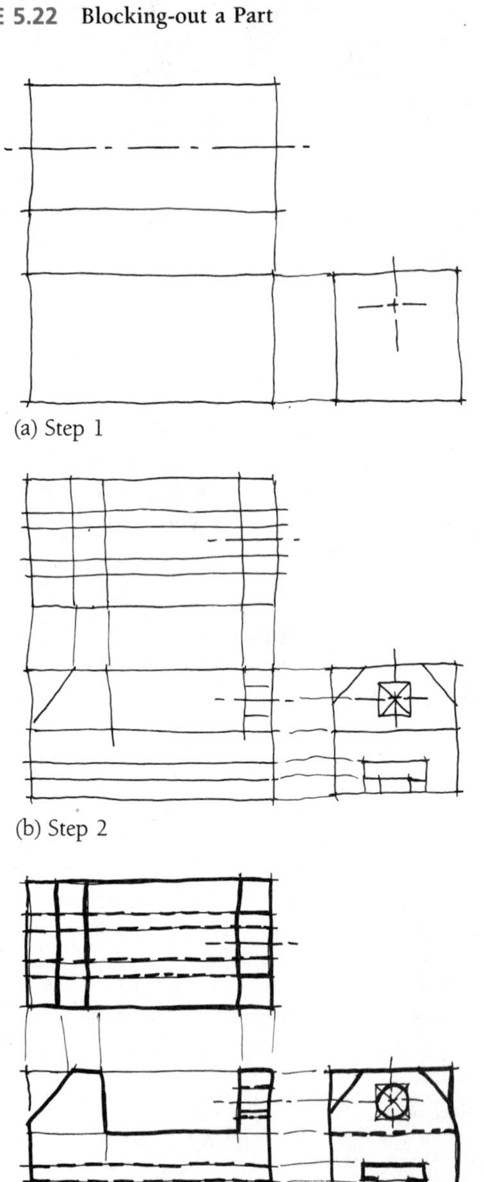

FIGURE 5.23 Blocking-out a Three-view Sketch (a) Block-out overall dimensions (b) Complete blocking-out all features (c) Darken all lines

usually obvious. You should use the minimum number of views necessary to describe the object completely. For example, only one or two views are needed for cylindrical parts because the diameter dimension will describe width and depth and features along the length are described in the longitudinal view (Fig. 5.22). Engineering sketches generally require at least two views.

5.5.3 Multiview Sketching

Figure 5.22 shows the three stages of sketching. The overall dimensions of the part were blocked out first in each view. Centerlines were added to establish circular or symmetrical aspects of the part. Next, the spring coils were drawn with construction lines. Finally, the lines were darkened.

Figure 5.23 is a multiview sketch of a part that required all three views. Each view is "in line" with its adjacent view, as are all the features of the part. Adjacent views of edges, holes, and other shapes are established by projecting lines between the views. Construction lines are extended view to view. Since alignment of the views is critical in multiview sketching, grid paper makes the sketching process easier and faster.

Because of the widespread use of computers in technical work, computer-aided design (CAD) is now used for many projects. However, sketching is and will continue to be the most effective and most used way to communicate graphic ideas. In fact, many companies now use a correctly dimensioned engineering sketch to speed the drafting stage of the design through the manufacturing cycle. This is called *simplified drafting* and has gained widespread acceptance in our highly competitive world.

You May Now Complete Exercises 5.1 Through 5.4 at This Time

5.6 Isometric Projection

Pictorial drawings are widely used for display illustrations and product literature. Isometric drawing is the most common pictorial technique.

Isometric projection is based on the theory that a cube representing the projection axes is rotated until its front face is 45° to the frontal plane and then is tipped forward or downward at an angle of 35° 16'. All three primary faces are displayed equally. In Figure 5.24 the part has been enclosed in a glass box and projected onto each of its corresponding surfaces. The viewing plane 1–2–3 is parallel to the projection plane. This is an isometric view. In true isometric projection, the three axes make equal angles with the projection plane and all three axes are equally foreshortened and make equal angles of 120°. A true isometric projection is about 81% of the size of an isometric drawing. Isometric drawing is used in industry.

Isometric drawing is commonly used in sketching. Isometric drawings are constructed along three axes, one vertical and the other two at 30° to the horizontal going both right

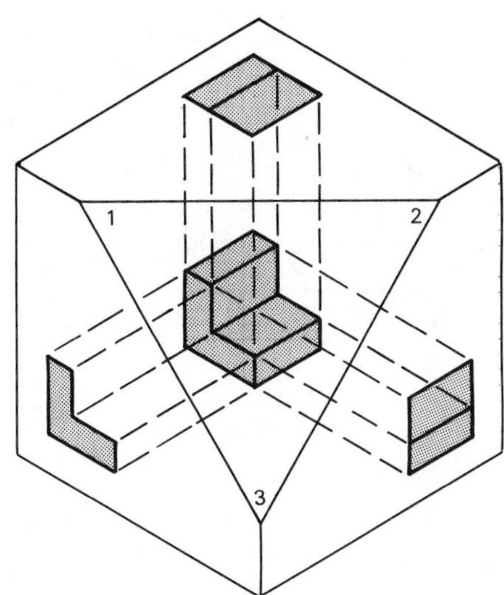

FIGURE 5.24 Isometric Projection

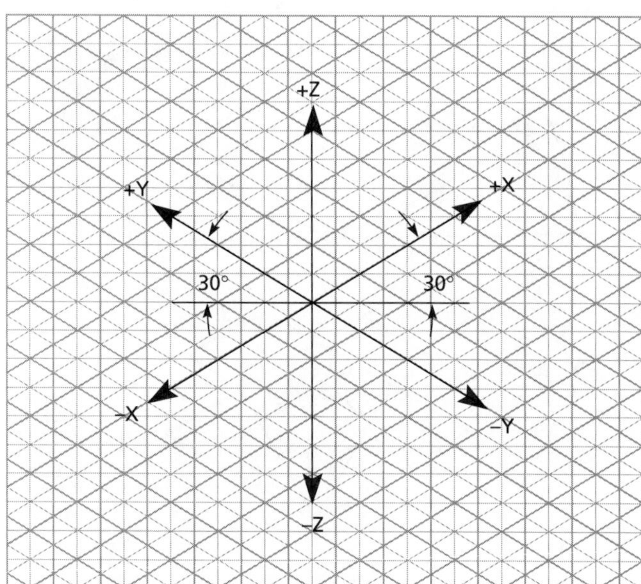

FIGURE 5.25 Isometric Axes

(a)

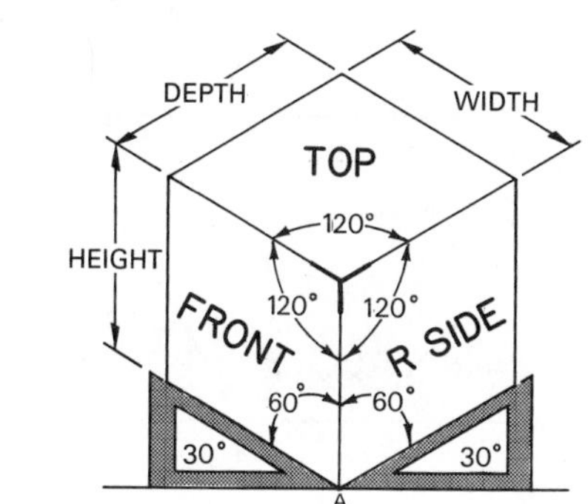

(b)

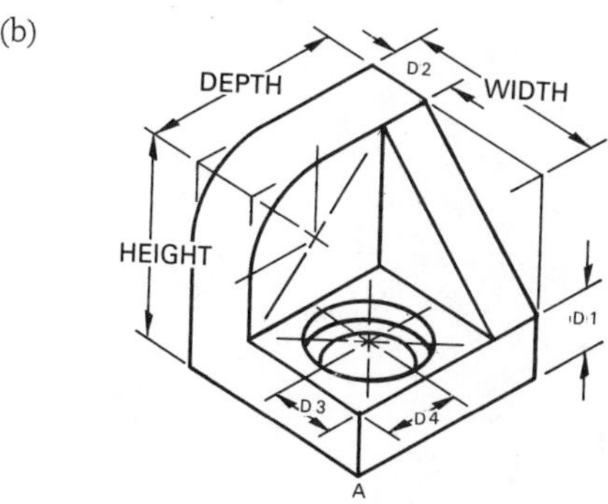

(c)

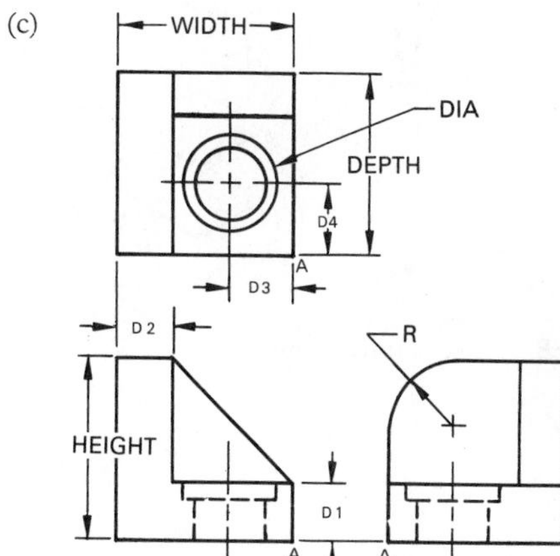

FIGURE 5.26 Isometric Projection

and left (**isometric axes**, Fig. 5.25). All lines in isometric drawings that are on or parallel to the three axes are drawn true length and are **isometric lines.** Lines not on or parallel to the axes are constructed with offset dimensions and are called **nonisometric lines.** Nonisometric lines are not true length.

5.6.1 Isometric Construction

Isometric construction using the box method is illustrated in Figure 5.26. The procedure for drawing an isometric box is shown in Figure 5.26(a) using 30° triangles. Starting at point A, the three axes are drawn. The edges of the box are constructed from the height, width and depth. *In an isometric drawing the dimensions are not foreshortened.*

After the part is boxed in, the remainder of the drawing is completed. Dividers (or a scale) are used to transfer dimensions shown in Figure 5.26(c) to the isometric view of Figure 5.26(b). All measurements are taken along isometric lines. Dimension D1 is measured along the vertical axis; dimensions D2, D3, and D4 are in the horizontal plane and are measured along or parallel to one of the receding axes. Locate the centerlines, then draw the circles and arcs.

5.6.2 Isometric Angles

Because of the distortion created by the isometric view, few angles appear as true angles. Angles appear larger or smaller than true size and must be established by **offset dimensions.** For example, the plane in Figure 5.27 has angles of 45° and

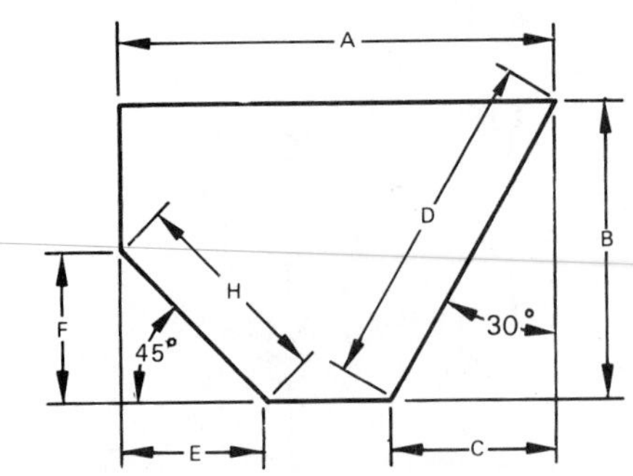

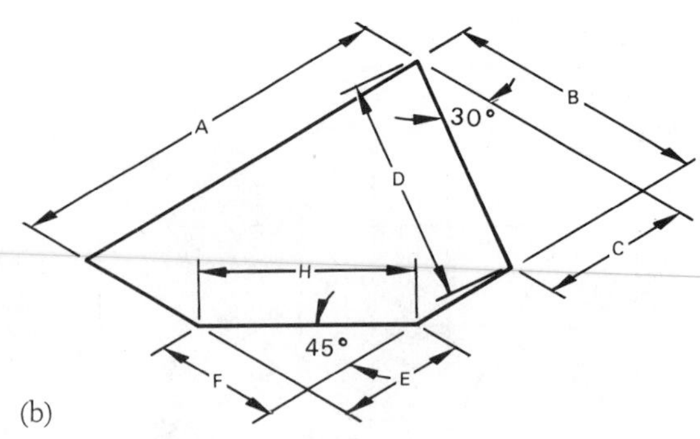

FIGURE 5.27 Isometric Angles (a) Orthographic (b) Isometric

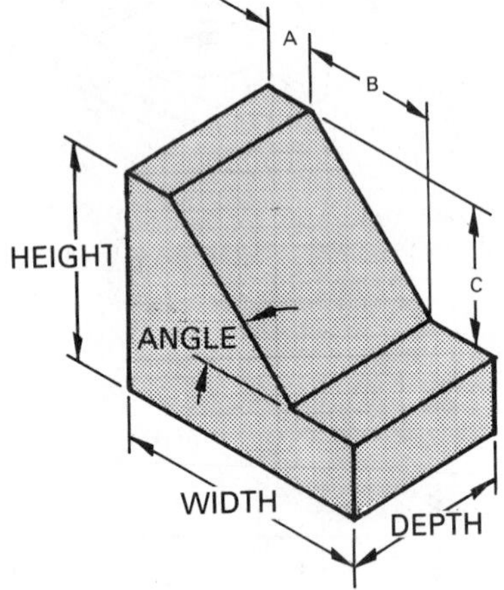

FIGURE 5.28 Offset Dimensions

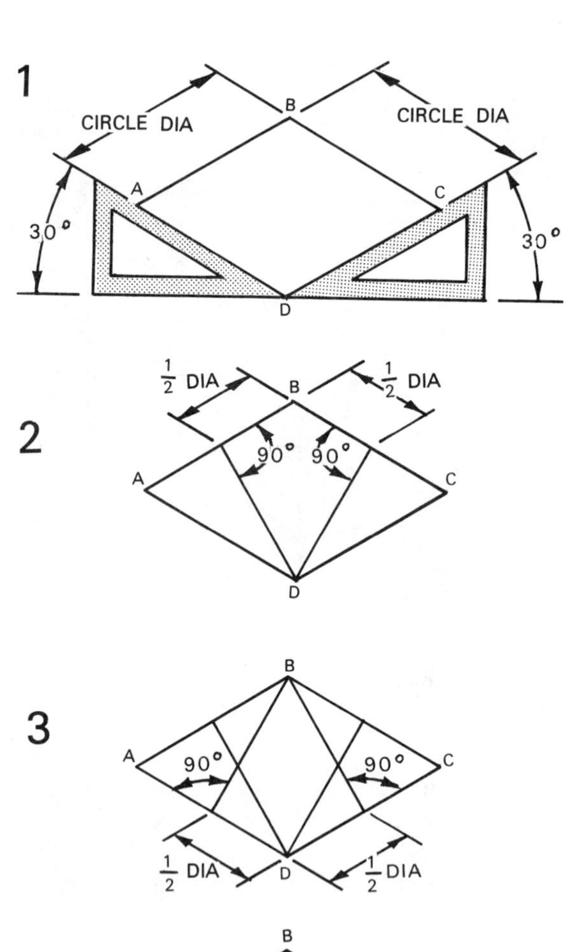

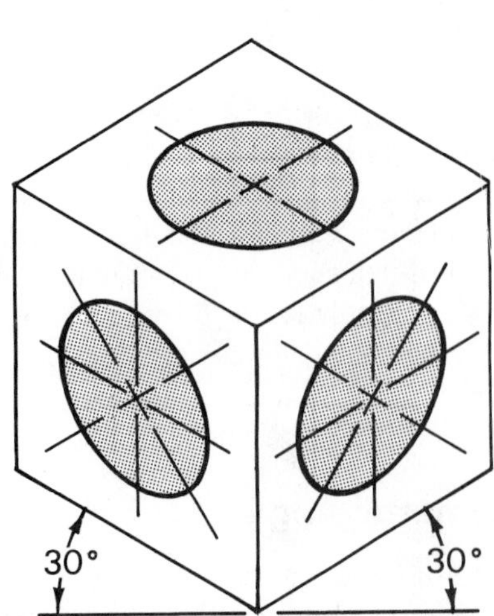

FIGURE 5.29 Isometric Ellipses

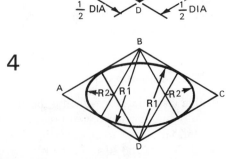

FIGURE 5.30 Ellipse Construction for Isometric Drawings

30°. Both angles are constructed from offset dimensions, measured along isometric lines from the top view of the plane [Fig. 5.27(a)]. The isometric view of the plane [Fig. 5.27(b)] is boxed in with true length dimensions A and B along the isometric axes. The 30° angle is constructed by transferring dimension C.

The part in Figure 5.28 has an angled surface. To draw the part in an isometric view, it is necessary to use dimensions A, B, and C because they can be taken along true length lines.

5.6.3 Isometric Circles and Arcs

Circles and circular arcs on isometric drawings appear to be elliptical (Fig. 5.29) unless they fall exactly on or parallel to the isometric viewing plane. Many methods are used to construct **isometric ellipses:** template, trammel, four-center, and point plotting. For sketches, freehand techniques are sufficient (you can also use a template). The **four-center method** (Fig. 5.30) does not create a perfect ellipse, but is accurate enough for most purposes. This method is used to draw circles or arcs on any isometric face (Fig. 5.31).

Isometric ellipses are easily sketched using this method (Fig. 5.32). Circles, arcs, or curves that do not lie in isometric planes, as in Figure 5.33, must be plotted with offset dimensions. A series of points is established along the curved outline. Offset dimensions for these points are transferred to the isometric drawing and are laid off along isometric lines.

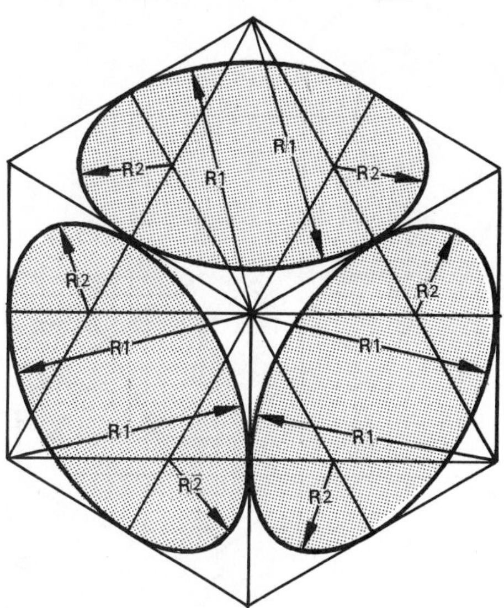

FIGURE 5.31 Isometric Ellipses and the Four-Center Method

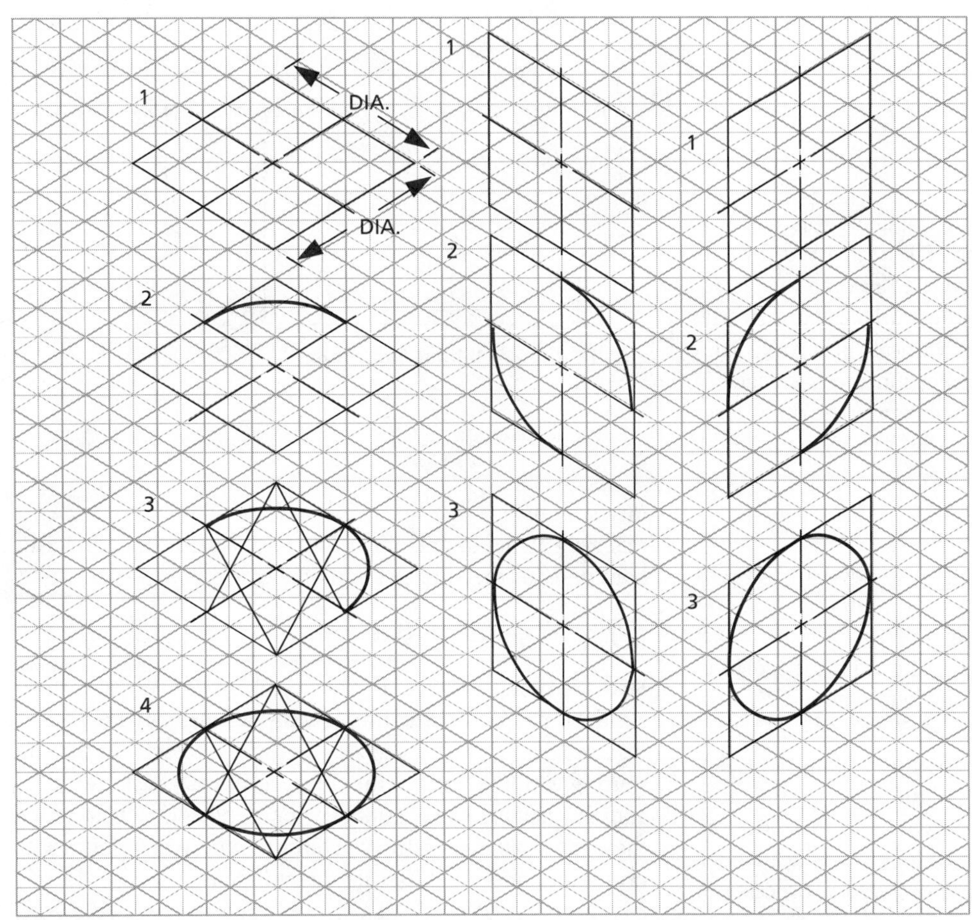

FIGURE 5.32 Sketched Isometric Arcs and Circles

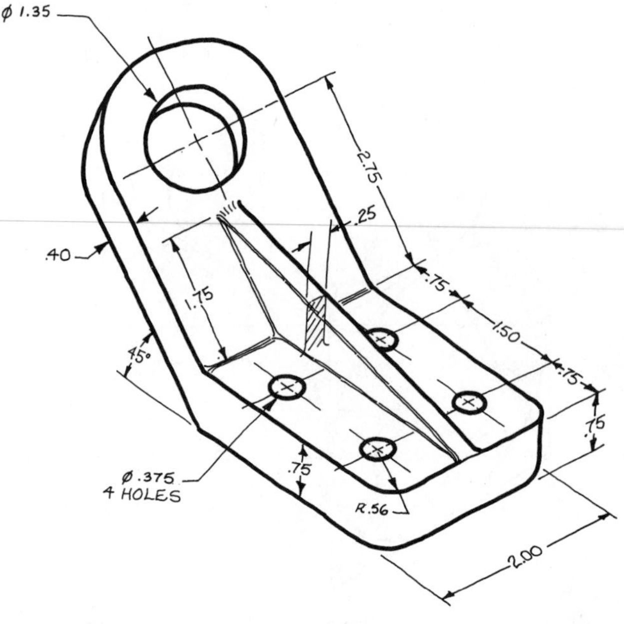

FIGURE 5.33 **Isometric Sketch of a Part**

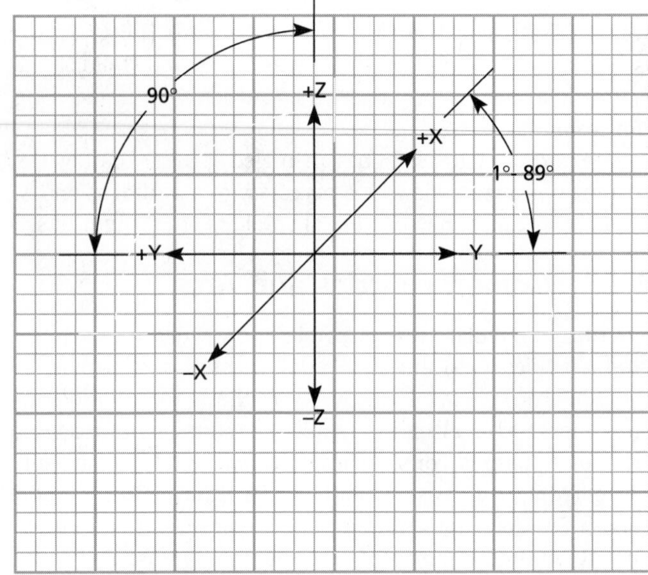

FIGURE 5.34 **Oblique Axes**

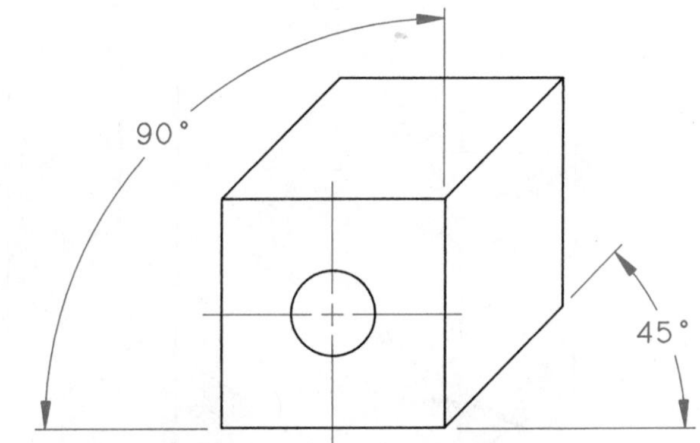

FIGURE 5.35 **Oblique Projection**

5.7 Oblique Projection

Oblique drawings are produced from parallel projectors that are angular to the projection plane. The primary difference between isometric and oblique is that a receding axis is required for oblique drawings. The other surface is drawn true shape and size. The three axes are vertical, horizontal, and receding (Fig. 5.34). In oblique projection the front face of the part is placed parallel to the image plane. The other faces of the part are on receding axes (1° to 89°). In Figure 5.35 the front face of the block and the diameter of the hole are drawn true shape and size. The most commonly used angle for the receding axis is 45°.

The two basic categories of oblique projection are **cavalier** and **cabinet** (Fig. 5.36). In a cavalier projection [Fig. 5.36(a)], receding lines are not foreshortened (full scale). In a cabinet projection [Fig. 5.36(b)] the receding lines have been foreshortened one-half their original length ($\frac{1}{2}$ scale). The most commonly used angles are 15°, 30°, 45°, 60°, and 75°.

Parts drawn with oblique projection are oriented so that the surface with the most curved features lies in the front plane. Circles and arcs are true projections in this position. Oblique projection is extremely useful for parts with parallel, curved, or irregular features. The construction process for slanted, inclined lines and planes is similar to that of isometric drawings. Locate each feature's end points along lines that are parallel to one of the axes; for slanted surfaces, locate both ends of the surface and connect the points.

5.7.1 Oblique Sketching

Start the oblique sketch as you would a multiview or isometric drawing; block out the overall dimensions (Fig. 5.37). Figure 5.38 shows the steps in this process. Block out the part starting with the front or the rear face and establish the width and the height. Next, establish the depth of each face and then *carve out* the features. Locate each circular form with centerlines and block out its three dimensions.

You May Complete Exercises 5.5 Through 5.8 at This Time

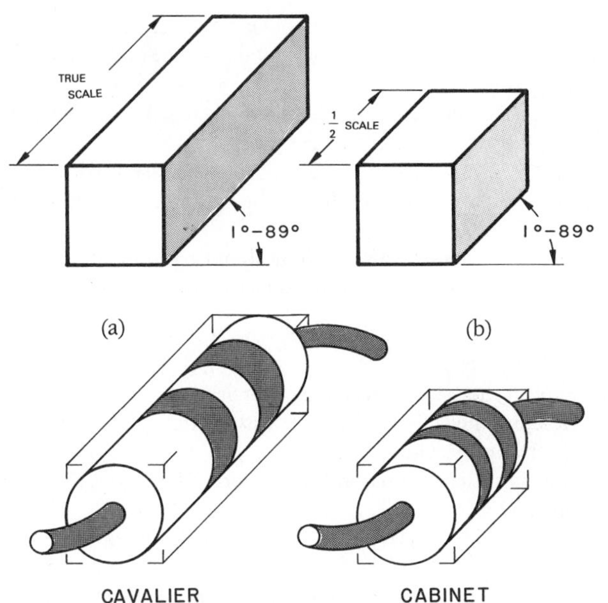

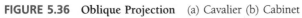

FIGURE 5.36 **Oblique Projection** (a) Cavalier (b) Cabinet

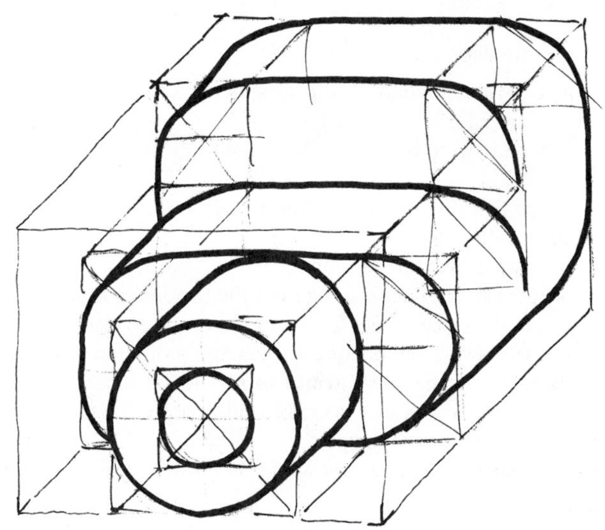

FIGURE 5.37 **Oblique Sketch**

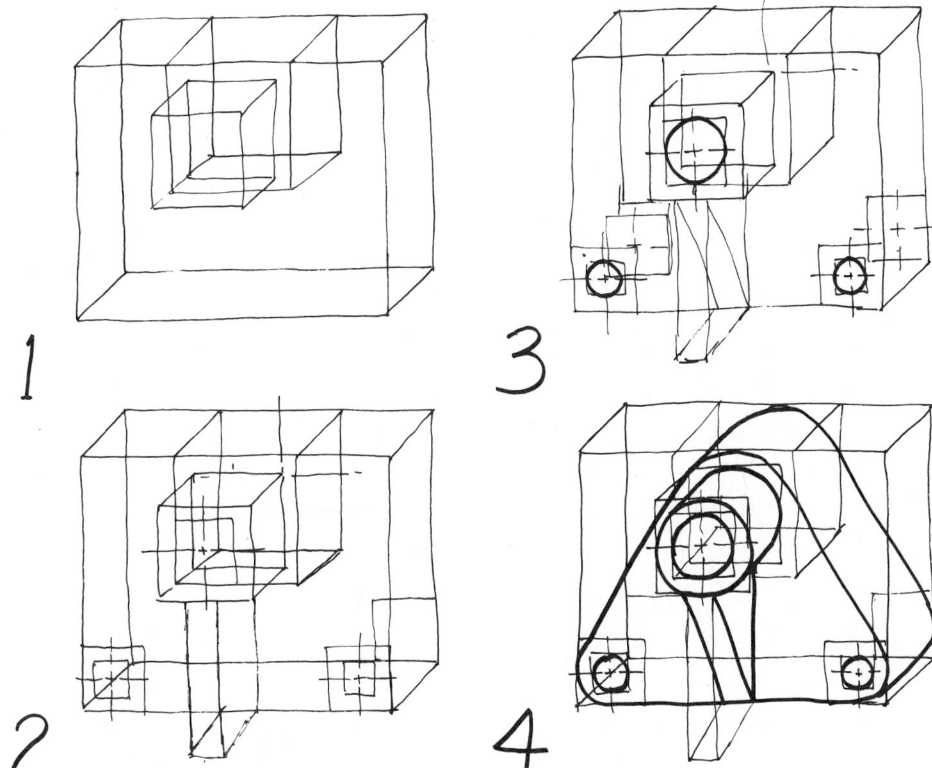

FIGURE 5.38 Step-by-Step Oblique Construction

QUIZ

True and False

1. The most commonly used angles for sketching oblique drawings are 15°, 20°, 25°, and 40°.
2. When drawing a circle, it is common practice to rotate the paper.
3. When blocking in a circle or an isometric ellipse, the sides are equal to the diameter.
4. 6H lead is best for sketching.
5. Sketch vertical lines starting from the bottom and move up.
6. Never show centerlines for round or curved portions of a part if it is drawn as an oblique or isometric projection.
7. Pictorial sketches are essential to the design process because they allow the designer to explore different possibilities, shapes, and orientations of the part.
8. Grid paper should be used whenever possible when sketching.

Fill in the Blanks

9. ________ or ________ lead is the best grade for sketching.
10. The pencil is held about ________ from the ________ when sketching.
11. For right-handed drafters, draw horizontal lines by moving the pencil from ________ to ________.
12. Circles are sketched by first drawing a ________.
13. ________ lines are used to lay out the outline of the part before darkening the lines.
14. ________ and ________ lines are sketched by moving the pencil from ________ to ________.
15. ________, ________, ________ sketches are used to represent the part pictorially during the ________ design stage of the project.
16. ________ lines are drawn vertical and receding at ________ degrees to the horizontal for isometric drawings.

Answer the Following

17. Explain the steps in sketching a circle.
18. Describe the difference between isometric, oblique, and perspective projections.
19. How is sketching used in conjunction with CAD in the design process?
20. How would you sketch an ellipse that lies in the horizontal plane?
21. What are the six standard views? Which views are most commonly represented on an orthographic drawing?
22. Why are parts always blocked out before darkening the lines?
23. How does the shape of the part help determine the use of isometric or oblique projection techniques?
24. Describe the process of sketching irregular curves.

EXERCISES

Transfer the given information to an "A" size sheet or .25 in. grid paper. Complete all views and solve for proper visibility, including centerlines, object lines, and hidden lines. Exercises that are not assigned by the instructor can be sketched in the text to provide practice and understanding for the preceding instructional material.

After Reading the Chapter Through Section 5.5.3, You May Complete the Following Exercises

Exercise 5.1 Sketch the one-view drawing.

Exercise 5.2 Sketch the two-view drawing.

Exercise 5.3 Sketch the circular part.

Exercise 5.4 Sketch the two-view section drawing.

These exercises can also be used for isometric and oblique problems after you have completed the chapter.

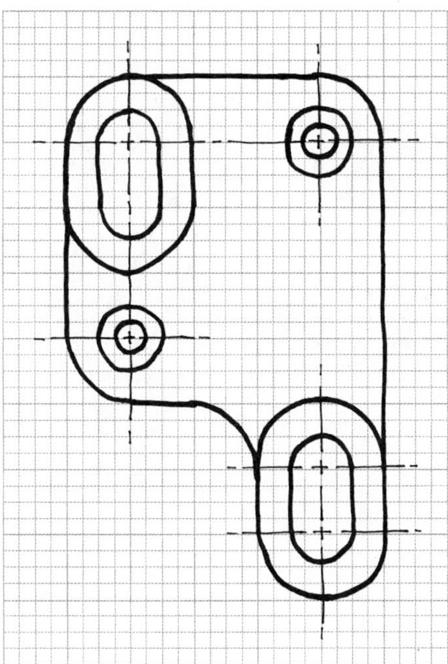

Exercise 5.1

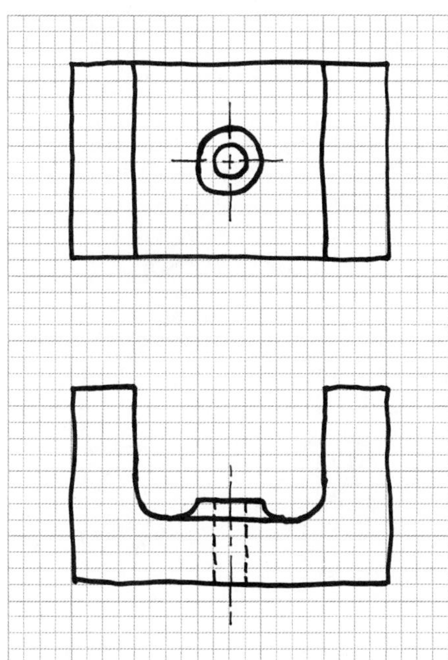

Exercise 5.2

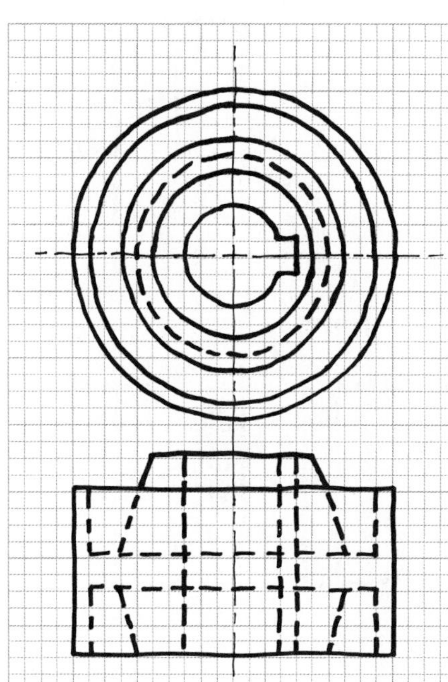

Exercise 5.3

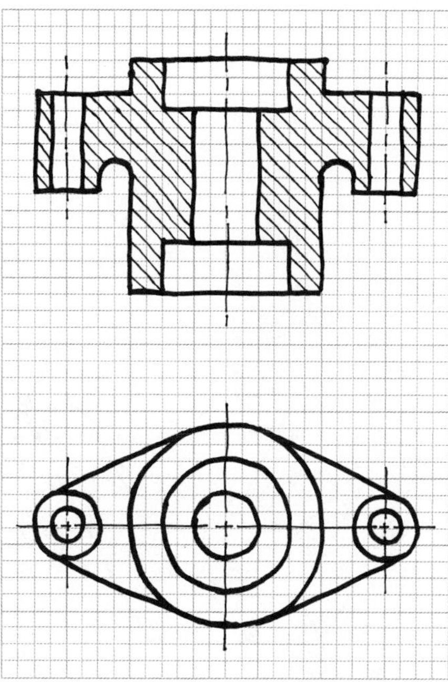

Exercise 5.4

After Reading the Chapter Through Section 5.7.1 You May Complete the Following Exercises

Exercise 5.5 Sketch the three-view part and complete an isometric sketch of the part on isometric grid paper.

Exercise 5.6 Sketch an isometric view of the part using isometric grid paper and complete a two-view drawing.

Exercise 5.7 Sketch an isometric view of the part on isometric grid paper. Also complete a three-view drawing.

Exercise 5.8 Sketch an oblique cabinet view of the part (use 45°) and complete a two-view drawing.

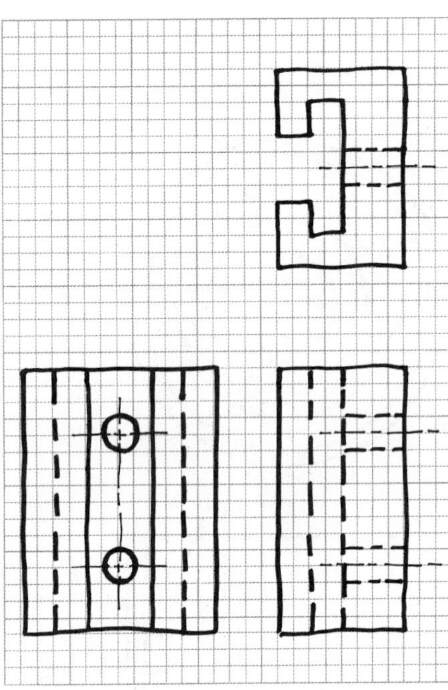

Exercise 5.5

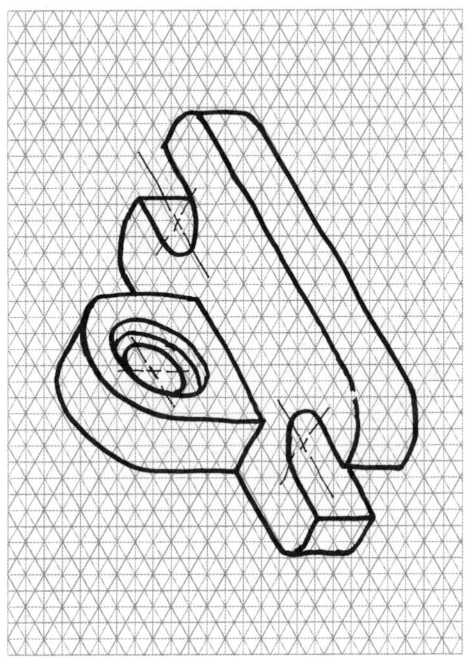

Exercise 5.7

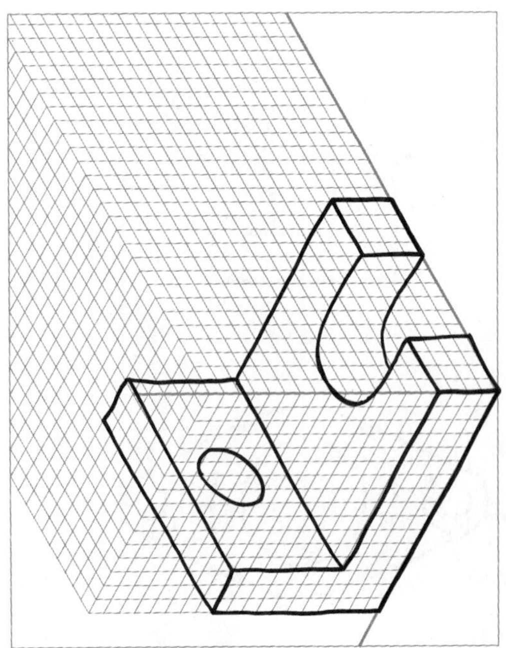

Exercise 5.6

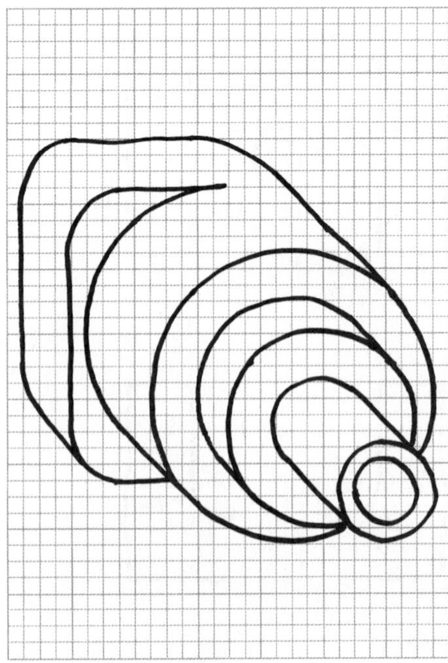

Exercise 5.8

PROBLEMS

Problems 5.1(A) through (H) Freehand sketch the assigned problems in multiview projection using grid paper. Draw the projects at two times the book scale. Problems can be either metric, inches, or decimal units. Two or three views may be required for a particular problem.

Problems 5.2(A) through (I) Same as Problem 5.1.

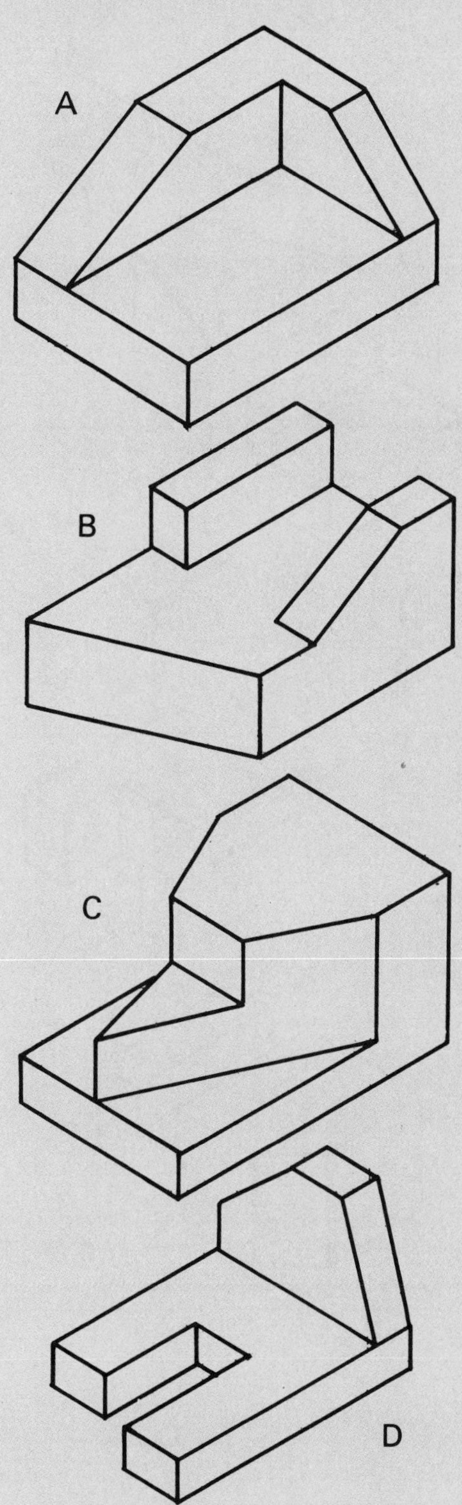

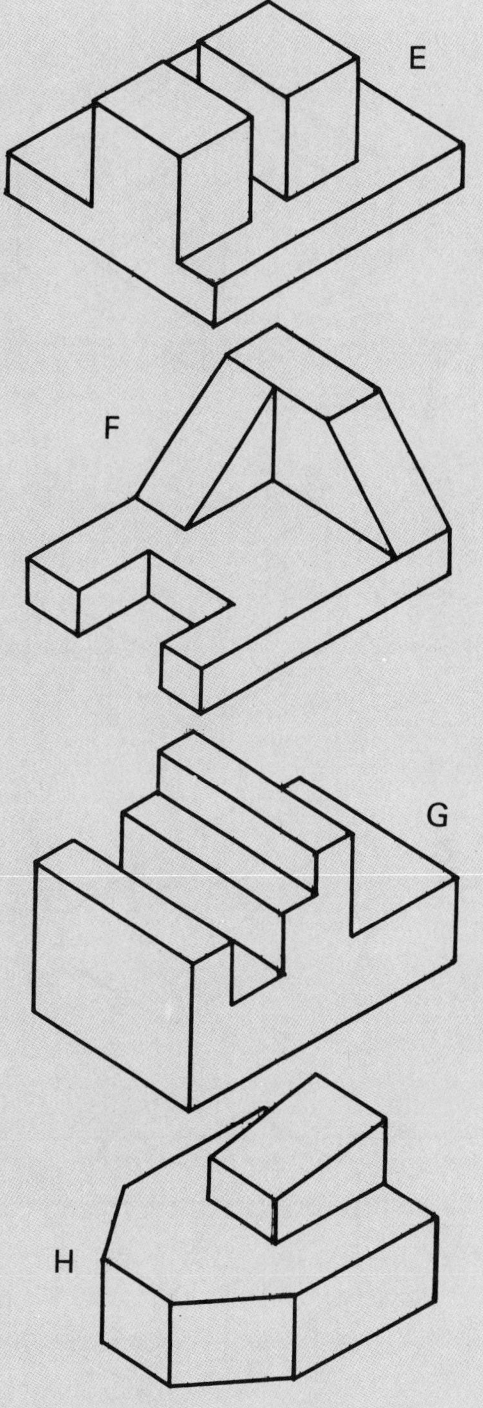

Problem 5.1(A) Through (H)

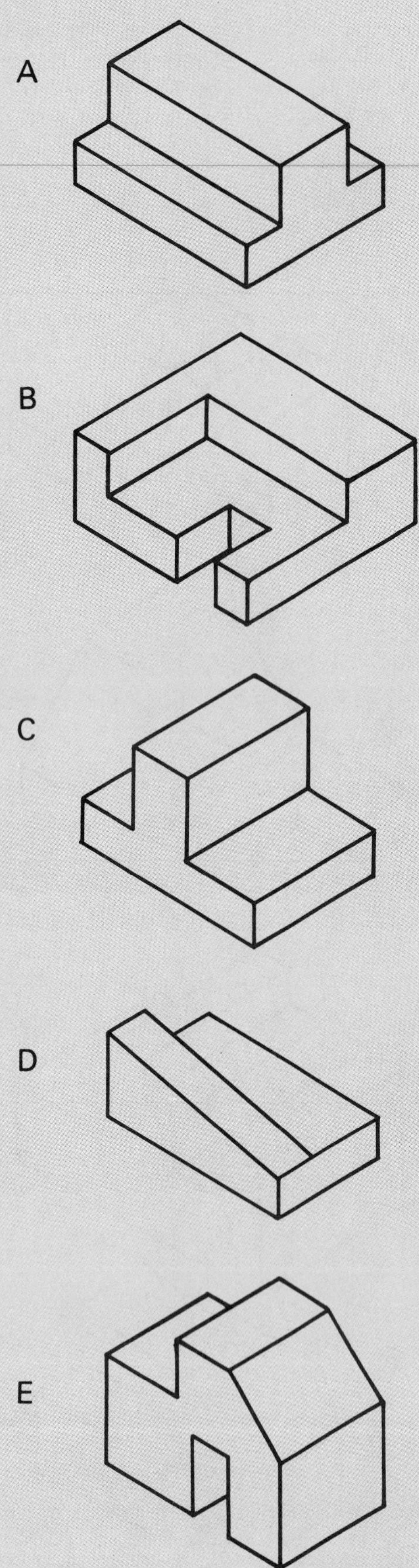

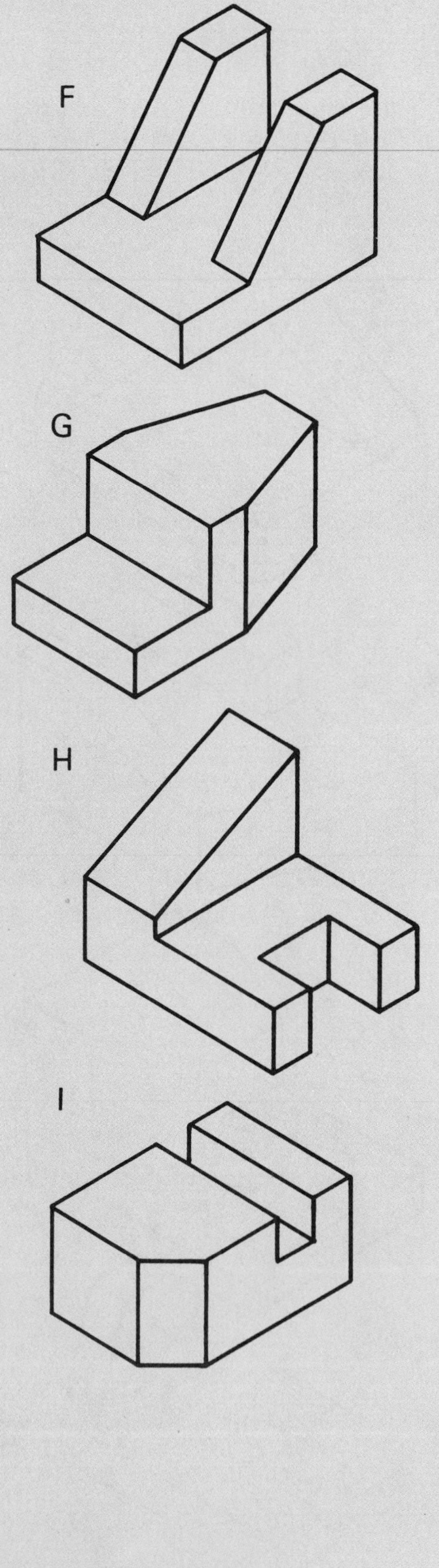

Problems 5.2(A) Through (I)

Problem 5.3(A) through (G) Using freehand sketching, draw each of the assigned problems. Use oblique projection. Be careful to choose the proper surface for the front face of the part. Draw at two times book scale.

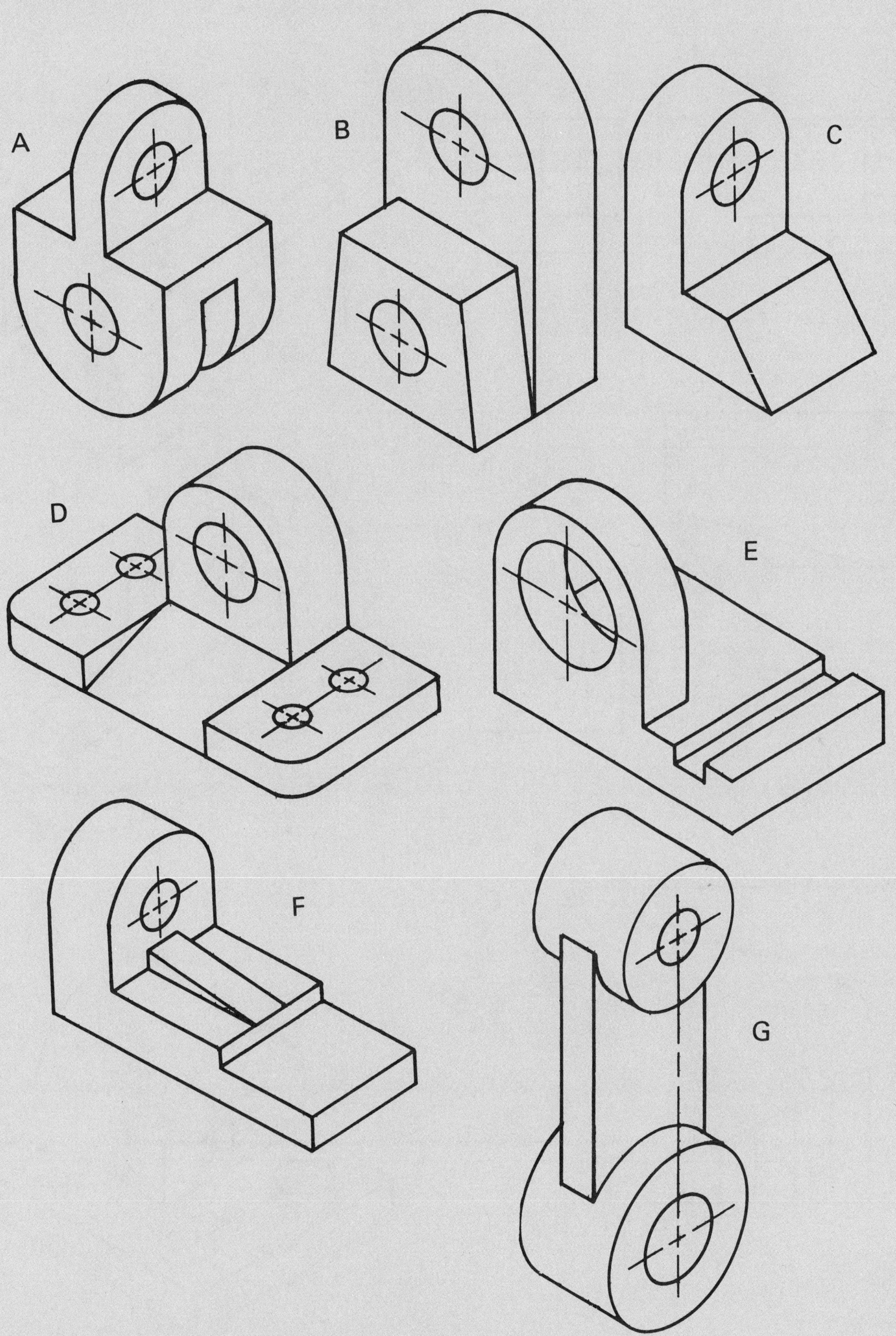

Problems 5.3(A) Through (G)

Problems 5.4(A) through (C) Complete the three views of each problem. On the same sheet sketch an isometric view. Draw at two times book scale.

Problems 5.4(D) through (F) Sketch three views of each problem. Draw at two times the book scale.

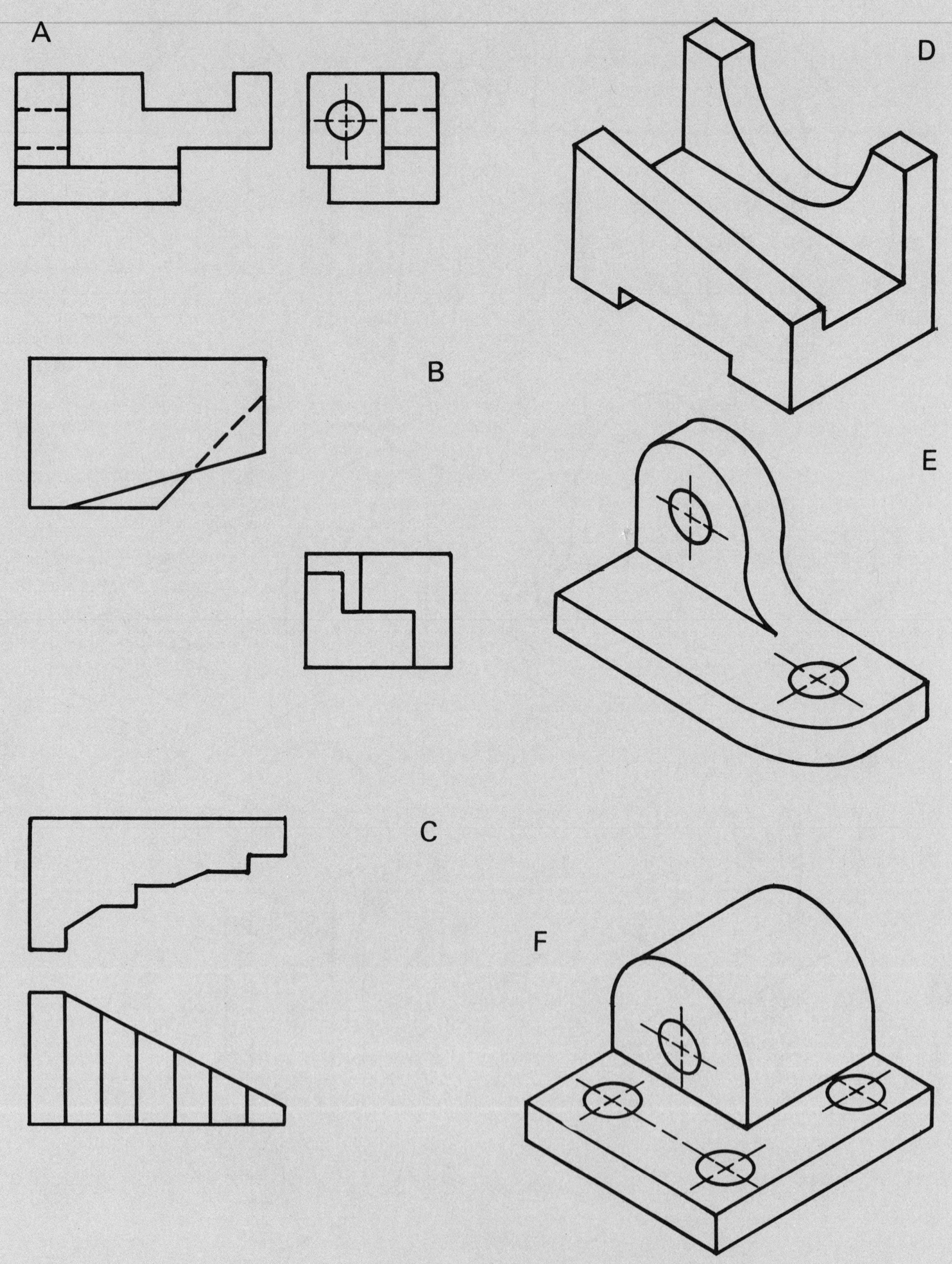

Problems 5.4(A) Through (F)

Problem 5.5 (A) through (J) Complete the given views and project a third view of each problem. Do an isometric sketch of each problem on a separate sheet of paper. Draw at three times the book scale. The isometric sketch will help solve for the three views of the part.

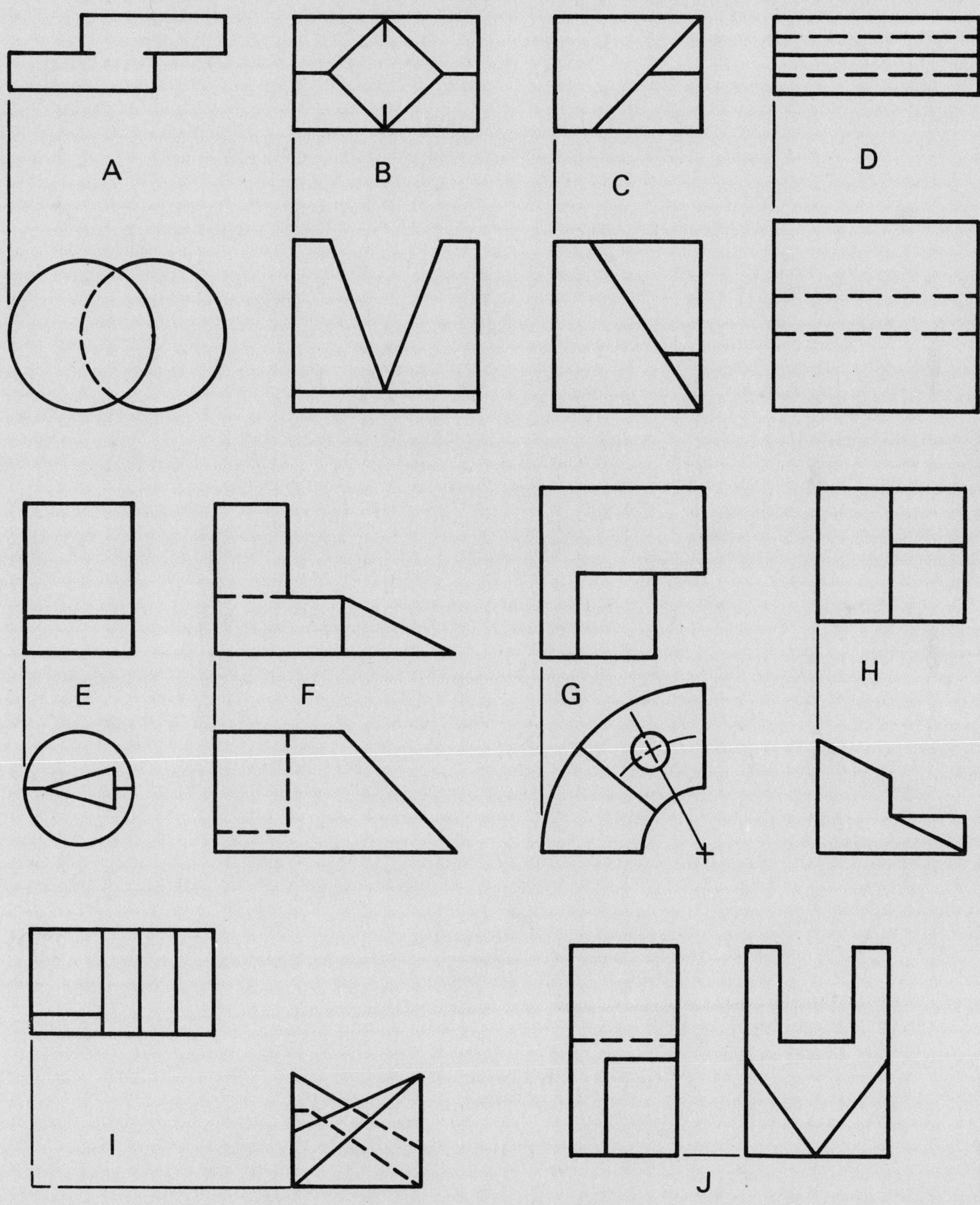

Problems 5.5(A) Through (J)

PART II

Projections, Views, Manufacturing, and Dimensioning

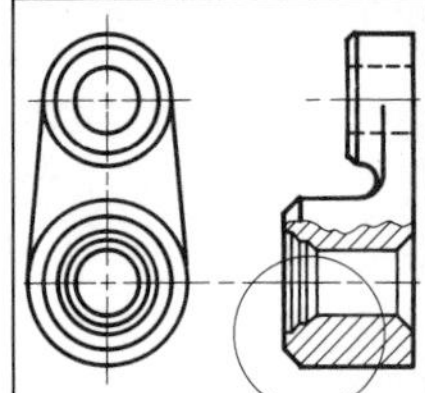

CHAPTER 6

Multiview Drawings

Learning Objectives

Upon completion of this chapter you will be able to accomplish the following:

1. Recognize the importance of orthographic projection in order to describe part features graphically.
2. Differentiate between first- and third-angle projection.
3. Identify the six standard views.
4. Demonstrate the ability to select a parts orientation and the number of views needed for complete part description.
5. Produce multiview drawings demonstrating standard line precedence.
6. Demonstrate familiarity with partial, revolved, and enlarged views.
7. Define methods of hole, fillet and round, tangent surface, runout, and thread representation.

6.1 Introduction

Multiview drawing using orthographic projection is the primary means of graphic communication in engineering work. These drawings are used to convey ideas, dimensions, shapes, and procedures for the manufacture of a part or the construction of a project. **Orthographic projection** is a procedure that is used to describe completely the shape and dimensions of a part with one or more views. Regardless of whether computer-aided design (CAD) systems or manual techniques are used, knowledge and understanding of multiview drawing based on orthographic projection is essential for the aspiring engineer, designer, or drafter.

There are two primary ways to explain orthographic projection: the **normal/natural method** and the **glass box method.** The normal or natural method is typical in mechanical and other engineering fields (Fig. 6.1). The glass box method is used in descriptive geometry and in teaching orthographic projection. This method requires you to imagine that the points, lines, and planes of the part are enclosed in a transparent "box" (Fig. 6.2). Views of the part are established on their corresponding glass box surfaces by using perpendicular projectors originating at each point of the part and extending to the related box surface. The box is hinged so that it can be unfolded onto one flat plane (the paper).

When the top, front, and side views are used, each view has a dimension in common with the other two views; the front view shows the height and width; the top view shows the depth and width; the side view shows the depth and height. The width dimension will vertically align the top and front views, and the height dimension will horizontally align the front and side views. The part is viewed perpendicular to

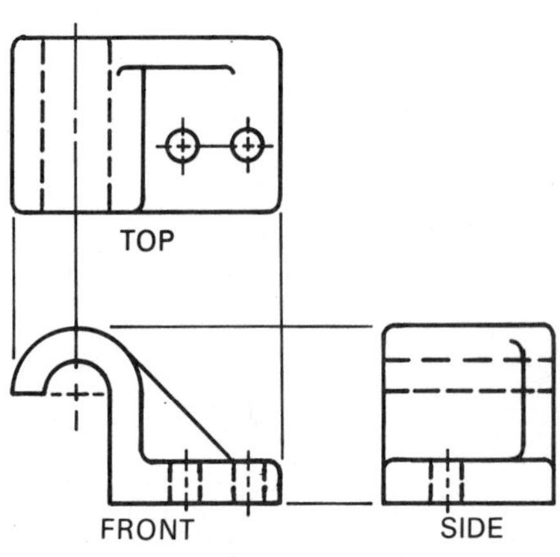

FIGURE 6.1 Three-View Drawing

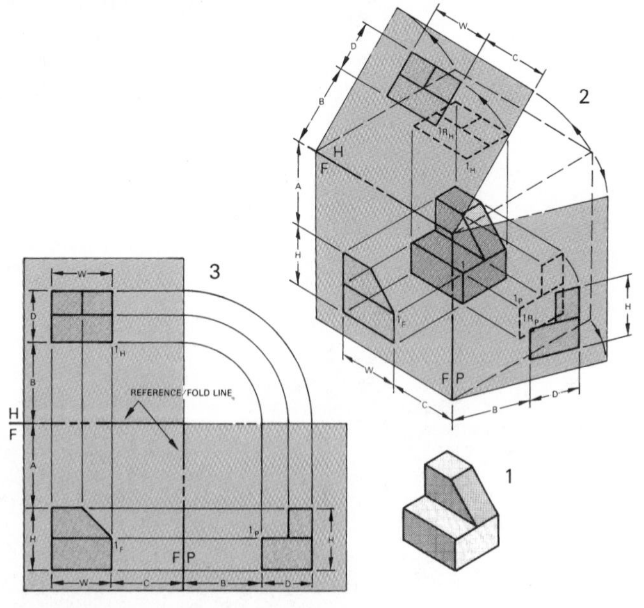

FIGURE 6.2 Multiview Drawing

each of its three primary surfaces with the position of the observer changed for each view. *See Color Plates 1, 2, 5, 6, 15, 23.*

6.2 Orthographic Projection

Orthographic projection is a drawing system using projectors from a part perpendicular to the desired planes of projection to produce the desired views. The figure outlined on one of the projection planes is called an **orthographic view.** An orthographic view shows the true size and shape of a surface when it is parallel to the projection plane (area ABCD in Fig. 6.3). If an area is inclined to the plane, the view of the area will be foreshortened (area BCEF in Fig. 6.3).

The glass box method of projection for a part is illustrated in its closed (folded) position and open (unfolded) position in Figure 6.4. The part is thought of as enclosed in the transparent box. The following concepts are used throughout the chapter and the text:

Lines	Dimensions
A = Vertical lines of sight	D = Depth
B = Horizontal lines of sight	H = Height
C = Projection lines	W = Width

Image Planes (Principal Projection Planes)

F = Front (frontal plane)
H = Top (horizontal plane)
P = Side (profile plane)

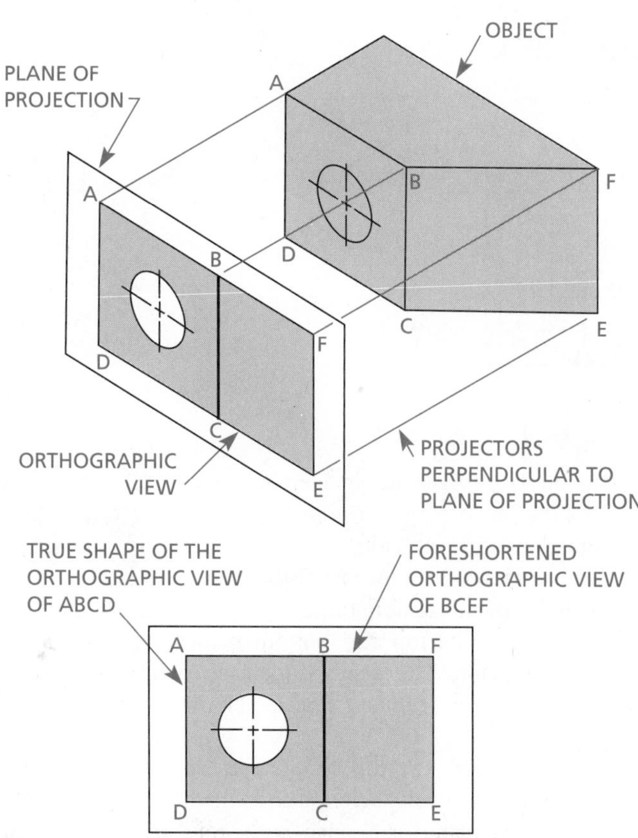

FIGURE 6.3 Third-Angle Orthographic Projection

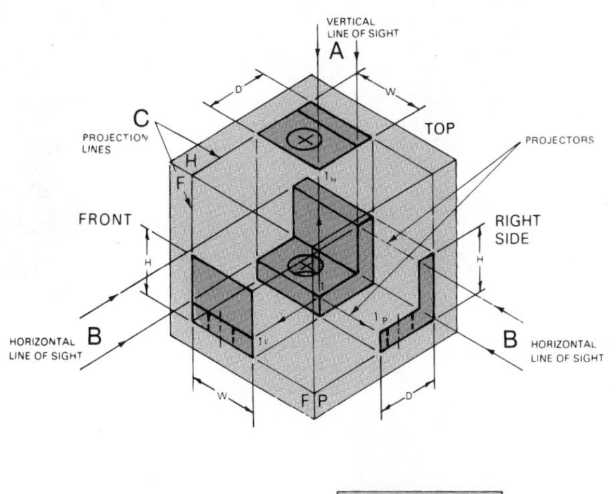

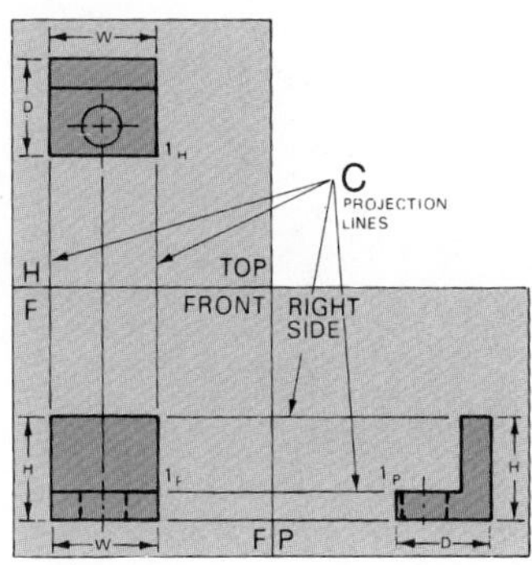

FIGURE 6.4 Orthographic Projection a Part

6.2.1 Line of Sight

The *line of sight represents the direction from which the part is viewed* (Fig. 6.4). The vertical lines of sight (A) and horizontal lines of sight (B) originate at infinity. The line of sight is always perpendicular to the image (projection) plane, represented by the surfaces of the glass box (top, front, and right side). Projection lines (C) connect the same point on the image plane from view to view, at right angles to the adjacent view.

A point is projected on the image plane where its line of sight pierces that image plane. In Figure 6.4, point 1, which represents a corner of the part, has been projected onto the three primary image planes. Where it intersects the horizontal plane, it is identified as 1H. Where it intersects the frontal plane, it is identified as lF. Where it intersects the profile plane, it is identified as 1P. The multiview drawing in Figure 6.4 shows the position of the unfolded image planes, which now lie in the same plane as the paper.

In Figure 6.5(a), the line of sight for each view is shown. These lines of sight establish the direction of viewing that the observer will take when completing the view. Figure 6.5(b) shows the three views properly aligned. In Figures 6.5(c), (d),

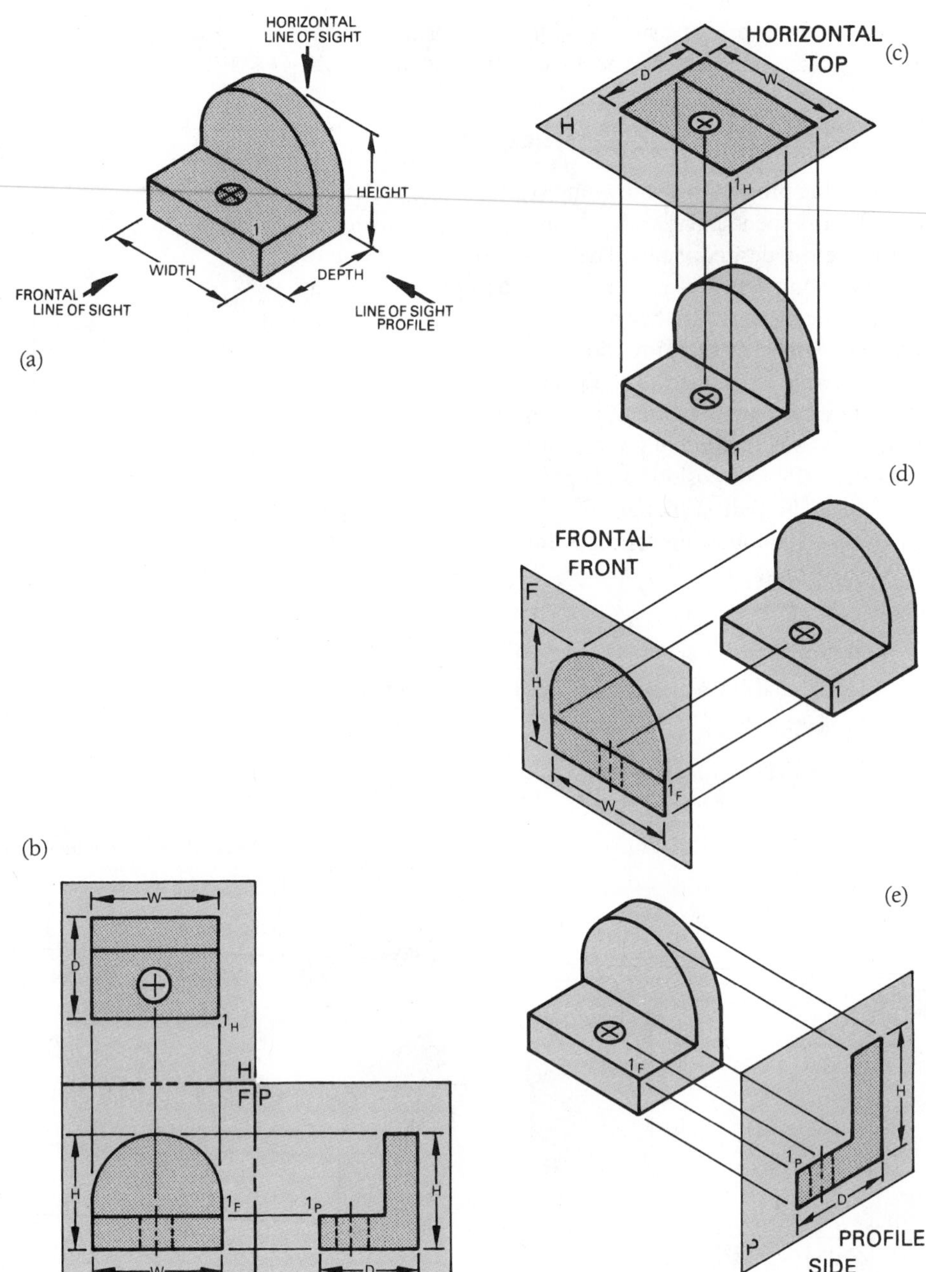

FIGURE 6.5 Line of Sight
(a) Lines of sight (b) Unfolded views (c) Top view (d) Front view (e) Right side view

and (e), the top, front, and side views, respectively, are analyzed separately. All points on each surface of the part are projected onto their corresponding image plane (view).

6.3 The Six Principal Views

When the glass box is opened, its six sides are revolved outward so that they lie in the plane of the paper. With the exception of the back plane, all are hinged to the front plane. The back plane is usually revolved from the left side view, but it can also be hinged to the right side view, as shown in Figure 6.6. Before it is revolved around its hinged fold line (reference line), each image plane is perpendicular to its adjacent image plane and parallel to the image plane across from it. A **fold line** is the line of intersection between any hinged (adjacent) image plane. The left side, front, right side, and back are all **elevation views.** In these views, the height dimension, elevation, and top, and bottom of the view can be determined and dimensioned. The top and bottom planes are both in the horizontal plane. The depth dimension, width dimension, and front and back can be established in these two horizontal planes.

In most cases, the top, front, and right side views are required. These are also referred to as the horizontal plane, H (top); frontal plane, F (front); and profile plane, P (side). These views are the three **principal projection planes.** The

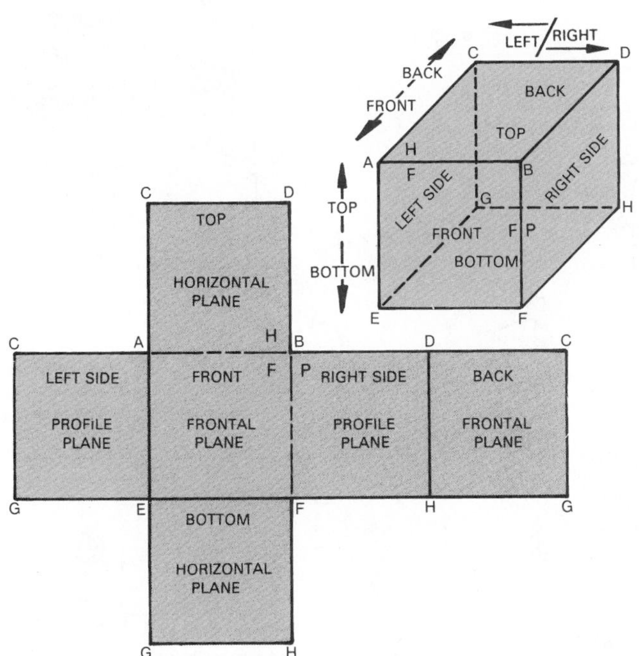

FIGURE 6.6 The Six Standard Views

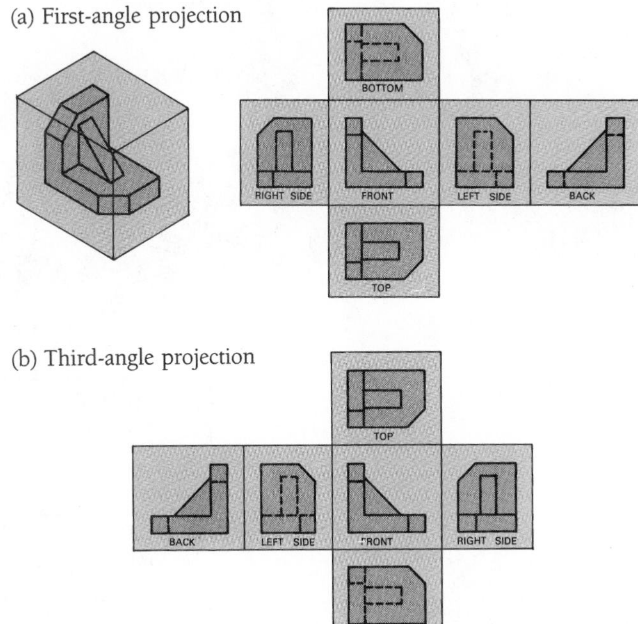

FIGURE 6.7 First- and Third-Angle Projection

top, front, and bottom are in line vertically; the left side, front, right side, and back are aligned horizontally.

When using directions to establish the location of a point or line, the top and bottom are shown in the frontal plane; the terms "above" and "below" are also used to describe directions in this plane. The horizontal view can be used to determine if a point is "in front of" or "in back of" a particular starting point or fold line. To locate a point to the right or left of a fold line or established point, the frontal or horizontal plane is used.

6.4 First- and Third-Angle Projection

The two types of orthographic projection used throughout the world are **first-angle** and **third-angle.** The six principal views of a part that have been presented are known as **third-angle orthographic projection.** This form of projection is used throughout the United States and Canada and is the primary form of projection in American industry. In third-angle projection, *the line of sight goes through the image plane to the part.* Projection lines are used to illustrate this projection from the part to where they intersect the image plane. Figure 6.7(b) illustrates third-angle projection and the normal procedure for unfolding the glass box.

First-angle orthographic projection is used in most foreign countries and on many American structural and architectural drawings [Figure 6.7(a)]. In this projection, the part is assumed to be in front of the image plane. Each view is formed by *projecting through the part and onto the image plane.* Figure 6.8 compares first- and third-angle projection.

6.4.1 ISO Projection Symbol

The internationally recognized projection symbols for first- and third-angle projection that are shown on drawings are shown in Figure 6.8. Symbols are required on drawings to be interchanged internationally. The symbol is normally placed to the left of the title block. This text uses third-angle projection exclusively.

6.5 Multiview Drawings

Multiview drawings represent the shape of a part using two or more views. These views, together with necessary notes and dimensions, are sufficient for the part to be fabricated without further information.

There are four basic types of drawings found in engineering work. The choice of which drawing is used is determined by the shape and complexity of the part. One-, two-, three-, and multiple-view drawings are found in industry.

1. **One-view drawings** (Fig. 6.9): Even though two adjacent views are considered the minimum requirement to describe three-dimensional parts, the third dimension of some parts (washers, shafts, bushings, spacers, sheet metal parts, etc.) may be specified by a note giving the thickness or diameter.
2. **Two-view drawings** (Fig. 6.10): Many parts may be described by showing only two views. These views must be aligned in a standard position that will clearly illustrate the part. In Figure 6.10, the side view was necessary to describe and dimension the part.
3. **Three-view Drawings** (Fig. 6.11): Most drawings consist of front, top, and side views arranged in their standard

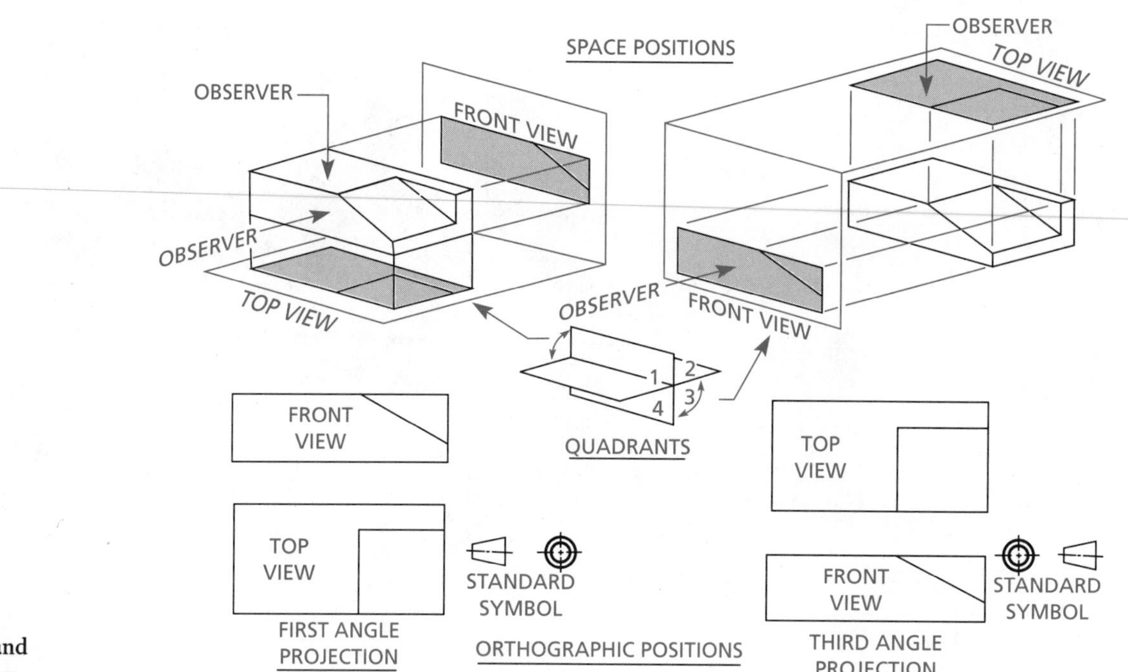

FIGURE 6.8 First- and Third-Angle Projection

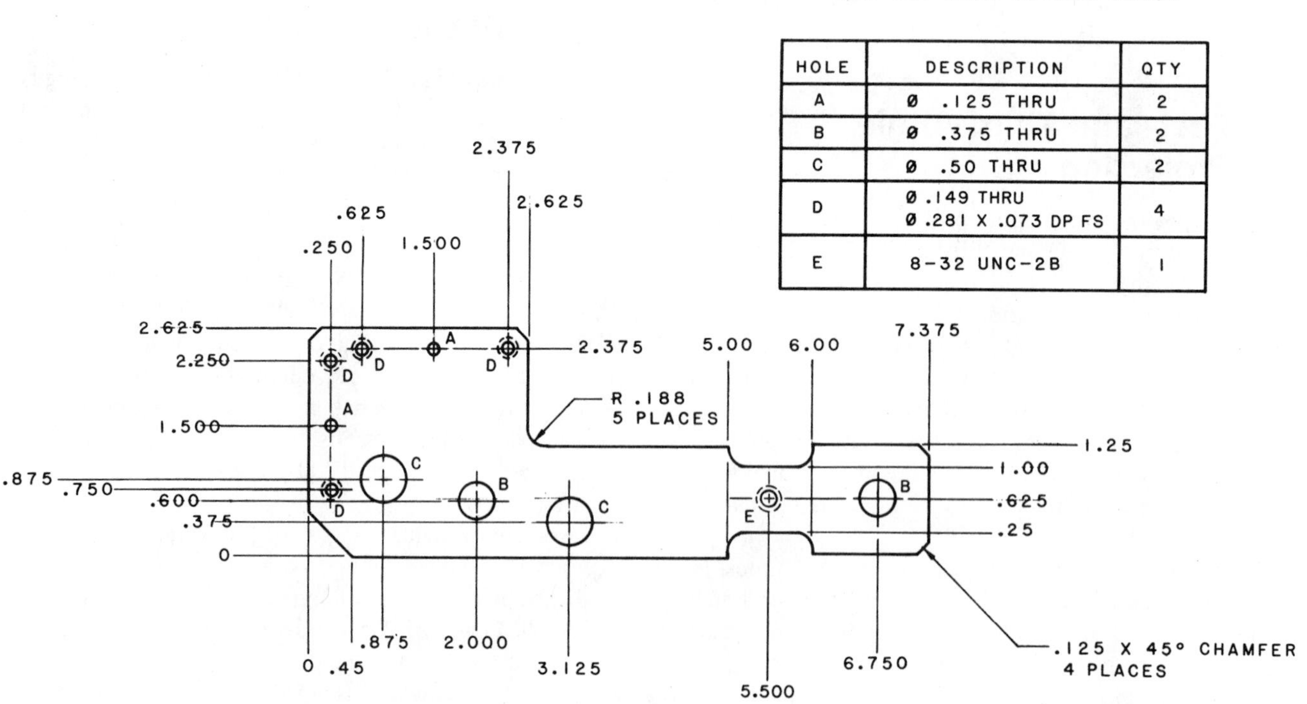

HOLE	DESCRIPTION	QTY
A	Ø .125 THRU	2
B	Ø .375 THRU	2
C	Ø .50 THRU	2
D	Ø .149 THRU Ø .281 X .073 DP FS	4
E	8-32 UNC-2B	1

FIGURE 6.9 One-View Detail of the Connector

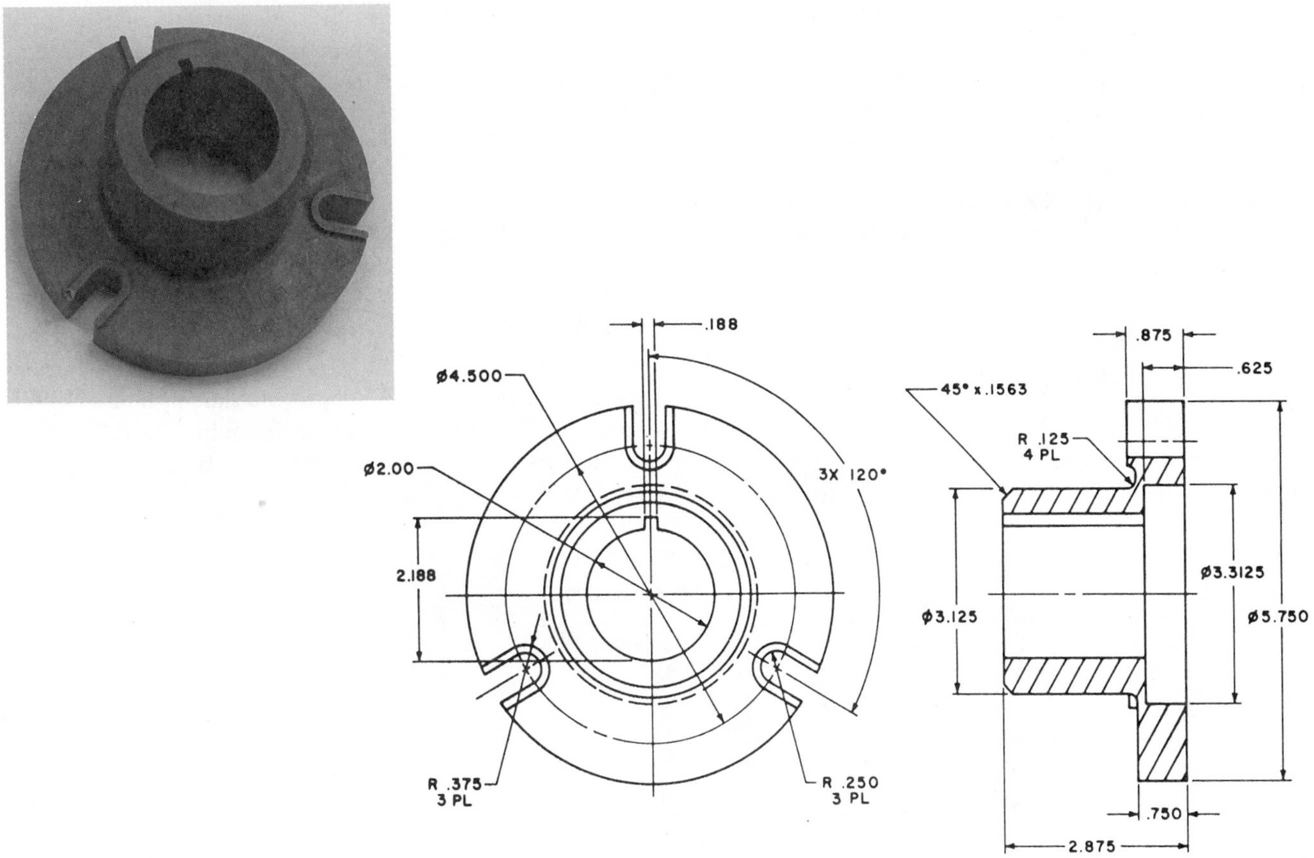

.188
Ø4.500
Ø2.00
3X 120°
2.188
R .375
3 PL
R .250
3 PL
.875
.625
45° x .1563
R .125
4 PL
Ø3.3125
Ø3.125
Ø5.750
.750
2.875

FIGURE 6.10 Two-View Detail of the Reel Post

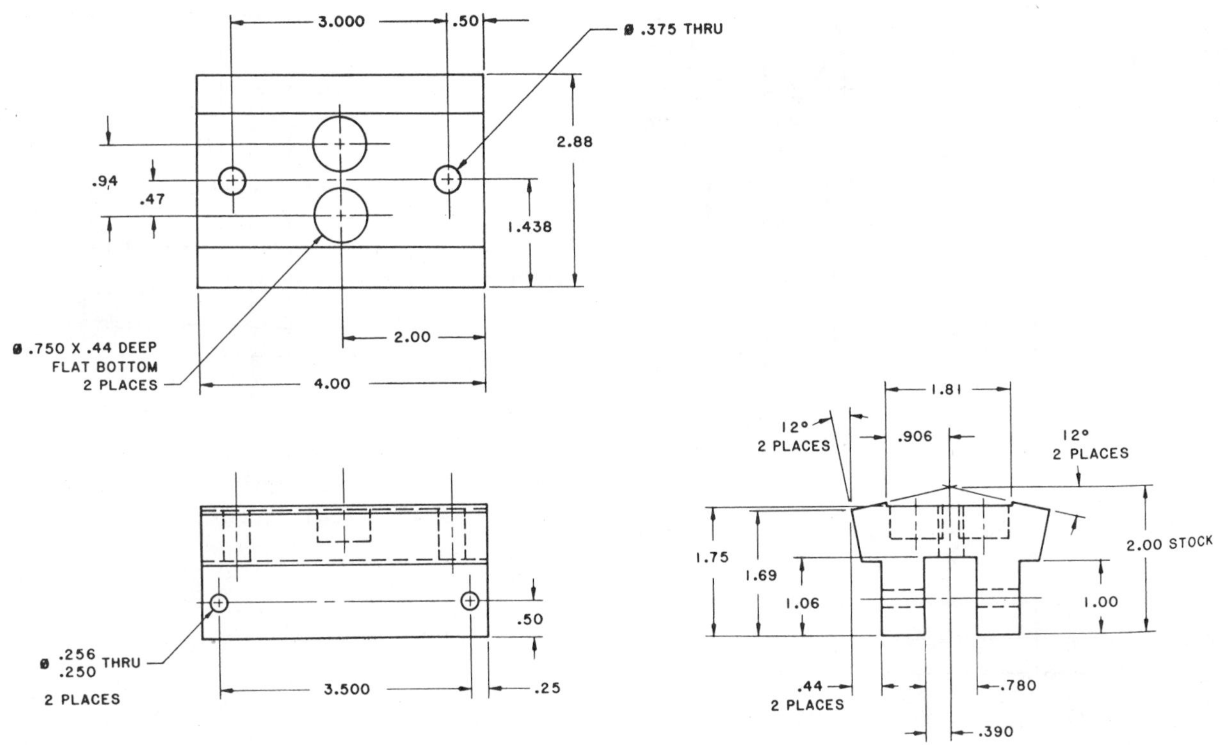

FIGURE 6.11 Three-View Detail of the Pad Mounting

positions. Any three adjacent views that best describe the shape of the part may be drawn. In Figure 6.11, each view of the part shows features that could not be graphically described in any of the other views. The holes show in the top and the front views, and the slot and angled surface in the right side view.

4. **Multiple-view and auxiliary view drawings** (Fig. 6.12): When a part cannot be defined graphically with one, two, or three views, a multiple-view drawing is required. The part shown in Figure 6.12 required four views to describe its configuration properly.

6.5.1 Choice and Orientation of Views on a Drawing

The first step in any drawing is to select the required views of the part to be drawn and dimensioned. Since dimensioning is not covered in this chapter, it is difficult to estimate the space needs of a part. Alternate positions of views can be used to conserve space, but they must be properly oriented to each other. For example, the right or left side might be placed adjacent to, and in alignment, with the top view. The rear view is sometimes placed in alignment with, and to the right of, the right side view. Before starting the drawing, you must analyze the configuration of the part and its view requirements.

A part is usually shown in a **natural** or **assembled position.** The minimum number of views necessary to describe the part is established first. Views are selected that will show the fewest hidden lines and convey maximum clarity. The top view is usually obvious or may be determined by a machining or fabrication process. The *front view should be the longest orientation* of the part.

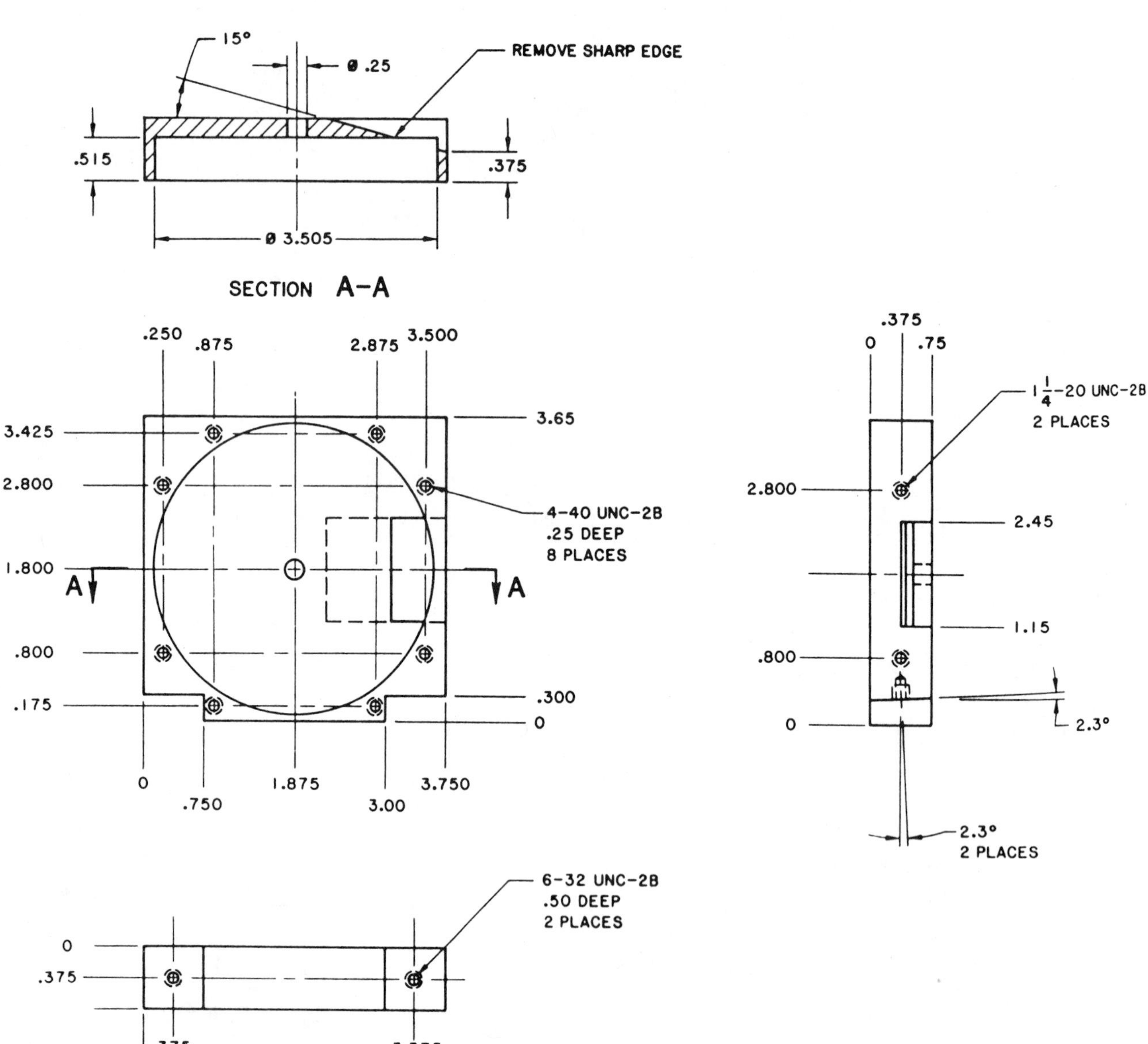

FIGURE 6.12 Top, Front, Back, and Side View of the Interface Bracket

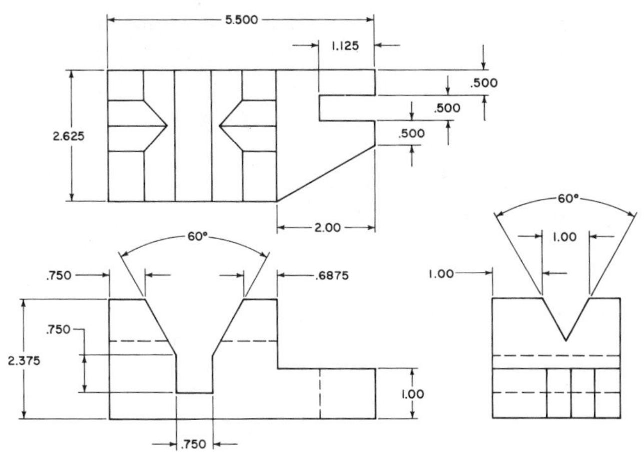

FIGURE 6.13 Three-View Detail of the Base Angle

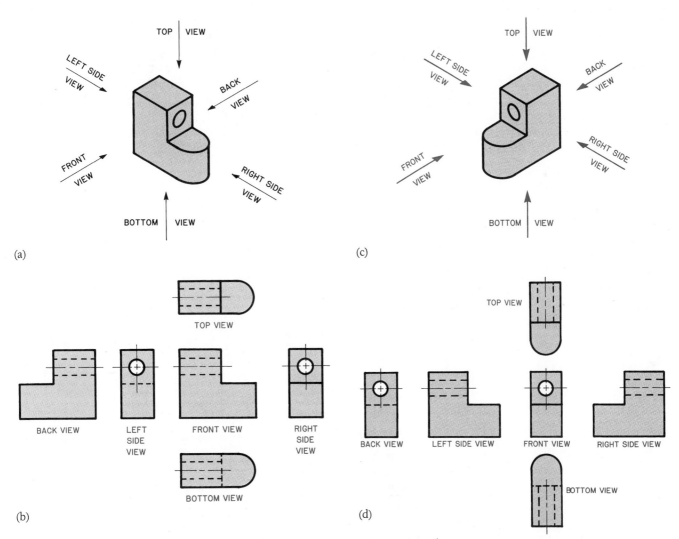

FIGURE 6.14 Views of a Part (a) Six standard views of a part (b) Six standard views of a part using third-angle projection (c) Alternative arrangement of a part in space (d) Alternative arrangement of views of a part

In Figure 6.13, the part required three views since it could not have been adequately described without all three views. The top view choice was obvious. The front view is the longest orientation and the right side view was required to describe the slot clearly.

6.5.2 Relationship of Views on a Drawing

The **relationship of views** on a drawing is determined by part orientation. In Figure 6.14(a), the six standard view directions of the part are labeled. In Figure 6.14(b), the views are laid out using third-angle projection. The placement and orientation of the top view determines that the front view will be the principal shape of the part. In Figure 6.14(c), the same part is shown differently, but not incorrectly. Here, the part has been turned so that the front view will not show the part's longest orientation. In fact, the side views show the longest orientation [Fig. 6.14(d)]. Although this orientation is not incorrect, it is less acceptable than that shown in Fig. 6.14(b).

6.5.3 Spacing Views

After the number of views are established, the next step in preparing a drawing is to establish the paper size. *The drawing must have space for views, dimensions, and notes.*

A simple method to roughly determine the sheet size is to add the dimensions of the part—add the width plus the depth. This is the total width of the views. Add extra space for separation of the views and a margin for each border. The height requirements of the drawing are determined by adding the height of the part to its depth and adding space for the separation of the views and the necessary margins. The paper size; A, B, C, D, E, or larger, is determined by these dimensions and, in industry, by company practice.

In Figure 6.15, the part has been laid out on the sheet using the preceding formula. The height, depth, distance between the lower border and the front view (A), space between the front and top views (B), and the space between the top view and the border (C) were added together to establish the height requirements of the drawing. The width, depth, space between the left border and the front view (D), the space between the front and the right side view (B), and the space between the right side view and the right border were added together to establish the width requirements. Remember, dimensions A, B, C, D, and E were determined by the space required for dimensioning.

The spacing requirements between the views are usually determined by the number of dimensions that will be placed in this space. In Figure 6.16, the shaded portion of the drawing shows the space between the top and front views and between the front and side views. If a number of dimensions must be placed between the top and front views, this area should be greater than that between the front and side views (unless a number of dimensions are also needed there).

The drawing is laid out by *blocking in the views with construction lines.* After the construction lines are drawn, the circles and radii are darkened. Each part requires careful individual consideration. There are no hard and fast rules for drawing layouts. You will, after practice, understand space requirements and adapt the drawing accordingly.

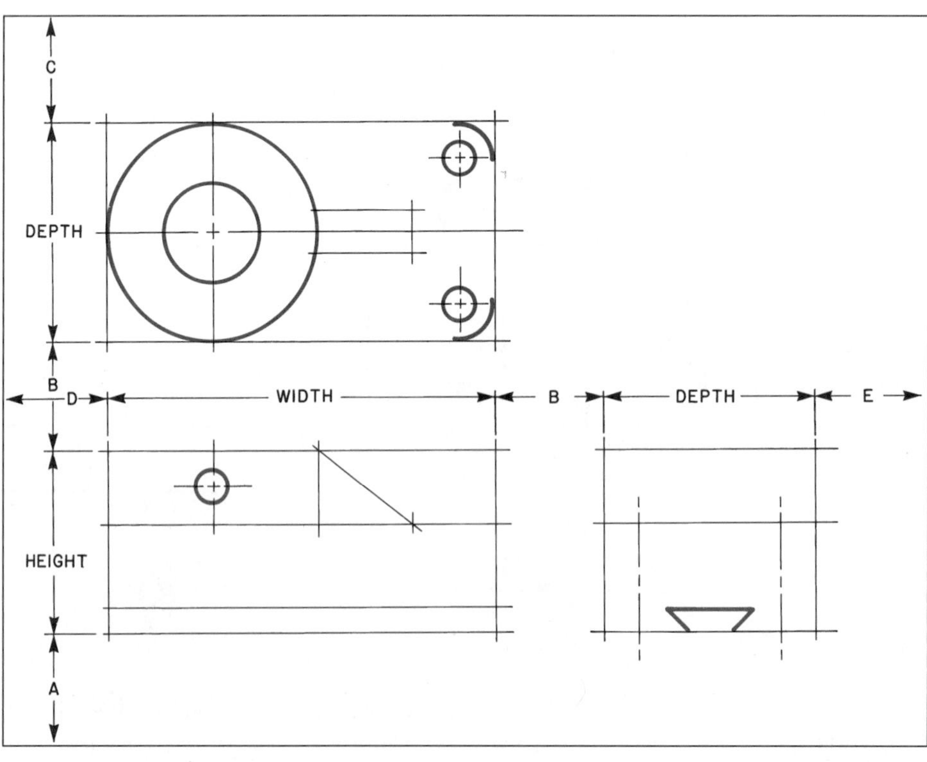

FIGURE 6.15 Laying Out a Drawing

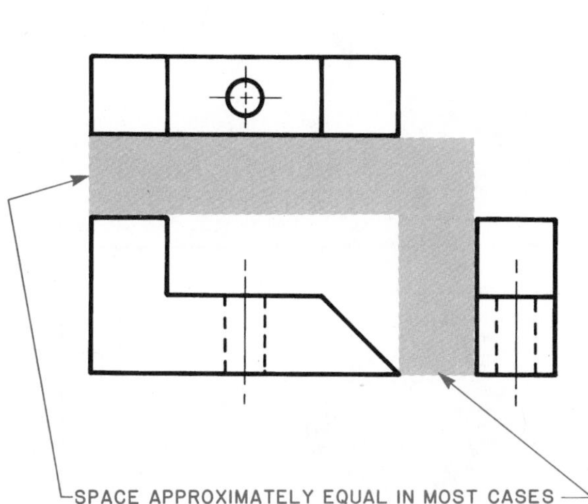

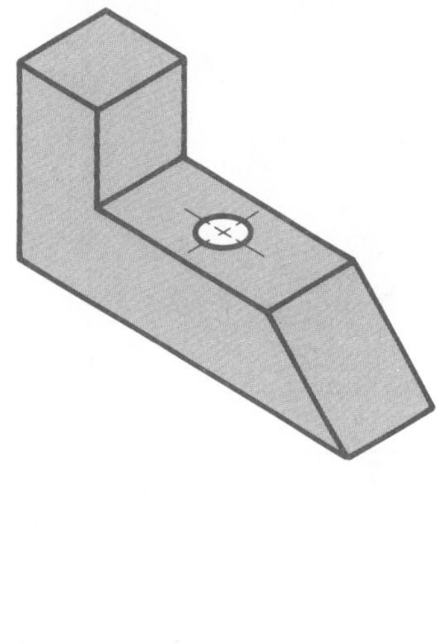

FIGURE 6.16 Spacing Views on a Drawing

6.5.4 Related and Adjacent Views

Two adjoining orthographic views aligned by projection lines are considered **adjacent views.** Two views that are adjacent to the same intermediate view are called **related views.** Each view shares one dimension with a related view and another dimension with an adjacent view.

6.5.5 Drawing Order

Whether the project is a one-, two-, three-, or multiview project, the same sequence of construction is used. Figure 6.17 provides a series of steps in the construction of a drawing:

1. Figure 6.17(a) shows an isometric view of the part to be drawn. Using the part's overall dimensions, establish the sheet size and format using the technique described. Determine the scale and dimensioning requirements at this time. Sketching possible view requirements and alternatives helps establish a well-planned drawing.
2. Using the part's overall dimensions, lay out the overall distances to establish the three views. Use the scale to measure and establish the dimensions with small construction lines as shown in Figure 6.17(b). Since dimensions are shared with adjacent views, it is necessary to scale only once for each of the three major dimensions. The width can be established in the top view and projected to the front view. The height can be established in the front view and projected to the side.
3. Using construction lines, connect the measured points to establish the outline of the part [Fig. 6.17(c)]. A drafting machine or a straightedge and triangles are used to draw these construction lines. Draw only construction lines that are necessary.
4. At this step [Fig. 6.17(d)], use your scale to measure all secondary details of the part and establish them on the drawing. Measure from the existing principal lines.
5. Draw all secondary features of the part. To avoid more measuring, project features to adjacent views where possible.
6. Centerlines and curved features of the part are established using construction lines. Fillets and circles of the part require centerlines for construction. All curved features are drawn with the aid of a template or compass. On projects where the primary shape of the part is curved or where there are prominent circular features, this is step 3 or 4.

   Check the drawing thoroughly before darkening any lines.
7. It is easier to match a straight line to a curve than a curve to a straight line, so circles, arcs, and fillets are the first features darkened on a drawing.
8. Darken the remaining lines. Match the line thickness of the curves and the straight lines.

After the drawing is complete, check it thoroughly. Fill in the title block as a last step. Since dimensioning is not presented here, this step has not been included in the preceding description.

6.5.6 Models for View Description and Reading a Drawing

Models made from plastic, metal, wood, clay, or soap may help you to visualize a part. By simply turning the model you can view the top, front, side, or any other view of the part. A number of illustrations in this chapter are accompanied by a photograph of the part that has been modeled. The drawing in Figure 6.18 provides views of the top, front, right side, and an isometric pictorial of the part.

Sketching the part pictorially also aids in understanding views. Normally, isometric or oblique sketching paper, with

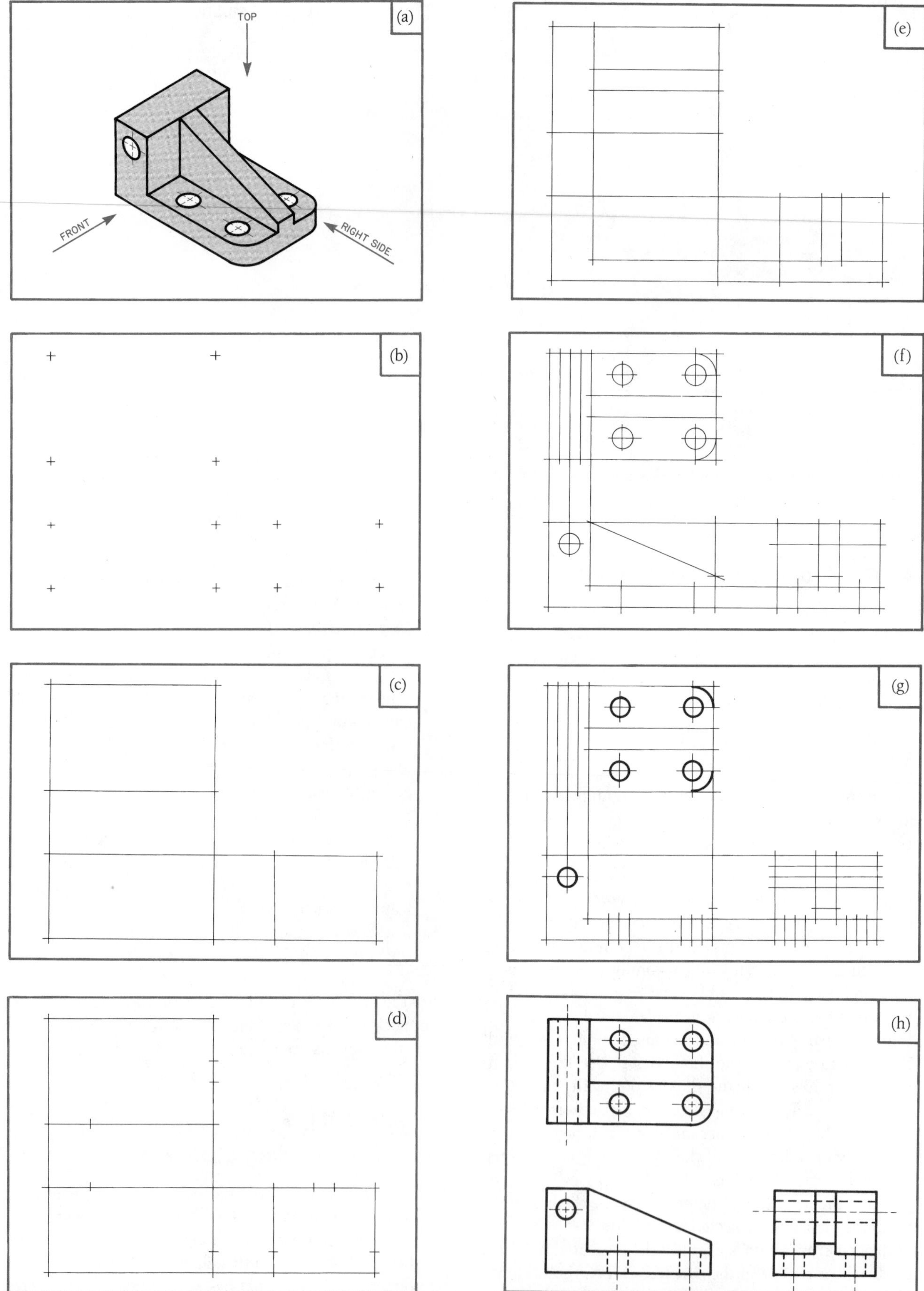

FIGURE 6.17 Steps in Constructing a Drawing (a) Isometric view of a part (b) Establish the overall dimensions of the part using a scale and space appropriately (c) Block-in the part using construction lines (d) Establish all the major features of the part (e) Block-in the secondary features using construction lines (f) Establish all holes and draw circles with construction lines using a compass or template (g) Darken arcs and circles (h) Darken drawing and remove construction lines

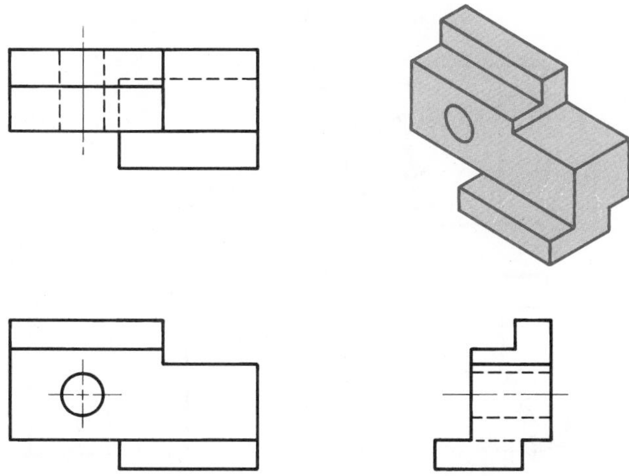

FIGURE 6.18 Three Standard Views of the Block

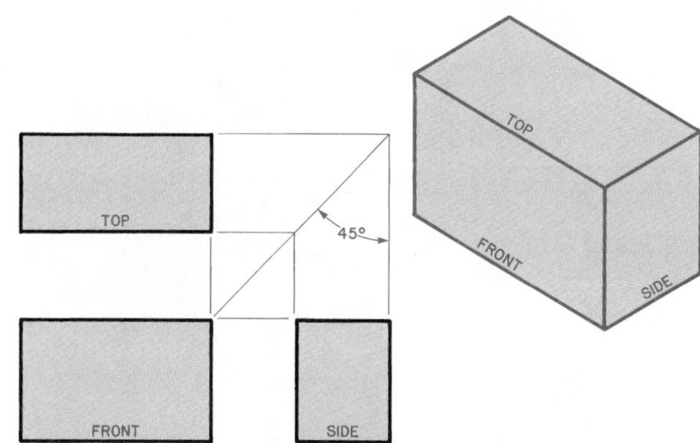

FIGURE 6.19 Establishing the Depth of a Part Using the Miter Line Method

preprinted grid lines, is used to "block out" the part before it is drawn in orthographic projection. Even with CAD systems, 3D sketching is an important part of the design, drafting, and visualization process.

6.6 View Projection Methods

There are four ways to project the third view of a part: the **miter method**, the **radius method**, the **divider method**, and the **scale method.** The miter method is used for learning how to project the third view and in understanding the relationship of the top and side views. Almost all industry drawings are completed by using the scale and the dividers to establish depth dimensions in the third view or by simply reading the third view.

6.6.1 Miter Lines for Transferring Depth Dimensions

The miter line method is a simple and straightforward procedure for establishing the depth dimensions of a three-dimensional part. After the front and the top views (or the front and side views) are drawn, the third view can be constructed. The **miter line**, a 45° line, is drawn from the upper right-hand corner of the front view of the part (Fig. 6.19). The upper edge line of the part, in the top view, is then extended until it intersects the miter line. The intersection point is used to establish the outside edge of the side view by drawing a vertical construction line through it. Since it is adjacent to the side view, the height of the part is projected from the front view. Other depth dimensions can now be extended to the miter line from the top view and then to the side view.

The drawing of the part in Figure 6.20 illustrates how each of the depth dimensions has been extended from the top view to the miter line and projected downward to establish the right side view. Height dimensions are projected directly from the front view.

6.6.2 Radius Method for Determining Depth

The radius method is shown in Figure 6.21. The upper right-hand corner of the front view is used to swing arcs R1 and R2 (90°) to establish the depth of the side view. The spacing between the front and top views and the front and side views is the same. Each feature in the top view is transferred to the side view using radii. Of course, the process could be reversed to transfer features from the side view to the top view.

6.6.3 Divider Method for Establishing the Depth Dimension

Since the divider method is quick and accurate, it is used for descriptive geometry problems and for engineering drawings. The third view can be placed at any distance from its adjacent projection. The divider method is shown in Figure 6.22.

Dividers are used to establish all depth dimensions in the third view. Most drafters will use a combination of dividers and scale measurements to draw the third view.

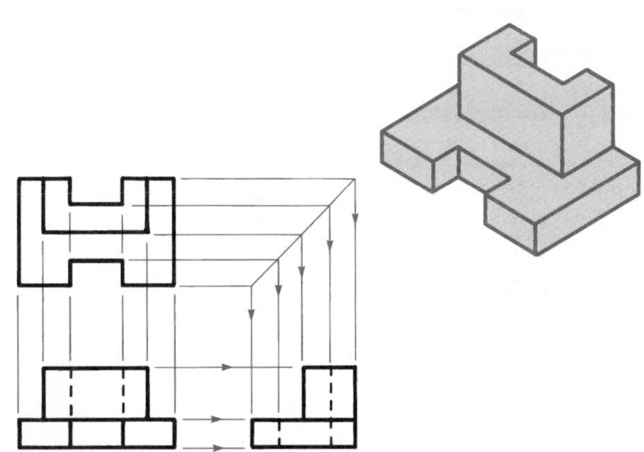

FIGURE 6.20 Miter Line Method of Projecting the Depth of the Third View

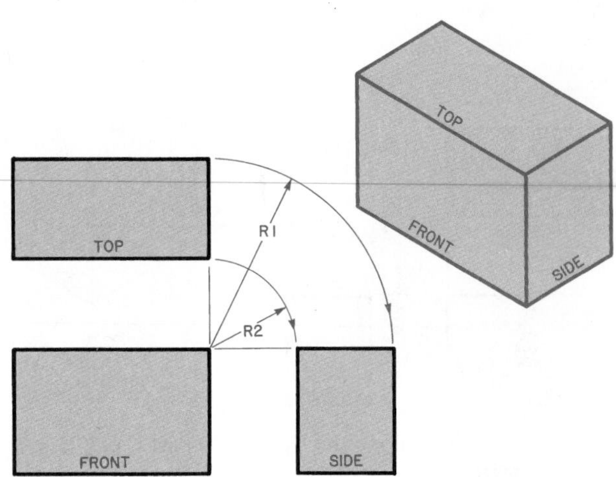

FIGURE 6.21 Radius Method of Projecting the Depth of the Third View

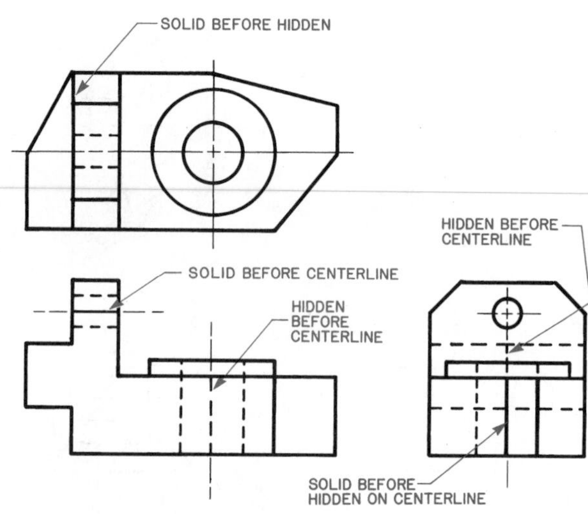

FIGURE 6.23 Precedence of Lines on a Drawing

6.6.4 Scale Method for Transferring Depth Dimensions

The scale method is commonly used in industry. The scale is used to measure the depth dimension of the part in the top or side views. Depth dimensions are then used to establish the third view. Many drafters use the dimensions of the part to construct each view (using the scale) without transferring dimensions. Although this method is acceptable, it does require the repetitious use of the scale and it takes longer. Measurements established once can be projected from adjacent views or transferred by dividers from related views. A minimum amount of scaling should be used in each view to increase efficiency and speed.

Although a typical drawing can be constructed with a combination of the four methods previously described, the use of the miter and radius methods is used for drawing simple parts and learning projection techniques. They are seldom used in industry.

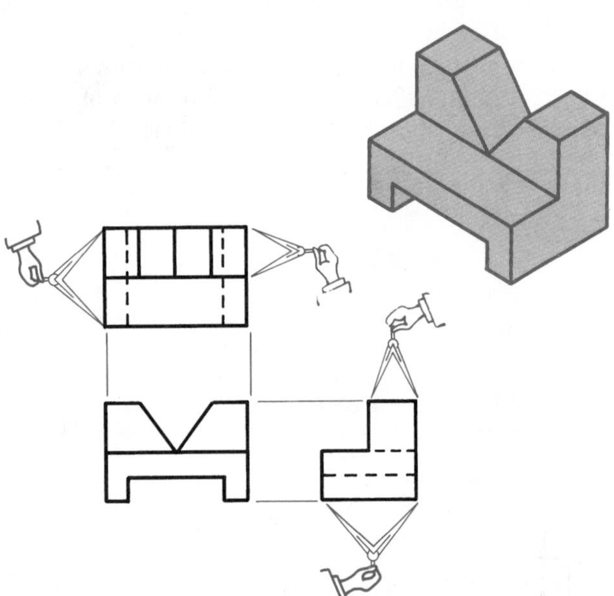

FIGURE 6.22 Transferring the Depth Dimension Using Dividers

6.6.5 Precedence of Lines on a Drawing

Since each view has so many features, they will at times interfere with one another. Because showing all the features in every view would only confuse the drawing, an order of **precedence of lines** has been established for engineering drawings. The most important lines are to be drawn and the less important ones are to be left off the drawing. Figure 6.23 shows the proper precedence of lines on a drawing.

All outside edges are **visible lines** and have precedence over all other lines. Visible edges are solid lines and always have precedence over **hidden lines. Dashed lines** represent hidden edges of the part and, therefore, have precedence over **centerlines**, which do not really exist as aspects of the parts geometry. **Dimension** and **extension lines** are always positioned so as to avoid coinciding with visible and hidden lines. The order of precedence of lines on a drawing is:

1. Visible (solid)
2. Hidden (dashed)
3. Cutting-plane or center line (depending on importance)
4. Break (solid)
5. Extension and dimension (solid-thin)
6. Section (crosshatch).

6.6.6 Interpreting Multiview Drawings

The use of numbers or letters to label the part's features may help develop understanding and visualization of three-dimensional parts. This method is also helpful in constructing views of complicated shapes. In Figure 6.24, each edge of the part, where it meets another edge, has been identified with a number or a letter. The ends of curved features are identified with letters and straight-line features are identified with

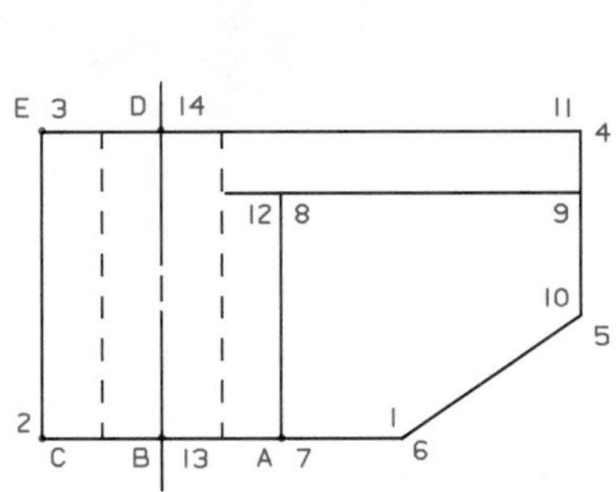
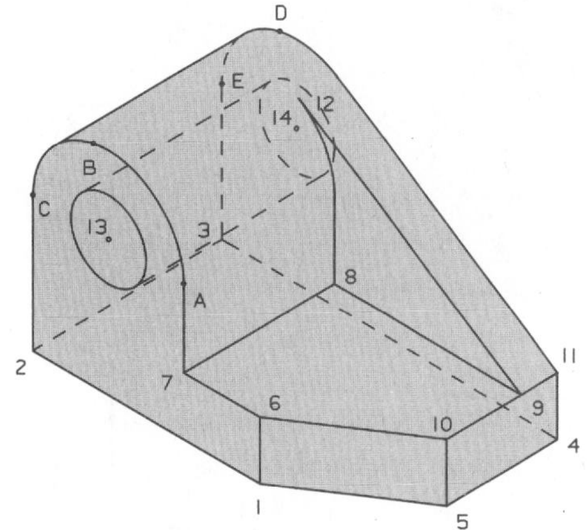
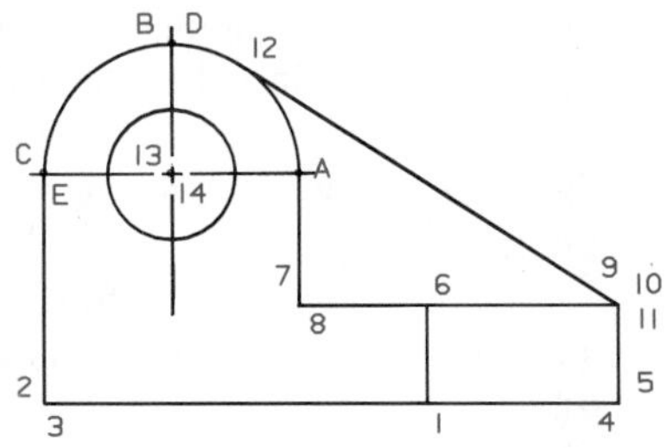
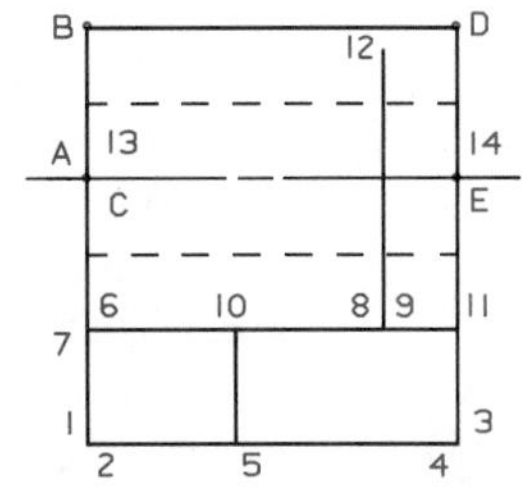

FIGURE 6.24 Labeling Points on a Part to Establish Features in Views

numbers. Each line can be seen in every view as true length, foreshortened, or as a point. Most lines, except for the angled lines 1–5 and 6–10 and line 9–12 will show as two numbered ends in two views and as a point view in another view. Line 13–14 is the centerline for the hole and for the curved surface. Projecting views and individual features of the part becomes a matter of locating points from view to view.

6.6.7 Hidden Lines in Views

Since every feature of a part is seen in each view as an edge or a surface, many features of the part are viewed as *hidden* features. These features that lie behind other features of a part are still represented. All features, lines, surfaces, and intersecting surfaces that cannot be seen directly as visible lines in a particular view are drawn with hidden lines.

The part in Figure 6.25 has two holes drilled through it. The holes must be represented in each view. The top view shows the holes as circles and visible. The front and left side views show the outside edges of the hole. Since they pass through the part and cannot be seen by the observer, the edges must be represented by hidden lines.

In Figure 6.26, the use of visible and hidden lines is shown. When constructing dashed and solid lines, the following drafting conventions for spacing must be maintained:

1. Do not leave a gap between a hidden line and a visible line (Fig. 6.27).
2. When a hidden line crosses a solid line, leave a gap (Fig. 6.28).
3. When a hidden line continues as a visible line, after crossing a visible line leave a gap (Fig. 6.29).

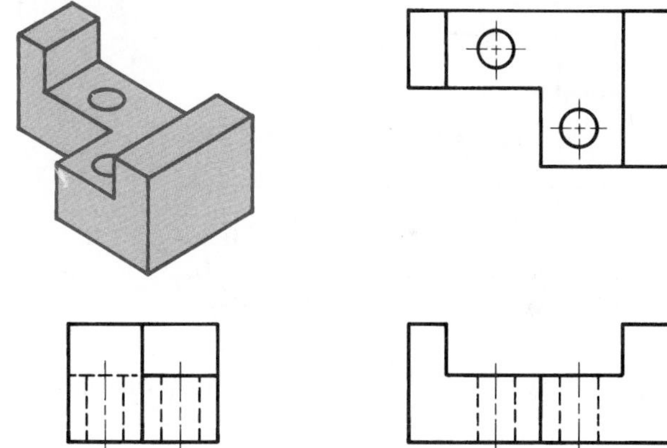

FIGURE 6.25 Views of the Holding Block

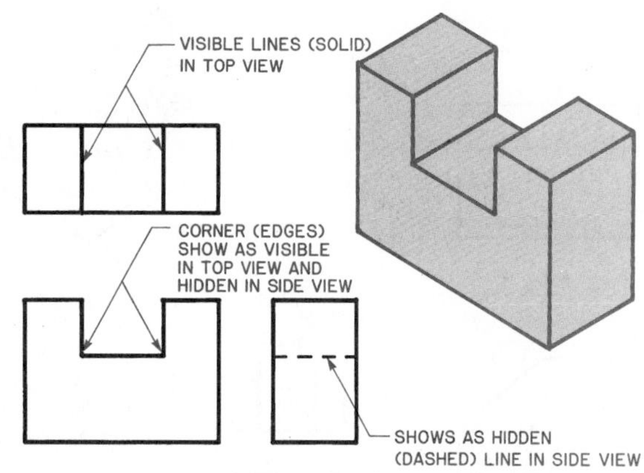

FIGURE 6.26 Solid (Visible) and Dashed (Hidden) Lines of a Part

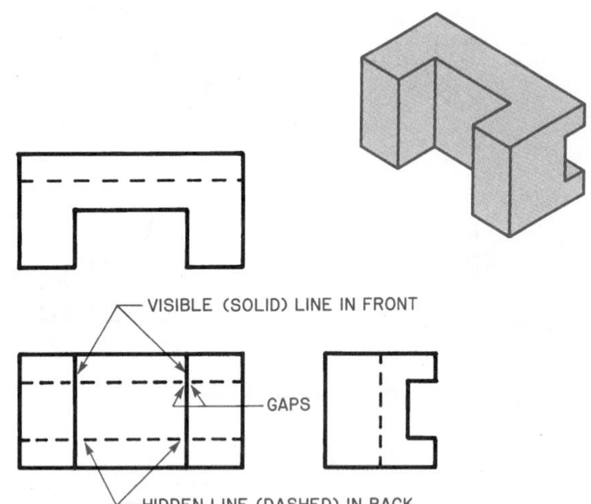

FIGURE 6.27 Drawing Dashed (Hidden) Lines

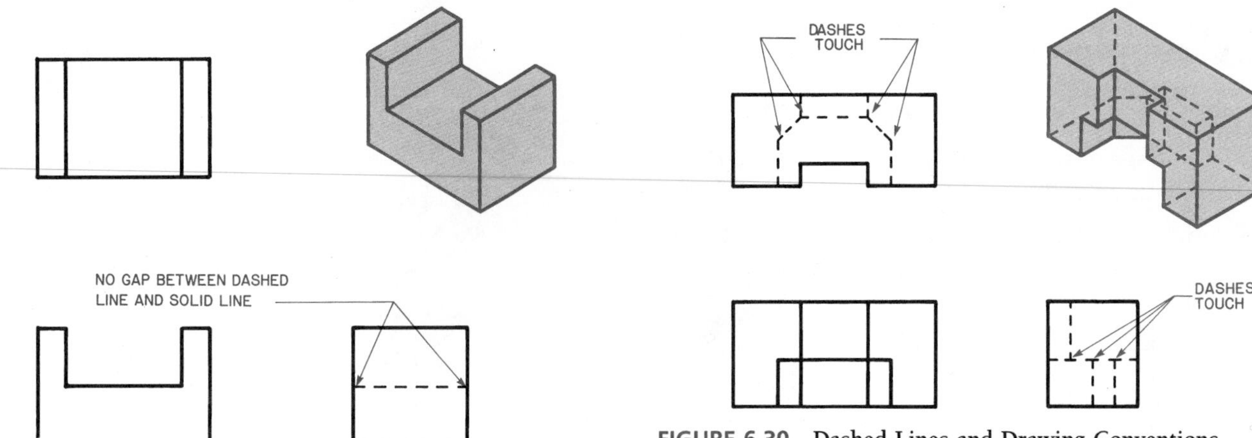

FIGURE 6.30 Dashed Lines and Drawing Conventions

4. Hidden lines that meet other hidden lines should not have gaps between them (Fig. 6.30). Hidden lines that establish corners always touch.
5. When a hidden line (or arc) meets a visible line (or arc) and is tangent to that line, leave a gap.
6. When hidden lines cross, draw the one that lies in front of the other as continuous and thru a space (between dashes) in the one behind it.

FIGURE 6.28 Visible and Hidden Lines on a Drawing

6.6.8 Curved Lines in Views

All curved features of a part are shown in each view. In most cases, a curved feature shows as a curved line or surface in only one view and as an edge in its adjacent view. The most common type of *curved feature* is the circle. Arcs and fillets are also widely used on parts. Circles, arcs, and fillets are really a part of a *curved surface*. A hole is a cylindrical surface too. Connected arcs and fillets are also portions of cylinders.

Holes and cylinders are both drawn the same way. In Figure 6.31, the part has both internal and external curved surfaces. The holes and the cylinder both show as circles in the top view and as straight lines in the front and side views. The hole shows as hidden features in these views and the cylindrical surface as visible lines. The outside arcs of the part also show as visible edge lines in the front and side views.

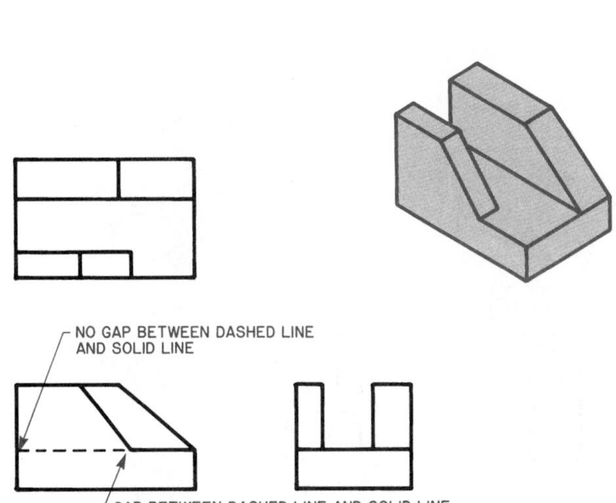

FIGURE 6.29 Drawing Dashed Lines

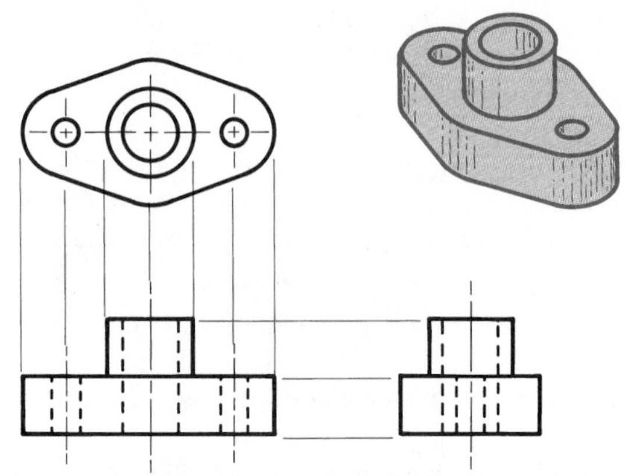

FIGURE 6.31 Curved Features in Views

Spheres, ellipsoids, or other similar shapes show as curved surfaces in more than one view. Unless cones and cylinders are part of angled surfaces, they show as curves in one view and straight lines in the other two views.

6.6.9 Use of Centerlines in Views

Curved features are established, located, and dimensioned using a centerline to position the feature in space. With the exception of fillets and rounds, all curves require **centerlines** to establish their curved features (Figs. 6.31 and 6.32). Centerlines for the end view of curved features are drawn as perpendicular crossing lines with short dashes at the center and as single centerlines in adjacent views. They are drawn to extend slightly beyond the boundaries of the part or curved feature and do not take precedence over visible or hidden lines.

Centerlines are also used on drawings where the part is *symmetrical about a centerline.* Cones, spheres, and other curved shapes require centerlines. When shown in adjacent views, they represent the *axis* of the curved surface.

You May Complete Exercises 6.1 Through 6.4 at This Time

6.6.10 Parallel Lines on Parts

Lines *parallel* in all three views are parallel lines and show as parallel in all views of the part. If the lines are shown from an end view, they appear as points. The part in Figure 6.33 is composed of parallel lines representing parallel and perpendicular surfaces.

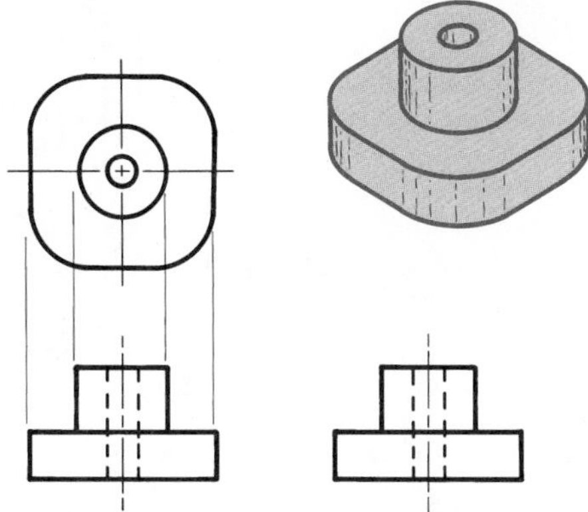

FIGURE 6.32 Curved Features

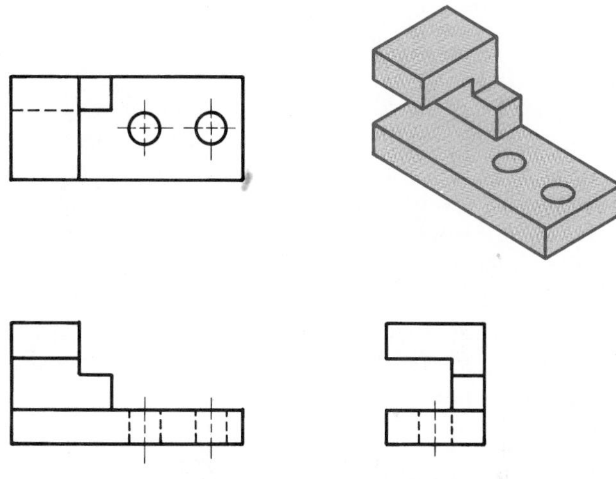

FIGURE 6.33 Parallel Edges

6.7 Drafting Conventions and Special Views

A variety of *drafting conventions* and procedures have been devised to enable the drafter to draw in a concise, clear, and speedy fashion.

As long as the geometry of a part is adequately described in another view, a **partial view** may be used. A partial view is a view where the dominate features, shape, and outline of the part are shown without the extra clutter of unneeded hidden lines. In Figure 6.34, the part has different shapes on each end. Since a top view is similar to the front view, it can be eliminated. Since they show only the visible lines of the corresponding end, the right and the left side views are partial views.

Hidden features on a partial view should include only those directly behind the visible shapes. In Figure 6.34, the cylinder's outside diameter (O.D.) lies directly behind the counterbored hole on each base plate. Since visible lines take precedence over hidden lines, this feature is not shown.

Enlarged views are increased size views of a complicated area of a part. This procedure is used extensively to provide sufficient space for dimensions. In Figure 6.35, **VIEW A** is an enlarged portion that clearly show the chamfers. The area to be enlarged is circled with a phantom line and the **view-letter**

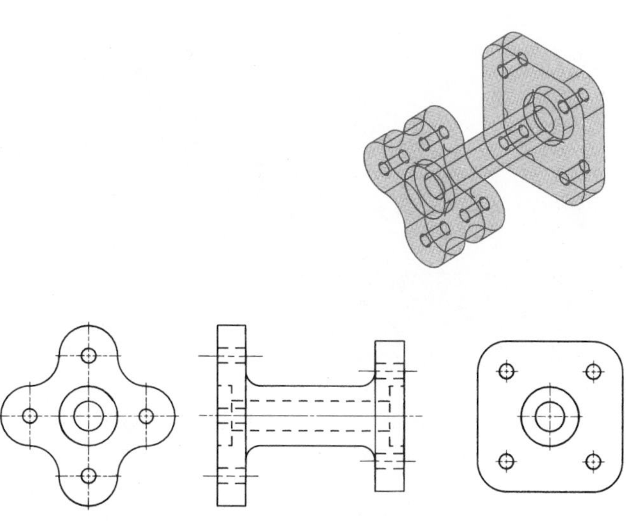

FIGURE 6.34 Partial Views

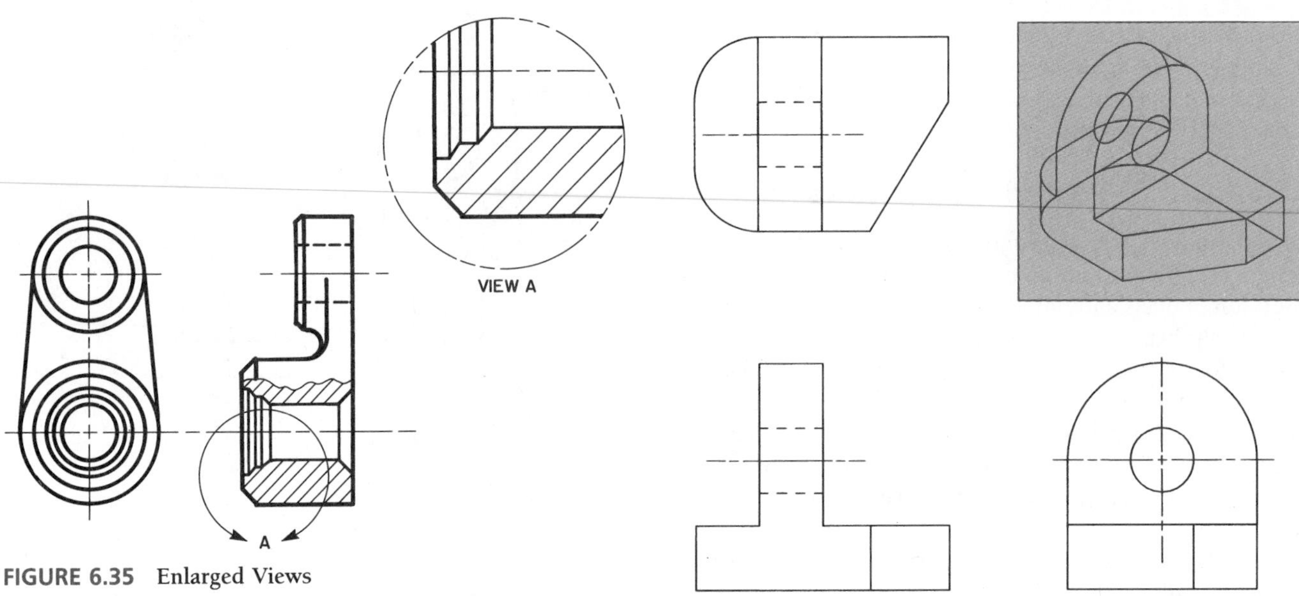

FIGURE 6.35 Enlarged Views

FIGURE 6.37 Surfaces on a Part

designation is positioned as in Figure 6.35. Phantom lines are drawn with the same thickness as centerlines and are similar except they consist of two short dashes and one long dash.

Rotated or **revolved views** are used where a true projection of the part would only confuse the reader of the drawing. The part in Figure 6.36 is an example of a part that is better when described with a rotated view. The detail of this part requires two views to describe its geometry and place dimensions adequately. The clevis portion of the arm was rotated parallel to the front view and projected as if it were a normal (true shape) view.

6.7.1 Surfaces and Edges on Multiple-View Drawings

It is easier to understand orthographic projection if you see parts as simple shapes, edges, lines, and points. *Surfaces* are created by combining lines, either combinations of straight lines or straight and curved lines. Surfaces, or areas, show **true shape/size** (TS) when they are parallel to the plane of projection and as **edges** (EV) when they are perpendicular to the plane of projection. A plane that appears true size/shape in a view is called a **normal surface**, and the view is a *normal view of a plane*.

Curved surfaces show as curved edges in views where they are perpendicular to the viewing plane and as plane shapes with straight sides in views where they are parallel to the viewing plane. When three surfaces come together, they meet at a corner (point). Most parts can be defined by establishing their corners (points in space). Figure 6.37 provides an excellent example. The part is composed of planer surfaces and curved surfaces. The hole shows as circular only in the side view and appears as an edge in the front and top views. The

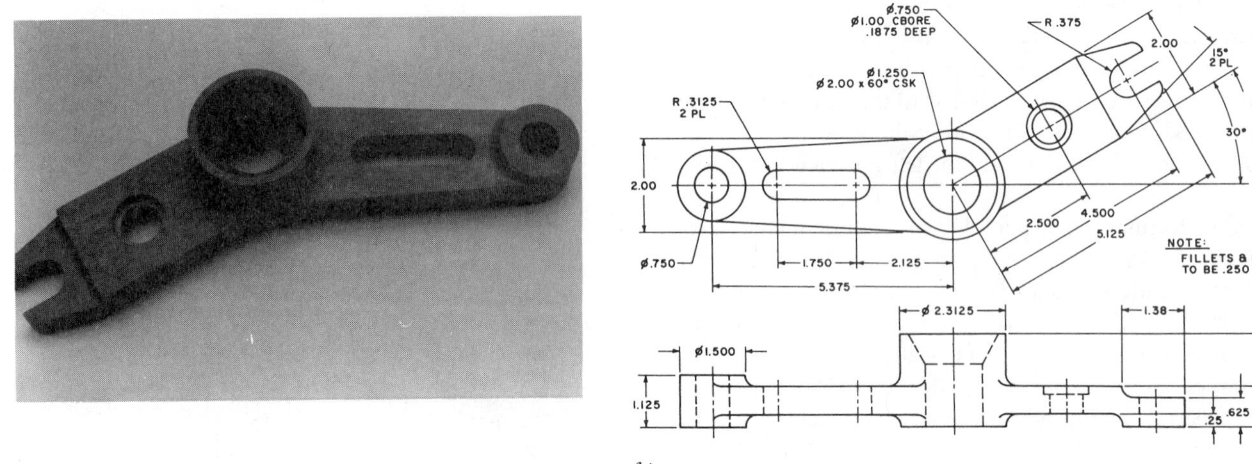

(a) (b)

FIGURE 6.36 Angle Frame (a) Photograph of angle frame (b) Drawing of angle frame

circular surface of the projected hole shows as a rectangle in the front and top views. All plane surfaces of the part show as true shape or as edges in their adjacent views. Since each of the curved surfaces is perpendicular to this plane of projection, the front view is all straight lines.

6.7.2 Reading a Drawing

Here are the mental steps required to read or interpret a drawing:

1. Study the total drawing by scanning all views and dimensions.
2. Visualize the shape of the part by making yourself the observer for each view.
3. Reduce the part to simple geometric shapes: planes, circles, surfaces, etc.
4. Study each view and feature as it relates to its adjacent and related projection. The depth, for instance, can be studied in the top and related side view.
5. If necessary, sketch a simple 3D pictorial of the part to clarify details.
6. Note each hole, tangent area, curved feature, or other special contours that distinguish the part.

To read the drawing for the part in Figure 6.38, first notice that three views were required to represent the geometry adequately. Most of the features are in the front and side views. The top view adds little, but does show the that the slot extends through the part. The front view shows the angled cut. This is the only surface that is a normal surface and, therefore, does not appear true shape in any view. The hole is described in the side view. The side view also shows that the slot extends the entire length of the part. A pictorial sketch would help to read this multiview.

(a)

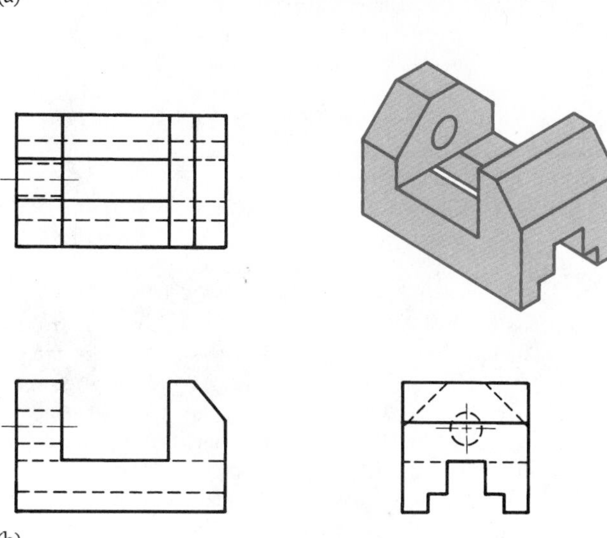

(b)

FIGURE 6.38 Three Views and Pictorial Illustration of the Guide (a) Photograph of part from left side orientation (b) Drawing of part

6.8 Visualization and Shape Description

The process of reading a drawing requires a certain level of skill at visualization. *Visualization is the process of converting a 2D drawing into a 3D image and a 3D part into a 2D orthographic projection.* Visualization is a skill that can be developed by studying a variety of drawings and parts.

In Figure 6.39, the part is shown with three views and a pictorial. Each of the views provides details of the part. The top view shows the depth and width of the part and visible lines representing the two removed portions, but you cannot tell the actual height of the block or the cutout height in the top view. The front view provides the height of the block and the cutout shapes. However, the small cutout on the right of the front view is still not completely clear (it could be an angled surface). Only the two side views can answer the remaining questions about its shape.

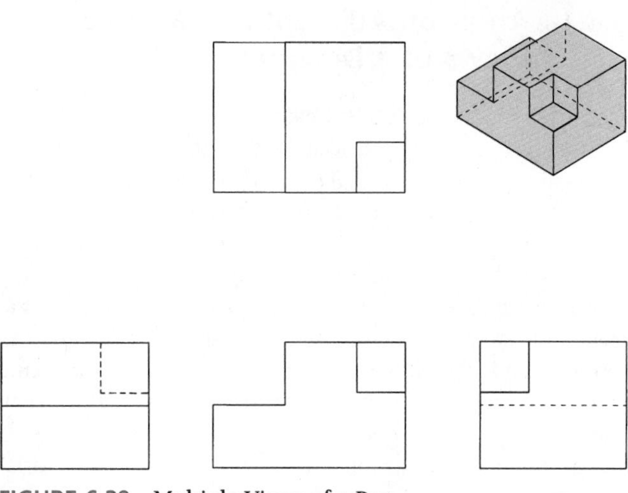

FIGURE 6.39 Multiple Views of a Part

ITEMS OF INTEREST

Evinrude

Who would have imagined that the son of an immigrant farmer, with only a third-grade education, would be responsible for the hours of pleasure experienced by people who fish and boat? Ole Evinrude was born in Norway and came to America with his parents to farm in Wisconsin. He wasn't a very good farmer, preferring to channel his energies into work on mechanical devices. At sixteen, he built his first project, a sailboat. He used this project to secure a job as a machinist in Madison. After several jobs in Chicago and Pittsburgh, he settled in Milwaukee working as a patternmaker. In his spare time, he "tinkered" with his idea of constructing a standard engine for the increasingly popular horseless carriage. The U.S. government became interested in this concept and contracted with him to produce fifty engines. As a result, he opened his own company.

The idea for the outboard motor was the result of being embarrassed during a summer picnic. His future wife asked him to row across the lake to get ice cream. On the return trip, the wind became so gusty that he was unable to row fast enough to keep the ice cream from melting. Ole was a large, strong man and was embarrassed over this inability to control his boat. The following Monday he began work on his outboard motor.

Evinrude introduced his 1.5-hp motor in 1907. It has remained essentially unchanged to this day. It has a horizontal cylinder with a vertical crankshaft, employing power direction changes with gears in a submerged lower unit. Ole was only thirty-two when he formed Evinrude Motor Company to produce the outboard motors.

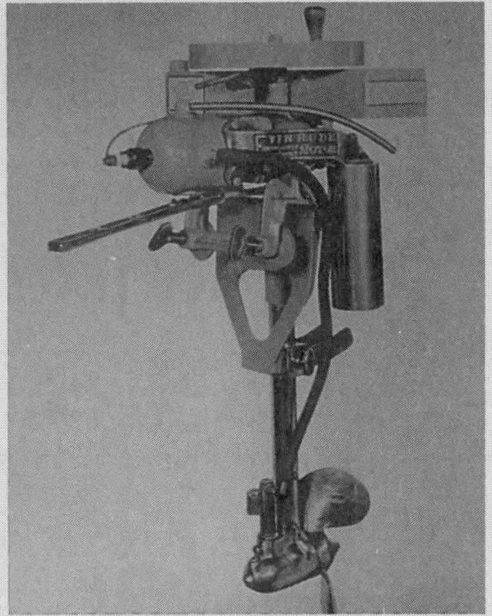

Evinrude's First Motor

The company was sold in 1914. Later, another company, Evinrude Light Twin Outboard (ELTO), produced the first practical twin-cylinder outboard. In it, many heavy engine parts were replaced with aluminum. Also, exhaust gases pass through the propeller hub.

Evinrude died in 1934. A few years, ago his original 1909 outboard motor was dedicated as a National Historic Mechanical Engineering Landmark. It was the first consumer product to be so recognized.

No doubt Evinrude spent many hours sketching his ideas. To manufacture those motors, many working and assembly drawings were also produced. Evinrude certainly had the genius to take an idea in his mind and make it into a valuable product. This is not so different from what we try to do today.

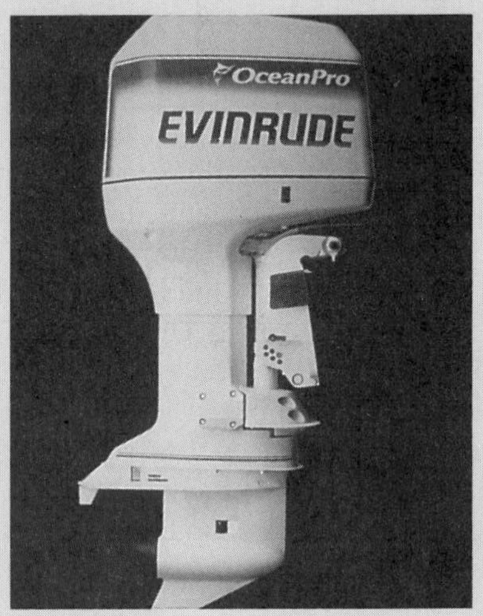

A Modern Outboard Motor

6.8.1 Areas on Adjacent and Related Views of a Drawing

Visualization can be used to examine a part by comparing surfaces and edges on adjacent and related views. *Adjacent areas cannot lie in the same plane.* If they did, they would not exist; they would not have a boundary between them,

The part in Figure 6.40 is a good example for studying areas. In each view, a surface or an edge is labeled. Surface A is shown true shape in the top view and as an edge in the front and side projections. Surface B is also true shape in the top view and, therefore, an edge view in the front and side views of the drawing. Surface C is true shape in the front view and will show as an edge view in the top and side views. If you cannot find it in the top and side projections, the pictorial view will help you locate surface C. Surface D is an angled surface; the slant angle is shown in the side view where it shows as an edge. The front and top projections of surface D are not true shape. Surface E is along the front of the part and is true shape in the front view. It shows as an edge in the top view and the side view. Surface F is at an angle and does not show in any view as true length. The top view shows this surface as an edge view and its angle to the part can be measured from the edge view of surface E. The side and front views of surface F show as foreshortened (not true shape). Surface G forms the right side of the part and shows as an edge in the top view and as true shape in the front and the side views. Surface H is an inclined surface and its slant angle can be measured in the front view as the angle it makes with surface B. Surface H is an edge in this view and shows foreshortened in the other two projections. Surface I is the top or highest surface on the part and shows as an edge in the front

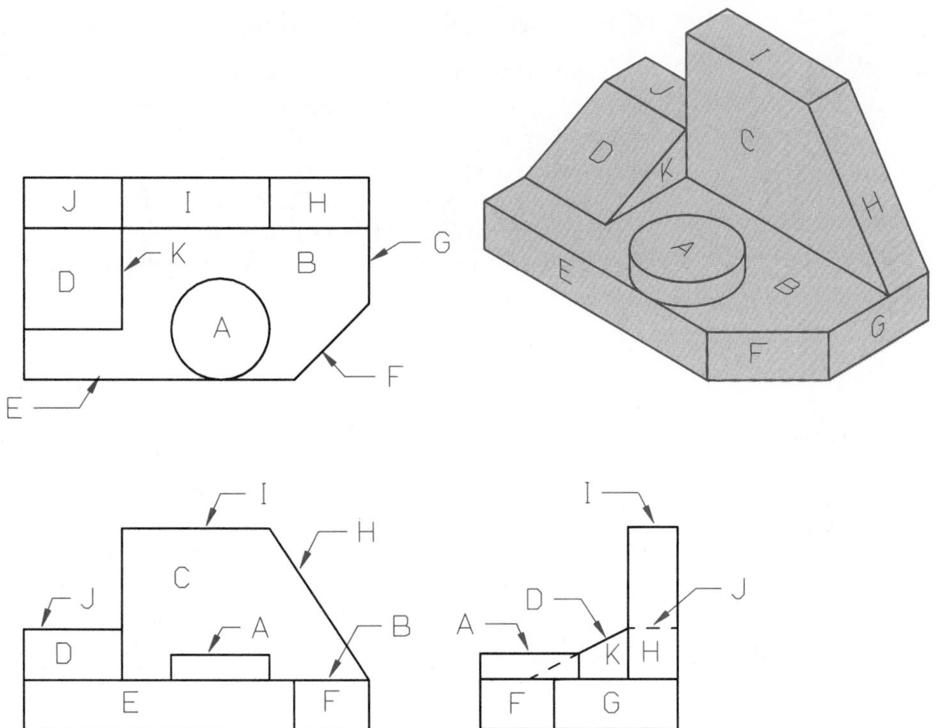

FIGURE 6.40 Related Surfaces and Edges

view, true shape in the top view, and as an edge in the side view. If a surface appears as an edge in the front view, it will also be an edge in the side projection. It will be true shape only in the top view. Surface J is parallel to surface I. Therefore, it also is true shape in the top view and an edge in the other two projections. Surface K is true shape in the side view and edge in the top and front views. Surfaces G and K are the only labeled surfaces that are true shape in the side view.

It is important to develop a sense of how each surface relates to another surface. Surface C, for instance, is parallel to surface E and perpendicular to surfaces I and B. Surface D is at an angle to surface B and surface C. Surface G is parallel to surface K and perpendicular to surface B and E. Parallelism, perpendicularlity, and angularity are important aspects of the visualization process.

6.8.2 Visualizing Similar Shapes of Surfaces

A simple rule of projection: *An area will project as a similar shape or as an edge in an adjacent view.* In Figure 6.41, the model of the part has an angled surface. The drawing of the part shows that the angled surface is a similar shape in the side and top views. It shows as an edge in the front view. Even though the top and side views show the surface as distorted, their outlines appear as similar shapes.

The shapes will have the same number of sides, and the sides of the areas are connected in the same sequence. Curved shapes may distort in related views, but they maintain similar shapes.

6.8.3 True Shape or Normal Surface of a Part

Surfaces that are parallel to a plane of projection are normal surfaces. They show true shape and each line, arc, circle, or other form that lies on this surface, or is parallel to it, will be true shape and true length/size. The part in Figure 6.42 illustrates this. The true shape surfaces (normal surfaces) are labeled in each of the three views of the part. The surfaces that are not normal to the projection plane are **inclined surfaces** and do not project as true shape in any given view.

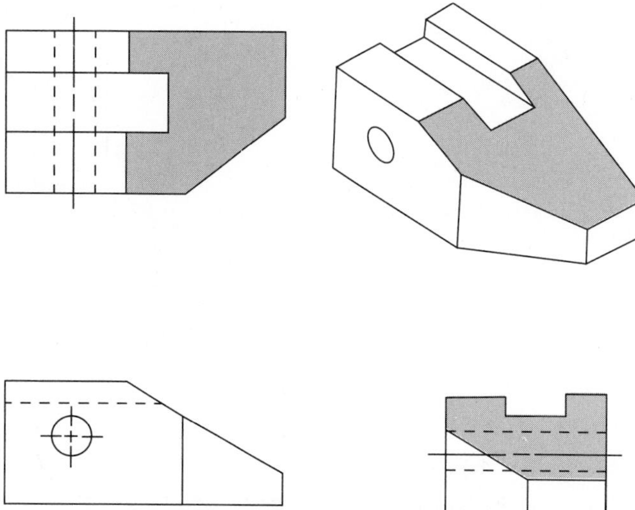

FIGURE 6.41 Angled Surfaces and Edge Views

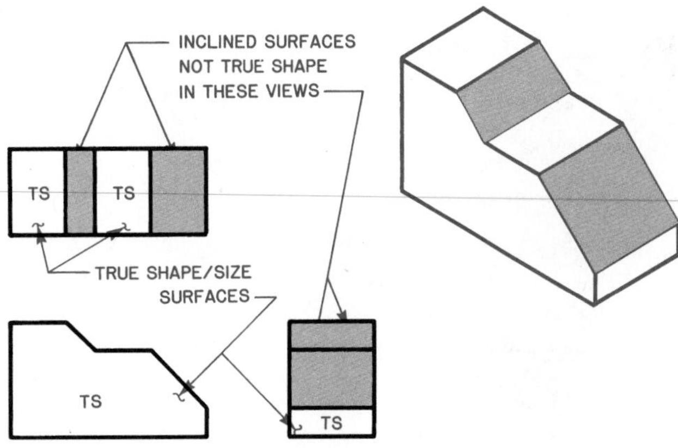

FIGURE 6.42 Inclined Surfaces and True Shape Surfaces

6.8.4 Edge Views and Edge Lines of a Surface

A surface projects as an edge in a view where the plane of projection is perpendicular to the surface. A line that shows as a point view is a normal edge; it is perpendicular to the projection plane.

Edges are always shown on views where the surfaces they represent are perpendicular to the adjacent view. In Figure 6.43, the front view of the part shows two perpendicular surfaces that project as edges in the top view. The surface that is at a slight angle and blends with its mating surfaces is not represented as an edge in the top view. The same drafting convention is used in the right side view and the left side view.

6.8.5 Angles on Multiview Drawings

In Figure 6.44, the part has two angled surfaces. The true angle of these surfaces is shown in the side view of the part where they show as edges. *Angles* can be measured only in views where they are in a normal plane.

Study Figure 6.45 carefully. The lower corner has been cut off at an angle to each of three surfaces. The angled surface is foreshortened in every view and is called an **oblique surface.**

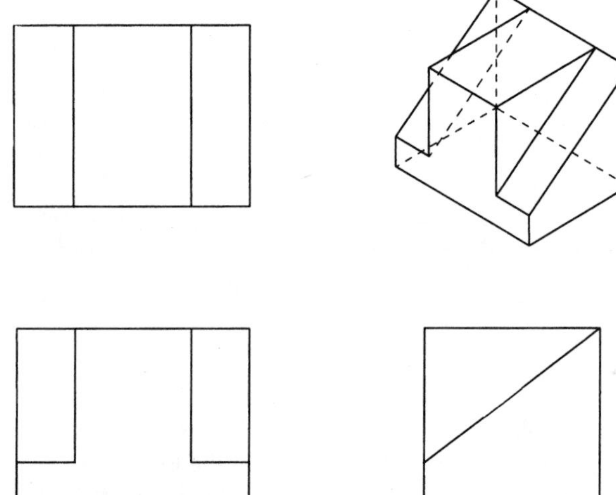

FIGURE 6.44 Inclined Surfaces of Parts

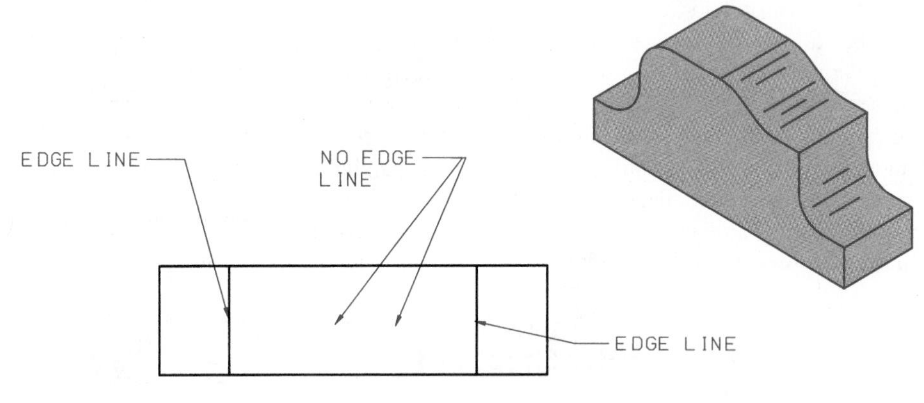

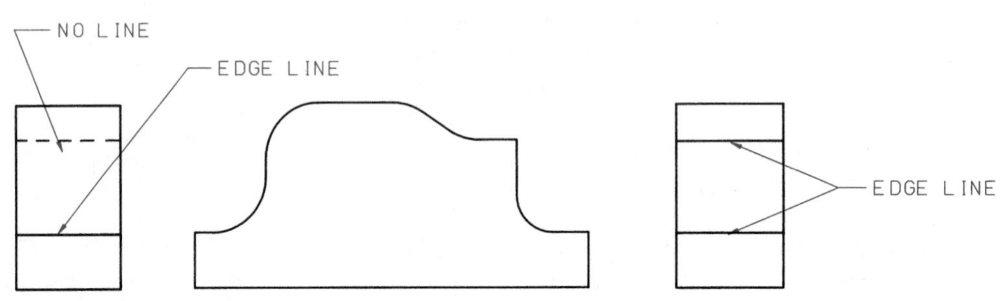

FIGURE 6.43 Curved Surfaces and Edge Lines

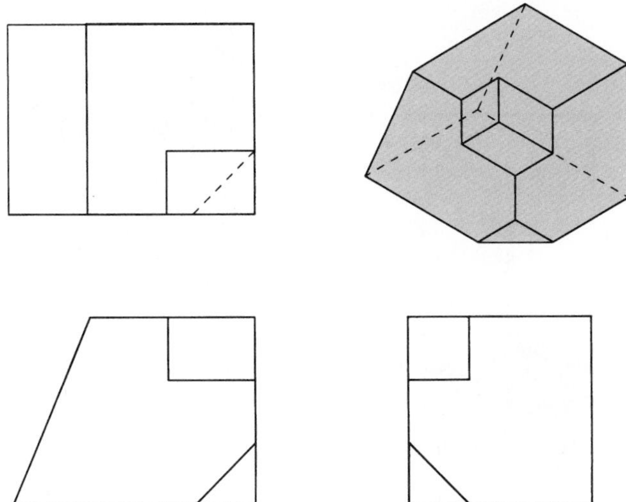

FIGURE 6.45 Inclined and Oblique Surfaces

6.8.6 Inclined Surfaces of a Part

An inclined surface shows as an edge in one view and foreshortened in the other two principal views. The edge view of the inclined surface shows the *true angle of the surface*. Figure 6.46 has three inclined surfaces. The angle that surface A and surface C make with the horizontal plane is shown in the front view where they each appear as edges. The *true angle* of surface B is measured in the side view where it appears as an edge. The other views of surface B show as foreshortened. The amount of foreshortening depends on the angle of the inclination. The greater the angle of incline to a view, the more the surface is foreshortened.

The part shown in Figure 6.47 has a number of angled surfaces, each represented by different shading. Each view shows the angle of two surfaces. The V cut in the top view shows two edges of surfaces that appear foreshortened in the front (and side) view. The angled surface on the front of the part is in the side view as an edge making a true angle with the base. The front view shows the edge lines of the two angled sides of the part.

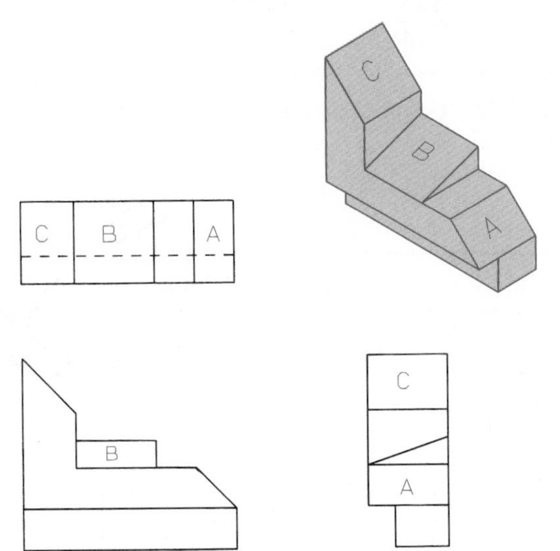

FIGURE 6.46 Inclined Surfaces in Adjacent and Related Views

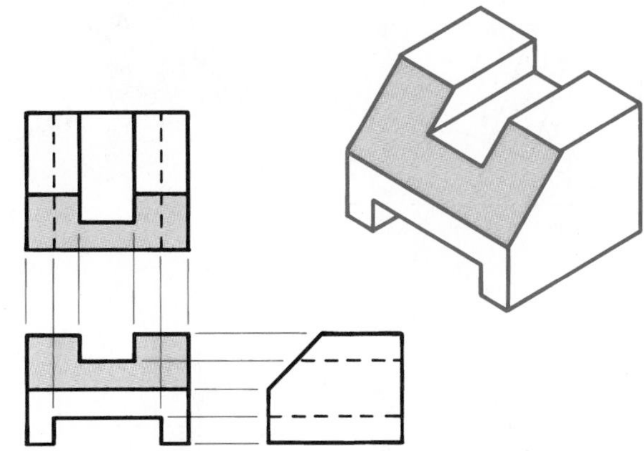

FIGURE 6.47 Surfaces Inclined in the Side View

6.8.7 Edge Views of Inclined Surfaces

The edge view of an inclined surface shows in a view where it forms a true angle in a normal plane. The adjacent and related views of the inclined surface always appear foreshortened. This concept is illustrated in Figure 6.48. The part has two angled surfaces: one inclined to the horizontal projection plane (top view); the other inclined to the frontal projection plane. The first inclined surface appears as an edge in the front view and its true angle with the horizontal plane can be measured here. The second inclined surface shows as an edge in the side view (hidden line) and foreshortened in the top and front views. The angle it makes with the frontal and the horizontal plane can be measured in only the side view.

Since many of its surfaces are at an angle to the standard projection planes, the part in Figure 6.49 is an example of a drawing that does not adequately describe the features of the part. When this happens, an **auxiliary view** showing the true

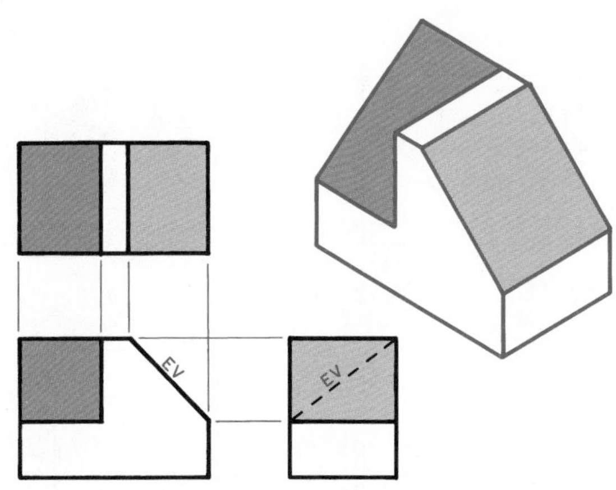

FIGURE 6.48 Edge Views and Inclined Surfaces

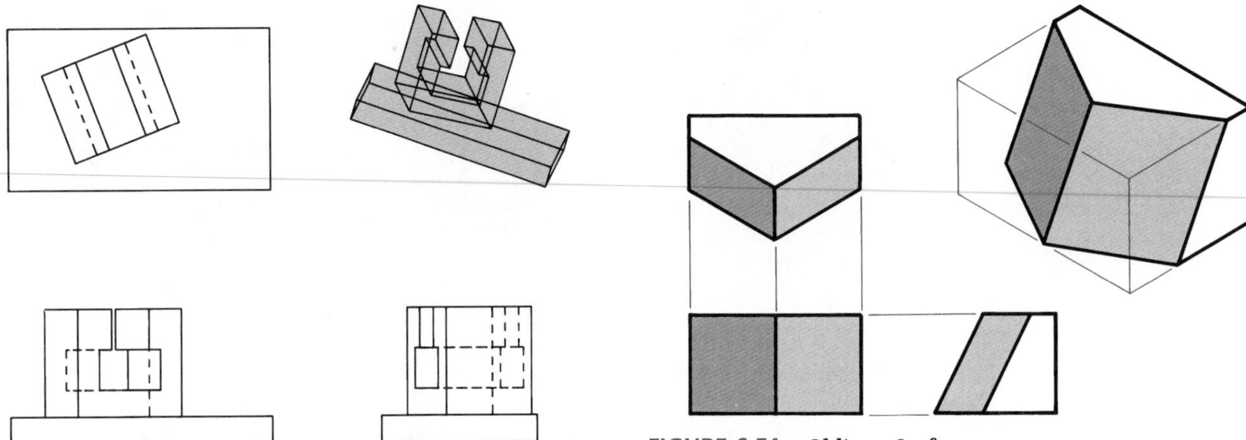

FIGURE 6.49 Distorted View of Surfaces

FIGURE 6.51 Oblique Surfaces

shape of the angled surface is necessary. The surfaces are at an angle to the frontal and profile projection planes. Nowhere do the vertical surfaces show as true shape.

6.8.8 Oblique Surfaces

Oblique surfaces are inclined to all three principal planes of projection, so in each view the surface appears foreshortened. Since it cannot appear as an edge, each view of the oblique plane always displays the same number of sides and has a similar shape. Figure 6.50 is an example of a part with an oblique surface. Since the surface has three sides, each view of the surface will have three sides. Each view shows the plane foreshortened. The true shape of an oblique plane is not shown in any of the principal projection planes. To establish a true shape view of an oblique surface, a secondary auxiliary view must be constructed.

In Figure 6.51, the part has two oblique surfaces. The intersecting line formed by the two oblique surfaces shows true length in the side view. This line is inclined to the base of the part, but since it shows as true length in one of the three principal planes of projection, it is not an oblique line. *An oblique line is inclined to all three principal planes of projection,* as in Figure 6.52 where the mating (intersection) line between the two oblique planes shows as foreshortened in all three views. This line is also known as an *oblique edge.*

6.8.9 Curved and Cylindrical Surfaces

Cylindrical shapes, as in Figure 6.53, show as true shape curves in views that are perpendicular to their surface. The

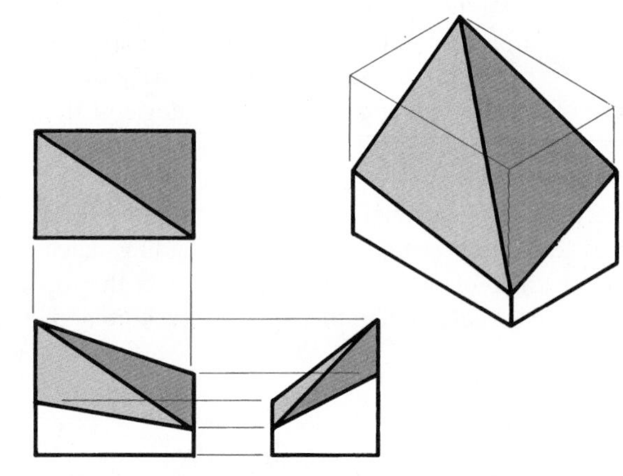

FIGURE 6.52 Oblique Surfaces

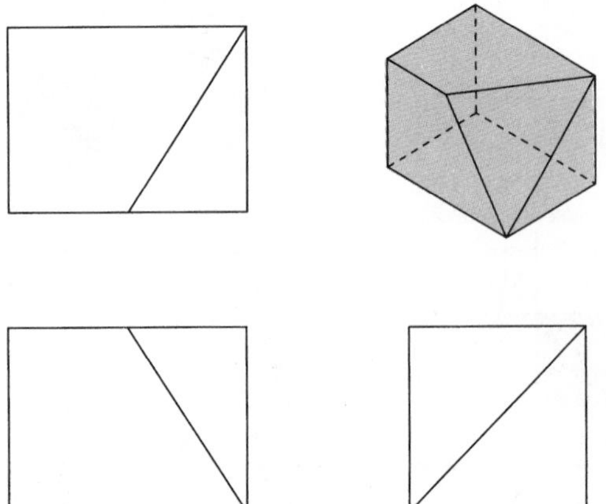

FIGURE 6.50 Oblique Surfaces

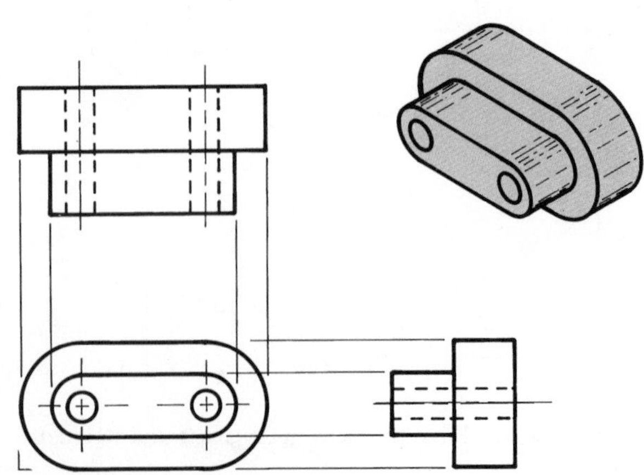

FIGURE 6.53 Cylindrical Features

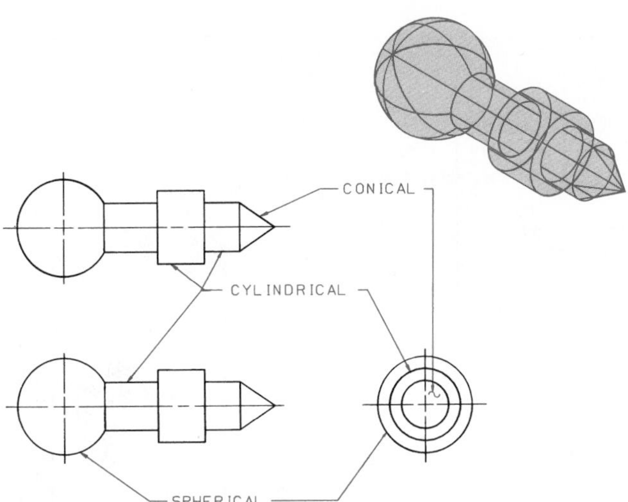

FIGURE 6.54 Representing Cylindrical, Conical, and Spherical Features

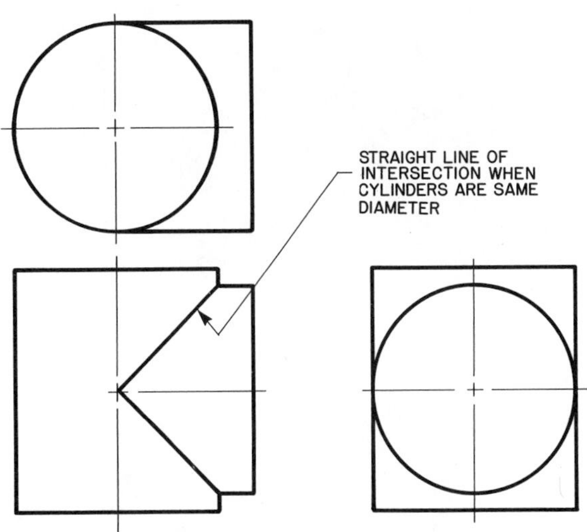

FIGURE 6.55 Intersecting Cylinders of the Same Diameter

front view of this part shows the true size/shape curve of the cylindrical surface. The side and the top view are parallel to the curved surface. In these views the cylindrical shape appears as an rectangle.

In Figure 6.54, the three types of curved surfaces are displayed. The **cylindrical surface** shows as a circle in one view and as a rectangle in the other two views. The **conical surface** appears as a circle in one view also, but its other two views show the surface as a triangle. The **spherical surface** shows as a circle in all three views; as would a ball when viewed from any direction.

For parts with curved features, *always provide at least one view where the curve appears as true shape.*

You May Complete Exercises 6.5 Through 6.8 at This Time

6.8.10 Intersection of Curved Surfaces

Where two cylindrical surfaces meet, a line of intersection must be constructed. When the line of intersection is manually derived, it must be represented according to established drafting conventions. Three situations are possible:

1. The two curved surfaces have the same diameter.
2. The two curved surfaces have different diameters.
3. One of the two curved surfaces is so small that it would be a waste of time to plot the line of intersection.

In Figure 6.55, the two curved surfaces have the same diameter. The line of intersection formed between them is a straight line. In Figure 6.56, the other two conditions are illustrated. The small-diameter cylindrical surface that intersects the vertical cylinder does not show a distinct enough

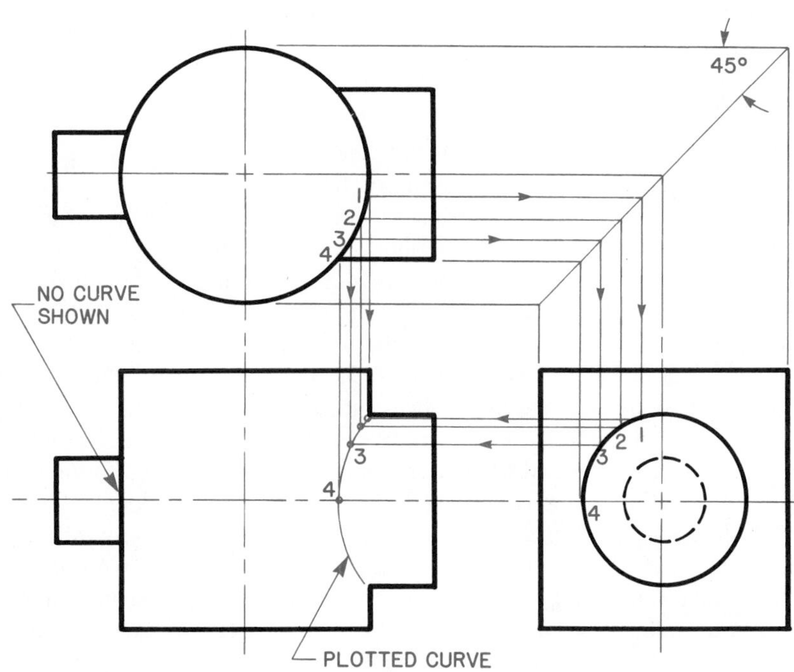

FIGURE 6.56 Intersection of Dissimilar-size Cylinders

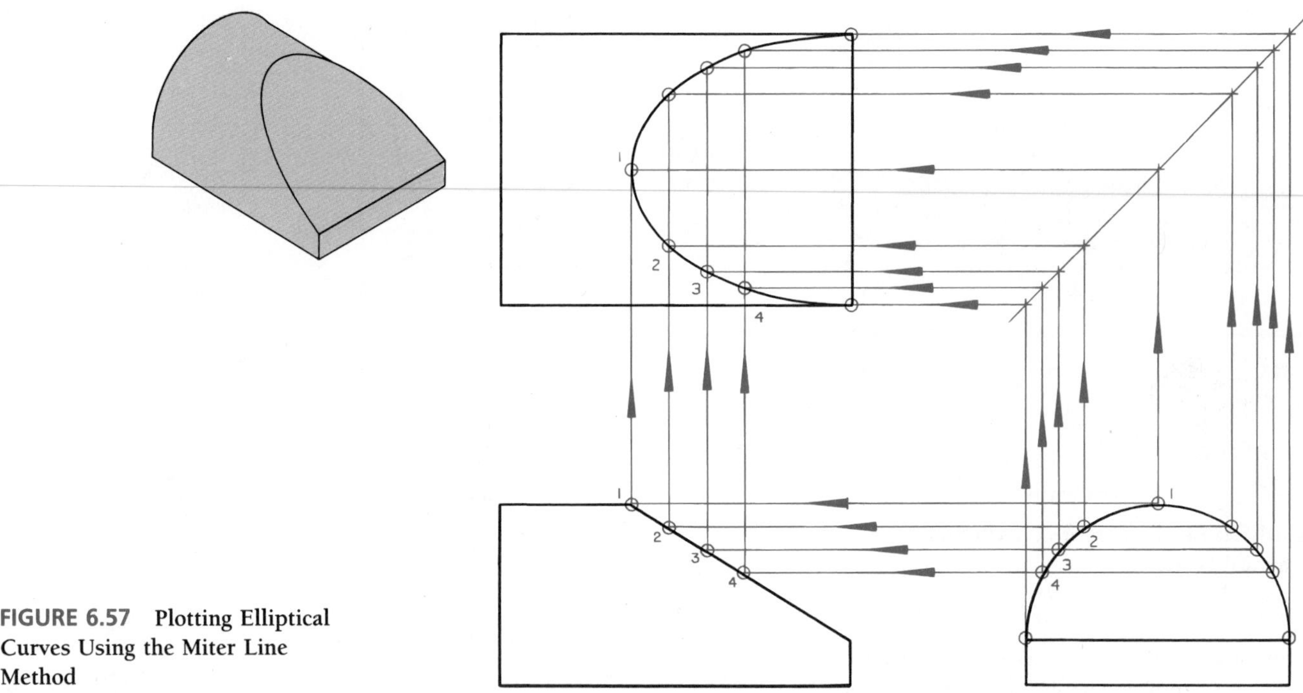

FIGURE 6.57 Plotting Elliptical Curves Using the Miter Line Method

line of intersection when it intersects the vertical cylinder. Therefore, it is accepted drafting practice to show the intersection as a straight line. Some drafters prefer to use an ellipse template and show a small curved intersection line. The right side of the intersecting cylinders shows a cylindrical surface large enough to be plotted. The *miter line* method can be used or transfer the points with dividers. Points are established on the curve of the cylinder in the top view, either randomly or evenly spaced as shown. The points are projected to the side view first. The side and the top views of each point are then projected to the front view. The intersection of related projection lines locates a point on the line of intersection. The points are connected with an irregular curve.

6.8.11 Plotting Elliptical Curves

Elliptical shapes are created by the intersection of planes and curved surfaces. In Figure 6.57, the curve is formed by the intersection of the curved surface and a flat plane surface (not shown). The resulting shape is a surface that is elliptical. This inclined surface does not appear as true shape in any of the given three principal views. To establish the line of intersection in the top view, the side view of the cylindrical surface has a series of points located along it, as in Figure 6.56. The greater the number of points used, the greater the accuracy of the plotted curve. The half-circle curve appears true shape. Each point on this curve is projected to the front view. The points are then transferred to the top view using the miter line method or with dividers. The intersection of related projection lines and transferred distances establishes points along the line of intersection. Connecting the points with a smooth curve completes the view.

6.8.12 Space Curves

Irregular-shaped surfaces (space curves), as shown in Figure 6.58, must be plotted. The curved surface of this part was cut by an inclined plane (not shown). The true shape of the inclined surface does not appear in any of the three views. To plot the resulting intersection, establish a number of points along the curve in the top view where the curve's edge line is shown. The more points that are used, the greater the accuracy of the plotted curve. Each point is projected to the front view. The points are now projected to the side view from the front view. The points are then transferred to the side view from the top view. The resulting series of points in the side view is connected using an irregular curve to establish a smooth curve.

6.8.13 Hole Representation

The part in Figure 6.59 has a number of curved features, including a through-hole and a counterbored hole. The diameter of the hole (.8125) is given for the two holes that are aligned. The counterbored hole has a diameter of .5625 for the through-hole and a counter bore diameter of .875 to a depth of .250. A machinist reading this drawing would be able to choose the proper equipment to machine these features. The machinist determines whether to use a drill, reamer, or boring tool. *A hole is always defined by its diameter.* Drills, reamers, bores, and other hole machining tools are described by their *diameter*.

Figure 6.60 provides a detailed explanation of how holes should be and should not be represented on drawings. The easiest hole callout gives the diameter symbol and the

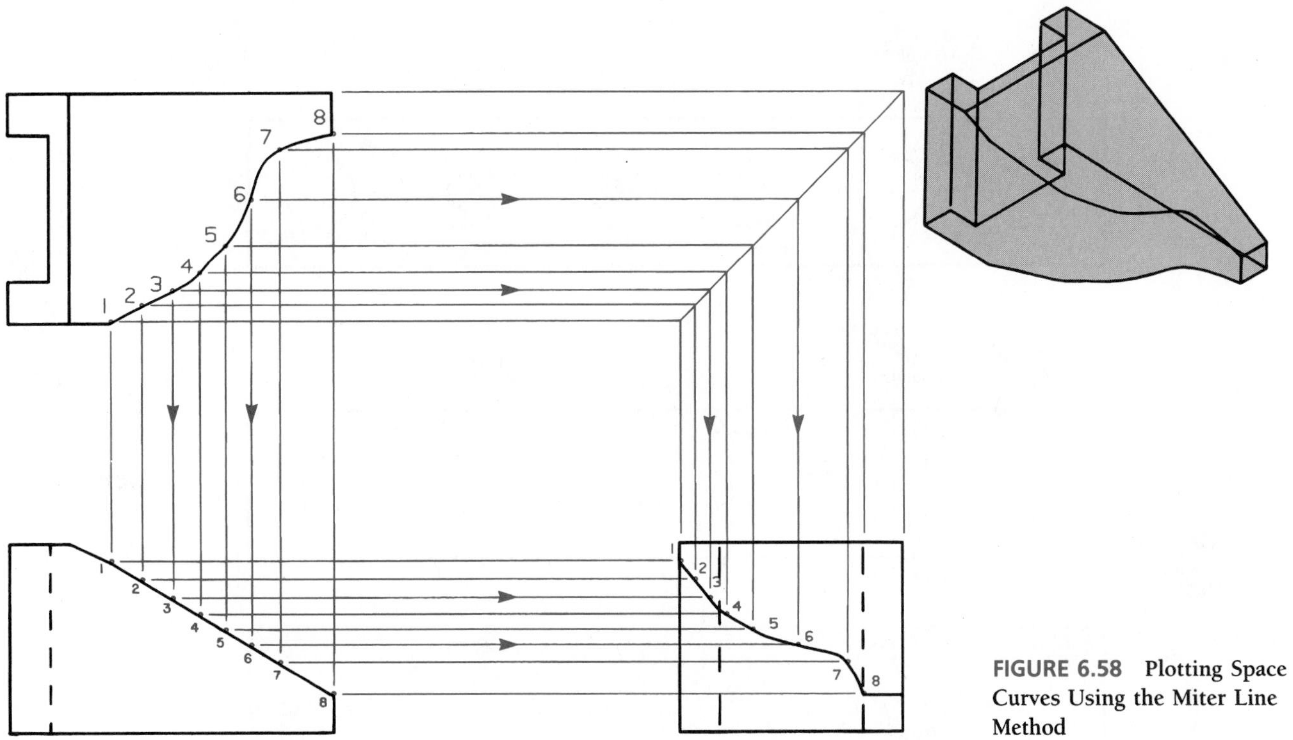

FIGURE 6.58 Plotting Space Curves Using the Miter Line Method

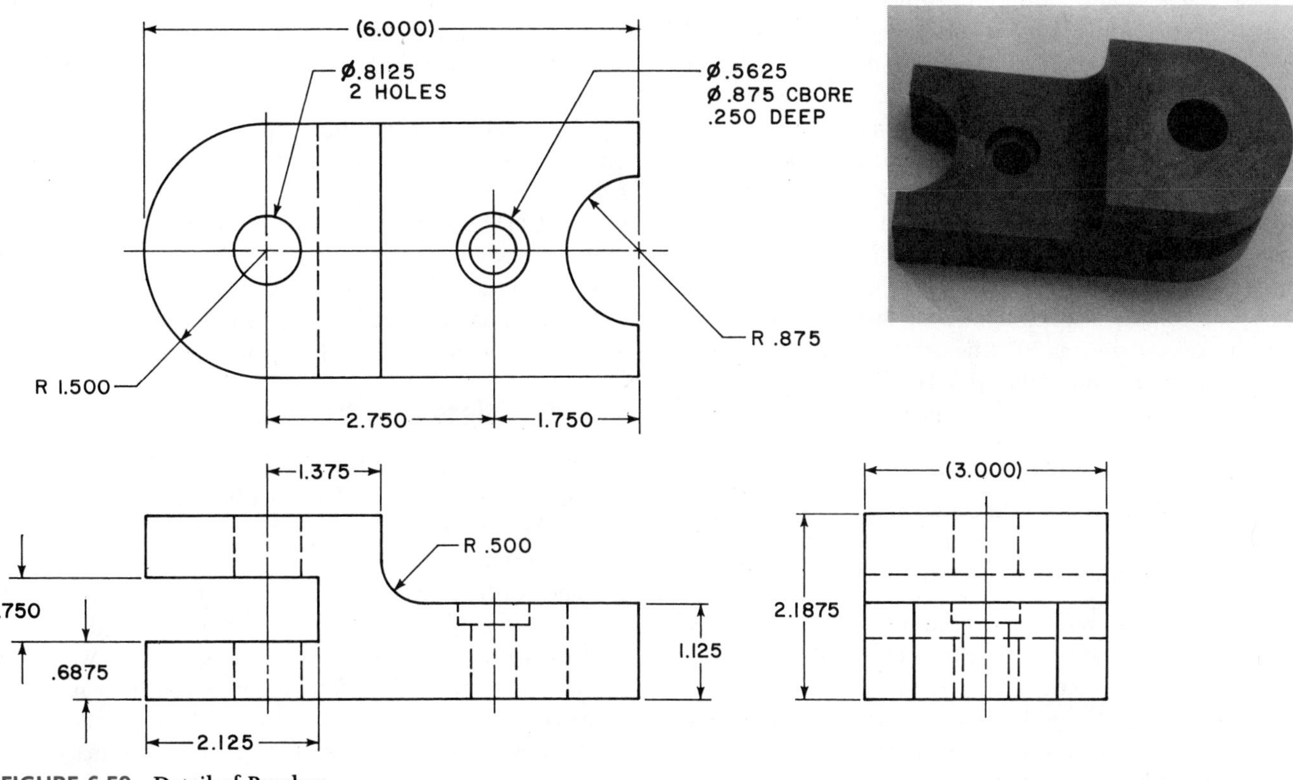

FIGURE 6.59 Detail of Breaker

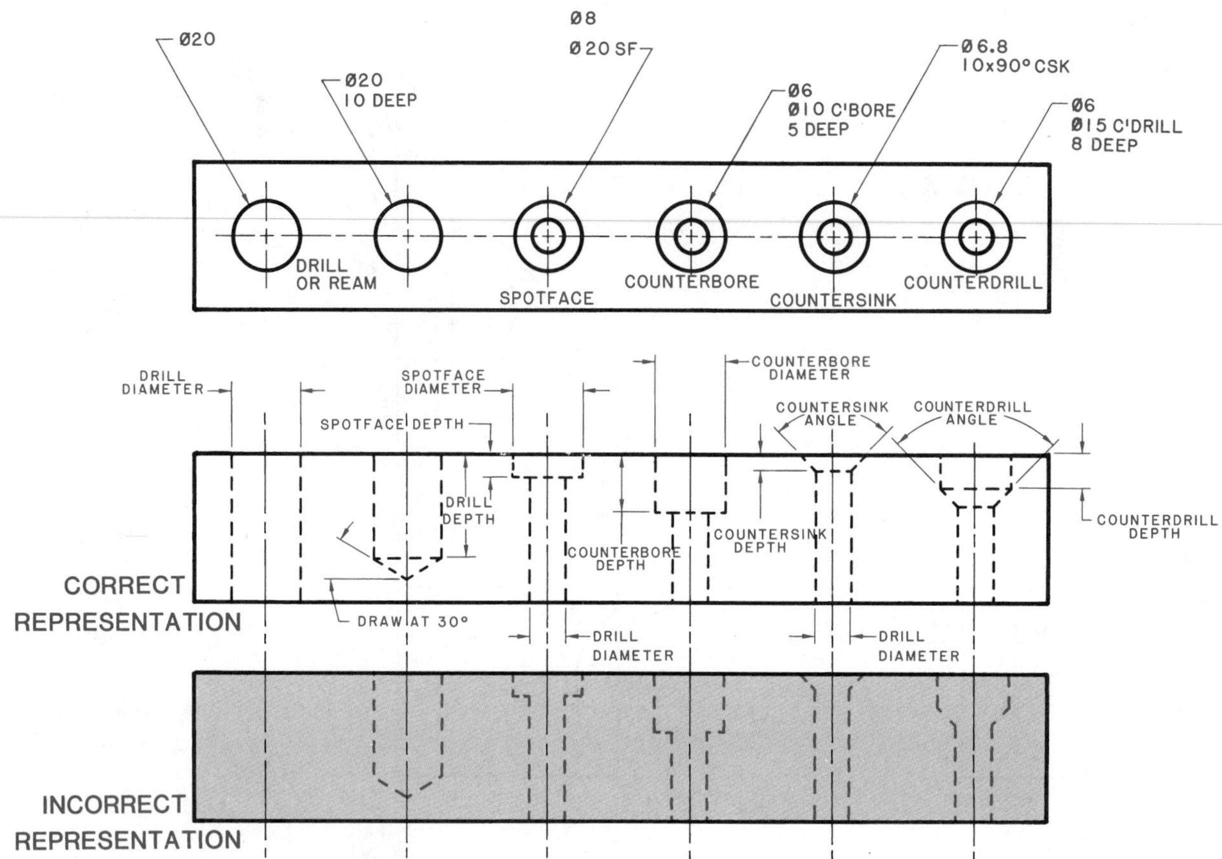

FIGURE 6.60 Types and Representation of Holes on Drawings

diameter value as in the **DRILL** or **REAM** callout. Unless the depth is given, the hole depth is understood to be through the part. When they completely penetrate the part, holes on drawings are sometimes noted with the word **THRU**. The word **THRU** is used in place of *through* on drawings.

A hole that does not go through the part is called a **blind hole.** It is shown in the depth view as two lines that represent the edges of the hole diameter and a centerline. A centerline is required for both blind and through-holes in every view in which they are shown. The bottom of the hole is a conical point. The conical shape is formed by the drill tip and, for convenience, is drawn at 30°. The depth of the blind hole is represented by the end of the cylindrical portion of the hole. The depth value is noted in the dimension under the diameter.

In Figure 6.60, the holes are depicted as through-holes. If they were blind holes, the drill depth would be stated under the diameter callout in the dimension. The following hole types are found on machined parts throughout industry:

- A **SPOTFACE** is a hole that has been drilled to the required depth and the upper part enlarged. The depth of the spotface is sometimes not noted. The spotface depth is drawn, depending on the part, .0625 in. (1.5 mm) to .125 in. (3 mm). Spotfacing is used to clean up the surface around the hole so that a bolt head or other item rests flush with the surface.
- A **COUNTERBORE** is similar to a spotface except the enlarged hole has a specific depth. The counterbore depth is specified in the callout dimension under the counterbore diameter.
- A **COUNTERSINK** is a hole that has been enlarged conically to a specified diameter and depth. The conical angle is drawn 90° for simplicity.
- A **COUNTERDRILL** is a countersink and a counterbore combined. The transition between the two diameters is a conical surface formed by the angle of the tip of the tool. Counterdrills are specified by their diameter and depth. The angle of the counterdrill is shown in the adjacent view.

6.8.14 Fillets and Rounds

Castings are rough parts that are usually machined along one or more of their surfaces. A casting will have curved intersections between mating surfaces because they cannot be accurately formed without these curved corners. Drawings of machined castings require the representation of these surfaces and their intersections. The intersection of two rough interior surfaces to form a rounded corner is called a **fillet.** Two rough exterior surfaces meet and form a corner called a **round.** In Figure 6.61, the part has a variety of rounds and fillets.

When two intersecting surfaces meet and one is machined, the corner becomes a sharp edge. If both surfaces are ma-

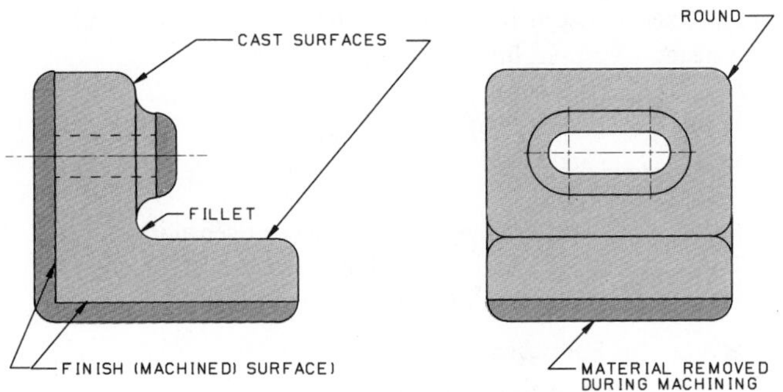

FIGURE 6.61 Fillets, Rounds, and Castings

chined, the corner is also a sharp edge. Rounds will show only when both mating exterior surfaces are unmachined. The material removed during machining is determined by casting dimensions and machining dimensions. Sometimes separate drawings are used. A *casting drawing* is for the foundry and a *machine drawing* is for the machine shop.

As a design requirement, fillets and rounds are used to reduce the possibility of failure of a joint. Sharp points are possible points of fracture. In many cases, the selection of the fillet radius is left to the patternmaker.

6.8.15 Tangent Surfaces

When a curved and a plane surface are tangent, a point of tangency may be required. In Figure 6.62, the cylindrical surfaces are connected by plane surfaces along the sides of the part. Since the cylindrical ends are different diameters, the tangent points of the cylinders and the planes will not fall along the centerline in the front view. Since the back surface is flush with the two diameters, tangent points at A fall along the centerline. Because the circles are staggered and of different diameters, the front view of the tangent points does not fall along the centerline. Tangent points B and C are determined by drawing construction lines perpendicular to the front edge and through the center of each cylindrical surface in the view where the diameter shows true shape. The intersection of this line and the circles' circumference determines the point of tangency (B and C).

6.8.16 Runouts and Edge Representation

After the point of tangency between a plane surface and a cylindrical surface has been determined, the runout can be drawn. **Runouts** are curves at the point of tangency. If the part is a casting, the runout will be a fillet at the tangent point, as in Figure 6.62. Points B and C are the points of tangency of the surface intersections, but they are also the transition point of the cast surfaces. The radius of the fillet is used to establish the runout; it is normally constructed with a template. Only 45° (one-eighth) of the curve is drawn for most situations.

You May Complete Exercises 6.9 Through 6.12 at This Time

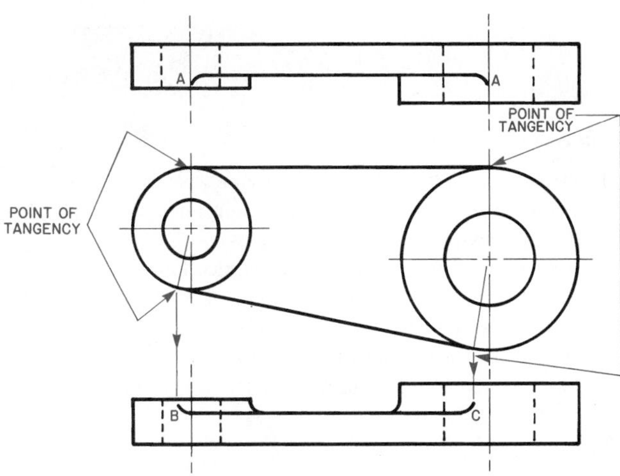

FIGURE 6.62 Runouts and Points of Tangency on Drawings

6.9 Opposite-Hand Parts

There are many industrial applications for parts that are the exact opposite of one another. These are called *opposite-hand parts* or *right-hand and left-hand parts*. Usually one drawing is used to describe both parts. Use a mirror to help you visualize this concept. The reflection in a mirror shows the opposite hand of the part. If a right-hand part was used, the mirror shows the left-hand view.

A car has many opposite-hand parts, both in the engine and on the body of the automobile. Do not confuse right-hand and left-hand parts with parts that are the same but are installed on both sides of an assembly. For instance, fenders and doors are right-hand and left-hand parts. But, the head lights, wheels, hubcaps, and head rests are not.

It is accepted practice to draw only one of the parts and to note on the drawing:

NOTE: RIGHT-HAND AND LEFT-HAND PART
REQUIRED.
RH PART SHOWN.

If there are differences between the two parts, it is normal practice to draw both parts. If the differences are minor, such as a hole size or the addition of a hole, then these differences may be established with a note or a callout:

.500 DIA THRU
LH PART ONLY

When both LH and RH parts are drawn, you can save time by tracing the completed side (or making a copy on an office copier), turning it over, and using it to draw the opposite side. A light table may be used to see through the drafting paper.

You May Complete Exercises 6.13 Through 6.16 at This Time

QUIZ

True or False

1. Partial views are used to save space and paper.
2. Centerlines, phantom lines, dimension lines, and leader lines are all drawn with the same thickness.
3. Centerlines take precedence over hidden lines.
4. The glass box method of projection is used for most drawings.
5. Adjacent and related views are the same.
6. Parallel lines are parallel in all views.
7. Most foreign countries use third-angle projection for their engineering drawings.
8. All orthographic projection is right-angle projection.

Fill in the Blanks

9. ________ view drawings are normally limited to thin, flat, or ________ round parts.
10. When the object is relatively simple, a ________ line is used to project the third view.
11. Dimensions can be transferred from the top to the side view using ________ lines, the ________ method or ________ .
12. ________ are considered to be a series of ________ in space having ________ but not ________ .
13. ________ are used to show round features of a part on drawings.
14. ________ lines always take precedence over hidden lines.
15. A ________ is a specific location in space.
16. ________ lines take precedence over visible lines.

Answer the Following

17. What is a fold line and how is it used?
18. What are the six standard views?
19. What is the difference between the *glass box method* and the *natural method*?
20. What is the image plane for projection?
21. Describe adjacent and related views.
22. Explain the difference between first- and third-angle projection.
23. What determines the spacing and choice of views for a drawing?
24. Describe the ISO projection symbol and its use.

EXERCISES

Exercises may be assigned as sketching, instrument, or CAD projects. Transfer the given information to an "A" size sheet of .25 in. grid paper. Complete all views and solve for proper visibility, including centerlines, object lines, and hidden lines. Exercises that are not assigned by the instructor can be sketched in the text to provide practice and understanding of the preceding instructional material.

After Reading the Chapter Through Section 6.6.9, You May Complete the Following Exercises

Exercise 6.1 Complete each of the given views and the third view, if required.

Exercise 6.2 Complete each of the given views and the third view, if required.

Exercise 6.3 Complete each of the given views and the third view, if required.

Exercise 6.4 Complete each of the given views and the third view, if required.

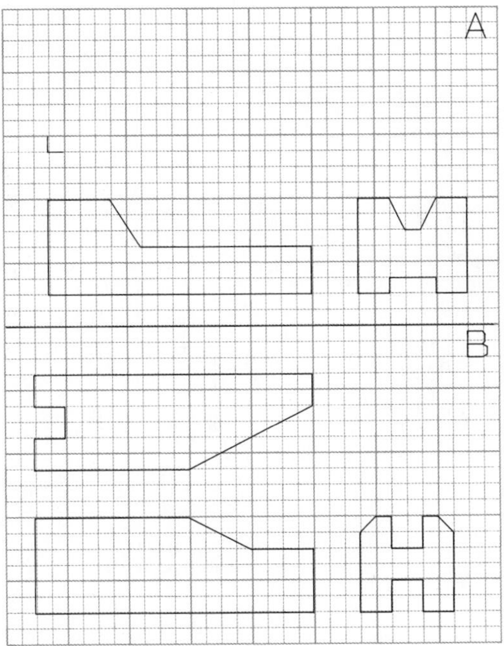

Exercise 6.1

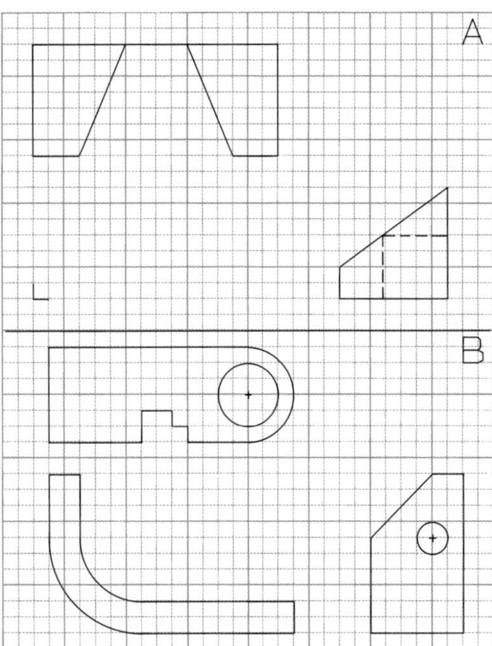

Exercise 6.3

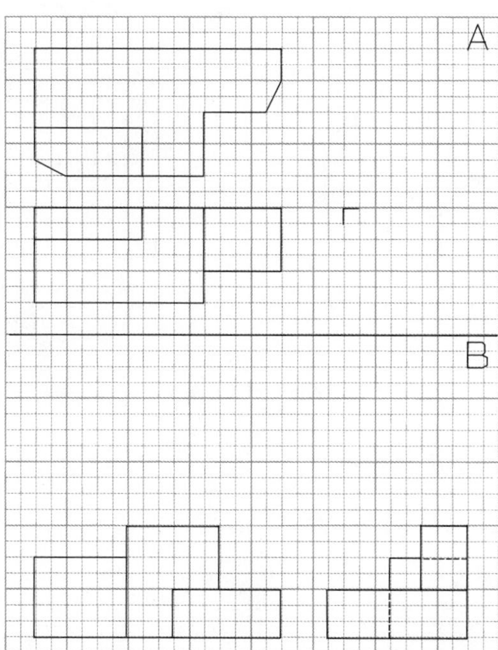

Exercise 6.2

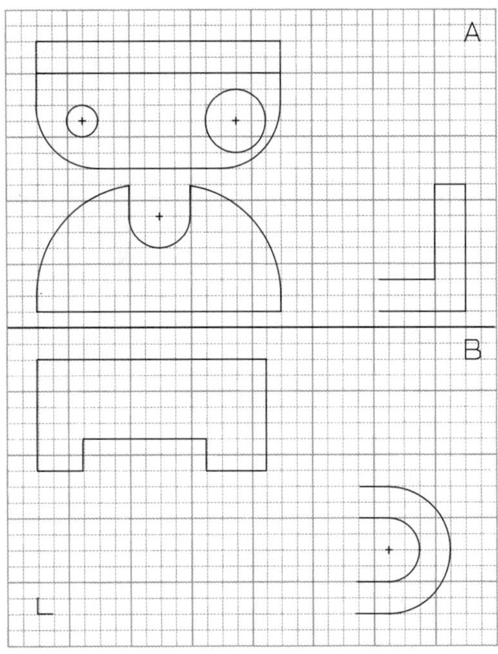

Exercise 6.4

After Reading the Chapter Through Section 6.8.9, You May Complete the Following Exercises

Exercise 6.5 Complete each of the given views and the third view, if required.

Exercise 6.6 Complete each of the given views and the third view, if required.

Exercise 6.7 Complete each of the given views and the third view, if required.

Exercise 6.8 Complete each of the given views and the third view, if required.

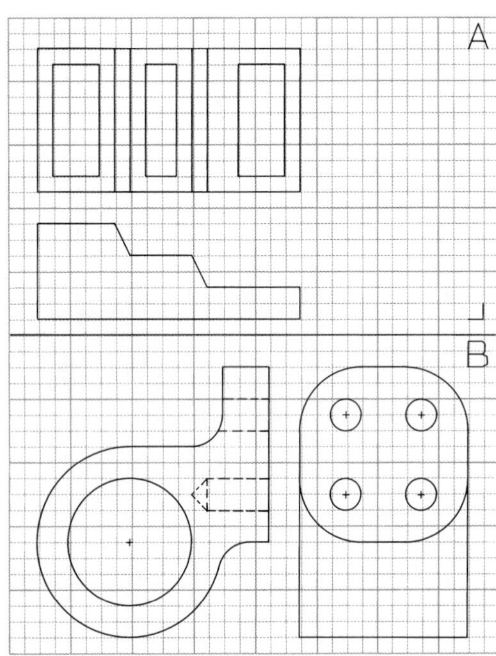

Exercise 6.5

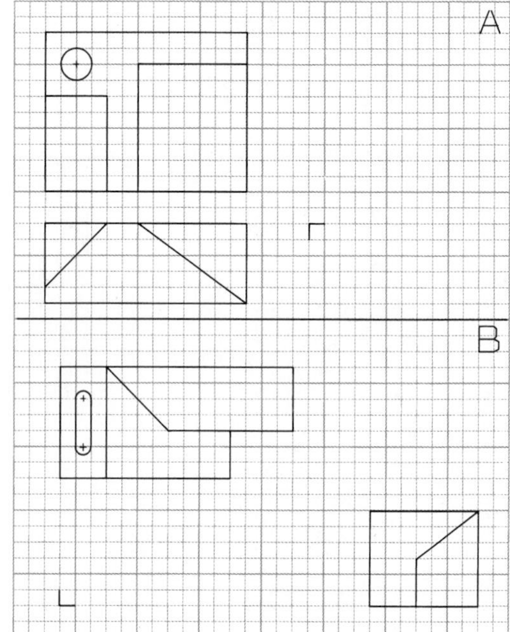

Exercise 6.7

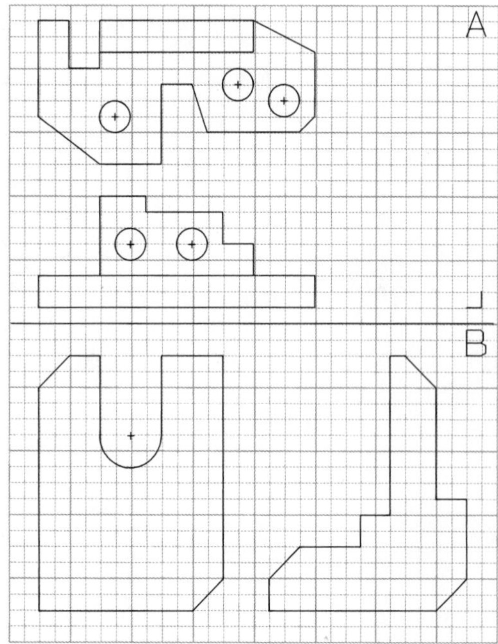

Exercise 6.6

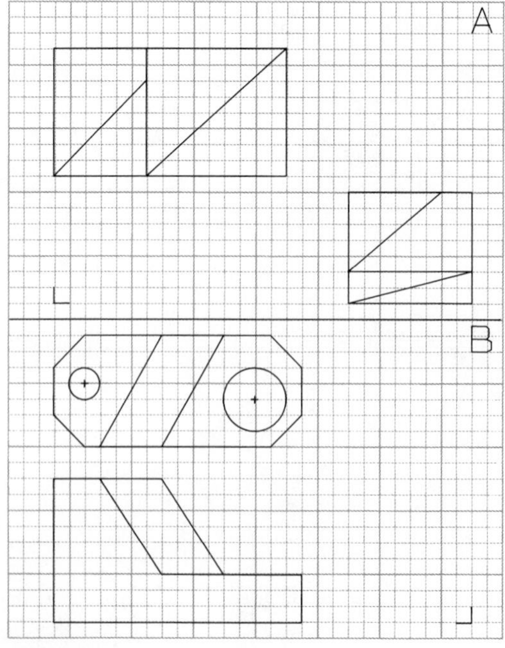

Exercise 6.8

After Reading the Chapter Through Section 6.8.16, You May Complete the Following Exercises

Exercise 6.9 Complete each of the given views and the third view, if required.

Exercise 6.10 Complete each of the given views and the third view, if required.

Exercise 6.11 Complete each of the given views and the third view, if required.

Exercise 6.12 Complete each of the given views and the third view, if required.

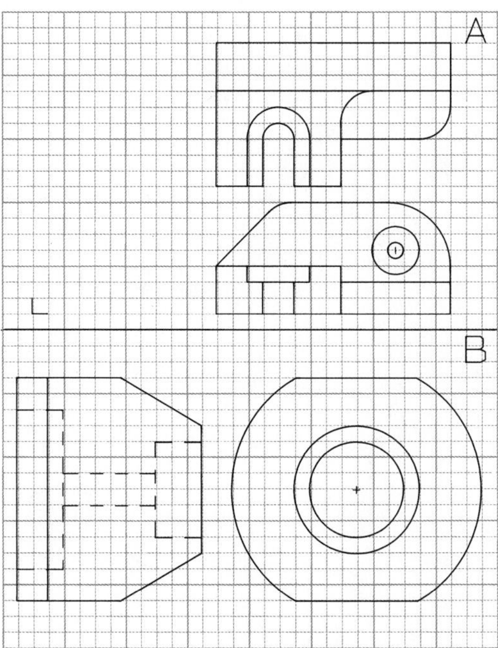

Exercise 6.9

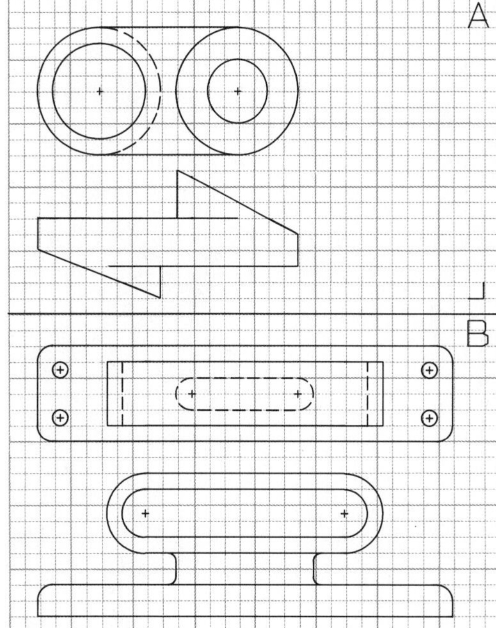

Exercise 6.11

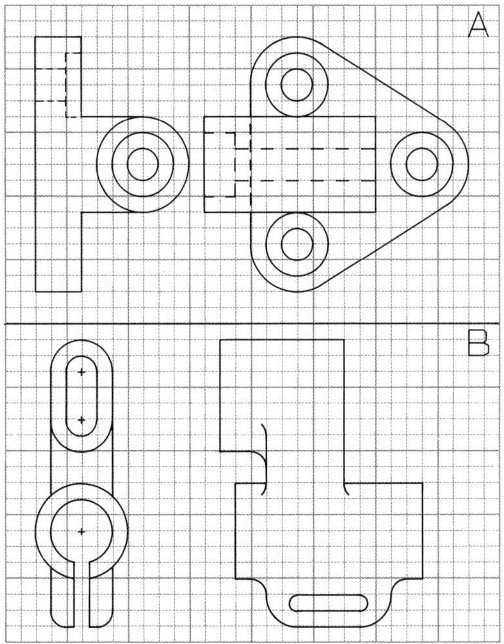

Exercise 6.10

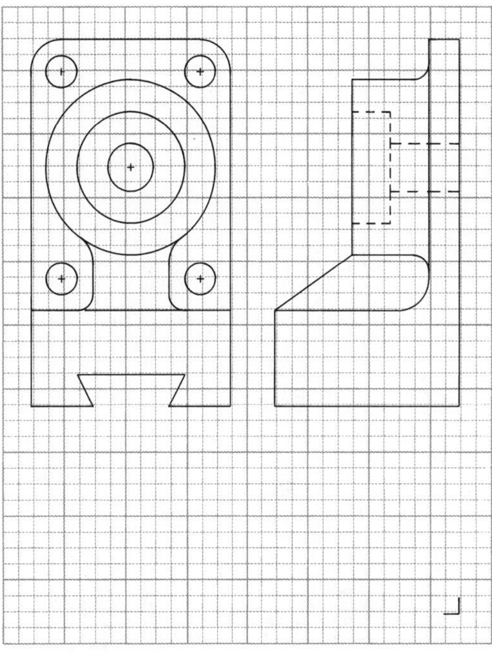

Exercise 6.12

After Reading the Chapter Through Section 6.9, You May Complete the Following Exercises

Exercise 6.13 Complete each of the given views and the third view, if required.

Exercise 6.14 Complete each of the given views and the third view, if required.

Exercise 6.15 Complete each of the given views and the third view, if required.

Exercise 6.16 Complete each of the given views and the third view, if required.

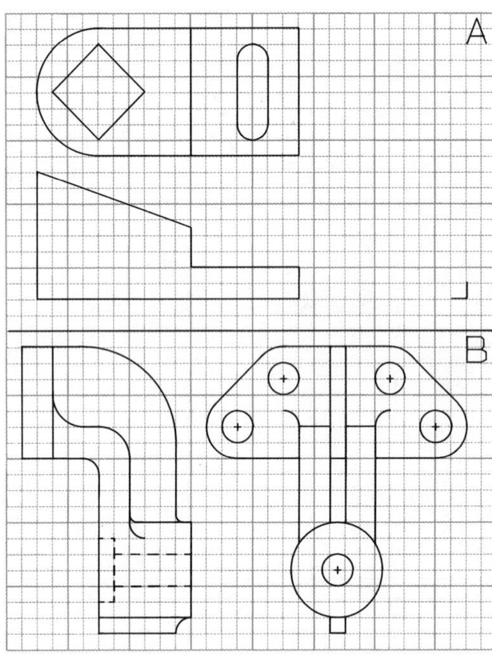

Exercise 6.13

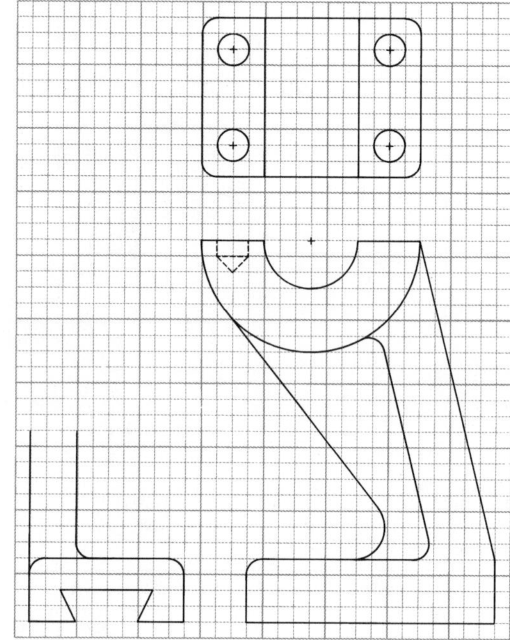

Exercise 6.15

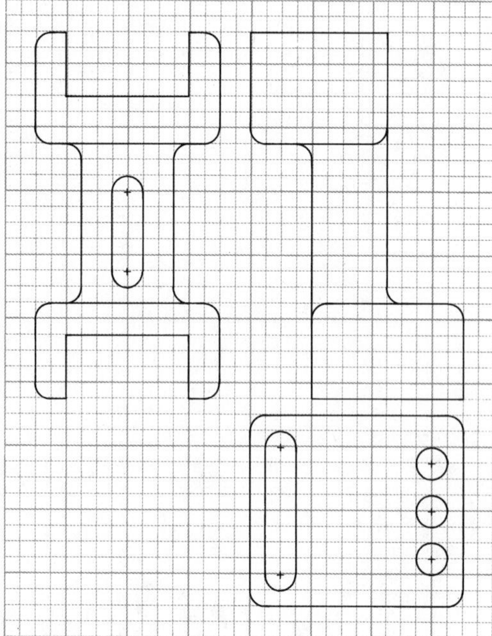

Exercise 6.14

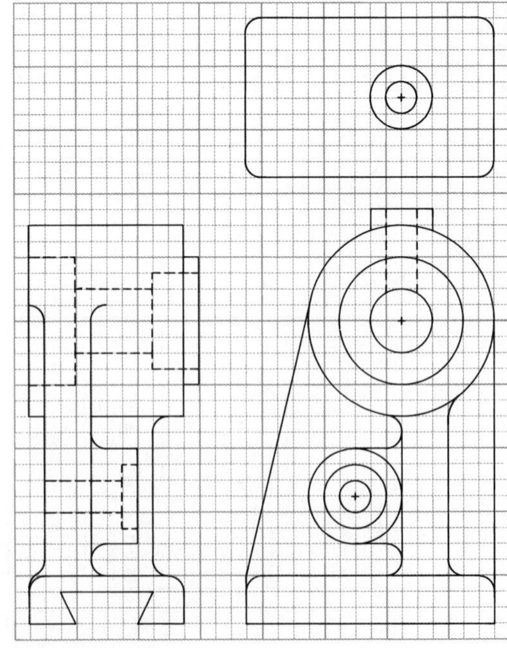

Exercise 6.16

PROBLEMS

Problems 6.1(A) through (K) Complete each of the problems on an "A" or "B" size sheet as required. Use one of the three scales provided in the lower-left corner of the page. Use dividers to take measurements from the drawing and set off on one of the scales to establish the part's dimensions. Round off dimensions where necessary. Solve for the missing view in each problem. All projects will have three views.

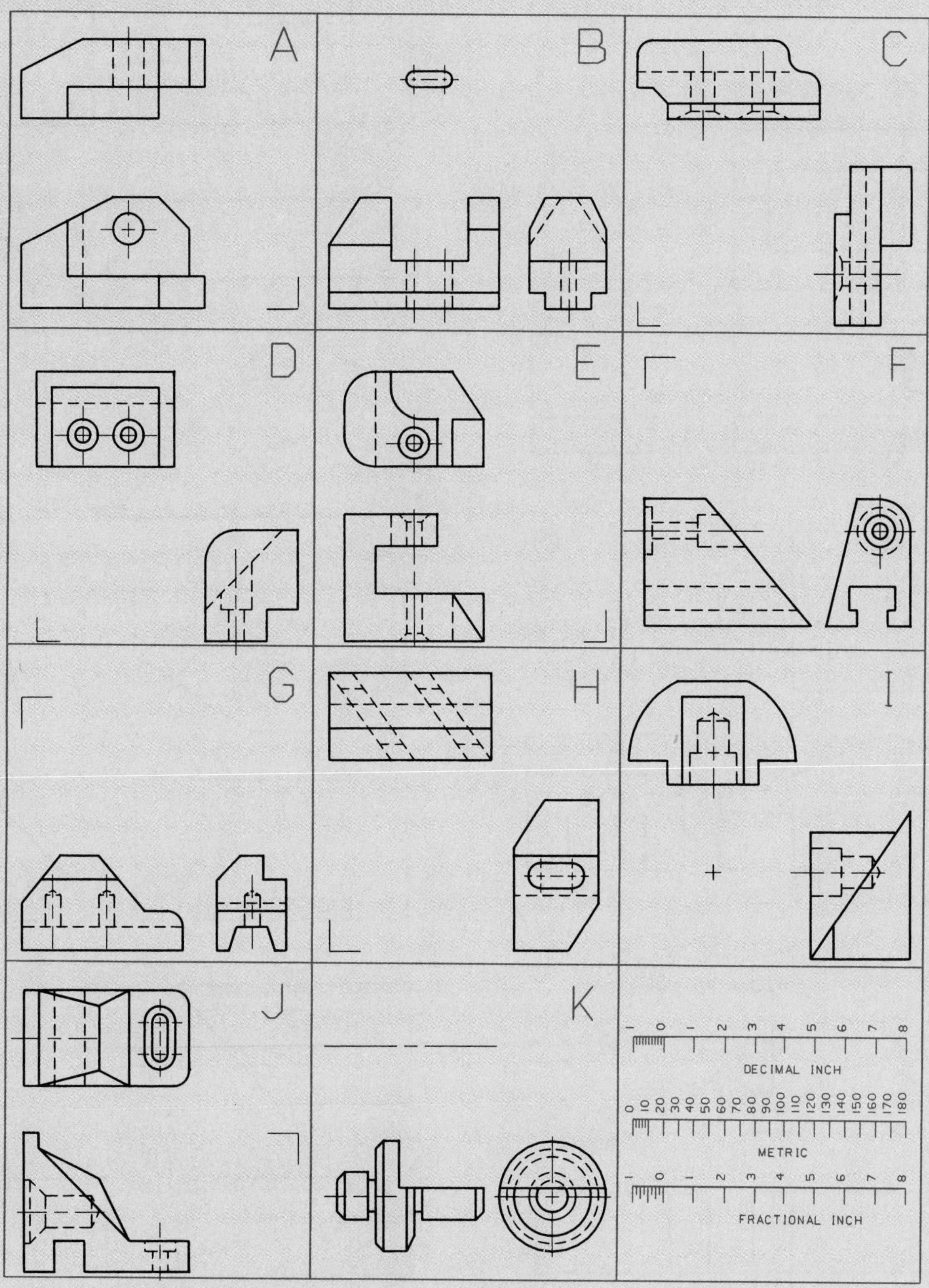

Problems 6.2 (A) through (G) Use the same directions as for Problem 6.1. In these problems, some of the given views are incomplete, although the outline of each of the three views is given. Complete the views as needed.

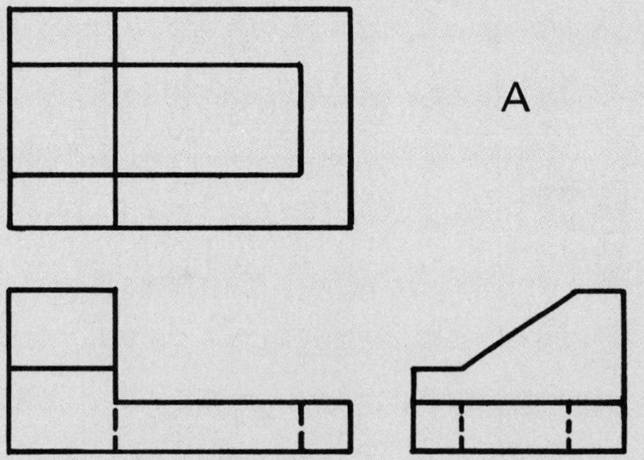

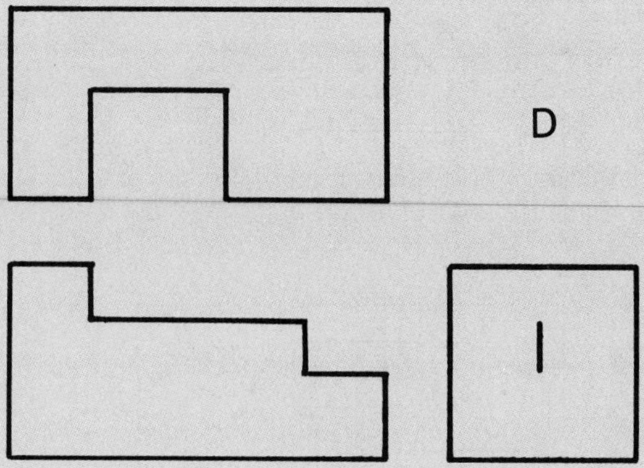

D

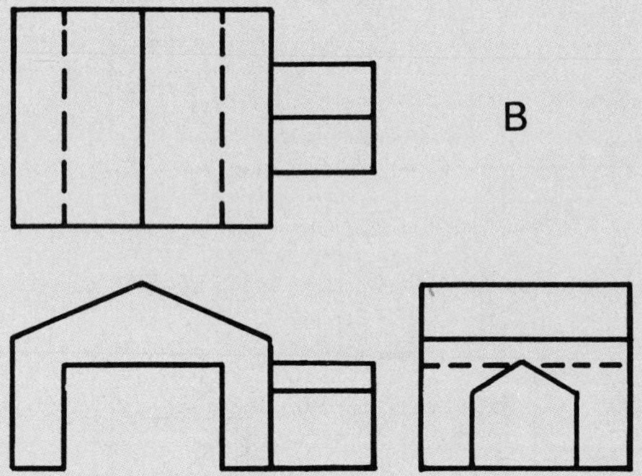

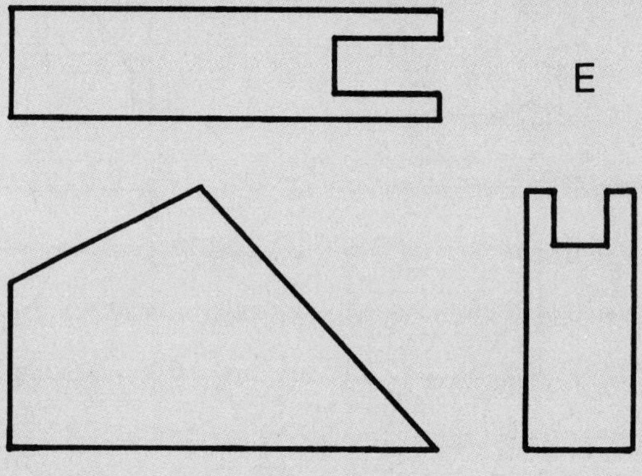

A

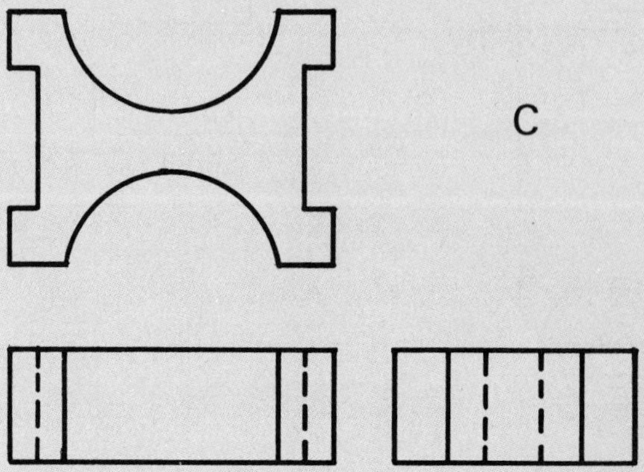

E

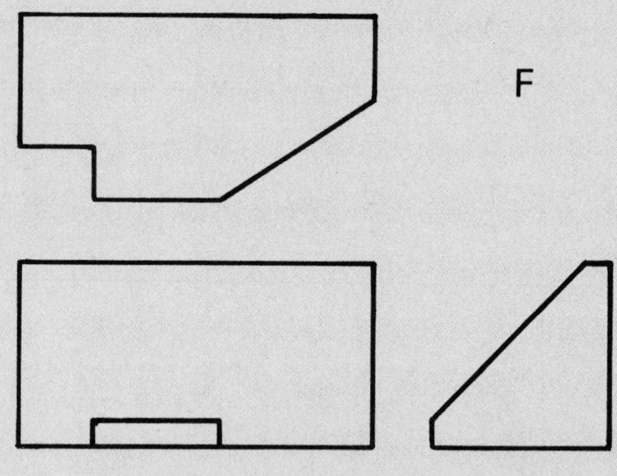

B

F

C

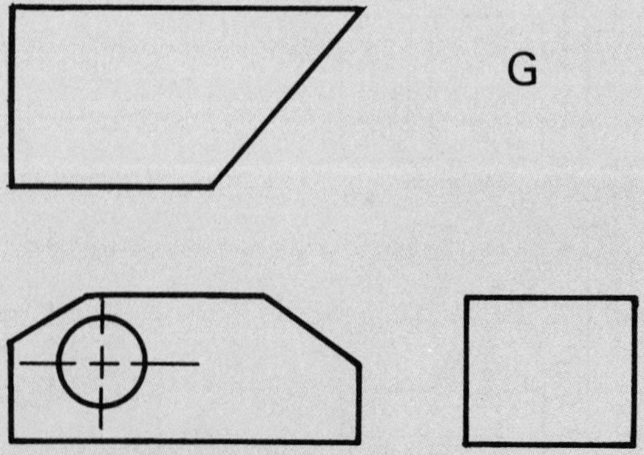

G

Problems 6.3 through 6.10 Draw enough views to describe the part graphically. These projects can be used later for dimensioning projects. Because of this, leave sufficient spacing between views to accommodate dimensions and notes.

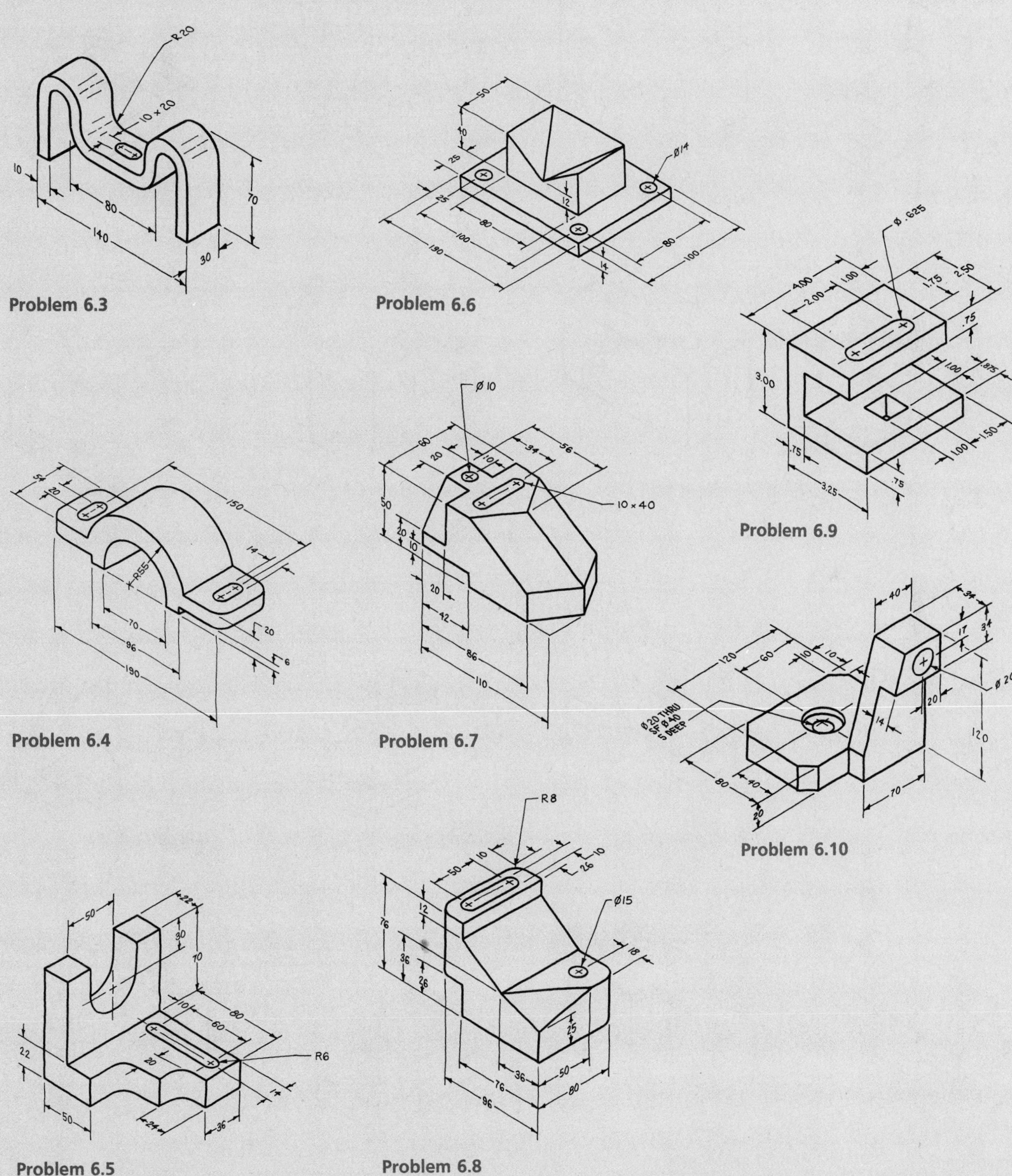

Problem 6.3

Problem 6.6

Problem 6.9

Problem 6.4

Problem 6.7

Problem 6.10

Problem 6.5

Problem 6.8

Problems 6.11 through 6.30 Draw three views of each of the given problems. Use an "A" size sheet for each project. Establish all dimensions by grid squares equaling 1 in. or 20 mm as assigned by the instructor.

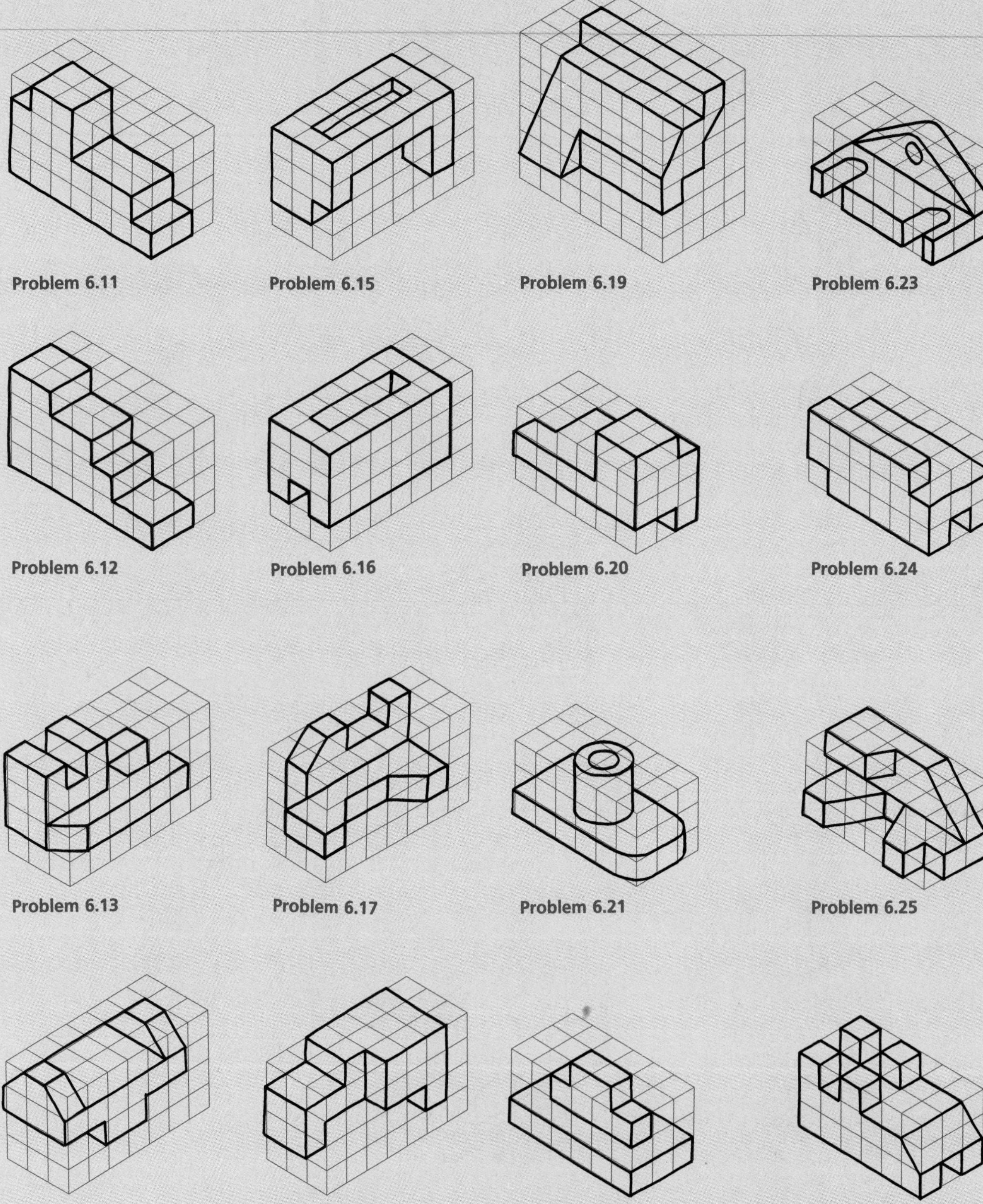

Problem 6.11

Problem 6.15

Problem 6.19

Problem 6.23

Problem 6.12

Problem 6.16

Problem 6.20

Problem 6.24

Problem 6.13

Problem 6.17

Problem 6.21

Problem 6.25

Problem 6.14

Problem 6.18

Problem 6.22

Problem 6.26

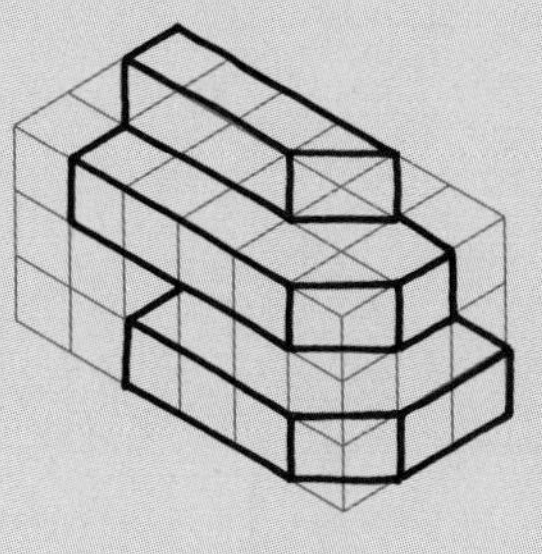
Problem 6.27

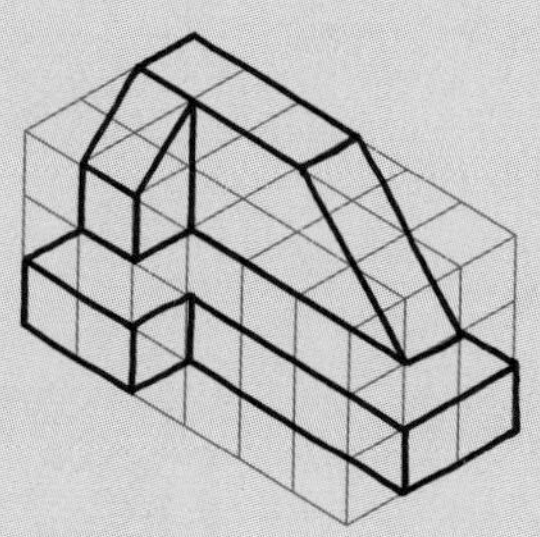
Problem 6.28

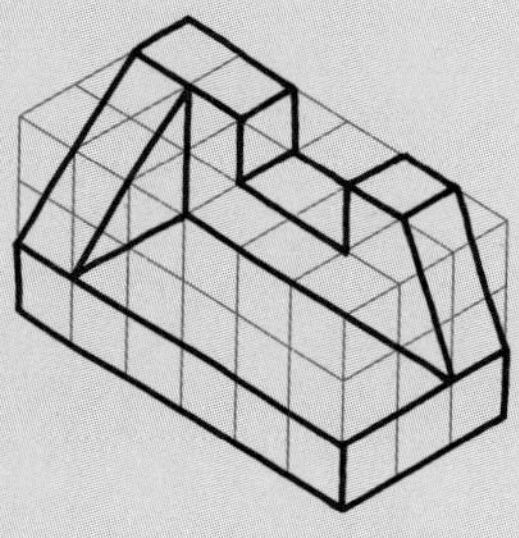
Problem 6.29

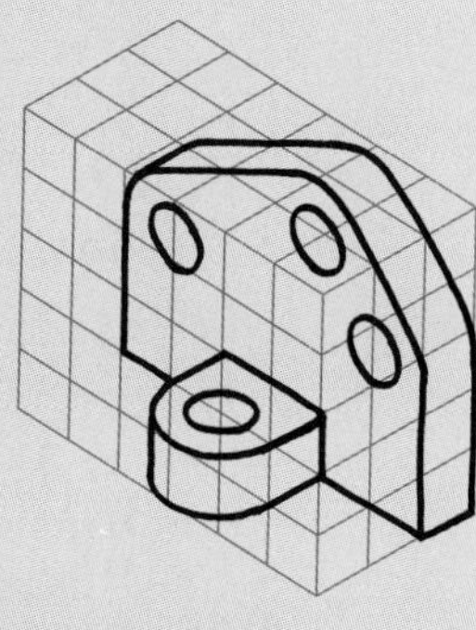
Problem 6.30

Problems 6.31 through 6.43 Draw, but do not dimension, each problem assigned by the instructor. Do not section any of the parts.

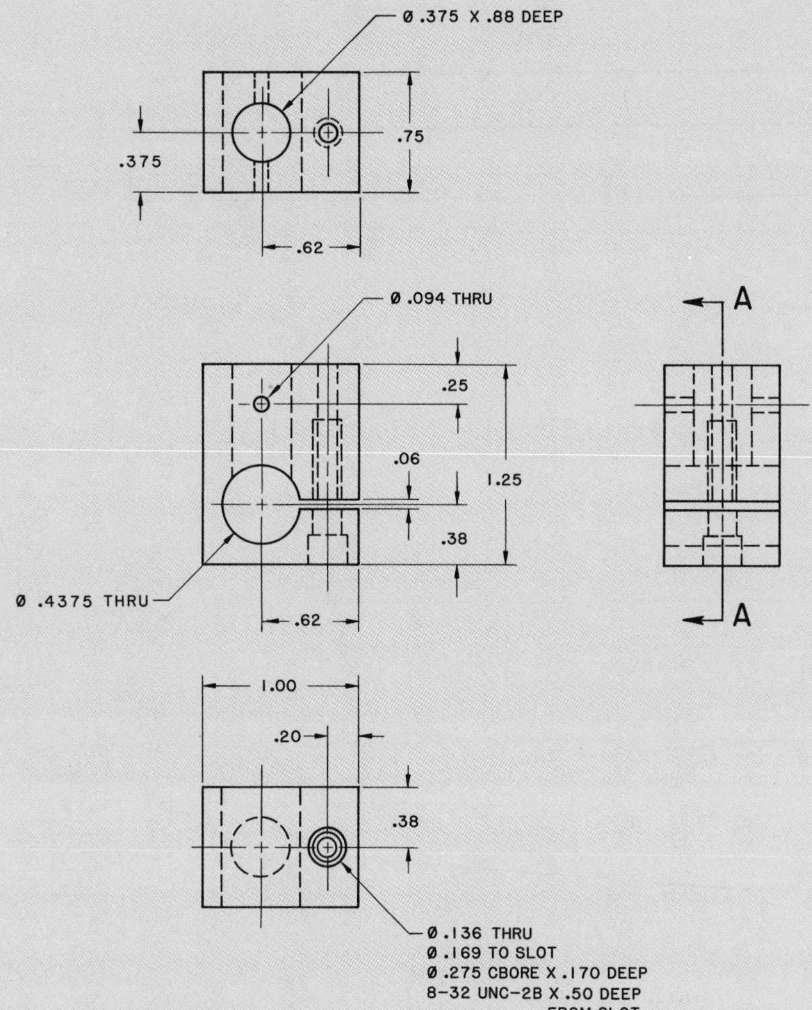

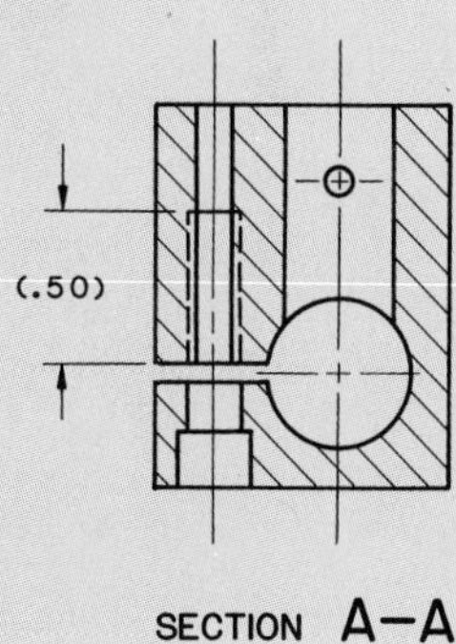

SECTION A–A

Problem 6.31

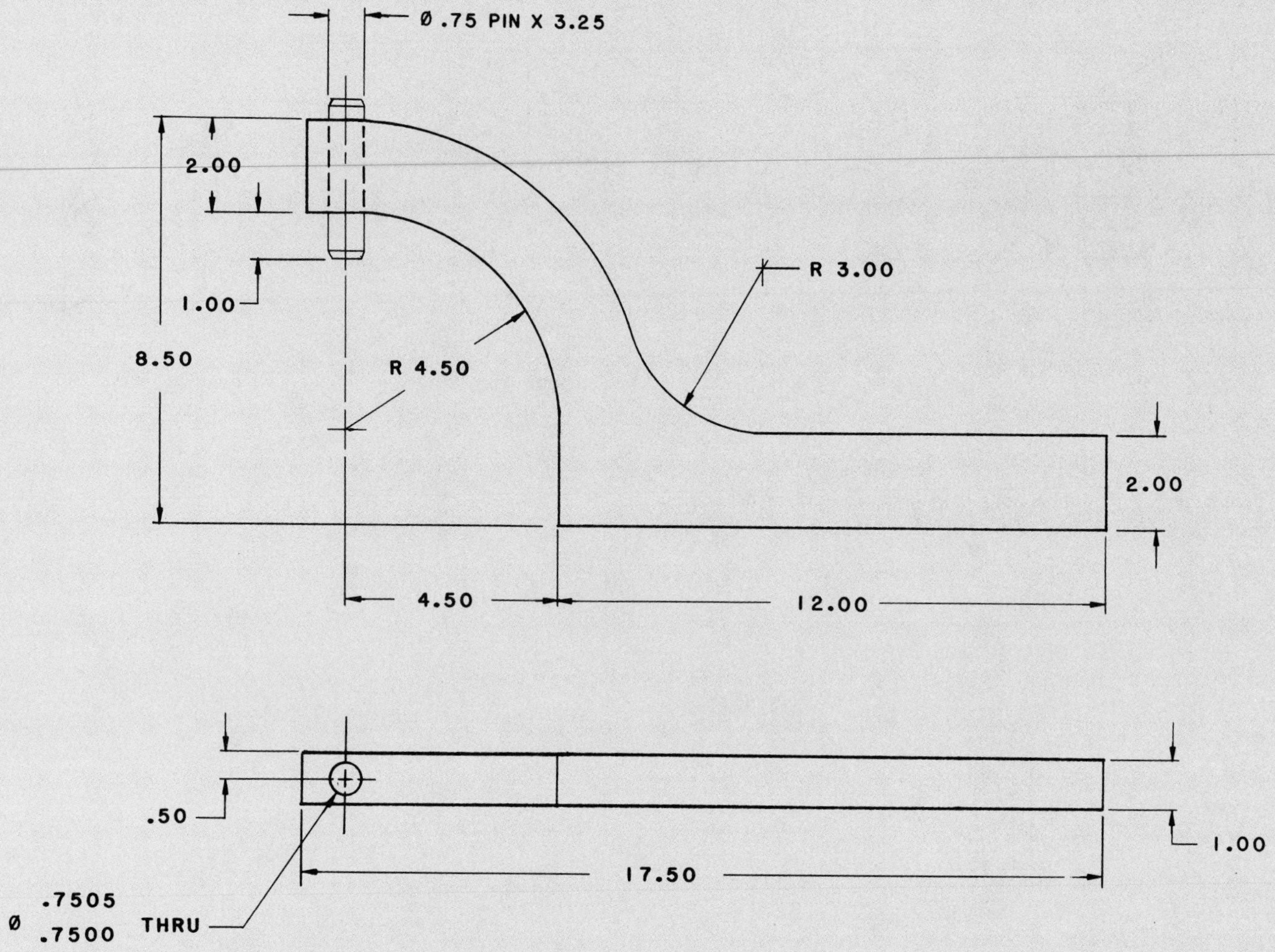

Ø .75 PIN X 3.25
2.00
1.00
8.50
R 4.50
R 3.00
2.00
4.50
12.00
.50
1.00
17.50
Ø .7505 .7500 THRU

Problem 6.32

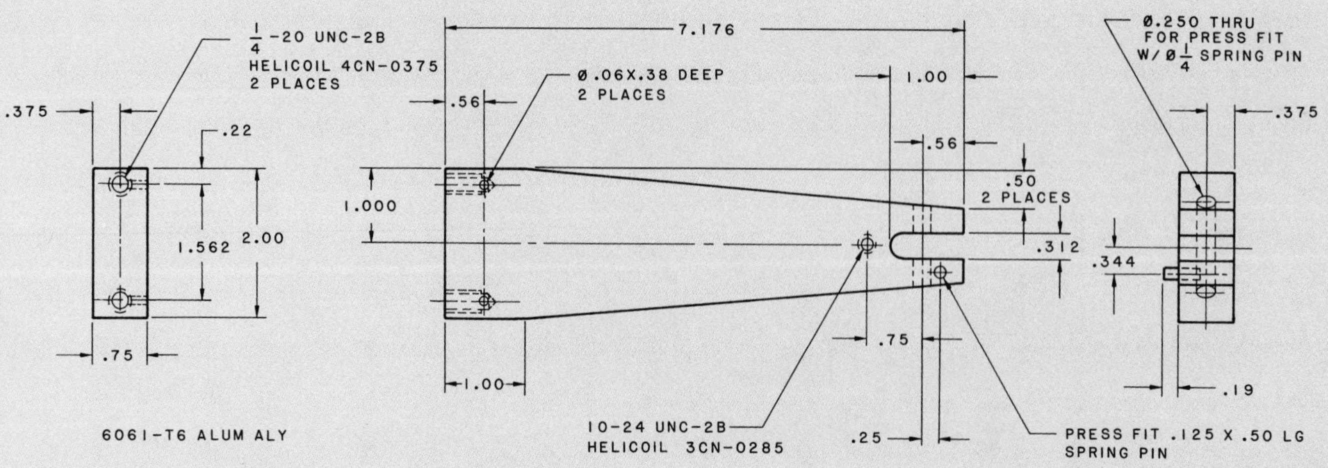

Problem 6.33

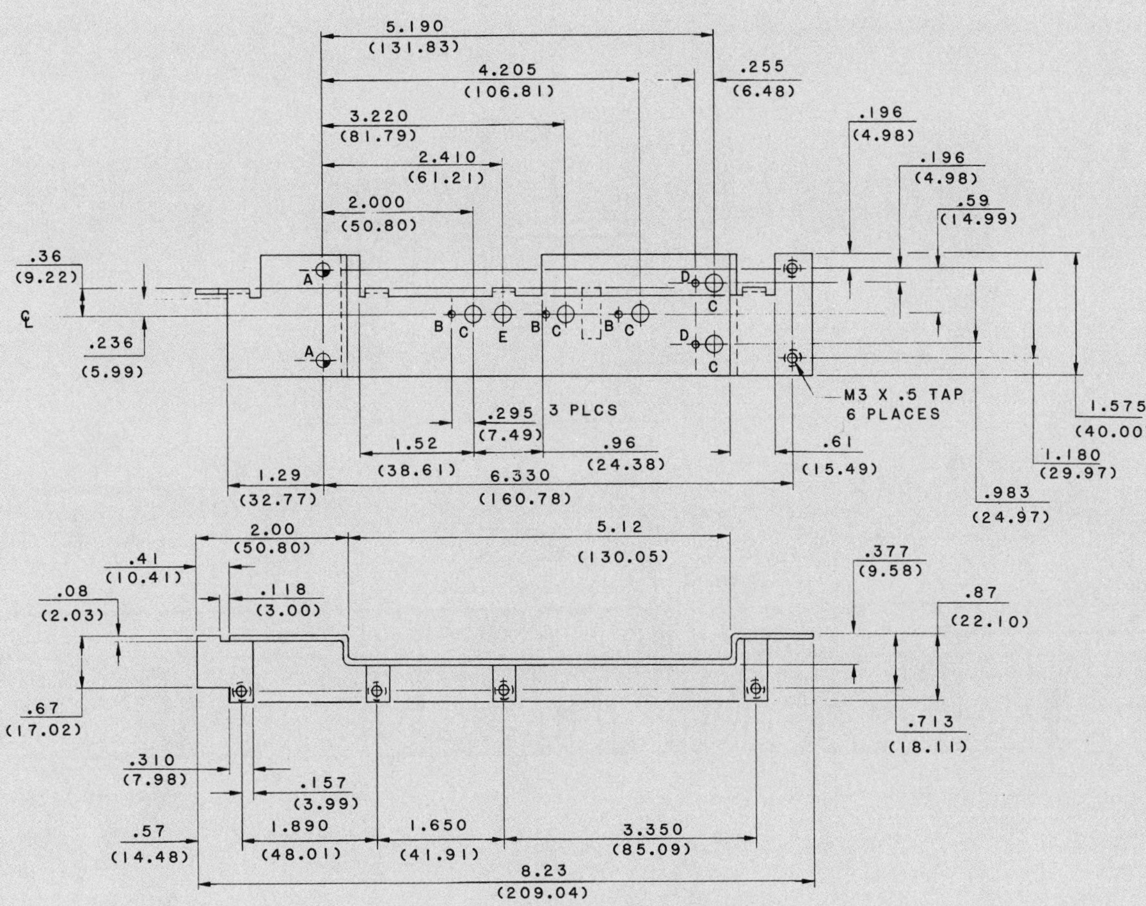

NOTES:

1. REMOVE ALL BURRS AND SHARP EDGES

2. BEND RADII .06 MAX

3. MARK PART NO. NEARSIDE

HOLE CHART		
CODE	DESCRIPTION±.005	QTY
A	.140 (3.56)	2
B	1.109 (2.78)	3
C	.219 (5.56)	5
D	.094 (2.38)	2
E	.188 (4.76)	1

SUBPANEL,PREV-2 BD

Problem 6.34

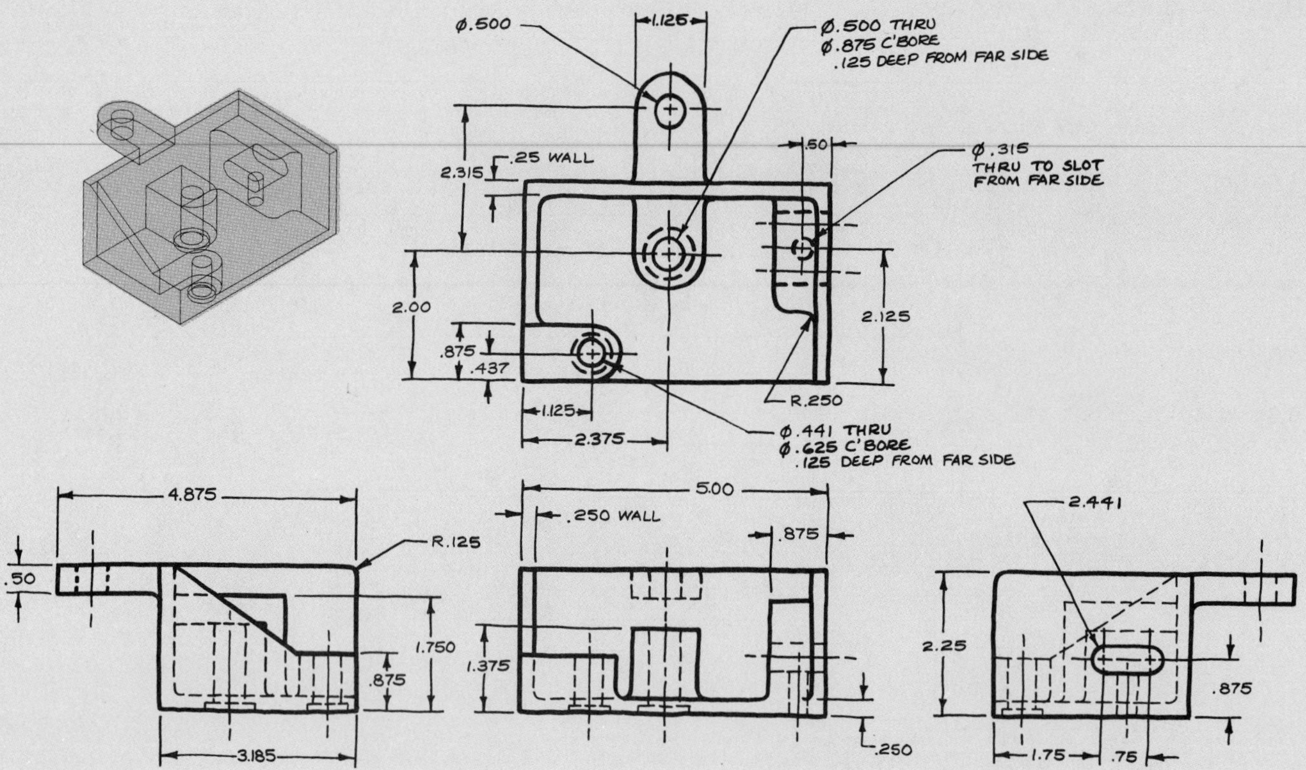

Problem 6.35

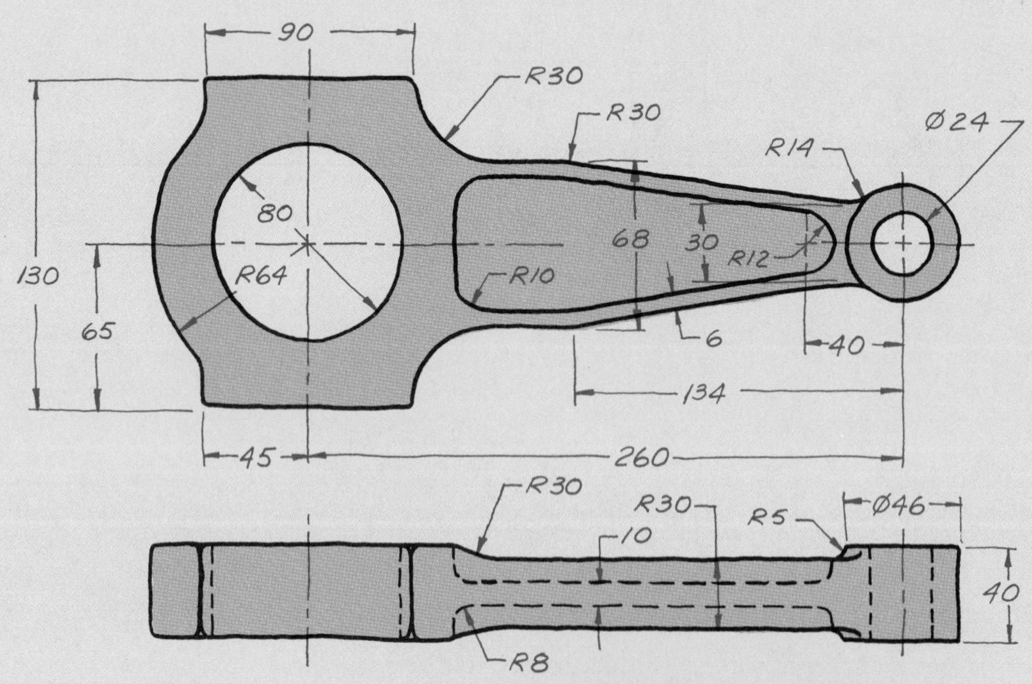

Problem 6.36

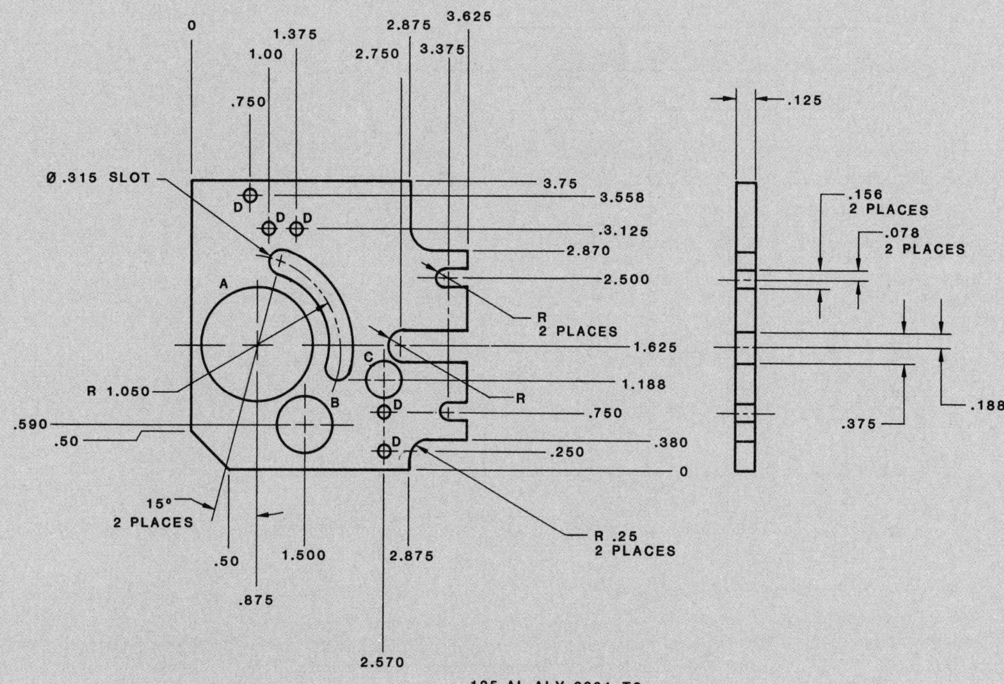

.125 AL ALY 6061-T6

ANODIZE, BLACK

HOLE	DESCRIPTION	QTY
A	Ø 1.552	1
B	Ø .688	1
C	Ø .500	1
D	Ø .149	5

Problem 6.37

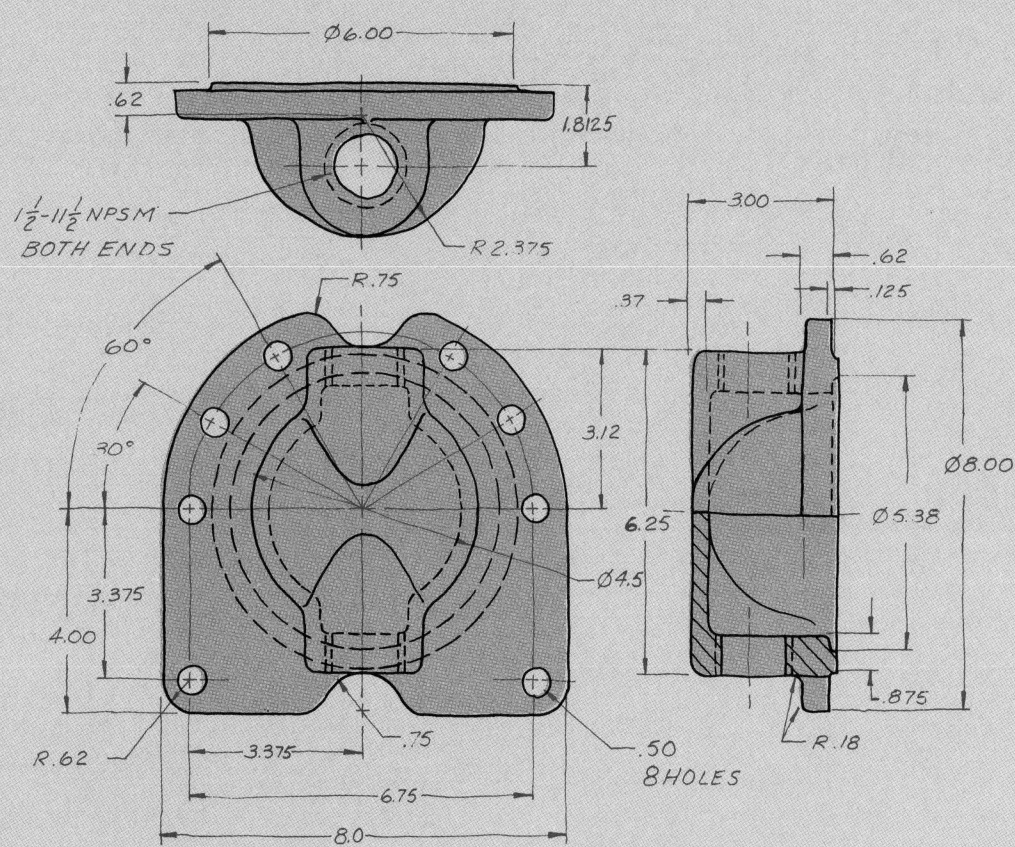

Problem 6.38

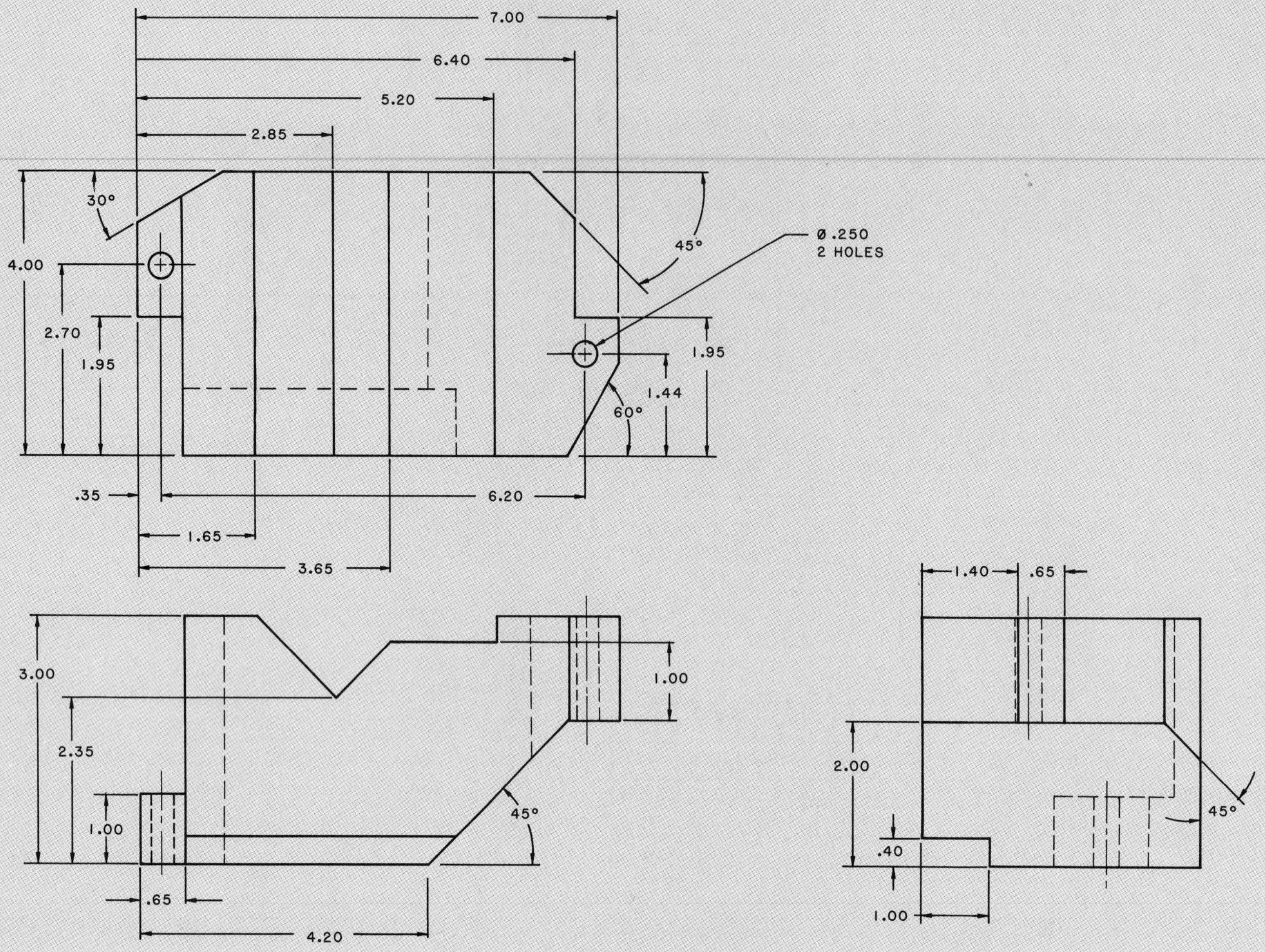

Problem 6.39

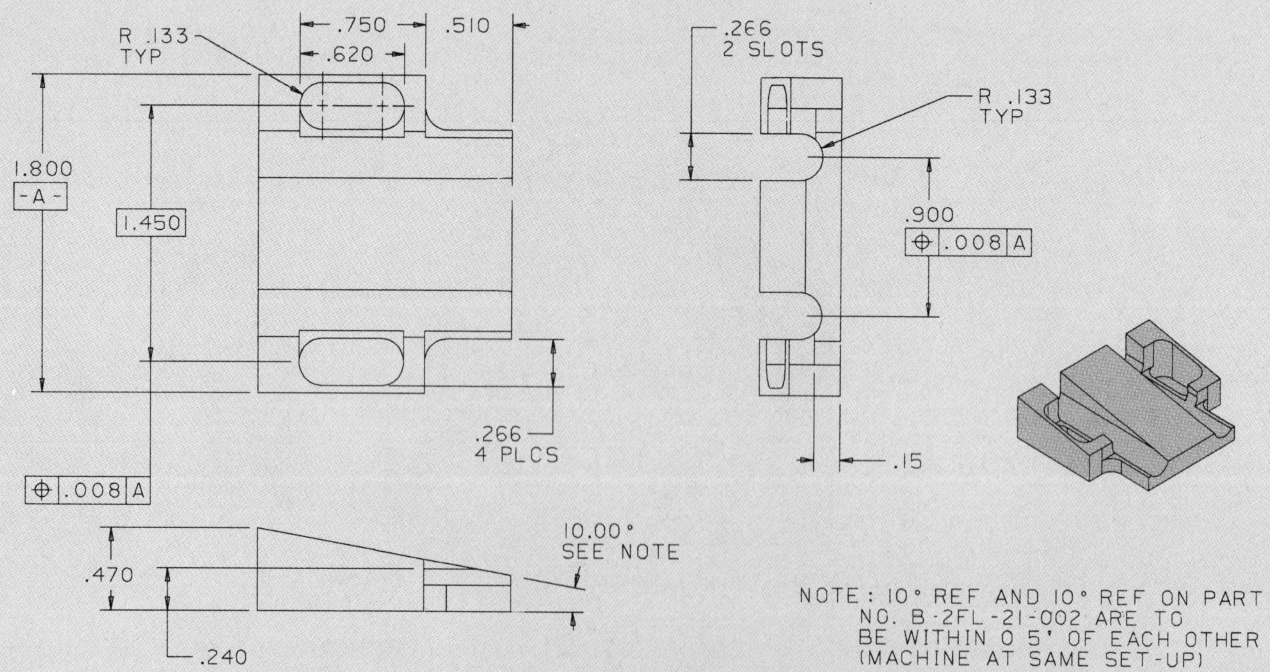

Problem 6.40

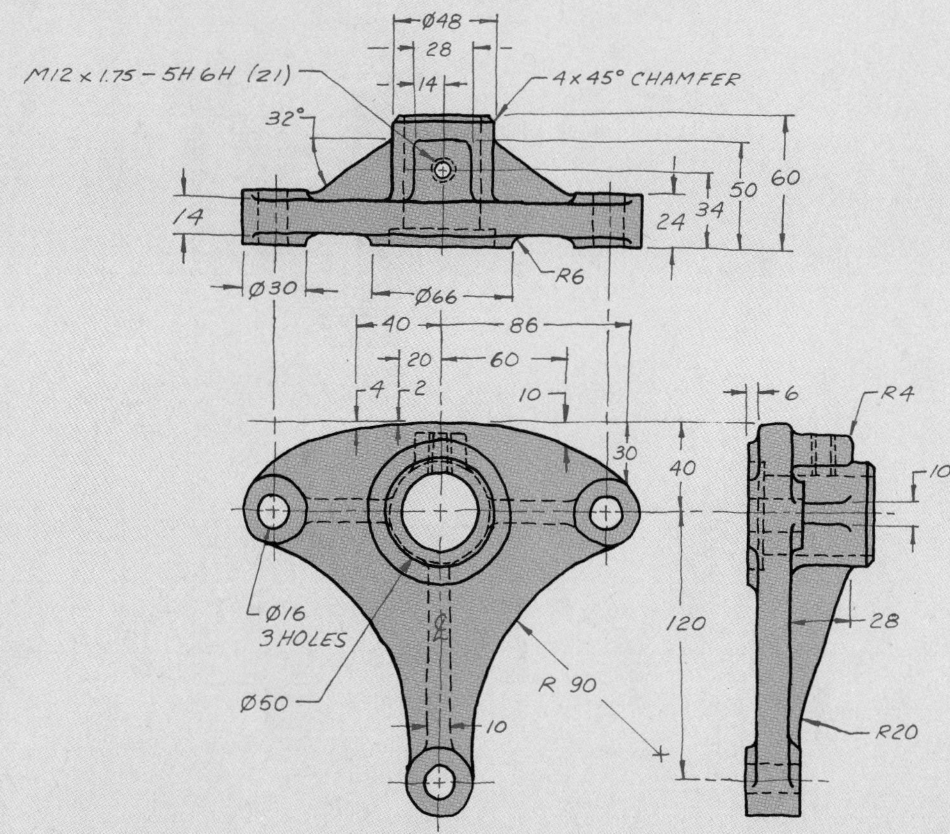

Problem 6.41

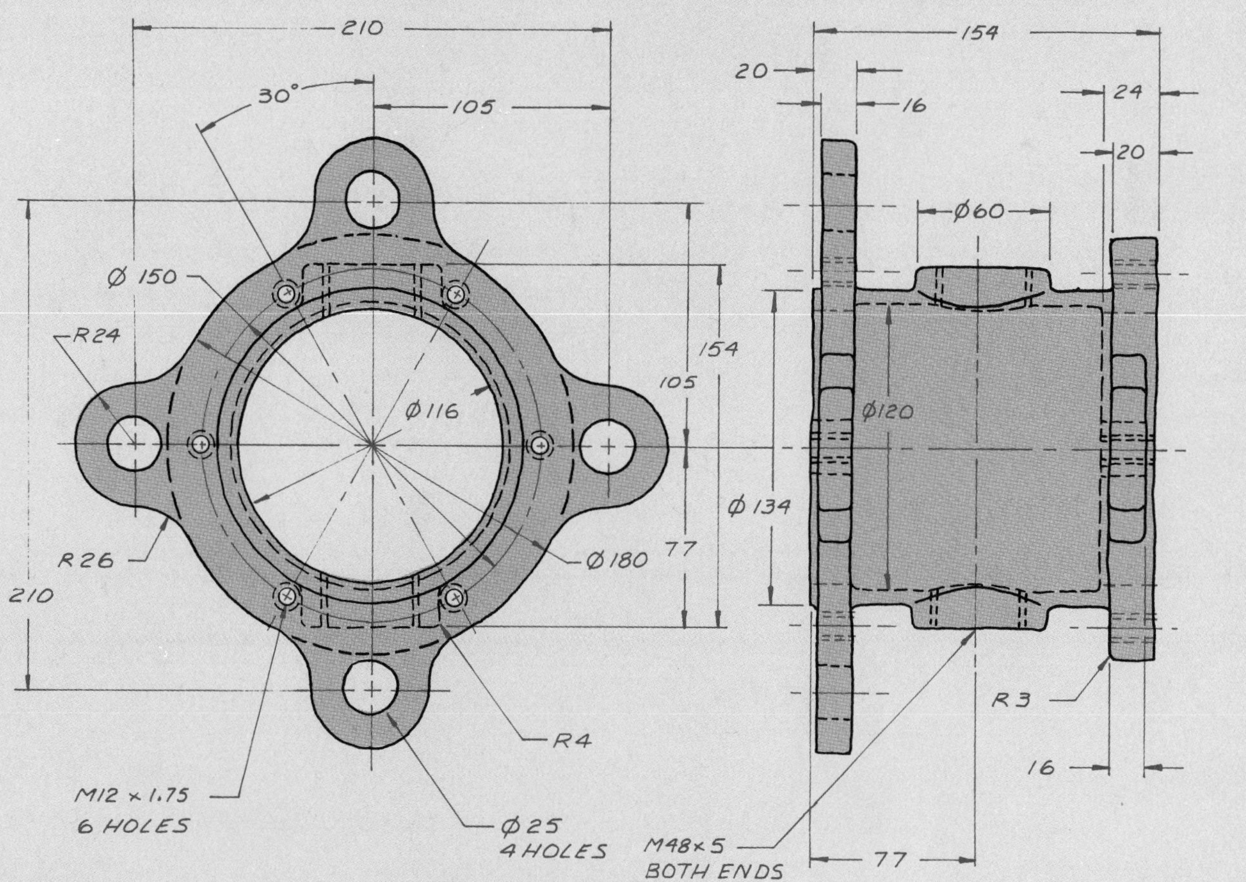

Problem 6.42

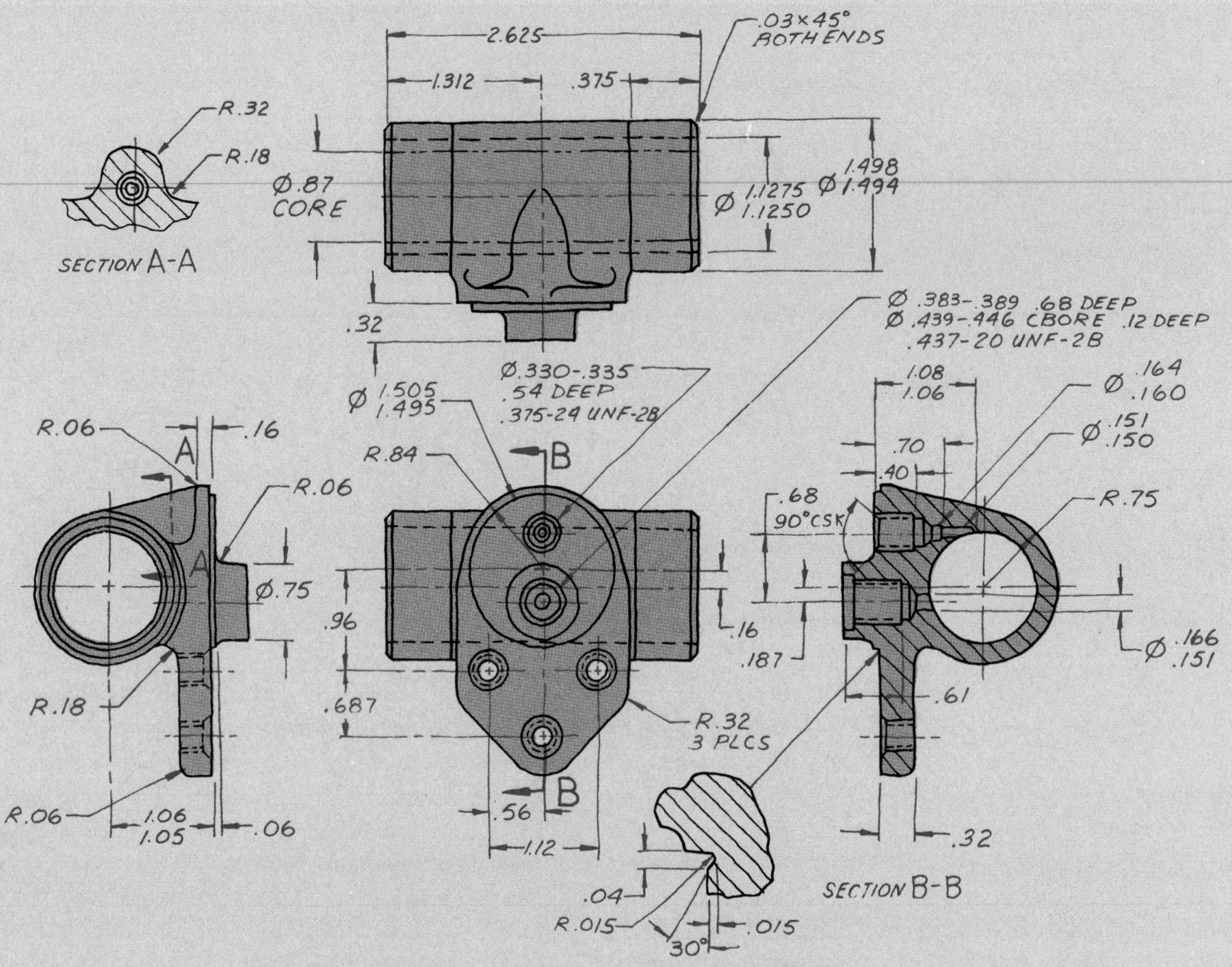

Problem 6.43

Problem 6.43 The three holes in the front view are .312-18 UNC with a countersink of .375 diameter and an angle of 90°. Use the Appendix to look up the top drill size. When dimensioning add the note using symbology.

Problem 6.44 Redraw Figure 6.9; show a side view and make the part 1.00 in. thick. Do not dimension.

Problem 6.45 Redraw Figure 6.10. Do not section or dimension.

Problem 6.46 Draw Figure 6.11. Do not dimension.

Problem 6.47 Draw Figure 6.12. Do not dimension.

Problem 6.48 Draw Figure 6.13. Do not dimension.

Problem 6.49 Draw Figure 6.36. Do not dimension.

Problem 6.50 Draw Figure 6.59. Show only the required views.

CHAPTER 7

Pictorials

Learning Objectives

Upon completion of this chapter you will be able to accomplish the following:

1. Recognize pictorial drawings as single-plane projections.
2. Develop an understanding of the ways in which pictorial drawings may be most useful.
3. Define and possess the ability to produce axonometric, oblique, and perspective drawings.
4. Understand and apply the functions of hidden lines, centerlines, and techniques for dimensioning on pictorials.
5. Develop familiarity with drafting conventions used to illustrate certain part features pictorially.

7.1 Introduction

Pictorial drawings are **single-plane projections** that present the three primary surfaces of the object to the viewer at the same time. Engineers and designers use pictorial sketches to refine and communicate 3D designs before they are formally drawn or modeled. Pictorial drawings are useful in design, construction, production, assembly, service, and sales. In this chapter, the pictorial drawings (Fig. 7.1) commonly used by the engineer, designer, drafter, and illustrator are described.

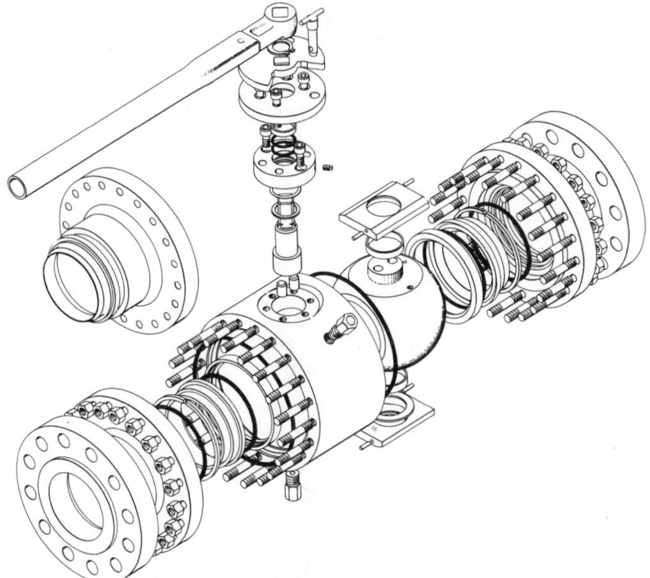

FIGURE 7.1 Technical Illustration of a Ball Valve Used for a Sales Catalog

The choice of pictorial drawing depends on its intended application since they are used in a variety of ways in industry:

- To explain complicated engineering drawings to people who are not trained to read multiview drawings.
- To help the designer work out problems in 3D space, such as clearances and interferences.
- To train new employees in the shop with illustrated training manuals.
- To speed up and clarify assembly processes.
- To transmit ideas, person to person, shop to shop, or sales to purchasing.
- To enhance visualization in education and business.
See Color Plates 1, 12, 13, 16, 36–38.

7.2 Types of Pictorial Drawing

Pictorial drawings may be divided into three groups because they differ from each other in the fundamental scheme of projection (Fig. 7.2): **axonometric**, **oblique**, and **perspective.** Each group is subdivided by varying the relationships between point of sight, plane of projection, and the object. In Figures 7.3, 7.4, and 7.5, the same assembly has been displayed using each group and subdivision. The four subdivisions of axonometric projection are illustrated in Figure 7.3: **isometric projection, isometric drawing, dimetric projection**, and **trimetric projection.** The three subdivisions of oblique projection are illustrated in Figure 7.4: **cavalier, cabinet**, and **general.** The three subdivisions of perspective projection are illustrated in Figure 7.5: **one-point, two-point** and **three-point** projections.

Regardless of the projection method chosen, the view of a

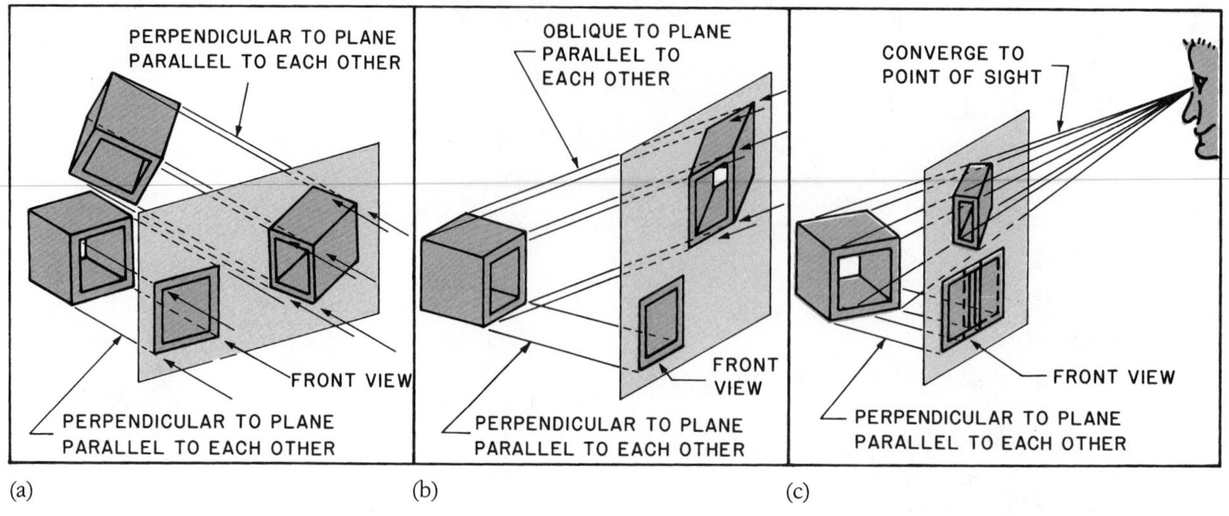

FIGURE 7.2 Kinds of Projection (a) Axonometric (b) Oblique (c) Perspective

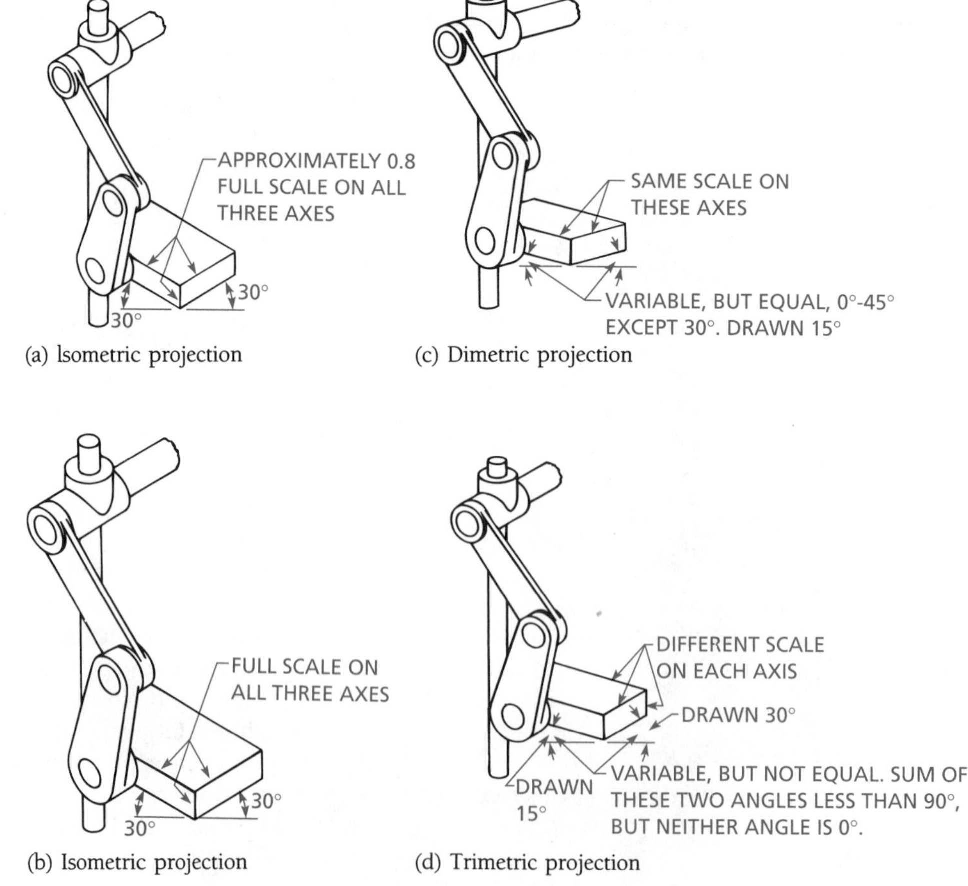

FIGURE 7.3 Types of Axonometric Projection

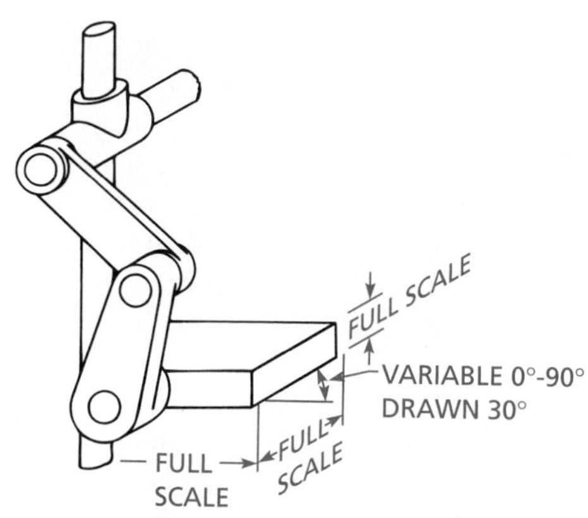

(a) Oblique projection (cavalier)

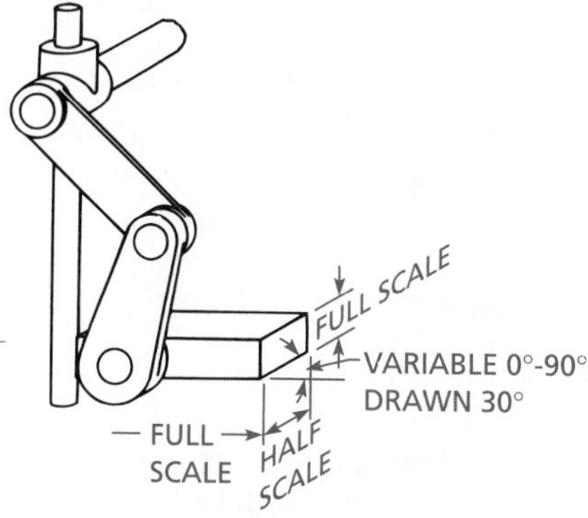

(b) Oblique projection (cabinet)

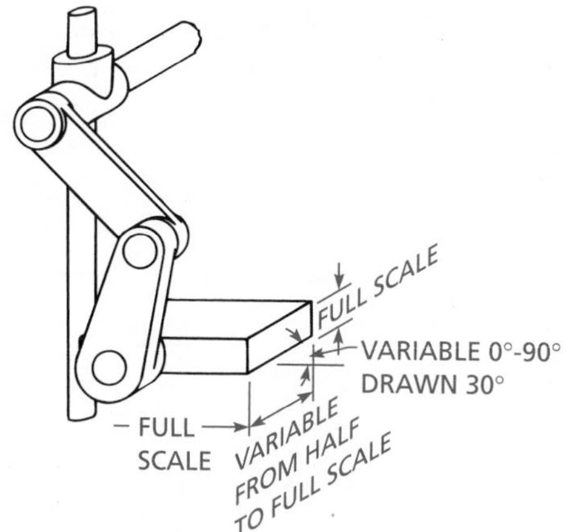

(c) Oblique projection (general)

FIGURE 7.4 Types of Oblique Projection

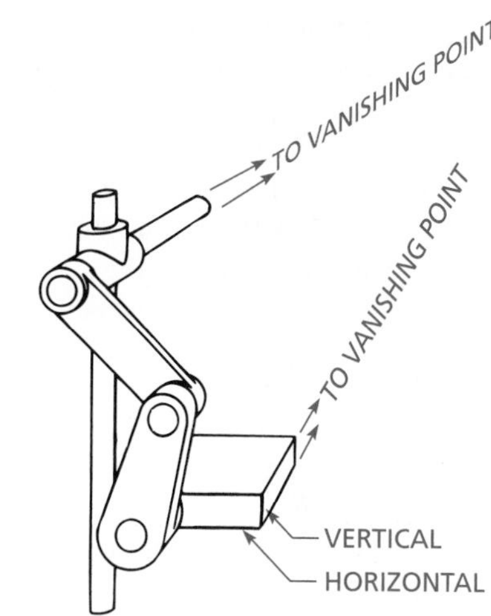

(a) Perspective (one-point)

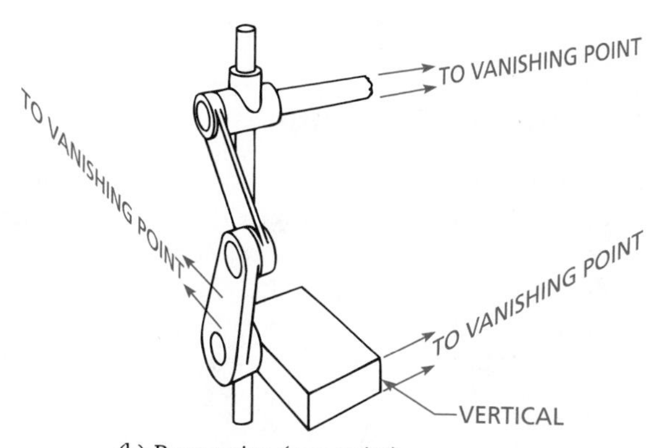

(b) Perspective (two-point)

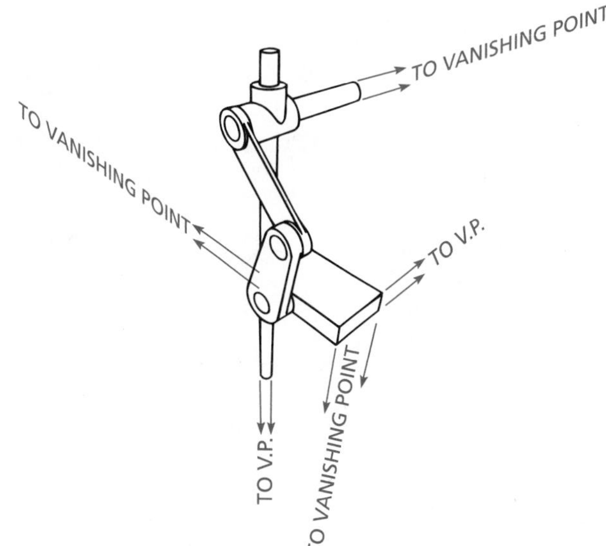

(c) Perspective (three-point)

FIGURE 7.5 Types of Perspective Projection

part that will give the greatest information possible is selected unless natural position or part relationships must take precedence.

7.3 Axonometric Projection

An **axonometric projection** is a projected view in which the lines of sight are perpendicular to the plane of projection, but in which the three faces of a rectangular object are all inclined to the plane of projection. The projections of the three principal axes may make any angle with each other except 90°. The three types of axonometric projections are isometric, dimetric, and trimetric. Isometric is the most common.

An **isometric projection** is a pictorial drawing in which the three principal faces and the three principal axes of the object are inclined equally to the plane of projection. The plane of projection is called the *isometric plane*. The three axes on the drawing make equal angles with each other [Fig. 7.6(a)], but may be placed in a variety of positions. A true orthographic projection of an object on the isometric plane is an isometric projection. The scales on all three axes are equal and foreshortened in a ratio of approximately 0.8 to 1.0. The term *axes* refers to the projections of the principal axes, unless otherwise stated.

A **dimetric projection** [Fig. 7.6(b)] is drawn with two axes making equal angles and the third axis at any selected angle. A **trimetric projection** uses three different scales (one for each axis) and has three different angles for the axes [Figure 7.6(c)]. Although trimetric projection is the most lifelike, it is the most time-consuming and difficult to draw. Most pictorials use isometric projection methods.

7.3.1 Isometric Drawing

The distances on each axis are measured *true length* with any standard scale in **isometric drawing.** This makes an isometric drawing larger than an isometric projection, which is normally 81% of the original in size. This is the isometric form most commonly used.

Both isometric projection and isometric drawing are based on the theory that a cube representing the projection axes is rotated until its front face is 45° to the horizontal plane and then tipped forward or downward at an angle of 35° 16′. The axes make equal angles of 120° [Figs. 7.7(a) and (b)] and the resulting rotation displays all three primary surfaces equally. Figures 7.7(d), (e), and (f) show the isometric cube in three different orientations. All three axes make equal angles with the projection plane and can be drawn easily using 30/60° triangles [Fig. 7.7(c)]. The three faces of the cube are identical in size and shape. The projected lengths of each edge are not foreshortened.

Because isometric drawings are constructed along the three axes, one vertical, and the other two at 30° to the horizontal, *each dimension is measured true length* along an axis. All lines in isometric drawings that are on or parallel to the three axes are drawn true length. Lines not on or parallel to the axes are constructed with offset dimensions.

The orientation of the axes determines what faces of the part are visible. The most typical orientation is shown in Figure 7.7(d), where the top, front, and side of the object are visible. Different arrangements are possible for the isometric axes as long as they remain at 120° to one another.

7.3.2 Isometric Construction

Isometric construction using the **box method** is illustrated in Figure 7.8. The three axes are drawn first, one vertical, one 30° receding to the right, and one at 30° receding to the left [Fig. 7.8(b)]. The edges of the box are constructed from the height, width, and depth dimensions transferred from the multiview drawing of the part [Fig. 7.8(a)]. If the distance is on or parallel to one of the axes, each measurement is full scale. Use a scale to mark dimensions or transfer them with dividers.

Isometric lines are true length lines that are parallel to or on one of the three axes. Lines that are not parallel to or on

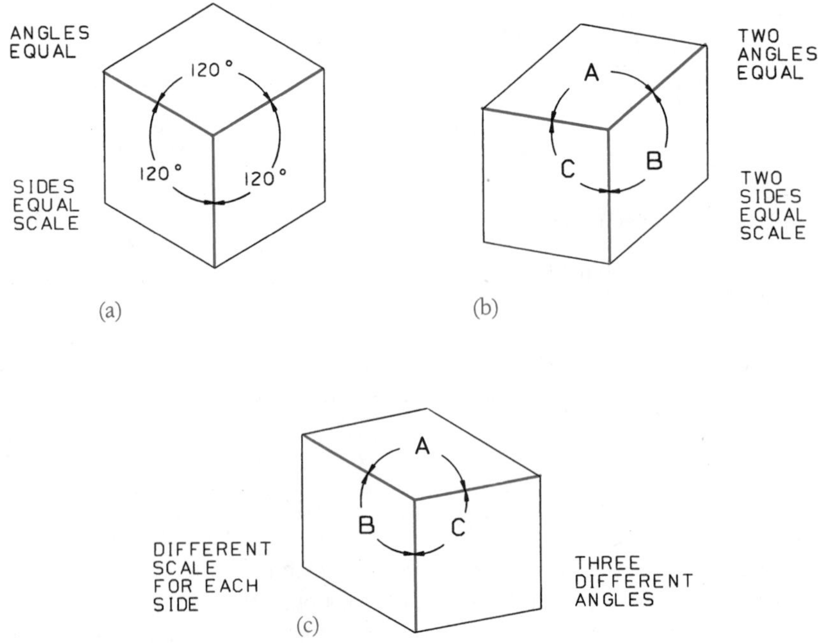

FIGURE 7.6 Axonometric Axes (a) Isometric (b) Dimetric (c) Trimetric

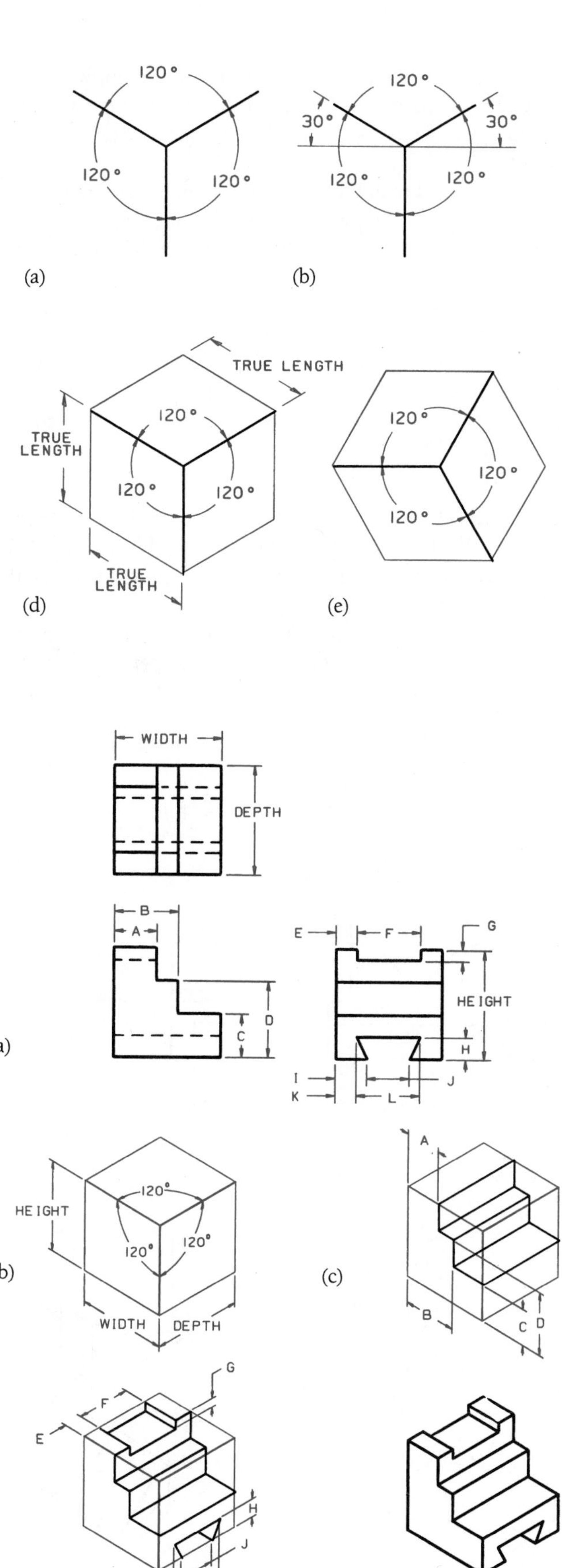

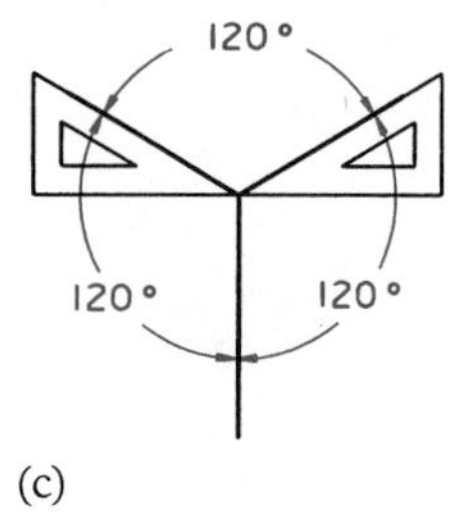

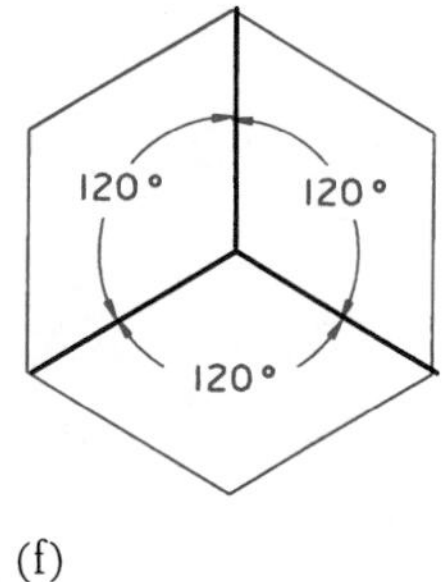

FIGURE 7.7 Isometric Axes
(a) Isometric axes
(b) 30° receding axes
(c) Using 30/60° triangles
(d) Isometric drawing (from top)
(e) Isometric drawing (turned on side)
(f) Isometric drawing (from bottom)

FIGURE 7.8 Isometric Construction Using the Box Method

an axis are called **nonisometric lines** and will not be true length on the isometric drawing. Nonisometeric lines are established from their end points located along isometric lines.

After the part is boxed in, the remainder of the drawing is completed. Dimensions A, B, C, and D are taken from the multiview drawing of the part in Figure 7.8(a) and transferred to the isometric box [Fig. 7.8(c)] to establish the step-like features of the part. All remaining features are established in the same manner [Fig. 7.8(d)]. After the part is complete, it is darkened [Fig. 7.8(e)].

One aid to the process of "blocking in" a part is to use isometric grids. In Figure 7.9, the part was drawn in three views on grid paper and then transferred to the isometric grid. Since the part's features all fell on grid lines, no measurements were necessary. Transferring the part from the three-view drawing to the isometric drawing simply involved counting grid lines.

7.3.3 Nonisometric Lines

The two lines that make the V-shaped feature in Figure 7.10 are not parallel to one of the three axes and are, therefore, **nonisometric lines.** Nonisometric lines cannot be scaled, but their end points are located using the box method and **offset dimensions.** Depending on their orientation, nonisometric lines may become longer or shorter on the isometric drawing. In this figure, the nonisometric lines are at the same angle, but slant from different directions. In the isometric view of Figure 7.10(d), the two lines now make different angles. One is longer than the original line; the other is shorter. This distortion is typical of nonisometric lines. Note that the nonisometric lines are established by locating their end points using

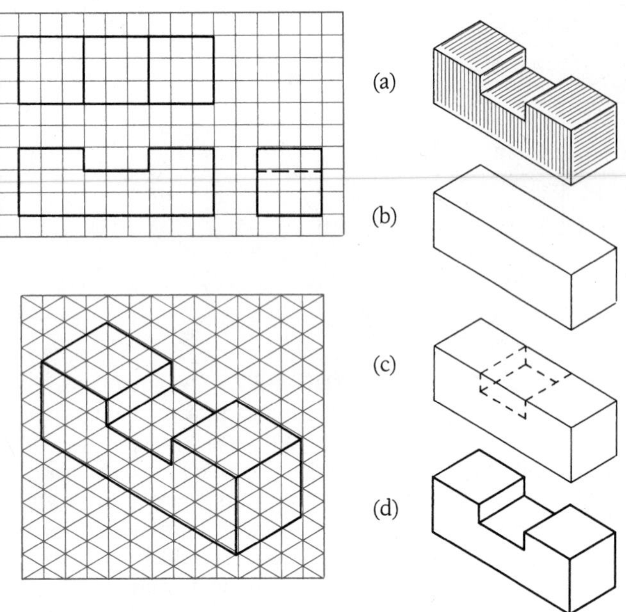

FIGURE 7.9 Box Construction and Grids for Isometric Drawings

offset dimensions [Fig. 7.10(d)]. The angled features are established using offset dimensions I and J.

7.3.4 Isometric Angles

The three major axes along an isometric cube are at 120° to one another, but, in reality, all axes of a cube are 90° or parallel to one another. *Because of the distortion created by the isometric view of the box, few angles appear as true angles.* Angles appear larger or smaller than true size on isometric drawings in relation to their position in the view. The lines that make an angle are nonisometric lines. Angles cannot be measured from the multiview drawing and must be drawn by locating their end points along isometric lines using offset dimensions.

The block in Figure 7.11 has an inclined surface. To draw the part in isometric, it was necessary to use dimensions A, B, and C. Points 1, 2, 3, and 4 are established in the isometric view with these dimensions.

7.3.5 Irregular Objects On Isometric Drawings

Any shape can be drawn in isometric with the box method and offset dimensions. In Figure 7.12, the pyramid has been drawn in isometric using this method. The isometric box is drawn with the three primary dimensions of the part [Fig. 7.12(b)], taken from the multiview drawing [Fig. 7.12(a)]. Using offset dimensions A and B, the base is established first (points 1, 2, and 3). Point 0 is located with offset dimensions C and D, as shown in Figure 7.12(c).

7.3.6 Circles and Arcs on Isometric Drawings

All circles and circular arcs on isometric drawings appear elliptical. Many methods are available to construct isometric ellipses: template, trammel, four center, and point plotting. A template should be used for instrument drawings whenever possible. For sketches, freehand techniques are normally sufficient.

If templates are not available, the trammel and point plotting methods are the most accurate, but are time-consuming and should be used only when the other methods are inadequate.

The four-center method, shown in Figure 7.13, does not create a perfect ellipse, but is accurate enough for most purposes. This method can be used to draw circles or portions of circles (arcs) on any isometric face/plane (Fig. 7.14). The following steps describe the construction of an isometric ellipse (Fig. 7.13):

1. Lines DA and DC are drawn along the two receding axes (at 30°). Line AB is parallel to DC, and line CB is parallel to AD. Each of the lines will be the same length as the diameter of the circle [Fig. 7.13(a)].
2. Construction lines are drawn from point D perpendicular to line AB at its midpoint and perpendicular to line CB at its midpoint [Fig. 7.13(b)].

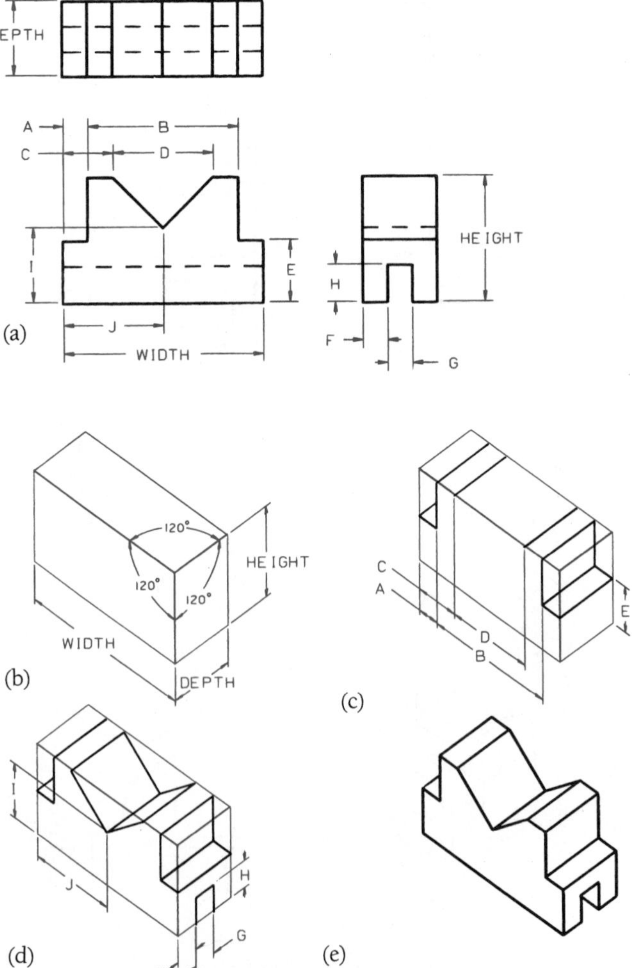

FIGURE 7.10 Nonisometric Lines in Isometric Drawing

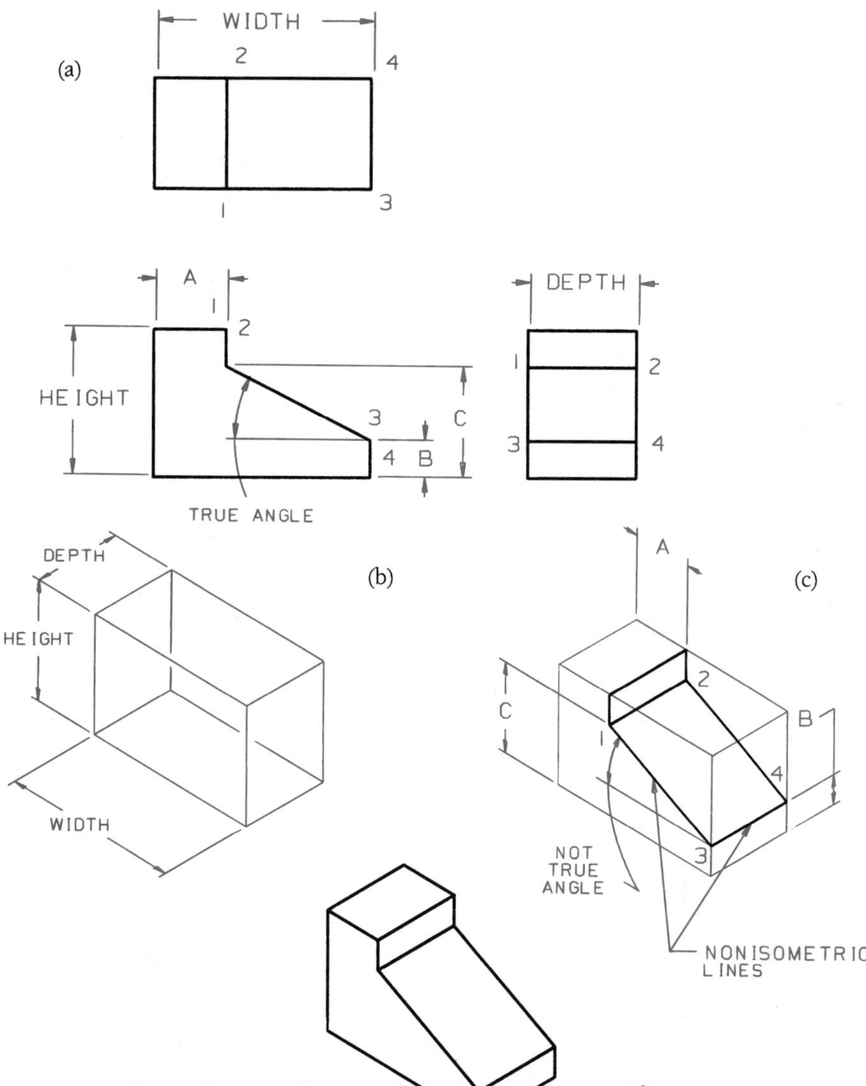

FIGURE 7.11 **Offset Dimensions for Isometric Construction**
(a) Three-view drawing
(b) Block-out overall dimensions
(c) Establish secondary features
(d) Completed isometric

3. Step two is repeated using point B and lines DA and DC [Fig. 7.13(c)].
4. The intersection of the construction lines is used to draw R1. The radius is equal to the distance from the intersection of the construction lines to one of the numbered points (1, 2, 3, or 4). The radius will thus be tangent to two edge lines (AB and AD, or CB and CD) [Fig. 7.13(d)].
5. Point D and point B are used to draw arc R2. Arc R2 originates at the intersection of the construction lines for both sides of the ellipse (B or D). Radius R2 will be tangent to two sides each (BA and BC, or DA and DC) [Fig. 7.13(e)].
6. All construction lines are erased and lines are darkened [Fig. 7.13(f)].

Arcs are required for parts that have fillets and rounds. The same procedure for construction is used for these cases. In Figure 7.15, the **round** has been constructed as a portion of an isometric ellipse. In Figure 7.15(a) the round is shown as a true shape. The ellipse is boxed in using two times the radius as each side for construction lines [Fig. 7.15(b)]. The radius of the bend is located to establish the tangent points [Fig. 7.15(c)]. Since one-quarter of the circle (ellipse) is being drawn, only one radius is necessary [Fig. 7.15(d)]. Unless the ellipse is an odd size or very large, a template is used for construction [Fig. 7.15(e)]. The major axis of the ellipse will be at 30° unless the curve falls in the top or bottom face of the part.

7.3.7 Using Offset Dimensions for Ellipse Construction

The **offset dimension** method locates a series of points along the curve of an ellipse. This method is more accurate than the four-center method, but is time-consuming and the quality of

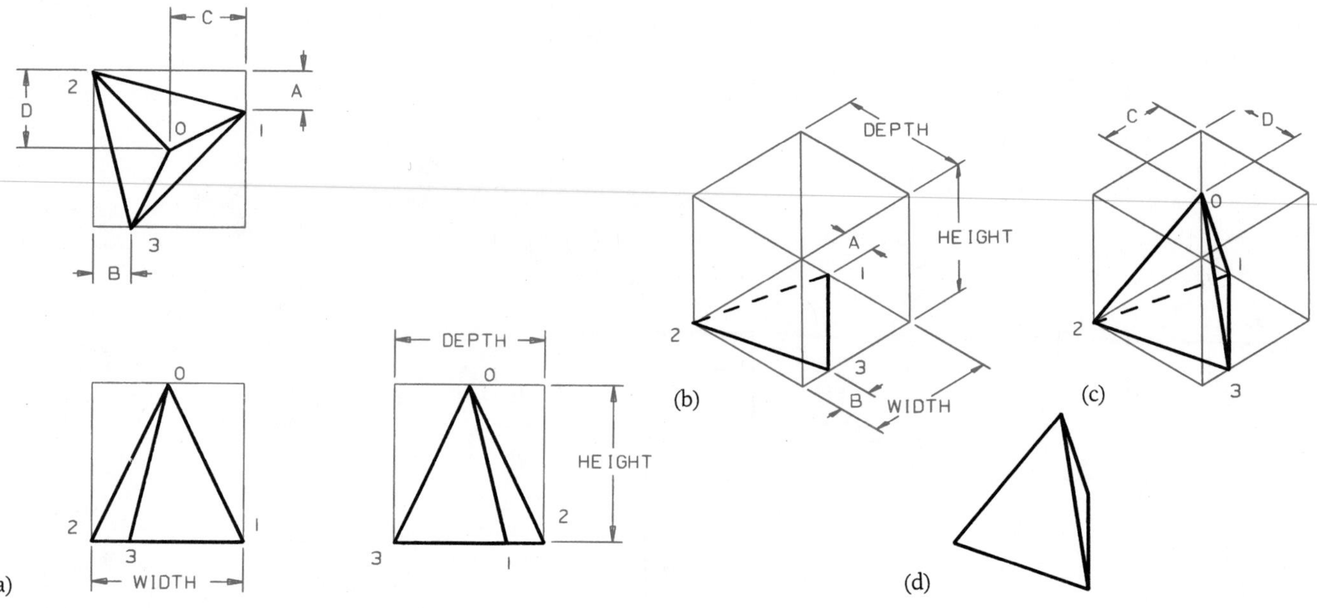

FIGURE 7.12 Construction of Irregular Objects in Isometric Drawing
(a) Three-view drawing
(b) Block-out height, width, depth, and base features
(c) Locate apex
(d) Darken in isometric

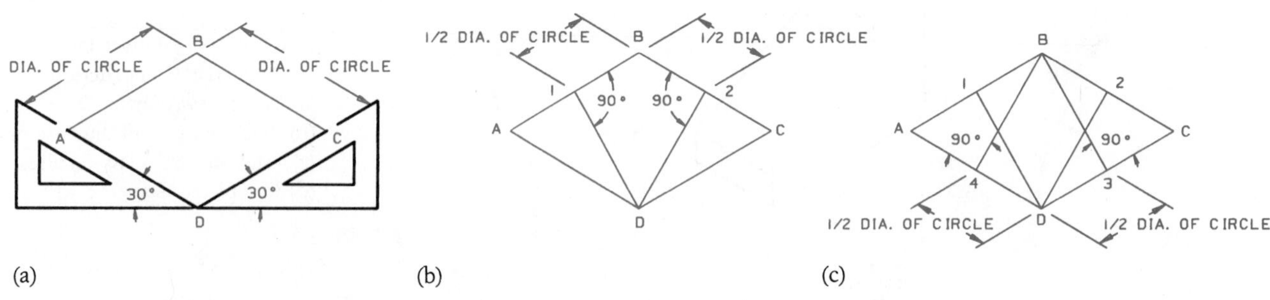

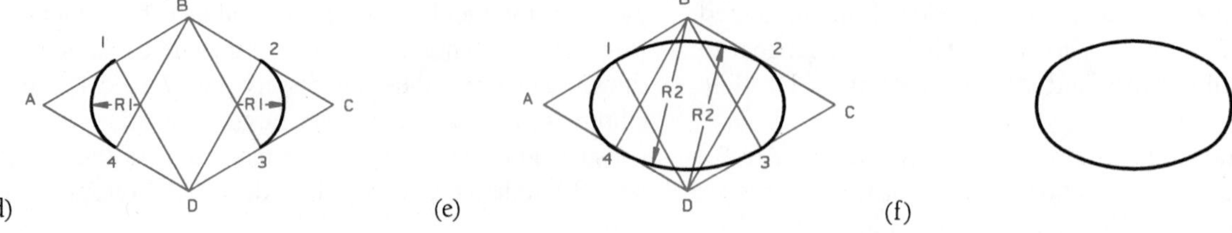

FIGURE 7.13 Drawing a Four-Center Ellipse
(a) Block-out overall size of ellipse
(b) Find midpoint of lines A-B and B-C
(c) Find midpoint of lines A-D and D-C
(d) Swing R1 arcs
(e) Swing R2 arcs
(f) Erase construction lines and darken ellipse

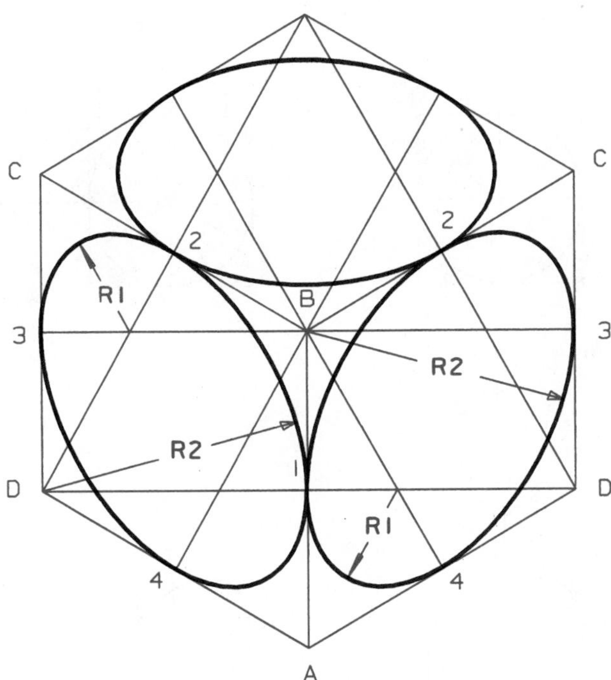

FIGURE 7.14 Four-Center Ellipses on Surfaces of an Isometric Cube

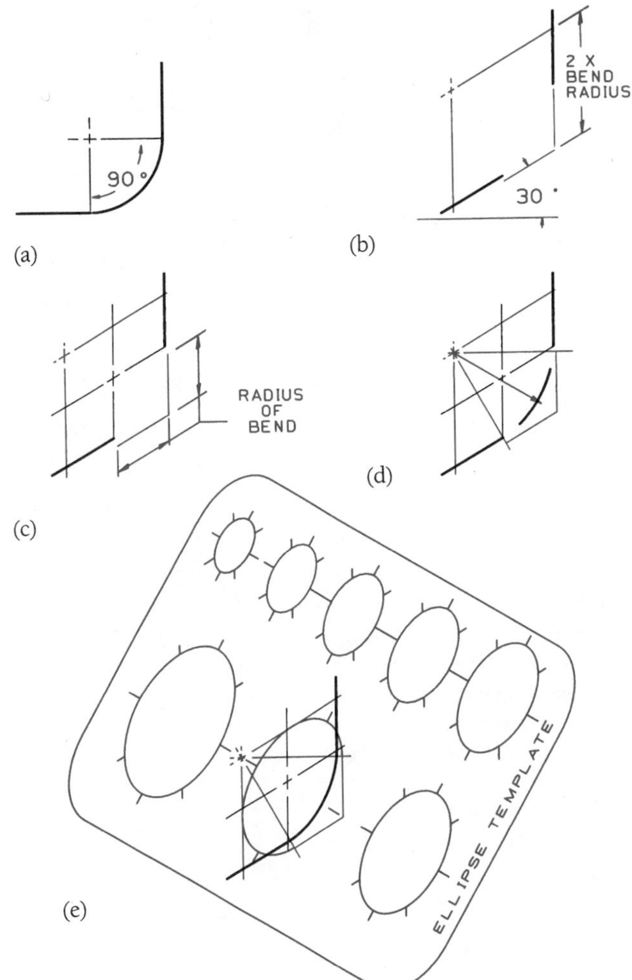

FIGURE 7.15 Construction of Arcs in Isometric Drawings

the finished curve depends on the drafter's skill. Two versions are described here.

The first method (Fig. 7.16) divides the circle evenly by drawing equally spaced construction lines emanating from the center of the circle to where they intersect the circle's circumference. The circle is drawn first along with its centerline [Fig. 7.16(a)]. Then, the circle is boxed in [Fig. 7.16(b)]. Equally spaced construction lines are drawn from the center of the circle to the circle's circumference [Fig. 7.16(c)]. The number of equally spaced lines depends on the desired accuracy level. The more points that are established on the circumference, the more accurate the ellipse. Here, a 30° spacing was used to establish 12 evenly spaced points along the circumference (points 1 through 12). The box shape is drawn in isometric [Fig. 7.16(d)], and each of the points is transferred from Figure 7.16(c) using offset dimensions. Dimensions D and C establish point 1. Dimensions A and B establish point 2. Points 3 and 12 are located at the tangent points of the circle and the box and are established on the isometric view at the intersection of the box and the centerline. To complete the ellipse, each of the four quadrants is drawn using the same method [Fig. 7.16(e)]. Because quadrants 1 and 3 are the same as are quadrants 2 and 4, only two quadrants need be established. The opposite side can be mirrored. The darkened finished ellipse is shown in Figure 7.16(f).

The second method (Fig. 7.17) is similar, but the points are arbitrarily fixed along the circumference of the circle and offset dimensions are taken [Fig. 7.17(c)]. The steps in Figures 7.17(a), (b), (d), (e), and (f) are the same as the first method. The points should be located to give a sufficient number of locations on the circumference to establish a smooth curve.

Circles, arcs, or curves that do not lie in isometric planes must be plotted with offset dimensions. A series of points must be established along the curved outline in this procedure. Offset dimensions for each point are transferred to the isometric drawing and marked off along isometric lines.

7.3.8 Curves on Isometric Drawings

A **space curve** can be constructed using offset dimensions and box construction. The methods used are not much different from those used to draw space curves on multiview drawings. The difference is in the use of receding axes drawn at 30°. Otherwise, all measurements are marked off the same. The offset method can be used for any shape. Points are located along the curve and their positions are transferred to the pictorial with dividers or a scale. A sufficient number of points is established to describe the curve accurately. After the points are located on the pictorial, a light curve is drawn freehand through the points, and an irregular curve is used to draw the curve with the appropriate line thickness.

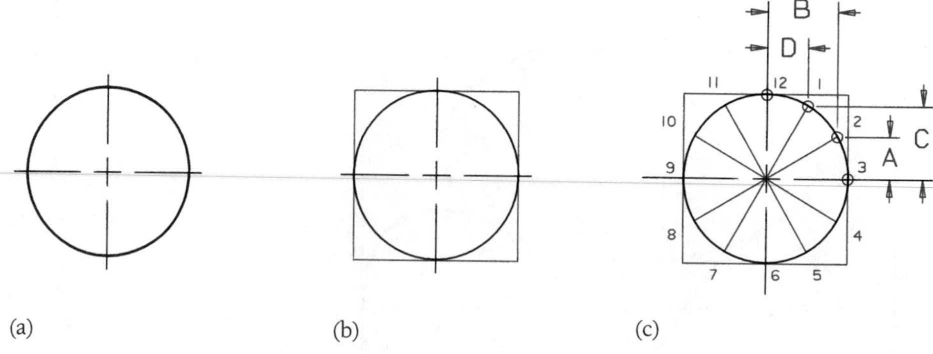

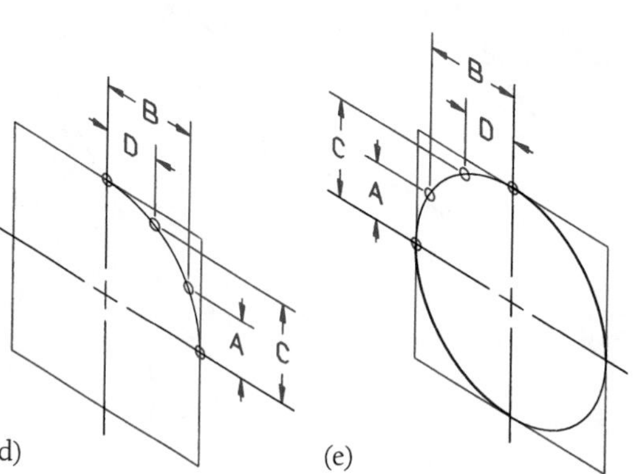

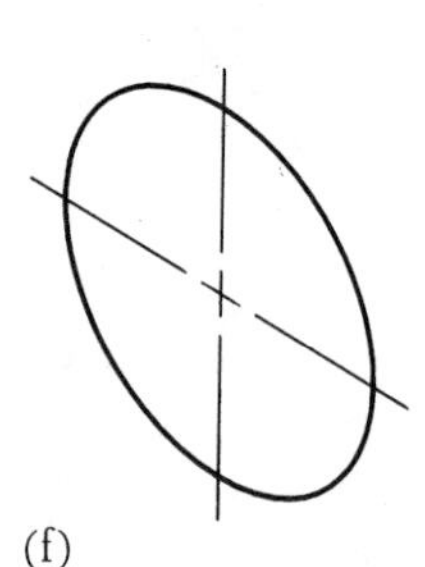

FIGURE 7.16 Construction of a True Ellipse Using Offset Dimensions (a) Draw circle and centerline (b) Draw square (c) Divide circle and establish points (d) Locate points along curve (e) Locate remaining points (f) Darken ellipse

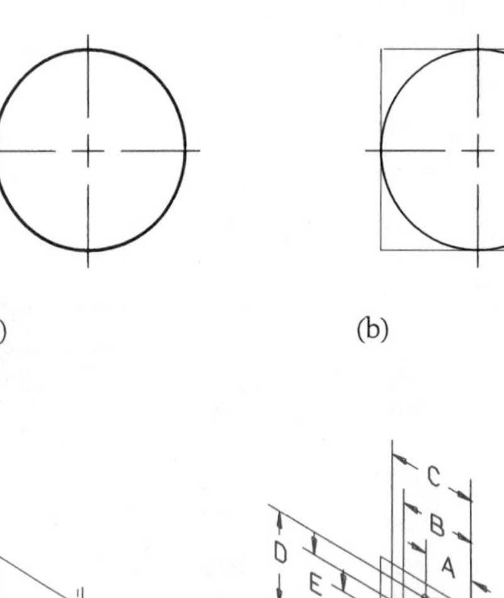

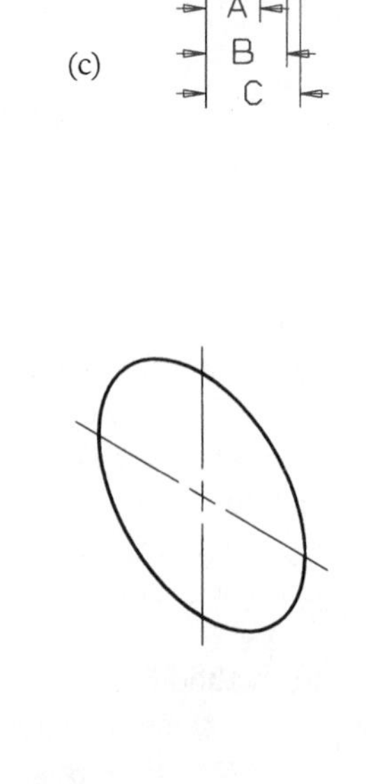

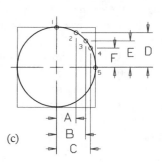

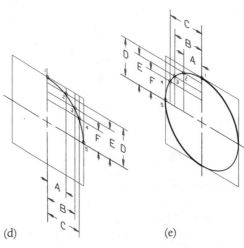

FIGURE 7.17 True Ellipse Construction
(a) Draw circle and centerline
(b) Draw square (c) Establish points along circumference of circle
(d) Locate points along curve
(e) Locate remaining points
(f) Darken ellipse

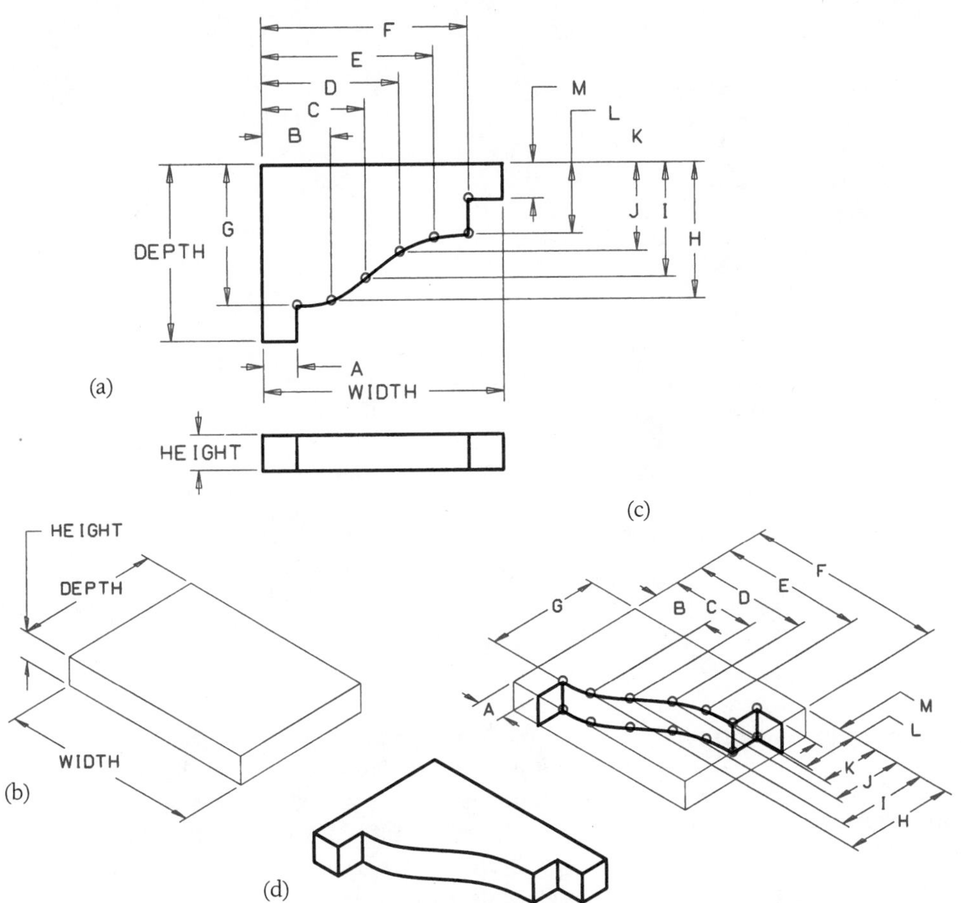

FIGURE 7.18 **Space Curves in Isometric Drawing** (a) Locate points along curve in two-view drawing (b) Block-out overall dimensions (c) Locate points (d) Erase construction lines and darken lines

In Figure 7.18(a), the top and the front view of the part are drawn first. Points are then established along the curve in the top view. Offset dimensions A through M are located from the edges of the part. The height, width, and depth are used to box in the part [Fig. 7.18(b)]. Using offset dimensions A through M, points 1 through 7 are located from the edges of the part [Fig. 7.18(c)]. All point dimensions are taken parallel to their corresponding axis (edge). Point 1 is located by dimensions A and G; point 2 by dimensions B and H; point 3 by dimensions C and I. Vertical lines are drawn from points 1 through 7 to establish the part's thickness [Fig. 7.18(c)]. The pictorial is completed by erasing the construction lines and darkening the outline [Fig. 7.18(d)].

7.3.9 Hidden Lines on Isometric Drawings

Hidden lines are omitted on most pictorial illustrations unless they are required for clarification of interior features. When an illustration requires hidden lines, a sectioned pictorial should be considered. In Figure 7.19, three variations of a pictorial drawing are shown. The part is shown with all hidden lines in Figure 7.19(a). In Figure 7.19(b), only visible lines are shown. In the last variation, the part is shown as it would appear modeled in a 3D CAD wireframe model with the hidden lines shown. CAD systems can display the part with or without hidden lines.

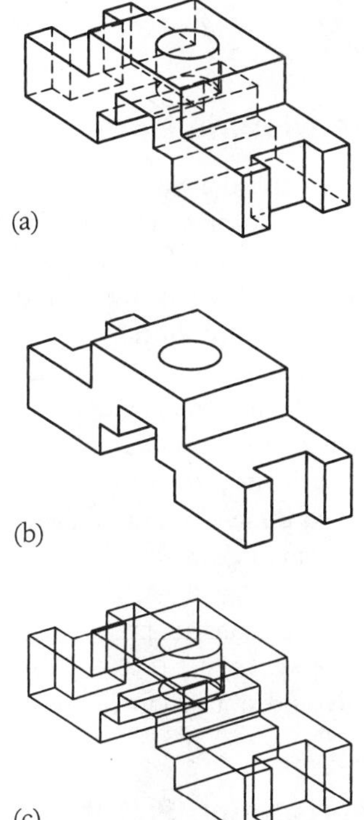

FIGURE 7.19 **Hidden Lines in Isometric Drawings** (a) Part with hidden lines (b) Part with hidden lines removed (c) Wireframe model of part of 3D CAD system

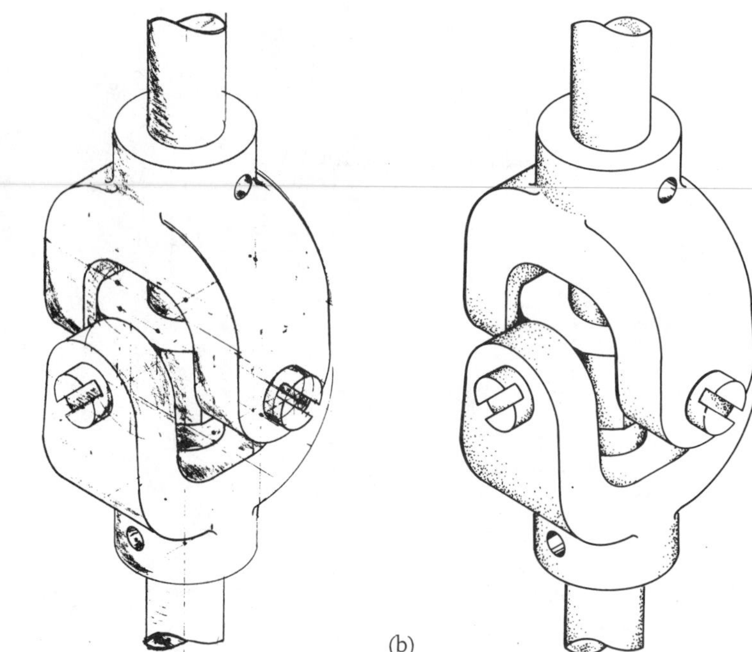

FIGURE 7.20 Pictorial Illustration Construction
(a) Construction of universal joint in an isometric drawing
(b) Pictorial illustration of universal joint

7.3.10 Centerlines on Isometric Drawings

Many pictorials do not show **centerlines** so the part looks more realistic. However, if dimensions are required on the drawing, centerlines are usually included.

During the construction of a pictorial, centerlines identifying the origin of a part's symmetrical or curved features are as necessary as any other construction line [Fig. 7.20(a)]. In most cases, they are erased after the pictorial is constructed [Fig. 7.20(b)].

7.3.11 Dimensioning Isometric Pictorials

Dimensions on isometric drawings can be either aligned or unidirectional. **Aligned dimensions** look pictorially correct but are harder to draw. **Unidirectional dimensions** and notes are easier to add to the illustration and are many times either typeset or labeled mechanically.

In Figure 7.21, the part shown is dimensioned using both methods. In aligned dimensioning, guidelines for the lettering must be drawn parallel to the item being dimensioned and in the isometric plane of the face being dimensioned. The arrowheads for aligned dimensions are drawn with their backs parallel to the extension line as shown in Figure 7.21(a) (upper right).

Unidirectional dimensions are positioned horizontally and are, therefore, easy to construct [Fig.7.21(b)]. Guidelines are drawn horizontally and dimensions are added with vertical lettering.

7.3.12 Dimensioning Pictorials

Any type of pictorial can be dimensioned. The rules of dimensioning used in conventional multiview drawings are applied whenever possible.

For all pictorials, the dimension lines, extension lines, and the dimension text (unless unidirectional lettering is used) should lie in the same plane as the line or feature being dimensioned [Fig. 7.21(a)]. Arrowheads should be long and narrow, with a ratio of 3:1, and should lie in the plane of the dimension and extension lines [Fig. 7.21(a)]. For unidirectional dimensioning, use vertical letters that can be read from the bottom of the sheet [Fig. 7.21(b)].

7.3.13 Dimetric Projection

An axonometric projection in which two sides and two of the axe s of a rectangular object make equal angles with the plane of the projection, while the third face and the third axes make a different angle, is called **dimetric projection** [Fig. 7.3]. Two of the angles on the drawing between the axes are equal, but the third angle is different. A variety of positions of the axes may be used. A dimetric projection may be constructed by scaling along the axes using two different scales under certain conditions. These scales change whenever the angles of the axes change. Most methods and rules for isometric drawings can also be applied to dimetric projection.

7.3.14 Trimetric Projection

An axonometric projection in which all three faces and three axes of a rectangular object make different angles with the plane of the projection is **trimetric projection** [Fig. 7.3]. The angles on a trimetric drawing between the axes are all different, and may be placed in a variety of positions. Under certain conditions, a trimetric drawing may be constructed by scaling along the three axes using three different scales. Each scale changes whenever the angle of the axis changes.

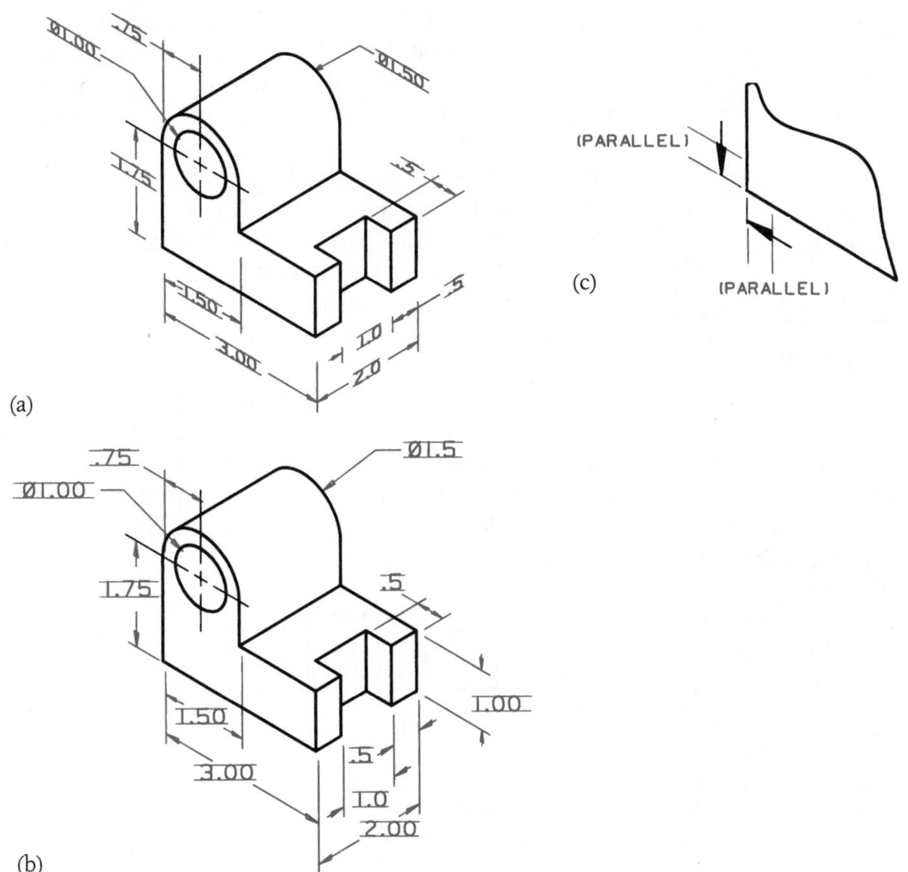

FIGURE 7.21 **Dimensioning Isometric Drawings**
(a) Aligned (b) Unidirectional
(c) Arrowhead construction

The choice of axes in axonometric projection depends on the part. To avoid the appearance of distortion on large flat areas, the angle which that face makes with the picture plane is increased. For more important faces where details must be shown more clearly, the angle between that face and the picture plane is decreased. Since the horizontal plane is less distorted and more detail can be seen in the vertical face, Figure 7.22(b) is clearer than (a).

You May Complete Exercises 7.1 Through 7.4 at This Time

7.4 Oblique Projection

A projected view in which the lines of sight are parallel to each other but inclined to the plane of projection is called an **oblique projection** (Fig. 7.4). The principal face is placed parallel to the plane of projection, making it and parallel faces show true shape. In all forms of oblique projection, the receding axis may be drawn in any direction (Fig. 7.23). Change the axis angle and choice of front face to obtain any orientation required to exhibit the part properly and clearly.

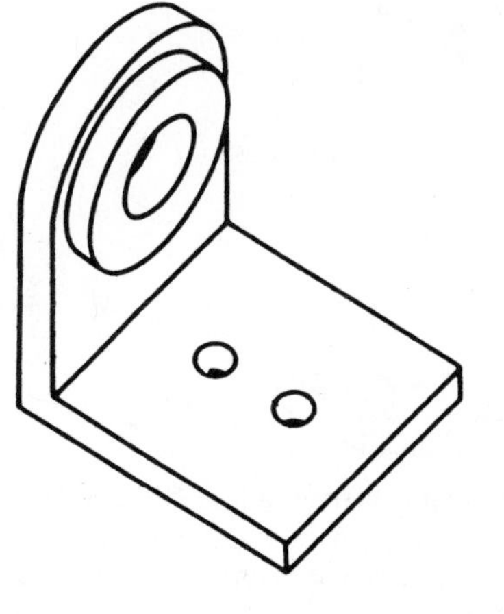

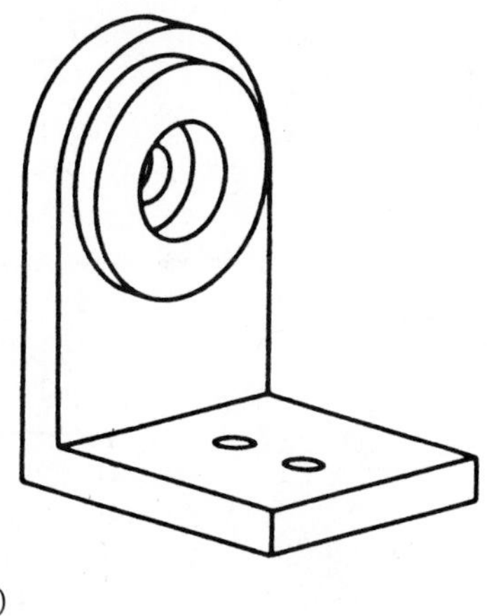

FIGURE 7.22 **Choice of Axonometric View Based on Part Distortion**

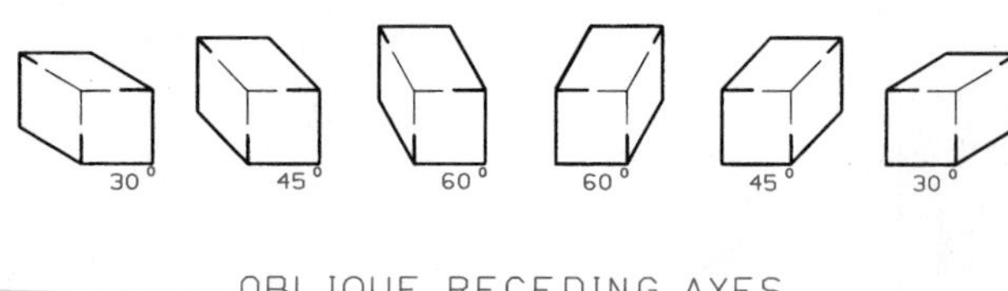

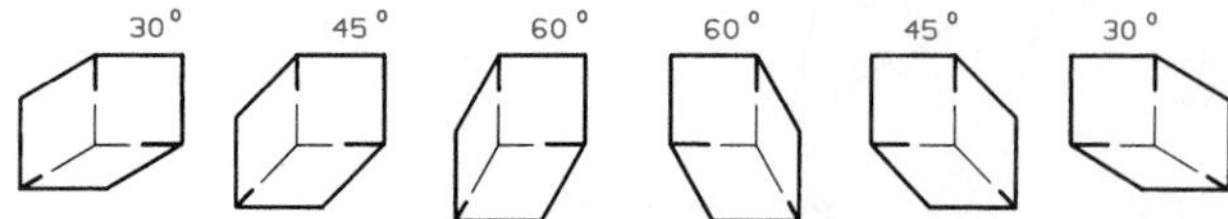

FIGURE 7.23 **Axis Choice for Oblique Projection**

Oblique drawings are similar to isometric drawings. However, they are produced from parallel projectors that are not perpendicular to the projection plane. The primary difference lies in the use of only one receding axis and the ability to draw one surface as true shape and size in the front plane.

There are three versions of oblique projection that differ only in the comparative scales used along the receding axis and the angle of the receding axes (Fig. 7.24). An oblique projection on which the lines of sight make an angle of 45° with the plane of projection is called a **cavalier projection** [Fig. 7.24(a)]. The front is drawn full scale and true shape. The same scale is used on all axes. Therefore, the receding faces are drawn full scale (but not true shape). An oblique projection in which the lines of sight make an angle of between 63° to 26° with the plane of projection is called a **cabinet projection** [Fig. 7.24(b)]. The scale on the receding axis is *one-half* of the scale on the other axes. An oblique projection in which the lines of sight make any angle other than 45° or 63° to 26° is called a **general oblique.** The scale on the receding axis should be between full scale and one-half scale of the horizontal and vertical axes [Fig. 7.24(c)]. The choice of the receding angle (1° to 89°) is determined by the shape of the object and the most descriptive view orientation.

The distortion often noticeable in oblique projection may be decreased by reducing the scale on the receding axis. Cyl inders and cones should have their axes on the receding axis to reduce distortion and to make it possible to draw circles.

Oblique projection is commonly used for objects that have a series of circles, curves, or irregular outlines in the same or parallel planes (Fig. 7.25). By placing curved outlines in the front face, they are drawn true shape and full scale without

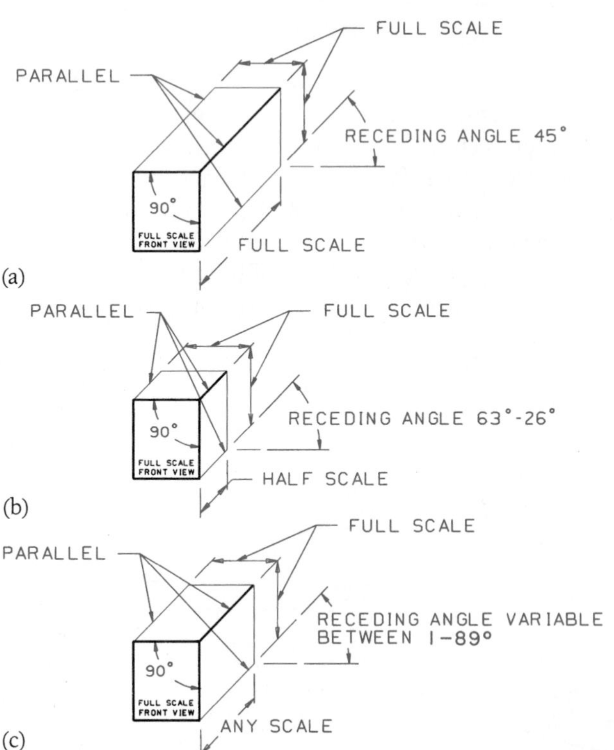

FIGURE 7.24 **Types of Oblique Projection**
(a) Cavalier projection (b) Cabinet projection (c) General oblique

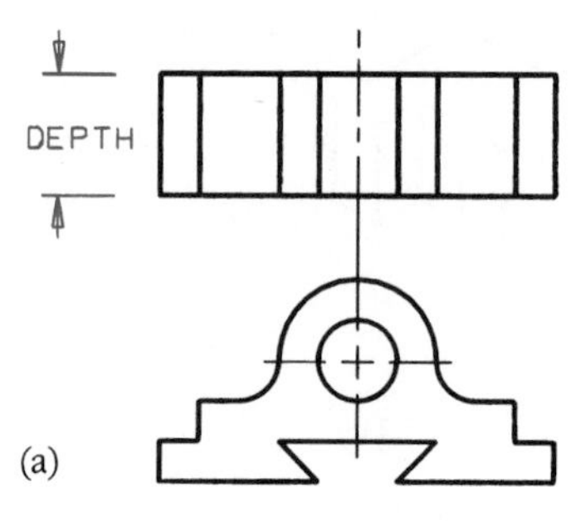

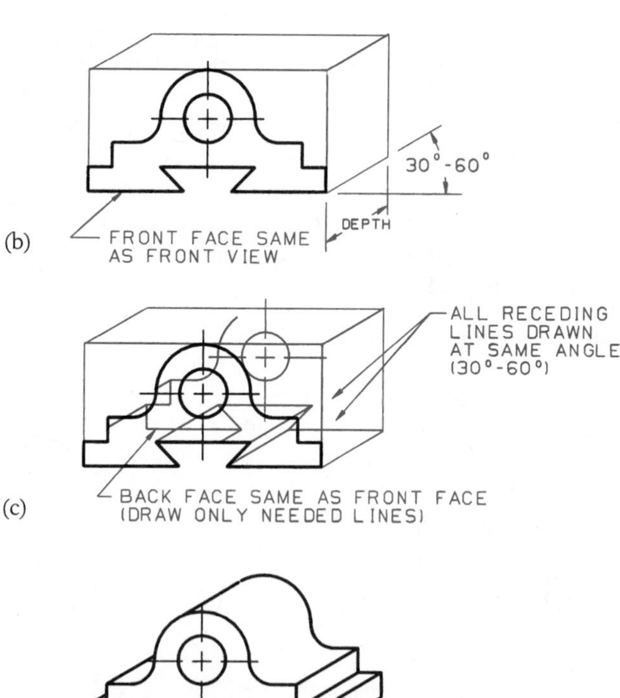

FIGURE 7.25 **Step-by-Step Construction of Oblique Drawing**
(a) Two-view drawing (b) Draw front face and establish receding angle (c) Establish rear (back) face and receding lines (d) Erase construction lines and darken

distortion. A standard circle template or a compass is used. The front face of an oblique projection is exactly the same as the front view of a part drawn in a multiview projection.

7.4.1 Oblique Construction

Objects drawn with oblique projection are oriented so that the surface with curved lines lies in the front plane bounded by the axes that are at 90°. All surfaces that lie on the front plane or that are parallel to it are drawn true shape and size.

Oblique construction is started by blocking in overall dimensions—establish the width and the height of the front and the rear face. Next, the depth of each face is established and then the part's edges are constructed. Circular features are then located and their dimensions blocked in. Last, the part is darkened. Care should be taken to ensure that the side of the part with the most curved edges is oriented to the front view. Figure 7.25 shows four steps in this process.

The front face of the part in Figure 7.25 (with all curved features) is drawn true shape and size, and all measurements on this front face are true length. The measurements are taken from the front view in Figure 7.25(a) and transferred to the view in Figure 7.25(b). The most commonly used angle for the receding axis is 45°. After the angle is determined, the rear face of the part is blocked in. The features of the rear face are then drawn [Fig. 7.25(c)]. The front and the rear faces are parallel and identical. Only the portions of it that show along the receding face must be drawn. The receding edge lines are drawn parallel to the receding axis between the corners and tangent to the curved feature on the top edge [Fig. 7.25(c)]. The part is completed by darkening in the visible edges [Fig. 7.25(d)].

7.4.2 Using Offset Measurements for Oblique Drawings

When the object to be drawn is placed so that the curved features do not fall in the front face or two or more faces have curved features, offset measurements must be used to establish the curves. After the feature's points are plotted, an irregular curve is used to draw the curve.

In Figure 7.26(a), the curved features are divided vertically and horizontally with construction lines. The intersection of vertical construction lines with related horizontal construction

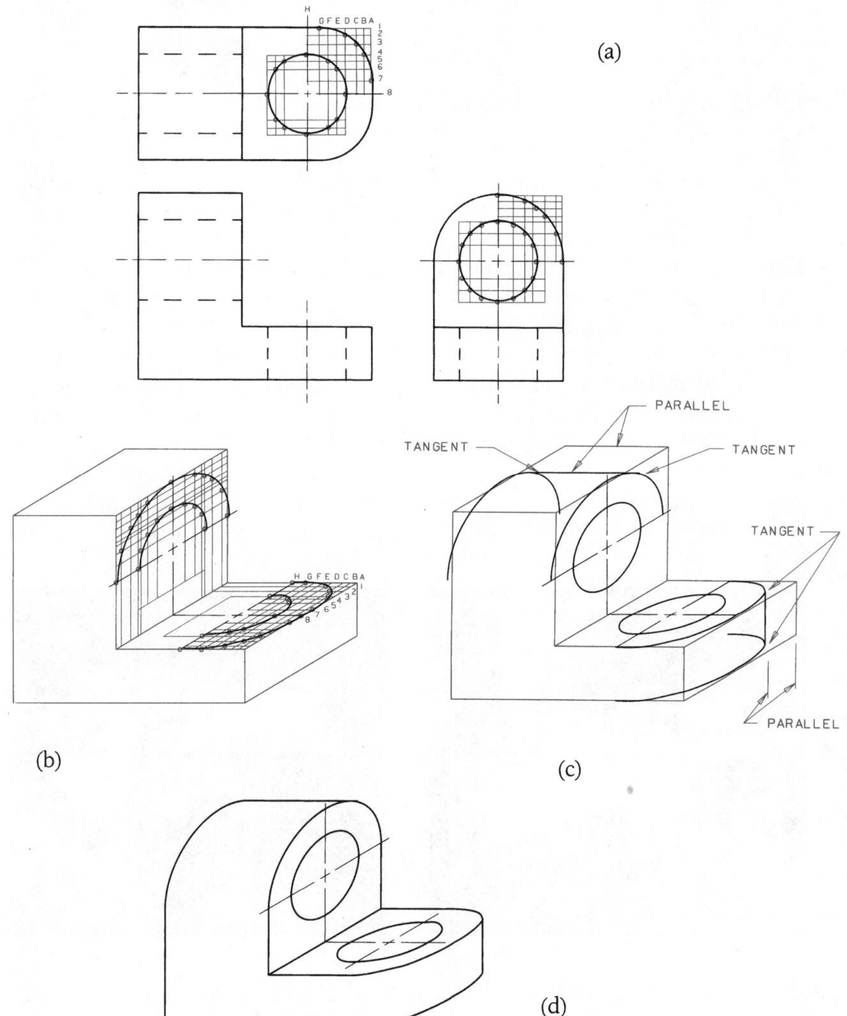

FIGURE 7.26 Offset Dimensions for Curves Not in the Front Face of a Part
(a) Locate points on three-view drawing (b) Locate points on curves (c) Establish tangent points (d) Erase construction lines and darken lines

ITEMS OF INTEREST

Pictorials

Pictorial drawings help the viewer to better visualize a part and gain a better understanding of its components and features. These drawings help bridge the gap between photograph and part.

Engineering drawings contain a wealth of information about a product: its size, shape, location of features, construction materials, and assembly specifications. However, engineering drawings (multiview projections) are not always easy to read for a nontechnical person untrained in those projection techniques. A more realistic looking 3D drawing (technical illustration) is produced for situations in which engineering drawings are not the most appropriate presentation, such as in marketing meetings and in maintenance documentation.

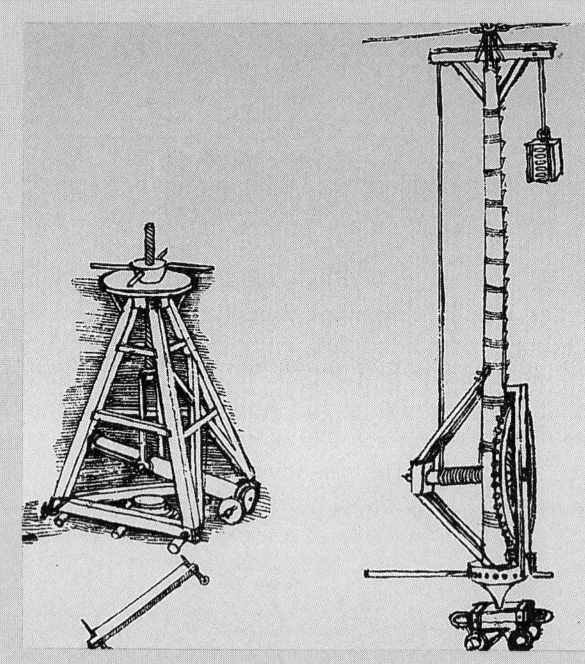

One of Leonardo's Mechanical Sketches

Early tries at technical illustration by the Egyptians and Greeks didn't really show all three dimensions in one view. Around 1500 A.D., pictorial drawings that showed all three dimensions in one view evolved. Leonardo da Vinci was the most famous of this group of inventors/ illustrators. His artistic ability and scientific foresight provided the means for true technical illustration. Techniques were further refined during the Industrial Revolution. After 1940, technical illustrations became popular design and development tools.

Today, technical illustrators produce pictorial drawings that aid in product development, assembly, marketing, illustration, repair, and maintenance. It is a fact that technical illustrations accelerate production, improve communications during development, and reduce product cost. Even people with limited technical knowledge can understand complex assemblies and interrelationships of parts and features. Indeed, pictorials prove to us "a picture is worth a thousand words."

The solid models on advanced CAD stations today are an extension of this same principle. Not only do these parts look three dimensional, they are 3D mathematically. The advanced renderings of parts on these sophisticated systems makes the parts look very real on the computer screen—in fact, they look almost as good as photographs. The 3D database can be used to make renderings and illustrations and can be used to generate a toolpath to machine the part. Improved visualization is one of the key factors driving the increasing popularity of these systems.

However, the lesson learned by mankind long ago is that we really need to be able to "see" a part to understand it. Leonardo da Vinci was a master of this long before the computer age.

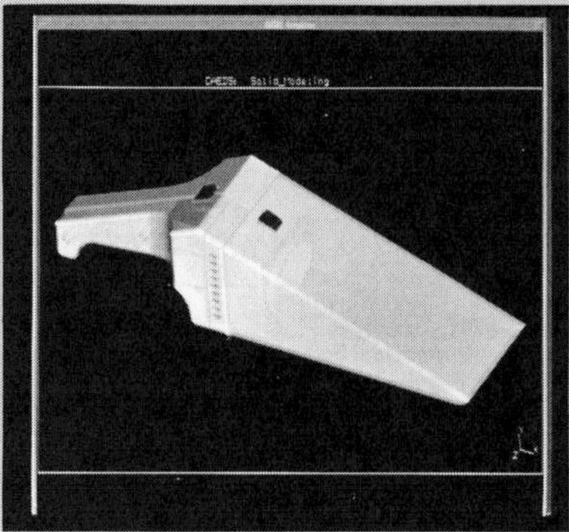

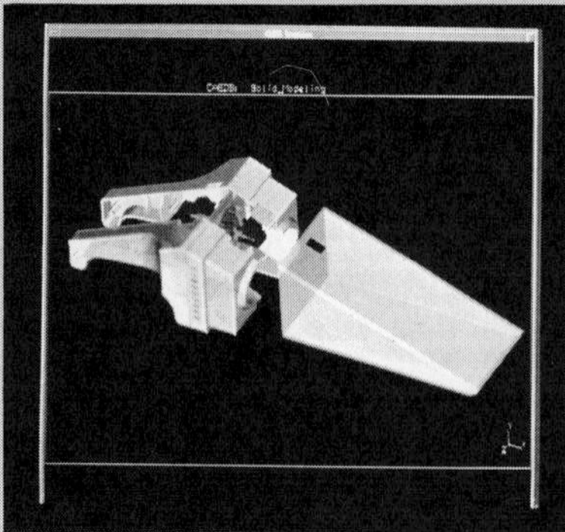

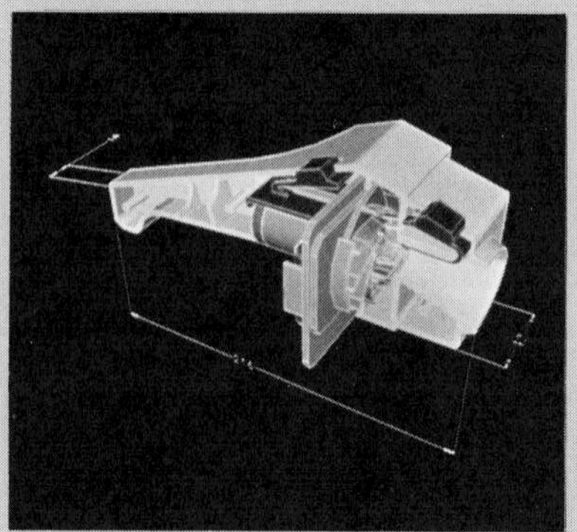

A 3D Solid Model

lines establishes points on the curved feature. The oblique projection is started by establishing a front face. However, the choice of front face here was made on the need to demonstrate this procedure. The construction lines are transferred to the oblique view and the points are plotted [Fig. 7.26(b)]. A smooth curve is drawn through the points using an irregular curve. Since the opposite portion can be mirrored, only half of each hole is plotted. The projection is completed by drawing the end lines parallel to the axes and tangent to the curves [Fig. 7.26(c)] and darkening in the lines [Fig. 7.26(d)]. If a cabinet drawing is used, the depth dimensions are halved and the same procedure is used in its construction.

The construction process for oblique projection with slanted, inclined lines and inclined planes is similar to that for isometric drawings. The end points are located along lines that are parallel to one of the axes.

7.5 Perspective Projection

A pictorial drawing made by the intersection of the picture plane with lines of sight converging from points on the object to the point of sight that is located at a finite distance from the picture plane is called a **perspective.** Figure 7.27 shows a perspective rendering of an array of radio telescope antennas. The use of perspective projection gives the illustration a "photo-like" realism. The observer is stationed at a fixed position relative to the object being drawn as with a photograph.

Perspective projection is used to provide illustrations that approximate how a particular object looks to the human eye or as a camera would record the object on film. Since a perspective drawing approximates how an object really looks,

FIGURE 7.27 Perspective Rendering of Radio Telescope Antennas

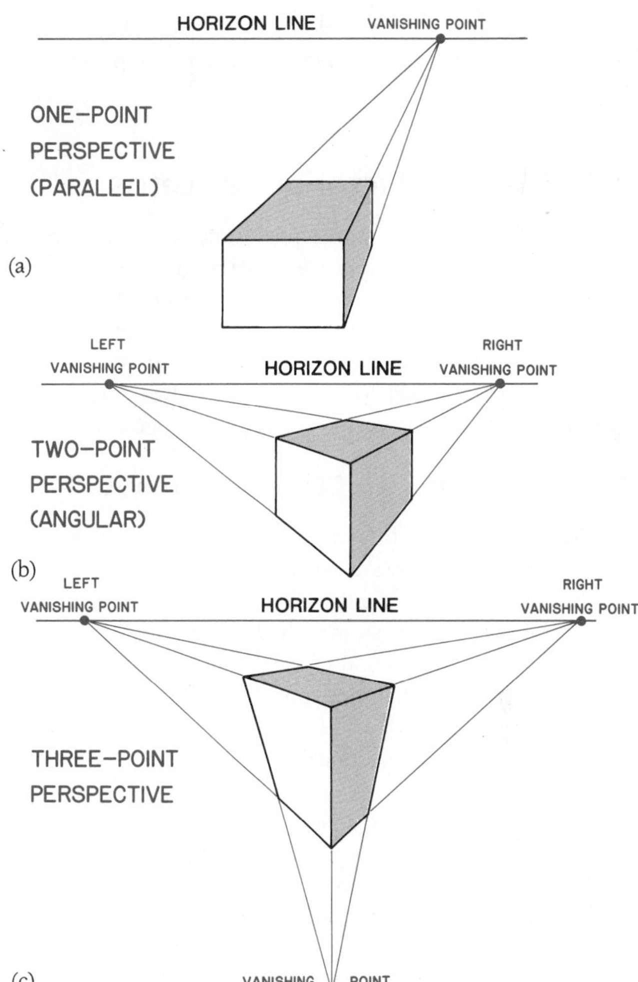

FIGURE 7.28 Types of Perspectives
(a) Parallel or one-point perspective (b) Angular or two-point perspective (c) Oblique or three-point perspective

it is not dimensionally correct and cannot be scaled. The only lines that can be scaled are those lines on the object that actually lie in the picture plane. Because of this distortion, perspective drawings are seldom found in engineering or design work. However, technical illustrations for advertisements, sales catalogs, technical manuals, and architectural renderings make extensive use of this form of pictorial projection. All lines in perspective drawings converge at one, two, or three points on the horizon (vanishing points) and, therefore, are not parallel, as in oblique and axonometric projection.

There are three basic types of perspective drawing: **parallel, angular** and **oblique** (Fig. 7.28). A perspective in which two of the principal axes of the object are parallel to the picture plane and the third is perpendicular to the plane is called **parallel perspective** or **one-point perspective** [Fig. 7.28(a)]. A perspective in which one axis of the object (usually the vertical axis) is parallel to the picture plane and the other two axes are inclined to it is called an **angular** or **two-point perspective** [Fig. 7.28(b)]. A perspective in which all three principal axes of the object are oblique to the plane or projection is called an **oblique** or **three-point perspective** [Fig. 7.28(c)].

7.6 General Pictorial Concerns

A variety of procedures and accepted drafting conventions for representing certain features are incorporated into pictorial illustrations including **sections** and **cutaways**, **breaks**, **fillets**, **rounds**, and **thread** representation. Shading and shadows are also added to pictorials to give a more lifelike representation of the part. Pictorial **assemblies** incorporate **exploded views** to show how a device fits together.

7.6.1 Sectional and Cutaway Views

Sectioned pictorials show the interior of a part or assembly. Section cutting planes are passed through centerlines and parallel to one of the principal faces of the part whenever possible. Figure 7.29(a) shows a half section of a part. Figure 7.29(b) shows a full section of a part. Section lines in a half section are drawn so that they appear to coincide if the planes were folded together [Fig. 7.29 (a)]. When a full section is used, the crosshatching should all be drawn in the same direction [Fig. 7.29(b)].

In assemblies, individual pieces are differentiated by using appropriate symbols and by changing the direction of the section lining. When a section plane passes through shafts, bolts, keys, pins, and solid round items, it is desirable to run the section around that item and show the entire bolt or shaft in the pictorial. Except for such cases, the section lines should show exactly what material has been cut. Figure 7.30(a) shows the first step in the creation of a section of a pictorial assembly. Each of the parts is blocked out. The assembly is then completed [Fig.7.30(b)].

7.6.2 Break Lines

For long parts, **break lines** may be used to shorten the length of the drawing. When the length of the part is beyond the size of the drawing format, and there are no features that must be displayed and dimensioned, you may shorten the drawing by using break lines. Position the break at a place on the part that does not interfere with the part. **Freehand breaks** are preferred as shown in Figure 7.31. Here, the preferred and acceptable methods for showing breaks are shown.

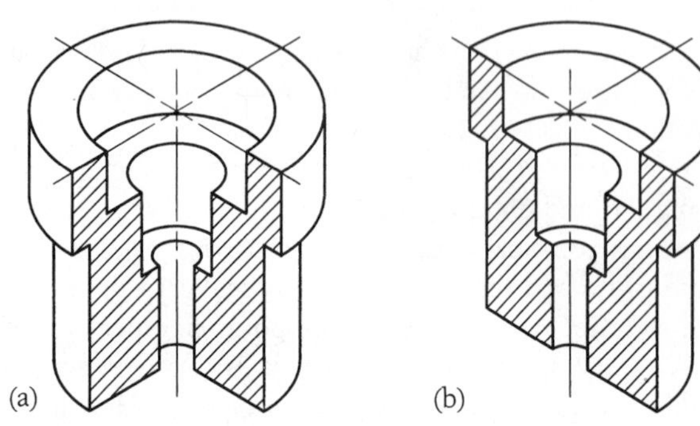

FIGURE 7.29 Half- and Full-Section Pictorials

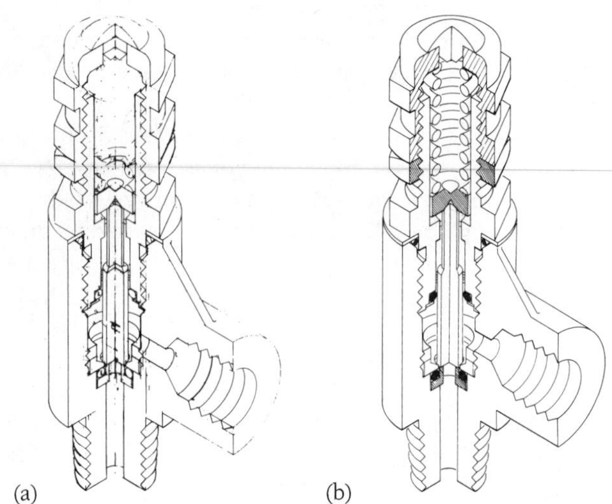

FIGURE 7.30 **Pictorial Construction** (a) Construction of needle valve in an isometric drawing using a half section (b) Completed pictorial of needle valve

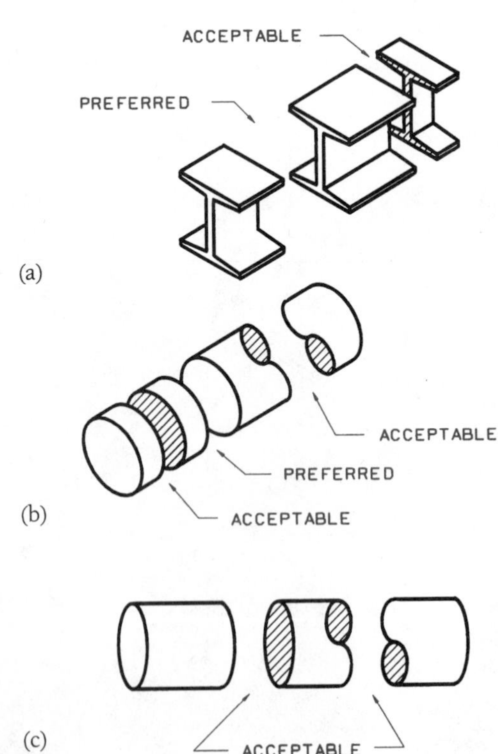

FIGURE 7.31 Break Lines for Pictorials

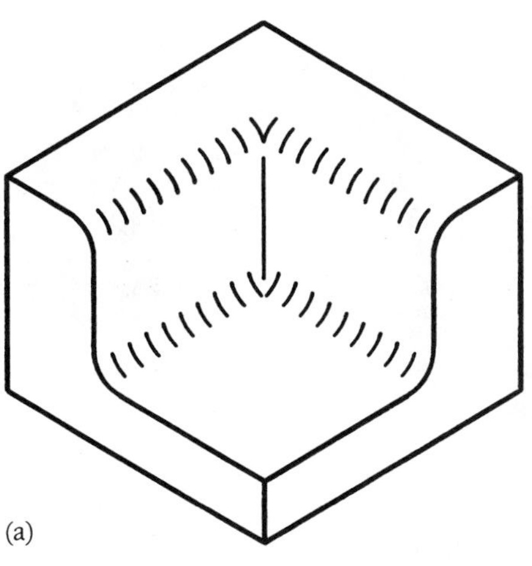

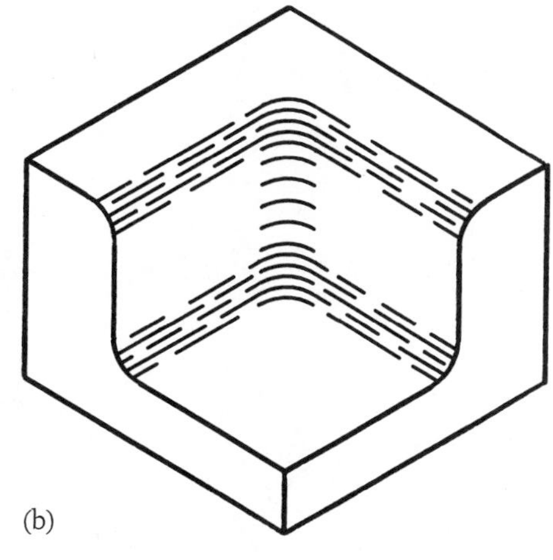

FIGURE 7.32 Fillets and Rounds in Pictorials (a) Curved highlighting (b) Straight highlighting

7.6.3 Fillets and Rounds

Fillets and **rounds** usually can be highlighted or can be shown as straight or curved lines representing the filleted and rounded edges of a part, as in Figure 7.32. Highlighting is drawn freehand.

7.6.4 Thread Representation

Threads may be represented by a series of ellipses or circles uniformly spaced along the centerline of the thread. Shading increases the effectiveness of the thread appearance, as in Figures 7.33 and 7.34. Threads should be evenly spaced, but it is not necessary to reproduce the actual pitch (distance between crests of the threads) or the exact number of threads.

7.6.5 Shading and Shadows

Many types of **shading** can be used on pictorials. The type of shading used depends on the purpose of the drawing and the type of pictorial. Smudge shading, as shown in Figure 7.33, is sometimes used on pictorial drawings. For catalog illustrations, some form of overall shading is generally preferred. Excellent results can usually be obtained with an airbrush (Fig. 7.34). For more artistic results, the pictorial may be shaded with shadows. When **shadows** are added to an illustration, they add depth and realism to the drawing. Choosing a proper light source can make the illustration look as if it is a photograph of an object.

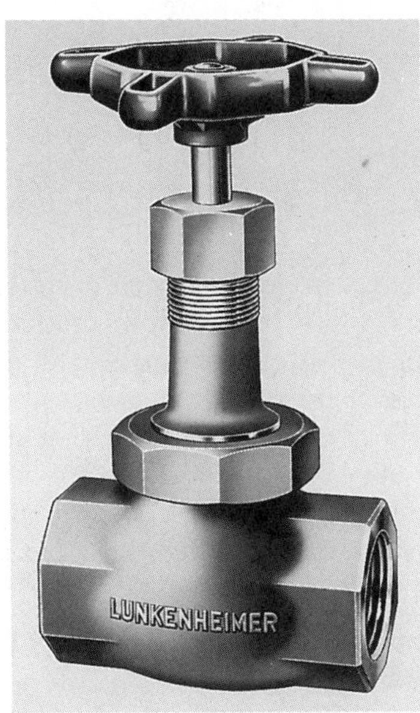

FIGURE 7.34 Airbrushed Shading in Pictorials

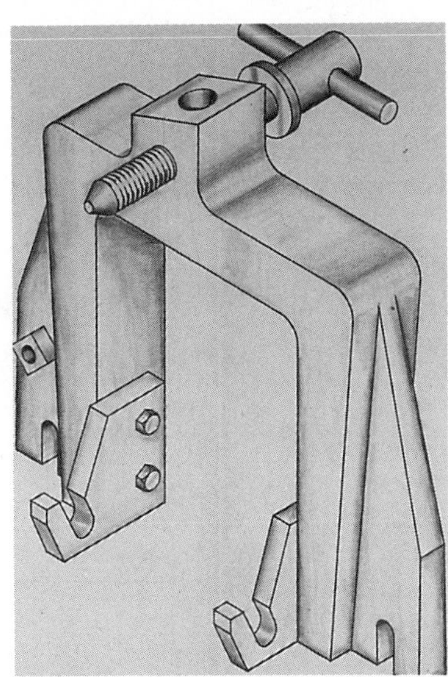

FIGURE 7.33 Shading Pictorials

7.6.6 Exploded Pictorial Views

A pictorial drawing showing the various parts of an assembly, separated, but in proper position and alignment for reassembly, is called an **exploded assembly** (Fig. 7.35). Exploded pictorials are used extensively in service manuals and as an

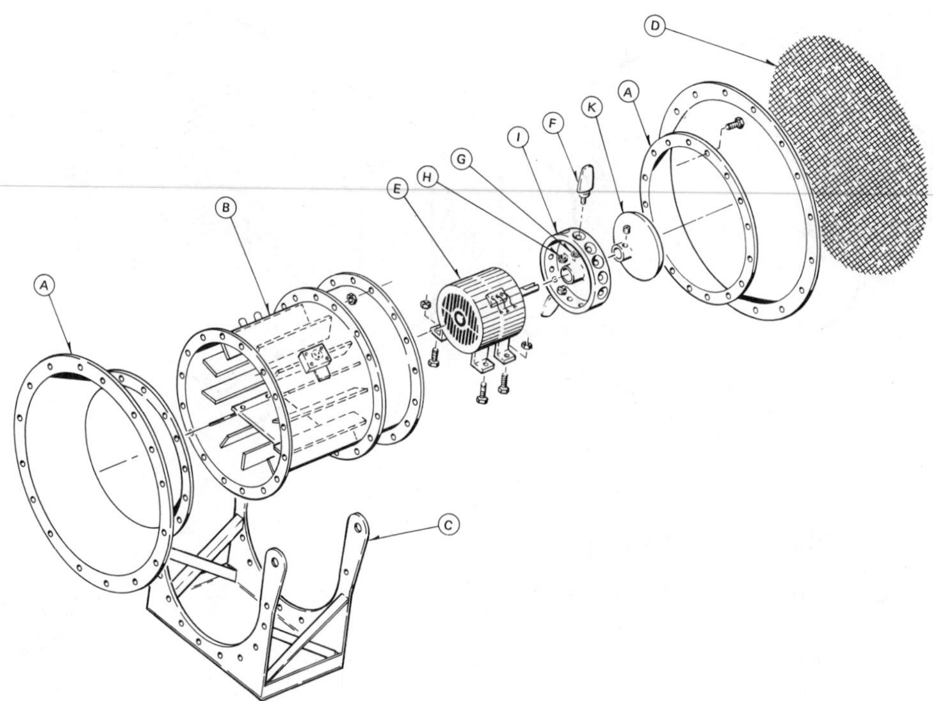

FIGURE 7.35 Exploded Pictorial

aid in assembling or erecting a machine or structure. Any type of pictorial drawing may be used for this purpose and shading may be as simple or as complete as desired. In Figure 7.35, the exploded assembly is drawn in isometric.

Each piece in an exploded assembly should be connected to its mating part by a centerline. If there is not sufficient room to extend the exploded pieces out from each other in one line, the piece can be moved. A jogged centerline still connects related pieces.

You May Complete Exercises 7.5 Through 7.8 at This Time

QUIZ

True or False

1. Curves that do not lie in isometric planes must be constructed with offset measurements.
2. A general oblique drawing is constructed using 45° for the receding axis.
3. An isometric drawing is constructed using true length measurements along all three axis lines.
4. A cavalier drawing is always foreshortened.
5. Centerlines are included on all pictorials.
6. Cabinet and cavalier drawings are types of perspective projection.
7. A trimetric projection uses different angles for all three axes.
8. Centerlines are always shown on pictorial drawings.

Fill in the Blanks

9. A ________ view shows the interior of the part or assembly.
10. ________ dimensioning is found on most pictorial drawings.
11. ________ projection approximates how a part will look to the human eye.
12. In true isometric projection all ________ ________ make equal ________ with the projection plane.
13. ________ lines are not parallel to or on one of the isometric axes.
14. A ________ oblique projection uses a half scale for all receding measurements.
15. The ________ ________ method of ellipse construction does not create an ellipse.
16. ________ features should be oriented so that they lie in the ________ face of the object when ________ projection is used on a drawing.
17. Pictorial drawings are ________ ________ ________ .
18. ________, ________, and ________ are the three general groups of pictorial projection.

Answer the Following

19. What is the difference between an isometric drawing and an isometric projection?
20. In what situation would an oblique drawing be used instead of an isometric drawing?
21. What are the three types of oblique projection? Describe each and how they differ.
22. Give four uses of pictorial drawings.
23. Describe the three types of perspective projection.
24. What are offset measurements and when are they used in the construction of a pictorial drawing?

EXERCISES

Exercises may be assigned as sketching, instrument, or CAD projects. Transfer the given information to an "A" size sheet of .25 in. grid paper. Complete all views and solve for proper visibility, including centerlines, object lines, and hidden lines. Exercises that are not assigned by the instructor can be sketched in the text to provide practice and understanding of the preceding instructional material.

After Reading the Chapter Through Section 7.3.14 You May Complete the Following Exercises

Exercise 7.1 Using the part provided in Exercise 6.5(B), draw an isometric pictorial.

Exercise 7.2 Using the part provided in Exercise 6.9(A), draw an isometric pictorial.

Exercise 7.3 Using the part provided in Exercise 6.10(B), draw an isometric pictorial.

Exercise 7.4 Using the part provided in Exercise 6.3(A), draw an isometric pictorial.

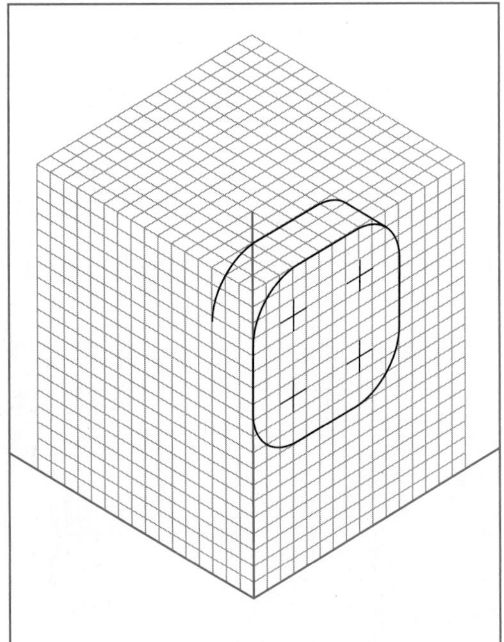

Exercise 7.1

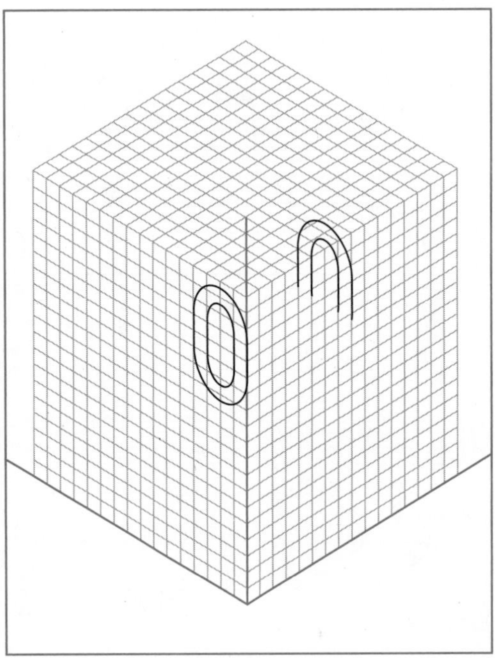

Exercise 7.3

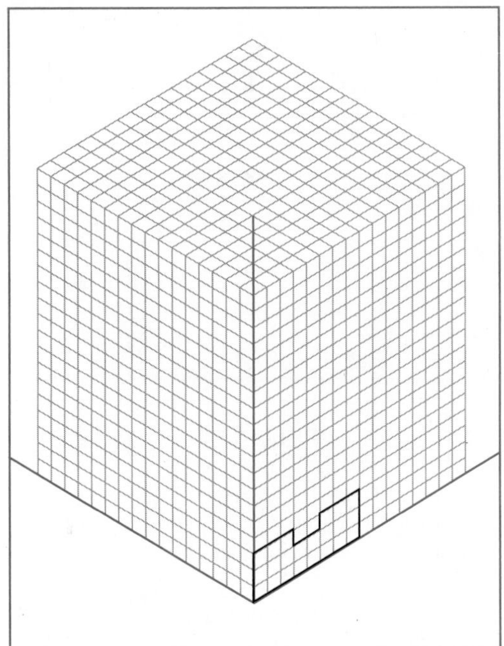

Exercise 7.2

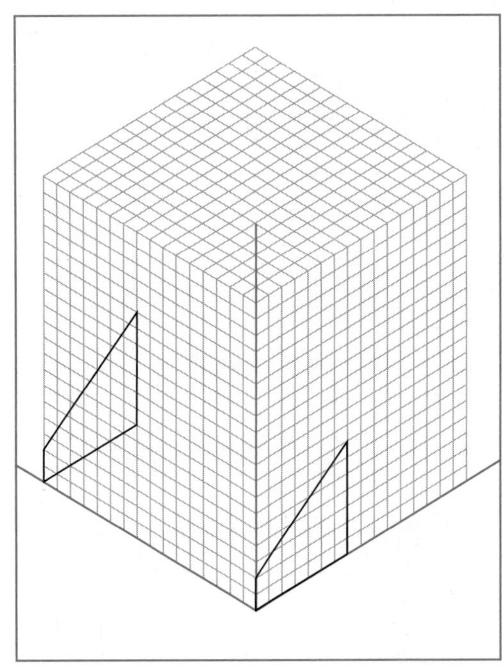

Exercise 7.4

After Reading the Chapter Through Section 7.6.6 You May Complete the Following Exercises

Exercise 7.5 Using the part provided in Exercise 6.13(B), draw an oblique pictorial.

Exercise 7.6 Using the part provided in Exercise 6.11(B), draw an oblique pictorial.

Exercise 7.7 Using the part provided in Exercise 6.4(B), draw an oblique pictorial.

Exercise 7.8 Using the part provided in Exercise 6.1(A), draw an oblique pictorial.

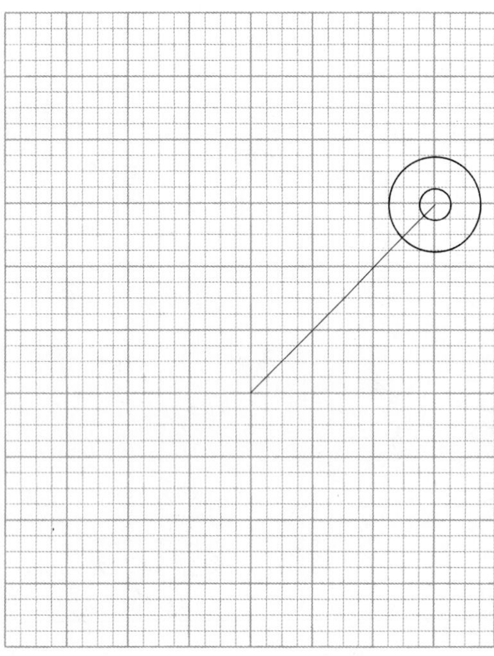

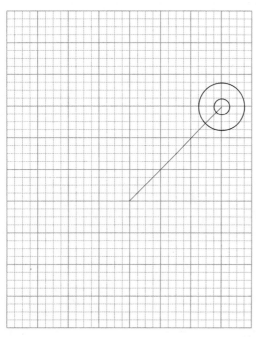

Exercise 7.5

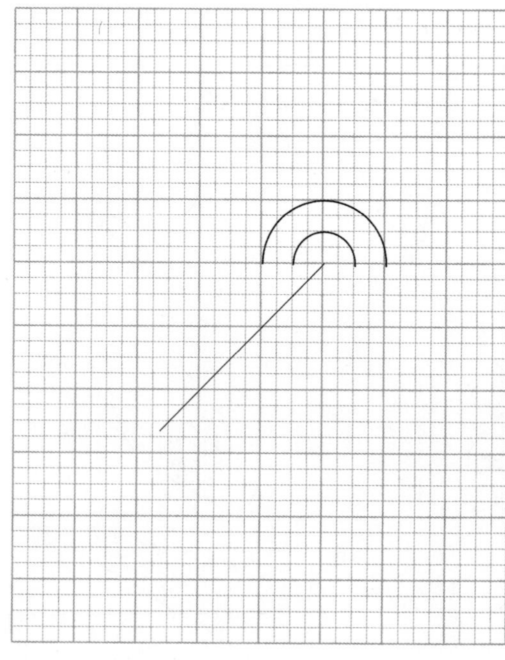

Exercise 7.7

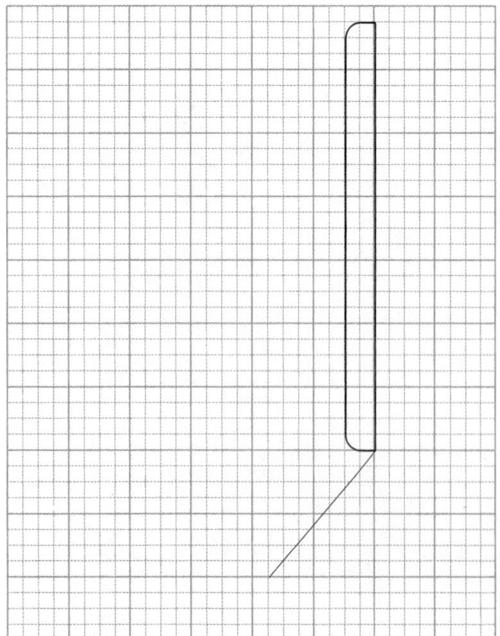

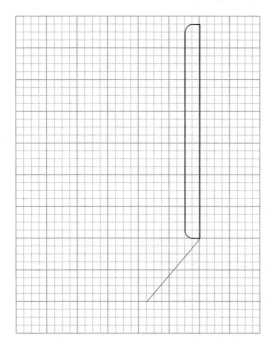

Exercise 7.6

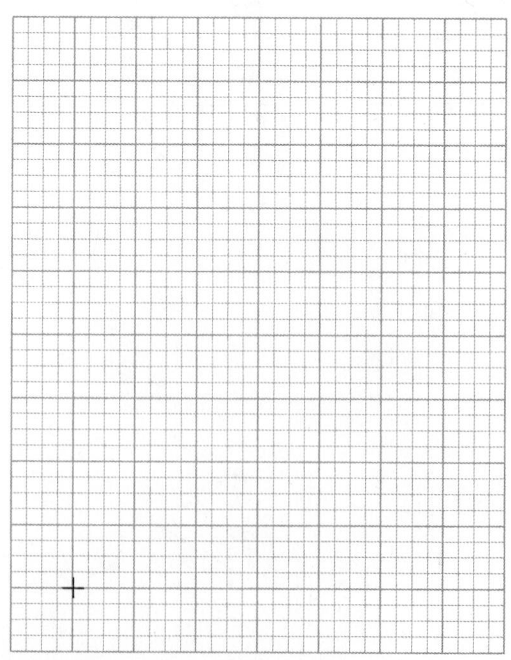

Exercise 7.8

PROBLEMS

Problems 7.1(A) through (K) These problems can be assigned as any of the three major types of pictorial projection. Unless assigned as a specific type of projection, these problems could be used o test the students' understanding of the suitability of projection types for a particular problem. The instructor can allow the student to determine the projection method based on the part's features and pictorial requirements as described by the instructor.

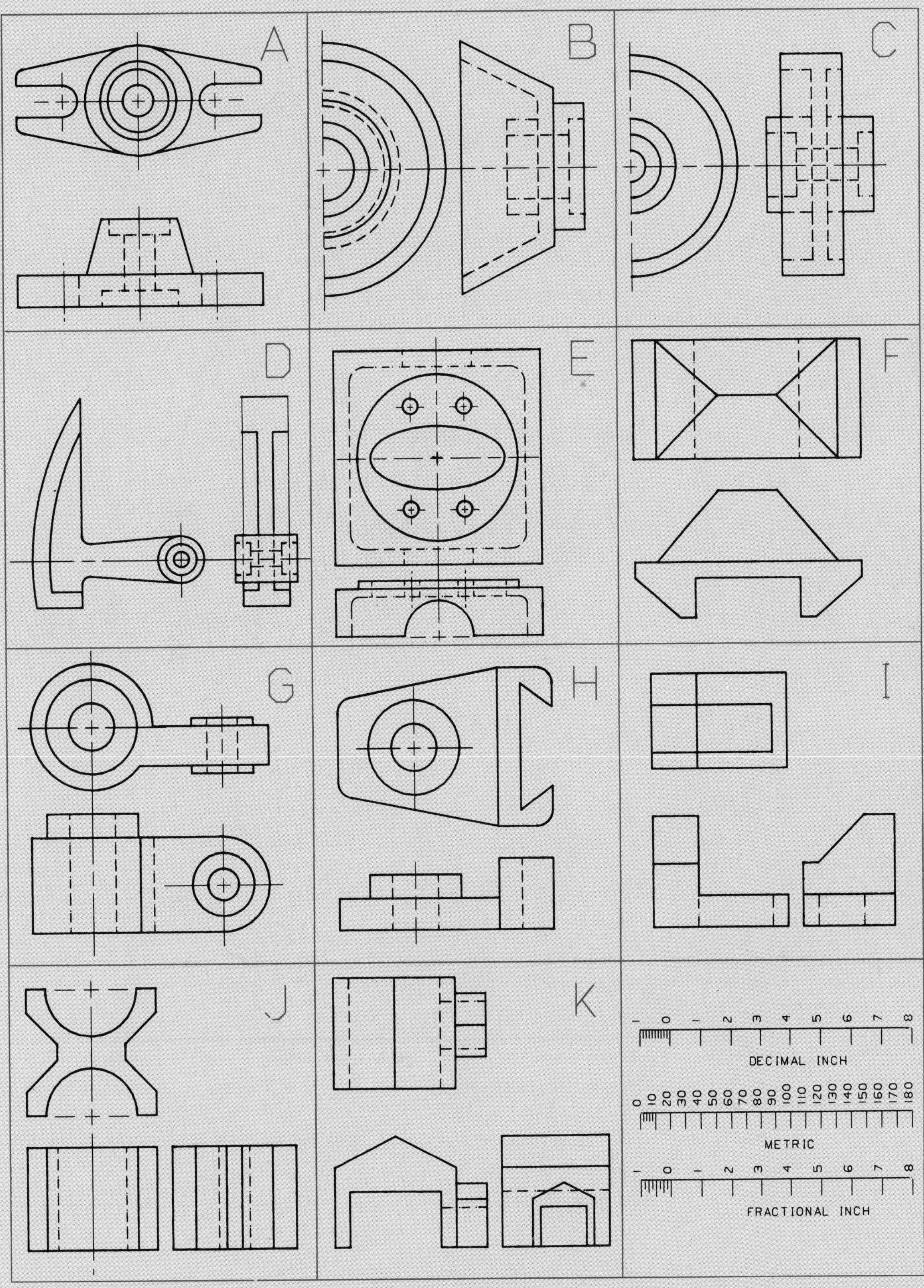

Problems 7.2 through 7.4 Created in SI. The scale provided is to be used when taking the part from the text and transferring it to the drawing board (or computer). The decimal scale can also be used if the problem as assigned will be done in decimal-inch units.

Problems 7.2(A) through (H) Meant for isometric projection, but can be drawn as any type of pictorial projection as assigned.

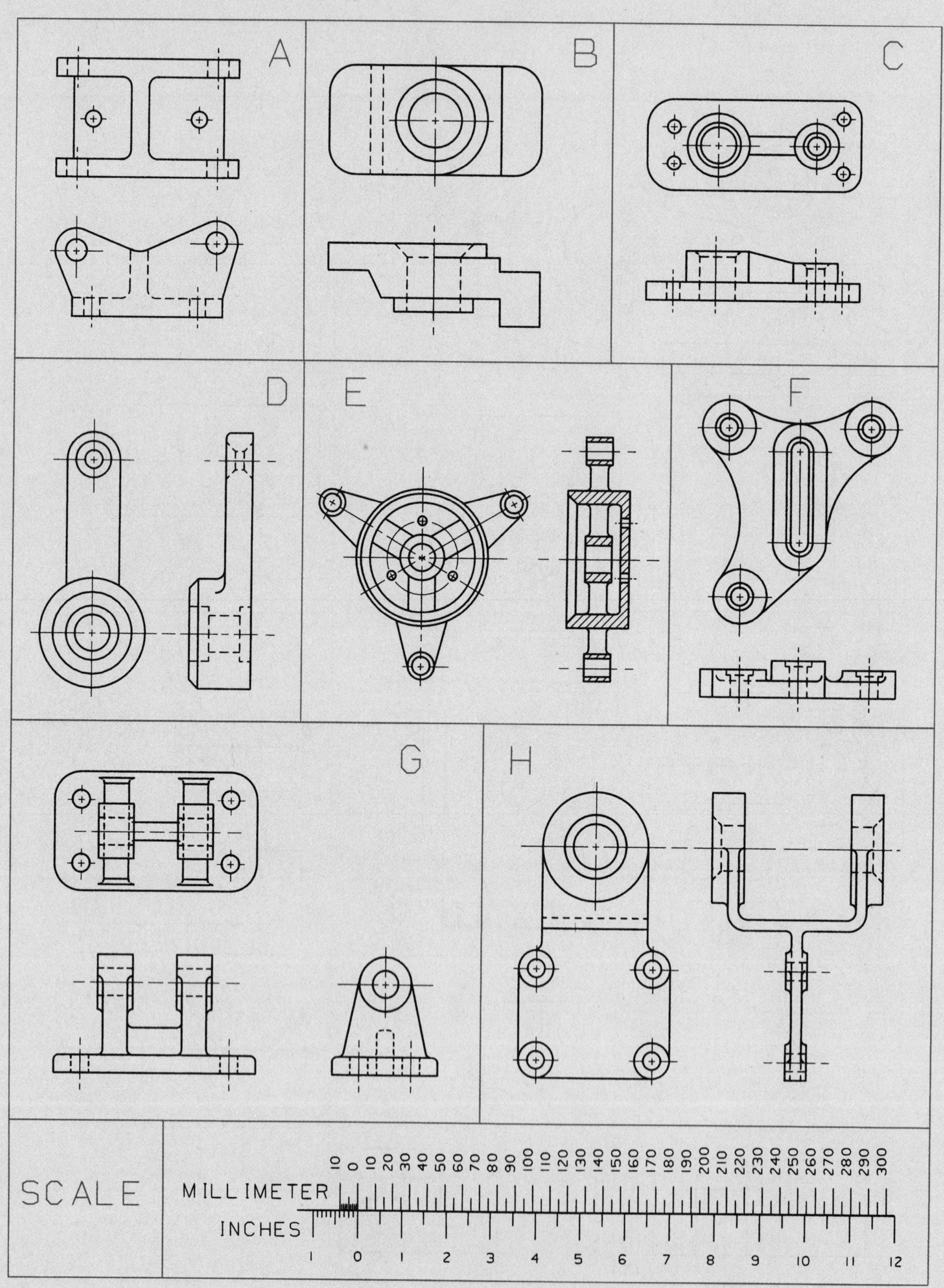

Problems 7.3(A) through (F) Designed as oblique projects, although the instructor can assign other methods of projection.

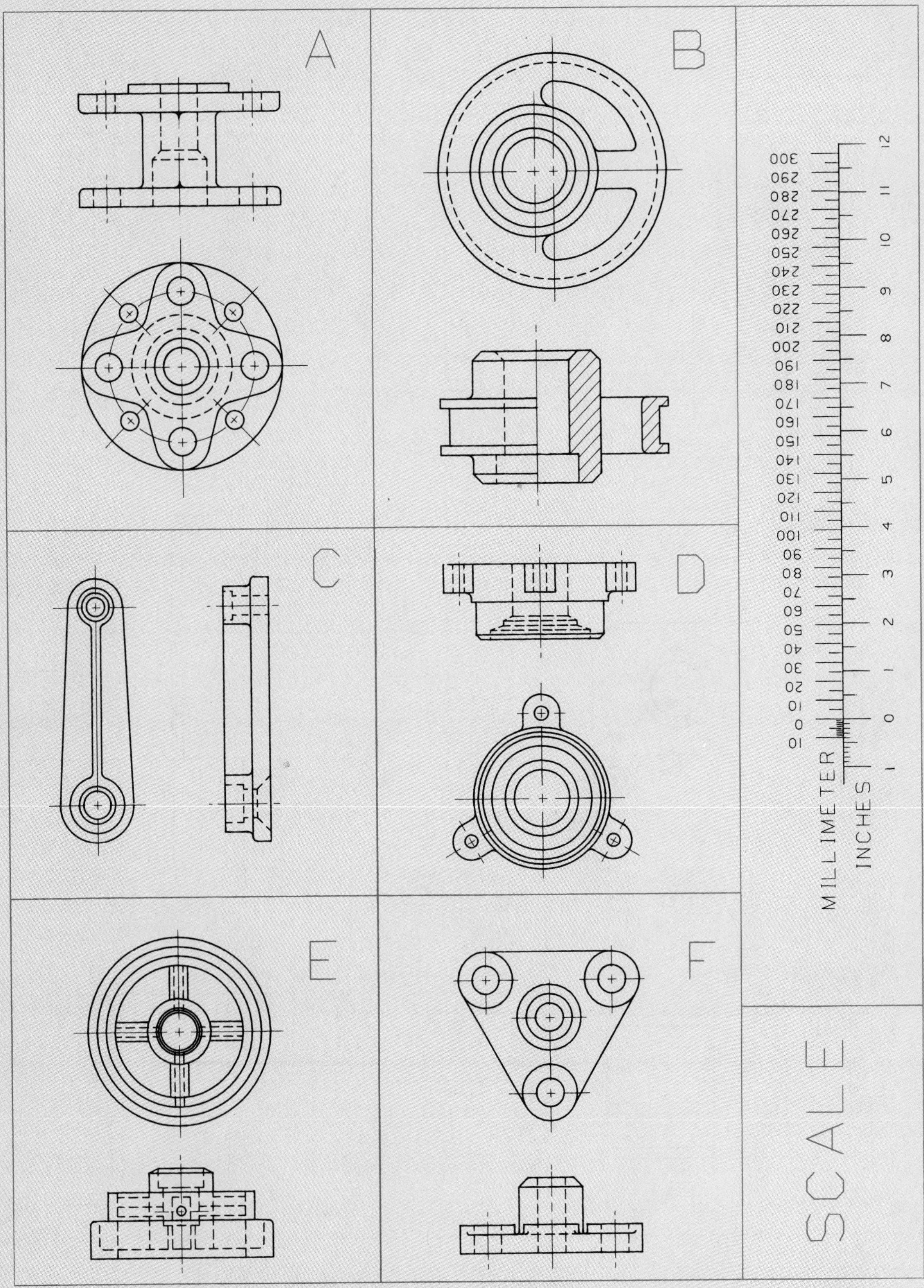

Problems 7.4(A) through (F) Use for any method of pictorial projection, although they are intended as perspective projection projects.

Instructor may assign any appropriate problems or figures found in Chapters 6, 8, or 11 as pictorial drawing projects, manual or CAD.

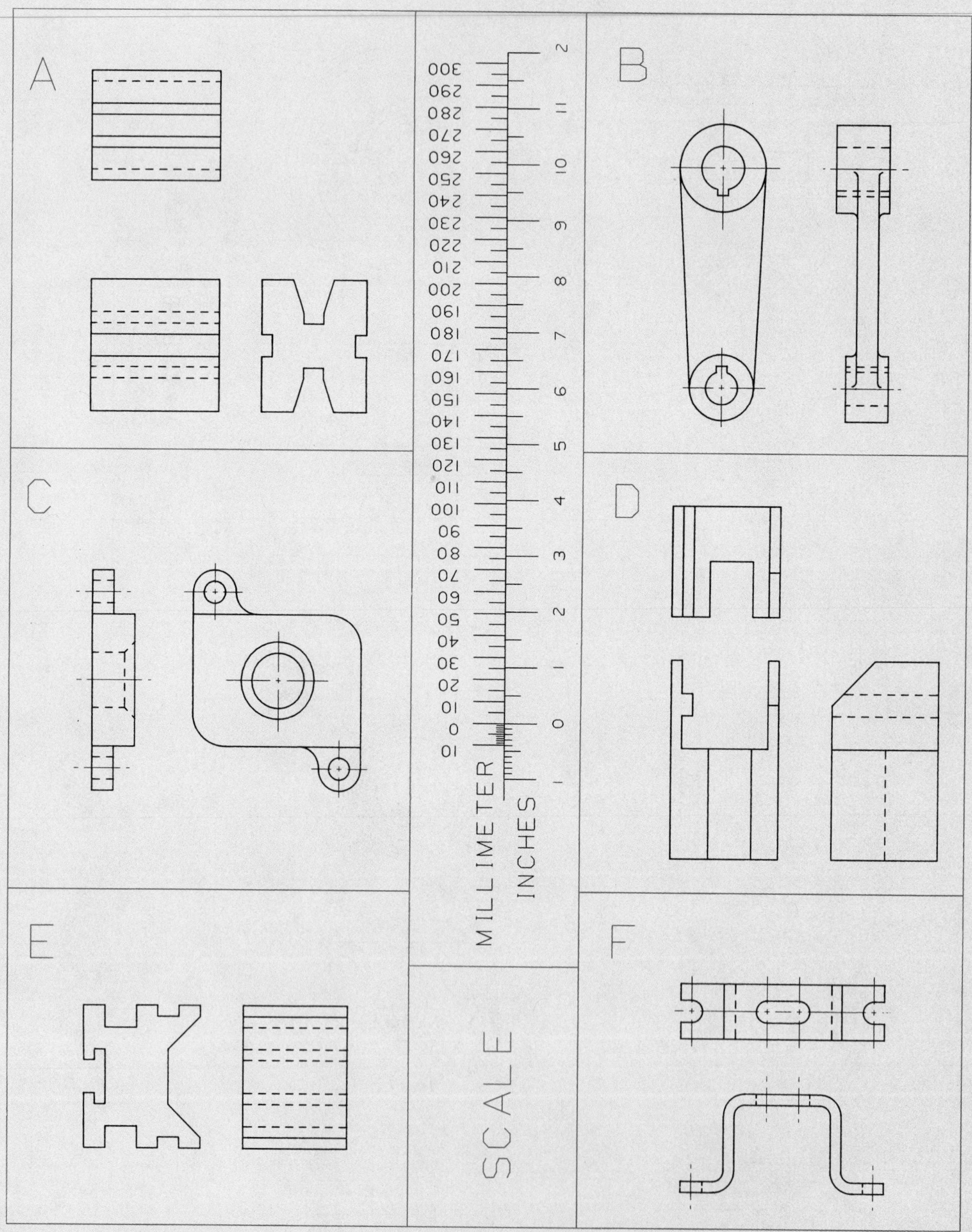

CHAPTER 8

Sectional Views

Learning Objectives

Upon completion of this chapter you will be able to accomplish the following:

1. Identify the need for sectional views in order to clarify interior part features.
2. Apply standard drafting conventions and line types to illustrate interior features.
3. Identify cutting planes and resulting views.
4. Differentiate between and produce full, half, offset, aligned, removed, revolved, broken-out, and assembly sections.

8.1 Introduction

Designers and drafters use **sectional views (sections)** to clarify and dimension the internal construction of a part when interior features cannot be clearly described by hidden lines in conventional views. For example, the valve in Figure 8.1 has a portion of its exterior body removed to allow a view of the disk, seating, and stem. Figure 8.2 shows the same valve in *full section*. This chapter presents different types of sections and discusses their variations when used on mechanical parts and assemblies. *(See Color Plates 25–27.)*

8.2 Sections

A sectional view is obtained by passing an imaginary **cutting plane** through the part, perpendicular to the **line of sight**, as in Figure 8.3 (**SECTION A–A**). The line of sight is the direction in which the part is viewed (Fig. 8.4). The portion of the part between the cutting plane and the observer is "removed." The part's exposed solid surfaces are indicated by **section lines** (uniformly spaced angular lines drawn in proportion to the size of the drawing).

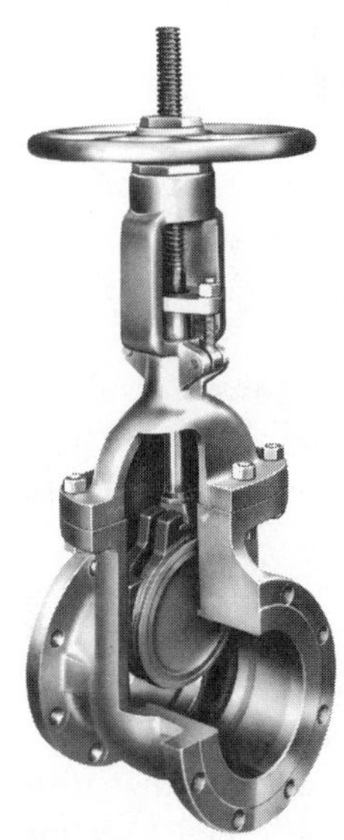

FIGURE 8.1 Sectioned Gate Valve

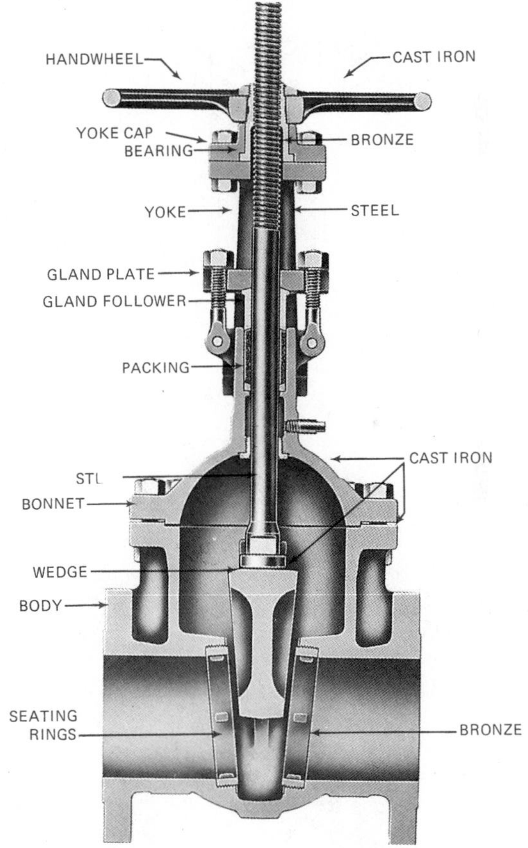

FIGURE 8.2 Front Section of a Gate Valve Showing the Stem and Disk

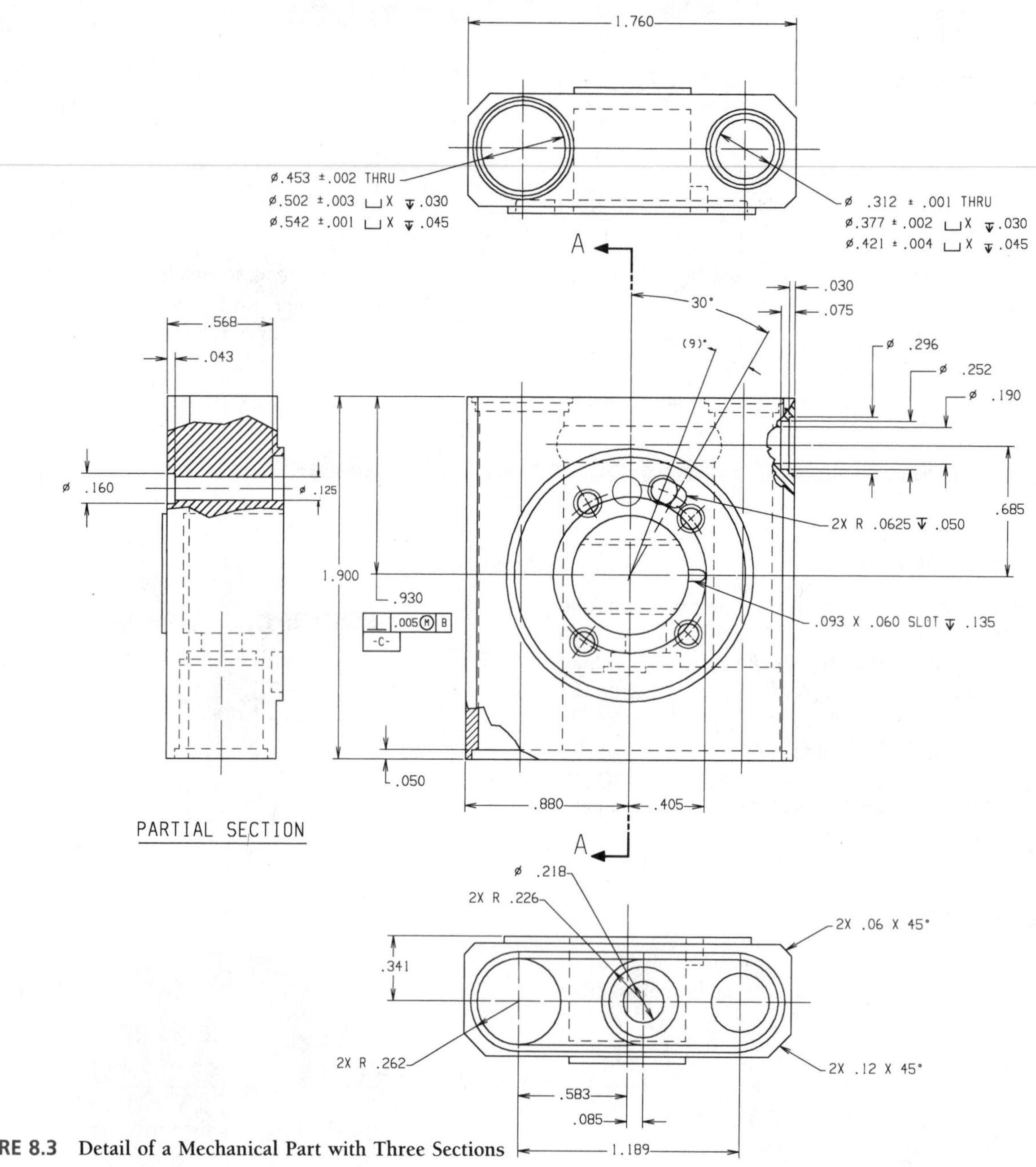

FIGURE 8.3 Detail of a Mechanical Part with Three Sections

In all section views on a drawing, section lines for the same part are identical in angle, spacing, and uniformity (Fig. 8.4). Construct section lines so that they are spaced clearly, are pleasing to look at, and will reduce and enlarge without distorting.

Many types of section views exist. Figure 8.3 shows a drawing of a complex part containing a full section (SECTION A–A), a partial section (left side), and a broken-out section (left corner of front view).

Sections are rotated 90° out of the plane of principal or auxiliary views from which they are taken, following customary projection rotation. A heavy line across or near the principal view indicates the plane of projection, with arrows to indicate the viewing direction line of sight (Fig. 8.4). This line is called a **cutting plane line** and it represents the edge of the imaginary cutting plane. When the plane of projection passes through the view, it is called the **cutting plane** and the resulting adjacent view is called a **section.** Each cutting plane

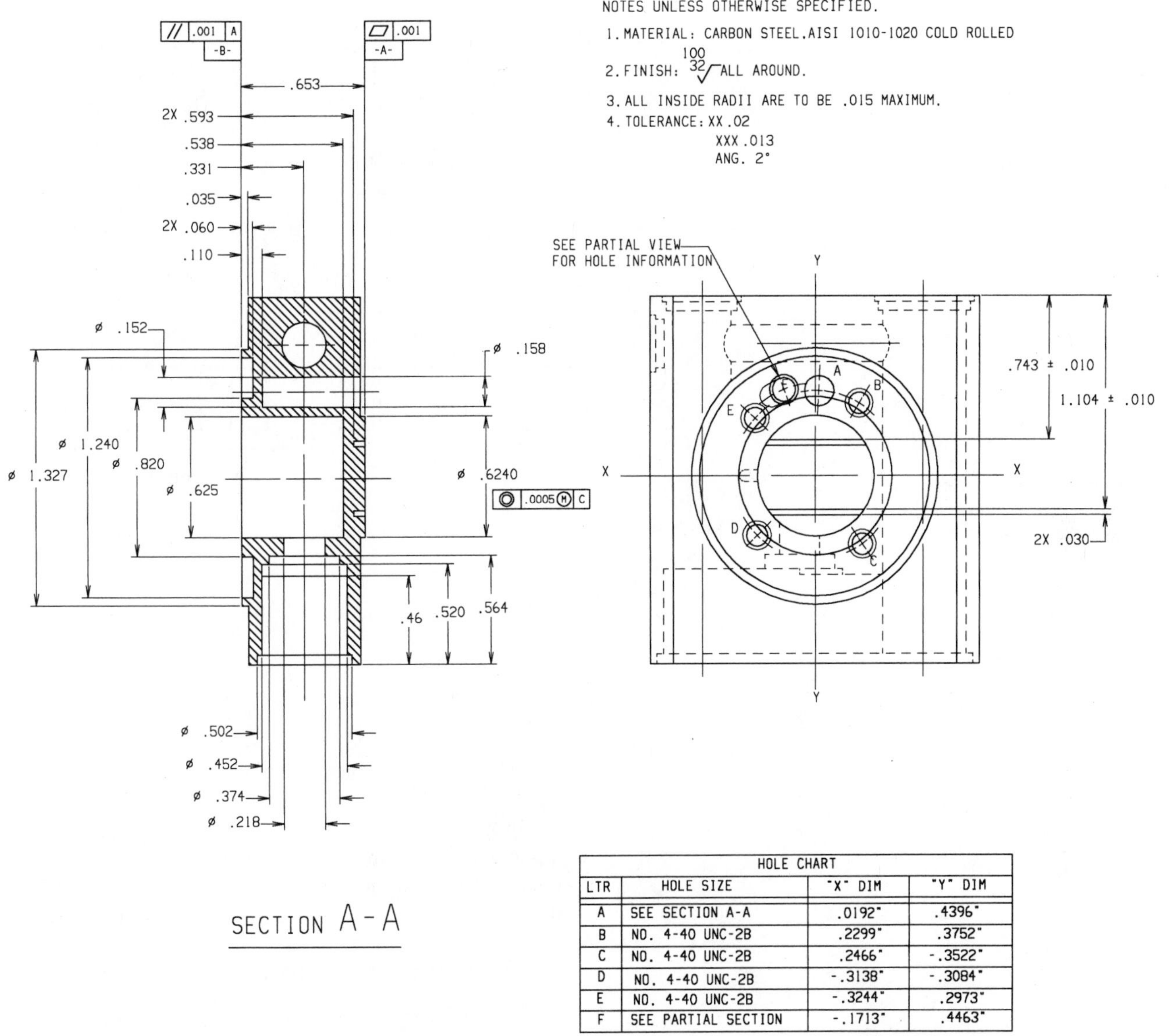

HOLE CHART			
LTR	HOLE SIZE	"X" DIM	"Y" DIM
A	SEE SECTION A-A	.0192"	.4396"
B	NO. 4-40 UNC-2B	.2299"	.3752"
C	NO. 4-40 UNC-2B	.2466"	-.3522"
D	NO. 4-40 UNC-2B	-.3138"	-.3084"
E	NO. 4-40 UNC-2B	-.3244"	.2973"
F	SEE PARTIAL SECTION	-.1713"	.4463"

FIGURE 8.3 Detail of a Mechanical Part with Three Sections—*Continued*

and corresponding view have **view identification letters** assigned to it (Fig. 8.4).

When cutting planes pass through solid portions of the part, these areas are shown by section lines in the adjacent section view. When the cutting plane passes through void areas such as a slot, hole, or other cutouts, the area is left without section lines in the adjacent section view (Fig. 8.4).

Since cutting planes are positioned to reveal interior details most effectively, selecting the proper location for the cutting plane is important. In Figure 8.5, the pictorial illustration of the section shows the cutting plane passing through the middle of the part. This is typically the most common location for the cutting plane.

8.2.1 Section Material Specification

Symbolic section lines are used on assembly illustrations for parts catalogs, display assemblies, promotional illustrations,

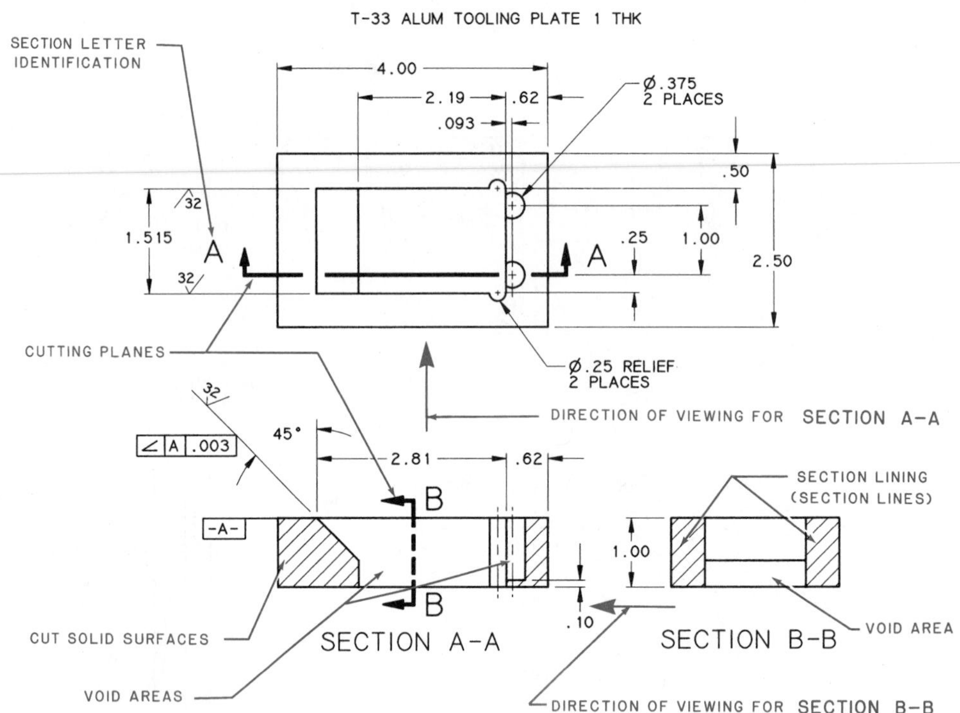

FIGURE 8.4 Three-View Drawing Using Sections as the Front and Side View

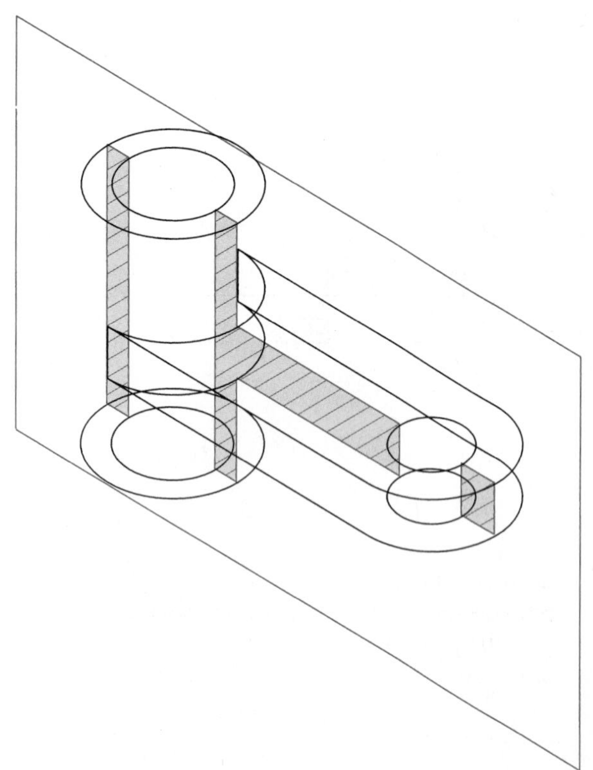

FIGURE 8.5 Sectioned 3D Part

and when it is desirable to distinguish between different materials.

Since it may not reduce and enlarge well, symbolic section lining is not recommended for drawings that will be microfilmed or put onto microfiche. Thus, the most common practice is to use the general-purpose symbol for all materials.

8.2.2 General-Purpose Section Lines

The symbol for cast iron [Fig. 8.6(a)] is considered the **general-purpose symbol.** General-purpose section lines do not distinguish between different materials and are normally drawn at an angle of 45°. Most drawings use general-purpose section lines [single lines at 45°, slanting from the lower left toward the upper right, and spaced evenly at about .10 in. (2.5 mm)]. Some drafters prefer to use $\frac{1}{8}$ in. (.125 in.) spacing when using decimal-inch measurements and 2.5 to 3.0 mm on drawings that use SI units. The exact material specification is given elsewhere on the drawing in note form or in the title block. An exception is made for parts made of wood, when it is necessary to show the direction of the grain.

Figure 8.7 shows measurements for the construction of general-purpose section lines. This figure includes examples of incorrect construction. The line thickness of section lines is thin (0.25 to 0.30 mm), sharp, and black. Section lines should not be too close [Fig. 8.7(d)] or they may merge and

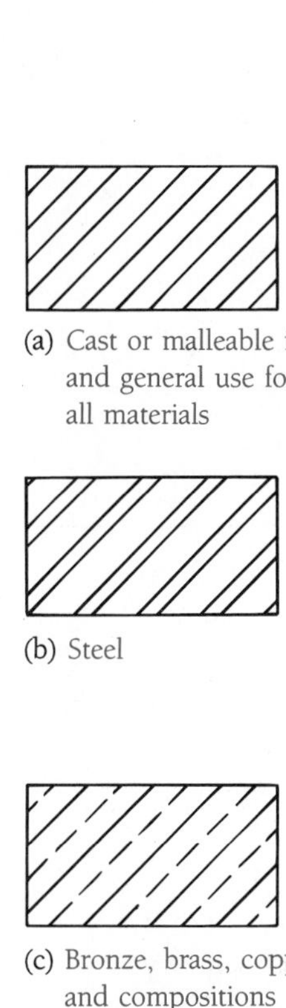

(a) Cast or malleable iron and general use for all materials

(b) Steel

(c) Bronze, brass, copper, and compositions

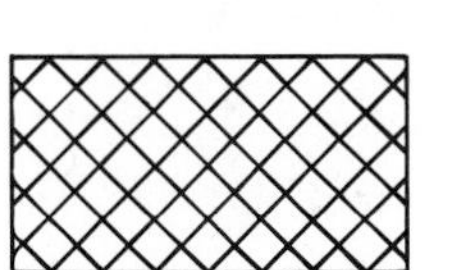

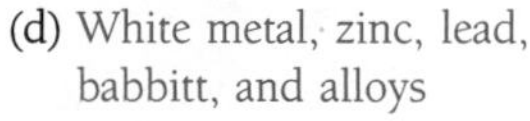

(d) White metal, zinc, lead, babbitt, and alloys

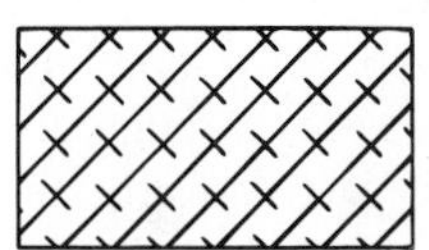

(e) Magnesium, aluminum, and aluminum alloys

(f) Rubber, plastic, and electrical insulation

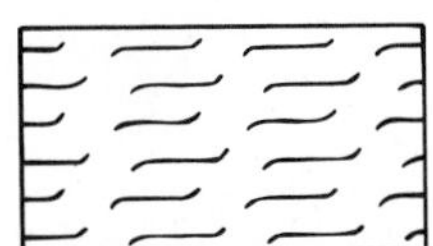

(g) Cork, felt, fabric, leather, and fiber

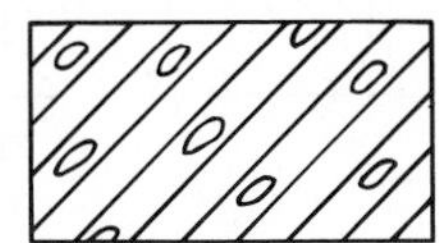

(h) Sound insulation

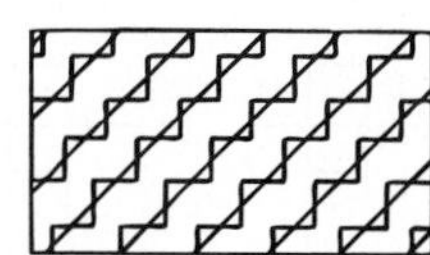

(i) Thermal insulation

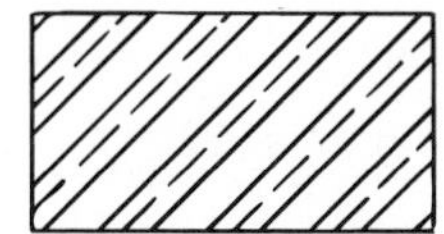

(j) Titanium and refractory material

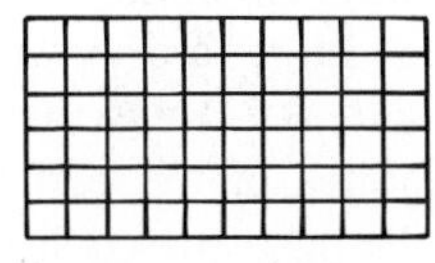

(k) Electric windings, electromagnets, resistance, etc.

(l) Concrete

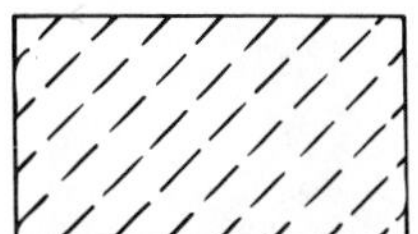

(m) Marble, slate, glass, porcelain, etc.

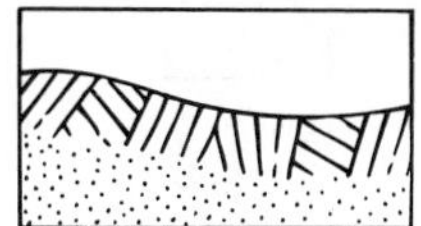

(n) Earth

(o) Rock

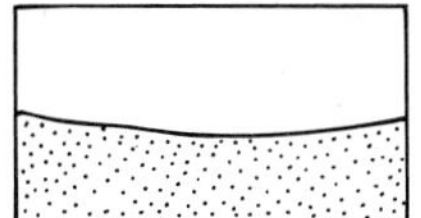

(p) Sand

(q) Water and other liquids

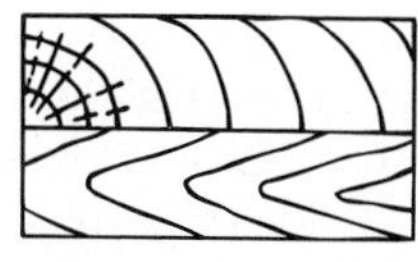

(r) Across grain / with grain } wood

FIGURE 8.6 Section Symbols for Material Specification

blot during reduction and reproduction. Section lines must be consistently spaced [Figs. 8.7(b) and (e)] and must end at visible object lines [Fig. 8.7(f)]. When the shape or position of a section area is such that the section lines would be parallel or perpendicular to a prominent visible line bounding the sectioned area, a different angle should be chosen [Fig. 8.7(g)].

8.2.3 Lines Behind the Cutting Plane

Hidden features behind the cutting planes are almost always omitted. In half sections, however, hidden lines are occasionally shown on the unsectioned half when needed for dimensioning or for clarity. The following rules apply when determining the precedence of lines on a section:

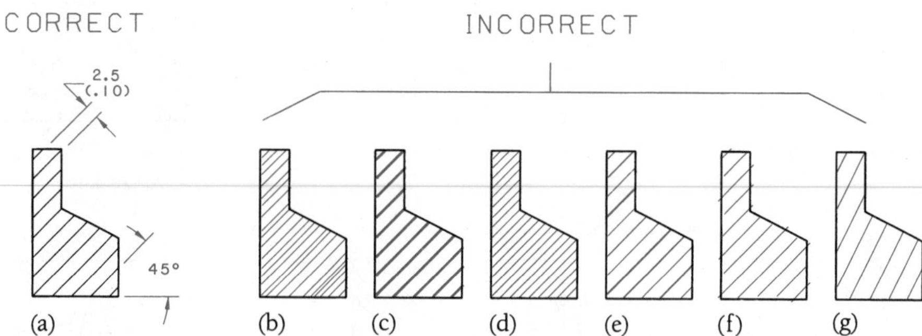

FIGURE 8.7 Section Lining (a) Correct example (b) Poor spacing (c) Thick lines (d) Close lines (e) Inconsistent lines (f) Lines not stopping at object lines (g) Lines perpendicular to object lines

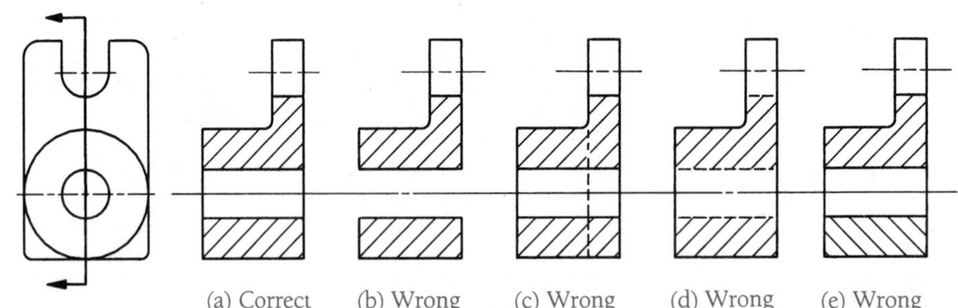

FIGURE 8.8 Hidden Lines in Sections (a) Correct example (b) through (e) Incorrect examples

1. Visible object lines take precedence over hidden lines and centerlines.
2. Hidden lines take precedence over centerlines.
3. Cutting plane lines take precedence over centerlines when locating a cutting plane. But, the cutting plane line can be omitted entirely if it falls along a centerline of symmetry for the part.

Figure 8.8 illustrates a few examples of line representation in sections. The correct procedure shows all visible lines as solid. Object lines on and behind the plane are still visible. The correct example shows all solid object lines [Fig. 8.8(a)]. Figure 8.8(b) does not show the back portion of the hole's edges (void area). Figure 8.8(c) shows a hidden line running through the section. This practice should be avoided because it complicates the drawing and does not add any clarity to the part definition. In some cases, it is acceptable practice to show hidden lines in sections, but, only where the part could not be properly defined.

Figure 8.8(d) incorrectly shows dashed interior lines representing the outline of the hole and the slot (void areas). The outline of a part should never be described using dashed lines. Section lines on the same part must run in the same direction, not opposing directions (on assemblies, matting parts that are sectioned have section lines with differing angles) [Fig. 8.8(e)].

Dimensions or other labeling should not be placed within sectioned areas of the drawing. When this is unavoidable, the section lines are omitted behind the label.

Some features are shown with double-spaced section lines as in Figure 8.9. Here, the cutting plane passes through distinct features of the part. To show the part correctly, the section lining is drawn at the same angle, but the *spacing is doubled.*

8.2.4 Sections as Views

The section view should appear on the same drawing sheet with the cutting plane view. Section views are projected directly from, and perpendicular, to the cutting plane, in

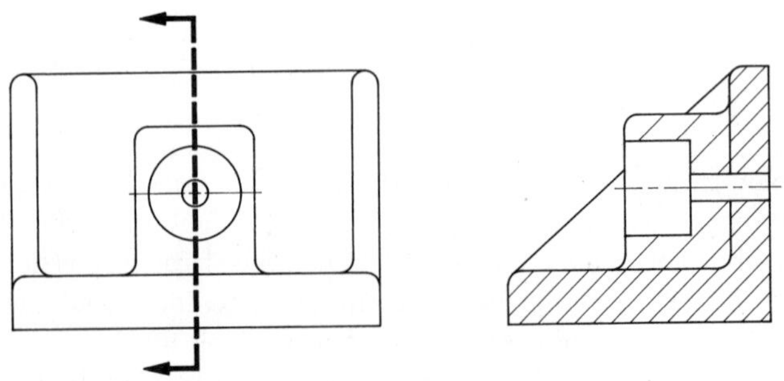

FIGURE 8.9 Double-spaced Section Lines

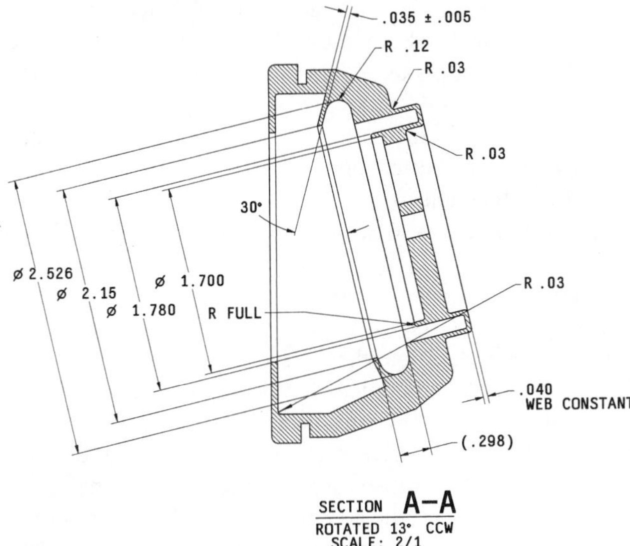

FIGURE 8.10 Rotated Sections

conformity with the standard arrangement of views. If, because of space limitations, this arrangement of views is impractical, the views should be clearly labeled.

The section view is placed so it is in direct projection with the principal view from which it is taken, behind and normal to the cutting plane. The view should not be rotated or shown on a different sheet than the cutting plane *unless* it is necessary due to the size of the view or the drawing space available. However, if rotation is necessary, specify the angle and the direction of rotation below the section label (Fig. 8.10). Here, **SECTION A – A** has been rotated counterclockwise (CCW) out of its normal position 13°. This section is also enlarged. In Figure 8.10, the section identification gives the following information:

SECTION A-A
ROTATED 13° CCW
SCALE: 2/1

The practice of using a section as a principal view is illustrated in Figure 8.4. **SECTION A – A** is the front view and **SECTION B – B** is the right side view of the part. *Avoid constructing a section through a section view.* This can lead to confusion and misinterpretation because it sometimes involves multiple plane rotations. Pass the cutting plane through an exterior view and not through a section view. In Figure 8.4, the cutting plane for **SECTION B – B** could have been drawn in the top view instead of through the front view (**SECTION A – A**).

8.2.5 Cutting Planes

The **cutting plane line** is shown on the view where the cutting plane appears as an edge (Figs. 8.4 and 8.11). The

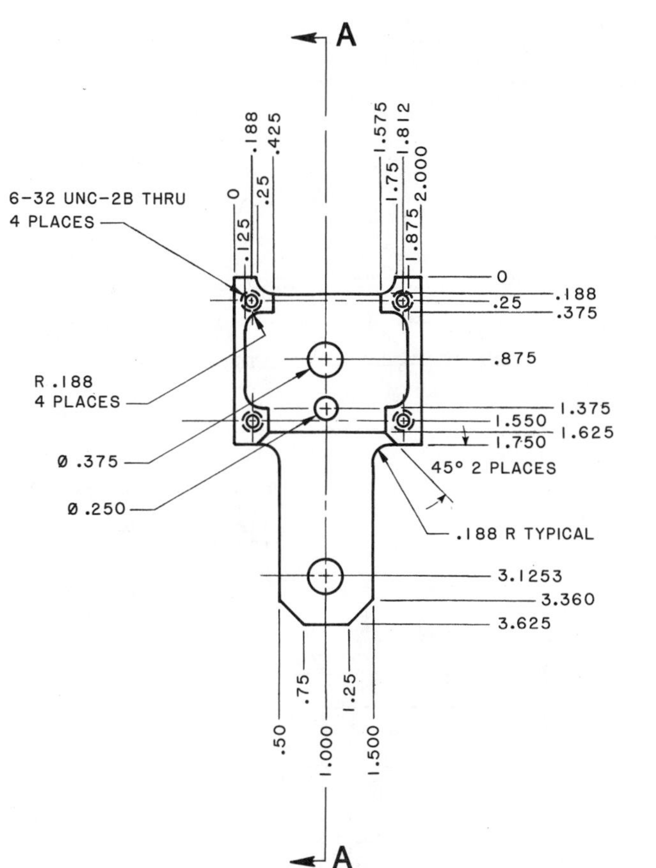

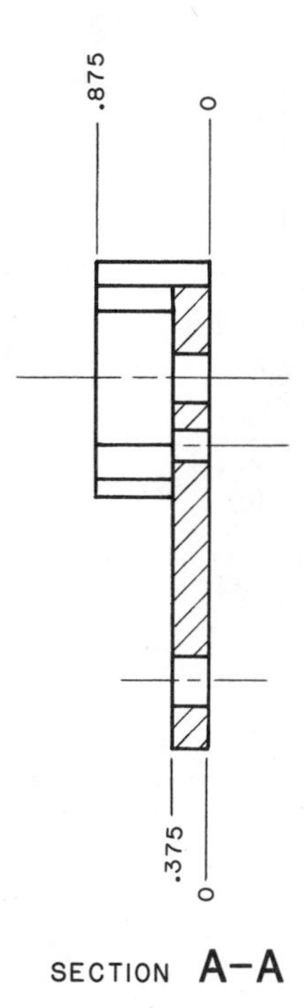

FIGURE 8.11 Section with Correct Cutting Plane Arrow Direction

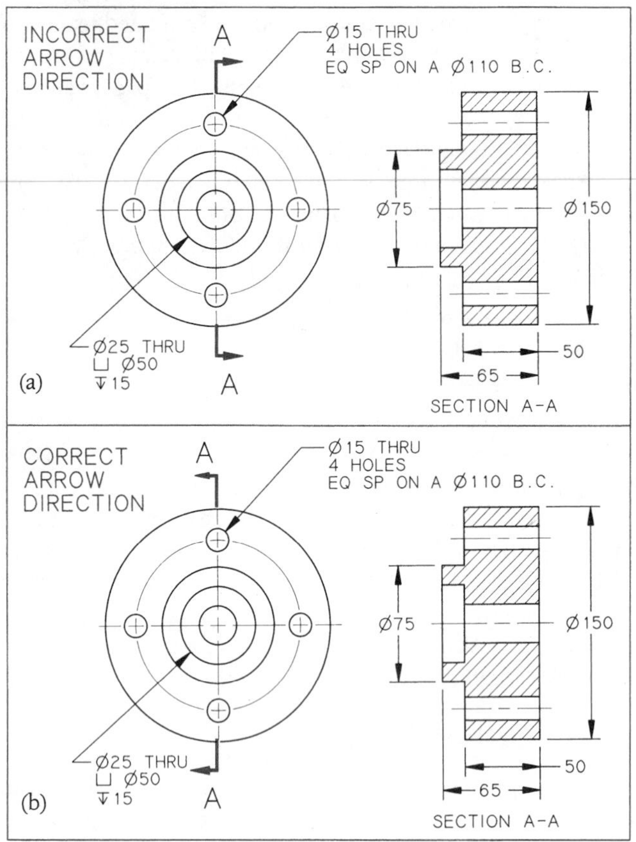

FIGURE 8.12 Arrow Direction on Sections

ends of the cutting plane line are turned 90° and terminated with large arrowheads to show the direction of sight (Fig. 8.4). The cutting plane arrows point away from the viewer and away from the section view. Figure 8.11 shows the proper direction of the cutting plane arrows. Figure 8.12(a) shows the incorrect direction for arrows and in Figure 8.12(b) the correct direction.

In simple sections, or when the location of the section is obvious, the cutting plane line is omitted. The cutting plane line and all identifying letters may be omitted *only* when the location of the cutting plane coincides with a centerline of symmetry (Fig. 8.13) or when the location is obvious. Figure 8.13 is an industry example of a welded pipe fabrication. The pipe and flange are separate pieces that are to be joined by welding. The pieces have section lines drawn at different angles so as to differentiate between the pipe and the flange.

Figure 8.14 shows the accepted sizes and line types to be used when constructing cutting plane lines. The first two examples in this figure follow the accepted ANSI standard. However, some companies use a solid line (third example) or portion of the cutting plane line, or just a portion of the cutting plane line—the bent ends and the arrows (Fig 8.11). The cutting plane line is always shown when the cutting plane is bent or offset or when the resulting section is not

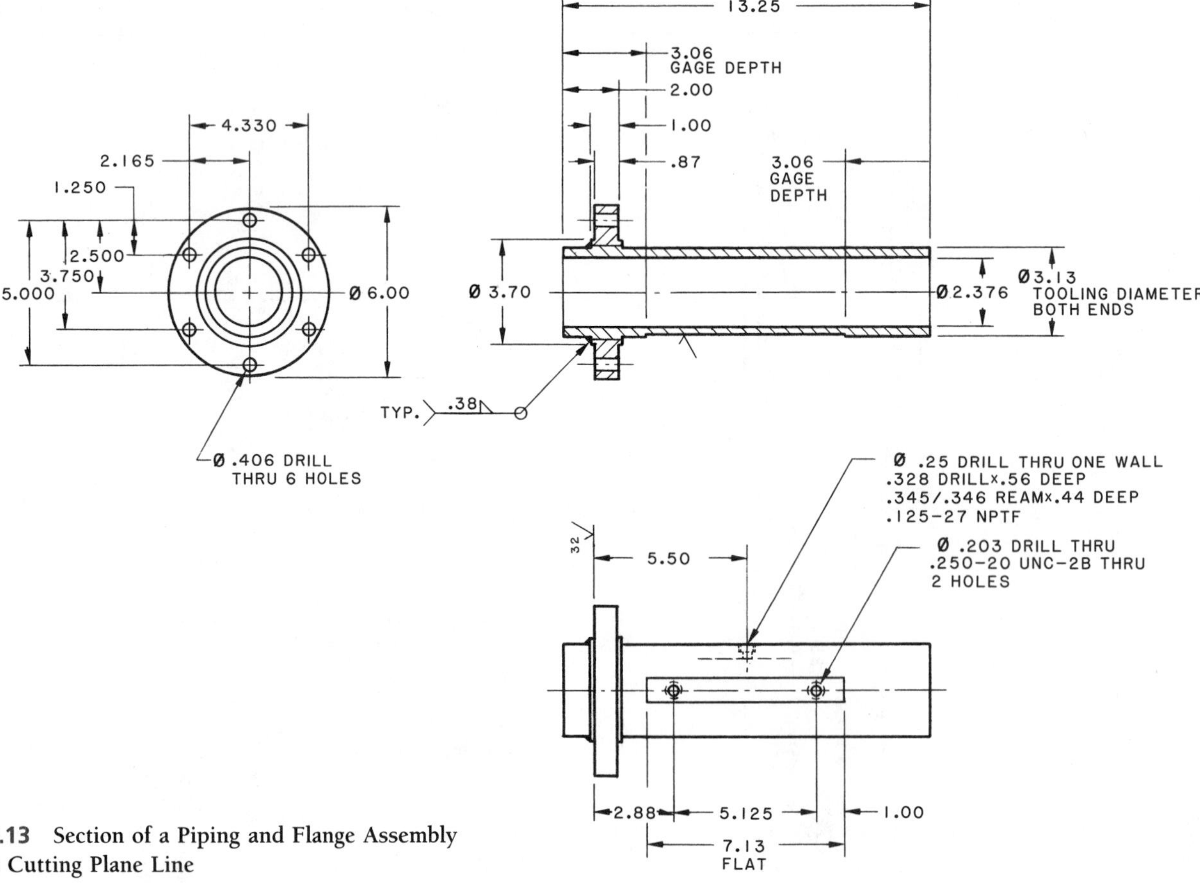

FIGURE 8.13 Section of a Piping and Flange Assembly Without a Cutting Plane Line

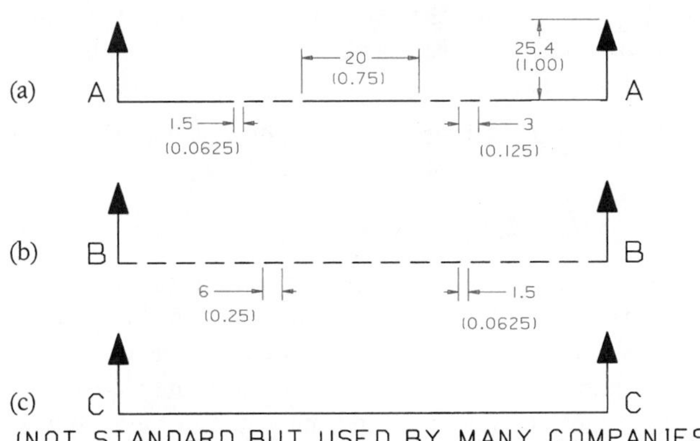

FIGURE 8.14 Dimensions for Drawing Cutting Planes and Arrows (a) Traditional method (b) Dashed method (c) Solid line method

symmetrical. Cutting plane lines are drawn with a 0.7 to 0.9 mm thickness. Border lines and cutting plane lines will be the thickest lines on the drawing.

8.2.6 Section Identification and Multiple Sections

To identify the cutting plane with its sectioned view, section identification letters (A, B, C, etc.) are placed adjacent or behind the arrowheads. The corresponding section views are identified by the same letters. For example, SECTION A–A, SECTION B–B, and SECTION C–C. If two or more sections appear on the same sheet, they are arranged in alphabetical order from left to right and/or top to bottom (Fig. 8.4).

Section letters are used in alphabetical order, excluding I, O, and Q. If all alphabet letters have been used, use double letters for additional section. For example, AA–AA, AB–AB, AC–AC, etc., in alphabetical order.

8.2.7 Conventional Representation

Conventional representation or accepted practice is any recognized practice of description or representation of a part that has been established in industry over time. Ordinarily, conventional representations involve simplifications to speed the drawing task.

For **outline sections**, limited section lines drawn adjacent only to the boundaries of the sectioned area are the preferred conventional representation for large sectioned areas because it eliminates the need to cover large areas with section lines. **Outline section lining** is used only where clarity is not sacrificed (Fig. 8.15).

Thin sections such as sheet metal, packing, and gaskets are drawn solid (filled). When drawing two or more thicknesses or layers, leave a narrow space between them to maintain their separate identities (Fig. 8.16).

8.3 Types of Sections

There are many types of sections used on technical drawings including the following:

1. **Full sections**
2. **Half sections**
3. **Offset sections**
4. **Aligned sections**
5. **Removed sections**
6. **Revolved sections**
7. **Broken-out sections**
8. **Assembly sections**
9. **Auxiliary sections**

A drawing may contain one or more of the sections listed above, as shown in Figure 8.3.

8.3.1 Full Sections

When the cutting plane extends through the entire part, in a straight line, usually on the centerline of symmetry, it results in a **full section** (Fig. 8.17). Full sections are the most common type of section view. In Fig. 8.17(a) a pictorial view of the part is given. In Fig. 8.17(b), a cutting plane is passed through the part and the sectioned area is shown. In Fig. 8.17(c), the line of sight is displayed and the part is split along

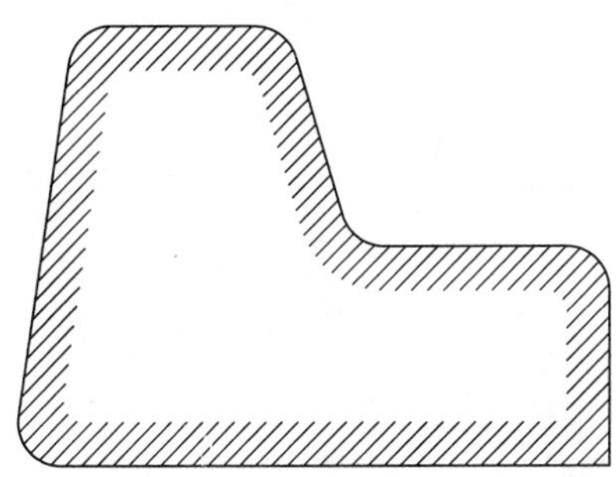

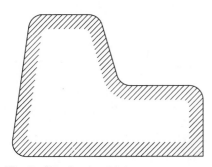

FIGURE 8.15 Outline Section Lining

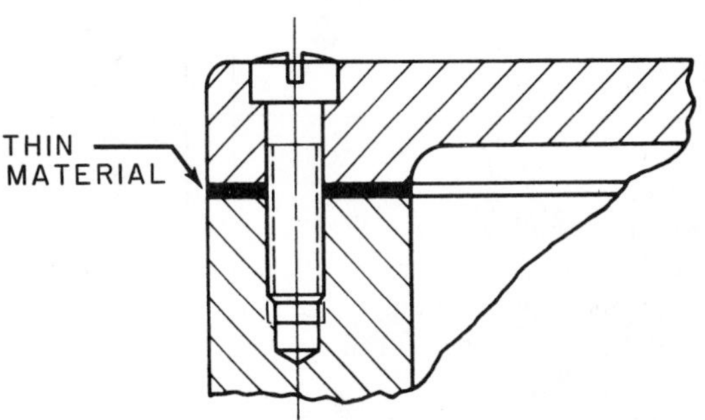

FIGURE 8.16 Thin Materials in Sections

the cutting plane. In Fig. 8.17(d) three views of the part are properly displayed: a front view, a left side view, and a full-section right side view. The cutting plane arrows point away from the section view.

8.3.2 Half Sections

The view of a symmetrical or cylindrical part that represents both the interior and the exterior features by showing one-fourth in section and the other three-fourths as an external view is known as a **half section.** Figure 8.18 is a half section obtained by passing two cutting planes at right angles to each other. The intersection line of the two cutting planes is coincidental with each axis of symmetry of the part. One-fourth of the part is "removed," and the interior is exposed. Figure 8.18(b) shows the part placed in the front and the top views with the front showing the half section. *The line that separates the sectioned half from the nonsectioned half is a centerline and not a visible solid line.*

You May Complete Exercises 8.1 Through 8.4 at This Time

8.3.3 Offset Sections

The cutting plane may be stepped or offset at right angles to pass through features not located in a straight line. **Offset sections** reduce the number of required sections for a complicated part. An offset section is drawn as if the offsets were in one plane. The offsets are not indicated in the sectioned view.

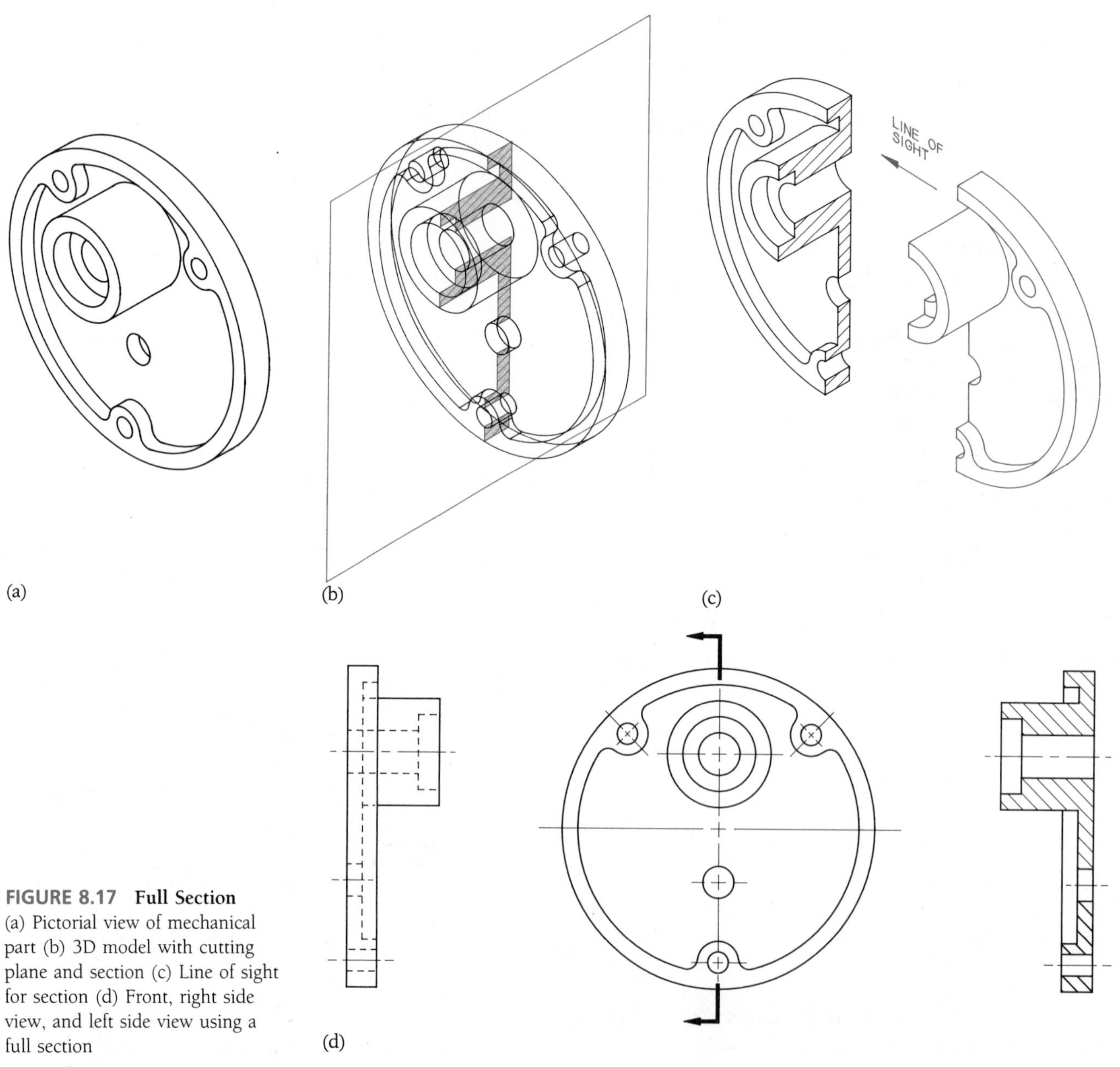

FIGURE 8.17 Full Section
(a) Pictorial view of mechanical part (b) 3D model with cutting plane and section (c) Line of sight for section (d) Front, right side view, and left side view using a full section

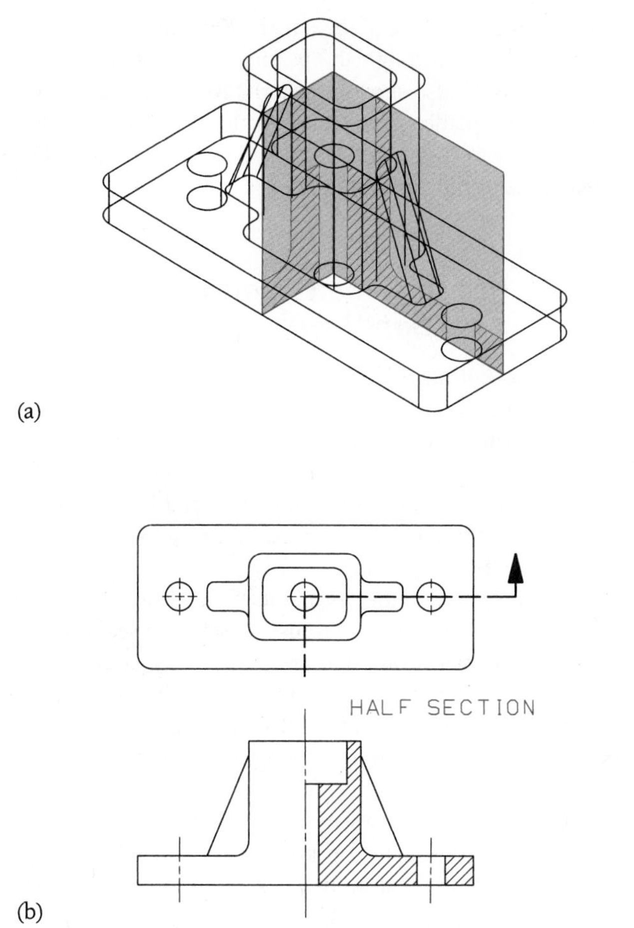

FIGURE 8.18 Half Sections (a) Pictorial illustration of a half section (b) Top view and front half section of a part

The part in Figure 8.19 has important features at three separate positions in the top view. The cutting plane is offset twice, once to pass through the hole and again to pass through the counterbored hole near the back of the part. No line is shown at the offset in the cutting plane line in the section view [Fig. 8.19(d)]. When changes in viewing direction are not obvious, place reference letters at each turning point of the cutting plane.

8.3.4 Aligned Sections

If the true projection of a part results in foreshortening or requires too much drafting time, inclined elements such as lugs, ribs, spokes, and arms are rotated into a plane perpendicular to the line of sight of the section. Section lines are normally omitted for rotated features. This type of section is called an **aligned section** (Fig. 8.20).

Aligned sections are the recommended conventional practice in industry. This convention speeds construction of the view, even though it is not a *true projection*. The true projection is completed only if it is important to establish clearance between features of a part or in an assembly of parts. Holes, slots, and similar features spaced around a bolt circle or a

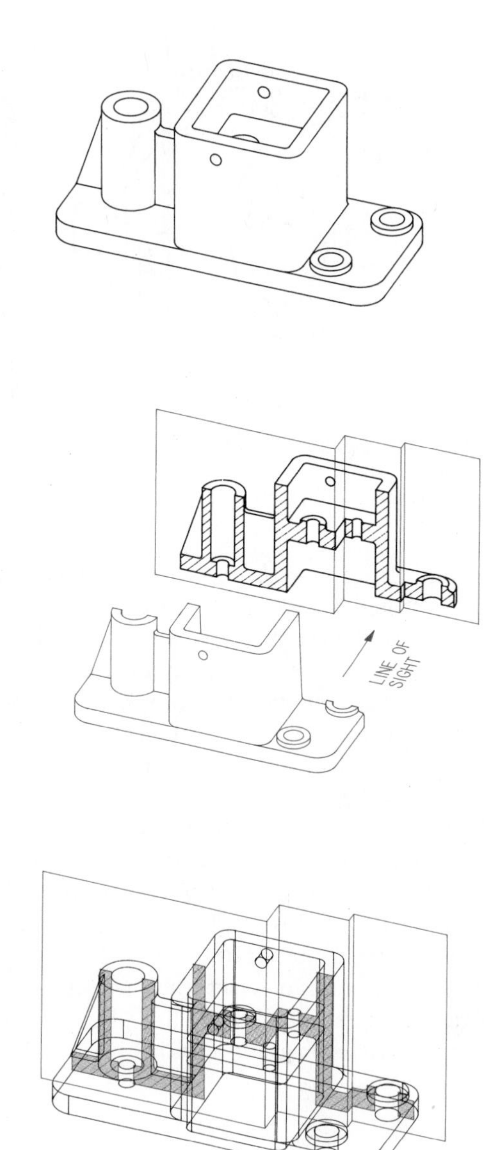

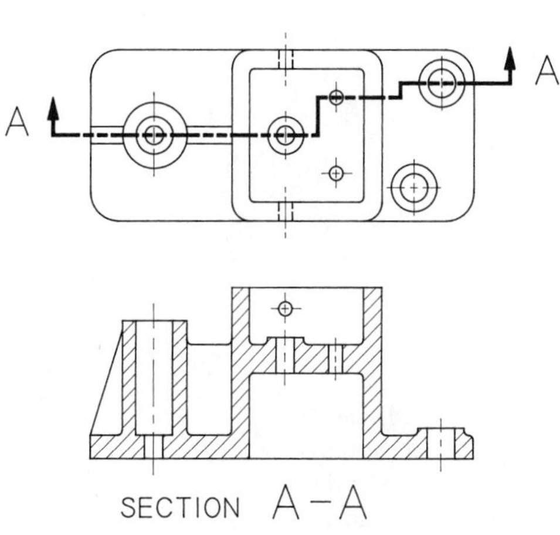

FIGURE 8.19 Multiple Bends in Offset Section

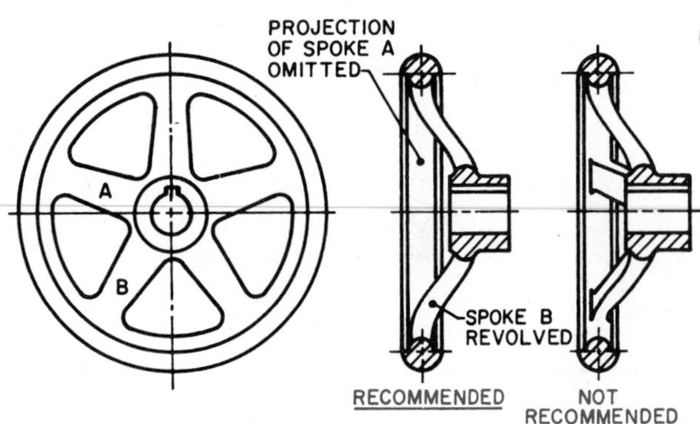

FIGURE 8.20 Spokes in Section

cylindrical flange may also be rotated to their true distance from the center axis and then projected to the section view.

Aligned sectioning preserves the feeling of symmetry, is easier to draw, and is more easily interpreted. The nonrecommended, foreshortened, true projected view of the part is provided to contrast the two methods in Figure 8.20. In this figure, the spokes of the wheel have been rotated to project as true shape in the right side view.

Another example of an aligned section is provided in Figure 8.21. Figure 8.21(a) shows the true front view projection of a part and in Figure 8.21(b) the rib has been rotated. Obviously, Figure 8.21(a) is less clear and a more complex projection than Figure 8.21(b).

When the features of a part lend themselves to an angular change in the direction of the cutting plane less than 90°, the section view is drawn as if the cutting plane and feature were rotated into the plane of the paper. In Figure 8.22, the cutting plane is drawn through the portion to be rotated.

Figure 8.22 also shows an *alternative way of sectioning a rib.* The cutting plane passes through the rib. Instead of leaving the rib area without section lines as is common practice, the area was **double sectioned.**

8.3.5 Nonsectioned Items in a Section View

When the cutting plane lies along the longitudinal axis of shafts, bolts, nuts, rods, rivets, keys, pins, screws, ball or roller bearings, gear teeth, ribs, and spokes, sectioning is not required except when internal construction must be shown. This convention is used mainly on assembly sections where more than one part is sectioned and a number of standard hardware items are found.

For shafts and other machine parts detailed as separate parts, it is normal practice to use broken-out sections for any internal construction that must be displayed. Sections through nuts, bolts, shafts, pins, and other solid machine elements that have no internal construction are not shown sectioned, even though the cutting plane passes through these features. These items are more easily recognized by their exterior. Figure 8.23 shows an example of a sectioned assembly. The shaft in this figure is unsectioned. Figure 8.24 shows another example of a sectional assembly. This shaft is also unsectioned.

When a cutting plane passes through a rib (Fig. 8.25), leave the rib portion of the section without section lining. Because ribs fall into the category of a *thin solid shape,* they are

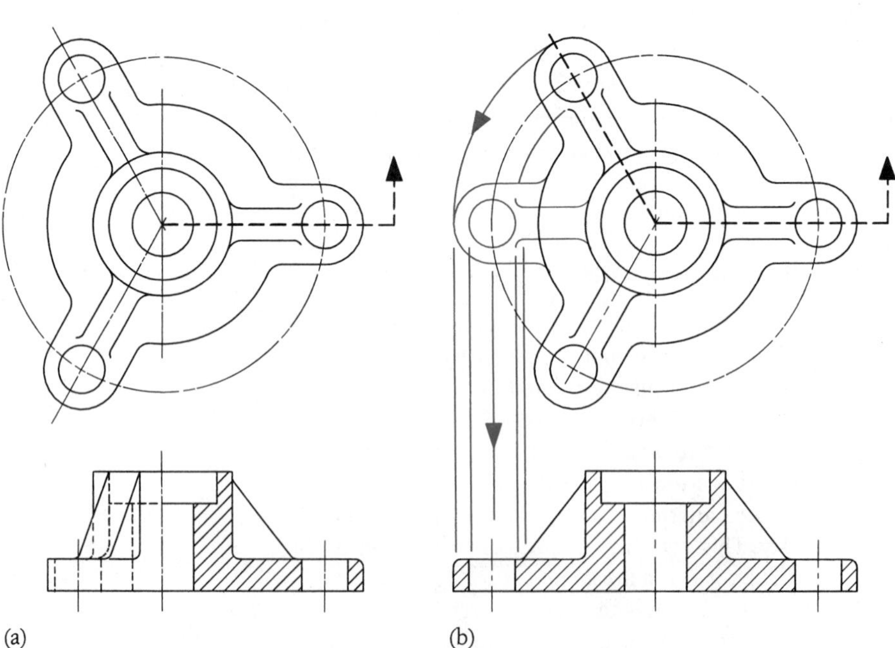

FIGURE 8.21 Full and Half Sections (a) Half section with true front projection of the part (b) Full section with aligned (rotated) frontal projection

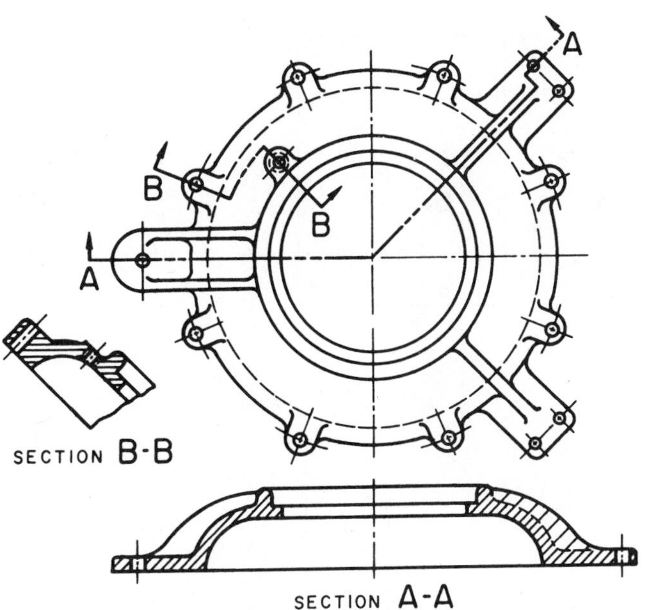

FIGURE 8.22 **Aligned Section Through a Rib** An alternative method of sectioning a rib with double-spaced section lines is also shown.

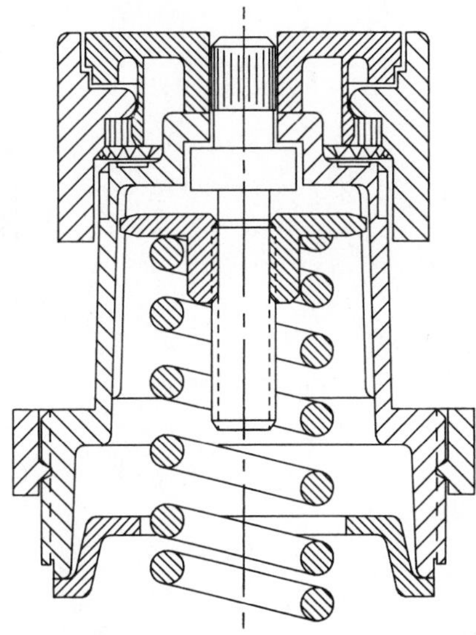

FIGURE 8.24 **Assembly and Solid Threaded Parts in Sections**

usually represented without section lining. Sectioning ribs gives the appearance of more mass than actually exists as in the (incorrect example) of Figure 8.25. Ribs are not sectioned when the cutting plane passes through them flat-wise, but are shown as visible edges. However, ribs are sectioned when the cutting plane passes perpendicular to them.

8.3.6 Removed Sections

Removed sections are used to show special or transitional details of a part, and are like revolved sections, except that they are placed outside the principal view. They may be drawn to a larger scale.

Removed sections that are symmetrical may be placed on centerlines extended from the imaginary cutting planes (Fig. 8.26). A removed section is usually not a direct projection from the view containing the cutting plane line; it is displaced from its normal projection position. In this case, formal identification is used.

If it is impractical to place a removed section on the same sheet with the regular views, clearly identify the sheet number and the drawing zone location of the cutting plane line. On the drawing where the cutting plane is shown, place a note that refers to the sheet and the zone where the removed section or section title is, along with a leader pointing to the cutting plane.

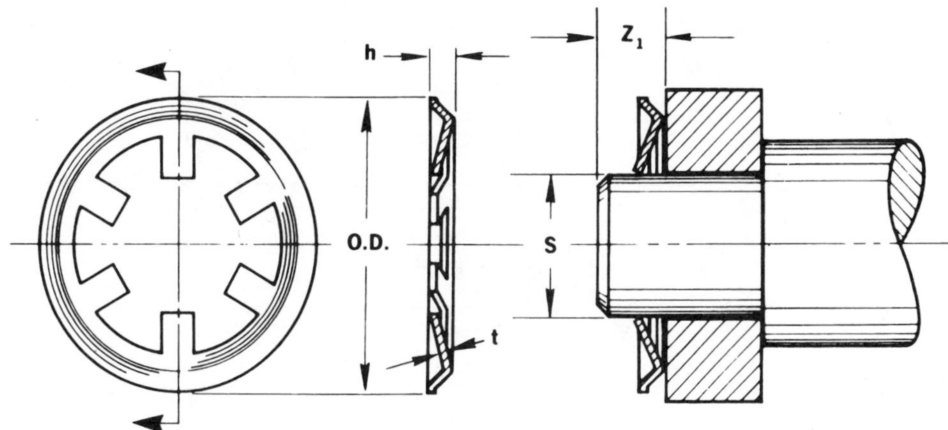

FIGURE 8.23 **Shafts and Solids in Section**

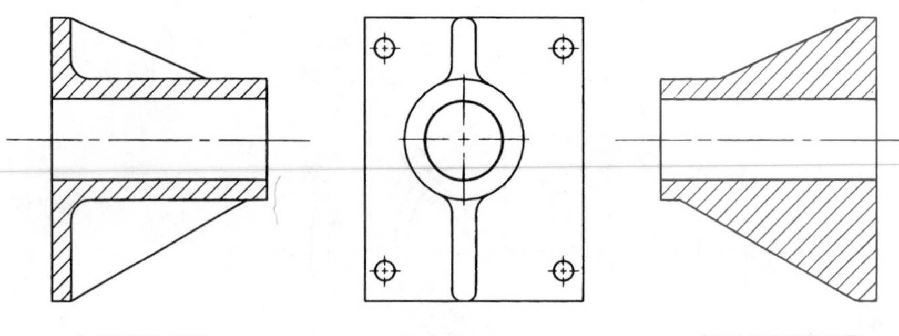

FIGURE 8.25 Ribs in Sectional Views

8.3.7 Revolved Sections

A **revolved section** is constructed by passing a cutting plane perpendicular to the axis of an elongated symmetrical feature such as a spoke, a beam, or an arm, and then revolving it in place through 90° into the plane of the drawing (Fig. 8.27). Visible lines extending on each side of the revolved section may be left in or removed and break lines used. Figure 8.28 uses both methods. The spoke section does not have the visible lines removed and broken, as does the wheel section. Cutting planes are not indicated on this type of section.

8.3.8 Broken-out Sections

When it is necessary to show only a portion of the object in section, the sectioned area is limited by a freehand *break line* and the section is called a **broken-out section** (Fig. 8.29). A cutting plane line is not indicated for this type of section. Broken-out sections are sometimes referred to as partial sections (Fig. 8.3.)

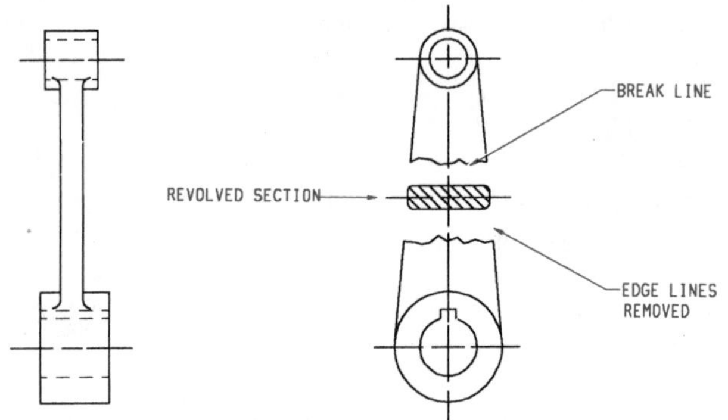

FIGURE 8.27 Revolved Section of an Arm

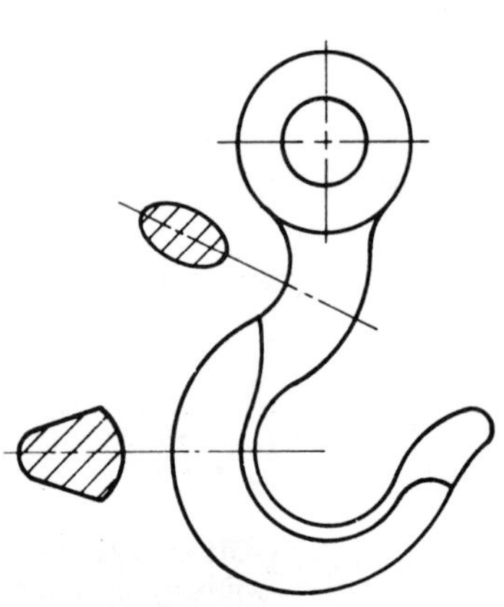

FIGURE 8.26 Removed Sections

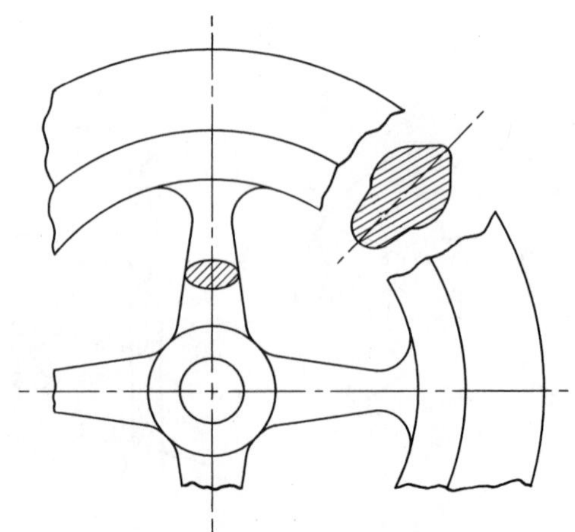

FIGURE 8.28 Revolved Section of a Spoke and Wheel

ITEMS OF INTEREST

Ultrasound

What do submarines, bats, whales, fish finders, and modern hospital technology have in common? They all depend on information gained by using ultrasonic waves.

Ultrasonic waves are vibrations similar to the sound waves that are audible to humans. They are measured by intensity, length, velocity, wave period, and frequency. The number of vibrations per unit time is the frequency of that wave. Waves with a frequency greater than 20,000 Hz are ultrasonic waves.

Ultrasonic waves are generated by passing an electric current through quartz or certain other materials. As an echo strikes the quartz, an electric current is produced, which in turn is used to produce a picture. This property of quartz is called *piezoelectricity.* The generating and receiving device is a *transducer.* In medical equipment, the transducer passes over the part of the body that is being examined.

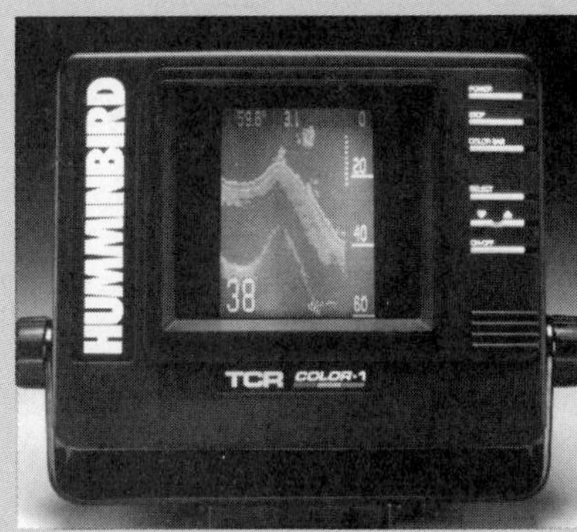

A Fish Finder

Since each tissue varies in density, these waves are reflected differently, producing different images. These images are displayed on a screen or recorded. Internal organs such as the heart and heart valves can be viewed in a static image or, by moving the transducer to different views, can be viewed in sequences in real time.

Recently, tiny ultrasonic transducers have been developed that can produce images from inside blood vessels and ducts. The transducer is rotated 360° to create a series of 2D cross sections. Computers are used to combine these 2D section images into a 3D image.

The sections produced with ultrasound equipment for medical applications are not unlike the sectional drawings used in mechanical drawing. Both types of sections show internal details. It doesn't really matter whether they are the internal features of a part or of an organ. The concepts are the same. Mechanical engineers also use ultrasound waves to check for internal defects, such as cracks and small holes (voids). Sectional views in medicine and in graphics are intended to show internal features that would not otherwise be visible. Sectional views are valuable for visualization regardless of the application involved.

8.3.9 Breaks and Sectioning

Conventional breaks are used to shorten a view of an elongated part (Fig. 8.30) and in broken-out sections. The type of break representation is determined by the material and the shape of the part. Solid and tubular rounds are shown in Figures 8.30(a) and (b). The break can be drawn with the aid of an ellipse template or constructed manually. Tubular

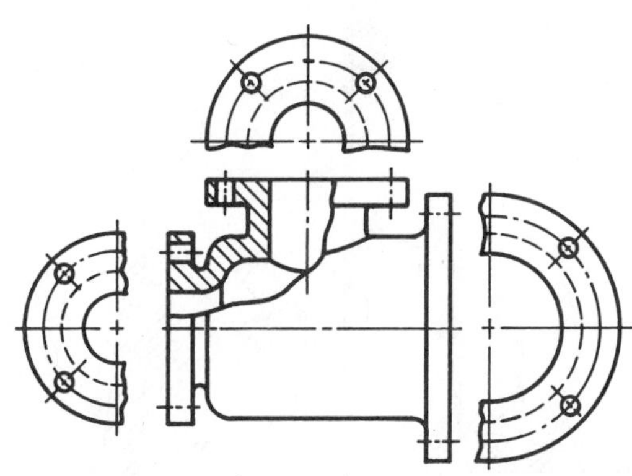

FIGURE 8.29 **Broken-out Section of a Pipe Fitting**

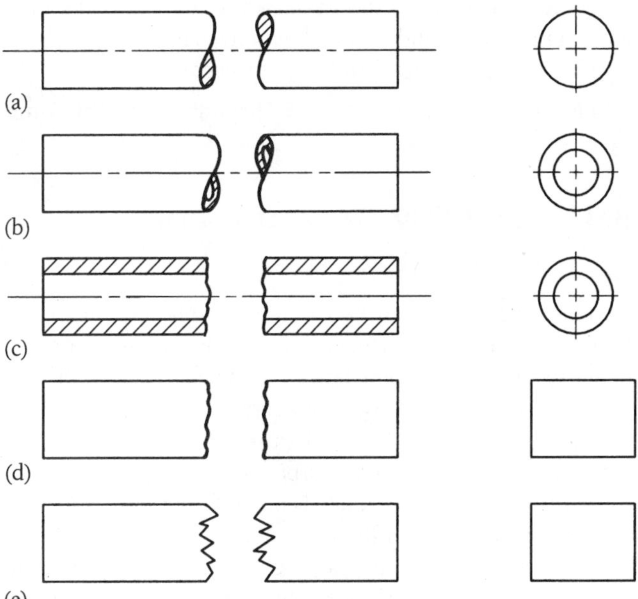

FIGURE 8.30 **Conventional Representation of Breaks in Elongated Parts** (a) Solid rod (b) Round tube (c) Sectioned tube (d) Rectangular bar (e) Wood

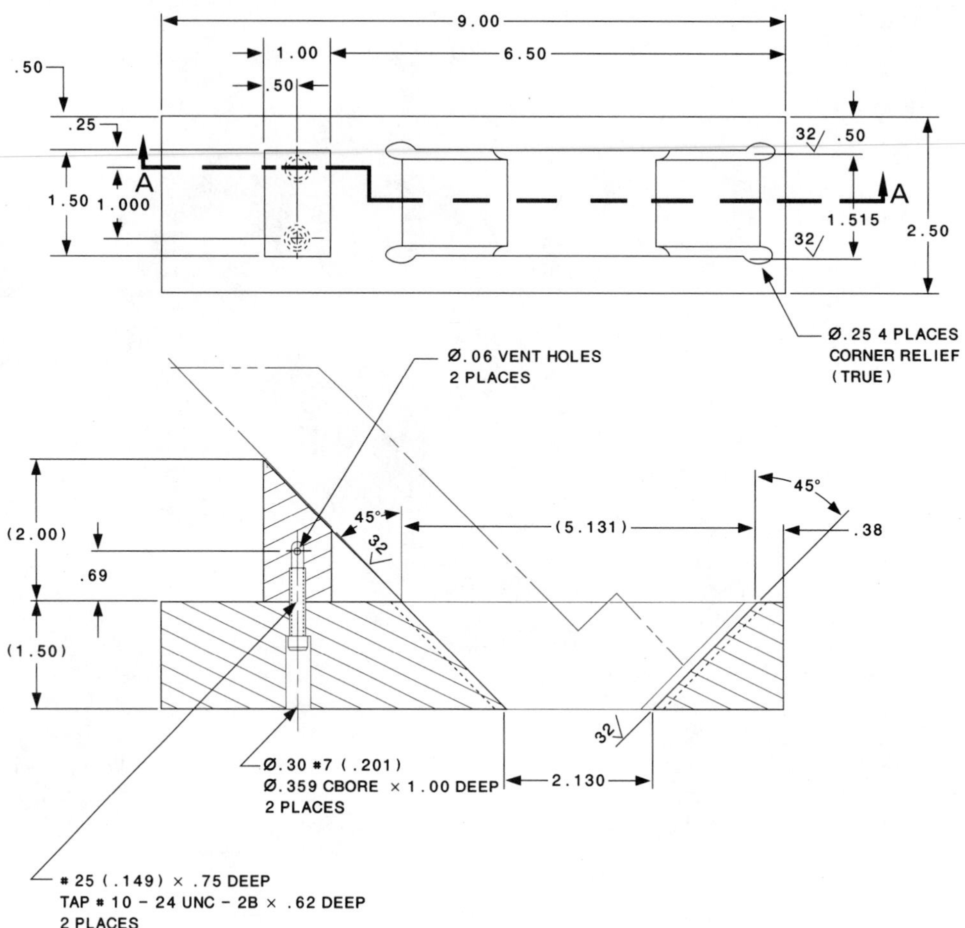

FIGURE 8.31 Assembly of Fixture with a Full Front Section View

shapes are sectioned as shown in Figure 8.30(c). Break lines for Figures 8.30(c) and (d) are drawn freehand. The break for wood is also drawn freehand but is jagged.

You May Complete Exercises 8.5 Through 8.8 at This Time

8.4 Assembly Drawings and Sectioning

Assembly sections show two or more mating parts in section (Figs. 8.23 and 8.24). General-purpose section lines are normally used on assembly drawings. When several adjacent parts are shown in a section view, the parts are sectioned as shown in the Figure 8.31. The fixture has its two major parts sectioned using the general-purpose sectioning symbol. Because the piece to be machined is not a portion of this fixture, it is shown in phantom lines and is not sectioned.

Figure 8.32 shows the jack assembly as a front section. Each individual part of the assembly has section lines running in different directions than adjoining parts. The threaded parts and other solid items are not sectioned.

You May Complete Exercises 8.9 Through 8.12 at This Time

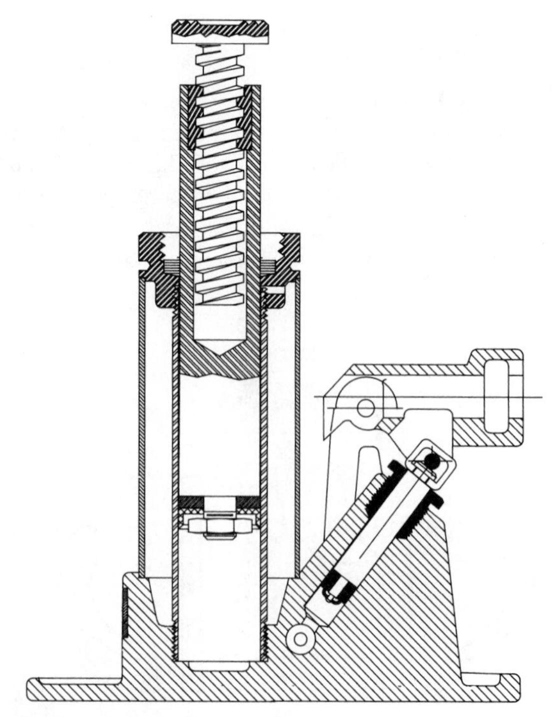

FIGURE 8.32 Assembly Section

QUIZ

True and False

1. Sections are used to describe the exterior of a part so that fewer views are required.
2. Section views are always rotated 90° as projections from existing views.
3. It is common conventional practice to show all hidden lines that fall behind the cutting plane.
4. The cutting plane arrows always point in the direction of sight.
5. Section lines should be drawn thick, black, and close together so as to be readily seen and identified.
6. Material-specific hatching symbols are used on all drawings.
7. Placing of dimensions within sectioned areas is a common and accepted practice.
8. Intersections in sections always show the true projection of the elements.

Fill in the Blanks

9. A section is an ________ cut taken through an ________ .
10. Section lining on assembly drawings should be drawn at ________ angles for each ________ .
11. A ________ taken through an existing ________ view should be avoided.
12. Section lettering for identification of sections and views should be used in ________ ________ .
13. Thin sections are always shown ________.
14. ________, ________, ________, and ________ are usually not shown sectioned.
15. The ________ - ________ symbol is used on most sectional drawings.
16. On simple drawings or where the section location is obvious it is common practice to ________ the ________ ________ ________ .

Answer the Following

17. What is the difference between a removed section and a revolved section?
18. When is a broken-out section likely to be used?
19. Describe the difference between a full section, a half section, and an external view.
20. What is an offset section and when is it used ?
21. What type of part features are rotated in aligned sections?
22. Define a cutting plane.
23. Name and describe three conventional practices used on sections.
24. Describe the difference between an aligned section and a revolved section.

EXERCISES

Exercises may be assigned as sketching, instrument or CAD projects. Transfer the given information to an "A" size sheet of .25 in. grid paper. Complete all views and solve for proper visibility, including centerlines, object lines, and hidden lines. Exercises that are not assigned by the instructor can be sketched in the text to provide practice and understanding for the preceding instructional material.

After Reading the Chapter Through Section 8.3.2 You May Complete the Following Exercises

Exercises 8.1(A) and (B) Draw the two views of the part and do a full section for the front view.

Exercises 8.2(A) and (B) Draw three views of the part. Construct a full front section.

Exercises 8.3(A) and (B) Section the appropriate views for each problem.

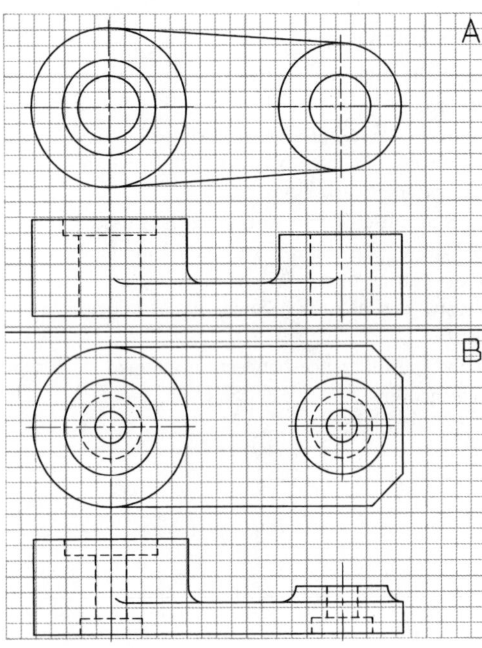

Exercise 8.1

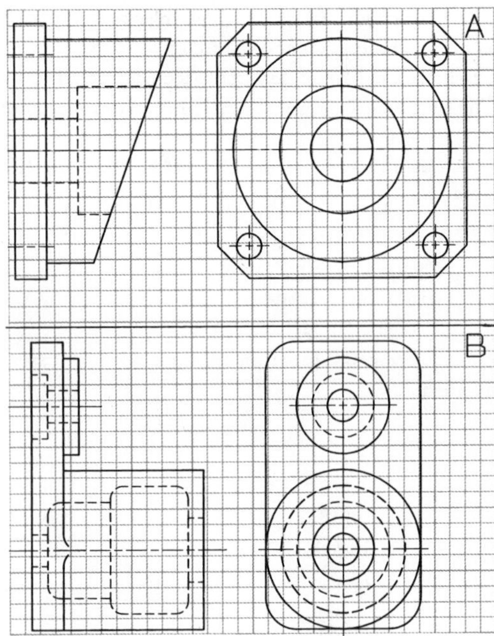

Exercise 8.3

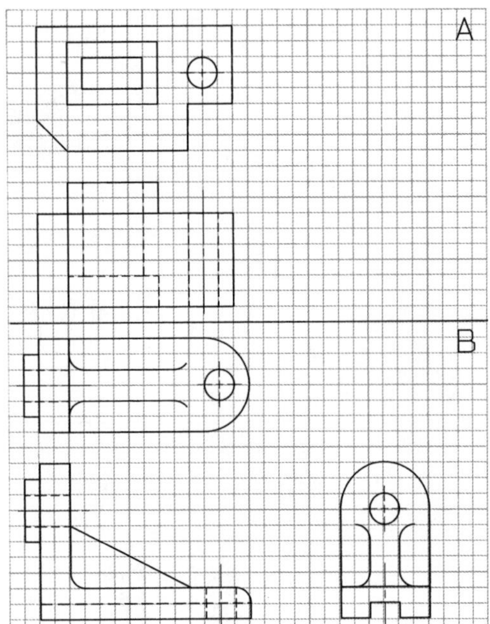

Exercise 8.2

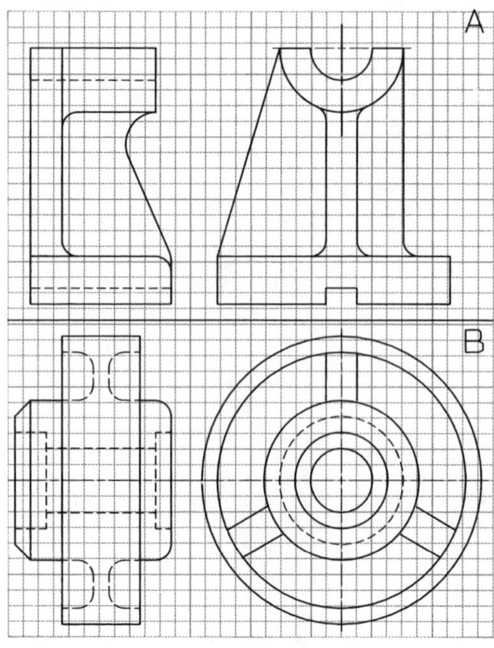

Exercise 8.4

Exercise 8.4(A) Draw a full left side section.

Exercise 8.4(B) Draw the two views. Construct a half section for the left side view.

After Reading the Chapter Through Section 8.3.9 You May Complete the Following Exercises

Exercises 8.5(A) and (B) Construct a full left side view section for each part.

Exercises 8.6(A) and (B) Draw half sections of the parts.

Exercises 8.7(A) and (B) Draw full sections of the parts.

Exercises 8.8(A) and (B) Draw half sections of the parts.

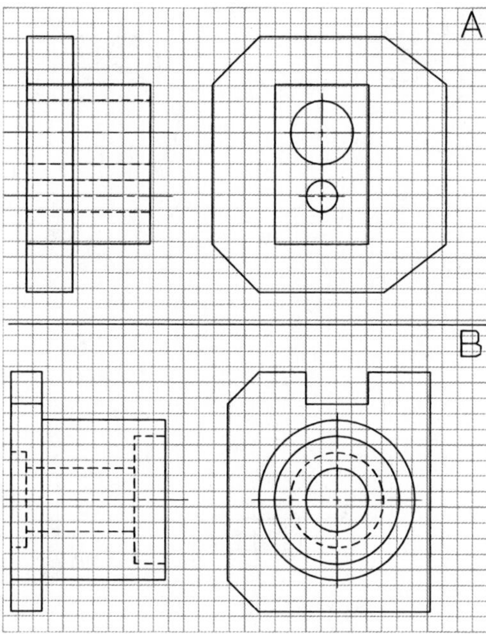

Exercise 8.5

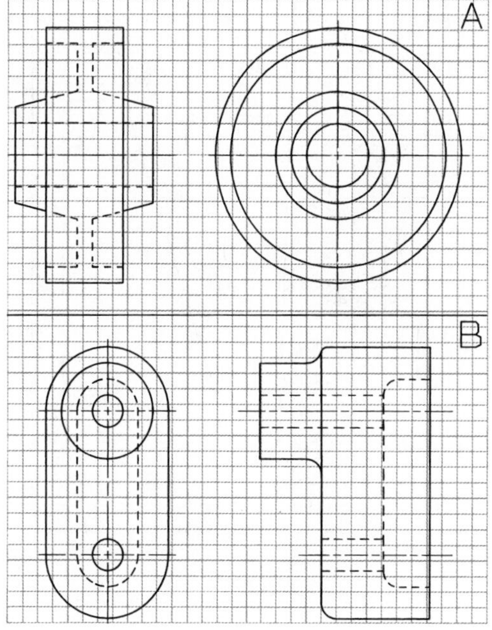

Exercise 8.7

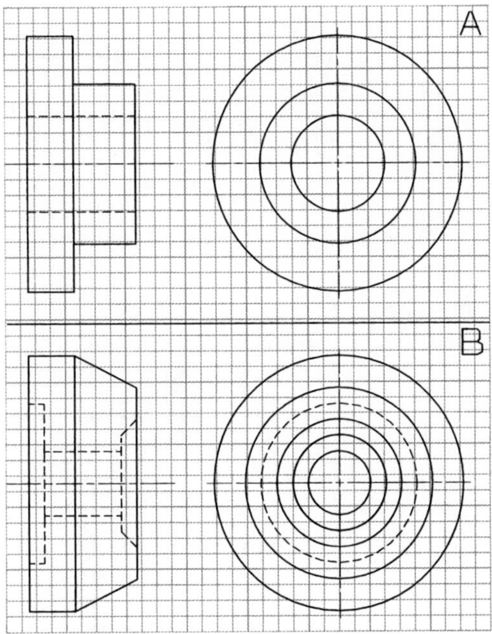

Exercise 8.6

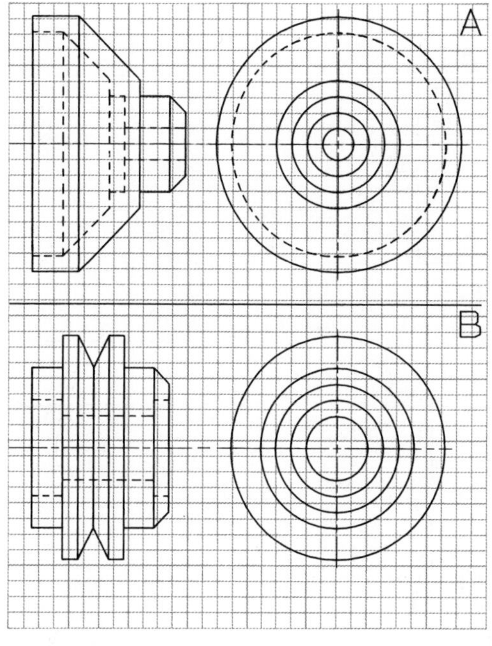

Exercise 8.8

After Reading the Chapter Through Section 8.4 You May Complete the Following Exercises

Exercise 8.9 Section the right side of the part.

Exercise 8.10 Section the whole part in the right side view and construct a partial (broken-out) section as required for hub in the front view (left). The right side view is an aligned view.

Exercise 8.11 Draw an offset section of the part. Pass the cutting plane through the two holes and the slot.

Exercise 8.12 Draw a complete full section of the assembly.

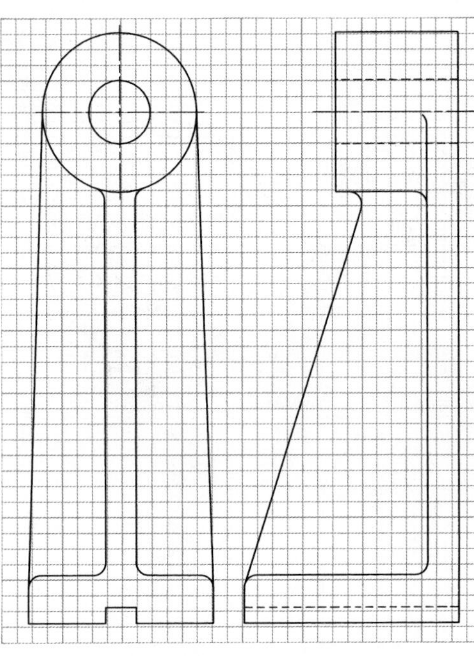

Exercise 8.9

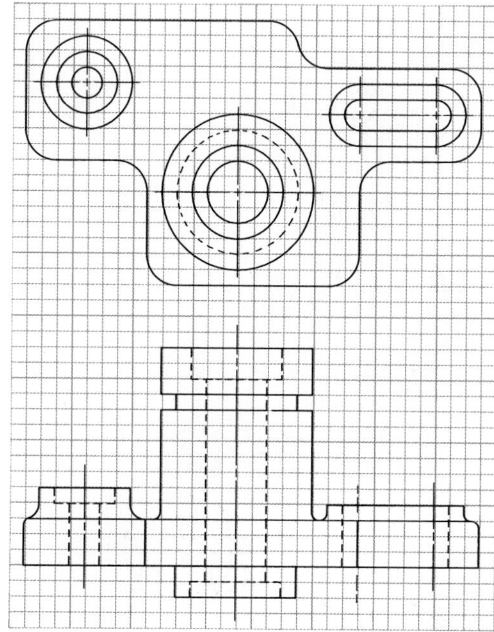

Exercise 8.11

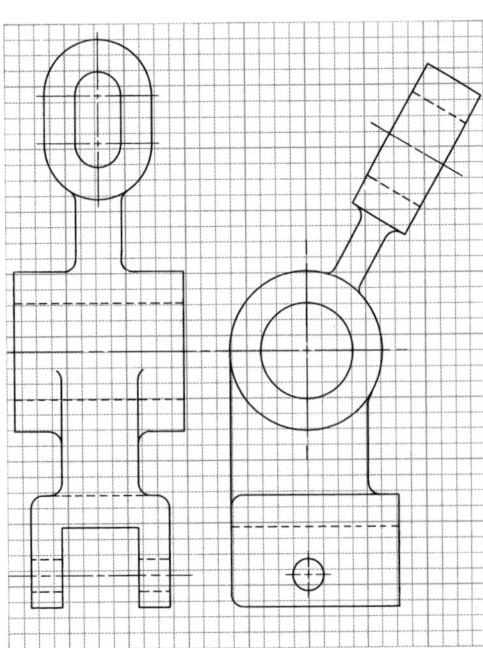

Exercise 8.10

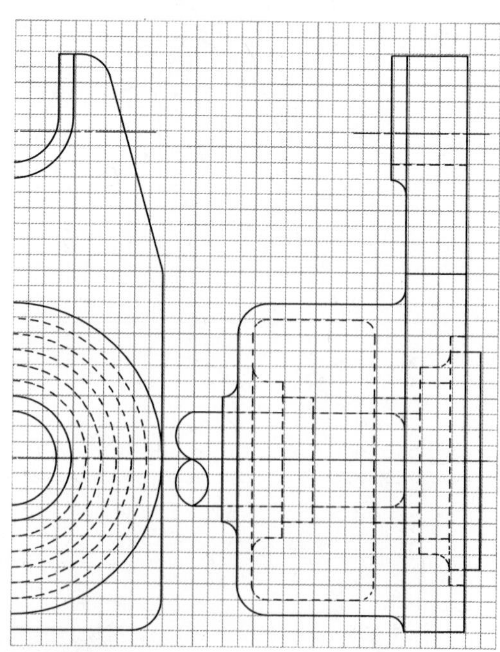

Exercise 8.12

PROBLEMS

To use these same projects for dimensioning, allow enough space between views and use an appropriate size sheet of paper when completing these problems. Complete all views and solve for proper visibility, including centerlines, object lines, and hidden lines. Do not dimension any of the following problems until you complete Chapter 11, Dimensioning, or are requested to do so by your instructor.

Problems 8.1(A) through (K) Using the scales provided, draw and section the appropriate views. Problems can be either metric, fraction-inches, or decimal-inch units. One, two, or three views may be required for a particular problem.

Problems 8.2(A) through (H) Same as Problem 8.1.

Problems 8.3 through 8.11 Establish the views and sections required to describe the part properly. Do not dimension the parts. Use half sections, broken-out sections, aligned sections, and revolved sections where useful to describe the part.

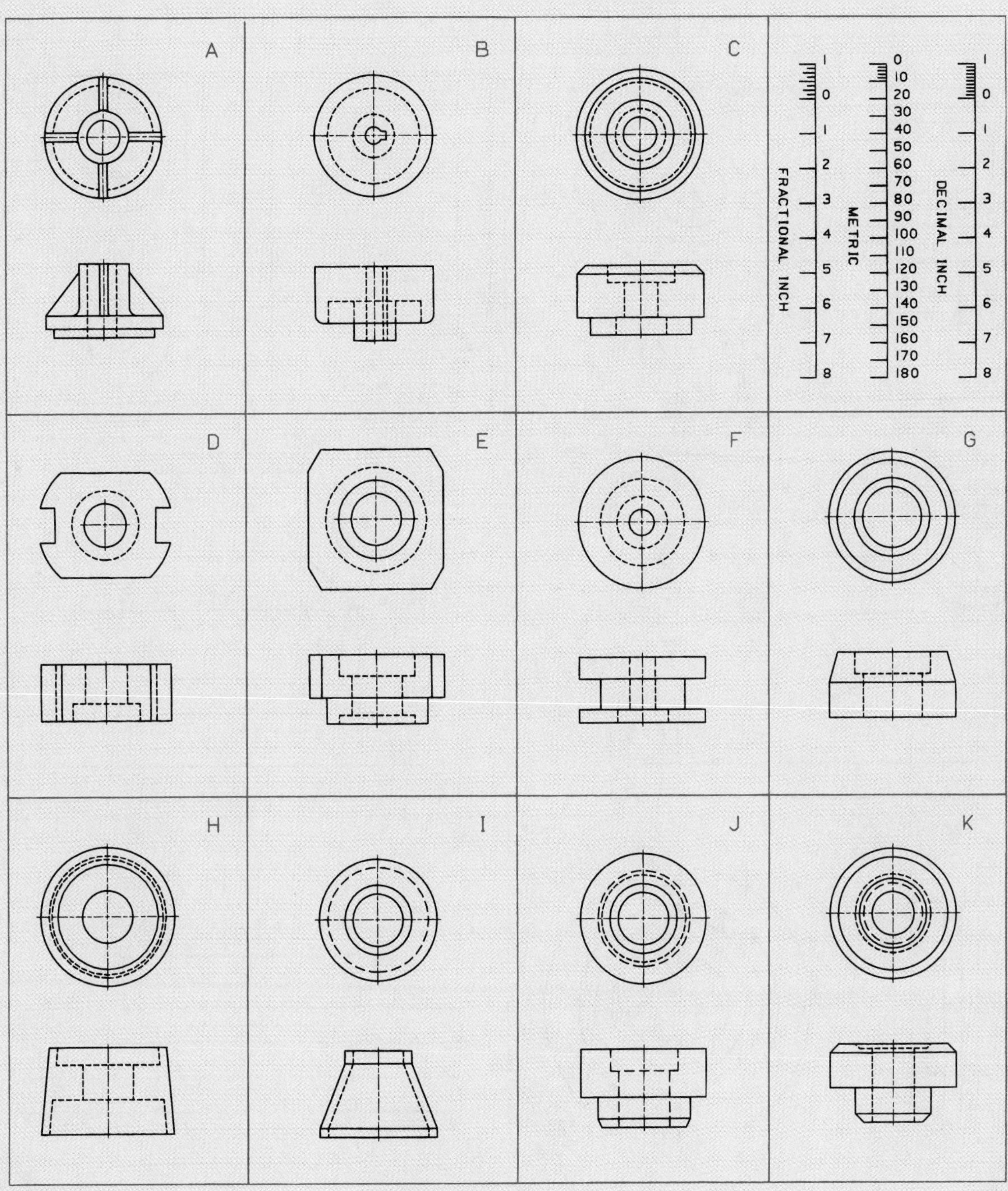

Problem 8.1 (A) Through (K)

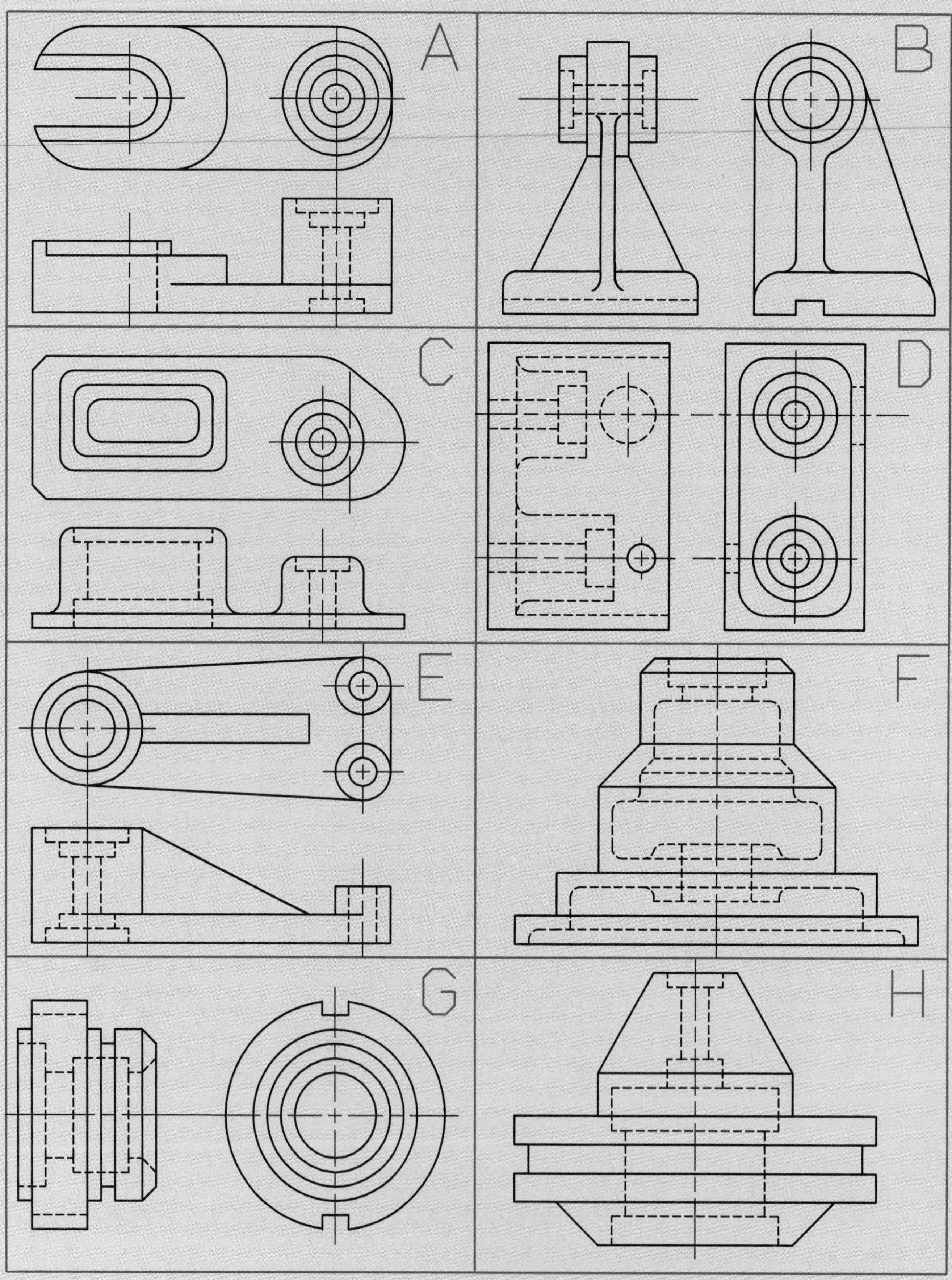

Problem 8.2

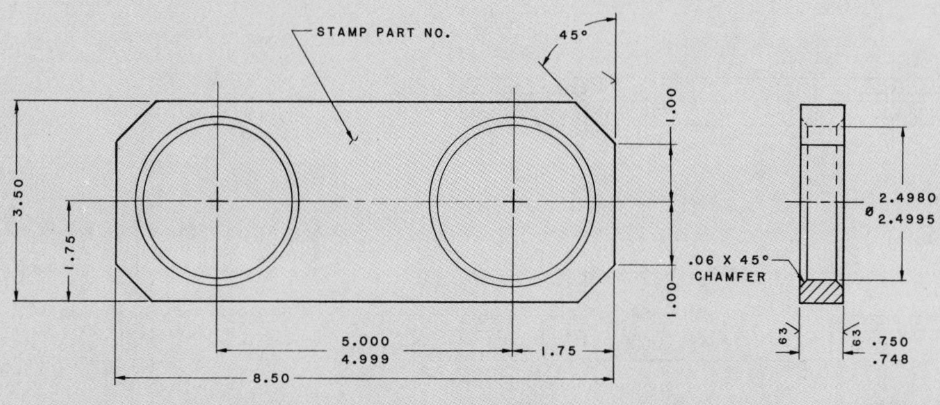
STAMP PART NO.
45°
1.00
3.50
1.75
Ø 2.4980
2.4995
.06 X 45°
CHAMFER
1.00
5.000
4.999
1.75
8.50
63
63
.750
.748

Problem 8.3

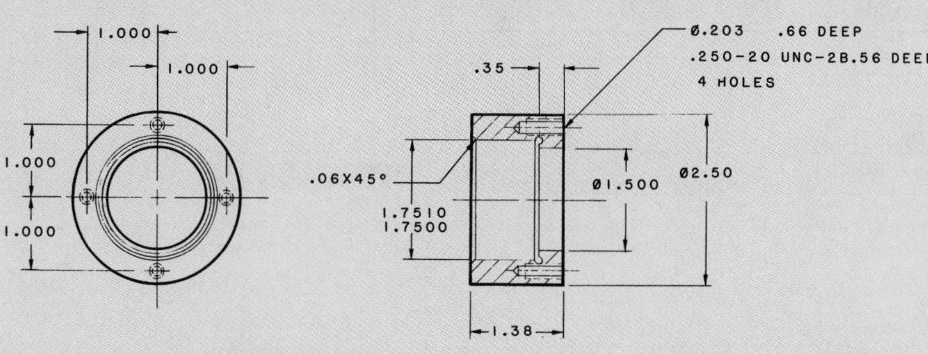
1.000
1.000
1.000
1.000
Ø.203 .66 DEEP
.250-20 UNC-2B.56 DEEP
4 HOLES
.35
.06X45°
Ø1.500
Ø2.50
1.7510
1.7500
1.38
HOUSING BEARING
GUIDE ROD

Problem 8.4

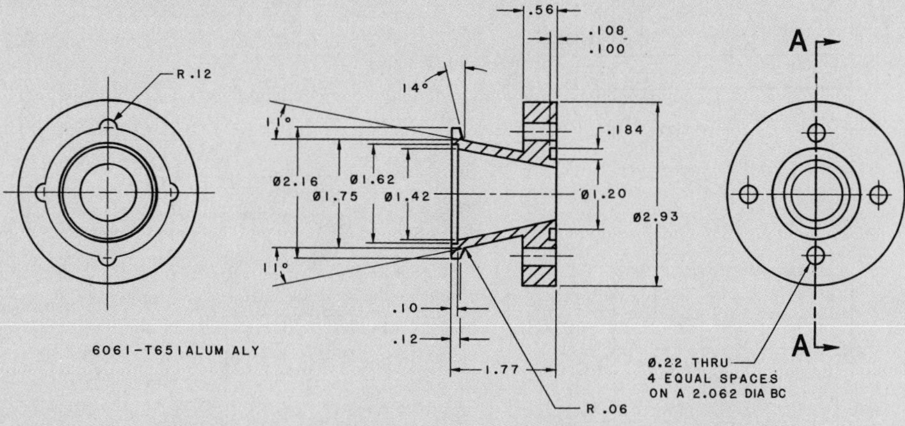
R.12
.56
.108
.100
14°
11°
.184
Ø2.16
Ø1.62
Ø1.75
Ø1.42
Ø1.20
Ø2.93
11°
.10
.12
1.77
R .06
6061-T651ALUM ALY
A
A
Ø.22 THRU
4 EQUAL SPACES
ON A 2.062 DIA BC
SECTION A–A

Problem 8.5

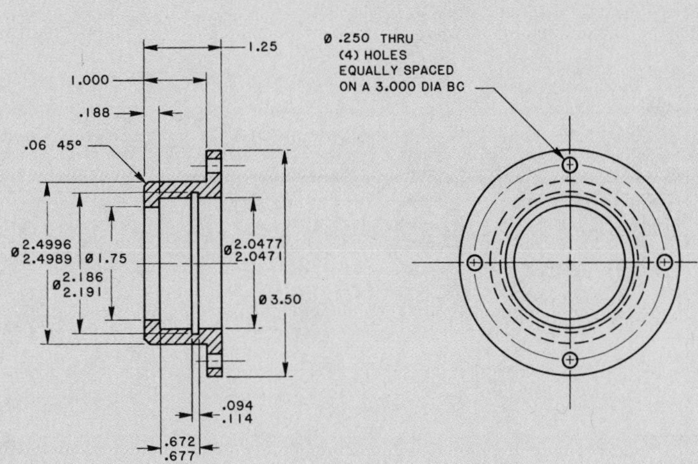
1.25
1.000
.188
.06 45°
Ø .250 THRU
(4) HOLES
EQUALLY SPACED
ON A 3.000 DIA BC
Ø 2.4996
2.4989
Ø 1.75
Ø 2.186
2.191
Ø 2.0477
2.0471
Ø 3.50
.094
.114
.672
.677

Problem 8.6

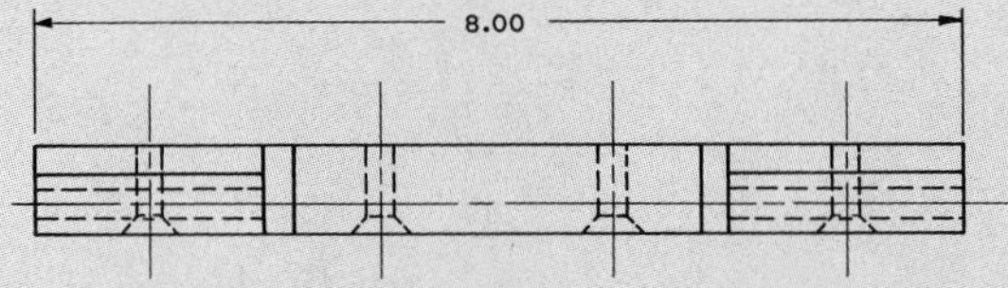

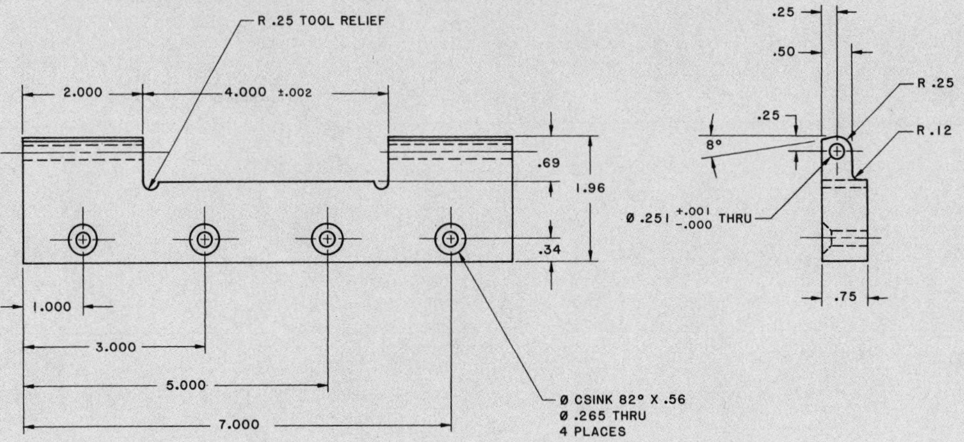

Problem 8.7

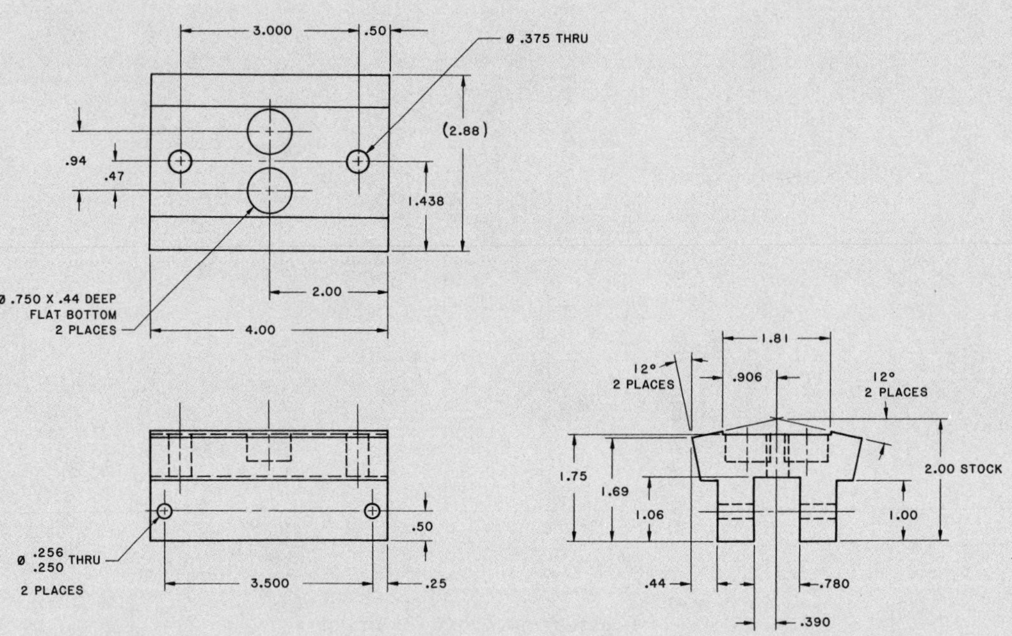

Problem 8.8

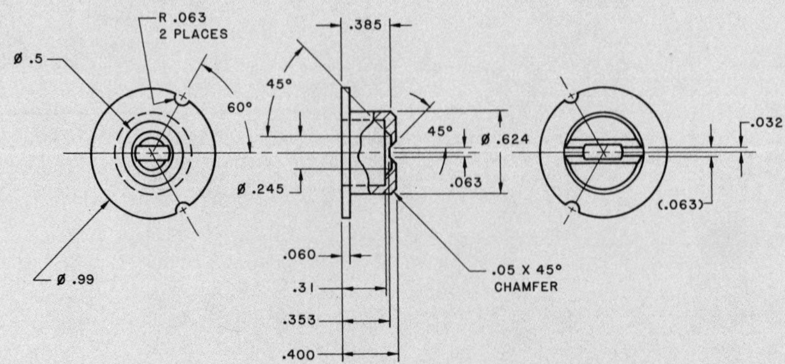

Problem 9.9

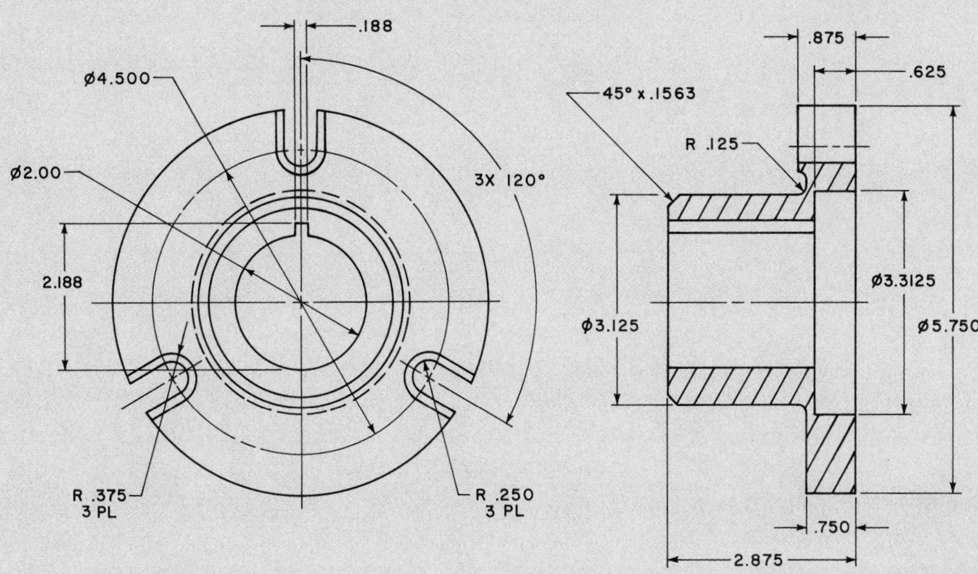

Problem 8.10

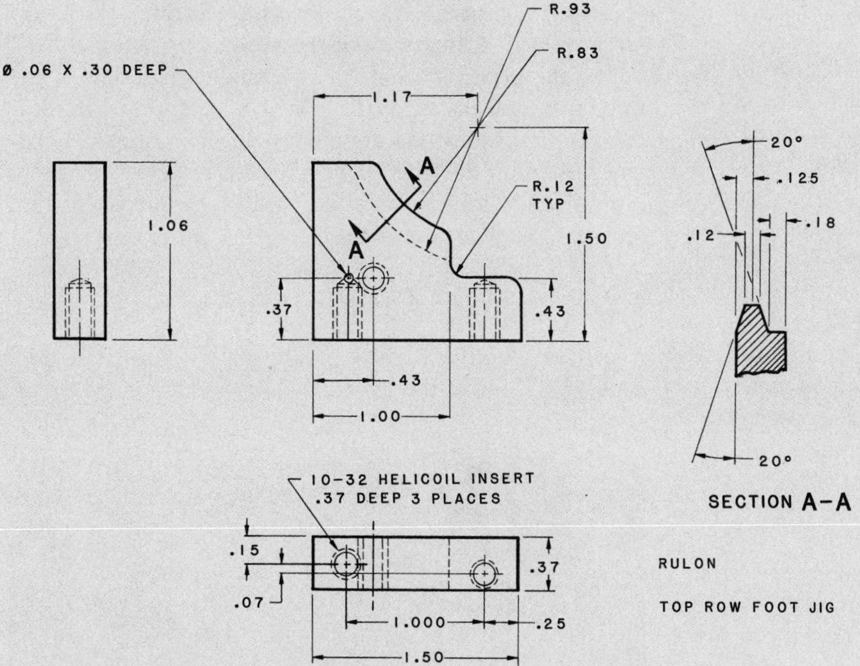

Problem 8.11

Problem 8.12 Redraw the part in Figure 8.4. Construct a top view, sectioned front view, and a left side view section.

Problem 8.13 Using Figure 8.11 construct a three-view drawing of the part. Provide a left side view section.

Problem 8.14 Redraw Figure 8.12.

Problem 8.15 Redraw the weldment shown in Figure 8.13.

Problem 8.16 Draw three views of the assembly shown in Figure 8.31. Section as needed.

CHAPTER 9

Auxiliary Views

Learning Objectives

Upon completion of this chapter you will be able to accomplish the following:

1. Identify the need for auxiliary views in order to show the actual shape, size, and relationship of a part feature that may not be parallel to any of the principal planes of projection.
2. Differentiate between and demonstrate ability to produce primary and secondary auxiliary views using the fold line method as well as the reference plane method.
3. Solve for the true shape of an angled surface using an auxiliary view.
4. Develop the ability to produce partial, broken, half, and sectional auxiliary views.

9.1 Introduction

Auxiliary views show the true shape/size of a feature, or the relationship of part features that are not parallel to any of the principal planes of projection. Many parts have inclined surfaces and features that cannot be adequately described using only principal views.

In Figure 9.1, the anchor has an inclined surface that cannot be shown true shape in a principal view. The detail of the part (Fig. 9.2) used an auxiliary view to describe the inclined surface and the hole true shape and size.

Auxiliary views are also used to dimension features that are distorted in principal views and graphically solve a variety of engineering problems, such as the interference between two parts or clearances between pieces of an assembly.

9.1.1 Selection and Alignment of Views

The proper selection of views, view orientation, and view alignment is determined by the features of a part and its assembled position. Usually, the front view is the primary view and the top view is obvious based on the position of the part in space. The choice of additional views is determined by the configuration of the part and the minimum number of views necessary to describe the part and show its dimensions.

As with all multiview drawings, auxiliary views are aligned with the views from which they are projected. In many cases, a centerline or a projection line continues between adjacent views to indicate the proper alignment (Fig. 9.3).

9.2 Auxiliary Views

Any view that lies in a projection plane other than the horizontal, frontal, or profile plane (or a plane parallel to one of these) is an **auxiliary view.**

Auxiliary views are classified by the view from which they are projected. **Primary auxiliary views** are projected from one of the principal views, are perpendicular to one of the three principal planes, and inclined to the other two. **Secondary auxiliary views** are projected from a primary auxiliary view and are inclined to all three principal planes of projection. **Successive auxiliary views** are projected from secondary auxiliary views.

9.2.1 Primary Auxiliary Views

A **primary auxiliary view** is one that is *adjacent* to and *aligned* with one of the principal views and are identified as

FIGURE 9.1 Anchor

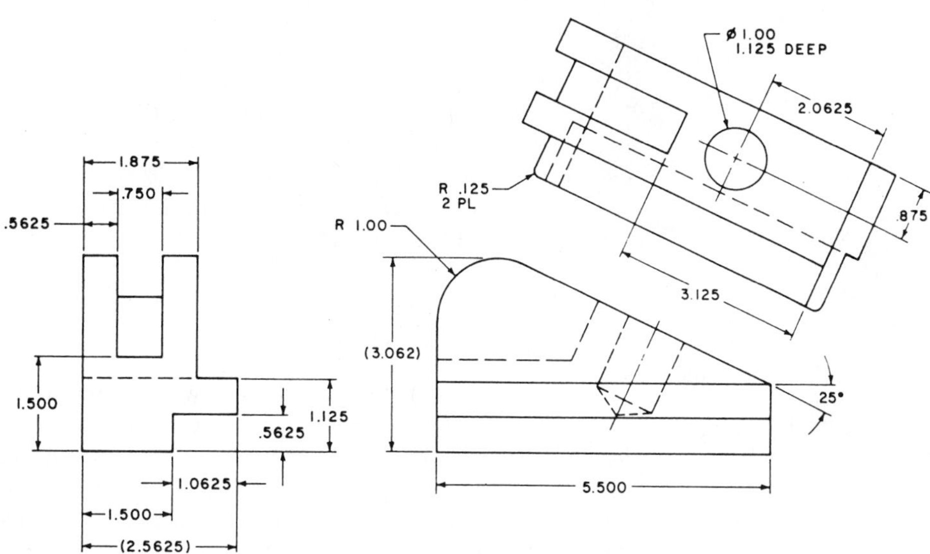

FIGURE 9.2 Detail of Anchor Showing Auxiliary View

front-adjacent, **top-adjacent**, or **side-adjacent** to indicate the principal view with which it is aligned. In industry, auxiliary views are used to show aspects of a mechanical part or portions of a system such as piping configurations that cannot be adequately represented in the three principal views. The machined block in Figure 9.4 required three primary auxiliary views to clarify the shape of the angled surfaces and the position of holes and slots.

Primary auxiliary views are divided into three types. Primary auxiliary views projected from the top (top-adjacent) view are **horizontal auxiliary views.** Primary auxiliary views projected from the front (front-adjacent) view are **frontal auxiliary views.** Primary auxiliary views projected from the side (side-adjacent) view are **profile auxiliary views.** These three types are represented in Figure 9.4 where auxiliary view A is projected from the top (horizontal) view, auxiliary view B is projected from the front (frontal) view, and auxiliary view C is projected from the side (profile) view. *Hidden lines that fall behind the true shape surface in an auxiliary view can normally be eliminated.* Each primary auxiliary view, besides being projected from one of the three principal views, will have one common dimension with at least one other principal view. The height (H) dimension in the front view is used to establish the limits of auxiliary view A by using dimension H. The depth (D) of the part can be found in the top view (and side view) and is used to establish the D dimension in auxiliary view B. Dimension A in auxiliary view C is taken from the view where the width of the slot is drawn true size (the front view).

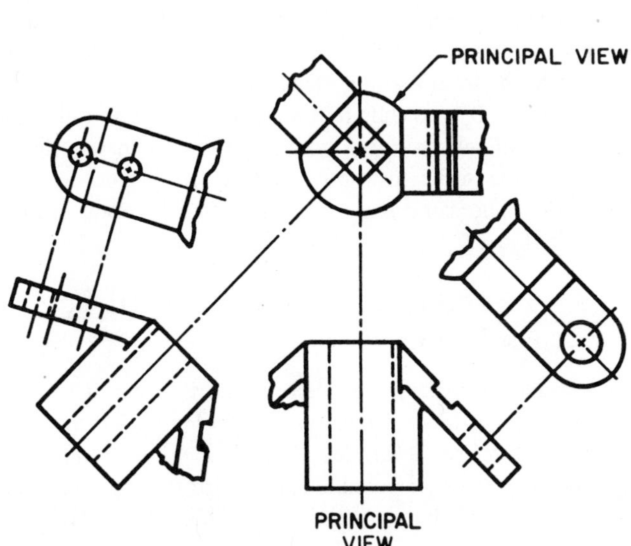

FIGURE 9.3 Principal and Auxiliary Views of a Part

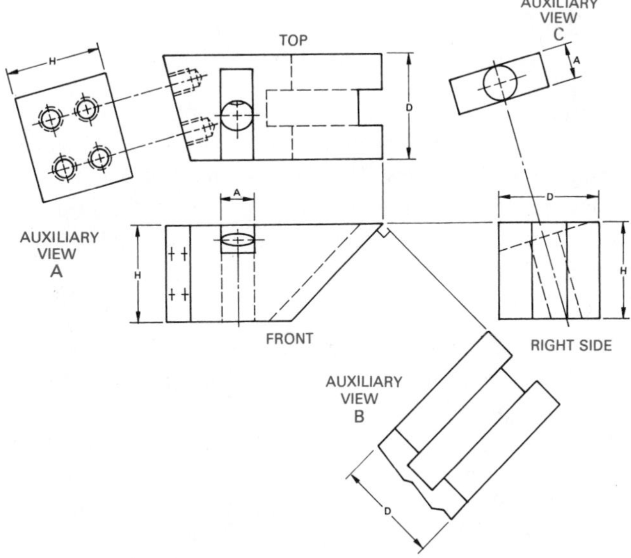

FIGURE 9.4 Auxiliary Views

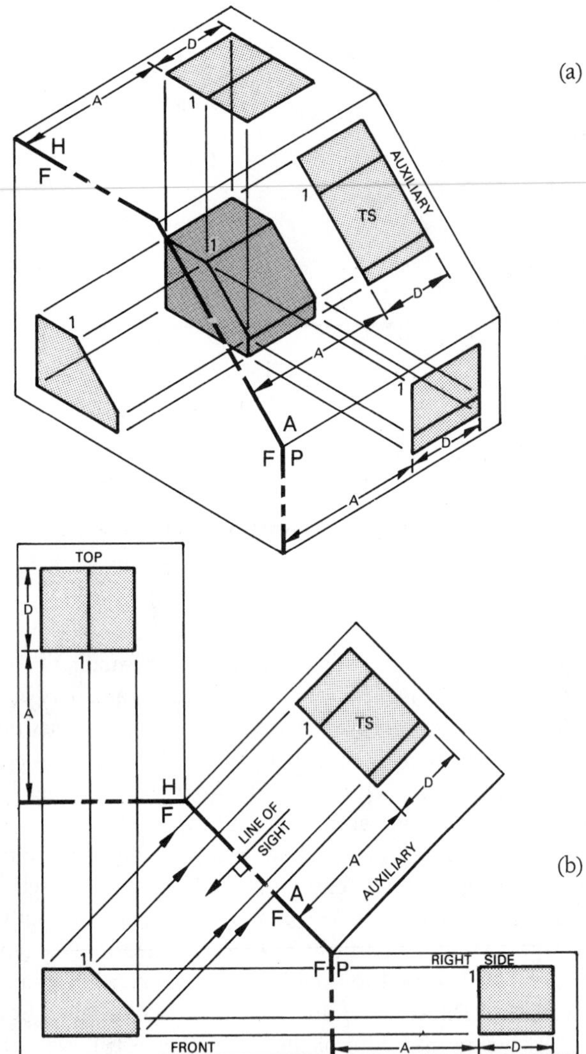

FIGURE 9.5 **Auxiliary View Projected from the Front View**
(a) The glass box method (b) The fold line method

9.2.2 Frontal Auxiliary Views (Fold Line Method)

The true shape of an inclined plane that appears as an edge in the front view must be projected from that view and is called a **frontal auxiliary view.** The **glass box method** is illustrated in Figure 9.5(a). The following steps (using the fold line method) describe the projection of the frontal auxiliary view shown in Figure 9.5(b):

1. The line of sight for a frontal auxiliary view is perpendicular to the inclined surface, which appears as an edge in the frontal view.
2. Fold line F/A is established perpendicular to the line of sight and parallel to the inclined surface (edge view).
3. Projectors are drawn from all points in the front view perpendicular to the fold line. Hidden lines were omitted in this example.
4. Measurements are taken (using dividers for speed and accuracy) from fold line H/F or P/F to establish the front face of the part in the auxiliary view. Dimension A is transferred from the top or side view to establish the distance from the F/A fold line to the front face of the part in the auxiliary view. The depth dimension (D) of the part is then transferred.

9.2.3 Horizontal Auxiliary Views (Fold Line Method)

The second type of primary auxiliary view is the **horizontal auxiliary view.** This auxiliary view is perpendicular to the horizontal plane and is inclined to the other two principal planes. The glass box method is shown in Figure 9.6(a). The auxiliary view is projected at a required viewing angle, and is not being used to solve for the true shape of an inclined surface. This does not show the true shape of a surface but provides a different viewing angle. The following steps describe the process of projecting a horizontal auxiliary view using the fold line method shown in Figure 9.6(b):

1. Establish a line of sight at a required angle of viewing; 45° was used here.
2. Fold line H/A is drawn perpendicular to the line of sight.
3. From each point in the top (horizontal) view, extend a projector parallel to the line of sight and perpendicular to the fold line. In this example, hidden lines are shown.
4. Dimension D is transferred from the side or front view to establish the distance from the H/A fold line to the top of the part. The height (H) dimension is then transferred to locate the bottom of the part. Visibility is determined and the view is completed.

9.2.4 Profile Auxiliary Views (Fold Line Method)

The third type of primary auxiliary view is the **profile auxiliary view.** In Figure 9.7, one of the surfaces of the part is inclined to the front and top view, and appears as an edge in the side view. By projecting an auxiliary view with a line of sight perpendicular to the edge view of the inclined surface, the true shape of the surface will be seen in the profile auxiliary view. The same basic steps are used to draw profile auxiliary views.

9.2.5 Secondary Auxiliary Views Using Fold Lines

A **secondary auxiliary view** is one that is adjacent to and aligned with a primary auxiliary view. In Figure 9.8 the part has *one surface that is inclined to all three principal planes of projection* and it is not possible to solve for the true shape of the surface in a primary auxiliary view. This type of surface is an **oblique surface.** Since all consecutive views of a part are at right angles, secondary auxiliary views are perpendicular to

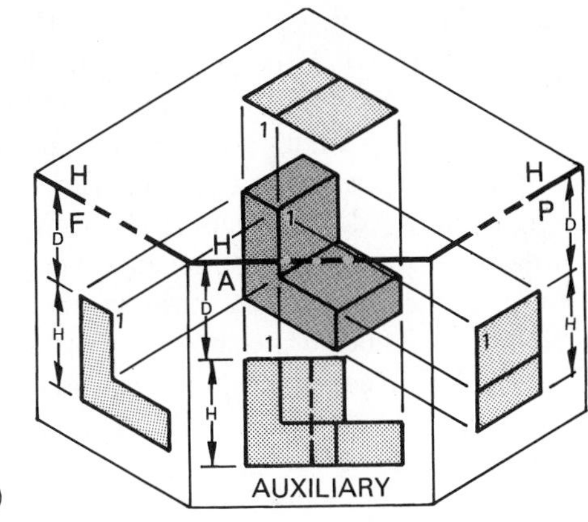

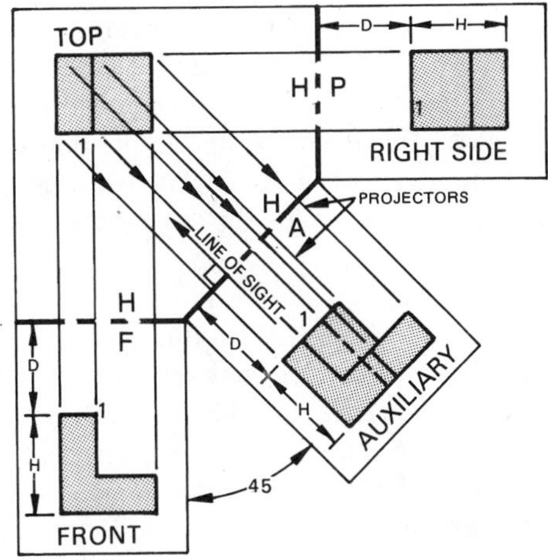

FIGURE 9.6 **Auxiliary View Projected from the Top View**
(a) The glass box method (b) The fold line method

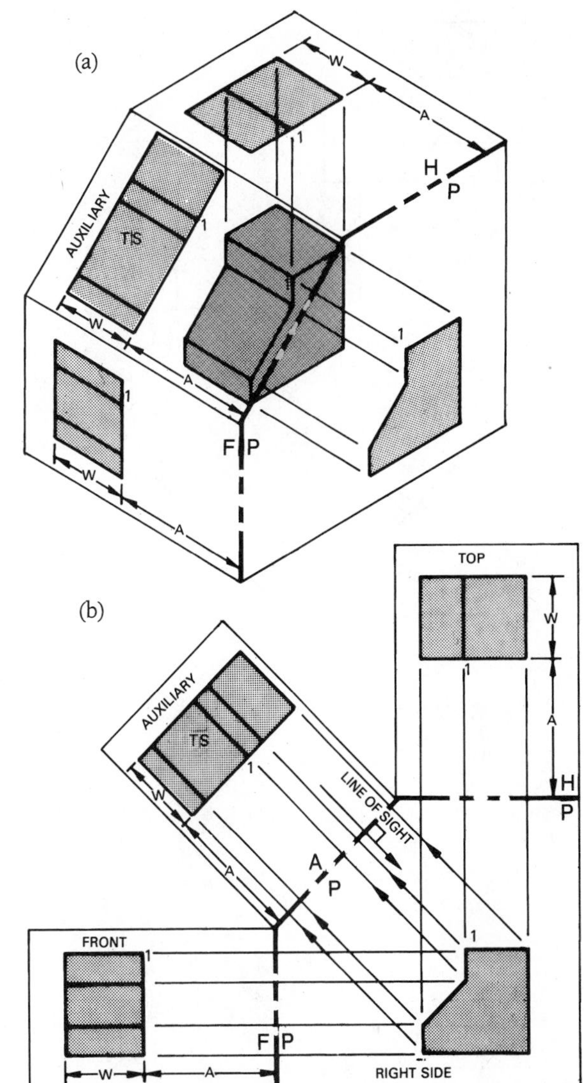

FIGURE 9.7 **Auxiliary View Projected from the Side View**
(a) The glass box method (b) The fold line method

primary auxiliary views. Views projected from a secondary auxiliary view are called **successive auxiliary views.** The following steps were used to draw the part in Figure 9.8:

1. Establish line of sight parallel to the true length (TL) line 1-2 in the front view.
2. Draw fold line F/A perpendicular to the line of sight and a convenient distance from the front view.
3. Complete the primary auxiliary view by transferring dimensions A, C, and D from the front view and draw the part.
4. Establish a line of sight perpendicular to the edge view (EV) of surface 1A-2A-3A in the primary auxiliary view.
5. Draw fold line A/B perpendicular to the line of sight and at a convenient distance from auxiliary view A.
6. Complete the secondary auxiliary view by transferring dimensions from the front view. Draw only plane 1-2-3, which will show true shape. Dimensions D and E establish points 3 and 2.

9.2.6 Adjacent Views

Each view and its preceding and following view are considered adjacent views. An **adjacent view** is any view that is aligned with another view by means of a direct projection. It is important to understand that principal views can also be adjacent views.

9.3 Auxiliary Views Using the Reference Plane Method

In drawing the auxiliary view, dimensions in one direction are projected into the auxiliary view from the adjacent view. The dimensions in the other direction are transferred to the auxiliary view by measurement. A **reference plane** can be used

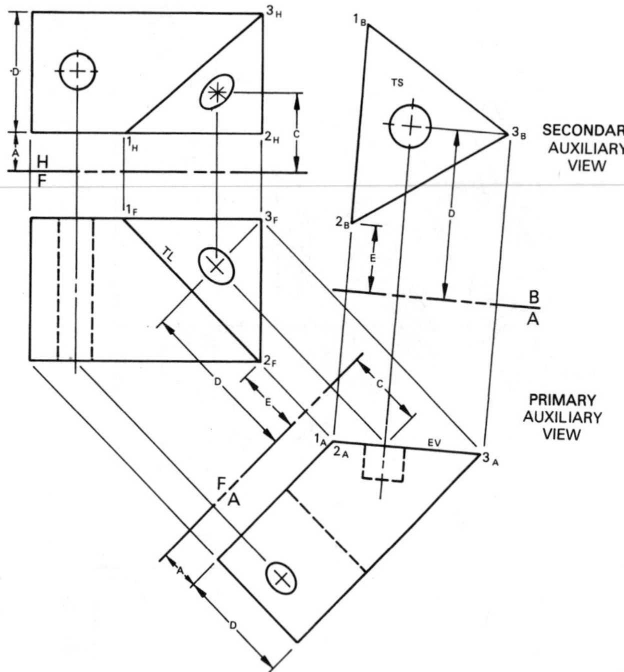

FIGURE 9.8 Secondary Auxiliary Views

instead of a fold line to transfer measurements. The reference plane is placed so that it is perpendicular to the inclined surface that is being drawn and is represented in this view by its edge (which shows as a line). All measurements are transferred from the edge view of the reference plane.

The reference line (edge) appears in the view where the inclined surface is shown as foreshortened. It will not appear in the view where the surface to be drawn is seen as an edge. In Figure 9.9, the reference plane shows as a plane surface in the front view and as an edge in the top view and the auxiliary view. Any convenient parallel position can be used to establish the reference plane.

9.3.1 Drawing an Auxiliary View

To draw the auxiliary view in Figure 9.9, first locate the reference plane for the auxiliary view so that it is parallel to the edge view of the inclined surface being drawn. Here, the plane was passed so that it coincided with the front surface of the part and shows as a reference line (edge) in the top and auxiliary views. The reference line is positioned a distance away from the edge view of the inclined surface that is equal to the depth of the part plus the space desired between the views.

Spacing the views can also be done as in drawing standard multiviews. The thickness [depth dimension (2.500) in Fig. 9.9] plus the space required between views determines the amount of space needed for the auxiliary view. Place the auxiliary view (the reference line edge) such that there is sufficient space for dimensions and notes.

After the extents of the view are projected (perpendicularly) from the adjacent view (the front view in Fig. 9.9), measurements to establish the thickness of the auxiliary view are then taken from the existing view. Measurements that can

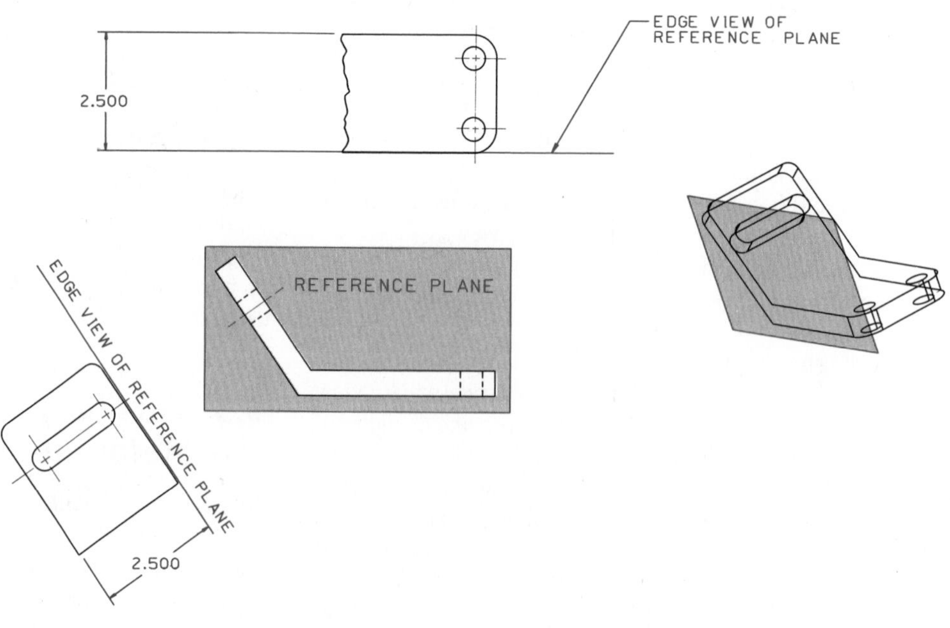

FIGURE 9.9 Auxiliary View Projection Using Reference Plane Method

ITEMS OF INTEREST

Hand Tools and Devices

There is evidence that hand tools and other devices were used by primates nearly a million years ago. It could be assumed that hand tools, after evolving with mankind for a million years or so, would now be specifically adapted for human use. In fact, this is not the case. Until recently, human biomechanical factors have been mostly ignored in the design of hand tools.

The human hand and wrist are complex structures of bones, nerves, ligaments, tendons, and arteries. Movement of the wrist occurs in two planes. The hand is flexed up and down in the first plane. Side to side movement or ulnar deviation occurs in the second plane. Continued use of tools that call for motions along these planes can injure the hand and wrist.

Recent studies have shown us how to design tools to avoid these types of motion. Using x rays of the wrist and computer-generated wire frame auxiliary models, designers are able to design tool configurations that use a relatively straight wrist motion. For example, by bending or rotating the handle of a pair of pliers or a hammer about 19°, grip strength is increased while fatigue is reduced. This bent handle design is now being used in softball bats and is approved for regulation play. Golf clubs and fishing rods also incorporate this technology into their designs.

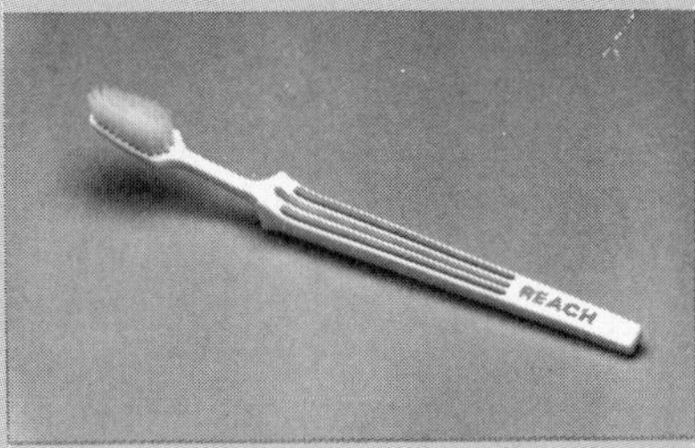

Ergonomically Designed Toothbrush

The most common hand-held device is probably the toothbrush. The only major development since it was introduced in 1780 was the use of nylon bristles. Nylon replaced hog hair in the 1930s.

Johnson and Johnson, Inc., designed a new toothbrush using human factor and time-motion research data. Prototypes were developed with different handle shapes and bristle head rotations. These prototypes were tested and the optimal features found from the testing were used in the final design. The result was the Reach toothbrush. The Reach incorporates a small bilevel bristle head into an angled, countered handle for easier handling, better gum stimulation, and better plaque removal.

The new toothbrush is one example of design with human factors. Future tool designs will require these research studies to produce hand tools specifically adapted for human use. Without using auxiliary views, projections, and models, none of this would have been possible.

be taken perpendicular to the reference line are transferred to the auxiliary view. At the same time, all dimensions that are parallel to the reference line are projected to the auxiliary view. This procedure (projection lines and transfer distances) is the same as for the fold line method. The line of projection for each point is always perpendicular to the reference line. All features can be established in the auxiliary view by projecting one point at a time.

The fold line or the reference line is always erased after the projection of the auxiliary view is completed (Fig. 9.10).

9.3.2 Secondary Auxiliary Views

For typical industry applications that require the detailing of an oblique surface, either the fold line or the reference plane method can be used to produce a secondary auxiliary view. The fold line method works the best for complicated parts that require the entire part to be projected.

In Figure 9.11, the part has a surface that is oblique with a slot positioned on it. To solve for the true shape view of the surface, a primary auxiliary view showing the surface as an edge was projected first. By projecting a view perpendicular to the edge view, a true shape secondary auxiliary view was established. The primary auxiliary view shows the true angle that the inclined surface makes with the base of the part.

9.4 Auxiliary View Conventions

Auxiliary views are aligned with the views from which they are projected. A centerline or projection line may continue between the adjacent views to indicate this alignment. In Figure 9.12, the centerline of the hole has been extended from the front view to the auxiliary view to show *alignment.* Only show hidden lines when they do not complicate the auxiliary view or where they are necessary to describe the part adequately.

You May Complete Exercises 9.1 Through 9.4 at This Time

9.4.1 Partial Auxiliary Views

To simplify a drawing, **partial auxiliary views** or **partial principal views** may be used to show only pertinent features not described by true projection in the principal or other views (Fig. 9.10). In Figure 9.13 the top view of the part is only partially shown; only the front view here is complete. In this example, the reference plane was passed through the center of the part, rather than along one of the edges. This method is frequently used where the part to be drawn is *symmetrical about its centerline* because passing the reference plane down the centerline of a part makes it easy to transfer the dimensions.

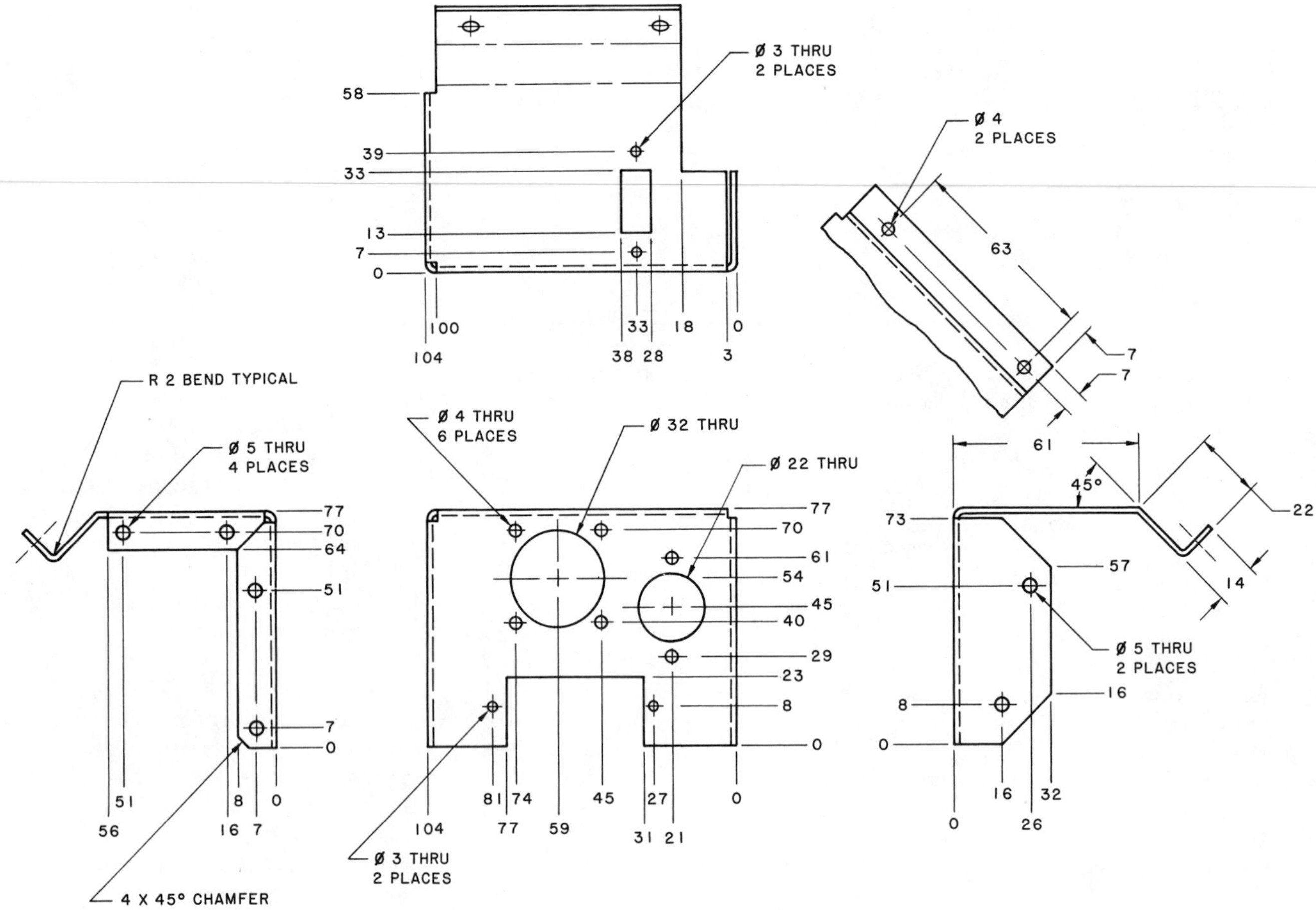

Ø 3 THRU
2 PLACES
Ø 4
2 PLACES
R 2 BEND TYPICAL
Ø 5 THRU
4 PLACES
Ø 4 THRU
6 PLACES
Ø 32 THRU
Ø 22 THRU
Ø 5 THRU
2 PLACES
Ø 3 THRU
2 PLACES
4 X 45° CHAMFER

FIGURE 9.10 Electronic Bracket Mount Detail

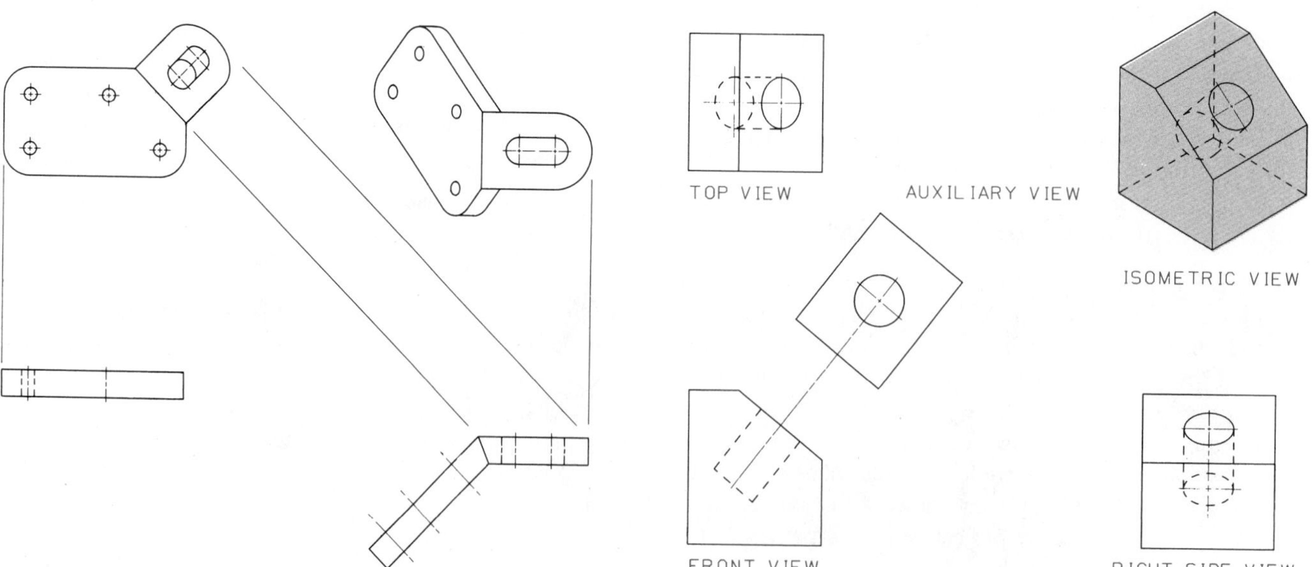

FIGURE 9.11 True Shape of an Oblique Surface by Auxiliary View

TOP VIEW
AUXILIARY VIEW
ISOMETRIC VIEW
FRONT VIEW
RIGHT SIDE VIEW

FIGURE 9.12 Front Adjacent Auxiliary View

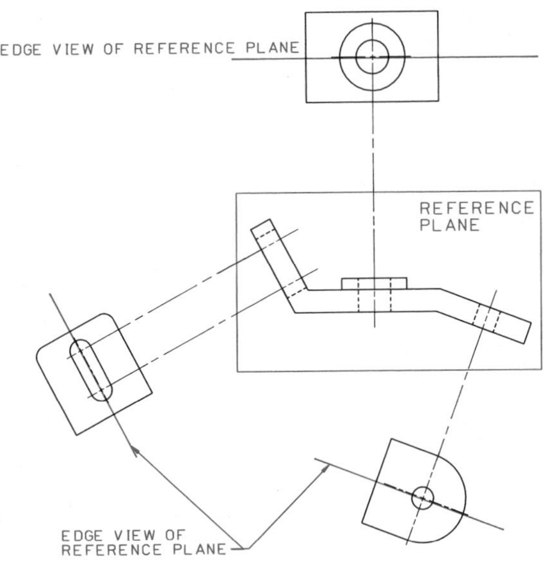

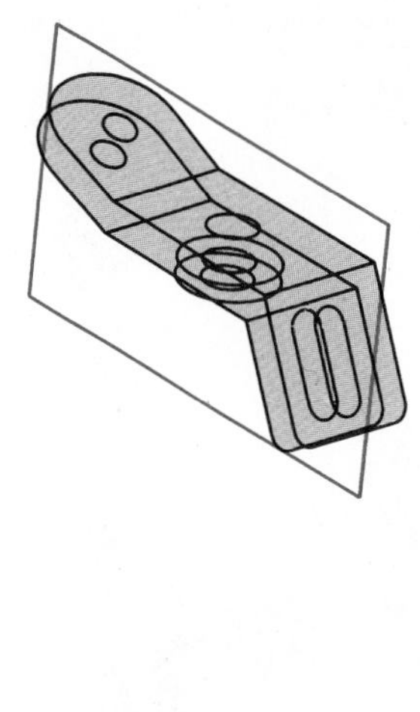

FIGURE 9.13 Reference Plane Method of Auxiliary View Projection

9.4.2 Broken and Half Auxiliary Views

In some situations only a portion of the auxiliary view is necessary. For example, in Figure 9.14, the base is partially shown in the auxiliary view using broken lines. In Figure 9.15, a **half top view** and a **half auxiliary view** are shown. Since the part is symmetrical about its centerline, little is gained by drawing a complete top view. The same is true for the auxiliary views.

9.4.3 Auxiliary Views of Curved Features

Circular and curved features are true size/shape in views where the line of sight is perpendicular to the edge view of the surface on which they lie. In the adjacent projection, the plane appears as an edge and parallel to the fold line. The length of the edge view line is equal to the circle's diameter.

When a circular plane is oblique it appears as an ellipse. An elliptical view of a circular plane along with each adjacent auxiliary view is plotted by locating a series of points along the outline of the circle in a true size view. These points are located in each adjacent view by projection and transferring distances to establish each individual point. The series of points is connected with a template or an irregular curve.

In Figure 9.16, a normal view (true size) of the circular plane is shown with its frontal edge view. Primary auxiliary view A forms a 30° angle with the line of sight; therefore, the edge view of the plane forms a 60° angle with the adjacent view (and fold line F/A). Auxiliary A shows the plane as a 30° ellipse. Secondary auxiliary view B is drawn by projection and transferring distances for each point. Auxiliary C is projected at a 70° angle to the edge view and shows as a 20° ellipse. Note that dimension D1 establishes points 3 and 7, and dimension D2 locates point 1 in auxiliary views A and C. Auxiliary B is a secondary auxiliary view.

A typical problem found in industry where this is technique is required is the location of a hole centered on a given surface. In Figure 9.17, plane 1-2-3-4 is given and a hole/circle of a specific size is to be drilled so that it is located in the exact center of an oblique plane. Primary and secondary auxiliary views were required. The following steps were used to complete the problem:

1. Line 1H-3H and line 2H-4H are horizontal lines (true length in the horizontal view). Therefore, a true length line need not be constructed to find the edge view. Draw H/A perpendicular to the horizontal lines and project auxiliary view A. Plane 1A-2A-3A-4A is an edge in this view [Fig. 9.17(a)].

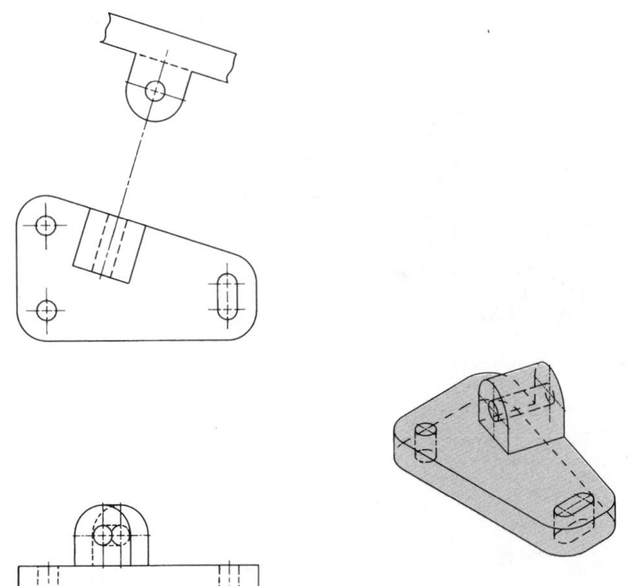

FIGURE 9.14 Broken Partial Auxiliary View

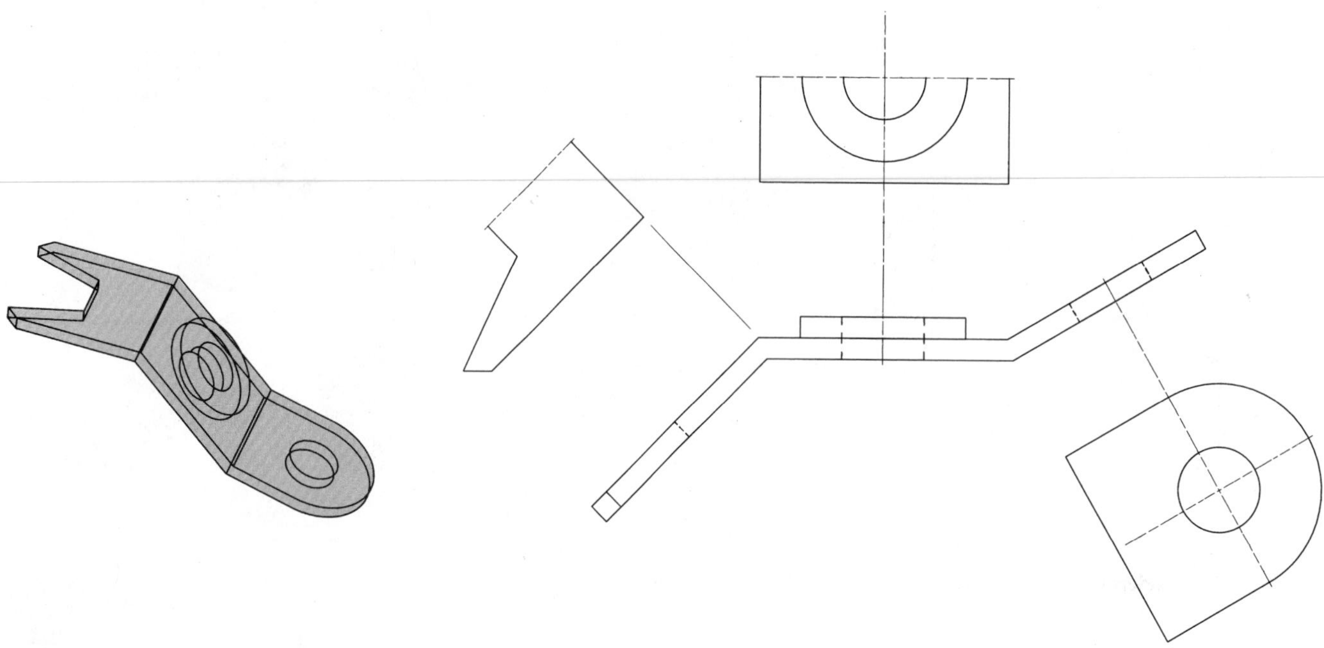

FIGURE 9.15 Half Views

2. Draw A/B parallel to the edge view of plane 1A-2A-3A-4A and project auxiliary B. This view shows the true size of the plane [Fig. 9.17(a)].
3. Locate the exact center of plane 1B-2B-3B-4B and draw the given circle [Fig. 9.17(a)].
4. To project the centered circle back to all previous views, a series of points needs to be located along its circumference. A simple method to locate points on the circle is to divide the circle evenly by drawing lines from the corners of the plane [Fig. 9.17(b)].
5. Locate each point in auxiliary A by projection where they fall on the edge of plane 1A-2A-3A-4A. Dimensions Dl, D2, and D3 are used to locate each point in the horizontal view by transferring them along their respective projection lines. Axis A (major diameter) and B (minor diameter) could also be used to locate and draw each view of the circle [Fig. 9.17(b)].
6. The frontal view of the circle is obtained by projection and transferring distances from auxiliary A (from H/A to each point on the edge view).
7. The points are then connected with a smooth curve using an irregular curve.

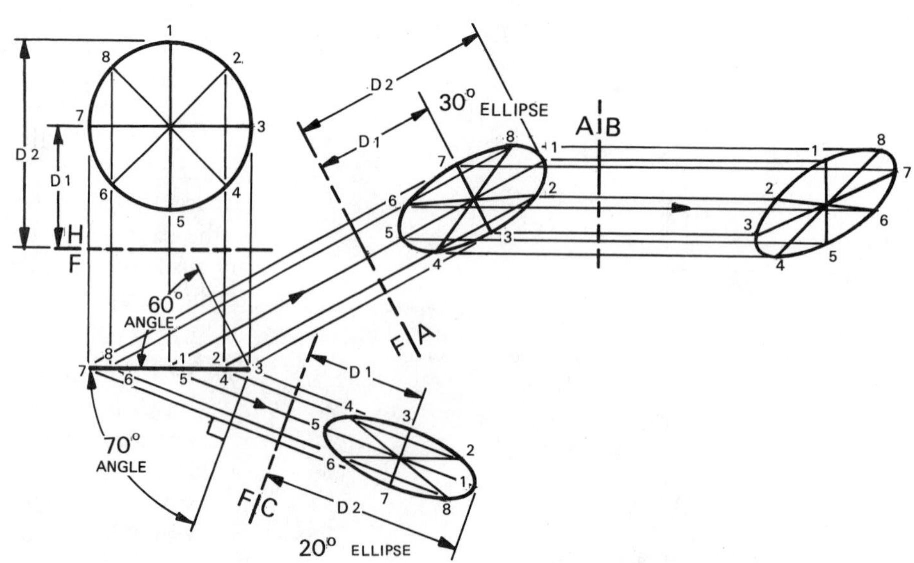

FIGURE 9.16 Plotting Points on Circular Planes

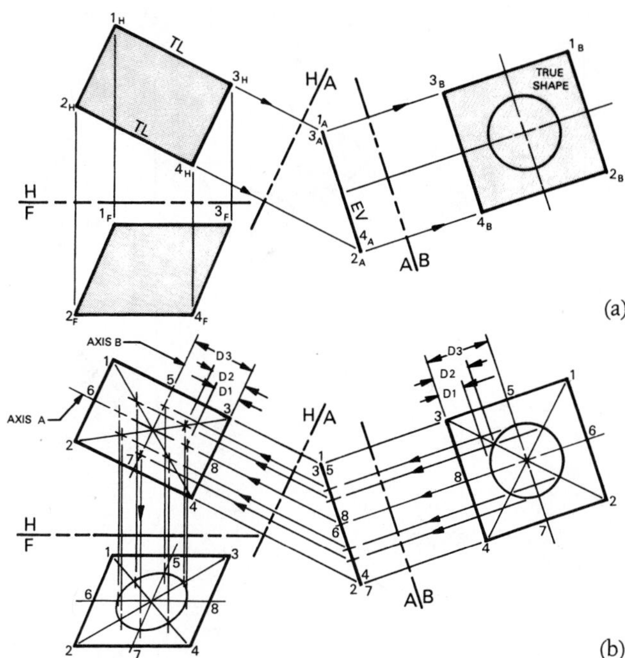

FIGURE 9.17 Curved Features and Auxiliary Views

9.4.4 Auxiliary Sections

Auxiliary views that are sections are called **auxiliary sections.** In Figure 9.18, the part has a section passed through the ribs since this feature cannot be adequately defined and detailed using the top, front, or partial side view. Section lining is drawn at a different angle than that of the lines of the view to avoid confusion.

9.4.5 Auxiliary Views and Dimensioning

The primary reason for projecting auxiliary views is to show and dimension the shape of a part that cannot be defined in one of the principal views. In Figure 9.19, the part has a surface that is inclined so it must be shown as true shape in order for it to be dimensioned. Dimensions were placed on each view where the part's features are shown true shape. The auxiliary view is used to dimension the slot and the holes.

You May Complete Exercises 9.5 Through 9.8 at This Time

QUIZ

True or False

1. Most auxiliary views are only partial projections.
2. Oblique, inclined, and otherwise distorted geometry is always shown on a view.
3. The top, front, and side views are always shown on a drawing when an auxiliary view is required to display inclined features for a part.
4. Auxiliary views may reduce the need for principal views of the part.
5. Auxiliary views are normally used for projecting a view to show the true shape of a surface that is inclined or oblique in the principal views.
6. A reference plane is always placed so that it is perpendicular to the inclined surface that is to be projected to an auxiliary view.
7. Auxiliary views are only used to display the true shape of the feature.
8. A third auxiliary view is one that is adjacent to the primary auxiliary view.

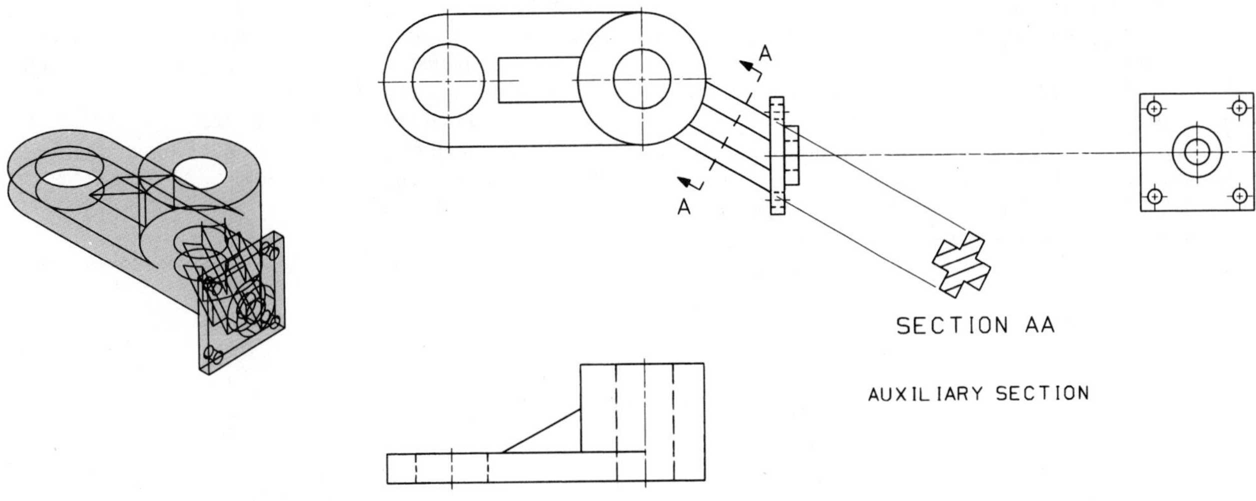

FIGURE 9.18 Auxiliary Section View

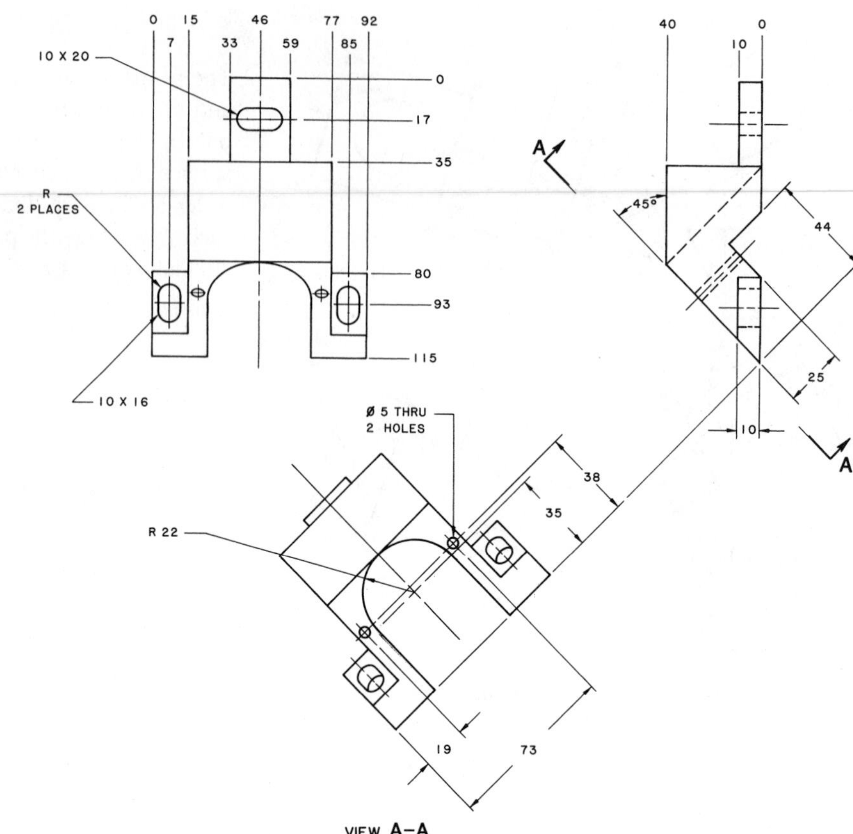

FIGURE 9.19 Auxiliary View and Dimensioning

VIEW A–A

Fill in the Blanks

8. A ________ auxiliary view is projected from one of the standard principal views.
9. A reference plane or a fold line can be established on a part to aid in the ________ of an ________ view.
10. ________ ________ views are projected from the front views.
11. The ________ ________ is normally passed through a prominent feature of the part so as to make projection of auxiliary views easier and quicker.
12. Half auxiliary views are normally used where the part is ______ about a ________ ________.
13. A ________ ________ view is adjacent and aligned with a secondary view.
14. A ________ ________ ________ is one that is adjacent to and aligned with one of the principal views.
15. ________ ________ that fall behind the true shape surface in an auxiliary view can be eliminated.
16. An ________ ________ is any view that is aligned with another view by means of a direct projection.

Answer the Following

17. What is the edge view of a plane and how is it used in the projection of a true shape view?
18. What is the primary purpose for an auxiliary view?
19. Compare the fold line method with the reference plane method.
20. Why are partial auxiliary views more common than complete auxiliary projections?
21. What is a fold line and how is it used?
22. What are half sections and why are they used?
23. What is a broken auxiliary view?
24. What is an adjacent view?

EXERCISES

Exercises may be assigned as sketching, instrument, or CAD projects. Transfer the given information to an "A" size sheet of .25 in. grid paper. Complete all views and solve for proper visibility, including centerlines, object lines, and hidden lines. Exercises that are not assigned by the instructor can be sketched in the text to provide practice and understanding for the preceding instructional material.

After Reading the Chapter Through Section 9.4, You May Complete the Following Exercises

Exercise 9.1 Draw the required views.

Exercise 9.2 Draw the required views.

Exercise 9.3 Draw the required views. Complete a full top view.

Exercise 9.4 Draw the three views as shown.

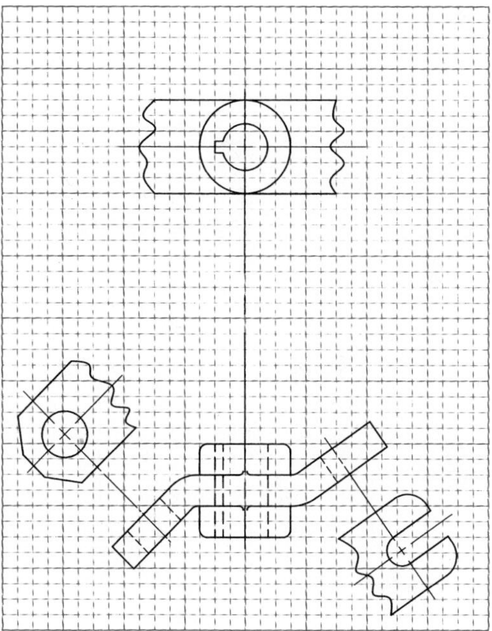

Exercise 9.1

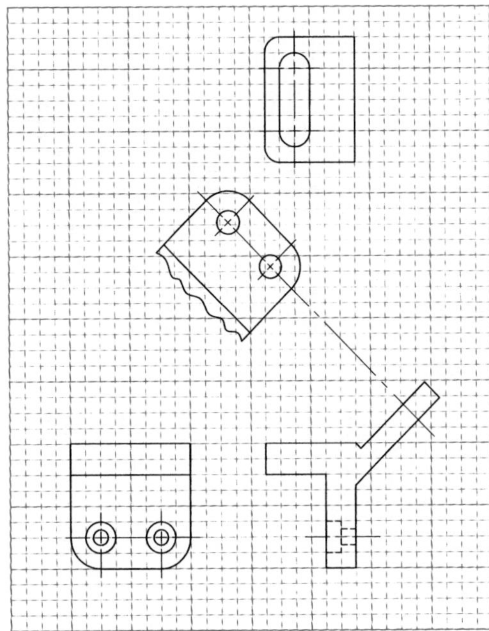

Exercise 9.3

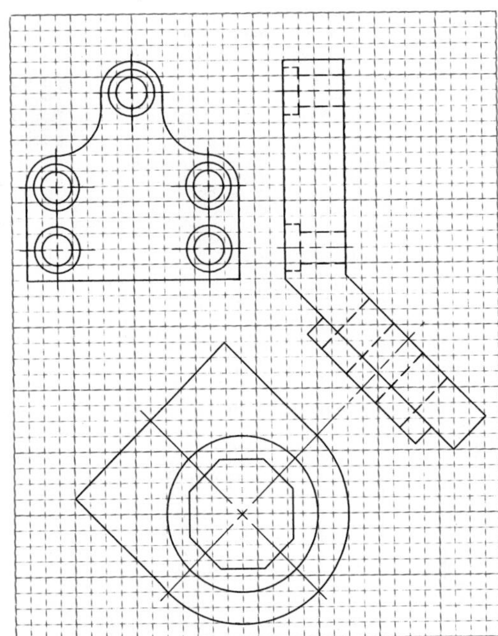

Exercise 9.2

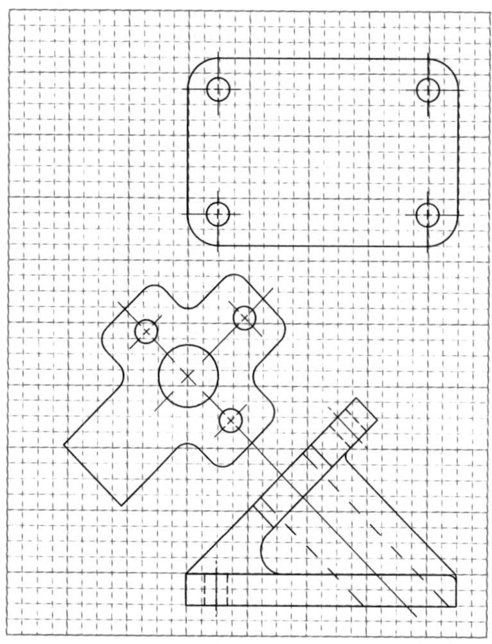

Exercise 9.4

After Reading the Chapter Through Section 9.4.5, You May Complete the Following Exercises

Exercise 9.5 Complete the required views and the auxiliary section.

Exercise 9.6 Draw the required views.

Exercise 9.7 Complete the required views and draw a full front view.

Exercise 9.8 Draw the required views. Project a secondary auxiliary view showing surface A or B as true shape/size.

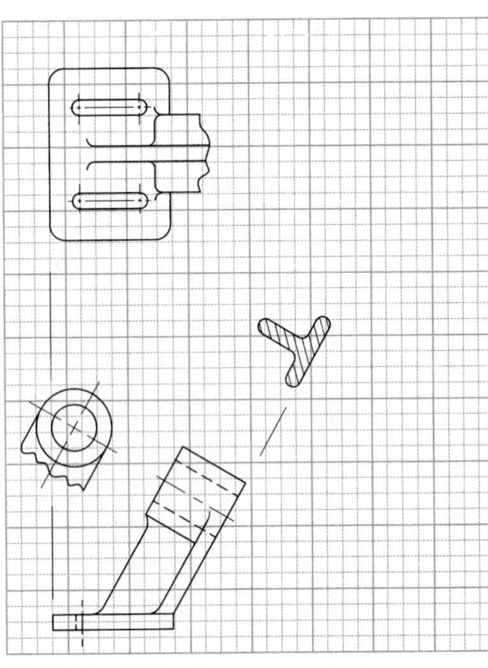

Exercise 9.5

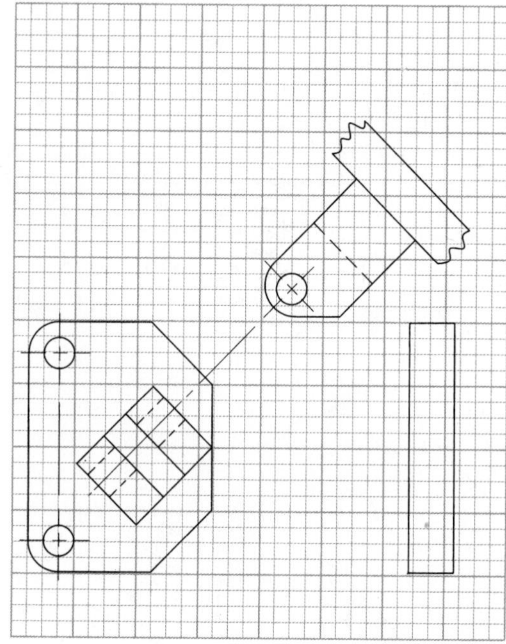

Exercise 9.7

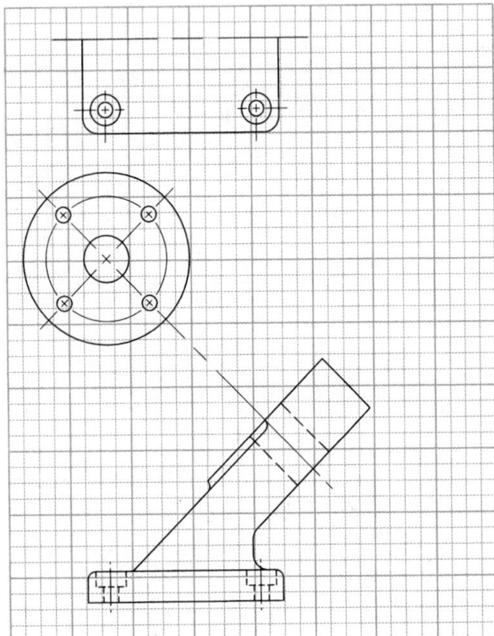

Exercise 9.6

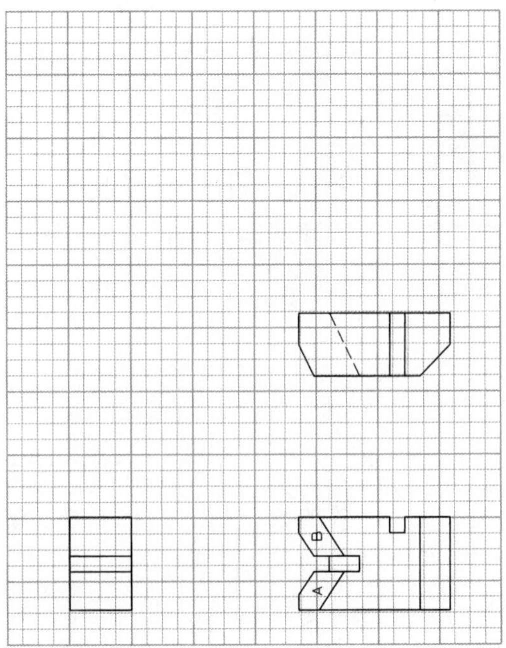

Exercise 9.8

PROBLEMS

Problems may be assigned as sketching, instrument or CAD projects. Use these projects for problems when completing the dimensioning chapter. Complete all views and solve for proper visibility, including centerlines, object lines, and hidden lines. When laying out these projects leave sufficient room to allow dimensioning. Instructor may assign projects to be dimensioned.

Problem 9.1 Draw the appropriate views of the part in order to describe completely each of its surfaces.

Problem 9.2 Draw the top, front, and auxiliary views of the part.

Problem 9.3 Draw the right side, top, and auxiliary views of the part in order to show each surface as true shape.

Problem 9.4 Draw the top, front, and auxiliary projections of the part.

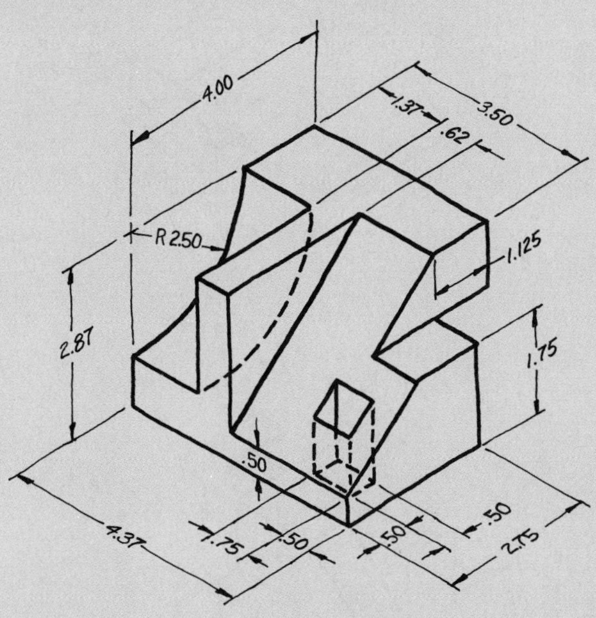

Problem 9.1

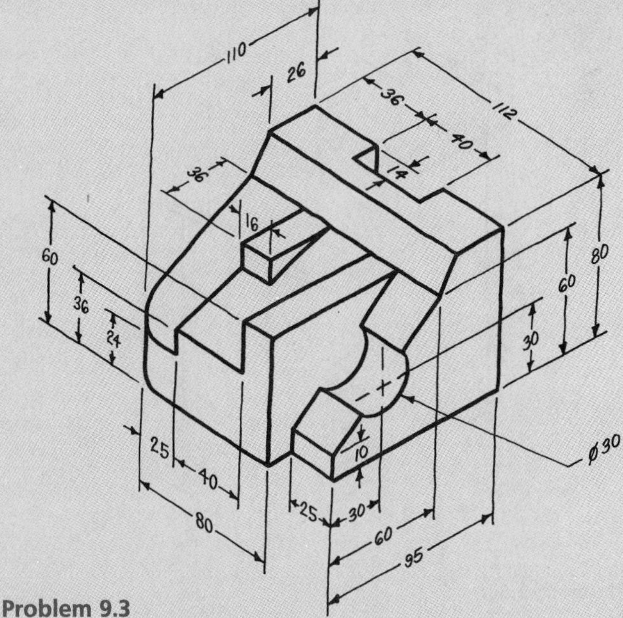

Problem 9.3

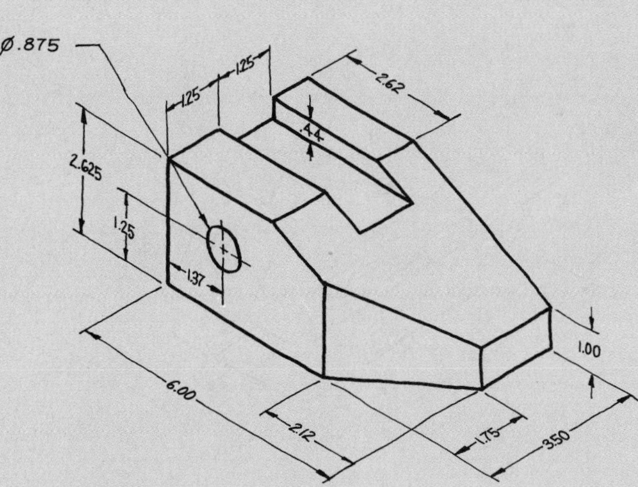

Problem 9.2

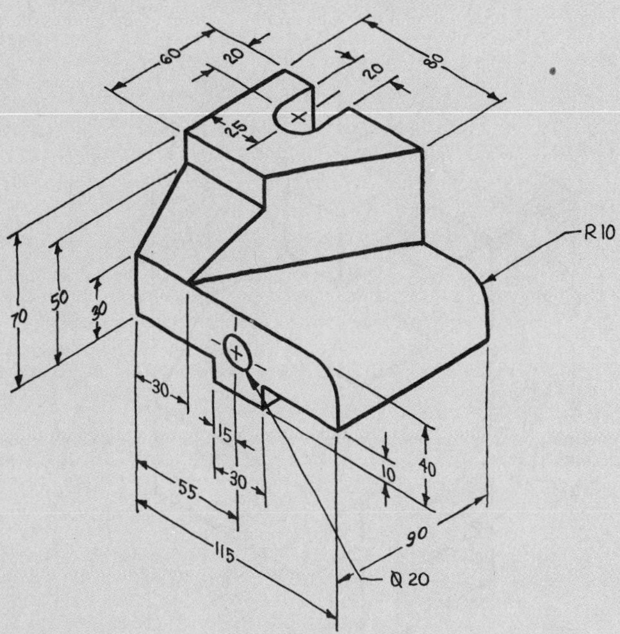

Problem 9.4

Problem 9.5 Draw the front and right side view of the part. Project a true shape view of the inclined surface. Position a 1.00 in. diameter hole in the middle of the surface and show in all views. The hole is to be .25 in. deep with a flat bottom.

Problem 9.6 Draw the top, front, side, and an auxiliary view projected from the top of the part. Center a 20 mm hole on the auxiliary view. The hole is 15 mm deep with a flat bottom.

Problem 9.7 Draw the appropriate views needed to describe the part completely.

Problem 9.8 Draw the top and front views and any auxiliary views needed to display the triangular surface's true shape.

Problem 9.9 Draw the views necessary to describe the part completely. The auxiliary projection should be a complete view.

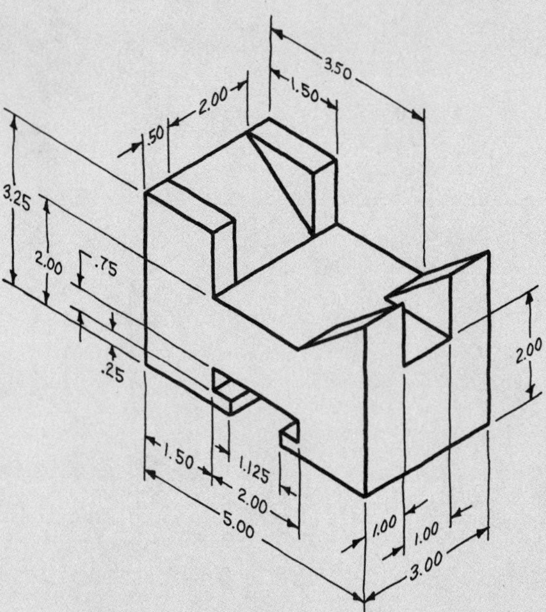

Problem 9.5

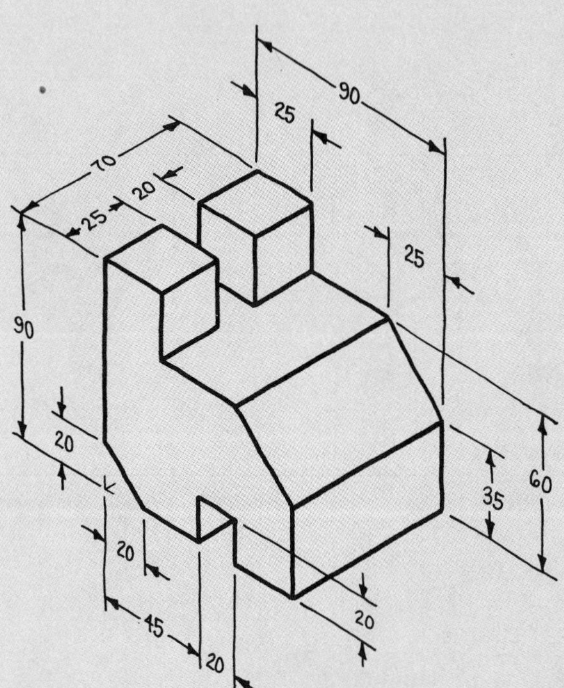

Problem 9.6

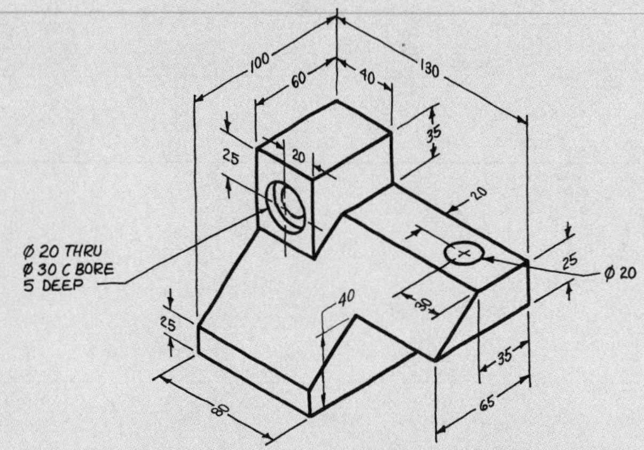

Problem 9.7

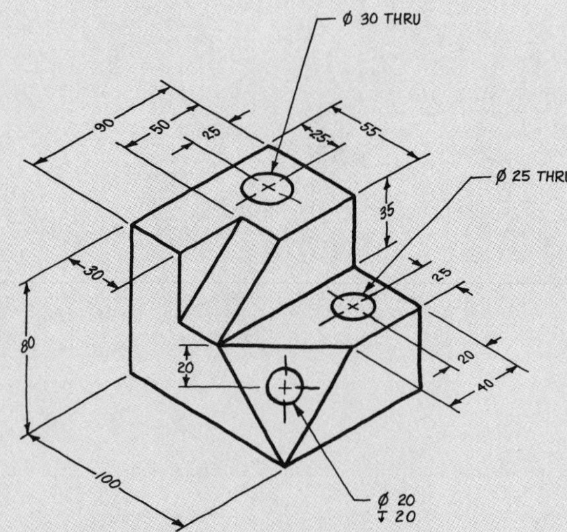

Problem 9.8

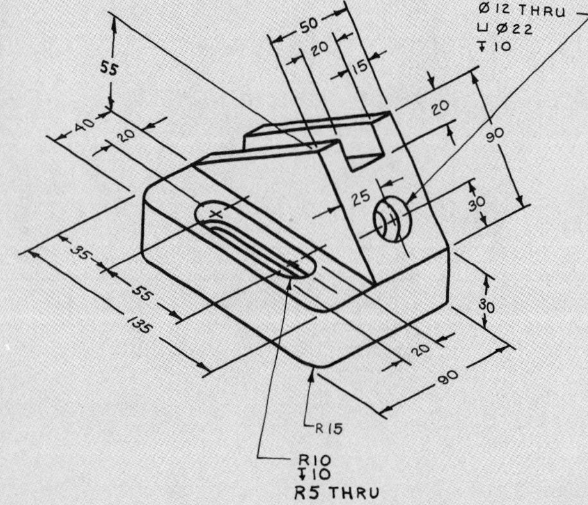

Problem 9.9

Problems 9.10 through 9.15 Draw the views required to detail each of the parts.

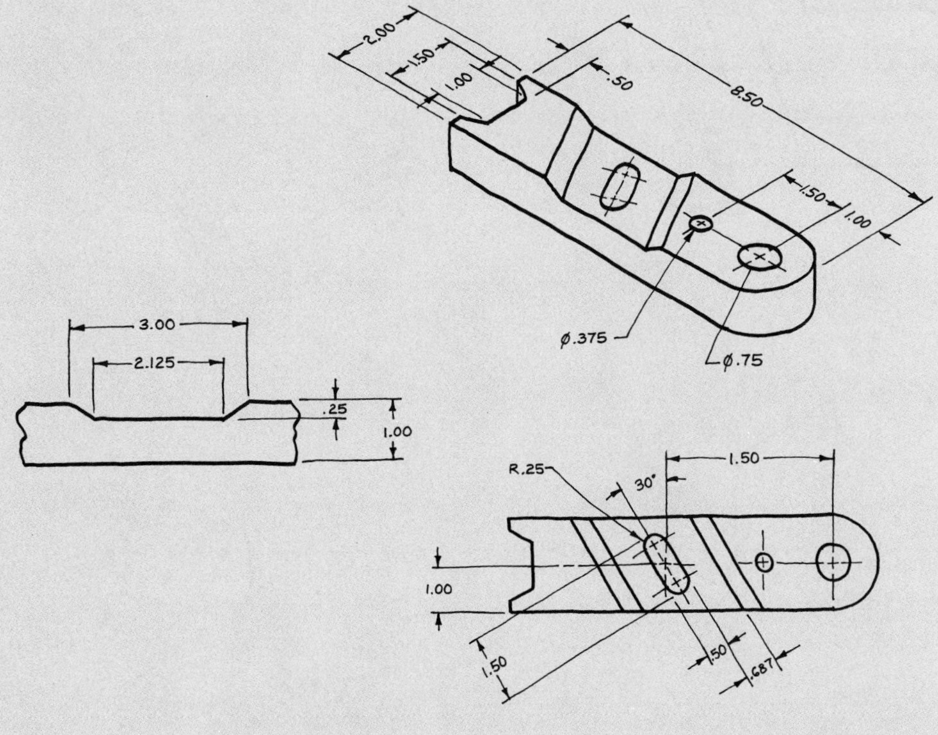

Problem 9.10

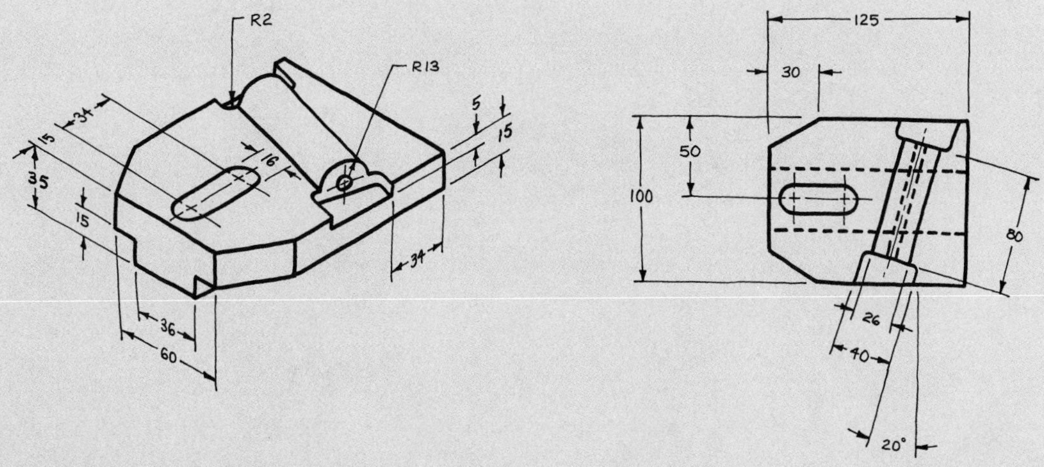

Problem 9.11

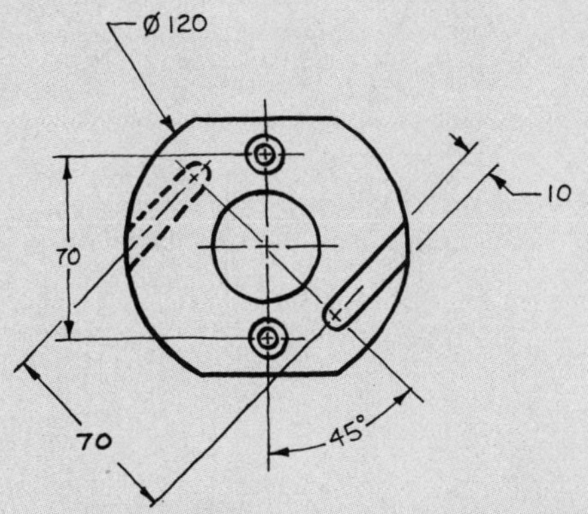

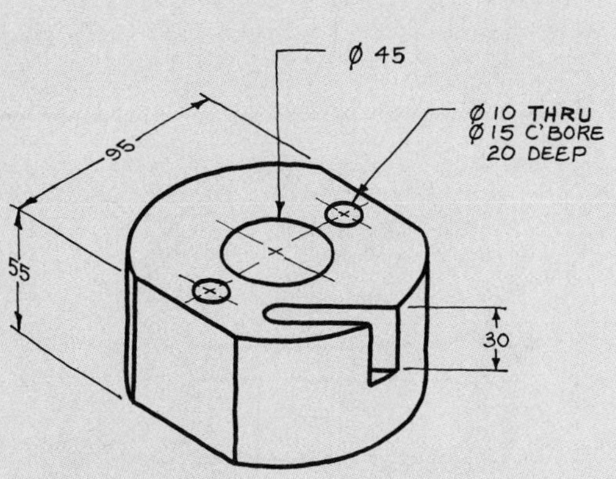

Problem 9.12

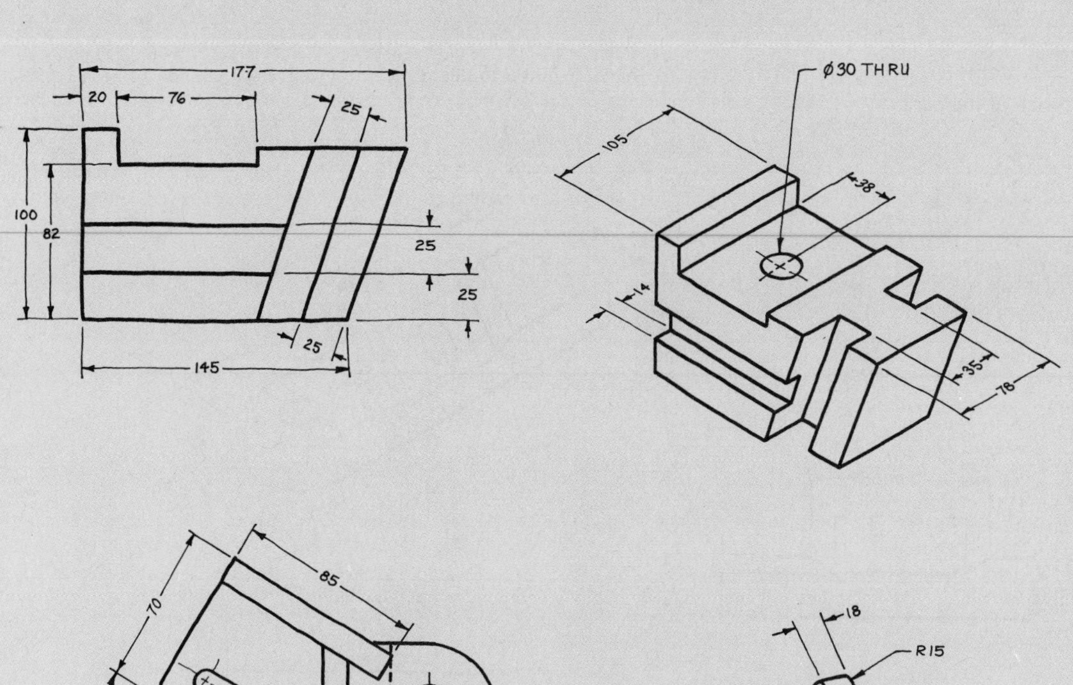

Problem 9.13

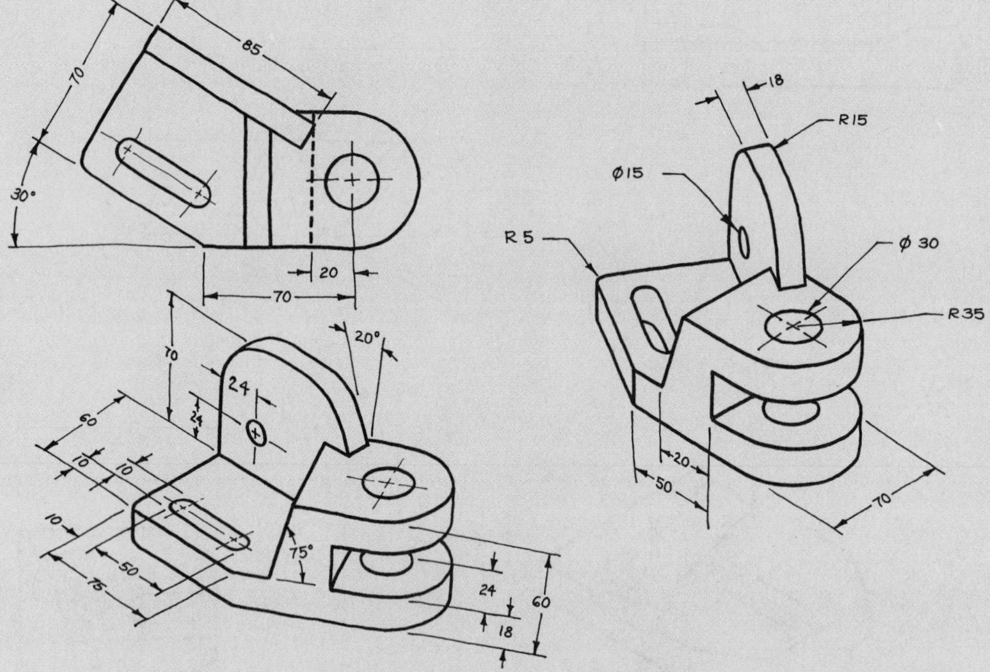

Problem 9.14

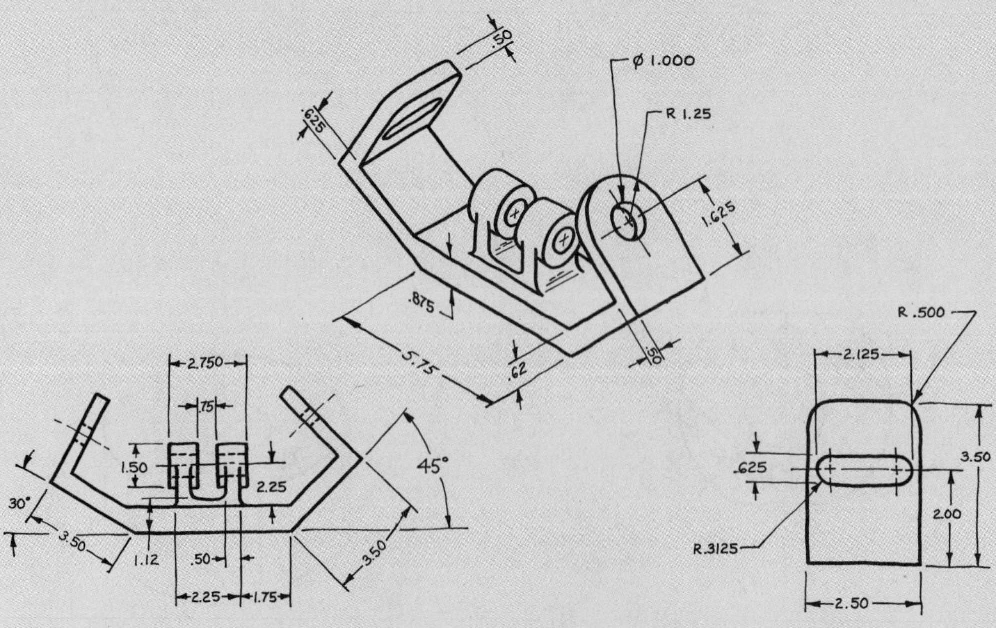

Problem 9.15

Problem 9.16 Draw the housing cover and complete the proper views to describe the cover.

(a)

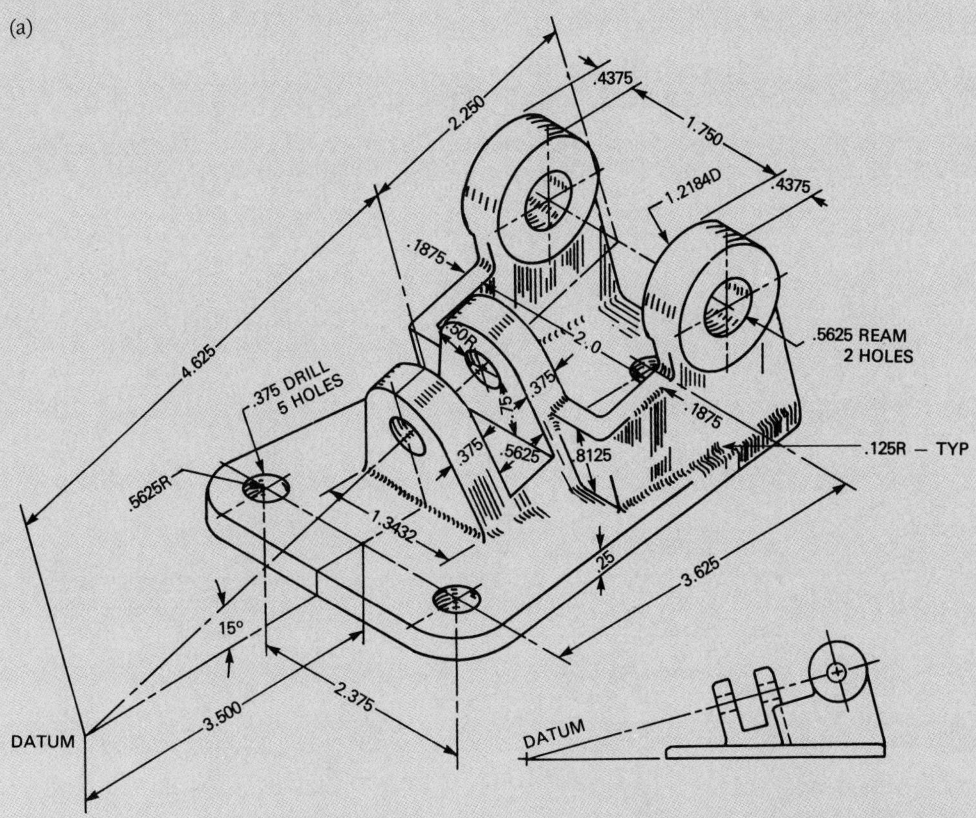

(b)

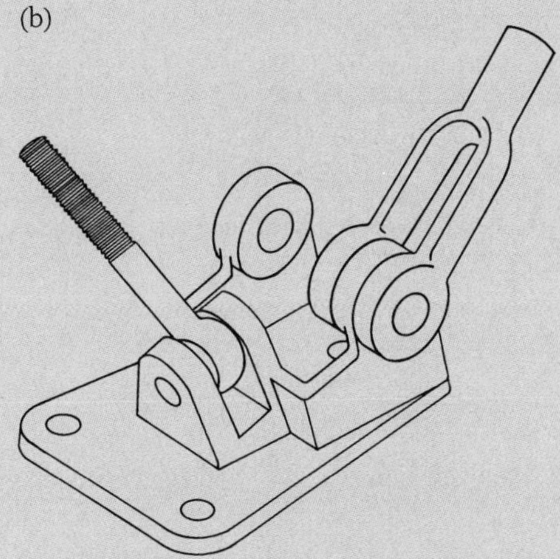

Problem 9.16

Problem 9.17 Draw the part and all required views.

Problem 9.18 Draw the part. Lay out the views required to describe the part completely. Do not use partial views.

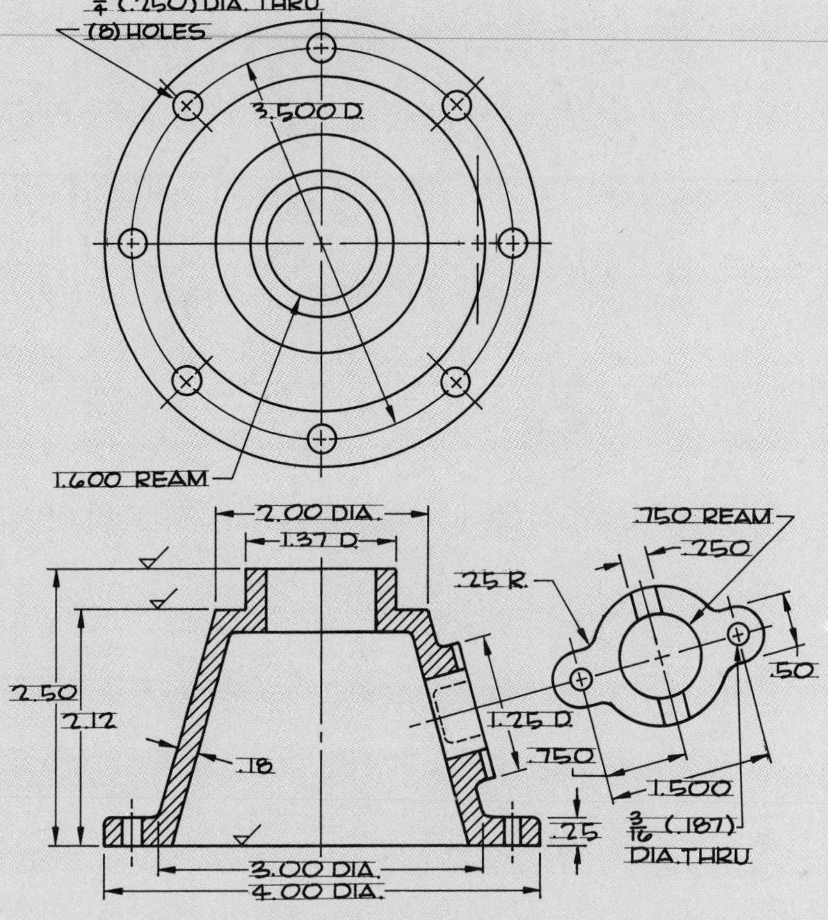

Problem 9.17

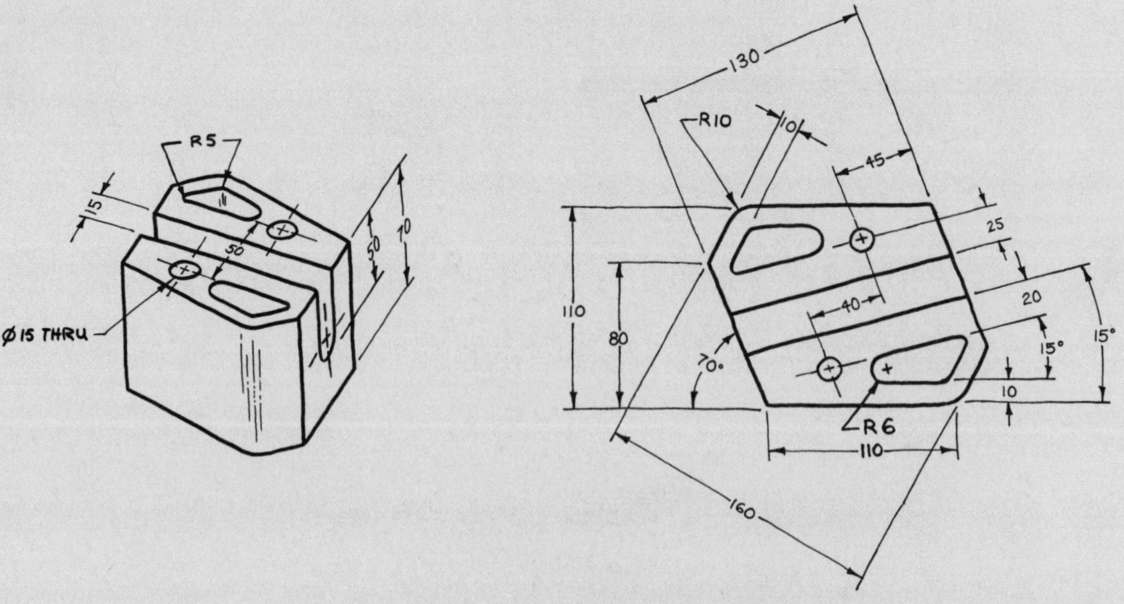

Problem 9.18

Problem 9.19 Draw the part and the required full views.

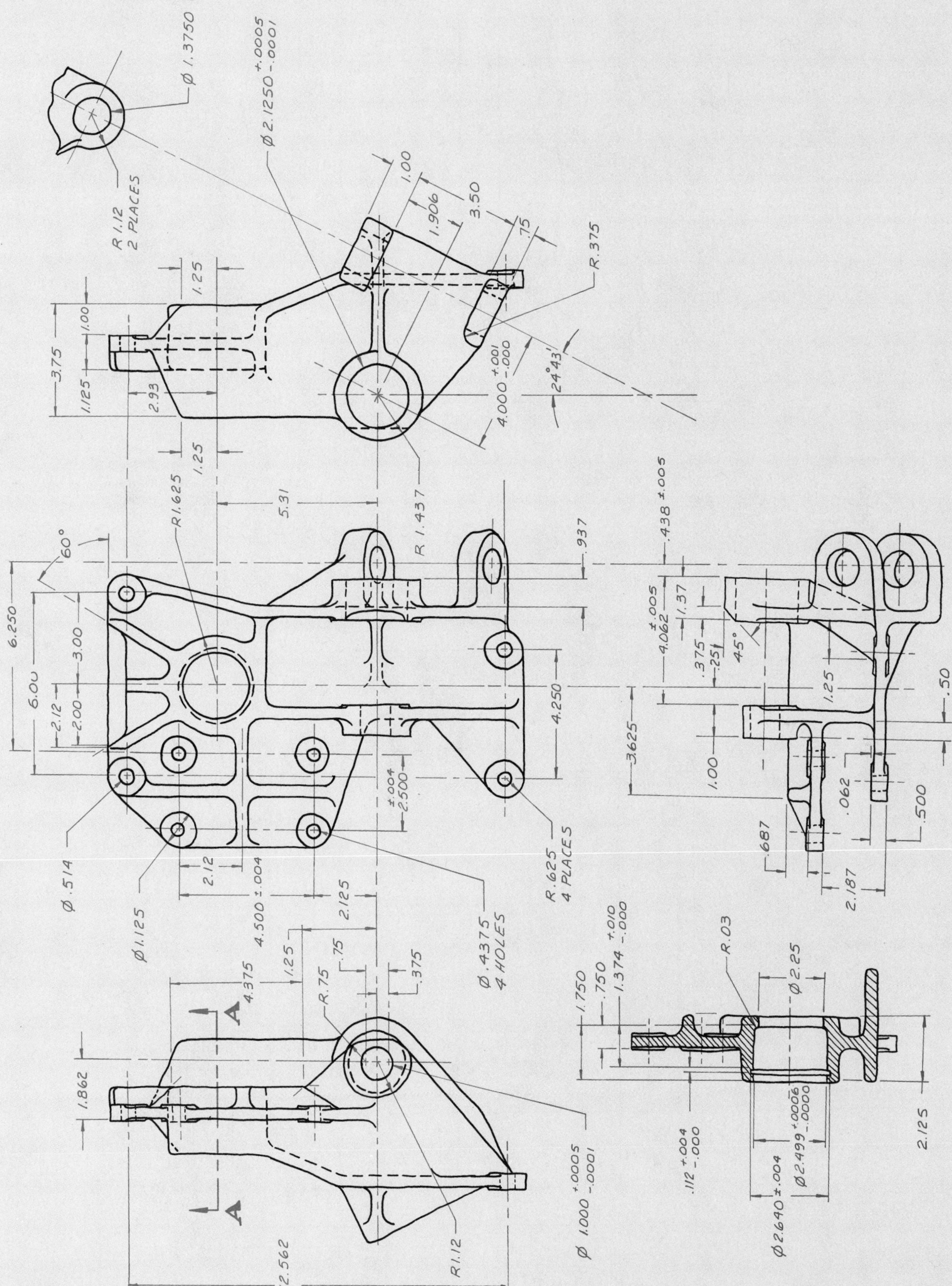

Problem 9.20 Draw the required full views for the part.

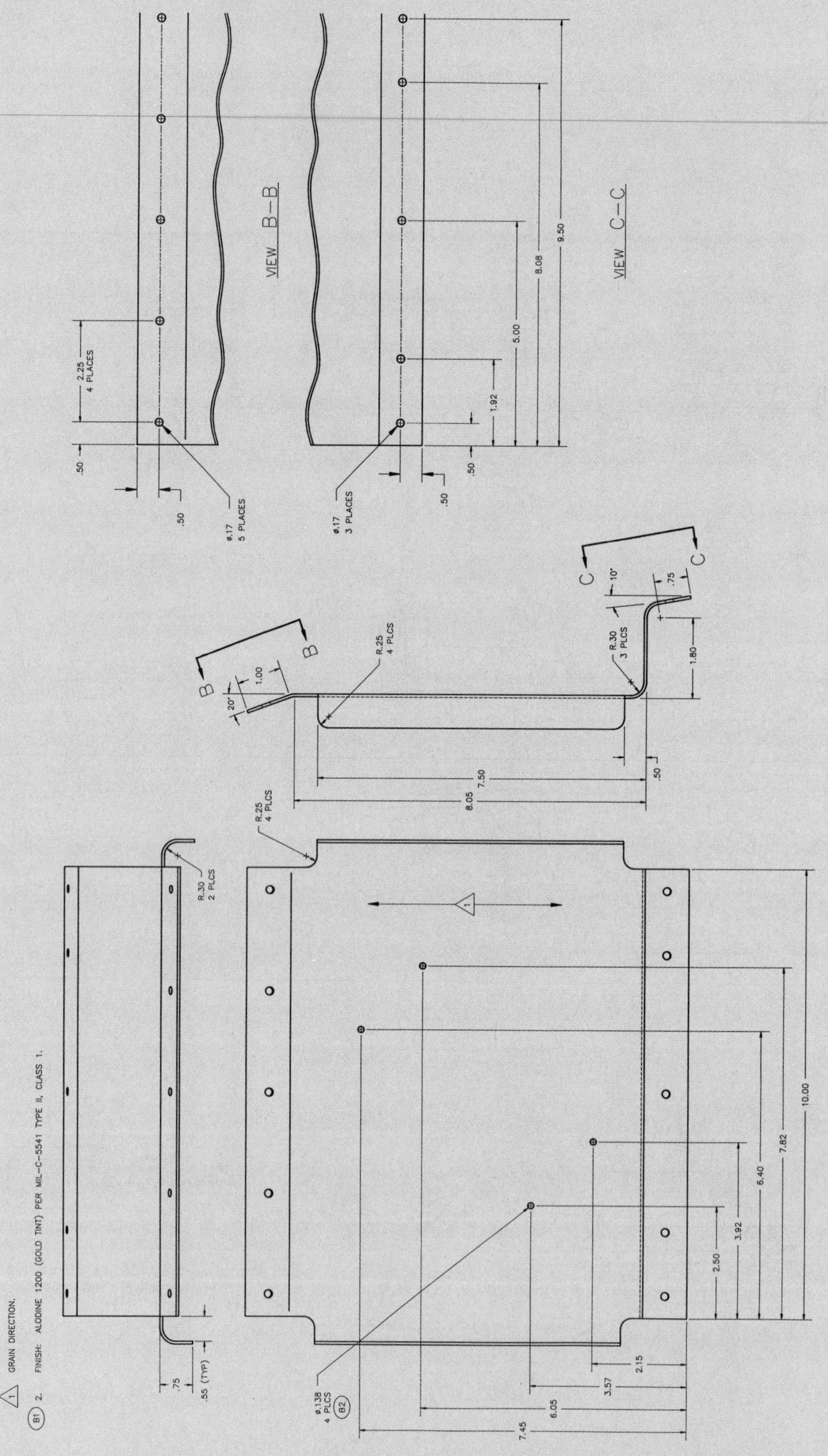

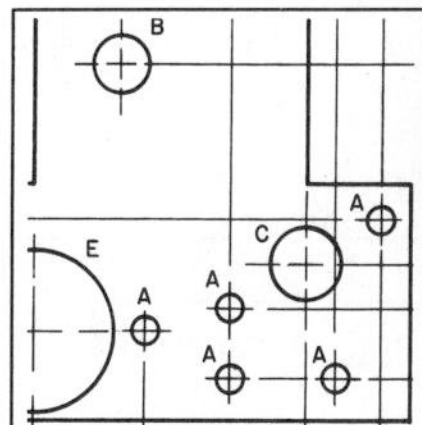

CHAPTER 10

Manufacturing Processes

Learning Objectives

Upon completion of this chapter you will be able to accomplish the following:

1. Identify the specific stages in the manufacturing process.
2. Demonstrate an understanding of materials used in the manufacturing process.
3. Develop an understanding of design for manufacturability (DFM) concepts.
4. Identify the basic types of machine tool operations.
5. Define the processes involved in the technology of materials forming.
6. Discern the difference between finishing techniques.
7. Demonstrate familiarity with the process of automated and computer-aided manufacturing.

10.1 Introduction

The purpose of any engineering drawing is to provide the information necessary to manufacture a part or system. To design and manufacture a part properly, the designer and drafter must understand manufacturing. This chapter describes basic manufacturing and production processes.

The engineering drawing shows the specific size and geometric shape of the part (Fig. 10.1). It also provides related information about material specifications, finish requirements, and required treatments, along with the revisions and releases made to the document. The drawing in Figure 10.1 shows the revisions in the upper right corner. The notes, in the lower left-hand corner, provide manufacturing with information about the part.

10.2 Manufacturing

Manufacturing is the process of coordinating workers, machines, tools, and materials to create a product. The primary purpose of manufacturing is to produce quality parts from raw materials and assemble related parts to create assemblies. Manufacturing steps include:

1. Materials selection and manufacturing methods
2. Assembly requirements
3. Production control
4. Planning and tooling requirements
5. Production and manufacturing of the product
6. Inspection and quality control.

Many companies have separate areas for product development, tooling and manufacturing, and facilities. **Product development** involves the conceptual work that is done in the development of a product. Producing and manufacturing a product requires new machines, tools, dies, jigs, and fixtures. Thus, **tool design** is very important for a successful product. **Facility designers** and **drafters** do building and plant upgrading, maintenance design, and new additions.

If you understand the cost, and the mechanical capabilities and limitations of basic processes, you can design the part with the manufacturing process in mind. The final product is what is manufactured and produced for sale. The drawing or computer-aided design (CAD) database is the starting point for the design-through-manufacture sequence.

10.2.1 Manufacturing Processes and Manufacturability

When the manufacturing department receives the engineering drawing, it is reviewed to ensure that all information necessary to make the part is provided. During this review, manufacturing engineers decide on tooling, machines, inspection, and the time needed to produce the part. New concepts, including the integration of manufacturing decisions into the beginning stages of design, are being implemented throughout industry today. This is called **design for manufacturability** (DFM). DFM is a company design philosophy. Because the manner in which a product is designed determines 70% to 90% of the total ongoing cost of the product, it makes sense to design in quality and manufacturability. **Concurrent engineering** is the effort to get design, engineering, manufacturing, and production to work in parallel rather than in sequence. The following considerations are important for a successful product, but they must be considered during the design stage when changes can be implemented most readily:

1. Material specification
2. Size and configuration
3. Production run (The number of parts needed; this greatly influences the production method.)

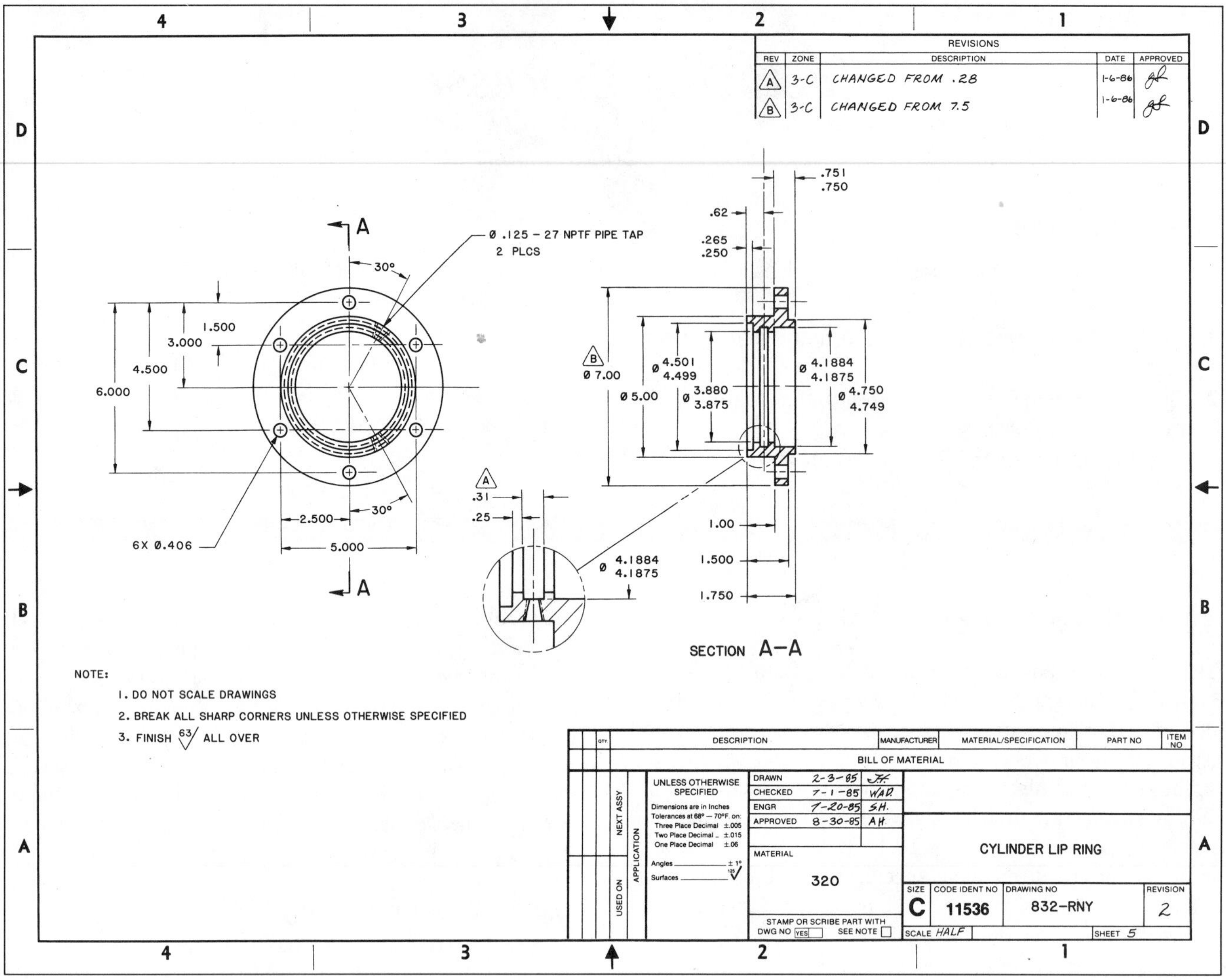

NOTE:
1. DO NOT SCALE DRAWINGS
2. BREAK ALL SHARP CORNERS UNLESS OTHERWISE SPECIFIED
3. FINISH 63/ ALL OVER

FIGURE 10.1 **Cylinder Lip Ring Detail Drawing**

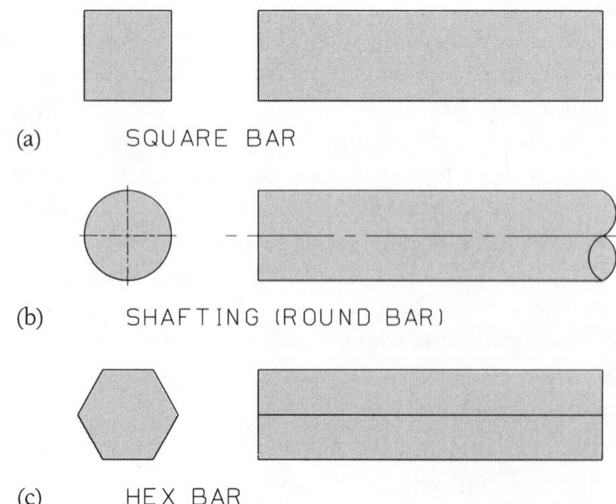

FIGURE 10.2 **Stock Forms** (a) Square bar (b) Shafting or round bar (c) Hex bar

4. The tolerances specified for the part
5. Machine and tooling operations.

The part is manufactured from a **stock piece** or other raw material. A variety of standard **stock forms** are available. **Bar stock** comes in square, round, and hexagonal shapes (Fig. 10.2). Figure 10.3 shows the available types of **structural shapes.** If a stock form is not used, then the part must be cast, extruded, or formed by means of other processes. The five basic families of processes are as follows:

1. **Molding** or **casting** into the proper configuration
2. Forming by **bending** into the required shape
3. **Cutting** or **sawing** into the proper size and shape
4. Pounding or **forging** into shape
5. **Fabrication** with fastening methods: *welding, riveting, bolting, screwing, adhering,* or *nailing* parts formed by the previously listed processes.

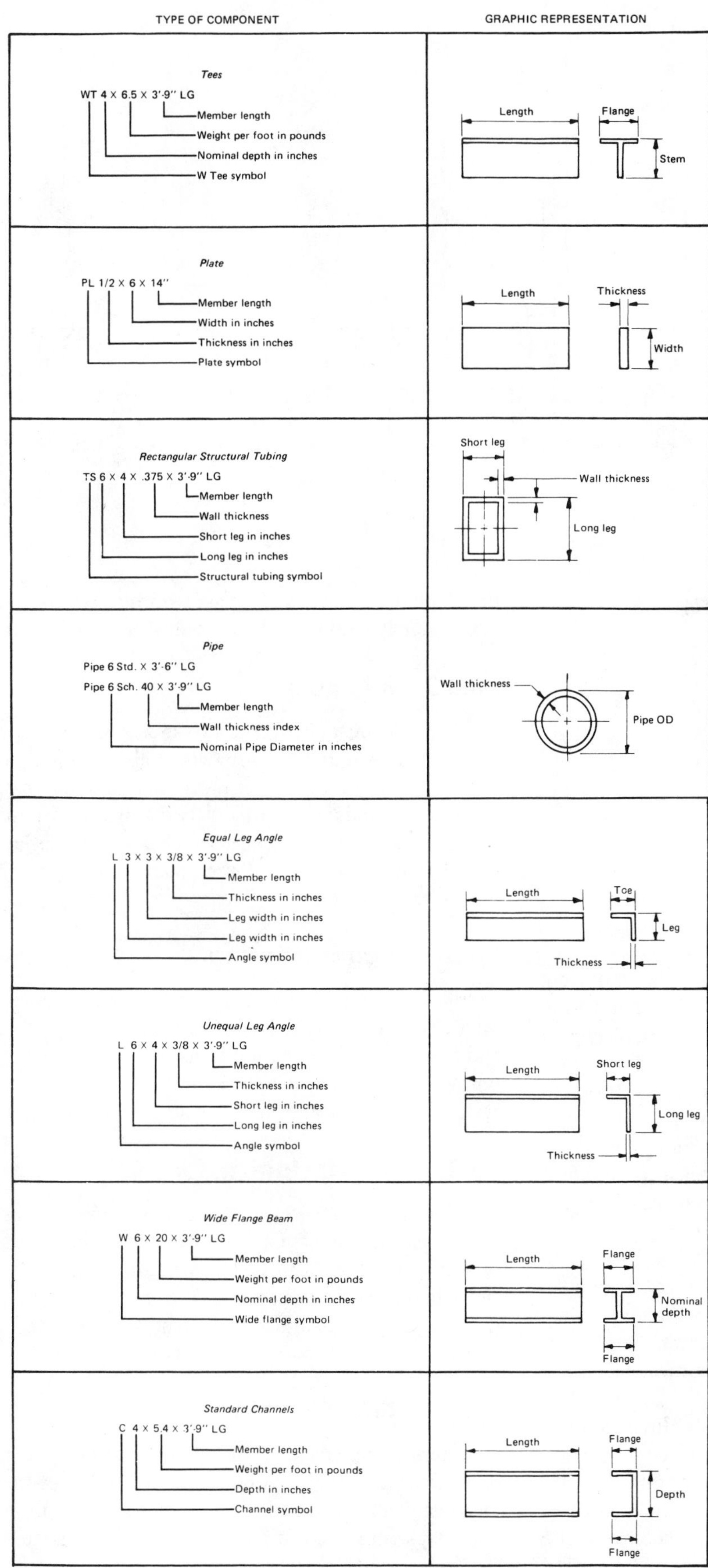

FIGURE 10.3 Structural Stock Forms

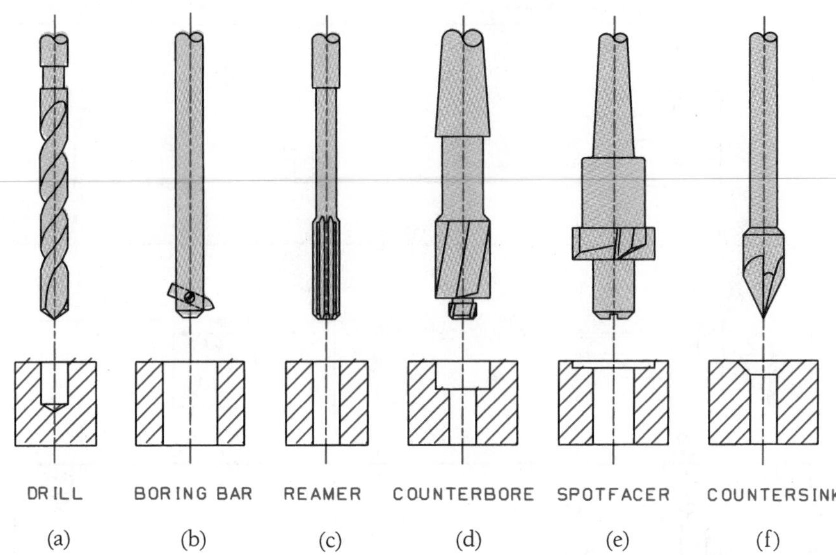

FIGURE 10.4 Machined Holes

10.3 Machine Tool Operations

Machine tools are machines that cut metal or form new material. Five basic processes are performed on machine tools:

1. Drilling (drilling, reaming, counterboring, countersinking, spotfacing)
2. Turning (lathe work)
3. Planing and shaping
4. Milling
5. Grinding.

10.3.1 Drilling

Drilling is one of the most common basic machine tool operations. Included in this process are drilling holes from under $\frac{1}{64}$ in. (0.4 mm) to more than 2 in. (50 mm). Machined holes can be made by counterboring, countersinking, spotfacing, spot or center drilling, and reaming (Fig. 10.4). In industrial drilling, the drill bit is a cutting tool held by a chuck and rotated by a large motor. The rotating tool is fed into the part at a controlled rate. The turning speed and feed rate of the drill are determined by the material and size of the hole.

Almost every machined part has drilled holes. The type of process required is determined by the tolerance of the hole. Drilling is also used to create rough holes before the boring, reaming, counterboring, countersinking, or tapping operations are performed. **Reamers** and **countersinks** (Fig. 10.4) are used after a hole has been drilled.

A tap drill is used when the hole will have a **thread** applied to it with a tapping tool. The tap drill must be the proper size to produce the minor diameter of the internal thread.

The part being machined must be held firmly in place during drilling. A **drill jig** was used to complete the precise drilling of the small part shown in Figure 10.5(a) (front). The drill jig has a hinged cover plate with integral drill bushings for accurately guiding and positioning the drill [Fig. 10.5(b)].

10.3.2 Reamers

When a hole must be precise, a **reamer** [Fig. 10.4(c)] is used. Reamers are required because twist drills make holes that are not accurately sized, are not precisely round, and have poor finishes. An undersized drill is used to remove most of the material, then the reamer finishes the hole. Reamers are made of tungsten carbide or tool steel. Many types of reamers are available.

10.3.3 Counterboring

A drilled hole must be made first before **counterboring** [Fig. 10.4(d)]. A counterbored hole is deeper than a spotfaced hole and has a specific dimension to its recessed depth. Counterboring is used so that socket head and fillister screws are seated with their heads flush or below the surface of the part.

10.3.4 Spotfacing

Spotfacing [Fig. 10.4(e)] is the same basic process as counterboring, but is done no more than $\frac{1}{8}$ in. (3 mm) deep. This process is used to clean up the area around the hole, especially if the part is made of a cast material. The spotface provides a smooth surface for fasteners (nuts, bolts, screws, rivets).

10.3.5 Countersinking

Countersinking creates a small chamfer or bevel at the edge of a hole [Fig. 10.4(f)]. A hole is drilled before countersinking. Countersinking makes it easier to insert dowel pins, bolts, taps, and reamers into the hole. Chamfers are usually 82° for flathead bolts.

(a)

(b)

FIGURE 10.5 Drill Jig (a) Drill jig with part (b) Drill jig with hinged top open

10.3.6 Center Drilling

Center drilling is required when the part is to be held between centers for machining on a lathe. Center drilling is also used to create an accurately located starting hole for a twist drill.

10.3.7 Taps and Dies

Taps and **dies** are used to machine internal and external threads. External threads on shafts are cut by a die; internal threads are cut by a tapping tool.

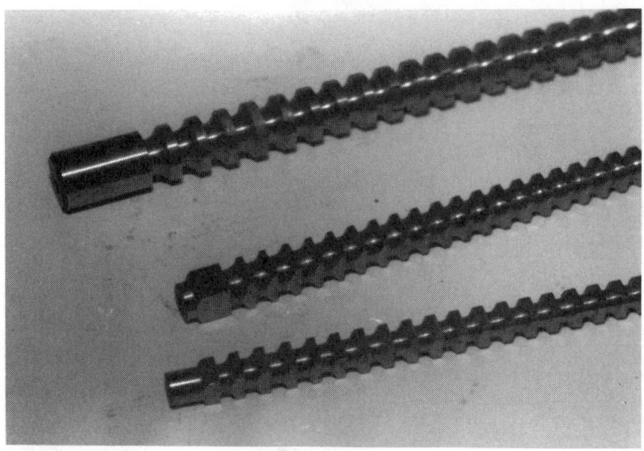

FIGURE 10.6 Broaches

10.3.8 Broaching

Broaches are used to create odd-shaped holes or openings (Fig. 10.6). A broaching machine is used to cut special features like keyseats and to form square, hexagonal, or odd-shaped holes after a drilled hole has been made. A broach is a long tool with a series of teeth or cutting edges that increase in size progressively so that each of the teeth removes only a small portion of the material as it is pulled or pushed through the part.

10.3.9 Boring

Boring is a machining process used to produce a wide range of precise tolerance holes and requires a milling machine, a lathe, or a special boring machine.

10.3.10 Turning Operations

Turning operations use the engine lathe (Fig. 10.7) , the turret lathe, and a variety of **boring machines.** The **vertical boring mill** is used for turning large parts that need round cuts and for facing and contouring.

The most common and versatile type of machine tool (found in every machine tool area) is the **engine lathe** (Fig. 10.8). The engine lathe is used to produce cylindrical part operations that include cutting threads, facing, tapering, parting, turning, and knurling (Fig. 10.9). A lathe is a machine that rotates the part rapidly while a stationary cutting tool is used to perform the operation. The machinist is setting up the part in Figure 10.7. A part is usually held in a lathe in a **chuck** [Fig. 10.10(a)]. The chuck is connected to the powered end of the machine [Fig. 10.10(b)]. Collets, face plates, drive plates, and other devices are also used to hold and drive the work piece in the lathe.

A lathe is also used for drilling, reaming, boring, counterboring, facing, threading, knurling, and polishing. Drilling, reaming, boring, and counterboring are done on the face of the part as it turns.

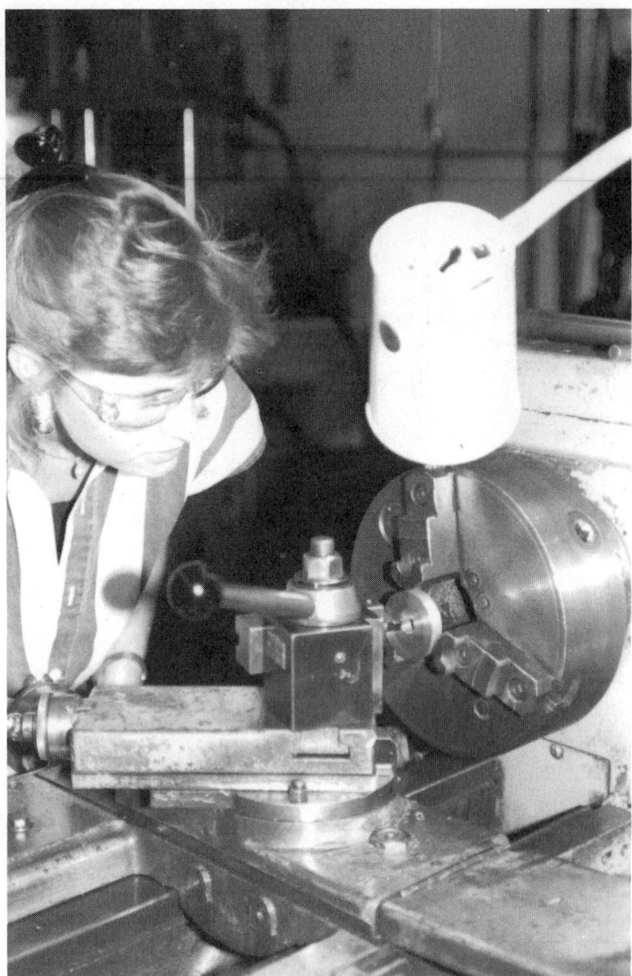

FIGURE 10.7 Machinist Setting Up a Part on a Lathe

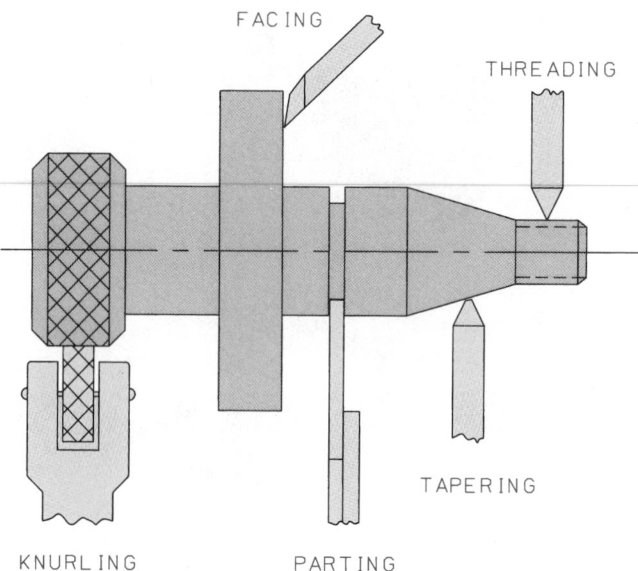

FIGURE 10.9 Lathe Processes

(a)

FIGURE 10.8 Part Being Machined on a Lathe

(b)

FIGURE 10.10 Lathe Chuck (a) Lathe chucks (b) Part held by a chuck on a lathe

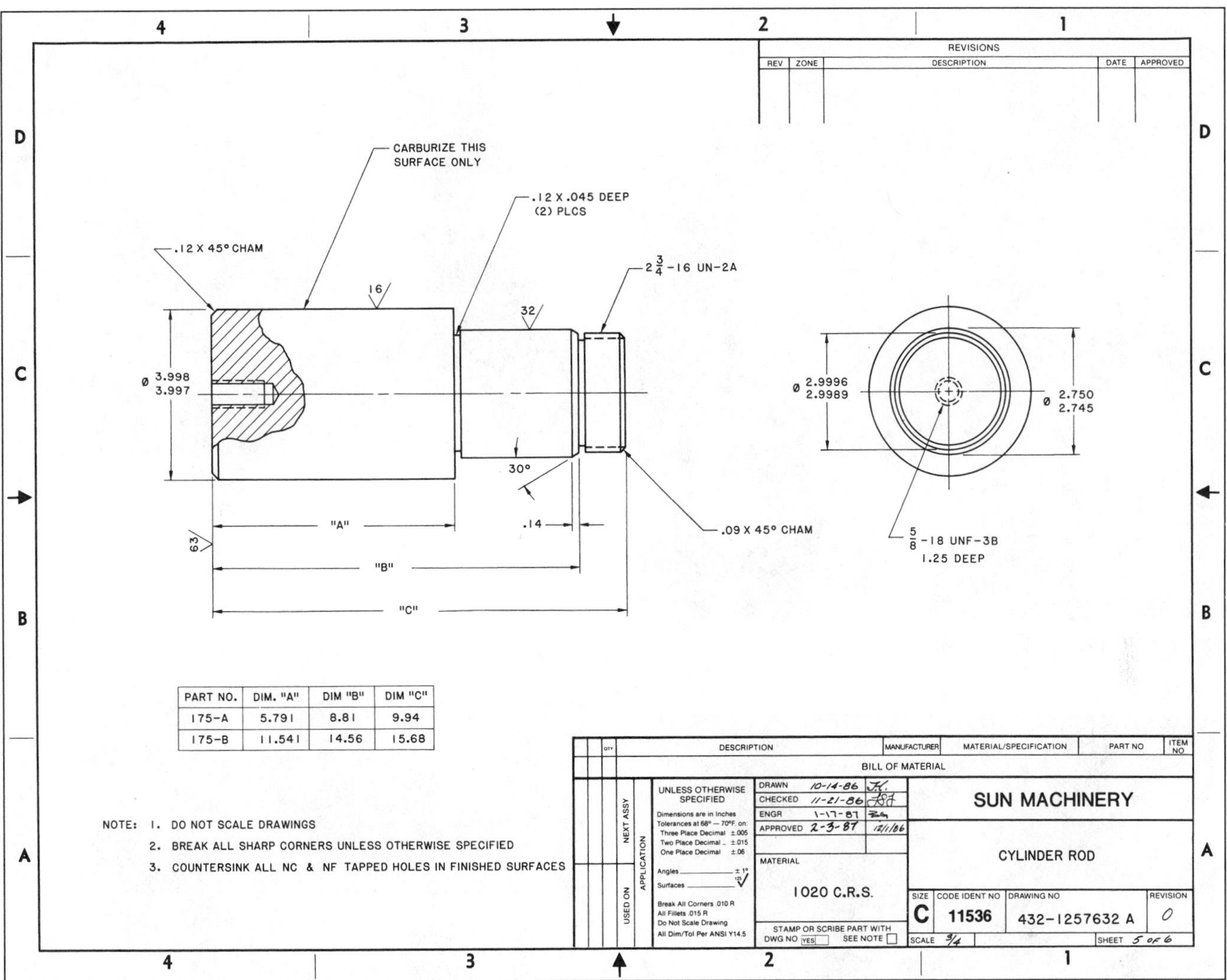

PART NO.	DIM. "A"	DIM "B"	DIM "C"
175-A	5.791	8.81	9.94
175-B	11.541	14.56	15.68

FIGURE 10.11 Cylinder Rod

Computer numerically controlled (CNC) engines lathes are also available. All of the functions are controlled by a programmed computer. Therefore, manual controls are limited on this type of machine. The **turret lathe** has a rotating multisided turret where a variety of cutting tools can be mounted. This allows the rapid changing of tools for small-to large-volume production.

Turning uses a lathe to reduce the outside diameter of a part. In this situation, the tool bit will travel parallel to the **Z** axis. **Facing** decreases the length of the part or flange and creates a flat surface (Fig. 10.9). **Threading** is done on an engine lathe with a single point tool (a slow process). Drilling and reaming can be done on a lathe but the hole location is limited to the center of the lathe's **Z** axis, in line with the tailstock center. **Knurling** (Fig. 10.9) is a pattern formed into the surface of a part, either for appearance or to provide a gripping surface. The pattern is either straight or diamond shaped.

The cylinder rod in Figure 10.11 is an example of a part that is produced on a lathe. Dimensions A, B, and C are given in three different sizes. This part is made from 1020 cold rolled steel (CRS) and requires a tapped hole in the large end. Chamfering, facing, parting, drilling, and threading are required to complete the part.

10.3.11 Milling Machines and Milling Cutters

A **milling machine** is one of the most important machines found in manufacturing. Milling machines are also one of the most accurate machines used. The typical milling machine has a table to which the part is securely fastened. Cutting is done by a rotary milling cutter with single or multiple cutting edges. One or more cutters are on each machine. Drilling, boring, reaming, slotting, facing, pocketing, and other types of cuts are made with this machine.

Milling machines are divided into two categories: vertical and horizontal. The classification depends on the orientation

FIGURE 10.12 Bridgeport Mill

(a)

(b)

FIGURE 10.13 **Milling Machine** (a) Horizontal mill (b) Horizontal mill machining a keyseat

(a)

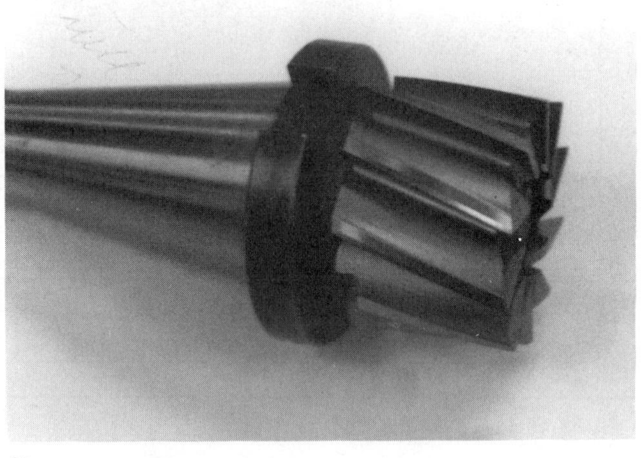

(b)

(c)

FIGURE 10.14 **Mills** (a) End mill and tool holder (b) Shell mill and holder (c) Face mill (slab type)

of the spindle. Figure 10.12 shows a vertical milling machine. The table is a flat surface with a variety of tee slots to insert clamping mechanisms that hold the part in place. Milling machines are also used to cut irregular surfaces, gears, slots, and keyways. Figure 10.13 shows a horizontal spindle milling machine with a side cutting mill, which is being used to cut a keyway. This is also referred to as a **slitting saw.** Cutters are held in place by **collet adapters, arbors,** and quick-change **holders.** Cutters fall into four basic categories:

1. End mills [Fig. 10.14(a)]
2. Shell mills [Fig. 10.14(b)]
3. Face mills [Fig. 10.14(c)]
4. Plane milling cutters, including side mills.

End mills are versatile cutters and are used for many types of machining work, especially where close tolerances must be maintained. The end mill in Figure 10.14(a) is held by a tool holder. An end mill is being used to mill the parts in Figures 10.15(a) and (b). End mills are also used for pocketing parts. **Shell mills** [Fig. 10.14 (b)] are used for simple facing or cutting steps that cannot be done by a face mill. **Face mills** are used for facing flat surfaces and are used primarily on horizontal milling machines. Face mills come with inserted teeth or are slab types. The face mill in Figure 10.14(c) is a **slab mill.**

10.3.12 Grinding

Grinding is also a cutting process except that the cutters used are grinding wheels made from irregular-shaped abrasive grit. This abrasive grit is used to cut or grind a part. The basic purpose of a grinding wheel is to provide a fine-finished surface and to maintain accurate size control.

Removing edges and corners from a part is done with stones and sandpaper or hand grinders. As an example, the note on the link pull detail in Figure 10.16 requires that the machinist "BREAK ALL SHARP CORNERS UNLESS OTHERWISE SPECIFIED."

Figure 10.17(a) shows a surface grinder that has a movable table. Grinding machines are divided into surface types: cylindrical, internal, and centerless. Vertical spindle surface grinders and surface grinders machine flat surfaces. Single-purpose abrasive machines such as abrasive cutoff machines and snagging grinders are commonly used. Figure 10.17(b) shows an outside diameter grinder (OD grinder) and Figure 10.17(c) shows a pedestal grinder.

10.3.13 Saws

Many types of **sawing machines** are found on the shop floor. The power saw (Fig. 10.18) is a band-saw cutoff machine that

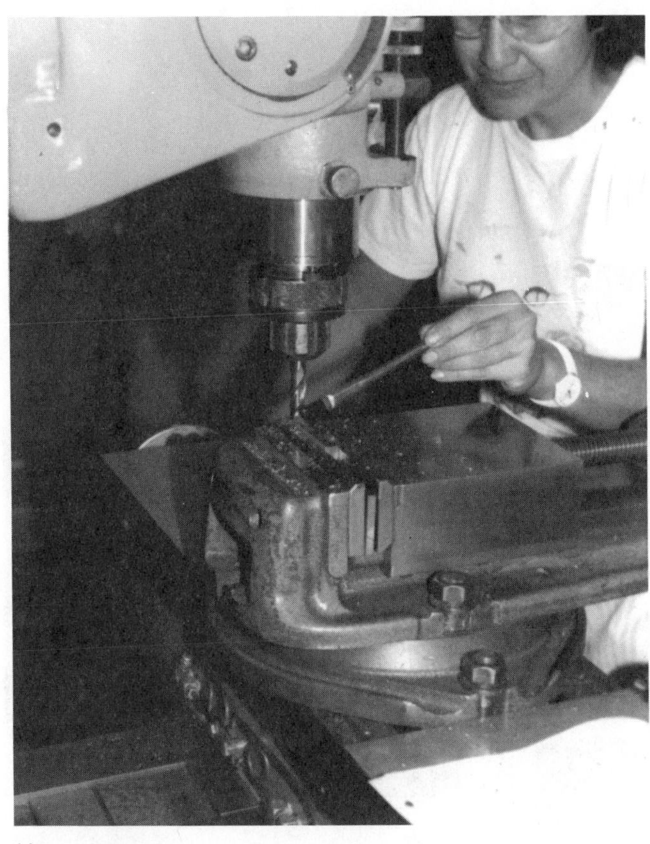

(a)

(b)

FIGURE 10.15 Machining a Part (a) Machinist adding lubricant to a part before milling (b) Part being machined on a vertical mill

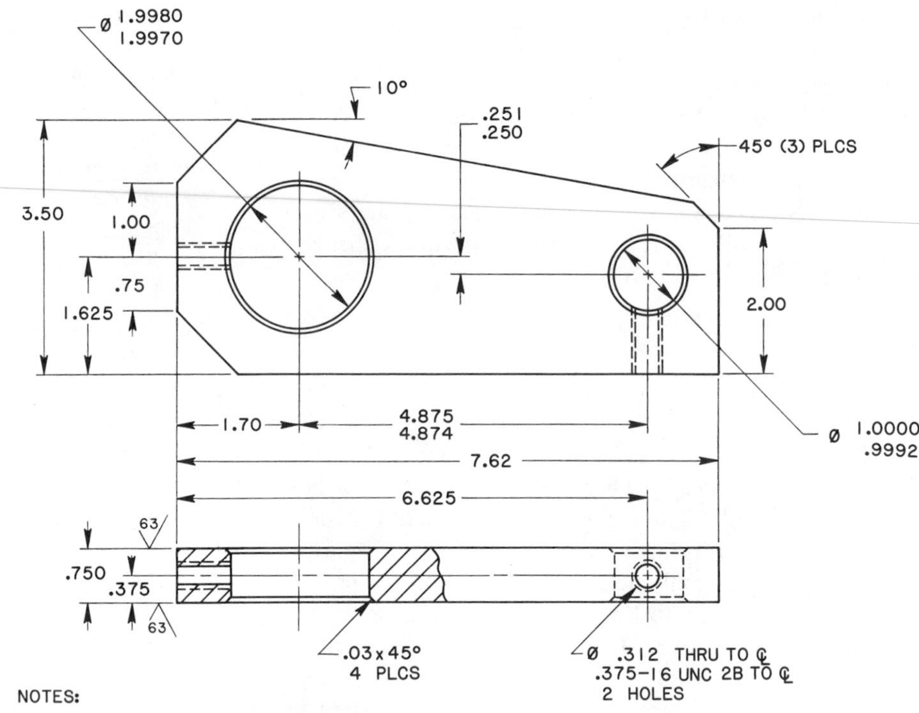

FIGURE 10.16 Pull Link Detail Drawing

(a)

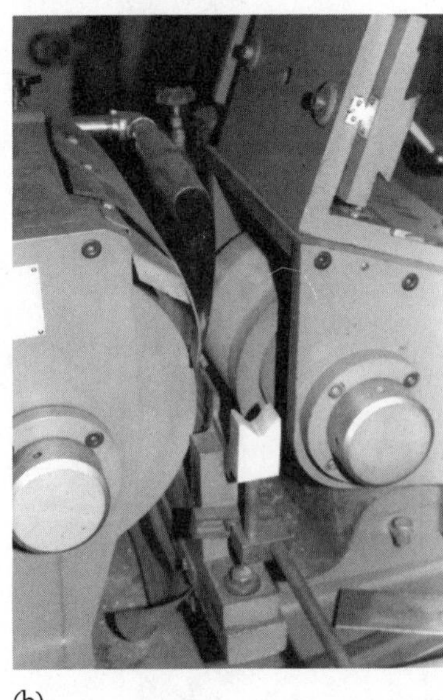

(b)

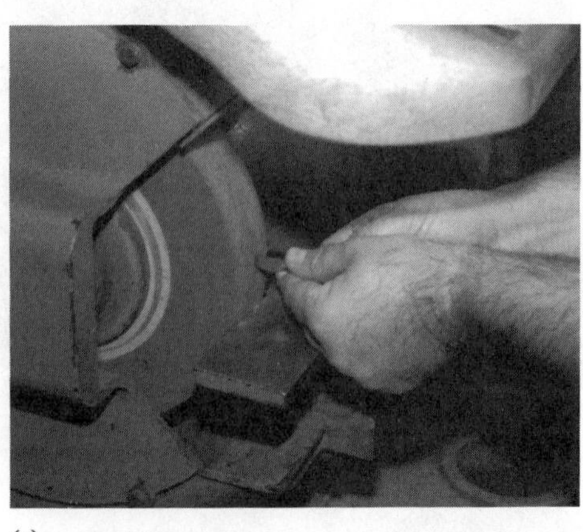

(c)

FIGURE 10.17 **Grinders** (a) Surface grinder (b) OD grinder (c) Pedestal grinder

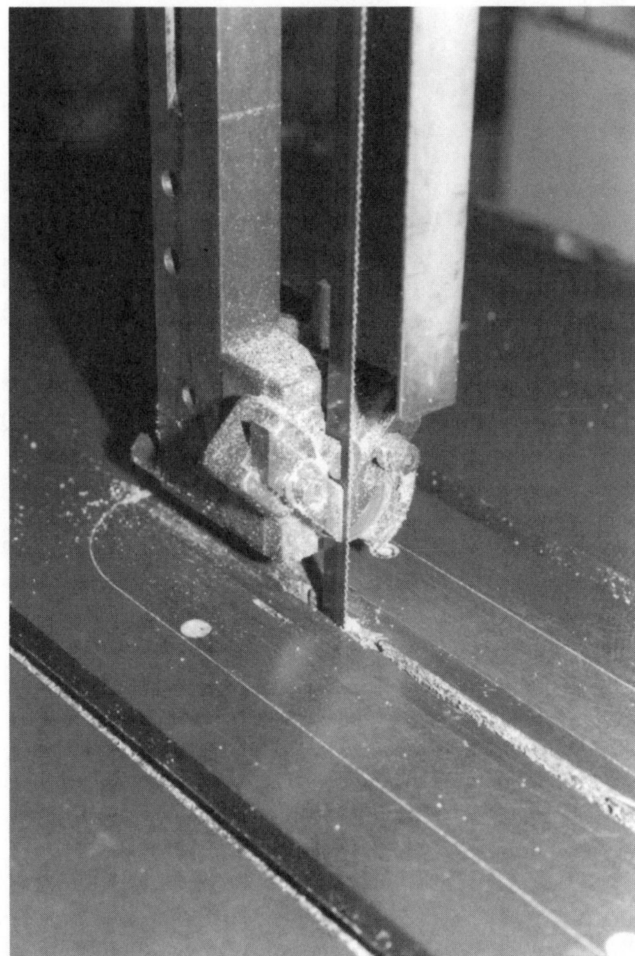

FIGURE 10.18 Metal Band Saw

uses a continuous band-saw blade. This type of saw is used to cut bar stock to length. These machines are used on thin material to cut irregular shapes, to make beveled cuts on tubing or solid stock, or to make slots or slits.

10.3.14 Shapers and Planers

Shapers and **planers** are limited to straight-line cuts. A shaper can handle relatively small parts; a planer is used on parts weighing up to several thousand pounds. Planers and shapers make facing cuts (both top and side), slotting, step cuts, and dove tails (both male and female). Both machines are used to create finished surfaces. Multiple pieces can be machined at the same time with a planer. The planer is used to machine large iron castings or steel weldments that weigh hundreds of pounds. Shapers come in both horizontal and vertical types. Vertical shapers are sometimes called slotters.

10.3.15 Hand-Held Measuring Devices

A variety of measuring tools are used in manufacturing. The **pocket steel ruler**, inside and outside **calipers**, micrometers, and vernier or dial calipers are used in manufacturing. **Vernier calipers** (Fig. 10.19) are used to measure both the inside

FIGURE 10.19 Vernier Calipers

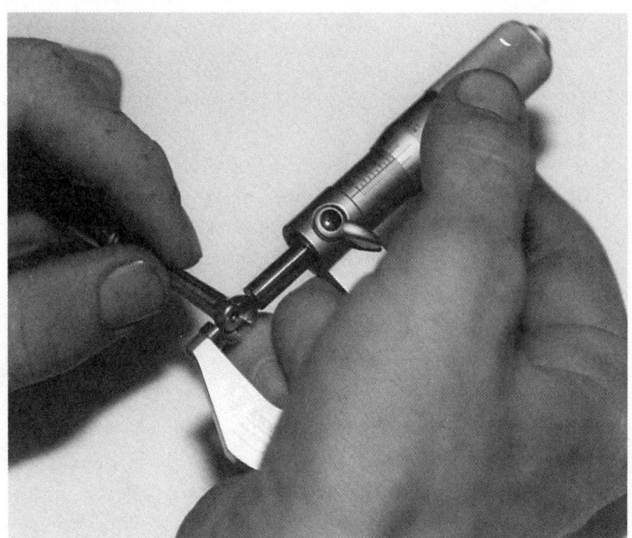

FIGURE 10.20 Measuring a Part with a Micrometer

and the outside of the part. Vernier calipers have a beam or bar marked in inches and hundredths or in centimeters and millimeters.

The **micrometer**, also referred to as the micrometer caliper, is available in inside and outside versions (Fig. 10.20). The micrometer is the most accurate of the precision hand-held measuring instruments. Digital versions of vernier calipers and micrometers are available in a variety of sizes.

10.4 Surface Texture Specification

A variety of standards have been developed by the American National Standards Institute (ANSI) and American Society of Mechanical Engineers (ASME) for specifying **surface textures.** The **surface roughness measurement** is important in machining. The finer the finish, the more expensive the machine process required. Processes such as milling, shaping,

ITEMS OF INTEREST

Computers in Manufacturing

Computer numerically controlled (CNC) machines have transformed manufacturing methods and techniques during the past twenty years. Today, they are integral parts, of flexible manufacturing systems (FMS) that can machine one part, a thousand parts or several different kinds of parts. Changes to the computer program controlling the system modify what part the system machines by redefining the sequence of events needed to complete the machining steps.

Numerical control (NC) began in 1947 with John Parson's experiments on producing aircraft components with three-axis curvature data to control machine tools. The U.S. Air Force awarded Parsons a contract in 1949 to build the first NC machine. The Massachusetts Institute of Technology took over the development contract in 1951 and produced the first machine in 1952. Refined industrial machines followed in 1955.

Early NC machines used either punched tape or punched cards to send commands to the machine. Most machines used punched tape and tape readers. The tape was fragile and broke easily in industrial settings and, if 1,000 parts were to be made, the reader read the tape 1,000 times. Because of the need to make NC more efficient, computer control was developed. The part programmer uses English-like commands to write the program and the computer does the work of translating the commands into machine code.

Distributed numerical control (DNC) allows control of a system of CNC machines by using a networked computer. By planning a network of computer control effectively, it is possible to control an entire factory. FMS systems are networked into the computer control scheme.

Numerical control was developed to increase productivity, increase quality, increase accuracy, reduce labor costs, and do jobs that were considered impossible or impractical. CNC machines require a large initial investment and have higher per hour operating costs than traditional machines tools, but the other advantages outweigh these disadvantages.

NC, CNC, and DNC machines will play increasingly important roles in automated and flexible manufacturing in the future. Today stand-alone or networked CNC machines are widely used in both large and small production shops. Because of the great advances in and great advantages of this technology, the "factory of the future" will rely on these machines to be the backbone of the machining processes.

Coordinate Measuring Machine (CMM)

Flexible Manufacturing System (FMS)

and turning can produce precise surface textures ranging from 125 to 32 microinches (μin.). Only a lathe can produce 8 μin. on a production basis. Grinding operations produce surface textures ranging from 64 to 4 μin. The Greek lowercase letter μ (microinch, μin; micrometer, μm) is used on the drawing.

The surface texture is a part of the design specifications. The surface texture value is used along with the **surface texture symbol** (Fig. 10.21). The surface texture symbol designates the waviness, lay, and classification of roughness. **Roughness** is the irregularity on the surface of the part. It is not the distance between the peaks and valleys of the

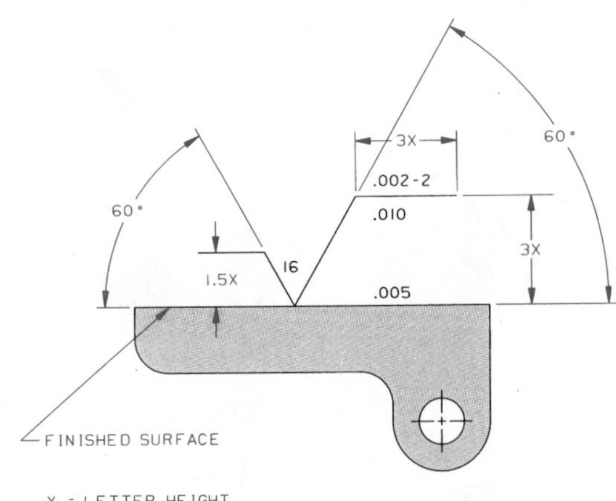

FIGURE 10.21 Surface Texture Symbol Specifications

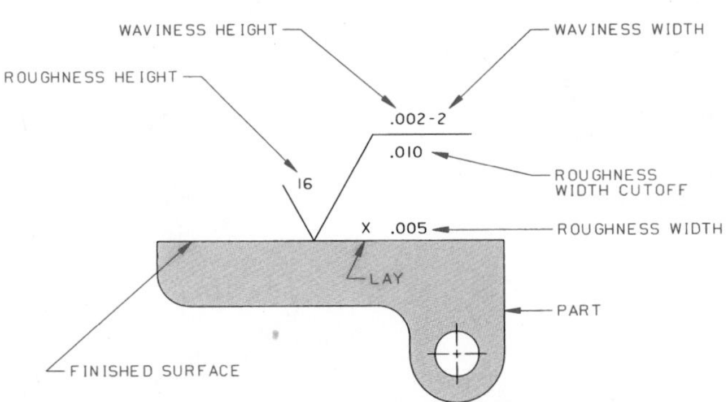

FIGURE 10.22 Surface Texture Symbol Description

roughness, but the average amount of irregularity above and below an assumed centerline. **Waviness** is the irregularity from the centerline. The **waviness height** is the peak-to-valley height of the roughness.

Roughness is caused by the action of the machining during the production process. The roughness height is designated above the V portion of the surface texture symbol (Fig. 10.21). The symbol is constructed with the measurements provided. Figure 10.22 defines each of the portions of this symbol and what they mean. *The symbol provides information on the waviness height, waviness width, roughness height, and width.* In Figure 10.23, a part is shown with an exaggerated measured profile. The **nominal centerline** or **profile of the part** is used to establish the surface roughness deviation.

A profilometer measures the **smoothness** of a surface texture roughness in microinches or micrometers. Surface texture is the deviation from the nominal centerline or nominal surface that forms the pattern of the surface and includes flaws, lay, waviness, and roughness. The direction of lay, roughness width cutoff, roughness width, waviness height, roughness height, and roughness width are shown in Figure 10.24.

The centerline or nominal surface line (Fig. 10.23) is a line about which the roughness is measured and is parallel to the

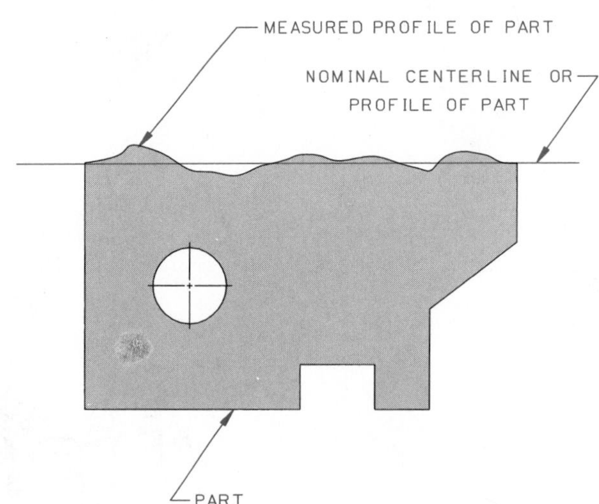

FIGURE 10.23 Nominal Center and Measured Profile of a Part's Surface

direction to the profile within the limits of the roughness width cutoff. The roughness consists of the finer irregularities in the surface texture including those that result from action in the production process, such as transverse feed marks and other irregularities. **Roughness height** (Fig. 10.24) is an average deviation expressed in microinches or micrometers measured normal to the centerline. **Roughness width** is the distance parallel to the nominal surface between successive peaks or ridges on the part. The nominal surface is the surface contour shape that is usually shown and dimensioned by the drafter. The **roughness width cutoff** is the distance over the surface on which the roughness measurement is made.

Waviness is caused by vibration of the machine during the machining process, heat treatment, or other processes applied to the part. The **waviness width** is rated as a measurement of spacing of successive wave peaks or wave valleys. The **lay** is the direction of surface pattern. **Flaws** are the irregularities including cracks, blowholes, checks, ridges, and scratches.

Figure 10.25 shows the removal of material by machining that will be specified by variations in the surface texture symbol: optional, required, prohibited and removal allowance. The preferred series of roughness height values is shown in Figure 10.26.

10.5 Production Processes

Production processes include casting, forging, bending, rolling, press work, injection molding, dies, electrical discharge machining, electrochemical machining, blow molding, and variations of other hot and cold processes.

Designing for automated production helps ensure a more efficient and cost-effective production process. *Design for manufacturability* ensures that the right process is chosen, existing factory resources are utilized, setup times are minimized, and tolerances are specified correctly. These

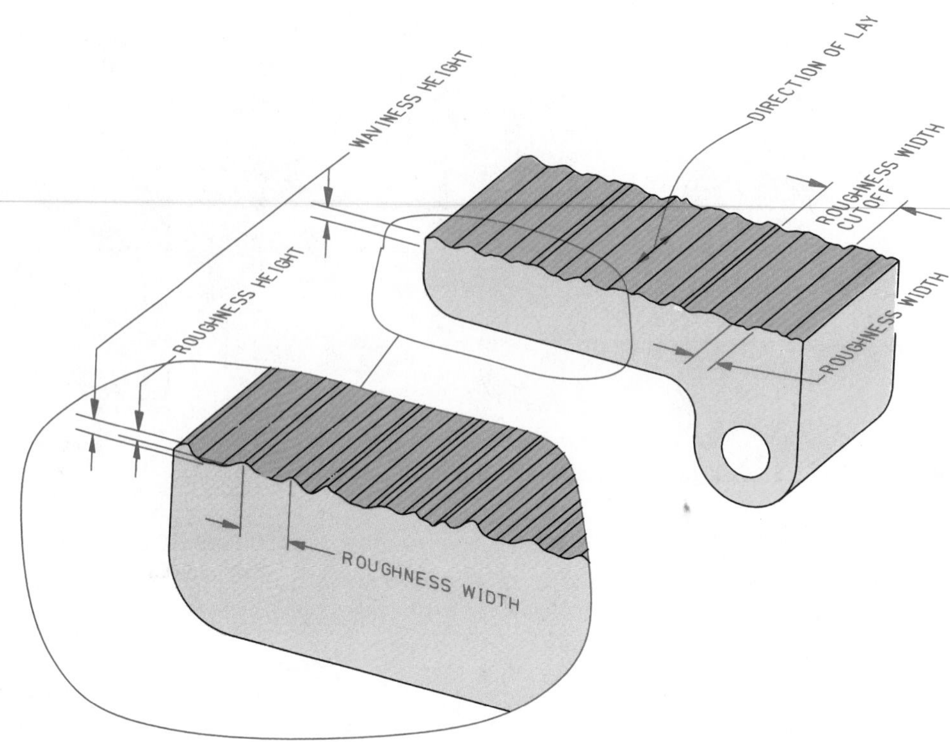

FIGURE 10.24 Surface Texture Terminology

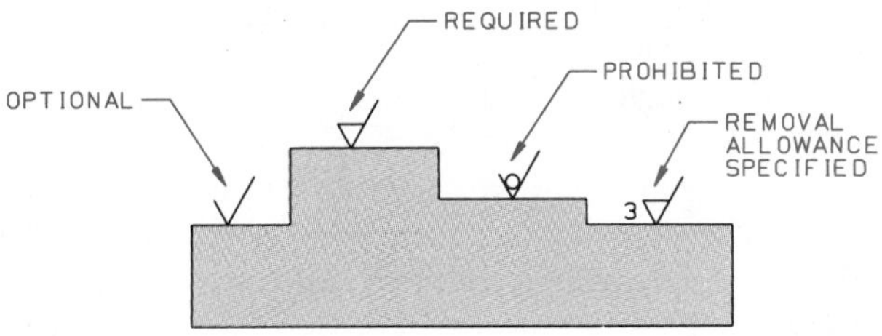

FIGURE 10.25 Surface Texture Symbols

ROUGHNESS HEIGHT RATING MICROMETERS	MICROINCHES	SURFACE DESCRIPTION	PROCESS
25.2	1000	VERY ROUGH	SAW AND TORCH CUTTING, FORGING OR SAND CASTING.
12.5	500	ROUGH MACHINING	HEAVY CUTS AND COARSE FEEDS IN TURNING, MILLING AND BORING.
6.3	250	COARSE	VERY COARSE SURFACE GRIND, RAPID FEEDS IN TURNING, PLANNING, MILLING, BORING AND FILING.
3.2	125	MEDIUM	MACHINE OPERATIONS WITH SHARP TOOLS, HIGH SPEEDS, FINE FEEDS AND LIGHT CUTS.
1.6	63	GOOD MACHINE FINISH	SHARP TOOLS, HIGH SPEEDS, EXTRA FINE FEEDS AND CUTS.
0.8	32	HIGH GRADE MACHINE FINISH	EXTREMELY FINE FEEDS AND CUTS ON LATHE, MILL AND SHAPERS REQUIRED. EASILY PRODUCED BY CENTERLESS, CYLINDRICAL AND SURFACE GRINDING.
0.4	16	HIGH QUALITY MACHINE FINISH	VERY SMOOTH REAMING OR FINE CYLINDRICAL OR SURFACE GRINDING,OR COARSE HONE OR LAPPING OF SURFACE.
0.2	8	VERY FINE MACHINE FINISH	FINE HONING AND LAPPING OF SURFACE.
0.05 0.1	2-4	EXTREMELY SMOOTH MACHINE FINISH	EXTRA FINE HONING AND LAPPING OF SURFACE.

FIGURE 10.26 Description of Roughness Height Values

procedures reduce labor costs and break down barriers between different areas (islands) of information in the company.

10.5.1 Casting

Casting is the process of forming parts to approximate rough sizes by introducing liquid material into a formed cavity called a mold and allowing the material to solidify by cooling. The mold is removed, leaving the solid, shaped part. Available casting methods include sand casting, mold casting, die casting, and investment casting. **Molding** and **die casting** are similar to casting except that the material used in the process is not liquid, but is softened to a plastic state and forced into the mold under high pressure.

Everyday items, from toys to electronic components, are cast. One common type of casting is **sand casting.** Figure

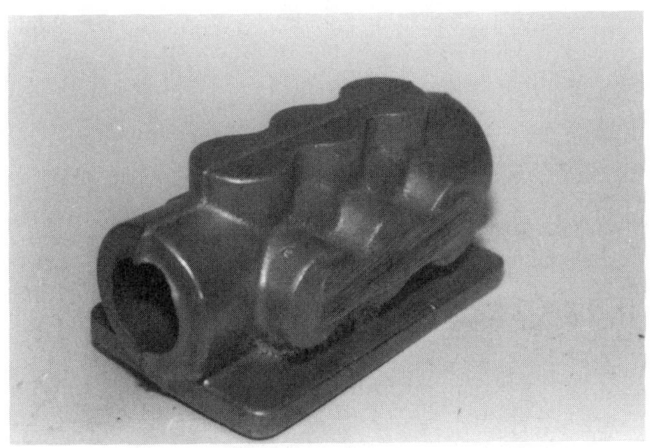

FIGURE 10.27 Sand-Cast Part Before Machining

FIGURE 10.28 Machined Sand-Cast Part

10.27 is a sand cast part before machining. Figure 10.28 shows a sand casting after machining. Casting is divided into two basic processes, **gravity** and **pressure.** Sand casting molds are formed by patterns. **Patterns** look like the cast part and are used to create a shape in the mold cavity. The wood pattern in Figure 10.29(a) is inserted into sand to create the proper configuration of the part (front). Figure 10.29(b) shows another example of a wood pattern. Here, the pattern is designed to shape six identical parts.

A drafter prepares a combination casting and machining drawing, although some companies do require separate drawings. The **casting detail** is used by the pattern maker and the **machining drawing** by the machinist. The **draft angle** is the angle of the **taper** of the part that makes it easier to withdraw the pattern from the mold. After the material hardens, the sand is removed from the casting (destroying the mold).

Tooling points on **three datum planes** (that are perpendicular) are used to locate dimensions on the casting. The planes are established by the tooling points on the casting.

Since it is not possible to cast sharp corners and angles accurately, the internal angles on a casting are filled with a material to eliminate sharp corners. Contoured surfaces that fill the sharp inside corners are **fillets** (Fig. 10.30). **Rounds** are the exterior corners that have been smoothed out to remove their sharp edges.

Aluminum, magnesium, zinc, copper, bronze, and brass as well as iron and steel are used to make castings. Designing a casting requires an understanding of how much the material will shrink during the cooling process. The dimensions shown on the casting drawing must reflect the **shrinkage allowance.**

Centrifugal casting is the process by which molten material is fed into a rotating mold. The rotation forces the molten material to fill the cavity or mold. Permanent molds are used for this process. **Die casting** is a permanent mold process that uses pressure to force the molten material into a metal die. **Injection molding** is also a type of permanent mold casting. Figure 10.31 shows an injection-molded part before and after machining. Injection molds are very similar to die casting molds. There are many other types of molding

(a)

(b)

FIGURE 10.29 **Casting Patterns** (a) Wood pattern and cast part (b) Wood pattern designed for casting multiple parts

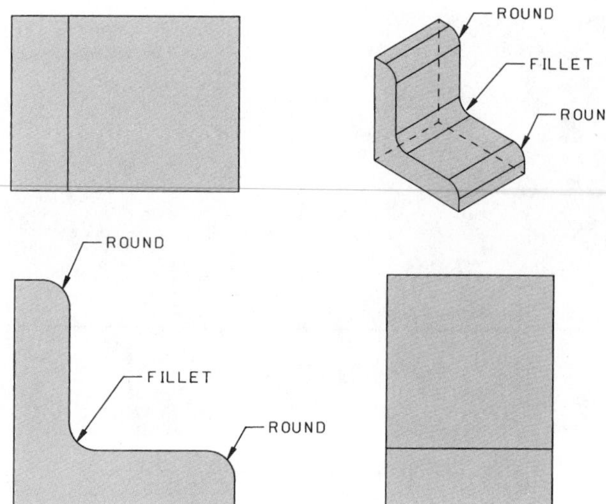

FIGURE 10.30 Fillets and Rounds

FIGURE 10.31 Injection Mold and Part

FIGURE 10.32 Extrusions

processes including blow molding, compression molding, transfer molding, layup molding, pressure molding, and vacuum molding.

10.5.2 Extruding

Materials forming processes use pressure to change the shape or the size of the material. This category of processes includes extruding, forging, stamping, punching, rolling, bending, and shearing.

Extrusion is a metalworking process used to produce long, straight semifinished products having constant cross sections (Fig. 10.32) such as bars, tubes, solid and hollow sections, wire, and strips. The metal is squeezed from a closed container through a die. **Cold extrusion** is also called **impact** or **cold forming** and is similar to cold forging.

Hot extrusion is used to make long and irregular-shaped parts. The billets and slugs are heated above their critical temperature, placed on a press, and squeezed through a die into the required shape.

10.5.3 Forging

Forging uses impact and pressure to form parts. Types include smith forging, upset forging, and drop forging. A forging is a metal part shaped to its desired form by hammering, pressing, or upsetting. The metal is usually heated to an elevated temperature. Forging without heating the material is known as cold forging.

A drop forging of a wrench before the part is cleaned is shown in Figure 10.33(a). The finished wrench is shown in Figure 10.33(b). In **drop forging,** the hot metal is forced into dies by means of drop hammers. The material itself is very hot, but not molten, and is forced into the die by pounding. This pounding force pushes the metal into the shape of the cavity of the die. Pounding does not create a very accurate part. Tolerances are large for this process and the dies are expensive; however, forging produces stronger parts than many manufacturing processes. Low carbon and low alloy steels and aluminum alloys are the most common materials used.

10.5.4 Stamping

In **stamping,** a punch and a die are used to cut or form sheet material. The assembled tool is called a die as is the cutting part of the tool. **Progressive dies** require that several operations take place in sequential order. Stamping includes cutting, parting, blanking, punching, piercing, perforation, trimming, slitting, shaving, forming, bending, coining, embossing, and drawing.

The part in Figures 10.34(a) and (b) was made by a progressive stamp that uses dies to cut or form the metal sheets into the desired form. Dies are assemblies that include a housing and the cutter.

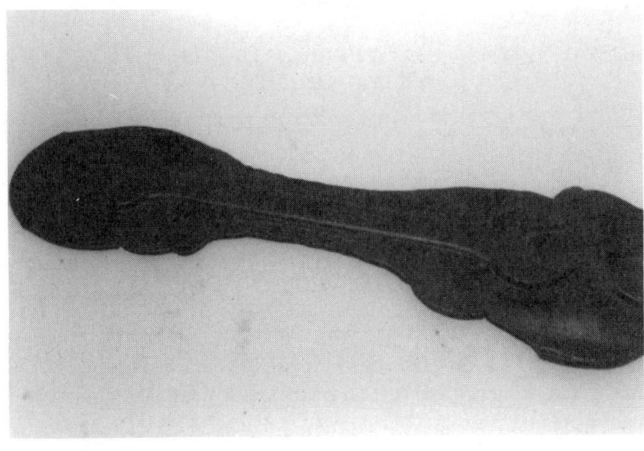

(a)

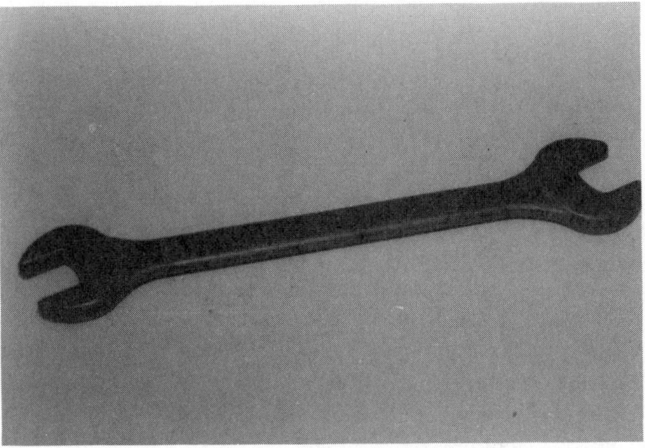

(b)

FIGURE 10.33 **Forging** (a) Forging of wrench (b) Wrench

(a)

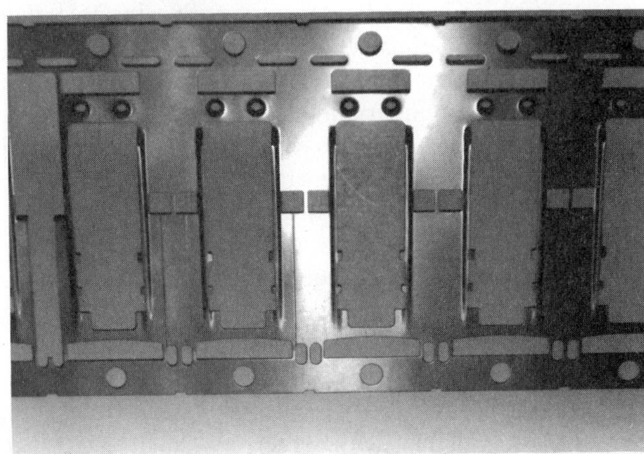

(b)

FIGURE 10.34 **Stamps** (a) Stamped part (b) Progressive stamping

10.5.5 Punching

Punching operations include shearing, cutting off, and blanking. **Shearing** is done along a straight line on a part. **Cutting** is performed on a part producing an edge other than a straight edge. **Blanking** produces parts with a punch and a die. Holes are also produced by punching in thin sheets of material. **Piercing** is similar to punching, except that no scrap is produced by the process. **Perforating** is a stamping operation performed on sheet to produce a hole pattern or decoration.

10.5.6 Electrical Discharge Machining and Electrochemical Machining

Electrical discharge machining (EDM) is a process that removes small particles of metal with an electrical spark. The material is vaporized by exposing the metal to sparks from a shaped electrode. The electrical discharge machine is a vertical spindle milling machine with a rectangular tank on the work table. The table can be moved along the **X** and **Y** axes or it can be numerically controlled. EDM was originally used stamping as a rough method for removing metal, but now it has been refined to do the precision work required in the electronics, aerospace, and toolmaking industries.

Electrochemical machining (ECM) has many of the same machining capabilities as EDM, but will machine a part much faster. ECM requires more electricity and is more expensive. This process uses electrolyte fluid and electric current to ionize and remove metal from the part.

10.6 Heat Treatment

Heat treatment is the process of applying heat to a material to change the properties of the material, but not the shape or size. Heat treatment can increase the strength and hardness, improve ductility, change the grain size and chemical composition, and improve the machinability of the part. Heat treatment is also used to relieve stresses, harden the part, and modify the electrical and magnetic properties of the material.

Heating metal just above its upper critical temperature for a specified period of time and controlled slow cooling in the furnace is called **annealing.** This results in a fully softened, stress-free part. Heating the metal just below the lower critical temperature and cooling by a predetermined method is called **process annealing.** Process annealing is often used on metals that have been work hardened. Process annealing softens the metal for further cold work.

The heating of metal above its lower critical temperature and **quenching** in water, oil, or air is called **hardening.** The resulting hardness is tested with the Rockwell hardness test. The **Rockwell hardness number** refers to the hardness of the steel. Although there are many different hardness scales, for steel, the higher the number, the harder it is.

Tempering is also called **drawing.** In this process, hardened steel is reheated to a predetermined temperature below its lower critical temperature and cooled at a specified rate. Tempering removes brittleness and toughens the steel (**tempered martensite**).

Heating the steel just below the upper critical temperature and then cooling the material in air is called **normalizing.** This improves the grain structure and removes the stresses. The **lower critical temperature** is the lowest temperature at which steel may be quenched to harden it. The **upper critical temperature** is the highest temperature at which steel can be quenched to attain the finest grain structure and the maximum hardness (martensite).

Typically, heat treatment requirements are listed in notes in a drawing or in the title block. Heat treatment is normally applied after the part has been machined, welded, or forged. To avoid problems during machining and heat treatment, consider the following during the design stage: balancing the areas of mass, avoiding sharp corners and internal recesses, and making hubs of gears, pulleys, and cutters a consistent thickness.

10.7 Automated Manufacturing Processes

Computer integrated manufacturing (CIM) is a system that links all information on the manufacturing floor, integrating design, production planning and control, and production processes. CIM encompasses both hardware and software. CAD/CAM is the integrator for computer-based applications in manufacturing, especially numerical control programming and robotics. CAD/CAM integration depends on a common engineering and manufacturing database (Fig. 10.35).

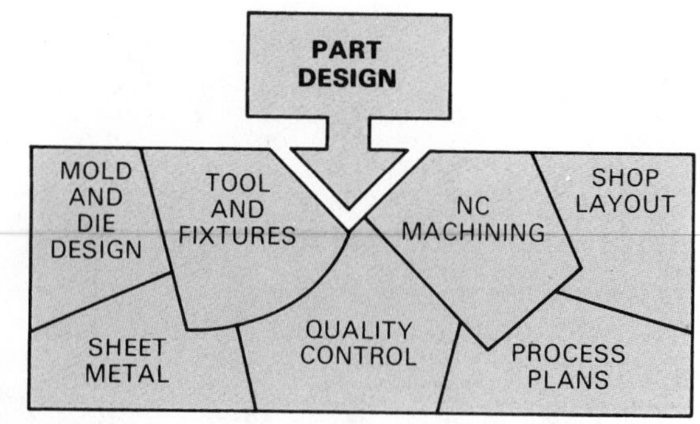

FIGURE 10.35 **Part Database** The part database created during the design phase is used by all groups associated with the manufacturing process.

10.7.1 Computer-Aided Manufacturing

CAD is a process in which a designer/drafter uses a computer to create or modify a design. *Computer-aided manufacturing* (CAM) *is a computer-based process to manage and control the operations in a manufacturing facility.* CAM includes numerical control (NC) for machining operations, tool and fixture design and setup, and integration of industrial robots into the manufacturing process.

The production process is computerized from the original graphics input through to the manufacture of the part on a numerically controlled machine. The NC programmer extracts accurate geometric data from the common database. The system serves all applications, promotes standardization, accumulates manufacturing information, and reduces redundancy and errors.

Using a CAD/CAM system, the engineer or designer applies the CAD features to create a model of the part. Then, using the information stored in the database, the manufacturing engineer applies the CAM capabilities. A CAD system may have a variety of specialized CAM capabilities including:

- Group technology
- Process planning
- Shop layout
- NC of machining operations
- NC postprocessing
- Sheet metal applications
- Tool and fixture design
- Mold design and testing
- Technical references and manufacturing documentation
- Quality control.

Use the following steps in CAM:

1. *Process planning:* The engineering drawing of the part to be tooled is interpreted in terms of the manufacturing processes.
2. *Part programming:* A part programmer plans the process for the portions of the job to be completed by numerical control. Part programmers are knowledgeable about the machining process and have been trained to program for numerical control. The two ways to program for NC are manual part programming and computer-assisted part programming.
3. *Verification:* The program is checked by plotting the tool movements on paper to discover errors in the program. The test of the part program is to make a trial part on the machine tool. A foam or plastic material is used for this test. CAD systems with CAD/CAM capabilities verify toolpaths and cutter motion on the display.
4. *Transfer media preparation:* Diskettes and direct com-

puter networks are now used as the transfer medium from computer to NC machine.

5. *Production:* Production involves ordering rough parts, specifying and preparing the tooling and any special fixturing that may be required, and setting up the NC machine. The NC system machines the part according to the instructions. In more automated operations, programmable robots change parts in NC operations.

10.7.2 Numerical Control

Numerical control (NC) is a form of programmable automation in which the process is controlled by numbers, letters, and symbols. In NC, the numbers form a program of instructions designed for a particular part or job. When the job changes, the program of instructions is changed. This capability to change the program for each new job makes NC flexible.

The two major types of NC machines used today are **point-to-point** and **continuous path.** Point-to-point machines operate on a series of programmed coordinates to locate the position of the tool. When the tool finishes at one point, it continues to the next point (position). This type of control is used for drilling and punching machines. Milling machines require continuous control of tool position and are continuous path. Complicated operations including contouring, angle surfaces, fillets, and radii can be programmed using continuous path The **toolpath** is the trace of the movement of the tip of a NC cutting tool that is used to guide or control machining equipment.

Figure 10.36 shows a CNC mill. In CNC, the design database is passed directly from the CAD/CAM system to the machine's computer. Figures 10.37(a) through (f) show a sequence of design, analysis, NC simulation, and manufacturing for a spindle. The menu shown in Figure 10.37(a) is used to manage the engineering data needed to model and test the part. The **solid model** in Figure 11.37(b) is used to assist the designer in creating, analyzing, and visually displaying the part. **Finite element analysis** packages are used for part analysis and generating a mesh of the geometry [Fig. 10.37(c)]. **Cutter paths** for NC machining are defined and modified by the part programmer [Fig. 10.37(d)]. Output from the postprocessor can then be used for NC machining of the actual part [Fig. 10.37(e)]. The finished assembly is displayed and reviewed [Fig. 10.37(f)].

10.7.3 Machining Operations

The types of parts that are produced on NC equipment from output generated by CAD/CAM systems include:

- Irregular or uniquely machined parts
- 2D parts created by point-to-point operations
- Lathe parts produced by turning operations
- 2½D parts that may require pocketing and profiling operations
- 3D parts produced by using all of the NC operations provided on the CAD/CAM system.

FIGURE 10.36 CNC Mill

Pocketing is completely removing material within a bounded area (Fig. 10.38). Machine pocket programs provide automatic pocketing on a CAD system by generating a toolpath to remove the material.

10.7.4 Tools and Fixtures

A **tool** is a piece of equipment that helps create a finished part. It may be anything that must be designed and/or made in order to manufacture the part. **Fixtures** are used to hold and locate parts of assemblies during machining or other manufacturing operations. The accuracy of the product being produced determines the precision with which a fixture is designed. To design and manufacture a finished part efficiently, product design engineers must work with tool and fixture designers as well as manufacturing engineers.

10.8 Robotics

Robotics is the integration of computer-controlled robots into the manufacturing process. Industrial robots are used to move, manipulate, position, weld, machine, and do a variety of other manufacturing processes. A **robot** is a reprogrammable,

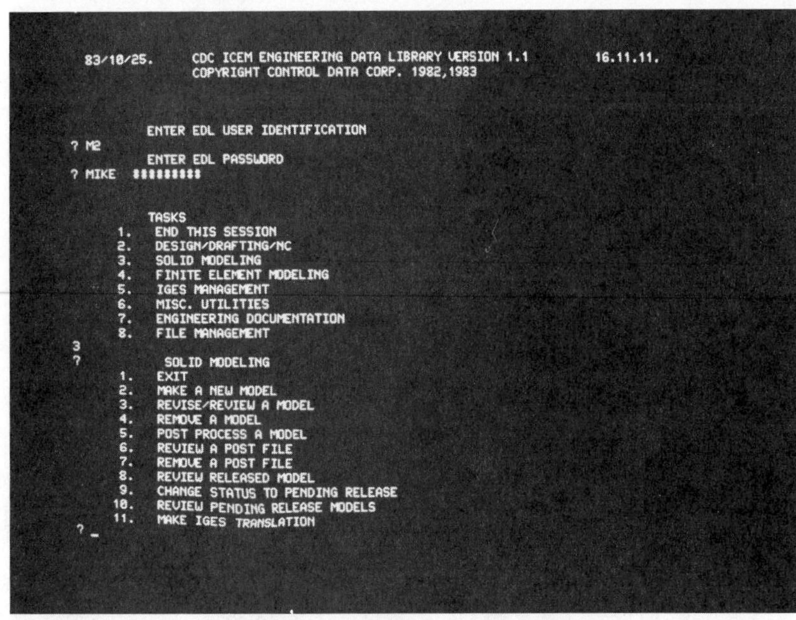

(a)

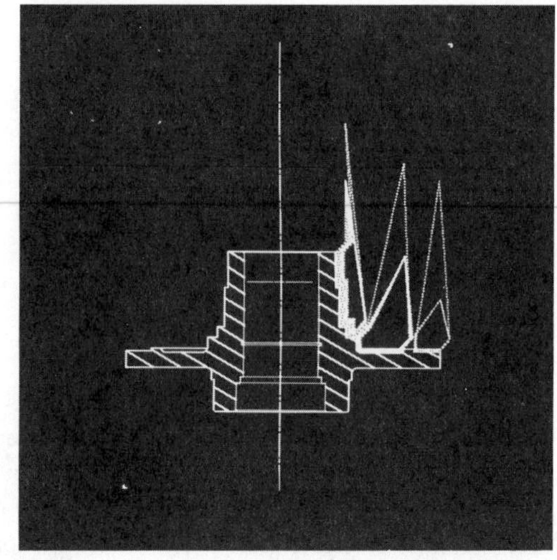

(d)

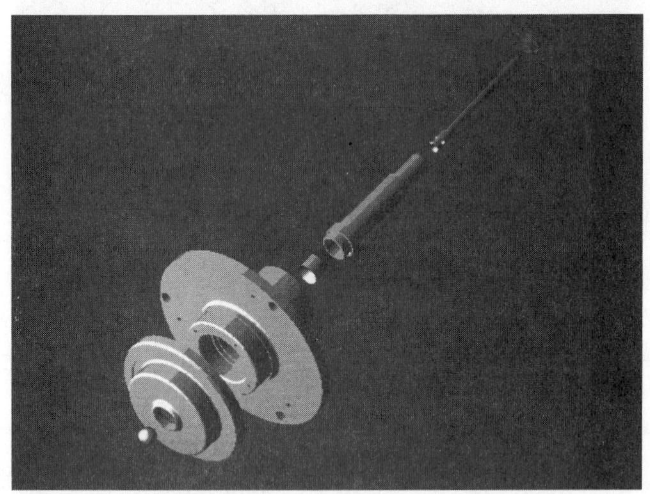

(b)

(e)

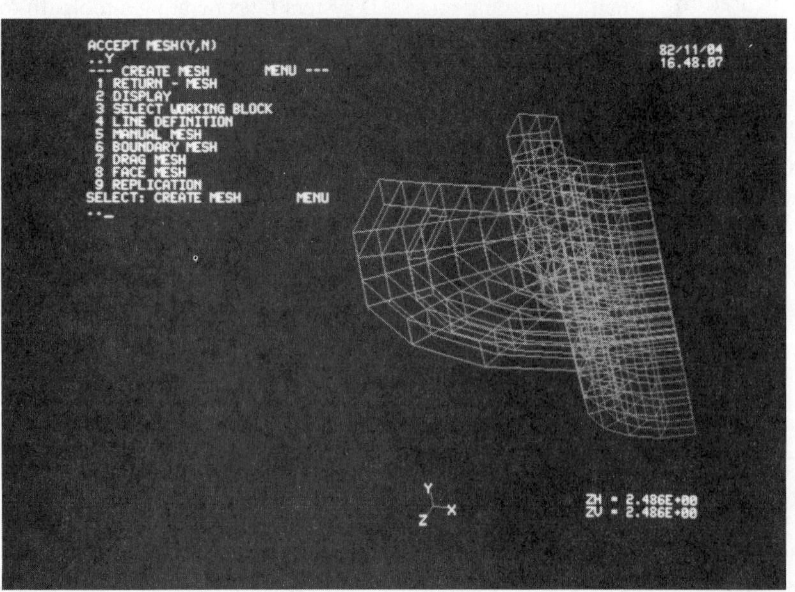

(c)

(f)

FIGURE 10.37 Sequence of Design, Analysis, NC Simulation, and Manufacturing for a Spindle

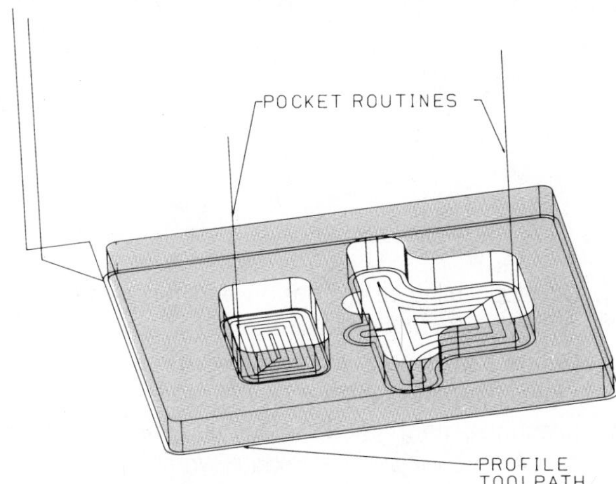

FIGURE 10.38 **Toolpaths** A variety of toolpaths can be created including profiling and pocketing.

multifunction manipulator designed to move material, parts (including the workpiece), tools, or specialized devices through variable, programmed motions for the performance of a variety of tasks. Robots are controlled by a microprocessor and are composed of a separate stand-alone computer station, the robot mechanism itself, and an electrical-hydraulic power unit. DFM includes part design with robotic manufacturing techniques designed into the part at the earliest stages of the project.

10.8.1 CAD/CAM Robotic Applications

The integration of CAD/CAM and robots results in increased productivity. The **robotic workcell** contains all the physical equipment needed to create a full functioning robot application. The equipment in the workcell must be arranged so that the **robot work envelope** includes all required devices and equipment for the assembly operation.

QUIZ

True or False

1. Patterns are made smaller than the real size of the part to allow for expansion of the metal (expansion-allowance).
2. The surface texture symbol is used for designating the classification of roughness, waviness, and lay.
3. Roughness is the distance between ridges or peaks on a surface.
4. Robots can be programmed.
5. NC is a form of programmable automation controlled by numbers, letters, or symbols.
6. Pocketing is the process of removing material from the outer boundaries of a part.
7. A true CAD/CAM system can create a common database, which is then used to derive part geometry for all areas of manufacturing and design.
8. Reamers are used to produce precise holes.

Fill in the Blanks

9. ________, ________, and ________ are done on a drill press.
10. ________, ________, ________, ________ and ________ are the five basic types of machining processes.
11. A ________ ________ is used to allow the cast part to be removed from the form more easily.
12. ________is the process of pouring molten metal into a mold.
13. Machining a continuous toolpath about a part is called ________.
14. The creation of a common ________ enables the part geometry to be used by many departments.
15. Toolpaths can be ________ and ________ on the display.
16. ________ is basically the same process as counterboring.

Answer the Following

17. Describe the difference between drilling, reaming, and boring.
18. From your own experience name five metal parts that have been cast.
19. Drilling is used before what types of basic tooling operations?
20. What are robots and how are they being used in industry?
21. Describe the components of the surface texture symbol.
22. What part does CAD play in the total process of CAM? How does the use of a common database effect the design-through-manufacturing process?
23. Explain the methods used to produce a precise hole.
24. What is a vertical milling machine?

CHAPTER 11

Dimensioning

Learning Objectives

Upon completion of this chapter you will be able to accomplish the following:

1. Analyze part features in terms of integral geometric shapes to facilitate concise dimensioning within prescribed tolerances.
2. Apply ANSI standards for dimensions and tolerances.
3. Apply angular, callout, overall, limited length, and area dimensions.
4. Develop ability to dimension and recognize standard symbols for curved features.
5. Define and dimension chamfers, threads, centerdrills, tapers, knurling, and keyways.
6. Recognize finish marks, general symbols and notes, and ANSI basic surface texture symbols.
7. Apply rectangular continuous coordinate dimensioning and polar coordinate dimensioning.

11.1 Introduction

The ability to analyze a part by recognizing that it is composed of simple geometric shapes enables you to understand what dimensions are required to manufacture that part. After all, the only real purpose of an engineering drawing is to convey information correctly so that the part can be manufactured correctly from the drawing.

Engineering drawings use dimensions and notes to convey this information. Knowledge of the methods and practices of dimensioning and tolerancing is essential to the drafter and designer. The multiview projections of a part provide a graphic representation of its shape (*shape description*). However, the drawing must also contain information that specifies size and other requirements.

Drawings are *annotated* with dimensions and notes. Dimensions must be provided between points, lines, or surfaces that are functionally related or to control relationships of other parts. Manufacturing personnel should not have to compute dimensions or guess intent. Each dimension on a drawing has a **tolerance**, implied or specified. The general tolerance given in the title block is called a **general** or **sheet tolerance**, as shown later in this chapter in Figure 11.62. Specific tolerances are provided with each appropriate dimension.

11.2 Dimensioning Standards

Uniform practices for stating and interpreting dimensioning and tolerancing requirements were established in **ANSI Y14.5M.** The Système Internationale d'Unites (SI), which is metric based, is used along with U.S. customary units in this chapter because SI units are expected to replace U.S. customary units on engineering drawings. Either type of unit can be used with equal results.

Some of the industry example drawings and problems in the text were completed before 1982 and, therefore, conform to earlier standards. You will be in contact with older standards in your career since companies sometimes continue to use older practices rather than face the expense of converting to the new standards. However, *use the most recent standard to complete the exercises in this text.*

11.2.1 Dimensioning Terms

The following terms are used throughout this chapter:

Dimension A numeric value expressed in appropriate units of measure and indicated on a drawing and in other documents along with lines, symbols, and notes to define size or geometric characteristics, or both, of a part or part feature. For example, 12.875 (in.), 25 (mm), etc.

Reference dimension A dimension, usually without tolerance, used for information only. It is considered auxiliary information and does not govern production or inspection operations. A reference dimension repeats a dimension or size already given or is derived from other values shown on the drawing or related drawings. Reference dimensions are enclosed within parentheses. For example, (23.50), (50), etc.

Datum An exact point, axis, or plane derived from the true geometric counterpart of a specified datum feature. A datum is the origin from which the location or geometric characteristics of the features of a part are established.

Feature The general term for a physical portion of a part (a surface, hole, slot, etc.).

Datum feature A geometric feature of a part used to establish a datum. For example, a point, line, surface, hole, etc.

Actual size The measured size of the feature.

Limits of size The specified maximum and minimum limits of a feature.

Tolerance The total amount by which a specific dimension is permitted to vary. The tolerance is the difference between the maximum and minimum limits.

11.2.2 Units of Measurement

The SI linear unit commonly shown on engineering drawings is the millimeter. The U.S. customary linear unit used on engineering drawings is the decimal inch. On drawings where all dimensions are either in millimeters or inches, individual identification of linear units is not required. However, the drawing must contain a note stating:

UNLESS OTHERWISE SPECIFIED, ALL DIMENSIONS ARE IN MILLIMETERS (or INCHES).

Dimensions are shown to as many decimal places as accuracy requires. The inch or millimeter symbol is omitted unless the dimension may be misunderstood or where feet and inches are used on construction drawings. When U.S. customary units are used, fractions and decimals are not mixed on the same drawing. If inch dimensions are shown on a millimeter-dimensioned drawing, the abbreviation "in." must follow the inch values. If millimeter dimensions are shown on an inch-dimensioned drawing, the symbol "mm" must follow the millimeter values.

Angular dimensions are expressed in either decimal parts of a degree or in degree, minutes, and seconds (° for degrees; ′ for minutes; ″ for seconds).

11.3 Types of Dimensioning

Decimal dimensioning is used on U.S. drawings except where certain commercial commodities are identified by standardized nominal designations such as pipe, steel, and lumber sizes.

Metric Dimensioning Use the following rules regarding the proper use of zeroes and decimal places in metric dimensioning (Fig. 11.1):

1. If the dimension is less than 1 mm, a zero precedes the decimal point.
2. If the dimension is a whole number, neither the decimal point nor a zero is shown.
3. If the dimension exceeds a whole number by a decimal fraction of 1 mm, the last digit to the right of the decimal point is not followed by a zero.
4. Neither commas nor spaces are used to separate digits into groups when specifying millimeter dimensions on drawings.

Decimal-Inch Dimensioning Use the following rules regarding the proper use of zeroes and decimal places in decimal-inch dimensioning (Fig. 11.2):

1. A zero is not used before the decimal point for values less than one in.

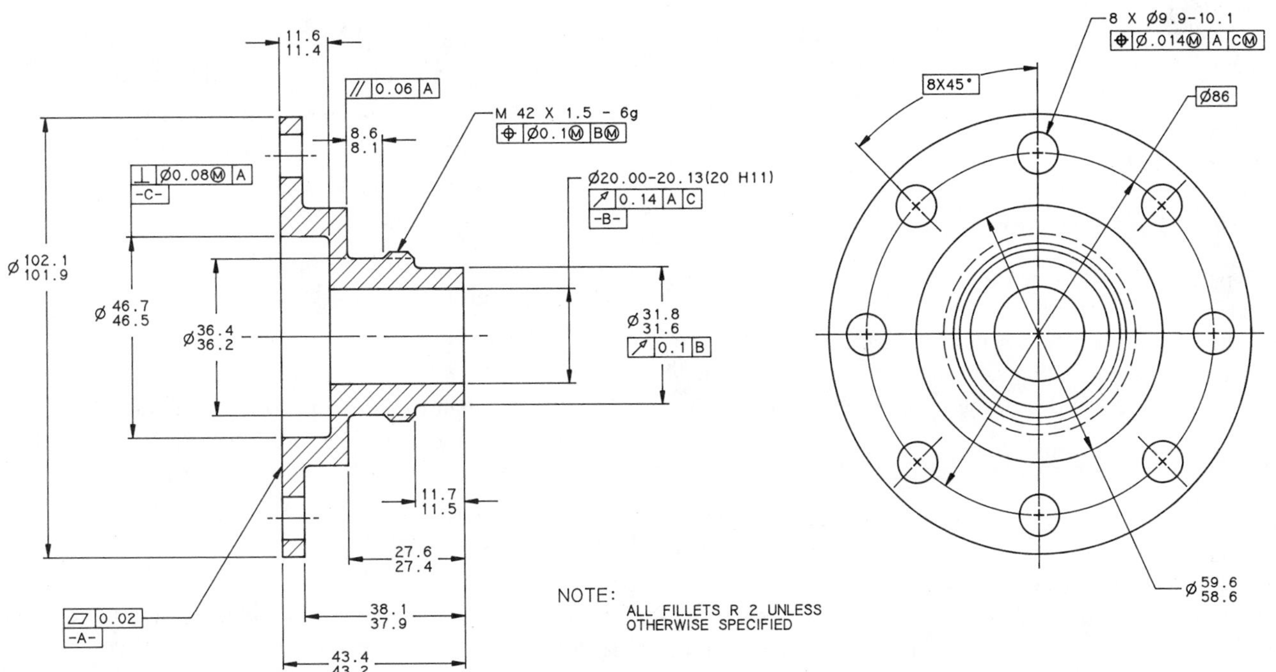

FIGURE 11.1 Geometric Tolerancing and Dimensioning Employed to Dimension a Mechanical Part

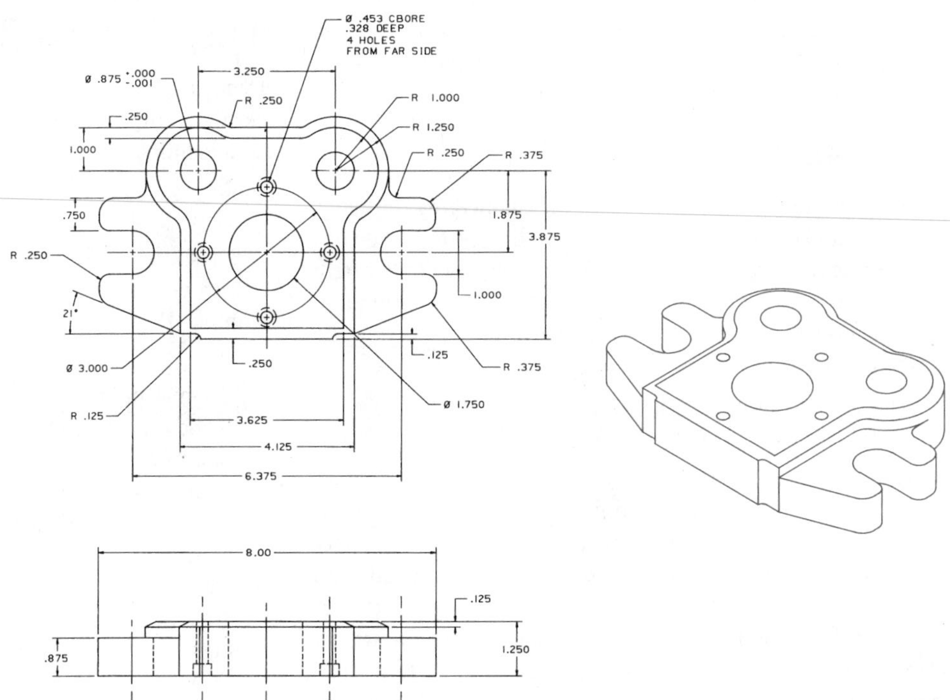

FIGURE 11.2 Mechanical Part Designed and Dimensioned in U.S. Standard Decimal-Inch Units

2. A tolerance is expressed to the same number of decimal places as its dimension. Zeros are added to the right of the decimal point where necessary for both the dimension and the tolerance.

Decimal Points (SI and U.S. units) The general rules regarding the proper use of decimal points in SI and U.S. units are:

1. Decimal points must be uniform, dense, and large enough to be clearly visible and to meet the reproduction requirements of ANSI Y14.2M. Decimal points are placed in line with the bottom of the associated digits.
2. When a dimension is 1 unit, always add a decimal and a zero (1.0).

11.3.1 Dual Dimensioning

Many parts designed in the United States are manufactured or traded in foreign countries. Therefore, some drawings use **dual dimensioning.** The top measurement, or first measurement when placed on the same line, is always the unit of measurement used to design the part. For example: $\frac{1.00}{25.4}$.

11.3.2 Dimensioning Numerals

Whole numbers in the inch system are normally shown to at least one decimal place (e.g., 1.0 or 2.0). This practice prevents dimensions from being "lost" on the drawing.

Common fraction dimensions are seldom used, except on construction drawings. Before the decimal-inch was adopted as a standard, firms did use common fractions for subdivisions of an inch. Some companies still use this system. Older drawings and sheet metal drawings also show common fractions.

Using decimals has many advantages. Decimals reduce arithmetic computation time. For example, it can take as much as five times longer to add a series of fractions than a series of decimals.

11.3.3 Rounding Off Decimal-Inch Measurements

ANSI has a standard method for rounding decimals:

1. If the last digit to be dropped is less than 5, there is no change in the preceding digits. For example:

   .47244 rounds to .4724
   .1562 rounds to .156

2. If the last digit to be dropped is greater than 5, the preceding digit is increased by 1. For example:

   .23437 rounds to .2344
   .55118 rounds to .5512

3. If the last digit to be dropped is 5 and followed by a zero, round the preceding digit to the nearest even number. For example:

   .98425 rounds to .9842
   .59055 rounds to .5906
   .19685 rounds to .1968

If precise calculation is required, values should be calculated to two places beyond the desired number of places; rounding should be based on the last two significant digits.

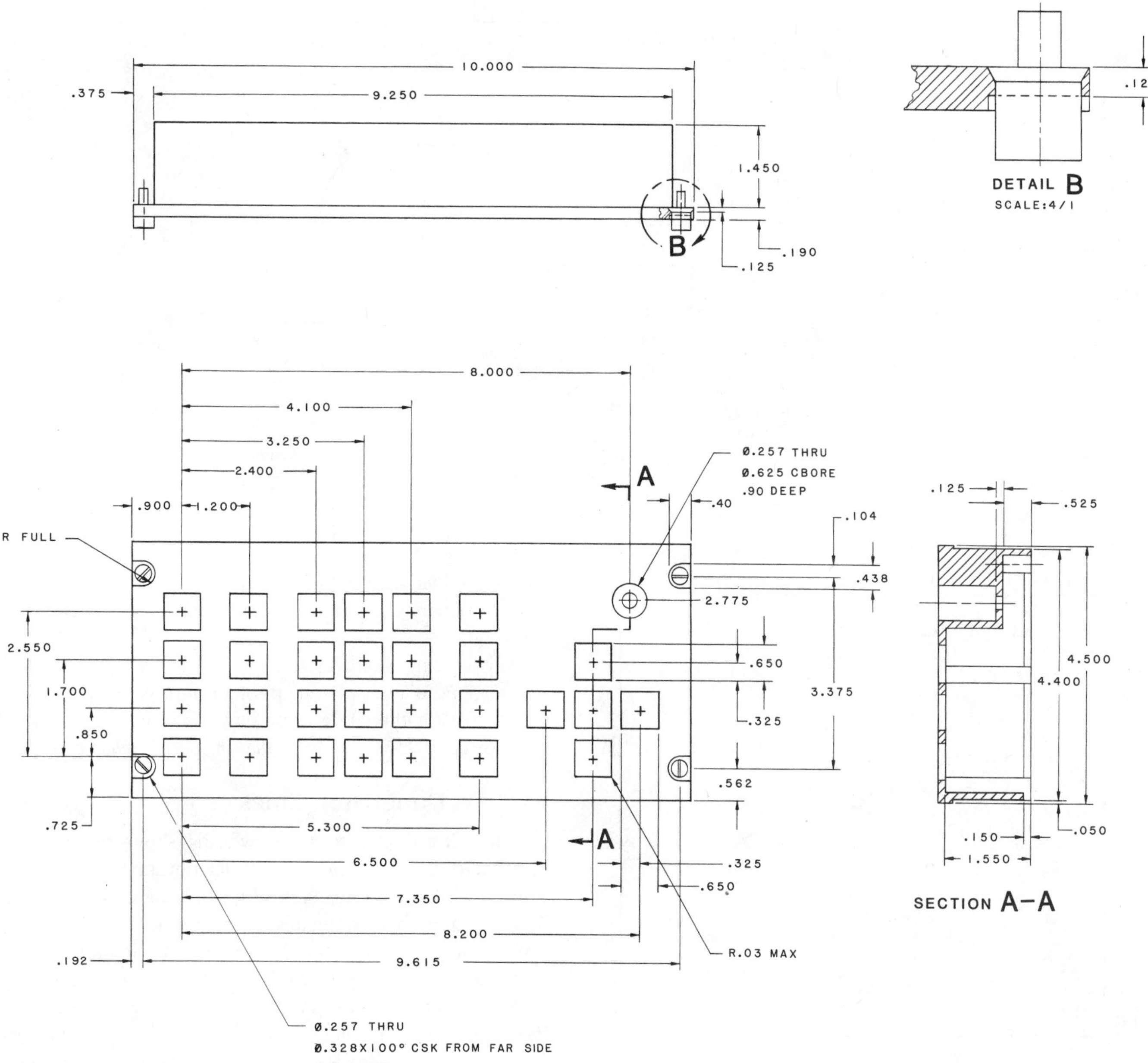

FIGURE 11.3 Panel Detail

11.3.4 Drawing Scale

Drawings should be drawn to a scale that is easy to read and interpret. Scales are constant within a given project where multiple drawings are needed. Scales are stated in the title block: 1:1 (full scale), 1:2 (half scale), 5:1, 10:1, and so on.

In some cases, such as when a portion of the drawing is enlarged, more than one scale is used on a drawing (Fig. 11.3). Here, **DETAIL B** is scaled at **4/1**. The predominant scale is shown in the scale area in the title block.

11.4 Dimensions

Dimensions use standard elements: dimension lines, extension lines, leaders, arrowheads, and dimension values. The types of dimensions include: vertical, horizontal, and aligned linear dimensions, angular dimensions, and callout dimensions using leaders for notes (Fig. 11.4). Figure 11.5 shows typical dimensions with examples in both decimal inches and millimeter measurements. When a dimension is small, the arrowheads and dimension line can go on the outside of the

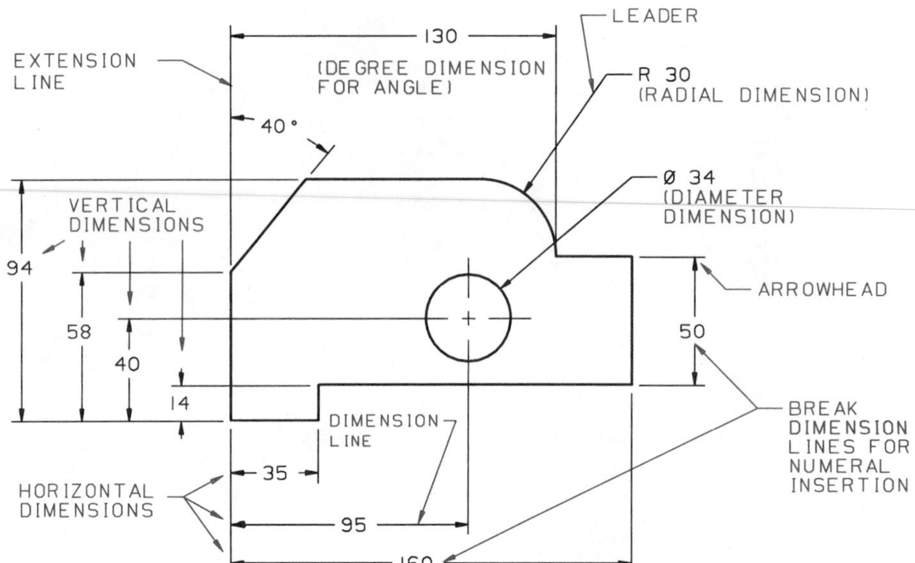

FIGURE 11.4 Dimension Elements

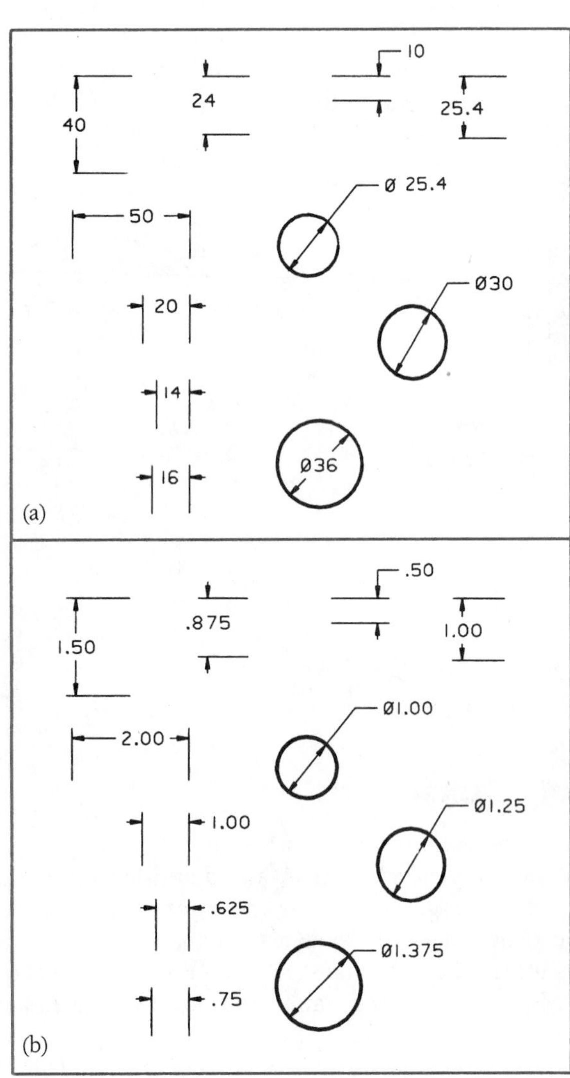

FIGURE 11.5 Millimeter and Inch Dimensions

extension lines with the value inside. Another method allows the dimension value to be placed outside of the extension lines.

Any drawing is only as good as its dimensioning. Accurate drawing and correct placement of all dimensioning elements is essential for the engineering drawing to transfer information correctly. *Dimension such that the part is made correctly.*

11.4.1 Dimension Lines

A **dimension line**, with its arrowheads, shows the direction and extent of a dimension. Numerals indicate the number of units of a measurement. Preferably, dimension lines are broken to insert these numerals, as shown in Figure 11.4. If horizontal dimension lines are not broken, the numerals are placed above and parallel to the dimension lines. *Do not use centerlines, phantom lines, an object line that represents the outline of a part or a continuation of any of these lines for dimension lines.* A dimension line is used as an extension line only where a simplified method of coordinate dimensioning is used to define curved outlines (as shown later in Fig. 11.39).

Crossing dimension lines should be avoided. Where it is unavoidable to cross them , the dimension lines are unbroken at the crossing point. The largest dimension always goes on the outside, farthest from the part's outline. This Figure 11.6 illustrates a number of rules:

1. Cross a dimension line only if it is unavoidable.
2. Do not place dimensions within the part outline unless there is no other place to show the dimension properly.
3. Place larger dimensions farthest from the part's outline.

Dimensions are usually placed outside the outline of the part. If directness of application makes it desirable or if extension lines or leader lines would be excessively long, dimensions may be placed within the outline of a view. On large, complex drawings, dimensions are sometimes placed within the out-

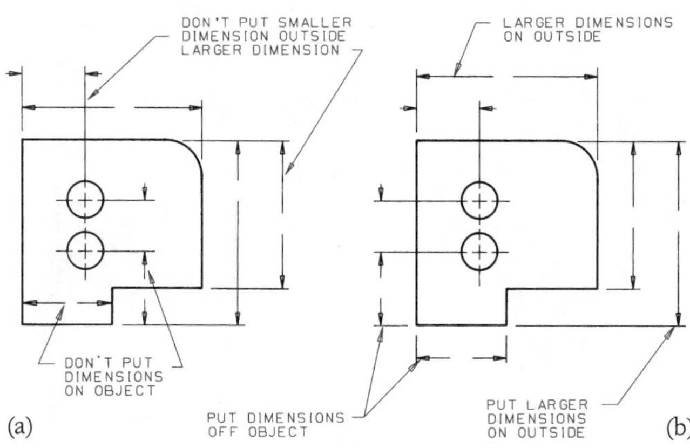

FIGURE 11.6 Dimensions and Extension Lines (a) Incorrect (b) Correct

line of the part, even when this contradicts the dimensioning rules. If it is necessary to place a dimensioning inside a part's outline that is sectioned, break the section lines around the dimension (Fig. 11.7).

11.4.2 Extension Lines

Extension lines indicate the extension of a surface or point to a location outside the part outline. Extension lines start with a short visible gap from the outline of the part and extend beyond the outermost related dimension line. Extension lines are drawn perpendicular to dimension lines. If space is limited, extension lines may be drawn at an oblique angle to illustrate clearly where they apply. If oblique lines are used, the dimension lines are shown in the direction in which they apply (Fig. 11.8).

Extension lines should not cross dimension lines. To minimize such crossings, the shortest dimension line is shown nearest the outline of the part (Fig. 11.9). If extension lines must cross other extension lines, dimension lines, or lines depicting features, they are not broken. However an extension line is broken where it crosses arrowheads or dimension lines close to arrowheads (Fig. 11.9).

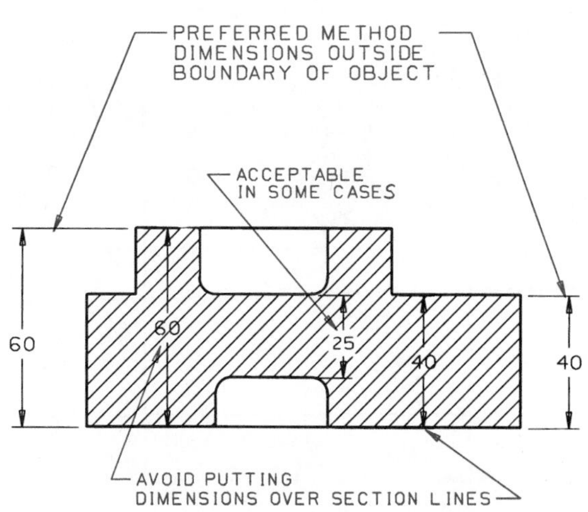

FIGURE 11.7 Dimensions on Section Lining

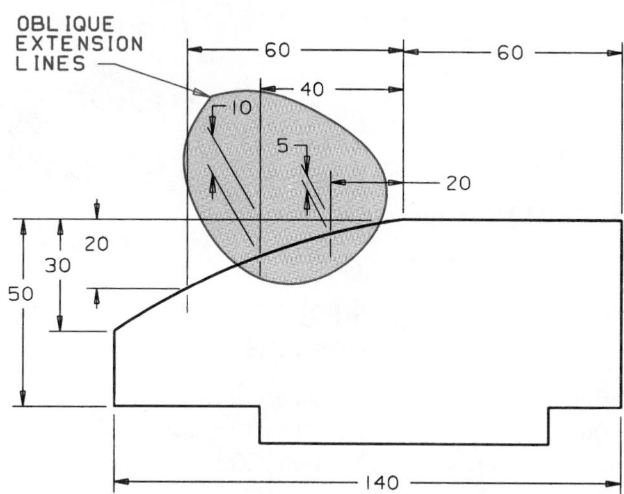

FIGURE 11.8 Oblique Dimensions

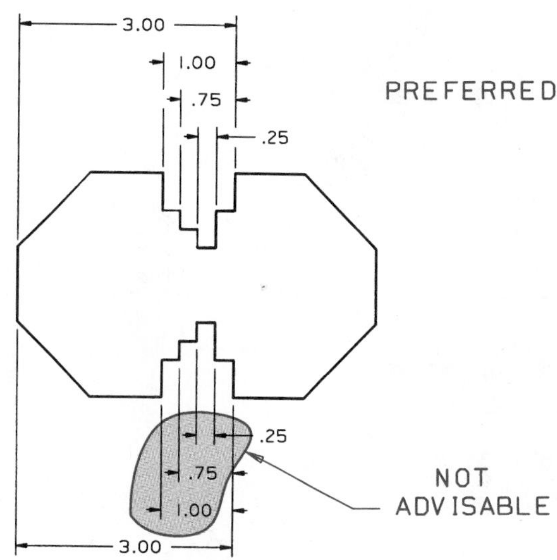

FIGURE 11.9 Breaks in Extension Lines

11.4.3 Drawing Dimension Arrowheads

The thickness of a dimension, leader or extension line is normally 0.3 to 0.35 mm. It is the thinnest line on the drawing (along with section lines) and must be drawn crisp and black.

The arrowhead used for dimensions is shown in Figure 11.10. The sides and back of the arrowhead are straight, not curved. An arrowhead is about three times as long as it is wide with a length approximately equal to the height of the lettering used on the drawing. Arrowheads are drawn completely filled. Other types of line terminators used throughout industry include open arrowheads, dots, and slashes. Keep arrowheads consistent and uniform.

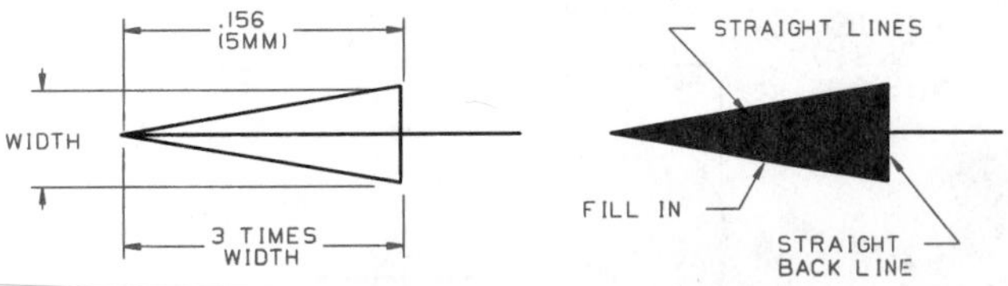

FIGURE 11.10 Arrowheads

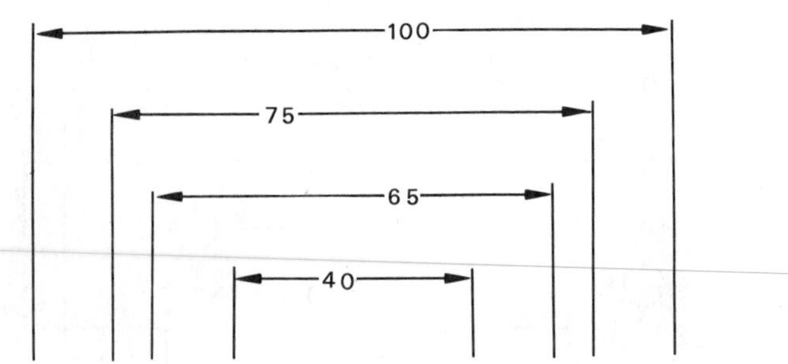

FIGURE 11.12 Staggered Dimensions

11.4.4 Drawing Dimension and Extension Lines

Dimension lines are aligned and grouped for a uniform appearance (Fig. 11.11). If there are several parallel dimension lines, the numerals should be staggered for easier reading (Fig. 11.12).

The minimum distance from the first dimension line to the part outline should be .375 in. (10 mm). The minimum spacing between parallel dimension lines should be .25 in. (6 mm) (Fig. 11.13). In general, .50 in. (12 mm) from the part and .375 to .50 in. (10 to 12 mm) between dimensions is suggested for large drawings and those that need to be greatly reduced. These spacings are intended as a guide when dimensioning, not as a rule. If the drawing meets the reproduction requirements of the accepted industry or military

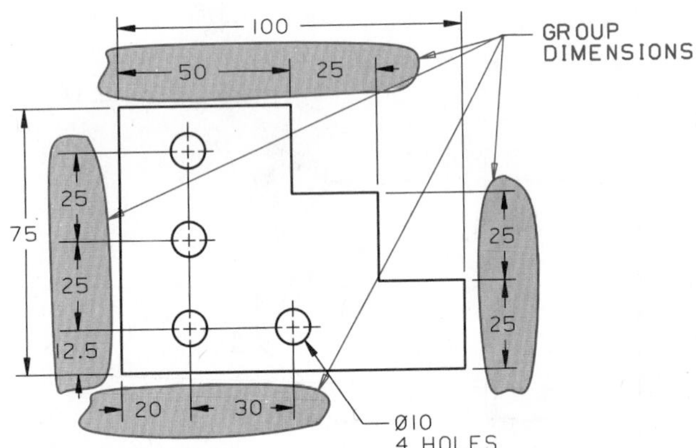

FIGURE 11.11 Grouping Dimensions

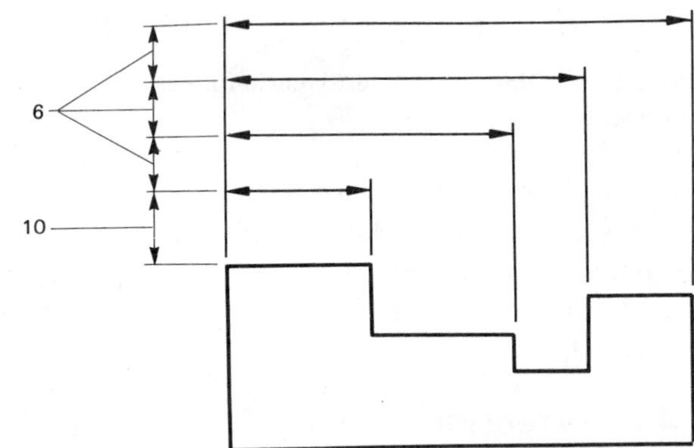

FIGURE 11.13 Setup and Spacing of Dimensions

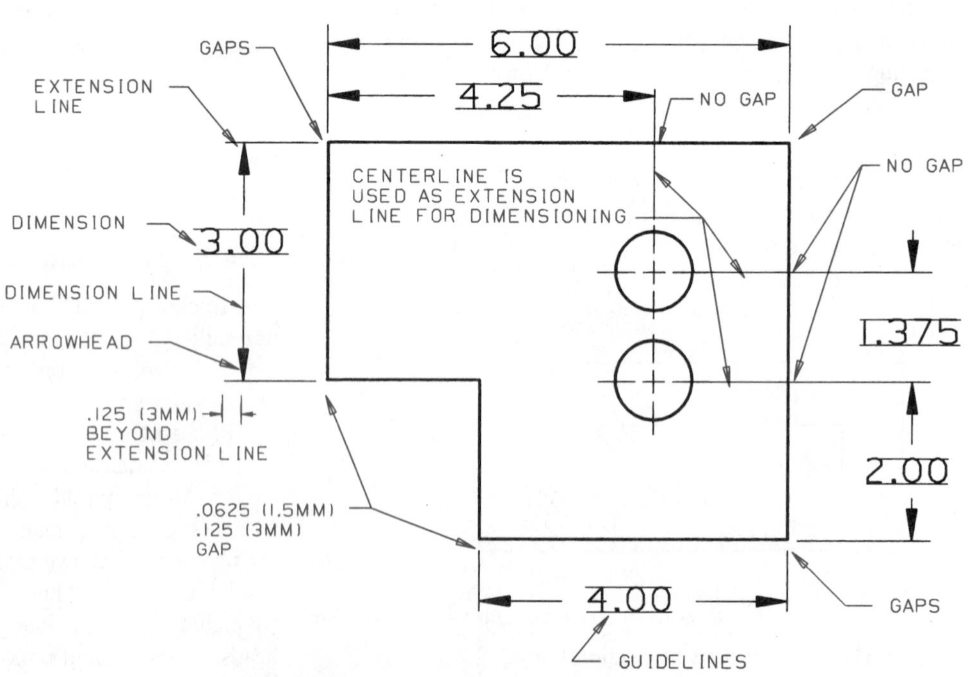

FIGURE 11.14 Gaps and Placement of Extension Lines

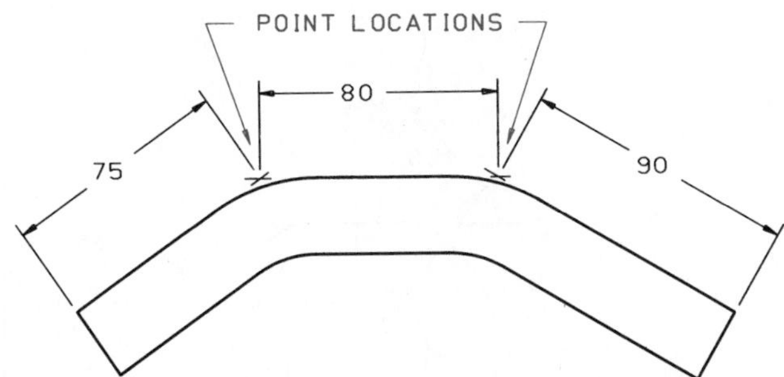

FIGURE 11.15 Point Locations Using Extension Lines

reproduction specification, these spacing requirements are not mandatory.

Extension lines should start about .06 in. (1.3 mm) from the part and end approximately .12 in. (2.5 mm) beyond the dimension line and arrowhead (Fig. 11.14). *Centerlines can be used as extension lines but not as dimension lines.*

All holes are dimensioned to their centerlines in two directions, except when the holes are arrayed in a circular pattern (as with a bolt circle). If a point is to be located only by extension lines, the extension lines (from the surfaces) pass through the point (Fig. 11.15).

Extension lines are not drawn to hidden lines on hidden features of the part. However, dimensioning to hidden features is acceptable in some circumstances (when a feature cannot be seen in another view).

For almost every dimensioning rule, there is an exception. The rules apply in 90% of the situations that you will encounter.

11.4.5 Lettering Dimensions

The preferred heights for lettering dimensions are shown in Figure 11.16. Dimension heights are standardized for each drawing size and reduction requirement. If reduction is not required, .125 in. (3 mm) height for lettering is acceptable. Follow the spacing of lettering in dimensions in Figure 11.16 to complete projects in this text.

Numerals that are placed parallel to dimension lines are called **aligned dimensions** (Fig. 11.17). Horizontal dimensions are readable from the bottom; vertical dimensions, from the right side of the drawing. Point-to-point dimensions of angled edges have the dimensions aligned (parallel) to the edge itself. Aligned dimensions are not accepted ANSI standard practice.

Unidirectional dimensioning (Fig. 11.18) places the dimension text parallel to the bottom of the drawing. This system is preferred since the drawing may be read and lettered without being turned.

For mechanical drawings, the dimension lines are broken to insert the measurement numerals. Piping, architecture,

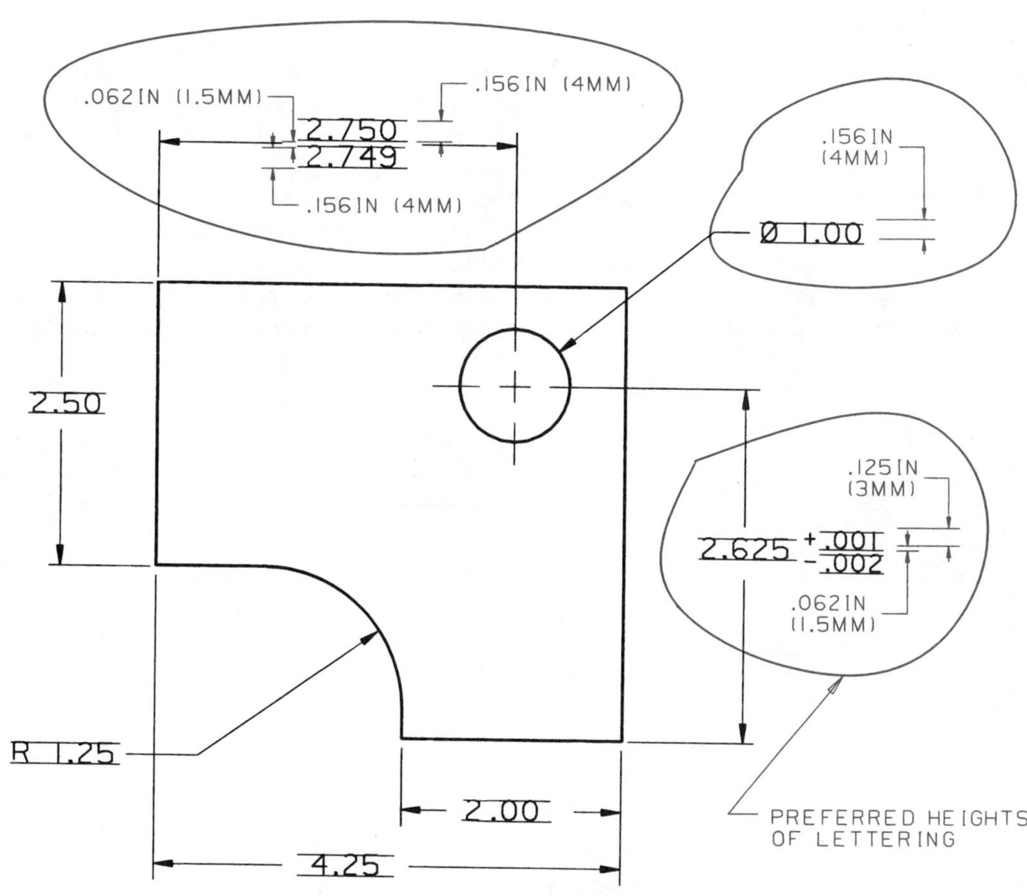

FIGURE 11.16 Preferred Lettering Height for Dimensions

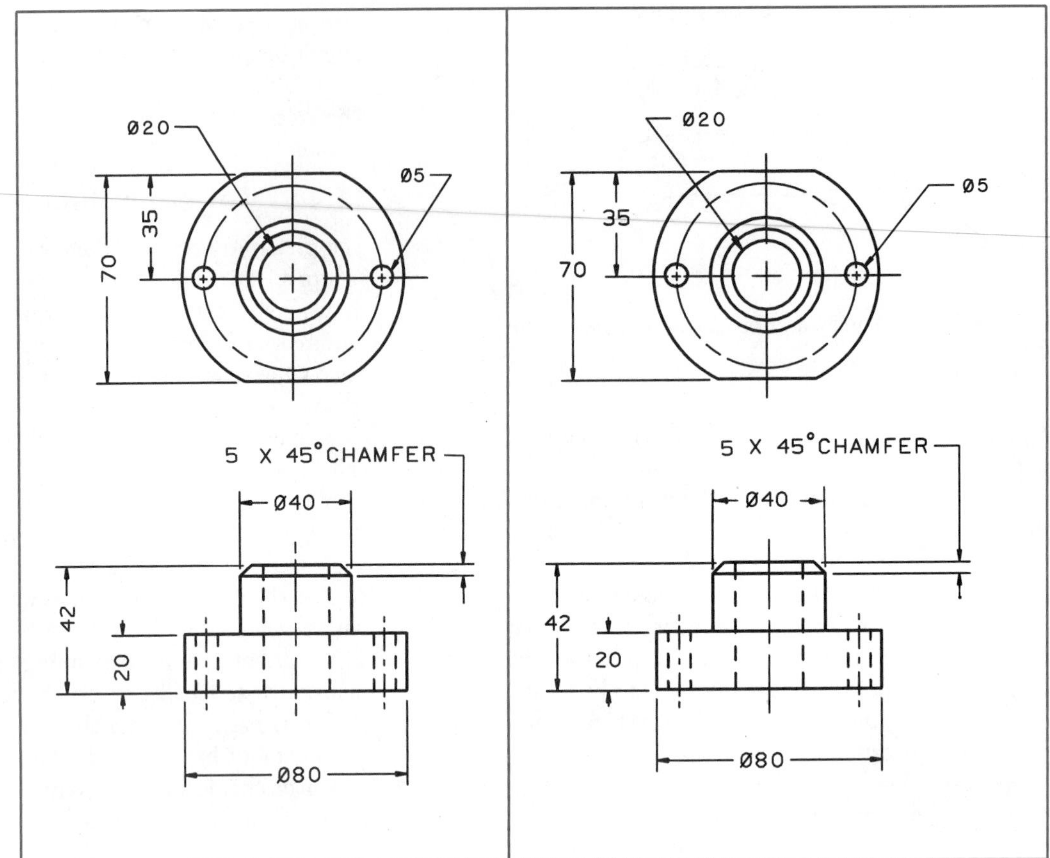

FIGURE 11.17
Dimensioning Methods
(a) Aligned dimensioning
(b) Unidirectional dimensioning

civil, structural, and other construction drawings do not normally break the dimension line, but place numerals above the dimension line.

11.4.6 Angular Dimensions

Size and location dimensions may be linear distances or angles. **Angular dimensions** are expressed in degrees, minutes, and seconds or as decimal equivalents of degrees. Decimal equivalents of degrees are shown on the part in Figure 11.19. If the angle is expressed in degrees and minutes, do not separate the numbers with a dash. If angles are less than 1°, precede the minute by 0°. For both unidirectional and aligned dimensioning, angular dimensions are placed to read horizontally between guidelines.

Angle dimensions should be avoided by locating the end points of inclined lines and planes. Because it is easier, quicker, and more reliable, coordinate dimensioning of angled features increases the accuracy during manufacturing.

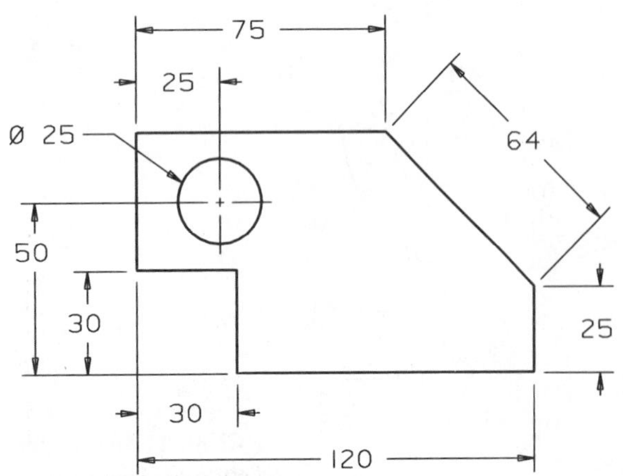

FIGURE 11.18 **Unidirectional Dimensions**

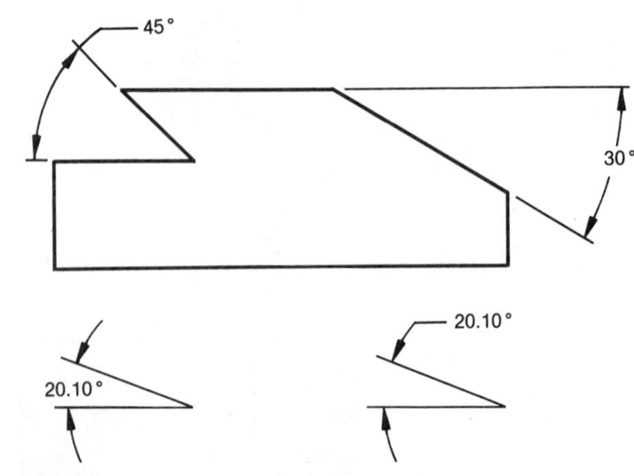

FIGURE 11.19 **Dimensioning Angles**

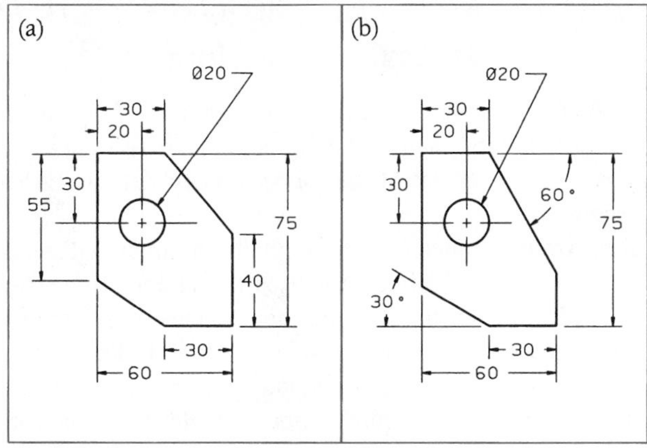

FIGURE 11.20 Square Versus Angular Dimensioning

The dimension line for an angle is drawn as an arc from a center at the intersection of the sides of the angle. A variety of methods are used to dimension angles (Fig. 11.19). The arrowheads terminate at the extensions of the two sides, inside or outside the extension lines.

Angles are used only where other forms of linear dimensions are unsuitable. In Figure 11.20, two methods of dimensioning a part are illustrated. One method uses angle dimensions and the other uses the offset method. Because it is easier for the machinist to locate the features of the part with linear measurements, the offset method is preferred.

11.4.7 Callout Dimensions and Notes Using Leaders

A **leader** is used to direct a dimension, note, or symbol to the intended feature on the drawing (Fig. 11.21). Leaders are used to point to a curved feature of a part or to reference a portion or surface. Most leaders are drawn at 45° to 60° to the horizontal (30° to the vertical) (Fig. 11.22). Leaders terminate in arrowheads. The dimension figure for a callout is placed at the end or the head of a short 6 mm (.25 in.) horizontal line.

Figure 11.23 shows the three most common uses of a leader: to call out a hole diameter, to call out a radius, and to reference a surface or part with a note. When a leader is used to dimension a circle or arc, it must point to or from (or

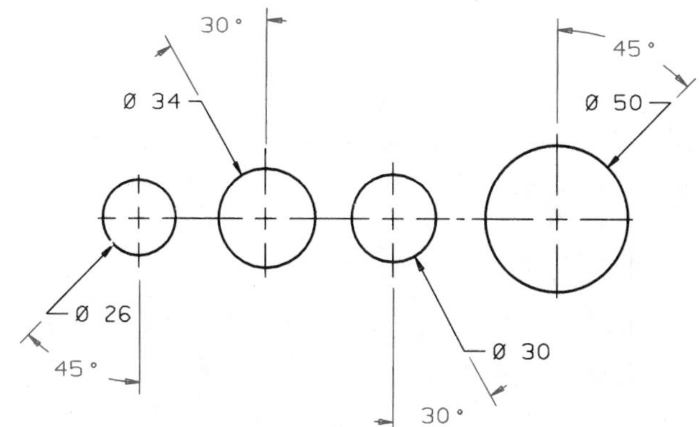

FIGURE 11.22 Leaders for Hole Callouts

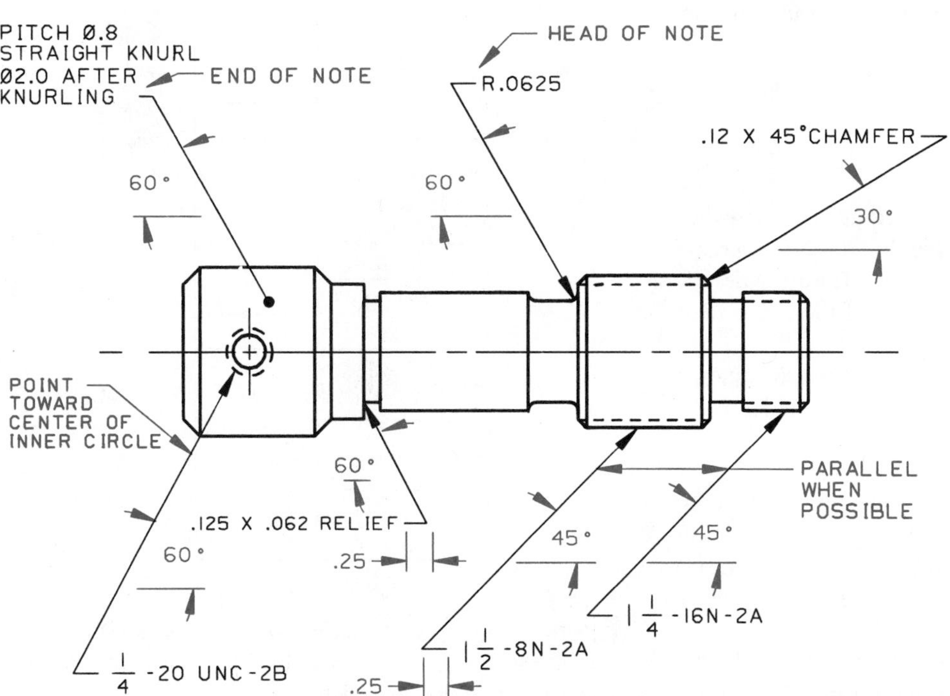

FIGURE 11.21 Leaders on Drawings

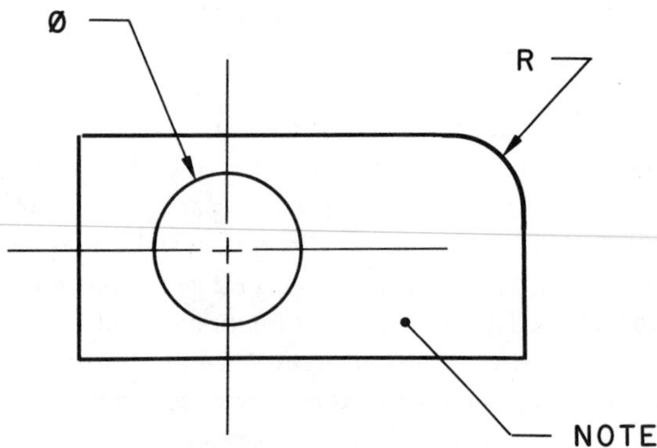

FIGURE 11.23 **Three Common Uses of Leaders**

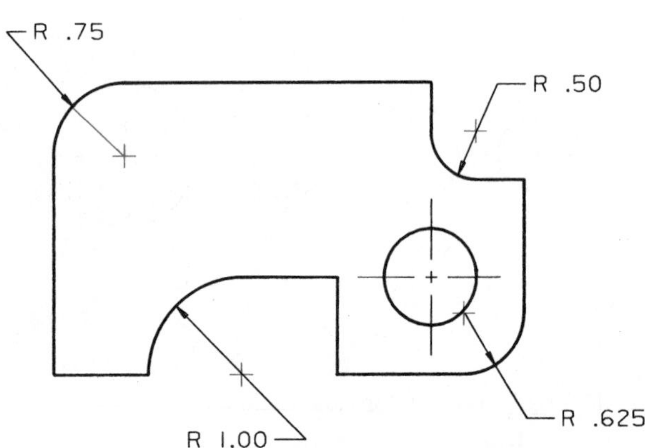

FIGURE 11.24 **Dimensioning Fillets and Arcs with Callouts**

through) the center of the circle or radius. The arrowhead points toward the center of the curve (Fig. 11.24). Therefore, *all leaders for radii and diameters are radial.* The arrowhead for a leader terminates at the circumference of the arc or diameter.

The crossing of dimension lines and extension lines by leaders should be kept to a minimum. The leader line is drawn at a different angle than object lines of the part or section lines on the drawing.

Leaders and their accompanying notes and callouts are kept outside dimension lines and away from the part being dimensioned. Leaders are placed on the drawing after the part is dimensioned. Although leaders can cross object, dimension, and extension lines, *never cross other leader lines.*

If too many leaders impair the legibility of the drawing, letters or symbols are used to identify features (Fig. 11.25).

11.4.8 Reference, Overall, and Not-to-Scale Dimensions

Reference dimensions are not used for manufacturing or inspection. Identify a reference dimension or reference data on drawings by enclosing the dimension or data *within parentheses.*

If an overall dimension is specified, one intermediate dimension is omitted or identified as a reference dimension (Fig. 11.26). When the intermediate dimension is more important than the overall dimension, the overall dimension is identified as a reference dimension.

To indicate that a feature is **not to scale**, the dimension should be underlined with a straight thick line or **NTS** (not to scale) should be added to the dimension.

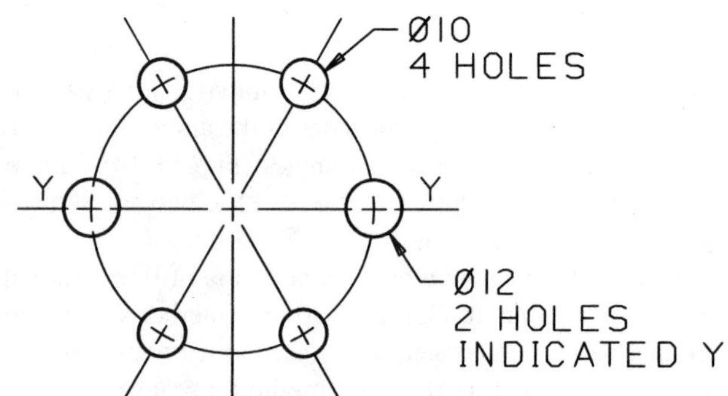

FIGURE 11.25 **Minimizing Leaders on Drawings**

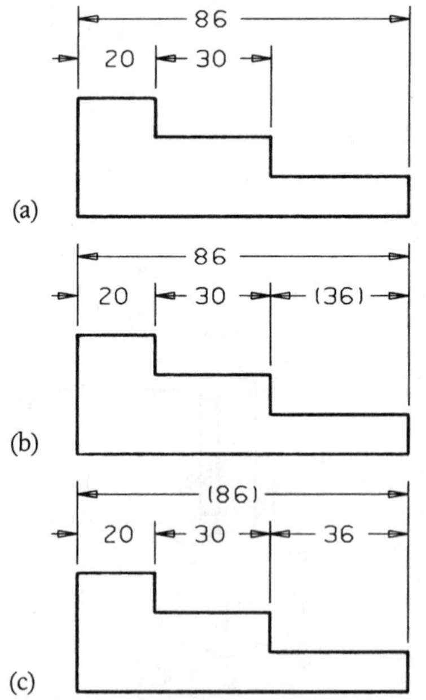

FIGURE 11.26 **Overall and Reference Dimensions**
(a) No reference dimension (b) Intermediate reference dimension
(c) Overall reference dimension

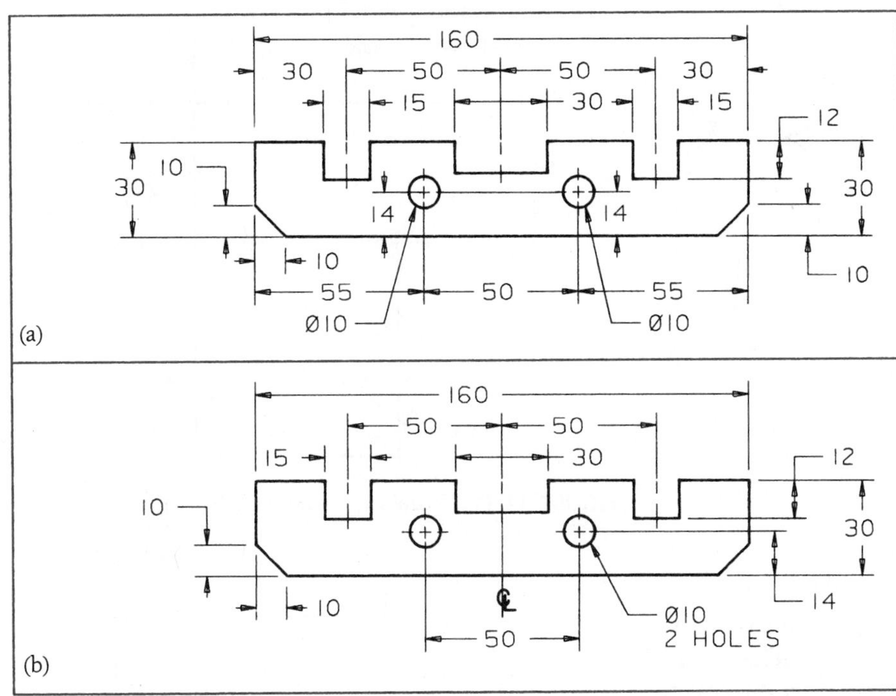

FIGURE 11.27
Overdimensioning a Part
(a) Overdimensioned (b) Correctly dimensioned

Only the dimensions required for manufacturing the part are on the drawing. In Figure 11.27, the overdimensioned part is shown at the top of the figure and the correctly dimensioned part below it.

11.4.9 Limited Length of Area Indicated

To indicate that a limited length or area of a surface is to receive additional treatment or consideration within limits specified on the drawing, the extent of these limits is indicated by use of a **chain line** (Fig. 11.28). In an appropriate view or section, a chain line is drawn parallel to the surface profile at a short distance from it. Dimensions are added for length and location {Fig. 11.28(a)]. For a surface of revolution such as a shaft, the indication is shown on one side only [Fig. 11.28(a)].

As long as the chain line clearly indicates the location extent of the *limited length,* dimensions may be omitted [Fig.11.28(b)]. When the *limited area* is shown on a direct view of the surface, the area is section-lined within the chain line boundary and dimensioned [Fig.11.28(c)].

11.5 Dimensioning Curved Features

Included in this section are methods of noting and dimensioning curved features such as radii, diameters, slots, counterdrills, countersinks, spotfaces, and counterbores. ANSI symbology is also covered.

11.5.1 Radius Dimensioning

Radius dimensions are used to call out slots, curves, arcs, rounds, and fillets. Each radius value on a radius dimension is preceded by the appropriate radius symbol R (Fig. 11.29). A radius dimension line uses one arrowhead, which points to the arc from the center. An arrowhead is not used at the radius center. The dimension line for any radius is an angular line extending radially through, from, or toward the center of

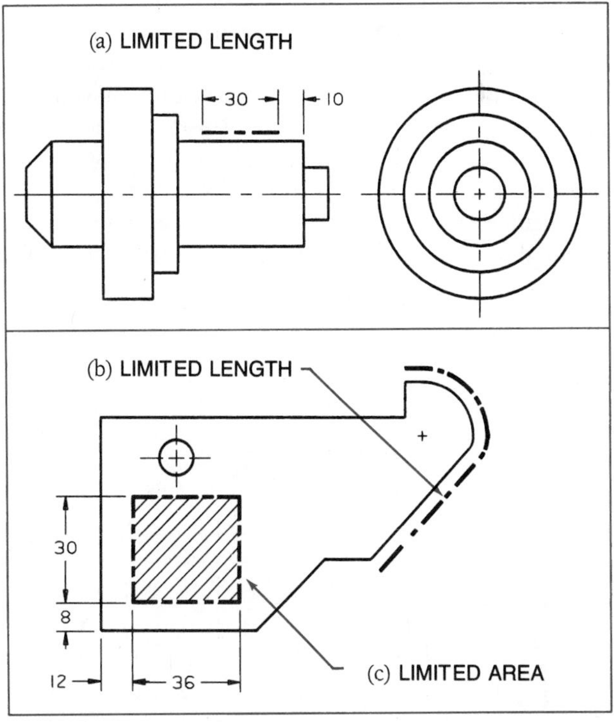

FIGURE 11.28 **Limited Length and Limited Area Indicators**
(a) and (b) Limited length (c) Limited area

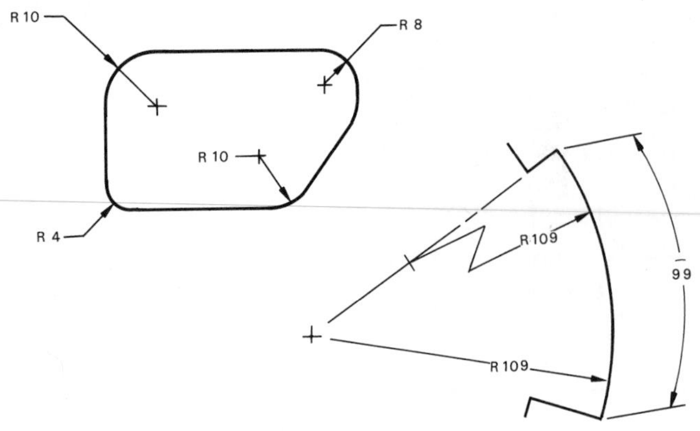

FIGURE 11.29 Dimensioning Radii and Arcs

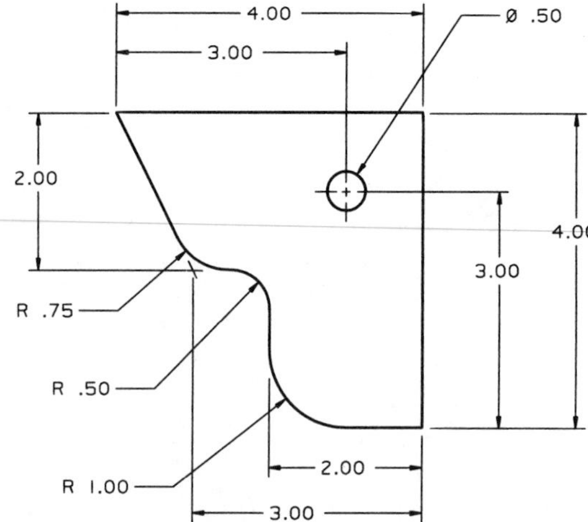

FIGURE 11.31 Radii with Unlocated Centers

the feature. Do not use horizontal or vertical lines when dimensioning arcs. Figure 11.29 illustrates the following:

1. If the location of the center is important and space permits, a dimension line is drawn from the radius center with the arrowhead touching the arc. Place the dimension between the arrowhead and the center.
2. When space is limited, extend the dimension line through the radius center.
3. When it is inconvenient to place the arrowhead between the radius center and the arc, place it outside the arc with a leader.
4. If the center of a radius is not located dimensionally, the center is not indicated.

To locate the center of a radius, draw a small cross at the center. Extension lines and dimension lines can be used to locate the center of an arc (Fig. 11.30). If the location of the center is unimportant, the drawing must clearly show that the arc location is controlled by other dimensioned features (Fig. 11.31). The center of a fillet or round is not located by dimensions.

Sometimes the center of an arc is moved on a drawing because there is a break or the center lies outside the drawing paper (Figs. 11.29 and 11.32). The new position is on a centerline of the arc, and the newly located "false" center leads to a *staggered dimension*. The portion of the dimension line touching the arc is a radial line drawn from the true center, whereas the staggered dimension is drawn parallel to the first radial line. When the radius dimension line is foreshortened and the center located by coordinate dimensions, the dimension line locating the center is foreshortened as well.

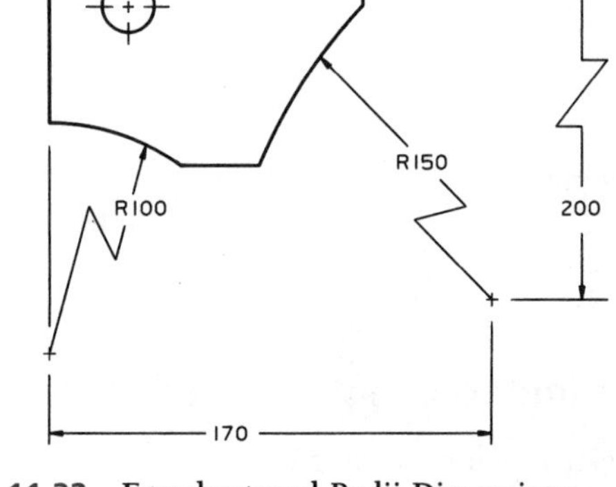

FIGURE 11.32 Foreshortened Radii Dimensions

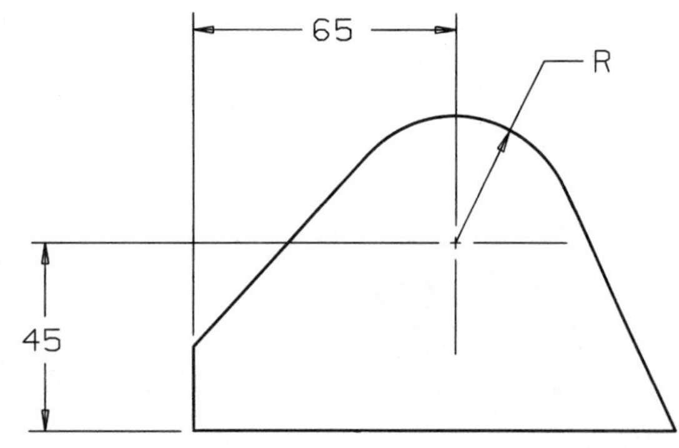

FIGURE 11.30 Radius with a Located Center

When a radius is dimensioned in a view that does not show the true shape of the radius, TRUE R is added before the radius dimension. A true shape view of the radius is shown and dimensioned whenever possible.

When a part has a number of radii of the same dimension, a note such as:

ALL RADII .75 UNLESS OTHERWISE NOTED

may be used instead of dimensioning each radius.

A **spherical surface** for a solid part is dimensioned by a radius dimension preceded by the symbol SR (Fig. 11.33).

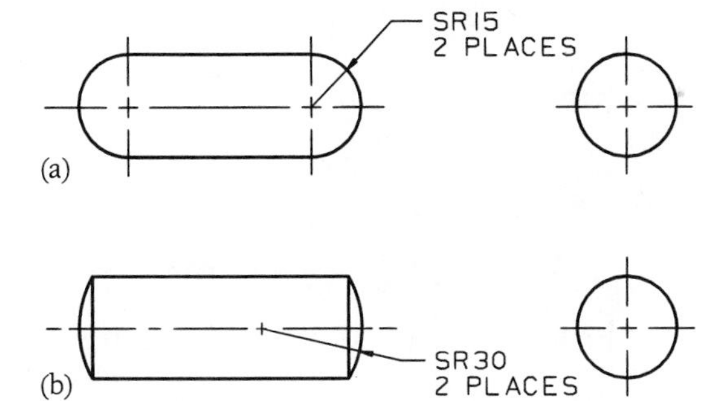

FIGURE 11.33 **Spherical Radius Dimensions** (a) Half sphere (b) Partial sphere

11.5.2 Detailing Chords, Arcs, and Rounded Ends

An angle measurement is the most common measurement used to dimension arcs and chords (Fig. 11.34). The arc dimension with the arc symbol and the chord dimension are used in applications such as nipple placement on large pressure vessels in piping design.

Overall dimensions are required for parts having rounded ends (Fig. 11.35). For the fully rounded ends of Figure 11.35(a), the radii are indicated but not dimensioned. For parts with partially rounded ends, the radii are dimensioned [Fig. 11.35(b)]. If corners are rounded, dimensions define the edges, and the arcs are tangent to the edge lines (Fig. 11.36).

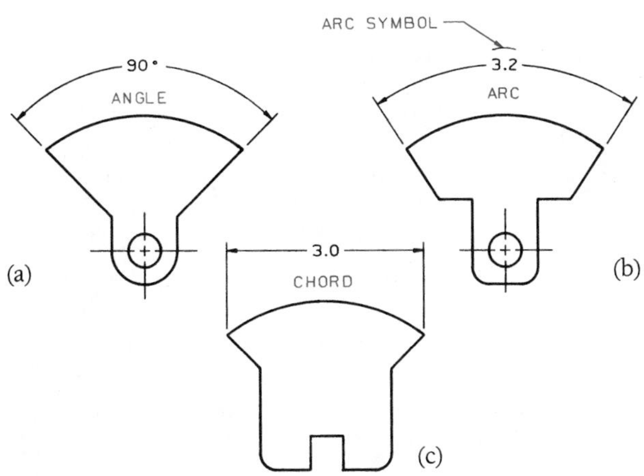

FIGURE 11.34 **Dimensioning Angles, Arcs and Chords** (a) Angle (b) Arc (c) Chord

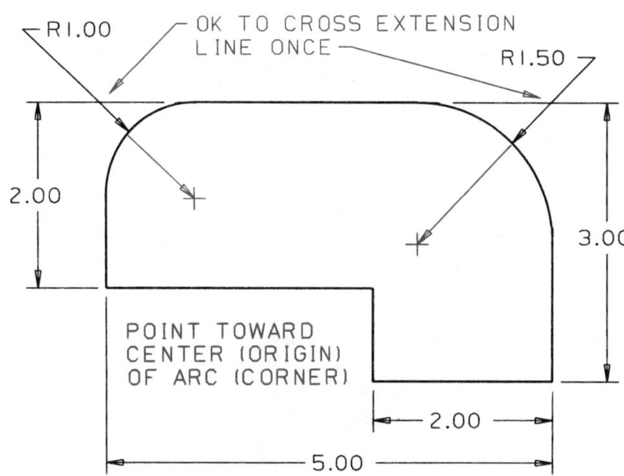

FIGURE 11.35 **Dimensioning Fully Rounded and Partially Rounded Ends**

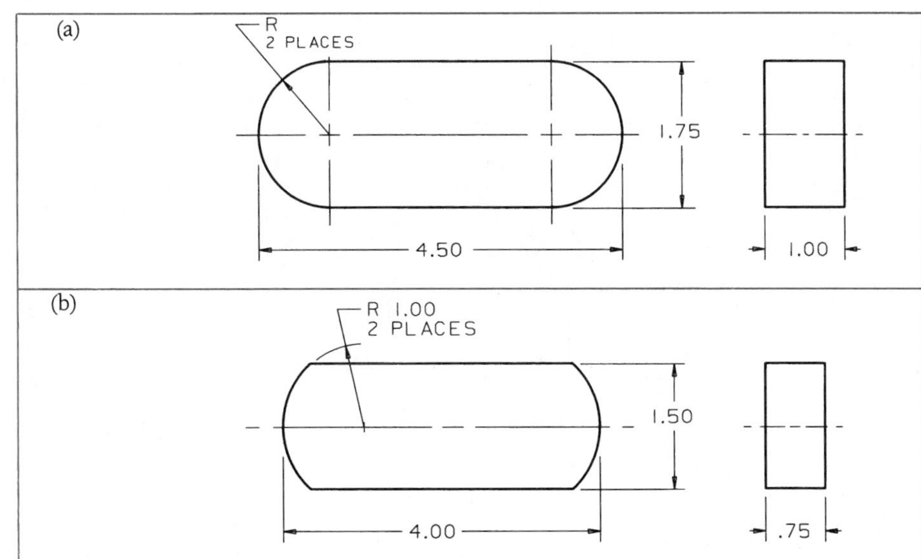

FIGURE 11.36 **Dimensioning Rounded Corners** (a) Fully rounded ends (b) Partially rounded ends

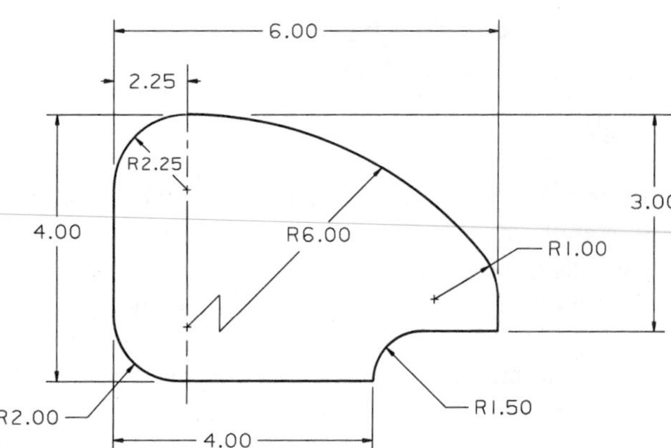

FIGURE 11.37 Dimensioning Circular Arc Outlines

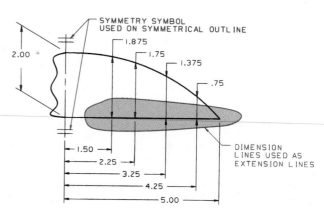

SYMMETRY SYMBOL USED ON SYMMETRICAL OUTLINE
2.00
1.875
1.75
1.375
.75
1.50
2.25
3.25
4.25
5.00
DIMENSION LINES USED AS EXTENSION LINES

FIGURE 11.39 Dimensioning Symmetrical Outlines Using Dimension Lines as Extension Lines

Radii dimensions use leaders to point to the arc. *The leader "aims" at the center point of the arc.* It is acceptable practice to cross one extension or one dimension line.

A curved outline composed of two or more arcs is dimensioned by the radii of all arcs and locates the necessary centers with coordinate dimensions (Fig. 11.37). Other radii are located on the basis of their points of tangency.

11.5.3 Irregular Outlines

Irregular outlines are dimensioned in Figures 11.38 and 11.39. Circular or noncircular objects are dimensioned by rectangular coordinates or an offset method. Coordinates are dimensioned from base or datum lines. If many coordinates are required to define an outline, the vertical and horizontal coordinate dimensions can be given in a table.

11.5.4 Symmetrical Outlines

Symmetrical outlines are dimensioned on one side of their *centerline of symmetry,* when only part of the outline can be conveniently shown (Fig. 11.39). Symmetry is indicated by applying symbols for part symmetry to the centerline.

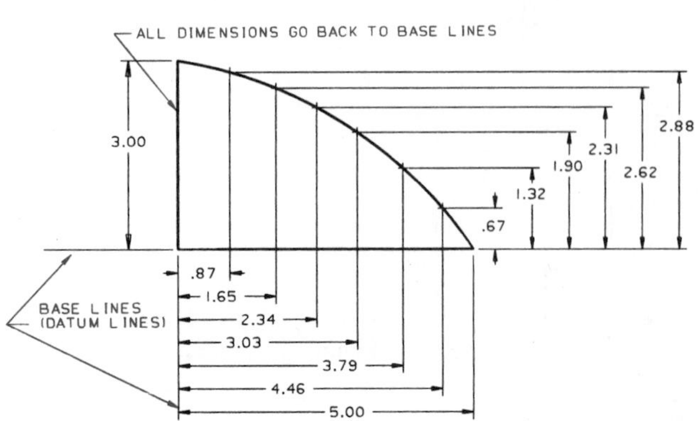

FIGURE 11.38 Coordinate Dimensioning of Curved Outlines

11.5.5 Diameter Dimensions

All diameter dimensions are preceded by the international symbol for diameter: a circle drawn the same size as the numerals, with a 60° slanted line passing through its center (∅). On some older U.S. standard unit drawings, the size of the diameter is called out with the abbreviation **DIA** after the numerals (**.375 DIA**). Some of the drawings in this text reflect this older practice.

The diameter symbol precedes all diametral values (Fig. 11.40). When the diameters of a number of concentric cylindrical features are specified, the diameter is dimensioned in a longitudinal view (Fig. 11.40).

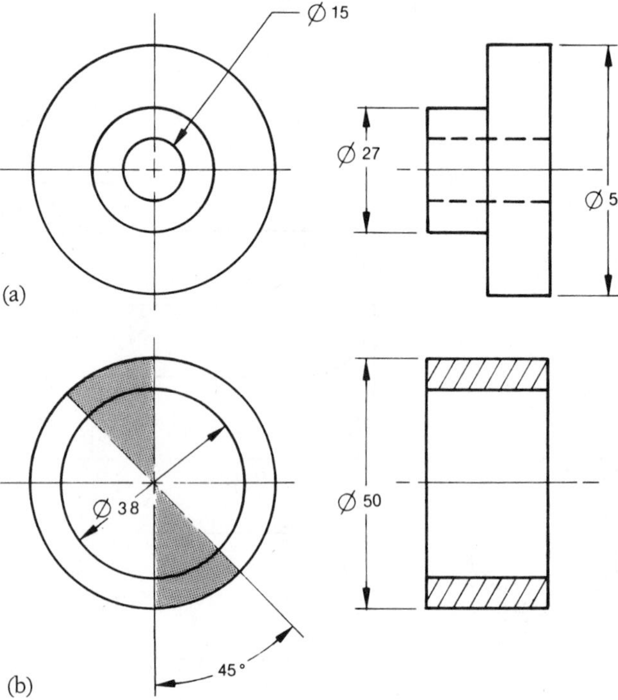

FIGURE 11.40 Dimensioning Diameters (a) Diameter callouts (b) Diameter dimensions and area to avoid

ITEMS OF INTEREST

The History of the Metric Standard

The metric system is a "standard" system of weights and measures based on the meter, a unit of length, and the kilogram, a unit of mass. But what does "standard" mean?

Noah was told to build his ark 300 cubits long. A cubit was the measured distance from the elbow to the extended finger. Some people have longer arms and fingers than others, so this is an interesting standard unit of length. On average, a cubit is about 18 inches, which means that Noah built an ark about 450 feet in length (as big as an ocean liner)!

The Cubit, An Ancient Unit of Measure

If you study the history of measurement, you will discover that standards were loosely defined and crudely measured. Many variations were found within a country. Charlemagne, emperor of the Holy Roman Empire from 800–814, used his foot as the standard length measurement. In Europe, the measure used was shorter than the English foot. The Chinese used a measure that was longer. King Henry stated that a yard was the distance from his nose to his outstretched middle finger of his right hand. An inch was the width of three barley corns laying side by side.

It was obvious to many that a "standard" system of measurement was desperately needed. In the 1790s Thomas Jefferson proposed a plan for the adoption of a standard system to Congress. Louis XVI of France tried to persuade the United States and Great Britain to cooperate in setting a standard. Although Great Britain and the United States did not join his effort, many other countries did. The end result of that project was the metric system. The standard was organized when a committee from the French Academy made its report to the National Assembly. It was adopted into French law and even though other countries adopted it, use of the system spread slowly.

Originally, the meter was one ten-millionth part of the distance from the North Pole to the Equator, passing through Paris. Later, they discovered the measurement was slightly short, so the defined it again. This time it was the distance between two marks on a platinum/iridium bar. The bar became known as the International Prototype Meter and was placed in the Bureau of Archives in Paris. A meter was later redefined as 1,650,763.73 wavelengths of the orange-red line of krypton 86. The International Bureau of Weights and Measures was formed in 1875 (Paris). The copies of the standards owned by the United States are housed in the National Institute of Standards and Technology, NIST, in Washington D.C. (formerly the National Bureau of Standards).

The metric system has been universally accepted—except in the United States and some of the British Commonwealth. Its use was legalized in the United States in 1866. In 1975, the U.S. Congress passed a bill allowing for voluntary conversion to the metric system. A special board, the U.S. Metric Board, was formed to implement this program.

The use of the metric system in the United States has increased consistently over the years. The automobile industry has been one proponent of this conversion. With cars assembled from parts manufactured all over the world, its seems very reasonable to agree on one standard measurement. However, there is an investment in the "English" system in the United States. The "English" system is now called the U.S. Customary System. New computer numerically controlled (CNC) machines can be used to cut a millimeter or an inch because of the way their motors are controlled. It seems reasonable to expect metric units to replace U.S. customary units in the United States and become the universal "standard." When is another question.

Holes in side views or section views are dimensioned when the hole cannot be adequately called out where it shows as a circle. If it is not included in a note, the depth of the hole can be dimensioned in the longitudinal view. The dimensions of a very large hole are shown by drawing the dimension line at an angle through the diameter (Fig. 11.40, lower left). For aligned dimensions placed inside the circular form, the area within the section should be avoided when the dimension runs through the diameter.

Holes should be called out with a leader and note. The leader points toward the center of the circle. *Solid round* shapes are dimensioned on the noncircular view.

You May Complete Exercises 11.1 Through 11.4 at This Time

11.5.6 Hole Depths and Diameter Dimensions

Holes dimensions are shown by pointing to the diameter with a leader and giving a note containing size and type. If the depth of the hole is not obvious or not dimensioned, the word THRU, implying drill through, follows the size specification.

A **blind hole** does not go through the part. The depth dimension of a blind hole is the depth of the full diameter from the surface of the part. If a blind hole is also counterbored or counterdrilled, the depth dimension is still from the outer surface.

A number of methods are used to call out hole diameter and depth:

Fraction-Inch		Decimal-Inch		Decimal-Inch (Symbology)
$\frac{1}{2}$ DIA THRU	or	.50 DIA THRU	or	Ø.50 THRU
$\frac{1}{2}$ DIA		.50 DIA		Ø.50
$\frac{3}{4}$ DEEP	or	.750 DEEP	or	↧.750

11.5.7 Dimensioning Slotted Holes

Figure 11.41 shows three methods for dimensioning slots. In Figure 11.41(a) the slot's centerlines are located between centers. A dimension from the edge of the part or other controlling feature is given as well. The slot width is given as an R (radius) pointing to the end of the slot arc. The R is accompanied by the note 2 PLACES. ANSI calls for R for a radius; however, many companies still use RAD. The method in Figure 11.41(b) uses a leader and a note stating the outside dimensions of the slot, 20 × 60. An R callout is also included. The slot also can be located from the part's edges [Fig. 11.41(c)] or its centerlines can be located from two controlling edges. The method in Figure 11.41(c) shows the dimensions of the slot on the view. A R callout is also used.

The choice of methods for dimensioning slots is determined by design and the required slot tolerance. If something fits into the slot, accurate tolerance and dimensions are required. The methods in Figures 11.41(a) and (c) are recommended when accuracy is important. The method in Figure 11.41(a) is used for milled slots and the method in Figure 11.41(c) is used for punched forms.

11.6 Dimensioning Features with Symbols

Geometric characteristics and other dimensional requirements can be established using standard symbols instead of traditional terms and abbreviations. These symbols must conform to ANSI Y14.2M [symbols denoting geometric characteristics (Fig. 11.42) and symbols identifying a basic dimension (Fig. 11.43)]. *A basic dimension is a numerical value*

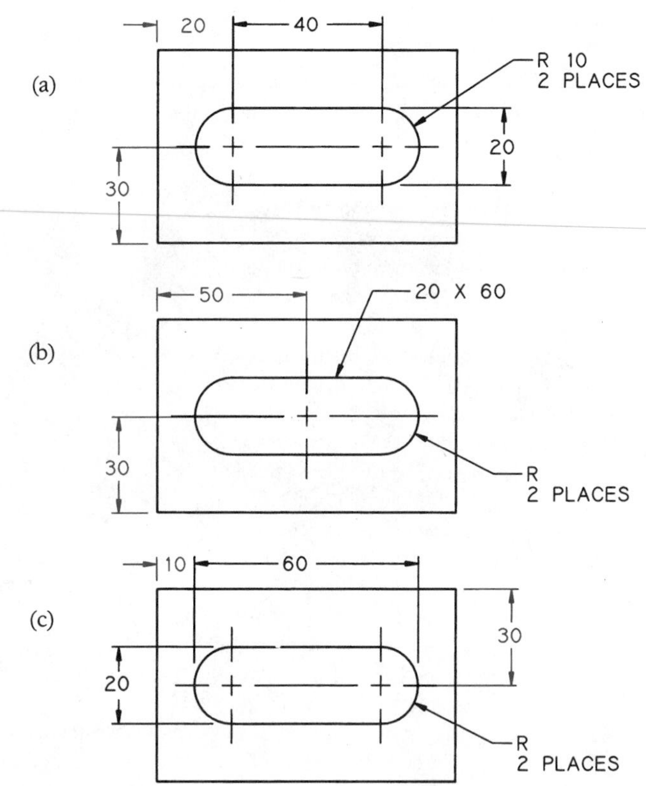

FIGURE 11.41 Dimensioning Slots (a) Locating centerlines of slot (b) Leader and note callout (c) Overall slot dimensions

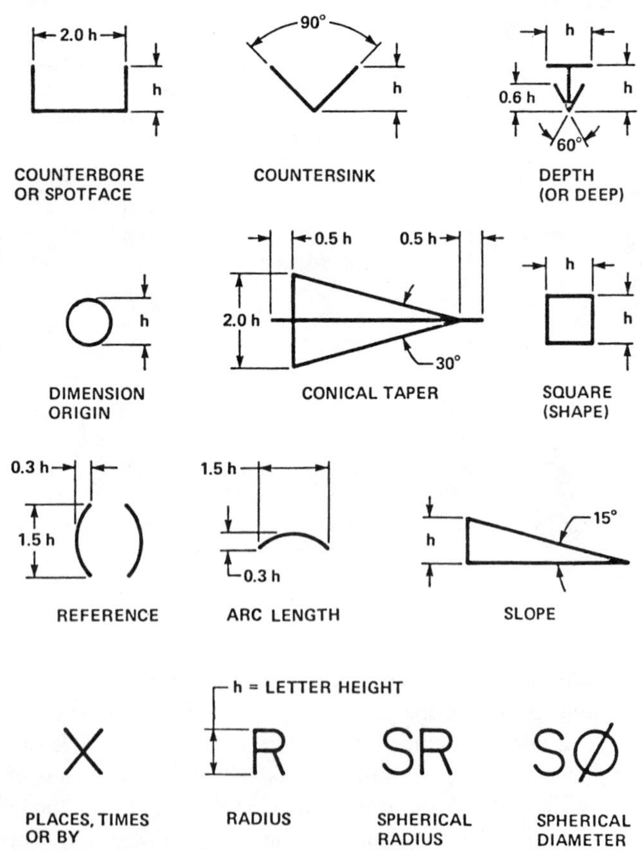

FIGURE 11.42 Form and Proportion of Dimensioning Symbols (ANSI)

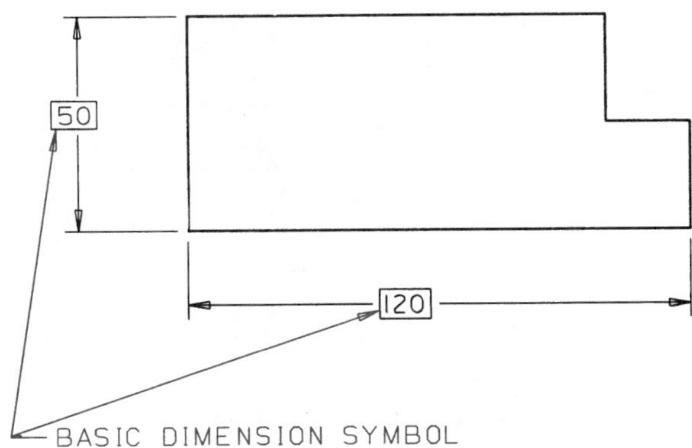

FIGURE 11.43 Basic Dimension Symbol

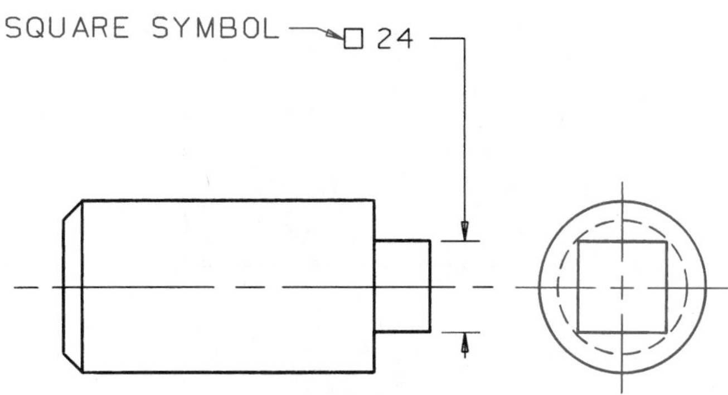

FIGURE 11.45 Square Dimensioning Symbol

describing the theoretically exact size, profile orientation, or location of a feature or datum target. Basic dimensions are the basis from which permissible variations are established by tolerances on other dimensions or in notes.

The symbols to indicate diameter, spherical diameter, radius, and spherical radius are shown in Fig. 11.44. Symbols precede the value of a dimension or tolerance given as diameter or radius. Reference dimensions (or reference data) are enclosed within parentheses. Symbology designates a variety of geometric features and dimensions including:

- The symbol to indicate a linear dimension is an arc length measured on a curved outline (Fig. 11.44). The symbol is placed above the dimension (Fig. 11.34).
- The symbol to indicate that a single dimension applies to a square shape precedes that dimension with the symbol for a square (Fig. 11.45).
- The symbol to indicate that a toleranced dimension between two features originates from one of these features is shown in Figure 11.46.
- The depth of a hole (Fig. 11.47), counterbore (Fig. 11.48), spotface (Fig. 11.49), countersink (Fig. 11.50) and counterdrill (Fig. 11.50) can be given symbolically. The symbol to indicate where a dimension applies to the depth of a feature precedes the dimension with the depth symbol.

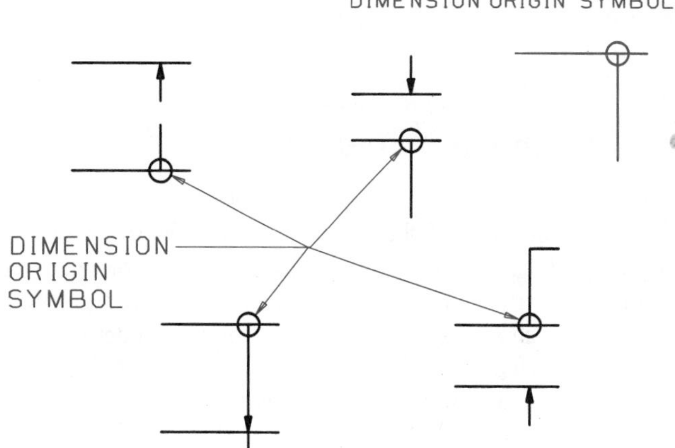

FIGURE 11.46 Dimension Origin Symbol

TERM	SYMBOL
AT MAXIMUM MATERIAL CONDITION	Ⓜ
REGARDLESS OF FEATURE SIZE	Ⓢ
AT LEAST MATERIAL CONDITION	Ⓛ
PROJECTED TOLERANCE ZONE	Ⓟ
DIAMETER	Ø
SPHERICAL DIAMETER	SØ
RADIUS	R
SPHERICAL RADIUS	SR
REFERENCE	( )
ARC LENGTH	⌒

FIGURE 11.44 Modifying Symbols for Dimensions (ANSI)

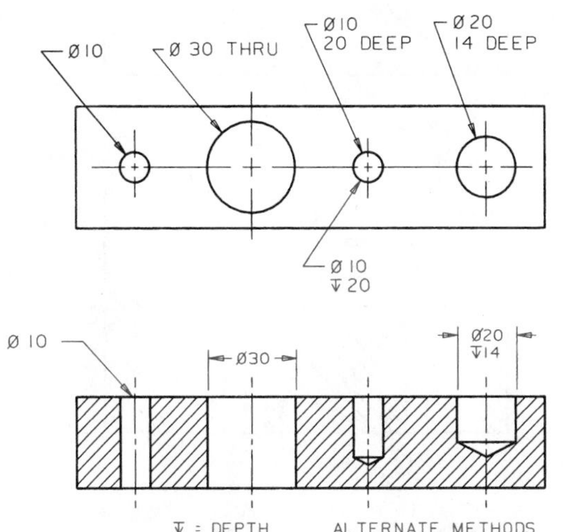

FIGURE 11.47 Using Symbols to Dimension Holes

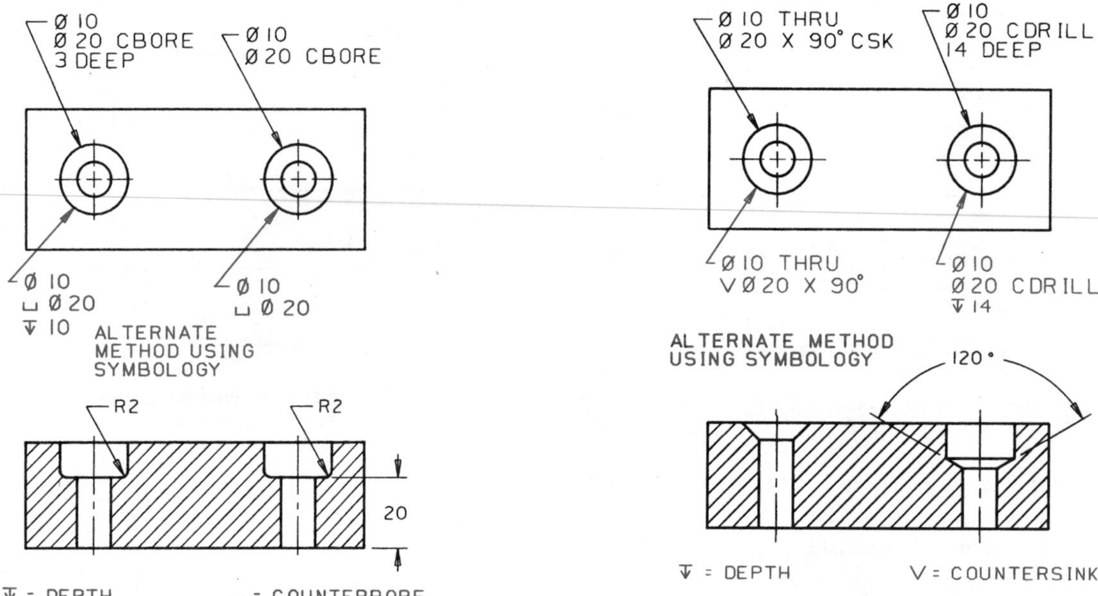

FIGURE 11.48 Dimensioning Counterbores

FIGURE 11.50 Dimensioning Counterdrills

11.6.1 Counterbored Holes

Counterbored holes are used extensively for socket head screws so the head of the screw is flush with or below the surface of the part. A counterbore is an enlarged hole, piloted from a smaller hole to maintain concentricity. Counterbored holes are machined to a square seat at a specified depth (Fig. 11.50). The depth is called out within the hole note as the distance from the upper surface (beginning surface) to the bottom of the counterbore. Either the symbol for the counterbore or the note **CBORE** is used. The depth symbol or the note **DEEP** can also be used. When the thickness of the remaining material has significance, it, rather than the depth, is dimensioned (Fig. 11.50).

FIGURE 11.49 Dimensioning Spotfaces

11.6.2 Spotfaced Holes

A **spotface (SF)** (Fig. 11.49) is a method of cleaning up and squaring a rough surface such as on a cast metal part. Material is removed so a screw head will seat flush against the surface. Its depth is usually not dimensioned.

The diameter of the spotfaced area is specified by the diameter symbol and a value. When a depth is required, either the depth or the remaining thickness of material may be specified. A spotface sometimes is specified by a note. If no depth or remaining thickness of material is specified, the spotfacing is the minimum depth necessary to clean up the surface. Figure 11.49 shows both methods for calling out the spotface.

11.6.3 Countersunk and Counterdrilled Holes

Countersinking (CSK) is a process that allows flathead screws to be positioned flush with the surface of the part (Fig. 11.50). The diameter and included angle of the countersink are specified.

For **counterdrilled holes (CDRILL)**, the diameter and depth of the counterdrill are given. Specifying the included angle of the counterdrill is optional (Fig. 11.50, lower right). The depth dimension is the depth of the full diameter of the counterdrill from the outer surface of the part. Symbology can also be used on these features. A counterdrilled hole differs from a counterbored hole in that the bottom of the counterdrilled hole is conical. Counterdrilled holes are created with a step drill or two drills of different diameters.

You May Complete Exercises 11.5 Through 11.8 at This Time

11.7 Dimensioning Special Features

Chamfers, threads, centerdrills, tapers, knurling, keyways, and other *geometric features* require specific, standardized dimensioning. These dimensions are based on the method used to machine them or on a standard purchased part mated with the feature.

11.7.1 Threads

Thread callouts are found on almost every mechanical drawing. Figure 11.1 shows a part designed with metric units. The callout M42 × 1.5 – 6G specifies a metric thread.

Nonmetric threads are classified according to the number of threads applied to a specific diameter. Unified (**UN**) thread is the standard type of thread for the United States. To specify screw threads, the nominal major diameter is given first, followed by the number of threads per inch, and the series designation. Finally, the class of fit between male and female threads is given, followed by an A for male threads or a B for female threads. For tapped holes, the complete note contains the tap drill diameter and depth of hole, followed by the thread specification and the length of the tapped threads. All threads are assumed to be right hand unless left hand is specified by LH following the class. A few examples of screw thread notations follow:

Decimal-Inch:

- .190-32 UNF-2A or #10-32 UNF-2A
- .250-20 UNC-2B or $\frac{1}{4}$-20 UNC-2B
- 2.000-16 UN-2A
- 2.500-10 UNS-2B

Metric:

- M6 × 1-4h6h
- M16 × L4-P2-4h6h

The thread type and size are given on the drawing and the machinist chooses the correct drill diameter. Specifying the drill and tapping requirements requires the diameter and depth. In this case, the tap drill size; its depth, the thread specification, and the depth of threads are provided, as in the examples that follow:

- .312 1.25 DEEP
  .375-16 UNC-2B, .88 DEEP (or $\frac{3}{8}$-16 UNC-2B)
  3 HOLES
- .422 1.25 DEEP
  .500-13 UNC-2B LH, 1.12 DEEP (or $\frac{1}{2}$-13 UNC-2B LH)
  2 HOLES

More information on methods of specifying and dimensioning screw threads is found in ANSI Y14.6.

11.7.2 Chamfers

Manual and automated assembly techniques both benefit from tapered features to help the parts engage. **Chamfers** are specified by dimensions or notes. It is not necessary to use the word CHAMFER when the meaning is obvious. If the chamfer is other than 45°, dimensions are used to show the direction of the slope. Figure 11.51 shows methods for dimensioning external chamfers. You can show chamfer dimensions by the chamfer angle and one leg, dimensioning both legs or pointing to the chamfer and giving the angle and one leg as a callout. Internal dimensions for chamfers (Fig. 11.52) are dimensioned by the included angle and the largest diameter. The metric method of dimensioning chamfers is also shown on this figure. For inch-unit drawings, the angle is sometimes given second and the leg first, for example .25 × 45°. This method is being replaced by ANSI Y14.5M. If chamfers are required for surfaces intersecting at other than right angles, the methods shown in Figure 11.53 are used.

11.7.3 Taper Dimensioning

Tapers are used on machines to align and hold machined parts that require simple and speedy assembly and disassembly. A round taper has a uniform increase in the diameter on a round part for a given length measured parallel to the axis of the workpiece (conical). Internal or external tapers are noted by taper per foot (**TPF**), taper per inch (**TPI**), or by degrees. TPF or TPI refer to the difference in diameters within 1 foot or 1 inch (Fig. 11.54(a)]. The difference is measured in inches. The *angles of taper* refer to the inclined angles with the part's centerline (axis) [Fig. 11.54(b)].

In Figure 11.54(a), the taper per foot, the length of the part, and the large diameter are given. In Figure 11.54(b), the diameter, length and the angle are given. In Figure 11.54(c)

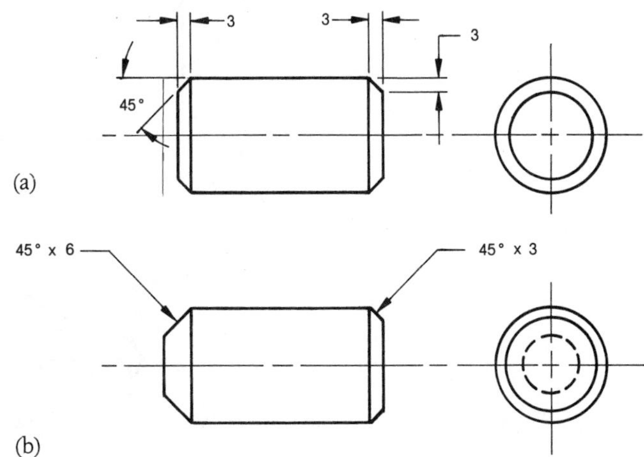

FIGURE 11.51 Dimensioning Chamfers (a) Dimensioning chamfer angle and one leg (b) Angle and one leg given in callout

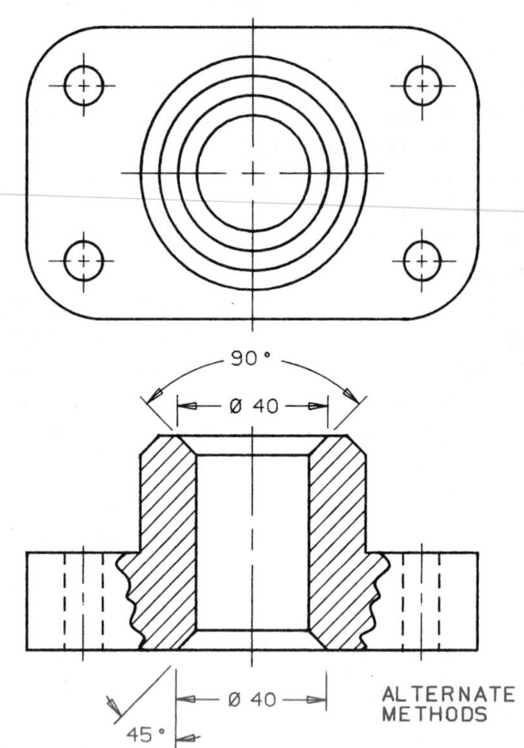

FIGURE 11.52 Internal Chamfers

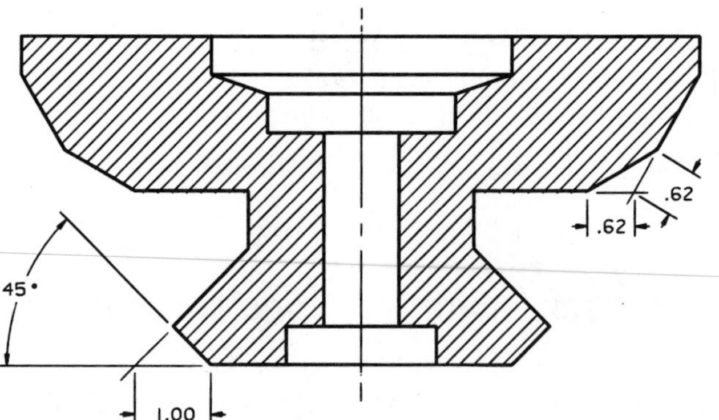

FIGURE 11.53 Dimensioning Chamfers Between Surfaces Not at 90°

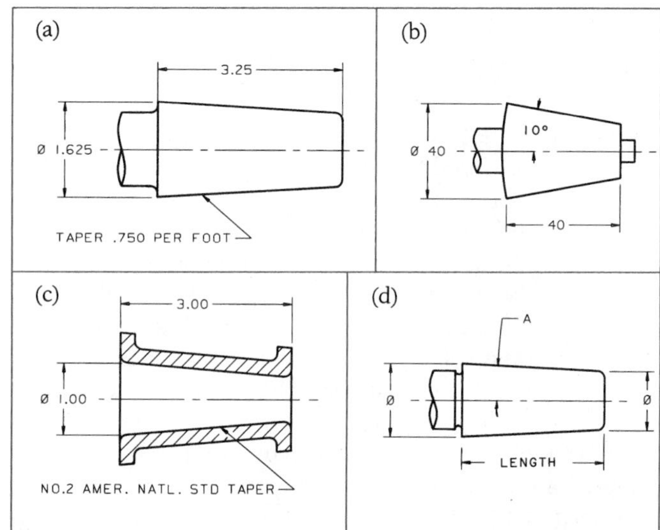

FIGURE 11.54 Dimensioning Tapers (a) Taper per foot (b) Taper angle, diameter, and length (c) National Standard Taper (d) Angle of taper, length, and diameter at both ends

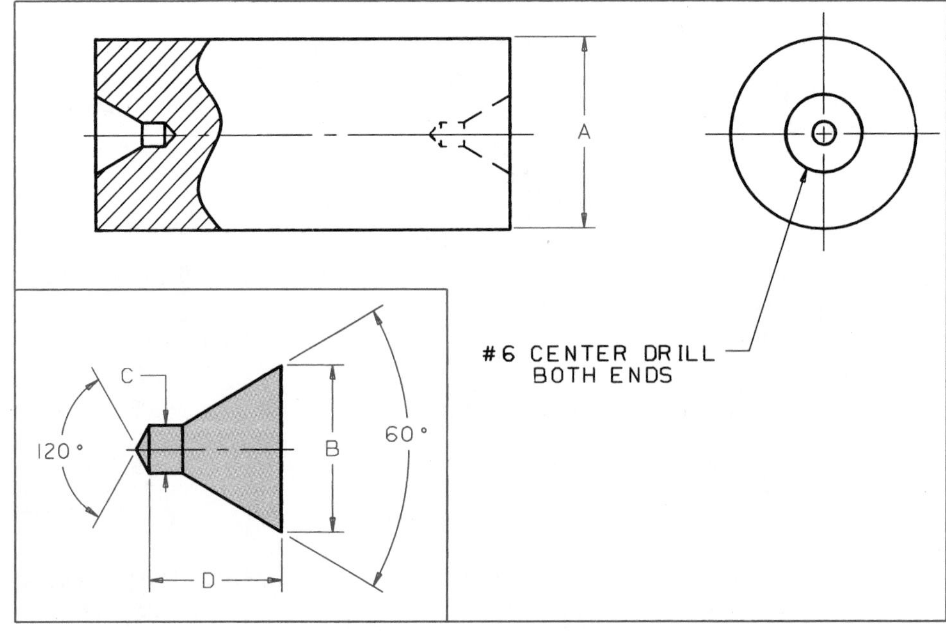

FIGURE 11.55 Center Holes and Center Drills

the length, diameter and the note **NO 2. AMER. NATL. STD TAPER** are given. In the last example [Fig. 11.54(d)], the two diameters, the length, and the angle are given.

11.7.4 Center Drill Dimensioning

When a part is held and turned between the centers of a lathe, a *center hole* is required on each end of the cylindrical workpiece. The center hole has a 60° angle that conforms to the center and a smaller drilled hole to clear the center's point. The center hole is made with a combination drill and countersink called a **center drill** (Fig. 11.55). In this figure, the A dimension is the workpiece diameter. The B dimension is the body diameter of the centerdrill. The C dimension is for the diameter of the drilled center hole, and the D dimension is the depth. The drill tip is drawn at 120°.

11.7.5 Keys and Keyseats

A **key** is a demountable machinery part. When assembled into keyseats, a key provides a positive means for transmitting torque between a shaft and hub. A **keyseat** is an axially located rectangular groove in a shaft or hub. Keyseats are dimensioned by width, depth, location, and, if required, length (Fig. 11.56). The depth is dimensioned from the opposite side of the shaft or hole.

11.7.6 Knurling

A **knurl** is a machined rough geometrical surface on a round metal part. Knurling is used to improve the grip or for press fitting the knurled part into a hole into a mating part. Knurling is also done for appearance.

Knurling is specified in terms of type, pitch, and diameter before and after knurling. When diameter control is not required, the diameter after knurling is omitted. If only a portion of a feature is to be knurled, axial dimensioning is necessary. Knurling can be either diamond or straight patterned and fine, medium, or coarse. Knurling is specified by a note that includes the type of knurl required, the pitch, the toleranced diameter of the feature prior to knurling, and the minimum acceptable diameter after knurling (Fig. 11.57).

11.8 Locating Features on a Drawing

The location of holes, slots, and machined features on a part is very important. During dimensioning, consider how the part is to be machined.

11.8.1 Geometric Analysis of a Part

Figure 11.58 shows simple geometric shapes and the dimensions required to describe the shapes. The machinist should have all necessary dimensions. *A drawing is never "scaled" to find a location or size that should have been described by a dimension.*

In Figure 11.59, a simple clamp is shown in three views with appropriate dimensioning. It is composed entirely of rectangular prisms. Each of these prisms must be sufficiently dimensioned so a machinist can make the part. *The three most important dimensions are also required: height, width, and depth.*

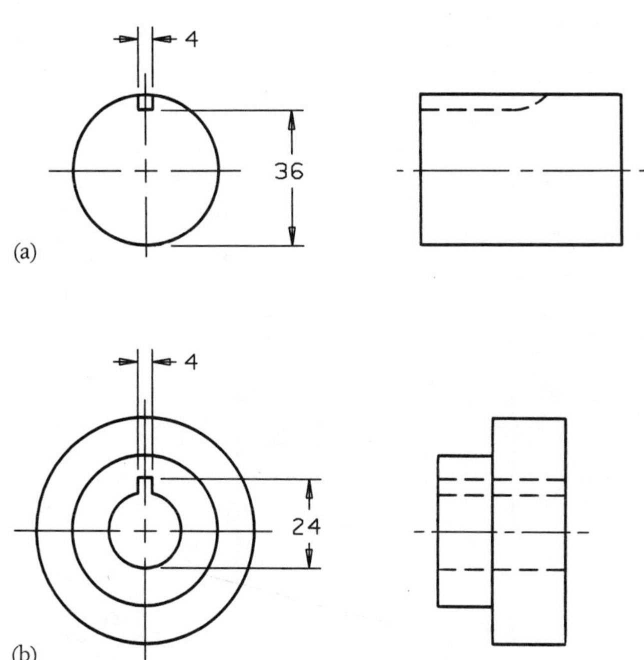

FIGURE 11.56 Dimensioning Keyseats (a) Shaft keyseat (b) Hub keyseat

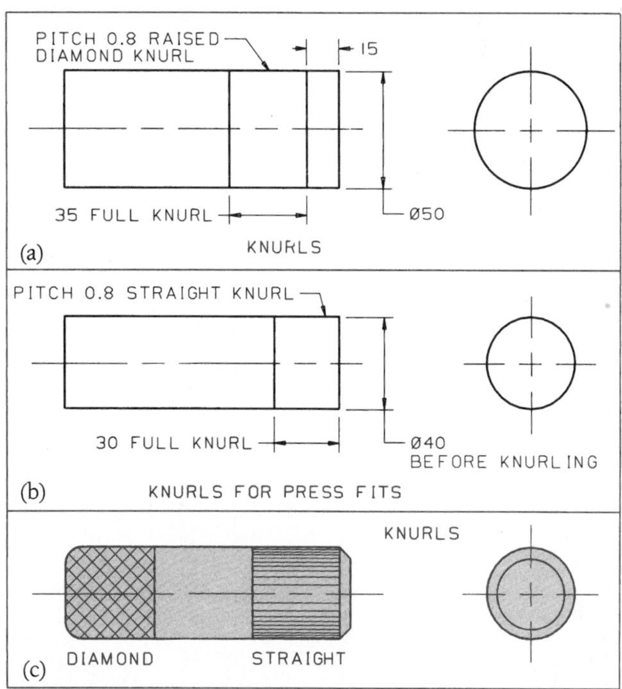

FIGURE 11.57 Dimensioning Knurling (a) Diamond knurl (b) Straight knurl (c) Knurling representation

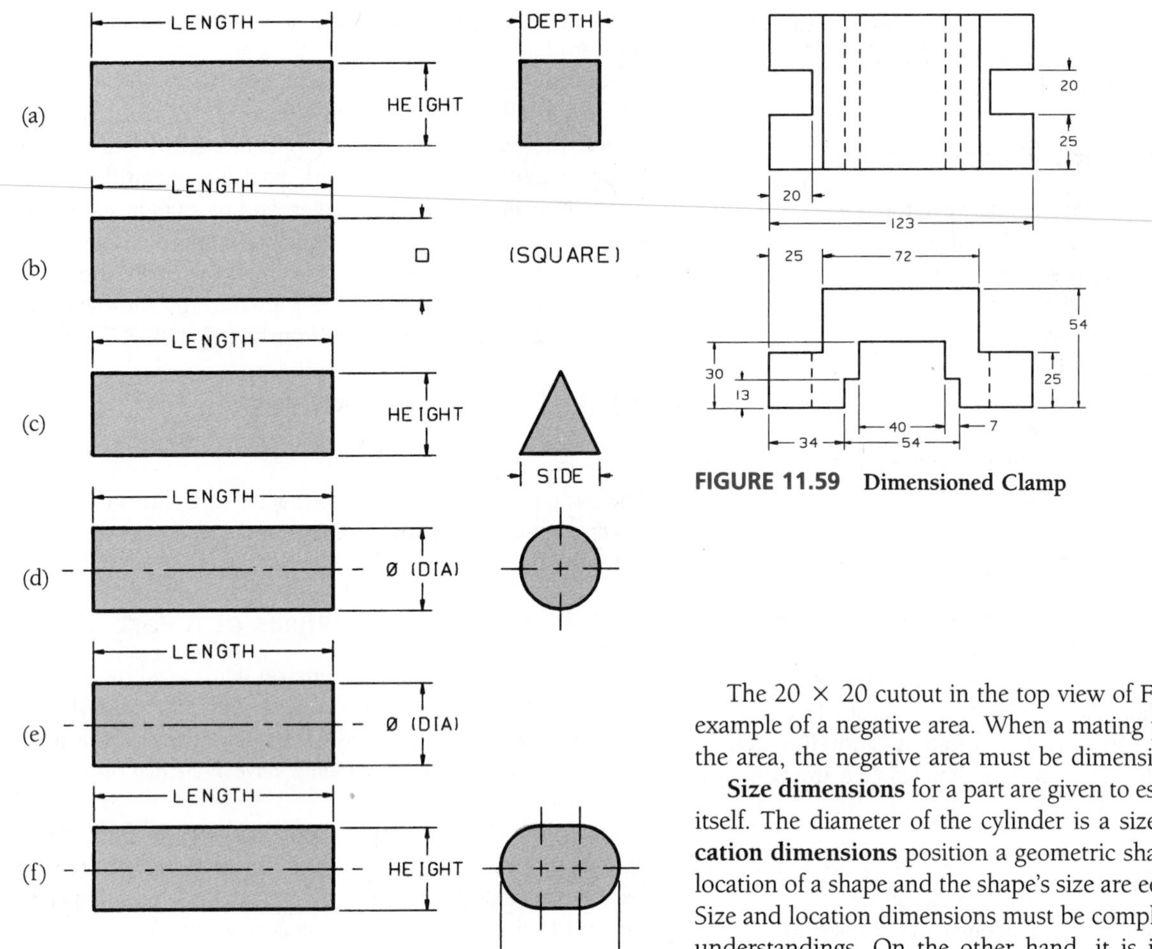

FIGURE 11.58 Dimensions for Common Geometric Shapes

FIGURE 11.59 Dimensioned Clamp

The 20 × 20 cutout in the top view of Figure 11.59 is an example of a negative area. When a mating part must fit into the area, the negative area must be dimensioned.

Size dimensions for a part are given to establish the shape itself. The diameter of the cylinder is a size dimension. **Location dimensions** position a geometric shape in space. The location of a shape and the shape's size are equally important. Size and location dimensions must be complete to avoid misunderstandings. On the other hand, it is important not to overdimension or to give two dimensions that locate the same feature.

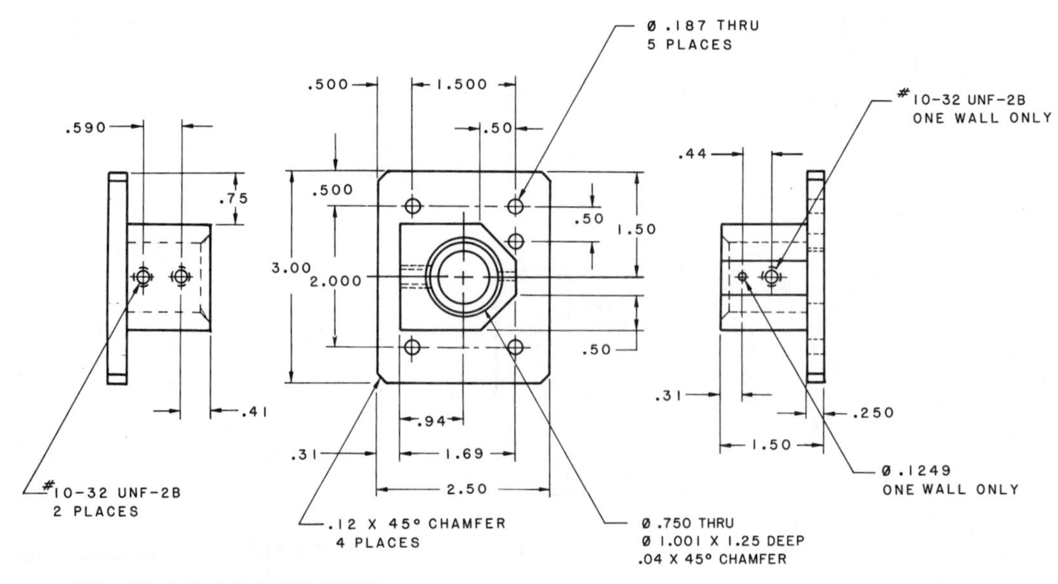

FIGURE 11.60 Base Mounting Detail

11.8.2 Mating Parts and Dimensions

In Figure 11.60, the part has features that obviously relate to a mating piece. The base of the part attaches to a mating part using the .187 clearance holes. A .750 diameter chamfered hole runs through the part. A shaft or other cylindrical items is to be inserted here during assembly and will be held in place by one or more screws entering the side of the part. In most cases, the part to be detailed will be accompanied by a description and or illustration of the assembly. From the assembly drawing, the use, location, orientation in space, and mating pieces are readily identified.

When dimensioning parts that must mate with other parts in an assembly, the related surfaces on each part should be dimensioned. Mating dimensions are also necessary to establish hole patterns used to secure one part to another.

11.8.3 Finish Marks and Machined Surfaces

Rough stock shapes, castings, and forgings have rough and unmachined surface textures. **Machined surfaces** are established on the drawing using finish marks. A **finish mark** is a symbol that tells the machinist the machining requirements for a surface. Machined surfaces must be established from a rough surface, in any direction (top, front, side). Many features on a casting or forging must be machined because these processes produce a part with every required geometric shape, but the surfaces are rough. All other machined surfaces or holes are established from that first machined surface. The symbol can be the traditional finish mark [Fig. 11.61(a)], the general symbol [Fig. 11.61(b)], or the ANSI recognized **basic surface texture symbol** [Fig. 11.61(c)].

The placements and measurements for construction of the three types of symbols are shown in Figure 11.61. Symbol templates are available for quick, easy, and accurate insertion of symbols on a drawing.

The general symbol establishes the surface to be machined without providing any details as to the quality or type of surface. The basic surface texture symbol, on the other hand, establishes a surface to be machined or how it can be altered to provide specifications for the lay, roughness, and waviness of a surface. This symbol is used whenever there is a need to control the surface irregularities of a part (Fig. 11.62).

You May Complete Exercises 11.9 Through 11.12 at This Time

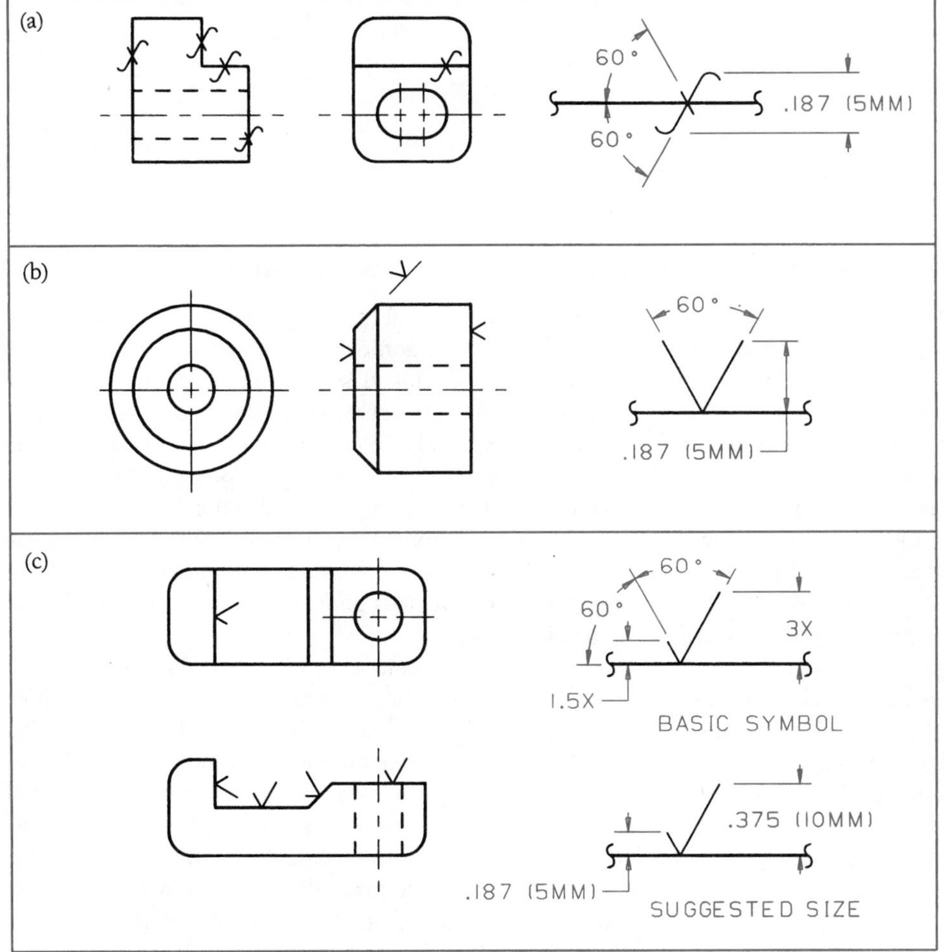

FIGURE 11.61 Finish Marks
(a) Traditional finish symbol
(b) Old general finish symbol
(c) New basic finish symbol

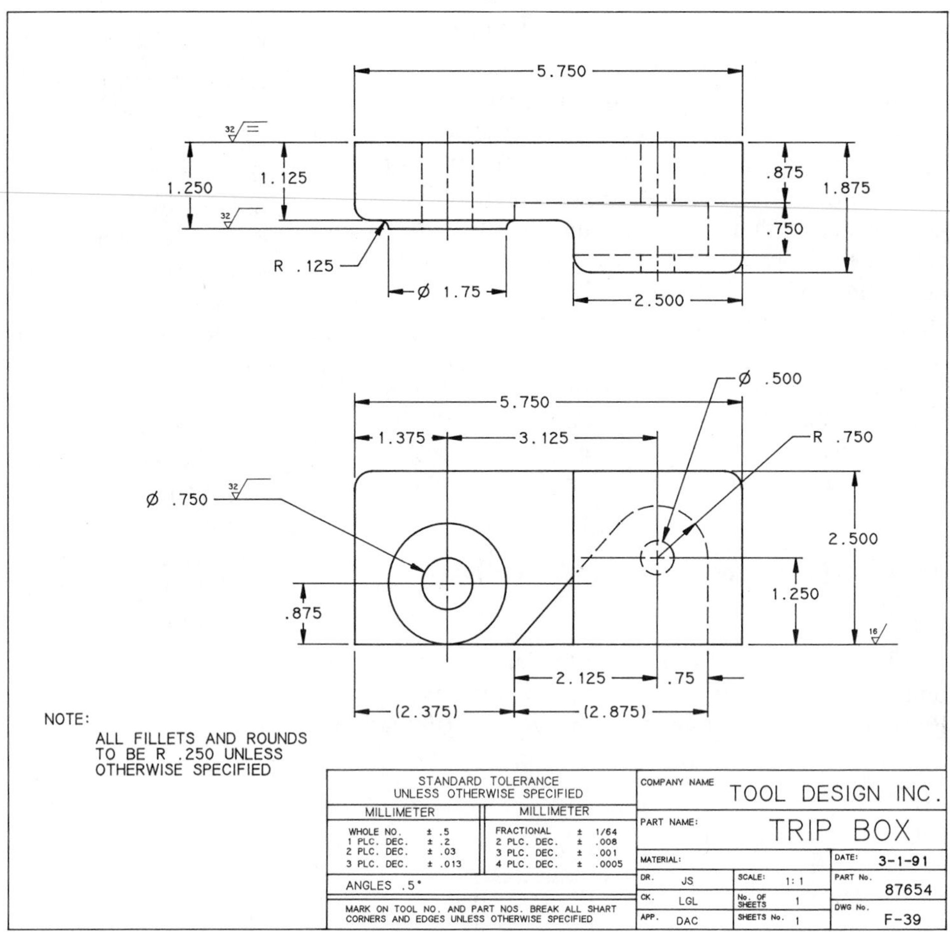

FIGURE 11.62 Finish Marks Used on Mechanical Detail

11.8.4 Locating Holes and Features on a View

Because machined surfaces are used to establish machined features such as holes and slots, it is important to locate them by dimensions from prominent features or surfaces (Fig. 11.63). The bracket arm is a cast part and has a number of machined surfaces and holes. The central hole is obviously the most important; therefore, all machined holes and slots are located from it. In the front view, the bottom surface was used to locate each of the height dimensions.

A general rule for dimensioning is to dimension from a rough surface to a finished surface once in each direction and between rough surfaces for all other nonmachined surfaces in each direction. Dimensions are given between all other machined surfaces and the first finished surface in each direction. Features such as holes and slots are dimensioned between each other and back to the prominent finished surface in each direction.

When the true shape of a feature or surface does not appear in one of the six standard views, an auxiliary view of that surface and feature should be projected to properly locate it (Fig. 11.64).

11.8.5 Locating Holes on a Part

Size dimensions for the part's features are established first, followed by the location of features such as holes. Holes are located from a machined surface. In most cases, holes are established in patterns and dimensioned accordingly. Figure 11.65 shows a detail of a connector. Because the part is thin (.25), only one view is needed. Each of the holes is located from the lower left corner. The 0,0 position (origin) is used to establish control surfaces from which all dimensions are taken. This part uses dimensions without dimension lines. This is called **rectangular coordinate dimensioning without dimension lines.**

Features that lie about a common center such as slots and holes can be dimensioned with angular dimensions [Fig. 11.66(a)] or with a note such as "equally spaced" [Fig. 11.66(b)]. **Offset dimensioning** takes each locating dimension along the axis of the part from its centerline. This method is preferred since it is easier to set up a machine to locate the holes for machining with rectangular coordinates.

Because each of the holes dimensioned in Figures 11.66 and 11.67 are taken from the center of the part, they are **dimensioned from a finished surface.**

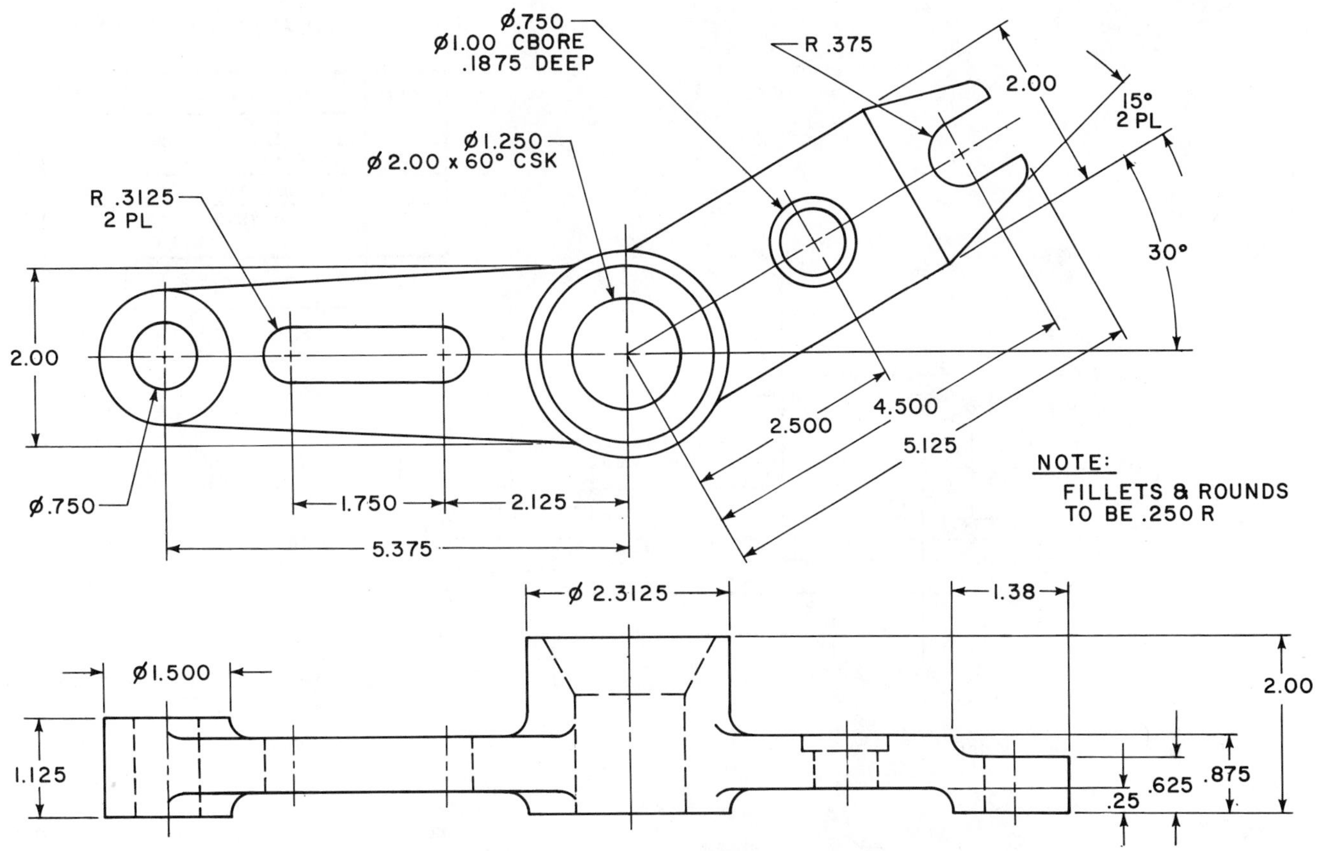

FIGURE 11.63 Dimensioning Machined Features

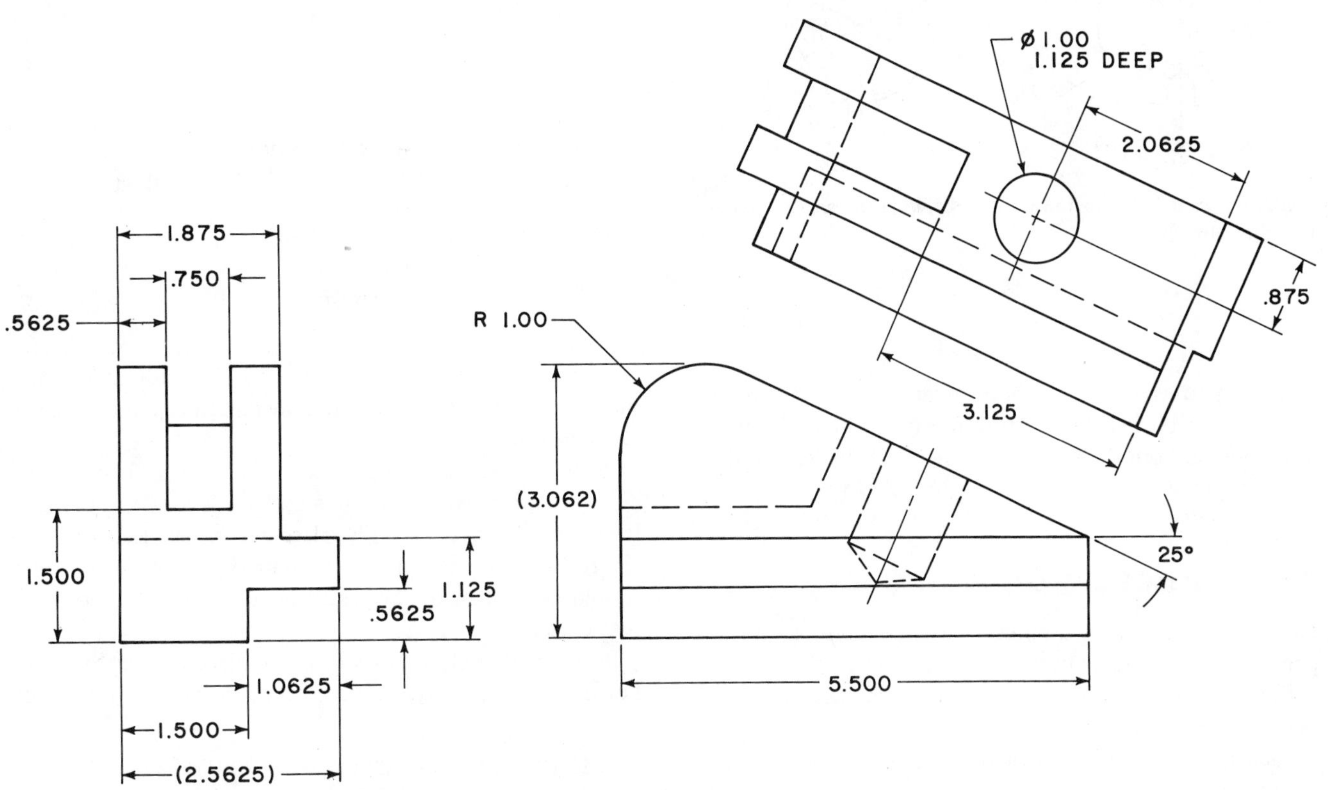

FIGURE 11.64 Anchor Detail

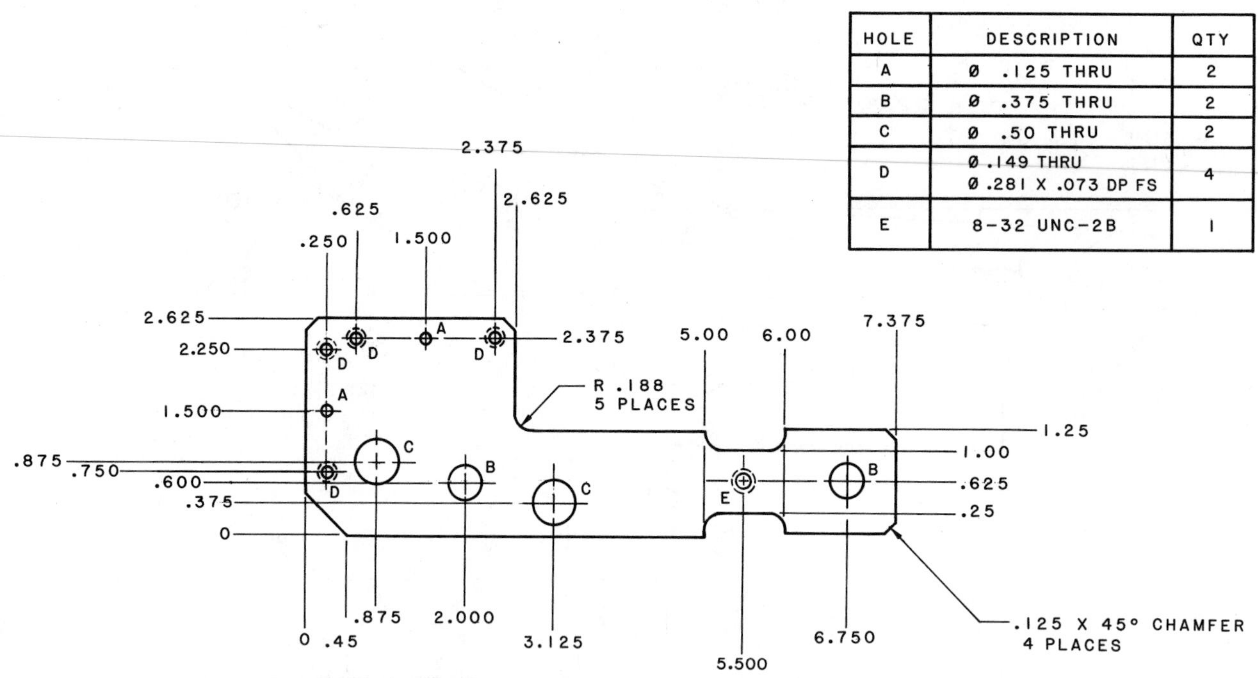

HOLE	DESCRIPTION	QTY
A	Ø .125 THRU	2
B	Ø .375 THRU	2
C	Ø .50 THRU	2
D	Ø .149 THRU Ø .281 X .073 DP FS	4
E	8-32 UNC-2B	1

FIGURE 11.65 Dimensioned Detail

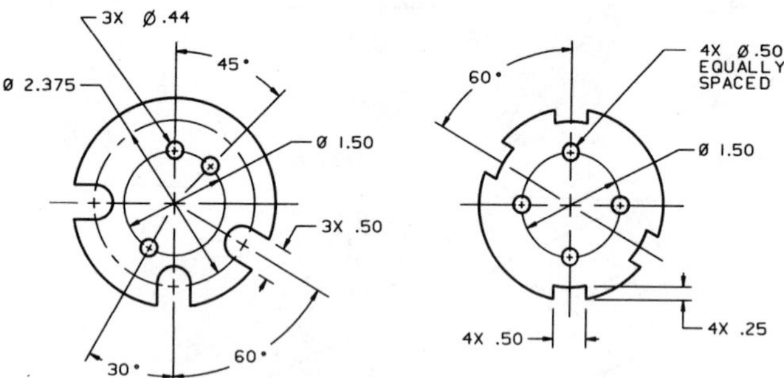

FIGURE 11.66 Dimensioning Repetitive Features (a) Three holes (b) Four holes

Hole patterns, a set of holes related to another mating part, are established by locating the center hole, or to the same hole in both directions. Dimensions between the holes are then given.

11.9 Notes on Drawings

Notes can be either **local notes**, as in the callout of a hole, or **general notes.** General notes are placed outside the geometry and beyond dimensions. Notes are one of the last items to be placed on the drawing.

General notes are located, according to ANSI standards, in the upper left-hand corner or lower left-hand of the drawing as in Figure 11.62.

Drawings completed to older ANSI standards have notes above the title block on the right side of the drawing. Some of the examples in the text reflect this older standard, as will many drawings encountered on the job. Many companies apply their own in-house standards, which may deviate from accepted ANSI standards.

This is an example of a typical general note:

NOTES:

1. MATL: .093 THK. ALUMINUM-5052.
2. FINISH: CLEAR ANODIZE- FRONT & REAR PANELS BRUSHED.
3. OPTIONAL RELIEF FOR BREAK.
4. MIN. BEND RAD. TYP. (4) PLACES.
5. SILK-SCREEN PER DWG. 18014-201

A variety of abbreviations are used in notes. Abbreviations should conform to ANSI YI.1. See the Appendixes for commonly used abbreviations on drawings. Keep the usage of abbreviations to a minimum.

Notes are lettered in uppercase except when they are long and detailed, in which case it is acceptable to use upper- and lowercase. This is the only place on mechanical drawings where lowercase lettering is permitted. The width of notes should be limited to the width of the parts list or revision block on the drawing.

Notes that apply to a view can be placed under the view. **Local notes** are placed away from the view outline (Fig. 11.68):

6-32 UNC-2B	4-40 UNC-2B	$\frac{1}{4}$-20 UNC-2B
.50 DEEP	.25 DEEP	2 PLACES
2 PLACES	8 PLACES	

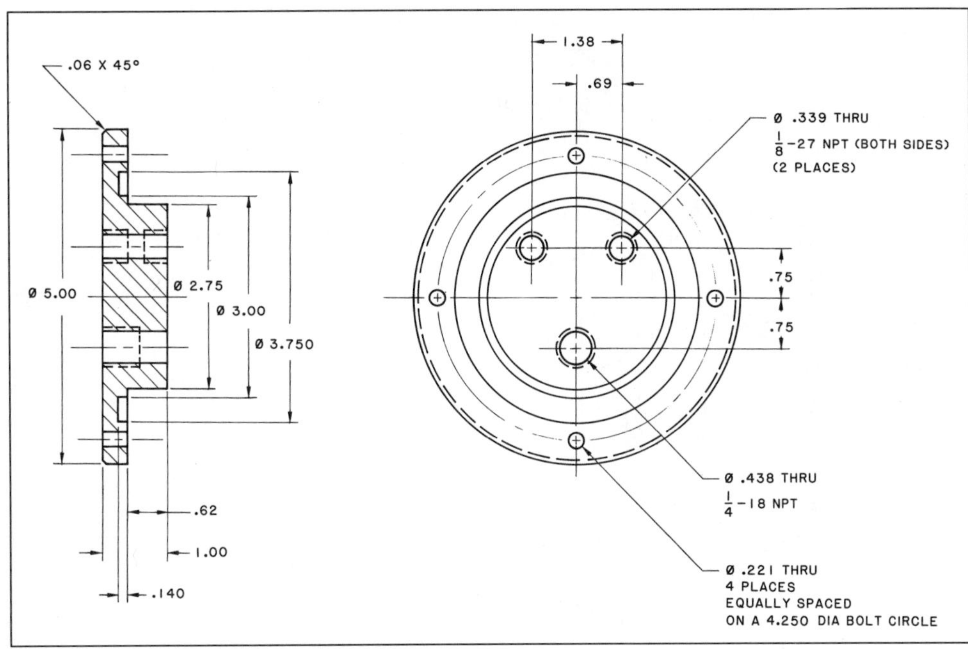

FIGURE 11.67 Square Dimensioning Holes

If a drawing is large and complicated, local notes are allowed within the view. However, avoid this practice when possible.

11.10 Locations of Features and Dimensioning Methods

To design a part, you must know how it will be manufactured. If the part to be manufactured does not require close tolerancing, simple dimensioning can be used. If the manufacturing method is automated, coordinate dimensioning is needed. Design for manufacturability (DFM) is an important factor in creating clear, precise, manufacturable, and successful parts and products.

Rectangular coordinate dimensions accurately locate features with respect to one another and, as a group or individually, from a datum or origin (Fig. 11.67). The features that establish the datum must be clearly identified on the drawing. Coordinate dimensioning is the most frequently used method of dimensioning in automation in the machine tool and manufacturing areas.

11.10.1 Rectangular Coordinate Dimensioning

Rectangular coordinate dimensioning locates features by dimensioning from two or three *mutually perpendicular planes.* The cylindrical part in Figure 11.69 is dimensioned using geometric tolerancing. The three mutually perpendicular planes are established from the center and bottom. This type of dimensioning establishes either **datum lines** (**X, Y,** and **Z** coordinate lines from which all dimensions are taken) or uses the centerlines of a symmetrical or circular shape.

In Figure 11.69, the hole pattern is dimensioned from the center of the piece. All dimensions are perpendicular to the center of the part. In Figure 11.70(a), the **X** and **Y** coordinates are used as datums/baselines. All dimensions are positioned rectangularly from the datum/baselines. The part is located in Quadrant 1 so all values are positive. In Figure 11.70(b) the circular part and its hole pattern are dimensioned from the center of the piece.

The four quadrants of the rectangular coordinate system are shown in Figure 11.71. The rectangular coordinate system has two perpendicular axes, **X** and **Y.** The plane formed by the **X** and the **Y** axes establishes the origin of the **Z** axis (Fig. 11.72). The intersection of the three axes is the origin and has a numeric value of zero (**X0,Y0,Z0**). The three reference planes are used to locate the part as in Figure 11.73. The **X** and the **Y** axes may also be established from the part as in Figure 11.74.

Figure 11.75 uses the rectangular coordinate method for dimensioning. Holes and curved features are dimensioned by locating center points from datum/baselines, indicated by zero coordinates. Dimensions are established so values can be easily entered when programming the part during CNC machining. This part also uses a hole chart. All holes are through the workpiece, so **X** and **Y** dimensions are required.

The part of Figure 11.75 is dimensioned using datum lines. The dimension lines are eliminated with only measurements and extension lines shown. This is called **rectangular coordinate dimensioning without dimension lines** (ordinate method). **Ordinate dimensioning** is one of the easiest and clearest ways to dimension a part. Dimensions are shown on extension lines without dimension lines or arrowheads. The baselines (or datum) are indicated as zero coordinates, or labeled **X, Y,** and **Z.**

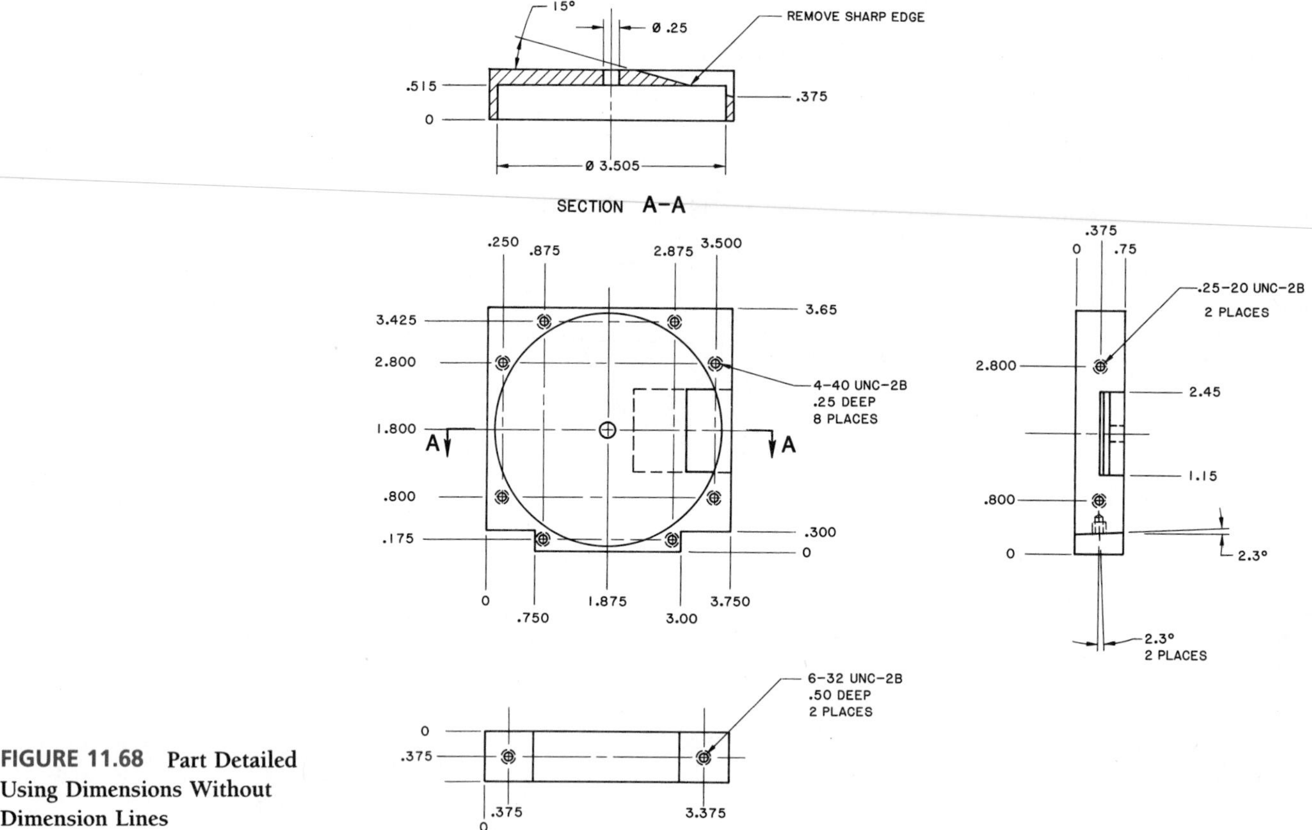

FIGURE 11.68 Part Detailed Using Dimensions Without Dimension Lines

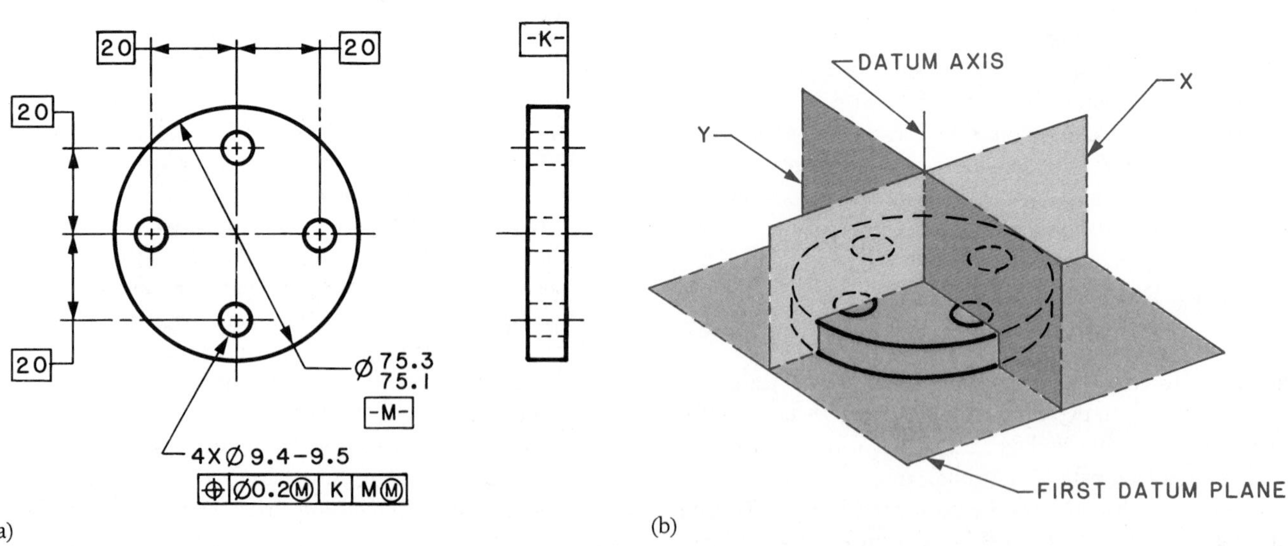

(a)

(b)

FIGURE 11.69 Part with Cylindrical Datum Feature (a) X and Y coordinates used as datums (b) Datum planes

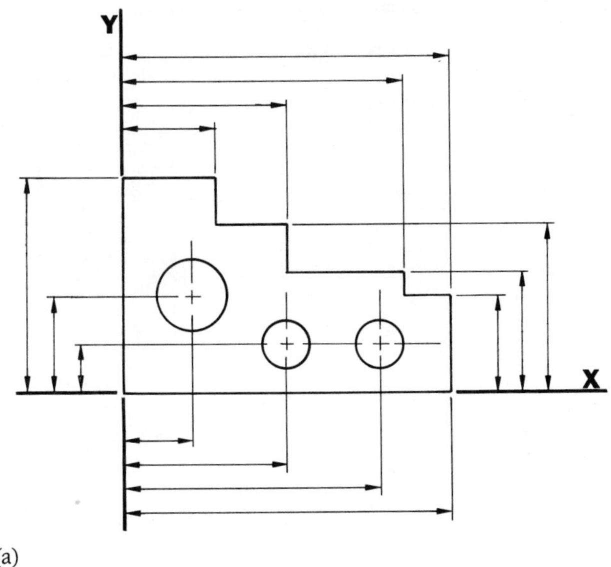

(a)

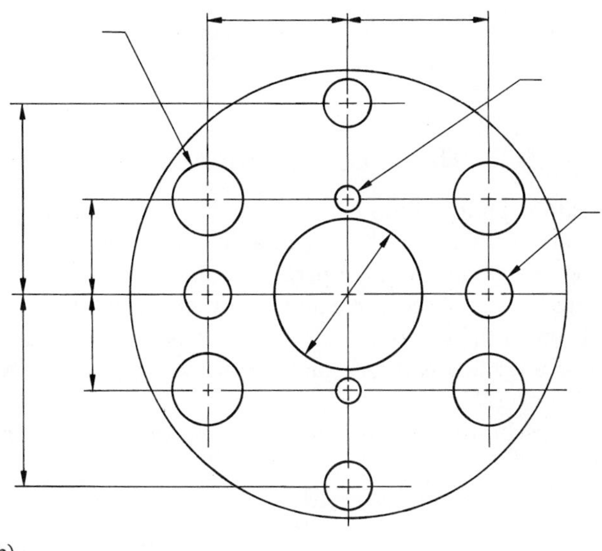

(b)

FIGURE 11.70 Dimensioning Methods (a) Datum line dimensioning (b) Rectangular coordinate dimensioning

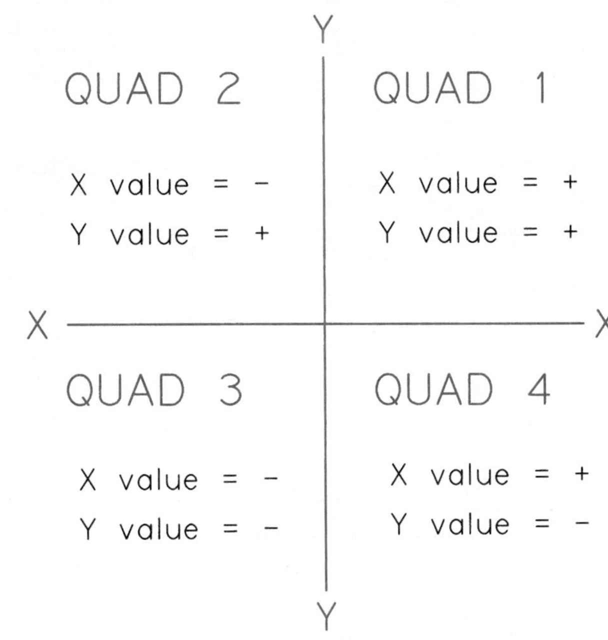

FIGURE 11.71 Quadrants

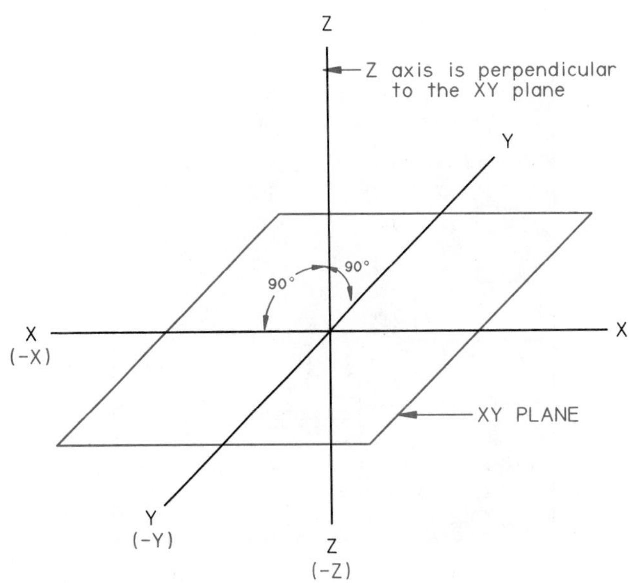

FIGURE 11.72 X,Y, Z Axes

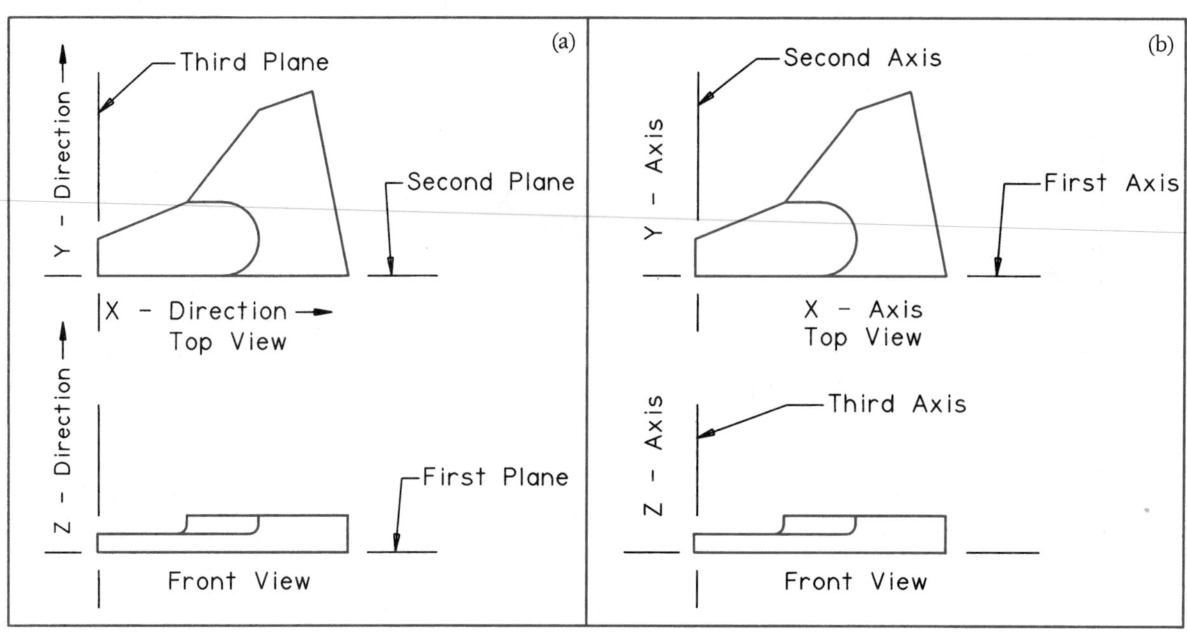

FIGURE 11.73 Reference Planes and Reference Axes

11.10.2 Polar Coordinate Dimensioning

When **polar coordinate dimensioning** is used to locate features, a linear and an angular dimension specify a distance from a fixed point at an angular direction from two or three mutually perpendicular planes. The fixed point is the intersection of these planes (Fig. 11.76). The holes are established with a radial value (R2 .62) and angles for each hole. The 0,0 position is in the lower left; the radial value and the angles are established from a location hole.

11.10.3 Datums and Tolerances

Datum points, lines, or **surfaces** are features that are assumed to be exact. They are baselines or references for locating other features of the part. A feature selected as a datum must be easily accessible and clearly identified. An artificial datum such as a construction hole or line edge is sometimes machined in a part for manufacturing and checking only. In Figure 11.77, the part is symmetrical about its vertical centerline. In this example, all dimensions are established from

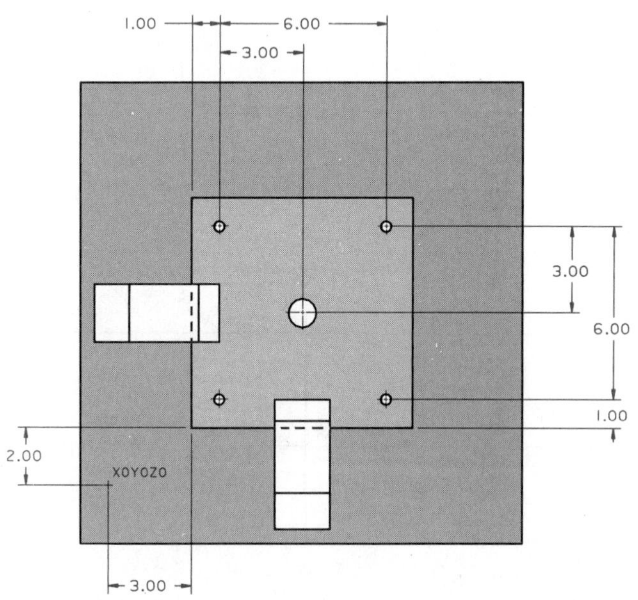

FIGURE 11.74 0,0 Position Established From the Workpiece
Work table and clamps shown with part.

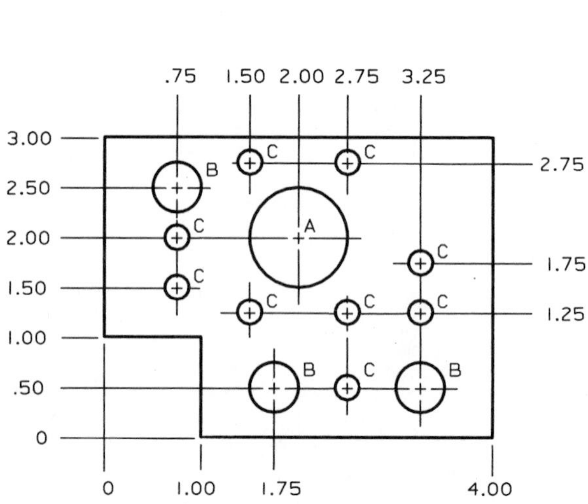

FIGURE 11.75 Hole Charts and Coordinate Dimensioning

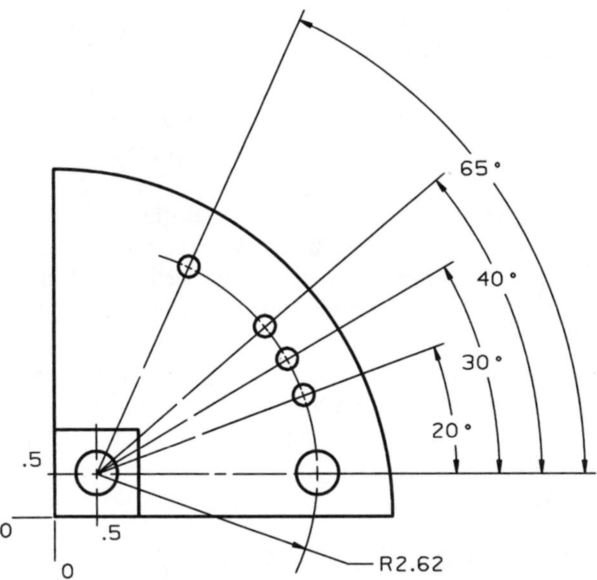

FIGURE 11.76 Polar Coordinate Dimensioning

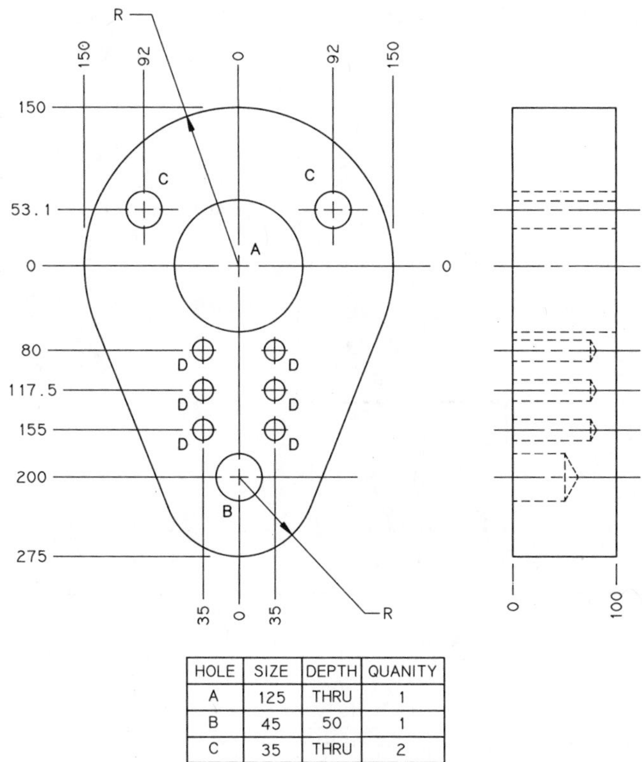

HOLE	SIZE	DEPTH	QUANITY
A	125	THRU	1
B	45	50	1
C	35	THRU	2
D	20	75	6

FIGURE 11.77 0,0 Position Established at Center of Part

the center hole. The lower left corner of the part is a large curve, which makes it inappropriate for establishing dimensions.

A datum surface must be more accurate than any location measured from that datum. It may be necessary to specify **form tolerances** for the datum surface to assure that locations can be accurately established. Mating parts use the same feature surface. When parts must match or mate, the related hole centers are used as the datum.

Dimensioning from a common base or datum reduces the overall accumulation of the tolerance. However, the tolerance on the distance between any two features, located with respect to a datum and not with respect to one another, is equal to the sum of their tolerances. Therefore, if it is important to control two features closely, the dimension is given directly from a datum or baseline.

11.10.4 Hole Charts and Tabular Dimensioning

Tabular dimensioning is a type of rectangular coordinate dimensioning in which dimensions from mutually perpendicular datums are listed in a table on the drawing (Fig. 11.78). This method is used on parts that require the location of a large number of similarly shaped features, such as multiple holes, slots, or hole patterns. The information is listed in tables. For automated tooling and programming CNC machines, providing **X** and **Y** dimensions and **Z** depths is the best method. Hole sizes are also given on a hole chart.

For complicated parts, or a part with a multitude of holes in one or more surfaces, **hole charts** simplify the drawing. In hole charts, the surface of hole entry and each hole are identified on the drawing. In Figure 11.78, the hole chart lists the **X** and **Y** position of each hole with the depth (**Z**).

The surfaces of hole entry are identified with the names of the principal views. The order of these views for hole charts is:

1. **top**
2. **front**
3. **right**
4. **left**
5. **bottom**
6. **rear**
7. **auxiliary view.**

The hole chart shows the surface of entry of each hole, the symbol number that identifies each hole, and the number of times each hole is used in this surface. It also gives the complete specification for each hole. Identical holes in a surface are shown by a single symbol number or letter. Hole charts are used for sheet metal details and drilling drawings for printed circuit boards. On parts with very complex hole patterns, the locating dimensions for the holes are shown in the chart as the **X** and **Y** positions in each view; this is called **rectangular dimensioning in tabular form** (Fig. 11.78).

In **X** and **Y** coordinate dimensioning, each hole has a separate identifying symbol. Group holes by giving diameters the

HOLE	FROM	X	Y	-Z
A1	X,Y	90	44	10
B1	"	26	150	30
B2	"		150	30
B3	"	26	26	30
C1	"	64	100	40
C2	"	40	76	40

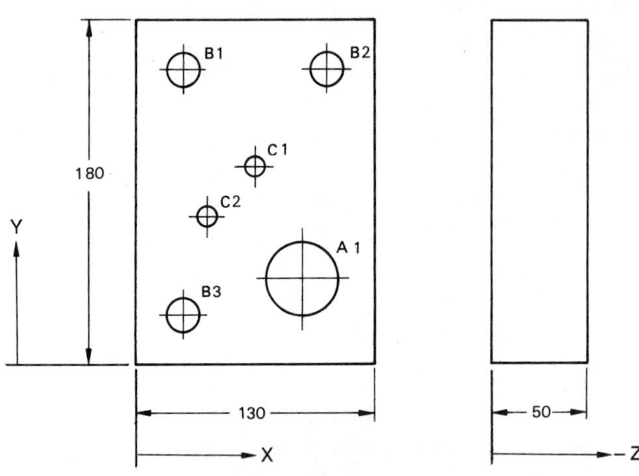

HOLE	DESC	QTY
A	Ø 40	1
B	Ø 20	3
C	Ø 10	2

FIGURE 11.78 Rectangular Coordinate Dimensioning in Tabular Form

same size and the same letter symbols or by numbering them consecutively. All holes are listed in the hole chart. Holes are normally listed alphabetically starting from the largest with the letter **A**. In another method of labeling holes for tabular dimensioning, each hole is numbered consecutively from number 1 without regard for size.

When the hole is completely through the part, **THRU** is used as the **Z** dimension. If more than one surface is to have holes called out, **X** and **Y** axes are established for each surface or view. The depth is specified for each hole, and the view is noted in the hole chart.

Tabular dimensioning is also found in many catalogs where a standard part has varying dimensions for size and length. Bolts, screws, keys, pipe fittings, valves, and other standard items have dimensions in tabular form (see Appendix C).

11.10.5 Repetitive Features or Dimensions

Repetitive features or dimensions are specified with an "X" following a numeral to indicate the "number of times" or "places" that a feature is required. Features such as holes and slots, which are repeated in a series or pattern, are specified by giving the required number of features and an X followed by the size dimension of the feature. A space is used between the X and the dimension (Fig. 11.79).

If it is difficult to distinguish between the dimension and the number of spaces, one space is dimensioned and identified as a reference (1.00 in Figure 11.79). *Reference dimensions* are enclosed in parentheses.

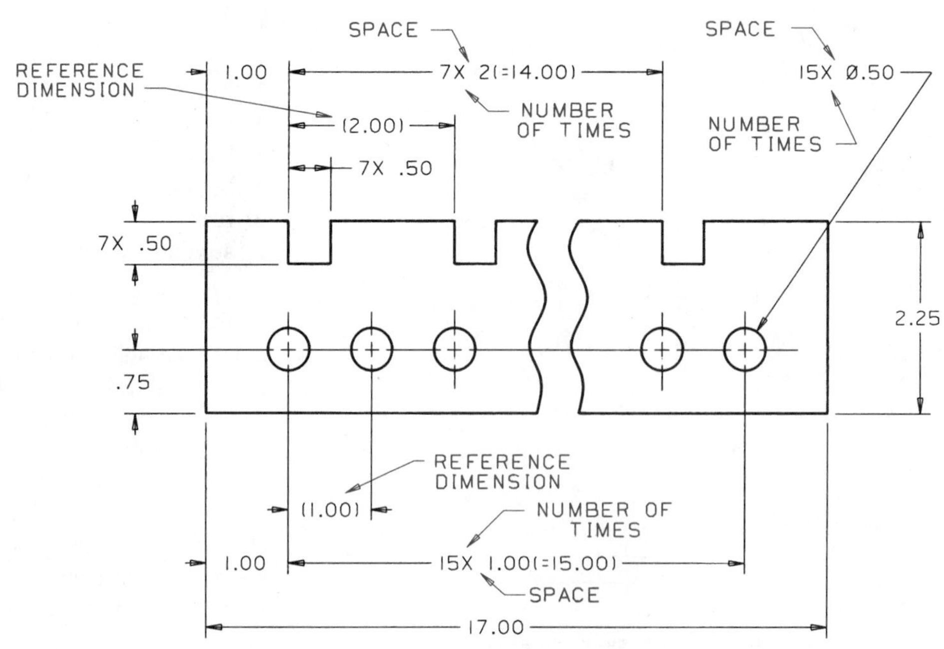

FIGURE 11.79 Repetitive Feature Dimensioning

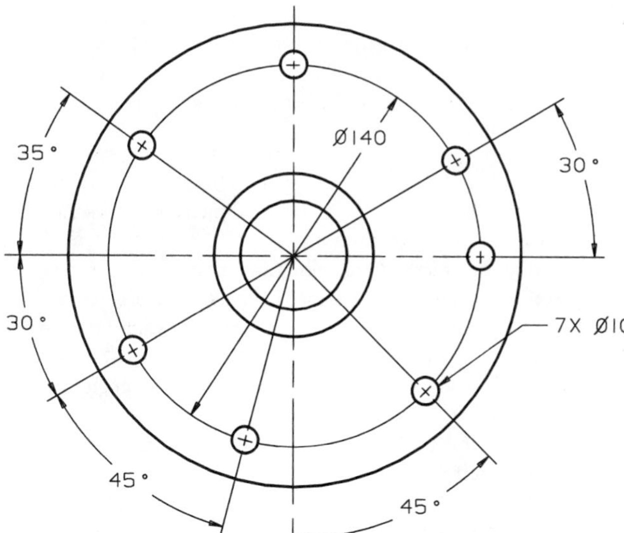

FIGURE 11.80 Dimensioning Repetitive Holes on a Common Center

The part in Figure 11.80 has repetitive features (holes) that are not equally spaced. Angle dimensions in degrees locate each hole from the vertical or horizontal centerline. A note gives the size and number of holes.

The notation X may also be used to indicate BY between coordinate dimensions. An example is when dimensioning chamfer, `50° X 45` or `30° X .250`. *The X is preceded and followed by a space.* If both these practices (BY and "number of features") are used on the same drawing, ensure that each usage is clear by providing proper spaces.

You May Complete Exercises 11.13 Through 11.16 at This Time

11.11 Basic Dimensioning Rules and Drawing Checklist

The following list is provided as guide to dimensioning and drawing. This list is by no means complete:

1. Give the dimensions that will be used to fabricate the part.
2. Make all figures legible; a misread dimension can result in an error in fabrication.
3. Do not crowd dimensions around the part. Allow space for the dimensions in their proper location by planning for dimensions in the layout stage of the project.
4. Do not dimension on the part unless it is absolutely unavoidable. Use extension lines and whenever possible keep figures off views. Place dimensions outside a view.
5. Use proper lettering techniques (with guidelines) for all lettering.
6. Dimension the views that show the characteristic shape and prominent features of each part.
7. Place dimension values so that they can be read from the bottom of the drawing unless it is for one of the construction engineering fields.
8. Do not use a part line as a dimension line. Object lines are used as extension lines only if unavoidable.
9. Dimension lines are located so they do not cross extension lines by placing the largest dimensions outside of smaller dimensions.
10. Never cross two dimension lines.
11. Place parallel dimensions equally spaced and the numerals staggered to avoid confusion on the drawing.
12. Give locating dimensions to the centers of circles that represent holes, cylindrical features, bosses, and slots.
13. Group related dimensions on the view where the contour of a feature is prominent.
14. Arrange a series of dimensions in a continuous line, i.e., chain dimensions.
15. Dimension from a machined (finished) surface, a centerline, or a datum (baseline) that is easily established during manufacturing.
16. Do not repeat dimensions for the same feature (double dimension).
17. Make dimensioning complete so that it is not necessary for manufacturing or inspection to add or subtract to obtain a needed dimension or manually scale the drawing.
18. Provide the diameter of a circle, never the radius.
19. Dimensions are required by the production method. Parts with radial ends will have diameters and center-to-center dimensions.
20. Dimension to limit the tolerance buildup and maintain ease of manufacture.
21. Dimension so that mating parts will fit in the worst case of tolerance buildup on the part.
22. When all dimensions are in inches, the inch symbol is generally omitted, except for construction drawings.
23. The radius of an arc should be provided and the abbreviation placed before the dimension.
24. If possible, avoid dimensioning to a hidden line.
25. Avoid dimensioning on sectioned areas of the part.
26. Use a note to establish repetitious features of a part, e.g., fillets with the same radius.

QUIZ

True or False

1. Dimensioning is not as important as a graphically correct drawing.
2. Holes should be called out with a note giving the radius of the hole and its depth.
3. Center drills are used to hold a workpiece between centers on a lathe.

4. Dual dimensioning is used on most drawings in the United States.
5. Simplified methods for showing threads should be used on metric drawings only.
6. The diameter symbol always follows the size dimension.
7. Symbols can be used when calling out counterbores, spotfaces, and counterdrills.
8. Leaders are always drawn radially from a curved feature when placing a local note.

Fill in the Blanks

9. Fractions are used on ________, ________, and ________ drawings in the United States.
10. Angles can be called out as ________ angles or ________, ________, and ________ .
11. SR is used to define ________ ________.
12. ________ dimensions are enclosed with parentheses.
13. Chamfers can be specified by ________ or ________ .
14. There are two types of knurling: ________ and ________ patterned.
15. ________ ________ ________ locates the features of a part by providing dimensions from two or three mutually perpendicular planes.
16. A ________ ________ does not go all the way through a part.

Answer the Following

17. What are the four types of linear dimensions? Describe the process of placing each on a part.
18. Describe four methods of calling out a taper.
19. Describe the process of geometric breakdown of a part.
20. Why are mating parts and mating dimensions important when dimensioning a part?
21. What is the difference between a radial and a diameter dimension and when should each be used?
22. What is a finish mark and why is it important when dimensioning a part?
23. Explain what NOTES are used for on a drawing. What is a local note and what is a general note?
24. Why are too many dimensions on a drawing a problem?

EXERCISES

Exercises may be assigned as sketching, instrument, or CAD projects. Transfer the given information to an "A" size sheet of .25 in. grid paper. Complete all views and solve for proper visibility, including centerlines, object lines, and hidden lines. Exercises that are not assigned by the instructor can be sketched in the text to provide an understanding of the preceding instructional material. Complete the views and add hidden lines where required.

After Reading the Chapter Through Section 11.5.5, You May Complete the Following Exercises

Exercise 11.1 Completely dimension the one-view part; it is .25 in. thick.

Exercise 11.2 Dimension the .125 in. thick aluminum plate.

Exercise 11.3 Dimension the two-view part as needed.

Exercise 11.4 Show the proper placement of all dimensions on appropriate views.

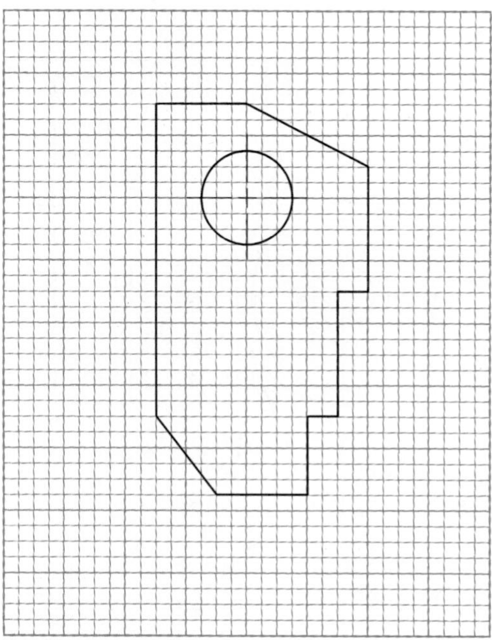

Exercise 11.1

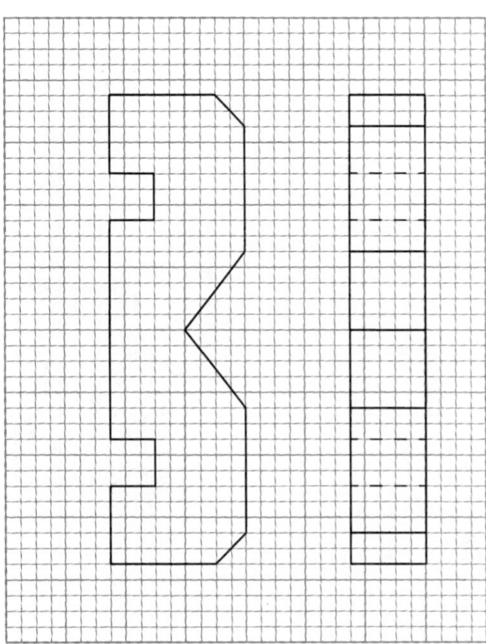

Exercise 11.3

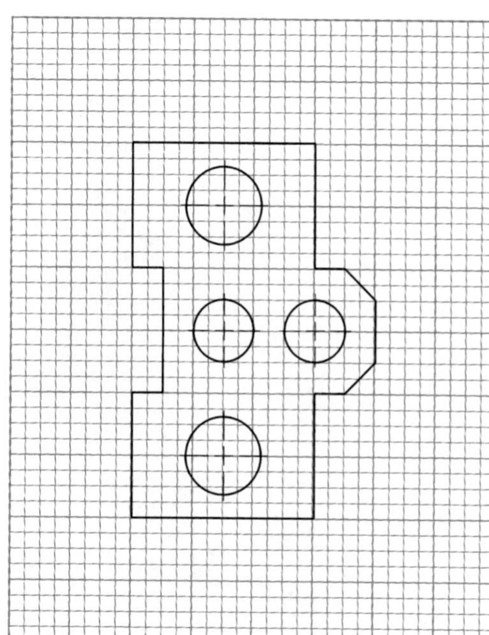

Exercise 11.2

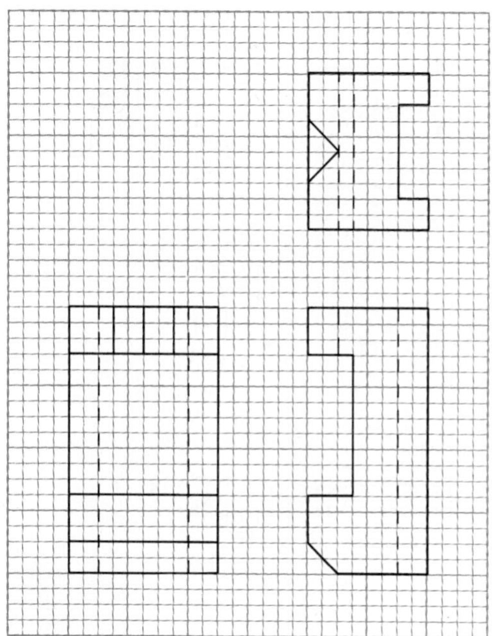

Exercise 11.4

After Reading the Chapter Through Section 11.6.3, You May Complete the Following Exercises

Exercise 11.5 Completely dimension the part as required. Use symbology to call out the spotfaced holes. Place finish marks on the machined faces and dimension accordingly.

Exercise 11.6 Completely dimension the part. The bottom surface is machined.

Exercise 11.7 Dimension the two-view part.

Exercise 11.8 Dimension the three views of the part. The bottom surface, the left side surface, and the boss are the only finished surfaces. Add appropriate fillets and rounds for the cast surfaces (at the top and around the boss). Put basic finish marks on machined surfaces.

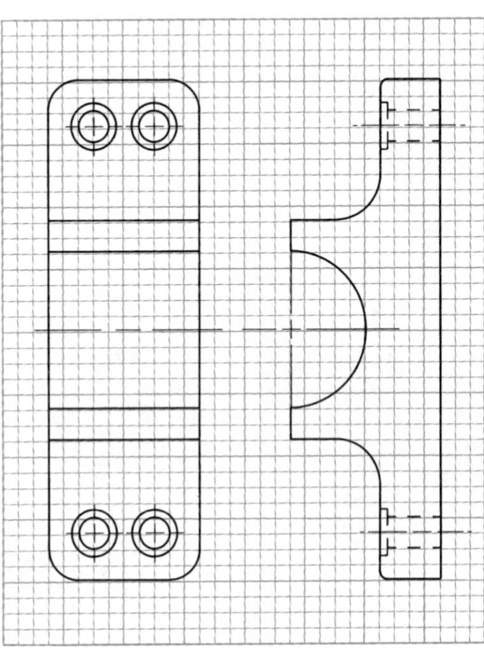

Exercise 11.5

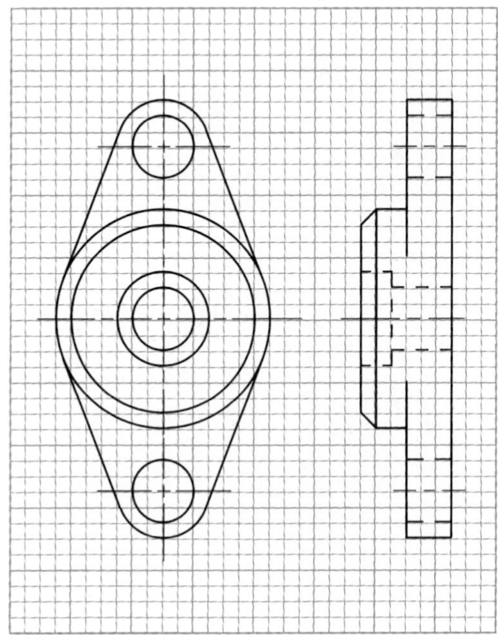

Exercise 11.7

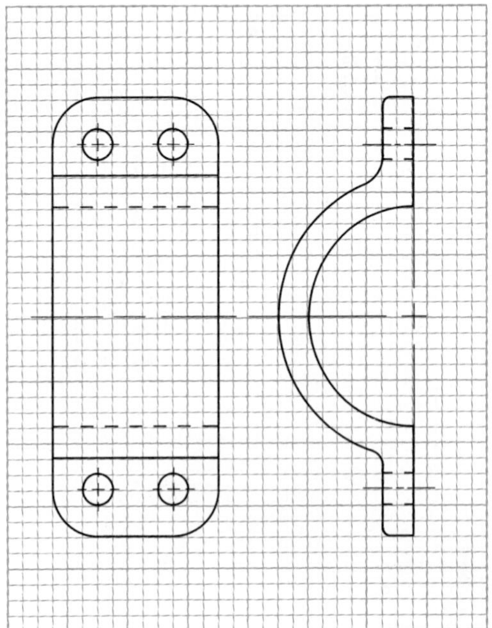

Exercise 11.6

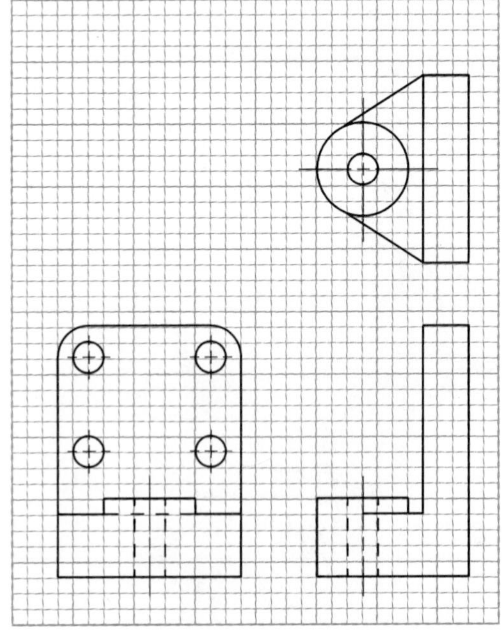

Exercise 11.8

After Reading the Chapter Through Section 11.8.3, You May Complete the Following Exercises

Exercise 11.9 Completely dimension the cast part. Add appropriate fillets and rounds to cast surfaces. Place finish marks on machine surfaces. The left surface and bottom surface are machined along with the upper U-shaped surface. All other surfaces are cast.

Exercise 11.10 Because of space limitations, dimension only the hole pattern and holes (call out the bolt circle), the slots, and the chamfer. Make sure all views are visually correct.

Exercise 11.11 Completely dimension the part. Complete the views for proper visibility. Call out the knurling with a note. The small hole goes through to the center hole only.

Exercise 11.12 Because of limited space, dimension only the hole pattern, slot, and the counterdrilled holes. Use symbology to call out the holes.

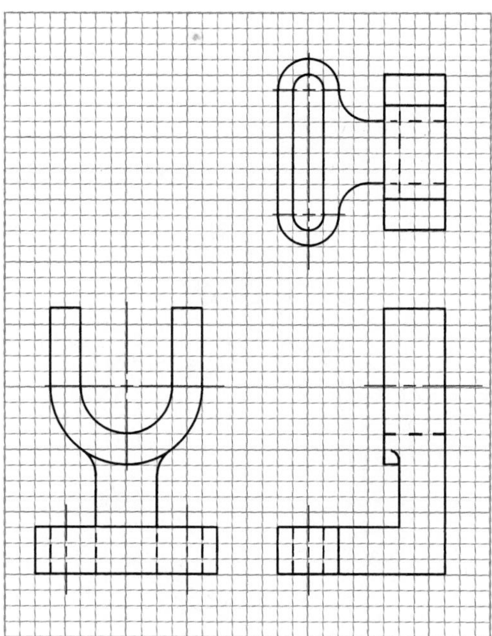

Exercise 11.9

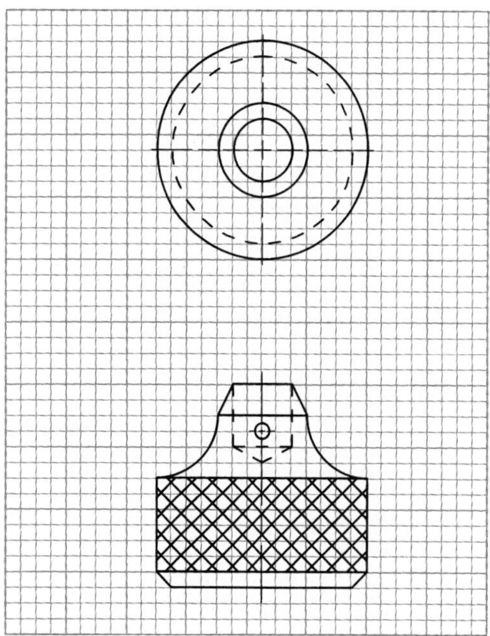

Exercise 11.11

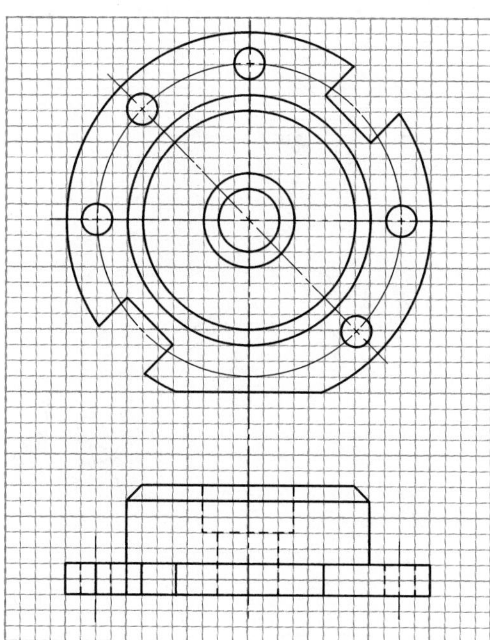

Exercise 11.10

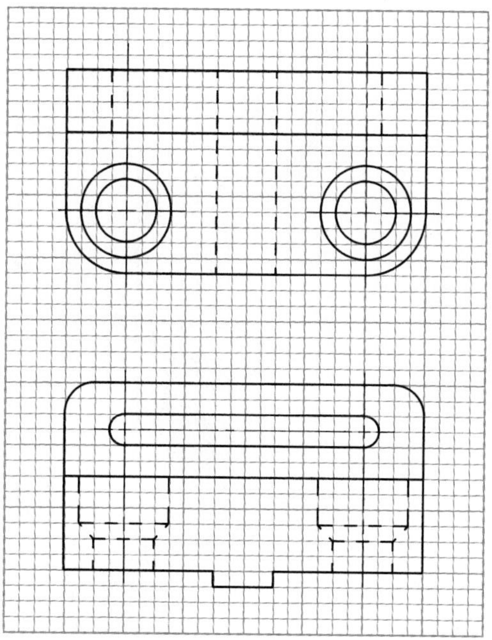

Exercise 11.12

After Reading the Chapter Through Section 11.10.5, You May Complete the Following Exercises

Exercise 11.13 Dimension the taper with a callout. Because of space limitations, dimension only the lateral length dimensions, not the diameter dimensions of the rest of the part.

Exercise 11.14 Place a #4 center hole for a centerdrill on both ends of the workpiece. Dimension the entire part and call out the centerdrill with a note. Complete the views.

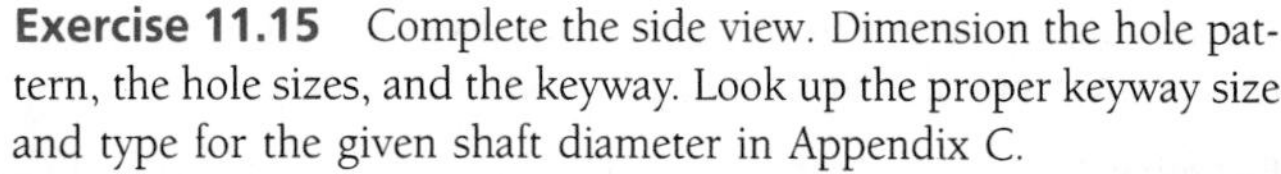

Exercise 11.15 Complete the side view. Dimension the hole pattern, the hole sizes, and the keyway. Look up the proper keyway size and type for the given shaft diameter in Appendix C.

Exercise 11.16 Complete the side view of the part. Dimension the holes with a callout using symbology. Call out the keyway based on the shaft diameter. Dimension only one "ear" of the part.

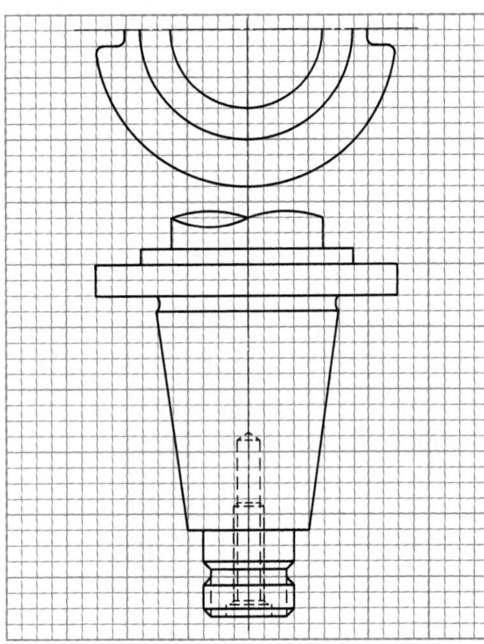

Exercise 11.13

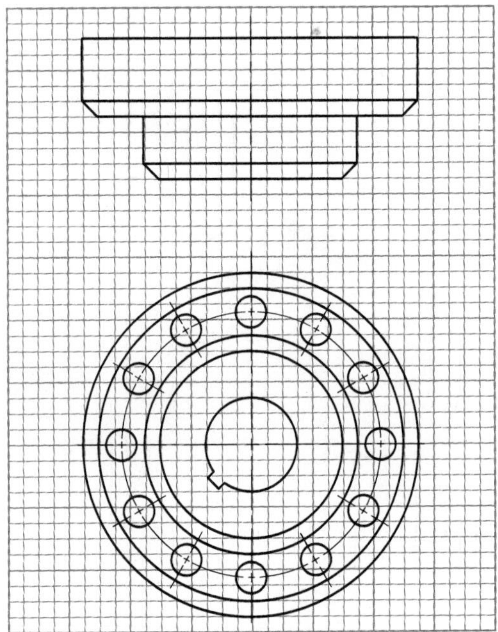

Exercise 11.15

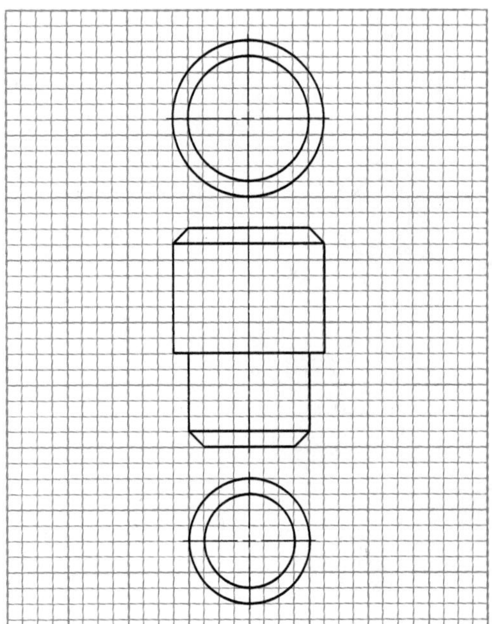

Exercise 11.14

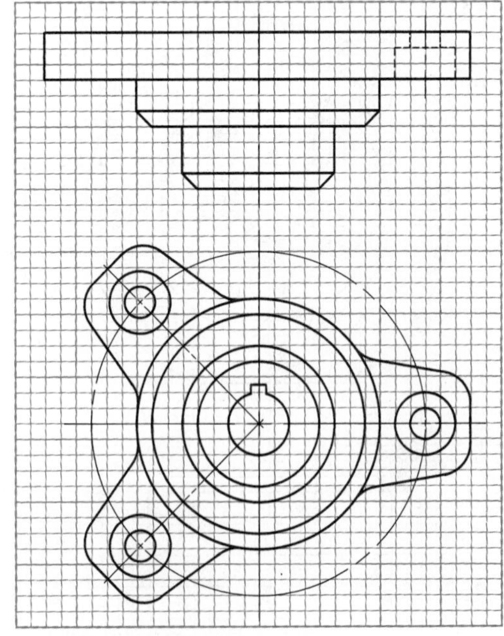

Exercise 11.16

PROBLEMS

Your instructor can assign any of the figures presented in this chapter as problems. For every part, redraw and dimension using the most recent ANSI standards.

CHAPTER 12

Geometric Dimensioning and Tolerancing

Learning Objectives

Upon completion of this chapter, you will be able to accomplish the following:

1. Become familiar with general and geometric tolerancing rules, symbology, and modifiers.
2. Recognize ANSI and ISO interpretations of angle of projection and limits of size.
3. Identify feature control frames and contents.
4. Develop an understanding of datums and datum systems.
5. Demonstrate skill in interpreting form, profile, orientation, location, and runout tolerances.
6. Recognize different types of fits.
7. Develop an understanding of standardized limits, fits, and respective calculations.

12.1 Introduction

Features of manufactured parts vary in size, form, orientation, or location. This variation in manufacturing is expected and, as long as it is understood and controlled, the part will perform as designed. You may have tried to assemble a consumer product sometime and found that holes did not line up between parts or that a hole for a bolt was not drilled perpendicular to the surface. The assembly process was no doubt frustrating and the resulting product may not have performed up to expectations without modifications. *Geometric dimensioning and tolerancing (GD&T) is a symbolic system of tolerancing to control the size, form, profile, orientation, location, and runout of a part according to geometry.* Cost-effective designs provide the largest allowable **tolerances** consistent with the function and interchangeability requirements of the design.

Even though technical drawings have been used to communicate engineering information for more than 6,000 years, the concept of tolerancing or holding variations within limits, has been used for only about 100 years. At one time, part variations were controlled by the worker rather than by engineering. It was not until the evolution of interchangeable manufacture that the *exact* size gave way to holding parts within *limits*. The *Taylor concept,* introduced in 1905 and still used today, introduced methods of limit gaging for holes and shafts. Increased production rates during World War II from larger factories that used a wider variety of suppliers created a high rate of scrap that sometimes hampered wartime requirements. Inadequacies in technical drawings for conveying this information became apparent. In 1945, the Gladman papers were published in Great Britain and these issues of drawing inadequacies were discussed at the first American, British, and Canadian Conference on the Unification of Engineering Standards.

Unfortunately, geometric tolerancing was used only partially in the 1950s and 1960s. In 1972, the International Organization for Standardization (ISO) established a separate subcommittee to develop dimensioning and tolerancing standards. During the 1970s and 1980s, GD&T was used extensively in industry and by the military. ANSI and ISO developed standards to ensure universal interpretation of tolerance requirements on drawings. The drawing in Figure 12.1 shows GD&T applied to a particular part. As product cycle times decrease and quality demands grow in the ever more competitive global marketplace of the 1990s, GD&T will play an increasingly important role in meeting those demands.

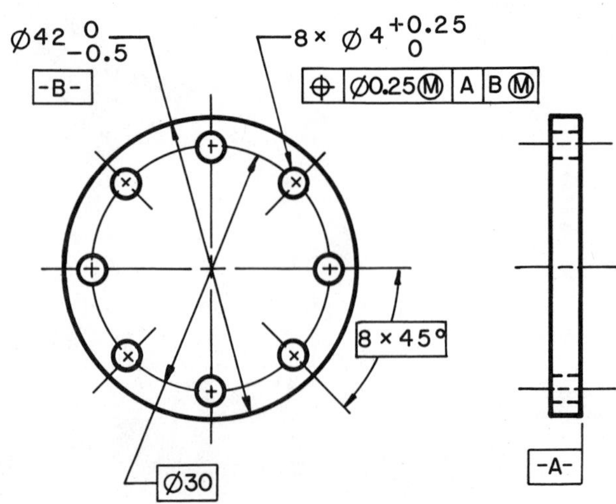

FIGURE 12.1 A Drawing That Uses Geometric Tolerancing

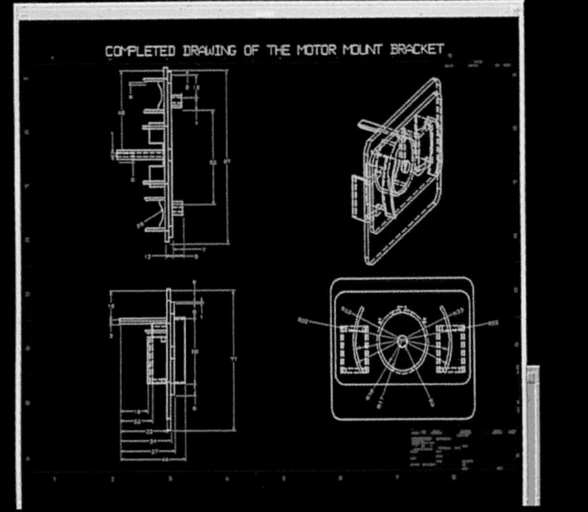

PLATE 1
A multiview drawing and a model of a motor mount bracket are shown in this plate.

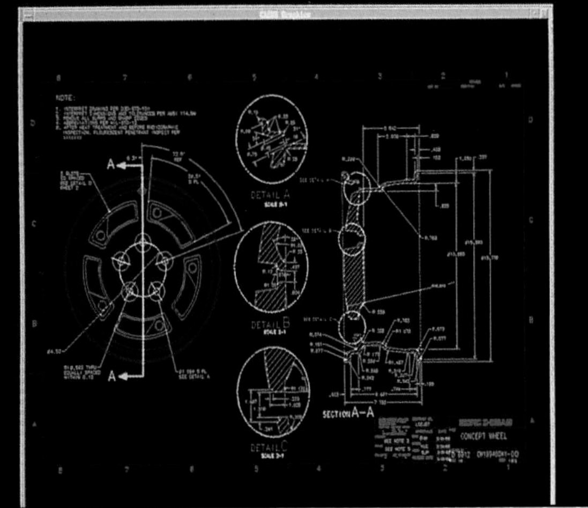

PLATE 2
This multiview drawing shows detail drawings of complex dimension regions in a part.

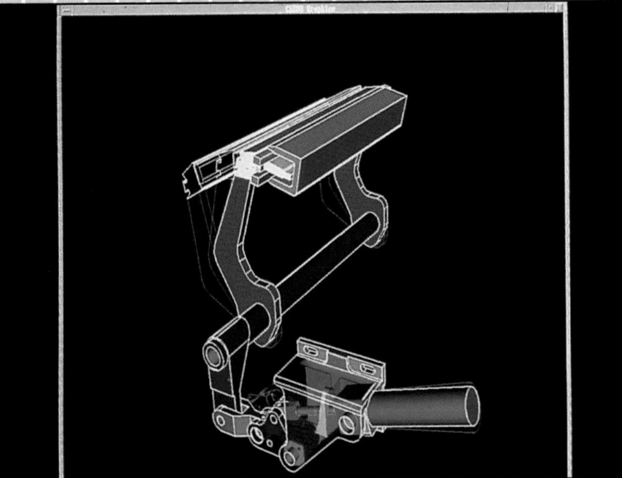

PLATE 3
The relationship and orientation of the components are shown in this solid model of a complicated mechanical assembly.

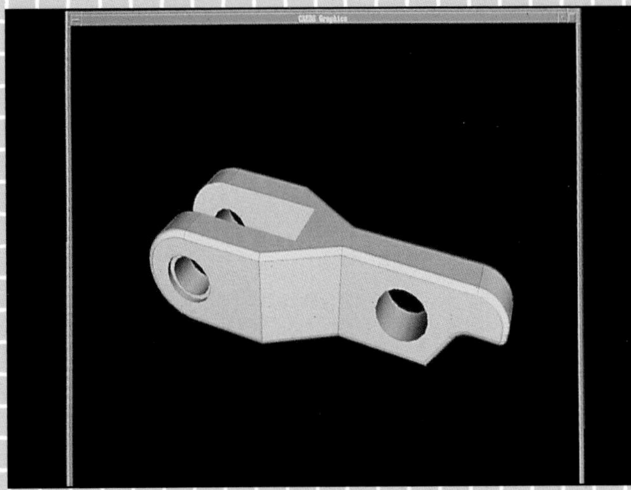

PLATE 4
This plate shows a solid model of a component in the assembly.

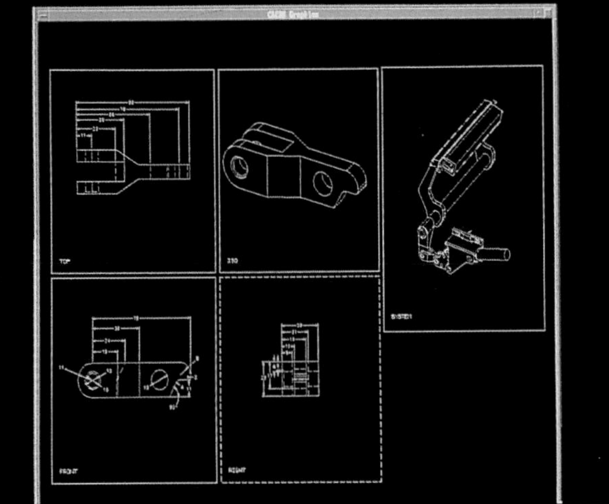

PLATE 5
A model of the part, the wire frame assembly and a multiview drawing of a component in the assembly are shown in this plate.

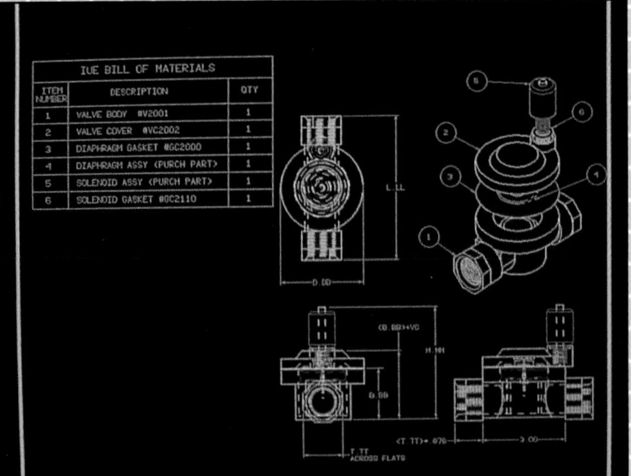

PLATE 6
A "Bill of Materials" list is shown in this multiview drawing.

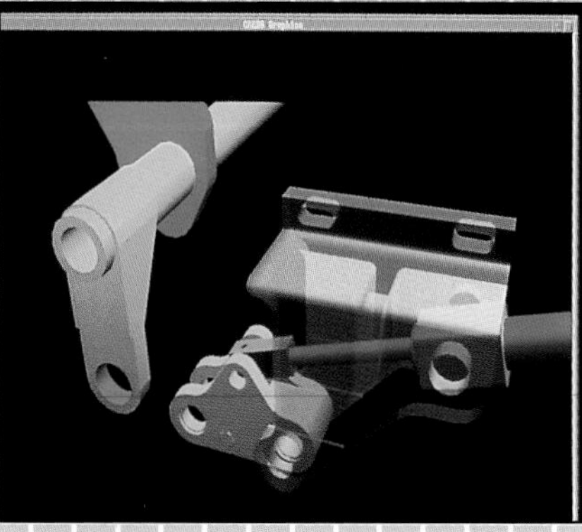

PLATE 7
A solid model of a complex mechanical assembly is shown in this plate.

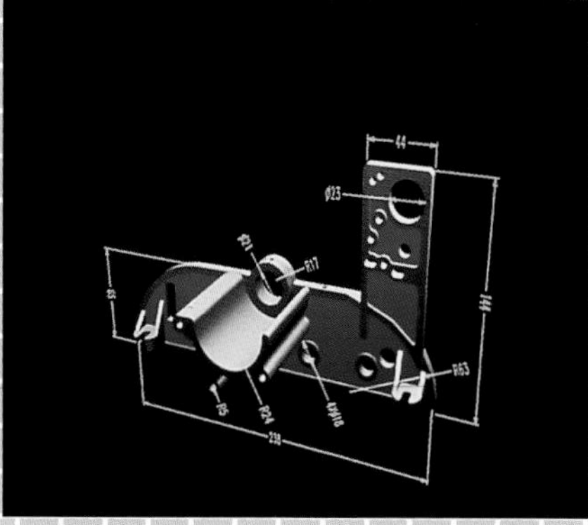

PLATE 8
This solid model of an assembly shows the relationship between parts and sizes.

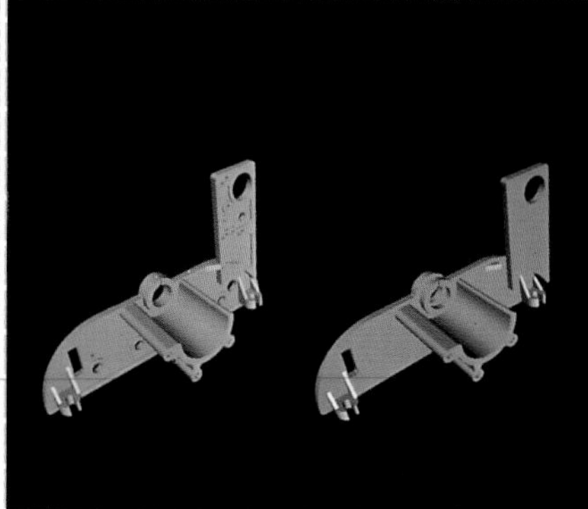

PLATE 9
Details and features of the mechanical part are shown in these solid models.

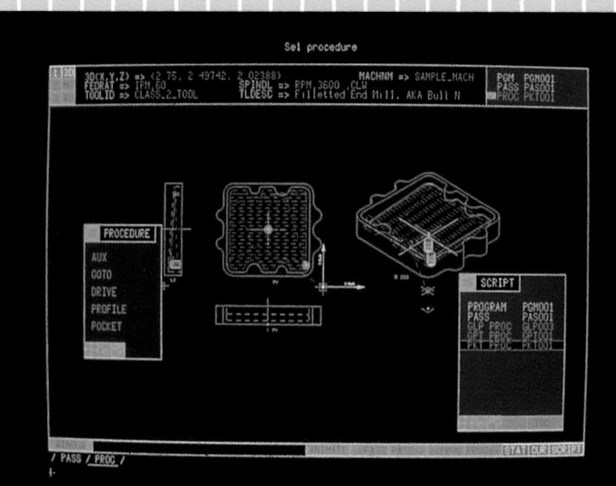

PLATE 10
The tool paths that are used to create a part in CNC milling are shown here.

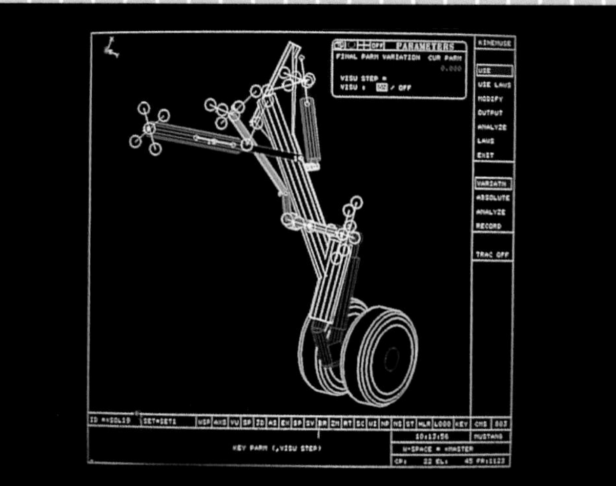

PLATE 11
This is a wire frame model of a mechanical assembly. Parameters in the model may be varied to produce different size components.

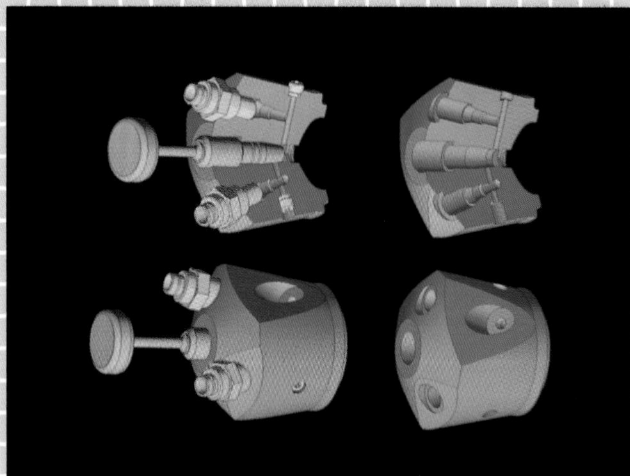

PLATE 12
Once the geometric model has been created, you can "explode" the valve housing model to show component parts, rotate the model to show different views, create a cross-section model, or check for interference with related parts.

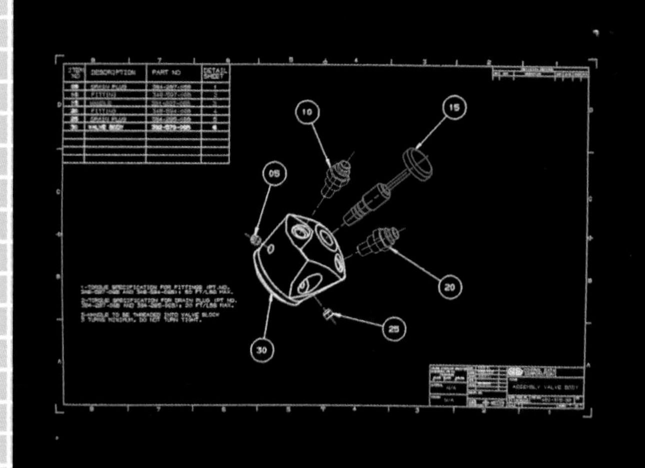

PLATE 13
Wire frame models may be created automatically from the solid model. Exploded views, assemblies, hidden line views, and design drawings can all be generated from the model as well.

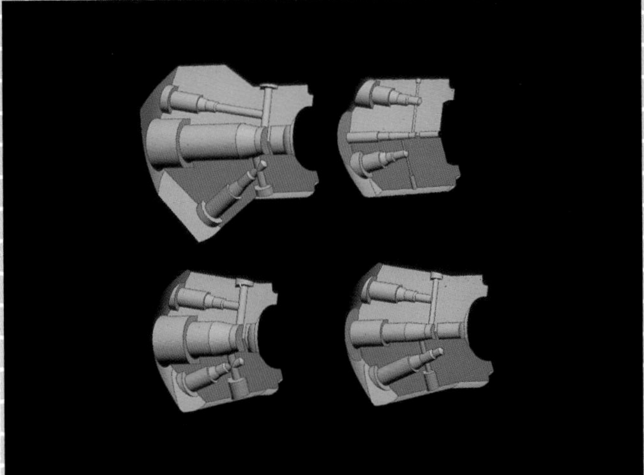

PLATE 14
Families of parts can be automatically generated by the solid modeler by manipulating dimension parameters.

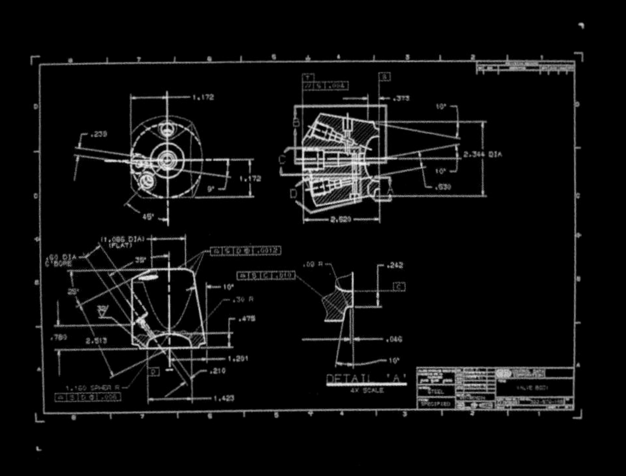

PLATE 15
Drafting and detail documentation for the part is produced from the solid model database.

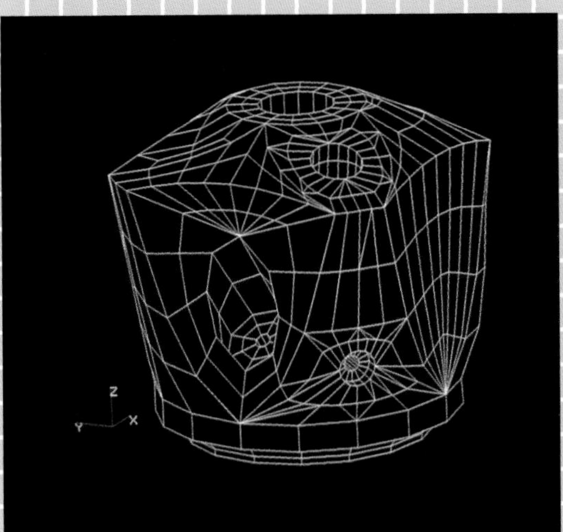

PLATE 16
A finite element model for analysis is created from the solid model database to ensure design requirements are met.

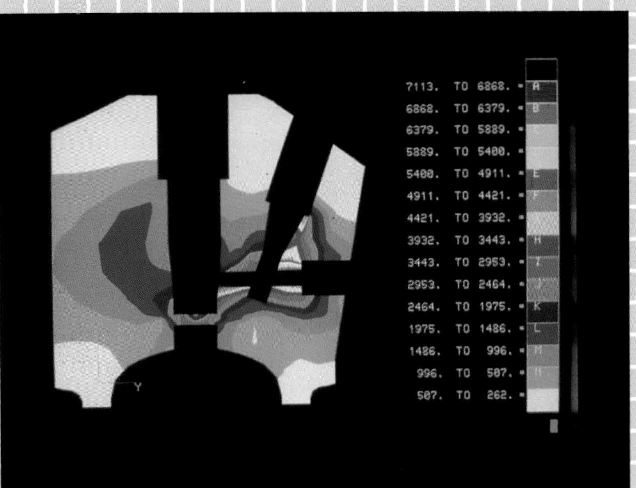

PLATE 17
Color schemes depict the results in a finite element analysis.

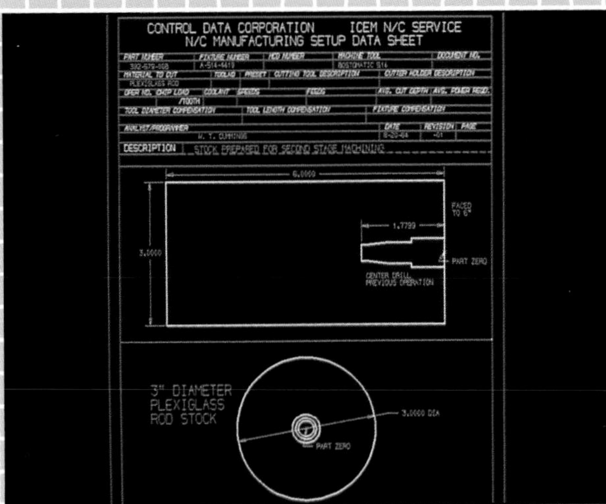

PLATE 18
Numeric control (NC) capability produces control tapes directly from the solid model design geometry. The NC output is used to machine the part.

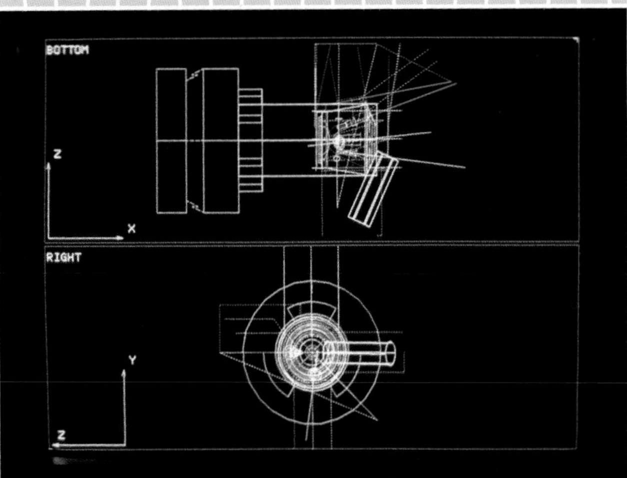

PLATE 19
Cutter paths for NC machining may be previewed before part machining to obtain optimum cutter path design.

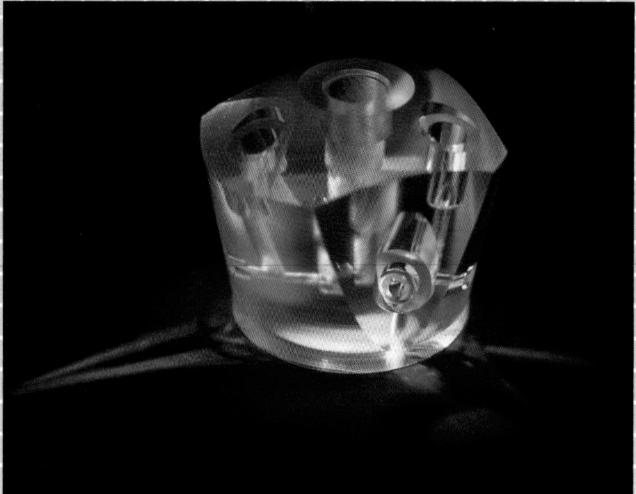

PLATE 20
The finished valve housing is shown in this plate.

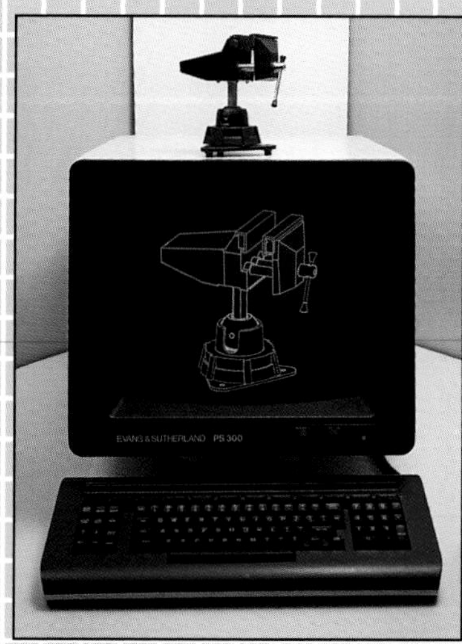

PLATE 21
This display screen shows a vise assembly.

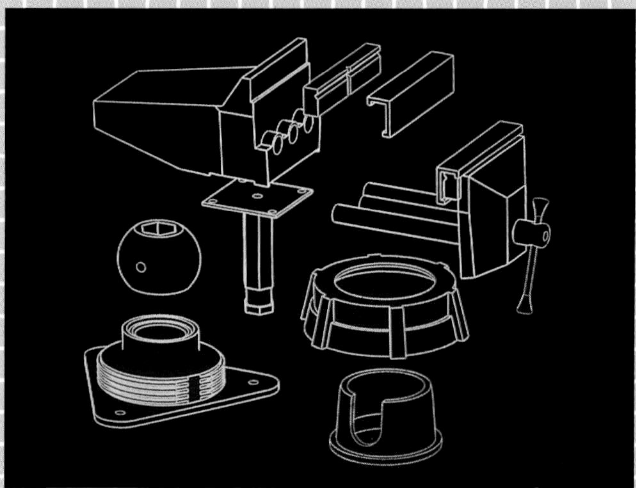

PLATE 22
An exploded view of the vise assembly is used to depict components of the assembly.

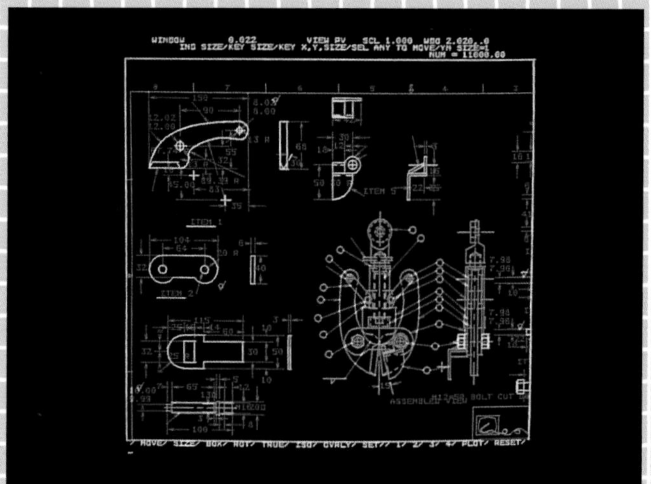

PLATE 23
An assembly and detail drawing of a mechanical part is shown here.

PLATE 24
This plate shows how a 2D concept may be used to create or modify a 3D model in a solid modeling design system.

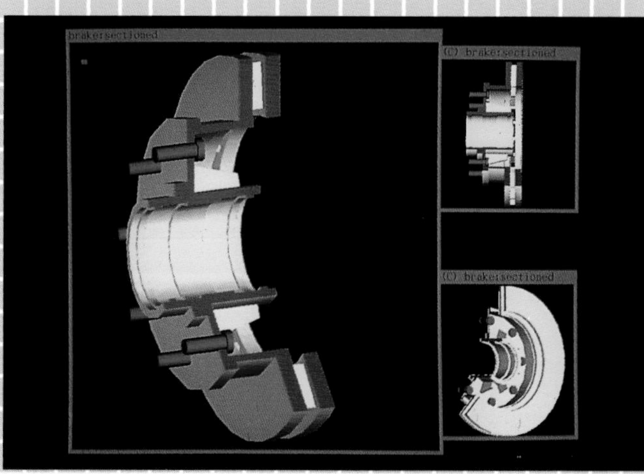

PLATE 25
AutoSolid® was used to create this design model.

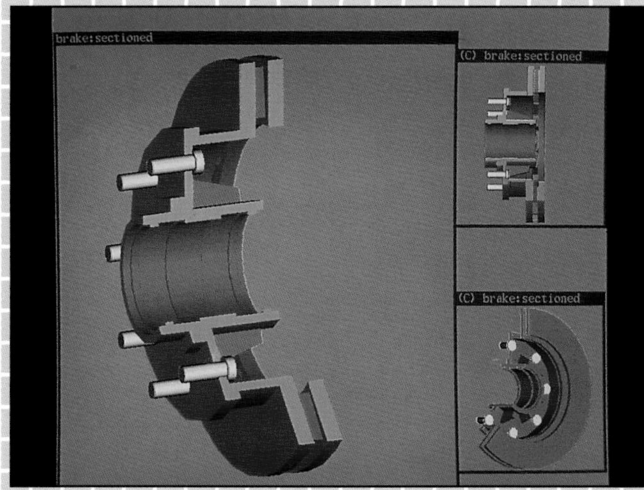

PLATE 26
A sectioned 3D solid model is illustrated in this plate.

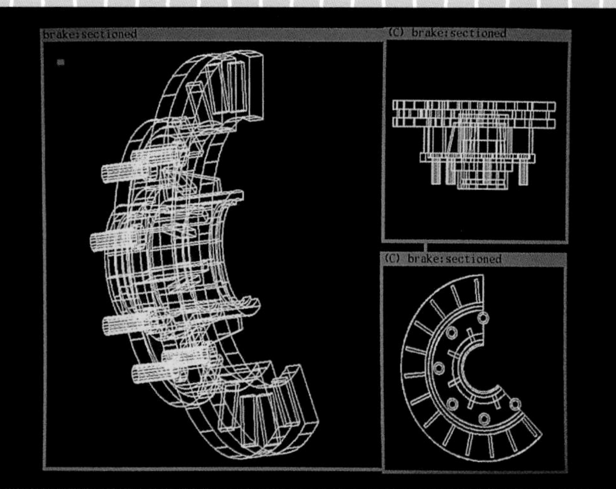

PLATE 27
This plate shows a wire frame display of a part.

PLATE 28
A solid model of an assembly is illustrated here.

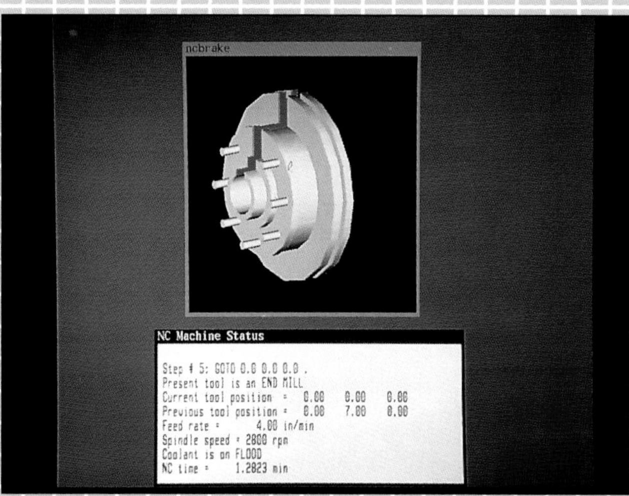

PLATE 29
The status of the NC machine may be displayed with the model in NC machining.

PLATE 30
A solid model design of a planetary gear system is depicted in this plate.

PLATE 31
AutoSolid® was used to shade this model of a gear design.

PLATE 32
This plate shows the interior view of a British rail car.

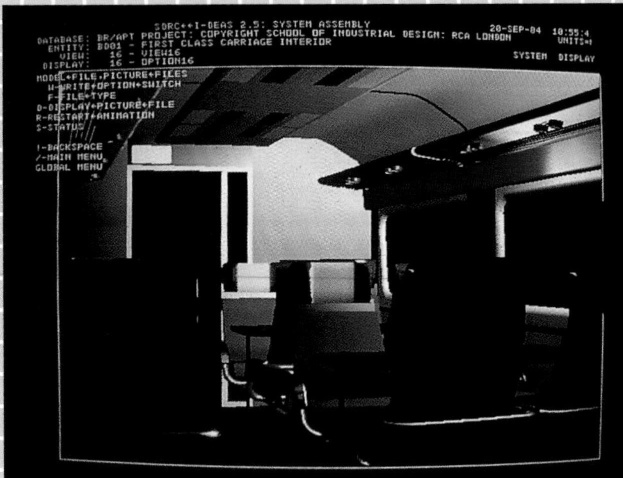

PLATE 33
The same rail car interior, but from a different angle than Plate 32, is shown here.

PLATE 34
An individual seating module design is shown.

PLATE 35
There are many color and fabric options for the seat. This plate shows the design with one choice of color and one choice of the fabric options.

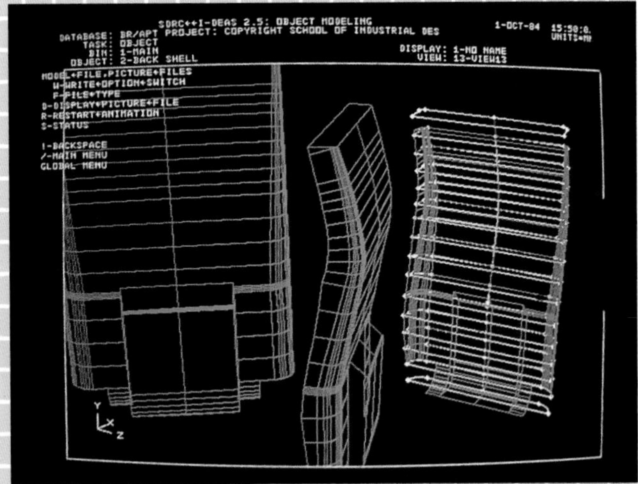

PLATE 36
The boundary display of the back support is shown here.

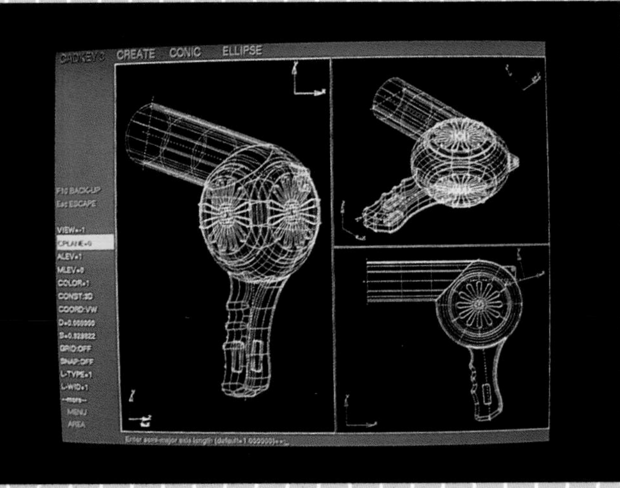

PLATE 37
This plate is a 3D wireframe design model of a hair dryer.

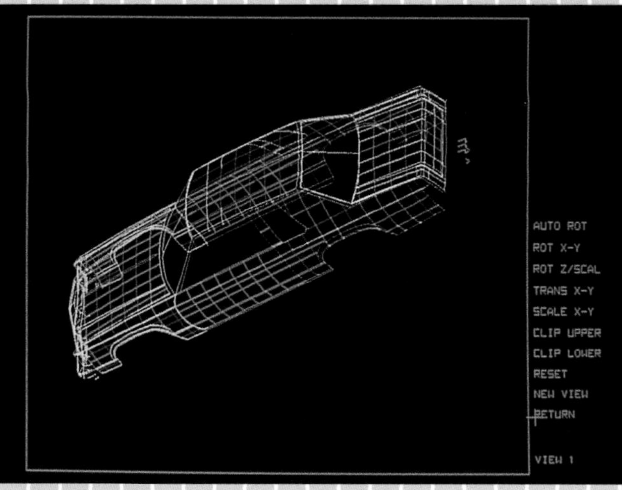

PLATE 38
A wireframe model of a car is shown here.

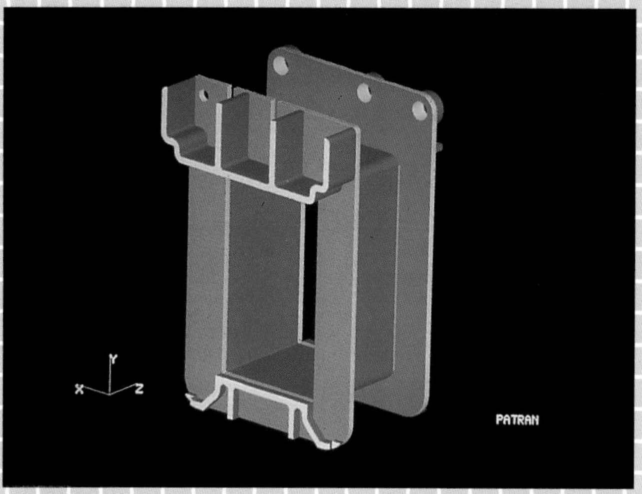

PLATE 39
A shaded model of a 3D part often shows the best visual display of the component.

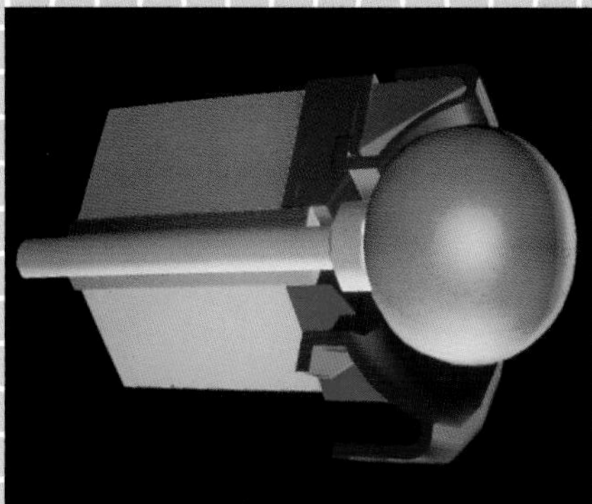

PLATE 40
The shaded model of a peach pitter assembly shows the components.

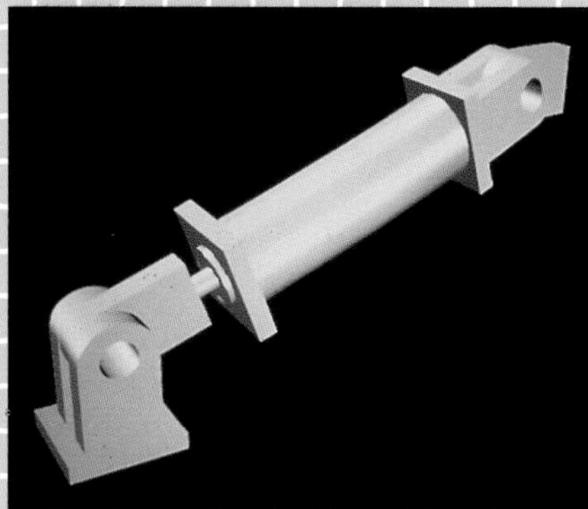

PLATE 41
Kinematic relationships that are essential for mechanical design are shown by a 3D modeler.

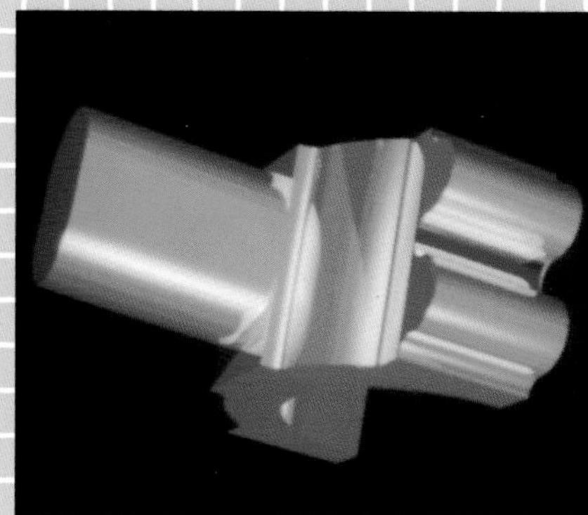

PLATE 42
This plate shows a solid model of a mechanical part.

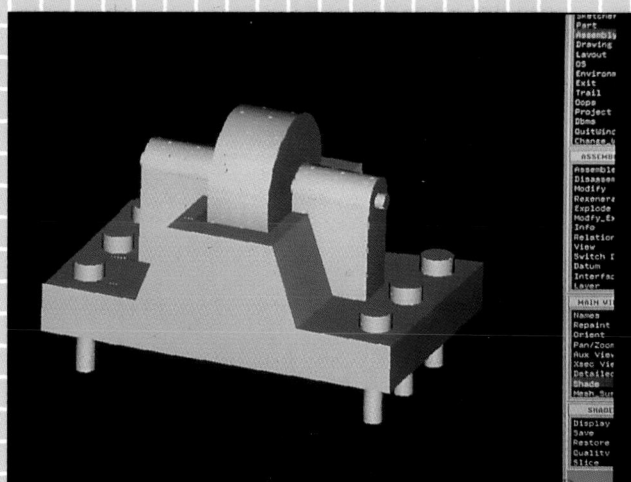

PLATE 43
A shaded 3D solid model of an assembly is shown.

PLATE 44
This display shows a solid model assembly.

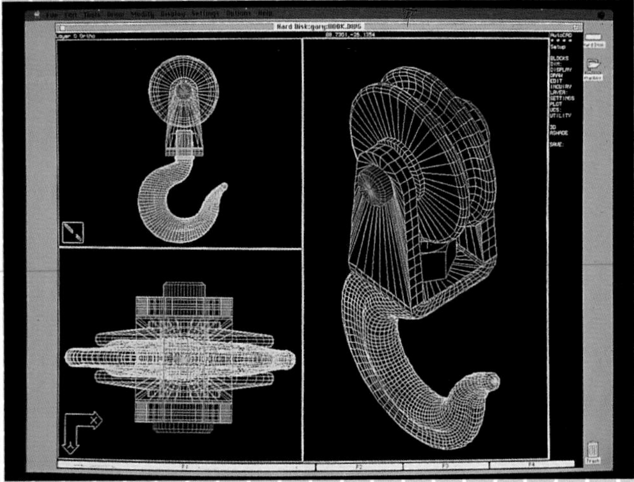

PLATE 45
Another 3D assembly model is shown in this plate.

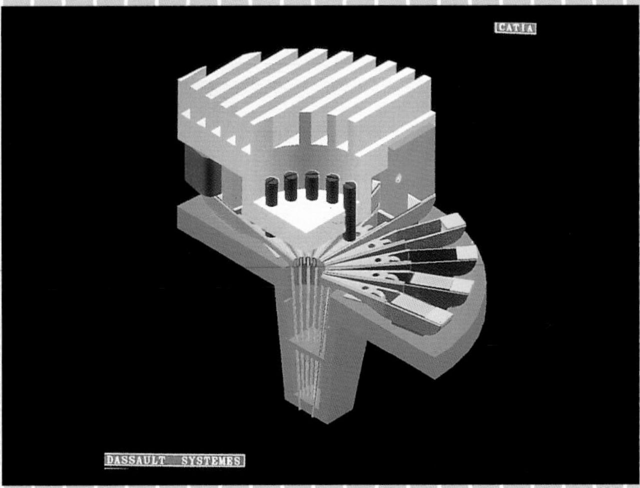

PLATE 46
Complex assemblies are easier to visualize and design in a solid modeler.

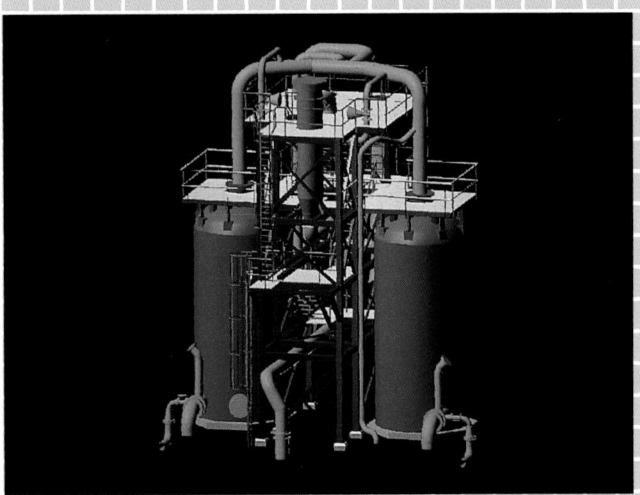

PLATE 47
Complex piping assemblies are easier to visualize as solid models.

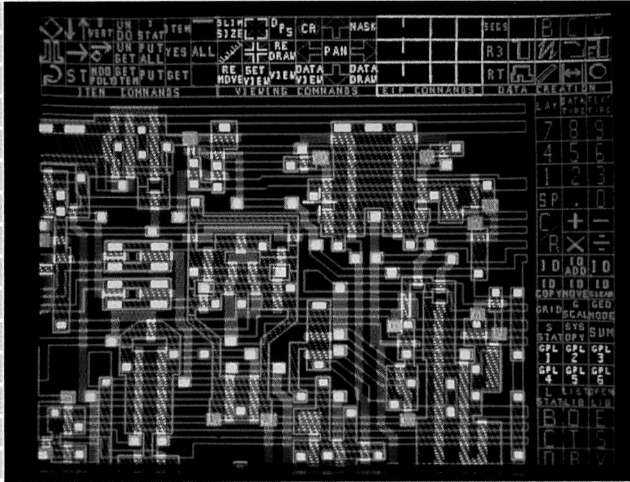

PLATE 48
The design of an integrated circuit (IC) is shown in this plate.

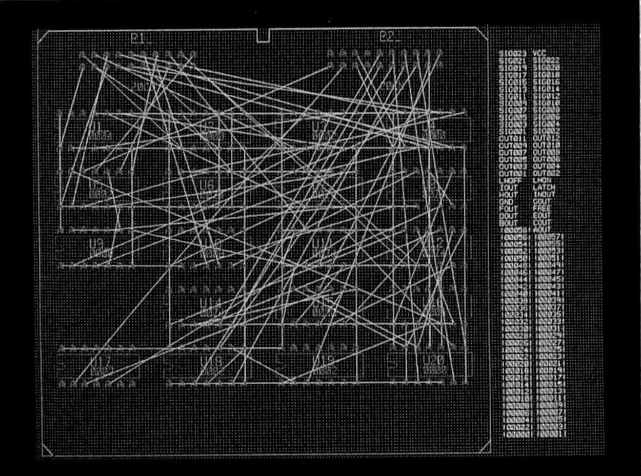

PLATE 49
Automatic routing of the traces in a circuit board simplifies the design of the board.

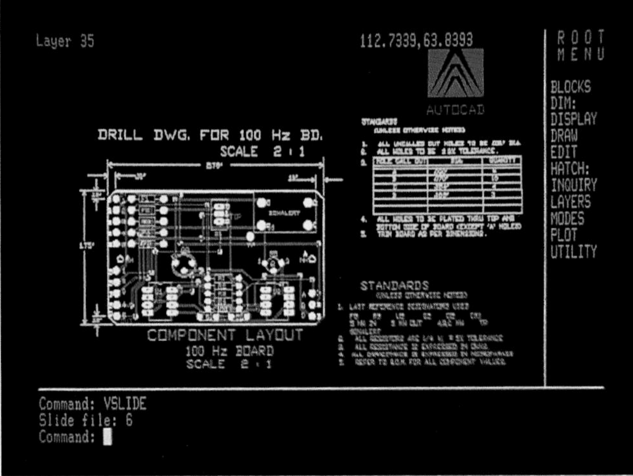

PLATE 50
A printed circuit board (PCB) assembly is shown here.

ITEMS OF INTEREST

Tolerancing and Its Role in Industry

Mass production of interchangeable parts played an important role in the Industrial Revolution. Much of the technology that we enjoy today also relies on interchangeable components. Automobiles and computer circuits are good examples of mass production and the importance of size control.

While it is impossible to make any part exactly the same size as another part, it is possible to keep component dimensions to a specific range of sizes. Geometrical relationships can also be specified. These dimension restrictions are specified with *tolerances*. Component function determines the degreee of tolerance. This process assures that parts made in one location are interchangeable with parts made in another location.

For example, Eagle Engine Manufacturing produces V-8 engines for top fuel dragsters. The Eagle engine can produce 3,000 hp and is designed to allow for different configurations. Cylinders are interchangeable and the head accommodates one to three spark plugs per cylinder. This means the engine can be configured for a dragster or a tractor. Specific parts for the engine were designed on a CAD system with tolerance capabilities to sixteen decimal places. The design was easily modified to fit another configuration with its tolerance specifications.

Producing components to specific tolerances makes it possible to mass produce goods and modify existing components to fit different needs. This system adds flexibility to the manufacturer, allowing the part to change quickly with market trends and technological advances. This kind of flexibility is essential to compete in the competitive world of today and tomorrow.

12.1.1 Terms Used in Geometric Dimensioning and Tolerancing

The following terms are used throughout the chapter:

Actual size The measured size.

Basic dimension The theoretically exact size, profile, orientation, or location of a feature or datum target. It is the basis from which permissible variations are established by tolerances on other dimensions, in notes, or in feature control frames.

Basic size The size to which limits or deviations are assigned. The basic size is the same for both members of a fit.

Clearance fit The relationship between assembled parts when clearance occurs under all tolerance conditions.

Datum A datum is the origin from which the location or geometric characteristics of features of a part are established.

Datum feature A geometric feature of a part that is used to establish a datum.

Datum target A specified point, line, or area on a part used to establish a datum.

Deviation The difference between the actual size and the corresponding basic size.

Interference fit The relationship between assembled parts when interference occurs under all tolerance conditions.

Lower deviation The difference between the minimum limit of size and the corresponding basic size.

Upper deviation The difference between the maximum limit of size and the corresponding basic size.

Feature The general term applied to a physical portion of a part, such as a surface, hole, or slot.

Feature of size A cylindrical or spherical surface, or a set of two parallel surfaces, each of which is associated with a size dimension.

Least Material Condition (LMC) The condition in which a feature of size contains the least amount of material within the stated limits of size; for example, the maximum hole diameter or the minimum shaft diameter.

Limits of size The specified maximum and minimum sizes.

Maximum Material Condition (MMC) The condition in which a feature of size contains the maximum amount of material within the stated limits of size, for example, the minimum hole diameter or the maximum shaft diameter.

Regardless of Feature Size (RFS) The geometric tolerance or datum reference applies at any increment of size of the feature within its size tolerance.

Tolerance The total amount by which a specific dimension is permitted to vary. The tolerance is the difference between the maximum and minimum limits.

Tolerance, bilateral A tolerance in which variation is permitted in both directions from the specified dimension.

Tolerance, geometric The general term applied to the category of tolerances used to control form, profile, orientation, location, and runout.

Tolerance, unilateral A tolerance in which variation is permitted in one direction from the specified dimension.

Tolerance zone An area representing the tolerance and its position in relation to the basic size not just size.

Transition fit The relationship between assembled parts when either a clearance or interference fit results.

True position The theoretically exact location of a feature established by basic dimensions.

Virtual condition The boundary generated by the collective effects of the specified MMC limit of size of a feature and any applicable geometric tolerances.

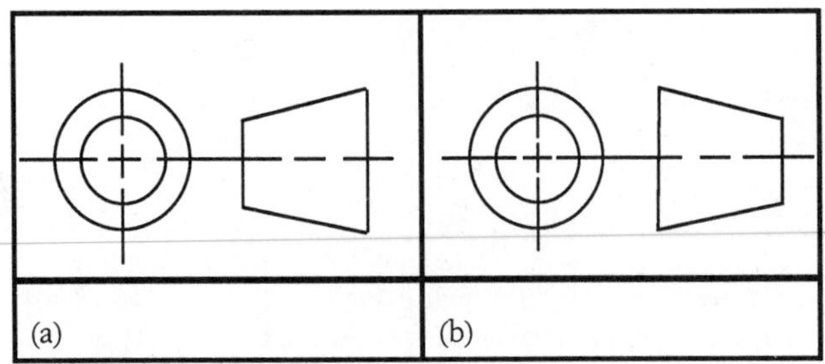

FIGURE 12.2 ANSI and ISO Orthographic Projection Symbols (a) Third-angle projection and (b) First-angle projection

12.2 Standards and Specifications

American National Standards Institute (ANSI) and ISO standards exist to ensure the universal interpretation of tolerance requirements. However, some companies do tailor these standards to meet the needs of their particular product requirements. Also, ANSI and ISO standards are not in complete agreement at this time. To avoid misinterpretation, a note such as "Interpret Drawing in Accordance with ANSI Y14.5M-1982" should be referenced directly on the drawing. The note should contain the standard, the revision, and the revision date.

Recall that ANSI standard drawings in the United States use third-angle projection [Fig. 12.2(a)] and that ISO drawings use first-angle projection [Fig. 12.2(b)]. Although the placement of the views is different, the resulting views are the same. However, limits of size are defined differently for the

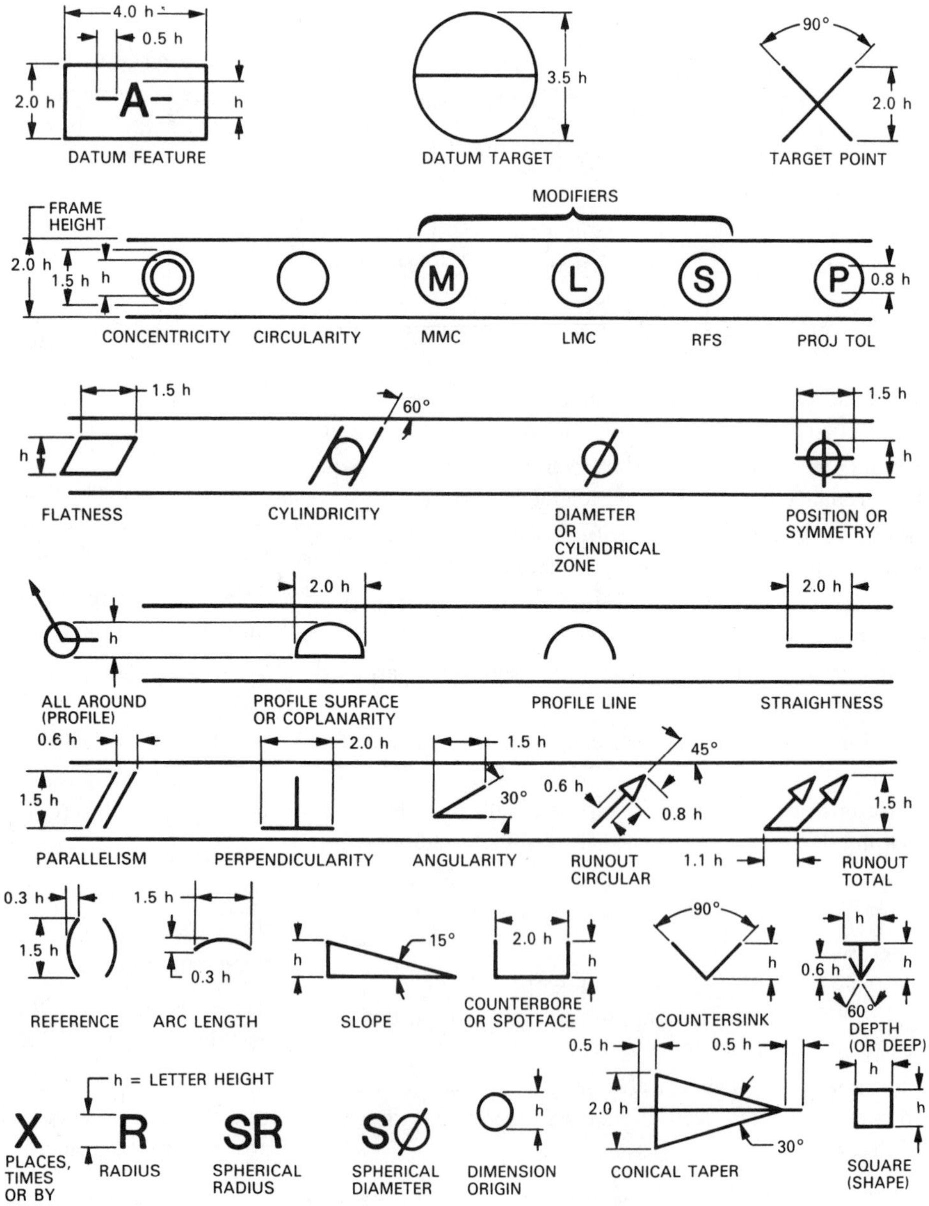

FIGURE 12.3 Geometric Tolerancing Symbols (sizes)

two standards. This book concentrates on ANSI standard tolerancing techniques. However, this chapter also describes ISO techniques.

12.3 Symbology

Geometric dimensioning and tolerancing is a **symbolic system** used to control economically function, interchangeability, size, form, profile, orientation, position, and runout of features or parts and to establish datums and other necessary tolerancing practices. This section describes the symbols for specifying geometric characteristics and other dimensional requirements on engineering drawings. Symbols should be of sufficient clarity to meet the legibility and reproducibility requirements of ANSI Y14.5M.

Symbols are preferred to notes because they use less space, overcome language barriers, and are less subject to interpretation. Most individual symbols not only represent an entire standardized engineering concept, but are combined in a **feature control frame** in various combinations to form complete engineering, production, and inspection quality specifications. The form and proportion of geometric tolerancing symbols are shown in Figure 12.3. The geometric characteristic symbols and the modifying symbols are further categorized in Figure 12.4.

Situations may arise where the desired geometric requirement cannot be completely conveyed by symbology. In such cases, a note may be used to describe the requirement, either separately or supplementing a geometric tolerance.

12.3.1 Geometric Characteristic Symbols

The symbols denoting geometric characteristics are described below:

Basic dimension symbols A basic dimension is identified by enclosing the dimension in a rectangle. The symbols used to identity a basic dimension are shown in Figures 12.3 and 12.4.

Datum feature symbol The datum feature symbol consists of a frame containing the datum identifying letter preceded and followed by a dash.

Letters of the alphabet (except I, O, and Q) are used as datum identifying letters. Each datum feature requiring identification is assigned a different letter. When datum features requiring identification on a drawing exceed single alpha lettering, the double alpha series is used: AA through AZ, BA through BZ, etc.

Datum target symbol The datum target symbol is a circle divided horizontally into two halves. The lower half contains a letter identifying the associated datum, followed by the target number, assigned sequentially starting with one, for each datum. If the datum target is an area, the area size may be entered in the upper half of the symbol. Otherwise, the upper half is left blank. A radial line attached to the symbol is directed to a target point (indicated by an "X"), target line, or target area.

FEATURE	TOLERANCE TYPE	SYMBOL	CHARACTERISTIC	
INDIVIDUAL (SINGLE)	FORM (SHAPE)	—	STRAIGHTNESS	
		⏥	FLATNESS	
		○	CIRCULARITY	
		⌭	CYLINDRICITY	
INDIVIDUAL OR RELATED	PROFILE (CONTOUR)	⌒	PROFILE OF A LINE	
		⌓	PROFILE OF A SURFACE	
RELATED	ORIENTATION (ATTITUDE)	∠	ANGULARITY	
		⊥	PERPENDICULARITY	
		//	PARALLELISM	
	LOCATION	⌖	POSITION	
		◎	CONCENTRICITY	
	RUNOUT	↗	CIRCULAR RUNOUT	
		⌰	TOTAL RUNOUT	
MODIFYING SYMBOLS		Ⓜ	MAXIMUM MATERIAL CONDITION-MMC	
		Ⓢ	REGARDLESS OF FEATURE SIZE-RFS	
		Ⓛ	LEAST MATERIAL CONDITION-LMC	
ADDITIONAL SYMBOLS		Ⓟ	PROJECTED TOLERANCE ZONE	
		Ø	DIAMETER (FACE OF DWG.)	
		SØ	SPHERICAL DIAMETER	
		R	RADIUS	
		SR	SPHERICAL RADIUS	
		( )	REFERENCE	
		⌒	ARC LENGTH	

FIGURE 12.4 Categories of Geometric Tolerancing Symbols

Material condition symbol The symbols used to indicate "maximum material condition," "regardless of feature size," and "least material condition" are shown in Figures 12.3 and 12.4.

Projected tolerance zone symbol The symbol used to indicate a projected tolerance zone is shown in Figures 12.3 and 12.4.

Diameter and radius symbols The symbols used to indicate diameter, spherical diameter, radius, and spherical radius are shown in Figure 12.3. The symbols precede the value of a dimension or tolerance given as a diameter or radius.

Reference symbol Reference dimensions or reference data are identified by enclosing the dimension or data within parentheses.

Arc length symbol The symbol used to indicate that a linear dimension is an arc length measured on a curved outline is shown in Figures 12.3 and 12.4. The symbol is placed above the dimension.

Counterbore or spotface symbol The symbol for indicating a counterbore or spotface is shown in Figure 12.3. The symbol precedes the dimension of the counterbore or spotface.

Countersink symbol The symbol for indicating a countersink is shown in Figure 12.3. The symbol precedes the dimensions of the countersink.

Depth symbol The symbol used to indicate that a dimension applies to the depth of a feature. It precedes that dimension (Fig. 12.3).

Square symbol The symbol used to indicate that a single dimension applies to a square shape. It precedes that dimension (Fig. 12.4).

Dimension origin symbol The symbol, a small circle placed at the origin, is used to indicate that a toleranced dimension between two features originates from one of these features.

Taper and slope symbols Symbols used for specifying taper and slope for conical and flat tapers are shown in Figure 12.3.

12.3.2 Modifiers

Modifiers stipulate whether a tolerance is to apply regardless of size or only at a specific size. The modifiers are shown in Figure 12.5. The following rules for modifiers are based on the size of features and the geometry involved:

1. Modifiers may be used only for features and/or datums that have a size tolerance. The MMC modifier may be used in conjunction with the straightness of a feature axis based on the cross-sectional size, flatness (by special note on features of size), datums of size used with profile tolerances, and all datums and features of size or orientation and position tolerances.
2. RFS and LMC are limited to use in conjunction with position tolerances. The RFS symbol is used exclusively in the United States. In other countries, unless otherwise specified, all tolerances automatically apply RFS.
3. Position, except in the case of a single-plane surface, requires a modifier for all features and datums (Fig. 12.6).
4. Circularity, cylindricity, runout, concentricity, straightness of element lines, and profile of a feature may not use the MMC modifier. The exception is for datums of size used in conjunction with profile. A special note is required to use the MMC modifier in conjunction with flatness:

(PERFECT FORM AT MMC NOT REQUIRED)

MODIFYING SYMBOLS		
symbol	abbreviation	meaning
Ⓜ	MMC	Maximum Material Condition
Ⓢ	RFS	Regardless of Feature Size
Ⓛ	LMC	Least Material Condition

FIGURE 12.5 Modifying Symbols

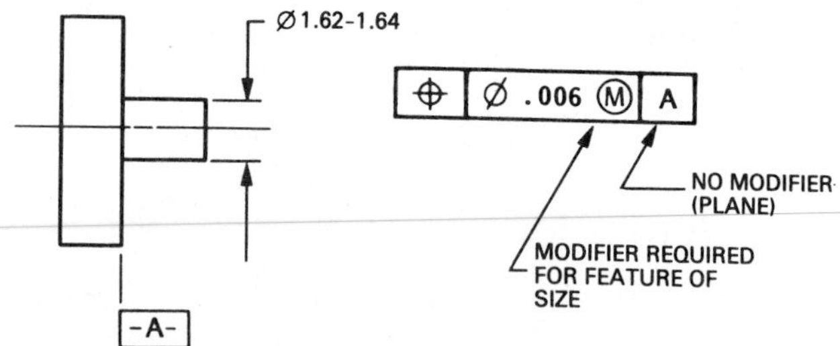

FIGURE 12.6 Single-Plane Surface, No Modifier

12.4 Feature Control Frame

Geometric tolerances are placed in a **feature control frame**, which contains a geometric characteristic symbol, the tolerance, modifiers, and datums (Fig. 12.7). The feature control frame consists of at least the first two compartments shown in Figure 12.8, but may contain three or more compartments. The *first compartment* contains the geometric characteristic

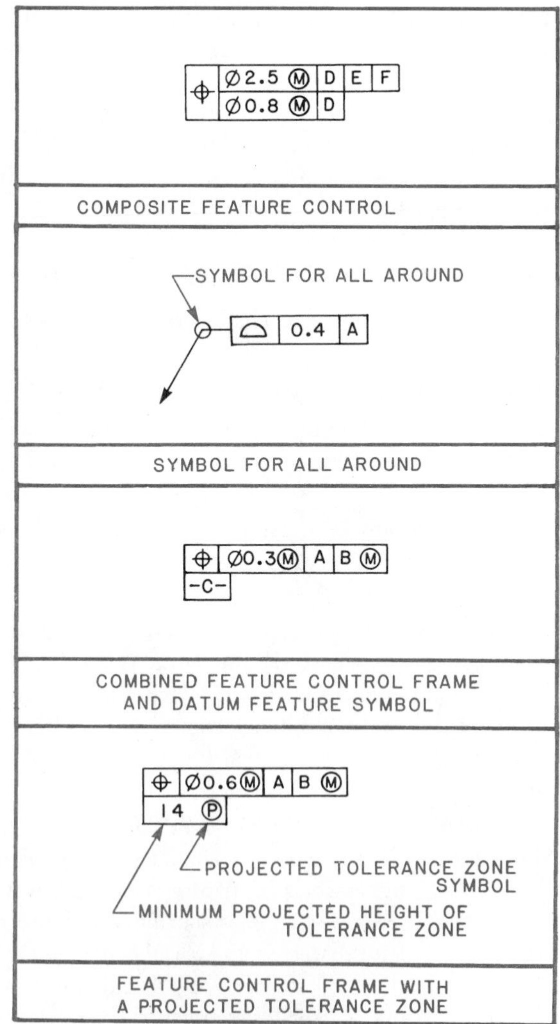

FIGURE 12.7 Feature Control Frames

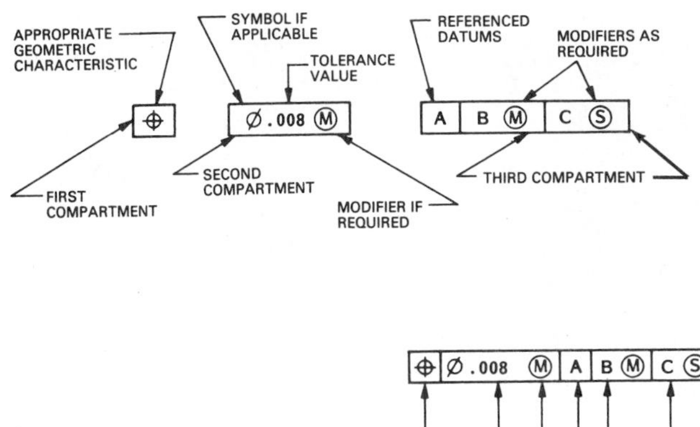

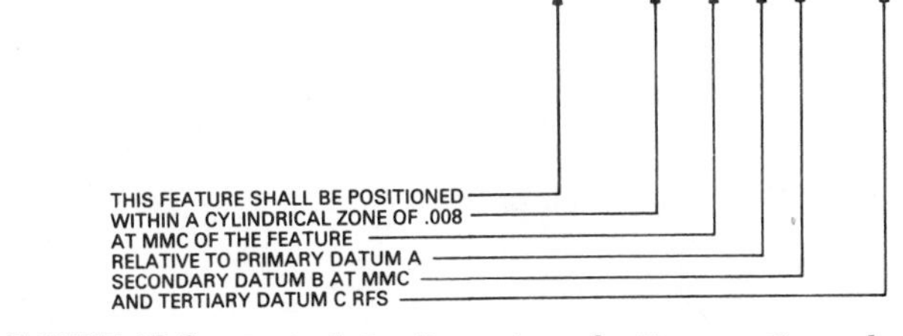

FIGURE 12.8 Typical Configuration of a Feature Control Frame

symbol (one of the 13 from Fig. 12.3). The *second compartment* may contain a zone shape symbol such as the diameter symbol indicating the diameter of a cylindrical zone, the tolerance, in inches or millimeters, and a modifier. The *third compartment* usually contains datums. This third compartment may have **separators** to order the datums.

Feature control frames are not repeated or referenced on a technical drawing. **Datum identification symbols,** -A-, may be repeated where it is essential to ensure the correct meaning.

12.4.1 Maximum Material Condition

In the **maximum material condition (MMC),** a feature or datum feature is at the tolerance limit, resulting in the part containing the most material (*weighing the most*). For an external feature, such as a pin or shaft, MMC is the maximum limit (Fig. 12.9). For an internal feature, such as a hole, MMC is the minimum limit (Fig. 12.10). Remember that a part weighs the most when the hole in it is the smallest size in the range. Figure 12.11 shows the MMC and least material condition in both an external and internal feature. The tightest fit between the two results when both features are at MMC. The loosest fit results when both features are at LMC.

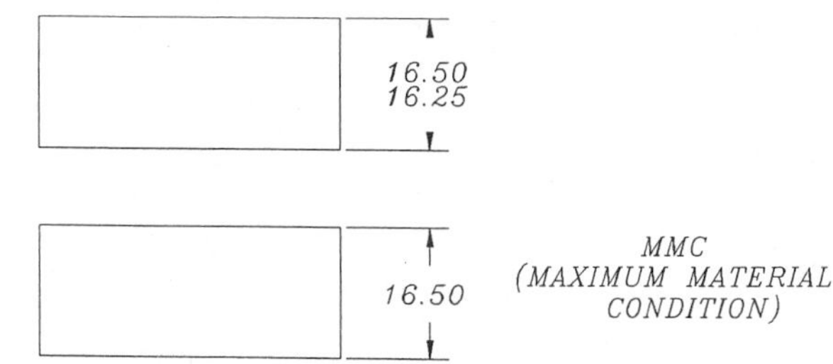

FIGURE 12.9 MMC of an External Feature

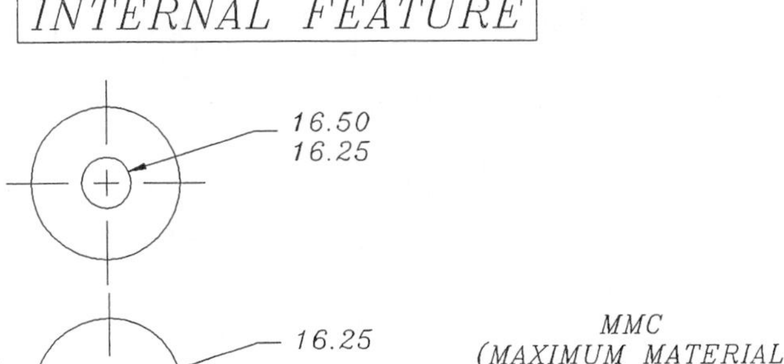

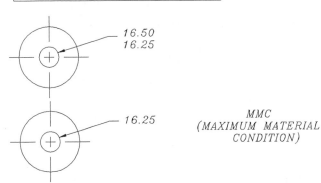

FIGURE 12.10 MMC of an Internal Feature

If the MMC modifying symbol is used in a feature control frame (Fig. 12.12), the specified tolerance applies only at MMC. In Figure 12.12, the perpendicularity tolerance is .004 when the feature is at MMC (.512). As the feature deviates from MMC, additional perpendicularity tolerance equal to the deviation is allowed. This is called the *bonus tolerance* (Fig. 12.13). In other words, as the male diameter decreases in size, the increase in perpendicularity results in the same fit to the mating part. The modifying symbol for MMC, specified in the feature control frame for the feature, datum, or both, works the same way for all geometric tolerances.

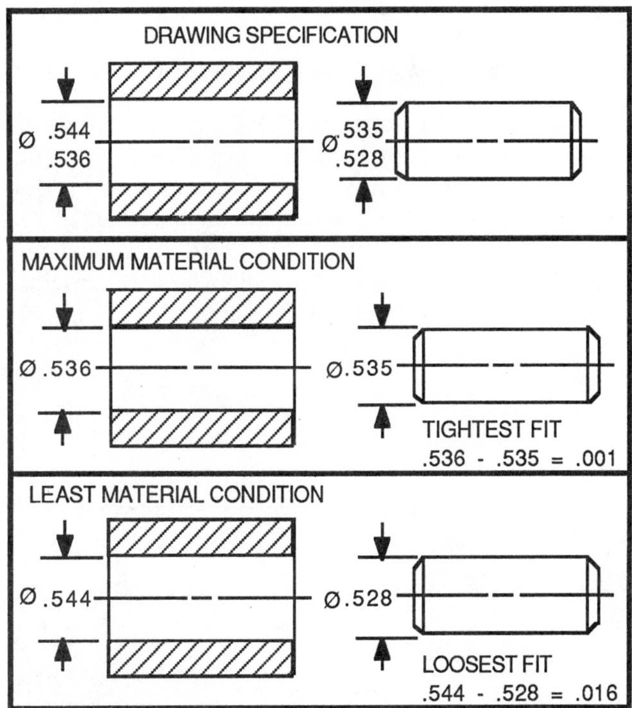

FIGURE 12.11 MMC and LMC Limits for External and Internal Features

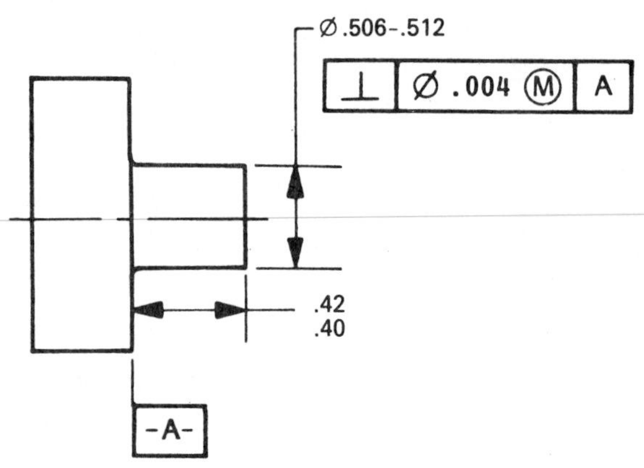

FIGURE 12.12 Perpendicularity When the Feature Is at MMC

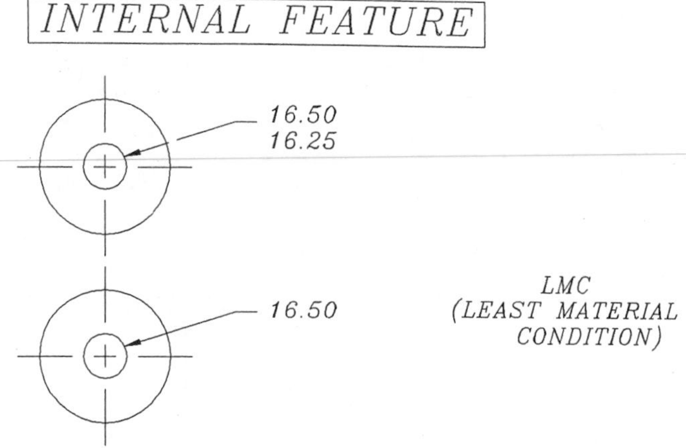

FIGURE 12.15 LMC of an Internal Feature

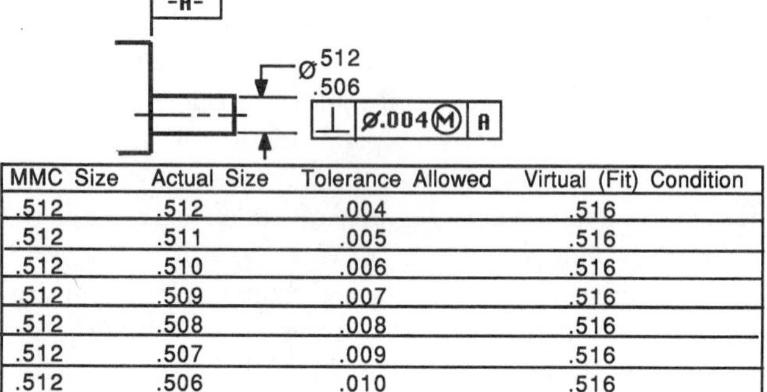

MMC Size	Actual Size	Tolerance Allowed	Virtual (Fit) Condition
.512	.512	.004	.516
.512	.511	.005	.516
.512	.510	.006	.516
.512	.509	.007	.516
.512	.508	.008	.516
.512	.507	.009	.516
.512	.506	.010	.516

FIGURE 12.13 Bonus Tolerance Addition to Geometric Tolerance at MMC

TABLE 12.1 Bonus Tolerance Addition to Geometric Tolerance at LMC

LMC Size	Actual Size	Tolerance Allowed	Minimum Bearing Area
.506	.506	.004	.502
.506	.507	.005	.502
.506	.508	.006	.502
.506	.509	.007	.502
.506	.510	.008	.502
.506	.511	.009	.502
.506	.512	.010	.502

12.4.2 Least Material Condition

In the **least material condition (LMC)**, a feature or datum feature is at the tolerance limit, resulting in the part containing the least material (*weighing the least*). For an external feature, such as a pin, LMC is the minimum limit (Fig. 12.14). For an internal feature, such as a hole, LMC is the maximum limit (Fig. 12.15). The modifying symbol for LMC, specified in the feature control frame for the feature, datum, or both, works the same way for all geometric tolerances. Table 12.1 shows the result of using Figure 12.11 as though LMC were specified in the feature control frame instead of MMC. LMC is generally used where minimum bearing areas, minimum wall thickness, or alignment of parts is the main concern, not fit.

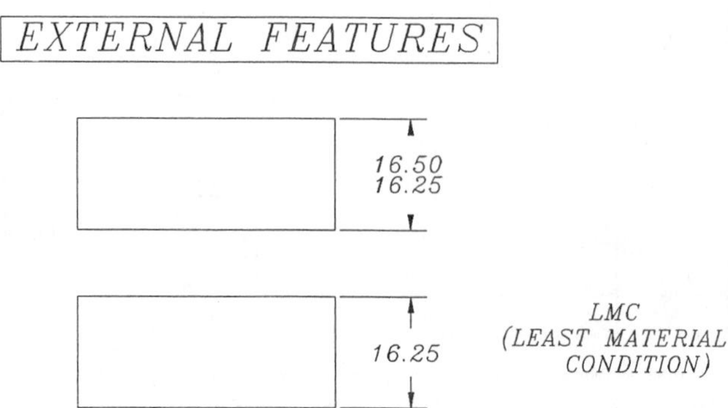

FIGURE 12.14 LMC of an External Feature

12.4.3 Regardless of Feature Size

The "regardless of feature size" (RFS) modifier is used only for features and datums of size and only with the **positional tolerance.** In the ISO standard, RFS applies to every geometric tolerance unless MMC or LMC is specified. In ANSI standards, RFS automatically applies to all geometric tolerances except position. MMC, RFS, or LMC must be specified for all features and datums of size for positional tolerance.

If the RFS modifier is placed after the feature tolerance in the second compartment, the tolerance must be met at all sizes.

12.4.4 Virtual Condition

The **virtual condition** (Fig. 12.16) is the condition resulting from the worst case effect of the size and geometric tolerance

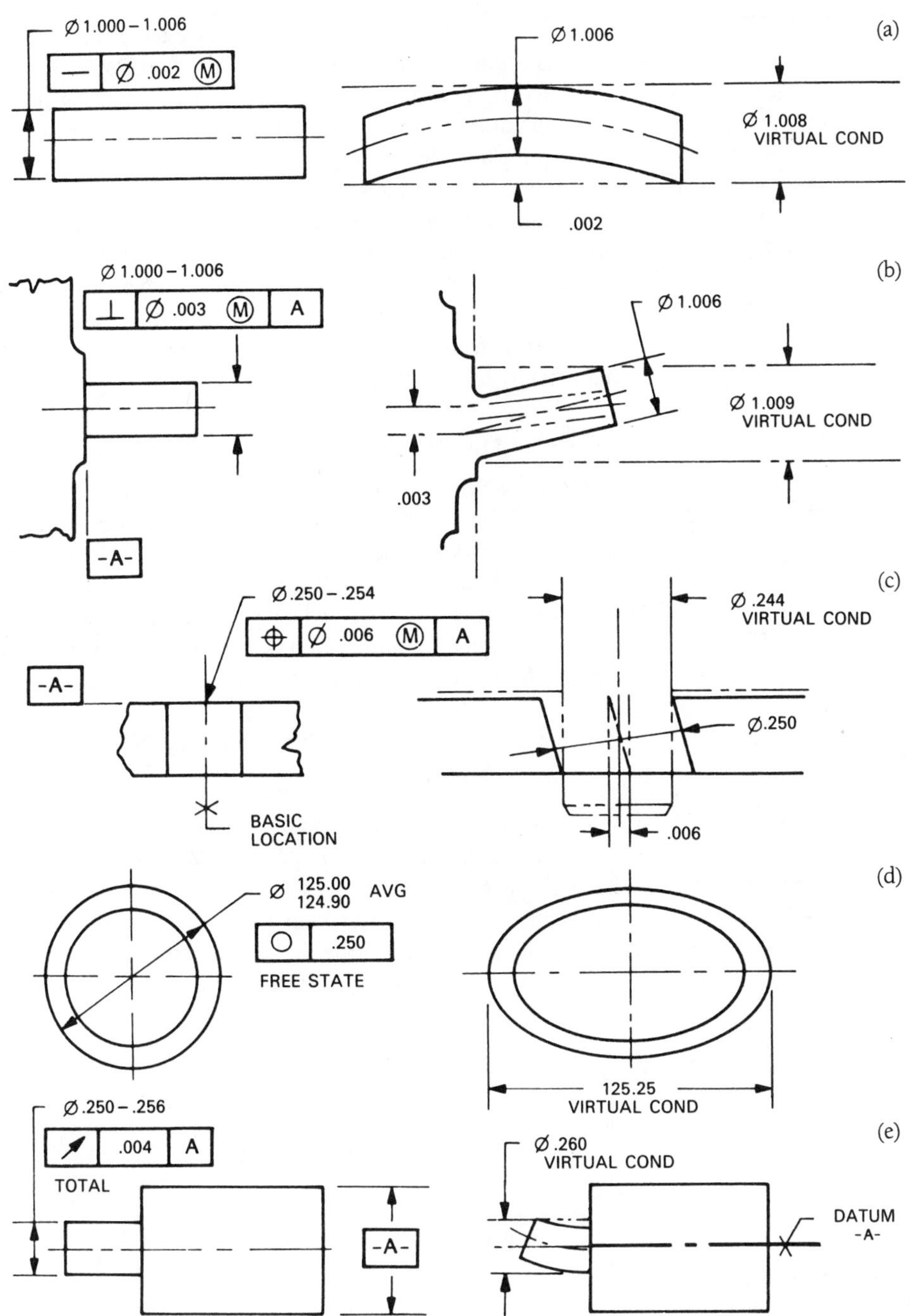

FIGURE 12.16 Examples of Virtual Condition

applied to the feature. The free assembly of components is dependent on the combined effect of the actual sizes of the part features and the errors of form, orientation, location, or runout. For example, if the axis is out of straight, the size of a shaft is virtually increased or the size of a hole is virtually decreased.

The formulas for determining the virtual condition are:

External Feature: Virtual Condition = MMC Size + Geometric Tolerance

Internal Feature: Virtual Condition = MMC Size − Geometric Tolerance

12.4.5 Angular Surfaces

If an **angular surface** is defined by the combination of a linear dimension and an angle, the surface must lie within a tolerance zone represented by two nonparallel planes (Fig. 12.17). The tolerance zone will be wider as the distance from the apex of the angle increases.

12.5 Datums and Datum Systems

Datums are theoretically exact geometric references derived from the datum feature. Figure 12.18 shows the primary datum

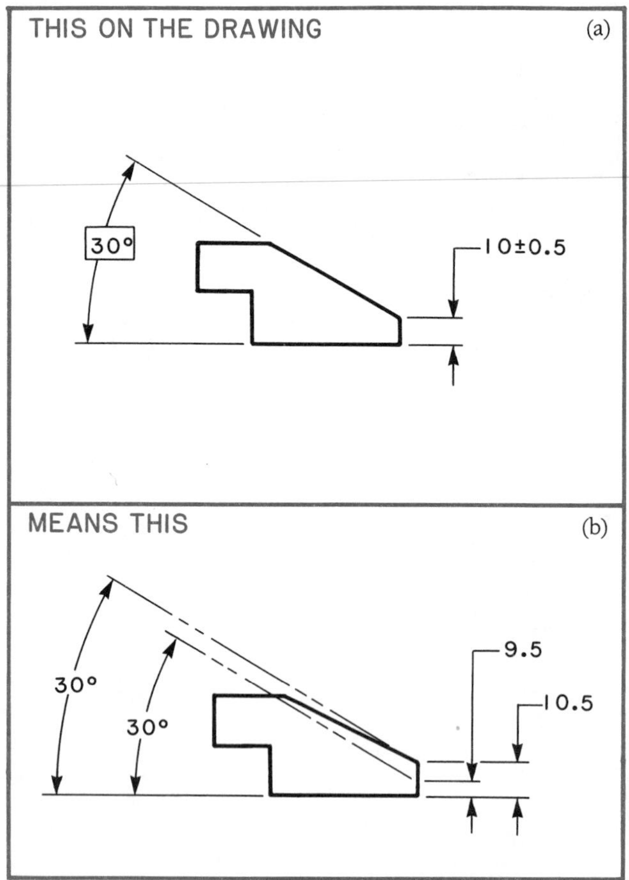

FIGURE 12.17 Tolerancing an Angular Surface Using Linear and Angular Dimensions

plane established on a surface by three area contact positions. Datums are not assumed to exist on the part itself, but are simulated by the more precisely made manufacturing or inspection equipment or a computerized mathematical model. A datum plane, for example, could be *simulated* from the datum feature by a surface plate (Fig. 12.19).

Datums are points, lines, and planes. Datums provide repeatable part and feature orientation for manufacturing and inspection consistent with the expected mating characteristics or orientation at assembly. Datums should be established from "hard" features on the part, such as one or two specific diameter(s) on a shaft (Fig. 12.20).

12.5.1 Applicability of Datums

A **datum** is a theoretically exact point, axis, or plane derived from the true geometric counterpart of a specified datum feature. A datum is the origin from which the location or geometric characteristics of features of a part are established. A **datum target** is a specified point, line, or area on a part used to establish a datum. Tolerances, as they relate to datums, are described according to the feature they locate.

12.5.2 Part and Feature Direction and Orientation

If a drawing contains two or more features, it is incomplete if one or more datums are not specified. Without datums, reliable engineering interchangeability is difficult or impossible; setup criteria for manufacturing and inspection are then arbitrary. Without datums, the design is compromised and the manufactured part or assembly may not function as intended.

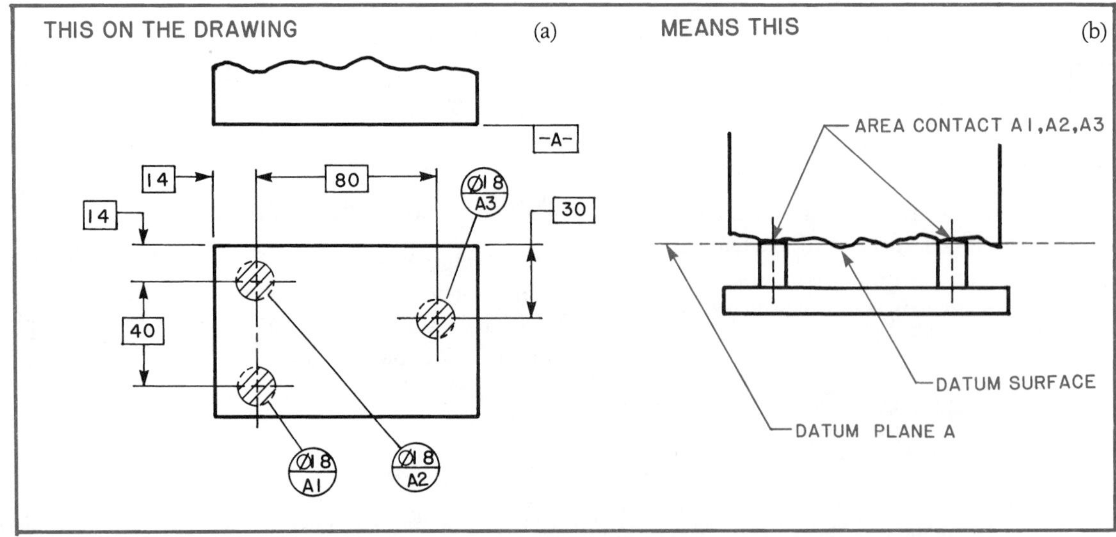

FIGURE 12.18 The Primary Datum Established by Three Area Contact Positions

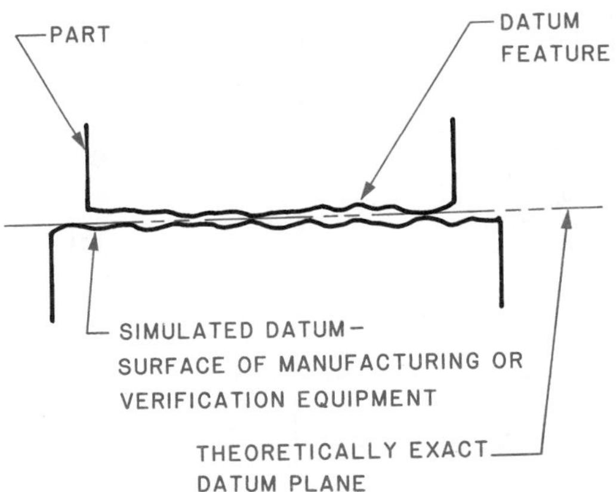

FIGURE 12.19 **Theoretical and Simulated Datum and Datum Plane**

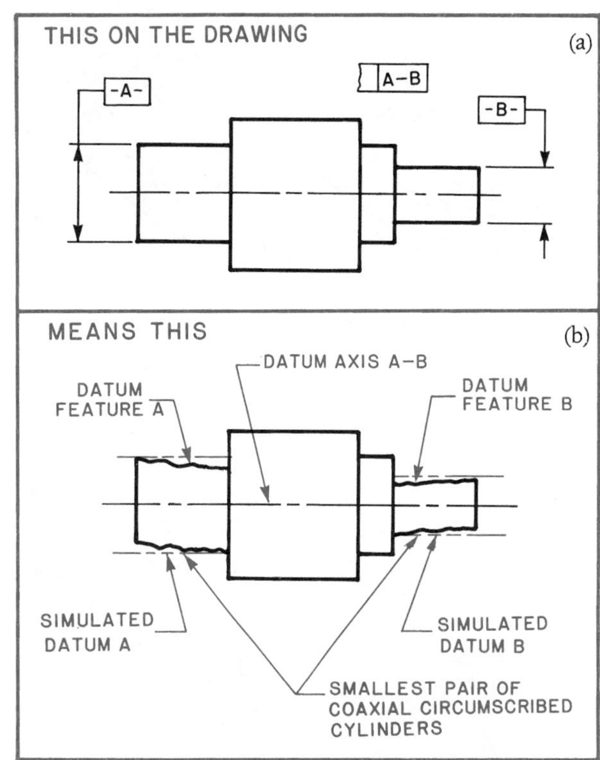

FIGURE 12.20 **Coaxial Datum Features**

12.5.3 Datum Reference Frame

Locations and measurements are taken relative to three mutually perpendicular planes (Fig. 12.21) called a **datum reference frame.** In inspection, a surface plate and two angle plates, perpendicular to it, may be used to simulate the datum reference frame. In manufacturing, the bed of the machine and clamps or other devices, along with the direction of machine movement, provide location relative to three mutually perpendicular planes.

12.5.4 Datum Features

Datum features are selected to ensure the orientation of the part and its associated features for interchangeability and to ensure functional relationships. If a functional datum feature is undesirable from a manufacturing or inspection standpoint, a nonfunctional feature with a precise toleranced relationship to the functional feature may be used, provided all design requirements are met.

12.5.5 Datum Precedence

The sequence of datums specified in the feature control frame determines the order in which the datum features contact the datum reference frames (Fig. 12.22):

1. The part primary datum feature is aligned with the primary datum.
2. While in full contact with the primary datum, the secondary datum feature is aligned with the secondary datum.
3. While in full contact with the primary datum and aligned to the secondary datum, the tertiary datum feature is pushed into contact with the tertiary datum.

The **primary datum** is established by full contact with of a minimum of three noncolinear points on the part (recall that three noncolinear points define a plane). The **secondary datum**

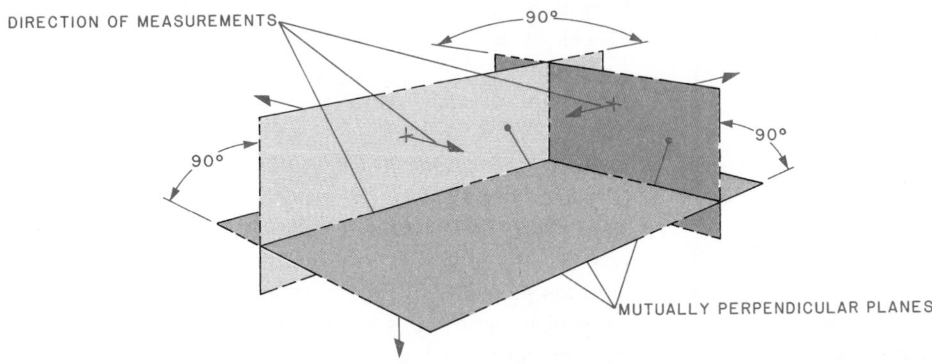

FIGURE 12.21 **Datum Reference Frame**

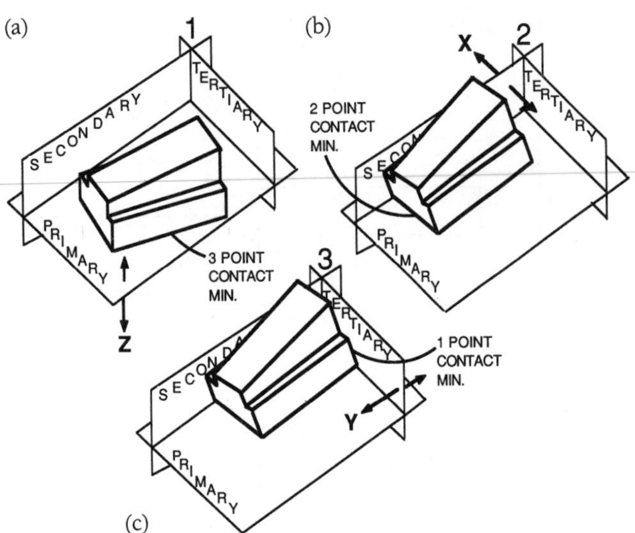

FIGURE 12.22 Datum Reference Frame—Datum Precedence

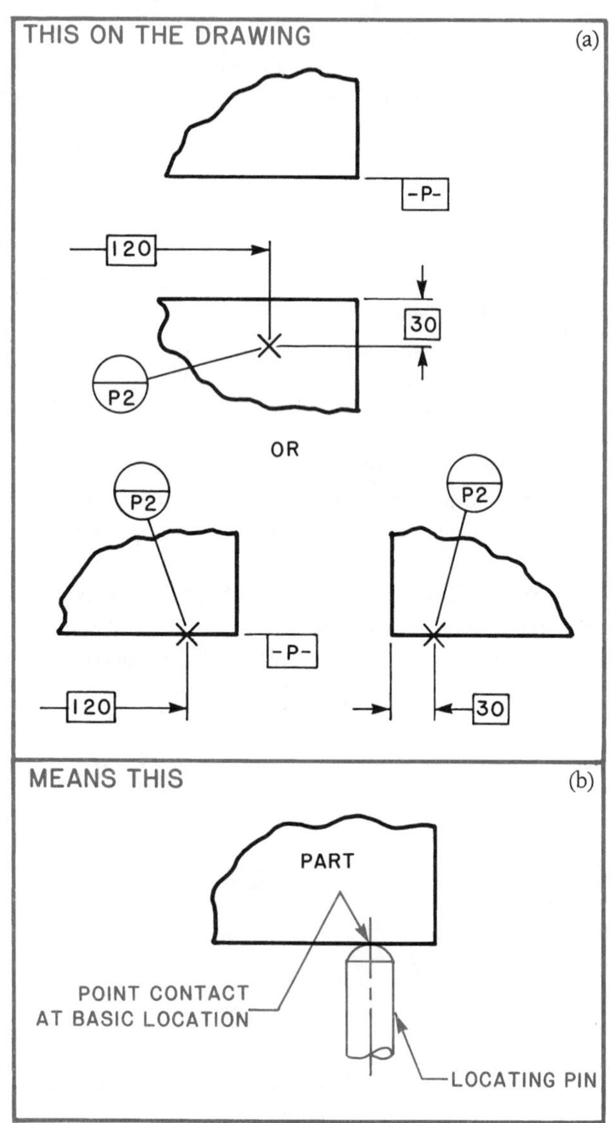

FIGURE 12.23 Datum Target Point

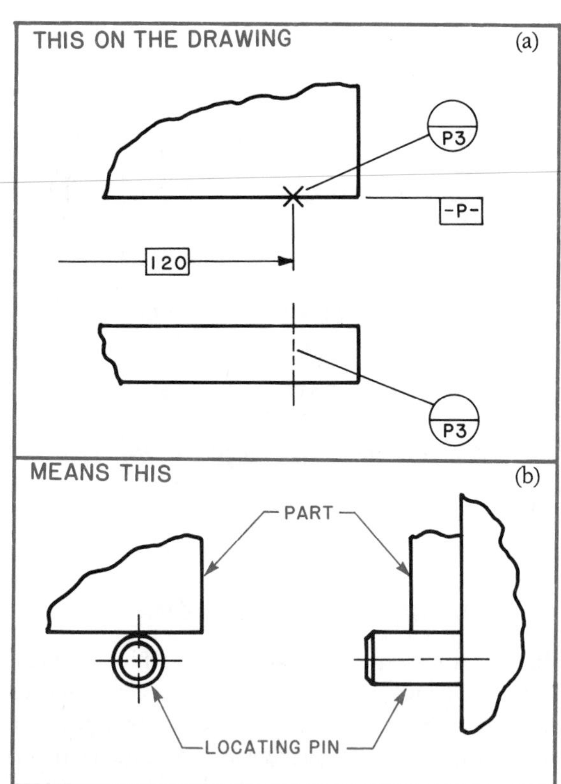

FIGURE 12.24 Datum Target Line

is perpendicular to the primary datum and is established by contacting a minimum of two points on the part (two points establish a line). The **tertiary datum** is perpendicular to the primary and the secondary datum and, therefore, needs only one point of contact on the part to establish it. In Figure 12.22, notice that the three directions of measurement, **X**, **Y**, and **Z**, are established on the part, as are their origin datums. Precise and repeatable measurements may now be made as the part is oriented and locked in position. Datums are specified on the drawing to ensure the intended datum reference frame.

12.5.6 Datum Targets

Datum targets are specific points (Fig. 12.23), lines (Fig. 12.24), or areas (Fig. 12.25) that are used where an entire surface may not be suitable as a datum feature. For example, the surfaces of castings and forgings are rough and difficult to use. If a limited portion of a feature is not a point, line, or local flat area (a portion of a cylindrical surface, for example) partial datums (Fig. 12.26) may be used instead of targets.

In Figure 12.27 datum targets A1, A2, and A3 are used to establish the primary datum. Datum targets B1 and B2 are used to establish a secondary plane perpendicular to the primary plane, and C1 is used to establish the tertiary plane perpendicular to the primary and secondary planes. The datum target identification symbol is shown in Figure 12.28.

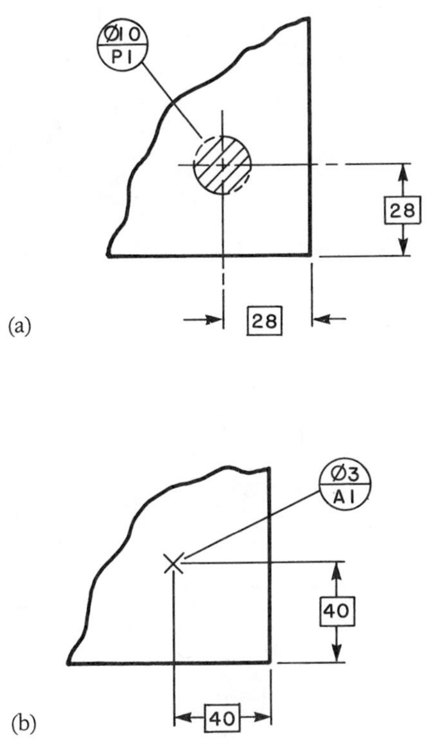

FIGURE 12.25 Datum Target Area

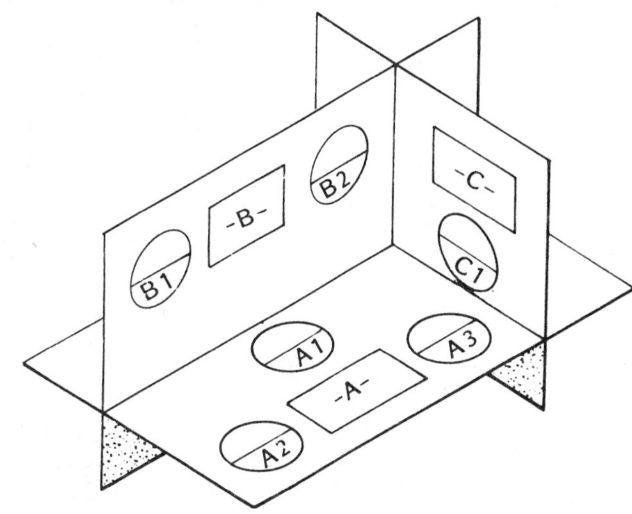

FIGURE 12.27 Datum Targets Used to Establish a Datum Reference Frame

12.5.7 Datum Target Depiction

Datum target points are depicted by a dense 90° "cross" at 45° ("X") to the centerline (Fig. 12.29), at two times letter height. The leader line from the datum target symbol does not terminate in an arrowhead. A solid leader line indicates the target is on the near side; a dashed leader line indicates the target is on the far side. The three mutually perpendicular planes from which to measure **X**, **Y**, and **Z** distances are locked in place and repeatable for each individual part. Datum targets may also be located on the drawing by dimensions (Fig. 12.30).

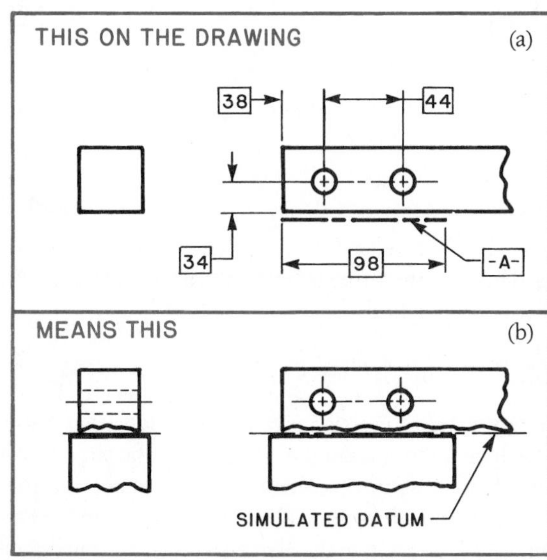

FIGURE 12.26 Partial Datums

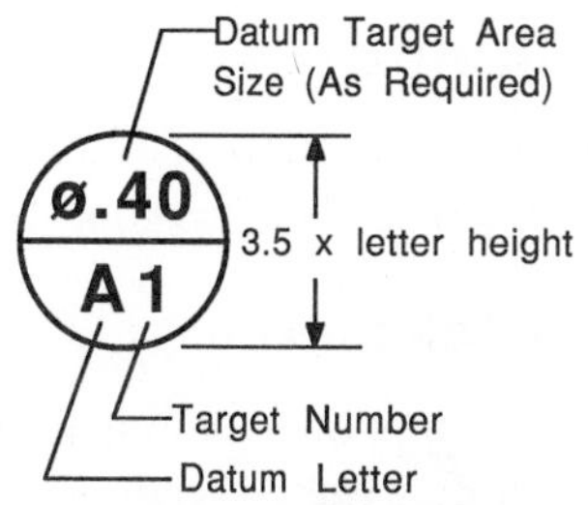

FIGURE 12.28 Datum Target Identification Symbol

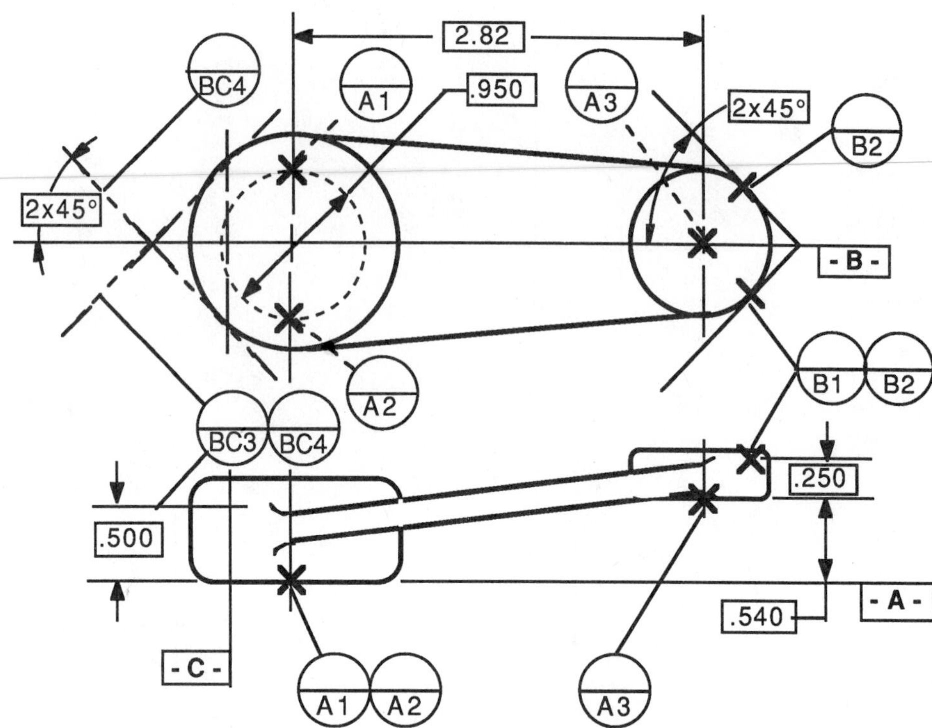

FIGURE 12.29 Datum Targets Showing "Step" and "Equalizing Dimensions"

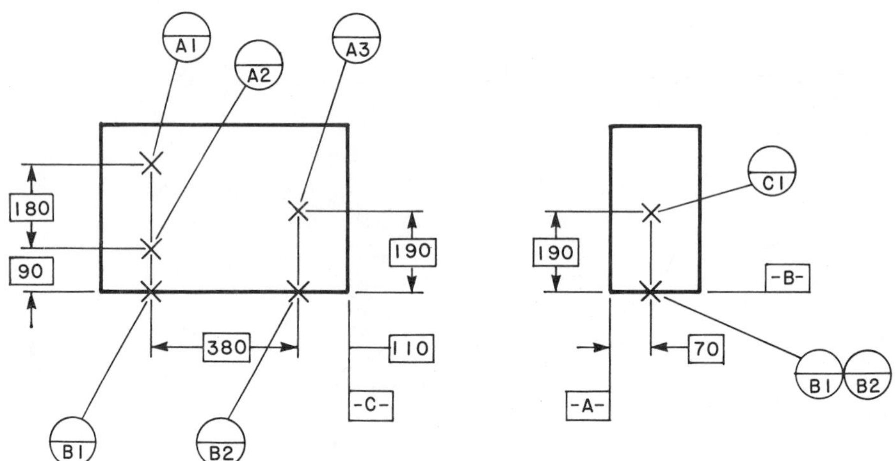

FIGURE 12.30 Dimensioning Datum Targets

12.6 Geometric Tolerance of Form, Profile, Orientation, Location, and Runout

Geometric tolerances of form, profile, orientation, location and runout are described in this section.

12.6.1 Form Tolerances

Form tolerances are applicable to individual features or elements of single features. These tolerances do not use datums because they are related to a perfect counterpart of themselves. These "pure form" tolerances are straightness, flatness, circularity, and cylindricity.

12.6.2 Straightness of Element Lines

The **straightness** tolerance specifies variation from a straight line. Each element line on the surface must be straight within the specified straightness tolerance. For element control, the leader from the feature control frame must be directed to the outline of the part where the element to be controlled appears

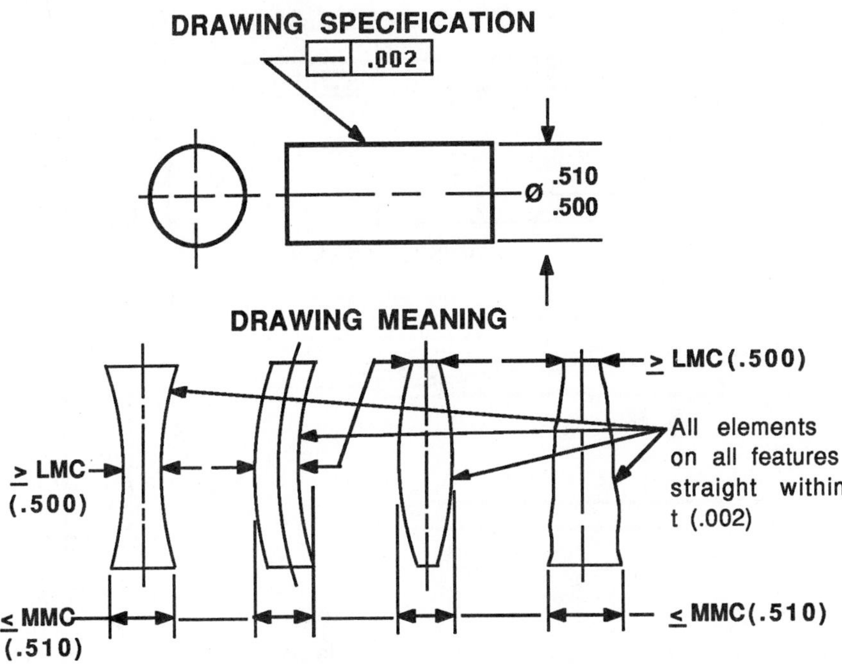

FIGURE 12.31 Straightness of Element Lines

as a straight line (Fig. 12.31). For a rectangular part, the view in which the leader is shown determines the direction of the indicator movement zone. In this case, each element on the surface is to be straight within the specified tolerance, and the feature must meet the size tolerance.

12.6.3 Straightness of an Axis

If straightness of an axis is specified, the leader from the feature control frame must be directed to the size dimension (Fig. 12.32), and a diameter symbol (⌀) must precede the tolerance in the feature control frame. An exception is made if the zone is not cylindrical.

12.6.4 Flatness

Flatness means that a surface has all elements in one plane. Flatness must be within the size tolerance, but has no orientation requirement. Therefore, it may be tilted in the size zone. A **flatness tolerance** specifies a tolerance zone defined by two parallel planes within which the surface must lie (Fig. 12.33). When a flatness tolerance is specified, the feature

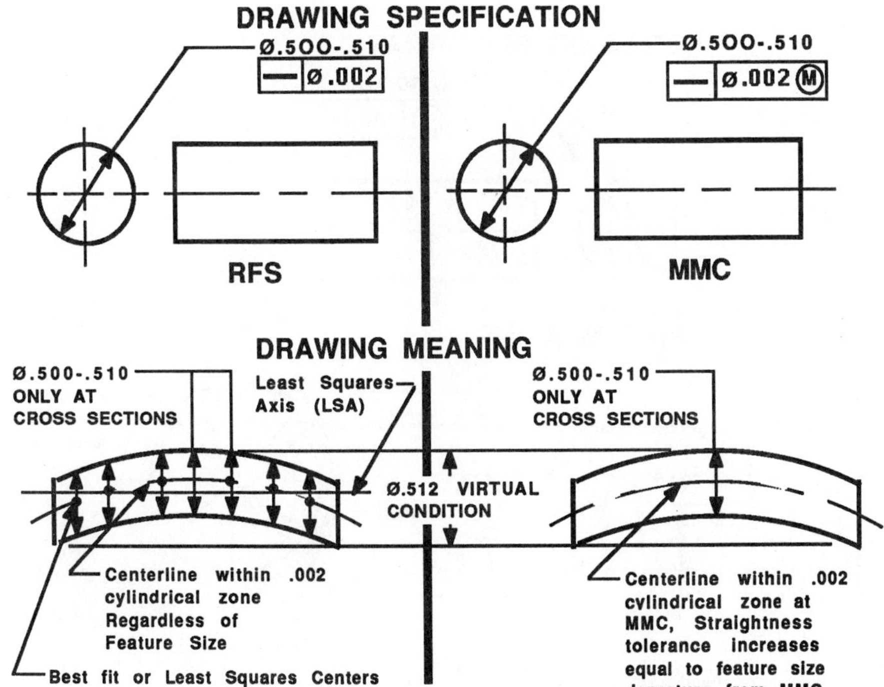

FIGURE 12.32 Straightness of the Axis

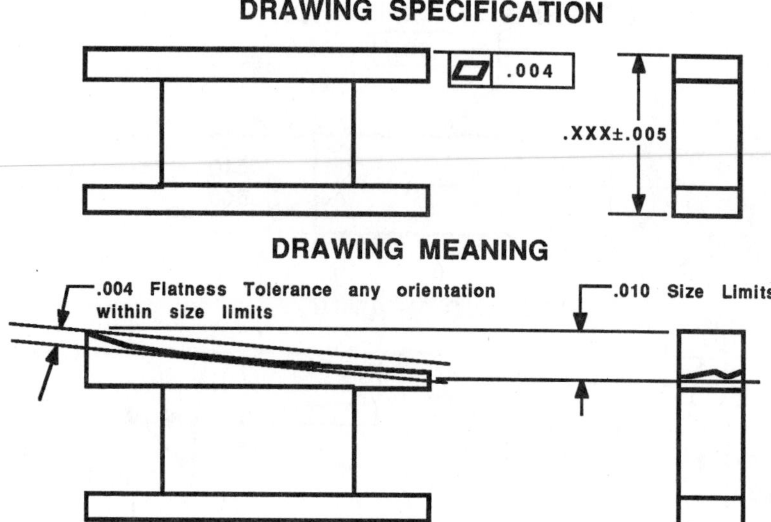

FIGURE 12.33 Flatness

control frame is attached to a leader directed to the surface or to an extension line of the surface. It is placed in a view where the surface elements to be controlled are represented by a line. If the considered surface is associated with a size dimension, the flatness must be less than the size tolerance.

12.6.5 Circularity

A **circularity tolerance** specifies a tolerance zone bounded by two concentric circles within which each circular element of the surface must lie, and applies independently at any plane (Fig. 12.34). The circularity tolerance must be less than the size tolerance, except for those parts subject to three-state variation.

12.6.6 Cylindricity

Cylindricity is a surface of revolution in which all points of the surface are equidistant from a common axis. A **cylindricity tolerance** specifies a tolerance zone bounded by two concentric cylinders within which the surface must lie (Fig. 12.35). In the case of cylindricity, unlike that of circularity, the tolerance applies simultaneously to the entire surface. The leader from the feature control frame may be directed to

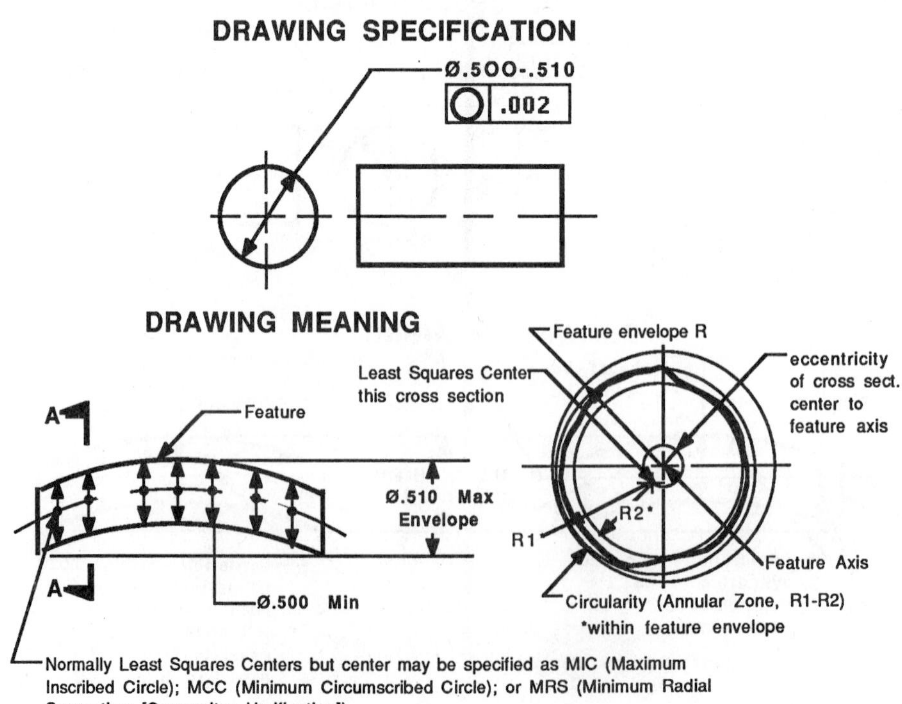

FIGURE 12.34 Circularity

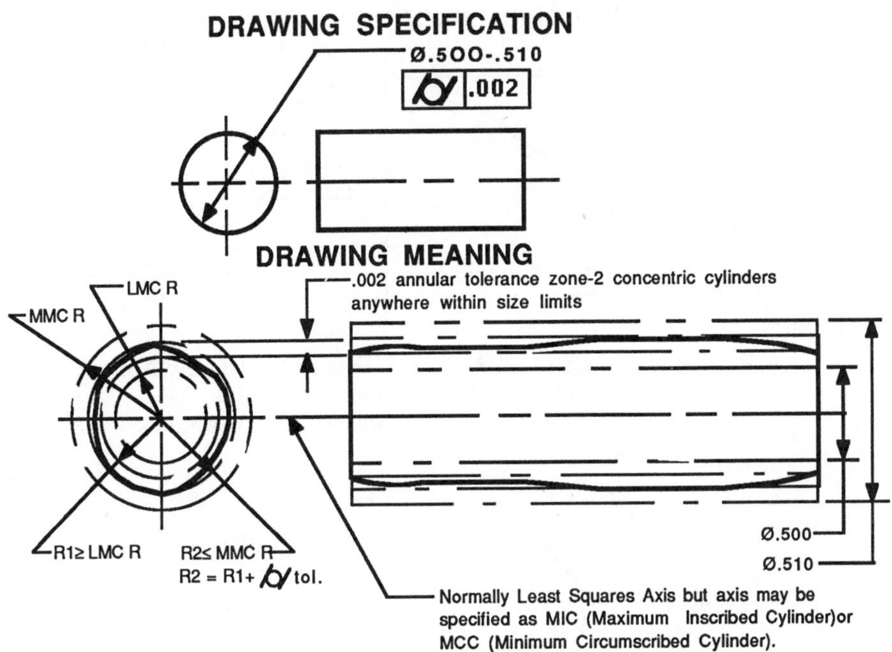

FIGURE 12.35 Cylindricity

either view. The cylindricity tolerance must be less than the size tolerance.

12.7 Profile Tolerances

A **profile** is the outline of a 2D part in a given plane. Profiles are formed by projecting a 3D figure onto a plane or by taking cross sections through the figure. The elements of a profile are straight lines, arcs, and other curved lines. If the drawing specifies individual tolerances for the elements or points of a profile, these elements or points must be individually verified. With profile tolerancing, the true profile may be defined by basic radii, basic angular dimensions, basic coordinate dimensions, formulas, or undimensioned drawings.

The **profile tolerance** specifies a uniform boundary along the true profile within which the elements of the surface must lie. It is used to control form or combinations of size, form, and orientation. Profile tolerances are specified as follows:

1. An appropriate view or section is drawn showing the desired basic profile.
2. Depending on design requirements, the tolerance may be divided bilaterally to both sides of the true profile or applied unilaterally to either side of the true profile. If an equal bilateral tolerance is intended, it is necessary to show only the feature control frame with a leader directed to the surface. For a unilateral tolerance, phantom lines are drawn parallel to the true profile to indicate that the line is extended to the feature control frame. If some segments of the profile are controlled by a profile tolerance and other segments by individually toleranced dimensions, the extent of the profile tolerance must be indicated.

12.7.1 Profile of a Line

A **profile of a line tolerance** specifies the limits of the boundaries of individual line elements of a surface (Fig. 12.36). The tolerance zone is two dimensional. Basic dimensions define the true profile. The shape of the tolerance zone is two parallel boundaries offset above and below the true profile.

12.7.2 Profile of a Surface

A **profile of a surface tolerance** specifies the limits of all surface elements at the same time (Fig. 12.37). If the surface is made up of one or more basic curves, arcs, straight lines, or other shapes, all are described by basic dimensions. Some segments of a surface may be controlled by profile tolerancing and other segments by different tolerances. Reference letters are used to define the extent of a controlled segment, such as FROM A TO B. Points A and B are directed to the appropriate location on the surface. If the relationship of features is to be zero at MMC, specify by placing a 0 in the control frame or use a note, for example, PERFECT ORIENTATION REQUIRED AT MMC. The orientation tolerance may then be equal to or less than the amount the feature deviates from MMC.

12.8 Orientation Tolerances

Orientation tolerances are parallelism, perpendicularity, and angularity. All require at least one datum specification. Some, such as perpendicularity, may use an additional (secondary) datum. Orientation tolerances may use element controls rather than surface requirements. In this case, *each element of each radial element* is specified below the feature control frame.

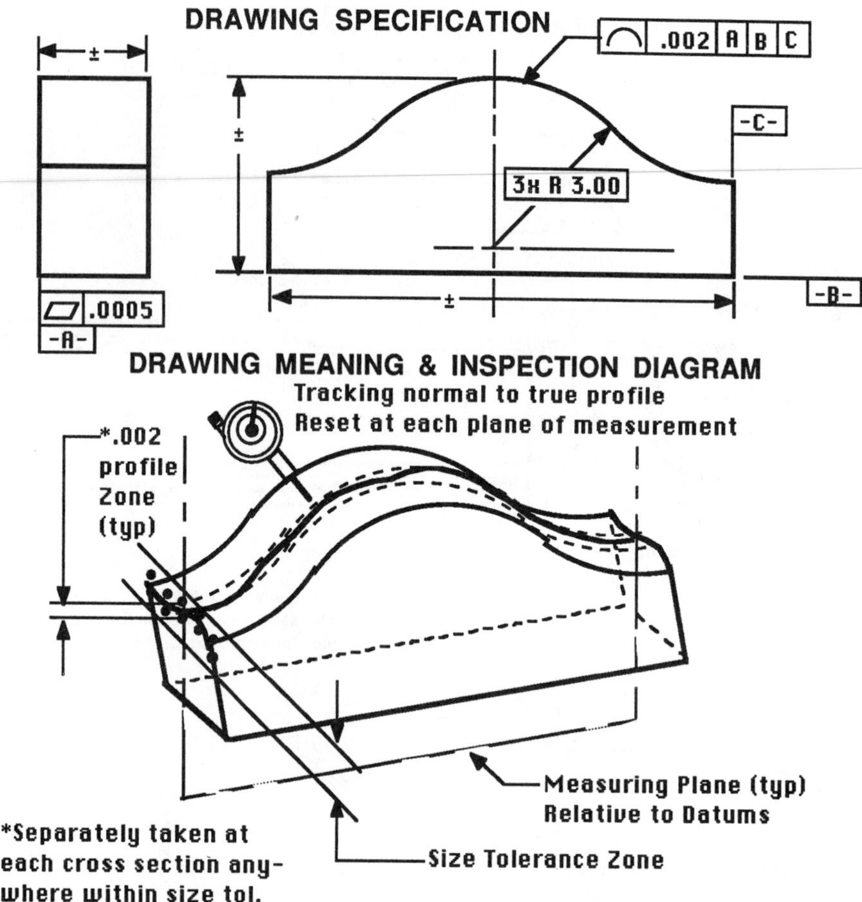

FIGURE 12.36 Profile of a Line

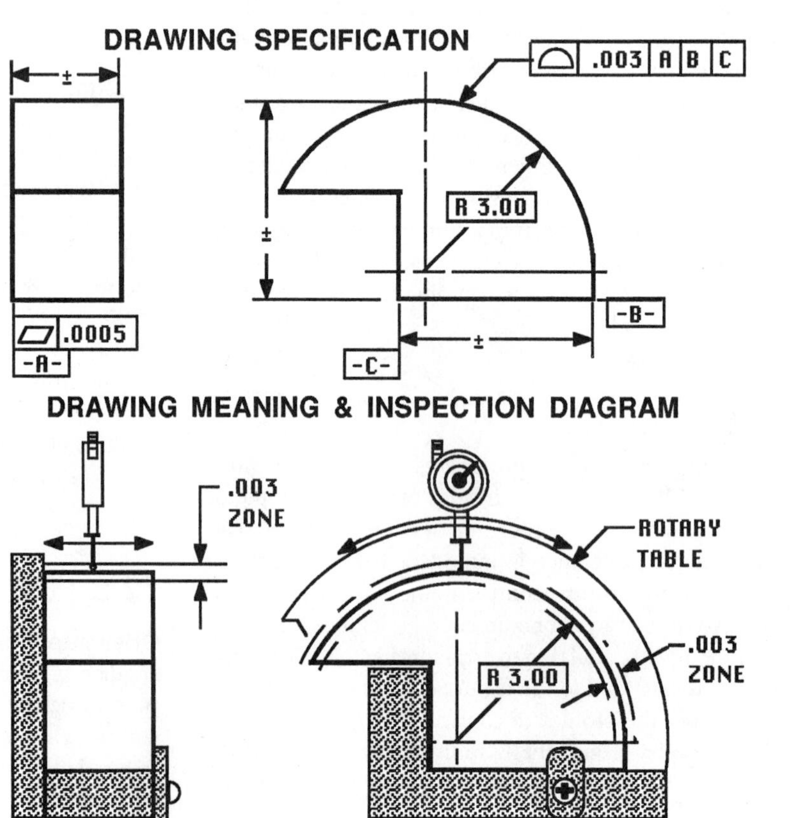

FIGURE 12.37 Profile of a Surface

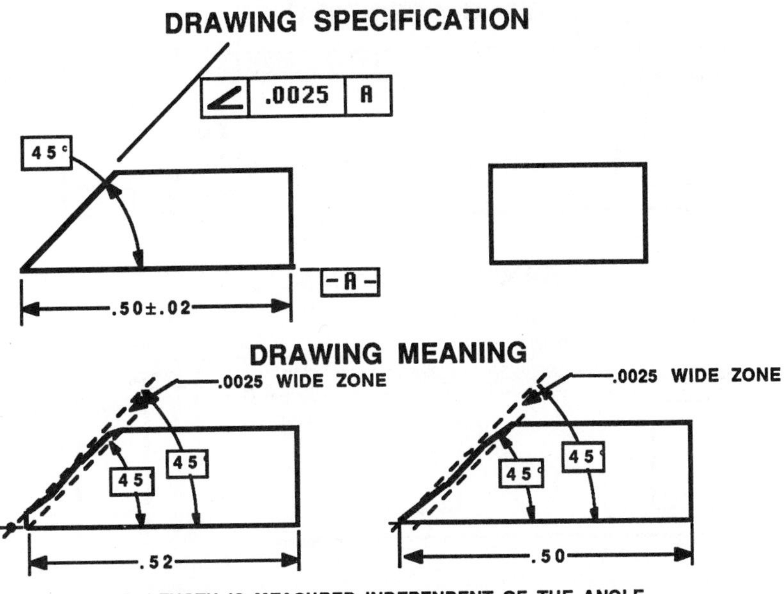

FIGURE 12.38 Angularity

Angularity, parallelism, perpendicularity, and in some instances profile are orientation tolerances applicable to related features. These tolerances control the orientation of features to one another. They are sometimes referred to as **attitude tolerances.** Relation to more than one datum feature should be considered if required to stabilize the tolerance zone in more than one direction. Note that angularity, perpendicularity, and parallelism, when applied to plane surfaces, control flatness if a flatness tolerance is not specified. Tolerance zones require an axis, or all elements of the surface, to fall within this zone.

12.8.1 Angularity

Angularity is when the surface or an axis is at a specific angle other than 90° from a datum plane or axis (Fig. 12.38). An **angularity tolerance** specifies one of the following:

1. A tolerance zone defined by two parallel planes at the specified basic angle from a datum plane or axis within which the surface of the considered feature must lie
2. A tolerance zone defined by two parallel planes at the specified basic angle from a datum plane or axis within which the axis of the feature must lie.

12.8.2 Perpendicularity

Perpendicularity is the condition of a surface, center plane, or axis at a right angle to a datum plane or axis (Fig. 12.39). A **perpendicularity tolerance** specifies one of the following:

1. A tolerance zone defined by two parallel planes perpendicular to a datum plane, or axis, within which the surface or center plane of the considered feature must lie

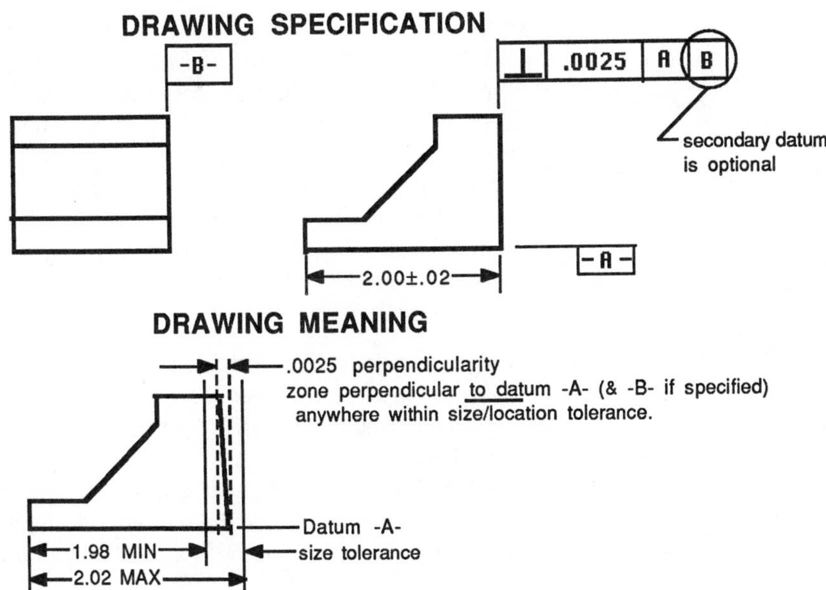

FIGURE 12.39 Perpendicularity

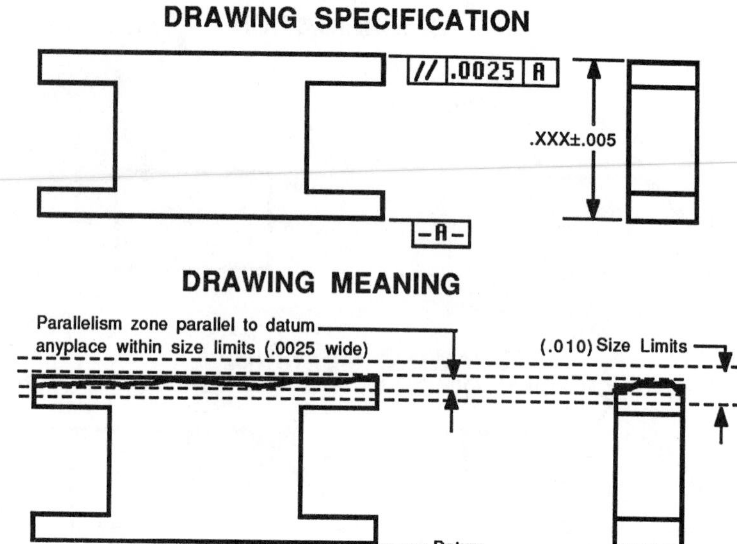

FIGURE 12.40 **Parallelism**

2. A tolerance zone defined by two parallel planes perpendicular to a datum axis within which the axis of the considered feature must lie
3. A cylindrical tolerance zone perpendicular to a datum plane within which the axis of the considered feature must lie
4. A tolerance zone defined by two parallel lines perpendicular to a datum plane, or axis, within which an element of the surface must lie.

12.8.3 Parallelism

Parallelism is the condition of a surface equidistant at all points from a datum plane or an axis equidistant at all points from a datum plane or an axis equidistant along its length to a datum axis (Fig. 12.40). A **parallelism tolerance** specifies one of the following:

1. A tolerance zone defined by two planes or lines parallel to a datum plane, or axis, within which the line elements of the surface or axis of the feature must lie
2. A cylindrical tolerance zone whose axis is parallel to a datum axis within which the axis of the feature must lie.

12.9 Runout Tolerances

Runout is a composite tolerance used to control the functional relationship of one or more features of a part to a datum axis. The types of features controlled by runout tolerances include those surfaces constructed around a datum axis and those constructed at right angles to a datum axis.

Runout tolerances control the composite form, orientation, and position relative to a datum axis. Each feature must be within its runout tolerance when the part is rotated about the datum axis. The tolerance specified for a controlled surface is the total tolerance or full indicator movement (FIM).

The two types of runout control are circular and total. The type used depends on design requirements and manufacturing considerations. Circular runout is normally a less complex requirement than total runout.

12.9.1 Circular Runout

Circular runout is the condition of a circular element on the surface with respect to a fixed point during one complete revolution of the part about the datum axis (Fig. 12.41). Circular runout controls circular elements of a surface. The tolerance is applied independently at any circular cross section as the part is rotated 360°. If applied to surfaces constructed around a datum axis, circular runout may be used to control the cumulative variations of circularity and coaxiality. If applied to surfaces at right angles to the datum axis, circular runout controls circular elements of a plane surface (wobble).

12.9.2 Total Runout

Total runout (Fig. 12.42) is the condition of a surface with respect to a perfect counterpart of itself, perfectly oriented and positioned. The indicator is moved across the feature, relative to the desired geometry, as the part is rotated about the datum axis.

Total runout provides composite control of all surface elements. The tolerance is applied simultaneously to all circular and profile measuring positions as the part is rotated 360°.

12.9.3 Position

Position is a total zone specification such as a diameter or total width centered on the basic location of the axis, center plane, or center point of a feature from the true position with respect to datum(s) (Fig. 12.43).

Locating a hole with rectangular coordinates and ± tolerances yields a square or rectangular zone. The worst case location for the axis of the mating features is at the diagonal.

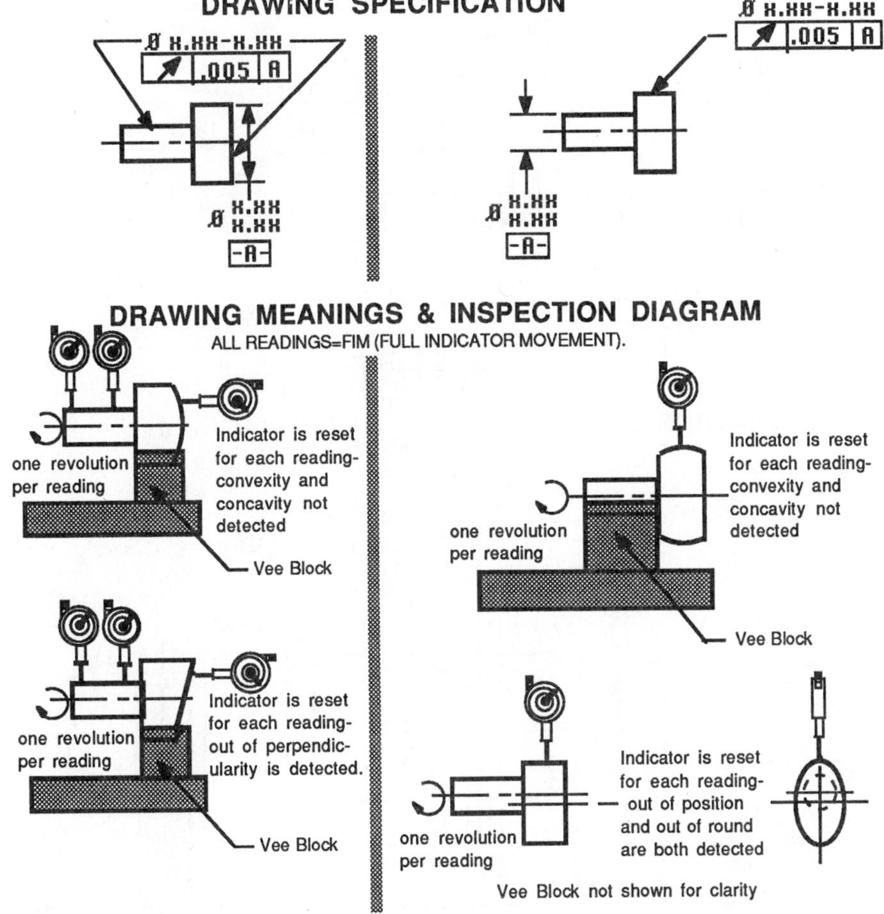

FIGURE 12.41 Circular Runout

Inscribing this square with a circle does not change the mating relationship, but it does yield a 58% greater area.

Another improvement in the positional tolerancing system is the change from a *chain* or feature-to-feature basis to a *basic grid* system. Tolerance accumulations are, therefore, avoided (Fig. 12.44). The grid for a pattern of zones is perfect in all respects. The locations of each of these zones are in perfect relationship to each other. A grid is established by placing basic dimensions between the features. Figure 12.45 shows examples of identifying basic dimensions for patterns.

12.9.4 Positional Patterns

Figures 12.46 shows how **patterns** are located on parts. The preference is for composite positional tolerancing, as shown in Figure 12.46. The use of ± dimensions to locate a pattern is not recommended.

12.10 Limits of Size

The **limits of size** of a feature describe the extent within which variations of geometric form are allowed. Where only a size tolerance is specified, the limits of size of an individual feature describe the extent to which variations in its geometric form, as well as size, are allowed.

The **actual size** of an individual feature at any cross section must be within the specified tolerance of size. The form of an individual feature is controlled by its limits of size.

The surface or surfaces of a feature must not extend beyond a boundary of perfect form at MMC. This boundary is the true geometric form represented by the drawing. No variation in form is permitted if the feature is produced at its MMC limit of size. Where the actual size of a feature has departed from MMC toward LMC, a variation in form is allowed equal to the amount of such departure. There is no requirement for a boundary of perfect form at LMC. Thus, a feature produced at its LMC limit of size is permitted to vary from true form to the maximum variation allowed by the boundary of perfect form at MMC. The control of geometric form by limits of size does not apply to the following:

1. Stock such as bars, sheets, tubing, structural shapes, and other items produced to established industry or government standards that prescribe limits for straightness, flatness, and other geometric characteristics
2. Parts subject to free-state variation in the unrestrained condition.

The limits of size do not control the orientation or location relationship between individual features. Features shown perpendicular, coaxial, or symmetrical to each other must be

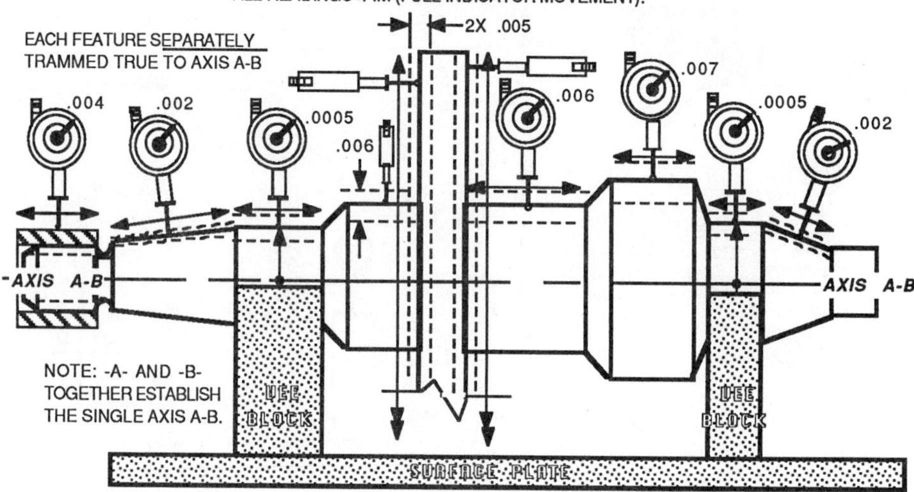
DRAWING SPECIFICATION
.005 A-B
.XXX-XX UNF-3B THD
.004 A-B
O.D.
10°
2X Ø .XX-.XX
.006 A-B (2X)
20°
.002 A-B
2X (BOTH TAPERS)
ANY 2 DIA.S RUNOUT TO A COMMON AXIS ARE RUNOUT TO EACH OTHER TO THE SUM OF THE RUNOUTS (-A- IS RUNOUT .001 TO -B-).
ALL DIAMETERS AND LENGTHS REQUIRE A SIZE TOLERANCE (NOT VERIFIED BY RUNOUT).
Ø .XXX-.XXX
.0005 A-B
-A-
Ø .XXX-.XXX
.007 A-B
-B-
DRAWING MEANING & INSPECTION DIAGRAM
ALL READINGS=FIM (FULL INDICATOR MOVEMENT).
2X .005
EACH FEATURE SEPARATELY TRAMMED TRUE TO AXIS A-B
.004
.002
.0005
.006
.007
.002
AXIS A-B
NOTE: -A- AND -B- TOGETHER ESTABLISH THE SINGLE AXIS A-B.
VEE BLOCK
SURFACE PLATE

FIGURE 12.42 Total Runout

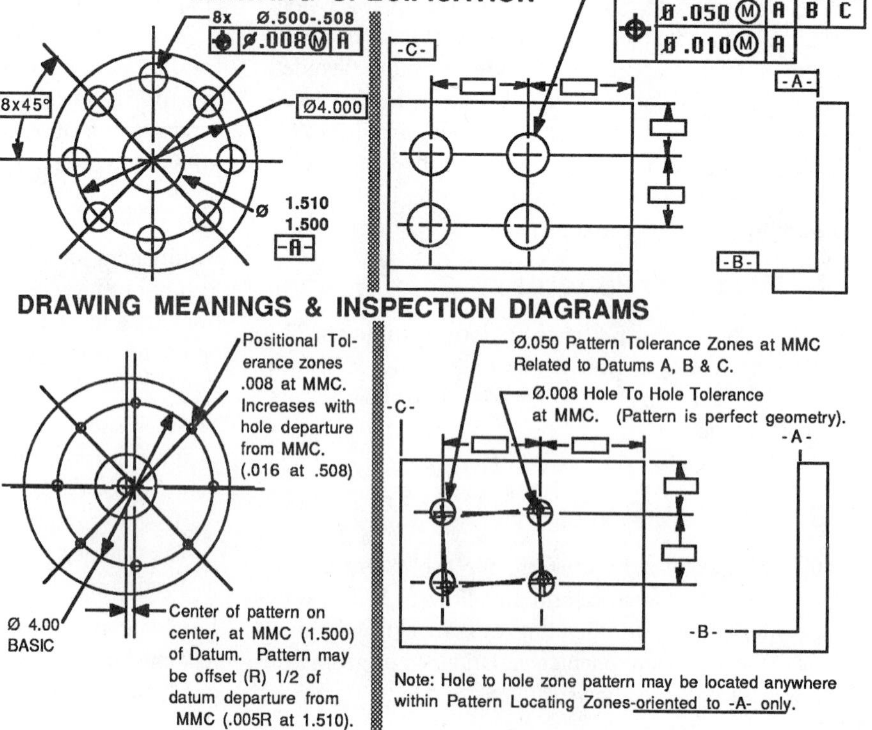
DRAWING SPECIFICATION
8x Ø.500-.508
Ø.008 Ⓜ A
8x45°
Ø4.000
Ø 1.510 1.500
-A-
-C-
Ø.375-.385
Ø.050 Ⓜ A B C
Ø.010 Ⓜ A
-A-
-B-
DRAWING MEANINGS & INSPECTION DIAGRAMS
Positional Tolerance zones .008 at MMC. Increases with hole departure from MMC. (.016 at .508)
Ø 4.00 BASIC
Center of pattern on center, at MMC (1.500) of Datum. Pattern may be offset (R) 1/2 of datum departure from MMC (.005R at 1.510).
Ø.050 Pattern Tolerance Zones at MMC Related to Datums A, B & C.
Ø.008 Hole To Hole Tolerance at MMC. (Pattern is perfect geometry).
-C-
-A-
-B-
Note: Hole to hole zone pattern may be located anywhere within Pattern Locating Zones-oriented to -A- only.

FIGURE 12.43 Position Location Tolerance

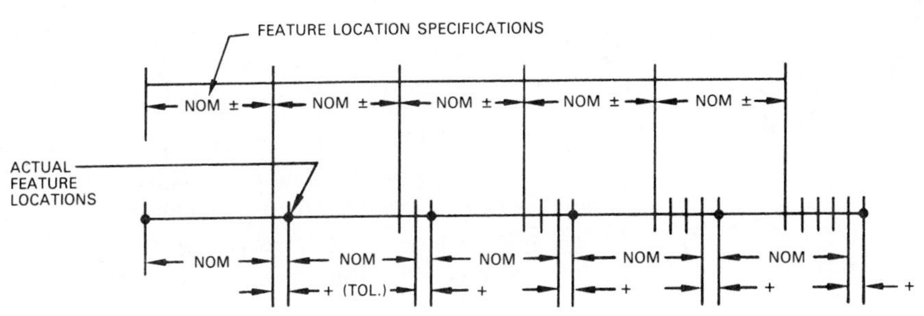

(a)

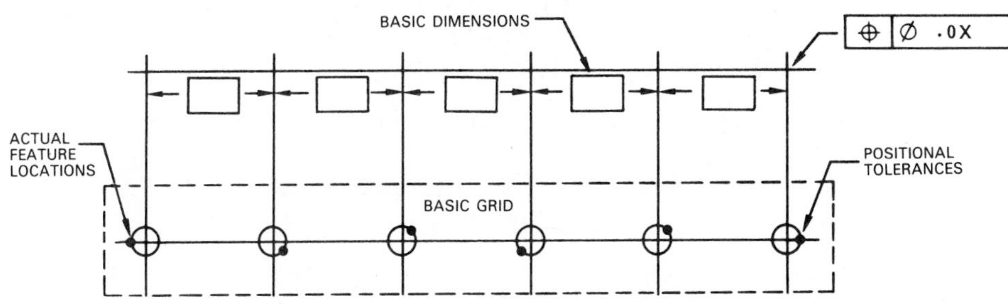

(b)

FIGURE 12.44 Conventional Chain Versus the Positional Grid System

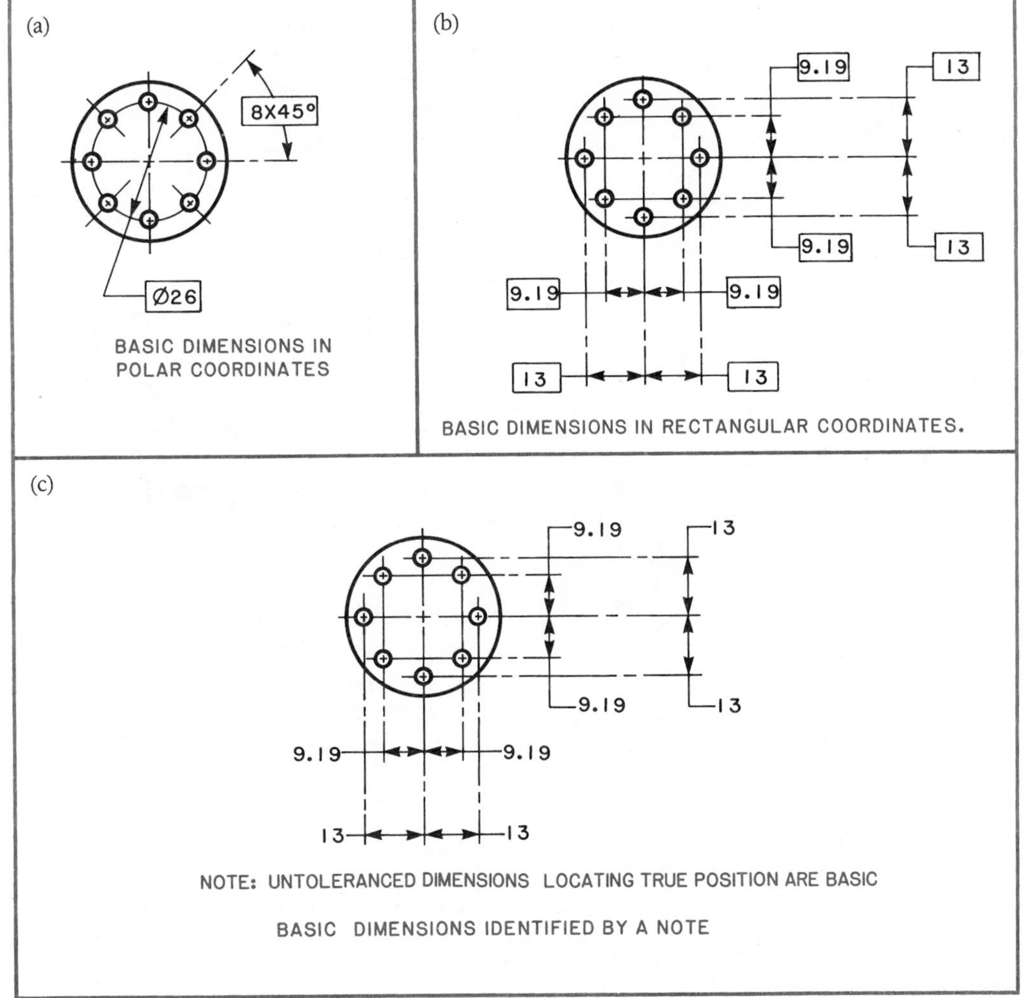

FIGURE 12.45 Identifying Basic Dimensions

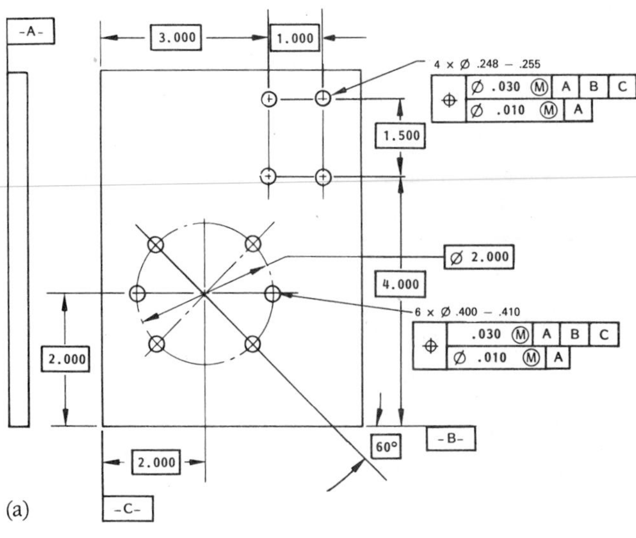

(a)

.030 DIAMETER AT MMC PATTERN-LOCATING ZONE (4 ZONES, BASICALLY RELATED AND BASICALLY ORIENTED TO THE DATUMS)

.010 DIAMETER AT MMC INTRAPATTERN TOLERANCE ZONES (4 ZONES, BASICALLY RELATED TO EACH OTHER)

FEATURE AXES MUST SIMULTANEOUSLY LIE WITHIN BOTH TOLERANCE ZONES

1.500

1.500

4.000

3.000

1.000

1.000

(b)

SQUARE PATTERN

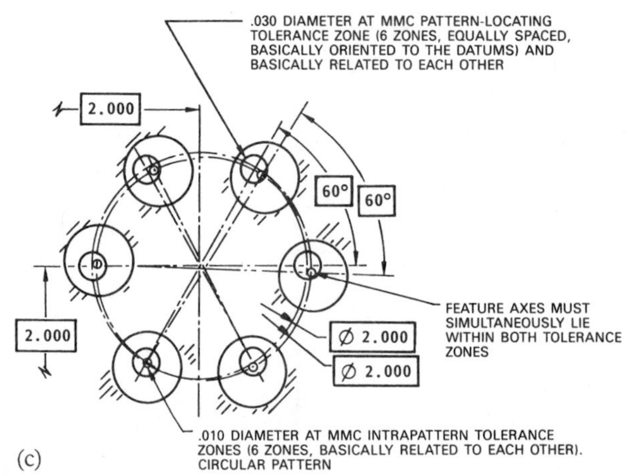

(c)

FIGURE 12.46 Composite Positional Tolerance to Datum Reference Frame

controlled for location or orientation. If it is necessary to establish a boundary of perfect form at MMC to control the relationship between features, use the following:

1. Specify a zero tolerance of orientation at MMC, including a datum reference (at MMC, if applicable), to control the angularity, perpendicularity, or parallelism or the feature.
2. Specify a zero positional tolerance at MMC, including a datum reference at MMC, to control coaxial or symmetrical features.
3. Indicate this control for the features involved with a note such as

   PERFECT ORIENTATION (OR COAXIALITY OR SYMMETRY) AT MMC
   REQUIRED FOR RELATED FEATURES.

4. Relate dimensions to a datum reference frame.

12.10.1 ISO Interpretation of Limits of Size

Where datums are specified in the ISO system measurements are made from the datums and made relative to them. Where datums are not specified, linear dimensions are intended to apply on a point-to-point basis or directly between the points indicated on the drawing. Unfortunately, caliper measurements float relative to one another and the exact shape is not known (Fig. 12.47). If the configuration is controlled, a form tolerance such as straightness or flatness is given.

Additionally, the direction of measurement can be a problem for a geometry that is not ideal. In Figure 12.48, the vertical measurements are not perpendicular to the horizontal

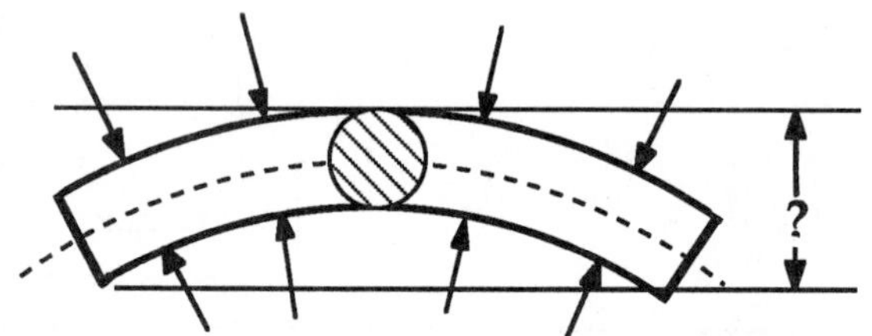

FIGURE 12.47 Caliper Measurements Do Not Measure Form Illustration shows a cylindrical part.

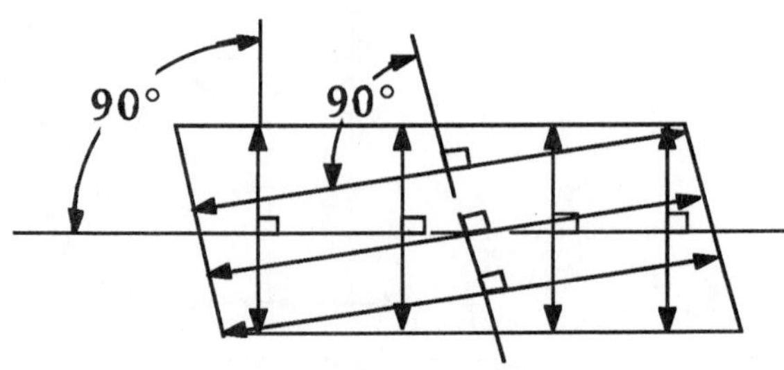

FIGURE 12.48 Measuring Orientation for Caliper Measurements Illustration shows a rectangular part.

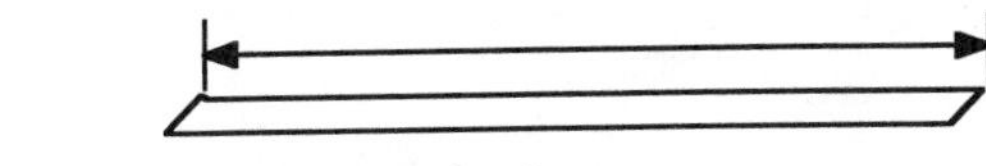

FIGURE 12.49 **Measuring Thin Parts**

ones. If the sides of a part are not parallel, finding the center plane to orient measurements is another problem.

For thin parts, the rule changes to making measuring parallel to the base (Fig. 12.49). Further, to make this system work, a rule of independence was devised:

> Every requirement on a drawing is intended to be applied independently, without reference to other dimensions, conditions, or characteristics, unless a particular relationship is specified.

This rule is voided, however, when "limits and fits" are specified. In that case, the Taylor principle is used (the basis of ANSI). Rule 1 of the ANSI standard regarding limits of size follows:

> *Rule 1:* The surface(s) of a feature shall not extend beyond a boundary (envelope)of perfect form at MMC. This boundary is the true geometric form represented by the drawing. There is no requirement for a boundary of perfect form at LMC (Fig.12.50).

This rule does not apply to stock materials that use established industry or government standards; to parts specified in the "free state" (nonrigid); and to those specifically excluded, such as straightness of the axis or where the note **PERFECT FORM AT MMC NOT REQUIRED** is specified.

This rule applies to individual features and, although they control the form of an individual feature of size, they do not

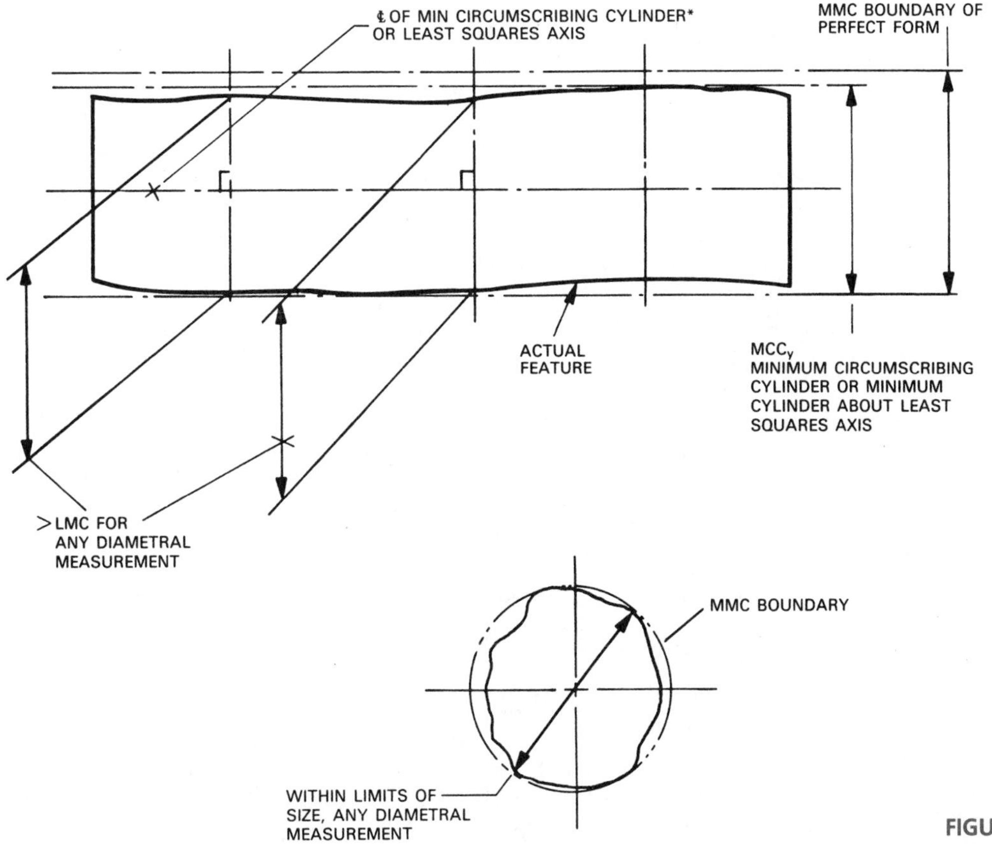

FIGURE 12.50 **Limits of Size Interpretation for Individual Features**
Continued

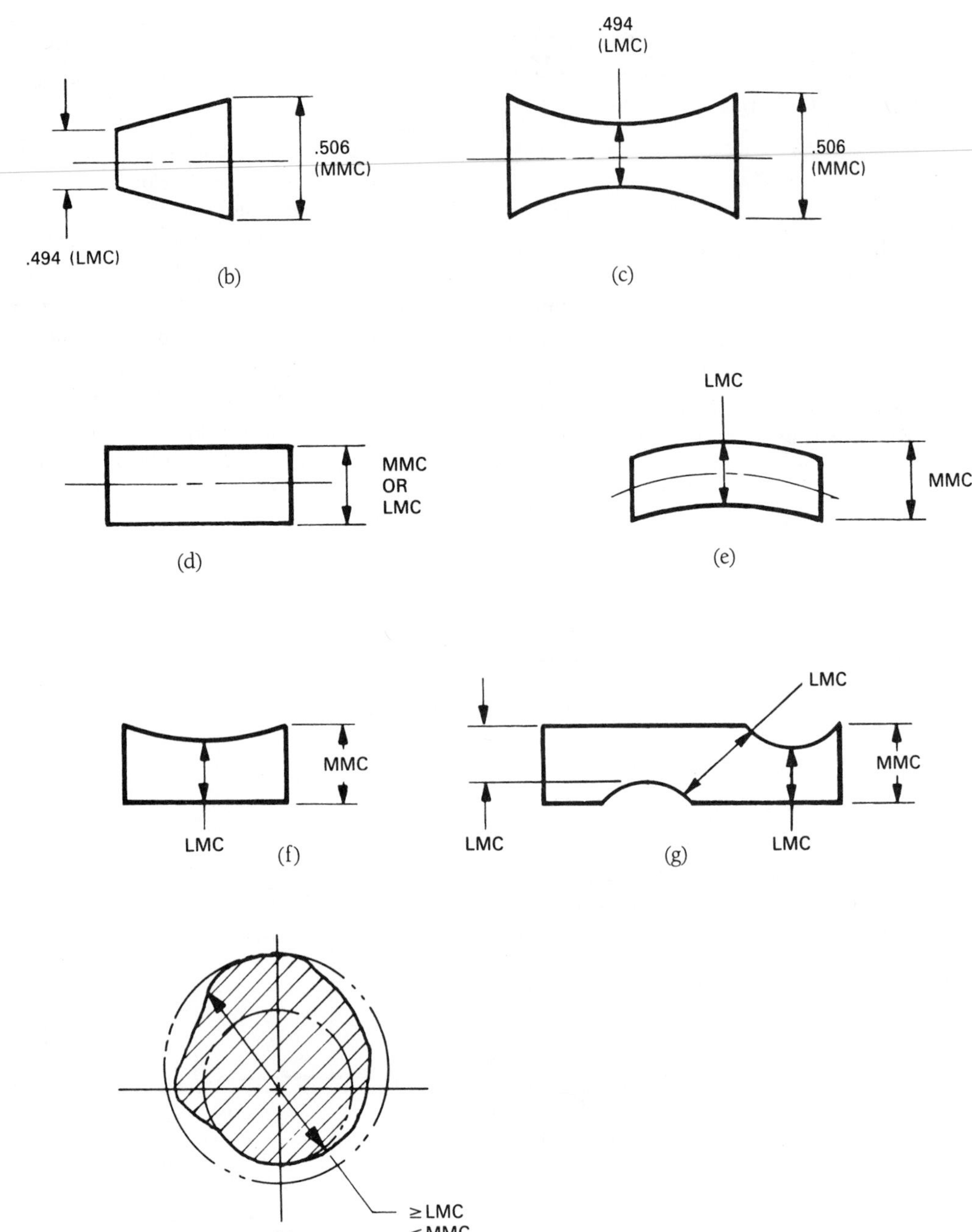

FIGURE 12.50 Limits of Size Interpretation for Individual Features— *Continued*

control the orientation, location, or runout of features to each other. These relationships are defined by tolerances, notes, or other specifications called out directly on the drawing.

Advances in metrology, especially where computerized mathematical models and algorithms are used such as on coordinate measuring machines (CMM), lend themselves to the ANSI version of limits of size. A ring gage made to the MMC size of a shaft, and as long as the shaft, may be used to verify a shaft for Rule 1 compliance. A micrometer or caliper is used to verify the LMC limit. A plug gage, made to the MMC size of a hole, and as long as the hole, may be used to verify a hole for Rule 1 compliance. An inside micrometer or caliper is used to verify the LMC limit.

12.10.2 U.S. Interpretation of Limits of Size

In the United States, the **limits of size** of a feature describe the extent within which variations of geometric form and size are allowed. This control applies solely to individual features of size. **Feature of size** refers to one cylindrical or spherical surface, or a set of two plane parallel surfaces, each of which is associated with a size dimension. Where only a **tolerance of size** is specified, the limits of size of an individual feature describe the extent to which variations in its geometric form and are allowed (Fig. 12.51). The **actual size** of an individual feature at any cross section is within the specified tolerance of size.

12.11 General Tolerancing Rules

Rules have been established to ensure uniform interpretation and avoid costly errors and misunderstandings. Study the following six rules carefully:

1. The system of indicating tolerances (whether size, location, or geometry) does not necessarily require the use of any particular method of production or quality.
2. Regardless of the number of places involved, all toleranced limits are considered to be absolute. Each limit is considered to be continued with trailing zeros. For example:

1.22		=	1.220 000 000 ...0
1.20			1.200 000 000 ...0
1.2		=	1.200 000 000 ...0
1.0			1.000 000 000 ...0
1.20	+.02	=	1.220 000 000 ...0
	−.00		1.200 000 000 ...0

This rule applies to all limits (plus or minus or limit dimensioned), including those where title block tolerances are applied.

3. All dimensions and tolerances are at 68°F (20°C) unless otherwise specified.
4. Surfaces drawn at 90° are subject to the title block tolerance specified for angles or by a note, such as

   PERFECT ORIENTATION REQUIRED AT MMC.

   This rule also applies to features that have a common centerline or axis of revolution. Where function or interchangeability is affected, the tolerances are to be specified.
5. Theoretical constructions such as centerlines or planes, shown at right angles, and from which features such as holes or pins are dimensioned, are considered to be at 90° basic. Variations in the inspection setup are subtracted from the allowable tolerances during the verification process.
6. The tolerance specified on the drawing is the total amount allowable, including manufacturing, inspection, and gaging variations. To assure rapid part acceptance, manufacturing does not usually use more than 90° of the available tolerance.

12.11.1 General Rules of Geometric Tolerancing

Use the following rules to apply geometric tolerancing to a part:

1. The surface(s) of a feature must not extend beyond a boundary (envelope) of perfect form at MMC. There is no requirement for a boundary of perfect form at LMC.
2. Position tolerance requirements for modifiers are specified in the feature control frame. A modifier, M, S, or L, is specified after the feature tolerance and after each datum for features and datums of size. No modifier is specified for a single-plane surface.
3. Requirements for modifiers of tolerances other than position are specified in the feature control frame. RFS applies, unless another modifier is specified, for all features and datums. The RFS modifier is usually not shown in the feature control frame. MMC is specified for features and datums of size where the design allows.

12.11.2 Setting Tolerances

The **nominal size** is often referred to as the **basic size** or **design size.** The nominal and associated tolerances have the same number of decimal places except in the metric system.

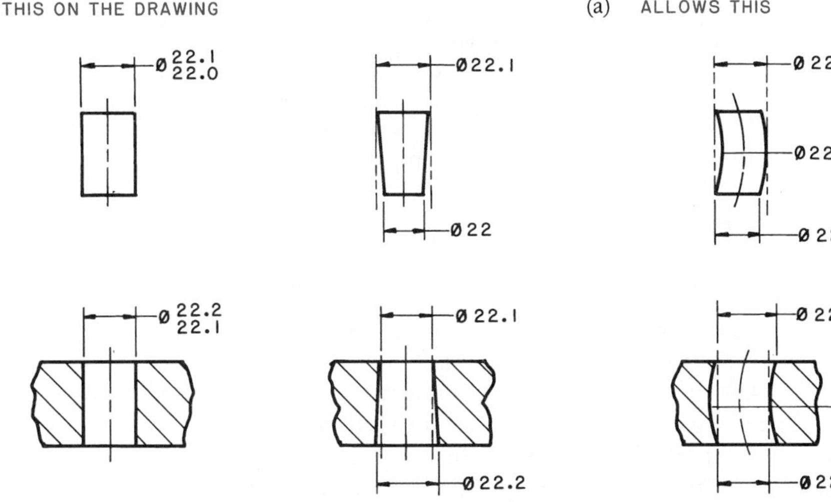

FIGURE 12.51 Tolerance Variations Allowed for Individual Features

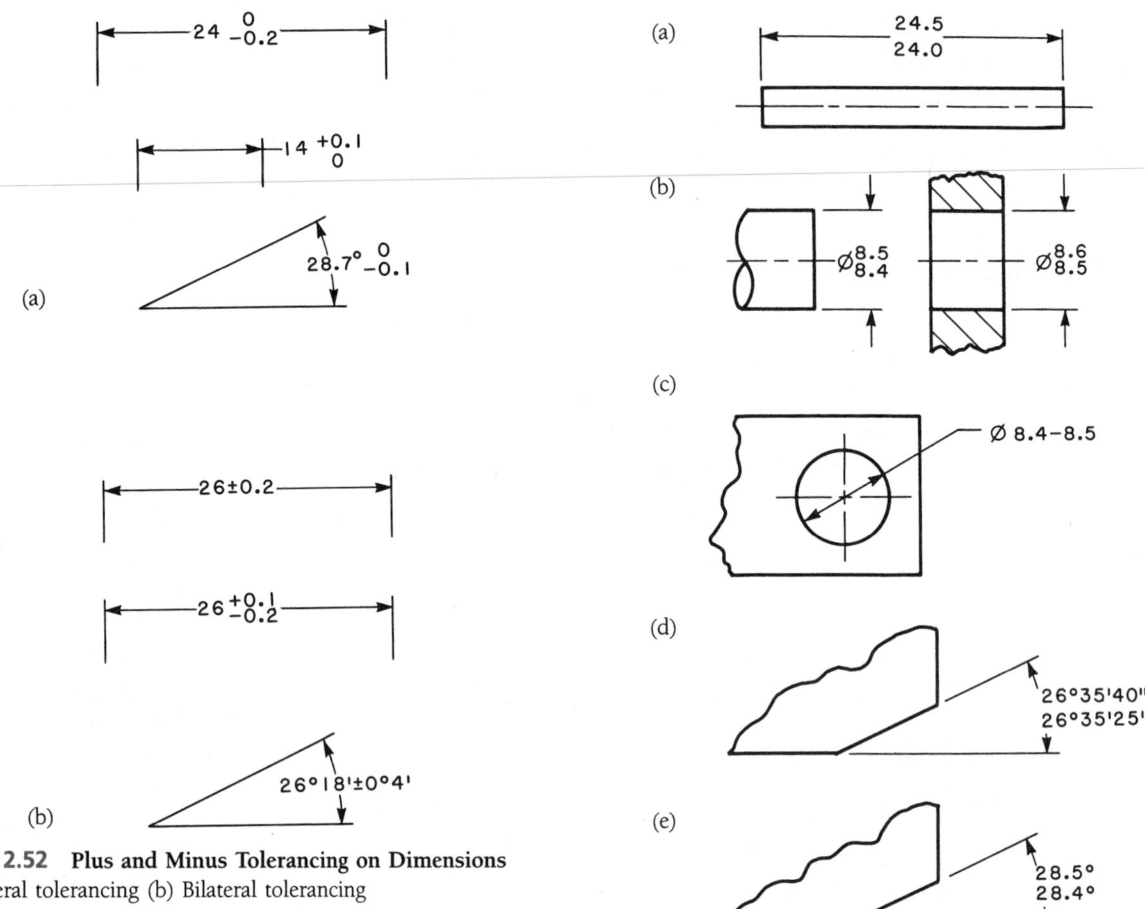

FIGURE 12.52 Plus and Minus Tolerancing on Dimensions
(a) Unilateral tolerancing (b) Bilateral tolerancing

FIGURE 12.53 Limit Dimensioning on Drawings

The "plus" value is shown above the "minus" value. European drawings also use +, + and −, and − tolerances.

In **unilateral tolerancing** [Fig. 12.52(a)], the tolerance is applied in one direction, the other value is zero:

$$1.200\ {}^{+.002}_{-.000} \qquad 1.200\ {}^{+.000}_{-.006}$$

In **bilateral tolerancing** [Fig. 12.52(b)], the tolerance is applied in both directions from the nominal:

$$.500 \pm .005 \quad \text{or} \quad 1.200\ {}^{+\ .002}_{-\ .005}$$

12.11.3 Limit Dimensioning

In **limit dimensioning,** the maximum value is placed above the minimum value (Fig. 12.53). In note form, the larger value is placed to the right of the lesser value, separated by a dash. Both limits have the same number of decimal places:

$$\begin{matrix} .750 \\ .748 \end{matrix} \quad \text{or } .748-.750$$

Even values are preferred. Although there is usually a trailing zero, the number of decimal places is minimized. Plus or minus and limit dimensions may appear on the same drawing. Generally, limit dimensions are used to specify the size of features; plus or minus dimensions are used to specify the location of features. Plus or minus dimensions, in bilateral form, are preferred for numerical control production, where the mean is used.

12.11.4 Title Block Tolerances

Title block tolerances are used where there is a uniformity in tolerances (Fig. 12.54). In Figure 12.55, the nominal dimension is given alone on the face of the drawing and a bilateral tolerance is shown in the title block. If larger tolerances are allowed for a particular feature, they should be specified. In European title blocks, tolerances are based on feature size. In the United States, they are based on the number of decimal places specified (Fig. 12.55).

12.11.5 Tolerance Accumulation

Figure 12.56 compares the tolerance values from three methods of dimensioning:

1. *Chain dimensioning:* The maximum variation between two features is equal to the tolerances on the intermediate distances. This method results in the *greatest tolerance accumulation.* In Figure 12.56(a), the tolerance accumulation between surfaces **X** and **Y** is ± .15.

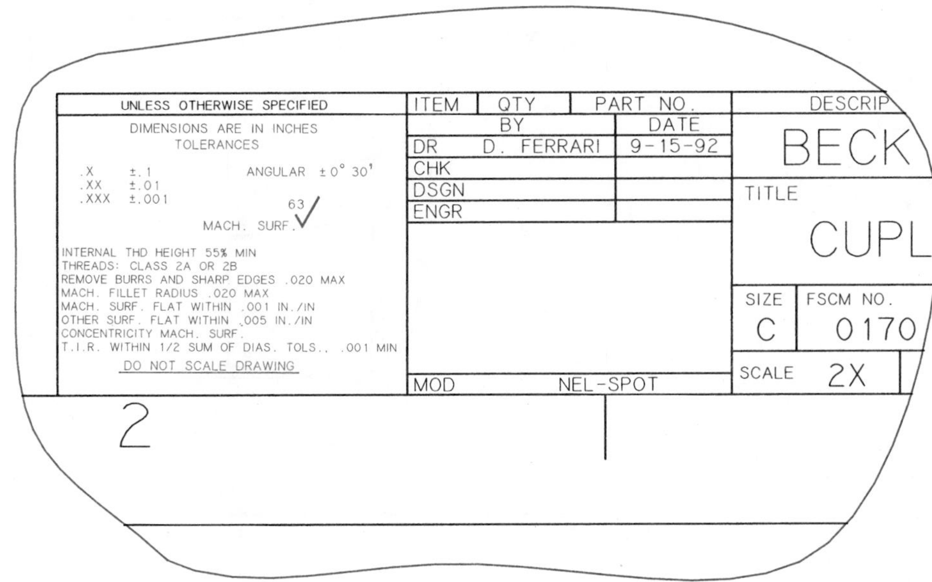

FIGURE 12.54 Title Block Tolerances

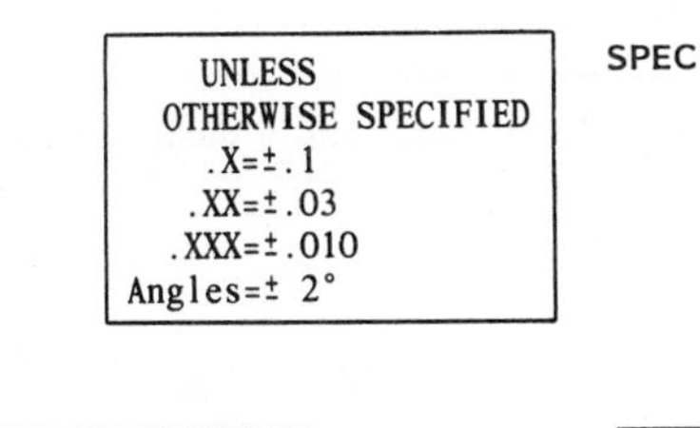

SPECIFICATION		MEANING
.5	=	.5±.1
.50	=	.50±.03
.500	=	.500±.010
12°	=	12° ±2°

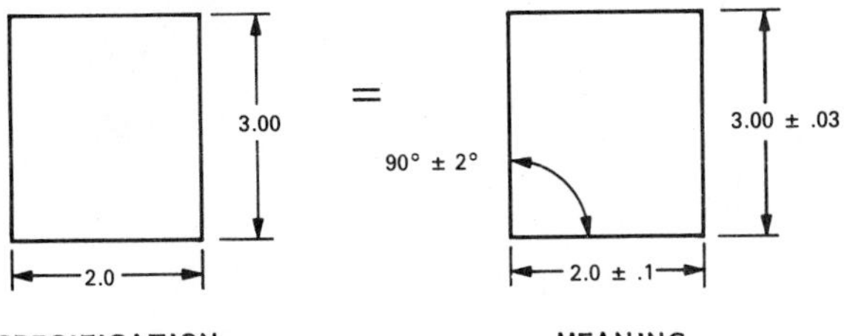

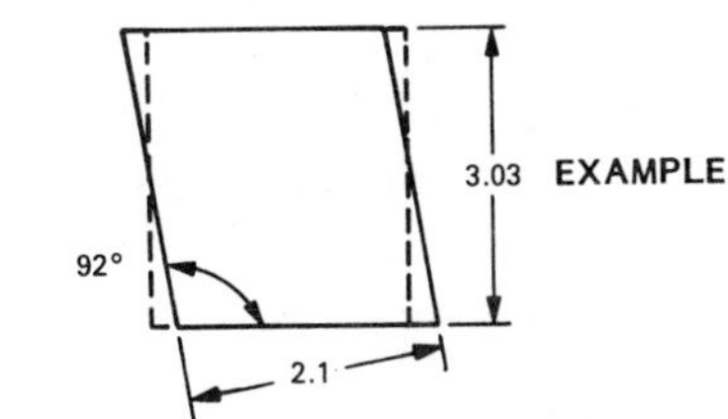

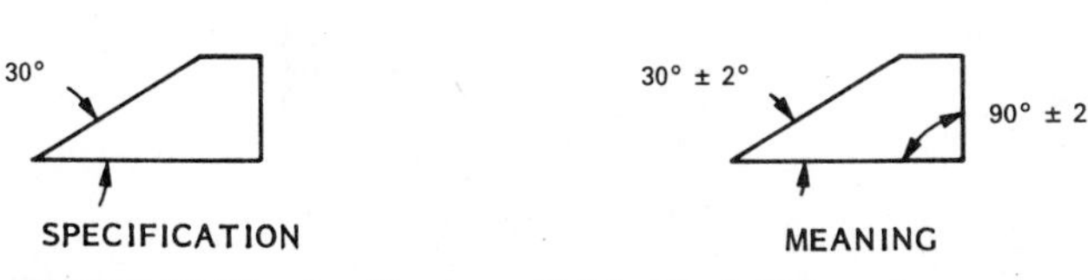

FIGURE 12.55 Application of Title Block Tolerance

2. *Baseline dimensioning:* The maximum variation between two features is equal to the sum of the tolerances on the two dimensions from their origin to the features. This results in a reduction of the tolerance accumulation. In Figure 12.56(b), the tolerance accumulation between surfaces **X** and **Y** is ± .1.
3. *Direct dimensioning:* The maximum variation between two features is controlled by the tolerance on the dimension between the features. This results in the *least tolerance.* In Figure 12.56(c), the tolerance between surfaces **X** and **Y** is ± .05.

12.11.6 Tolerances for Flat and Conical Tapers

Taper is defined as the ratio of the difference in the diameters of two sections, perpendicular to the axis, of a cone, to the distance between these sections. A **conical taper** is specified by one of the following methods:

1. A basic taper and a basic diameter (Fig. 12.57)
2. A size tolerance combined with a profile of a surface tolerance applied to the taper
3. A toleranced diameter at both ends or a taper and a toleranced length.

A **flat taper** is defined by specifying a toleranced slope and a toleranced height at one end (Fig. 12.58). **Slope** is defined as the inclination of a surface expressed as a ratio of the difference in the heights at each end, above and at right angles to a baseline, to the distance between those heights. Flat and conical tapers are toleranced as shown in Figure 12.59.

12.11.7 Single Limits: Min and Max

The unspecified limit in minimum dimensions (the maximum limit) approaches infinity. Therefore, overall lengths are not

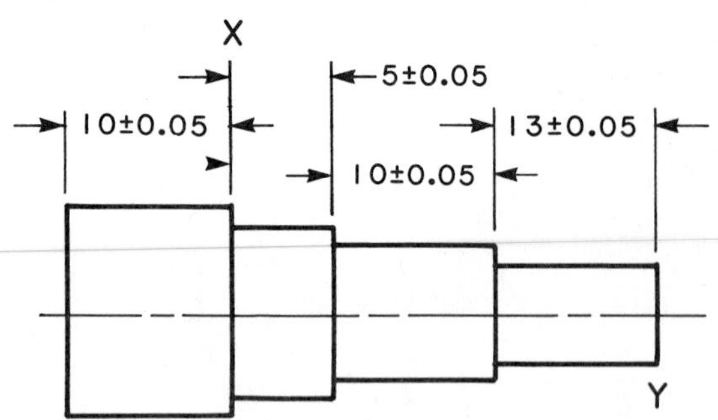

(a) Chain dimensioning—greatest tolerance accumulation between X and Y

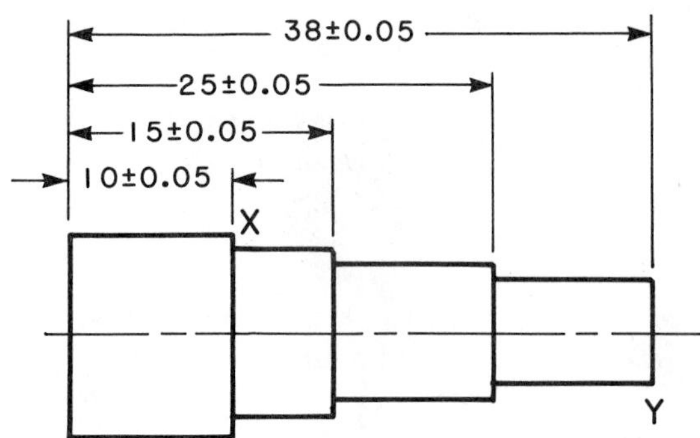

(b) Base line dimensioning—lesser tolerance accumulation between X and Y

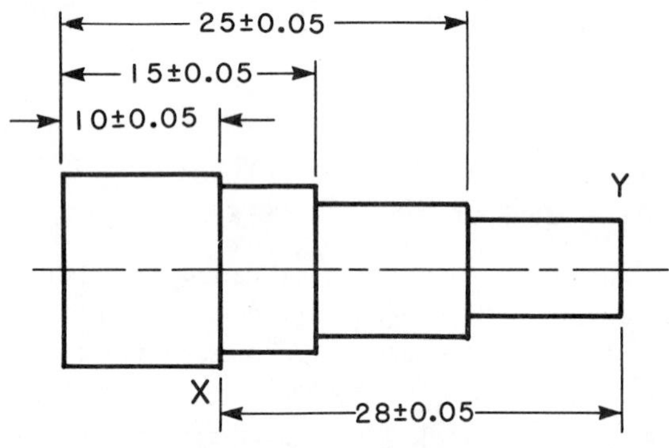

(c) Direct dimensioning—least tolerance between X and Y

FIGURE 12.56 Tolerance Accumulations

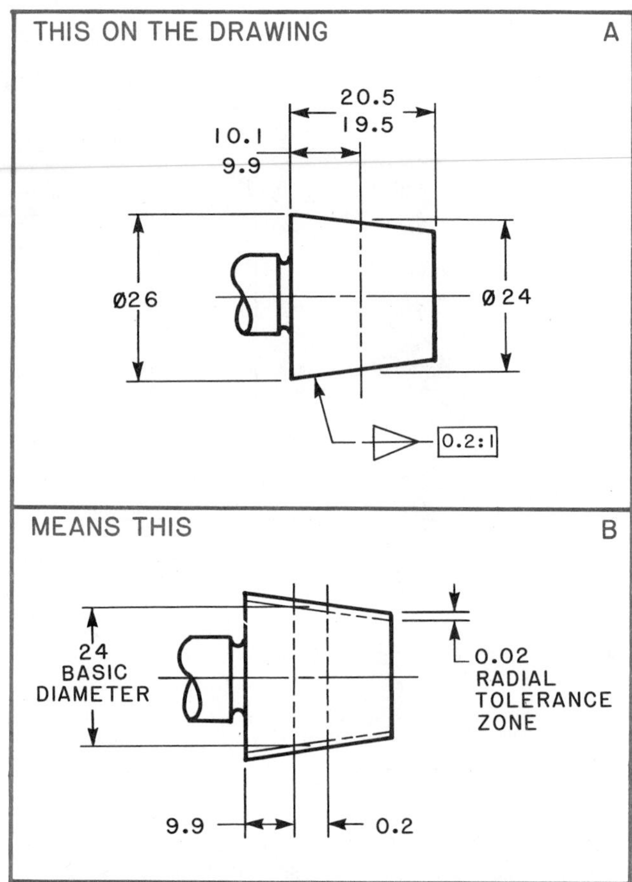

FIGURE 12.57 Specifying a Basic Taper and a Basic Diameter

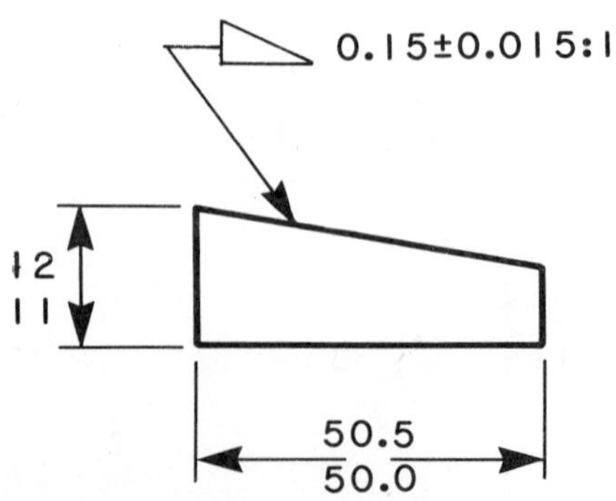

FIGURE 12.58 Slope Designation on a Drawing

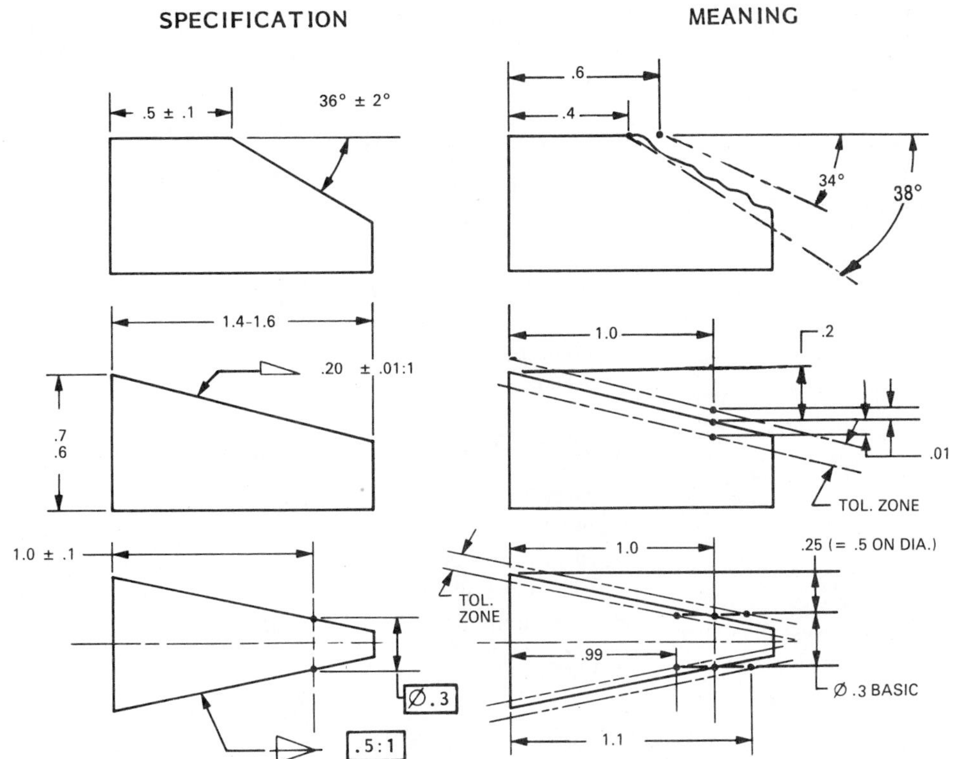

FIGURE 12.59 Flat and Conical Taper Tolerance Zones

specified as minimums. The unspecified value in maximum dimensions (the minimum limit) approaches zero.

12.11.8 Reference Dimensions

Reference dimensions, as discussed in Chapter 11, are specified by enclosing the dimension in parentheses, for example, (.500). No tolerance is given. Reference dimensions are not intended to govern production or inspection; they are informational, not controlling.

12.11.9 Functional Dimensions

Tolerances accumulate (Fig. 12.56). An example of how to control this situation is shown in Figure 12.60. The most important function of the firing pin is to project far enough to detonate the primer, but not far enough to pierce the primer. Also, the point must be fully below the bolt face, in the retracted position, to prevent premature detonation in the cartridge. This function is controlled by dimension A, a direct dimension from the point face to the interface with the bolt in the full forward position. Dimension B is established similarly. Dimensions that affect function should be dimensioned directly to avoid tolerance buildup. Conversely, dimension C1 was replaced by (C) during a producibility team review. This dimension must be long enough to so that the hammer will drive the pin to its full forward position, but the length is not critical because the pin has plenty of overtravel. Dimension (C), for ease of manufacture, was taken to the end of the spherical surface that is contacted by the hammer. The tapered section does not have to be accurate, but it must be located. This is accomplished by dimension d.

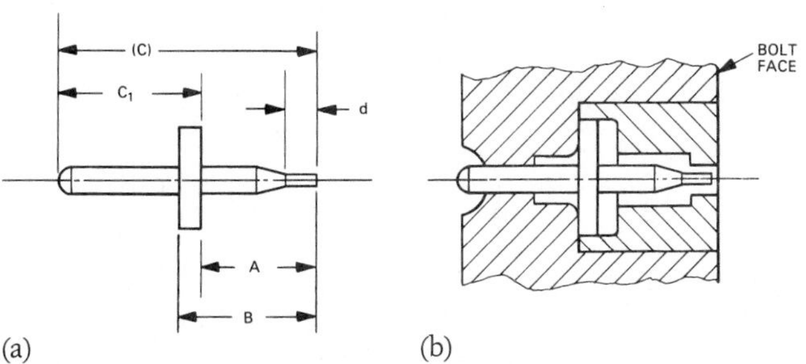

FIGURE 12.60 Functional Dimensioning: Firing Pin Assembly (a) Firing pin (b) Firing pin assembly

12.11.10 Nonmandatory Dimensions

If practical, the finished part is defined without specifying the manufacturing method. For example, only the diameter of a hole is given without indicating whether it should be drilled, reamed, punched, or made by any other operation. If manufacturing, processing, verification, or environmental information is essential to the definition, it is specified on the drawing. The affected dimensions are identified as "NON-MANDATORY (MFG DATA)." This allows improved or superior methods to be used at the discretion of the manufacturing or quality control departments.

12.11.11 Coordination, Interface Control, and Correlation Dimensions/Tolerances

On large projects or programs, dimensions and tolerances are agreed on by all parties. A coordination drawing has agreement on function, mating, shipping, equipment, removal, etc. These dimensions are then "flagged" on the hardware drawings. Therefore, the design is protected from inadvertent changes. Interface control and correlation drawings are handled the same way. The only difference is that coordination drawings are prepared for the total system, whereas correlation and interface control drawings are prepared for major subsystems. Dimensions on these drawings are flagged to be in compliance with the coordination drawings.

12.12 Limits and Fits

Production and inspection benefit from the use of standard limits. ANSI B4.2 and ISO 286 describe these systems. The ISO system has more than 500 possible tolerance zones for holes and shafts; ANSI has about 150. Many products may be standardized by using the system of limits and fits: drills, reamers, clevis pins, bushings, keys, keyways, gages, and bolts. **Renard preferred numbers** are used in the metric system with limits and fits to maximize standardization. The tables presented are not restricted to the preferred numbers. Sizes in design are often determined by factors other than cost such as mechanical and thermal stress or weight.

The three types of **fits** are **clearance**, where there is always clearance; **transition**, where there may be clearance or interference; and **interference**, where there is always interference (Fig. 12.61). The assumption used in "limits and fits" is the Taylor principle that each individual feature is in perfect form at MMC.

Limits and directly applied tolerance values are specified as follows:

1. **Limit dimensioning:** The high limit (maximum value) is placed above the low limit (minimum value). When expressed in a single line, the low limit precedes the high limit and a dash separates the two values.
2. **Plus and minus tolerancing:** The dimension is given first and is followed by a plus and minus expression of tolerance.

12.12.1 Single Limits

For **single limits,** MIN or MAX is placed after a dimension where other elements of the design definitely determine the other unspecified limit (depth of holes, length of threads, corner radii, chamfers, etc.). Single limits are used where the intent is clear and the unspecified limit can be zero or approach infinity without interfering with the designed function of the part.

12.12.2 Tolerance Expression

The conventions used for the number of decimal places carried are different for metric and inch drawings. For millimeter dimensions use the following rules. For **unilateral tolerancing**, when the plus or minus value is nil, a single zero is shown without a plus or minus sign:

Example:

$$32^{\;0}_{-0.02} \quad \text{or} \quad 32^{+.02}_{\;0}$$

Where **bilateral tolerancing** is used, both the plus and minus values have the same number of decimal places, using zeros where necessary:

Example:

$$32^{+.25}_{-.10} \quad \text{not} \quad 32^{+.25}_{-.1}$$

Where **limit dimensioning** is used and either the maximum or minimum value has digits following a decimal point, the other value has zeros:

Example:

$$\begin{matrix} 25.45 \\ 25.00 \end{matrix} \quad \text{not} \quad \begin{matrix} 25.45 \\ 25 \end{matrix}$$

For **inch dimensions,** both limit dimensions or the plus and minus tolerance and its dimension are expressed with the same number of decimal places:

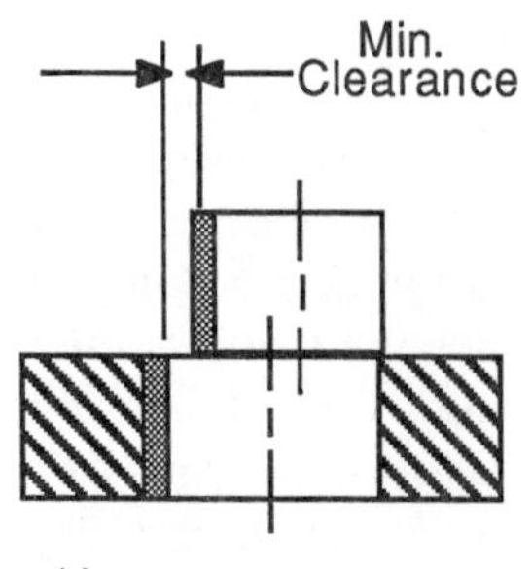

(a)

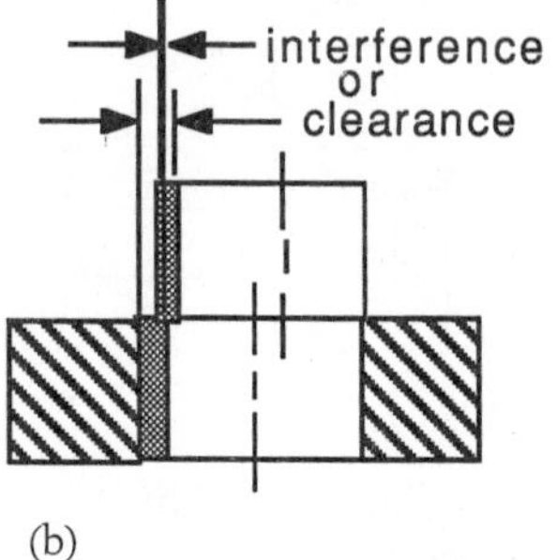

(b)

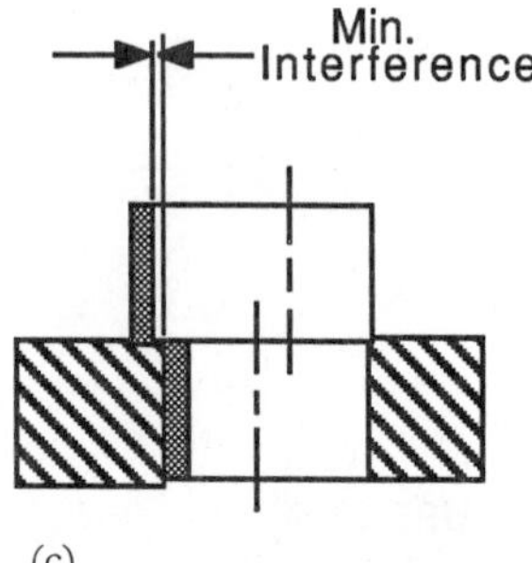

(c)

FIGURE 12.61 Basic Types of Fits (a) Clearance fit (b) Transition fit (c) Interference fit

Examples:

.5 + .005	not	.50 + .005
+.005 −.000	not	+.005 0
25.0 + .2	not	25 + .2

12.12.3 Preferred Metric Fits

For metric application of limits and fits, the tolerance may be indicated by a basic size and tolerance symbol. See ANSI B4.2 for complete information on this system. The preferred metric fits are defined as follows:

Loose Running H11/c11	Suitable for wide commercial tolerances or allowances on external members
Free Running H9/d9	Not suitable for use where accuracy is essential, but good for large temperature variations, high running speeds, or heavy journal pressures
Close Running H8/f7	Suitable for running on accurate machines and for accurate location at moderate speeds and journal pressures
Sliding Fit H7/g6	Not intended to run freely, but to move and turn freely and to locate accurately
Locational Clearance H7/h6	Provides snug fit for locating stationary parts, but can be freely assembled and disassembled
Locational Transition H7/k6	Suitable for accurate location; a compromise between clearance and interference
Locational Transition H7/n6	For more accurate location where greater interference is permissible
Locational Interference H7/p6	Suitable for parts requiring rigidity and alignment with prime accuracy of location, but without special bore pressure required
Medium Drive H7/s6	Suitable for ordinary steel parts or shrink fits on light sections, tightest fit usable with cast iron
Force Fit H7/u6	Suitable for parts that may be highly stressed or for shrink fits where the heavy pressing forces required may be impractical

12.12.4 Preferred Inch Fits

There are three general groups of fits: running and sliding fits, locational fits, and force fits. **Running** and **sliding** fits provide similar running performance, with a suitable lubrication allowance, throughout the range of sizes. The first ten preferences for inch fits are as follows: RC 4, RC 7, RC 9, LC 2, LC 5, LT3, LT6, LN 2, FN2, and FN4. Running and sliding fits are defined as follows:

RC 1 Close sliding fits are intended for the accurate location of parts that must assemble without perceptible play.

RC 2 Sliding fits are intended for accurate location, but with greater maximum clearance than class RC 1. Parts made to this fit move and turn easily, but are not intended to run freely; in the larger sizes, they may seize with small temperature changes.

RC 3 Precision running fits are about the closest fits that can be expected to run freely, and they are intended for precision work at slow speeds and light journal pressures, but are not suitable where appreciable temperature differences are likely to be encountered.

RC 4 Close running fits are intended chiefly for running fits on accurate machinery with moderate surface speeds and journal pressures, where accurate location and minimum play are desired.

RC 5 and RC 6 Medium running fits are intended for higher running speeds, or heavy journal pressures, or both.

RC 7 Free running fits are intended for use where accuracy is not essential, or where large temperature variations are likely to be encountered, or under both these conditions.

RC 8 and RC 9 Loose running fits are intended for use where wide commercial tolerances may be necessary, together with an allowance, on the external member.

12.12.5 Locational Fits

Locational fits are fits intended to determine only the location of the mating parts; they may provide rigid or accurate location, as with interference fits, or provide some freedom of location, as with clearance fits. They are divided into three groups: **clearance fits (LC)**, **transition fits (LT)**, and **interference fits (LN)**.

LC Locational clearance fits are intended for parts that are usually stationary, but which can be freely assembled or disassembled. They range from snug fits for parts requiring accuracy of location, through the medium clearance fits for parts such as spigots, to the looser fastener fits where freedom of assembly is the prime consideration.

LT Locational transition fits are a compromise between clearance and interference fits, for applications where accuracy of location is important, but either a small amount of clearance or interference is permissible.

LN Locational interference fits are used where accuracy of location is of prime importance and for parts requiring rigidity and alignment with no special requirements for bore pressure. Such fits are not intended for parts designed to transmit frictional loads (these are covered by force fits).

12.12.6 Force or Shrink Fits

Force or shrink fits are a special type of interference fit, normally characterized by maintenance of constant bore pressures throughout the range of sizes. The interference varies almost directly with diameter, and the difference between its minimum and maximum value is small so as to maintain the resulting pressures within reasonable limits.

FN 1 Light drive fits are those requiring light assembly pressures, and they produce more or less permanent assemblies. They are suitable for thin sections or long fits, or in cast-iron external members.

FN 2 Medium drive fits are suitable for ordinary steel parts or for shrink fits on light sections. They are the tightest fits that can be used with high-grade cast-iron external members.

FN 3 Heavy drive fits are suitable for heavier steel parts or for shrink fits in medium sections.

FN 4 and FN 5 Force fits are suitable for parts that can be highly stressed or for shrink fits where the heavy pressing forces required are impractical.

12.12.7 Preferred Tolerance Zones

A profile tolerance may be applied to an entire surface or to individual profiles taken at various cross sections through the part. **Preferred tolerance zones** are shown in Figure 12.62.

A **positional tolerance** defines a zone within which the center, axis, or center plane of a feature of size is permitted to vary from true position. Basic dimensions establish the true

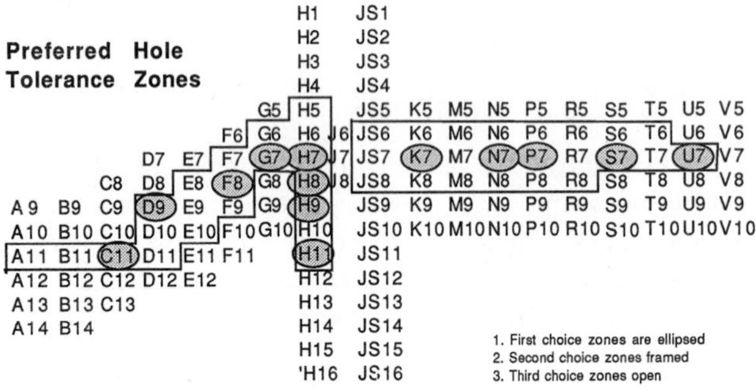

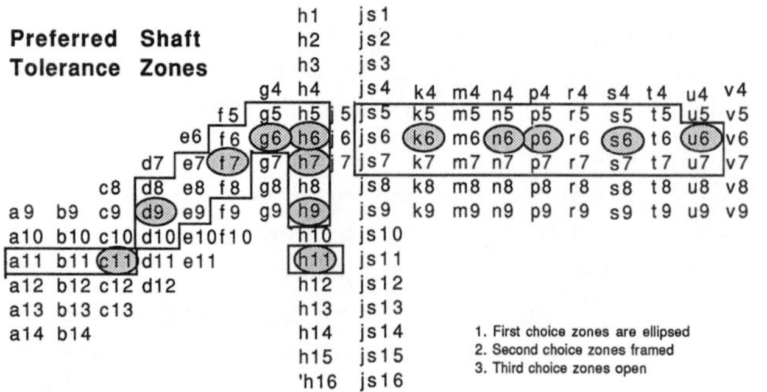

FIGURE 12.62 Preferred (Standardized) Tolerance Zones

position from specified datum features and between interrelated features. A positional tolerance is indicated by the position symbol, a tolerance, and appropriate datum references placed in a feature control frame.

12.12.8 Metric Preferred Sizes

Metric **preferred sizes** are based on the Renard series of preferred numbers. The first choice is rounded from the R10 series where succeeding numbers each increase by 25%. The second choice is rounded from the R20 series, which has 12% increments. The rationale for first choice sizes is the selection of every second number in the series such as 1, 1.6, 2.5, etc. This series is rounded from the R5 series of preferred numbers in which the increments are 60%. Preferred sizes from 1 to 300 are given in metric (Table 12.2) and .01 to 20.00 in inches (Table 12.3). The **hole basis system** is the preferred system for selecting standard tools and gages.

12.12.9 Standardized Tolerances

Standardized **metric tolerances** are given in Table 12.4. International tolerance (IT) grade values are used. The basis for these is the tolerance unit *i*, which is defined as follows: *i equals .45 times the cube root of D plus .001 times D (where D equals the nominal dimension in millimeters).*

Standardized **inch tolerances** are given in Table 12.5. The equivalent to IT values are based on the following formula (in inches): *i equals .052 times the cube root of D plus .001 times D (where D equals the nominal dimension in inches).*

IT grades for manufacturing processes are shown in Figure 12.63. Production costs may be reduced by using limits to dimensions or grades for which gaging equipment is available. Doing so allows production personnel to apply "Go" (MMC) and "Not Go" (LMC) gages to the inspection of small parts.

12.13 Calculating Limits and Fits

Metric tolerancing makes extensive use of limits and fits and uses the symbology extensively. In the United States, the symbology is supplemented by the limits, or the limits are specified and the symbology referenced on drawings to prevent misinterpretation. Tables 12.6 and 12.7 are metric and inch shaft position tables. These are used in conjunction with Tables 12. 4 and 12.5 to calculate limits.

12.13.1 Calculating Metric Hole Limits

The hole basis system places the low limit of a hole at exactly the basic or design size. The high limit is calculated by adding the IT value (Table 12.4) to it. For example:

$$\varnothing 80\ \text{H7} = 80 + 0.03, -0$$

where .03 comes from Table 12.4.

TABLE 12.2 Preferred Metric Basic Sizes (B.S.4318)

Choice			Choice			Choice		
1st	*2nd*	*3rd*	*1st*	*2nd*	*3rd*	*1st*	*2nd*	*3rd*
1					23			122
	1.1				24	125		
1.2			25					128
		1.3			26	130		
	1.4			28				132
		1.5	30				135	
1.6				32				138
		1.7			34	140		
	1.8		35					142
		1.9			36		145	
2				38				148
		2.1	40			150		
	2.2			42				152
		2.4			44		155	
2.5			45					158
		2.6			46	160		
	2.8			48				162
3			50				165	
		3.2		52				168
	3.5				54	170		
		3.8	55				175	
4					56			178
		4.2		58		180		
	4.5		60					182
		4.8		62			185	
5					64			188
		5.2	65			190		
	5.5				66			192
		5.8		68			195	
6			70					198
		6.2		72		200		
	6.5			74				205
		6.8	75				210	
	7				76			215
		7.5		78		220		
8			80					225
		8.5			82		230	
	9			85				235
		9.5			88	240		
10			90					245
	11				92		250	
12				95				255
		13			98	260		
	14		100					265
		15			102		270	
16				105				275
		17	110					285
	18				108		290	
		19			112			295
20				115		300		
		21			118			
	22		120					

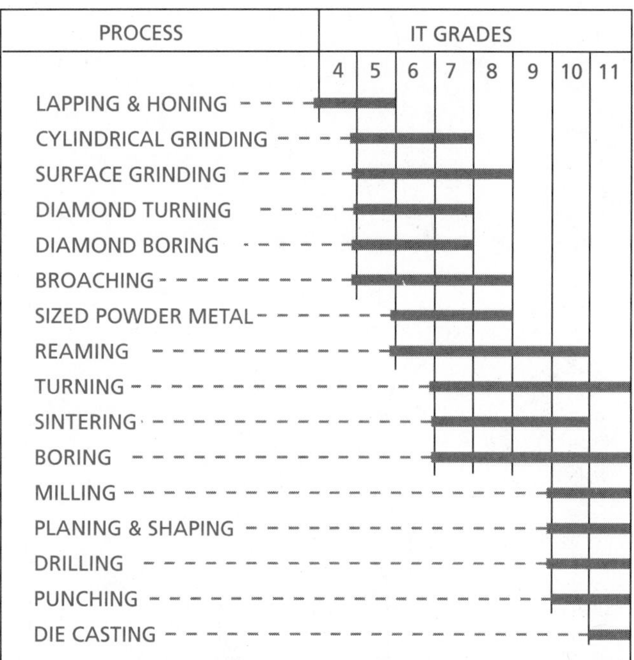

FIGURE 12.63 IT Grades for Manufacturing Processes

12.13.2 Calculating Inch Hole Limits

The low limit of the hole is at the basic or design size. The high limit is calculated by adding the IT value (Table 12.5) to it. For example:

$$\varnothing 3.1500\text{ H7} = 3.1500 + 0.0012, - .0000$$

where .0012 comes from Table 12.6.

12.13.3 Calculating Metric Shaft Limits

When using a hole basis system, unless the shaft is at the *h* position (see figure at the bottom right of Table 12.6), the value in Table 12.6 must be added algebraically to calculate the upper limit of a shaft; subtracting the IT value (Table 12.4) from the upper limit yields the lower limit.

Example:

$$\varnothing 80\text{ k6} = 80 + .021, + .002 \text{ or } 80.002 - 80.021$$
$$(.021 - .019) = .002$$

where .021 is from Table 12.6 and .019 from Table 12.4.

Example:

$$\varnothing 80\text{ c11} = .80 - .150, - 340 \text{ or } 79.660 - 79.850$$
$$(- .150 - .190) = - .340$$

where − .150 is from Table 12.6 and .190 from Table 12.4.

12.13.4 Calculating Inch Shaft Limits

When using a hole basis system, unless the shaft is at the *h* position (see figure at the bottom right of Table 12.7), the value in Table 12.7 must be added algebraically to arrive at

TABLE 12.3 Preferred Basic Sizes in Inches

Decimal			Fractional							
0.010	2.00	8.50	1/64	0.015625	2¼	2.2500	9½	9.5000
0.012	2.20	9.00	1/32	0.03125	2½	2.5000	10	10.0000
0.016	2.40	9.50	1/16	0.0625	2¾	2.7500	10½	10.5000
0.020	2.60	10.00	3/32	0.09375	3	3.0000	11	11.0000
0.025	2.80	10.50	1/8	0.1250	3¼	3.2500	11½	11.5000
0.032	3.00	11.00	5/32	0.15625	3½	3.5000	12	12.0000
0.040	3.20	11.50	3/16	0.1875	3¾	3.7500	12½	12.5000
0.05	3.40	12.00	¼	0.2500	4	4.0000	13	13.0000
0.06	3.60	12.50	5/16	0.3125	4¼	4.2500	13½	13.5000
0.08	3.80	13.00	3/8	0.3750	4½	4.5000	14	14.0000
0.10	4.00	13.50	7/16	0.4375	4¾	4.7500	14½	14.5000
0.12	4.20	14.00	½	0.5000	5	5.0000	15	15.0000
0.16	4.40	14.50	9/16	0.5625	5¼	5.2500	15½	15.5000
0.20	4.60	15.00	5/8	0.6250	5½	5.5000	16	16.0000
0.24	4.80	15.50	11/16	0.6875	5¾	5.7500	16½	16.5000
0.30	5.00	16.00	¾	0.7500	6	6.0000	17	17.0000
0.40	5.20	16.50	7/8	0.8750	6½	6.5000	17½	17.5000
0.50	5.40	17.00	1	1.0000	7	7.0000	18	18.0000
0.60	5.60	17.50	1¼	1.2500	7½	7.5000	18½	18.5000
0.80	5.80	18.00	1½	1.5000	8	8.0000	19	19.0000
1.00	6.00	18.50	1¾	1.7500	8½	8.5000	19½	19.5000
1.20	6.50	19.00	2	2.000	9	9.0000	20	20.0000
1.40	7.00	19.50	...	.....	...	.....	....	.....
1.60	7.50	20.00						
1.80	8.00	....	All dimensions are given in inches.					

TABLE 12.4 Metric Table-Standard Tolerances

Nominal[1] sizes		IT Tolerance Grades												IT = ISO Series of Tolerances					
Over mm	*Up to and including* mm	*IT01*	*IT0*	*IT1*	*IT2*	*IT3*	*IT4*	*IT5*	*IT6*[3]	*IT7*	*IT8*	*IT9*	*IT10*	*IT11*	*IT12*	*IT13*	*IT14*[2]	*IT15*[2]	*IT16*[2]
—	3	0·3	0·5	0·8	1·2	2	3	4	6	10	14	25	40	60	100	140	250	400	600
3	6	0·4	o·6	1	1·5	2·5	4	5	8	12	18	30	48	75	120	180	300	480	750
6	10	0·4	0·6	1	1·5	2·5	4	6	9	15	22	36	58	90	150	220	360	580	900
10	18	0·5	0·8	1·2	2	3	5	8	11	18	27	43	70	110	180	270	430	700	1100
18	30	0·6	1	1·5	2·5	4	6	9	13	21	33	52	84	130	210	330	520	840	1300
30	50	0·6	1	1·5	2·5	4	7	11	16	25	39	62	100	160	250	390	620	1000	1600
50	80	0·8	1·2	2	3	5	8	13	19	30	46	74	120	190	300	460	740	1200	1900
80	120	1	1·5	2·5	4	6	10	15	22	35	54	87	140	220	350	540	870	1400	2200
120	180	1·2	2	3·5	5	8	12	18	25	40	63	100	160	250	400	630	1000	1600	2500
180	250	2	3	4·5	7	10	14	20	29	46	72	115	185	290	460	720	1150	1850	2900
250	315	2·5	4	6	8	12	16	23	32	52	81	130	210	320	520	810	1300	2100	3200
315	400	3	5	7	9	13	18	25	36	57	89	140	230	360	570	890	1400	2300	3600
400	500	4	6	8	10	15	20	27	40	63	97	155	250	400	630	970	1550	2500	4000
500	630	—	—	—	—	—	—	—	44	70	110	175	280	440	700	1100	1750	2800	4400
630	800	—	—	—	—	—	—	—	50	80	125	200	320	500	800	1250	2000	3200	5000
800	1000	—	—	—	—	—	—	—	56	90	140	230	360	560	900	1400	2300	3600	5600
1000	1250	—	—	—	—	—	—	—	66	105	165	260	420	660	1050	1650	2600	4200	6600
1250	1600	—	—	—	—	—	—	—	78	125	195	310	500	780	1250	1950	3100	5000	7800
1600	2000	—	—	—	—	—	—	—	92	150	230	370	600	920	1500	2300	3700	6000	9200
2000	2500	—	—	—	—	—	—	—	110	175	280	440	700	1100	1750	2800	4400	7000	11000
2500	3150	—	—	—	—	—	—	—	135	210	330	540	860	1350	2100	3300	5400	8600	13500

Tolerance unit 0.001 mm

[1]STANDARD TOLERANCE IN MICRONS (1μ = 0.001 mm)

[2]Not applicable to sizes below 1 mm

[3]Not recommended for fits in sizes above 500 mm

ISO TOLERANCE GRADE 6 IN ABBREVIATED FORM IS IT6

TABLE 12.5 Inch Values-Standard Tolerances

IT Grade	01	0	1	2	3	4	5	6	7	8	9	10	11	12	13	14*	15*	16*
≤ 0.12	0.012	0.02	0.03	0.05	0.08	0.12	0.15	0.25	0.4	0.6	1.0	1.6	2.5	4.0	6.0	10.0	16.0	25.0
> 0.12 to 0.24	0.015	0.025	0.04	0.06	0.10	0.15	0.2	0.3	0.5	0.7	1.2	1.8	3.0	5.0	7.0	12.0	18.0	30.0
> 0.24 to 0.40	0.015	0.025	0.04	0.06	0.10	0.15	0.25	0.4	0.6	0.9	1.4	2.2	3.5	6.0	9.0	14.0	22.0	35.0
> 0.40 to 0.71	0.02	0.03	0.05	0.08	0.12	0.2	0.3	0.4	0.7	1.0	1.6	2.8	4.0	7.0	10.0	16.0	28.0	40.0
> 0.71 to 1.19	0.025	0.04	0.06	0.10	0.15	0.25	0.4	0.5	0.8	1.2	2.0	3.5	5.0	8.0	12.0	20.0	35.0	50.0
> 1.19 to 1.97	0.025	0.04	0.06	0.10	0.15	0.3	0.4	0.6	1.0	1.6	2.5	4.0	6.0	10.0	16.0	25.0	40.0	60.0
> 1.97 to 3.15	0.03	0.05	0.08	0.12	0.2	0.3	0.5	0.7	1.2	1.8	3.0	4.5	7.0	12.0	18.0	30.0	45.0	70.0
> 3.15 to 4.73	0.04	0.06	0.1	0.15	0.25	0.4	0.6	0.9	1.4	2.2	3.5	5.0	9.0	14.0	22.0	35.0	50.0	90.0
> 4.73 to 7.09	0.05	0.08	0.12	0.2	0.3	0.5	0.7	1.0	1.6	2.5	4.0	6.0	10.0	16.0	25.0	40.0	60.0	100.0
> 7.09 to 9.85	0.08	0.12	0.2	0.3	0.4	0.6	0.8	1.2	1.8	2.8	4.5	7.0	12.0	18.0	28.0	45.0	70.0	120.0
> 9.85 to 12.41	0.10	0.15	0.25	0.3	0.5	0.6	0.9	1.2	2.0	3.0	5.0	8.0	12.0	20.0	30.0	50.0	80.0	120.0
> 12.41 to 15.75	0.12	0.2	0.3	0.4	0.5	0.7	1.0	1.4	2.2	3.5	6.0	9.0	14.0	22.0	35.0	60.0	90.0	140.0
> 15.75 to 19.69	0.15	0.25	0.3	0.4	0.6	0.8	1.0	1.6	2.5	4.0	6.0	10.0	16.0	25.0	40.0	60.0	100.0	160.0

Table shows standard tolerances in 0.001 inches for diameter steps in inches.

*Up to .04 in, grades 14 to 16 are not provided.

TABLE 12.6 Metric Tolerance Zone Position—Standard Fits

*****************TABLE 5*****************

METRIC-TOLERANCE ZONE POSITION TABLE-SHAFT UPPER LIMITS FROM ZERO LINE (BASIC SIZE) SEE GRAPHIC AT LOWER RIGHT.

Loose, free & close running, sliding, locational, drive and force fits

Over-To	c11	d9	f7	g6	h6	k6	n6	p6	s6	u6
≤3	-60	-20	-6	-2	0	6	10	12	20	24
3 to 6	-70	-30	-10	-4	0	9	16	20	27	31
6 to 10	-80	-40	-13	-5	0	10	19	24	32	37
10 to 14	-95	-50	-16	-6	0	12	23	29	39	44
14 to 18	-95	-50	-16	-6	0	12	23	29	39	44
18 to 24	-110	-65	-20	-7	0	15	28	35	48	54
24 to 30	-110	-65	-20	-7	0	15	28	35	48	61
30 to 40	-120	-80	-25	-9	0	18	33	42	59	76
40 to 50	-130	-80	-25	-9	0	18	33	42	59	86
50 to 65	-140	-100	-30	-10	0	21	39	51	72	106
65 to 80	-150	-100	-30	-10	0	21	39	51	78	121
80 to 100	-170	-120	-36	-12	0	25	45	59	93	146
100 to 120	-180	-120	-36	-12	0	25	45	59	101	166
120 to 140	-200	-145	-43	-14	0	28	52	68	117	195
140 to 160	-210	-145	-43	-14	0	28	52	68	125	215
160 to 180	-230	-145	-43	-14	0	28	52	68	133	235
180 to 200	-240	-170	-50	-15	0	33	60	79	151	265
200 to 225	-260	-170	-50	-15	0	33	60	79	159	287
225 to 250	-280	-170	-50	-15	0	33	60	79	169	313
250 to 280	-300	-190	-56	-17	0	36	66	88	190	347

1. Basic hole system (unilateral hole basis) employed (table covers shafts).
2. Values are in thousandths of a millimeter (microns); sizes in mm.
3. Values represent the upper limit (relative to the zero line) of shafts
4. The selected fits indicated are recommended in ANSI B4.1. these are somewhat similar to those in (UK standard) BS 4500.
5. Add value from table algebraically to basic size for upper shaft limit

Clearance Fits

FIT	HOLE	SHAFT
loose	H11	c11
free	H9	d9
close	H8	f7
sliding	H7	g6
locationa	H7	h6

Transition Fits

FIT	HOLE	SHAFT
locationa	H7	k6
locationa	H7	n6

Interference Fits

FIT	HOLE	SHAFT
locationa	H7	p6
med. drive	H7	s6
force	H7	u6

USING THE TABLE ABOVE

Ø30 f7 (Shaft)

zero line 30.00

Basic size "-20 (.02) from above " table to get upper limit

IT7 f 29.980 29.959

IT7 (other table)=.021

30.000-.020=29.980
29.980-.021=29.959

Note:
•Upper case letters represent holes/bores. (e.g. H7).
•Lower case letters represent shafts (e.g. f6).
•The letter (location symbol) represents the position/distance to the "zero line"/Basic size.
•H and h are on the zero line.
• The number (quality no.) represents the tolerance grade. Higher no.s yield coarser fits.
•*See tolerance grade (IT) table for limits not shown in these tables.
*Add IT to hole for max limit (Basic size is min hole)
*Subtract IT from shaft limit calc. from table for min.

METHODS OF INDICATING
1. Ø30 f7
2. Ø29.980 +.000, -.021 (Ø30 f7)
3. Ø 29.959 - 29.980 (Ø30 f7)

INDICATING FITS
1. Ø.30 H8/f7 (hole first)
2. Ø30 H8 (30.000-30.033) f7 (29.959 - 29.980)

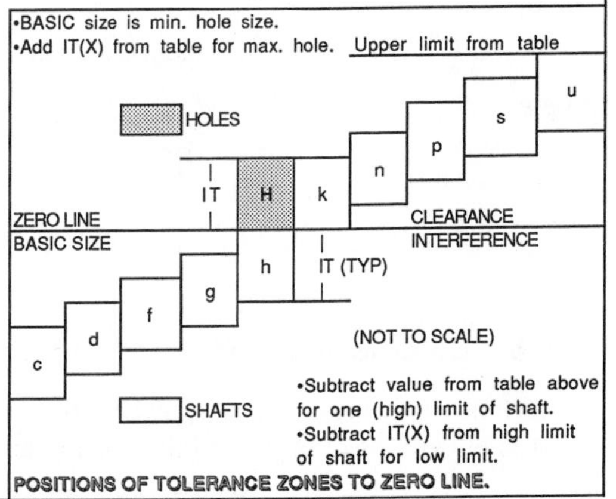

TABLE 12.7 Inch Tolerance Zone Position—Standard Fits

******************TABLE 6******************

INCH -TOLERANCE ZONE POSITION TABLE-SHAFT UPPER LIMITS
FROM ZERO LINE (BASIC SIZE) SEE GRAPHIC ON METRIC TABLE PAGE.

RUNNING, SLIDING, CLEARANCE TRANSITION AND INTERFERENCE LOCATIONAL FITS

Over-To	c9&c10	d8	d9	e7&E8	e9	f6,f7	f8	g4,g5	g6	h5,6,7&9	js6	js7	k6	k7	n5	n6	n7	p6	r6	Sp-10	Sp-11	Sp-12
0-.12	-2.50	-1.0	-1.0	-.6	-.6	-.3	-.3	-.10	-.10	0	0.1	0.2	-	-	0.45	0.5	0.65	0.65	0.75	-4.0	-4.0	-5
.12-.24	-2.80	-1.2	-1.2	-.8	-.8	-.4	-.4	-.15	-.15	0	0.15	0.25	-	-	0.5	0.6	0.8	0.8	0.9	-4.5	-4.5	-6
.24-.40	-3.00	-1.6	-1.6	-1.0	-1.0	-.5	-.5	-.20	-.20	0	0.2	0.3	0.5	0.7	0.65	0.8	1	1	1.2	-5.0	-5.0	-7
.40-.71	-3.50	-2.0	-2.0	-1.2	-1.2	-.6	-.6	-.25	-.25	0	0.2	0.35	0.5	0.8	0.8	0.9	1.2	1.1	1.4	-6.0	-6.0	-8
.71-1.19	-4.50	-2.5	-2.5	-1.6	-1.6	-.8	-.8	-.30	-.30	0	0.25	0.4	0.6	0.9	1	1.1	1.4	1.3	1.7	-7.0	-7.0	-10
1.19-1.97	-5.00	-3.0	-3.0	-2.0	-2.0	-1.0	-1.0	-.40	-.40	0	0.3	0.5	0.7	1.1	1.1	1.3	1.7	1.6	2	-8.0	-8.0	-12
1.97-3.15	-6.00	-4.0	-4.0	-2.5	-2.5	-1.2	-1.2	-.40	-.40	0	0.3	0.6	0.8	1.3	1.3	1.5	2	2.1	2.3	-9.0	-10.0	-14
3.15-4.73	-7.00	-5.0	-5.0	-3.0	-3.0	-1.4	-1.4	-.50	-.50	0	0.4	0.7	1	1.5	1.6	1.9	2.4	2.5	2.9	-10.0	-11.0	-16
4.73-7.09	-8.00	-6.0	-6.0	-3.5	-3.5	-1.6	-1.6	-.60	-.60	0	0.5	0.8	1.1	1.7	1.9	2.2	2.8	2.8	3.5	-12.0	-12.0	-18
7.09-9.85	-10.00	-7.0	-7.0	-4.0	-4.0	-2.0	-2.0	-.60	-.60	0	0.6	0.9	1.4	2	2.2	2.6	3.2	3.2	4.2	-15.0	-16.0	-22
9.85-12.41	-12.00	-8.0	-7.0	-5.0	-4.5	-2.5	2.2	-.80	-.70	0	0.6	1	1.4	2.2	2.3	2.6	3.4	3.4	4.7	-18.0	-20.0	-28
12.41-15.75	-14.00	-10.0	-8.0	6	-5.0	-3.0	2.5	-1.00	-.70	0	0.7	1	1.6	2.4	2.6	3	3.8	3.9	5.9	-22.0	-22.0	-30

1. Basic hole system employed (table covers shafts).
2. Values are in thousandths of an inch
3. Values represent the upper limit (relative to the zero line) of shafts
4. Values indicated "Sp-X" are not used in the ISO (International ISO 286) syst
5. <u>Add value from table algebraically to basic size for upper shaft limit.</u>

FORCE AND SHRINK FITS

Over-To	s6	t6	u6	x7	Sp-5
0-.12	0.85	-	0.95	1.3	0.5
.12-.24	1	-	1.2	1.7	0.6
.24-.40	1.4	-	1.6	2	0.75
0.4-.56	1.6	-	1.8	2.3	0.8
.56-.71	1.6	-	1.8	2.5	0.9
.71-.95	1.9	-	2.1	3	1.1
.95-1.19	1.9	2.1	2.3	3.3	1.2
1.19-1.58	2.4	2.6	3.1	4	1.3
1.58-1.97	2.4	2.8	3.4	5	1.4
1.97-2.56	2.7	3.2	4.2	6.2	1.8
2.56-3.15	2.9	3.7	4.7	7.2	1.9
3.15-3.94	3.7	4.4	5.9	8.4	2.4
3.94-4.73	3.9	4.9	6.9	9.4	2.6
4.73-5.52	4.5	6	8	11.6	2.9
5.52-6.30	5	6	8	13.6	3.2
6.30-7.09	5.5	7	9	13.6	3.5
7.09-7.88	6.2	8.2	10.2	15.8	3.8
7.88-8.86	6.2	8.2	11.2	17.8	4.3
8.86-9.85	7.2	9.2	13.2	17.8	4.3
9.85-11.03	7.2	10.2	13.2	20	4.9
11.03-12.41	8.2	10.2	15.2	22	4.9
12.41-13.98	9.4	11.4	17.4	24.2	5.5
13.98-15.75	9.4	13.4	19.4	27.2	6.1
515.75-17.72	10.6	13.6	21.6	30.5	7
17.72-19.69	11.6	15.6	23.6	32.5	7

Running and Sliding Fits

FIT	HOLE	SHAFT
RC 1	H5	g4
RC 2	H6	g5
RC 3	H7	f6
RC 4	H8	f7
RC 5	H8	e7
RC 6	H9	e8
RC 7	H9	d8
RC 8	H10	c9
RC 9	H11	Sp-10

Clearance Locational Fits

FIT	HOLE	SHAFT
LC 1	H6	h5
LC 2	H7	h6
LC 3	H8	h7
LC 4	H10	h9
LC 5	H7	g6
LC 6	H9	f8
LC 7	H10	e9
LC 8	H10	d9
LC 9	H11	c10
LC 10	H12	Sp11
LC 11	H13	Sp12

Clearance Locational Fits

FIT	HOLE	SHAFT
LT 1	H7	js6
LT 2	H8	js7
LT 3	H7	k6
LT 4	H8	k7
LT 5	H7	n6
LT 6	H7	n7

Interference Locational Fits

FIT	HOLE	SHAFT
LN 1	H6	n5
LN 2	H7	p6
LN 3	H7	r6

Force and Shrink Fits

FIT	HOLE	SHAFT
FN 1	H6	Sp5
FN 2	H7	s6
FN 3	H7	t6
FN 4	H7	u6
FN 5	H8	x7

<u>Note:</u>
- Upper case letters represent holes/bores (e.g. H7).
- Lower case letters represent shafts (e.g. f6)..
- The letter (location symbol) represents the position/distance to the "zero line"/Basic size.
- H and h are on the zero line.
- The number (quality no.) represents the tolerance grade. Higher no.s yield coarser fits.
- "See tolerance grade (IT) table for limits not shown in these tables

"Add IT(x) for hole limit (from 0). (Basic hole is min size.)
"Subtract IT from max shaft limit calculated from basic size and value in table for other limit.

METHODS OF INDICATING
1. Ø.30 f7
2. Ø.2995 +.000, -.0006 (Ø.30f7)
3. Ø.2989-.2995 (Ø.30f7)

INDICATING FITS
1. Ø.30 H8/f7 (hole first)
2. Ø.30 H8 (.3000-.3009)
f7 (.2989-.2995)

IT Grade (7)
Location on zero line (H)
Ø30 H7
Basic Size (Ø30) **Hole Designation**

IT Grade (7)
Location to zero line (f)
Ø30 f7
Basic Size (Ø30) **Shaft Designation**

the upper limit of a shaft; subtracting the IT value (Table 12.6) from the upper limit yields the lower limit.

Example:

Ø 3.15 k6 = 3.1500 − .0008, − .0015 or 3.1485 − 3.14992 (− .0008 − .0007) = − .0015

where − .0008 is from Table 12.7 and .0007 from Table 12.6.

Example:

Ø 12.00 d9 = 12.00 − .007, − .012 or 11.988 − 11.993 (− .007, − .005) = − .012

where − .007 is from Table 12.7 and .005 from Table 12.6.

12.13.5 Calculating Fits

Once the limits are computed, the **fit calculation** is simple. A fit consists of two sets of limits and the maximum and minimum clearance. Except for interference, force, or shrink fits, the shaft is given a tolerance grade one number less than the hole (Fig. 12.64). Using the example in Figure 12.49 of Ø 80 H7/k6 (LT3, clearance locational fit; see Table 12.6) the calculated limits are:

HOLE 80.000-80.030
SHAFT 80.002-80.021

Subtracting the largest (MMC) shaft from the smallest (MMC) hole is 80.000 − 80.012 = −.012 (minimum clearance or maximum interference). Subtracting the smallest (LMC) shaft from the largest (LMC) hole is 80.030 − 80.002 = .028 (maximum clearance or minimum interference).

Tables for the most popular fits are in Appendix C.

You May Complete Exercises 12.1 Through 12.8 at This Time

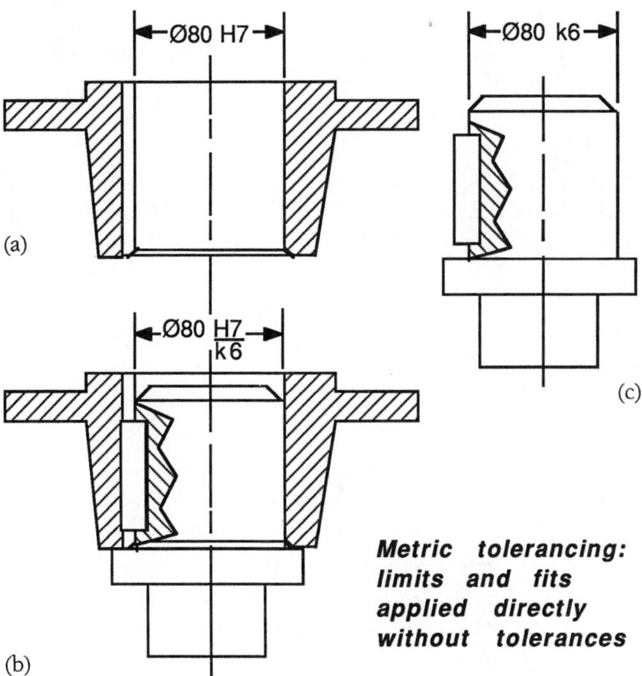

FIGURE 12.64 Metric Symbology for Limits and Fits

QUIZ

True or False

1. Bilateral tolerancing allows variation in both the positive and the negative direction.
2. In a feature control frame, a separator is not required between the geometric characteristic and the tolerance.
3. A tolerance of 1.2-1.4 allows a feature of 1.202 to be accepted.
4. A tertiary datum may be established by a single point of contact.
5. The four basic types of fit are clearance, interference, transition, and running.
6. The ANSI system of limits and fits has been standardized at about 500 fits.
7. Tolerances are understood to apply at 68°F unless otherwise specified.
8. A datum is identified, located, and established in a feature control frame.

Fill in the Blanks

9. A theoretically exact dimension is a ________ dimension.
10. A datum may be a ________, a ________, or a ________.
11. The abbreviations for the modifiers are ________, ________, and ________.
12. The surfaces of a feature must not extend beyond a ________ ________ ________ at MMC.
13. ________, ________, and ________ are the basic types of fits.
14. In ________ and ________ tolerancing, the ________ is given first and is followed by a ________ and ________ expression of tolerance.
15. The classes of fits are arranged into three groups: ________ and ________ fits, ________ fits, and ________ fits.
16. ________ means producing parts relative to standard measures and forms.

Answer the Following

17. What is the difference between circularity and cylindricity?
18. What is a tolerance zone?
19. What does the following statement mean?: "Every manufactured part varies in size, but as long as it is understood and controlled, the part will function as designed."
20. Define the three datum planes and how they are located.
21. How is perpendicularity defined?
22. What is title block tolerance and how is it applied to part features?
23. Explain the MMC of an internal feature.
24. What does RFS mean?

EXERCISES

Exercises may be assigned as sketching, instrument, or CAD projects. Transfer the given information to an "A" size sheet of .25 in. grid paper. Exercises that are not assigned by the instructor can be sketched in the test to provide practice and understanding of the preceding instructional material. Complete all views by showing all hidden lines, centerlines, tolerances, datums, and dimensions.

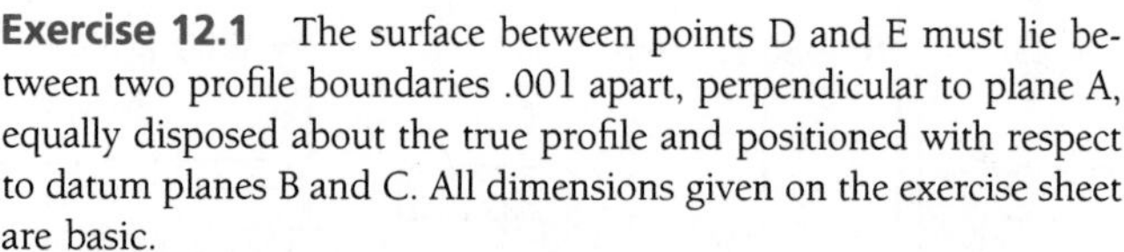

Exercise 12.1 The surface between points D and E must lie between two profile boundaries .001 apart, perpendicular to plane A, equally disposed about the true profile and positioned with respect to datum planes B and C. All dimensions given on the exercise sheet are basic.

Exercise 12.2 Specify different tolerances for each segment of the profile; between A and B use .005, between B and C use .004, and between C and D use .002. Complete the views and dimension the part.

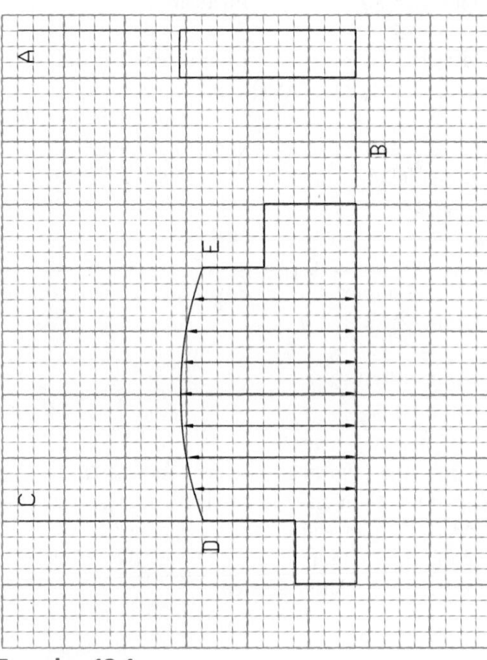

Exercise 12.1

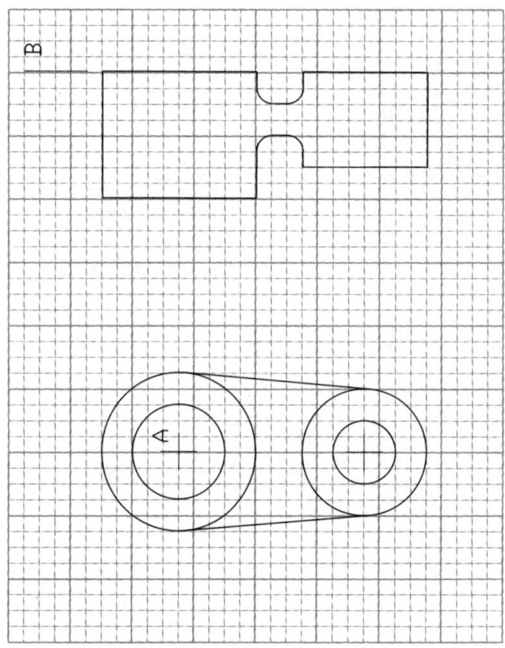

Exercise 12.3

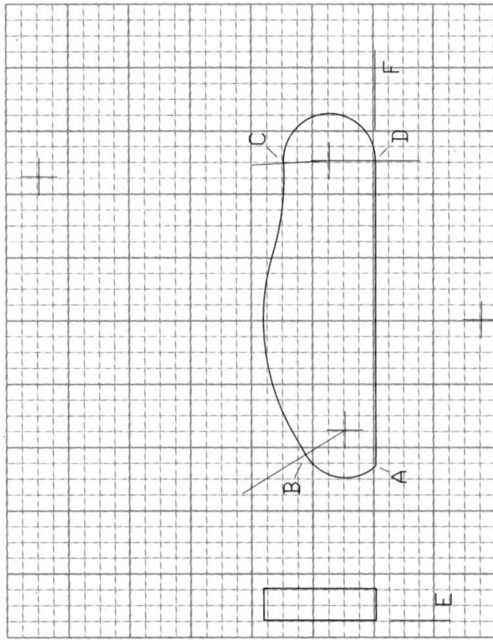

Exercise 12.2

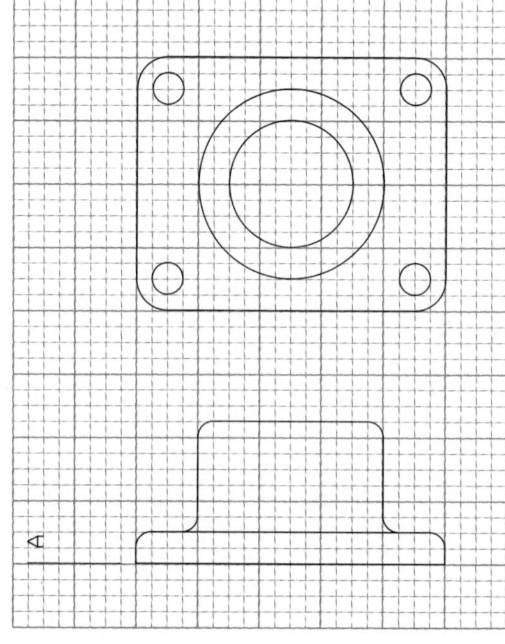

Exercise 12.4

Exercise 12.3 The two holes are to be parallel with a maximum tolerance of .002. Complete the views and dimension the part.

Exercise 12.4 The large hole at MMC is to be perpendicular to datum plane A with a maximum tolerance of .004. Complete the views and dimension the part.

Exercise 12.5 Use positional tolerancing for the part. The holes are to be at a depth of 1/2 of the part thickness. Show the bolt circle diameter as a reference dimension.

Exercise 12.6 Use positional tolerancing for the part. The large hole is to be toleranced at .016 at MMC from datums A, B, and C. The small holes are to be toleranced at .001 at MMC from the large hole. Dimensions for the hole pattern are basic.

Exercise 12.7 Dimension the part using composite positional tolerancing. All hole patterns are to be basic. The bolt circle hole pattern is 6 × 60° basic.

Exercise 12.8 Dimension the part using positional tolerancing. Use symbology to call out the counterbored holes. The counterbores are to have a .001 diameter tolerance zone at MMC in relation to datum B.

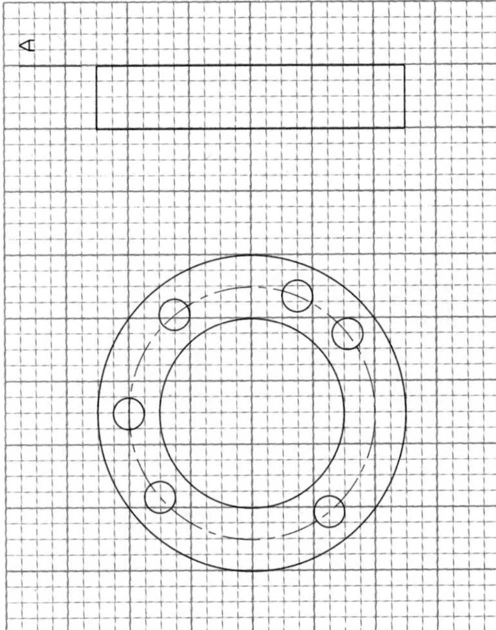

Exercise 12.5

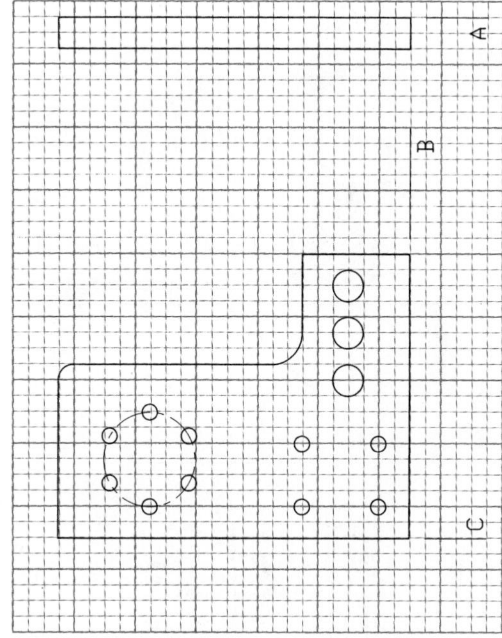

Exercise 12.7

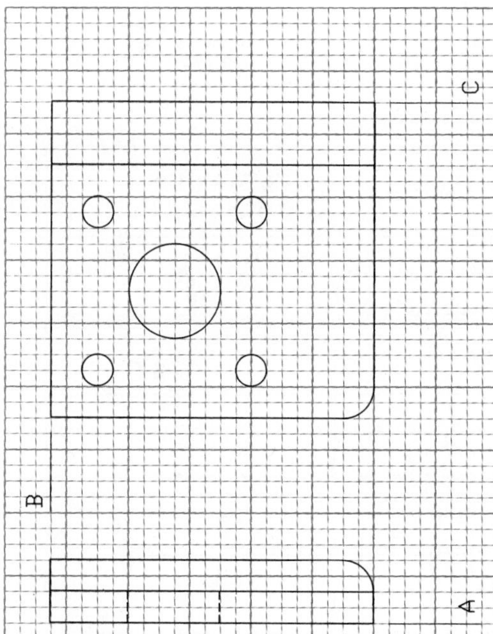

Exercise 12.6

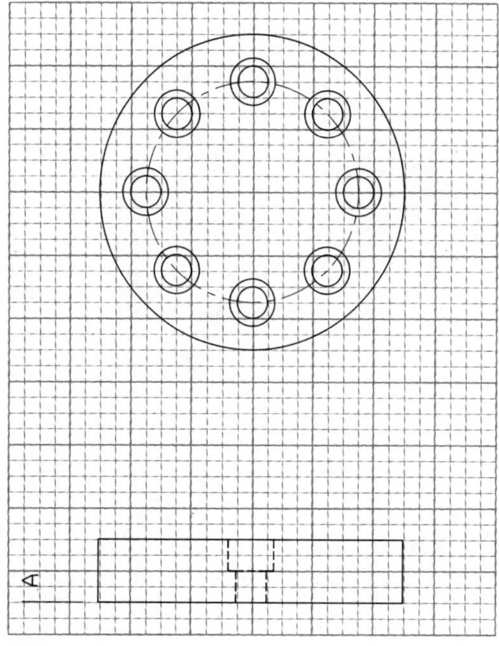

Exercise 12.8

PROBLEMS

The concepts and methods presented in the chapter should be applied for drawing projects throughout the text. The problems presented here are meant to introduce and familiarize you with simple geometric tolerancing situations. On problems that do not have units, use either metric or decimal-inch scales and transfer directly from the text using a 2× scale.

Problem 12.1 Complete the table for each of the four cases. Use Tables 12.1 through 12.7 and the Appendix C to check the results.

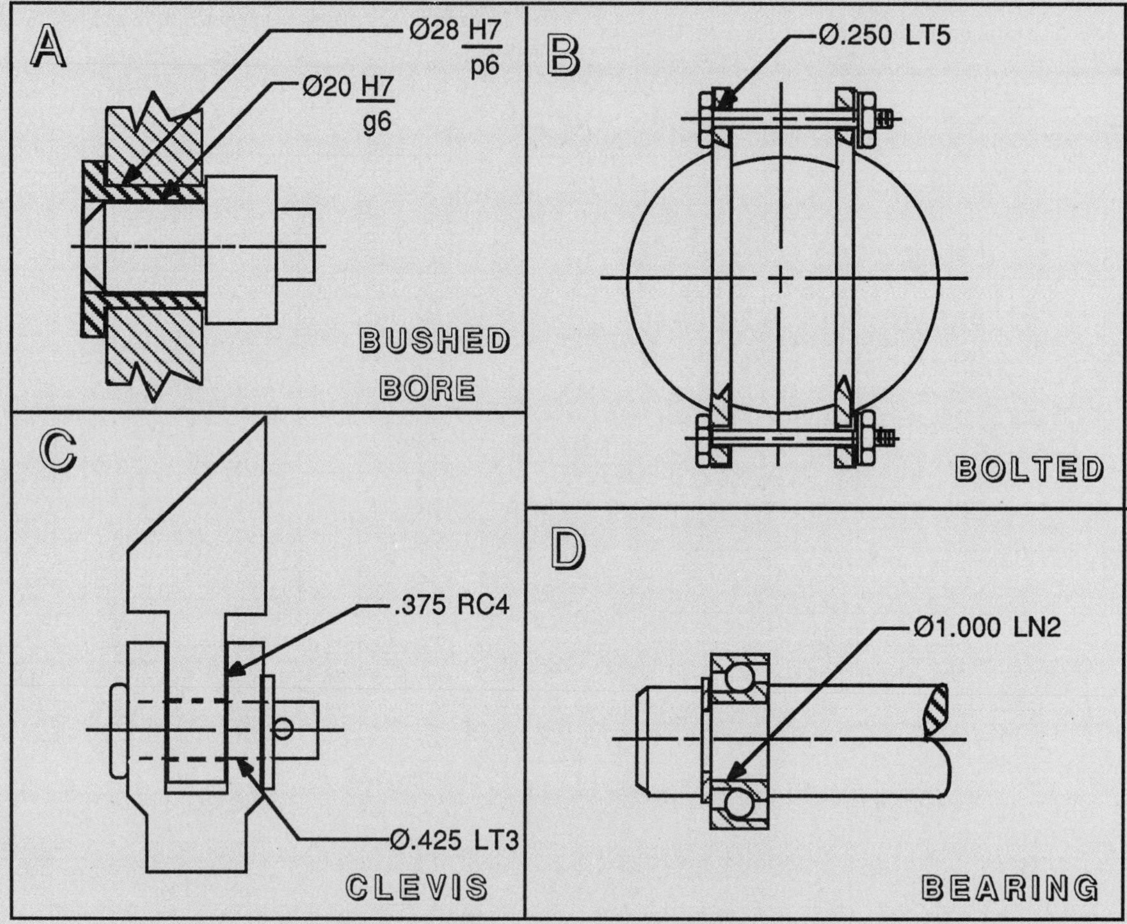

FIT IDENT	Ø or width mm or in.	FIT CALLOUT	FEATURE	SIZE LIMITS	CLEARANCE/ INTERFERENCE
A	Ø28 mm	H7/p6	MALE		
			FEMALE		
	Ø20 mm	H7/g6	MALE		
			FEMALE		
B	Ø.250 in.	LT5 (H7/n6)	MALE		
			FEMALE		
C	.375 in.	RC4 (H8/f7)	MALE		
			FEMALE		
	.425 in.	LT3 (H7/k6)	MALE		
			FEMALE		
D	1.00 in.	LN2 (H7/p6)	MALE		
			FEMALE		

Problem 12.2 Follow the instructions at the bottom of the figure.

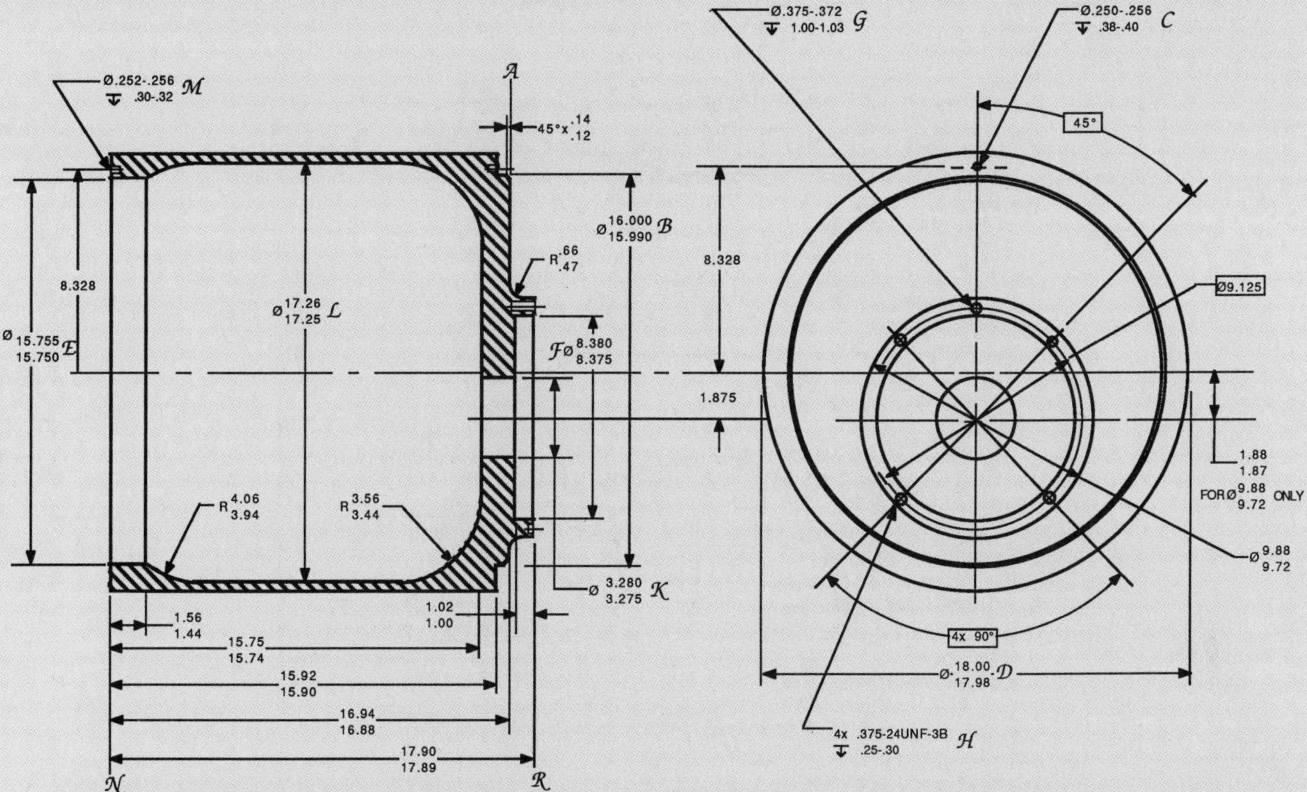

1. Establish a Datum Reference Frame using features A, B and C.
2. A is flat to .001, B is perpendicular to A within .010 at MMC. C is Positioned within .005 at MMC to A and B at MMC
3. Feature D has a Positional tol. of .020 at MMC to A and B at MMC.
4. E has an .008Circ. Runout to A and B 5. F is positioned within .004 at MMC to R., B at MMC and C at MMC.
6. G is positioned to .005 at MMC to R., F at MMC and C at MMC.
7. H has a Composite Positional Tol. of .020 at MMC to R., F at MMC and G at MMC, and positioned to .006 at RFS to R, the zone is projected .510.
8. K is positioned to .010 at MMC to R and F at MMC. 9. L has a Circ. Runout of .025 to D 10. M is Positioned to .008 at MMC to N, E at MMC and C at MMC. 11. N and R are parallel within .005 & .003 to A respectively.

Problems 12.3 and 12.4 Redraw each of the frames and note the parts of each.

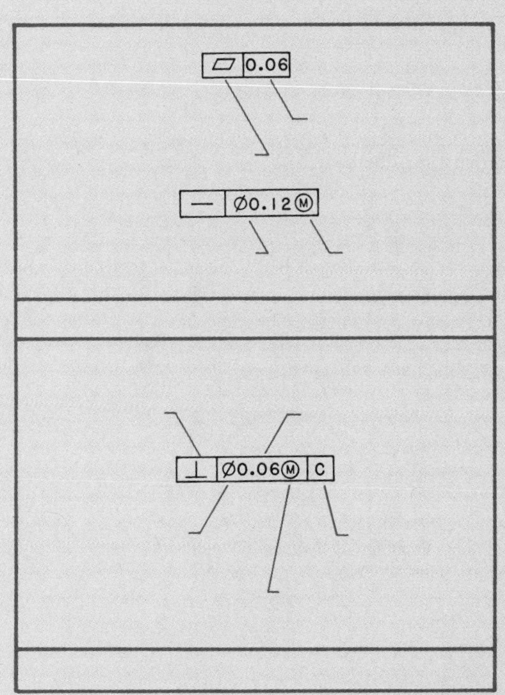

Problem 12.3

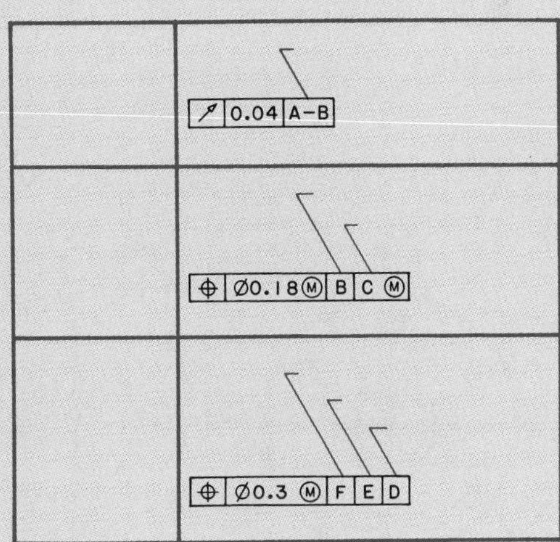

Problem 12.4

Problem 12.5 The surfaces all around the part outline must lie between parallel boundaries .06 (1.5 mm) apart and perpendicular to datum -A- .

Problem 12.6 Draw and dimension the part.

Problem 12.7 Complete the part's dimensions and tolerancing.

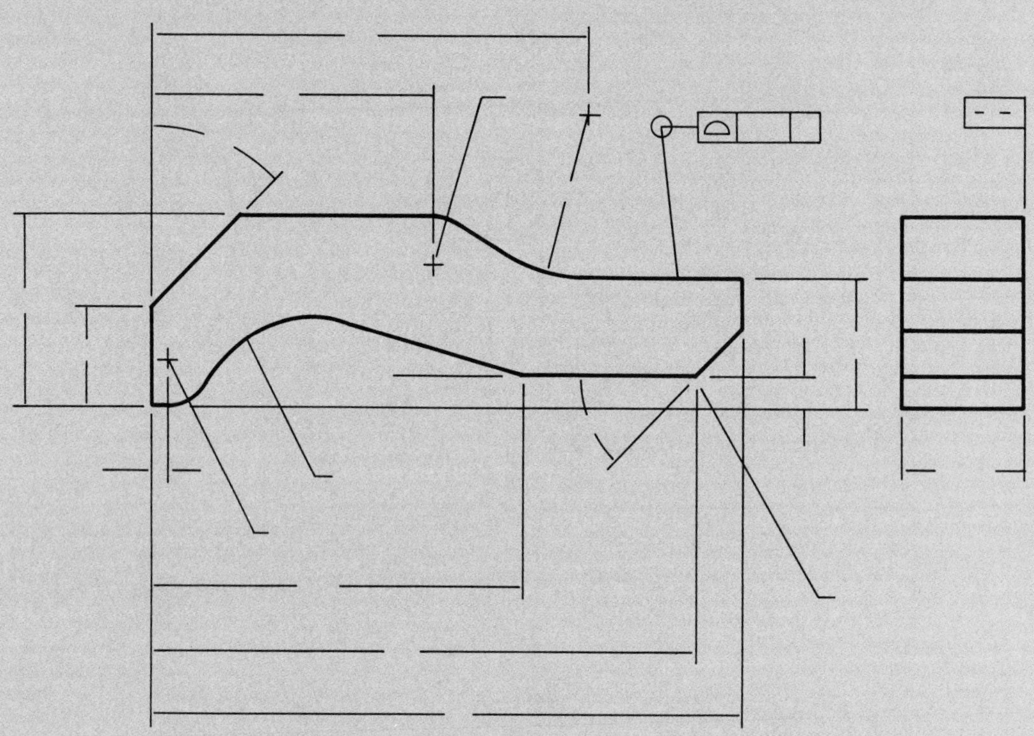

Problem 12.5

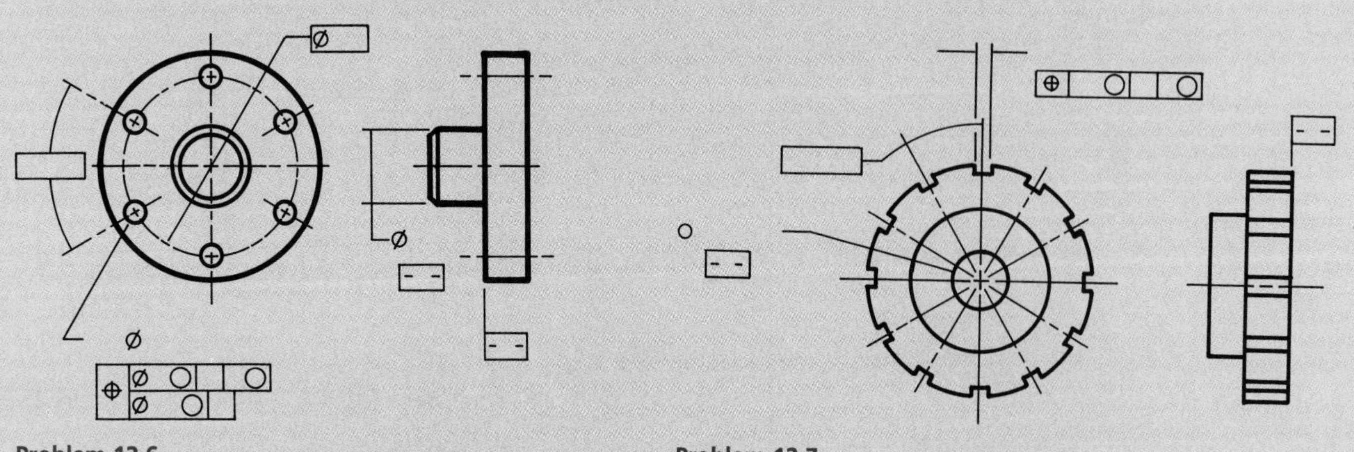

Problem 12.6

Problem 12.7

Problem 12.8 Draw the part and complete the dimensioning.

Problem 12.9 Complete the part as required.

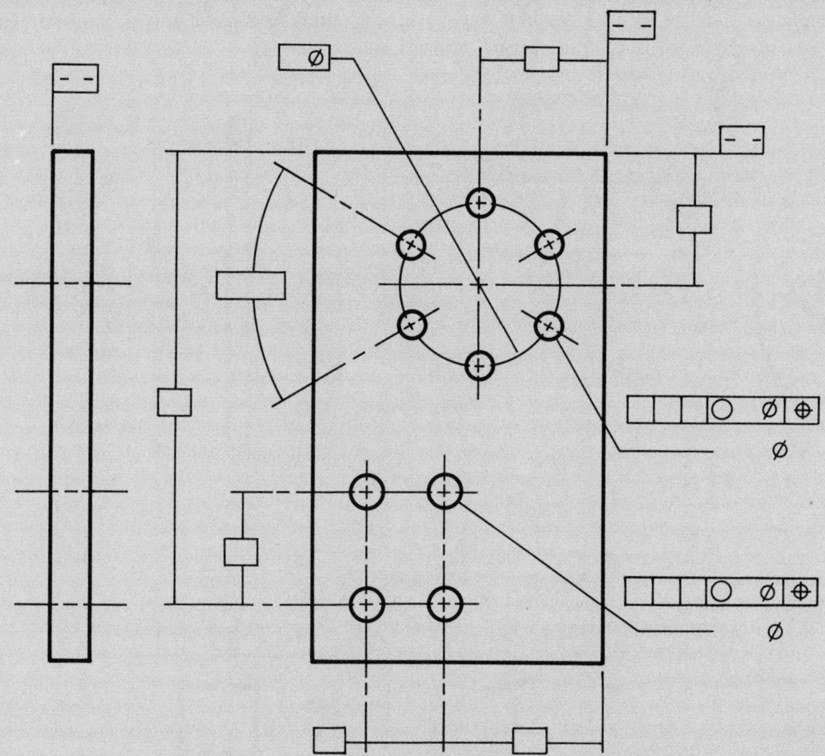

Problem 12.8

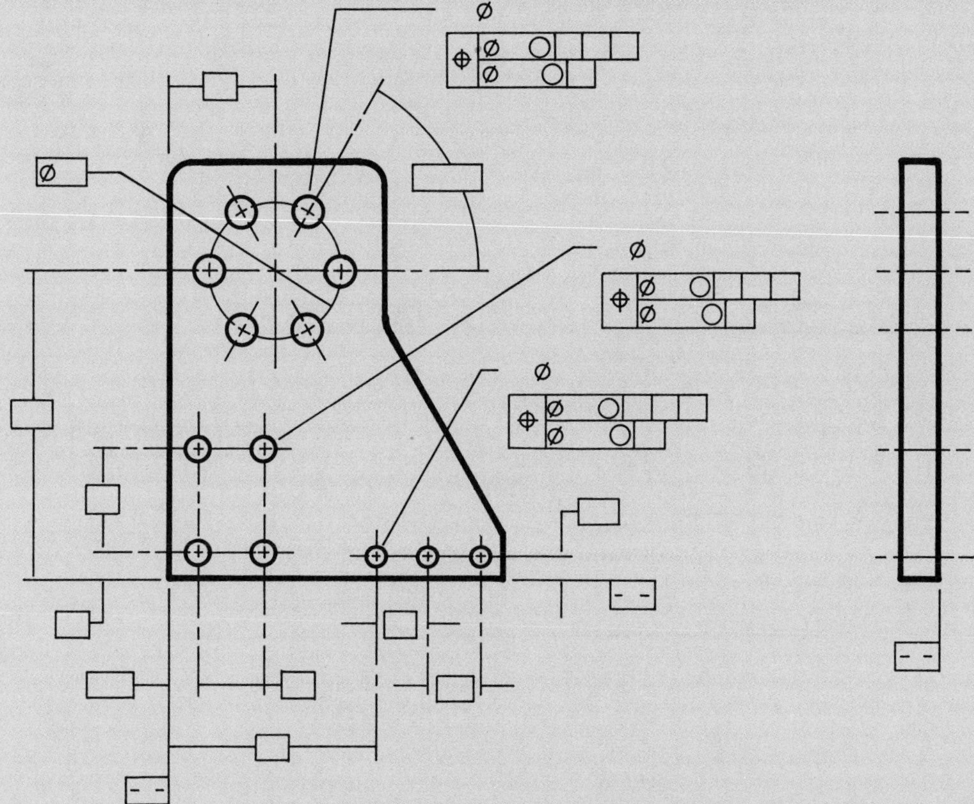

Problem 12.9

Problem 12.10 Draw and dimension the part as required.

Problem 12.11 Complete the portion of the part as shown.

Problem 12.12 Complete the project as required. Specify the different profile tolerances between the segments: .12 (3 mm) between A and B, .10 (2.5 mm) between B and C, and .05 (1.2 mm) between C and D.

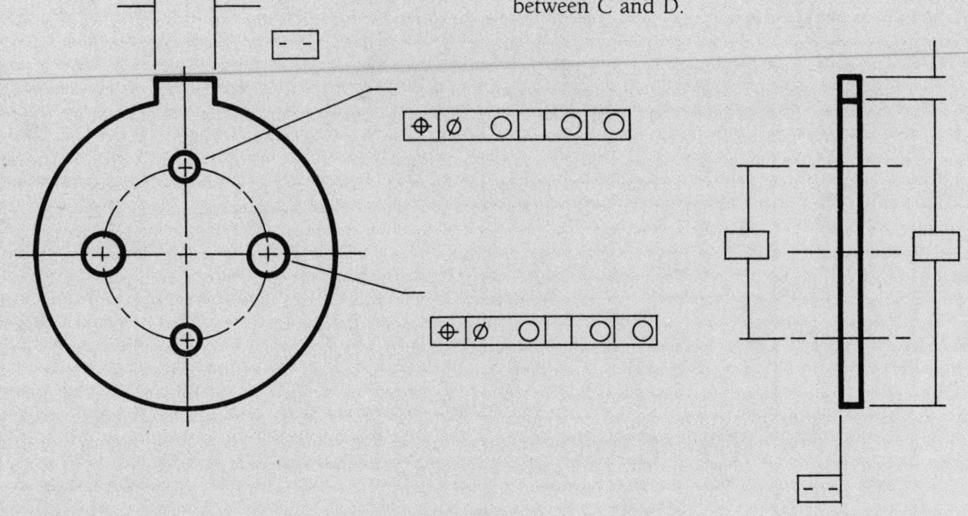

Problem 12.10

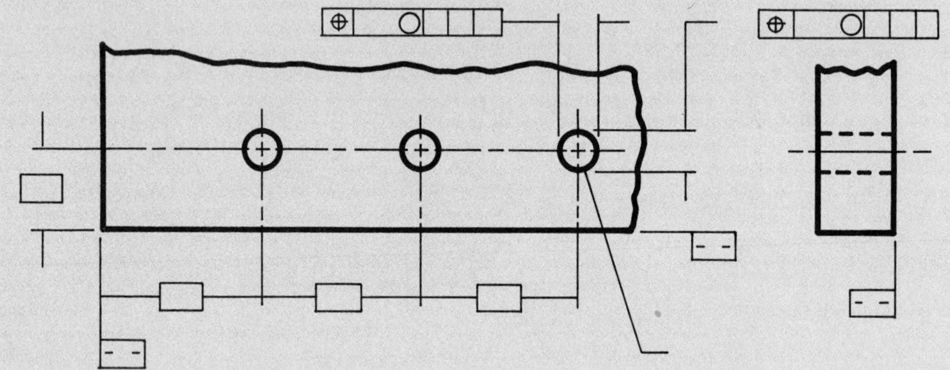

Problem 12.11

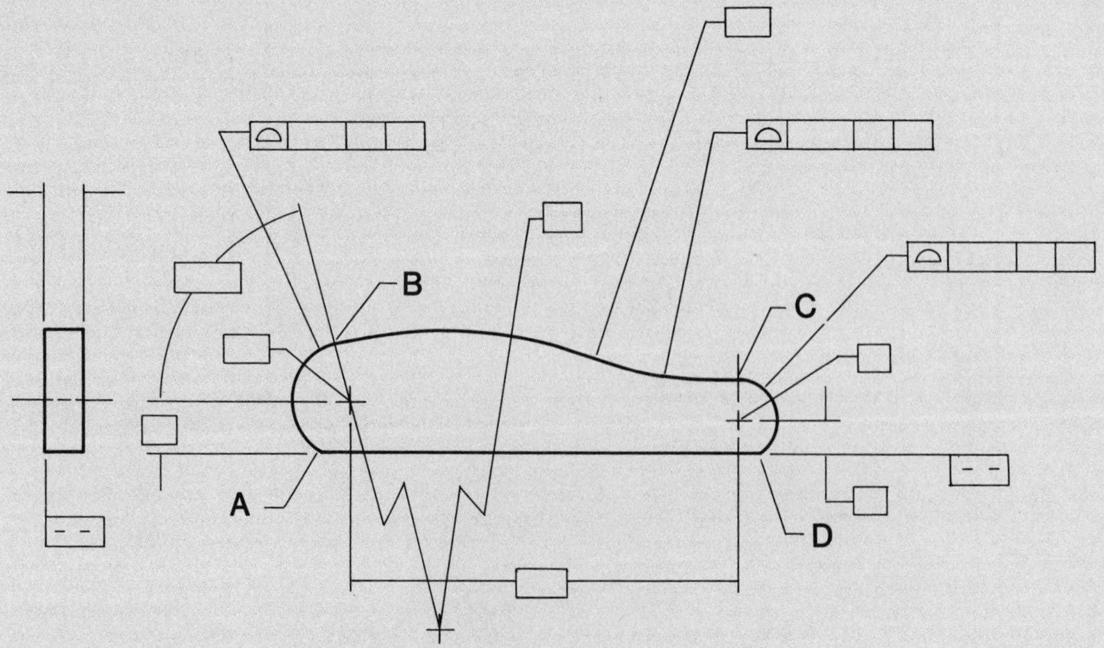

Problem 12.12

Problem 12.13 Redraw the rotor and show all dimensions.

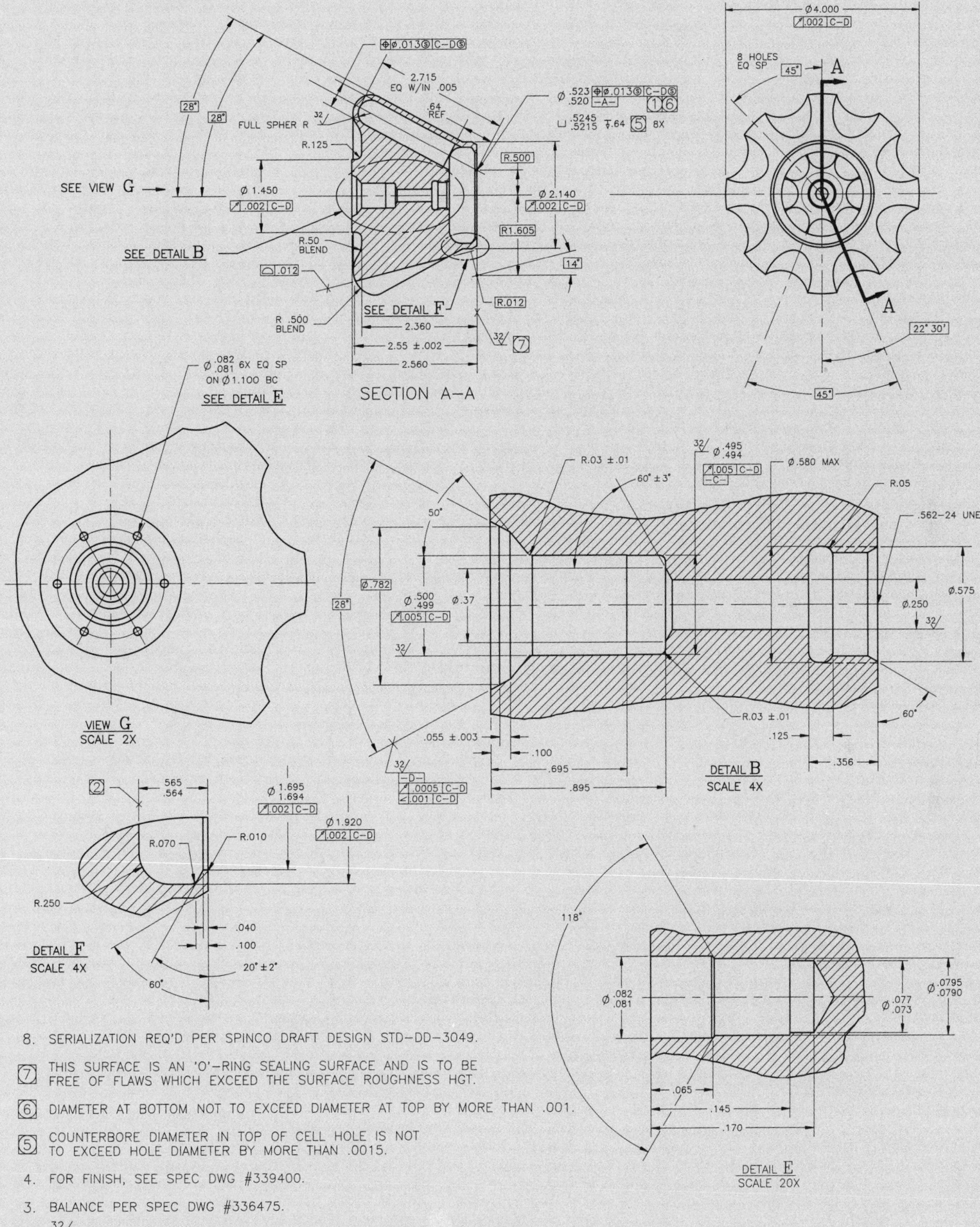

8. SERIALIZATION REQ'D PER SPINCO DRAFT DESIGN STD–DD–3049.

7. THIS SURFACE IS AN 'O'–RING SEALING SURFACE AND IS TO BE FREE OF FLAWS WHICH EXCEED THE SURFACE ROUGHNESS HGT.

6. DIAMETER AT BOTTOM NOT TO EXCEED DIAMETER AT TOP BY MORE THAN .001.

5. COUNTERBORE DIAMETER IN TOP OF CELL HOLE IS NOT TO EXCEED HOLE DIAMETER BY MORE THAN .0015.

4. FOR FINISH, SEE SPEC DWG #339400.

3. BALANCE PER SPEC DWG #336475.

2. 32√ PRIOR TO MARKING. THIS SURFACE IS AN 'O'–RING SEALING SURFACE AND IS TO BE FREE OF FLAWS WHICH EXCEED THE SURFACE ROUGHNESS HGT.

1. TRUE POSITION OF DIAMETER -A- TO BE WITHIN THE CONICAL ZONE AS SHOWN. THE MAX TRUE POSITION VARIATION BETWEEN CAVITIES TO BE .004.

NOTE: (UNLESS OTHERWISE SPECIFIED)

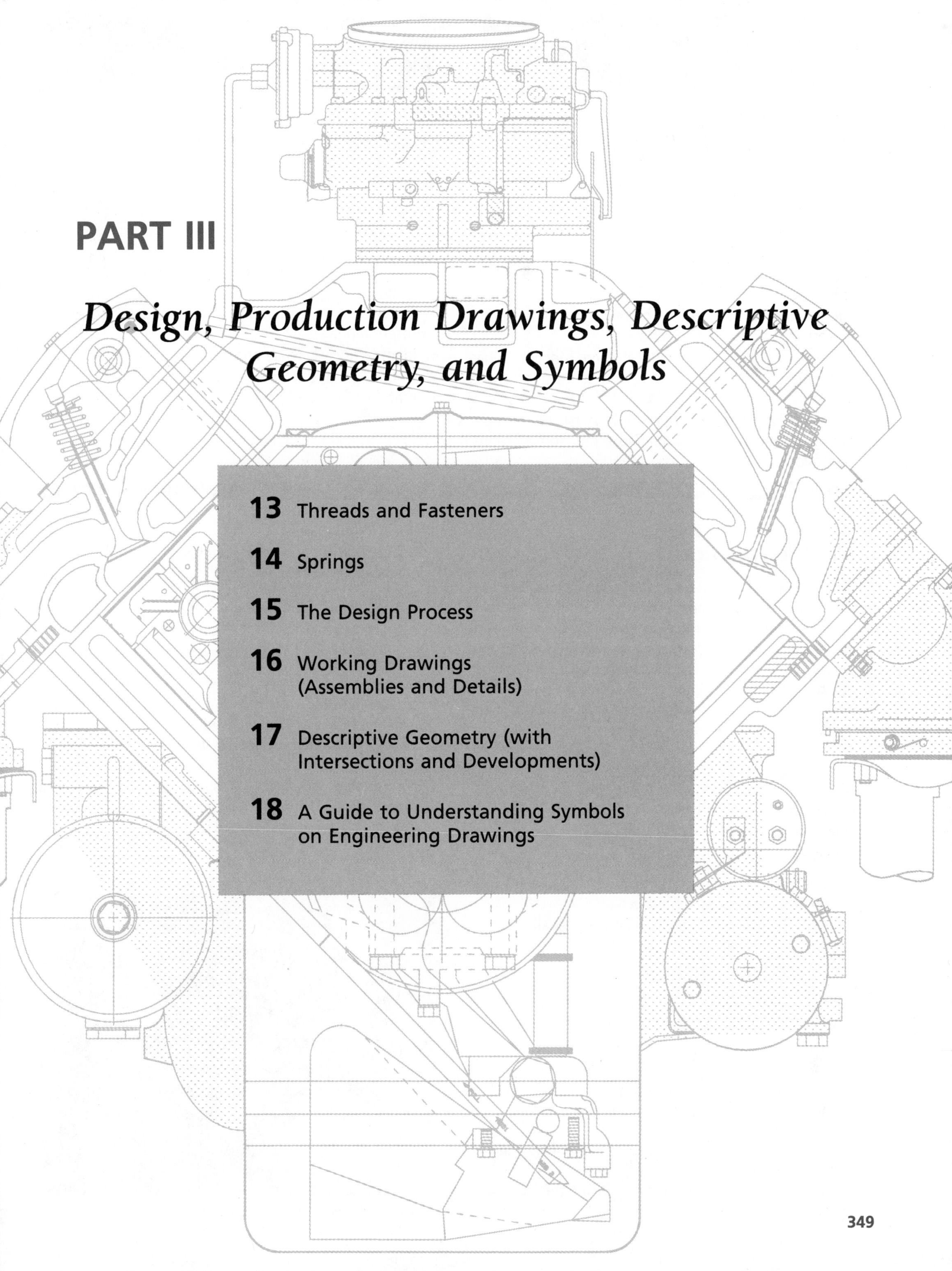

PART III

Design, Production Drawings, Descriptive Geometry, and Symbols

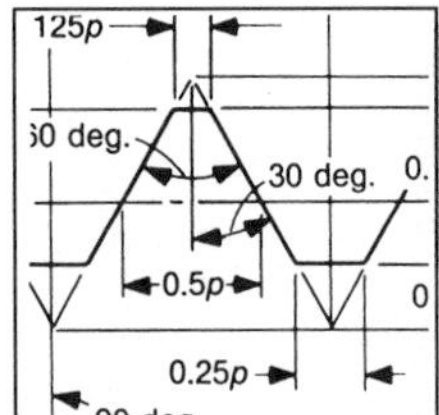

CHAPTER 13

Threads and Fasteners

Learning Objectives

Upon completion of this chapter you will be able to accomplish the following:

1. Identify the variables, requirements, and considerations necessary for fastener selection.
2. Develop an understanding of thread function while recognizing standard thread forms, series, terms, and parts.
3. Differentiate between and produce ANSI standard detailed, schematic, and simplified screw thread representations.
4. Identify and compare Acme, buttress, metric, and pipe threads.
5. Develop an understanding of bolt, nut, and screw representation.
6. Identify quick-release and semipermanent pins.
7. Define key and keyseat variations and design considerations.

13.1 Introduction

Fasteners (Fig.13.1) are used to join components in an assembly. They are interchangeable, readily available as standard parts, and manufactured to specific requirements to maintain a high degree of precision and quality. Most, but not all fasteners have threads. There are more than a million types of fasteners. This chapter presents the common types of fasteners, covers thread specifications, and discusses non-threaded types of fasteners.

13.1.1 Fastener Selection

There are many factors to consider when selecting the proper fastener. Design for manufacturability (DFM) concepts are considered at this stage including:

- Use off-the-shelf readily available standard fasteners
- Use the minimum number of fasteners
- Use fewer large fasteners rather than many small fasteners
- Avoid separate washers
- Design for automated assembly
- Design for drop-in assembly
- Eliminate separate fasteners by design (for example, snap fits).

The selection of the proper fastener for a project involves:

- Assembly requirements for assembly and disassembly during manufacturing, shipping, installation, service, and maintenance
- Conditions of operation: temperature, vibration, movement, corrosion, and impact
- Quantity of fasteners required to secure the parts adequately
- Variety of fasteners on the assembly
- Function of the fasteners in the assembly: location and fastening

13.2 Screw Threads

A **thread** is a helical or spiral groove formed on the outside (external) or inside (internal) surface of a cylinder. **Screw threads** support and transfer loads and transmit power. A variety of thread styles are used in the valve shown in Figure 13.2.

Threads on round parts such as shafts or bolts are **external threads** (Fig. 13.3) and threads on interior surfaces of a cylindrical hole are **internal threads** (Fig. 13.4). A die is used

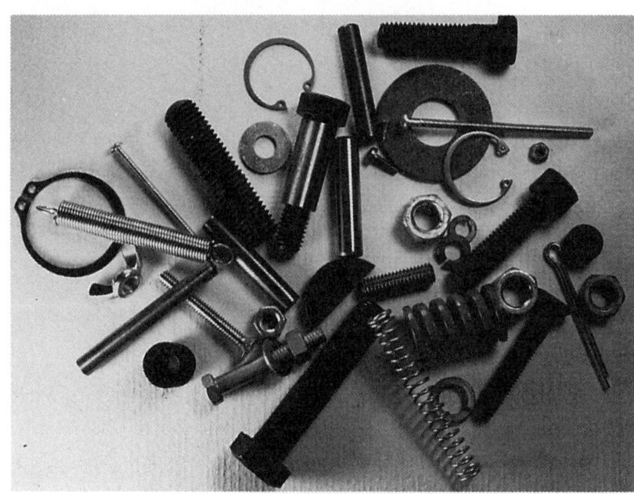

FIGURE 13.1 **Fasteners and Threads**

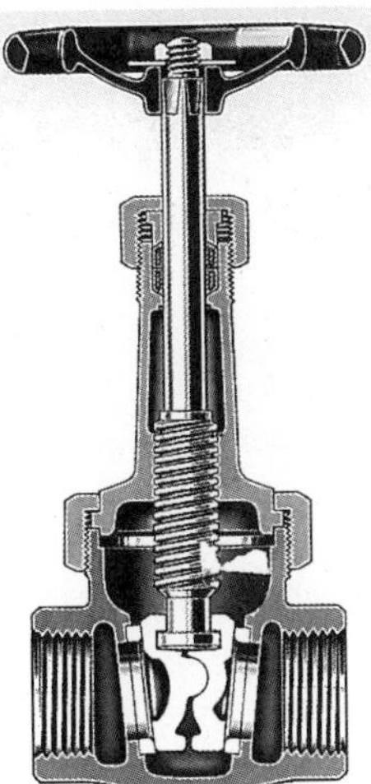

FIGURE 13.2 **UNC, Acme, and NPT Threads Used in the Design of this Rising Stem Gate Valve**

to cut external threads and a tap is used to cut internal threads.

The different forms of threads are selected based on the requirements of the design. Eight standard styles are presented in Figure 13.5: (a) American National thread form, (b) the British Standard (Whitworth) thread form, (c) the worm thread form, (d) the square thread form, (e) the sharp V thread form, (f) the knuckle thread form, (g) the Acme thread form, and (h) the buttress thread form. The ISO metric thread form is shown in Figure 13.6, and the Unified National (UN) thread form is shown in Figure 13.7.

The Acme and square thread forms are used to transmit power. Acme threads (Fig.13.2) are used to move the valve stem up and down to open and close the valve. Worm threads are similar to Acme threads and are also used to transmit power.

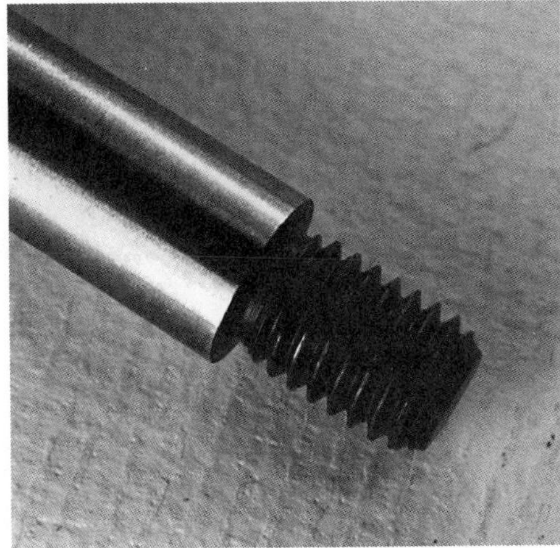

FIGURE 13.3 **External Threads**

FIGURE 13.4 **Internal Threads**

The knuckle thread form is used for sheet metal products such as the base of a light bulb, bottle and jar tops, and plastic bottles and caps. Buttress threads are used in high stress designs and to transmit power along the axis in one direction.

ISO metric threads (Fig.13.6) are the internationally recognized standard for thread forms. The ISO thread is very similar to the UN thread form except that its thread depth is not as great. The ISO thread has the same basic profile as the UN thread form.

The **Unified National thread** form (Fig.13.7) is used in the United States and is practically identical to the obsolete American National thread form. In fact, threads manufactured to either form are functionally interchangeable. American National threads are designated as N, NC, NF, NEF, or NS. Unified National threads are designated similarly: UN, UNC, UNF, UNEF, UNS, or UNM.

13.2.1 Thread Terms

The following terms are used throughout the chapter:

Class of thread An alphanumerical designation to indicate the standard grade of tolerance and allowance specified for a thread.

Crest The top surface joining the two sides of the thread.

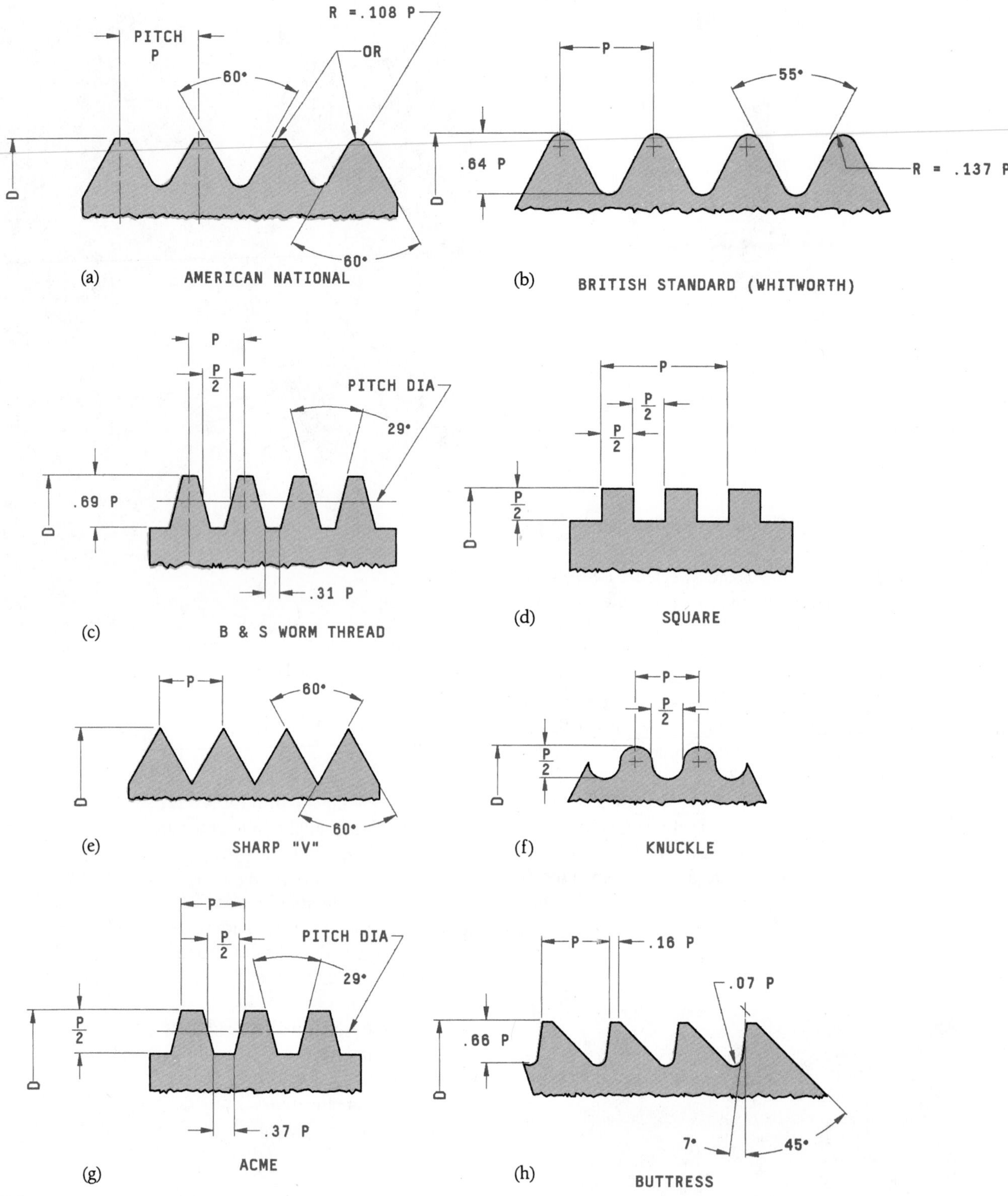

FIGURE 13.5 Standard Thread Forms

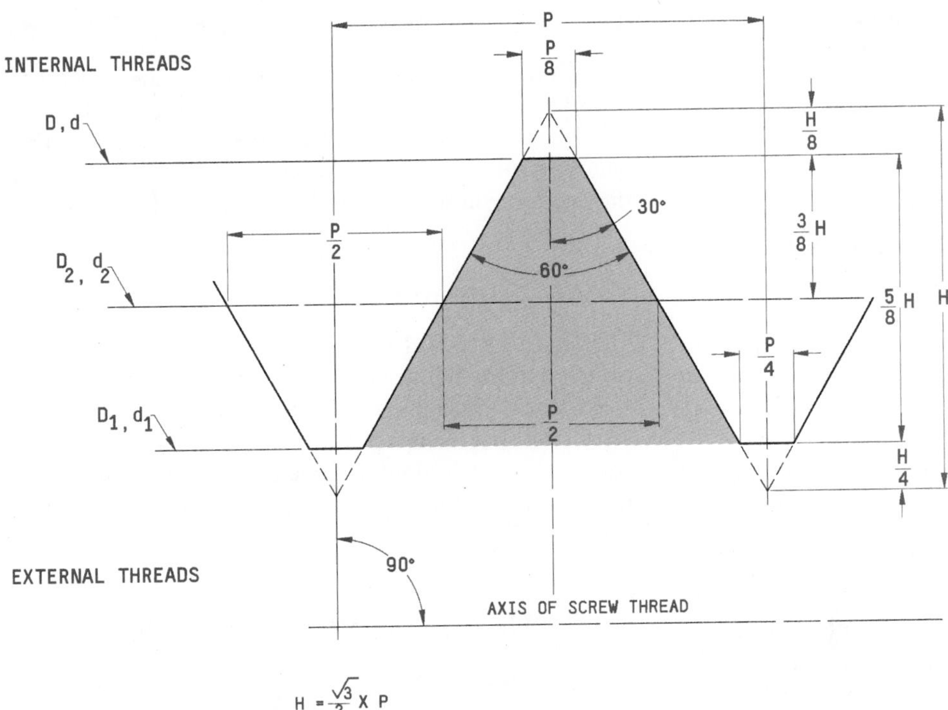

FIGURE 13.6 Basic M Thread Profile (ISO 68 Basic Profile)

Depth of thread engagement The radial distance, crest to crest, by which the thread forms overlap between two assembled mating threads.

Major diameter The major diameter is that of the major cylinder-distance across the crests of the thread.

Minor diameter The minor diameter is that of the minor cylinder-root diameter of the thread.

Nominal size The designation that is used for general identification of a thread based on the major diameter.

Pitch The axial distance from a point on one screw thread to the corresponding point on the next screw thread. Pitch is equal to the lead divided by the number of thread starts.

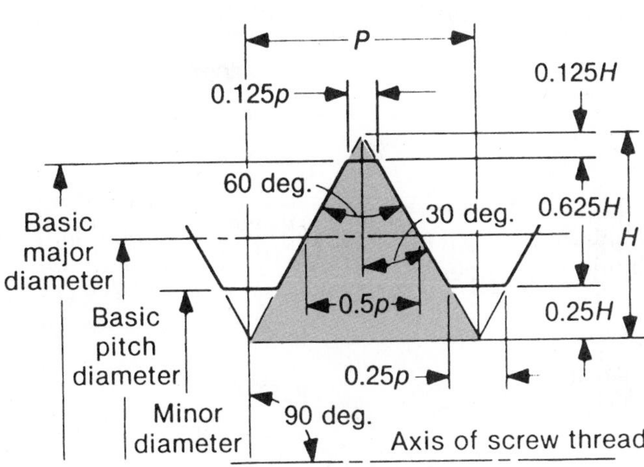

FIGURE 13.7 Basic Profile for UN and UNR Screw Threads

Profile of thread The contour of a screw thread ridge and groove delineated by a cutting plane passing through the thread axis. Also called form of thread.

Root The bottom surface joining the two sides of the thread.

Root diameter The diameter of an imaginary cylinder bounding the bottom of the roots of a screw thread (minor diameter of the thread).

Thread designations A capital letter abbreviation of names used to designate various thread forms and thread series.

Thread series Groups of diameter/pitch combinations distinguished from each other by the number of threads per unit of measurement.

13.2.2 Thread Parts

The configuration of the thread in an axial plane is the **thread form** (profile). The three parts that make the form of a thread are the **crest**, the **root**, and the **flank** (Fig.13.7). The crest of a thread is at the top, the root is on the bottom, and the flank joins them. The **fundamental triangle** (shaded part of Fig.13.7) is the triangle formed when the thread profile is extended to a sharp V at both crests and roots. The height of the fundamental triangle (*H*) is the distance between the crest and the root diameters (for Unified threads, *H* equals 0.866025 times the thread pitch).

A thread having full form at both crests and roots is a complete or full form thread. When either the crest or root is not fully formed, it is an **incomplete thread.** Such threads occur at the ends of externally threaded fasteners that are

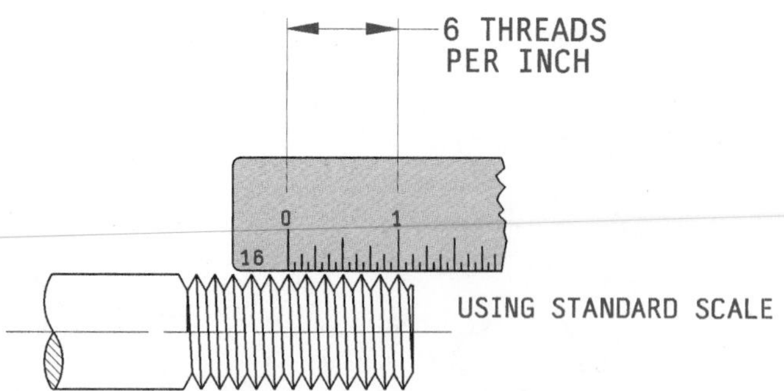

FIGURE 13.8 Using a Scale to Measure Threads per Inch

pointed (conical), at thread runouts where the threaded length blends into the unthreaded shank, and at the countersinks on the faces of nuts and tapped holes.

Thread pitch (*P*) is the distance, measured parallel to the thread axis, between corresponding points on adjacent threads. Unified screw threads are designated in **threads per inch**, which is the number of complete threads occurring in one inch of threaded length. Thread pitch is the reciprocal of threads per inch. The standard inch scale can be placed along the threads when a screw thread pitch gage is unavailable (Fig.13.8). Counting the number of threads in one inch will give the threads-per-inch measurement.

The **pitch diameter** is the diameter of a theoretical cylinder that passes through the threads so that the widths of the thread ridges and thread widths would each equal one-half of the thread pitch (Fig.13.9).

The combination of allowances and tolerances in mating threads is called the **fit** and is a measure of tightness or looseness between them. A **clearance fit** is one that always provides a free running assembly. An **interference fit** is one that always results in a positive interference between the threads.

When assembling externally threaded fasteners into internally threaded nuts or tapped holes, the axial distance of contact of the fully formed threads is the **length of thread engagement** (Fig.13.10). The distance these threads overlap in a radial direction is the **depth of thread engagement.**

13.2.3 Right-hand and Left-hand Threads

Unless otherwise specified threads are right-hand. A left-hand thread turns counterclockwise to advance (Fig. 13.11). Figure 13.12 shows a turnbuckle that is designed with both right-hand and left-hand threads. When the buckle is turned in one direction, it will pull both rods together, thus tightening the connection.

13.2.4 Thread Lead

The **lead** of a thread is the axial distance it travels in one complete turn (the axial distance between two consecutive crest points). Since the lead is the axial distance a point will advance in one complete turn, **single threads** have a lead equal to the pitch, **double threads** have a lead equal to two times the pitch, and **triple threads** have a lead of three times the pitch (Fig. 13.13).

13.3 Unified National Thread Series

Thread series are groups of diameter pitch combinations that differ by the number of threads per inch. For fasteners, the popular thread series are Unified coarse, fine, and 8-pitch. The two general series classifications are *standard* and *special.*

The standard series consists of three series with graded pitches (coarse, fine, and extra-fine) and eight series with constant pitches (4, 6, 8, 12, 16, 20, 28, and 32 threads per inch).

13.3.1 Constant-Pitch Thread Series Applications

The various constant-pitch series (UN/UNR) with 4, 6, 8, 12, 16, 20, 28, and 32 threads per inch offer a comprehensive range of diameter-pitch combinations where the threads in the coarse, fine, and extra-fine series do not meet the particular requirements of the design.

The **8-thread series** (8UN) is a uniform-pitch series for large diameters or as a compromise between coarse and fine thread series. Although originally intended for high-pressure-joint bolts and nuts, it is now widely used as a

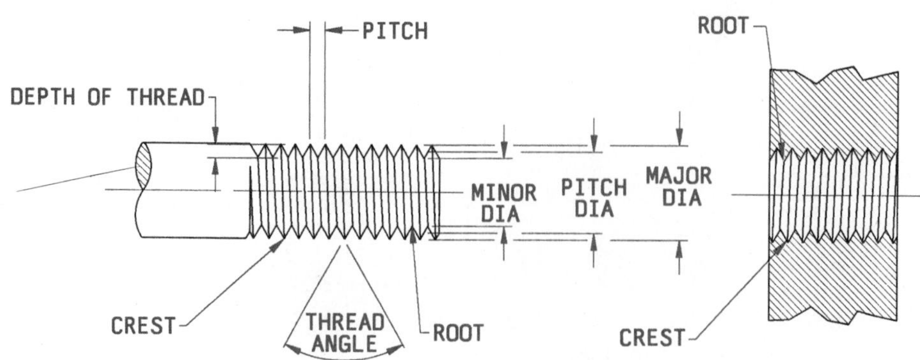

FIGURE 13.9 Unified National Thread Terminology

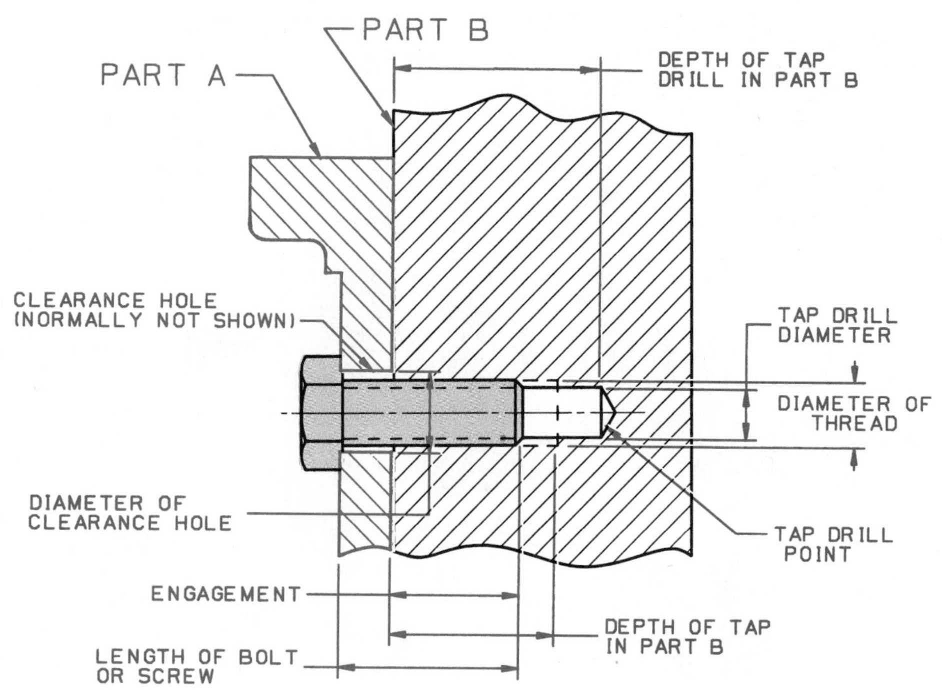

FIGURE 13.10 Thread Engagement

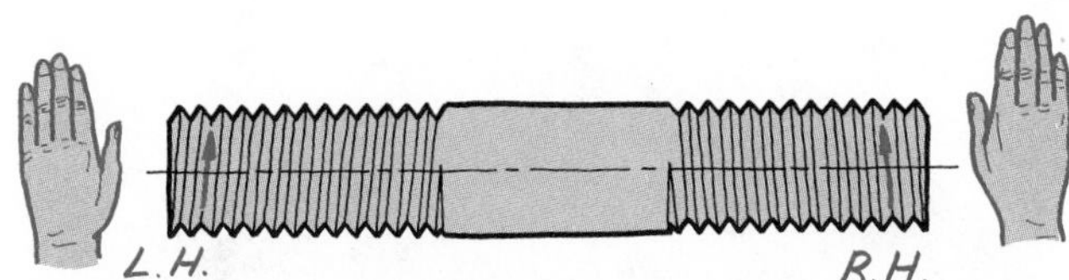

FIGURE 13.11 Right-Hand and Left-Hand Threads

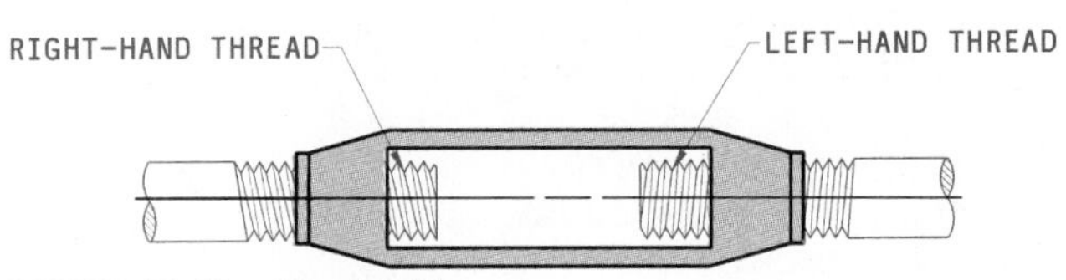

FIGURE 13.12 Turnbuckle

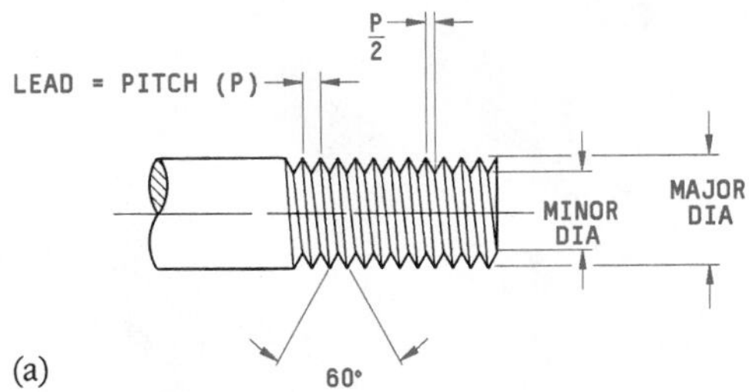

(a) SINGLE THREAD

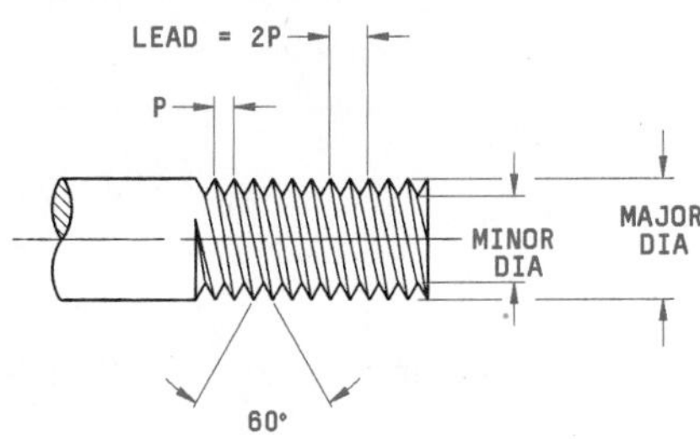

(b) DOUBLE THREAD

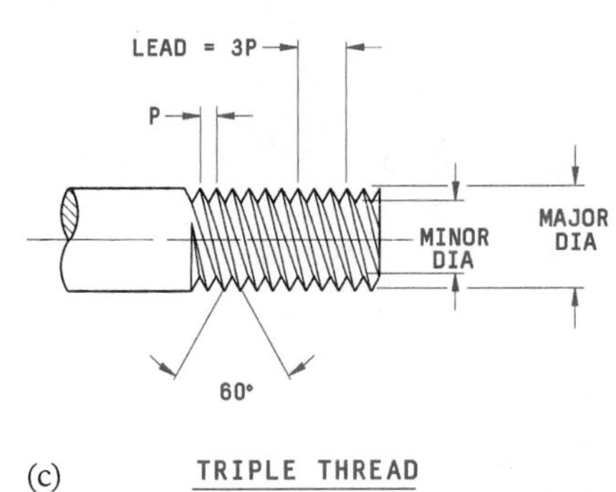

(c) TRIPLE THREAD

FIGURE 13.13 Single, Double, and Triple Threads

ITEMS OF INTEREST

Fasteners

How a product is fastened together is important to both the manufacturer and the customer or user of the product. We have all complained about the difficulty and cost to replace some minor component in an assembled product. Obviously, if rapid and easy disassembly were considered in the beginning in the design stage of the product, everyone would save time and money.

A fastener is any kind of device or method that is used to hold parts together. The permanent fastener choices are soldering, brazing, riveting, welding, and adhesives. Removable fasteners include nuts and bolts, screws, studs, pins, rings, or keys. Snap fits can also be designed into the part itself, eliminating the need for separate fasteners.

The choice of a suitable material for the fastener is also important. Because new materials, like carbon fiber composites, are being used, the choice of fastener material is becoming increasingly complex. Also, fasteners used on assemblies (for instance, aircraft and automobiles) must function in all weather conditions without deteriorating in a reasonable amount of time.

One of the most popular removable fasteners is the screw. Archimedes, the Greek mathematician, first used the idea in a screw conveyor to raise water. The threads on a screw provide a fast and easy method of fastening two parts together. However, screws are not the method of choice in automated assembly because of the complex motion required for insertion. Standards are being established to unify screw threads throughout the world. These standards would cut the costs of parts, reduce paperwork, simplify the inventory process, and improve quality control.

The selection of the proper fastening method and material is crucial for the product to be an economic success. The cost of the fastener itself is small compared to the costs associated with that fastener over the lifetime of the assembly. Every designer and drafter in industry knows how complicated the proper selection of a fastener can be, particularly when design for disassembly might be as, or, more important than design for assembly.

Industrial Fasteners

Various Fasteners

substitute for the coarse thread series for diameters larger than one inch.

The **12-thread series** (12UN) is a uniform-pitch series for large diameters requiring threads of medium-fine pitch. Although originally intended for boiler applications, it is now used as a continuation of the fine-thread series for diameters larger than $1\frac{1}{2}$ inches.

The **16-thread series** (16UN) is a uniform-pitch series for large diameters requiring fine-pitch threads. It is suitable for adjusting collars and retaining-nuts and also serves as a continuation of the extra-fine thread series for diameters larger than $1\frac{11}{16}$ inch.

13.4 Screw Thread Selection

The first consideration in determining screw thread selection is the *length of thread engagement* required between threaded components (Fig.13.10). For fastening applications, the lengths of engagement are derived from thread formulas based on the basic major diameter, nominal size of the thread, and the material of the internal threaded part. The basic diameter of the thread is *D*. To determine the optimum strength of steel screws, the length of engagement in mating materials should equal *D* for steel; 1.50 × *D* for cast iron, brass, bronze, or zinc; 2.00 × *D* for forged aluminum; 2.50 × *D* for cast aluminum and forged magnesium; and 3.00 × *D* for cast magnesium or plastic.

Thread form is the second consideration. Normally, the choice is limited to UNC, UNF, or SI metric for fasteners. Other thread forms such as square, Acme, buttress, knuckle, and worm are used for special applications.

Thread series is the third consideration. The Unified Screw Thread Standard Series gives preference to the coarse and fine thread series.

The **class of thread fit** is the fourth consideration. The class of threads determines the degree of looseness or tightness between mating threads.

13.4.1 Thread Form

There are dozens of screw thread forms. However, for inch series mechanical fasteners, only three have significance: UN, UNR, and UNJ. All are 60° symmetrical threads with essentially the same profile. The principal difference between them is the contour at the root of the external thread. For metric fasteners, SI metric threads are designated.

UNR applies only to external threads. The difference between UN and UNR threads, in addition to designation, is that a flat or optional rounded root contour is specified for UN threads, while only a rounded root contour is specified for UNR threads. The design of UNJ threads developed from a search for an optimum thread form. This thread has root radius limits of 0.150 to 0.180 times thread pitch.

13.5 Standard Thread Fits

Thread fit is a measure of looseness or tightness between mating threads. **Classes of fit** are specific combinations of allowances and tolerances applied to external and internal threads.

For unified inch screw threads there are three thread classes for external threads, 1A, 2A, and 3A, and three for internal threads, 1B, 2B, and 3B. All are clearance fits, which means they assemble without interference. *The higher the class number, the tighter the fit.* The designator "A" denotes an external thread, "B" denotes an internal thread. The mating of class 1A and 1B threads provides the loosest fit, the rating of class 3A with 3B the tightest.

Additionally, there is a class 5 thread fit. Class 5 is an interference fit, which means that the external and internal threads are toleranced so that a positive interference occurs when they are assembled. Class 5 interference fits are standard only for coarse thread series in sizes 1 in. and smaller.

The requirements of screw-thread fits are determined by use and should be specified by indicating the proper classes for the components. For example, a Class 2A external thread should be used with a Class 2B internal thread. When choosing a class fit for threads, no tighter thread fit should be selected than the function of the parts requires.

Classes 1A and 1B are very loosely toleranced threads, with an allowance applied to the external thread. These classes are ideally suited when quick and easy assembly and disassembly are a prime design consideration. They are standard only for coarse and fine thread series in sizes $\frac{1}{4}$ in. and larger. They are rarely specified for mechanical fasteners.

Classes 2A and 2B are by far the most popular thread classes specified for inch series mechanical fasteners. Approximately 90% of all commercial and industrial fasteners produced in North America have this class of thread fit.

Classes 3A and 3B are suited for closely toleranced fasteners such as socket cap screws and set screws, and other high-strength fasteners. Classes 3A and 3B have restrictive tolerances and no allowance.

13.6 Thread Representation

On working drawings, threads are seldom drawn as they would actually appear; instead notes and specifications are used. The American National Standards Institute (ANSI) recognizes three conventions for representing screw threads on drawings: **detailed** (Fig. 13.14), **schematic** (Fig.13.15), and **simplified** (Fig.13.16) representations.

The detailed representation is an approximation of the actual appearance of screw threads. Minor modification includes showing the thread profile as a sharp V where the actual thread has flat crests and roots. Also, the normal helices are shown as straight lines connecting the thread, crest to crest and root to root. The detailed conventional representation is limited to where the basic diameter is over more than

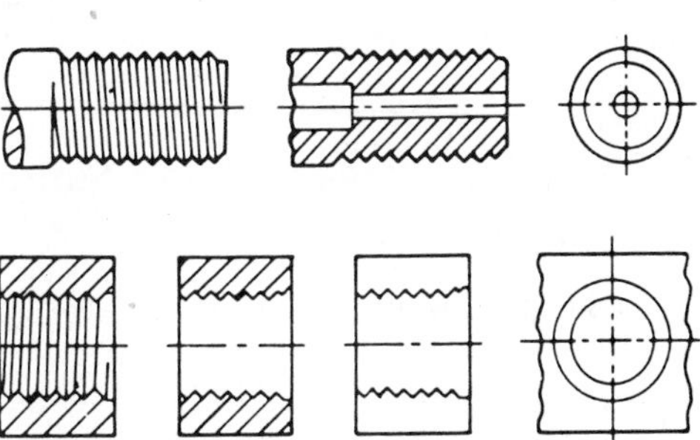

FIGURE 13.14 Detailed Thread Representation

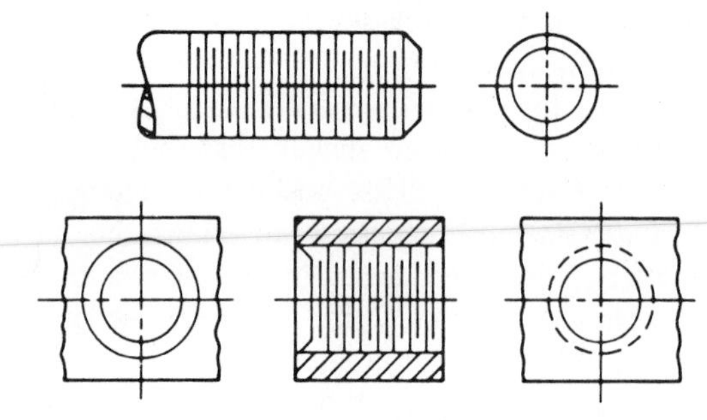

FIGURE 13.15 Schematic Thread Representation

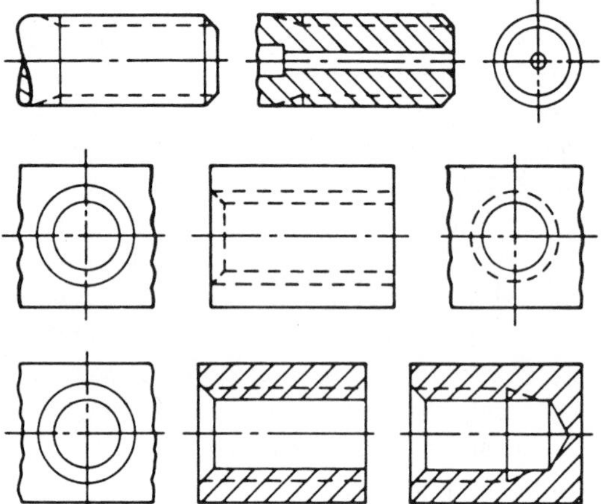

FIGURE 13.16 Simplified Thread Representation

1 in. and where detail or relation of component parts could be confused by less realistic thread representation. When internal holes are drawn using the detailed method, the lines representing the threads are sometimes omitted.

The **simplified method** showing internal threads and the simplified method to represent internal threads in a section are shown in Figure 13.17. Figure 13.18(b) shows the simplified method to represent external threads.

The **schematic method** is only used for external threads [Fig.13.18(a)] or sectioned internal threads (Fig.13.19). The schematic method is not used to represent internal nonsectioned threads.

Figure 13.20 shows the **detailed method** of representing internal threads in a section. Notice that the detailed method can be drawn with or without the lines of the threads. External threads drawn with detailed representation are shown in Figure 13.18(c).

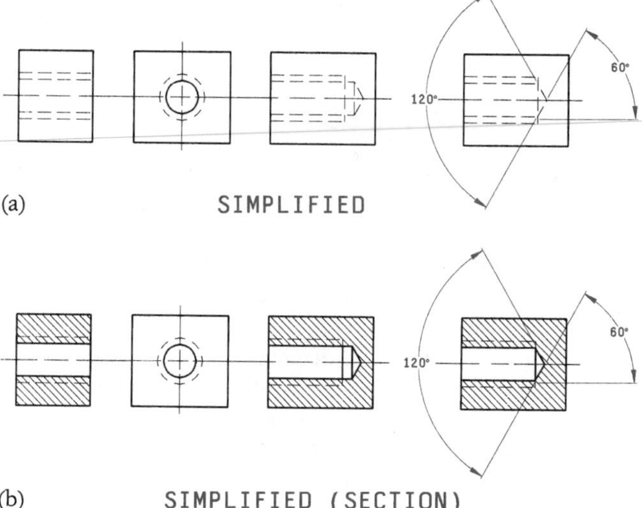

FIGURE 13.17 Internal Simplified Thread Representation

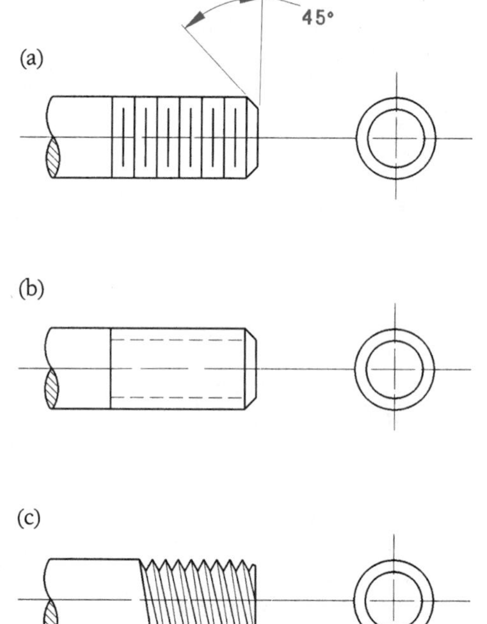

FIGURE 13.18 External Thread Representation
(a) Schematic (b) Simplified (c) Detailed

13.6.1 Drawing Threads Using Simplified Representation

The simplified method is drawn by following the steps shown in Figure 13.21. Both internal and external threads are drawn with this method. Here, external threads are being constructed. The diameter of the screw is drawn and its end is established [Fig. 13.21(a)]. The pitch (*P*) is measured as shown in Figure 13.21(b). Lines are drawn at 60° through the pitch measurements. The thread depth is where the 60° lines cross. The thread depth is used to draw the chamfered end. The chamfer is drawn at 45° and the threaded length is established [Fig. 13.21(c)]. The thread depth is used to draw the dashed lines that represent the minor diameter of the thread [Fig. 13.21(d)].

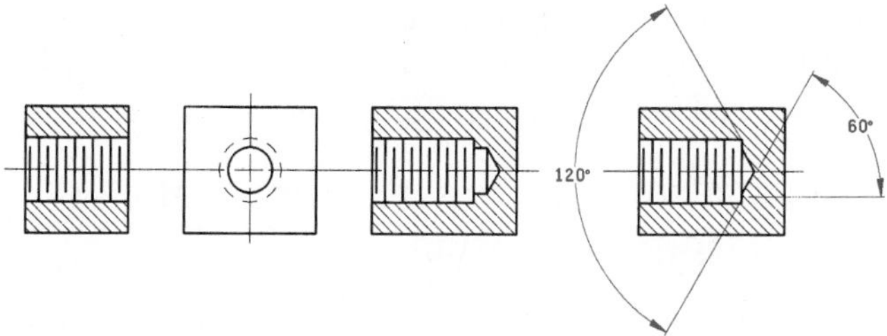

FIGURE 13.19 Internal Schematic Thread Representation

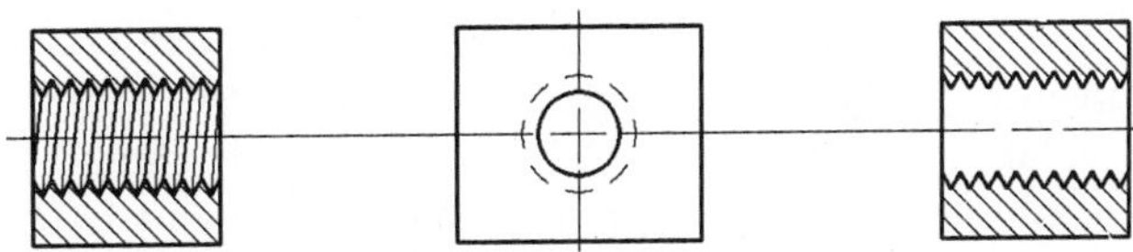

FIGURE 13.20 Internal Detailed Thread Representation

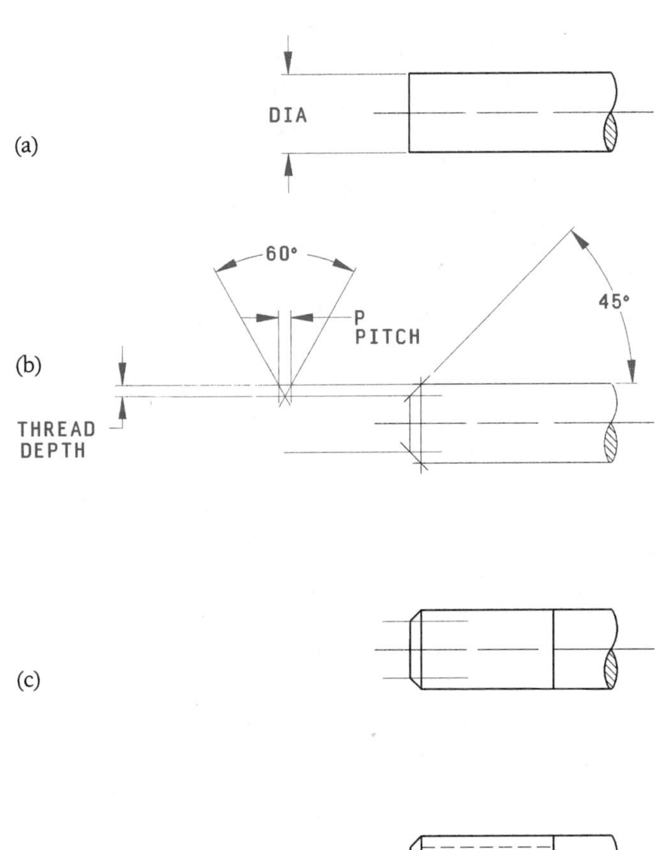

FIGURE 13.21 Drawing Threads Using Simplified Thread Representation

13.6.2 Drawing Threads Using Schematic Representation

Schematic representation is almost as effective as the detailed representation and is much easier to draw. The alternating lines, symbolic of the thread roots and crests, are usually drawn perpendicular to the axis of the thread or sometimes slanted to the approximate angle of the thread helix. This construction should not be used for internal threads or sections of external threads.

Drawing schematic threads is similar to the simplified method. The screw diameter and end are drawn first [Fig.13.22(a)]. The chamfer is completed using 45° and the thread depth [Fig. 13.22(b)]. The pitch (*P*) is used to establish the spacing of the thread crests [Fig. 13.22(c)]. The root lines are drawn up to the thread depth [Fig. 13.22(d)]. The thread is completed by darkening in the lines [Fig. 13.22(e)]. This method is called the **uniform-line method.** The slope-line representation is shown in [Fig. 13.22(f)]. The slope

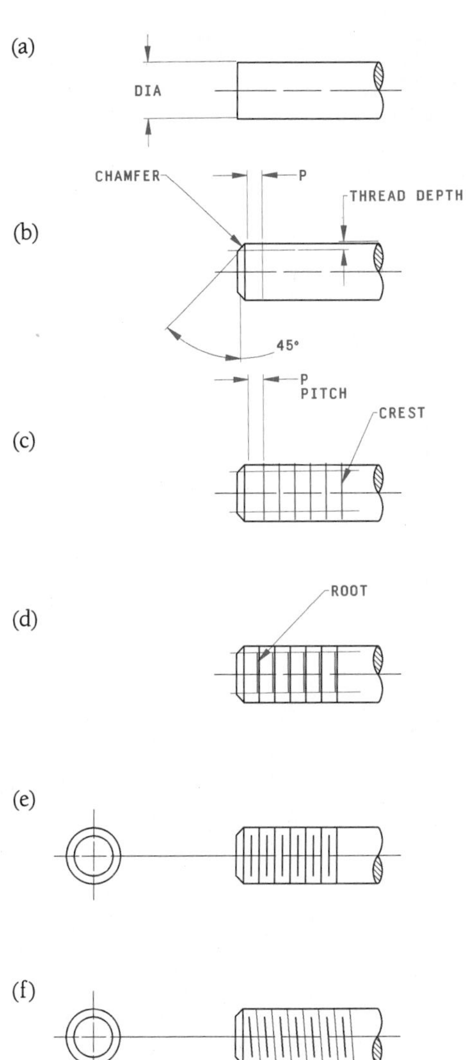

FIGURE 13.22 Drawing Threads Using Schematic Thread Representation

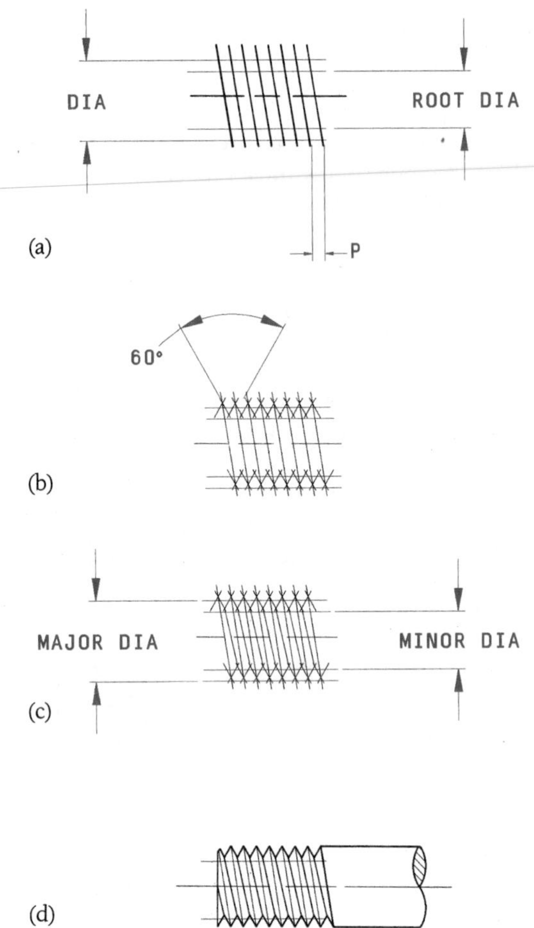

FIGURE 13.23 Drawing Threads Using Detailed Thread Representation

angle is equal to one-half of the pitch. In actual industrial practice, drafters draw the screw diameter, construct a 45° chamfer, and use the chamfer depth to locate the thread root.

13.6.3 Drawing Threads Using Detailed Representation

The detailed thread representation is drawn only when a mechanical advantage must be calculated or analyzed graphically (or for illustrations). Figure 13.23 shows four steps in drawing threads with detailed representation. Step (a) is the same as for simplified and schematic thread representation. The diameter of the screw thread is layed out and one half of the pitch is measured. Using the pitch (*P*), the top and the bottom lines of the shaft are divided along its length into the required number of threads. The sloped lines (crest lines) are drawn with an angle of one-half the pitch. Draw the threads as sharp V's at 60° as in Figure 13.23(b). The ends of the root lines will be established where the thread lines cross at the root. The root lines are drawn by connecting the roots [Fig. 13.23(c)]. In Figure 13.23(d), the threads are darkened.

13.6.4 How to Draw Acme Threads

A step-by-step procedure for drawing Acme threads is given in Figure 13.24. The Acme thread has a depth equal to one-half of its pitch. The drawing is begun by drawing the shaft diameter (major diameter), the minor diameter, and the pitch diameters with construction lines [Fig. 13.24(a)]. The pitch diameter lines are divided into segments equaling one-half the pitch [Fig. 13.24(b)]. The angle of the thread profile is one-half of 29° ($14\frac{1}{2}$°). Usually, 15° is used to simplify the procedure. The 15° lines are drawn through the half pitch distances established along the pitch diameter lines [Fig. 13.24(c)]. The angled lines will fall between the major diameter and the minor diameter [Fig.13.24(d)]. The crests are completed and the root lines are then drawn [Fig. 13.24(e)]. The ends of the threads are completed, the construction lines

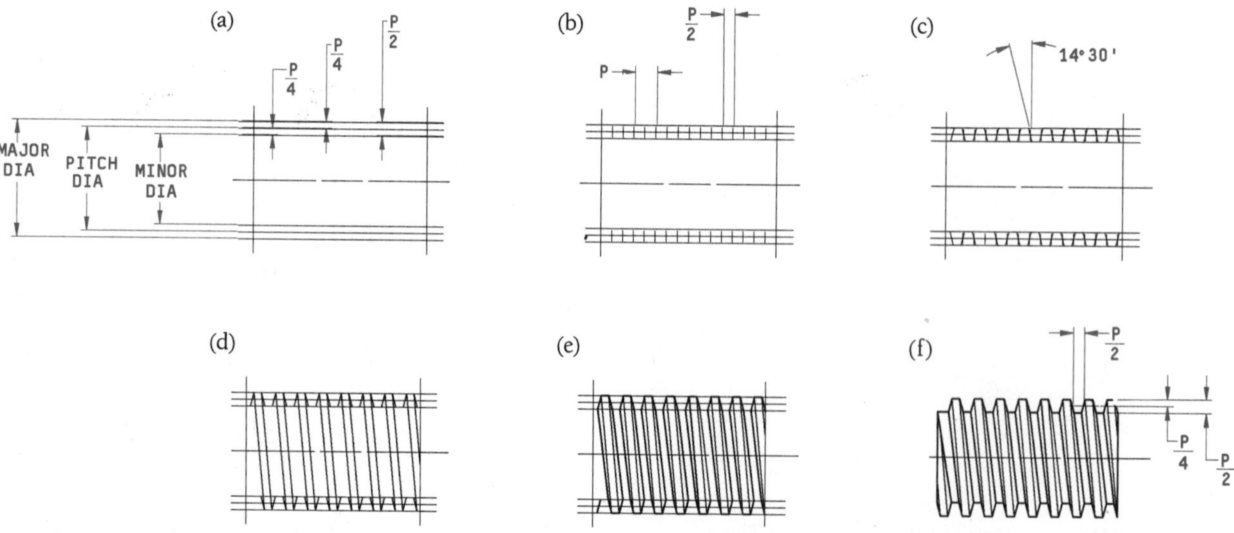

FIGURE 13.24 Drawing Acme Threads Using Detailed Thread Representation

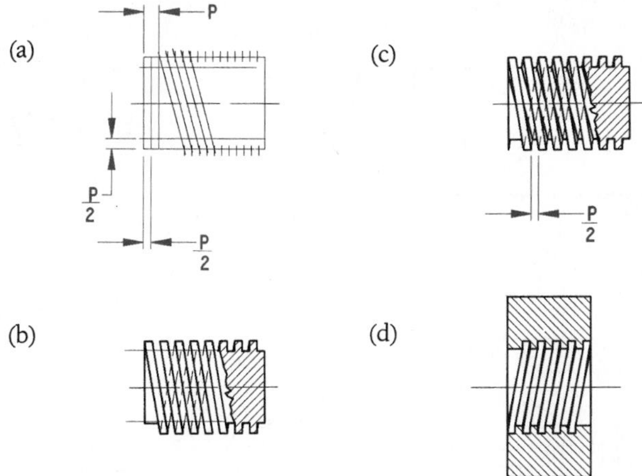

FIGURE 13.25 Detailed Acme Threads

erased, and the drawing is darkened [Fig. 13.24(f)]. Detailed Acme threads are shown in Figure 13.25.

13.6.5 Tap Drills

Threaded holes are drilled first and then tapped (Fig.13.26). The tapping tool extends far enough into the hole to thread the required length of full threads. The tap drill, therefore, must extend beyond the required thread depth. The major diameter represents the outside diameter of the thread and the minor diameter represents the tap drill diameter.

Figure 13.27 shows how to represent tapped holes using the simplified method. The drilled hole is drawn accurately with its diameter and depth as shown. The drill tip is 118°, but for simplicity it is drawn at 130° (30° from the horizontal). The tap drill is represented the same for holes. For drilled and tapped holes the depth of the full thread is accurately drawn. The tap drill is drawn 3× the pitch below the threaded portion. This distance includes a number of incomplete threads created by the chamfer end of the tapping tool. Although normally drawn at 3× the pitch, this distance is determined by the drill size as is whether to use a bottoming tap or a plug tap. In some cases, the thread will extend to the bottom of the drilled hole, or a **thread relief** will be required.

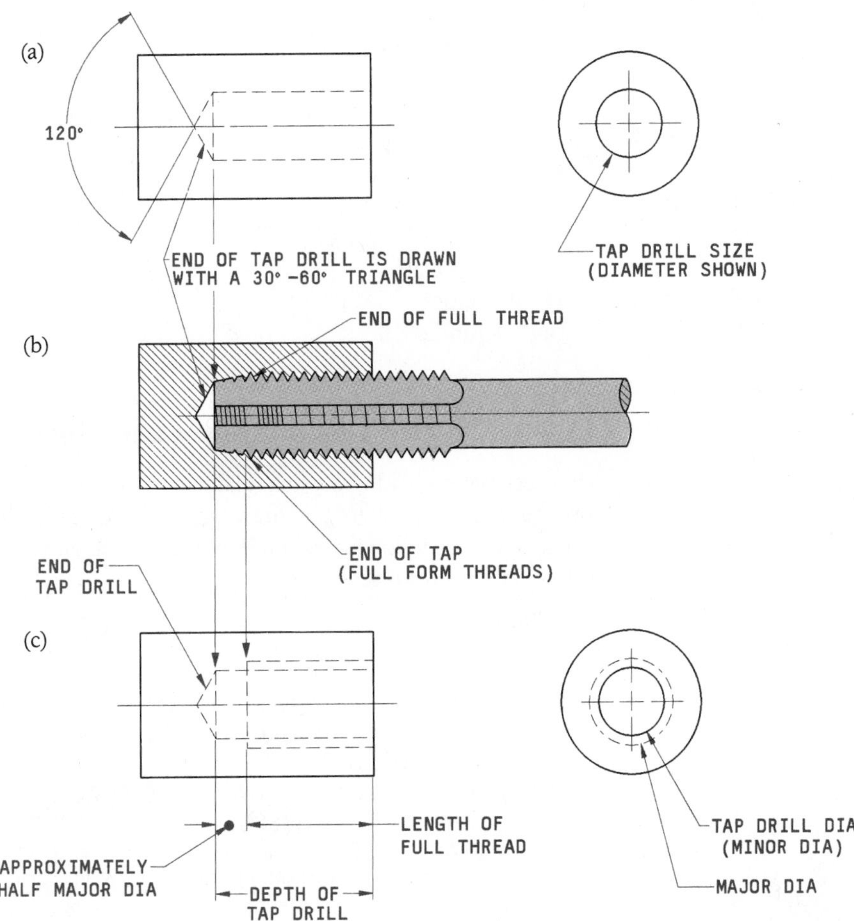

FIGURE 13.26 Blind Holes and Taps

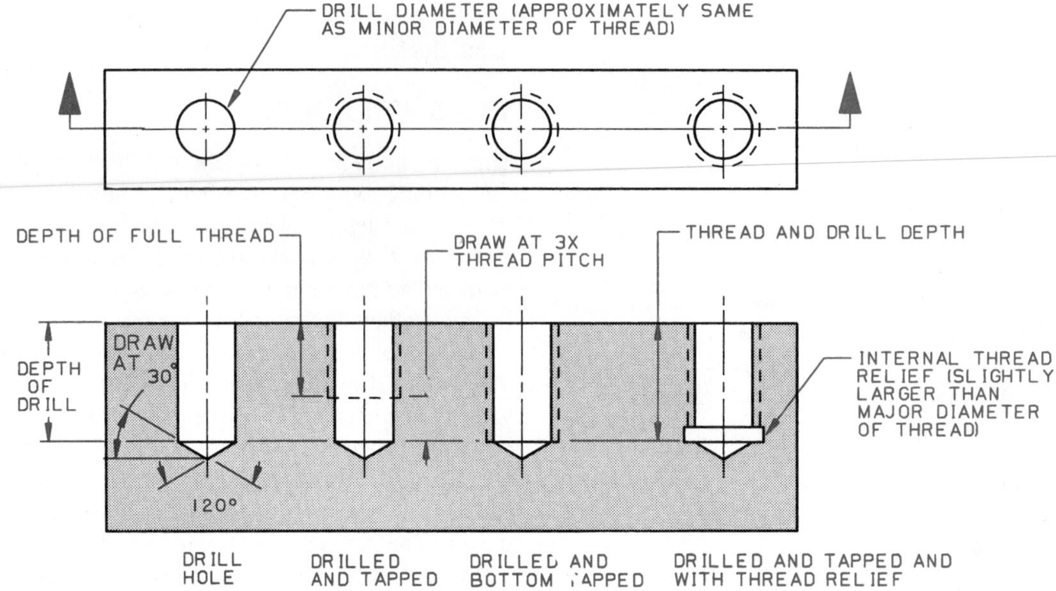

FIGURE 13.27 Drilling and Tapping Holes

An internal thread relief is slightly larger than the major diameter of the thread (Fig.13.27). The circular views of the threaded holes show the tap drill as a solid line and the major thread (major diameter) as a dashed line.

13.7 Designating Threads and Thread Notes

The thread designation includes the nominal diameter, the number of threads per inch (or the pitch and lead), the letter symbol of the thread series, the number and letter of the thread class, and any qualifying information. The thread length, the hole size, and the chamfer or countersink may be included in the note or dimensioned on the drawing of the part.

The series symbols and the class numbers identify the controlling thread standard and define the details of thread design, dimensions, and tolerances not specifically covered on the drawing. Series, class, and dimensional letters in a thread designation are shown as follows:

A	external, American, aeronautical
B	internal
C	coupling, coarse, or centralizing
EXT	external
EF	extra fine
F	fine, fuel and oil
G	general purpose, gas, pitch allowance
H	house
I	intermediate
INT	internal
J	controlled radius root
L	lead, locknut
LE	length of engagement
LH	left-hand (absence of LH indicates RH)
M	metric, mechanical, microscope, miniature
MOD	modified
N	national
O	outlet, objective
P	pipe, pitch
R	railing, rounded root, American National Class 1 allowance
RH	right-hand
S	straight
SE	special engagement
SPL	special
T	taper
UN	unified

13.7.1 Thread Designation Examples

The designation and the pitch diameter limits are in note form and referenced to the drawing of the thread with a leader line. The following example illustrates the elements of a designation of the screw thread:

.250-20 UNC-2A

where

.250 = NOMINAL DIAMETER IN DECIMAL FORM
20 = NUMBER OF THREADS PER INCH OF PITCH AND LEAD
UNC = THREAD FORM, SERIES, AND TOLERANCE FORMULATION SYMBOL
2 = CLASS NUMBER
A = INTERNAL OR EXTERNAL SYMBOL (A IS EXTERNAL)

Thread sizes are shown as decimal callouts except for fractional sizes. When specifying decimal sizes, a minimum of three or maximum of four decimal places, omitting any zero in the fourth decimal place, should be shown as the nominal size:

1.000-8 UNC-2A

$1\frac{3}{4}$ - 8 UN-2A

Numbered sizes may also be shown; the decimal equivalent should be in parentheses:

No. 10(.190)-32 UNF-2A

Unless otherwise specified, threads are right-hand; a left-hand thread shall be designated **LH** as:

$\frac{1}{4}$ - 20 UNC-3A-LH

13.8 Acme Threads

There are four classes of general-purpose Acme threads and five classes of centralizing Acme threads. The general-purpose Acme threads have clearances on all diameters for free movement and may be used in assemblies where both internal and external members are supported to prevent movement.

There is only one class of stub Acme thread for general usage. It is the Class 2G (general-purpose) thread, which uses two threads with modified thread depths. Stub Acme threads are used for power applications.

When designating Acme threads, the designation covers the nominal size, the number of threads per inch, the thread form symbol, and the thread class symbol:

1.750-4 ACME-2G

where

1.750 = NOMINAL DECIMAL SIZE
4 = NUMBER OF THREADS PER INCH
ACME = THREAD FORM AND SERIES SYMBOL
2G = THREAD CLASS SYMBOL

13.9 Buttress Threads

Buttress threads are used for high-stress applications where the stress is along its axis in only one direction. The buttress thread is designated **butt** or **push-butt.** Since the design of most components having buttress threads is so special, no diameter pitch series is recommended. The two classes of buttress threads are Class 2 (standard grade) and Class 3 (precision grade).

When only the designation "butt" is used, the thread is a "pull"-type buttress with the clearance flank angle of 45° leading and the pressure flange angle of 7° following. In thread designations on drawings and in specifications, the designation should be shown as:

2.500-8 BUTT-2A-LH

where

2.500 = NOMINAL SIZE (BASIC MAJOR DIAMETER IN INCHES)
8 = THREADS PER INCH (TPI)
BUTT = BUTTRESS FORM OF THREAD, PULL-TYPE
2 = CLASS 2 (MEDIUM) THREAD
A = EXTERNAL THREAD
LH = LEFT HAND

13.10 Metric Threads

A wide variety of threaded fasteners are manufactured using metric threads. This section contains general metric standards for a 60° symmetrical screw thread with a basic ISO 68, profile designated "M".

The simplified, schematic, and detail methods of thread representation also apply to metric screw thread drawing practices. The following additional definitions are used for metric threads: **Bolt thread (external thread)** is used in ISO metric thread standards to describe all external threads. All symbols associated with external threads are designated with lowercase letters.

Nut thread (internal thread) is used in ISO metric thread standards to describe all internal threads. All symbols associated with internal threads are designated with uppercase letters.

13.10.1 Metric Classes of Fit

There are two recognized classes of thread fit. One is for general-purpose applications and contains tolerance classes 6H/6g; the other is used where closer thread fits are required and contains tolerance classes 6H/4g to 6g.

The **tolerance grade** is indicated by a number. The system provides for a series of tolerance grades for each of the four screw thread parameters: minor diameter, internal thread (4,5,6,7,8); major diameter, external thread (4,6,8); pitch diameter, internal thread (4,5,6,7,8); and pitch diameter, external thread (3,4,5,6,7,8,9).

The **tolerance position** is the allowance and is indicated by a letter. A capital letter is used for internal threads and a lowercase letter for external threads. The system provides a series of tolerance positions for internal and external threads:

Internal threads G, H
External threads g, h

The tolerance grade is given first, followed by the tolerance position: 4g or 5H. To designate the tolerance class, the grade and position of the pitch diameter is shown first followed by the major diameter (external thread) or the minor diameter (internal thread) 4g6g for an external thread and 5H6H for an internal thread. If the two grades and positions are identical, it is not necessary to repeat the symbols. Therefore, 4g alone, stands for 4g4g, and 5H alone means for 5H5H.

13.10.2 Designation of Metric Screw Threads

Metric screw threads are identified by the letter "M" for the thread form profile, followed by the nominal diameter size and the pitch expressed in millimeters, separated by the × sign and followed by the tolerance class separated by the dash (-) from the pitch.

The simplified international practice for designating coarse pitch *M* profile metric screw threads is to leave off the pitch. Thus, a **M14 × 2** thread is designated just **M14.** However, to prevent misunderstanding, it is mandatory to use the value for pitch in all designations shown on drawings.

The thread acceptability gaging system requirements of ANSI B1.3M may be added to the thread size designation. The numbers are shown in parentheses: (22), (21). The following is an example of a close tolerance external thread designation:

M8 × 1.25-4g6g (22)

Two examples of internal thread designation are:

M6 × 1-6H (21)

M6 × 1-5H6H (21)

Unless otherwise specified in the designation, the screw thread helix is right-hand. When a left-hand thread is specified, the tolerance class designation is followed by a dash and LH. The following example is of a left-hand external thread with a M profile:

M6 × 1-4g6g-LH

where

M = METRIC THREAD SYMBOL, ISO 68 METRIC THREAD FORM
6 = NOMINAL SIZE IN MILLIMETERS
1 = PITCH IN MILLIMETERS
4g6g = TOLERANCE CLASS
4g = MAJOR DIAMETER TOLERANCE SYMBOL (4 = TOLERANCE POSITION; G = TOLERANCE GRADE)
6g = PITCH DIAMTER TOLERANCE SYMBOL (6 = TOLERANCE POSITION; G = TOLERANCE GRADE)
LH = LEFT HAND

A fit between *mating threads* is indicated by the internal thread tolerance class, followed by the external thread tolerance class separated by a slash:

M6 × 1-6H/6g

M6 × 1-6H/4g6g

13.11 Dimensioning Threads

The thread length dimensioned on the drawing should be the gaging length or the length of threads having full form. That is, the incomplete threads are outside or beyond the length specified.

Should there be reason to control or limit the number of incomplete threads on parts having a full-body diameter shank, the overall thread length, including the vanish (runout or incomplete) threads, are represented and dimensioned on the drawing, in addition to the full thread length (Fig.13.28). All representation of fully formed threads should indicate the *thread runout* (incomplete threads) as shown in the figure. Overall thread length should be represented and dimensioned on the drawing, and should include the thread runout.

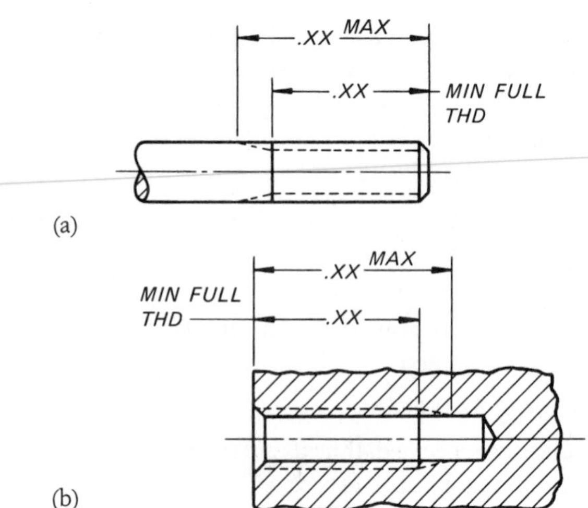

FIGURE 13.28 Dimensioning Thread Length

13.11.1 Thread Chamfers

If required, thread **chamfers** or **countersinks** should be specified on the drawing. It is preferable to specify the chamfer by length and diameter to avoid confusion. Figure 13.29 shows three methods of dimensioning an external chamfer. The chamfer length should be 0.75 to 1.25 times the pitch, rounded off to a two- or three-place decimal. When a callout cannot properly or clearly designate an internal threaded hole, the depth, size, and countersink (chamfer) are dimensioned (Fig.13.30). If the chamfer and minor diameter are very close to being the same, the minor diameter of a thread may be eliminated to improve clarity. On end views of countersunk threaded holes where countersunk diameters and the major diameters of threads are close to being the same, the major diameter may be eliminated for clarity.

13.11.2 Threads on Drawings

Holes are located by their centers. Leaders have the arrowheads pointing toward the center in the circular views. When the circular view is not available, the arrow of the leader line should touch the axial centerline of the hole. Figure 13.31 shows an example of UN thread callouts (they are not meant to be equivalents in the example).

The full depth of the drilled hole for **blind tapped holes** should be specified on the drawing (Figs.13.27 and 13.30). Blind holes do not go all the way through the part. If the wall

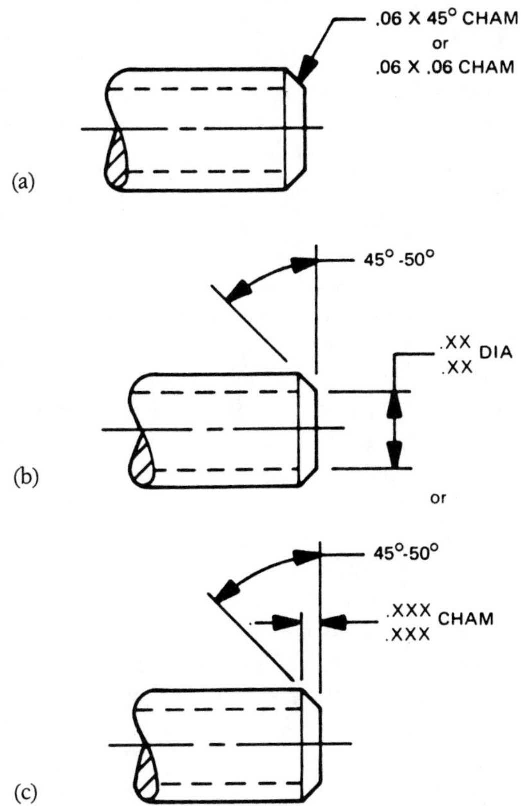

FIGURE 13.29 Dimensioning Chamfers at the End of External Threads

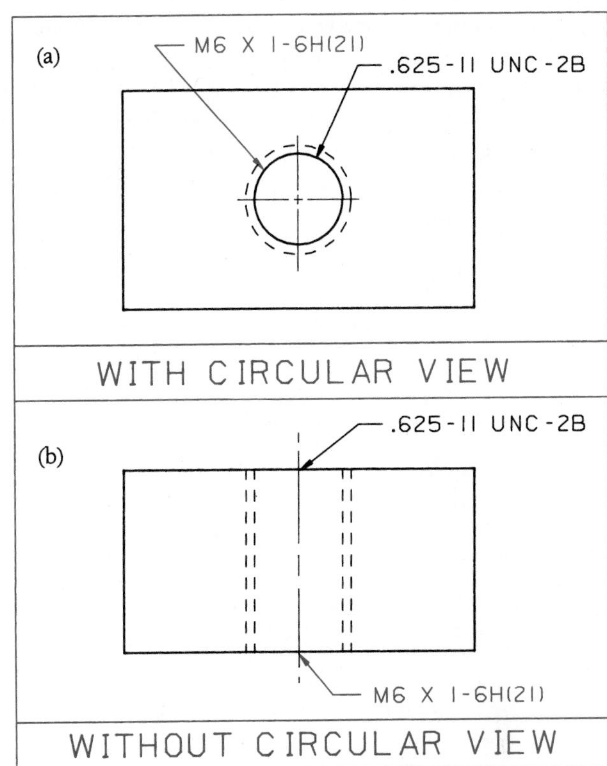

FIGURE 13.31 Calling Out Threads on Drawings

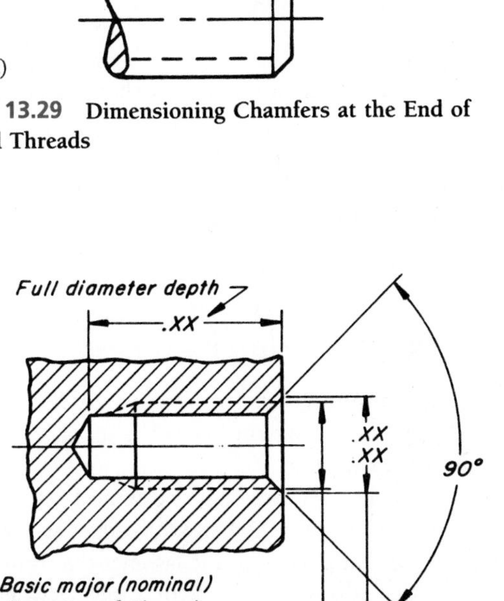

FIGURE 13.30 Dimensioning Countersink, Drill Depth, and Size on Internal Threaded Holes

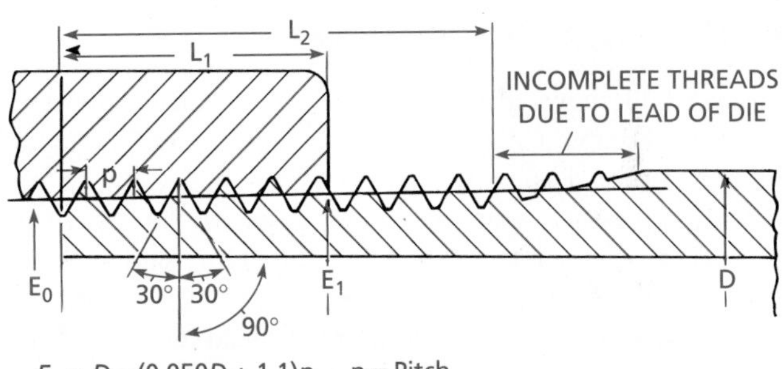

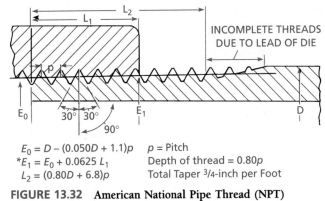

$E_0 = D - (0.050D + 1.1)p$ p = Pitch
$^*E_1 = E_0 + 0.0625\ L_1$ Depth of thread = $0.80p$
$L_2 = (0.80D + 6.8)p$ Total Taper 3/4-inch per Foot

FIGURE 13.32 American National Pipe Thread (NPT)

Nominal Pipe Size	D	TPI	P	E0	E1	L1	L2
.750	.840	14	.071	.758	.778	.320	.533
3.000	3.500	8	.125	3.340	3.388	.766	1.200

at the drill point is the limiting consideration in addition to, or instead of, the full diameter depth, the drill point depth or the *wall thickness* may be dimensioned or stated in a note. In some cases, the depth may be specified as a minimum full diameter depth and the note "DO NOT BREAK THRU" should be added. Hole size limits should be shown on the drawing.

13.12 Pipe Threads

The American National Standard taper pipe thread is tapered $\frac{1}{16}$ in. per inch ($\frac{3}{4}$ in. per ft) to ensure a tight joint at the fitting (Fig. 13.32). The crest of the thread is flattened, and the root is filled so that the depth of the thread is equal to 80% of the pitch. The number of threads per inch for a given nominal diameter can be obtained in Appendix C.

Pipe threads are designated in established trade sizes that signify a nominal diameter only. The designation of tapered threads includes the nominal size, the number of threads per inch, the thread form, and thread series symbols as shown in the following examples:

6-8 NPT		**.125-27 NPT**
	Explanation	
6 =	nominal pipe diameter in inches	= .125
8 =	number of threads per inch	= 27
N =	American Standard National thread	= N
P =	pipe	= P
T =	taper	= T

.750-14 NPSL		**12-8 NPTR**
	Explanation	
.750 =	nominal pipe diameter in inches	= 12
14 =	number of threads per inch	= 8
N =	American National Standard thread	= N
P =	pipe	= P
S =	straight	
	taper	= T
L =	locknuts & locknut pipe threads	
	rail fittings	= R

13.12.1 Drawing Pipe Threads

Figure 13.33 shows a male (external) and female (internal) pipe thread drawn using simplified representation. The taper on a pipe thread is so slight that it does not show up on drawings unless it is exaggerated. It is drawn to $\frac{1}{8}$ in. taper per inch. The ANSI recommendation for representing pipe threads is the same as for all other threads. The simplified form is the most commonly used, the detailed form the least.

You May Complete Exercises 13.1 Through 13.4 at This Time

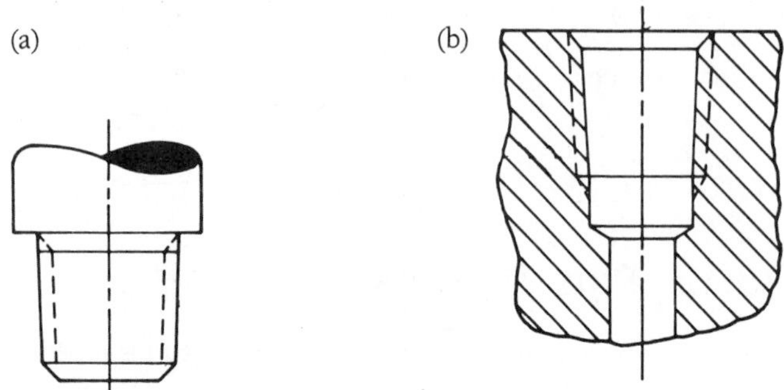

FIGURE 13.33 Internal and External Pipe Thread Representation

13.13 Fasteners

Basic industrial **fasteners** (Fig.13.34) include square and hex bolts, cap screws, carriage bolts, machine screws, plow bolts, lag screws, studs, nuts, and rivets. Other fasteners have also been standardized over the years as to type, style, usage, properties, dimensions, and tolerances.

Semipermanent assembly fasteners include bolts, screws, studs, nuts, washers, snap rings, nails, and pins. Rivets are considered **permanent fasteners.** Fastener selection is made by considering strength, appearance, durability, corrosion resistance, materials to be joined, total cost of assembly parts, and assembly and disassembling labor involved or machines and power tools required. Whenever possible, design for automated assembly using common standard parts.

The *installed cost* is far more important than the initial cost of the fastener. For example, a rivet is much cheaper than the high-strength bolt, nut, and washer that replaced it, but the greater holding power and the lower installed cost of the high-strength bolting system has for all practical purposes displaced riveting as standard fastening for structural joints.

When designating fasteners on a drawing, provide the following:

- product name
- nominal or actual size in fractions, decimal equivalent, or metric units
- thread specification, if appropriate
- length in fractions, decimal equivalent, or metric units
- material and protective coating, if applicable
- finish, where required.

13.13.1 Representing Fasteners

In general, a template is used for construction of standard fasteners. When a CAD system is used, a standard library of parts is normally available.

Figure 13.35 shows a sketch of two typical fasteners. The *head styles* shown here are the **hex** and **socket** varieties. The *bearing surface* is that portion of the fastener that is in contact with the part that is being fastened (or a washer when one is used). The *point* is at the opposite end from the head and is normally chamfered. The *threaded* part of the *body* extends from the point toward the bearing surface. Some fasteners are completely threaded (the whole body) and some are partially threaded.

13.13.2 Studs

Studs are fasteners with no head but with threads at both ends of the shank. Studs come in *continuous threaded* types and *double-ended* varieties. In most applications, the stud is screwed into a workpiece on one end and a nut is used on the other end (Fig. 13.36). In other applications, the stud has a nut on both ends and is used to secure two pieces. In Figure

Description	Military Reference	Description	Military Reference	Description	Military Reference
–1– Pan head	MS 35204 thru MS 35219 and MS 35221 thru MS 35236	–12– Socket head cap screw	MS 35455 thru MS 35461	–23– Flat washer	MS 15795
80° to 82° –2– 82° Flat head	MS 35188 thru MS 35203 and MS 35237 thru MS 35251 and MS 35262	–13– Set screw	AN 565	–24– Lockwasher-spring	MS 35337 MS 35338 MS 35339 MS 35340
99° to 101° –3– 100° Flat head	AN 507	–14– Self-locking	Plastic pellet can be applied to all types of screws	–25– Lockwasher-ext. tooth	MS 35335
–4– Fillister head	MS 35361 and MS 35366	–15– Hex nut	MS 35649 MS 35650 MS 35690	–26– Lockwasher-int. tooth	MS 35333 MS 35334
–5– Drilled fillister head	MS 35263 thru MS 35278	–16– Self-locking nut (non-metallic collar)	Can be supplied with fibre or plastic collar. All sizes and material	–27– Lockwasher-csk. tooth	MS 35336 MS 35790
–6– Slotted hex head	Made to order in 1020 Bright. 1035 Heat Treat and Alloy Steel	–17– Self-locking nut (deflected beam)	Can be supplied in Steel, Brass, Stainless - all sizes	–28– Spring pin	MS 9047 MS 9048 MS 171401
–7– Tapping screw-Type 1	AN 504 AN 506	–18– Clinch nut	Supplied to order for special applications	–29– Grooved pin	MS 35671 thru MS 35679
–8– Tapping screw-Type 23	AN 504 AN 506	–19– Clinch nut	Supplied with fibre locking collar in various shank lengths	–30– Taper pin	AN 385
–9– Tapping screw-Type 25	AN 530 AN 531	–20– Self-locking nut	Made with Nylon pellets in standard and special sizes	–31– Weld stud	Supplied with welding nibs under and top of head
–10– Drive screw	AN 535	–21– Semi-tubular	MS 20450	–32– Weld nut (self locating)	Supplied with standard thread sizes
–11– Sems	Supplied with all types of heads, also with Internal and External Lockwashers	–22– Shoulder	Made to specifications in steel and brass	–33– Weld nut	Supplied with standard thread sizes

FIGURE 13.34 Industrial Fasteners

13.37 the cover plate for the check valve has eight studs and 16 nuts.

Continuous thread studs are threaded from end to end and are often used for flange bolting with two nuts. Continuous threaded studs come in two types: Type 1 and Type 2. Type 1 is for general purpose and Type 2 is for pressure piping. If a stud is to be inserted into a tapped hole (Fig.13.36), it is recommended that it be held in place by jamming it against the bottom of the hole. A class 5 fit is recommended for such service. The thread engagement should be $1\frac{1}{4}$ times the diameter of the stud for steel, $1\frac{1}{2}$ times for cast iron, and $2\frac{1}{2}$ times for softer materials.

Double-ended studs come in four types: Type 1 is unfinished, Type 2 is finished and has an undersize body, Type 3 is full bodied and finished, and Type 4 is finished and is close-body, milled to specifications.

A typical stud application is shown in Figure 13.38 of the swivel-heel clamp assembly. Here, two studs are used in the design of this tooling component. Studs are designated on drawings as shown below:

For Type 1 continuous:

CONTINUOUS THREAD STUD, $\frac{1}{2}$ - 13 × 8, ASTM A307, ZINC PLATED

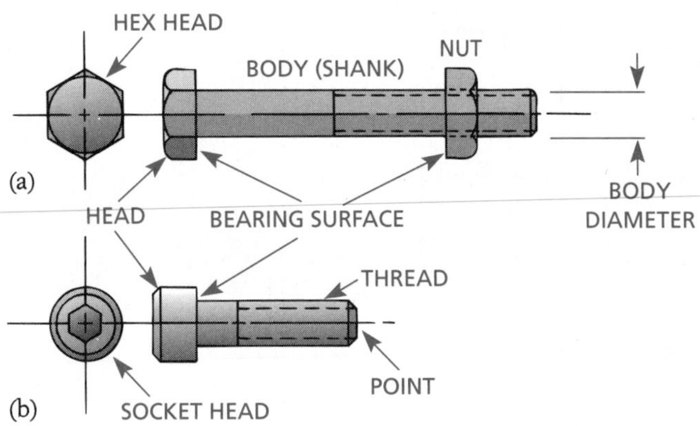

FIGURE 13.35 **Bolt and Screw Terminology**
(a) Bolt (b) Screw

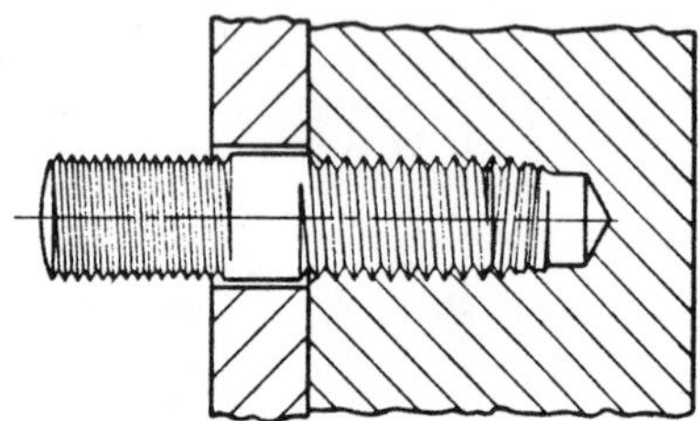

FIGURE 13.36 **Threaded Stud in Blind Hole**

FIGURE 13.37 **Check Valve**

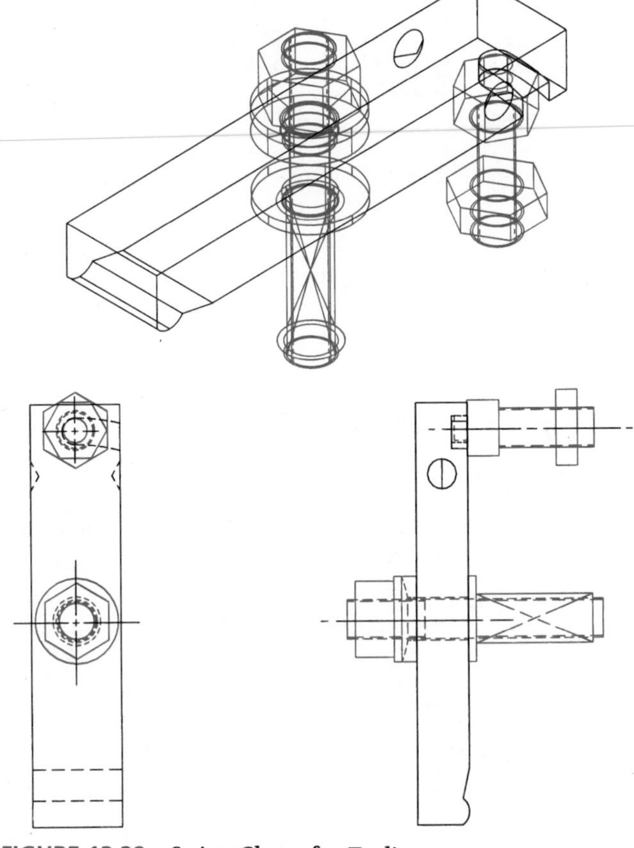

FIGURE 13.38 **Swing Clamp for Tooling**

For Type 2 continuous:

ANSI/ASME B16.5 STUD BOLT, .875-9 × 12, ASTM A 354, GRADE BD

For metric continuous:

CONTINUOUS THREAD STUD, M24 × 3 × 200, ASTM F568 CLASS 8.8, ZINC PHOSPHATE AND OIL

For double-ended:

TYPE 4 DOUBLE END STUD $\frac{3}{4}$ - 10 × 8.50, ASTM A499, CADMIUM PLATED

TYPE 2 DOUBLE END STUD, M10 × 1.5 × 90, ASTM F568 CLASS 9.8, ZINC PLATED

13.13.3 Bolts

A **bolt** is a device with a head on one end of a shank and a thread on the other end. Designed for insertion through holes in assembly parts, it is mated with a tapped nut. The diameter of all bolts is measured as the outside or major diameter of the thread; the length of a headed bolt is measured from under the head to the end of the bolt. The length of a bolt with a countersunk (flathead) head is the overall length. The point (tip) of a bolt is always included in the measured length.

Figure 13.39 illustrates the common types of bolts available. **Hexagon bolts** (Fig.13.40) can be used in a threaded

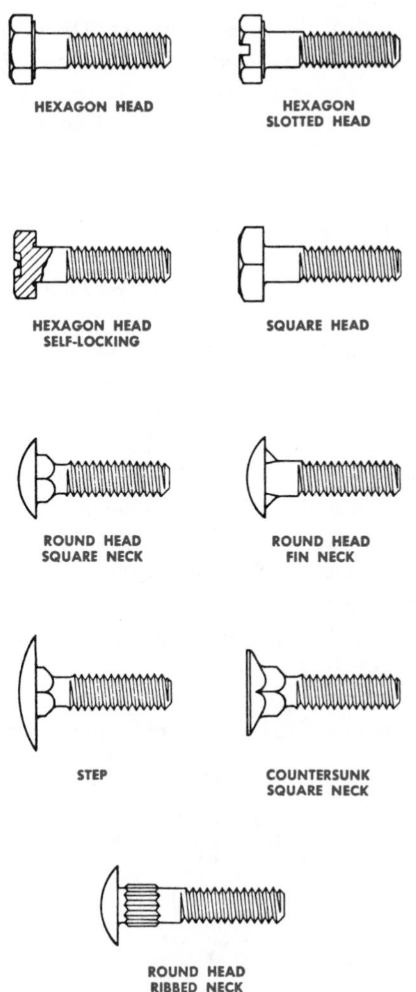

FIGURE 13.39 Bolt Head Types

hole or with a nut. A typical application of a hexagon bolt is shown in Figure 13.41. Hexagon bolts are available with either plain or slotted heads, and also come in metric sizes.

Square-head (Fig.13.42) and **round-head** (Fig.13.43) bolts are usually made of low carbon steel and are referred to as "black bolts." They are available in an unfinished style and with coarse threads. Square-head bolts are adequate for heavy machinery, conveyors, and fixtures. Round-head bolts have various shaped necks under the head that are embedded in wood or metal and act as a locking device. **Countersunk bolts** are shown in Figure 13.44. Bolts are designated on drawings as shown in the following examples:

$\frac{1}{2}$-13 × 3 $\frac{1}{2}$ HEX CAP SCREW, SAE GRADE 8 STEEL

.625-11 × 2 ROUND HEAD SQUARE NECK BOLT, STEEL

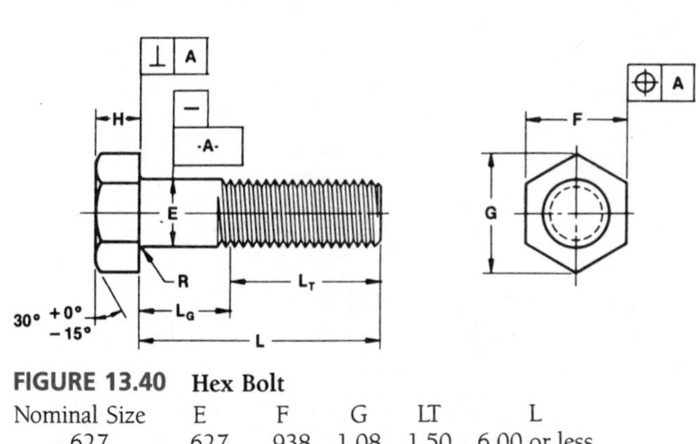

FIGURE 13.40 Hex Bolt

Nominal Size	E	F	G	LT	L
.627	.627	.938	1.08	1.50	6.00 or less
1.000	1.000	1.500	1.73	2.25	6.00 or less

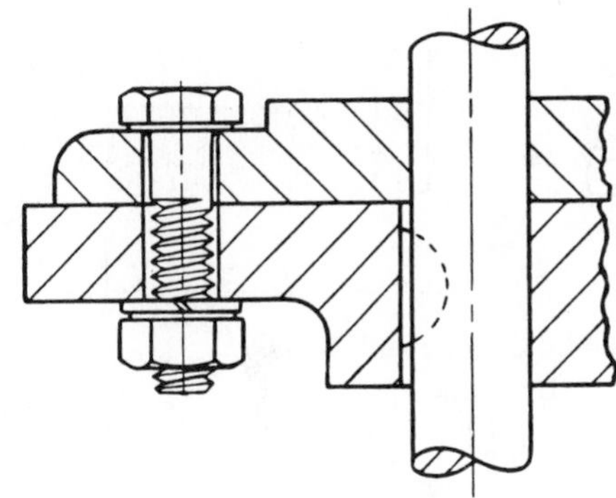

FIGURE 13.41 Hex Bolt Used on Assembly

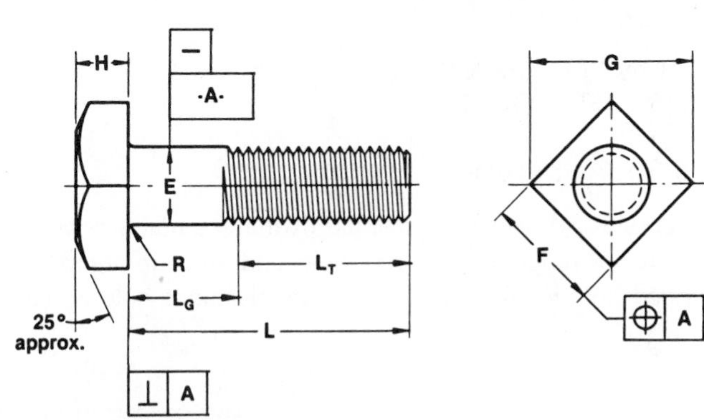

FIGURE 13.42 Square-Head Bolt

Nominal Size	E	F	G	LT	L
.875	.875	1.312	1.856	2.00	6.00 or less
1.500	1.50	2.250	3.182	3.25	6.00 or less

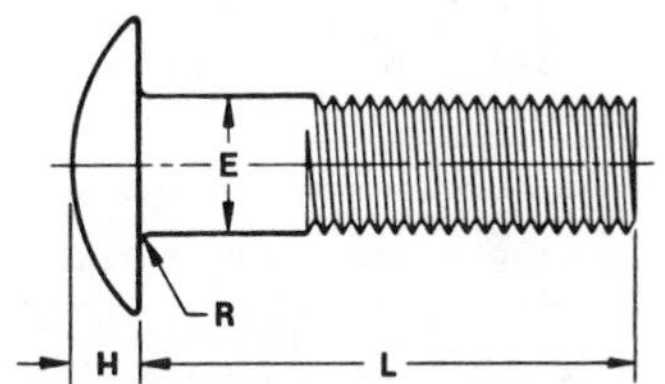

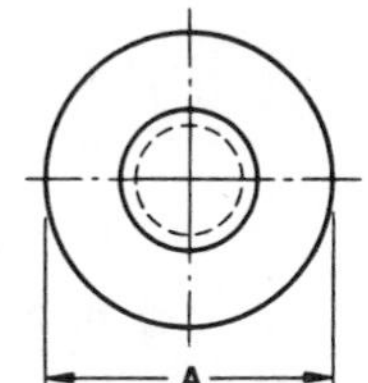

FIGURE 13.43 Round-Head Bolts

Nominal Size	A	E	H
.312	.719	.312	.176
.625	1.344	.625	.344

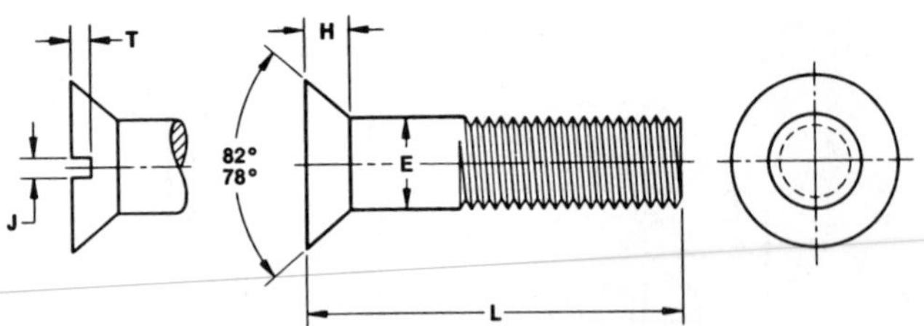

FIGURE 13.44 Countersunk Bolts

Nominal Size	Head Dia	E	H	J	T
.750	1.16	.750	.40	.14	.17
1.25	2.33	1.250	.67	.22	.29

For metric:

HEX BOLT, M20 × 2.5 × 160, CLASS 4.6, ZINC PLATED

HEAVY HEX STRUCTURAL BOLT, M22 × 2.5 × 160, ASTM A325M

13.13.4 Screws

A **screw** is a threaded fastener used without a nut. Screws are inserted through a clearance hole and into an internally tapped hole in the mating part. The clearance hole is only slightly larger than the screw diameter. Therefore, it is not shown on the assembly drawing. Only the body diameter of the screw is shown, as is the outside diameter of the threads. When sectioning assemblies, screws and other fasteners are not sectioned.

13.13.5 Machine Screws

Machine screws differ from cap screws mainly in range of basic diameters, head shapes, and driver provisions. Machine screws are so named because they are completely machined from bar stock. They are usually restricted to light assemblies such as instrument panel mountings, moldings, and clip fasteners. The size selection is determined by the tightness required of the parts to be fastened. Machine screws can be assembled into a nut or a threaded hole in a functional part. Figure 13.45 shows various screw head shapes available as standard parts. Screw selection is made by considering design needs such as surface condition, appearance, size of hole, cover clearance, driving provisions, and expected environmental exposure.

Machine screws come in either fine or course thread and are normally confined to light assembly applications. Machine

PAN Low large diameter with high outer edges for maximum driving power. With slotted or Phillips recess for machine screws. Available plain for driving screws.	FLAT UNDERCUT Standard 82° flat head with lower 1/3 of countersink removed for production of short screws. Permits flush assembles in thin stock.
TRUSS Similar to round head, except with shallower head. Has a larger diameter. Good for covering large diameter clearance holes in sheet metal. For machine screws and tapping screws.	100° FLAT, 100° Has larger head than 82° design. Use with thin metals, soft plastics, etc. Slotted or Phillips driving recess.
BINDER Undercut binds and eliminates fraying of wire in electrical work. For machine screws, slotted or Phillips driving recess.	FLAT TRIM Same as 82° flat head except depth of countersink has been reduced. Phillips driving recess only.
ROUND Used for general-purpose service. Used for bolts, machine screws, tapping screws and drive screws. With slotted or Phillips driving recess.	OVAL Like standard flat head. Has outer surface rounded for added attractiveness. Slotted, Phillips or clutch driving recess.
ROUND WASHER Has integral washer for bearing surface. Covers large bearing area than round or truss head. For tapping screw only; with slotted or Phillips driving recess.	OVAL UNDERCUT Similar to flat undercut. Has outer surface rounded for appearance. With slotted or Phillips driving recess.
FLAT FILLISTER Same as standard fillister but without oval top. Used in counter bored holes that require a flush screw. With slot only for machine screws.	OVAL TRIM Same as oval head except depth of countersink is less. Phillips driving recess only.
FILLISTER Smaller diameter than round head, higher, deeper slot. Used in counterbored holes. Slotted or Phillips driving recess. Machine screws and tapping screws.	ROUND COUNTERSUNK For bolts only. Similar to 82° flat head but with no driving recess.
HEXAGON Head with square , sharp corners, and ample bearing surface for wrench tightening. Used for machine screws and bolts.	SQUARE (SET-SCREW) Square, sharp corners can be tightened to higher torque with wrench than any other set-screw head.
HEXAGON WASHER Same as Hexagon except with added washer section at base to protect work surface against wrench disfigurement. For machine screws and tapping screws.	SQUARE (BOLT) Square, sharp corners, generous bearing surface for wrench tightening.
82° FLAT, 82° Use where flush surface is desired. Slotted, clutch, Phillips, or hexagon-socket driving recess.	SQUARE COUNTERSUNK For use on plow bolts, which are used on farm machinery and heavy construction equipment.

FIGURE 13.45 Machine Screw Head Styles

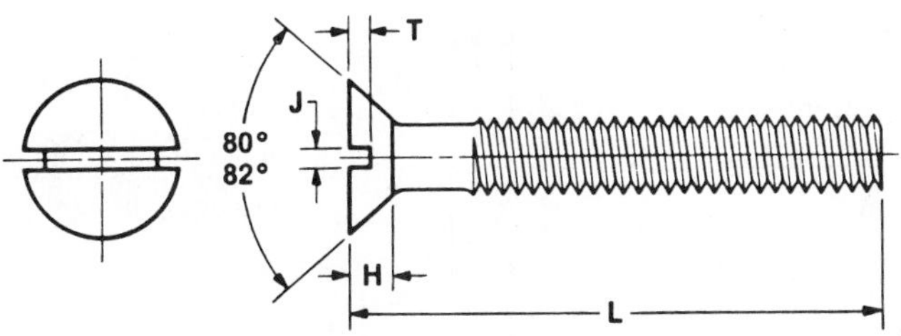

FIGURE 13.46 Slotted Flat Countersunk Head Machine Screw

Nominal Size	Head Dia	H	J	T
#5(.125)	.25	.075	.04	.03
.500	.875	.223	.10	.10

screw sizes are divided into two categories: fractional sizes and numbered sizes. Numbered sizes are confined to those below $\frac{1}{4}$ diameter. Fractional sizes range between $\frac{1}{4}$ and $\frac{3}{4}$ inch. Number 0 has a diameter of .06 inches; .013 inches is added to each numbered size above Number 0. Figure 13.46 shows a slotted flat countersunk head machine screw. Machine screws 2 inches and under in length come fully threaded. All lengths greater than 2 inches have $1\frac{3}{4}$ inch thread. Machine screws are called out the same as bolts:

.25-20 × 1.5 SLOTTED PAN HEAD MACHINE SCREW, STEEL, ZINC PLATED

6-32 × 1.50 SLOTTED FLAT COUNTERSUNK HEAD MACHINE SCREW

For metric:

M8 × 1.25 × 30 SLOTTED PAN HEAD MACHINE SCREW, CLASS 4.8 STEEL, ZINC PLATED

M4 × 0.7 × 40 RECESSED PAN HEAD MACHINE SCREW, BRASS

13.13.6 Cap Screws

Cap screws are similar to machine screws except that there are fewer head styles available. Cap screws have their heads cold-formed from smaller diameter stock. Cap screws are for applications that require closer tolerances and greater holding power per diameter. Figure 13.47 shows three examples of cap screws. Cap screws are finished and are more expensive than similar size bolts and machines screws. Cap screws come in course, fine, or special threads. Cap screws 1 inch in diameter and under have a class 3A thread; those greater than 1 inch in diameter have a class 2A thread.

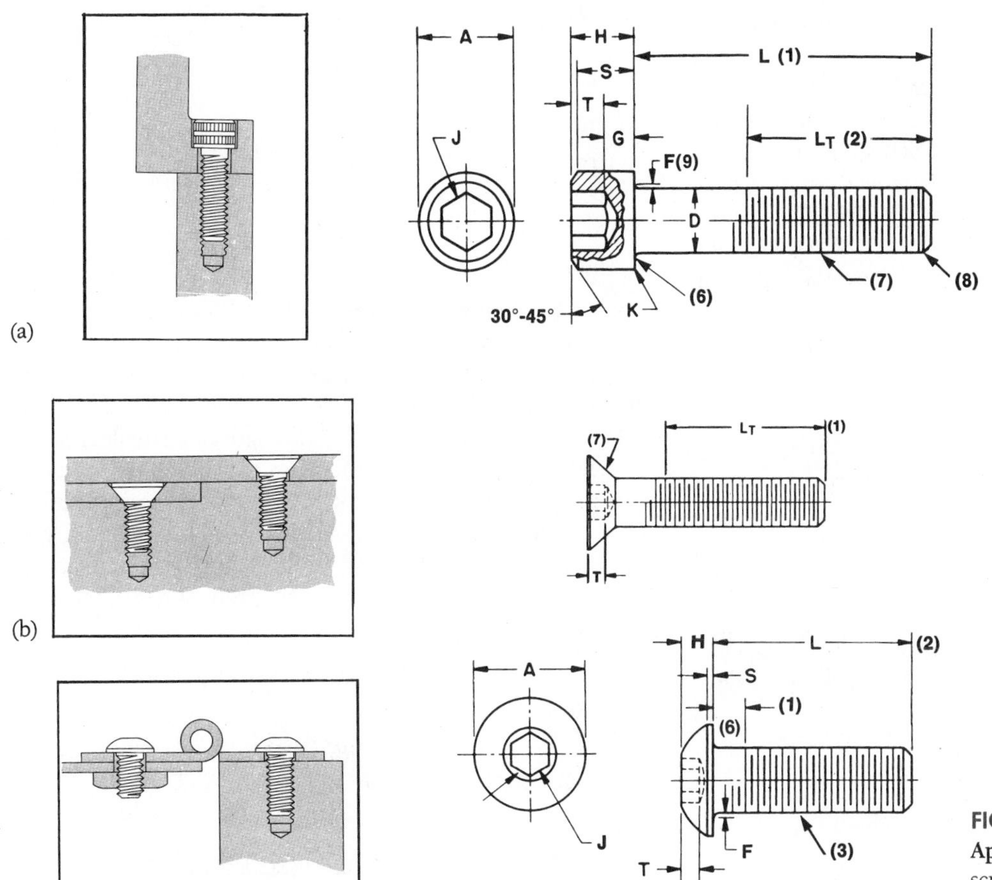

FIGURE 13.47 Cap Screw Applications (a) Socket head cap screw (b) Flat head cap screw (c) Round head cap screw

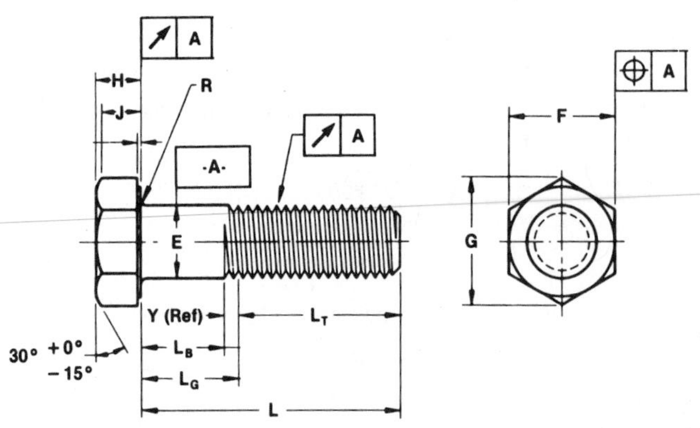

FIGURE 13.48 Hex Cap Screws

Nominal Size	E	F	G	H	J	LT	L
.500	.500	.750	.86	.32	.21	1.25	6.00 or less
.75	.750	1.125	1.29	.48	.32	1.75	6.00 or less

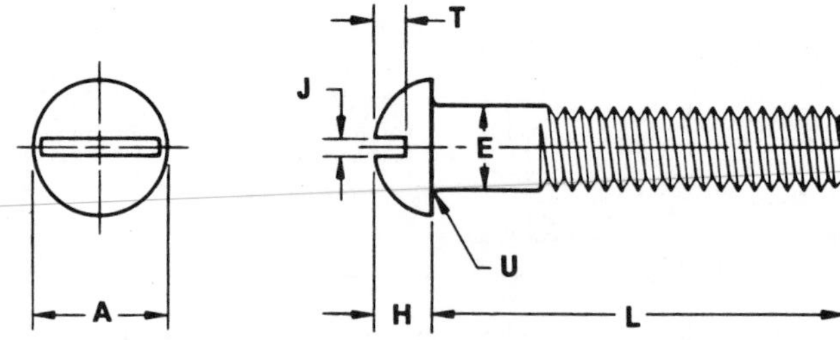

FIGURE 13.50 Slotted Fillister Head Cap Screws

Nominal Size	A	E	H	J	O	T
.312	.437	.312	.20	.08	.25	.11
.562	.812	.562	.37	.11	.46	.21

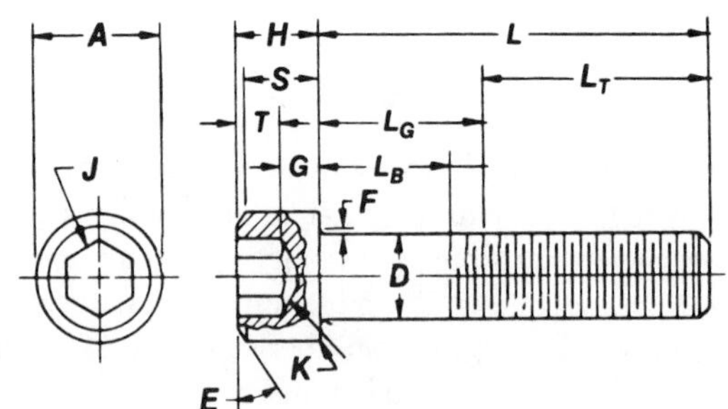

FIGURE 13.51 Socket Head Cap Screws

Nominal Size	A	D	H	J	LT
.375	.56	.375	.372	.312	1.25
1.000	1.50	1.000	1.000	.750	2.50

Cap screws are available in steel, brass, bronze, aluminum, and titanium. Steel hex head cap screws (Fig.13.48) are available in diameters from $\frac{1}{4}$ to 3 inches and have their strength indicated on their hex head by a geometric symbol. Slotted head cap screws come in round (Fig.13.49), fillister (Fig.13.50), or flat heads.

Socket head cap screws (Fig.13.51) are used throughout industry for precision, high-strength fastening and where the head of the screw must be flush or below the part's surface. A clearance hole for the head is counterbored into the part (Fig.13.47). Socket head cap screws are also made with socket button heads and socket flat heads.

The metric format for designating fasteners can be abbreviated. For example, **SOCKET HEAD SHOULDER SCREW** becomes **SHSS.** American standard fasteners can also have abbreviated designations. For example, **HEXAGON HEAD CAP SCREW** can be abbreviated **HEX HD CAP SCR.** When designating cap screws on your drawing, use the following format:

.138-32 × 1.00 HEXAGON SOCKET HEAD CAP SCREW, ALLOY STEEL, CADMIUM PLATED

$\frac{1}{4}$ - 28 × 1.75 HEXAGON SOCKET FLAT COUNTERSUNK HEAD CAP SCREW, ALLOY STEEL

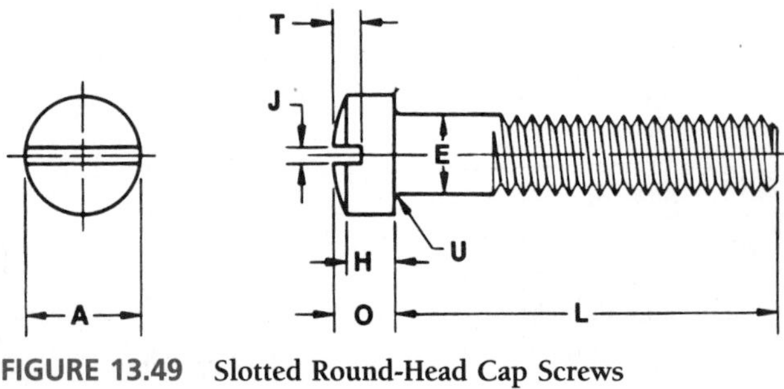

FIGURE 13.49 Slotted Round-Head Cap Screws

Nominal Size	A	E	H	J	T
.250	.437	.250	.19	.07	.11
.500	.812	.500	.35	.10	.21

For metric:

B18.3.1M-M6 × 1 × 20 HEXAGON SOCKET HEAD CAP SCREW

IFI-535 - 6 × 1 × 8 SOCKET COUNTERSUNK HEAD CAP SCREW, ZINC PLATED

Socket head shoulder screws are used for location and fastening by combining the features of dowels and screws, and for applications requiring a pivot. This type of screw has an enlarged, toleranced, unthreaded portion of the screw body called a *shoulder* (Fig.13.52). The length of a shoulder screw is measured from under its head to the end of its shoulder. The threaded portion is not included in the length specification. When designating a shoulder screw on a drawing, give the nominal size or basic shoulder diameter in fractions or decimal equivalent, shoulder length, product name, material, and finish as shown here:

$\frac{1}{4}$ × 1.250 HEX SOCKET HEAD SHOULDER SCREW, ALLOY STEEL

1.25 × 4.25 HEX SOCKET HEAD SHOULDER SCREW, ALLOY STEEL, PHOSPHATE COATED

For metric:

B18.3.3M-8 × 25 SOCKET HEAD SHOULDER SCREW

B18.3.3M-10 × 50 SHSS, ZINC PLATED

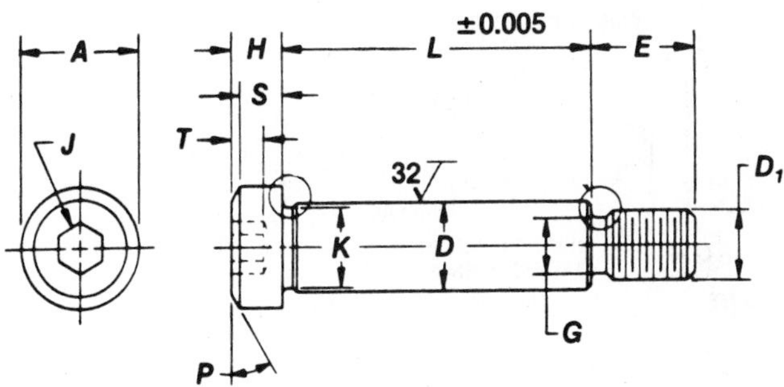

FIGURE 13.52 Hexagon Socket Head Cap Shoulder Screw

Nominal Size	A	D	D1	E	G	H	J	K
.500	.75	.500	.375	.625	.30	.31	.25	.47
1.000	1.31	.998	.750	1.000	.63	.625	.50	.97

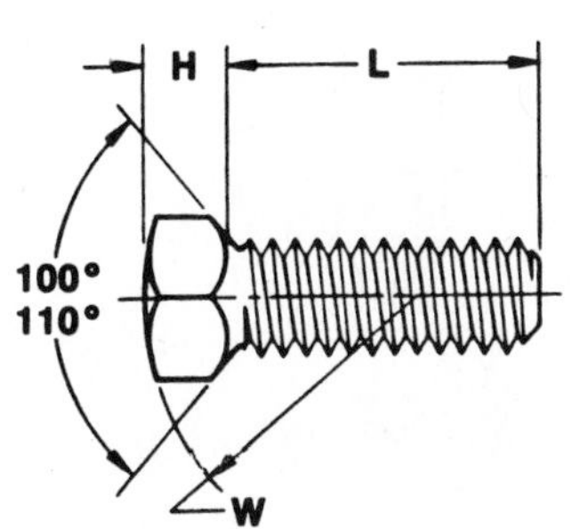

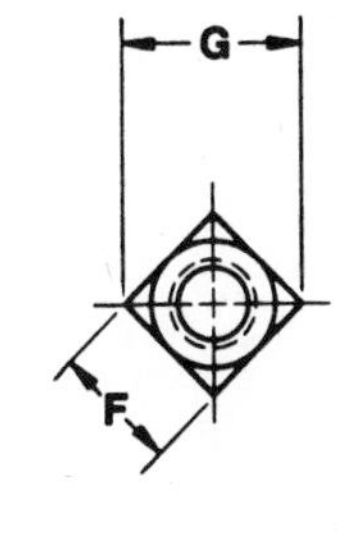

FIGURE 13.54 Square Head Set Screws

Nominal Size	F	G	H	W
.250	.250	.35	.19	.62
.500	.500	.70	.38	1.25

13.13.7 Set Screws

There are three types of **set screws:** slotted, socket, and square head. In a set screw, there are three types of holding power: torsional (resistance to rotation), axial (resistance to lateral movement), and vibrational. Set screws are used for a variety of applications, such as securing components to shafts (Fig. 13.53).

Set screws are available in number sizes from 0 to 12 and in fractional sizes from $\frac{1}{4}$ to 2 inch. Metric set screws come in nominal diameters of 1.6, 2, 2.5, 3, 4, 5, 6, 8, 10, 12, 16, 20, and 24 millimeters.

The size of a set screw is an important factor in holding power. A rough rule of thumb is that the set screw diameter should be 25% of the shaft diameter. When more than one set screw is used, it should be placed near and in line with the first one. If the second set screw must be in the same location as the first, it should be staggered at an angle of 60°.

Square-head set screws protrude above the surface of the part (Fig.13.54). Headless types disappear below the work surface when tightened. **Socket set screws** have spline or hex sockets (Fig.13.55). Slotted set screws are tightened with screw drivers (Fig.13.56). Figure 13.56 also shows examples of six standard point forms available for both socket and slotted set screws. The *cone point* is used where two parts must be joined in a permanent position relative to each other. The *cup point* is used for applications that require rapid assembly. The *oval point* is used in applications similar to the cup point. The *flat point* is used where fine adjustments are needed. Since they penetrate a mating hole drilled in the shaft, the half dog and full dog points have the greatest holding power.

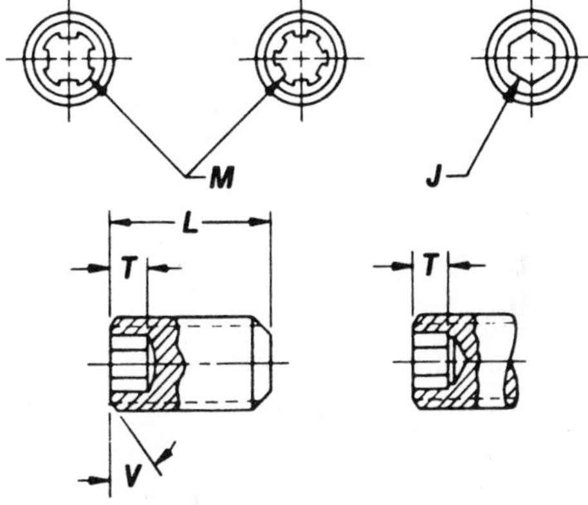

FIGURE 13.55 Socket Head Set Screws

Nominal Size	J	M	T
.250	.125	.14	.13
.375	.188	.21	.18

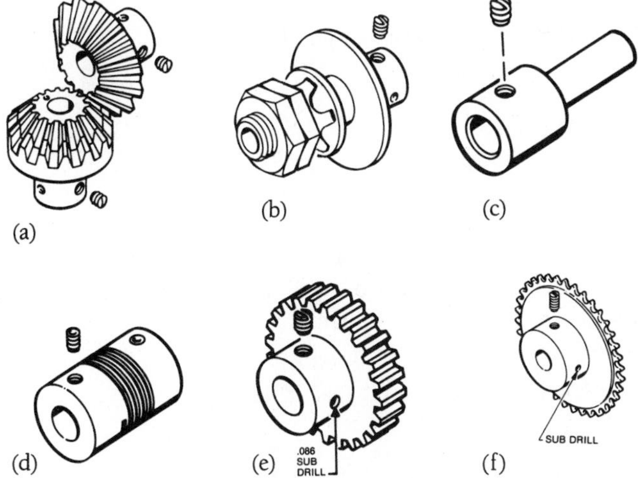

FIGURE 13.53 Set Screws in Use

A set screw is designated on a drawing by giving the nominal size, threads per inch, length, product name, point style, material, and protective coating (if needed):

$\frac{1}{4}$-20 × .375 HEXAGON SOCKET SET SCREW, CUP POINT, ALLOY STEEL

.250-20 × .50 SLOTTED HEADLESS SET SCREW, HALF DOG POINT, STEEL

For metric:

B18.3.6M-10 × 1.5 CUP POINT SOCKET SET SCREW, ZINC PLATED

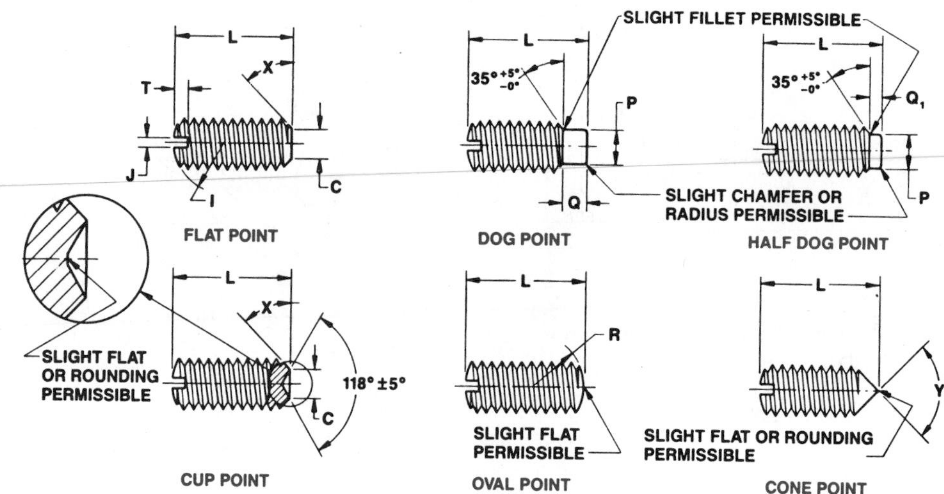

FIGURE 13.56 Slotted Headless Set Screws

Nominal Size	J	P	Q	Ql	T
.250	.04	.15	.13	.06	.06
.375	.06	.25	.19	.09	.09

13.14 Nuts

Many types of nuts are available to satisfy specific design and functional requirements. Lock nuts, swivel nuts, hex nuts, flange nuts, coupling nuts, square nuts, slotted nuts, and jam nuts are just a few of the types used in industry. Most nuts are either hex head or square-head varieties. Nuts are identified by the size of bolt they fit, not by their outside dimensions.

Flange nuts incorporate a washer into the nut that increases the bearing area of the nut. **Hexagon nuts** are available as unfinished, plain, slotted, regular, heavy, and jam types. Semifinished hex nuts are available in plain, slotted, jam, thick plain, thick slotted, and castle varieties. Semifinished nuts have one side machined on the bearing side of the nut. Heavy nuts are .125 inches wider across the flats on the hexagon. **Slotted nuts** (Fig. 13.57) have slots for use with cotter pins, which prevents the nut from coming off or untightening. Regular hex nuts (Fig.13.58) are thinner than their size designations. A $\frac{1}{2}$ inch regular hex nut is actually $\frac{7}{16}$ inch thick, and a $\frac{1}{2}$ inch heavy hex nut is $\frac{31}{64}$ inch thick. Metric nuts are also thinner than their designated size. An M6 × 1 metric hex nut, Style 2 is 5.70 mm thick. There are two types of metric nuts, Style 1 and Style 2. The nominal size, threads per inch, product name, material, and protective finish are given to designate a hex nut on a drawing:

$\frac{1}{2}$ - 13 HEX NUT, STEEL, ZINC PLATED

.750-20 HEX NUT, SAE J995 GRADE 5, CORROSION RESISTANT STEEL

For metric:

HEX NUT, STYLE 2, M20 × 2.5, ASTM A563 CLASS 9, ZINC PLATED

HEAVY HEX NUT, M30 × 3.5, ASTM A563M CLASS 105, HOT DIP GALVANIZED

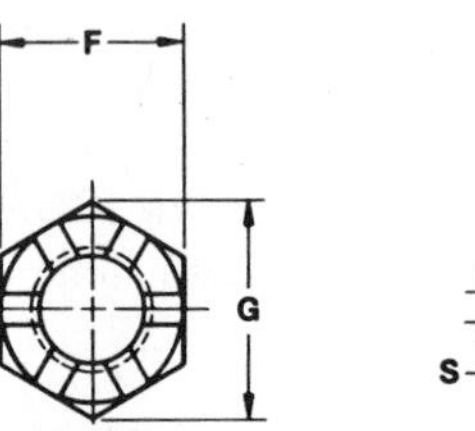
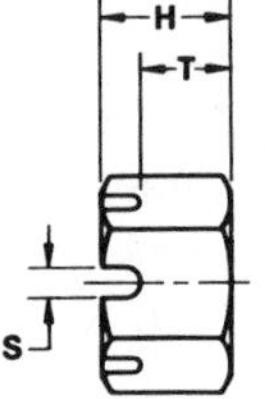
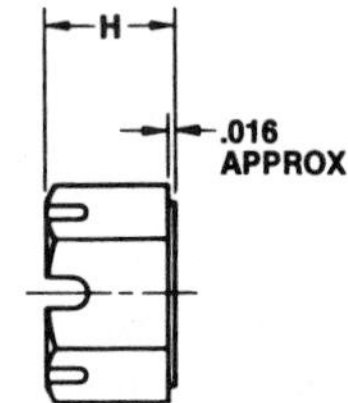

FIGURE 13.57 Hex Slotted Nuts

Nominal Size	F	G	H	S
.500	.75	.86	.56	.18
1.000	1.50	1.72	1.018	.30

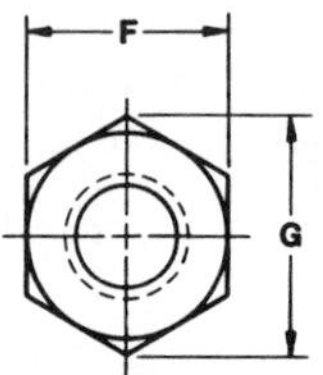
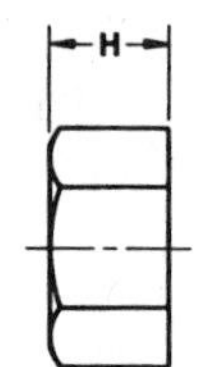
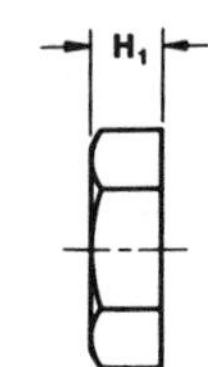

FIGURE 13.58 Hex Flat and Jam Nuts

Nominal Size	F	G	H	H1
.500	.75	.86	.43	.31
.750	1.12	1.29	.66	.44

Jam nuts are thin hex nuts and are used where height is restricted, or as a means of locking the working nut, if assembled as in Figure 13.59. Jam nuts are designated the same as hex nuts:

.500-16 HEX JAM NUT, STEEL, ZINC PLATED

For metric:

HEX JAM NUT, M10 × 1.5, ASTM A563M CLASS 04, ZINC PLATED

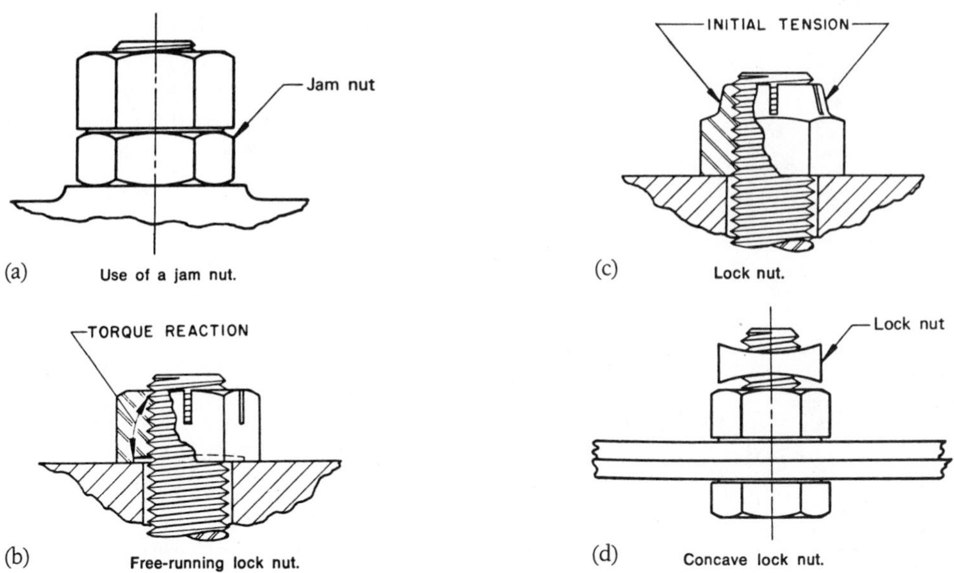

FIGURE 13.59 Lock and Jam Nut Applications

Because they must be installed with an open-ended wrench and not a socket wrench, **square nuts** are less common. Square nuts are designated on drawings the same way as hex nuts:

1.000-8 SQUARE NUT, STEEL

13.15 Standard Bolt, Nut, and Screw Representation

Bolts, screws, and nuts should be drawn with the aid of a template. When a template is not available, use the fastener's dimensions for drawing the part. A simplified method is also acceptable (Fig.13.60). These dimensions are acceptable when constructing bolts and screws. The most important dimensions on fasteners are their diameter and length, which must be accurately constructed because they affect clearances.

Figure 13.60 shows some approximate dimensions that can be used to draw fasteners. Although they do not correspond exactly to the fastener's actual dimensions, it is standard practice to simplify the constructions. In this figure, *the basic sizes of each part of a fastener are given relative to the diameter dimension.* Each dimension is a fraction of the diameter. Chamfered end points are normally drawn at 45°. When drawing the end view of a slotted fastener, the slots are drawn at 45° not at 90 or 180°. The head of hex head bolts and nuts

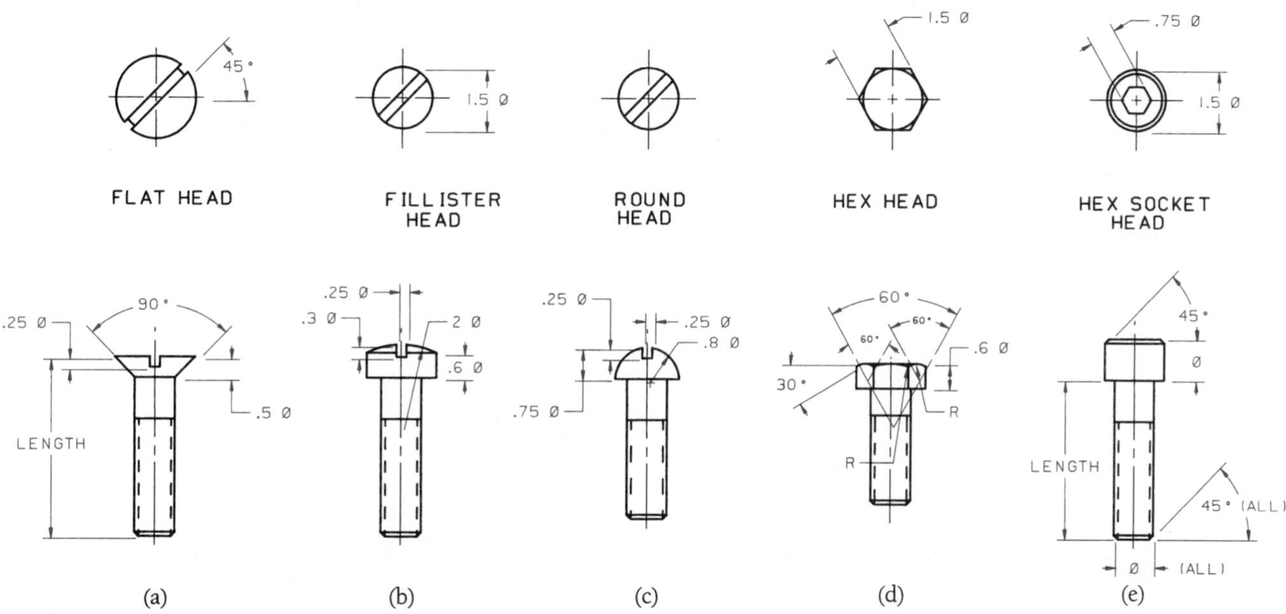

FIGURE 13.60 Approximate Sizes for Drawing Screws

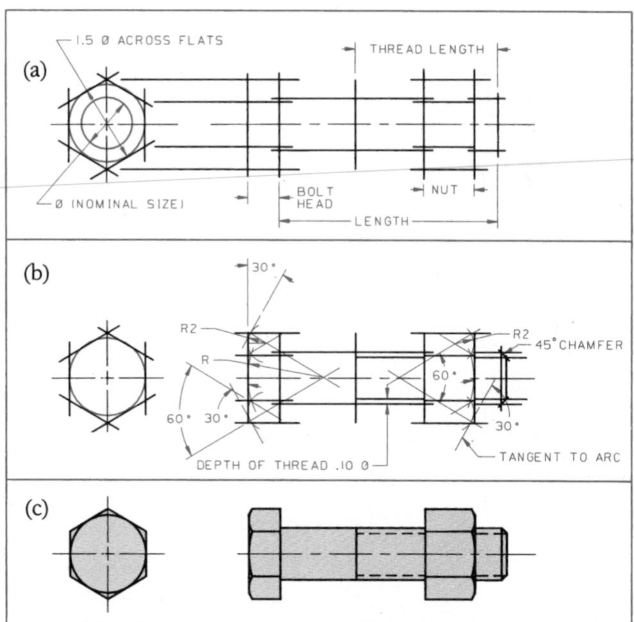

FIGURE 13.61 Drawing a Hex Bolt and Nut Without a Template

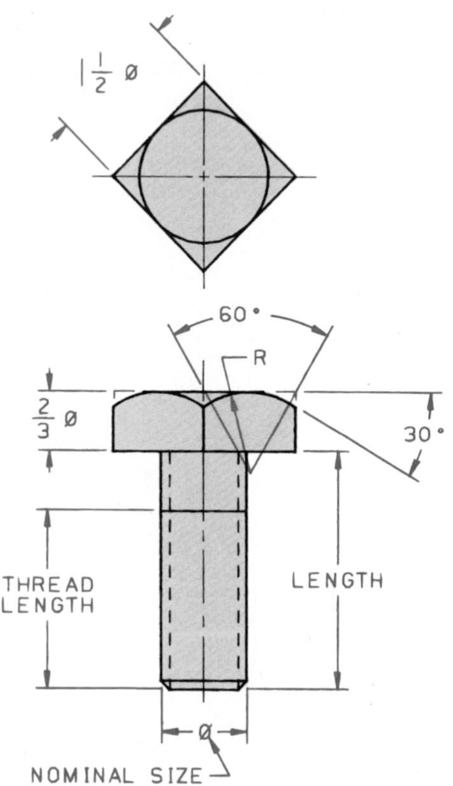

FIGURE 13.62 Drawing a Square-Head Bolt Without a Template

is drawn so that three surfaces are visible from an elevation view. The depth of the hex on a hex socket head cap screw is not drawn. Figure 13.61 shows three steps in the construction of a hex head bolt and nut. Figure 13.62 shows the dimensions used to construct a square-head bolt.

13.16 Washers

Washers are used in conjunction with threaded fasteners. The three basic types of washers are *plain, spring lock,* and *tooth lock.* Plain washers spread the bearing area of the fastener head or nut and are normally used with soft metals. Spring washers maintain tension on the nut or bolt head, and tooth lock washers provide teeth that dig into the fastener and the part to prevent the fastener from loosening. **Plain washers** are flat and ring shaped (Fig.13.63). Washers are designated on drawings by providing the product name and type, size (ID), material, and finish:

TYPE A PLAIN WASHER, $1\frac{1}{2}$, STEEL, CADMIUM PLATED

TYPE B PLAIN WASHER, NO. 12, STEEL

For metric:

PLAIN WASHER, 6MM, NARROW, SOFT, STEEL, ZINC PLATED

Spring lock washers are split on one side and are helical in shape. They have the dual function of acting as a spring take-up to compensate for developed looseness and a loss of tension between component parts of an assembly and as a hardened thrust bearing to aid in assembly and disassembly of bolted fastenings. Lock washer (Fig.13.64) sizes are selected by the nominal bolt or screw sizes. Figure 13.65 shows two commonly used types of **tooth lock washers:** Type A

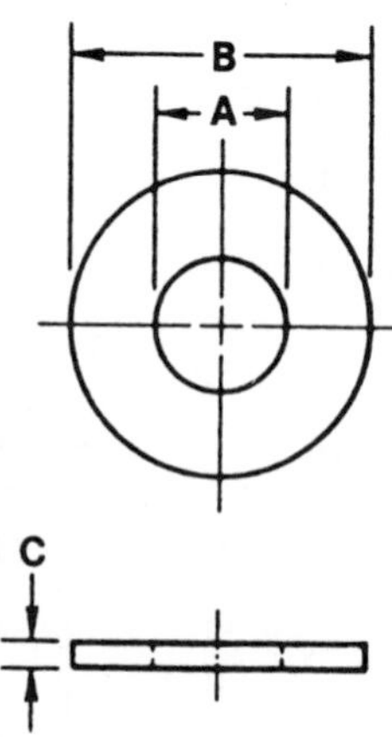

FIGURE 13.63 Plain Washers

Nominal Size	A	B	C
.500	.531	1.06	.09
1.000	1.062	2.50	.16

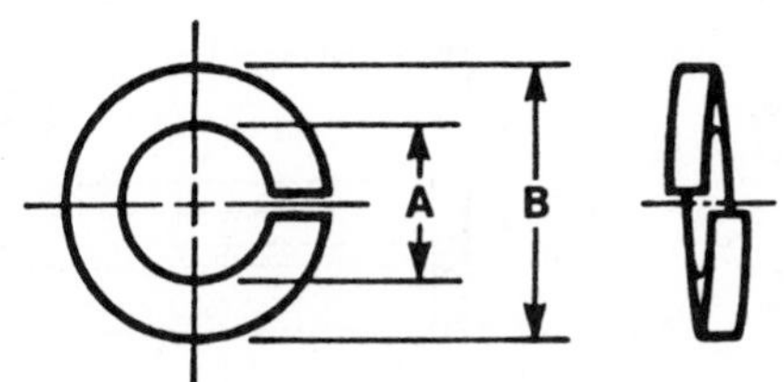

FIGURE 13.64 Spring Lock Washers

Nominal Size	A	B	Width
.500	.518	.87	.125
.625	.65	1.07	.156

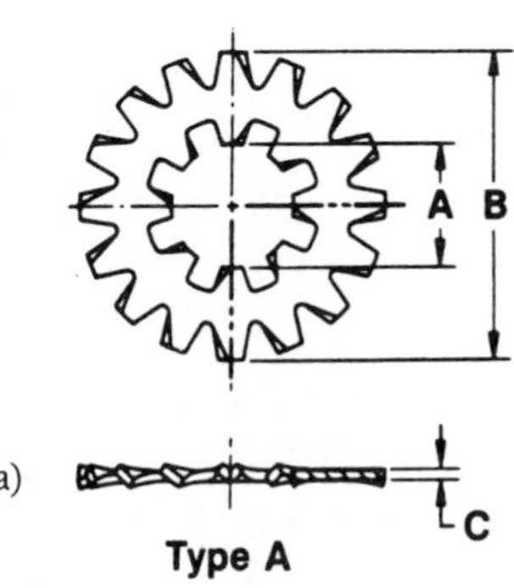

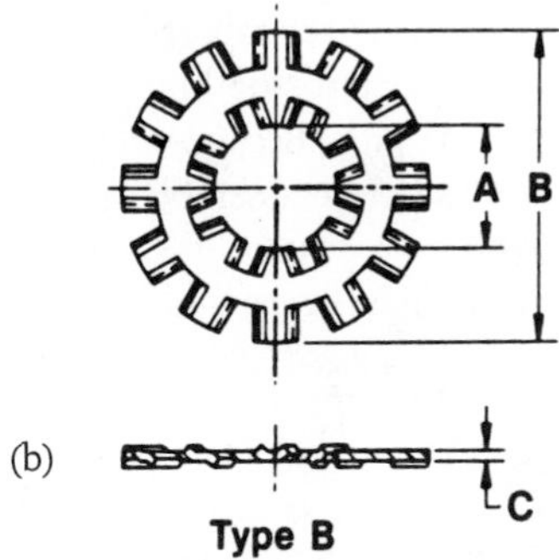

FIGURE 13.65 Tooth Lock Washers

Nominal Size	A	B	C
#10(.190)	.20	.76	.04
.500	.53	1.41	.06

and Type B. Both are internal-external types. Designate lock washers on drawings as:

HELICAL SPRING LOCK WASHER, .125 REGULAR, CORROSION RESISTANT STEEL, CADMIUM PLATED

HELICAL SPRING LOCK WASHER, $\frac{3}{8}$ EXTRA DUTY, STEEL, PHOSPHATE COATED

INTERNAL-EXTERNAL TOOTH LOCK WASHER, NO. 10 (0.760 O.D.), TYPE A, STEEL, CADMIUM PLATED

EXTERNAL TOOTH LOCK WASHER, .625, TYPE B, STEEL, PHOSPHATE COATED

For metric:

4MM INTERNAL TOOTH, TYPE A

You May Complete Exercises 13.5 Through 13.8 at This Time

13.17 Machine Pins

Standard machine pins are used throughout industry where there is a need for the assembly and alignment of mating parts, and for attaching gears, cams, collars, pulleys, sprockets, and other mechanical parts to shafts. Three types of pins are used to secure a gear to a shaft in Figure 13.66: (a) a straight pin, (b) taper pin, and (c) a spring pin. Most of the pin types have metric-sized standard equivalents. Eight common types of pins are found in industry and are recognized as American National standards:

- **straight**
- **tapered**
- **spring**
- **grooved**
- **dowel**
- **cotter**
- **clevis**
- **push-pull**

Pins can be either quick release or semipermanent. **Quick release** pins include the cotter, clevis, push-pull, and positive locking varieties. Dowel, tapered, straight, grooved, and spring pins are semipermanent types because they all require some form of pressure to insert.

13.17.1 Straight Pins

Straight pins (Fig.13.67) are somewhat difficult to align during assembly and must be a precise fit to make them secure.

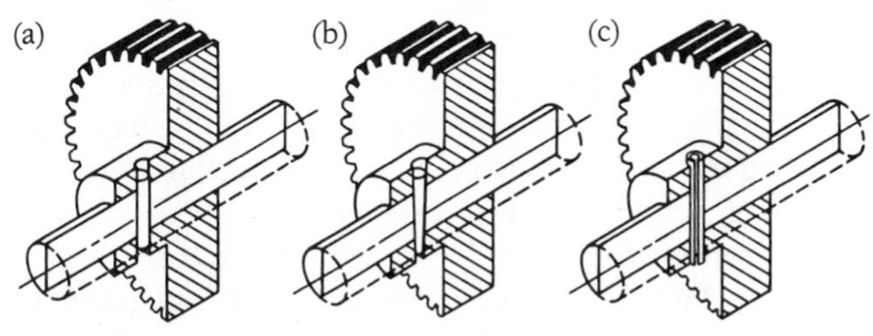

FIGURE 13.66 Pinning Applications

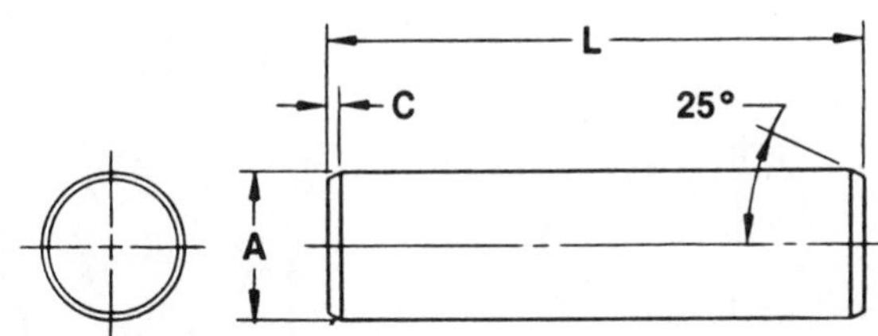

FIGURE 13.67 Straight Pins

Nominal Size	A	C
.250	.2500	.025
.375	.3750	.040

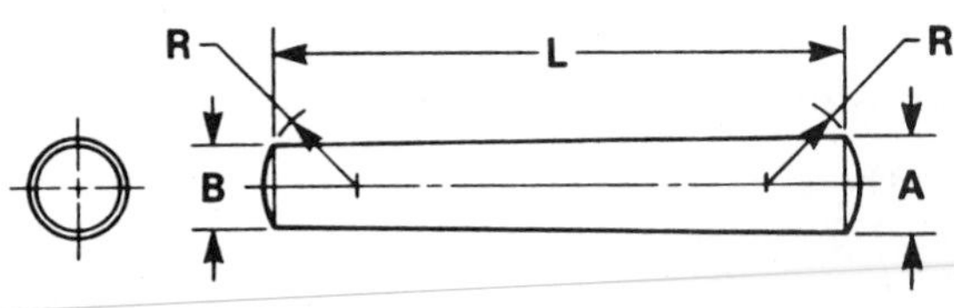

FIGURE 13.68 Taper Pins

Nominal Sizes	A	R
#4(.2500)	.2500	.26
#8(.4920)	.4920	.50

To designate a pin on a drawing the product name, nominal size, length, material, and the finish (if required) are given:

PIN, CHAMFERED STRAIGHT, $\frac{5}{16}$ × 2, STEEL

13.17.2 Tapered Pins

Tapered pins (Fig. 13.68) can be used for ease of assembly and disassembly. Taper pins fall out more easily than dowels. Taper pins come in sizes from $\frac{1}{16}$ to $1\frac{1}{2}$ inches and are normally steel. Taper pins are called out by a number, from 0 (small diameter) to 14 (large diameter), and by their length requirement. The large end of a taper pin is constant for a particular size pin, but the small end changes according to the length. Taper pins have a taper of $\frac{1}{4}$ inch per foot.

Step drilling or tapered reaming is required for taper holes. The information contained Figure 13.69 should be provided on all taper details. Taper pins are designated as:

PIN, TAPER (COMMERCIAL CLASS) NO. 2 × 1 $\frac{1}{4}$, STEEL

13.17.3 Spring Pins

Since the spring force retains the pin in the hole, **spring pins** (rolled pins) reduce the possibility of falling out during operation. The hole for a spring pin is drilled slightly smaller than the pin. Spring pins are reusable and can be repeatedly removed without distortion or losing their locking efficiency.

Spring pins come in two basic styles. One type has a slot throughout its length (Fig. 13.70) and the other is shaped in the form of a coil (Fig. 13.71). Spring pins are designated on drawings as:

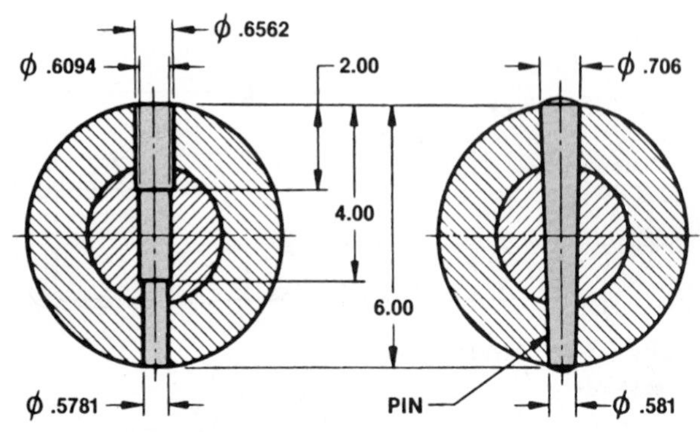

FIGURE 13.69 Dimensioning Taper Pins

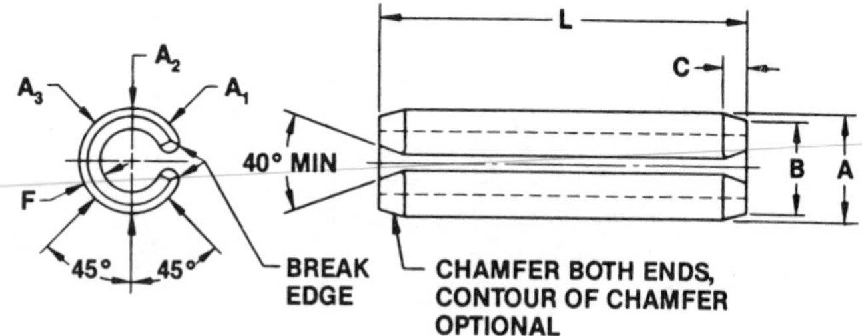

FIGURE 13.70 Slotted Spring Pins

Nominal Size	A	B	C
.375	.39	.36	.09
.500	.521	.48	.11

PIN, COILED SPRING, $\frac{1}{2}$ × 2 $\frac{1}{4}$, STANDARD DUTY, STEEL, ZINC PLATED

PIN, SLOTTED SPRING, .250 × .75, AISI 420 CORROSION RESISTANT STEEL

For metric:

PIN, COILED SPRING, 10 × 40, HEAVY DUTY, STAINLESS STEEL, PHOSPHATE COATED

PIN, SLOTTED SPRING, 20 × 60, STANDARD DUTY, CHROME-NICKEL AUSTENITIC STAINLESS STEEL, CADMIUM PLATED

13.17.4 Grooved Pins

Grooved pins (Fig. 13.72) are tapered or straight with longitudinal grooves pressed into the body. The pin will deform when pressed into the part. Because they hold securely even after repeated removal and reassembly, grooved pins are used in situations where repeated disassembly is required. Grooved pins are designated as:

PIN, TYPE B GROOVED, $\frac{5}{16}$ × 2, CORROSION RESISTANT STEEL

13.17.5 Dowel Pins

Dowel pins (Fig. 13.73) are heat-treated and precision-ground pins. Dowels are used to precisely align mating parts or to retain parts in a fixed position; they are not used as fasteners. Since the dowels are press fit, holes for dowels are reamed and not drilled. The dowel is slightly larger than the

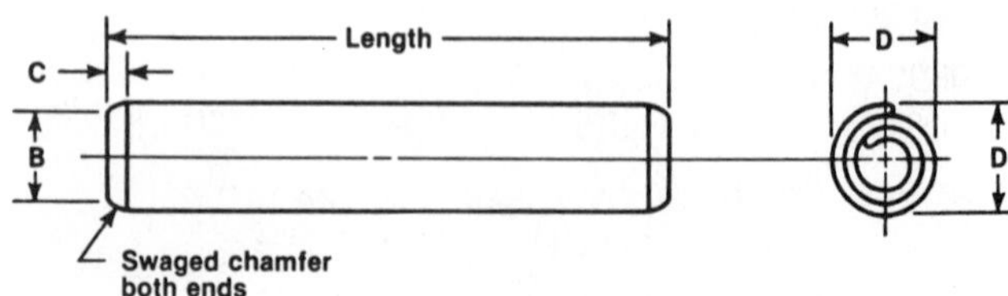

FIGURE 13.71 Metric Coiled Spring Pins

Nominal Size	B	C	D
10	9.75	2.5	10.80
20	19.6	4.5	21.10

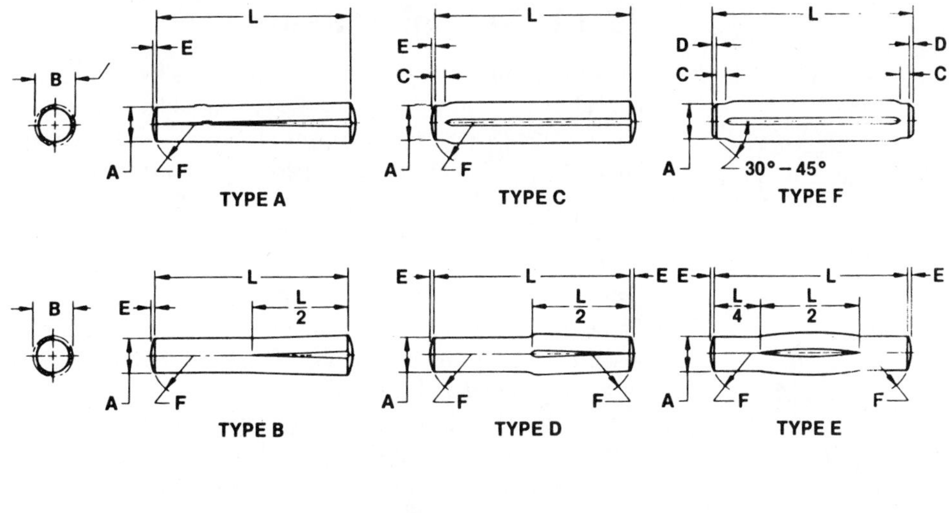

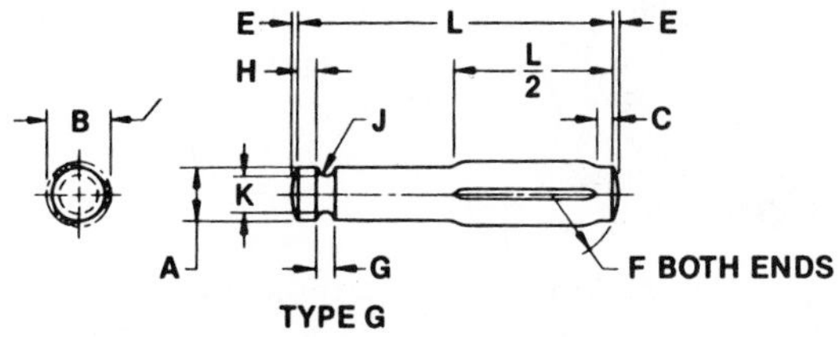

FIGURE 13.72 Grooved Pins

hole and the dowel pin is forced ("press fit") into the hole to provide accurate alignment between mating parts. Dowels provide alignment and the screws are used for fastening. A general rule is to use dowels that are close to the same diameter as the screws. The length of the dowel should be $1\frac{1}{2}$ to 2 times its diameter in each plate or part to be doweled. Dowel pins are designated as:

PIN, HARDENED GROUND PRODUCTION DOWEL, .500 × 1.75, STEEL, PHOSPHATE COATED

PIN, UNHARDENED GROUND DOWEL, $\frac{3}{4} \times 1\frac{1}{2}$, STEEL

For metric:

PIN DOWEL, 16 × 70, STAINLESS STEEL

13.17.6 Clevis Pins and Cotter Pins

Clevis pins (Fig. 13.74) are used with cotter pins to retain parts on a shaft or lock a nut and bolt. **Cotter pins** (Fig. 13.75) are used with clevis pins to retain parts on a shaft or to lock a slotted nut and bolt. Cotter pins are used where quick and easy assembly and disassembly are required. Clevis pins and cotter pins are designated as:

PIN, CLEVIS, .438 × 1.19, STEEL, CADMIUM PLATED

PIN, CLEVIS, $\frac{1}{4} \times 0.77$, STEEL

PIN, COTTER, $\frac{1}{8} \times 1\frac{1}{2}$, EXTENDED PRONG TYPE, STEEL, ZINC PLATED

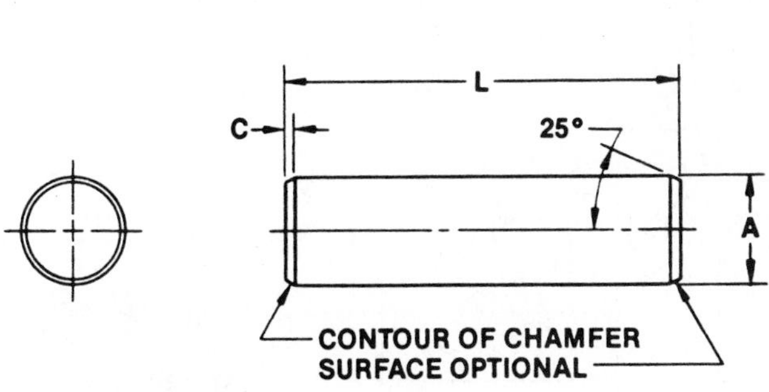

FIGURE 13.73 Dowel Pins

Nominal Size	A	C
.375	.371	.04
.500	.496	.04

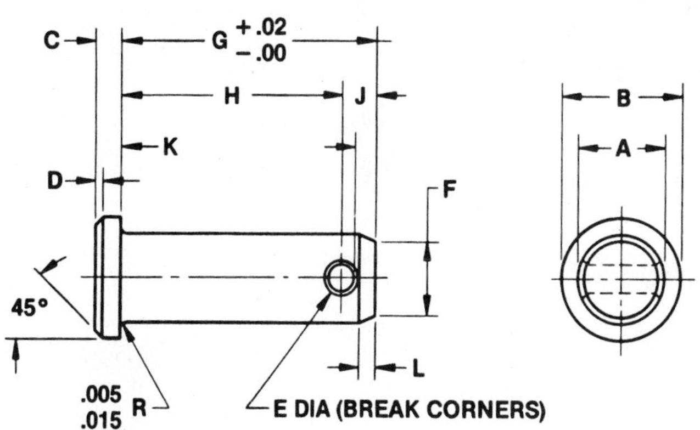

FIGURE 13.74 Clevis Pins

Nominal Size	A	B	C	D	F	G	H	J	L	pin size
.375	.37	.51	.13	.03	.33	1.06	.95	.12	.07	.093
.500	.49	.63	.16	.04	.44	1.36	1.22	.15	.08	1.250

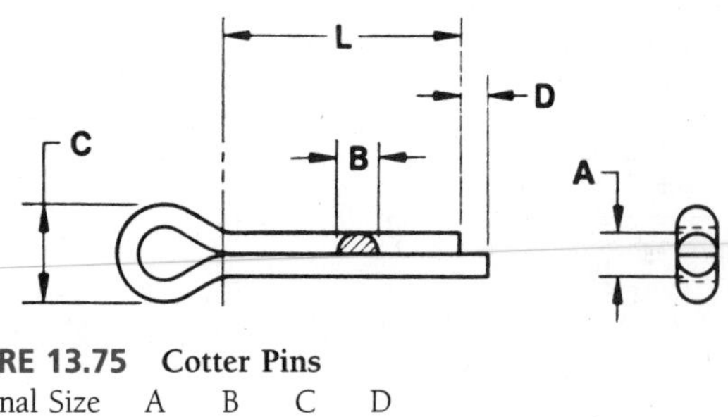

FIGURE 13.75 Cotter Pins

Nominal Size	A	B	C	D
.135	.12	.12	.25	.06
.188	.17	.17	.38	.09

13.18 Rivets

Figure 13.76 shows four types of typical riveted joints: single-riveted lap, double-riveted lap, single-riveted butt, and double-riveted butt. The most common types of rivets are solid, tubular, split, and blind rivets. Solid rivets are used in assembling parts not to be taken apart.

Solid rivets are shown on drawings as in Figure 13.77. If plans, elevations, or sections show the conventional signs for the head of the shop rivets or field rivets, the corresponding lengthwise view of the rivet fastenings is normally omitted.

Rivets are available in a variety of end points. The choice of head and end point is determined by the application. The hole size and type will be determined by the rivet choice. Figures 13.78 through 13.81 show four standard types of rivets. Designate rivets on drawings as:

.146 × .500 SEMI-TUBULAR, OVAL HEAD, STEEL, CADMIUM PLATED

$\frac{1}{4} \times 1\frac{1}{4}$ FLAT HEAD SMALL SOLID RIVET, STEEL, ZINC PLATED

13.19 Retaining Rings

Retaining rings are semipermanent fasteners found on many assemblies. Retaining rings are used as shoulders that can be located along a shaft (or pin) or in a recessed hole to keep the components of an assembly properly positioned, as shown in Figure 13.82. Many different styles of retaining rings are available (Fig.13.83).

Retaining rings can easily be installed in machined grooves, internally in housings or externally on shafts or pins. Some styles of retaining rings do not require grooves but have a self-locking spring-type action. The two types of retaining rings are *internal* and *external.*

Radially assembled rings are designed to be snapped directly onto a shaft. Axially assembled rings require special tools to expand (for external rings) or to contract (for internal rings) the ring to slide over a shaft (external) or slip into a grooved housing (internal) while installing.

13.20 Collars

A **collar** (Fig.13.84) is a ring installed over a shaft and positioned adjacent to a machine element such as a pulley, gear, or sprocket. A collar is held in position, in most cases, by a set screw. The advantage of a collar lies in that axial location can be established virtually anywhere along the shaft to allow adjustment of the position at the time of assembly. Typical collar applications include:

1. Spacer on a machine shaft
2. Thrust collar on pillow block
3. Hub or plate on a sprocket
4. Adjustment for torsion spring
5. Clutch part
6. Locating a gear or cam on a shaft.

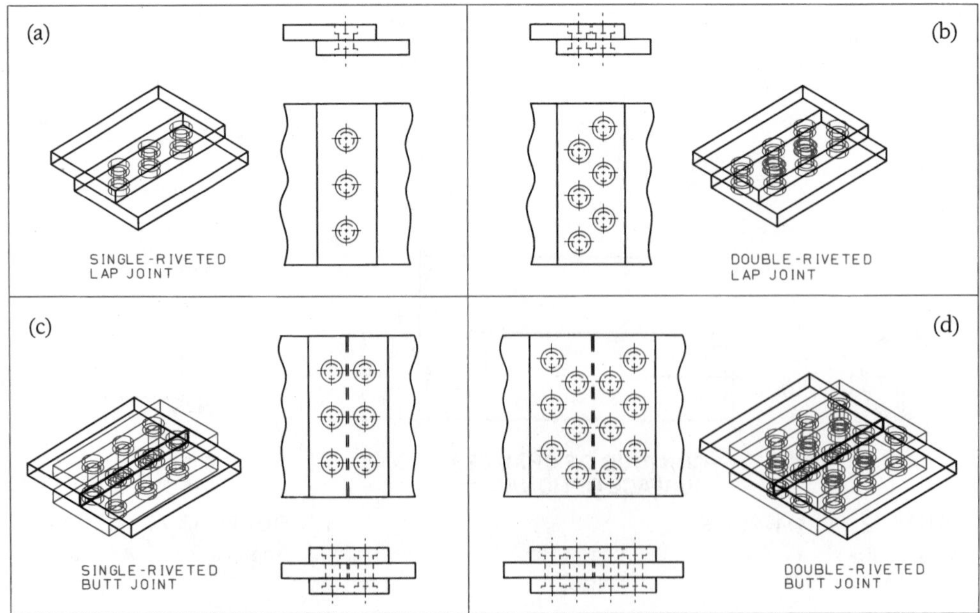

FIGURE 13.76 Riveted Joints

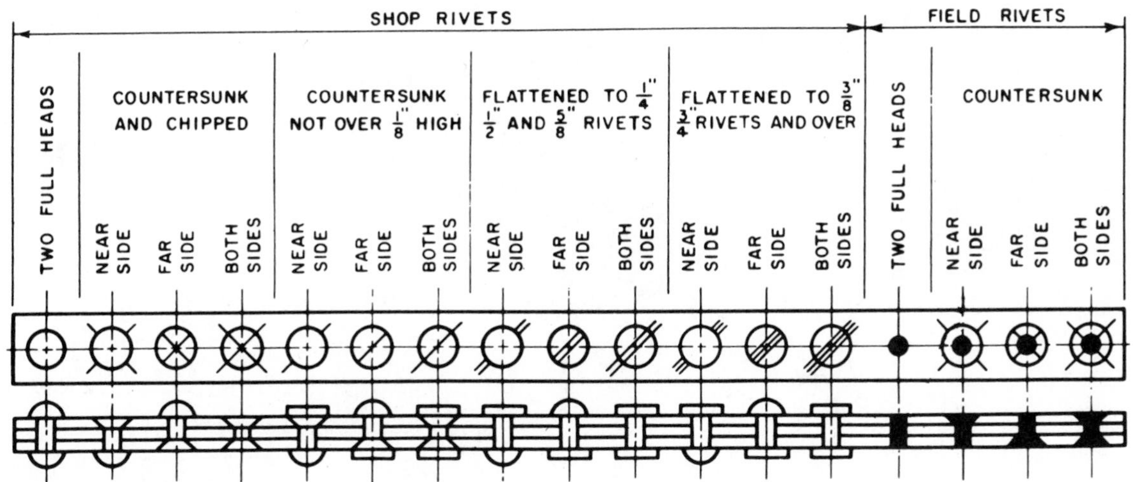

FIGURE 13.77 Drawing Conventions for Solid Rivets

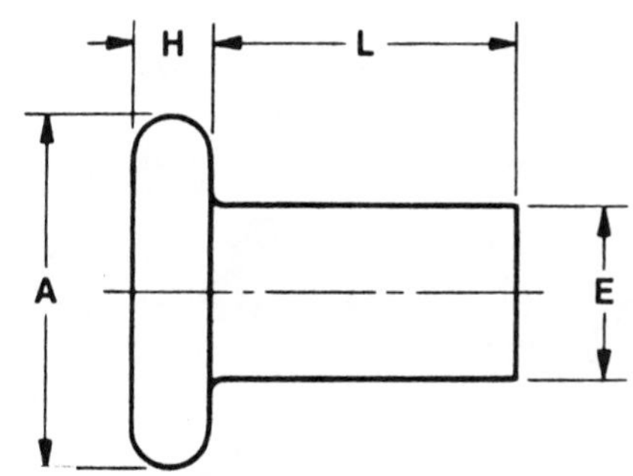

FIGURE 13.78 Flat Head Rivets

Nominal Size	A	E	H
.125	.125	.25	.04
.250	.250	.50	.09

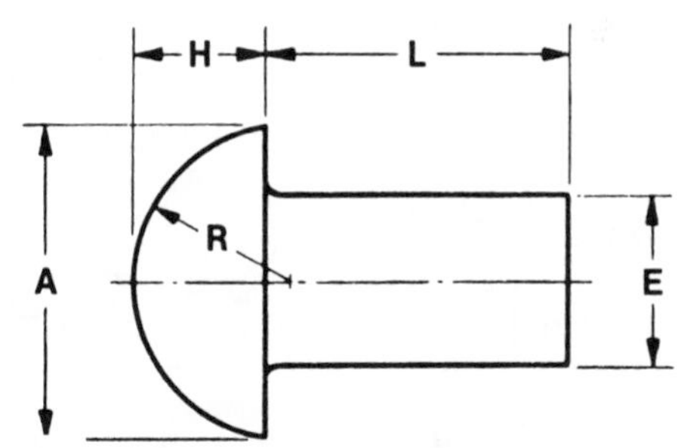

FIGURE 13.80 Button Head Rivets

Nominal Size	A	E	H	R
.094	.18	.094	.07	.08
.281	.51	.281		

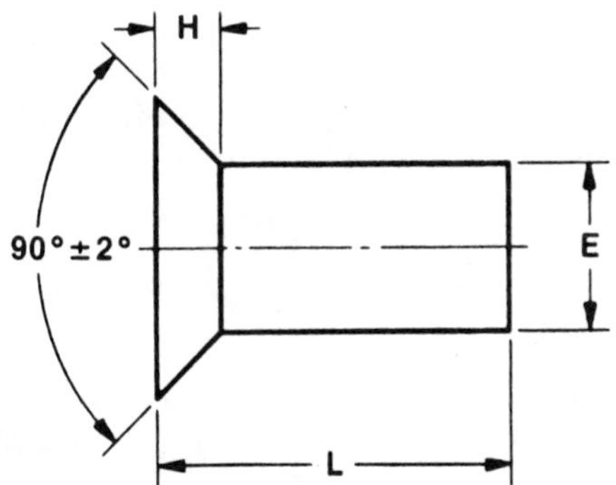

FIGURE 13.79 Flat Countersunk Head Rivets

Nominal Size	Head Dia.	E	H
.156	.29	.15	.06
.312	.58	.31	.13

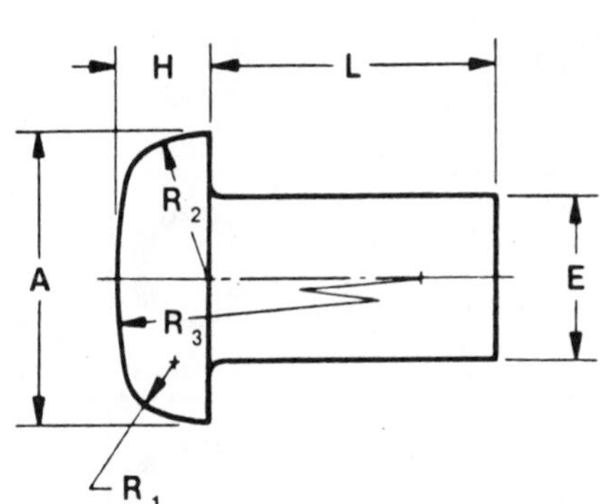

FIGURE 13.81 Pan Head Rivets

Nominal Size	A	E	H	R1	R2	R3
.188	.33	.187	.11	.05	.15	.64
.312	.55	.312	.18	.09	.26	1.07

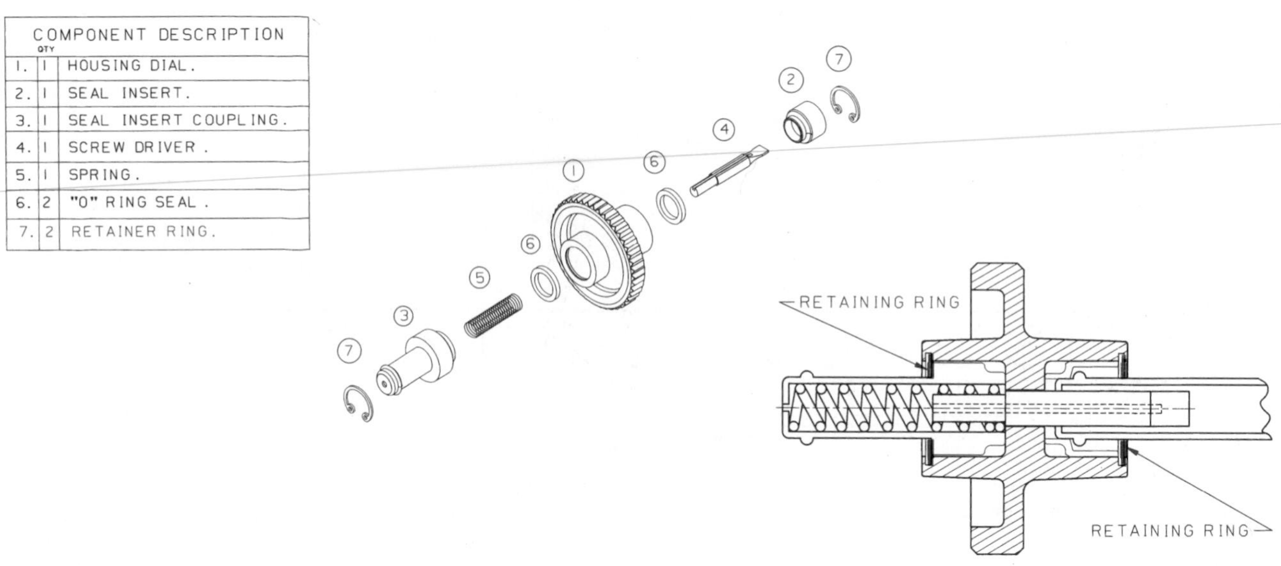

	QTY	COMPONENT DESCRIPTION
1.	1	HOUSING DIAL.
2.	1	SEAL INSERT.
3.	1	SEAL INSERT COUPLING.
4.	1	SCREW DRIVER.
5.	1	SPRING.
6.	2	"O" RING SEAL.
7.	2	RETAINER RING.

FIGURE 13.82 Internal Retaining Rings Used on an Assembly

Type	Style	No.	Use	Size Range (in.)	Size Range (mm)
INTERNAL	BASIC	N5000	For housings and bores	.250—10.0 in.	6.4—254.0 mm.
EXTERNAL	BOWED	5101	For shafts and pins	.188—1.750 in.	4.8—44.4 mm.
EXTERNAL	REINFORCED	5115	For shafts and pins	.094—1.0 in.	•
EXTERNAL	TRIANGULAR NUT	5300	For threaded parts	6-32 and 8-32 10-24 and 10-32 1/4-20 and 1/4-28	
INTERNAL	BOWED	N5001	For housings and bores	.250—1.750 in.	6.4—44.4 mm.
EXTERNAL	BEVELED	5102	For shafts and pins	1.0—10.0 in.	25.4—254.0 mm.
EXTERNAL	BOWED E-RING	5131	For shafts and pins	.110—1.375 in.	2.8—34.9 mm.
EXTERNAL	KLIPRING	5304 T-5304	For shafts and pins	.156—1.000 in.	4.0—25.4 mm.
INTERNAL	BEVELED	N5002	For housings and bores	1.0—10.0 in.	25.4—254.0 mm.
EXTERNAL	CRESCENT®	5103	For shafts and pins	.125—2.0 in.	3.2—50.8 mm.
EXTERNAL	E-RING	5133	For shafts and pins	040—1.375 in.	1.0—34.9 mm.
EXTERNAL	TRIANGULAR	5305	For shafts and pins	.062—.438 in.	•
INTERNAL	CIRCULAR	5005	For housings and bores	.312—2.0 in.	•
EXTERNAL	CIRCULAR	5105	For shafts and pins	.094—1.0 in.	•
EXTERNAL	PRONG-LOCK®	5139	For shafts and pins	.092—.438 in.	•
EXTERNAL	GRIPRING®	5555	For shafts and pins	.079—.750 in.	2.0—19.0 mm.
INTERNAL	INVERTED	5008	For housings and bores	.750—4.0 in.	19.0—101.6 mm.
EXTERNAL	INTERLOCKING	5107	For shafts and pins	.469—3.375 in.	11.9—85.7 mm.
EXTERNAL	REINFORCED E-RING	5144	For shafts and pins	.094—.562 in.	2.4—14.3 mm.
EXTERNAL	HIGH-STRENGTH	5560	For shafts and pins	.101—.328 in.	•
EXTERNAL	BASIC	5100	For shafts and pins	.125—10.0 in.	3.2—254.0 mm.
EXTERNAL	INVERTED	5108	For shafts and pins	.500—4.0 in.	12.7—101.6 mm.
EXTERNAL	HEAVY-DUTY	5160	For shafts and pins	.394—2.0 in.	10.0—50.8 mm.
EXTERNAL	PERMANENT SHOULDER	5590	For shafts and pins	.250–.750	6.4–19.0 mm.

FIGURE 13.83 Retaining Ring Styles

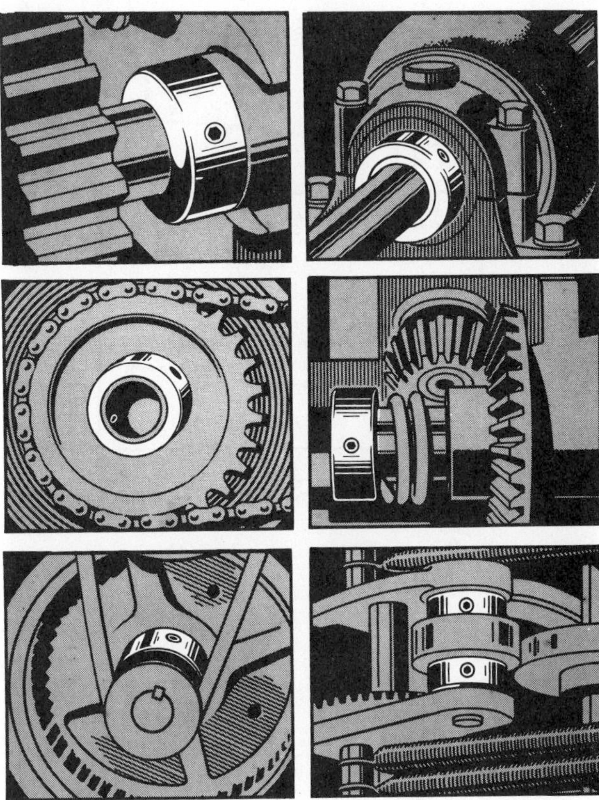

FIGURE 13.84 Coupling Applications

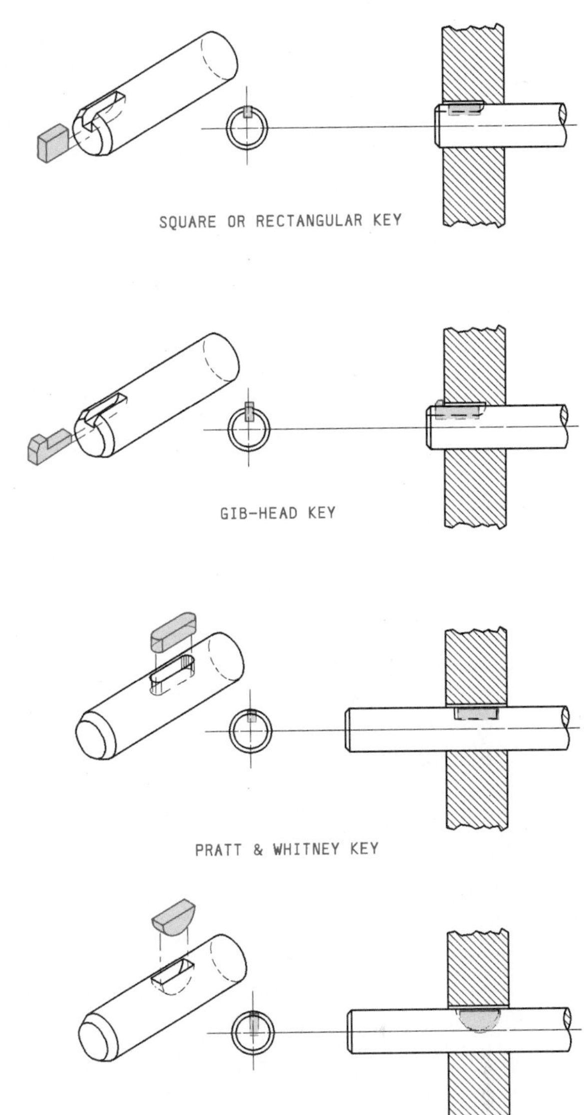

FIGURE 13.85 Types of Keys

13.21 Keys and Keyseats

A **key** is a machine component used to assemble a shaft and the hub of a power-transmitting element (gear, sprocket, pulley) to transmit torque. Keys are removable to facilitate assembly and disassembly of the shaft and components. A key is installed in an axial groove machined into the shaft, called a **keyseat** (Fig.13.85). A similar groove in the hub of the power-transmitting element is usually called a keyway but is more properly called a keyseat.

Square keys (the width and the height are equal) are preferred on shaft sizes up to 6.50 inches in diameter. Square keys (Fig.13.86) are sunk halfway into the shaft and extend halfway into the hub of the assembly. Above 6.50 inches in diameter, rectangular keys are recommended. The rectangular key (flat key) is recommended for larger shafts and is used for smaller shafts where the shorter height is acceptable for the design requirements.

The **taper key** (Fig.13.86) permits the key to be inserted from the end of the shaft after the hub is in position. If the opposite end of the key is not accessible to be driven out, the gib head key provides the means of extracting the key. On both the plain taper and the gib head key, the taper is $\frac{1}{8}$ inch per foot. The cross-sectional dimensions of the key, W and H, are the same as those used for parallel keys, with the height, H, measured at the position specified in Figure 13.86.

13.21.1 Key Size Versus Shaft Diameter

For a stepped shaft (one that has multiple diameters), the size of a key is determined by the diameter of the shaft at the point of location of the key, regardless of the number of different diameters on the shaft. Sizes and dimensions for keys are found in tables in the *Machinery's Handbook* and in ANSI B17.1. Figure 13.87 shows the preferred dimensions for parallel keys as a function of the shaft diameter. The width is normally one-fourth of the diameter of the shaft.

13.21.2 Woodruff Keys

Woodruff keys, which are almost in the shape of a half circle, are used where relatively light loads are transmitted. One advantage of Woodruff keys is that they cannot change their axial location on a shaft because they are retained in a pocket.

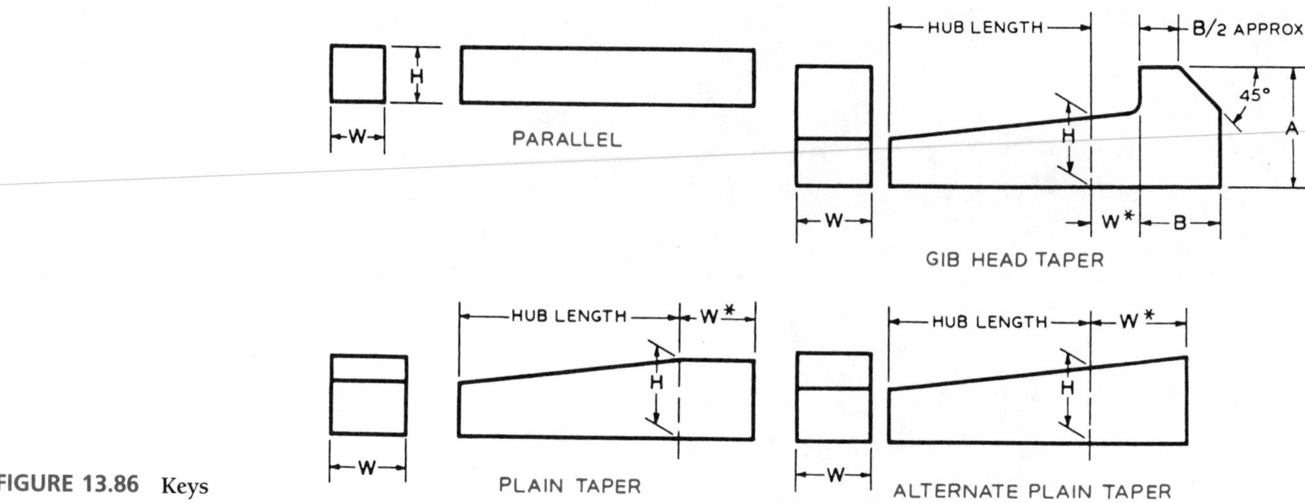

FIGURE 13.86 Keys

Woodruff keys can be either the full radius or the flat bottom type and come in two styles (Figs.13.88 and 13.89).

13.21.3 Design of Keys and Keyseats

The key and keyseat are designed after the shaft diameter is determined. Then, with shaft diameter as a guide, the size of

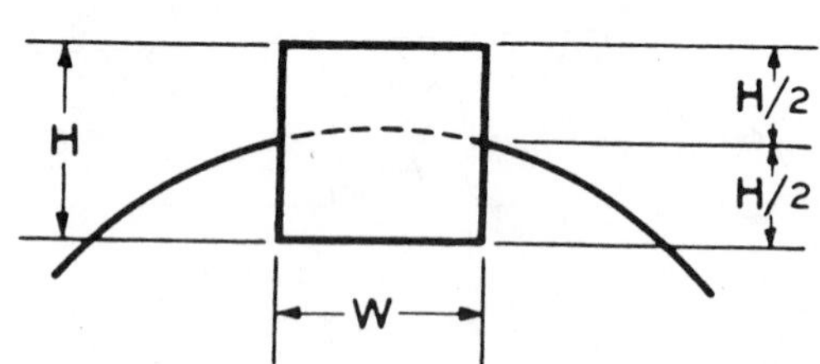

FIGURE 13.87 Key Sizes for Square Keys

Nominal Shaft Dia.	H	W
.875.1.25	.25	.25
1.750.2.25	.50	.50

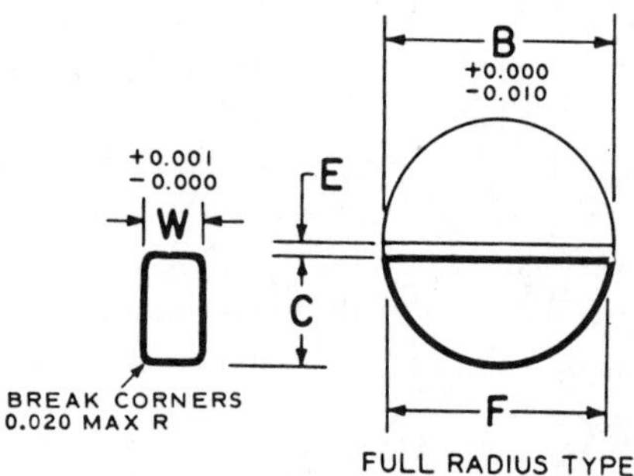

FIGURE 13.88 Full-Radius Woodruff Keys

Key #	W × B	C	D	E	F
403	.125 × .375	.17	.17	1/64	.37
806	.250 × .7501	.31	.30	1/16	.74

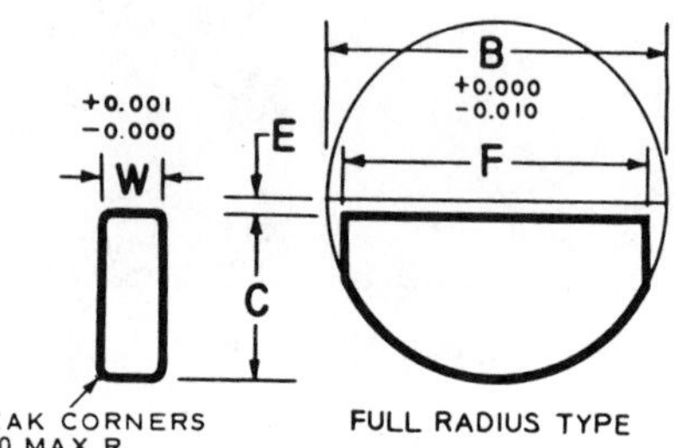

FIGURE 13.89 Woodruff Keys

Key#	W × B	C	D	E	F
817.1	.250 × 2.125	.40	.39	21/32	1.38
1217.1	.375 × 2.125	.40	.29	21/32	1.38

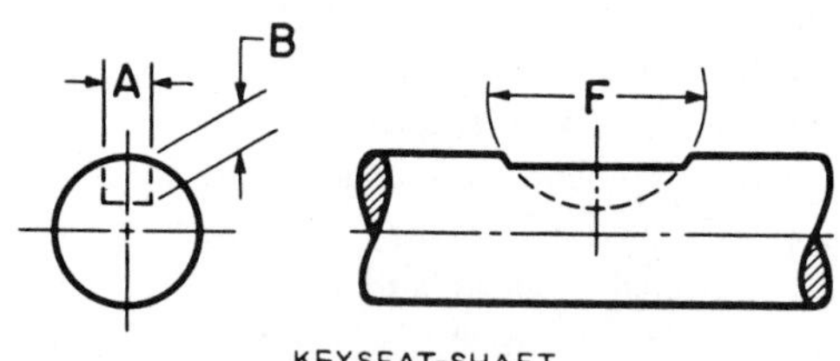

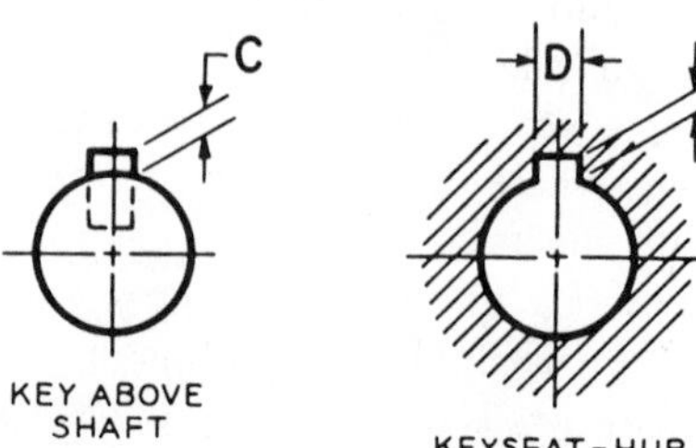

FIGURE 13.90 Keyseat Dimensions

Key#	Nominal Size	A	B	C	D	E	F
403	.125 × .375	.12	.10	.06	.12	.06	.375
806	.250 × .750	.24	.18	.12	.25	.13	.750

CHORDAL HEIGHT

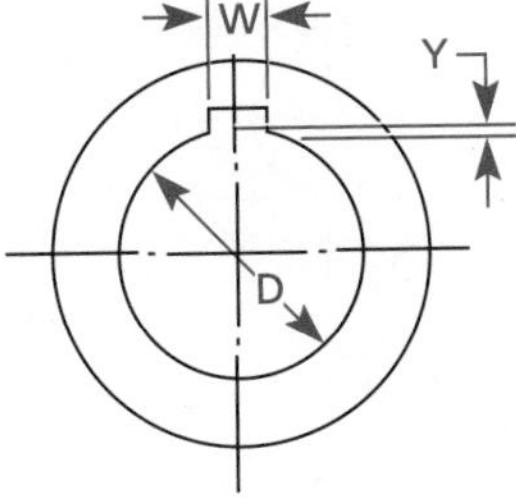

The chordal height Y is determined from the following formula:

$$Y = \frac{D - \sqrt{D^2 - W^2}}{2}$$

The distance from the bottom of the shaft keyseat to the opposite side of the shaft is specified by dimension S. The following formula may be used for calculating this dimension:

$$S = D - Y - \frac{H}{2} = \frac{D - H + \sqrt{D^2 - W^2}}{2}$$

DEPTH OF SHAFT KEYSEAT

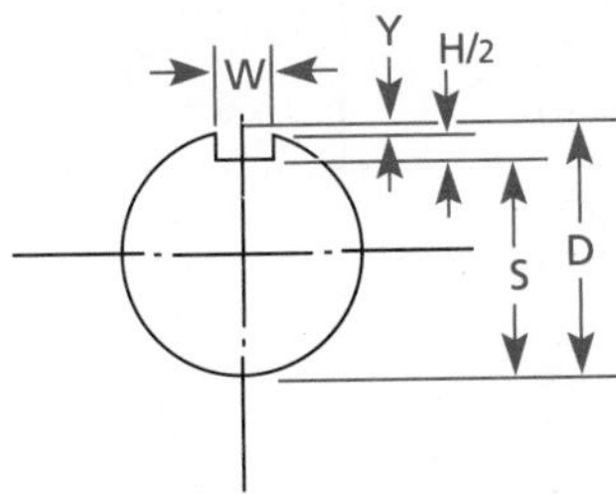

The distance from the bottom of the hub keyseat to the opposite side of the hub bore is specified by dimension T. For taper keyseats, T is measured at the deeper end. The following formula may be used for calculating this dimension:

$$T = D - Y + \frac{H}{2} + C = \frac{D + H + \sqrt{D^2 - W^2}}{2} + C$$

DEPTH OF HUB KEYSEAT

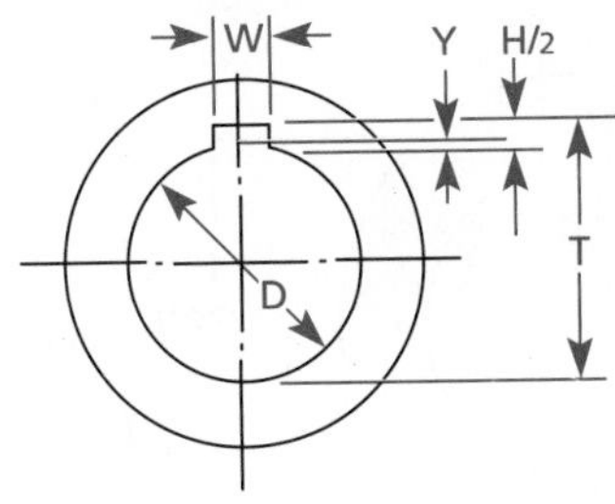

Symbols
C = Allowance
+ 0.005 inch clearance for parallel keys
− 0.020 inch interference for taper keys
D = Nominal shaft or bore diameter, inches
H = Nominal key height, inches
W = Nominal key width, inches
Y = Chordal height, inches

FIGURE 13.91 Calculating Keyseats

the key is selected from ANSI B17.1 or ANSI B17.2. The only remaining variables are the length of the key and its material. One of these can be specified, and the requirements for the other can then be computed. Typically, the length of a key is specified to be the hub length of the element in which it is installed to provide for good alignment and stable operation. Figure 13.90 shows keyseat dimensions for Woodruff keys. Keys are designed to fail before the shaft or hub fails, thus resulting in a lower cost for replacement.

If rectangular and square keys are used, keyseats in the shaft and the hub are designed so that exactly one-half of the height of the key is in the shaft keyseat and the other half is in the hub keyseat. Figure 13.91 shows the resulting geometry. The distance Y is the radial distance from the theoretical top of the shaft, before the keyseat is machined, to the top edge of the finished keyseat to produce a keyseat depth of exactly $H/2$. To assist in machining and inspecting the shaft or the hub, the dimensions S and T can be computed and shown on the part drawings. The equations are given in Figure 13.91. Tabulated values (Fig. 13.92) of Y, S, and T are available in the standard and in the *Machinery's Handbook*. Standard key sizes are also listed in Appendix C.

13.21.4 Dimensioning Keyseats

Keyseats (Fig. 13.93) are dimensioned by giving the width, depth, location, and, if required, length. For shafts, the width of the keyseat, the distance from the bottom of the shaft to the bottom of the keyseat, and the length are given. For the hub, give the width of the keyseat and the distance from the bottom of the shaft hole to the top of the keyseat.

When designating keys on drawings, the key number or size, length, and product name are given:

$\frac{1}{2} \times 3$ SQUARE KEY

NO. 403 WOODRUFF KEY

$\frac{1}{4} \times 1\frac{1}{2}$ SQUARE GIB HEAD KEY

$1\frac{1}{4} \times 4$ SQUARE PLAIN TAPER KEY

NO. 8 PRATT & WHITNEY KEY

$\frac{1}{8} \times \frac{3}{32} \times \frac{3}{4}$ RECTANGULAR KEY

You May Complete Exercises 13.9 Through 13.12 at This Time

NOMINAL SHAFT DIAMETER	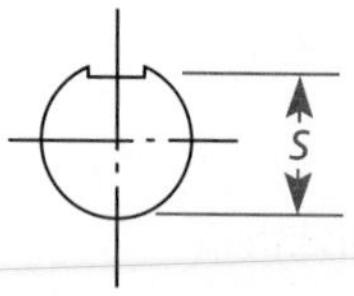 PARALLEL AND TAPER		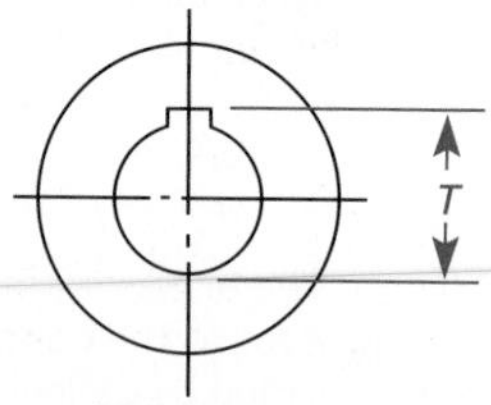 PARALLEL		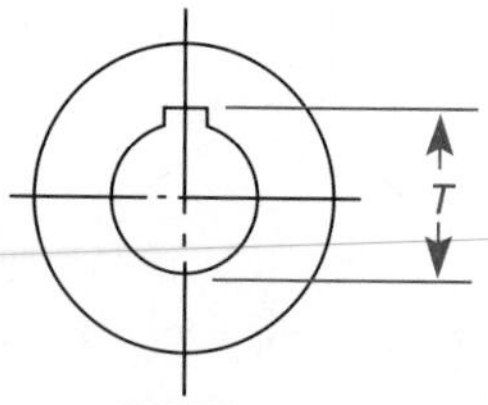 TAPER	
	SQUARE	RECTANGULAR	SQUARE	RECTANGULAR	SQUARE	RECTANGULAR
	S	*S*	*T*	*T*	*T*	*T*
1/2	0.430	0.445	0.560	0.544	0.535	0.519
9/16	0.493	0.509	0.623	0.607	0.598	0.582
5/8	0.517	0.548	0.709	0.678	0.684	0.653
11/16	0.581	0.612	0.773	0.742	0.748	0.717
3/4	0.644	0.676	0.837	0.806	0.812	0.781
13/16	0.708	0.739	0.900	0.869	0.875	0.844
7/8	0.771	0.802	0.964	0.932	0.939	0.907
15/16	0.796	0.827	1.051	1.019	1.026	0.994
1	0.859	0.890	1.114	1.083	1.089	1.058
1-1/16	0.923	0.954	1.178	1.146	1.153	1.121

FIGURE 13.92 Shaft Diameter and Keyseat Dimensions

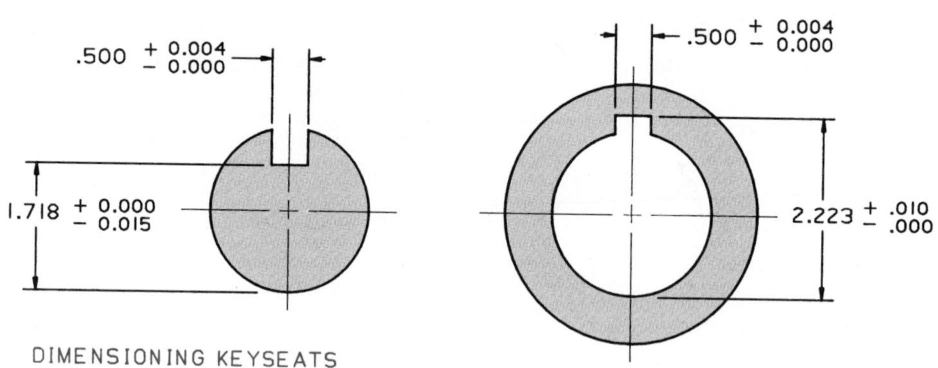

FIGURE 13.93 Dimensioning Keyseats

QUIZ

True or False

1. M5 × 1.50-6g is a designation for an American National thread.
2. UNC means United National thread form.
3. The B symbol for threads indicates that the thread is external.
4. Dowel pins are used to align and locate parts.
5. Studs are fasteners that secure two left-handed parts together.
6. Carriage bolts have a square body under the head.
7. The 13 in ".500-13 UNC-2B" means the number of threads per foot.
8. Taper pins are tapered $\frac{1}{8}$ inch per foot.

Fill In The Blanks

9. ________ , ________ , and ________ are types of keys.
10. ________ , ________ , and ________ are primarily for power transmission.
11. ________pins are used to retain parts such as slotted nuts and ________ pins.
12. ________ threads are used in place of the old ________ threads.
13. ________ ________ rings are installed on ________ machined grooves in housings.
14. A ________ key is shaped similar to a half circle.
15. Basic industrial fasteners include ________ and ________ bolts, ________ ________, carriage bolts, studs, ________ , and ________ .
16. ________ are fasteners with no head but with threads at both ends of the shank.

Answer the Following

17. Describe how an axially assembled external retaining ring might be used in a design.
18. What is a set screw and describe some of its possible design functions?
19. Name and describe three types of locating pins.
20. What are the meanings of UN, UNF, and UNC?
21. What is the difference between class 1, 2, and 5 threads?
22. List four considerations in the selection of a fastener.
23. Describe two types of studs.
24. What is the difference between a right-hand and a left-hand thread?

EXERCISES

Exercises may be assigned as sketching or instrument projects. Transfer the given information to an "A" size sheet of .25 in. grid paper. Complete all views and solve for proper visibility, including centerlines, object lines, and hidden lines. Exercises that are not assigned by the instructor can be sketched in the text to provide practice and understanding for the preceding instructional material. Dimensions for fasteners used on exercises can be located in figures throughout the chapter and in Appendix C.

After Reading the Chapter Through Section 13.12, You May Complete the Following Exercises

Exercise 13.1 Complete the three parts using the appropriate threads. Use detailed thread representation.

Exercise 13.2 Draw the detailed representation of the Acme threads.

Exercise 13.3(A) Calculate the normal engagement, effective thread, and pipe end. Draw the flange and pipe as shown. The NPT pipe thread has a $\frac{3}{4}$ inch per foot taper. The pipe has a 3 in. nominal size.

Exercise 13.3(B) Complete the pipe plug and flange. The plug has a standard NPT $\frac{3}{4}$ in. nominal pipe thread with a $\frac{3}{4}$ inch per foot taper. Calculate and draw the effective thread, normal thread engagement, and length. Use simplified thread representation. Complete the end views showing only details that are visible and the threads.

Exercise 13.4 Draw the threaded shaft as shown. Include all chamfers, reliefs, and threads. Use schematic thread representation.

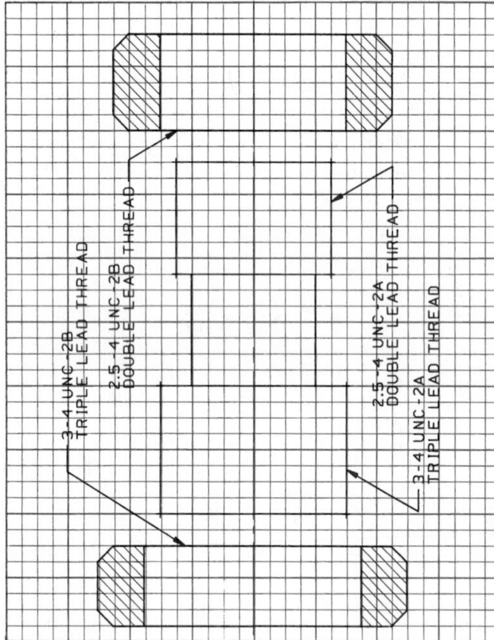

Exercise 13.1

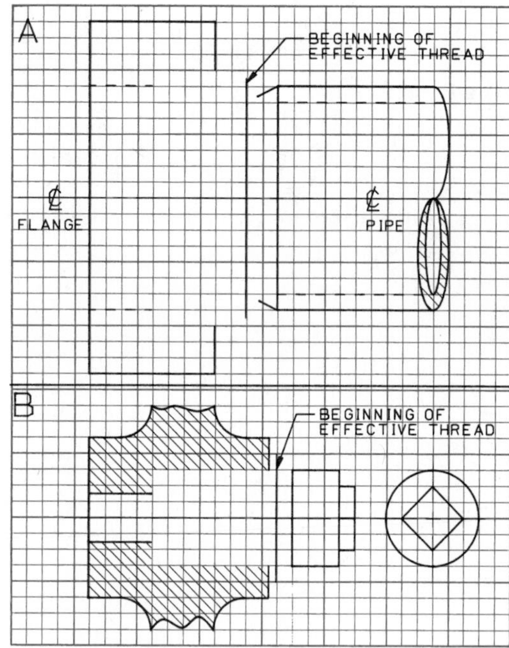

Exercise 13.3

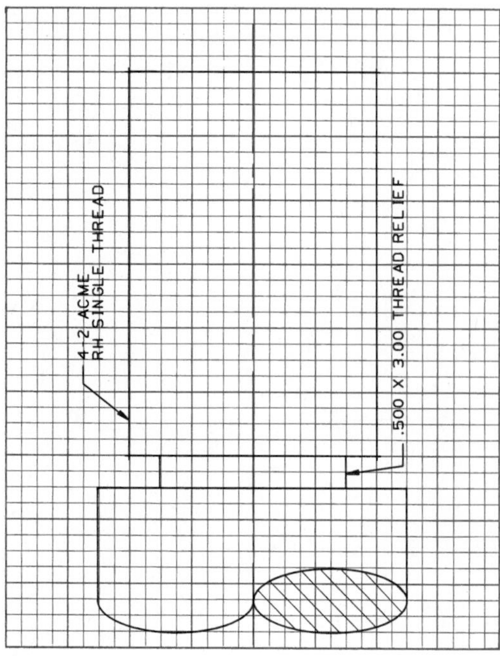

Exercise 13.2

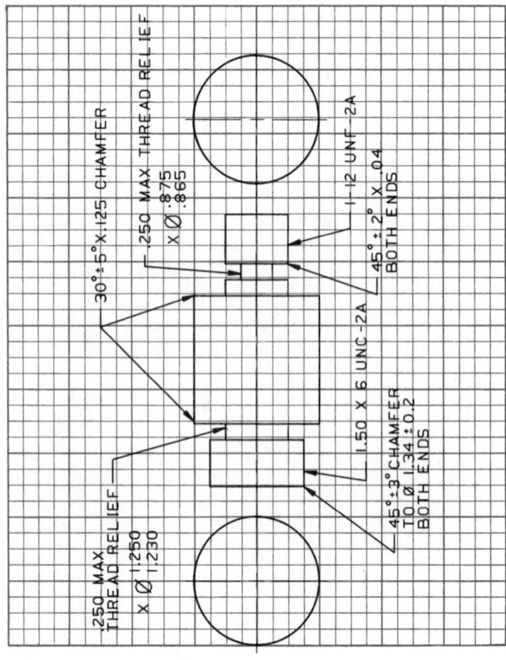

Exercise 13.4

After Reading the Chapter Through Section 13.16, You May Complete the Following Exercises

Exercise 13.5(A) Using detailed representation draw a 1.50-6 UNC-2A × 4 square head bolt. Draw only the axial (side) view in the given space.

Exercise 13.5(B) Draw a detailed representation of the 1-8 UNC-2A × 3 hex socket head cap screw. Show the side and end views in the given space and label the drawing correctly.

Exercise 13.6(A) Fasten the rest block to the plate with four $\frac{3}{8}$ in. diameter hex socket head cap screws (S) and two $\frac{3}{8}$ in. diameter steel dowel pins (D). Calculate the screw and dowel lengths. The plate will have threaded through-holes. The rest block will have clearance through holes for the screws to pass through. Calculate the screw's length of engagement and counterbore the block so that the screw's head will be flush with the top surface. Calculate the counterbore diameter and depth. The dowel will be press fit (interference fit) into the block and the plate. Calculate the ream hole diameter for the dowels.

Exercise 13.6(B) Dimension the hole pattern for the screws and the dowels. Call out the proper clearance hole size for the drilled clearance holes, screws and the reamed holes for the dowels.

Exercise 13.7 Draw the moving shaft pivot as shown. Use two hex socket head shoulder screws. Show the screws in both views. The screws have different diameters and lengths.

Exercise 13.8 Read the section on set screws and complete the exercise as described. Determine the proper diameter and length of the set screws as per shaft diameter. There are two hex socket set screws required for each shaft. They are installed 90° to one another. Use a cup point for the small set screws and a dog point for the two larger set screws.

A
TOP OF BOLT HEAD
B
TOP OF SCREW HEAD

Exercise 13.5

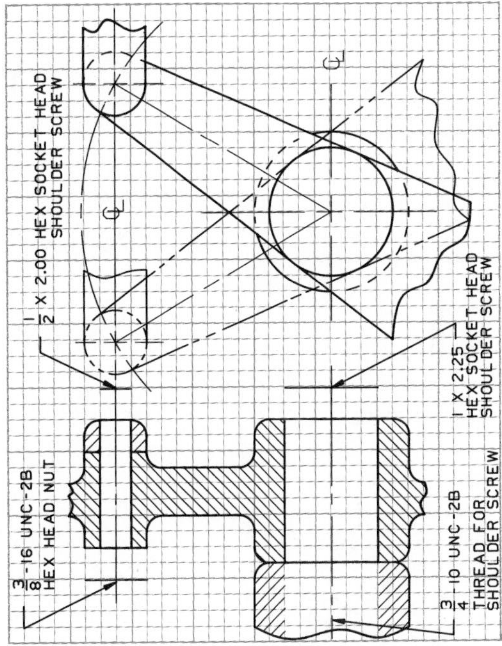

Exercise 13.7

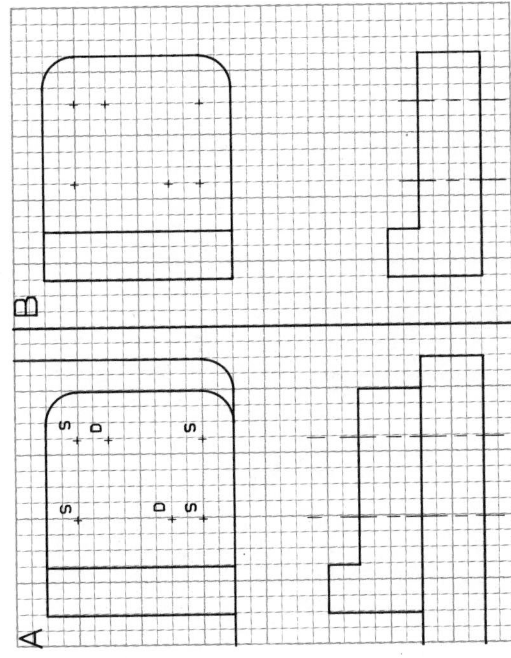

Exercise 13.6

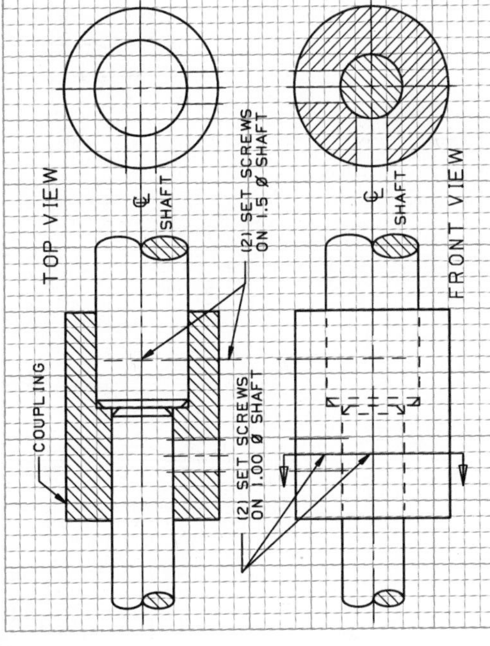

Exercise 13.8

After Reading the Chapter Through Section 13.21.4, You May Complete the Following Exercises

Exercise 13.9(A) Attach the collar to the shaft using one of the two following types of spring pins:
Pin, Coiled Spring, 10 × 100, Metric, Steel
Pin, Slotted Spring, .500 × 4.00, ANSI 302

Exercise 13.9(B) Attach the collar to the shaft with a tapered pin. Call out the hole size and dimensions for tapered holes and use the following pin: Pin, Tapered, No. 8 × 2.50, Steel.

Exercise 13.10 Fasten the hitch at C with a .500 in. clevis pin. Use two .500 in. plain washers above and below the hitch and plate. Show a .135 in. diameter cotter pin to secure the clevis pin. Fasten the plates at B with three .625 in. diameter hex bolts. Use lock washers on both sides and hex flat nuts on the bottom. Show fasteners in both views. You will need to determine the length of the bolts and the clevis pin based on the fastening requirements. Call out the clevis pin, cotter pin, washers, nuts, and bolts on a separate parts list and attach to the drawing.

Exercise 13.11(A) Calculate the size and length of a Woodruff key or a square key (ask your instructor). Draw the key in the view provided. The shaft has a 2.00 in. diameter.

Exercise 13.11(B) Secure the shaft to the sprocket using a tapered gib key or a taper key. Determine the size and length of a key for the 1.50 in. diameter shaft.

Exercise 13.12(A) Draw and dimension the shaft and a basic 5100 external retaining ring. See Appendix C or manufacturing catalogs for the ring dimensions.

Exercise 13.12(B) Same as Exercise 12(A) except use a N5000 basic 3 in. internal retaining ring for the housing. Draw and dimension completely.

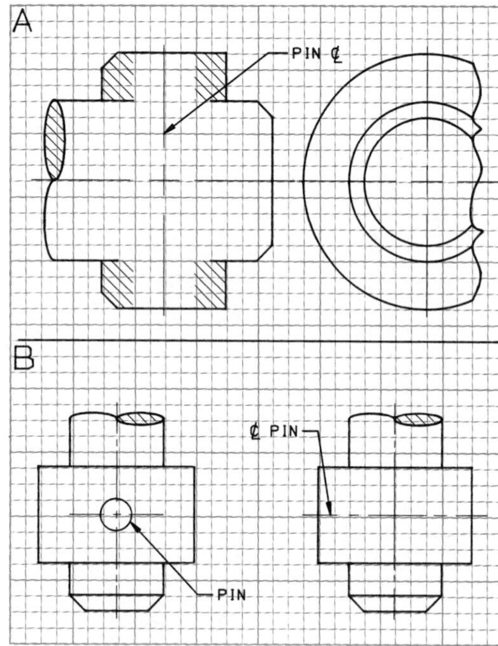

Exercise 13.9

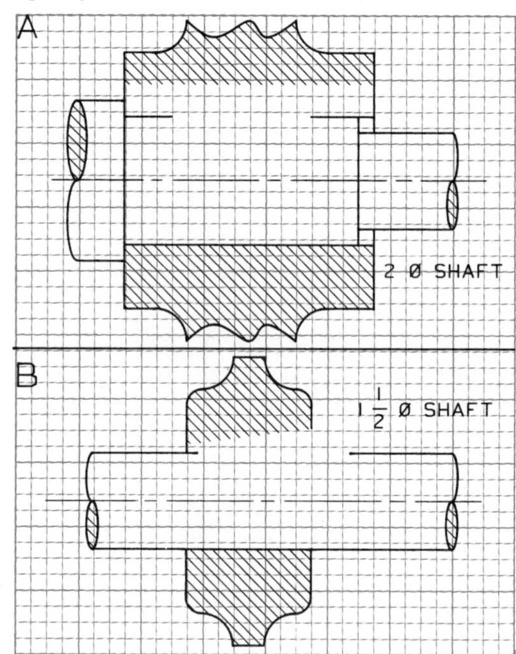

Exercise 13.11

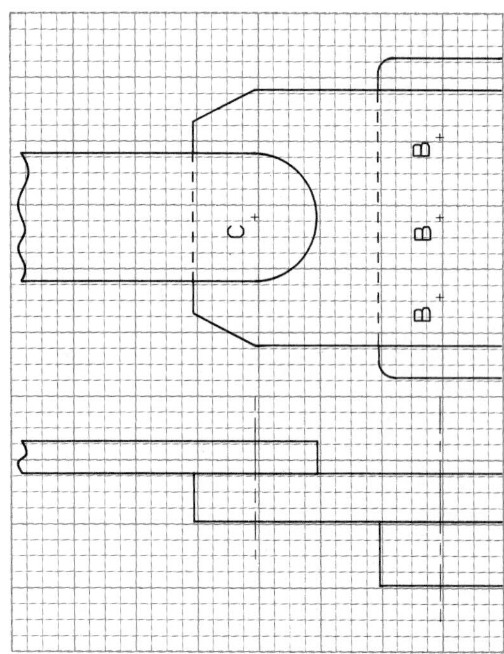

Exercise 13.10

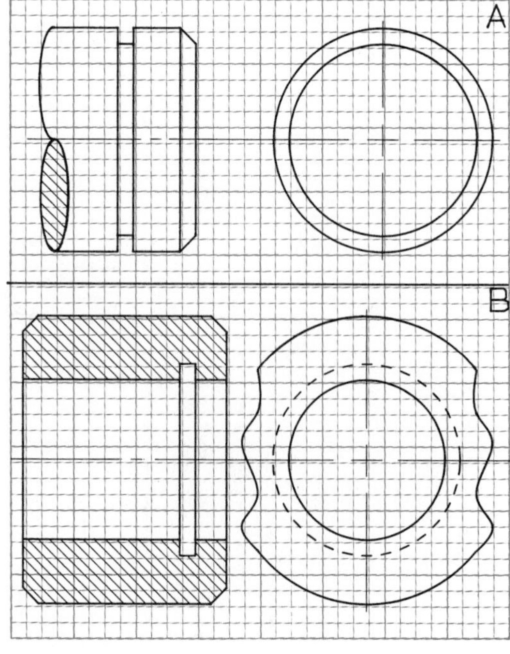

Exercise 13.12

PROBLEMS

Problem 13.1 Draw a 1.00 in. pitch thread (two times size) of the following thread types: Acme, square, and a UNC.

Problem 13.2 Draw 3-2 Acme thread with a length of 5.00 inches using detailed thread representation.

Problem 13.3 Draw a 1.25 × 4.50 hex socket head shoulder screw full size. Show length view and end view of head. Use schematic method to display threads.

Problem 13.4 Fasten a 1.25 in. plate to an aluminum casting (3.00 inches thick) using a .500-13 UNC socket head cap screw. Calculate and show the screw in two views. Dimension and call out tap drill, clearance hole, and tap size.

Problem 13.5 Connect a 1.50 in. and a 1.375 in. plate with a 1.00-in. socket head shoulder screw and appropriate nut. Show in two views and call out all hole sizes.

Problem 13.6 Bolt together two 1.50 in. thick steel plates with two 1.25-12 UNF hex head bolts. Use lock washers on both ends and the appropriate nut. Show in section. Construct a small parts lists for the hardware.

Problem 13.7 Fasten a 4.0 × 4.0 × 4.0 × 2.00 in. thick steel angle plate to a steel part using four .375-16 UNC socket head caps screws and two .375 inch diameter dowels. Design the bolt pattern and calculate all fastener sizes. Dimension and call out all fasteners. Counterbore the plate so that the screw heads will be below the surface.

Problem 13.8 Draw a 50 mm diameter shaft and a 74 mm wide collar (O.D. 100 mm/I.D. 51 mm). Fasten the collar to the shaft with two appropriately sized socket set screws with a dog point. Show in two views.

Problem 13.9 Draw a 2.50 in. diameter shaft and a 4.00 in. wide collar (O.D. 5.00 in./I.D. 2.51 in.). Connect the two parts with a square key 2.00 inches long. Calculate the key size and show in two views. Dimension the views as required.

Problem 13.10 Same as Problem 13.9 but use a Woodruff key and keyseat. Dimension the views as required.

Problem 13.11 Connect two sheets of .125 inch thick aluminum with a .125 inch diameter button head rivet. Show in two views at 2 times size.

Problem 13.12 Using a butt joint connect two .500 inch thick sheets (6.00 inches wide) of steel with twelve 1.125 inch diameter pan head rivets. Use double rivets on each side of the joint. Show in two views and dimension completely.

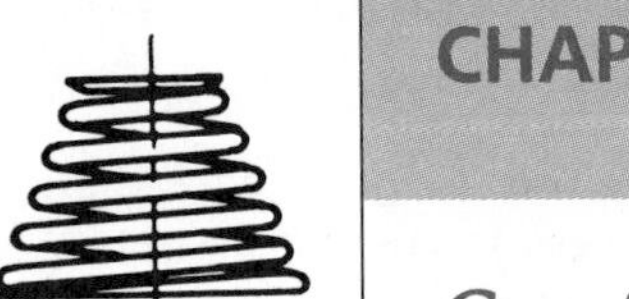

CHAPTER 14

Springs

Learning Objectives

Upon completion of this chapter, you will be able to accomplish the following:

1. Develop an understanding of the purposes for and uses of springs in mechanical assemblies.
2. Identify the various types of springs used in mechanical assemblies.
3. Differentiate between left-hand and right-hand springs.
4. Produce drawings of basic spring types.

14.1 Introduction

A **mechanical spring** is an elastic body whose mechanical function is to store energy when deflected by a force and to return the equivalent amount of energy upon being released. Helical springs (Fig. 14.1) are similar to threads in that they are spiral shaped.

Most springs are represented by their centerline and phantom lines defining their outside diameter. Seldom are springs drawn pictorially (coils drawn). However, at the end of the chapter, a step-by-step procedure for drawing spring coils is provided.

A number of requirements are applicable to all spring drawings, including material specifications and inspection notes. Since most springs are standard configurations and sizes, specifications and notes are more important than the drawing itself. Material specifications are designated in a general note on the drawing.

Springs are produced according to specific standards and specifications. ANSI recognizes six types of springs:

1. Compression—helical, cylindrical, volute, coned disk (Belleville)
2. Extension—helical
3. Garter—helical
4. Torsion—helical, torsion bar, spiral
5. Flat—cantilever
6. Constant force—flat

14.1.1 Spring Terms

The following terms are used throughout this section and on drawings of mechanical springs:

Coils, active The number of coils used in computing the total deflection of a spring. Those coils that are free to deflect under load.

Deflection, total The movement of a spring from its free position to maximum operating position. In a compression spring, it is the deflection from the free length to the solid (compressed) length.

Force The force exerted on a spring to reproduce or modify motion, or to maintain a force system in equilibrium.

Helix The spiral form (open or closed) of compression, extension, and torsion springs.

Length, free The overall length of a spring in the unloaded position.

Length, solid The overall length of a compression spring when all coils are fully compressed.

Load The force applied to a spring that causes deflection.

Cylindrical
Right-Hand Helix

Convex
Right-Hand Helix

Cylindrical With Coned End
Left-Hand Helix

Concave
Right-Hand Helix

Conical
Right-Hand Helix

FIGURE 14.1 **Helical Compression Spring Forms**

Pitch The distance from center to center of the wire in adjacent active coils (recommended practice is to specify number of active coils rather than pitch).

Set Permanent distortion of the spring when stressed beyond its elastic limits.

Total number of coils Number of active coils n plus the coils forming the ends.

14.2 Right-hand and Left-hand Springs

If dictated by design requirements, the direction of helix is specified as "LEFT-HAND" (LH) or "RIGHT-HAND" (RH). Otherwise, the direction of helix is specified as "OPTIONAL." Usually, the direction is not important, except when a plug is screwed into the end or when one spring fits inside another. In the case of the latter, one spring is designated left-hand and the other spring right-hand. Figure 14.2 shows how the coils look for right-hand and left-hand springs. Look at the back of your hands; the spring will coil to the left or right.

14.3 Compression Springs

A **compression spring** is an open-coil helical spring that resists a compressive force applied along the axis. Compression springs are coiled as a constant-diameter cylinder. Other common forms of compression springs such as conical, tapered, concave, convex, or various combinations of these are used as required by the application (Fig. 14.1). While square, rectangular, or special-section wire may have to be specified, round wire is predominant in compression springs. Figure 14.3 shows the recommended way to specify compression springs.

There are four basic types of compression spring ends, as shown in (Fig. 14.4). The particular type of ends specified affect the pitch, solid height, number of active and total coils, free length, and seating characteristics of the spring. The type of ends are specified on the drawing and dimensioned as required.

Depending on the application of the compression spring, the following requirements are specified:

TO WORK OVER ________ MAX DIAMETER ROD
TO WORK IN ________ MIN DIAMETER BORE
ID (with tolerance)________
OD (with tolerance)________

14.4 Extension Springs

Extension springs (Fig. 14.5) absorb and store energy by resisting a pulling force. Various types of ends are used to attach the extension spring to the source of the force. Most extension springs are wound with an initial tension, which holds the coils tightly together. The load necessary to overcome the internal force and just start coil separation is the same as the initial tension.

14.5 Helical Extension Springs

A **helical extension spring** is a close-wound spring, with or without initial tension, or an open-wound spring that resists an axial force trying to elongate the spring. Extension springs are formed or fitted with ends that are used for attaching the spring to an assembly. Guidelines for specifying dimensional and force data on engineering drawings showing helical extension springs (Fig. 14.6) are similar to those established for helical compression springs. Usually, all coils in an extension spring are active. Exceptions are those with plug ends and those with end coils coned over swivel hooks. The total number of coils required is specified.

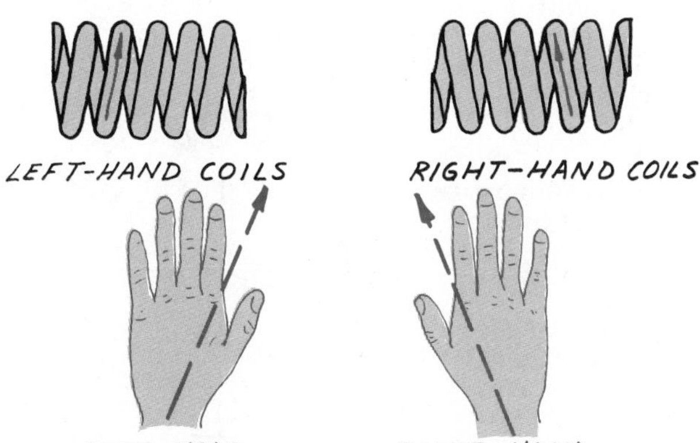

FIGURE 14.2 Left-Hand and Right-Hand Springs

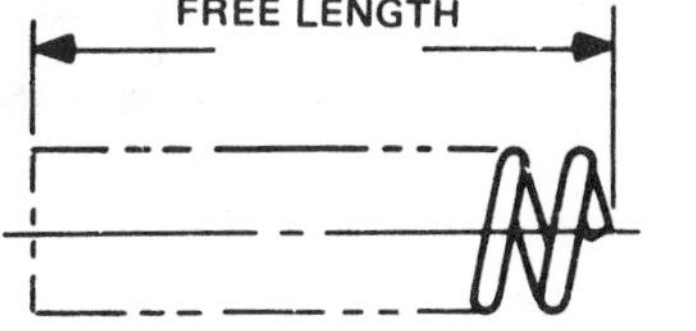

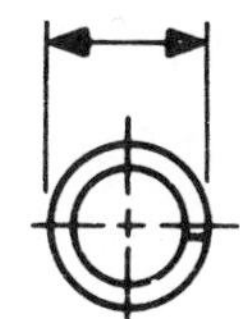

SPRING DATA

MATERIAL SPECIFICATION ....
WIRE DIAMETER ....
DIRECTION OF HELIX ....
TOTAL COILS ....

FIGURE 14.3 Drawing Requirements for Helical Compression Springs

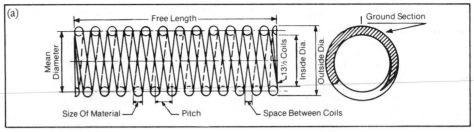

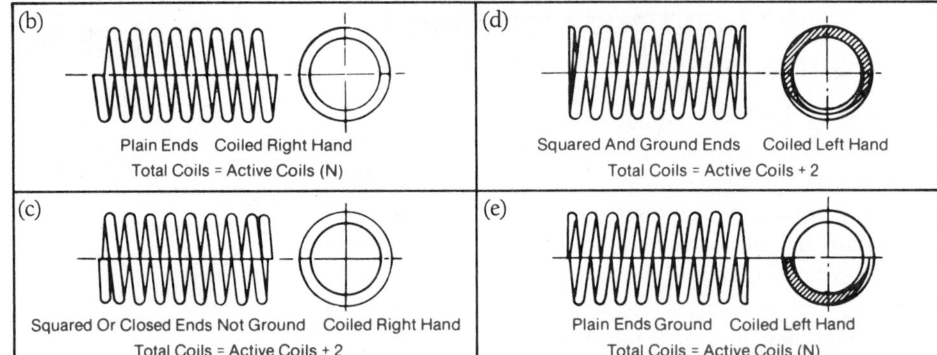

FIGURE 14.4 End Finishes for Compression Springs

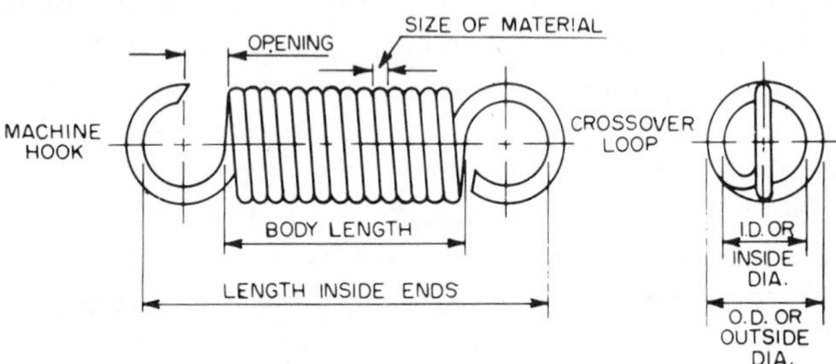

FIGURE 14.5 Extension Springs

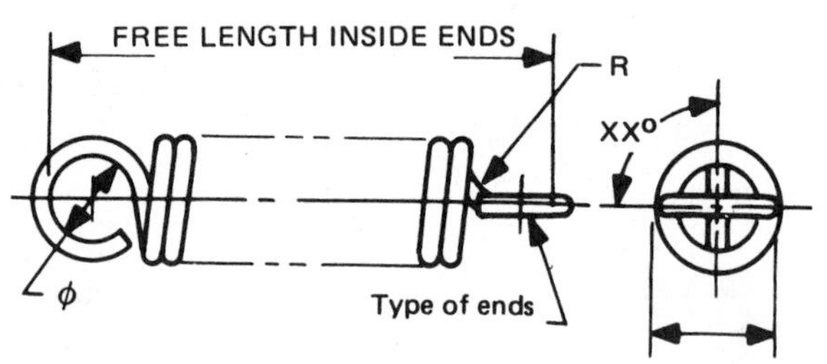

SPRING DATA

MATERIAL SPECIFICATION . . . .
WIRE DIAMETER . . . .
DIRECTION OF HELIX . . . .
TOTAL COILS . . . .
EXTENDED LENGTH WITHOUT PERMANENT SET . . . .
RELATIVE POSITION OF ENDS . . . .
INITIAL TENSION . . . .
FORCE AT OPERATING LENGTH OF______

FIGURE 14.6 Drawing Requirements for Helical Extension Springs

14.6 Garter Extension Springs

A **garter spring** (Fig. 14.7) is a long, close-coiled extension spring with its ends joined to form a ring. Garter springs are used in mechanical seals on shafting, to hold round segments together, as a belt, and as a holding device. The diameter over which the spring is to function is specified. For example, a shaft diameter may be used, although other than an actual shaft may be involved.

14.7 Helical Torsion Springs

Helical torsion springs (Fig. 14.8) are springs that resist a force or exert a turning force in a plane at right angles to the axis of the coil. The wire itself is subjected to bending stresses

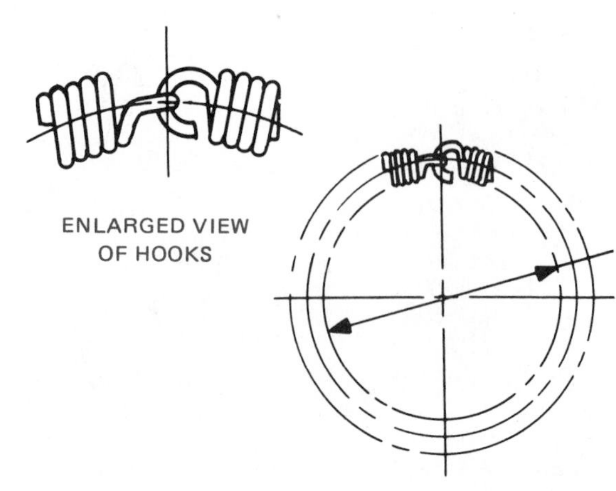

FIGURE 14.7 Drawing Requirements for Garter Springs

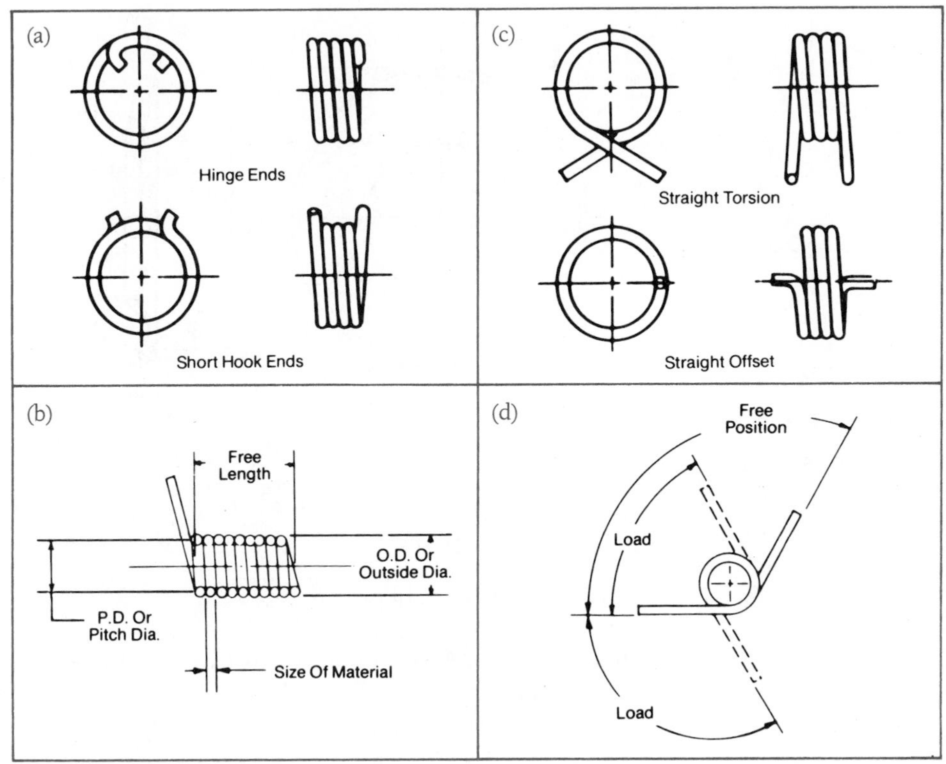

FIGURE 14.8 Drawing Requirements for Helical Torsion Springs

rather than torsional stresses. Usually, all coils in a torsion spring are active. The total number of coils required and the length in the free position are specified. The helix of a torsion spring is important. Either "LEFT-HAND" or "RIGHT-HAND" is specified.

14.8 Spiral Torsion Springs

Spiral torsion springs (Fig. 14.9), made of rectangular section material, are wound flat, with an increasing space between the coils. A spiral torsion spring is made by winding flat spring material on itself in the form of a spiral. It is designed to wind up and exert a force in a rotating direction around the spring axis. This force may be delivered as torque or it may be converted into a push or pull force.

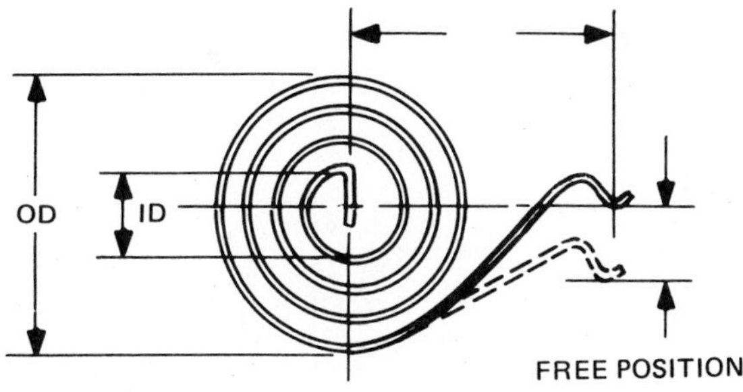

SPRING DATA

MATERIAL SPECIFICATION ....
MATERIAL SIZE ....
OUTSIDE DIAMETER ....
INSIDE DIAMETER ....
DEVELOPED LENGTH OF MATERIAL ....
ACTIVE LENGTH OF MATERIAL ....
NUMBER OF COILS IN FREE POSITION ....
TORQUE AT FINAL POSITION ....
MAXIMUM DEFLECTION BEYOND FINAL POSITION WITHOUT SET .....
TYPE OF ENDS ....

FIGURE 14.9 Drawing Requirements for Spiral Torsion Springs

14.9 Spring Washers

Because of trends toward miniaturization and greater compactness of design, **spring washers** are used more often today. They have space and weight advantages over conventional wire springs and are often more economical. Their applications include keeping fasteners secure, distributing loads, absorbing vibrations, compensating for temperature changes, eliminating side and end play, and controlling end pressure. Figure 14.10 shows a finger spring washer used for preloading ball bearings.

A **coned disk (Belleville) spring** (Fig. 14.11) is a spring

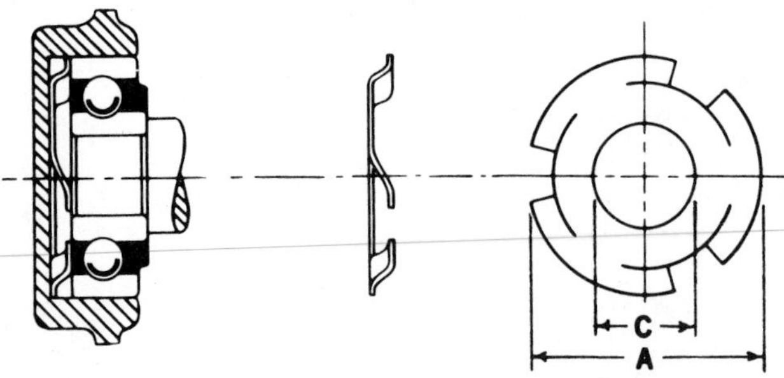

FIGURE 14.10 **Finger Spring Washer Installed in a Bearing Housing**

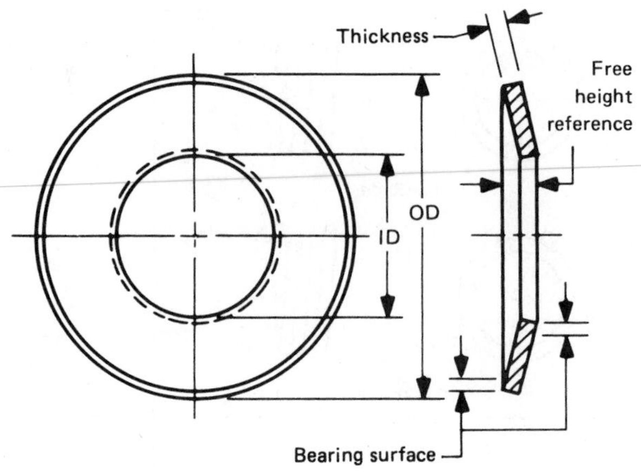

SPRING DATA
MATERIAL SPECIFICATION ....
THICKNESS OF MATERIAL ....
FREE HEIGHT ....
FORCE AT COMPRESSED HEIGHT OF____
(Special data) ....

FIGURE 14.11 **Drawing Requirements for Coned Disk (Belleville) Springs**

washer in the form of the frustrum of a cone. It has constant material thickness and is used as a compression spring.

14.10 Flat Springs

The term **flat springs** covers a wide range of springs or stampings fabricated from flat strip material, which, when deflected by an external load, releases stored energy. Only a small portion of a complex-shaped stamping may actually be functioning as a spring. Leaf springs used on the rear of cars and vans are examples of flat springs.

A flat spring includes all springs made of flat strip or bar stock that deflects as a cantilever or as a simple beam. Figure 14.12 is an example of a detail drawing of a flat spring. A pictorial view shows the part in its finished state and gives the bending angle in degrees. The dimensioned view is of the flat (unfolded) part.

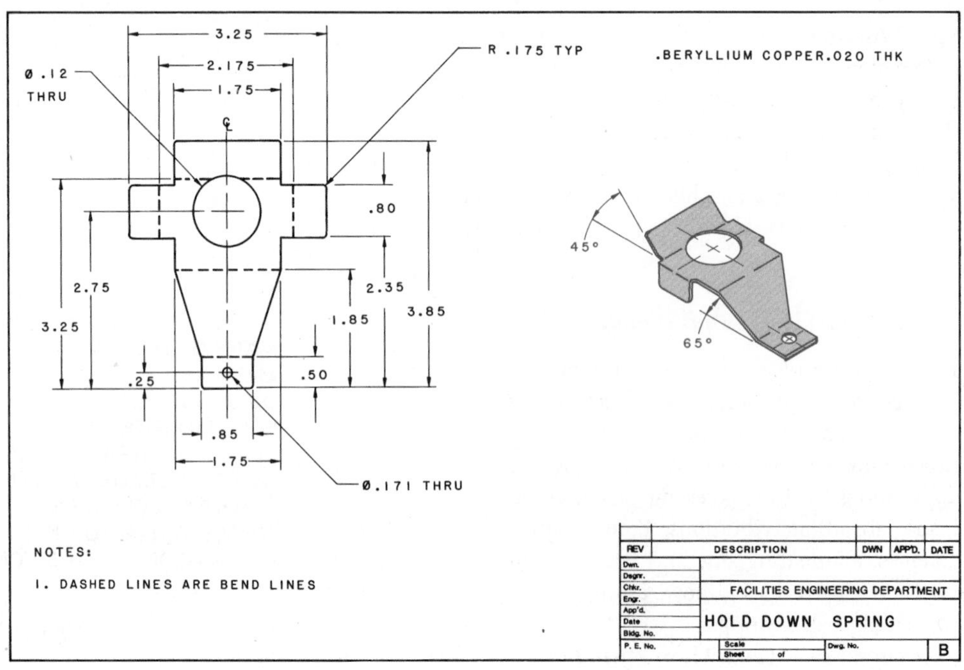

FIGURE 14.12 **Hold Down Spring**

ITEMS OF INTEREST

Why Change ANSI Standards?

Ever since you began your study of technical or engineering drawing, you have been learning how to produce drawings according to standards that have been established by the American National Standards Institute (ANSI). You may not have realized that these established standards change and evolve to match the needs of the ever-changing technology involved in manufacturing and production. At first thought it might seem odd to you that an established standard could or should change. After all, the very purpose of a standard seems to oppose the idea of changing it. However, standards exist to assist manufacturers and to make the production of parts and assemblies more efficient. If you think carefully about the quick evolution of modern technology and the new worldwide marketing and manufacturing environment, it seems reasonable to expect standards to change.

In 1935 the American Standards Association (the predecessor to ANSI) published the first recognized standard for engineering drawings in the United States—"American Drawing and Drafting Room Practices." The document was 18 pages long and the entire subject of tolerancing was covered in two paragraphs. It was clear in that era that the assembly line manufacturing process created largely by Henry Ford and his Model T replaced the old "fit and file to size" craftsman type manufacturing process forever. This mass assembly process created a pressing need for different shops to be able to produce the same parts. The exchange of drawings for those parts became critical and a group was formed to create a standard way to communicate manufacturing and engineering details graphically. It took years for the group to publish the standards document. Whose standard is best is, and will continue to be, a difficult question for any group of individuals charged with creating a standard.

World War II provided the motivation to continue to improve engineering drawings and mass production techniques. Scrap rates were too high and assemblies were hampered by the limitations of the plus/minus tolerance system. It became apparent that geometry and not just variation in size controlled many assemblies. How many times have you drilled a hole with a hand drill and realized that

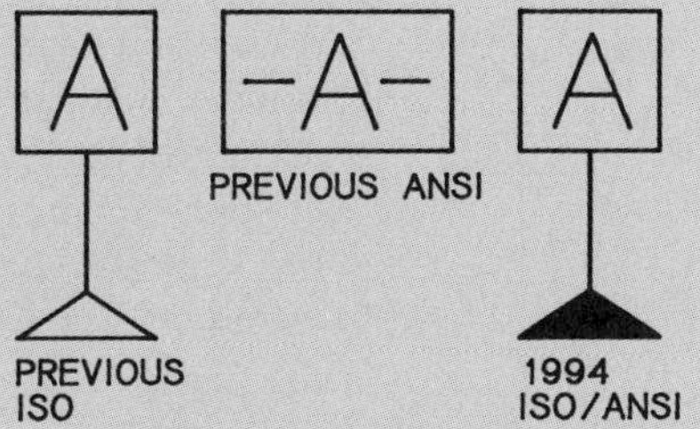

Datum-Identifying Symbols

the hole was the correct size but it was of no use to the assembly because the axis of the hole wasn't perpendicular to the right plane? The U.S. Army published an Ordinance Manual on Dimensioning and Tolerancing in 1945 that used symbols to specify form and position tolerances. Unfortunately, the American Standard Association's "American Standard Drawing and Drafting Room Practice," second edition published in 1946, lacked a comprehensive section on tolerancing. To make matters worse, the Society of Automotive Engineers published its own standards in 1946 and in 1952. After the war, there were *three different standards* for tolerancing engineering drawings. It wasn't until 1966 after years of debate that ANSI published the first unified standard on tolerancing and dimensioning—ANSI Y14.5. The standard was updated in 1973 to replace notes with symbols for all tolerancing, and the current version was published in 1982. As you may have guessed, a new version is expected soon—in 1993 or 1994. If that all seems a bit complicated, remember that the rest of the world has been developing their own standards and formed the International Organization for Standardization (ISO).

Evolving technology and the need to compete in world class manufacturing seem to be the drivers for the next rounds of ANSI standards revisions. CAD and CAM have become key components in manufacturing since the last ANSI Y14.5 revision. Producing a part with a CNC machine becomes easier when the part is dimensioned with that production technique in mind. The use of decimals has replaced fractions for that same reason. Unfortunately, anyone who has used a CAD system will tell you that keeping to ANSI standards has never seemed to worry the makers of CAD systems very much. The outcome of that dilemma has not been resolved. However, competing in world class manufacturing is very important to all American manufacturers. Parts for any one assembly will more than likely be produced in a variety of shops across the world. To be part of that network, it seems that ANSI standards and ISO standards must be compatible so that engineering drawings convey the same information worldwide. Conveying information quickly and correctly is a must to effectively compete in today's world class manufacturing environment.

The men and women charged with making ANSI and ISO standards have a difficult and complex job. Each document they produce must contain compromises. As technology and the world environment changes, standards must evolve with them. As the world economy becomes more unified, more world unified standards will certainly follow. The ability to change and adapt seems more important to our continued success than ever.

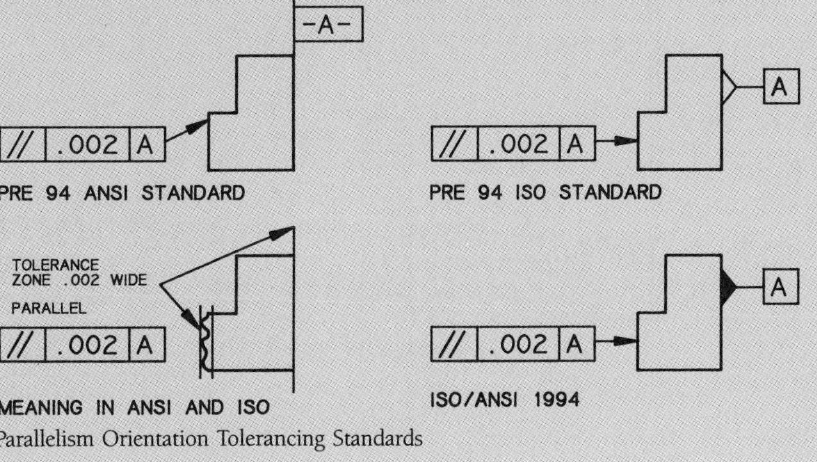

Parallelism Orientation Tolerancing Standards

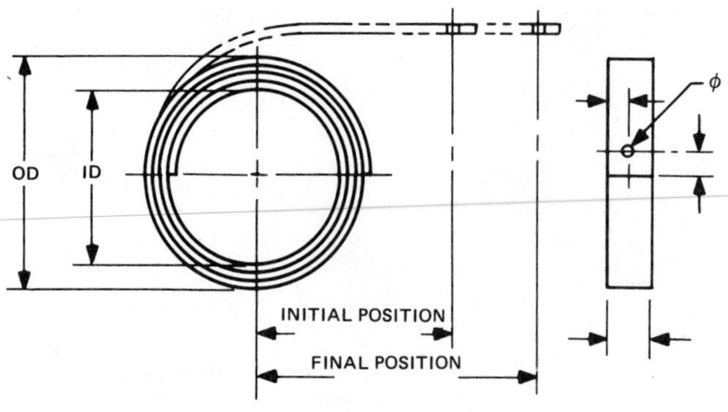

SPRING DATA

MATERIAL SPECIFICATION	....
MATERIAL SIZE	....
ACTIVE LENGTH	....
NUMBER OF COILS	....
FORCE	....
FITS OVER	....

FIGURE 14.13 Drawing Requirements for Constant Force Springs

14.11 Constant Force Springs

A **constant force spring** (Fig. 14.13) is a strip of flat spring material that has been wound to a given curvature so that, in its relaxed condition, it is in the form of a tightly wound coil or spiral. A constant force is obtained when the outer end of the spring is extended tangent to the coiled body of the spring. A constant torque is obtained when the outer end of the spring is attached to another spool and wound in either the reverse or same direction as it is originally wound. Because the material used for this type of spring is thin and the number of coils would be difficult to show in actual form, it is acceptable to exaggerate the thickness of the material and to show only enough coils to depict a coiled constant force spring.

14.12 Drawing Springs

Springs are drawn using simplified methods, except when the spring must be pictorially correct for dimensioning. Even when these situations occur, it is normal practice to show only a limited number of coils and use the simplified method for the remaining coils. The simplified method of representing springs uses phantom lines to establish the springs outside diameter, and a centerline to locate its axis. Figure 14.14 shows an industrial detail of a torsion spring. The ends are drawn true, and the coils are shown with phantom lines.

In Figure 14.15, six active coils were required along with plain open ends. The following steps are used to draw the coils of a *compression spring:*

1. Lay out the free length (overall length), coil centerline, and the outside diameter of the spring. These dimensions are blocked-in with construction lines. The *mean diameter* is drawn as shown in the side view (end view) of the spring. The mean diameter equals the outside diameter of the coil minus the wire diameter.

   One coil diameter (wire diameter) is drawn in the side view (Fig. 14.15). The inside diameter and the outside diameter of the coil are drawn in the side view (end view).

   The front view of the spring is divided into even spaces based on the total number of coils. Each of the coil cross-section diameters is lightly drawn along the top and bottom of the coil length, at the appropriate divisions.
2. Lightly draw the coil winding (left- or right-hand) as shown. The appropriate end style is then constructed. The plain open end is used in this example.
3. Darken the coil, using appropriate line weights, and dimensioned accordingly. (Dimensions were not shown in this example; refer to previous examples throughout the chapter.)

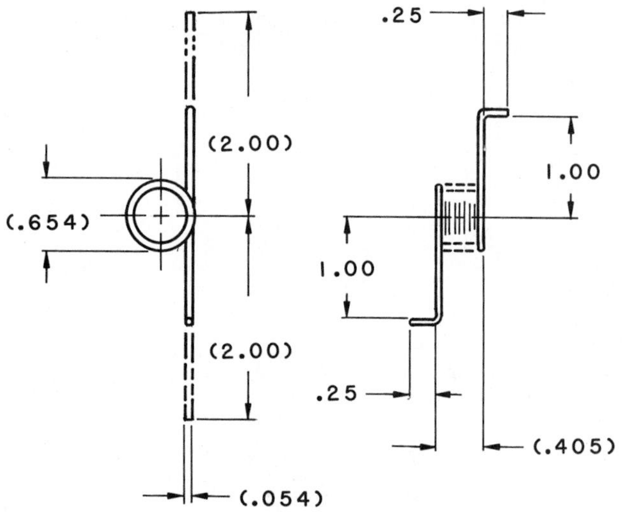

FIGURE 14.14 Torsion Spring Detail

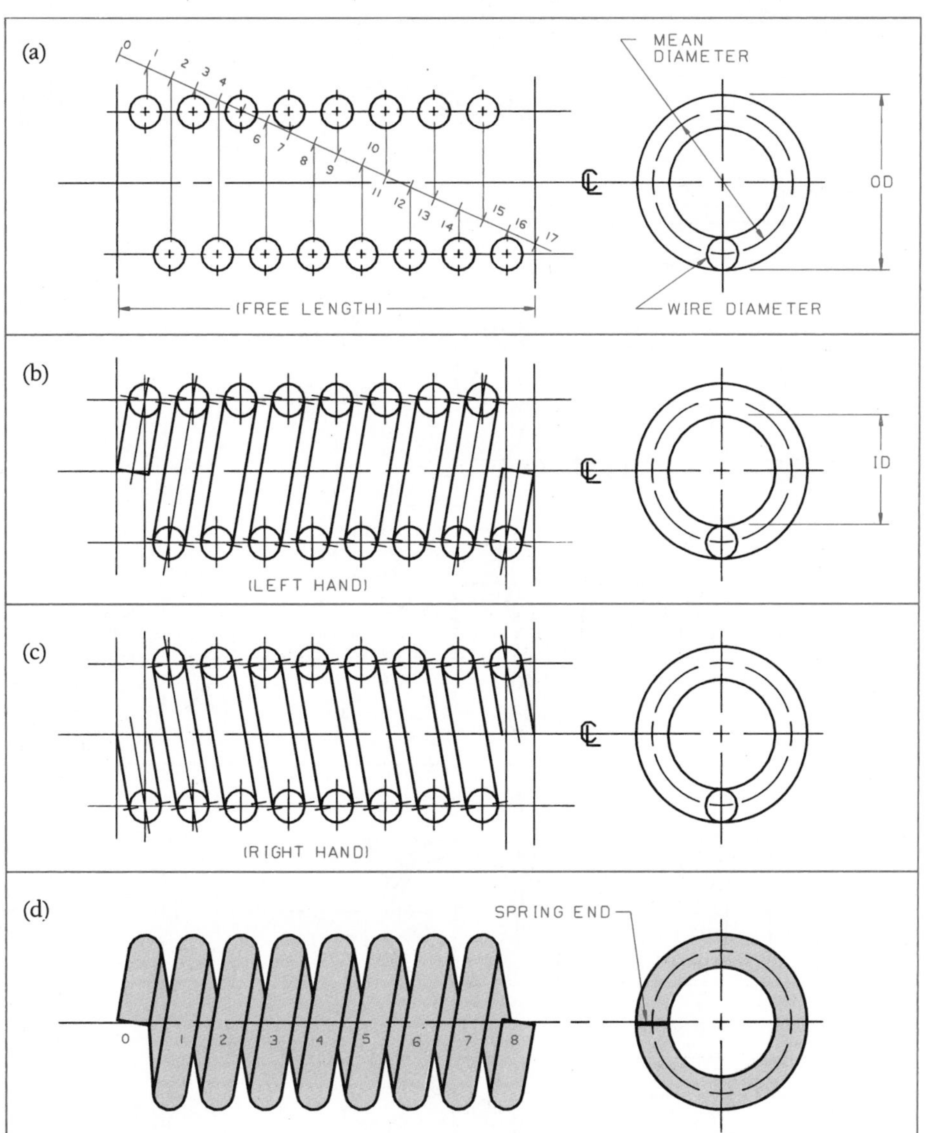

FIGURE 14.15 Drawing a Compression Spring

Drawing an *extension spring* is similar to constructing a compression spring except that the coils are solid in the relaxed (unloaded) position. In other words, the coils touch. The following steps were used to draw a full-loop-over-center extension spring [Figs. 14.16(a) through (e)]:

1. Draw centerlines and the outside and inside diameter. Then draw the end loops (they will be the same as the end view) at the required length and complete the construction.
2. Using a circle template and the appropriate diameter (wire size), draw the wire diameters on the top and the bottom.
3. Extend a construction line from the end of the edge of the wire diameter on the lower left to the edge of the upper left diameter. Draw lines parallel to the first construction line along the total length of the spring coils.
4. Draw circles that represent the wire diameters along the upper portion of the coil length as shown. Then adjust the spring end as shown. The spring ends are established by a 30° construction line extended from the coil end diameter.
5. Complete the coil and end visibility carefully. Use appropriate line weights to darken and complete the drawing. Add dimensions.

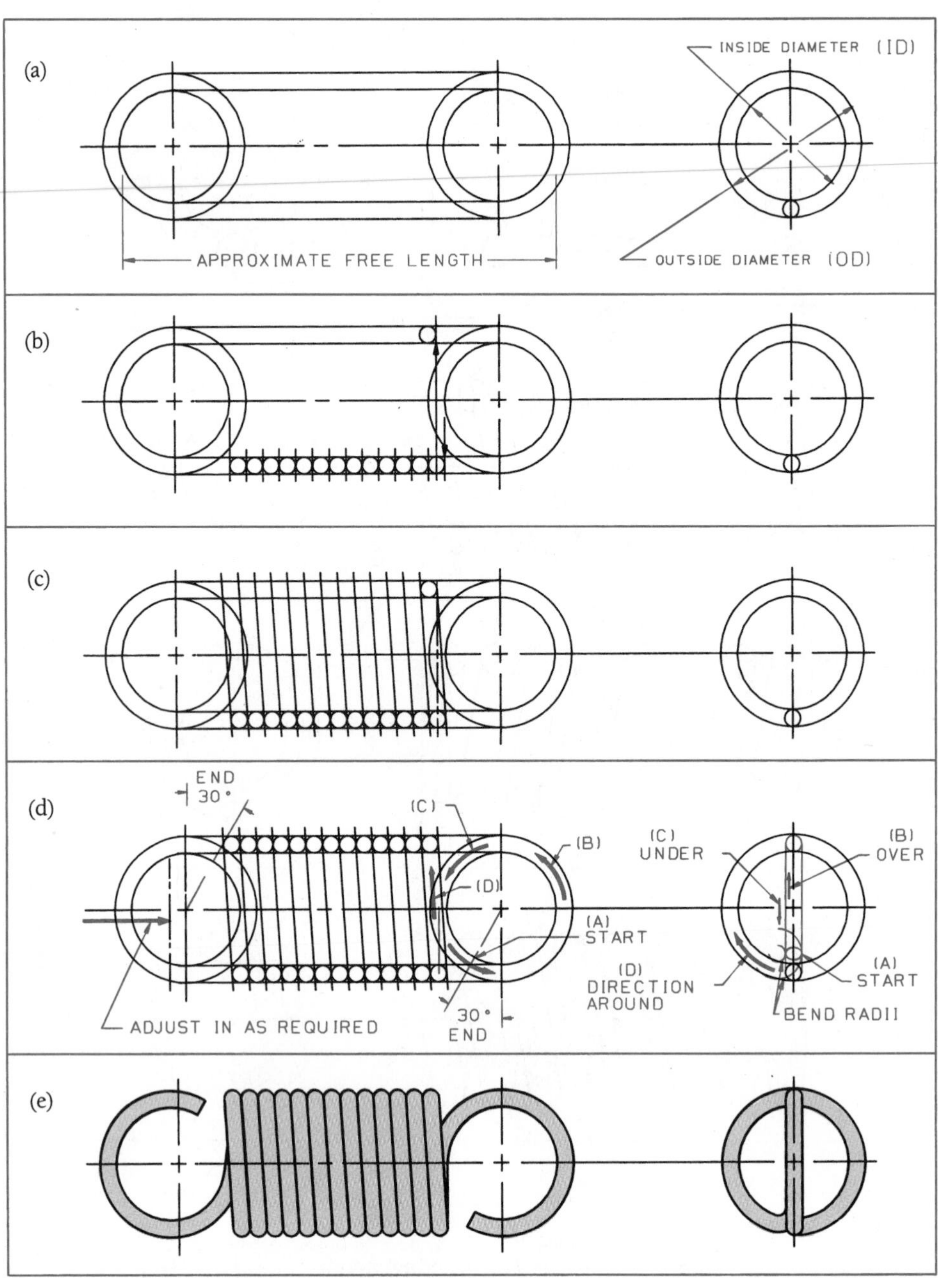

FIGURE 14.16 Drawing an Extension Spring

QUIZ

True or False

1. Spring washers should not be used in applications where weight and space are the prime considerations.
2. The solid length of a spring is the overall length of a spring in the unloaded position.
3. Usually, all coils in a torsion spring are active.
4. A Belleville spring is a coned disk spring.
5. Set is the permanent distortion of the spring when stressed beyond its elastic limits.
6. A garter spring may not be used to hold round segments together.
7. There is really only one basic type of compression spring end.
8. The force that is applied to a spring that causes deflection is known as load.

Fill in the Blanks

9. ________ , ________ , and ________ are three types of end configurations used on extension springs.
10. A ________ ________ ________ is a spring washer in the form of the frustrum of a cone.
11. A ________ is the spiral form of compression, extension, and torsion springs.
12. A ________ ________ is an open-coil helical spring that resists a compressive force along the axis.
13. The ________ ________ is the movement of a spring from its free position to maximum operating position.
14. ________ ________ absorb and store energy by resisting a pulling force.
15. A ________ ________ spring is made by winding flat spring material on itself in the form of a spiral.
16. A ________ ________ spring is a strip of flat spring material that has been wound to a given curvature so it is in the form of a tightly wound coil.

Answer the Following

17. What is the difference between the free length and the solid length of a spring?
18. What is the difference between the terms, "active number of coils" and "total number of coils"?
19. Describe the difference between a left-hand spring and a right-hand spring.
20. What is a compression spring?
21. Describe how "pitch" is defined for springs.
22. What is the difference between a helical extension spring and a garter extension spring?
23. Describe the basic function of an extension spring.
24. What is a spring washer?

EXERCISES

Exercises may be assigned as sketching or instrument projects. Transfer the given information to an "A" size sheet of .25 in. grid paper. Complete all views and solve for proper visibility, including centerlines, object lines, and hidden lines. Exercises that are not assigned by the instructor can be sketched in the text to provide practice and understanding of the preceding instructional material. Dimensions for fasteners used on exercises can be located in figures throughout the chapter and in Appendix C.

Exercise 14.1 Using detailed representation, draw the compression spring as shown. List all pertinent specifications on the drawing. The spring is steel, has a wire diameter of .250 inches, is left-hand wound, with square ends, and has eight active coils and ten total coils.

Exercise 14.2 Draw all coils for the compression spring. The spring is right-hand wound, has a wire diameter of .187 inches, with plain ends, and has a total of eighteen active coils (also eighteen total coils). List all controlling specifications.

Exercise 14.3 Complete the helical extension spring. The spring is to be right-hand wound, has a .250 inch wire diameter, and comes with round ends as shown. Draw all coils. List all specifications.

Exercise 14.4 Complete the helical torsion spring using a wire diameter of .200 inches and seventeen coils. The spring is left-hand wound. Draw all coils. List all specifications.

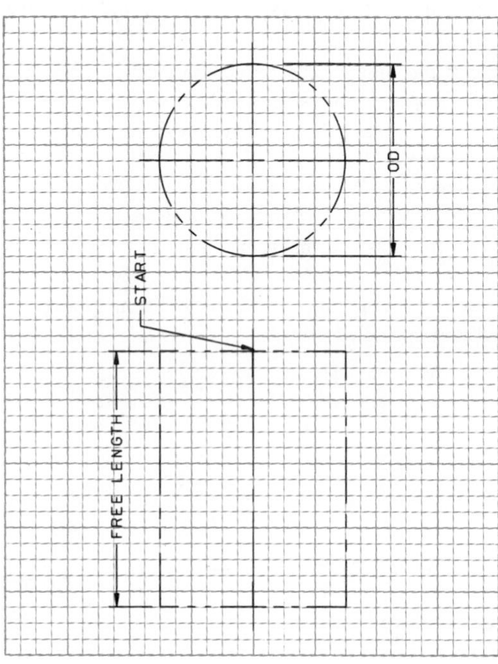

Exercise 14.1

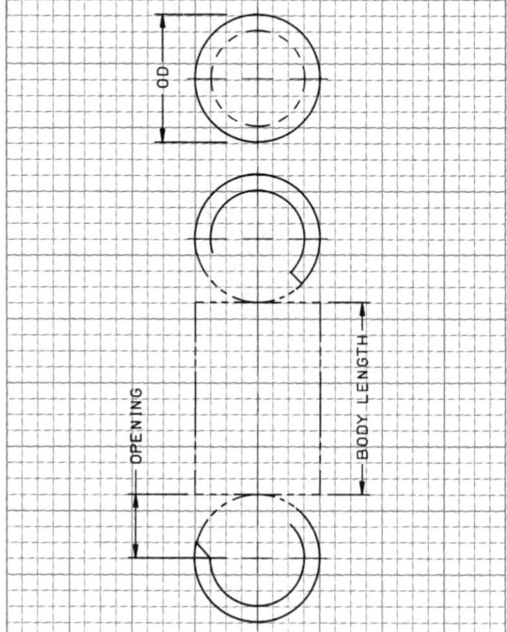

Exercise 14.3

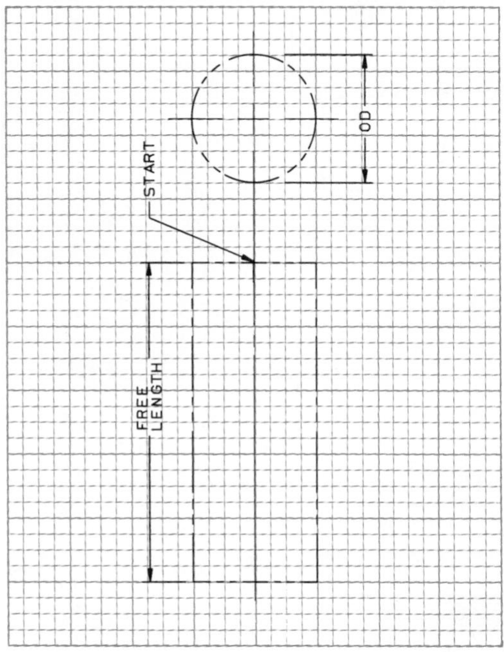

Exercise 14.2

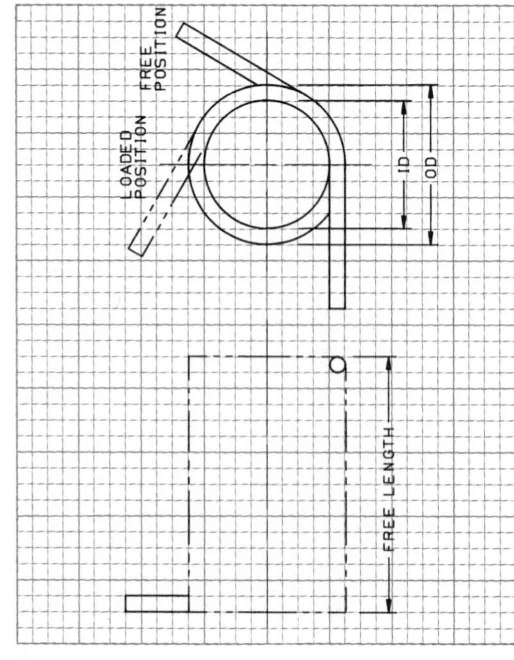

Exercise 14.4

PROBLEMS

Problem 14.1 Draw a detailed representation of a compression spring. List all specifications on the drawing. The spring is steel, has a wire diameter of .200 inches, is right-hand wound, with square ends, and has ten active coils and twelve total coils. Use the same OD.

Problem 14.2 Draw a compression spring showing five coils at each end and the remainder with phantom lines. The spring is left-hand wound, has a wire diameter of .125 inches, comes with plain ends, and has a total of twenty active coils (also twenty total coils). List all controlling specifications. Use the same OD.

Problem 14.3 Draw a helical extension spring. The spring is to be left-hand wound, has a .200 inch wire diameter, and comes with round ends. Draw all coils and list the specifications and use the same OD.

Problem 14.4 Construct a helical torsion spring with a wire diameter of .187 inches and fifteen coils. The spring is right-hand wound. Draw all coils. List all specifications. Use the same OD.

Problem 14.5 Design and detail an extension spring with a full loop over center on the right end and a long hook over center on the left end. The spring is right-hand wound and has a free length of 180 mm with a 6 mm wire size. There are fourteen total coils. The coil length is 80 mm with an O.D. of 50 mm. Show all dimensions.

Problem 14.6 Design and detail an extension spring with the following specifications:

Approximate free length = 1700 mm
Winding = Left-hand (special)
Wire size = 5 mm
O.D. = 50 mm
Ends = Full loop over center for both

Problem 14.7 Draw and dimension a compression spring with plain closed ends and a wire diameter of 10 mm. The spring will be left-hand wound, with an O.D. of 48 mm. The free length is 160 mm. Calculate the solid length. There are ten total coils (eight are active).

Problem 14.8 Design and detail a compression spring with the following specifications:

Free length = 4.00 inch
Coils = 6 total; 3 active
Wire size = .50 inch
Ends = Closed ground
O.D. = 3.75 inch
Winding = Left-hand
Solid length = (calculate)

Problem 14.9 Design and detail a compression spring with the following specifications:

Free length = 190 mm
Coils = 14 total; 12 active
Wire size = 6 mm
Ends = Plain open
O.D. = 60 mm
Winding = Right-hand
Solid length = (calculate)

Problem 14.10 Design and detail a compression spring with the following specifications:

Free length = 5.00 inch
Coils = 7 total; 5 active
Wire size = .375 inch
Ends = Ground open
O.D. = 3.00 inch
Winding = Left-hand
Solid length = (calculate)

Problem 14.11 Design and detail (draw 2 × size) a torsion spring with the following specifications:

Free length = .875 inch
Coils = 5
Wire size = .125 inch
Ends = Straight and turned to follow radial lines to center of spring and extend .375 inch from outside diameter of spring
O.D. = 1.375 inch
Winding = Left-hand

Problem 14.12 Design and detail a torsion spring with the following specifications:

Free length = 50 mm
Coils = 10
Wire size = 6 mm
Ends = As assigned by instructor
O.D. = 70 mm
Winding = Right-hand

CHAPTER 15

The Design Process

Learning Objectives

Upon completion of this chapter you will be able to accomplish the following:

1. Recognize the options of material, instrumentation, manufacturing technology, and fabrication personnel involved in the design process.
2. Interpret the criteria for product and manufacturing engineering that result in design for manufacturability.
3. Identify and define design parameters and considerations.
4. Analyze and utilize the stages involved in the design process while recognizing their flexibility.
5. Develop an understanding of critical-path scheduling and just-in-time production concepts.

15.1 Introduction

The **design process** is an organized, interactive engineering activity that results in a well-defined concept and a specific plan to turn that concept into reality. The design process is a logical and planned sequence used by an individual or a team to develop a solution to a specific problem. The stages described in this chapter are not necessarily applicable to all design situations. They can, however, be considered a guide to the design process.

Although the end product is specified in the form of drawings, computer images, sketches, and engineering specifications, design involves more than producing a drawing and having the part made. Design is an interactive process with planned steps, input from several people, and checkpoints. Each design involves choices of materials, instrumentation, manufacturing processes, and fabrication. **Design for manufacturability** (DFM) is a philosophy that integrates the manufacturability of a product with its design.

Few people understand the complexity of products or the amount of effort required to bring them to market. The automobile engine in Figure 15.1 is an example of a complex assembly. We seldom think of the complexity of the engine and the time devoted to its design. Product description and development include design, drafting, analysis, and manufacturing (Fig. 15.2).

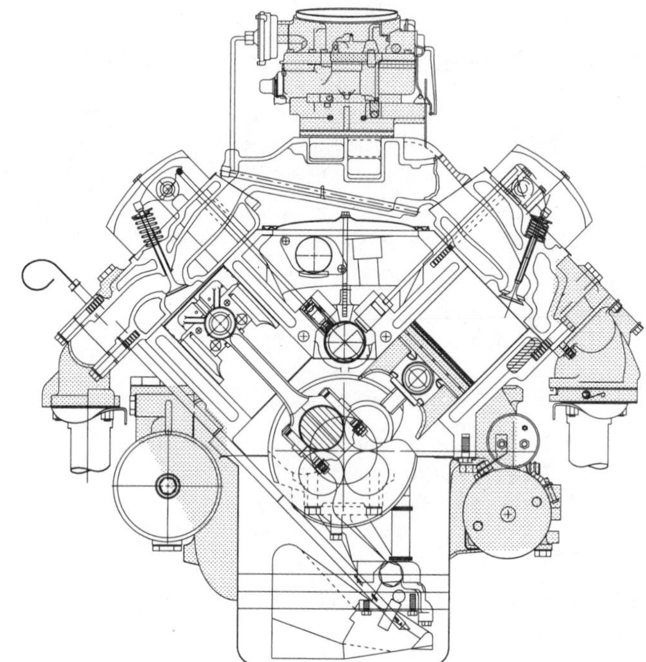

FIGURE 15.1 Automobile Engine

It is impossible to describe how to design every item. It has been said that you cannot teach design. However, a thorough presentation of design concepts leading to the understanding of the conceptualization process involved in design and mastery of the stages involved in the design process will lay a solid foundation for anyone aspiring to become a designer.

There are two main divisions of engineering design: **system design** and **product design.** Although system design is an important field, a majority of this text is devoted to product and mechanical design and drafting techniques. Therefore, this chapter is primarily a detailed analysis of the design process as it relates to product development and mechanical design. *(See Color Plates 3-5, 7, 12-20, 32-36, 46.)*

15.2 Design Engineering

The design effort encompasses both product engineering and manufacturing engineering. To design and produce a product efficiently, the manufacturability of the part must considered during design. Therefore, the product engineering and the

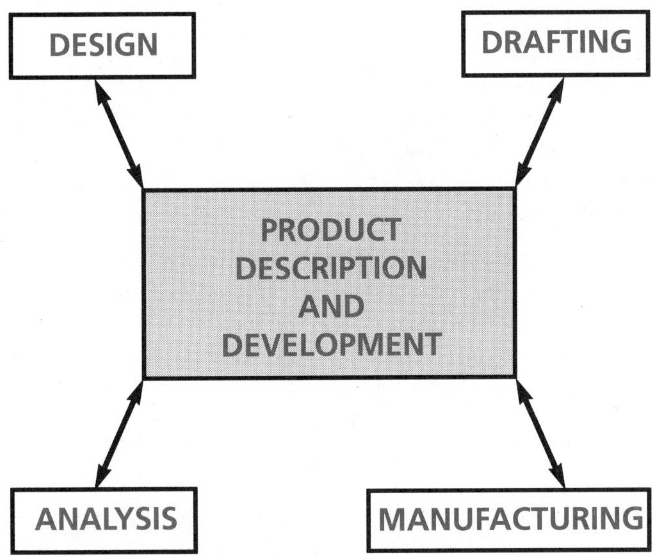

FIGURE 15.2 Product Description and Development

FIGURE 15.3 Trackman

manufacturing engineering of a successful project integrate the following activities:

Product Engineering
- Product description
- Specifications
- Models
  - Test
  - Prototype
  - Fit and function
  - Presentation
- Analysis
  - Stress/strain
  - Fatigue/corrosion
  - Movement/kinematics
  - Load-forces/dynamics
  - Heat-energy/thermodynamics
- Layout and detail drawings
- Redesign

Manufacturing Engineering
- Production method
- Costs
- Quantity
- Tooling
  - Dies
  - Tools
  - Jigs and fixtures
- Robot workcells
- Material management and movement
- Ordering
- Production planning
- Manpower requirements
- Testing
- Inspection
- Quality control
- Distribution
  - Packaging
  - Shipping
  - Storing/stacking
- Facility management

The designer must consider multiple factors and make decisions based on compromises. Seldom is a design everything the designer wanted when the project started. The true test of a successful design is if the design is functional and manufacturable.

Designs are functional when they satisfy a need and are available to the public in some form or quantity that is cost effective and profitable for the company. Before a design is accepted, it must be tested and researched thoroughly. After the development and testing of a product or mechanical design, the design data are released to the factory for production.

15.2.1 The Designer and Designing

Designers use their education and experience to invent new products, to create new systems, or to improve existing products or systems or add innovations. The process of design involves creativity and the ability to discover new solutions to existing problems or invent new products to fill a need. Being a designer has less to do with natural talent than with cultivating an eye for detail, accumulating knowledge, and gathering experience with design.

Designing is an intellectual activity for which there are no hard rules. Expertise in design is acquired through time and experience. Designers keep their minds open to new concepts and learn from co-workers, journals, magazines, and past failures. They are well informed and choose the best features from several approaches for their designs.

Product design (Fig. 15.3) of a new item involves the creation of commercially profitable, useful, or desirable devices. **Mechanical design** includes a wide range of industrial products (Fig. 15.4) and tools used in manufacturing (jigs and fixtures, dies, molds). The system designer uses existing standard parts combined in a unique functional manner to satisfy a need or an industrial requirement.

During the design process, the designer uses many different types of documents and consults with a variety of specialists (Fig. 15.5). During the design process, the designer has many things to consider:

- Geometric arrangement of the components or design configuration
- Effects of motion, forces, heat, and environment

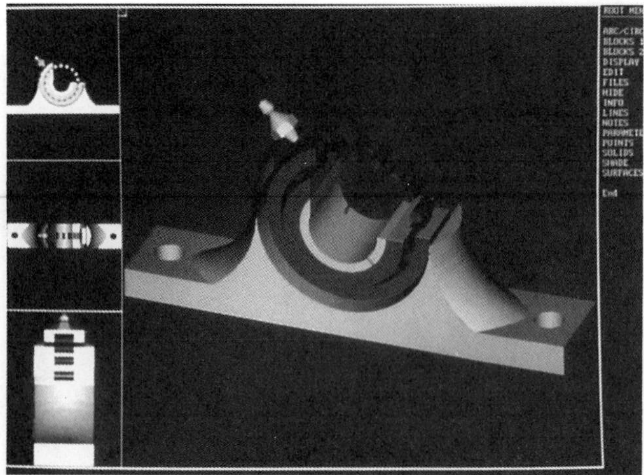

FIGURE 15.4 Bearing Design

- Human capabilities, limitations, and requirements (human factors)
- Manufacturing and production processes
- Material selection

The designer must have basic creative instincts, an inquisitive mind, and the ability to communicate. Successful designer traits include the following:

- Intuition
- Good communication skills—written, verbal, graphic
- Open mind to problem-solving
- Inquiring mind
- Understanding of fundamental principles of design
- Ability to integrate and balance several ideas and solutions
- Ability to do self-evaluation
- Visualization skills
- Mathematical skills

The space program (Fig. 15.6) is an excellent example of designers creating new designs for exploration in unique, new environments. The program required designers who were not captive to preconceived notions and were not afraid to push the frontiers of knowledge.

15.3 Product and Industrial Design

Product design is done by an industrial designer working in conjunction with engineering, manufacturing, and marketing. Products are mass produced for the consumer, educational, or industrial markets. Figure 15.7 is an example of an **industrial product.** Here, function is more important than visual appeal. The oscilloscope is sold to industry, military, and educational markets. The gas generator (Fig. 15.8) is a product for both individual consumer and industrial markets.

The rollerball pointing device used for computer input (Fig. 15.3) is an example of a consumer product.

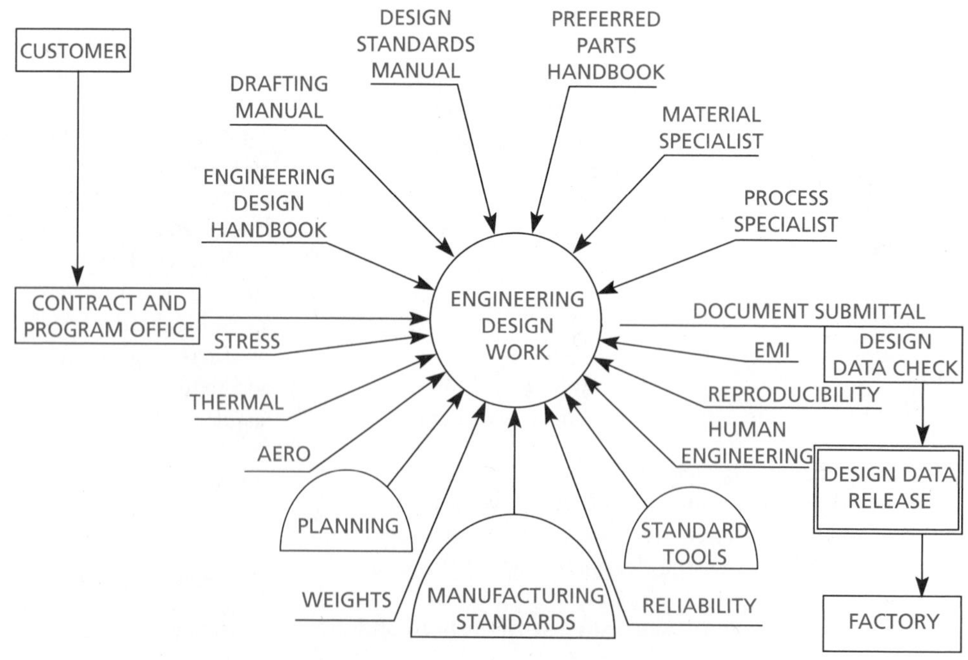

FIGURE 15.5 Engineering Design Flow Diagram

FIGURE 15.6 Apollo Lunar Surface Experiment

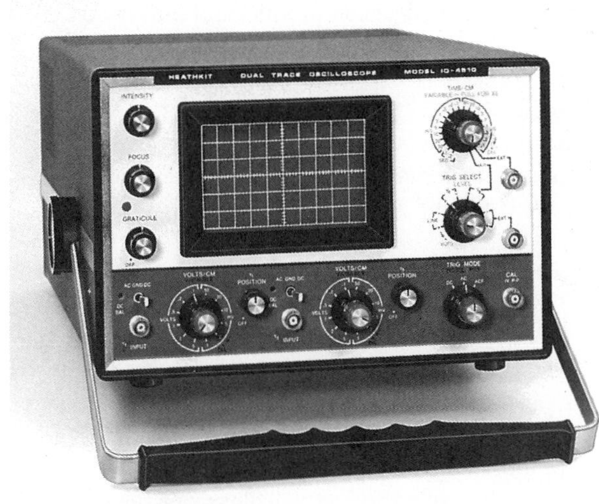

FIGURE 15.7 Oscilloscope

FIGURE 15.8 10-kW Gas Generator Mockup

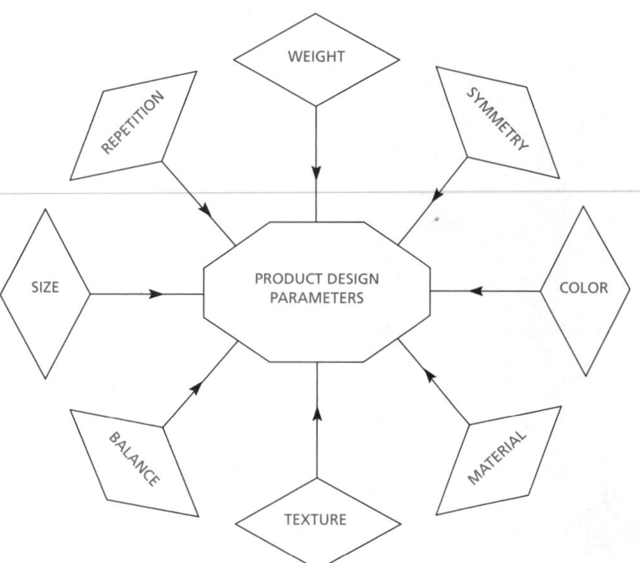

FIGURE 15.9 **Product Design Parameters**

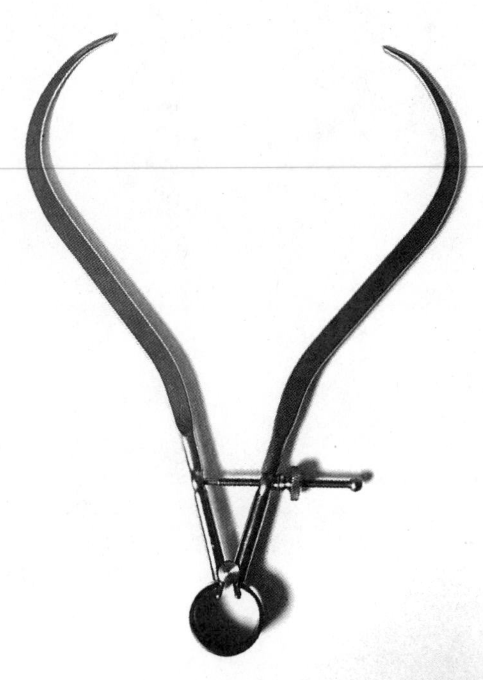

FIGURE 15.10 **Calipers**

15.3.1 Product Design Parameters

Product design parameters (Fig. 15.9) determine manufacturing and production methods:

- Weight
- Material
- Symmetry
- Size
- Texture
- Color
- Repetition
- Balance

Each design parameter affects the other parameters. For example, the weight of the product is influenced by the material selection and its size, the color is influenced by the material and the material is influenced by the texture. If, for example, the texture of the product is more important than all the other factors, the texture determines the material choice.

The shape of the product must be considered with regard to its symmetry, proportion, repetition, and balance. Geometric proportions that are pleasing to the eye and appear balanced are used to provide a repetitive shape to the part. A reliable, completely functional, long-lasting product will not necessarily sell if it is poorly proportioned and unappealing in color and shape. A good designer determines the proper mix of these factors based on their relative importance to the project.

15.3.2 Design Considerations

A careful and systematic overview of **design considerations** will lead to a successful product. The calipers (Fig. 15.10) were designed for a very specific use. Calipers are used by designers, engineers, machinists, and others involved in manufacturing. Calipers must be accurate, lightweight, sturdy, simple to use, easy to handle, unbreakable, have a long life, and must not be affected by the environment. The **function** of this product is the foremost concern. The function determines the material choice; nonmetals would not allow for the accuracy required of the finished product. The requirements for corrosion resistance and strength limit the material choices.

Design considerations include the following:

- Function
- Constraints
- Materials
- Appearance
- Environmental effects on product and product on environment
- Product life
- Reliability
- Safety requirements
- Standardization and interchangeability of components
- Maintenance and service requirements
- Costs

Function The required **functions** of a product should be defined early in the design process. Unnecessary functions or features should be eliminated if they do not increase the value of the product. The designer determines if the product could be used for more than one purpose or if the product could be less complicated.

The function of a product will always be the primary consideration. If the product does not function properly, it is unmarketable. A basic rule is to *minimize the complexity of a design*. The simpler the product, the easier and cheaper it is to manufacture. Strength requirements are influenced by function too. The designer should try to reduce rotational stress, bending, and complex motion. Interferences of moving parts are also important aspects of function.

Constraints *A basic rule of design is design for simplicity within the constraints.* First, define the **constraints** of the project. Are the size, weight, and volume of the product adequate? The projected cost of the product is also a constraint. A "widget" that performs 25 different functions, but costs five times more than the consumer will pay, is not a well-designed product.

The size of a product is influenced by the user of the item. For example, a child's hand is smaller than an adult's hand. If the item will be used by only one sex, then research at the marketing level defines many of the constraints.

Materials The selection of the **material** used for the design is a complex process. The following is a partial list of *materials properties* that must be considered during design:

Strength A measure of a material's capacity to resist different types of forces.

Elasticity The stiffness of a material and its capacity to deflect under load.

Ductility Ability of a metal to deform before fracturing.

Fatigue When a material fails after many load cycles.

Bearing characteristics The suitability of a material to be used as an element resting on another part and in motion.

Hardness and brittleness A characteristic of a material to shatter before deforming.

Damping Ability of a material to dissipate energy caused by vibration.

Temperature Effective range where the material properties will be suitable.

Toughness Ability of a material to absorb energy before fracturing.

Resilience or elasticity Ability of a material to store energy when permanently deforming.

Wearing Ability of a material to withstand rubbing motion causing removal of material.

Corrosion Ability of a material to resist deterioration caused by a reaction to the environment.

Toxicity The possibility of producing a poisonous effect. Material safety data sheets (MSDS) are now required documentation for products.

Machinability The relative ability of a material to be machined.

Forgeability Ability of a material to be forged.

Formability Ability of a material to be formed.

Castability Ability of a material to be cast.

Weldability Ease with which a material can be welded.

The material that best suits the design and manufacturing requirements is sometimes a compromise. **Availability** is a factor in material selection. An exotic material may be the best choice, but if it is hard to procure, another material may be better for production.

Appearance **Appearance** is one of the most important considerations in consumer product design. Sometimes, for the successful marketing of a product, the function and other characteristics will be less important than the appearance.

Environment The effect the product has on the environment is more important now than at any other time in history. Industrial history is full of how profit took precedence over the environment.

The effects of the environment on the product are also a concern for the designer. For example, consider the environmental effects on the tractor in Figure 15.11. Operating conditions such as temperature variation, dust and dirt, vibration, and moisture level must be considered during design. *Designing any product starts with an understanding of when, how, by whom, and where a product is to be used.*

The oscilloscope (Fig. 15.7) was designed to be carried. The electronics inside must be shielded properly by the packaging. Heat is allowed to escape from the package by vents in the sheet metal on the sides of the package.

The space shuttle was designed to withstand heat (thousand of degrees) while reentering the atmosphere. The shuttle tiles are made from ceramics that can withstand this type of repeated thermal shock.

Product Life The **operation life** of a product is its time of operation before it fails. The **shelf life** of a product is the period of time it can be in storage and still operate correctly. The designer influences the life of a product by choice of

FIGURE 15.11 Case Tractor

material, features, manufacturing methods, and assembly methods. Often, it is the intention of the designer to have the product wear out after a given life. The tractor in Figure 15.11 is an example of a product that is designed to be maintained over a long period of use.

The Viking lander (Fig. 15.12) had to have extremely long life in order to operate effectively in a hostile environment. The original parts had to be reliable and provide error-free operation for an extended period while operating in a severe environment.

Reliability The reliability of a product is influenced by its complexity and sensitivity to the environment. **Reliability** is a product's ability to function properly each time it operates. Each product is designed to have adequate reliability to last the average expected life. The higher the quality of components, the longer its life and the higher its reliability. Parts designed for the military and for space exploration must have the highest reliability.

Safety **Safety** involves the safe, correct performance of a product in service. Some dangerous products are designed as fail-safe so as to prevent any injury or harm to the environment. **Fail-safe** means that a product incorporates features for automatically counteracting the effect of an anticipated source of breakdown. Products can be dangerous when they do not perform correctly, when they are operated incorrectly, or when there is insufficient protection for the operator.

Standardization and Interchangeability By using standard, off-the-shelf items, the cost of the product can be reduced. Systems design is the assembly of standard components in unique configurations to accomplish a specific task such as the production of power. The use of standard parts and previously designed parts is important to DFM because it saves time and reduces costs.

The ability of a unit to use similar parts or have different components that can be substituted is called **interchangeability.** This reduces the production costs and eliminates possible shortage problems and possible delays in production.

Maintenance *Design for simplicity in disassembly and maintenance. Design for recycleability* whenever the product is to be a throwaway.

Products that require repair, service, and maintenance are designed to be disassembled at specific intervals of operation or stages of wear. *Design for disassembly* by providing clearance for tools and hands during maintenance and consider repair procedures. For example, the valve in Figure 15.13 is designed to allow removal of the handwheel, stem, bonnet, and disk to replace the composition ring. This service is completed without disassembling the pipeline.

Costs The number of parts produced influence the **cost** of the product. In general, the greater the quantity manufactured, the lower the overall unit cost. If you purchase each part separately for an automobile, it costs about 100 times as much as you paid for it new. Most consumer products can be made much cheaper by producing them in large quantities. Industry must hold down costs while increasing quality.

The design, production, and marketing costs of a product must be estimated early in the design process to bring the product to market and make a fair return on the investment. *Designing in quality instead of inspecting out problems will ensure a greater profit and a better product.*

FIGURE 15.12 **Viking Lander**

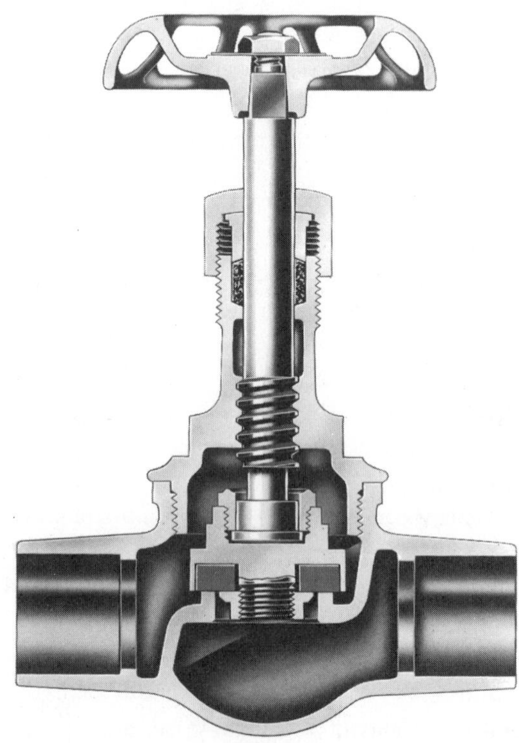

FIGURE 15.13 **Composition Disk Globe Valve**

FIGURE 15.14 **Product Design Optimization**

15.3.3 Product Design Optimization

The optimum product is created when all factors are properly analyzed and balanced. Figure 15.14 shows the 12 major influences that, when properly considered, will yield a superior product. This process is called **product optimization.** Each of the 12 factors affects the success of the product. A good designer factors in each and develops an optimum product.

15.3.4 The Design Tree

The **design tree** (Fig. 15.15) is useful in analyzing a particular project and in illustrating the decision process during the crucial initial design or redesigning phase of the project. Although the physical configuration, materials, manufacturing methods, assembly procedures, and equipment costs may be altered and balanced between selections, there are stages in the design process of a new product in which equal function, quality, and performance levels can be obtained. The goal is to maximize the advantages of the product while maintaining the critical specifications.

The design tree starts with the trunk (product idea). The two main branches are the material selection and the physical configuration. The material selection limb splits at metal and nonmetal. What grades and types of metals or nonmetals should be considered? The physical shape of the product is determined by selecting its size, shape, and features. The branching process can continue to include possible modifications and enhancements.

15.3.5 Product Design Example

A sphere is considered the perfect geometric shape. However, if a sphere is penetrated or truncated by some other geometric shape, it appears either ellipsoidal or flattened, as Figure 15.16(a) shows.

Figure 15.16(b) is an enlargement of the sphere in outline form. Notice that the surface seems flattened at those places where the penetrating cylinder passes through it. How can the sphere be made to look like a sphere instead of an ellipsoid? The dashed lines show the sphere as it would be without the penetrating cylinder.

Designers overcome this optical illusion by subtle changes in shape. To create the illusion that the flattened sides are not flattened, the designer first changes the shape of the sphere around the flattened area. Four radii are layed out. The designer "builds up" the sphere and thus negates the flattened look.

When a sphere is truncated, as in this design for a coffee percolator [Fig.15.16(b)], the same process is used. The drawing on the left shows the squat, flattened look the percolator would have if the designer did not modify the basic shape. On the right, the designer subtly uses the ellipsoidal shape shown here by the solid lines to give the finished product a spherical shape.

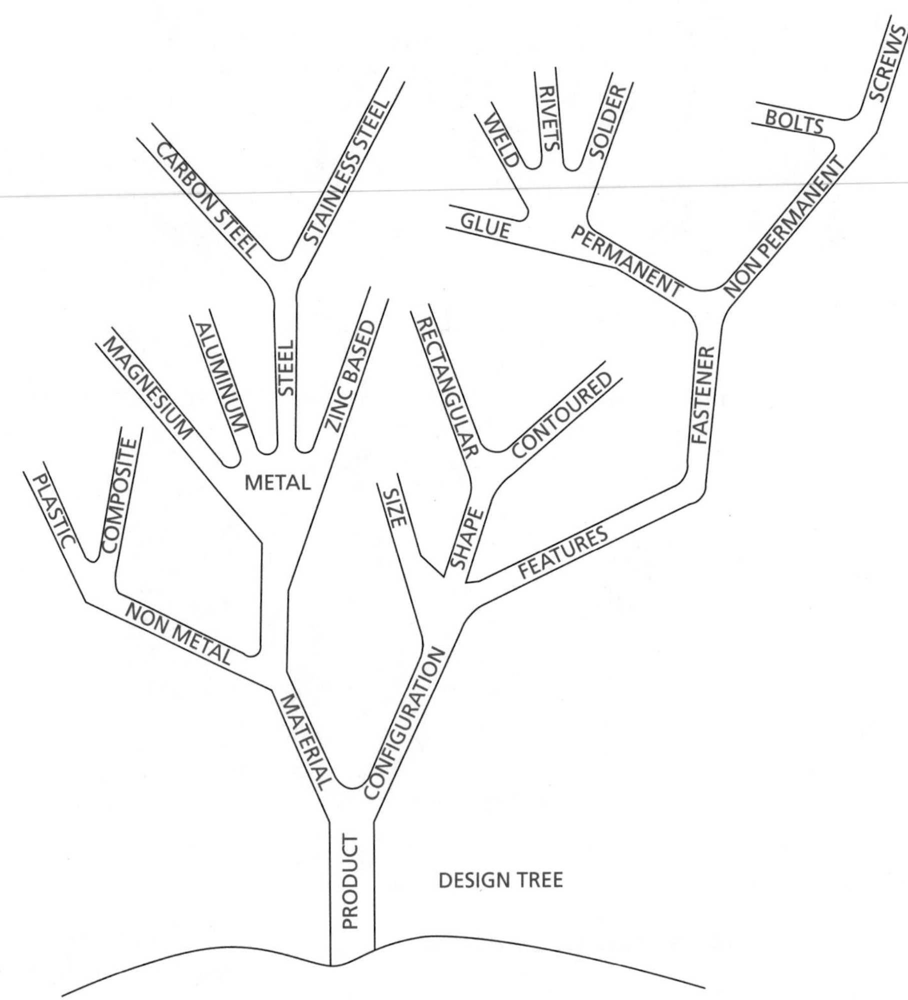

FIGURE 15.15 Design Tree

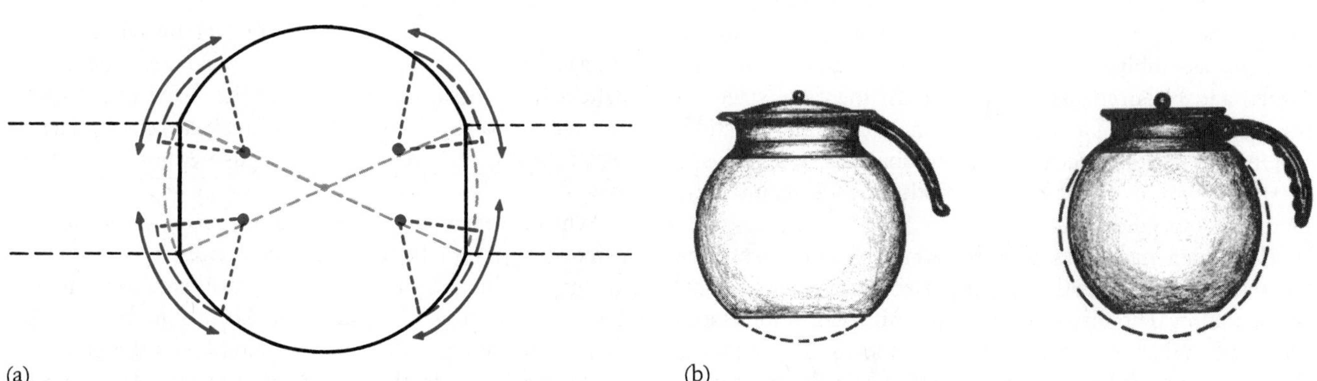

(a) (b)

FIGURE 15.16 **Product Design** (a) The intersected sphere in design (b) Spheres in product design

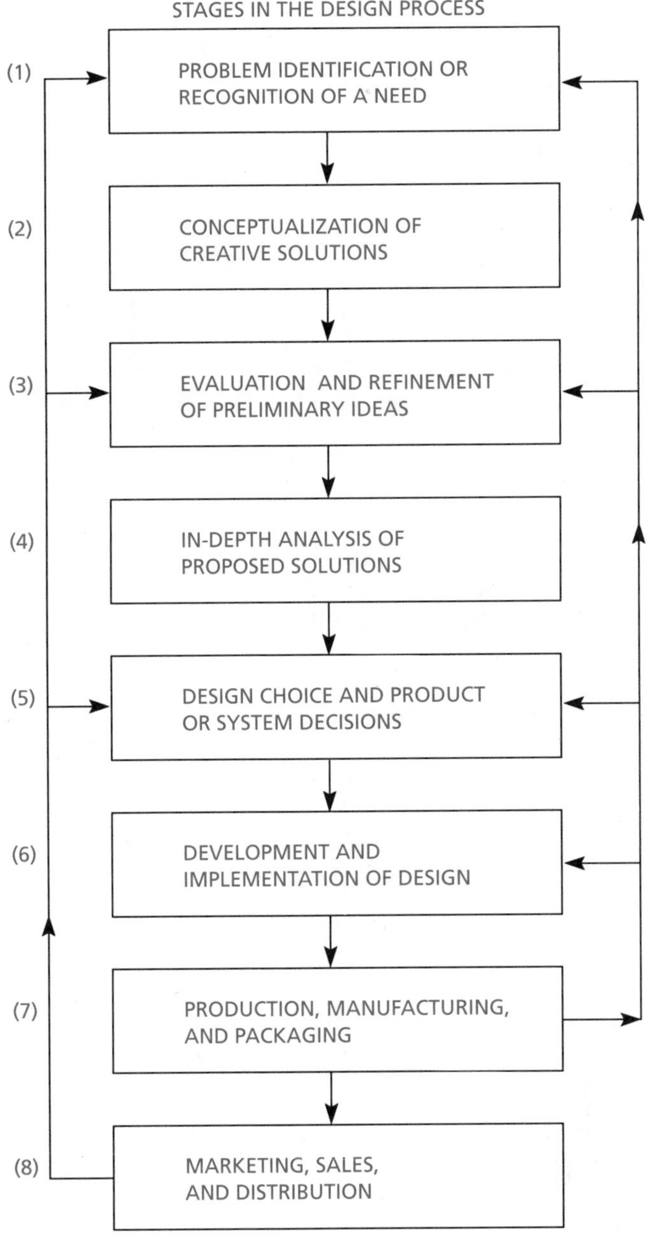

FIGURE 15.17 **Stages in the Design Process**

15.4 The Design Process

The **design process** begins when a customer expresses a need for a product. For simplicity, the design process has been separated into eight individual stages (Fig. 15.17). The actual design process is more flexible and is not as linear as described here. Design for manufacturability means that the eight stages are not separate, but are an integrated whole. Many of the eight stages represented here are performed simultaneously by the design/manufacturing team. The stages do not always flow in a straight line. Sometimes, the stages are different or there are more or fewer stages. Manufacturing always comes after the engineering and analysis, but manufacturing decisions are integrated into the preliminary design and engineering stage.

The flow diagram of Figure 15.17 shows various stages in the design process. The general flow is from top to bottom as the arrows beneath each box indicate. The flow lines and arrows on the left side of the illustration flow from the bottom up. For instance, the marketing and sales division is consulted and influences the problem identification (1), evaluation (3), and the design choice (5). On the right side of the figure the flow lines and arrows also point upward. Here, manufacturing (7) affects almost every other stage in the process.

15.4.1 Problem Identification or Recognition of a Need

The design process starts with the **identification** of a problem (Fig. 15.18), an observed need, or a potential new idea for a product or system. This stage requires that the designer or design team be thoroughly aquainted with the problem or need. You should attempt to answer the following at this stage of the project:

- *Who* needs it?
- *What* is needed?
- *Where* is it needed?
- *When* is it needed?
- *Why* is it needed?
- *How many* are needed?

History, Background, and Existing Information The **background** of the project is presented to the design team by a concerned party: the company management, an outside client, or a company inventor who has a new idea. The marketing department may be called on to do a survey on the potential for a particular product. For example, a computer company that has a well-received product line servicing the

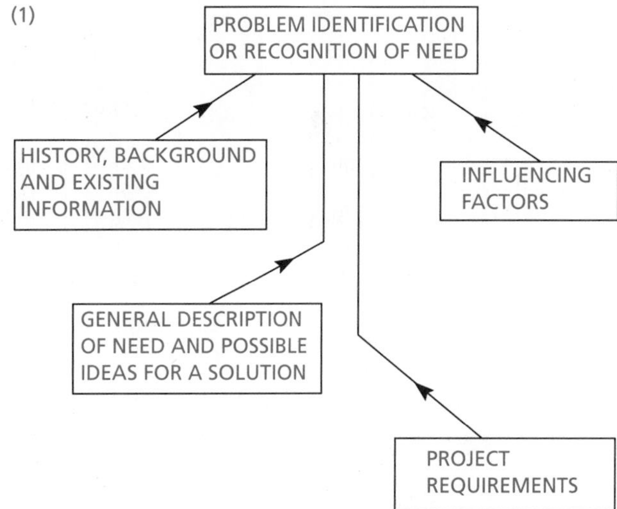

FIGURE 15.18 **Stage 1 in the Design Process**

(a)

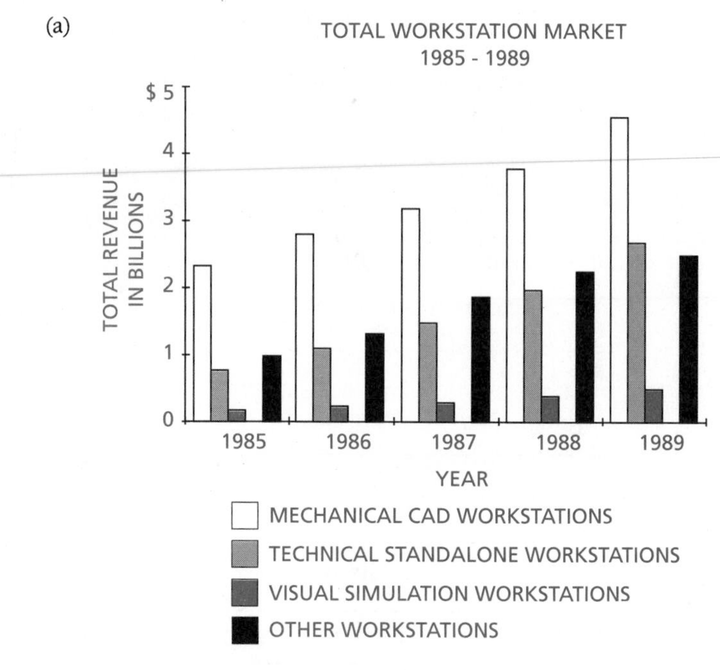

(b)

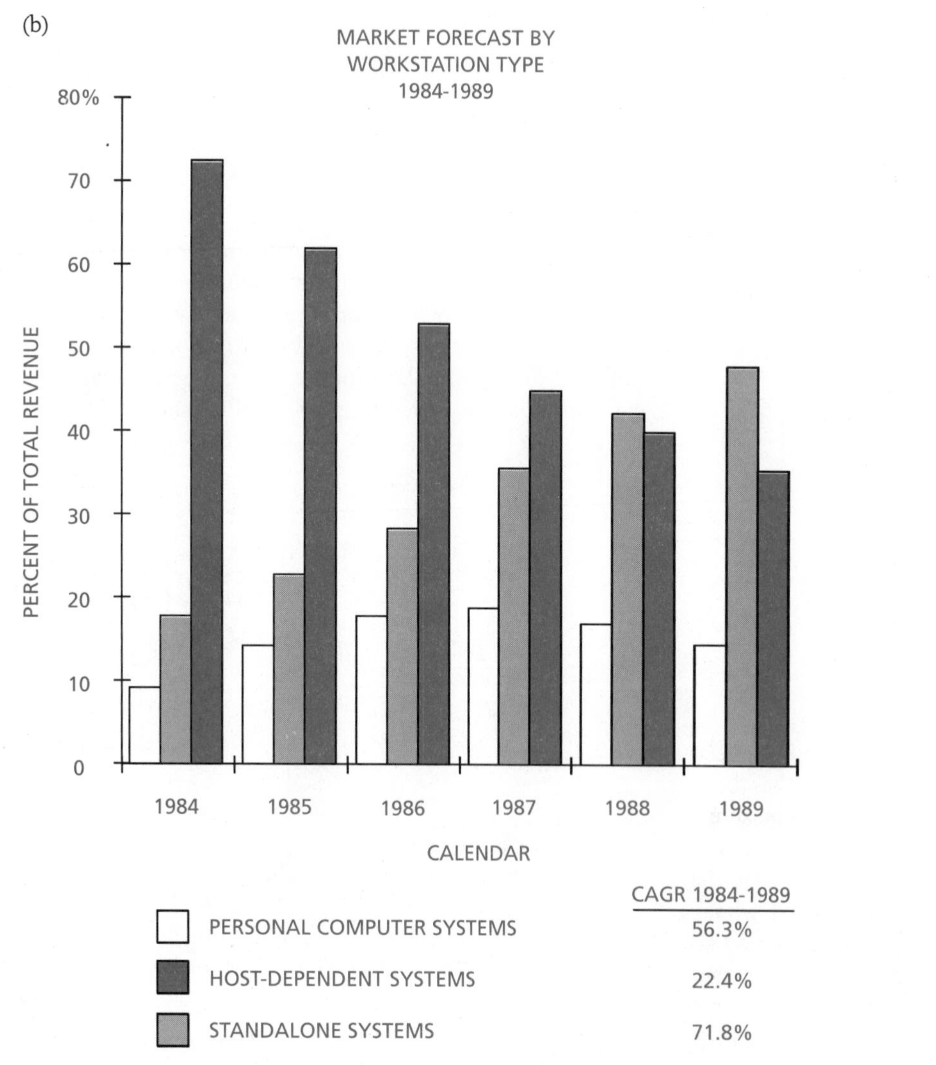

FIGURE 15.19 **Example of a Market Analysis** (a) Graph depicting workstation market data (b) Market forecast data in graph form

private sector with personal computers wishes to enter the engineering workstation market. Surveys may be needed to determine the total sales of computers in that sector and create a forecast for the future. The company's management and the design team then discuss the potential for their firm to be successful in this market. Figure 15.19(a) is an example of an analysis of the past sales in the engineering workstation market. The forecast for workstation type is shown in Figure 15.19(b).

A number of questions can be asked at this stage of the process:

- What exists now?
- How was the problem faced in the past?
- Is this a new problem?

The answers to these questions may not be complete at this stage and, in fact, they may create more questions.

FIGURE 15.20 48″ Gate Valve for Transalaska Pipeline

General Description of a Need and Possible Ideas for a Solution The **general description** and possible solutions to the problem come from a series of meetings conducted with the design team. For example, a new product line will affect manufacturing equipment, facility space, manpower requirements, shipping, and storage areas of the firm. The size of the product alone may necessitate a total retooling of the facility. The job may also be a one-time opportunity and have little continued sales after the project is complete. The economics of the project is of primary concern. How will it affect existing product lines? Is it worth the risk? Will the company turn a profit for its effort? Will the company be able to sustain a continued presence in this market after the project is complete?

Influencing Factors All products and systems have an effect on the users of the product or system. The cost of a project determines its feasibility. All factors that may influence the total cost and the economic feasibility of the design must be considered before the project is initiated.

Systems design is influenced by the environmental constraints imposed on it by the government and special interest groups. The design of a power plant, chemical facility, hydroelectric plant, bridge, housing complex, or building is defined by the acceptable effect it has on the environment. The valve in Figure 15.20 is on the Alaskan pipeline (Fig. 15.21). Environmental impact reports were a major part of the design effort and were needed to convince the public that the pipeline was feasible and safe. The conditions under which the pipeline and valves operate stretched the limits of pipeline technology.

The economics of a particular solution must be understood. Can the product or system make money? Is the existing budget adequate? Will the project involve new markets or replace an existing one? The marketing and manufacturing departments have considerable input that must be integrated into the total economic analysis.

FIGURE 15.21 Gate Valve Being Installed on Transalaska Pipeline

Project Requirements Basic parameters are identified at this stage. All ideas and suggestions are recorded as notes and rough sketches. The project's size, shape, color, material, and general configuration are discussed. No decisions are made at this point. If the product is to be efficiently produced, design for manufacturability must be integrated into the project from the onset.

15.4.2 Conceptualization of Creative Solutions

Before the design process goes any further, **creative possibilities** for a solution to the design problem should be purposely investigated. Background information and research of pertinent data help the designer see the range of answers to the design problem.

This stage (Fig. 15.22) includes researching every available source of information. Each of the design elements listed in Stage 1 is analyzed thoroughly. Even at this stage, there should be no attempt to find a final solution. Data from outside sources are integrated into the design process during this stage. Former solutions to the same or a similar problem can be discussed and expanded.

The proper research of all existing information on the subject is very important. Since the research process helps build a professional "database" that can be tapped for other projects, a new or inexperienced designer benefits from any existing information. Oddly enough, it is not the lack of background sources but the overabundance of information that is a problem. The designer must differentiate between what is useful and important and what is nonessential. The following list provides some sources for acquiring information:

- Textbooks
- Periodicals, technical magazines
- Library research
- Engineering standards
- Technical reports
- Published papers presented at conferences
- Manufacturing specifications
- Catalogs of parts
- Patents
- Handbooks
- Previous designs in the company
- Coworkers , other designers and engineers

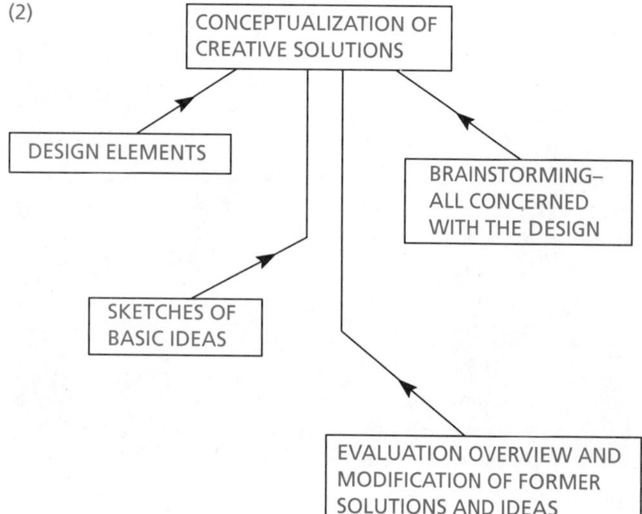

FIGURE 15.22 Stage 2 in the Design Process

Design Elements The **elements of design** must be identified in this stage of the project. The identification and classification of design elements help to clarify and divide important elements from minor concerns. Listing the design elements helps to clear any misconceptions regarding the project. The following is a list of steps to take to help identify the design elements:

1. Define the *basic design problems* relevant to the solution.
2. Define the *secondary design problems* that are not the designer's concern but must be solved.
3. Identify *perceived problems* that are not really important.
4. Identify *obstructions* to the design.
5. Find and discuss all *hidden difficulties.*
6. Scrutinize any hindrances to the design that are really not important.

Sketches and Layout of Basic Ideas Although a few rough sketches may have been made during Stage 1, more developed sketches and pictorial layouts (Fig. 15.23) are used to define any preliminary ideas. All notes and preliminary sketches should be kept on file. All those concerned with the project should bring the notes and sketches to the next meeting at which a short brainstorming session may introduce creative solutions to the problem.

Brainstorming **Brainstorming** is a group problem-solving technique that elicits a spontaneous contribution of ideas from all members of the group. No idea is rejected at this point, and all members of the group are considered equal. Ideas are not explored in depth. All suggestions are recorded and used later. Creative solutions are given as much merit as practical or obvious solutions. Multiple products or variations of one product or solution should be thoroughly investigated before the next stage.

Review and Modification After the brainstorming session, all notes, sketches, surveys, marketing analysis, and research data are reviewed. Any ideas that show no merit are filed now. Many possible answers to the problem are still considered, but a basic or general idea of the direction of the project is sought.

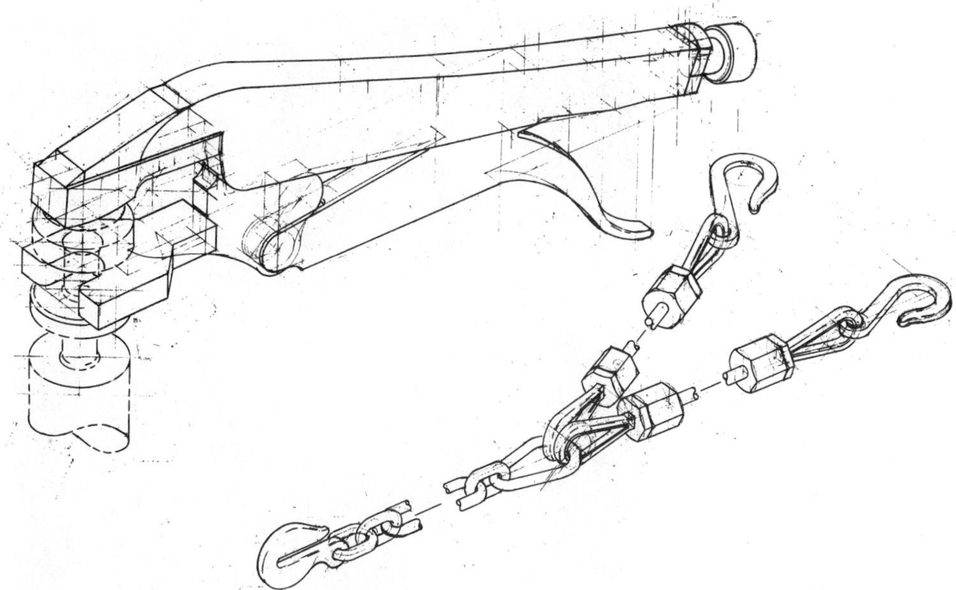

FIGURE 15.23 Pictorial Layout

15.4.3 Evaluation and Refinement of Ideas

The **evaluation** of possible solutions and their **refinement** into an end product (Fig. 15.24) is done in this stage. Refinement of a design is more restrictive at this point. More than one solution is still pursued, but the basic parameters of the project have been used to control the breadth of the design effort. An analysis of the project is completed at this point.

Creative Choices and Alternative Size and Shape Before a formal proposal is formulated, a number of possibilities for the project should be sketched and evaluated to determine size and shape. The basic parameters determined here help define the engineering and scientific analysis needed and what must be understood before the final decision on the design is made.

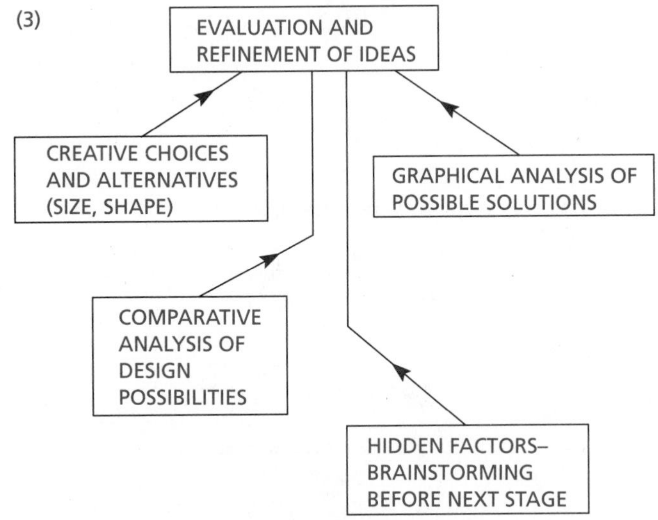

(3)

EVALUATION AND REFINEMENT OF IDEAS

CREATIVE CHOICES AND ALTERNATIVES (SIZE, SHAPE)

GRAPHICAL ANALYSIS OF POSSIBLE SOLUTIONS

COMPARATIVE ANALYSIS OF DESIGN POSSIBILITIES

HIDDEN FACTORS–BRAINSTORMING BEFORE NEXT STAGE

FIGURE 15.24 Stage 3 in the Design Process

Comparative Analysis of Design Possibilities To be successful, the analysis of a problem must include both its requirements and limitations. Any analysis of the data gathered to this time must include the possibility of a compromise solution to the problem. From this analysis, a decision must be made to proceed with any graphical analysis that may be needed.

Graphical Analysis of Possible Solutions **Engineering analysis** includes the use of graphics to define a number of possible solutions and analyze them. Figure 15.25 is an example of descriptive geometry being used to find a clearance between a pipeline and a fixed point in space.

Human Engineering and Graphical Analysis Product design includes consideration of **human engineering** requirements. Human engineering involves analyzing how people are affected by the performance of specific tasks and consideration of the man-machine interface. The first concern is referred to as **ergonomics** and the second as **human factors.**

Ergonomics is the study of physiological responses to physically demanding work; environmental stress caused by temperature, noise, and lighting; motor skills for assembly; and visual-monitoring tasks. Human factors is the modeling of the human body in a work-related setting. Human factors data are useful in the design of a factory workstation that provides a comfortable environment, thereby increasing work output and decreasing job stress. Three views of a man at work are shown in Figure 15.26. The seating and standing heights of a typical male, the comfortable reaches from both positions, and the maximum and minimum working area are defined graphically.

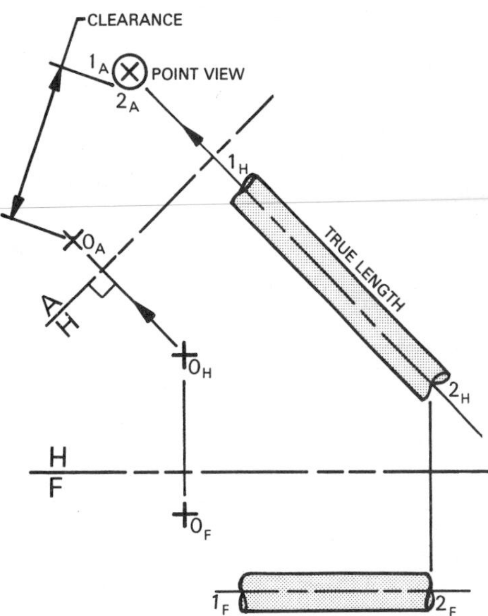

FIGURE 15.25 Descriptive Geometry Problem Solving for the Shortest Distance Between Pipe and a Fixed Point

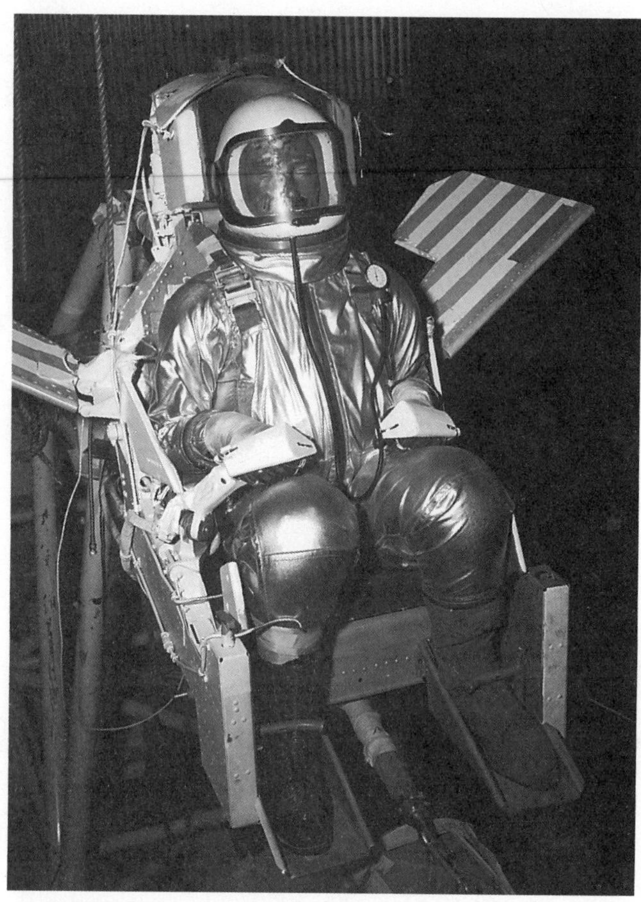

FIGURE 15.27 Human Factors "Dummy"

The interaction of workers with their tools, equipment, or workstation is the focus of human engineering design. The space program spawned the first intensive study of human physiology of man in the healthy state. Many of the tasks associated with space travel required intensive study of adverse environments. The restrictive work environment of a spaceship required designs incorporating the findings of human factors and ergonomic research.

The human factors dummy in Figure 15.27 was used to study and design an ejection seat for an airplane. The dummy is attired in a pressurization suit and is fitted into an ejection seat for engineering tests. Notice the foot clamps, arm guards,

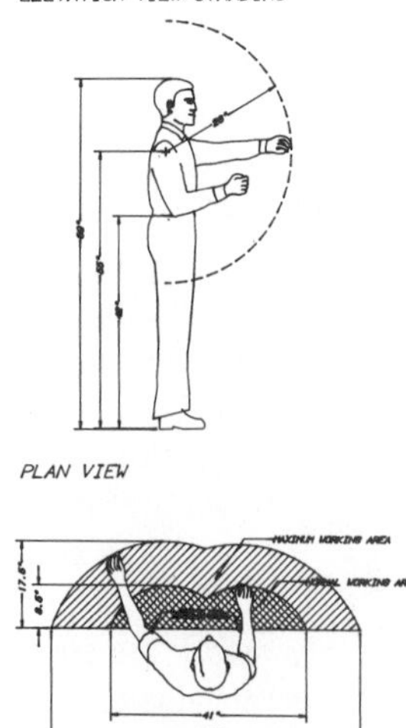

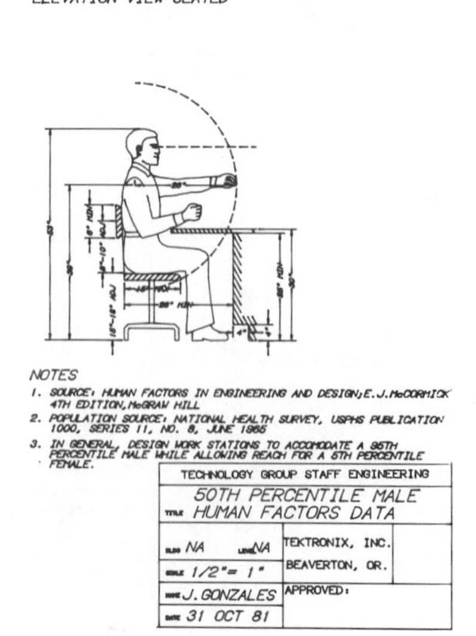

FIGURE 15.26 Male Human Factors Data

and stabilizing fins on the seat. The seat was designed by engineers to have a stable supersonic ejection with maximum protection for the pilot.

Typically, the operation of a system or product must incorporate the following:

1. Minimize the possibility of injury caused by the improper use of the product or system. Designs must incorporate safety features that make normal usage error-free. It is important to anticipate misuses of the product.
2. The design should be as efficient as possible. Limiting user fatigue and stress due to repetition is essential to proper design.
3. Systems or products should be designed with physical attractiveness, operation ease, and error-free operation.
4. The product or system should be designed with a positive, efficient, and functional user interface.
5. Products and systems must be designed to prevent catastrophic failures and must fail in a relatively safe mode at the end of their useful life.

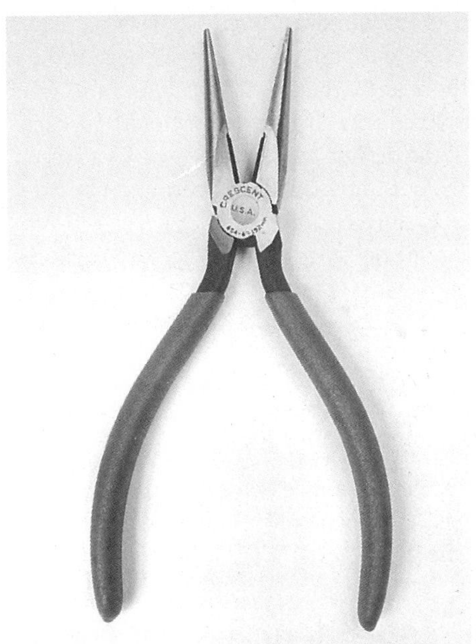

FIGURE 15.28 Needle-Nose Pliers

Product Design and Human Factors The mechanical factors, anthropometric (human) dimensions, anatomical considerations (body and limb rotational and movement characteristics), ergonomic factors, and the work environment must also be considered in the design of consumer products. Hand tools must be designed to be strong, functional, easy to carry, safe to operate, easy to store, compact, insulated from electrical shock, and slip resistant.

The design of the needle-nose pliers in Figure 15.28 required the designer to align the center of gravity of the pliers with the grasping hand so that the user does not have to overcome rotational movement or torque of the tool. The tool handle is oriented so that the user's wrist remains in the most comfortable and natural position while applying force. The sketches in Figure 15.29 show variations in designs for handle orientation. The handle must be long enough so that the user's grip includes all fingers and provides proper leverage during operation. The maximum handle spread must not be so great that someone with a small hand could not fully open the tool's jaws during operation.

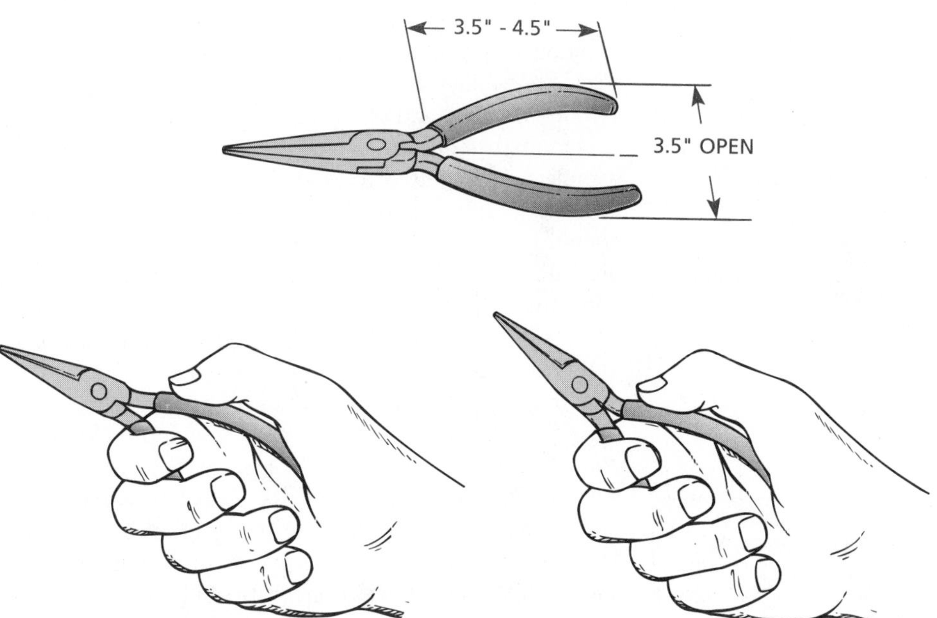

FIGURE 15.29 Design Studies of Pliers Handles

Hidden Factors What possible factors may have been overlooked in the design? Are any aspects of the design suspect? If these and other pertinent questions are not satisfactorily answered, the design team should go back to the beginning of the design process and review each step. Solutions to any problems must be solved here and not later when they could prove more costly.

A variety of questions should be openly discussed at this point:

- What could go wrong?
- Will the product work?
- Will it sell?
- Will the company lose money?

Business managers might determine that although the product will sell, the company, with its current assets, cannot produce the item.

15.4.4 In-Depth Analysis of Proposed Solutions

The **analysis and evaluation** (Fig. 15.30) of possible design solutions is normally done by a thorough investigation of the data. The use of graphs, charts, and diagrams can greatly improve this analysis and help in communicating the data to others involved in the project. Data can be categorized into three divisions: **survey data**, data gathered by the marketing department evaluating possible acceptance by the public; **design data**, data gathered by analyzing the performance characteristics of the test model; and **comparison data**, data used to balance two or more design solutions against each other based on material, manufacturability, or exclusive design features.

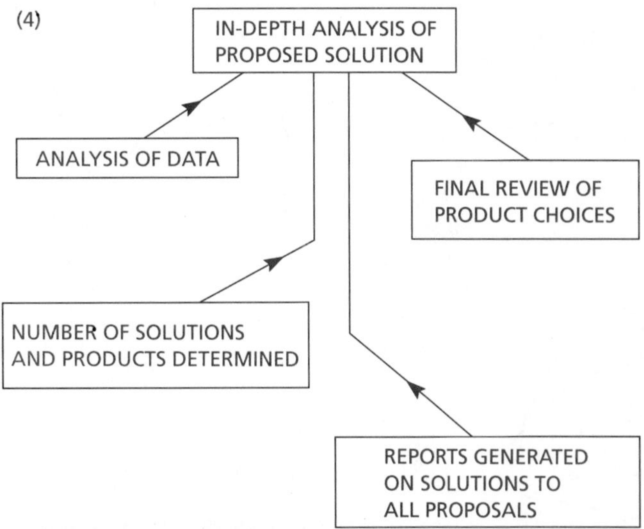

FIGURE 15.30 Stage 4 in the Design Process

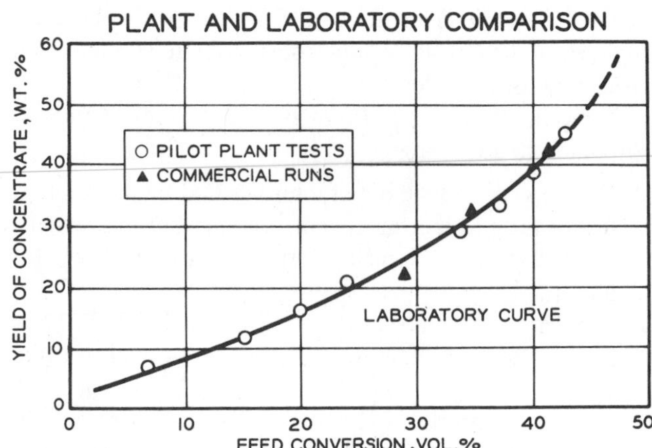

FIGURE 15.31 Graph Layout for Publications

Analysis of Data **Marketing data** are useful in identifying features and capabilities desired by the public. Marketing data will, at times, drive the project design. The performance characteristics of a possible design solution are determined by analyzing the data generated from hardware testing.

Graphs and Charts Information about the fundamentals of drafting and design is sometimes displayed by **graphs, charts,** and **diagrams.** Graphs contain the following information: axes or scale lines, major divisions, title, designators on axes, units, one or more curves, and captions or notes (Fig. 15.31). A graph should be easily understood without extensive explanation.

Number of Solutions and Products Determined The process of creating a new design always gives rise to more than one solution. Often the different solutions involve compromises between cost, reliability, and time. Design for manufacturability concepts and procedures help ensure a design that is functional, cost effective, and timely. Each possible solution may produce a marketable item. A record of each design possibility should be kept, including all sketches, written descriptions, and other data. It is appropriate to investigate more than one solution at a time. If a team is working on a project, it may need to be divided into subunits, each investigating different solutions. A variety of different designs should be created and developed to the point at which they are sufficiently defined. They should then be evaluated and compared during the next stage of the project. Duplication of competing teams should be minimized by the sharing of data and resources.

In Figure 15.32, the model of the hovercraft is one of many design solutions being investigated by a ship building company. The surface-effect ship in Figure 15.33 is a more detailed design alternative. The Coast Guard hovercraft, shown in Figure 15.34, is a full-scale test prototype. The design products require a complete analysis of design alternatives and a comparison of performance capabilities before production.

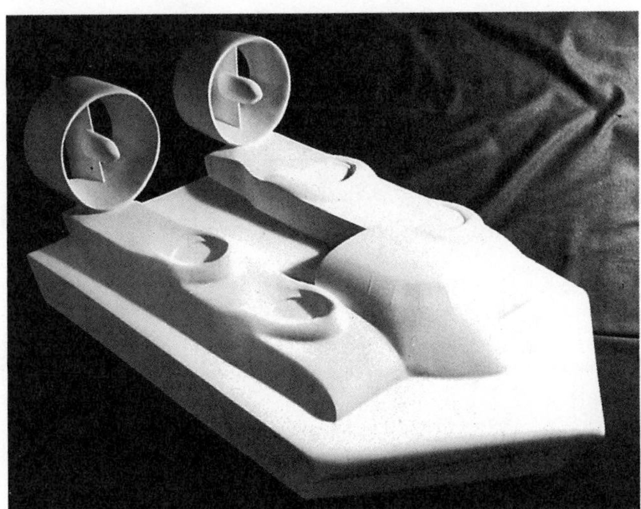

FIGURE 15.32 Model of Hovercraft Produced by Carving Artfoam

FIGURE 15.33 Surface-effect Ship Model

FIGURE 15.34 Coast Guard Hovercraft Test Model

Reports Generated on Solutions to All Proposals

Technical reports containing design data on each possible solution are generated at this point in the process. All factors, both positive and negative, must be clearly defined before the project is developed further. Properly presented reports consist of complete graphic descriptions and are well written.

Final Review of Project Choices The design department must evaluate each of the possible solutions before submitting their findings and suggestions to management during Stage 5. Two or more design solutions should be prepared by the design team. During a team meeting, each of the designs should be compared. The merits of each design solution should be clearly presented, as well as any drawbacks to the design. Each of the features incorporated into the design should be compared to those of other possible solutions. A list should be prepared that ranks the relative importance of each feature.

Materials, manufacturing, and facility requirements should be discussed at this meeting to balance each solution's effectiveness against others based on cost and company capabilities. At this point, select the best designs and prepare for the next stage of the project.

15.4.5 Design Choice and Product Decisions

After all the data have been gathered on the design problem and the remaining possible solutions have been clearly defined, the last step in the decision process is addressed (Fig. 15.35). Management has the final say on which project design solution to pursue. The designers, engineers, and other company personnel involved in the project present their findings and design choices to management for a decision.

Design Decision A complete technical report on each solution must be submitted at this juncture by the design

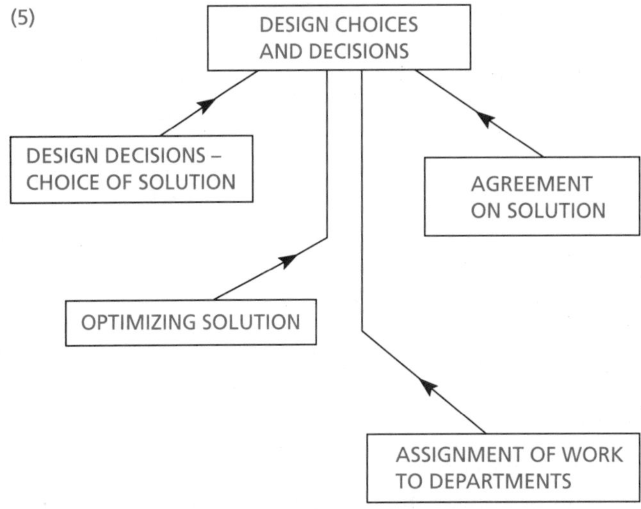

FIGURE 15.35 Stage 5 in the Design Process

ITEMS OF INTEREST

Ballpoint Pens

Before the invention of the fountain pen, man used the sharpened shaft of a feather as a standard writing instrument. The earliest form of a fountain pen was found in a 6,000-year-old Egyptian tomb. It was a reed into which ink was poured and was used to write on parchment and papyrus. In 1884, the fountain pen, as we now know it, was invented. Because of the leaks and breaks in the nibs, you could always spot users of these pens from their ink-stained hands. It became obvious—there had to be a better way to write than this.

People began experimenting using a ball bearing point. A Hungarian, Ladisloa Biro (a journalist), and his brother Gerry (a chemist) worked on the problem of the ink on the ball point clotting and not running smoothly. When the war broke out, the Biro brothers emigrated to Argentina and in 1943 formed a new company to perfect and produce the pen.

They finally worked out a formula for the ink. In 1944 the tube for the ink and the ball bearing point were combined to form a new pen. An Englishman spotted one of the new pens and knew about the problem the wartime aircraft crews were having with the fountain pen leaking at high altitudes. An instant market was available for the new pen!

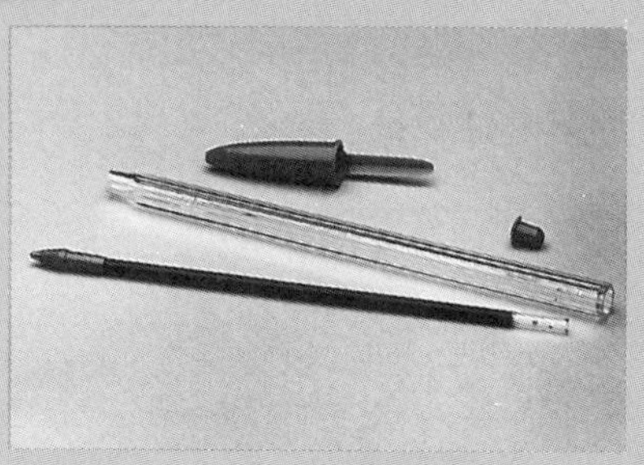

Bic Pen

After the war, the Biro brothers' invention (which had been sold earlier to their financial backer) was sold in the United States as Eversharp and in Europe as Bic. The Biro name became a household word, but the inventors disappeared into obscurity. A medium-point Bic pen produces more than two miles of writing. The French company sells more than 12 million ballpoint pens per year.

The ballpoint pen is a simple design. A precision ground ball, usually stainless steel and $\frac{1}{32}$ in. in diameter, is the main component. It is ground to an accuracy of a few millionths of an inch. The ball can also be a tungsten and carbon compound almost as hard as diamond.

The ball is fitted into a housing that allows rotation in all directions. The tip of the housing is bent over the edge of the ball so it does not fall out of the pen. Ink is fed through a narrow tube from the reservoir to the ball housing. If the reservoir is not open to the air, a small hole is made in the housing to prevent a partial vacuum from forming and preventing the ink from flowing. The ink is distributed evenly around the ball by ridges inside the housing.

Ballpoint pens remain the most popular, inexpensive, and efficient writing tool of today. The Biro brothers gave us another important invention we hardly notice, but would not want to live without.

group. The report consists of a design proposal and a timetable for completion of the project. All data regarding the design solutions must accompany the proposal, including cost analysis, time studies, capital requirements (personal, equipment, and facility), design layouts, and other materials that may help in the decision process.

The management decision team, which include members from the design, manufacturing, and marketing departments, evaluates the merits of each solution submitted by the design department. The evaluators of competing designs consider the following:

Design Comparison

- Capability to satisfy the original statement of the design intent and project definition
- Cost, manufacturability, and reliability
- Design requirements for precision, operating efficiency/flexibility, maintenance projections, and environmental impact
- Material and manufacturing processes
- Configuration and complexity on manufacturing costs

The management team, after each of the above is satisfactorily investigated, can reach a decision on the design choice or choices. The design team then proceeds to the next stage, in which the selected design is optimized.

Optimizing Solution The design team further develops and refines the selected design. The selection of materials, processes, and other design requirements can be further refined at this stage. Each feature and capability of the design should be analyzed and evaluated. Any changes should be made at this point so that during Stage 6 (development and implementation) there are no drastic changes in the design. Any new desirable features or capabilities should be evaluated. A list of concerns and considerations of the design should include the following:

Design Refinements

- Should any new or desired features be added at this time?
- Should the effective life of the design be extended or decreased?
- Are there any aspects of the design's appearance that should be changed?
- Based on the design's potential configuration and operating conditions, what materials are acceptable?
- Is the design manufacturable?
- Can the design be more flexible and interchangeable without cost increases?
- What are the basic cost parameters for the design?
- What are the tolerance requirements?

- Have the stress factors and alignment problems that may be encountered been determined?
- Has the need for study models, prototypes, and test models for motion, stress, or other design studies been established?

Agreement on Solution A consensus must be reached on the acceptability of the solution. There should be complete agreement as to the direction and choice of management's decision. All basic features and design requirements are now solidified into an accepted solution.

Assignment of Work The project schedule is determined after the final design is chosen and the project launch is approved by management. Several methods of project scheduling are used in industry. One method is the **project evaluation and review technique** (PERT). This method is used to coordinate the many activities associated with a successful design project. All departments must be coordinated efficiently to move the project from design through production in a smooth and error-free environment.

The success of the project depends on this coordination. At the project launch, each department is given specific work assignments and time requirements. The **critical-path method** of scheduling project work assignments is used in conjunction with PERT to control the project.

15.4.6 Development and Implementation of Design

The **development and implementation** stage of the design process includes the drawing, modeling, testing, analysis, and refinement of the design (Fig. 15.36). The actual documentation of the project can be done manually or on a CAD system.

Physical models are used for design and testing (Fig. 15.37). Actual full-scale prototypes, developed from design drawings, are used to test for strength and design flaws. The rotor mount in Figure 15.38 failed during the test of a prototype design. Other models are used for testing aspects of the design, such as the all-brass antenna model on the destroyer that was used for reflectivity studies (Fig. 15.39).

The design and development of a project takes many forms. The space shuttle started as a design concept and proposal [Fig. 15.40(a)]. This pictorial rendering is quite unlike the final vehicle. The original designs were drawn, modeled [Figs. 15.40(b) and (c)], and tested [Figs. 15.40(d) and (e)] many times before the design was refined [Fig. 15.40(f)] and brought to completion [Fig. 15.40(g)].

Stage 6 in the design process is the heart of all design efforts. When most of the work is done, the project is brought to fruition. All previous steps must have been done effectively if this step is to be successful.

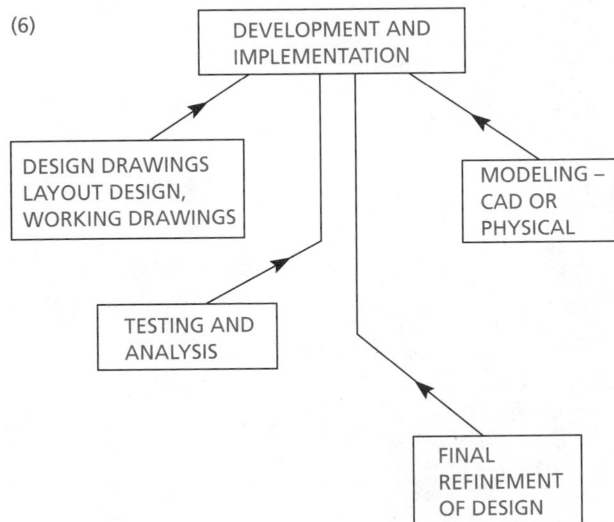

FIGURE 15.36 Stage 6 in the Design Process

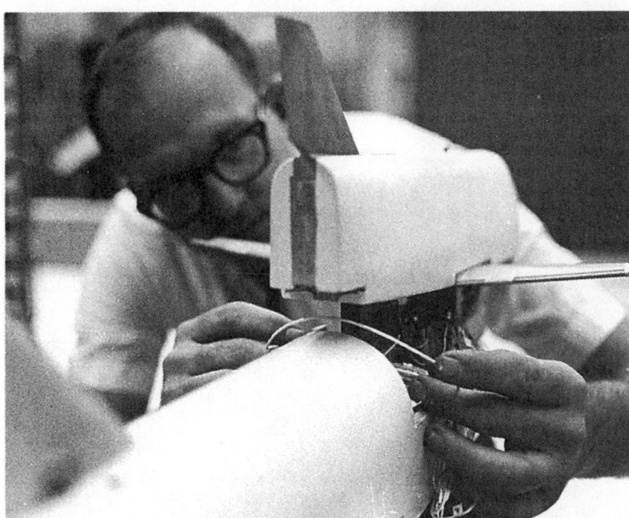

FIGURE 15.37 Shaker Test Being Prepared for the Space Shuttle Using $\frac{1}{100}$ Scale Model

FIGURE 15.38 Rotor Test Failure

FIGURE 15.39 All-Brass Model of Destroyer Used for Reflectivity Studies

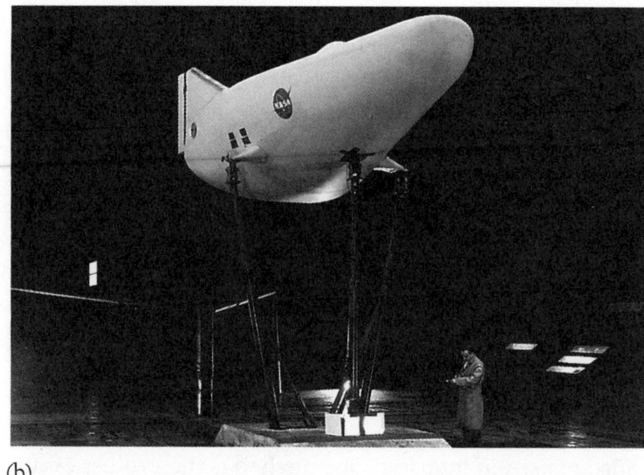

(b)

(c)

(a)

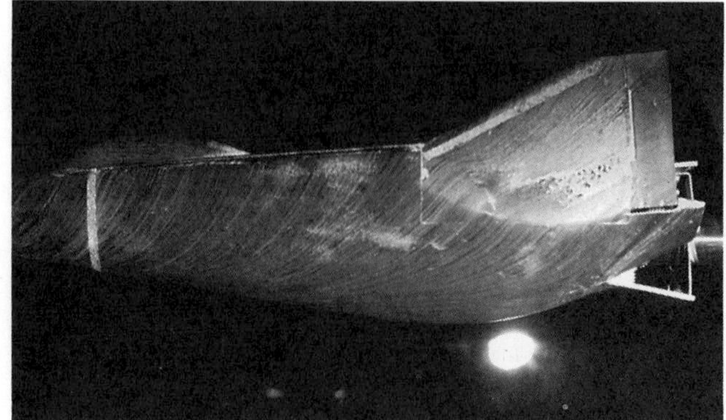

(d)

FIGURE 15.40 Development of Space Shuttle Design
(a) Space shuttle concept
(b) Wind tunnel study of original space shuttle concept
(c) Space shuttle concept with astronaut window
(d) Flow studies in wind tunnel using oil
(e) Full-scale test prototype of reentry spacecraft shown here before being dropped from a B-52 in 1966
(f) Space shuttle model being tested in wind tunnel in 1975
(g) Illustration of the final space shuttle design in 1979

(e)

(f)

(g)

FIGURE 15.40—*Continued*

FIGURE 15.41 Tractor-Trailer Models

Design Layout and Working Drawings At this step in the design process, all preliminary work must be complete and no major changes in the design should be implemented. The acceptance of the project by all parties is essential.

Technical drawings generated manually or on a CAD system are the primary means of communicating design and manufacturing information at this time. Drawings are also required before the construction of physical models: test, prototype, mock-ups, presentation, process, system, or product.

Physical Models **Models** are used throughout the industry as scaled representations of systems designs and for refinement and testing of product designs. A **systems model** shows an installation. **Product models** are used in many stages of the design process to establish scale, appearance, and function of a product. Scale models of tractor-trailer designs are shown in Figure 15.41.

The designer can request a model at almost any stage of the design process. The type of models requested depends on the product or system and the availability of modeling facilities. At times, outside vendors are called on to complete a model.

Engineering systems models are a design tool that can eliminate unnecessary problems, bad design, inefficient planning, and other expensive, time-loss situations. They are used throughout the petrochemical, nuclear, and conventional power-generation industries. They are also used in food and beverage processing, pulp and paper manufacture, pharmaceutical processing, and other systems.

When working with a three-dimensional model (Fig. 15.42), the designer can visualize the design sequences and operations that are necessary for the project. Models are most advantageous if they are used as a working tool, from the beginning stages of a project through the entire design phase. The preliminary models may not resemble the final design model at all.

Mock-up models are full-size models. They are used primarily to determine the appearance of the product. The size, configuration, color, and artistic considerations can be

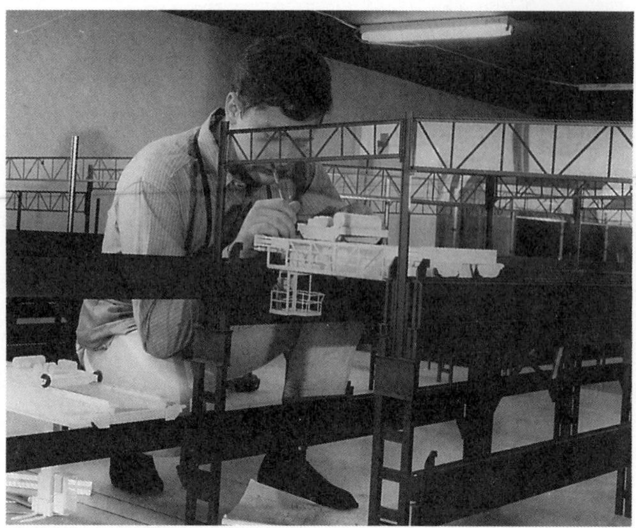

FIGURE 15.42 Designer Modeling Turbine Crane for a Power Plant

determined with the mock-up. This type of model is not often used for movement or operation design.

Product models are used more often in mechanical engineering to help design various parts of machinery or other mechanical devices. In some cases, these models are built to a scale larger than that of the project itself.

Prototype models are similar to product models, but are sometimes working simulations of the product [Fig. 15.40(e)]. Some prototype models are mock-ups of the eventual product. 3D prototypes are sometimes required in the design of mechanisms to test their performance and capabilities. A prototype will often be close to the same configuration as the finished product. Prototypes are tested under the typical operating conditions of the proposed product.

FIGURE 15.43 Display Model of Offshore Buoy

Presentation models, such as the one in Figure 15.43, are created to display a project, product, or design to the general public or for sales.

Testing and Analysis One of the primary responsibilities of the designer is to create a design that will withstand the stresses under which the product, part, or system will function. If the design is to be operated safely, the proper design for strength becomes the major concern. After the product or system is designed, it must undergo testing and analysis. Lab testing for strength of materials and finite element analysis using a CAD system are both used throughout industry to determine the adequacy of the design.

15.4.7 Production, Manufacturing, and Packaging

At this point (Fig. 15.44), the design must be presented to all interested, involved, and essential parties. The configuration of the design is presented in drawings, renderings, and possibly a prototype or presentation model. All aspects of the design are discussed at this meeting. The limitations and capabilities of the design, the new or improved features, problems in the design process of the product, and validation of all research, scientific, and engineering aspects of the project must be available. Graphs, charts, notes, sketches, models (of rejected design alternatives), and design calculations to establish the solution may all be used during the meeting. The production department will require much of this information to establish manufacturing requirements and a production schedule.

Detail drawings of each aspect of the product, assemblies, and any other graphic documentation are complete at this

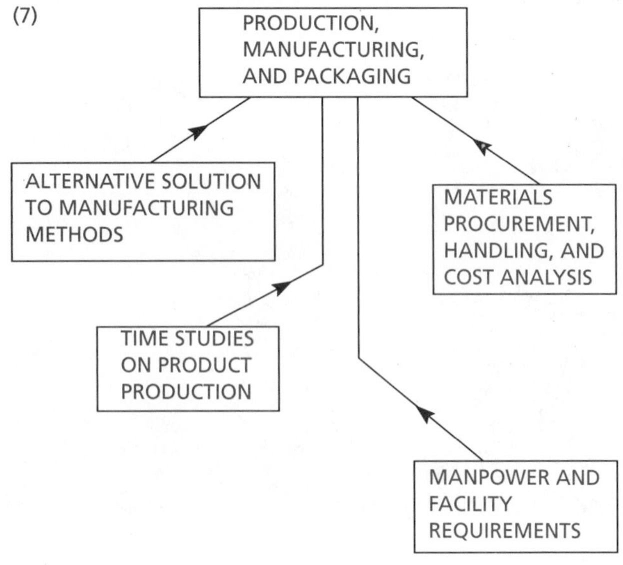

FIGURE 15.44 Stage 7 in the Design Process

stage. Communication has been continuous throughout the design process, and although this is the sixth step in a formal explanation of the design process, the flow of information has been back and forth between these steps throughout the process.

Manufacturing Considerations in Design The design of a particular part or product usually determines the material and the manufacturing process. Failure to understand the materials and the manufacturing options will doom the project from the start. The stress, vibration, environmental operating conditions, tolerance requirements, and surface finish are determined during previous stages in the design process. Therefore, by the time the materials and the manufacturing methods are selected, they are almost defined by default. The decisions by the designers at this stage are merely refinements.

Of course, the selection of materials in Stage 5 determines many of the manufacturing methods used in the production of the part. Each of the factors in the following list is critical to the manufacturability of a particular material and must be considered during this design stage:

- Size limitations
- Configuration
- Thermal characteristics
- Tolerance requirements
- Hardness
- Weight limits
- Required ultimate strength
- Elasticity
- Surface texture roughness
- Precision

A variety of manufacturing processes are available. Regardless of the process or the material, the designer should design for the most efficient and cost-effective process. The following processes are used in manufacturing and were covered in more detail in Chapter 10, Manufacturing Processes:

- Machining
  - Drilling
  - Boring
  - Milling
  - Planing
  - Reaming
  - Broaching
  - Turning
- Welding
- Casting
- Forging
- Forming
- Stamping
- Extruding
- Bending

Assembly of the part is partially determined by the material selection. The estimated maximum number of parts and the minimum run is also a factor in the selection of the process. Manufacturing assembly processes include:

- Brazing
- Riveting
- Bolting
- Welding
- Gluing

The following is a partial list of concerns and suggestions for the designer. If these conditions are met, the chance of a successful, manufacturable part is greatly increased.

1. Design for standard machines and processes.
2. Design within the cost-effective limits of available and effective manufacturing procedures.
3. Design to limit the number of manufacturing processes.
4. Design to permit efficient production in acceptable quantities and within time requirements.
5. Design for the most cost-effective process that will deliver a product meeting the design parameters.
6. Design for assembly ease.

Although the preceding discussion centers on the designers, responsibilities and the design requirements, it should be understood that the actual selection of manufacturing processes are made by the manufacturing engineer in conjunction with the product designer.

Alternative Solutions to Manufacturing Methods **Automated manufacturing** involves the design of a product so that it can be readily manufactured, fabricated, assembled, handled, tested, quality controlled, packaged, stored, and shipped by automated methods. Designs have always been concerned with materials and manufacturing methods. Increased productivity requirements, brought on by foreign competition and profitability margins, have made automated manufacturing methods essential to the survival of a company. Figure 15.45 shows the influences of automation on the design process.

Material considerations, processes, automated technology, and other factors can be assembled to provide a review list for the optimum design. Some aspects of automated manufacturing design include designing for simplicity. The more complex a part, the more difficult it is to manufacture automatically. A simple part will require less manipulation by the robot.

Designing with automated manufacturing in mind takes more initial design time, but you are rewarded with a more efficient, cost-effective, and better product. The following checklist can be used to maximize the results of your design effort:

1. Incorporate every design aid available to reduce manufacturing costs, without adversely influencing the product's essential features.

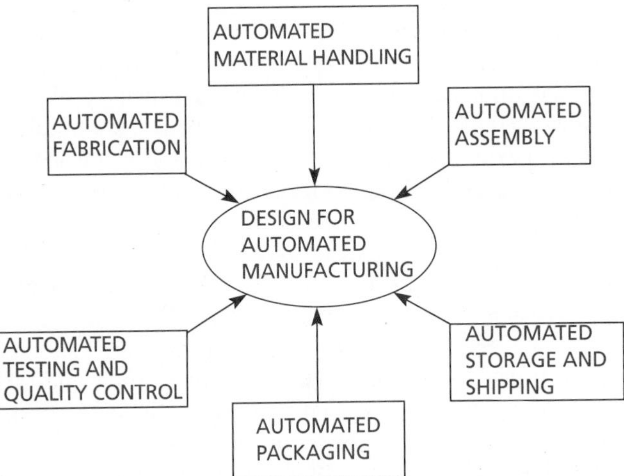

FIGURE 15.45 Design for Automated Manufacturing

2. Understand the basic capabilities and limitations of your in-house production and outside vendors' capabilities pertaining to the part's manufacture and materials.
3. Determine the manufacturing methods, whether manually produced, automated production, or a combination of the two, early in the design process so as to maximize the successful creation of the product.
4. Use a design review process that maximizes the effect of any design rules created in the design process. Be willing to review the results and redefine any design requirements based on new knowledge and automated capabilities. Keep an open mind about the material, process, and automated manufacturing methods.
5. Keep up-to-date on new and developing automated processes, machinery, time and production studies, advanced materials, and new technology.

Time Studies on Product Production The rate of production has a direct effect on the profitability of a product. **Time studies** are conducted to optimize a product's manufacturing cycle. Material handling, production elapsed time (manufacturing and assembly), and part removal (and transportation) are all considered in this study.

Materials Procurement, Handling, and Cost Analysis **Availability** of a material or standard part will influence its selection as much as design requirements. **Material procurement and handling** are essential parts of the total design effort. If the specified standard part is temporarily out of stock or not available, a product can be delayed in the manufacturing stage.

The increased cost of substituting one material for another at the manufacturing stage can completely destroy the profitability of the product. The introduction of **just-in-time** manufacturing, where parts arrive at the manufacturing station exactly when needed in the assembly or production process, requires extensive coordination of all departments.

Manpower and Facility Requirements A company's facility and work force will at times determine when and where the part or product is made and assembled. The design process includes decisions based on available space, machinery, and, trained personnel. The implementation of automated methods has been influenced by a lack of highly trained personnel. The use of robots in the manufacturing stage will also affect personnel, facility, and equipment considerations.

Industrial Packaging The field of packaging design includes design of boxes to hold consumer items such as perfume, electronic products, food products, and household items. It also includes industrial packaging for mechanical and electronic designs (sheet metal enclosures). The artistic design of boxes for consumer items are considered under Stage 8. Here, the discussion is confined to sheet metal enclosures required for mechanical and electronic systems. The electronic meter in Figure 15.46(a) is composed of electronic components housed in a sheet metal enclosure [Fig. 15.46(b)].

Space requirements, safety, function, operation, service, and environmental conditions are all factors in designing the enclosure. Most of these requirements are known in the early

(a)

(b)

FIGURE 15.46 Example of Industrial Packaging (a) Meter face (b) Meter panel and package

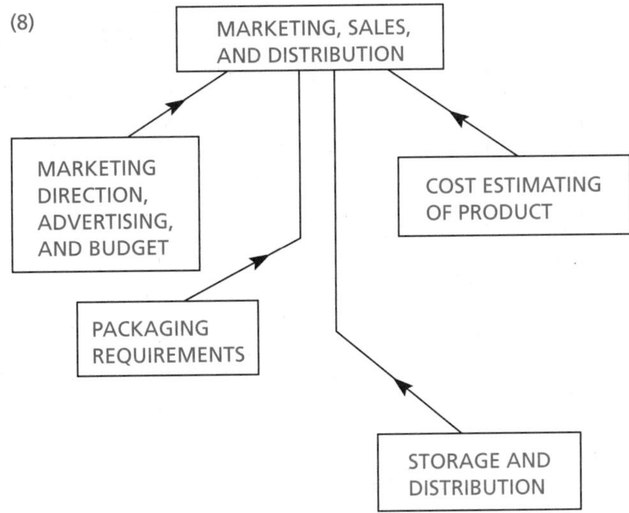

FIGURE 15.47 Stage 8 in the Design Process

portion of the design process so as to provide sufficient leeway in the packaging.

15.4.8 Marketing, Sales, and Distribution

Marketing, sales, and distribution are essential aspects of the total design effort (Fig. 15.47). The product or system may never exist without a thorough economic analysis. The costs of advertising, marketing, packaging, shipping, and distribution greatly affect the cost to bring the product to market. The input of the sales and marketing department, the packaging department, and the shipping and storage facility makes the important connection to the world of business and also affects profit margin.

Advertising, Budget, and Marketing Direction The marketing department is responsible for an accurate product survey before the product is designed. A poor design decision with respect to customer needs on the part of the designers would leave the marketing department with the job of selling an unusable or undesirable product. Sometimes the advertising campaign can create a need for a product in the minds of the public or sell an inferior product with slick advertising. Marketing will help determine many functional requirements of the product before the final design is accepted. Surveys that help determine the need, size, color, shape, feel, and acceptable cost can be completed well before the engineers and designers are through with their work.

Cost Estimating of Product The **cost analysis** of the product includes the expenses generated by the engineering, design, manufacturing, sales, and shipping departments. Each department must submit a detailed cost estimate for the labor hours, materials, and overhead cost for each stage of the design development process in which it is involved. Design costs include modeling and drafting; sales costs include advertising and marketing; packaging costs include art design and box design. Since they include personnel, equipment, facility, and material considerations, manufacturing costs may be the largest single cost of the product. Cost estimating includes how long it takes to generate an acceptable return and how many items must be sold before the product is considered an economic success.

Packaging Requirements The design of a product does not end with the product itself. Without proper presentation, the product may fail to achieve the required sales to be successful. **Packaging** is almost as important as the product, especially in the world of mass marketing and international sales. Packaging also includes new concepts of "green" packaging or environmental packaging, which includes the use of biodegradable packaging and the elimination of overpackaging. Too much packaging creates unnecessary amounts of waste.

A simple box is not so simple. As a matter of fact, it can be rather complex. Size must be considered, along with shape, text, color, art, and competition from other boxes. Most packaging for consumer products must be designed for appearance as well as function.

Although packaging is normally part of Stage 7, most of the design input about the appearance comes from the advertising and marketing departments. The actual design of the dies and patterns is the domain of the manufacturing group. The marketing department, in conjunction with the packaging designers assigned to the manufacturing department, must agree on a box design that is functional, attractive, and can be produced within cost restraints.

Software programs have been designed specifically for the packaging industry. Software is available that offers a library of parametric designs for creating standard box configurations.

Once the individual box design is complete [Fig. 15.48(a)], copies are *nested* together [Fig. 15.48(b)], mirroring and duplicating the images and interlocking them like jigsaw puzzle pieces to produce as many boxes as possible from a single sheet of cardboard. The software performs automatic bridging, a technique in which gaps are left during the process of cutting the plywood so that it does not fall apart. The die manufacturer then inserts steel blades into the plywood, bridging the gaps with blades so that the boxes will fall out like shapes from a cookie cutter.

The ability to prepare the entire design [Fig.15.48 (c)], including the structure, graphics, and machine codes, on a single computer not only speeds and simplifies the prototyping phase, it also eliminates expensive mistakes.

Shipping and Distribution **Just-in-time manufacturing** is a process in which the component parts of a system design arrive at the assembly line station at the time of installation. This process requires careful control of materials, equipment, fabrication processes, and subassembly transportation to the assembly site at the appropriate time. **Field fabrication** of parts of the system also depends on the timely arrival of appropriate materials. The planners of any project have most of the responsibility for this stage of the project. The designers and procurement department must be in constant communication at almost every stage to prevent shortfalls or overstocking of supplies. It is this coordination between all involved departments that makes a project meet both time constraints and cost estimates.

Shipping and distribution of a product includes storage and warehousing of the product. The timely arrival of a consumer product on the market greatly affects its sales performance.

The shipping department must be aware of the product's size, weight, and any factors that would influence the method of shipment. If the product is to be stored at the facility, will there be enough warehouse space? Should the warehouse be automated to handle the product with robotic systems? How many of the product should be available at any one time?

The quantity of items to be handled and stored is also an important consideration. *Design for stacking* whenever possible. If the product is to be shipped to the general public, contracting with outside shipping sources is also done.

Illustrations for Presentation, Sales, Advertising, and Catalogs Pictorial illustrations are used in this stage of the design process to present the product, system, or concept to a nontechnical or purchasing audience. Renderings of products and concepts provide a realistic illustration of the proposed item [Figs. 15.40(a) and (g)]. Renderings and models help introduce a product, concept, or system design to the general public or an interested potential customer.

15.5 General Problem Instructions and Design Process Summary

The list of design projects provided at the end of this chapter are intended as a guide to both instructor and student, indicating some possibilities and alternatives for further work in design. Since the lists are hardly definitive and should not be used to restrict the student's creativity, the student may wish to choose, with the approval of the instructor, an unlisted design project in which he or she is personally interested. The major objective and importance of the project should be for the student to do some original thinking in the use of design principles and of DFM concepts. Because they outline a project from the abstract to actuality, the following eight steps in the design procedure should be carefully considered.

Design Process Summary

1. *Identification:* Defining the design objective.
   a. Making a list of known facts and existing information.
   b. Ask the following questions: What? Why? Where? Whom? How? When?
2. *Conceptualization:* Brainstorming, creative solutions.
   a. How many ways can it be solved?
   b. Thought starters, make a list of values.
   c. Similarities; environmental requirements.

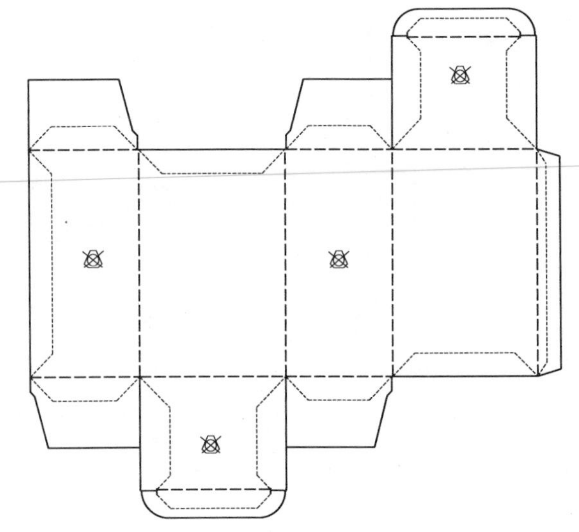

(a)

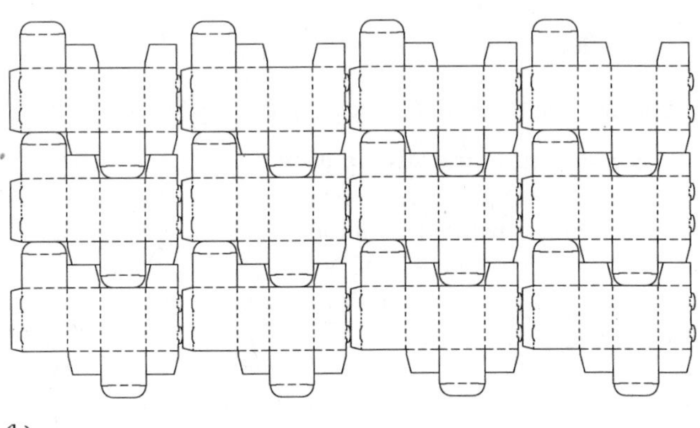

(b)

(c)

FIGURE 15.48 Packaging Design Example (a) Die design using Ovation CAD/CAM (b) Nested design layout (c) Package graphics

   d. Checklist, brainstorming, material options.
   e. Is there a simpler way?
3. *Evaluation:* Application, functional requirements, synthesis.
   a. What makes the design good: economy, simplicity, reliability, durability, usefulness, attractiveness, manufacturability, easy to promote in sales, easy to service?
   b. What are the alternatives to the design?
4. *Decision:* Design optimization.
   a. What materials should be used?
   b. Should the parts be interchangeable?
   c. Should we use standard parts?
   d. Is it an economical manufacturing process?
   e. How easy is it to operate?
5. *Development:* Implementation of design.
   a. Create working drawings and details.
   b. Model the part: CAD, physical, types, number of models.
   c. Check the design.
   d. Test and analyze: modeling, debugging.
   e. Improve and redesign for aesthetic or functional refinements.
6. *Production:* Manufacturing, packaging, handling the product.
   a. Facility needs.
   b. Personnel requirements.
   c. Materials and processes for manufacturing.
   d. Packaging design.
   e. Material handling and product handling.
7. *Marketing:* Sales and distribution.
   a. Staff training.
   b. Servicing the product.
   c. Low maintenance costs, customer acceptance.
   d. Sales strategy on how to present.
   e. Product and main features.
   f. Distribution of the product. Who? How?

Problem solutions may be completely freehand, a combination of freehand and instrument drawings, or all instrument-prepared accurate design layouts. The method used is dependent on the project size, the facility capabilities, and the class requirements. A final report may include a whole or partial design, idea sketches, layouts, details, assemblies, exploded pictorials, patent drawings, graphical solutions, investigation reports, sales promotion suggestions, proposed newspaper announcements, graphs, charts, and a model. The laboratory sessions should be devoted mainly to the development of ideas by graphical documentation, unless a model shop is available. Depending on the size and complexity of the project, the instructor may choose to form design teams of two to four students.

QUIZ

True or False

1. The design process is a series of eight rules that should be followed exactly.
2. All designs developed by a company's design team eventually get manufactured or produced.
3. Function is the most determining factor in design.
4. Human engineering is the study of people and how they engineer/design projects in industry.
5. The project cost and the profit margin are two of the most important factors in any design.
6. Reliability is the length of time a product will operate properly.
7. Human factors plays an important role in all product design.
8. Off-the-shelf means a mechanical item that is too large for shelf storage.

Fill in the Blanks

9. ________ design uses standard components arranged in a unique configuration.
10. ________ is the first stage of the design process.
11. Durability, life, quality, economy, and simplicity are all factors in product ________ ________ .
12. The ________ ________ can be used to describe and determine the material and configuration requirements of a design.
13. ________ design and ________ design are the two main divisions of design projects.
14. The effects of motion, heat, ________ , and environment are some of a designer's concerns.
15. Design parameters include weight, size, ________ , ________ , ________ , color, ________ , and ________ .
16. Machinability, castability, ________ , ________ and ________ are all factors in the selection of a product's material.

Answer the Following

17. Compare systems design with product design. Describe both and explain their differences.
18. Name ten factors that influence design at the onset of the project.
19. Describe the use of models in the design process.
20. How are graphs and charts used in the design process?
21. What is human engineering and how do human factors and ergonomics influence design?
22. What is DFM and how does it affect the design process?
23. List six sources of information available to the designer.
24. What is critical-path scheduling?

PROBLEMS

Problem 15.1 *Play Yard Toy.* Design a child's backyard play toy. Include a slide, ladder, swing, tunnel, rope climb, and tire walk. Design for safety, strength, and creative play situations. Make sure that all aspects of the design are sized for the child.

Problem 15.2 *Small Tools.* Do a short survey on the variety of small tools, such as screwdrivers, pliers, and saws. Attempt to redesign an existing product to better suit the needs of the user. Improve the design; attempt to find creative and aesthetically pleasing alternatives.

Problem 15.3 *Packaging Machine.* Take a small household product and design a packaging machine to load and box the item.

Problem 15.4 *Garbage Dumper.* Design a garbage can and a lift device that could empty the can into the dump truck. Expand on this idea by designing stackable recycling containers for glass, cans, plastic, and paper products.

Problem 15.5 *Fishing Rod Holder.* Create a series of fishing rod holders that will accommodate all sizes and shapes of handles. One version should be for a boat-mounted holder and the other a shore fishing model.

Problem 15.6 *Face Guard.* Design a face guard for baseball hitters to use while batting. Do a study on the typical human face and how to protect it without reducing visibility or inhibiting the batter.

Problem 15.7 *Trailer Hitch.* Design an interchangeable hitch that could be used for all size balls and every class of hitch, if possible. Research should include the ball sizes, weight limits, classifications, and materials of existing hitches.

DESIGNING FOR THE PHYSICALLY CHALLENGED (Problems)

Most colleges have a department devoted to the physically challenged. An excellent design project would include the research and analysis of how the life of a physically challenged person is determined by the man-made items found in daily life. Contact your physically challenged program and request to interview a number of wheelchair-bound persons. Other types of physical limitations could also be studied. Design projects for the blind, hearing impaired, or seeing impaired could also be considered.

Since the physically challenged use many of the existing facilities, this is an excellent area in which to research, analyze, and design products and fixtures. Do some research and tabulate your findings such that the number of fixtures and the number of users of a facility are displayed on a graph. Determine the number of users of a particular facility. Chart the information, as in Figure 15.49. Figure 15.50 is an example of a sketch used to determine the proper heights of restroom facilities for the physically challenged. Your problems should have rough sketches, development layouts, and accurate assemblies and details.

Many products will require studies involving human factors and ergonomics. The student should research the project by consulting *Human Factors Design Handbook,* Woodson, Wesley E., McGraw-Hill Book Company, and *Ergonomic Design for People at Work,* Eastman Kodak Company, Lifetime Learning Publications. Both of these volumes contain pertinent data regarding designs for the physically limited.

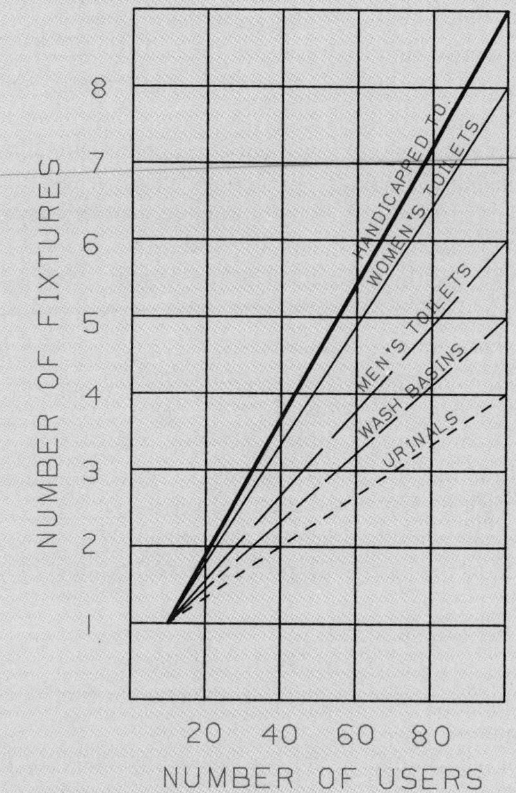

FIGURE 15.49 Graphical Analysis of Research Data on Facility Use of Restroom

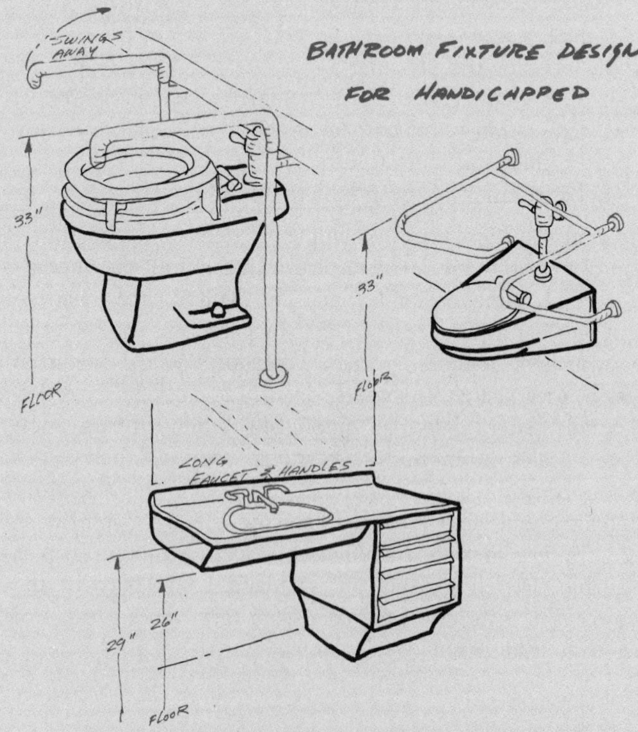

FIGURE 15.50 Bathroom Fixture Design for Handicapped

Problem 15.8 *Bathroom or Restroom Layout.* The bathroom or restroom configuration includes a number of appliances. Washbasins, commodes, mirrors, tubs, showers, faucets, lights, dryers, hand towel dispensers, soap dispensers, cabinets, and floor space must all be considered. Figure 15.50 shows some design alternatives for commodes and a sink. In addition to taking notes about the facility, bring along a tape measure and roughly determine the optimum height and size for a well-designed restroom or bathroom.

One way of researching the needs and capabilities of a physically challenged person is to imagine yourself in their place. If possible, borrow a wheelchair and spend some time taking notes as to where and what difficulties are encountered during a typical day. If a wheelchair is unavailable, then use a chair with wheels (such as a computer chair) and move around your home while thus confined.

When researching a restroom or a bathroom, take notice of each detail. First, do what you would normally do in that facility. On entering, is there enough room for your wheelchair? Is there sufficient clearance for your hands, for the chair, for your extended feet? How difficult is it to get on and off the commode? Can you reach the basin and the faucets? How difficult is it to perform simple grooming procedures? Are the switches and the electrical plugs accessible? Combing and drying hair, applying makeup, and washing the face and hands should all be investigated.

Problem 15.9 *Kitchen Design.* Complete an analysis of the kitchen area as described for the bathroom. Can the stove and oven be safely operated by a wheelchair-bound person? Are the freezer and refrigerator accessible? Could someone get a beer or pop from the refrigerator's top shelf? Are the small appliances (microwave, blender, toaster) easily operated from the chair? Could the dishes be washed by hand from a wheelchair? Measure the kitchen cabinets and the placement of each appliance, major and minor. A rolling chair can be used to do the research in this area. Design special cabinets to hold small appliances. Redesign a refrigerator and freezer to accomodate a physically limited person. Sketch a variety of design improvements for the layout of the total kitchen.

Problem 15.10 *Car Jack.* Design a jack that could be easily used by a physically limited person who has limited use of their legs. The jack must be easy to operate and a convenient size. Possibilities include the incorporation of the jack into the car design itself.

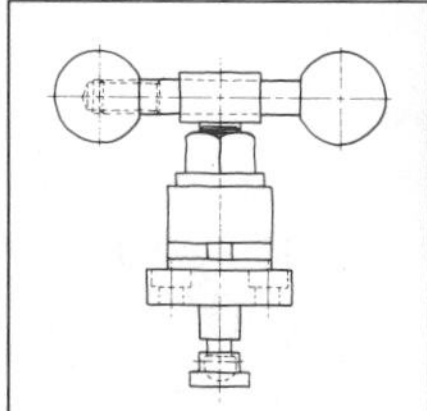

CHAPTER **16**

Working Drawings (Assemblies and Details)

Learning Objectives

Upon completion of this chapter, you will be able to accomplish the following:

1. Convey engineering information for production while recognizing basic types of assembly and drawing categories.
2. Produce assembly and detail drawings with appropriate view selection and dimensions.
3. Compile parts lists and supply information for notes and pre-printed drawing sheets.
4. Identify and apply simplification and checking procedures.
5. Develop an understanding of drawing reproduction and storage methods.

16.1 Introduction

The primary purpose of a drawing is to convey engineering requirements to produce a finished product or part (Fig. 16.1). **Working drawings** include **assembly drawings** and **detail drawings** of a project. Drawings are required for all product design, production requirements, and manufacturing specifications. Each of the parts in Figure 16.2 required working drawings: detail drawings for each separate piece and an assembly drawing. A set of drawings for a system, tool, or product contains sufficient engineering information such that the following functions may be performed:

- Order material
- Plan manufacturing operations, tooling, and manufacturing facilities
- Process material
- Inspect and control product quality and reliability
- Assemble
- Test and model
- Package, box, and ship
- Determine cost
- Catalog
- Install and service
- Conduct final acceptance test
- Make alterations
- Record for duplication, repair, or replacement

See Color Plates 3, 5, 7, 9, 12-20, 21-23, and 32-36.

16.2 Assembly Design Considerations

The way the product will be assembled must be considered early in the design phase. This is one of the many reasons why the design of the product and the process must occur simul-

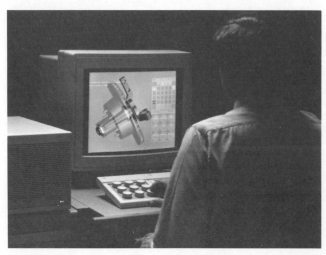

FIGURE 16.1 Solid Model of an Assembly

FIGURE 16.2 Disassembled Boiler Feed Pump

taneously. The optimal method of assembly can be chosen early in the design process if the product and the production-assembly process are designed at the same time. There are five basic types of assembly:

1. **Manual assembly** is performed manually with only the assistance of hand tools.
2. **Semiautomatic assembly** includes a combination of manual and automated processes. The operator loads the product manually, the machine then performs one or more assembly operations, and the operator then unloads the product.
3. **Adaptive assembly** involves programming the system to adapt itself automatically to certain variations based on sensors.
4. **Automatic assembly** performs all operations automatically without human intervention or decisions. Automatic assembly places constraints on the design for both the assembly of the product and the orientation, presentation, and gripping of the parts.
5. **Flexible assembly** uses flexible manufacturing equipment, which builds families of related products or subassemblies on the same setup or with quick, automated setup changes built into the process.

16.3 Categories of Drawings

Engineering drawings can be classified in three broad categories:

- Layout and design study drawings
- Engineering and production detail drawings
- Assembly drawings

16.3.1 Layout and Design Study Drawings

Layout and design drawings depict proportions, dimensions, materials, and the relationship of parts in new or modified designs. Usually, they precede production drawings, are drawn to scale, and are useful in preparing and checking part drawings and assemblies. When practical, layouts should be prepared in a form suitable for use in manufacturing, especially in situations where prototype or developmental models of equipment are required.

16.3.2 Engineering and Production Detail Drawings

Detail drawings provide a complete engineering definition of the finished system, assembly, or part (Fig. 16.3). This includes design data references, laboratory instructions, and engineering specifications. These drawings are prepared for the shop floor and are used in the production of the desired product.

Individual parts of an assembly are machined separately during the manufacturing stage of the project. Parts are also called **workpieces.** A typical workpiece is detailed so the fixture designer can produce tools for holding and locating the workpiece during the machining operations.

Engineering production drawings are not prepared to accommodate a particular method of manufacture. Engineering production drawings are prepared so that they can be used without additional explanation. Manufacturing or processing instructions may also be provided. But this information is for reference purposes only unless such data are vital to the end definition and engineering control of the product.

16.3.3 Assembly Drawings

An **assembly drawing** defines the complete end item requirements and establishes item identification for the assembled configuration of two or more pieces, parts, subordinate assemblies, or any combination joined to form an assembly (see Fig. 16.4 on p. 438). Assembly drawings always include a **parts list** or **bill of materials.** The sleeve valve in Figure 16.5 required an assembly drawing to show the proper relationship of the parts. Color Plates 3, 5, 7, 9, 12-20, 21-23, and 32-36 show various types of assembly drawings.

16.4 Assembly Drawings and the Design Process

In most cases, the assembly is conceived as a whole (layout assembly) and broken into individual pieces (detail drawings) for manufacturing and later assembly. This is called the **top-down** approach to design.

Assemblies are seldom designed from the bottom-up, unless the individual parts are standard items that are to be assembled into a unit. **Bottom-up design** means that each component of a unit is designed separately and then put together in an assembly.

The sleeve valve in Figure 16.5 comes in four pieces (see p. 439). The design is *limited to the minimum number of parts.* Each piece fits inside the other. Since each piece was toleranced in relationship to the housing, the product was *designed as a unit.* The housing and end cap hold the other parts together. The bolts and nuts needed for assembly are not shown in the figure.

After the design and layout of the project are complete, a detail drafter pulls separate parts from the assembly and draws them on individual sheets. This process is called **detailing.** Details include appropriate views and dimensions.

Standard items that can be purchased off the shelf are shown on the assembly drawing and listed in the parts list. They do not require a separate detail unless they are modified in some way. Standard parts include bolts, screws, nuts, retaining rings, dowels, pins, springs, gears, bearings, clamps, and purchased subassemblies.

As the product is designed and detailed, the manufacturing department begins to determine the tooling requirements. **Tool designers** create appropriate fixtures to hold the individual parts during manufacturing. Machining involves the use of fixtures to locate, hold, and position the part for accurate, economical, and efficient production. Jigs and fixtures

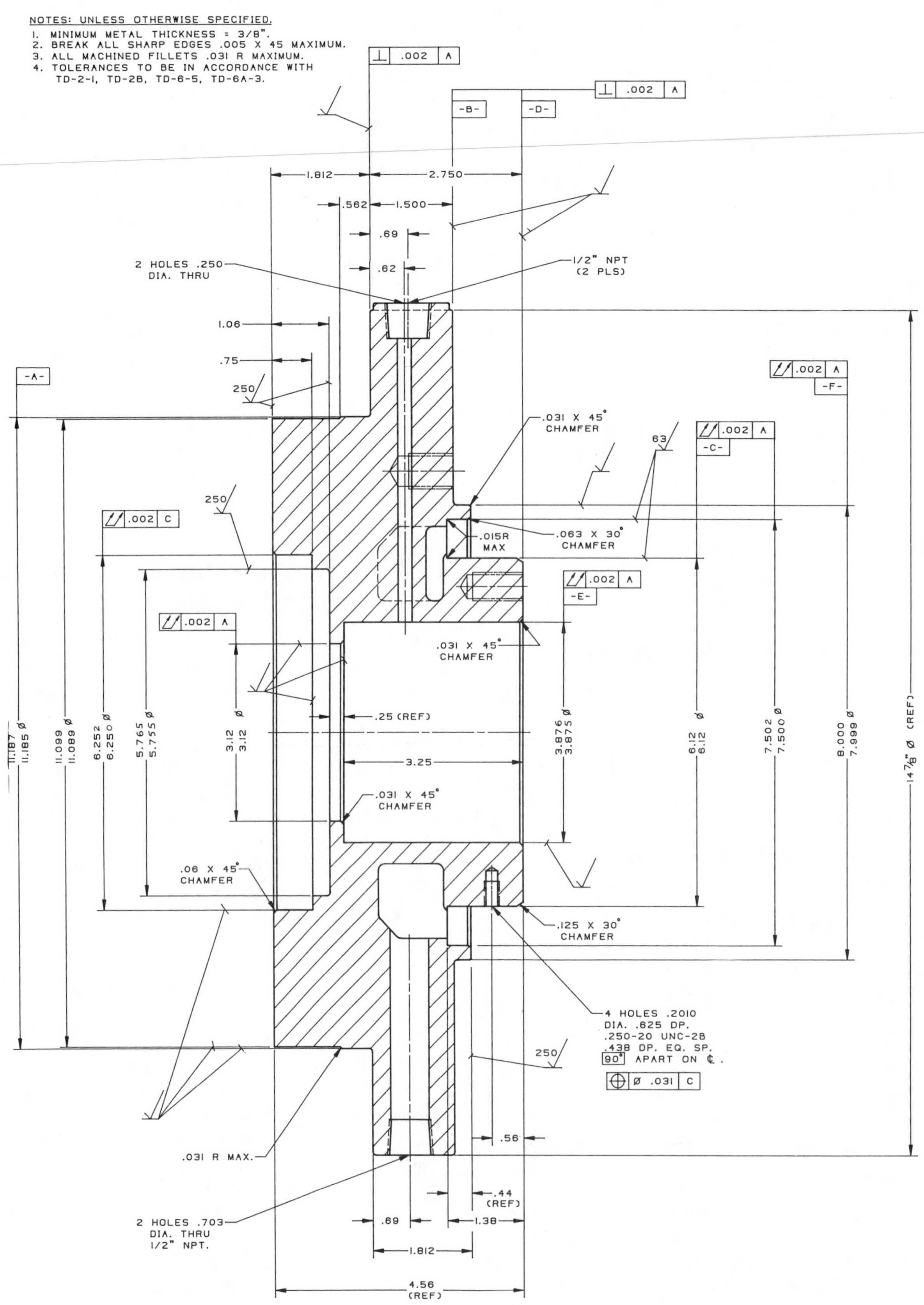

(a)

FIGURE 16.3 Cover Detail

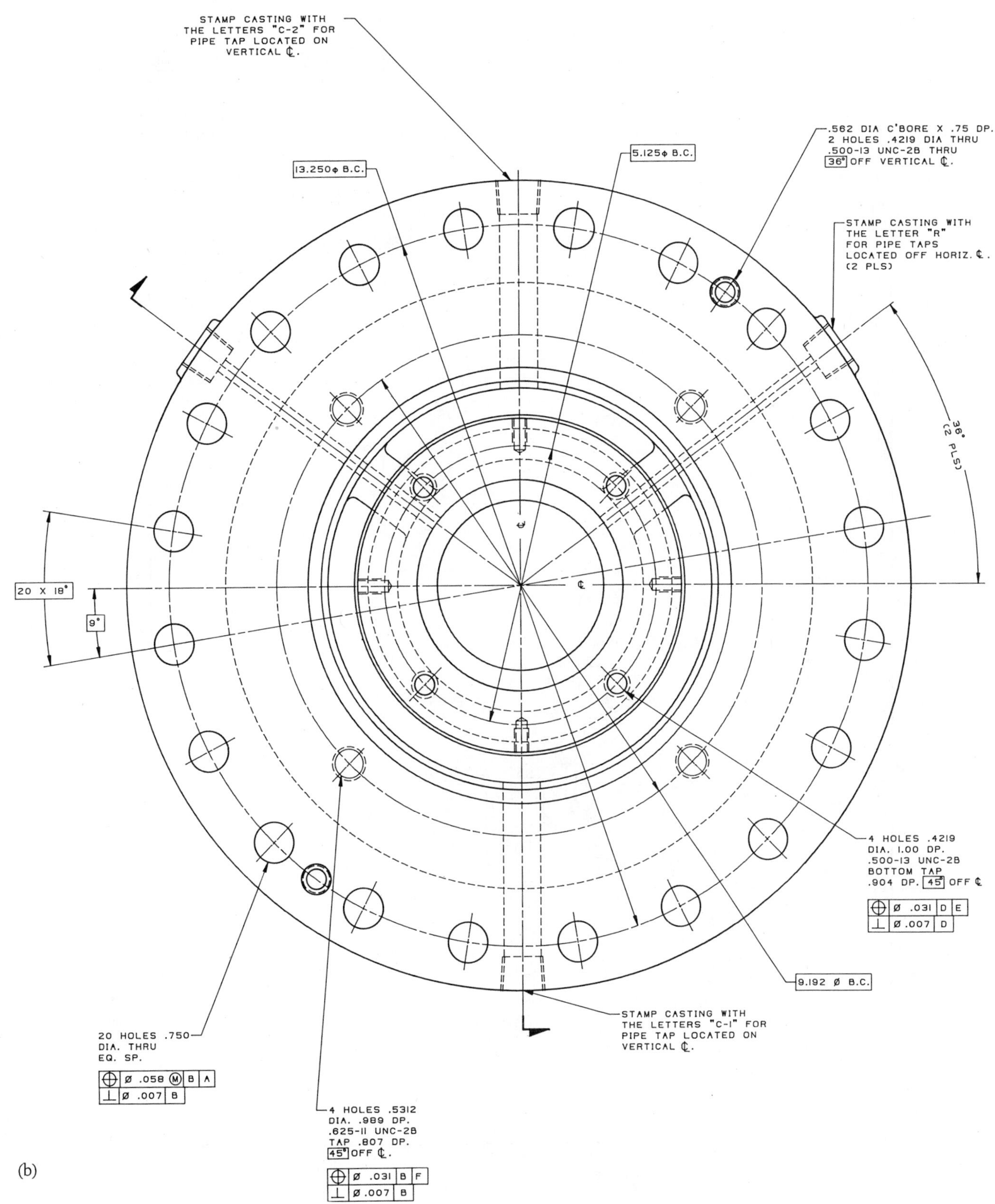
STAMP CASTING WITH
THE LETTERS "C-2" FOR
PIPE TAP LOCATED ON
VERTICAL ℄.
.562 DIA C'BORE X .75 DP.
2 HOLES .4219 DIA THRU
.500-13 UNC-2B THRU
36° OFF VERTICAL ℄.
5.125⌀ B.C.
13.250⌀ B.C.
STAMP CASTING WITH
THE LETTER "R"
FOR PIPE TAPS
LOCATED OFF HORIZ. ℄.
(2 PLS)
36°
(2 PLS)
20 X 18°
9°
℄
4 HOLES .4219
DIA. 1.00 DP.
.500-13 UNC-2B
BOTTOM TAP
.904 DP. 45° OFF ℄
Ø .031 D E
Ø .007 D
9.192 Ø B.C.
STAMP CASTING WITH
THE LETTERS "C-1" FOR
PIPE TAP LOCATED ON
VERTICAL ℄.
20 HOLES .750
DIA. THRU
EQ. SP.
Ø .058 M B A
Ø .007 B
4 HOLES .5312
DIA. .989 DP.
.625-11 UNC-2B
TAP .807 DP.
45° OFF ℄.
Ø .031 B F
Ø .007 B
(b)

FIGURE 16.3 Cover Detail—*Continued*

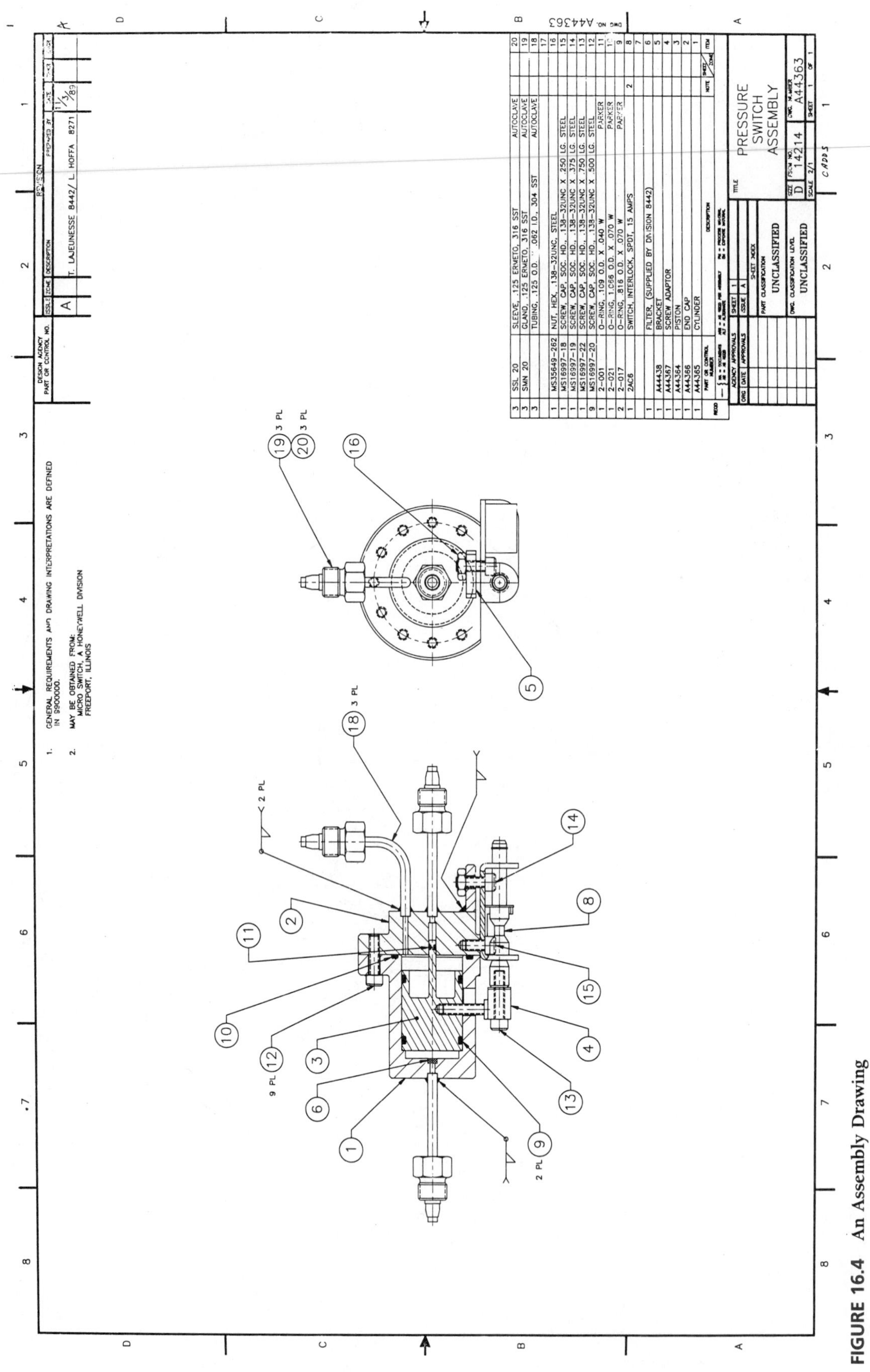

FIGURE 16.4 An Assembly Drawing

FIGURE 16.5 Disassembled Grove Flexflo Flexible Rubber Sleeve Valve

are also assemblies and are designed, laid out, and detailed similar to product assemblies.

16.4.1 Assembly Drawing Types

An **assembly drawing** may contain the detailed design requirements for one or more parts used in the assembly. It is prepared for each group of items that is to be joined to form an assembly and reflects either:

- A logical level in the assembly or disassembly sequence
- A functional unit
- A stocked standard off-the-shelf purchased item

The assembly may be shown on the same drawing sheet on which the details are shown or on a separate sheet. If the assembly is shown on a separate sheet, it will be sheet number one and the details shown on sheet number two, etc. All sheets bear the same drawing number.

An assembly drawing may define either a **separable** or **inseparable** assembly. **Welded assemblies** are inseparable assemblies. Figure 16.6 (see p. 440) is an example of a welded assembly. The parts list is a material stock list and simply lists the stock steel sizes of each piece. On most weldments, the assembly is considered a detail of a part and each piece is dimensioned on the drawing.

16.4.2 Mechanical Assemblies and Drawing Requirements

A mechanical assembly may have parts made from a sand mold, a permanent mold or die castings (see Fig. 16.7 on p. 441), rolled, extruded, or pressed shape forgings, plates, bars, sheet metal, or a combination of any two or more of these processes. These parts may be assembled into a complete unit by welding, brazing, soldering, riveting, bolting or other fastening methods. After assembly, additional work, such as a machining, may be necessary to complete the item.

The function of an assembly drawing is to provide a complete specification for joining, in proper relationship, two or more detail parts or subassemblies to form an assembly. This type of drawing usually includes a graphic layout of component parts, necessary notes, and, if a separate list or bill of material is not used, a tabulated list of parts. It should show the spatial relation of each part or subassembly, the method of fastening, and the type of fasteners.

When necessary, the assembly drawing indicates subsequent operations to form the completed item. For example, heat treatment, machining dimensions, and finishes are specified here. If a mechanical assembly is made entirely from cut shapes with sufficient information for cutting each piece, the information to fasten and finish them might be given on a single drawing.

16.4.3 Detail Drawings

Detail drawings are fully dimensioned, accurately laid-out engineering drawings of individual parts (see Fig. 16.8 on p. 442). All information needed to manufacture and produce the part is included. Adequate view description, correct dimensioning and tolerancing, accurate notes, and material designation are shown on the detail drawing. Components may appear on the assembly drawings or on separate details, or be established by written description.

16.4.4 Assembly Drawing Considerations

Assembly drawings for production parts may also be created from the detail drawings *after they have been approved by the checker*. This procedure gives a final check of the detail drawings for space clearances, limits, and satisfactory function in assembly.

Assembly drawings for jigs and fixtures are produced from the designer's sketches. The final assembly is broken into individual components that are detailed separately. The checker checks the assembly and the details in the final stage.

Product design, production volume, and facilities are the factors that determine the need for an assembly drawing. The quality of the finished product depends on the effective attachment methods, regardless of the quality of the individual parts.

Welded, soldered, or brazed parts that have a particular relation to one another and parts that are permanently assembled are shown in their assembled positions. Necessary dimensions and specifications are included for size control and other conditions. Parts that are pressed and line reamed in place, parts that are secured with pins, bushings, and similar assemblies, and parts that are machined after assembly require an assembly to drawing show these operations and specifications for assembly control. Parts that need a surface finish applied after assembly may require an assembly drawing to ensure proper overall finish. The drain assembly in Figure 16.9 (see p. 443) is an example of an assembled device.

16.4.5 View Selection and Dimensions on Assembly Drawings

Views are chosen to depict the assembly in its natural position in space, to define clearly how the parts fit together, and

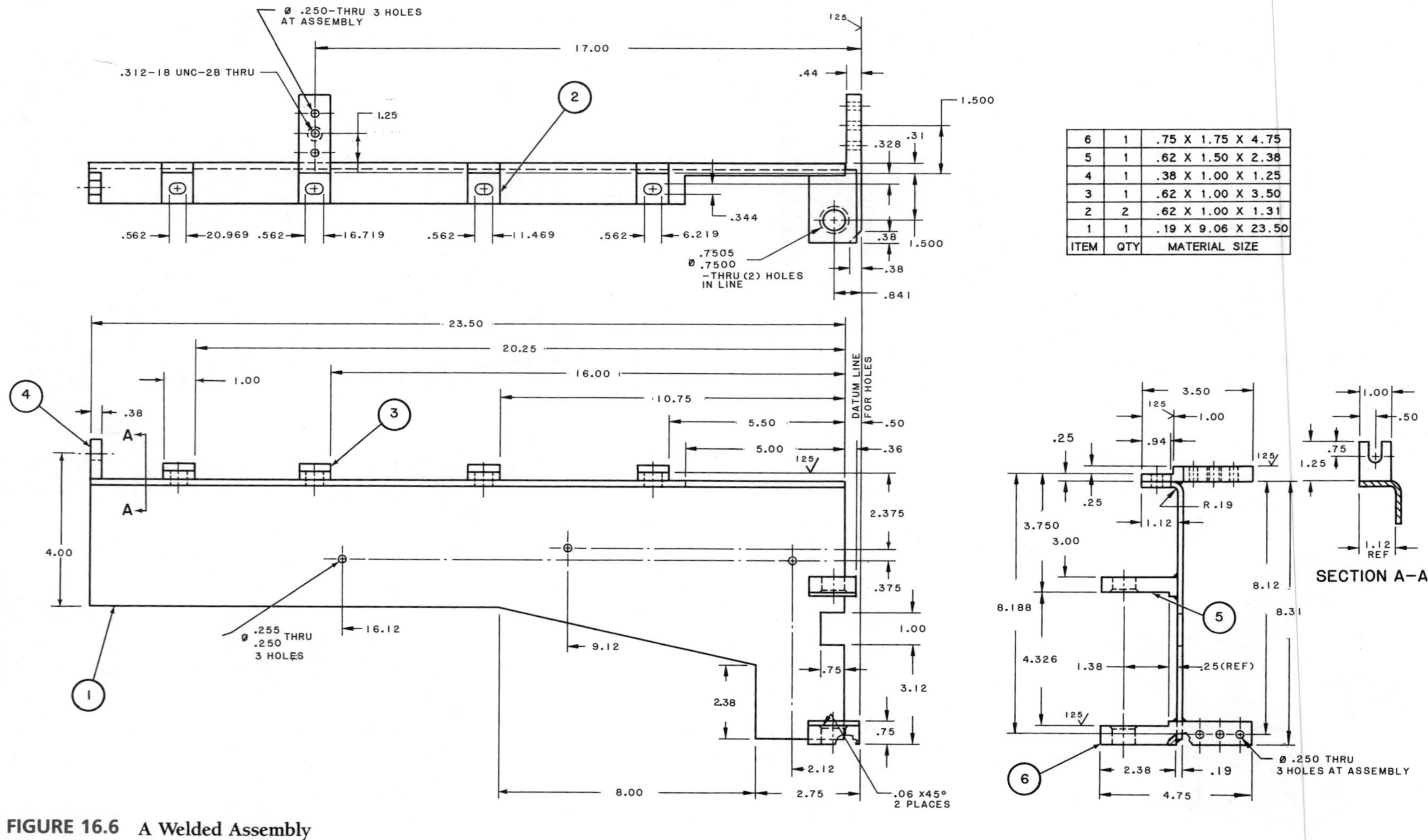

ITEM	QTY	MATERIAL SIZE
6	1	.75 X 1.75 X 4.75
5	1	.62 X 1.50 X 2.38
4	1	.38 X 1.00 X 1.25
3	1	.62 X 1.00 X 3.50
2	2	.62 X 1.00 X 1.31
1	1	.19 X 9.06 X 23.50

FIGURE 16.6 A Welded Assembly

FIGURE 16.7 Die Cast Parts

to describe the functional relationship of the parts. The minimum number of views needed to define the assembly should be used. Often only one view is required.

Dimensions on assembly drawings are confined to setup dimensions, dimensions needed for assembly, dimensions required for machining after assembly, and clearance dimensions. Overall dimensions (height, width, and depth) may be included on the assembly for packaging assistance. When necessary, open and closed positions of movable parts on the assembly are given.

16.4.6 Hidden, Crosshatching, and Phantom Lines on Assembly Drawings

Hidden lines should not be shown on assembly drawings, especially if their use would confuse the reader. Instead, section views are used to show the relationship of internal parts. Conventional section lining may be used. Material symbols are optional on assembly drawing sections. The section lines are drawn at angles to the object outlines and should be drawn at a different angle for each adjacent part.

If there is some doubt, the note FRONT, to indicate the forward operating position, should be added to the detail drawing of a part to indicate the position of the part in the assembly. Showing a part before an assembly process operation by using phantom lines and after the operation by using object lines is also an accepted procedure.

16.5 Assembly Drawings and Parts Lists

An **assembly** is a combination of two or more parts joined in one working unit. A **subassembly** is an assembly of parts that aid in producing a larger assembly. The purpose of an assembly drawing is to show the spatial relation of each part to the others and to identify all parts in the assembly by a number for each unique part. Assembly drawings include a list of each part of the assembly. This list is called a **parts list.** The parts list must be keyed to the drawing so that individual parts are clearly identified on the assembly. This is done by *ballooning* the drawing (Fig. 16.4).

Assembly drawings consist of two parts: an assembly delineation drawing and a parts list, which is integral or separate.

16.5.1 Parts Identification on an Assembly

Ballooning is the process of identifying each part in the assembly. Each part has a circle with a number inside it and a leader extending from the balloon to the piece and ending with an arrowhead. Balloons are either placed in a line down the middle of the drawing, horizontally or vertically, or they are scattered throughout the drawing. The choice of methods depends on the complexity of the assembly. Leader lines from balloons should not cross and can be straight lines or curved. Balloon circles are drawn anywhere from .5 to .75 in. in diameter (12 to 20 mm) with a template on assemblies.

In some cases, the parts of an assembly are called out on the drawing without the aid of numbered balloons. In Figure 16.10, the shutter assembly has been displayed pictorially. Each of the parts is described in notes that have leaders pointing to the part.

16.5.2 Parts Lists

The assembly drawing must have a complete parts list. Each parts list includes individual part numbers, the name and description of each part, and the material and quantity of all items required for one complete assembly (see Fig. 16.11 on p. 444).

On tool and die drawings, the parts list is sometimes called a **stock list.** Here, material allowance is added for purchase information. Die drawings also have a stock list on the assembly drawing sheet.

The assembly parts list is placed above the title block when on the same sheet as the drawing. The part numbers are arranged to read upward so that new parts can be added if needed. The precise method of listing parts varies with companies. The parts or stock list sometimes use the vertical line divisions of the revision record (block at top right of the drawing form). Spacing of horizontal divisions is uniform in both revision blocks and the parts list.

Parts List Heading Arrangement When an integral parts list is included on a drawing sheet the heading PARTS LIST is placed on the bottom of the list and the part numbers read upward. Figure 16.12 (see p. 445) is an example of a company title block with a parts list. The quantity, item number (balloon number), part number, and description are shown here.

When the parts list is separate from the part drawing, the heading PARTS LIST is at the top and the list is constructed from the top down. The following describes the four basic columns (Fig.16.13) used on a parts list:

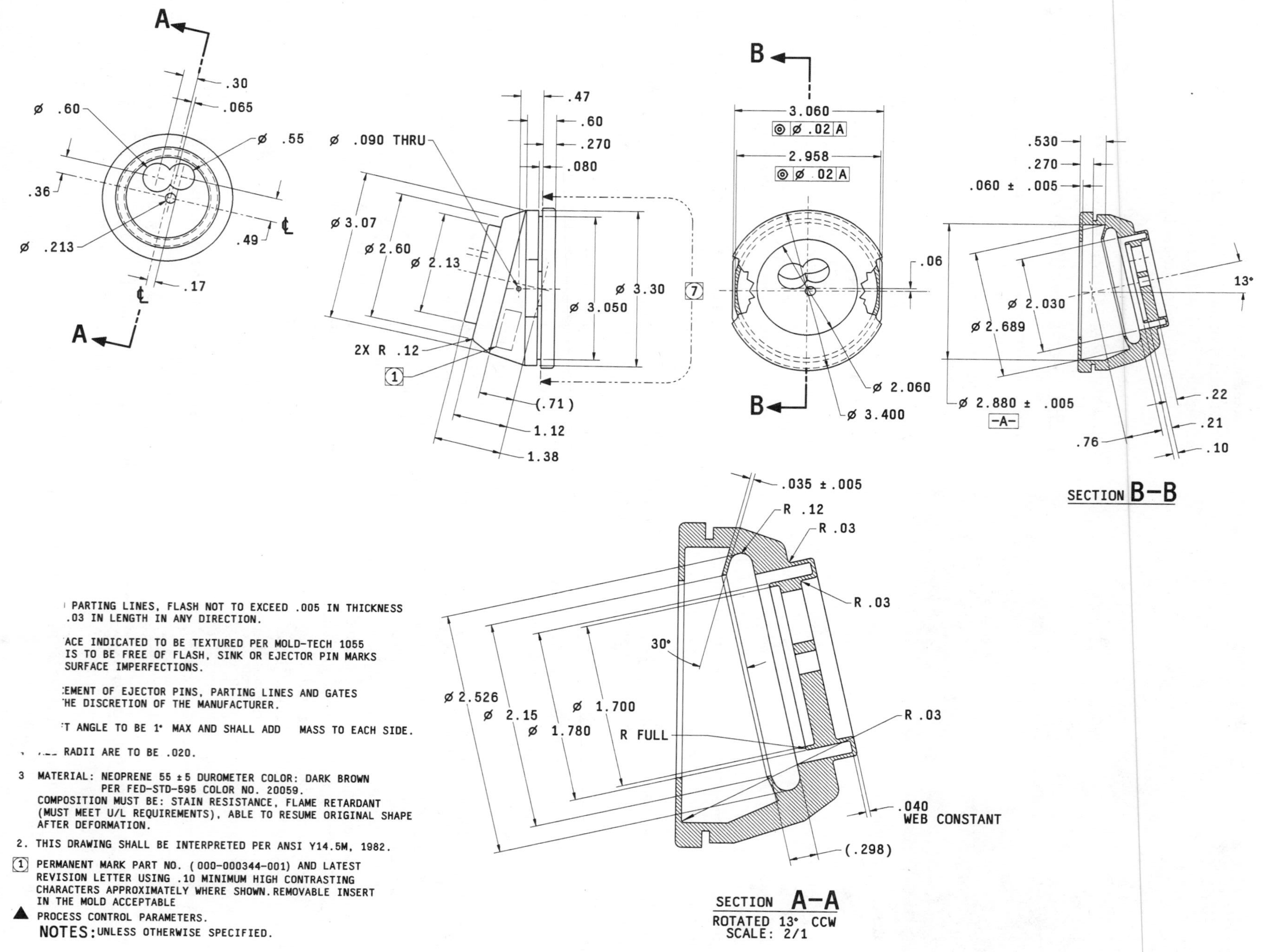

FIGURE 16.8 Acoustic Microphone Cup

FIGURE 16.9 Multiport Drain

1. *Quantity required:* The number entered in this column denotes the quantity, volume, length, or other unit of measure required to complete one of the items to which the column applies. When this number applies to other than quantity, the unit of measure is entered in this column or in an optional unit of measure column.
2. *FSCM:* The Federal Supply Code for Manufacturer's number assigned to the originating design activity whose part or identifying number appears in column 3 is shown in this column. Notice that many company title blocks do not include this entry.
3. *Part or identifying number:* The identifying number for each item on the parts list is shown in this column.
4. *Nonmenclature or description:* The assigned noun or name of the item whose identifying number appears in the parts number column appears in this column.

16.5.3 Drawing Sheets

Drafting sheets are preprinted sheets with the border, title block (Fig. 16.14), and revision block preprinted on polyester film, vellum, bond, or other type of paper. You must know what each aspect of the drawing sheet means and what information must be added before the project is complete. This understanding is also important when reading existing drawings. The following list describes each part of a typical drawing sheet as shown in Figures 16.15(a) and (b):

1. Ancillary drawing number. Permits the drafter to file print copies so that, when folded correctly, all drawing numbers will appear in the upper left corner.
2. Sheet number for multiple sheet drawings.
3. Ancillary revision identification.
4. Revision identification symbol.
5. Description of the revision or the identification of the change authorization document.
6. Issue date of the revised drawing.
7. Required approval signature for revisions.
8. Microfilm alignment arrowheads.
9. DSJ, distribution key or code, if used.
10. Company name and address. Must agree with FSCM number for companies with multidivisions and departments.

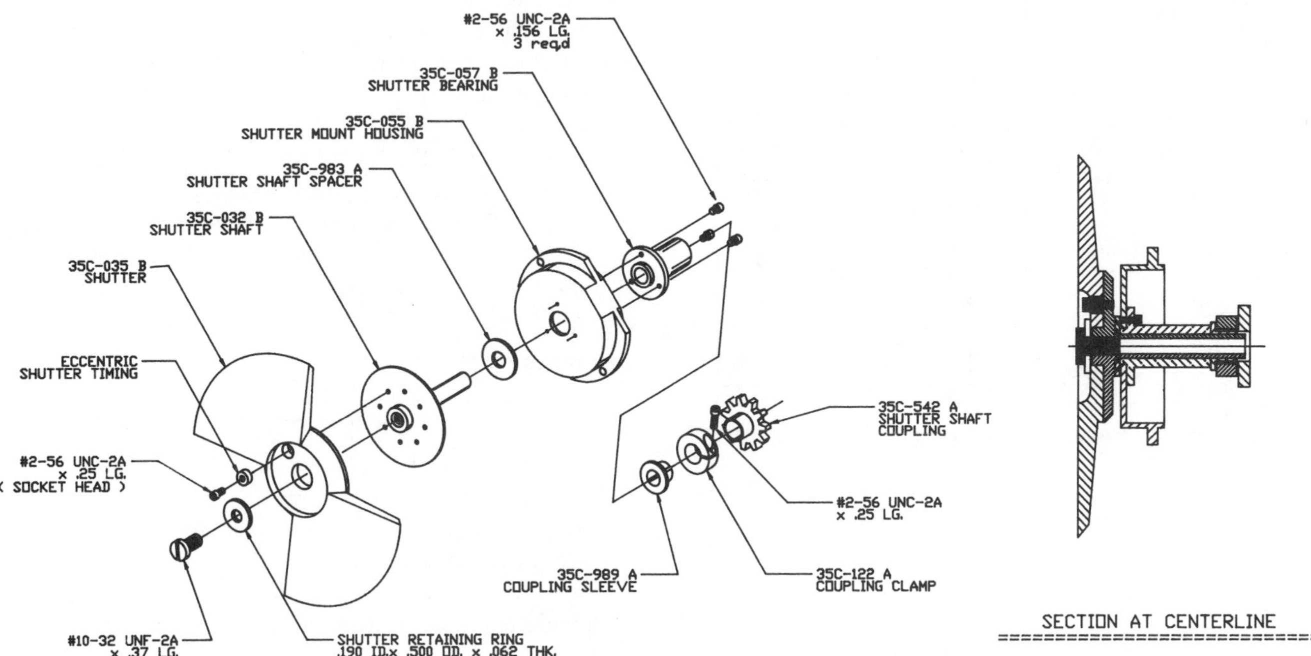

FIGURE 16.10 UltraCam Shutter Assembly

ITEM	QTY	PART NO.	DESCRIPTION
55	1	664-359239	CABLE, ASSY - MONOCHRONOMETER, YEL
54	1	664-359238	CABLE, ASSY - MONOCHRONOMETER, GRN
53	1	664-359237	CABLE, ASSY - MONOCHRONOMETER, RED
52			
51			
50	1	693-349584	BOARD, P.W. - U.V. SCANNER
49	1	301-961283	SPRING, CPRSN-.063 OD X 1 LG CRES
48			
47			
46	12	165-359506	WASHER, FLT .127 TFL-
45			
44	1	125-362041	PIN, DOWEL -.1553 DIA X .450L CRES
43	4	125-811591	PIN, DOWEL -.1251 DIA X .375L CRES
42	4	125-824305	PIN, DOWEL -.0626 DIA X .312L CRES
41			
40			
39	3	130-961281	RING, RETAINING - EXT, .073 ID
38	2	105-828447	NUT, HEX 2-56 S-BK
37			
36			
35	6	101-827620	SCREW, CAP 4-40 X 1" S-HXSO
34	2	101-961013	SCREW, CAP 2-56 X 1-3/8 S-HXSO
33	2	101-803947	SCREW, CAP 2-56 X .875 S-HXSO
32	1	101-961282	SCREW, CAP 2-56 X .750 S-HXSO
31	2	101-961201	SCREW, MACH 2-56 X 1" FL-S-SL
30	2	101-828409	SCREW, MACH 2-56 X .375 P-S-BK-SL
29	3	101-828408	SCREW, MACH 2-56 X .25 P-S-BK-SL
28			
27			
26	2	201-361974	GEAR, (MODIFIED) 42 TEETH
25	1	201-359899	GEAR, (MODIFIED) 84 TEETH
24	3	201-349033	GEAR, (MODIFIED) 132 TEETH
23			
22			
21	1	150-359263	BUSHING, MIRROR HOUSING
20	6	145-863274	BEARING
19			
18			
17	1	201-356344	RACK, GEAR U.V. MODIFICATION
16	1	520-348987	GRATING
15	1	333-349032	PLATFORM, GRATING
14	1	223-356345	SHAFT, CROSSOVER U.V. MODIFICATION
13	4	223-349019	SHAFT, GEAR
12			
11	1	548-361969	ASSY. FILTER
10			
9	1	499-348983	SPUD
8	1	105-348982	NUT, RETAINER
7	1	178-348984	EXTENSION TUBE #1, U.V. SCANNER
6	1	178-349027	EXTENSION TUBE #2, U.V. SCANNER
5	1	178-347010	EXTENSION TUBE #3, U.V. SCANNER
4	1	299-348981	HOUSING, MIRROR
3			
2	1	299-349023	DIRECTION CHAMBER U.V. LIGHT (LEFT)
1	1	299-348989	DIRECTION CHAMBER U.V. LIGHT (RIGHT)

E SIZE | 223-356367 DWG. NO.

BY	DATE
DR *D.M.DUARTE*	1/18/90
CHK	
DSGN	
ENGR	

BECKMAN — BECKMAN INSTRUMENTS, INC. SPINCO DIVISION 1050 PAGE MILL ROAD PALO ALTO, CALIFORNIA 94304

TITLE MONOCHRONOMETER ASSEMBLY

E SIZE | CODE IDENT NO. 07978 | DWG NO. 223-356367

MOD L10-A | SCALE 1/2 | 1ST USE 355899 | SHEET 1 OF 1

FIGURE 16.11 Monochrometer Assembly Parts List

11. Drawing title.
12. Assigned drawing number.
13. Weight record. Should indicate whether it is actual, estimated, or calculated, when required, and if it is gross (before machining) or net (after machining).
14. FSCM number. If required for identification of the company or design activity whose drawing number is used.
15. Predominant scale of the drawing.
16. Drawing size letter designation.
17. The signature of the drafter and the date the drawing was started.
18. The signature and date of the responsible person who checked the drawing.
19. The signature and date of the responsible engineer to signify the approval of the design by engineering.
20. Signature of responsible issuing person and initial date of issue.
21. Notes.
22. Approval by an activity other than those described previously.
23. The appropriate surface texture designation that applies.
24. The general tolerances that apply to the overall document.
25. The appropriate material specification that should include type, grade, class, or other classifications as applicable.
26. Zones: letters vertically (bottom up, starting with A) and numbers horizontally (right to left, starting with one).

16.6 Notes

Notes are used on drawings (see Fig. 16.16 on p. 448) to supply information that cannot be presented in any other descriptive way. A standard method of applying, placing, and revising notes on engineering drawings is used to maintain company-wide uniformity. Although there are standard formats, placements, and sequences for notes on the drawing, each company will have their own **company standards.**

General notes (Fig. 16.16) are those that apply to the total drawing and, if placed on the drawing at each point of application, would be repetitive and time-consuming to apply. **Local notes** are those that apply to a specific portion, surface, or dimension on a drawing. The following rules can be used as a guide when putting notes on a drawing:

Use notes to:

- Clarify features that can be more accurately defined by words than by graphical delineation and dimensions.
- Give instructions for applying special treatments.
- Give instructions for utilization of specific processes.
- Describe instructions to supplement standard symbols.
- Provide additional information to the drawing document or to its use.
- Provide notes so the part can be made correctly the first time.

2	13	140-862005-606	STUD,SELF-CLINCH-FH, NO. 10-32 X.750 LG
2	12	140-862005-206	STUD,SELF-CLINCH-FH, NO.4-40 X .750
4	11	104-044042-014	FASTENER,SELF-CLINCHING, NO. 10-32
26	10	140-862005-404	STUD SELF-CLINCH-FH, NO.8-32 X .500
22	9	104-044042-002	FASTENER,SELF-CLINCHING, NO. 8-32
4	8	106-044316-006	WASHER EXTERNAL TOOTH, NO. 6
8	7	102-044729-003	LOCKNUT, HEX, NO. 2-56
2	6	140-017322-011	FASTENER,SELF-CLINCHING SS NO. 4-40
8	5	102-044629-001	NUT, SADDLE
7	4	104-045364-001	NUT, HEX JAM, NO. 1/4-20
6	3	104-044356-003	INSERT,THREADED STAINLESS STEEL,NO. 8-32 X.248 LG
3	2	140-021009-001	NUT-HEX, NO. 8 LIGHT
1	1	000-012345-051	COVER,CONTROL,FREQUENCY PANEL (13)
-001	ITEM NO.	VERSATEC PART NO.	DESCRIPTION
QUANTITY PER VERSION			PARTS LIST

PROPRIETARY
The contents of this document are PROPRIETARY TO VERSATEC INC. and are not to be disclosed to others or used for purposes other than intended without the written approval of Versatec

MATERIAL:
SEE ABOVE P/L DATA BASE B.O.M AVAILABLE.

UNLESS OTHERWISE SPECIFIED
DIMENSIONS ARE IN INCHES. ALL PARTS TO BE DEBURRED AND EDGES BROKEN .010 MAXIMUM

TOL.	1PLC.	2PLC.	3PLC.	ANG.
±	.1	.03	.010	1°--

THIRD ANGLE PROJECTION

FINISH: [7]

SIGNATURE	DATE
DRN. VALENZUELA	9-10-90
CHK.	- -
APPV.	- -
APPV.	- -
APPV.	- -
DO NOT SCALE DRAWING	

VERSATEC A XEROX COMPANY — SANTA CLARA CALIFORNIA 95051

TITLE: COVER,CONTROL, FREQUENCY PANEL

SIZE	CODE IDENT.	DRAWING NO.	REV.
E	50804	000-012345	17

SCALE: 1/2 , AS NOTED — SHEET 1 OF 1

2 1 A

FIGURE 16.12 Cleaning Kit Box Title Block and Parts List

QTY REQD	FSCM	PART OR IDENTIFYING NO.	NOMENCLATURE OR DESCRIPTION
PARTS LIST			

(a) COLUMNAR ARRANGEMENT FOR INTEGRAL PARTS LIST

PARTS LIST			
QTY REQD	FSCM	PART OR IDENTIFYING NO.	NOMENCLATURE OR DESCRIPTION

(b) COLUMNAR ARRANGEMENT FOR SEPARATE PARTS LIST

FIGURE 16.13 Parts List Arrangement (a) Columnar arrangement for integral parts list (b) Arrangement for separate parts list

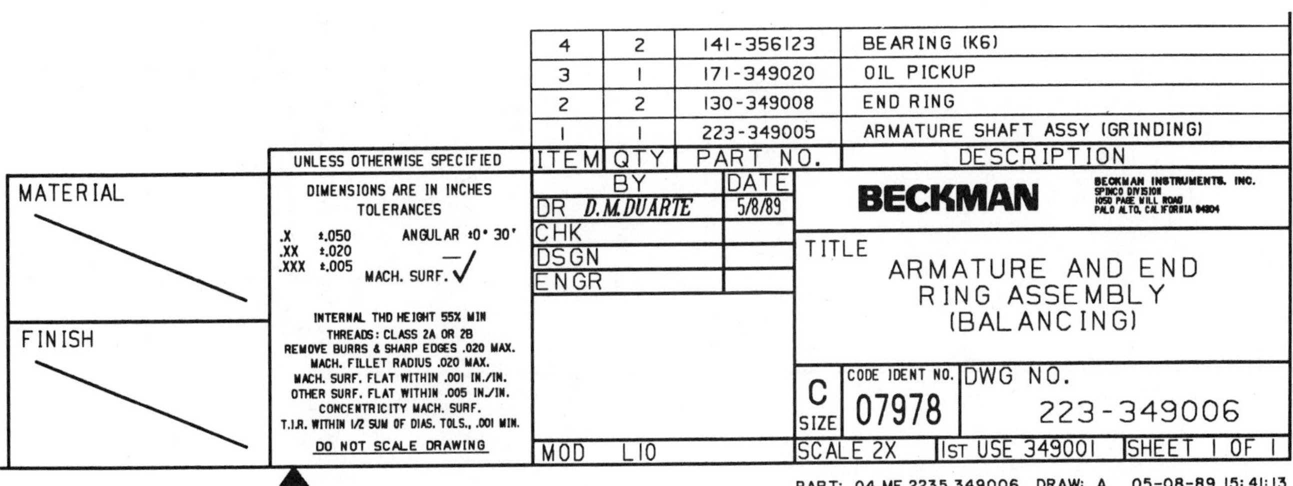

4 2 141-356123 BEARING (K6)
3 1 171-349020 OIL PICKUP
2 2 130-349008 END RING
1 1 223-349005 ARMATURE SHAFT ASSY (GRINDING)
ITEM QTY PART NO. DESCRIPTION
MATERIAL
FINISH
UNLESS OTHERWISE SPECIFIED
DIMENSIONS ARE IN INCHES
TOLERANCES
.X ±.050
.XX ±.020
.XXX ±.005
ANGULAR ±0° 30'
MACH. SURF.
INTERNAL THD HEIGHT 55% MIN
THREADS: CLASS 2A OR 2B
REMOVE BURRS & SHARP EDGES .020 MAX.
MACH. FILLET RADIUS .020 MAX.
MACH. SURF. FLAT WITHIN .001 IN./IN.
OTHER SURF. FLAT WITHIN .005 IN./IN.
CONCENTRICITY MACH. SURF.
T.I.R. WITHIN 1/2 SUM OF DIAS. TOLS., .001 MIN.
DO NOT SCALE DRAWING
BY DATE
DR D.M.DUARTE 5/8/89
CHK
DSGN
ENGR
BECKMAN
TITLE
ARMATURE AND END RING ASSEMBLY (BALANCING)
C SIZE
CODE IDENT NO. 07978
DWG NO. 223-349006
MOD L10
SCALE 2X
IST USE 349001
SHEET 1 OF 1
PART: 04.ME.2235.349006 DRAW: A 05-08-89 15:41:13

FIGURE 16.14 Armature and End Ring Assembly Title Block and Parts List

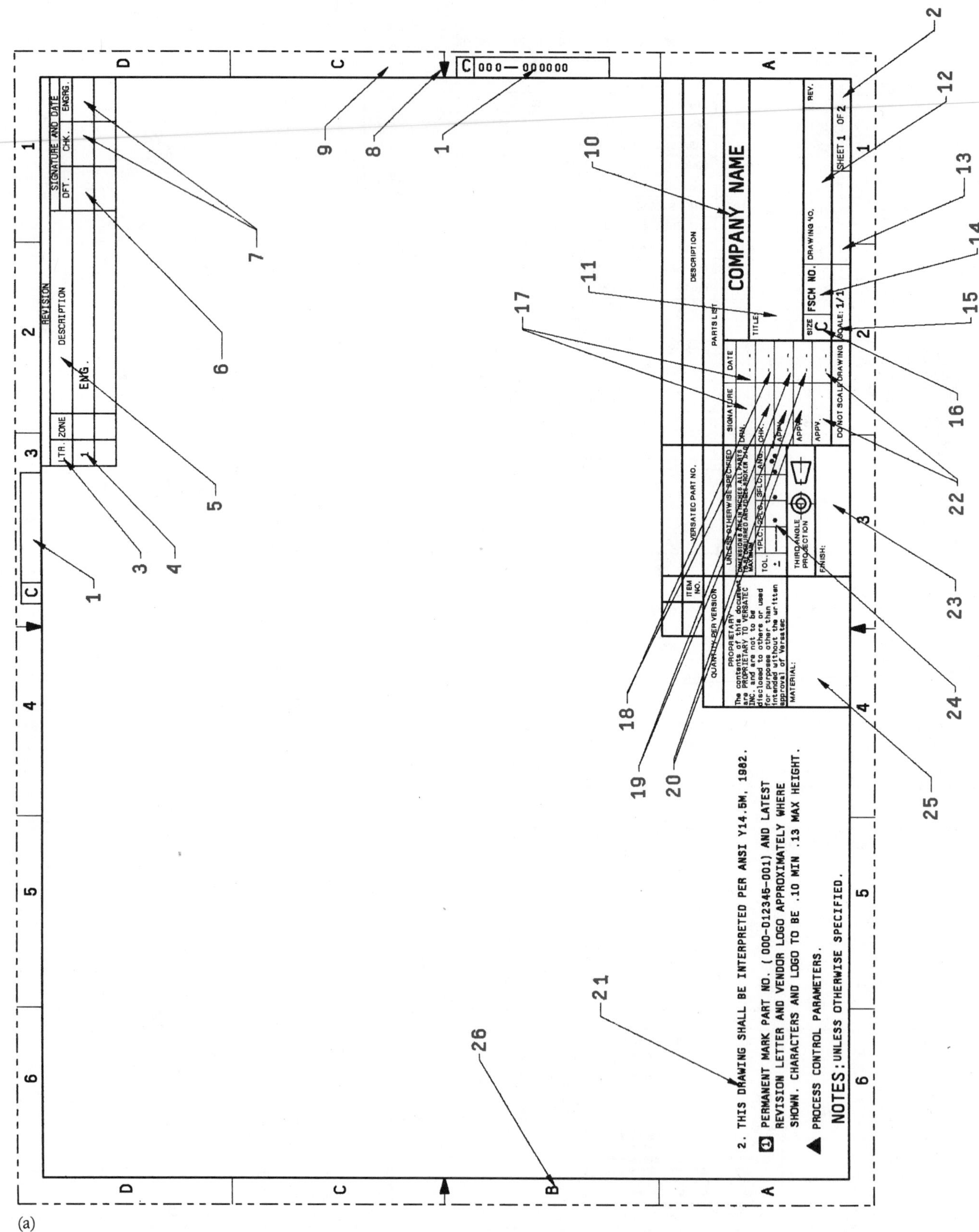

(a)

FIGURE 16.15 Standard Drawing Sheet Format

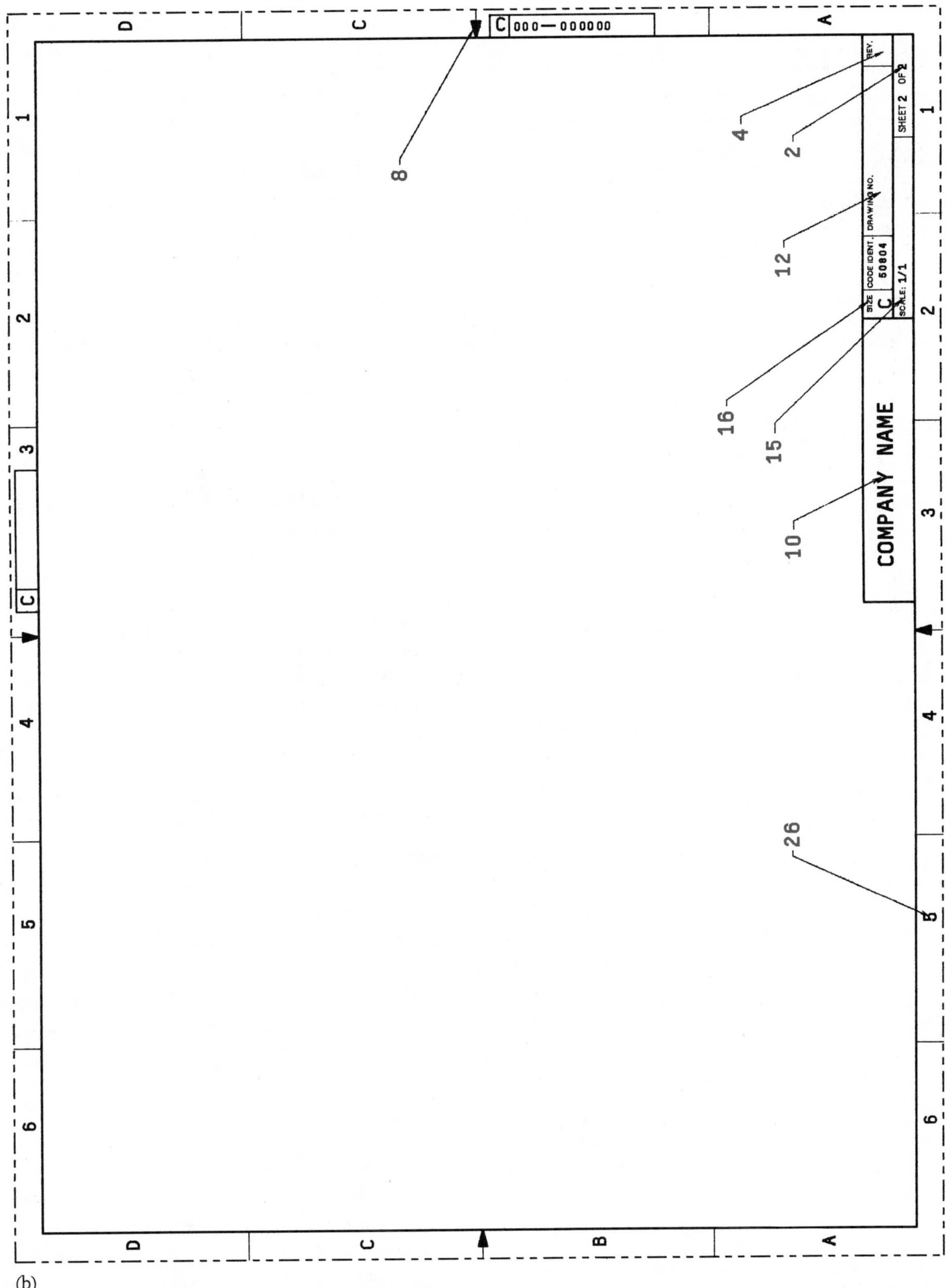

(b)

FIGURE 16.15 Standard Drawing Sheet Format—*Continued*

A

4 ASSEMBLY METHOD AT VENDOR OPTION, WITH MANUFACTURING ENGINEERING APPROVAL.

[3] FOR CARTON SPECIFICATION REFER TO ITEM (2) .

2. THIS DRAWING SHALL BE INTERPRETED PER ANSI Y14.5M, 1982.

1. PKG & IDENTIFY WITH P/N (000-000345-001) AND LATEST REVISION LETTER.

▲ CRITICAL FUNCTIONALITY PARAMETER.

NOTES: UNLESS OTHERWISE SPECIFIED.

FIGURE 16.16 General Notes

Notes should be:

- Clear and concise.
- In the present tense.
- Positioned parallel to the bottom edge of the drawing.
- Carefully composed to relay one message; capable of one interpretation.
- Preprinted on appliqués or, if the drawings are computerized, entered in a standard library for repeated use.

Notes should not:

- Be underlined on the drawing.
- Contain abbreviations other than the most commonly understood shop trade terms.
- Duplicate information on an associated parts list or shop practices reference document.
- Contain dimensions that are already documented elsewhere on the drawing.
- Describe complex processes. This kind of information should be documented either in a specification or process document.
- Reference information that is given elsewhere in the product documentation.

16.6.1 General Notes

General notes on drawing sizes "B" through "F" are placed in the upper left-hand or lower left-hand corner of the sheet. Some companies construct their note sequence from the bottom-up, as in Figure 16.17; others from the top-down (Fig. 16.18). Notes at the top of a sheet are numbered top-down; those at the bottom part of a sheet are numbered bottom-up. The width of the general note column should not exceed 6 to 8 inches (150 to 200 mm).

16.6.2 Local Notes

The placement of **local notes** on the drawing must be outside the outline of the part. Position the note close to the item that is being referenced.

Fabrication operations such as **BEND**, **DRILL**, **PUNCH**, **TAP OR BORE** are not shown on the drawing. This permits manufacturing to determine the type of operation required to produce the part within the required tolerances. Features such as **SPOTFACE**, **COUNTERBORE**, **COUNTERSINK**, **UNDERCUT**, and **THREAD** may be used in notes.

16.7 Revision of Engineering Drawings

The **revision block** (Fig. 16.19), located in the upper right-hand corner of the drawing, is used to record changes to the drawing. Revisions initiate a substantial number of change documents in all functions of a business. Therefore, the need for accuracy and completeness in the revision process should not be underestimated.

16.7.1 Revision Terminology

The following terms are used to describe the process of revising drawings and are found in revision blocks on drawings:

Added A new feature or view introduced to the document.

Approval An endorsement attesting to a revision made on a drawing or parts list.

Change A specific alteration made as part of a revision on a drawing. A revision may include one or more changes.

Deleted A feature or view removed from the document.

Obsolete (inactive, canceled) A condition in which the

A

[7] PAINT YELLOWISH GREY (CODE BAP) AND REIDENTIFY PAINTED PART AS BAP-012345-001, PER PAINT SPECIFICATION DWG 000-014123.

6 NOTE DELETED.

5 NOTE DELETED.

4 BEND RELIEF ⌀ .125 OPTIONAL.

3 ALL BEND RADII ARE TO BE MINIMUM.

2. THIS DRAWING SHALL BE INTERPRETED PER ANSI Y14.5M, 1982.

[1] PERMANENTLY MARK PART NO. (000-012345-001) AND LATEST REVISION LETTER AND VENDOR LOGO APPROXIMATELY WHERE SHOWN. CHARACTERS AND LOGO TO BE .10 MIN .13 MAX HEIGHT.

▲ PROCESS CONTROL PARAMETERS.

NOTES: UNLESS OTHERWISE SPECIFIED.

[13] MATERIAL: GALVANEAL ZINC PRE-PLATED, CARBON STEEL,AISI COLD ROLLED 14GA (.075 THICK).

[12] AREAS INDICATED ARE TO BE MASKED AS SHOWN.

11 HOLE DIMENSIONS APPLY TO UNPAINTED PART.

[10] FEATURE TO BE FREE OF PAINT.

[9] FEATURE TO BE SINGLE PUNCHED SUCH THAT FEATURE SIZE SURFACE IS FREE OF BURRS/PROTRUSIONS GREATER THAN .002.

[8] PAINT SPECIFICATION TEXTURE APPLIES TO THESE SURFACES ONLY.

8 7 6 5

FIGURE 16.17 Notes Listed from Bottom-up

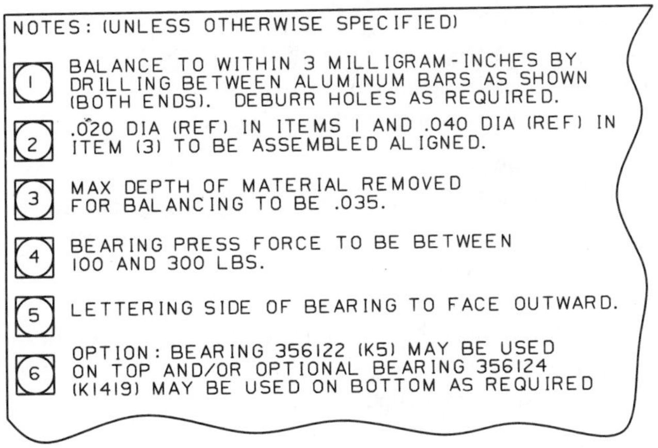

NOTES: (UNLESS OTHERWISE SPECIFIED)

1. BALANCE TO WITHIN 3 MILLIGRAM-INCHES BY DRILLING BETWEEN ALUMINUM BARS AS SHOWN (BOTH ENDS). DEBURR HOLES AS REQUIRED.
2. .020 DIA (REF) IN ITEMS 1 AND .040 DIA (REF) IN ITEM (3) TO BE ASSEMBLED ALIGNED.
3. MAX DEPTH OF MATERIAL REMOVED FOR BALANCING TO BE .035.
4. BEARING PRESS FORCE TO BE BETWEEN 100 AND 300 LBS.
5. LETTERING SIDE OF BEARING TO FACE OUTWARD.
6. OPTION: BEARING 356122 (K5) MAY BE USED ON TOP AND/OR OPTIONAL BEARING 356124 (K1419) MAY BE USED ON BOTTOM AS REQUIRED

FIGURE 16.18 Notes Listed from Top-down

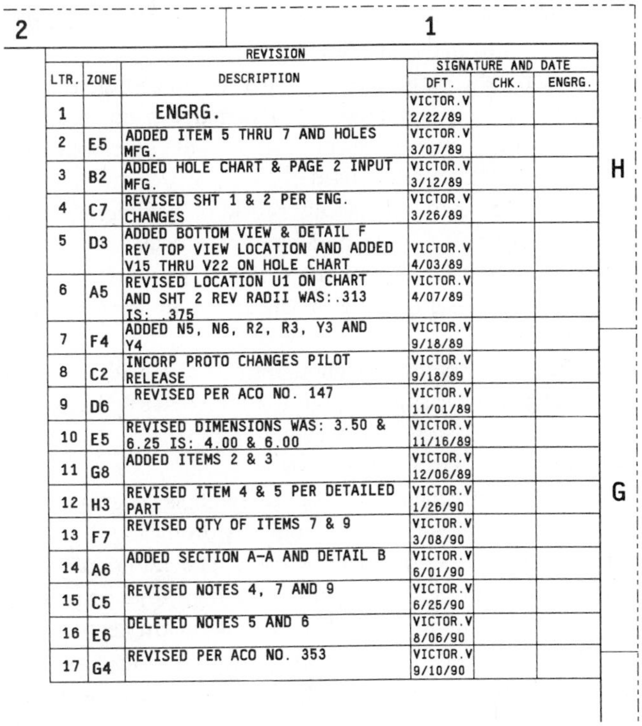

2 1

REVISION					
			SIGNATURE AND DATE		
LTR.	ZONE	DESCRIPTION	DFT.	CHK.	ENGRG.
1		ENGRG.	VICTOR.V 2/22/89		
2	E5	ADDED ITEM 5 THRU 7 AND HOLES MFG.	VICTOR.V 3/07/89		
3	B2	ADDED HOLE CHART & PAGE 2 INPUT MFG.	VICTOR.V 3/12/89		
4	C7	REVISED SHT 1 & 2 PER ENG. CHANGES	VICTOR.V 3/26/89		
5	D3	ADDED BOTTOM VIEW & DETAIL F REV TOP VIEW LOCATION AND ADDED V15 THRU V22 ON HOLE CHART	VICTOR.V 4/03/89		
6	A5	REVISED LOCATION U1 ON CHART AND SHT 2 REV RADII WAS: .313 IS: .375	VICTOR.V 4/07/89		
7	F4	ADDED N5, N6, R2, R3, Y3 AND Y4	VICTOR.V 9/18/89		
8	C2	INCORP PROTO CHANGES PILOT RELEASE	VICTOR.V 9/18/89		
9	D6	REVISED PER ACO NO. 147	VICTOR.V 11/01/89		
10	E5	REVISED DIMENSIONS WAS: 3.50 & 6.25 IS: 4.00 & 6.00	VICTOR.V 11/16/89		
11	G8	ADDED ITEMS 2 & 3	VICTOR.V 12/06/89		
12	H3	REVISED ITEM 4 & 5 PER DETAILED PART	VICTOR.V 1/26/90		
13	F7	REVISED QTY OF ITEMS 7 & 9	VICTOR.V 3/08/90		
14	A6	ADDED SECTION A-A AND DETAIL B	VICTOR.V 6/01/90		
15	C5	REVISED NOTES 4, 7 AND 9	VICTOR.V 6/25/90		
16	E6	DELETED NOTES 5 AND 6	VICTOR.V 8/06/90		
17	G4	REVISED PER ACO NO. 353	VICTOR.V 9/10/90		

H G

FIGURE 16.19 Revision Block on Format

drawing has been discontinued by the design activity. The words *inactive* or *canceled* may be used.

Redrawn A new original drawing with the same drawing number that has been substituted for a previous drawing.

Revision (revised) One or more changes to a drawing, made after distribution or release, according to an established revision procedure.

Revision designation Alphabetic, numeric, or alphanumeric characters that identify a revision.

16.7.2 Revising Drawings

Revisions are made by erasure, crossing out, addition, or redrawing. When evaluating the method to be used to revise a drawing, first give consideration to achieving and maintaining the best possible quality, legibility, and reproducibility by the most economical means. Unless otherwise specified, use the most recently approved graphics symbols, designations and letter symbols, abbreviations, and drawing practices. The exception to this is the use of geometric and position tolerance symbols that may be different from the latest issue of ANSI Y14.5. If use of the latest symbol is desired, an explanatory note should be provided on the drawing. Superseded symbology on the drawing should remain unchanged, provided the interpretation is clear.

16.7.3 Incorporating Changes

Dimensional changes entered on a drawing are made to the same scale as the portion of the drawing undergoing revision. If the drawing is not to scale, and the pictorial portion of the drawing is made to proportion, all dimensional changes are made to the proportions of the delineation affected.

When information is added to a drawing, the additions must match the lettering style and line weight of the existing drawing as closely as possible.

16.7.4 Simplifying the Drafting Process and Saving Time

Saving time on a project may mean bringing it to market ahead of the competition. Overdrawn and detailed designs add time and cost to a project. The following list can be used to check for simplicity:

1. Use text description wherever possible to eliminate drawing completely.
2. Use text description wherever practical to eliminate projected views.
3. Eliminate views where the shape can be given by description, e.g., HEX, SQ, DIA.
4. Show partial views of symmetrical objects.
5. Avoid the use of elaborate, pictorial, or repetitive detail.
6. When necessary to detail threads, do not show them over the full length of the stud, bolt, or tapped hole.
7. Eliminate detail of nuts, bolt heads, and other standard hardware. Show outlines when it is necessary to show position.
8. Reduce detail of parts on assembly drawings. Simply show the part position.
9. Avoid the use of unnecessary hidden lines that do not add clarification.
10. Sectioning should be used only when it is necessary for the clarity of the drawing.
11. Simplify graphics for holes and tapped holes by use of symbols.
12. Omit views with no dimensional or written instruction.

13. Within limits, a small drawing is usually easier and quicker to make than a large one.
14. When two parts are only slightly different, complete graphical representation of both parts is not required. The note "same as except ______" or "otherwise same as ______" may be used.
15. Drawings made to modify stock of commercial parts should be as plain as possible. Avoid detail.
16. Use standard abbreviations wherever possible.
17. When necessary, enlarge small details on larger parts for clarity.
18. Draw small parts large enough to avoid crowding so that they may be easily read, but not unnecessarily large so as to waste space on the drawing.
19. Do not duplicate dimensions.
20. Substitute recognized standard symbols to greatly simplify the drawing of commonly used objects.
21. Eliminate repetitive data by use of general notes.
22. When drafting, use as much freehand drawing as the work permits in preference to using instruments.
23. Where practical, use geometric symbols instead of notes.
24. Where acceptable, use rectangular coordinate or tabular dimensioning instead of dimension lines.

16.8 A Checklist for Drafters and Designers

A drawing should be checked after it has been completed. Compare it to the following questions:

Readability

1. Is the drawing easy to read?
2. Are the part outlines distinct from dimension lines?
3. Is the lettering neat and clear?
4. Is all of the information on the drawing?
5. Will the drawing make a good print?
6. Have all the rules of standard drafting practice been followed?
7. Is the nomenclature correct? Will everyone understand it the same way?
8. Is the drawing title truly descriptive?

Completeness

9. Are all necessary views given?
10. Are some views unnecessary?

Notes

11. Are the general notes properly located?
12. Are any exceptions to the general notes clearly pointed out?
13. Are any notes crowded or hard to find?
14. Could any of the notes be misunderstood?
15. If a specially purchased item is required, is procurement information given?
16. If special procedures are required in manufacture or assembly, have they been noted on the drawing?

Parts List

17. Does the parts list agree with the drawing?
18. Have overall dimensions been given?
19. Are standard parts specified correctly?

Dimensioning

20. Are out-of-scale dimensions (if any) clearly marked?
21. Is it necessary to leave a dimension out of scale?
22. Are all dimensions given?
23. Are there any duplicate dimensions?
24. Are dimensions kept well away from the outline of the part?
25. Is the scale designated?

Tolerances

26. Have all tolerances given been carefully considered?
27. Are all tolerances to the maximum possible?
28. Are any tolerances too large? Too small?
29. Has the drawing been checked for possible tolerance stackups?

Finishes

30. Are all machine finishes given and do they conform to applicable specifications?
31. Are all paint and plating finishes specified?

Processes

32. Is heat treatment needed?
33. Have standard manufacturing processes been followed?
34. Can the part be simply and economically produced?

Materials and Parts

35. Are standard or purchased parts used to the maximum extent?
36. Are all special or reworked parts noted?

Assembly

37. Have you made sure there are no mechanical interferences?
38. Will parts assemble without difficulty?
39. Does the work agree with associated mechanisms?
40. Are all parts properly numbered or designated?

Cost

41. Could the function have been accomplished at less expense with the same results?
42. Could the design have contained fewer parts?
43. Have you given thought as to how this would be built?

Reliability

44. Have you checked the design for possible failure?
45. Have you considered safety factors?

16.9 Reproduction and Storage of Drawings

The last step in the design detailing process is outputting the drawings of the project. Reproduction of drawings involves a process called whiteprinting.

16.9.1 Whiteprinting

Whiteprinters (Fig. 16.20) are used to make copies of drawings. The whiteprinter is still referred to as a "blueprint" machine by many people. As a drafter you can expect to run prints of drawing projects.

Regardless of the type of paper, the whiteprinter makes a "positive" image of the drawing on whiteprint paper. The process depends on the transmission of light through the drawing paper and onto the developing surface. The lines, lettering, or other graphics block the light in order for a positive image to be developed on the print. Because of the necessity to block light, your drawing should have high-quality, crisp, dark lines. Use H or HB grade lead and press sufficiently hard to create a dark crisp drawing.

16.10 Jigs, Fixtures, Dies, and Tooling

After the part has been designed, it is necessary to establish the tools to produce the part. Jigs, fixtures, dies, and other tooling require intense design work. Changes in production methods, the development of new manufacturing products, and, in particular, the perfecting of economical production methods involve many design problems with tooling, dies, jigs, and fixtures. Making drawings requires special experience and is done either by a company division maintained for that purpose or by independent tool specialists.

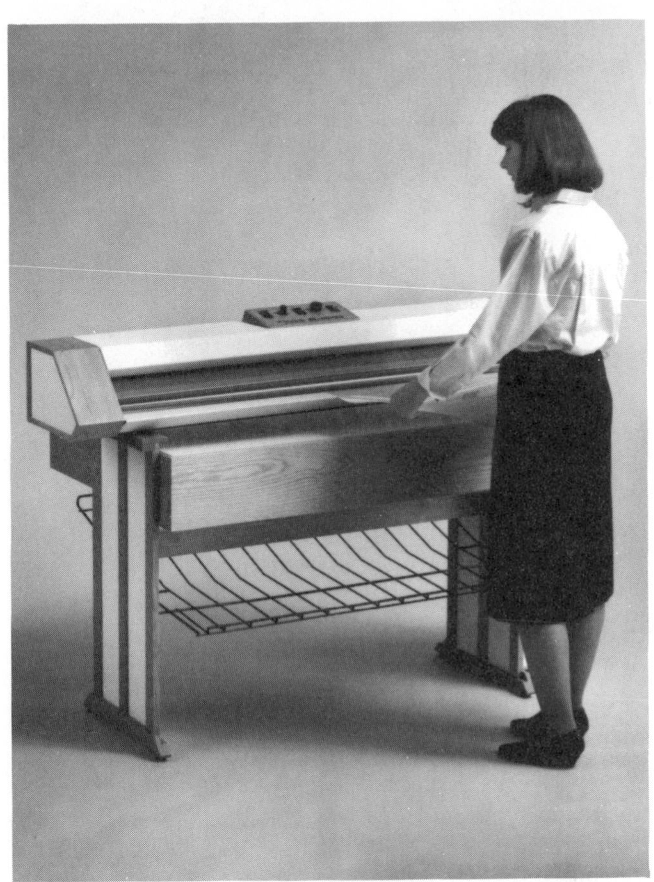

FIGURE 16.20 Running Prints

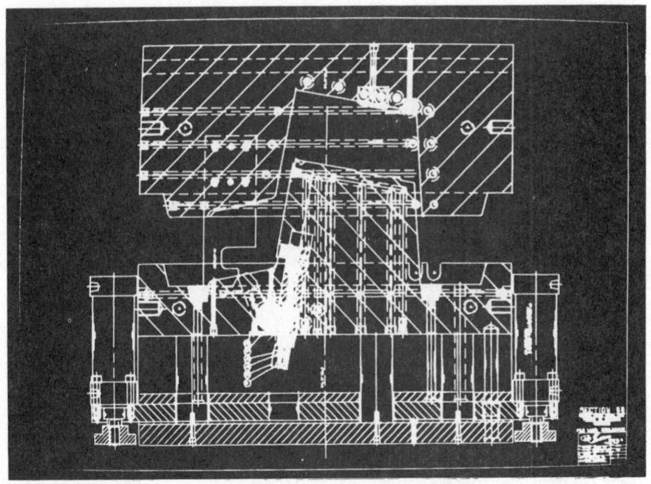

FIGURE 16.21 Mold Design on a CAD System

In general, a **tool** is a piece of equipment that helps create a finished part. It may be anything that must be designed or made in order to manufacture the part. The following is a list of tools found in industry:

1. *Molds:* Used to form a variety of parts for consumer, industrial, and medical applications (Fig. 16.21).
2. *Dies:* Used to forge, cast, extrude, and stamp materials in various physical states (solid through fluid).
3. *Tooling:* The individual component of a mold or a die. Tooling might include a cavity, nest, core, punch, bushing, slide, or sleeve.

FIGURE 16.22 Tooling Fixture Designed on a CAD System

ITEMS OF INTEREST

America's First Automobile

Who produced America's first automobile? One might be tempted to name Henry Ford or maybe even Thomas Edison because of his electric car. However, the first car manufactured in the United States was the creation of Charles Duryea in Springfield, Massachusetts, in 1895.

Duryea was born on a farm in 1861 at a time when people relied on mechanical devices to accomplish their farm work. At seventeen, he began to cultivate his mechanical aptitude by assembling discarded farm parts into bicycles. Later on, he sold bicycles built from parts manufactured to his specifications.

He first saw a gasoline engine while he was displaying bicycles at the Ohio State Fair. The engine was much larger than could possibly be used in an automobile, but he knew that smaller engines were possible. He also knew that a German, Karl Benz, had recently patented the first automobile. He decided to build and patent the first American automobile.

After many years of thinking about his horseless carriage, Duryea and his brother finally built their first car in 1892. It had a gasoline-powered internal combustion engine with an electric ignition. The engine was attached to a converted horse buggy.

In 1893, the Duryeas produced a prototype called the *buggaut* and established themselves as the makers of the first successful American automobile. By 1895, they offered an improved 700-pound version for $2,000.

Charles Duryea continued work on his dream machine and obtained nineteen patents, one of which was the first automobile patent issued to an American manufacturer. In 1896, the Duryea Motor Wagon company produced thirteen cars of the same design. They were the first manufacturers to produce many copies of a single design. The car won several races against domestic and foreign automobiles.

Modern Internal Combustion Engine

4. *Fixtures:* Used to hold and locate parts of assemblies during machining or other manufacturing operations (Fig. 16.22). The accuracy of the product being produced determines the precision with which the fixture is designed. Figure 16.23 shows a fixture for holding a part while machining.

Once the tool fixture is designed, a detail drawing provides a geometric description of each part of the fixture.

16.11 Working Drawing Example for a Consumer Product

The recalibratible electronic level shown in Figure 16.24 is capable of reading all angles through 360°, rather than only traditional level and plumb measurements. The "SmartLevel" is designed to increase the speed, accuracy, and efficiency of standard leveling tasks and also to provide a tool for tasks requiring direct measurement of angles, slopes, grades, and pitches.

A SmartLevel is composed of two components: an electronic sensor module that contains the functions and display readout for the tool, and an ergonomically-designed rail that is available in lengths of 2, 4, and 6 feet.

16.11.1 Working Drawings for the SmartLevel

Figures 16.25(a) through (l) (see pages 454 through 464) shows assembly drawings (from the sensor module), a computer-generated parts list, and individual details for the SmartLevel. This product has parts made of teak, aluminum, ABS, high impact polycarbonate, and silicone rubber. The electronic drawings are not included in this set.

This is an example of the type of drawings found in industry for a small consumer product. Since these drawings were done manually, they do not look perfect. The choice of views, assembly pictorial views, dimensioning on the details, the revisions, the title block information, and the notes of each drawing (material designations, production and manufacturing requirements, and finishes) should be examined carefully.

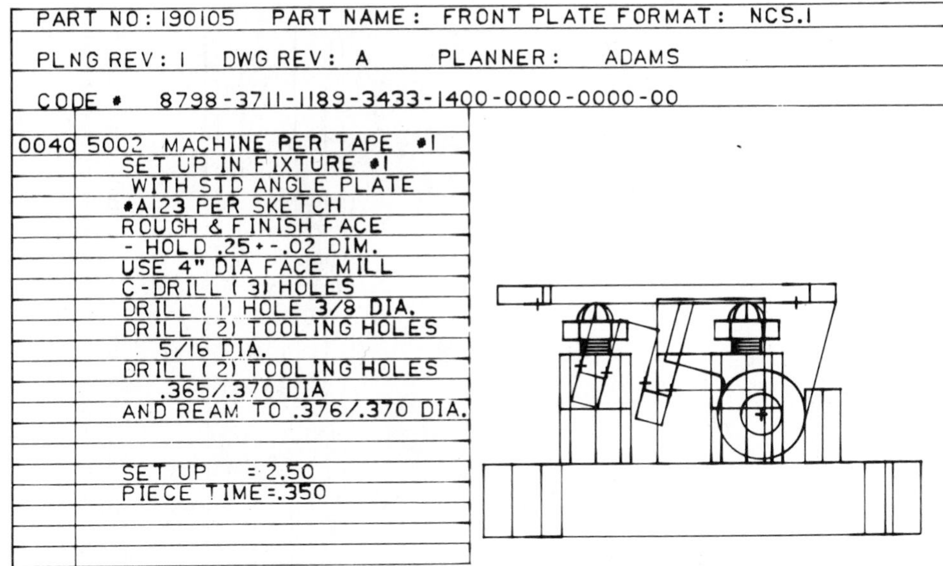

PART NO: 190105 PART NAME: FRONT PLATE FORMAT: NCS.I
PLNG REV: I DWG REV: A PLANNER: ADAMS
CODE • 8798-3711-1189-3433-1400-0000-0000-00
0040 5002 MACHINE PER TAPE •I
SET UP IN FIXTURE •I
WITH STD ANGLE PLATE
•A123 PER SKETCH
ROUGH & FINISH FACE
- HOLD .25+-.02 DIM.
USE 4" DIA FACE MILL
C-DRILL (3) HOLES
DRILL (1) HOLE 3/8 DIA.
DRILL (2) TOOLING HOLES
5/16 DIA.
DRILL (2) TOOLING HOLES
.365/.370 DIA
AND REAM TO .376/.370 DIA.
SET UP = 2.50
PIECE TIME=.350

FIGURE 16.23 Planning Machine Operations on a CAD System

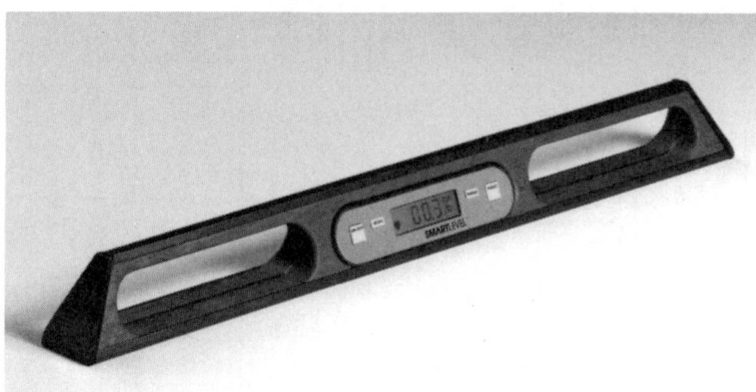

FIGURE 16.24 SmartLevel

QUIZ

True or False

1. The revision block is one of the most important things you should look at when reading an existing drawing for the first time.
2. Dimensions are given on all assemblies to locate and define the parts geometry.
3. Sections are used on assemblies to provide a convenient way of clearly displaying the unit's geometry.
4. A parts list and a bill of materials are two distinct and different aspects of an assembly drawing.
5. Since they will normally differ for each project, general notes are seldom standardized.
6. Hidden lines are shown whenever possible on the assembly drawing to clarify each part's geometry.
7. A parts list that is on the assembly drawing is listed from the top-down and is placed below the revision block.
8. Ballooning is the process of calling out each part of an assembly by providing a circle attached to a leader, which points to the piece on the drawing.

Fill in the Blanks

9. A ________ provides prints of a drawing that have white lines with a blue background.
10. ________ are used to hold and locate a part during machining operations.
11. Layout drawings are used to establish and depict ________, ________, and the relationship of ________.
12. ________ is the process of taking individual pieces of an assembly and redrawing them with sufficient views and ________.
13. Working drawings are used to order ________, plan ________ operations, determine ________, and assemble the unit.
14. Welded assemblies are considered ________ assemblies.
15. Assemblies are fastened by one or more of the processes including ________, ________, ________, ________, and ________.
16. ________ on assemblies are provided for setup, assembly, and clearance.

Answer the Following

17. What are jigs, fixtures, and dies?
18. Why are notes required for most drawings? What is the difference between local and general notes?
19. What is the function of the assembly drawing?
20. Explain the difference between the concept of bottom-up design versus the top-down design approach.
21. Describe a parts list and each of its major headings.
22. How are views selected for an assembly drawing?
23. Explain the difference between a separable and an inseparable assembly.
24. Name three ways in which you could simplify a drawing.

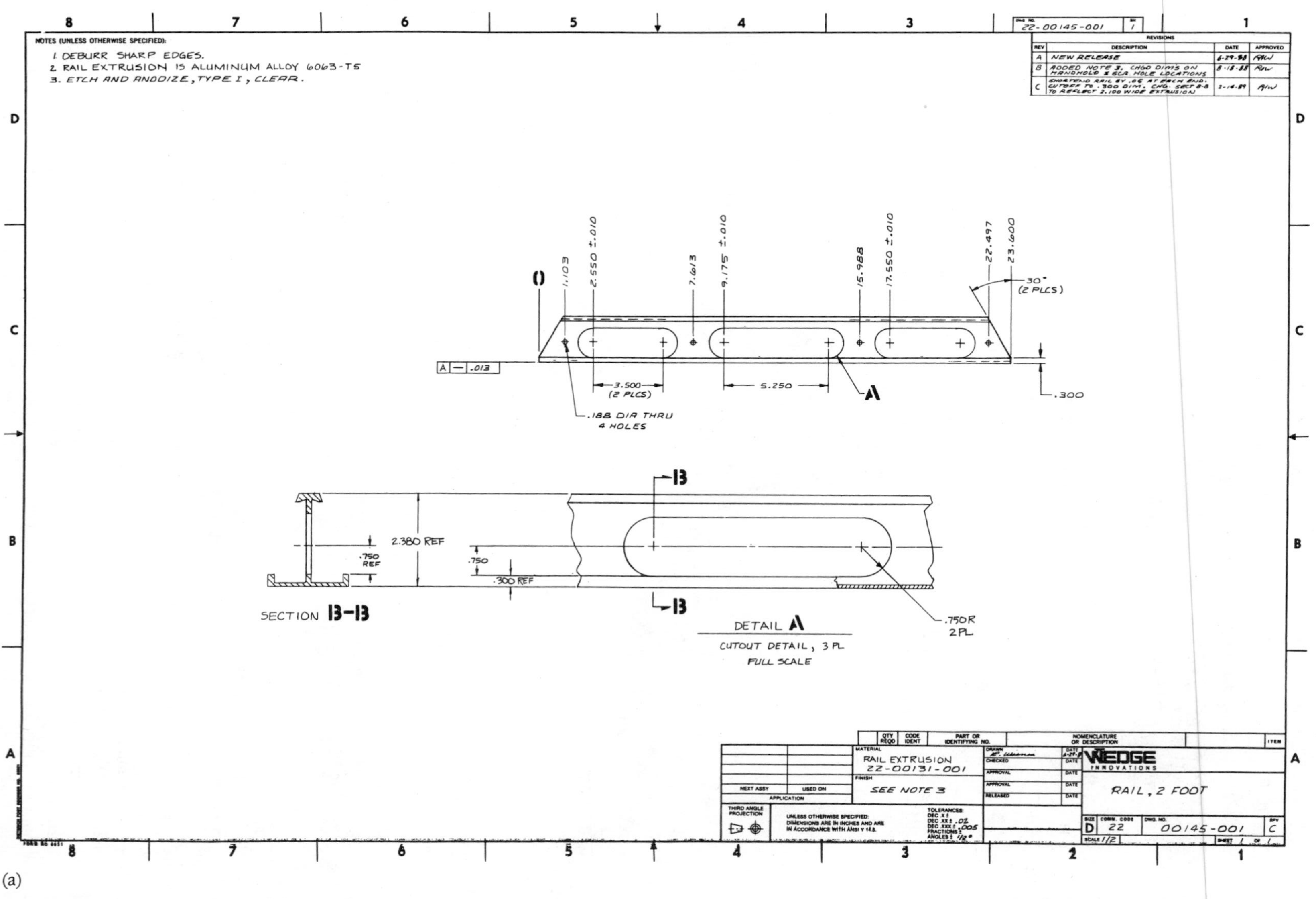

(a)

FIGURE 16.25 SmartLevel Production Drawings (a) Rail detail (b) Rail extrusion (c) Wood insert (d) End cap (e) Module assembly (f) Computer-generated parts list for the model assembly (g) Front panel assembly (h) Parts list for front panel assembly (i) Housing detail (j) Housing detail sections (k) Front panel detail (l) Keytop detail

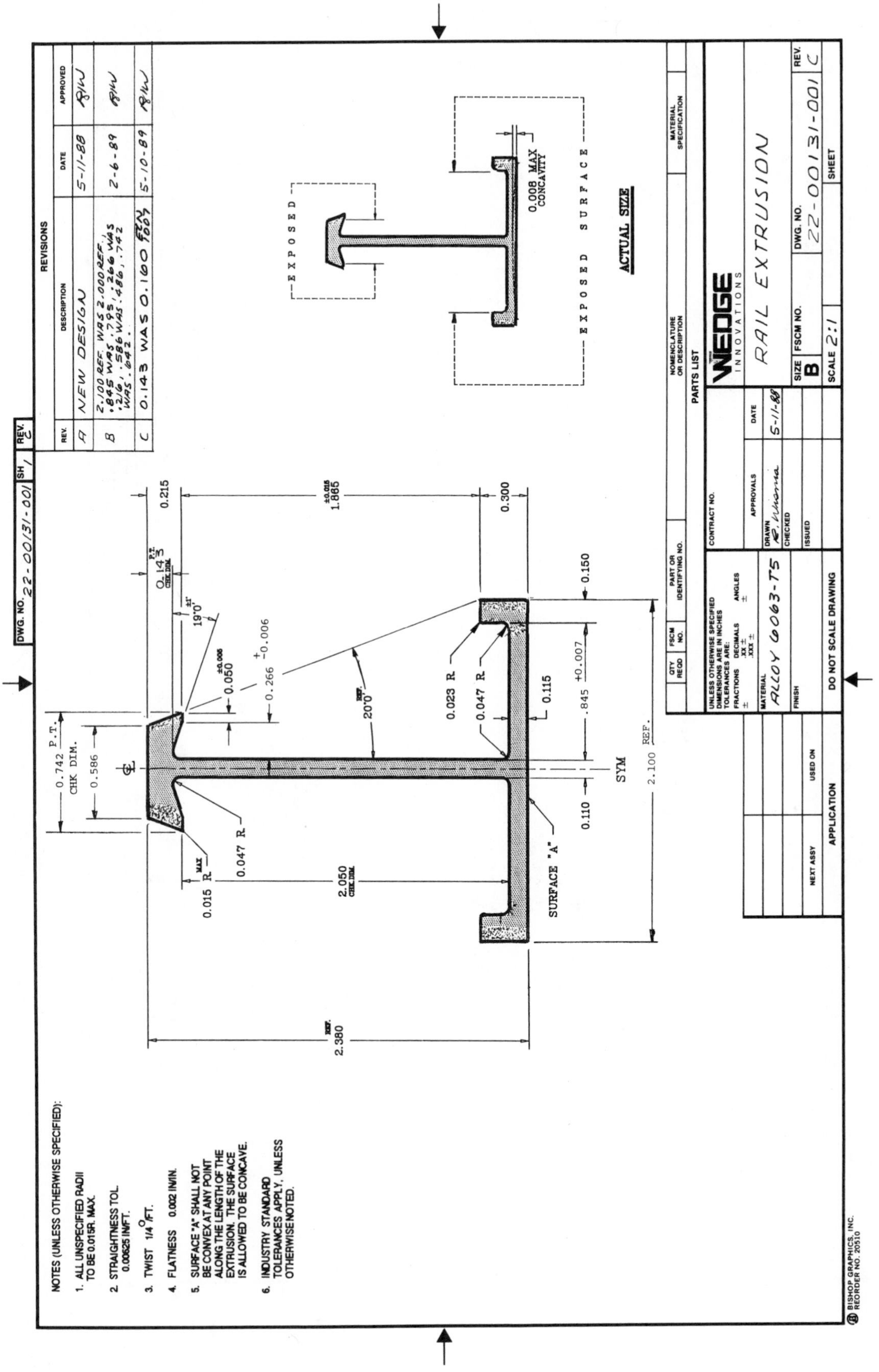

DWG. NO. 22-00131-001 SH 1 REV. C

NOTES (UNLESS OTHERWISE SPECIFIED):

1. ALL UNSPECIFIED RADII TO BE 0.015R. MAX.
2. STRAIGHTNESS TOL. 0.00625 IN/FT.
3. TWIST 1/4° /FT.
4. FLATNESS 0.002 IN/IN.
5. SURFACE "A" SHALL NOT BE CONVEX AT ANY POINT ALONG THE LENGTH OF THE EXTRUSION. THE SURFACE IS ALLOWED TO BE CONCAVE.
6. INDUSTRY STANDARD TOLERANCES APPLY, UNLESS OTHERWISE NOTED.

REVISIONS			
REV.	DESCRIPTION	DATE	APPROVED
A	NEW DESIGN	5-11-88	RJW
B	2.100 REF. WAS 2.000 REF., .845 WAS .795, .266 WAS .216, .586 WAS .486, .742 WAS .642.	2-6-89	RJW
C	0.143 WAS 0.160 ECN 1007	5-10-89	RJW

0.742 P.T. CHK DIM.
0.586
℄
0.143 P.T. CHK DIM.
0.215
19°0' ±1°
0.050 ±0.006
0.266 +/-0.006
0.015 R MAX
0.047 R
2.380 REF.
2.050 CHK DIM.
1.865 ±0.015
20°0' REF.
0.023 R
0.047 R
0.300
SURFACE "A"
0.115
0.110
.845 +0.007
0.150
SYM
2.100 REF.

EXPOSED
EXPOSED SURFACE
0.008 MAX CONCAVITY
ACTUAL SIZE

QTY REQD	FSCM NO.	PART OR IDENTIFYING NO.	NOMENCLATURE OR DESCRIPTION	MATERIAL SPECIFICATION

PARTS LIST

UNLESS OTHERWISE SPECIFIED DIMENSIONS ARE IN INCHES TOLERANCES ARE: FRACTIONS ± DECIMALS .XX ± .XXX ± ANGLES ±

MATERIAL ALLOY 6063-T5

FINISH

DO NOT SCALE DRAWING

NEXT ASSY | USED ON

APPLICATION

CONTRACT NO.

APPROVALS	DATE
DRAWN R. Wissma	5-11-88
CHECKED	
ISSUED	

WEDGE INNOVATIONS

RAIL EXTRUSION

SIZE B | FSCM NO. | DWG. NO. 22-00131-001 | REV. C

SCALE 2:1 | SHEET

BISHOP GRAPHICS, INC. REORDER NO. 20510

(b)

FIGURE 16.25 SmartLevel Production Drawings—*Continued* (b) Rail extrusion

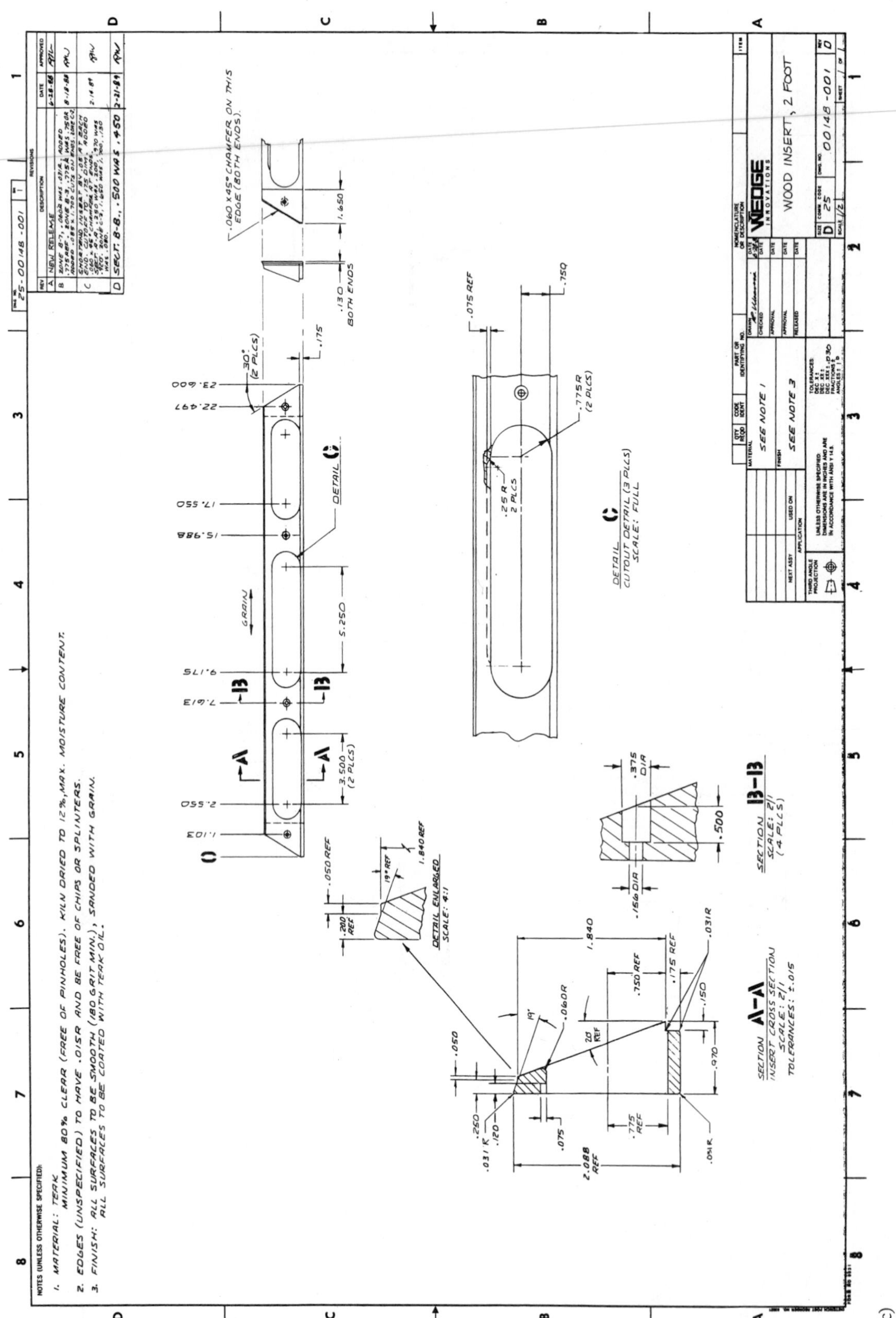

(c)

FIGURE 16.25 SmartLevel Production Drawings—*Continued* (c) Wood insert

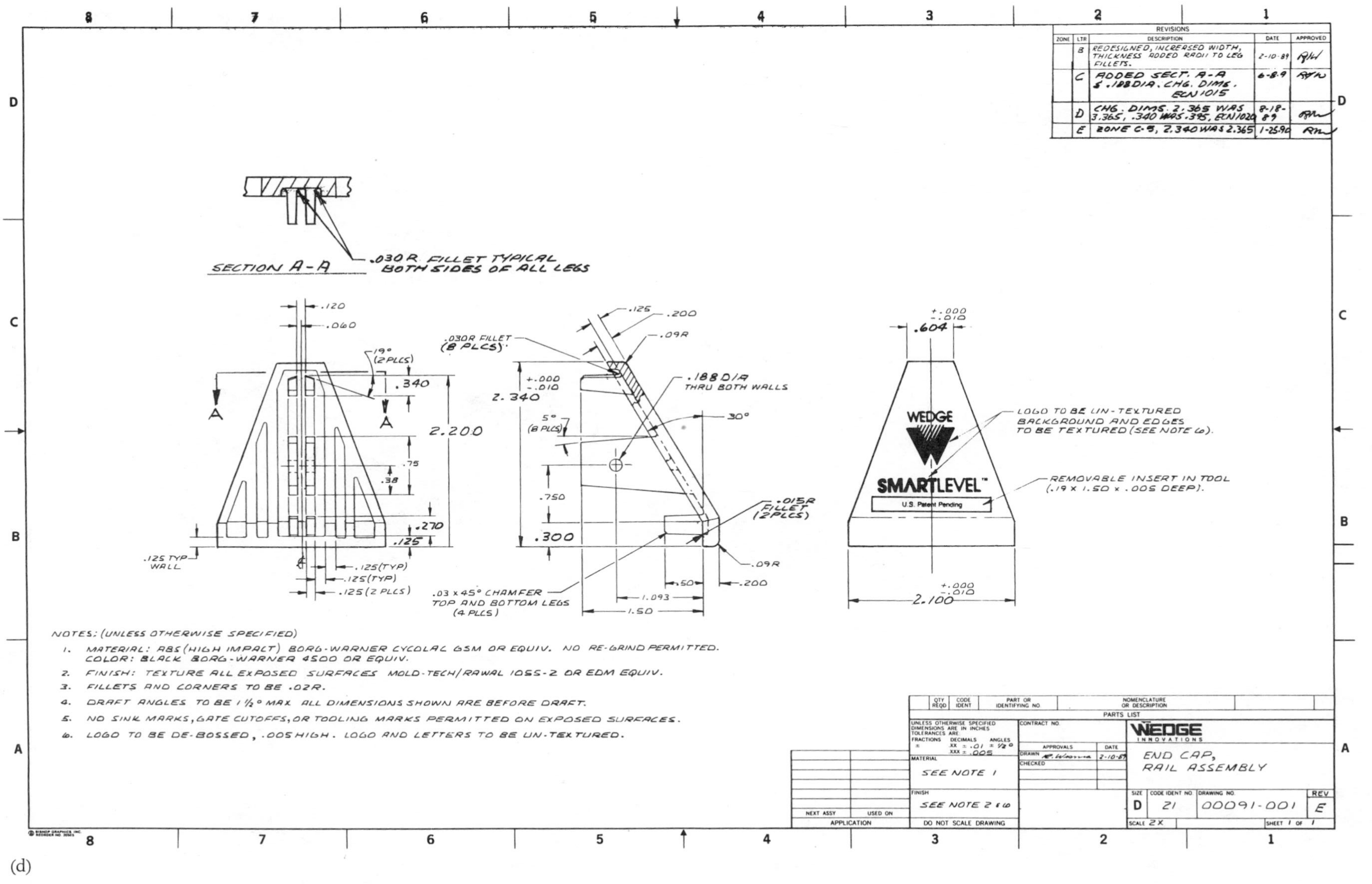

(d)

FIGURE 16.25 SmartLevel Production Drawings—*Continued* (d) End cap

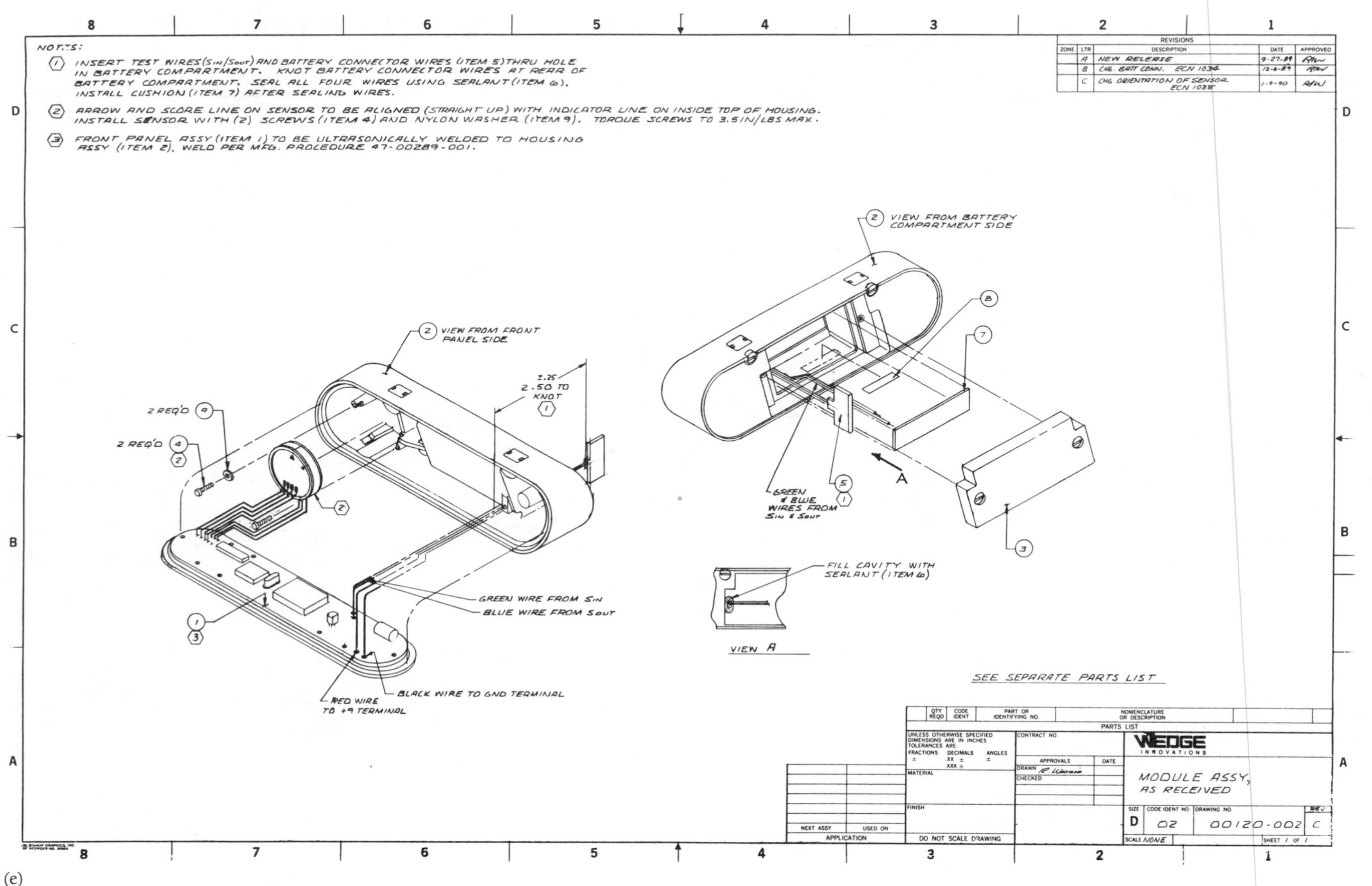

(e)

FIGURE 16.25 SmartLevel Production Drawings—*Continued* (e) Module assembly

PARENT PART NUMBER : 02-00120-002 REV. B, PL120.2
PARTS LIST : MODULE ASS, AS REQUIRED DATE : 12-5-90

ITEM	PART NUMBER	DESCRIPTION	REF. DESIGN	QTY	U/M
1	03-00121-001	FRONT PANEL ASSY		1	EA
2	03-00122-001	HOUSING ASSY		1	EA
3	03-00123-001	BATTERY CIVER ASSY		1	EA
4	81-00124-007	SCR, THRD CUTTING (TYPE 25) PH, CR, #2X1/2		2	EA
5	54-00399-001	CONNECTOR, BETTERY, MOLDED		1	EA
6	51-00141-001	SEALANT, RTV		A/R	OZ
7	23-00140-001	CUSHION, BATTERY		1	EA
8	26-00291-001	LABEL, SERIAL NUMBER		1	EA
9	82-00321-001	WASHER, NYLON, #1		2	EA

(f)

PARENT PART NUMBER : 03-00121-001 REV. F, FR.PNLASSY
PARTS LIST : FRONT PANEL ASSY DATE : 6-20-89

ITEM	PART NUMBER	DESCRIPTION	REF. DESIGN	QTY	U/M
1	21-00112-001	FRONT PANEL (SILKSCREENED)		1	EA
2	23-00113-001	KEYTOP, LEFT HAND		1	EA
3	23-00113-002	KEYTOP, RIGHT HAND		1	EA
4	62-00250-001	LCD, 15 PIN		1	EA
5	23-00030-001	ELASTOMETRIC CONNECTOR STRIP, ACTIVE		1	EA
6	23-00030-002	ELASTOMETRIC CONNECTOR STRIP, DUMMY		1	EA
7	03-00183-001	PCB/SENSOR ASSY		1	EA
8	81-00124-002	SCR, THRD CUTTING (TYPE 25), PH, CR, #2X3/16		8	EA
9	81-00124-003	SCR, THRD CUTTING (TYPE 25), PH, CR, #2X1/4		2	EA
10	21-00223-001	PRESSURE BAR		1	EA
11	24-00246-001	INSULATOR, KEYPAD, RIGHT HAND		1	EA
12	24-00246-002	INSULATOR, KEYPAD, LEFT HAND		1	EA

(g)

FIGURE 16.25 **SmartLevel Production Drawings—*Continued*** (f) Computer-generated parts list for the model assembly (g) Parts list for front panel assembly

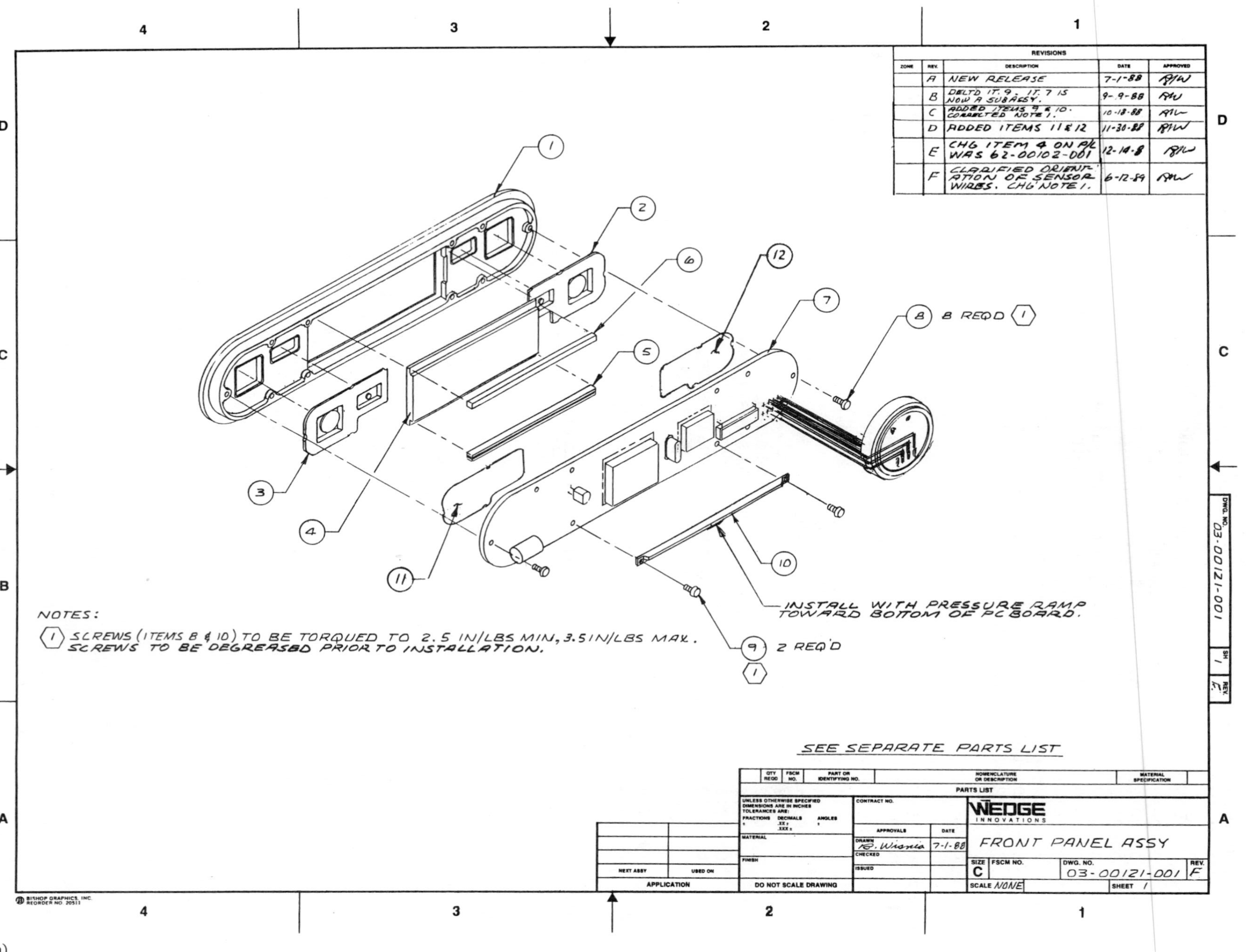

(h)

FIGURE 16.25 SmartLevel Production Drawings—*Continued* (h) Front panel assembly

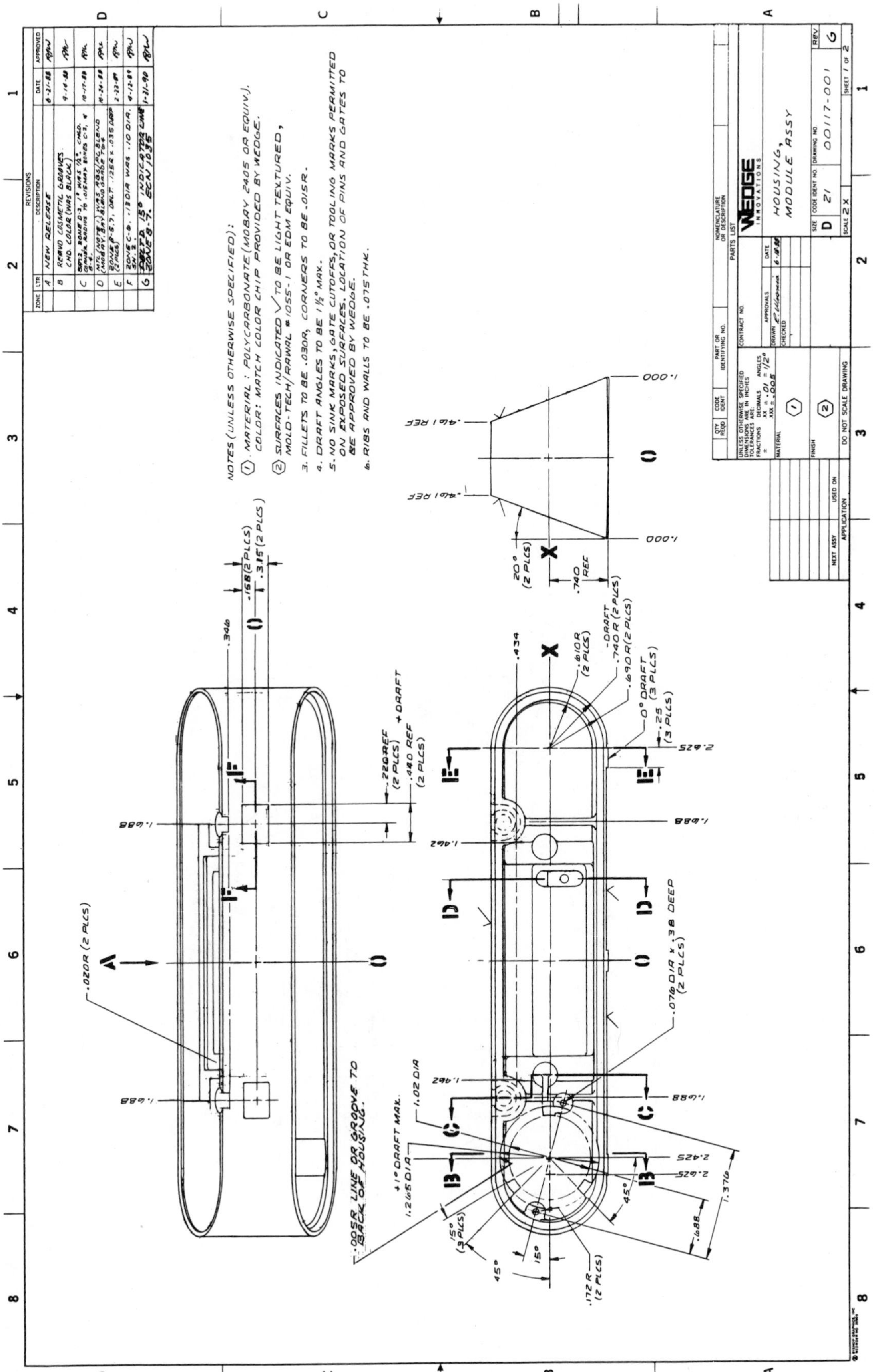

(i)

FIGURE 16.25 SmartLevel Production Drawings—*Continued* (i) Housing detail

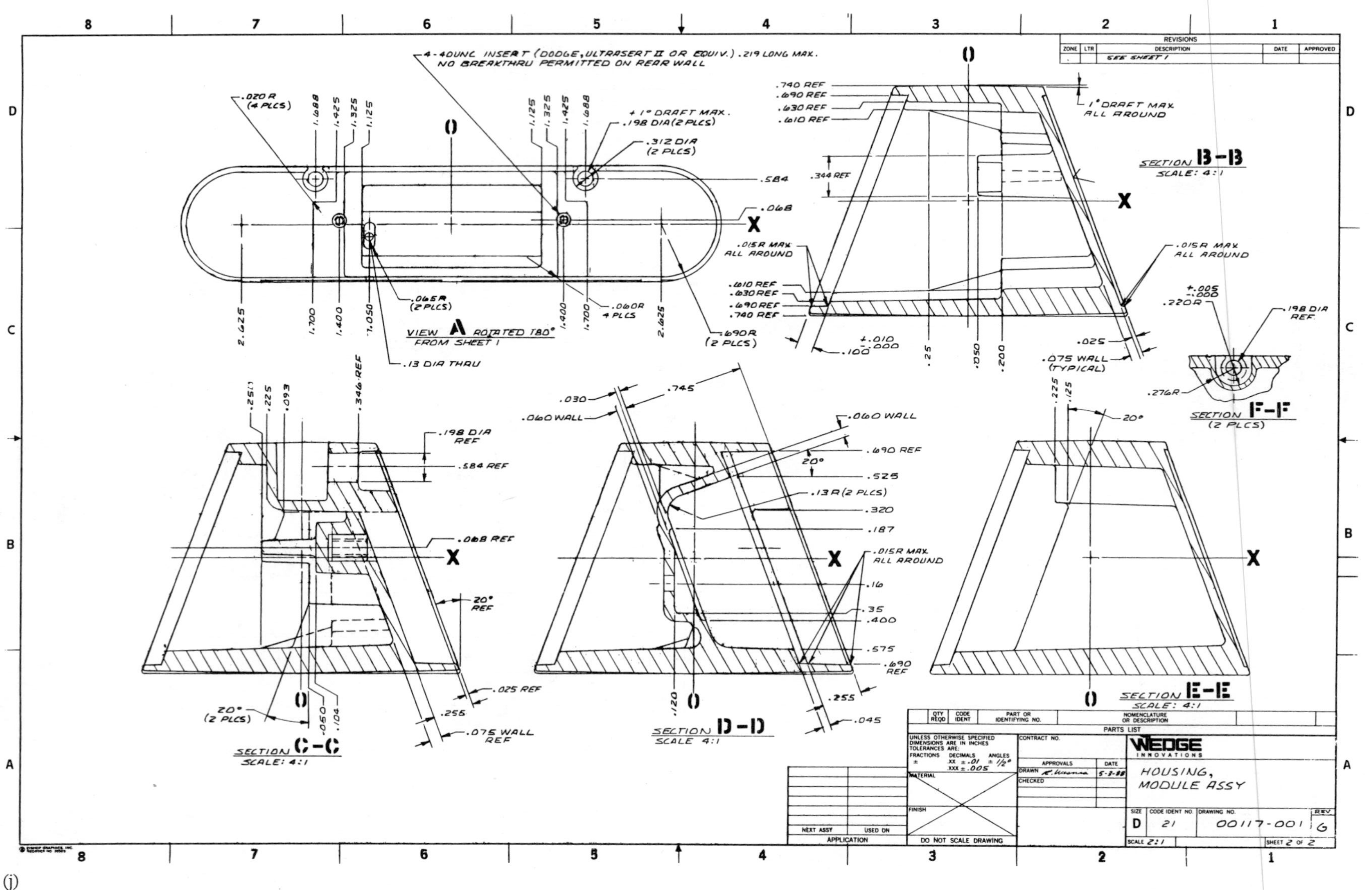

(j)

FIGURE 16.25 SmartLevel Production Drawings—*Continued* (j) Housing detail sections

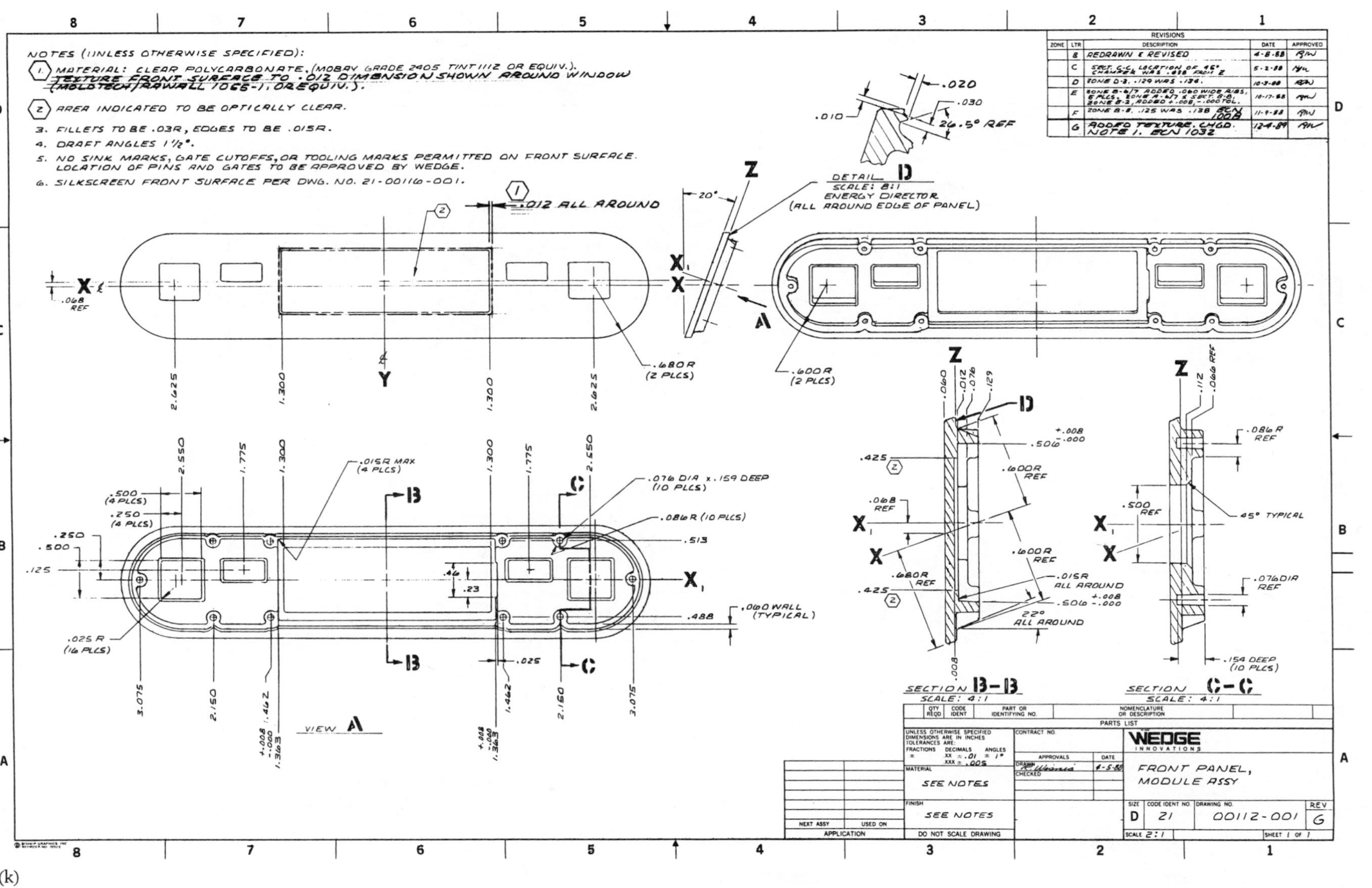

(k)

FIGURE 16.25 SmartLevel Production Drawings—*Continued* (k) Front panel detail

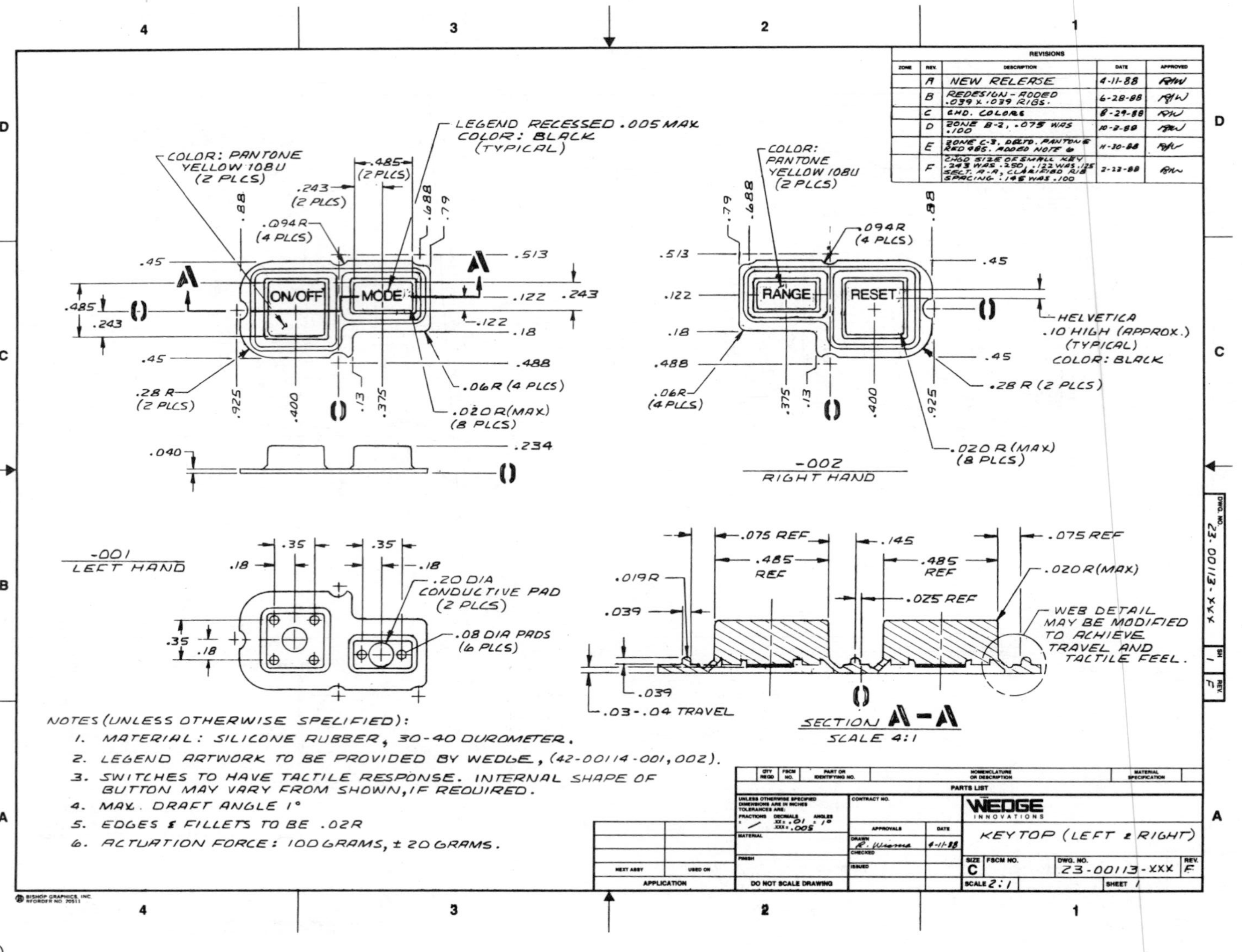

(l)

FIGURE 16.25 SmartLevel Production Drawings—*Continued* (l) Keytop detail

EXERCISES

Exercises may be assigned as sketching, instrument, or CAD projects. Transfer the given information to an "A" size sheet of .25 in. grid paper. Exercises that are not assigned by the instructor can be sketched in the text to provide practice and understanding for the preceding instructional material. Draw the drawing format, title block, and other standard information for the exercise. If using AutoCAD or another system that provides standard formats, use them instead of the one provided here or in the worksheets.

Exercise 16.1 Do a complcte parts list for Problem 16.27 or 16.22.

Exercise 16.2 Do a complete parts list for Problem 16.28 or 16.23.

Exercise 16.3 Do a complete parts list for Problem 16.29 or 16.24.

Exercise 16.4 Do a complete parts list for Problem 16.30 or 16.25.

ITEM	QTY	PART NO.	DESCRIPTION

Exercise 16.1

ITEM	QTY	PART NO.	DESCRIPTION

Exercise 16.2

ITEM	QTY	PART NO.	DESCRIPTION

Exercise 16.3

ITEM	QTY	PART NO.	DESCRIPTION

Exercise 16.4

PROBLEMS

Detail Drawings

For detail drawings use appropriate ANSI standards for dimensioning and tolerancing for all problems. Complete a rough sketch of the part before drawing it with instruments. Choose an appropriate size drawing sheet for each project.

For all drawings, use ANSI or ISO standard sheet sizes, parts list format, revision block, and title blocks as shown in this chapter. The dimensions given for individual parts are in most cases for construction of the part's geometry only. With the exception of a few of the projects shown in the chapter body, the problems given here are by no means meant to represent the correct way of dimensioning the part. The given dimensions will enable you to draw the part. *It will be your responsibility to select the proper views and place the dimensions, notes, and other information on the drawing.* Show all finish marks and use symbology wherever possible.

All pictorial drawings are to be converted to multiview details with appropriate view selection and proper dimensioning methods. In most cases, the drawings presented here as projects can be drawn full scale. If a project prohibits full-scale rendering, a reduced scale can be used.

Decimal-inch projects and metric drawings are provided. You may convert any of the projects to the other measurement system. In many cases, you will find that converting the dimensions will give odd and inappropriate sizes. You may redesign any of the parts using even and logical sizes for that measurement system. As an example, a 1.00 in. measurement converts to 25.4 mm. It is acceptable to change the metric dimension into 25 or 24 mm. The same is true for standard parts such as screws, nuts, washers, and other off-the-shelf items; look up the closest standard size before ordering the item (placing on the parts list).

Problem 16.1 Draw and detail the adjustable guide.

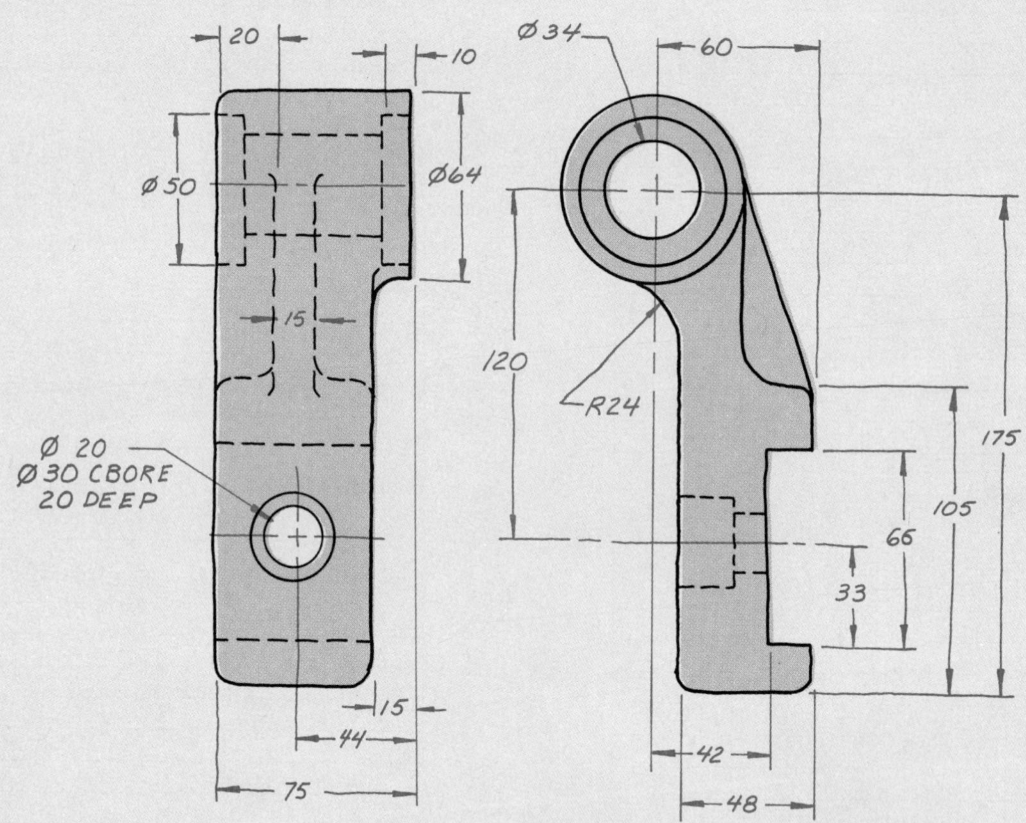

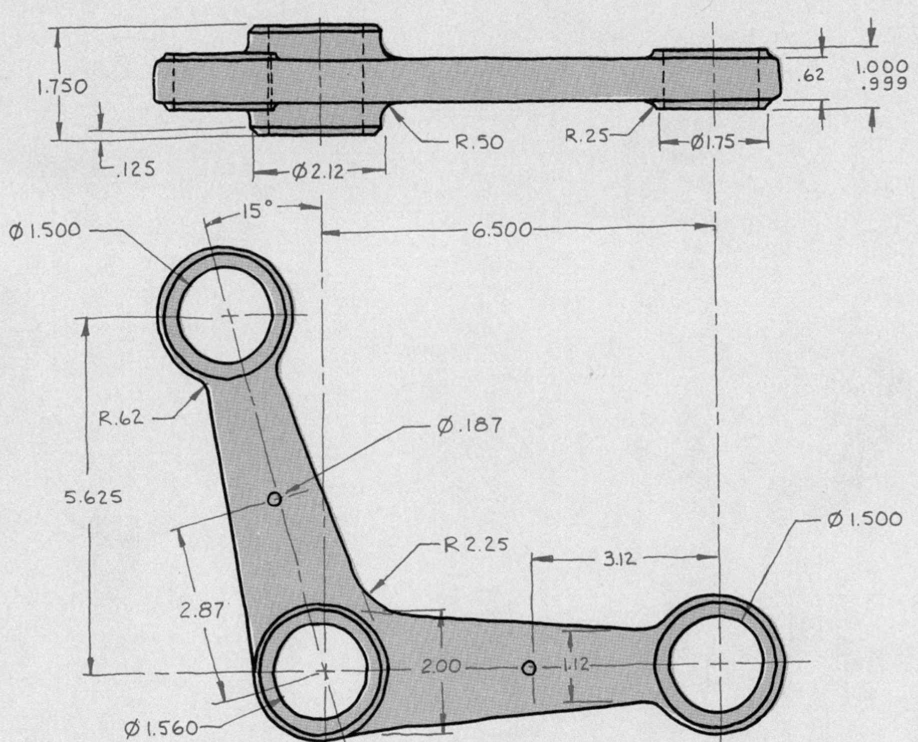

Problem 16.2 Do a detail drawing of the crank arm.

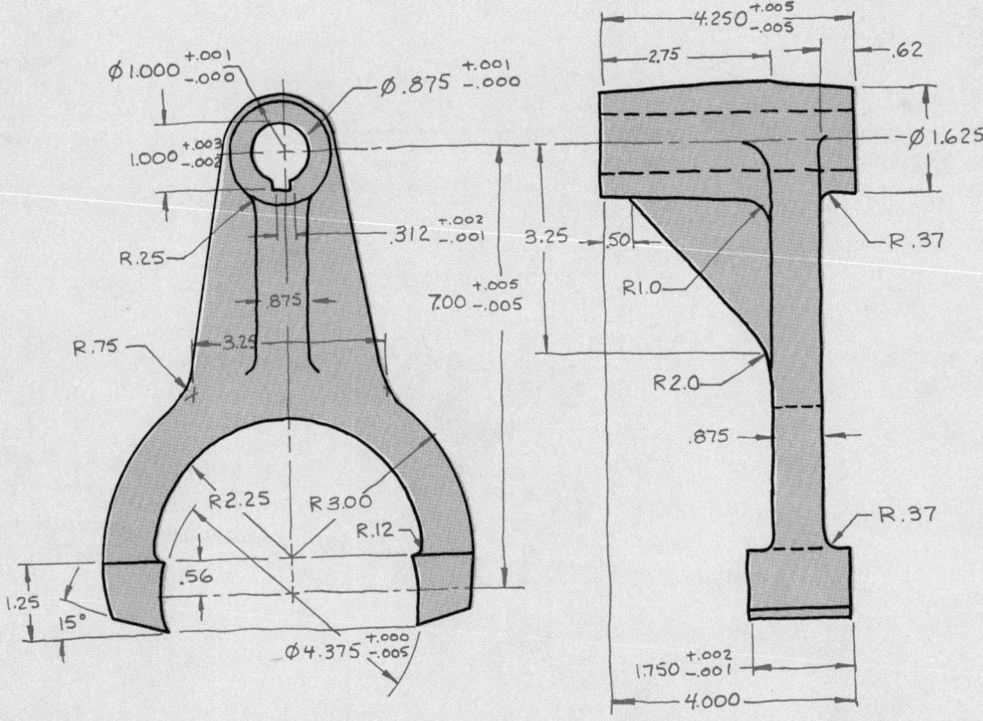

Problem 16.3 Draw and detail the shifter fork.

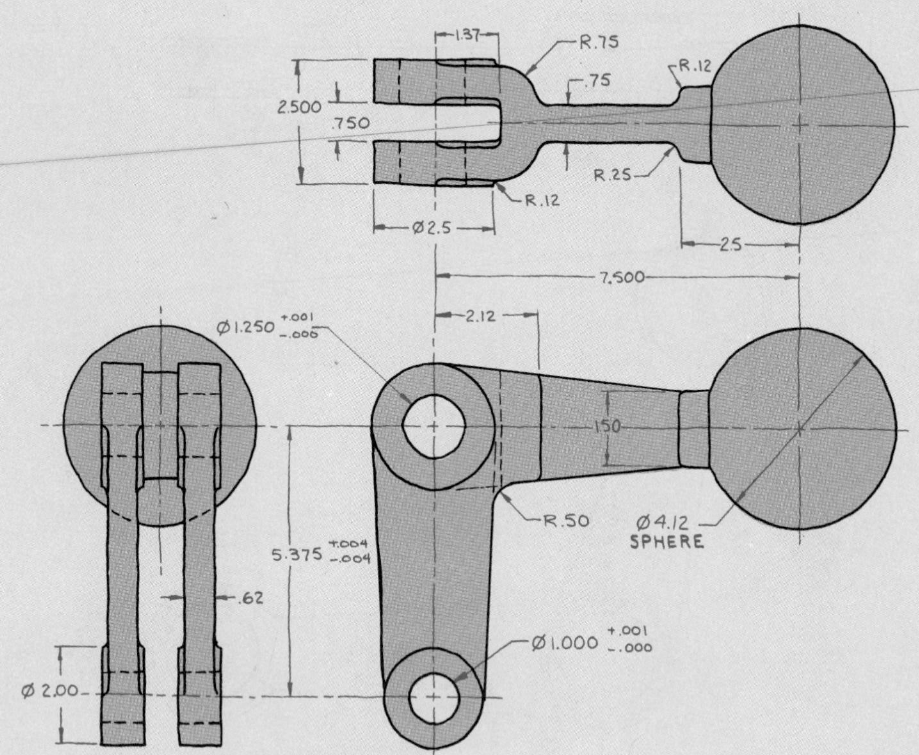

Problem 16.4 Draw and detail the flywheel.

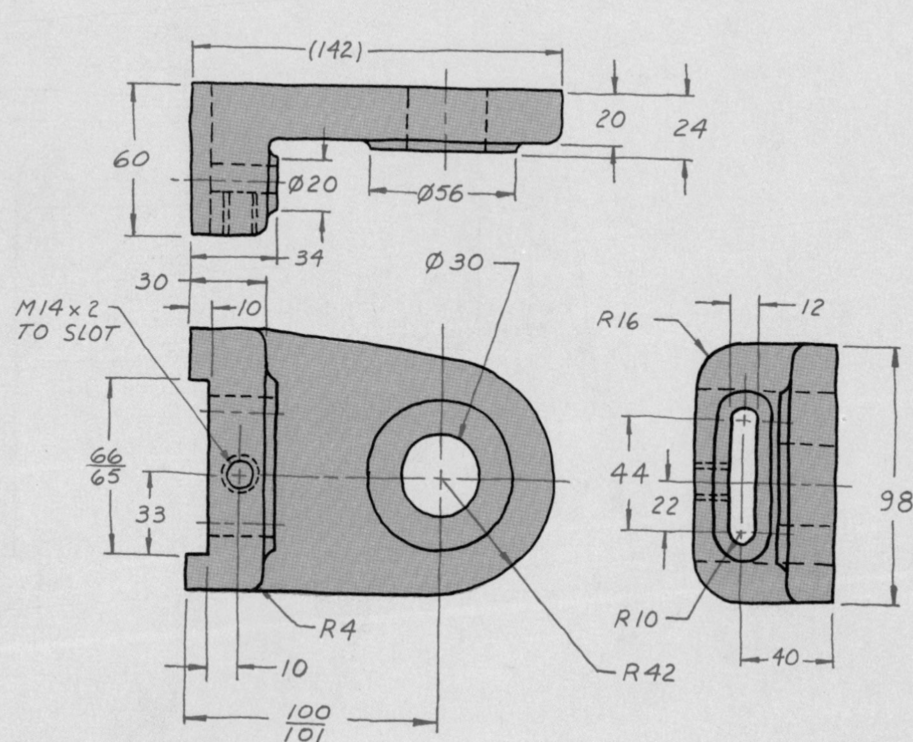

Problem 16.5 Do a detail drawing of the bearing adjustment.

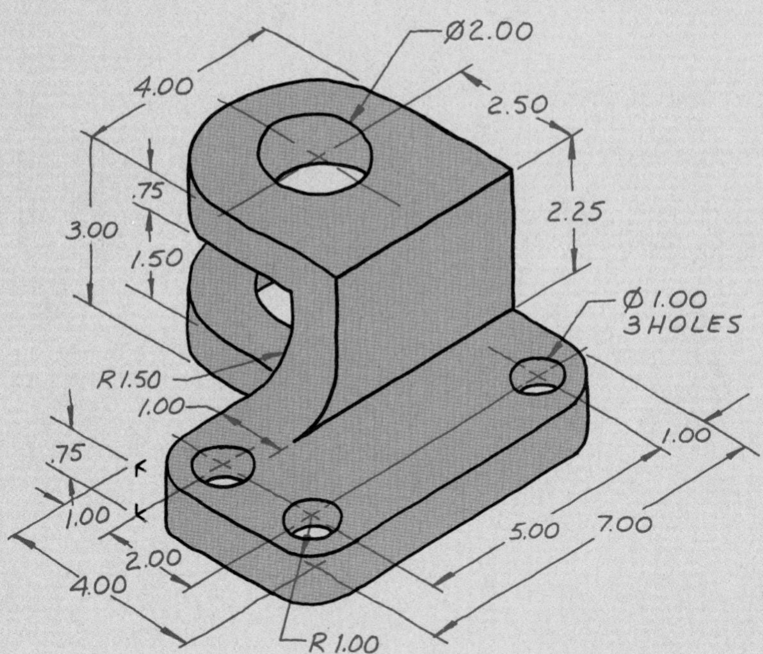

Problem 16.6 Draw and detail the guide bracket.

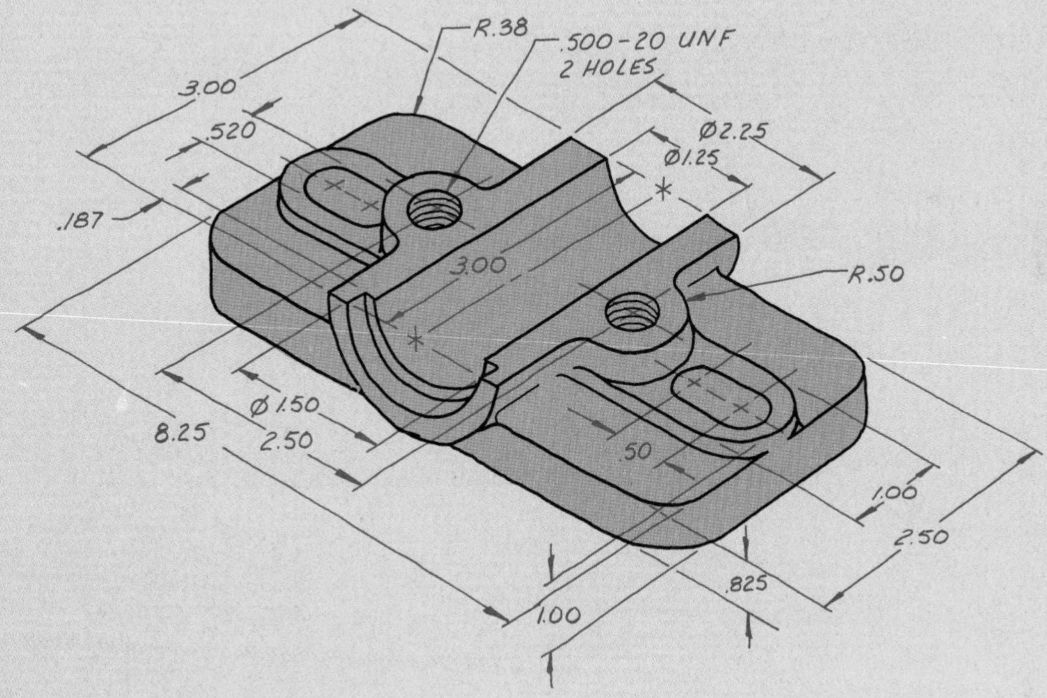

Problem 16.7 Detail the journal bearing housing.

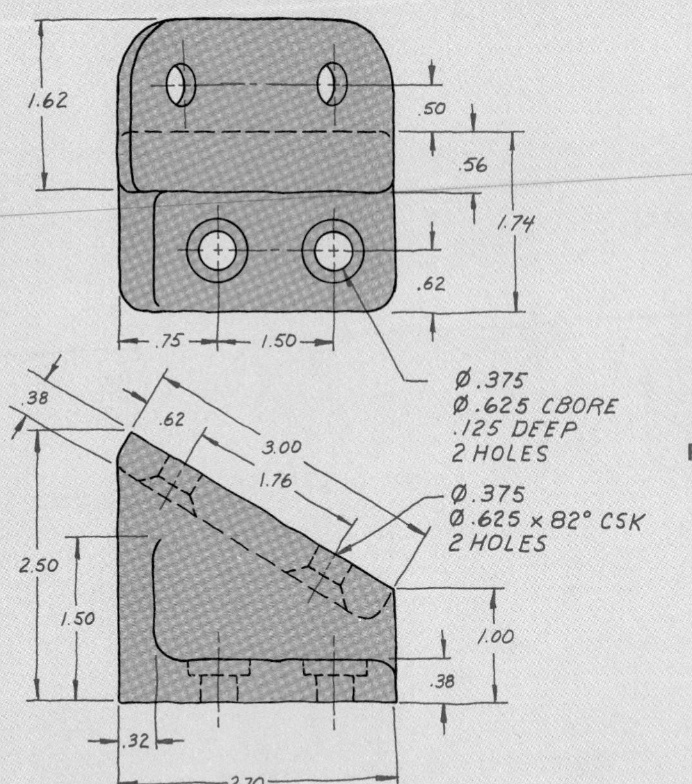

NOTE: ALL FILLETS AND ROUNDS R .18

Problem 16.8 Draw and detail the offset bracket.

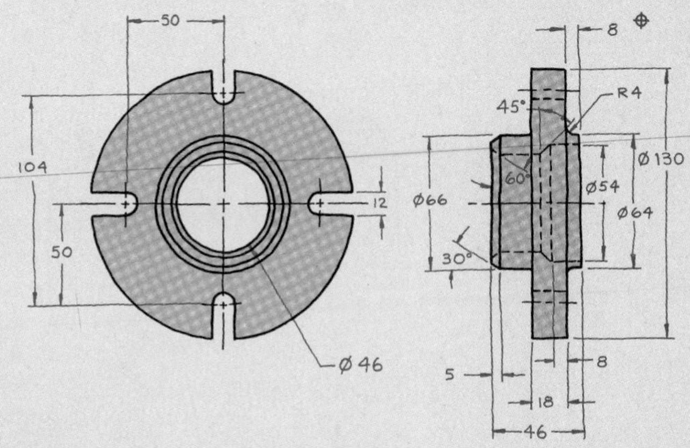

Problem 16.9 Complete a detail of the thrust bearing cap.

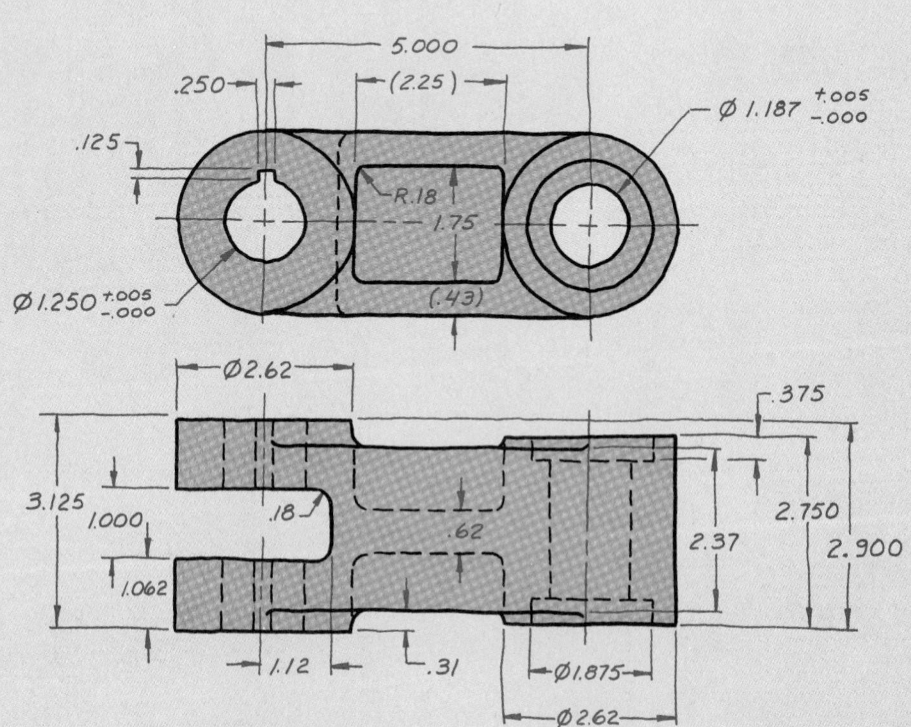

Problem 16.10 Detail the anchor bracket.

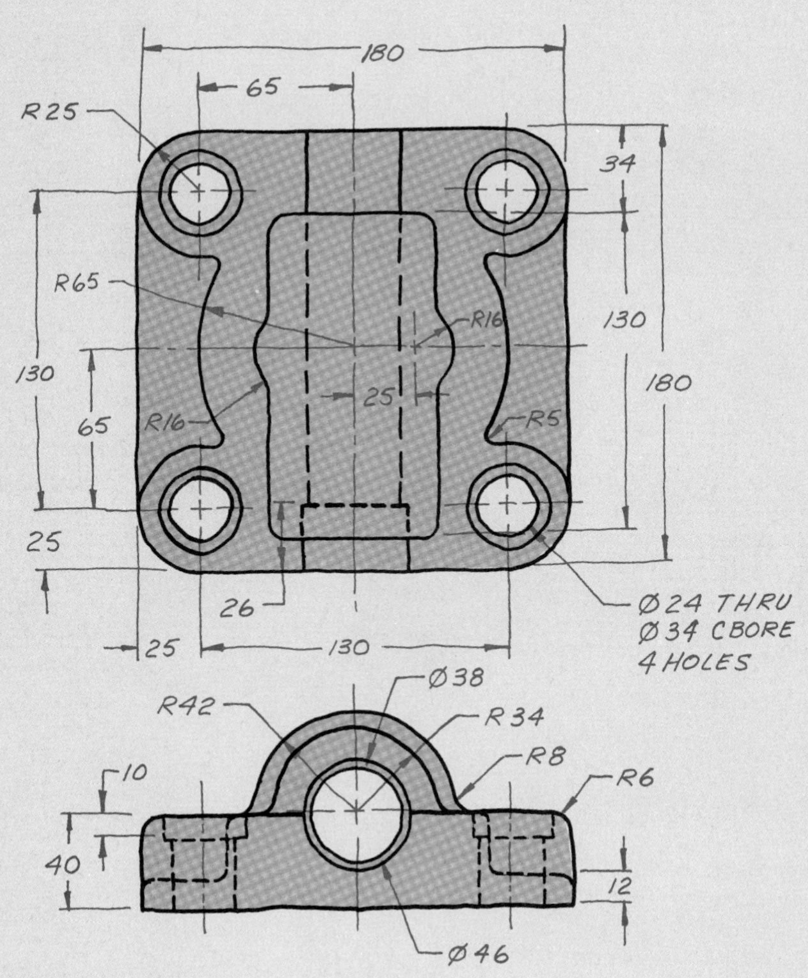

Problem 16.11 Draw and detail the bracket.

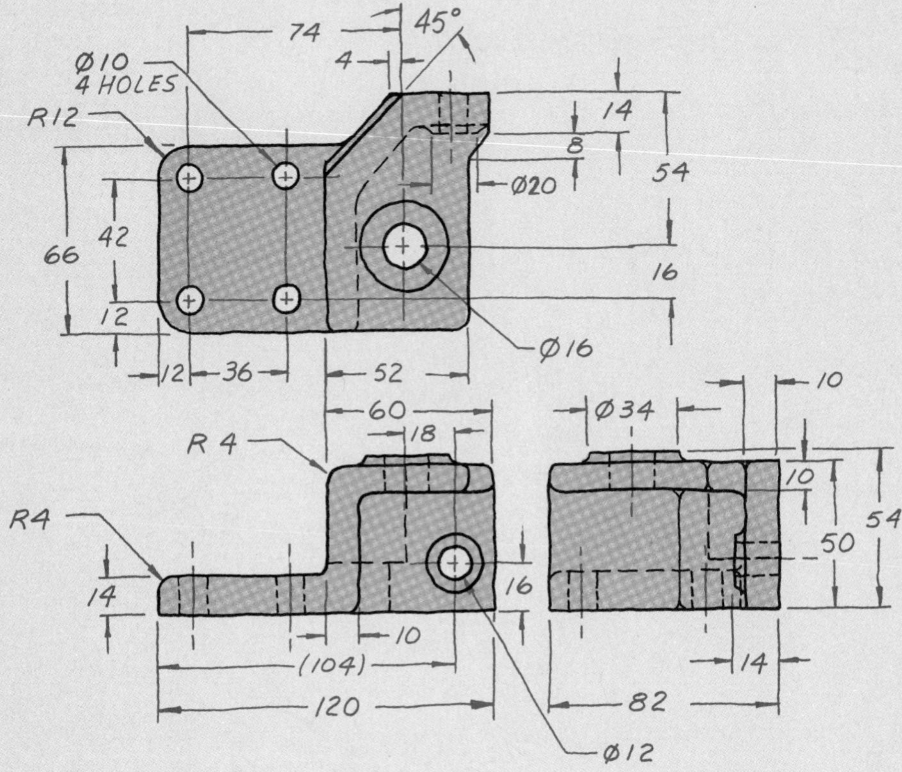

Problem 16.12 Draw and detail the part.

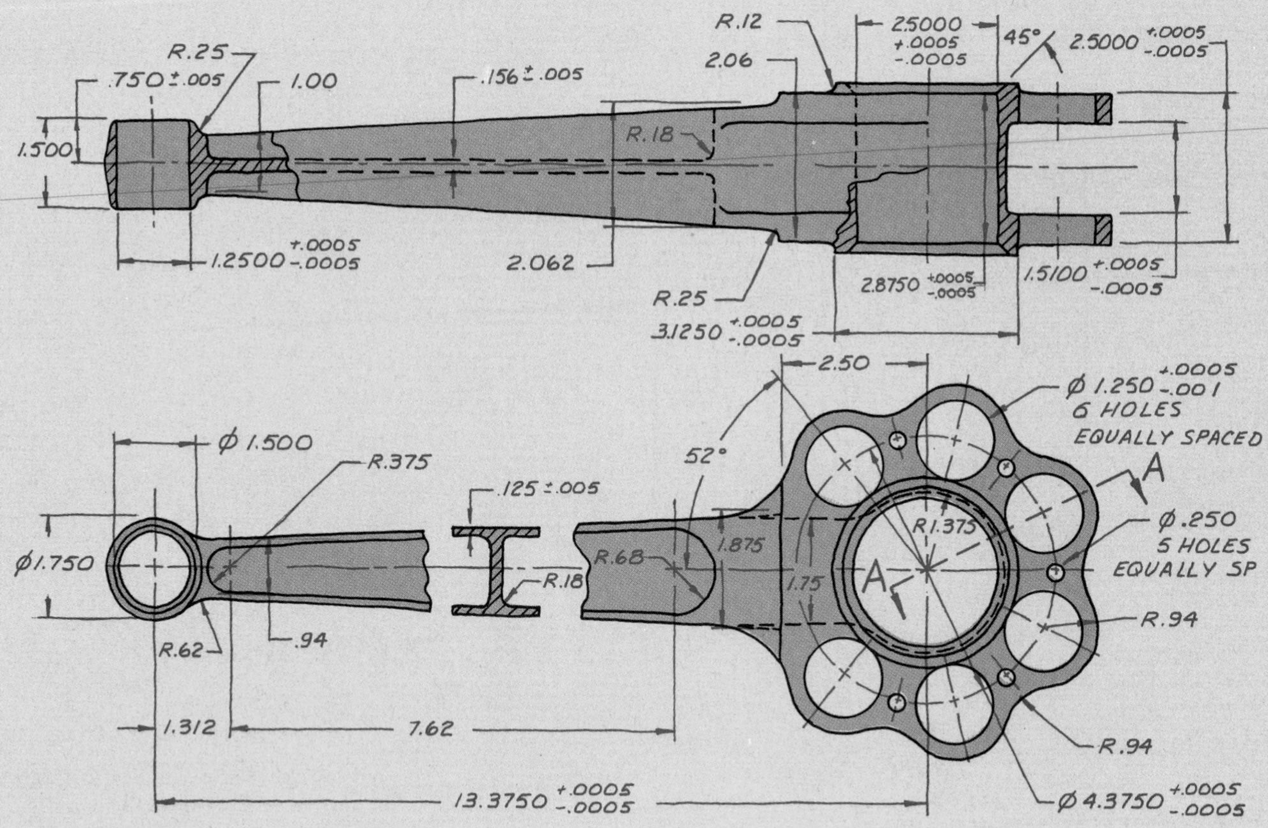

Problem 16.13 Draw and detail the master connecting rod.

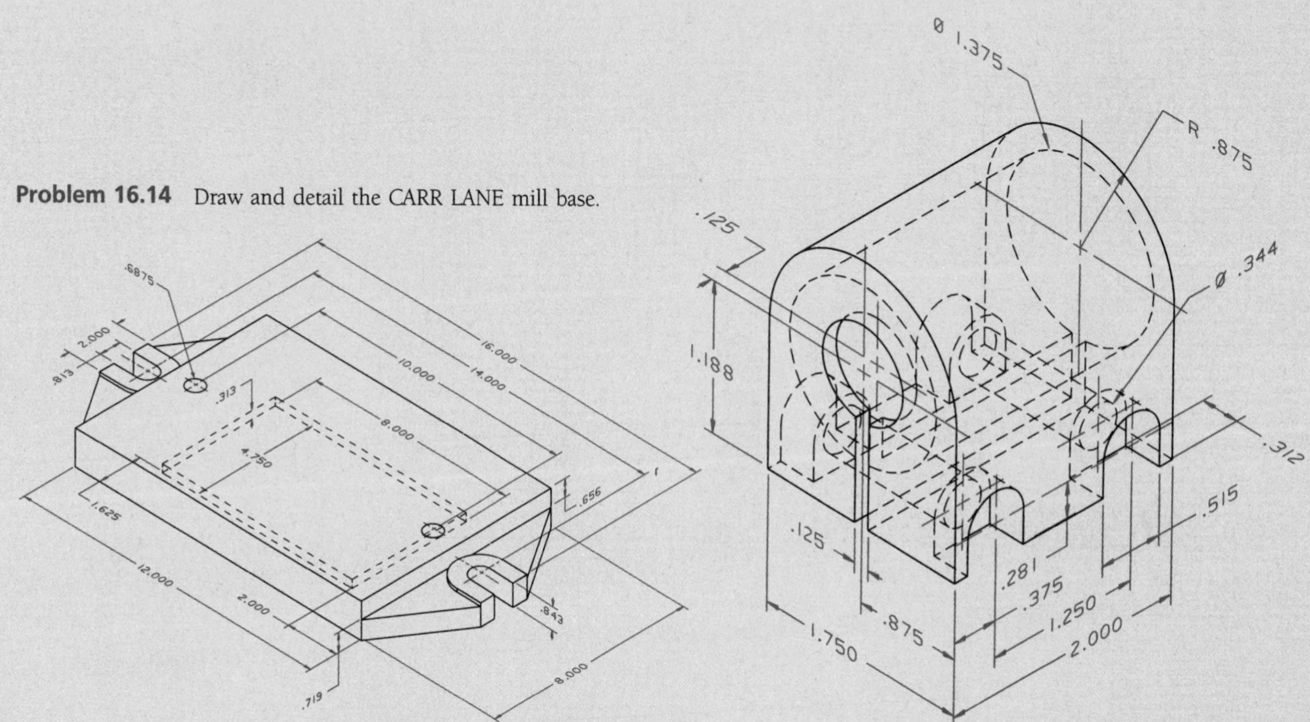

Problem 16.14 Draw and detail the CARR LANE mill base.

Problem 16.15 Draw and detail the CARR LANE cylinder mount.

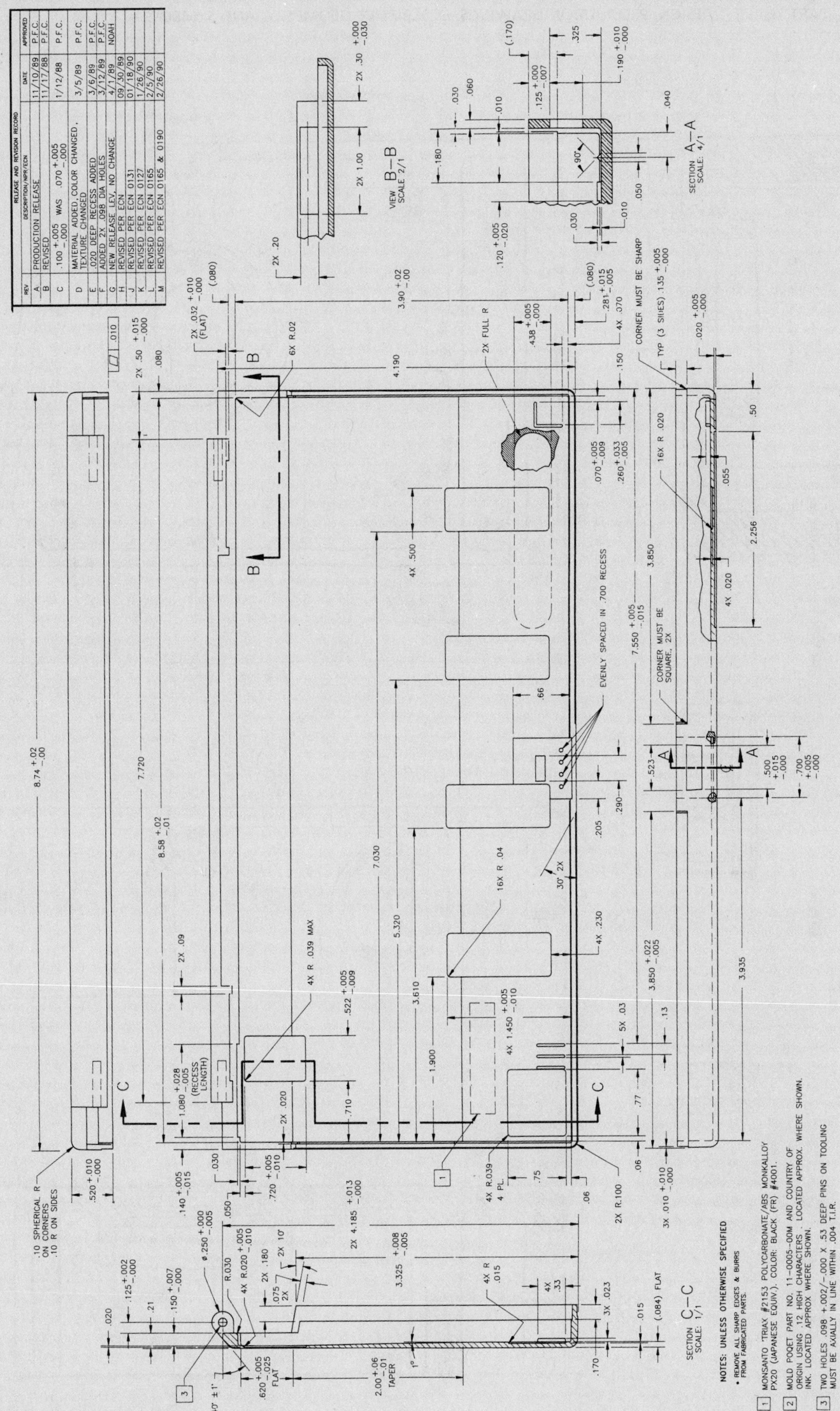

Problem 16.16 Draw and detail the part.

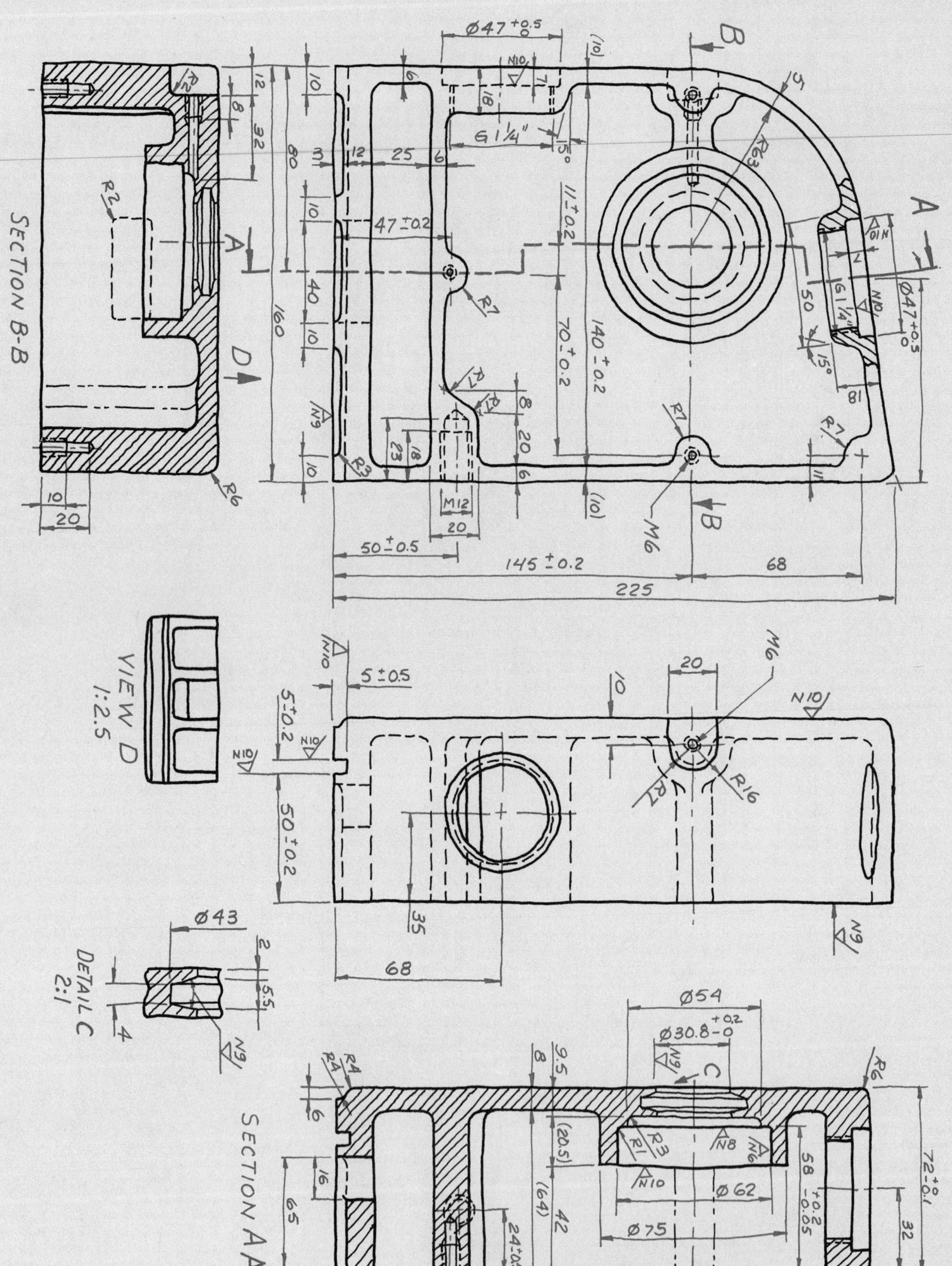

Problem 16.17 Draw and detail the casting.

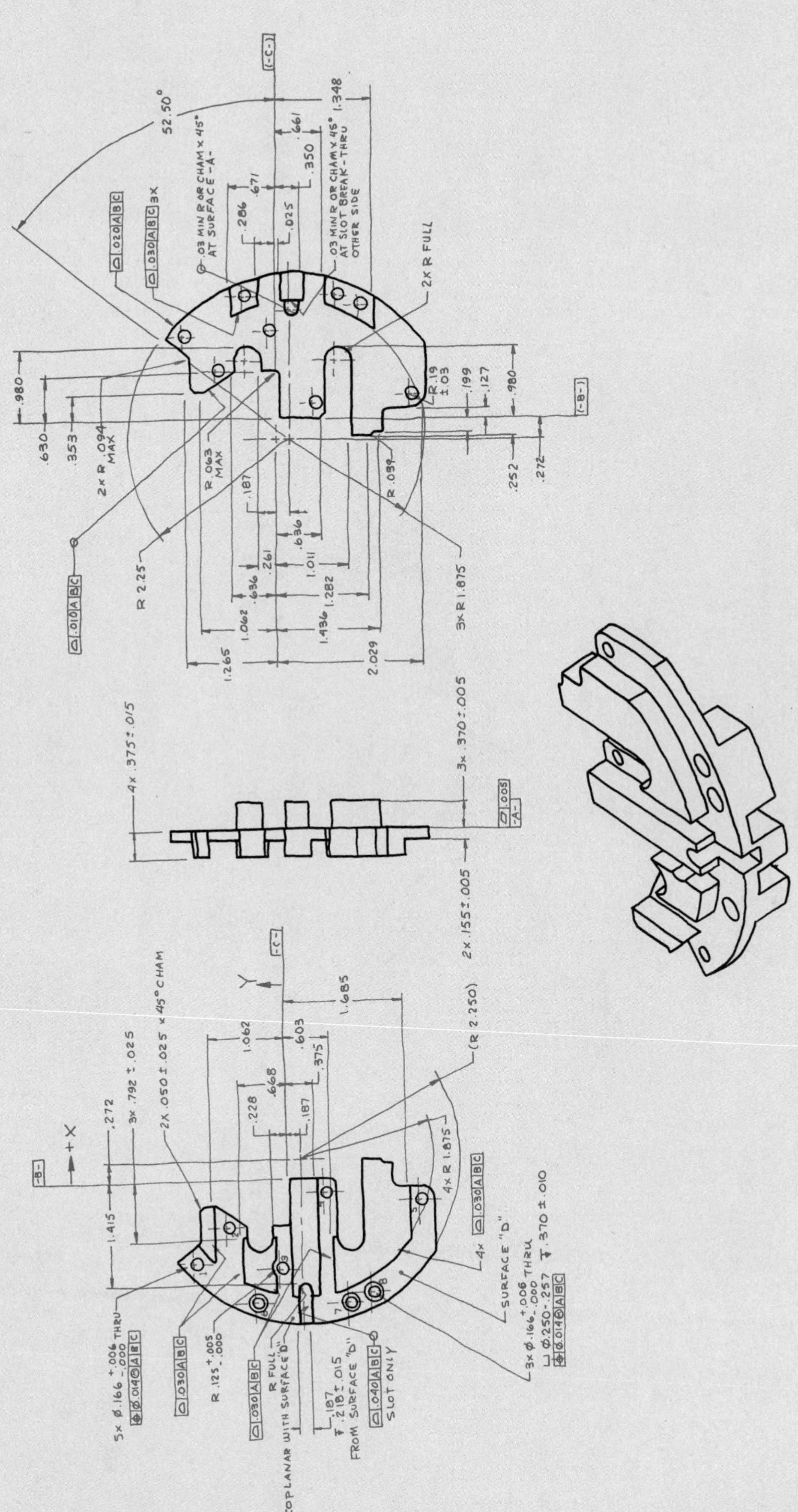

Problem 16.18 Draw and detail the safety shield.

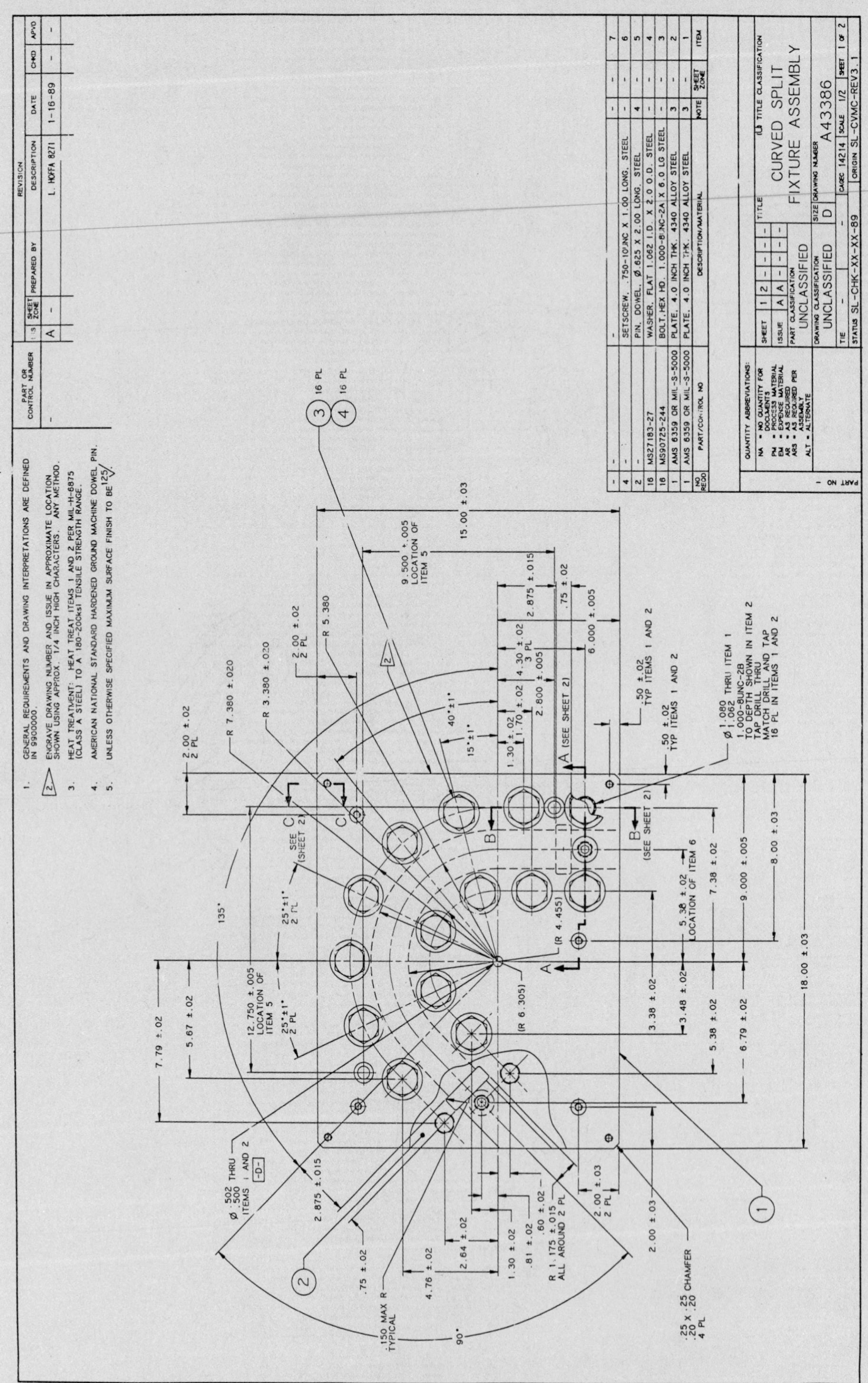

Problem 16.19 (a) Draw and detail the fixture assembly.

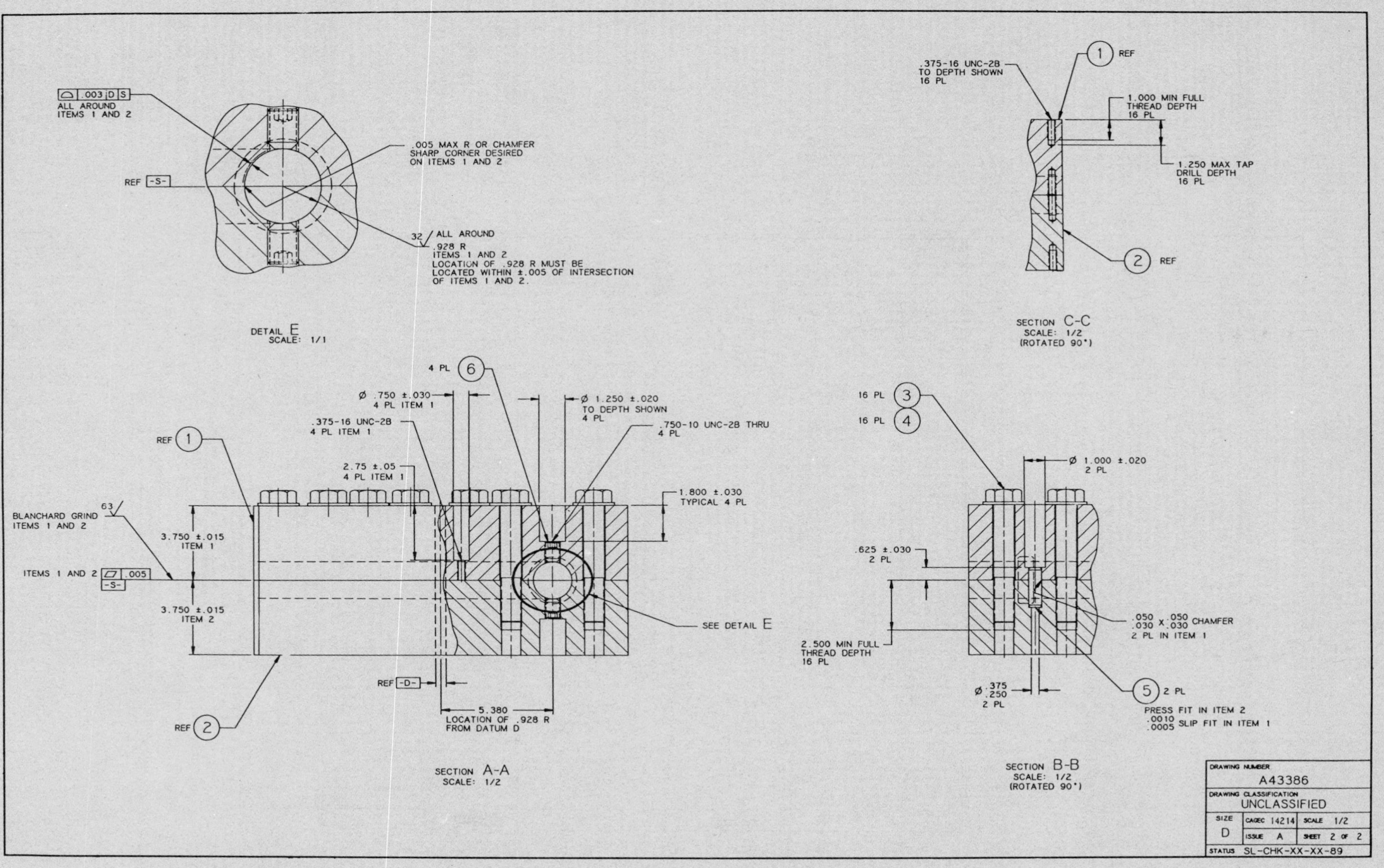

Problem 16.19 (b) Redraw the chassis.

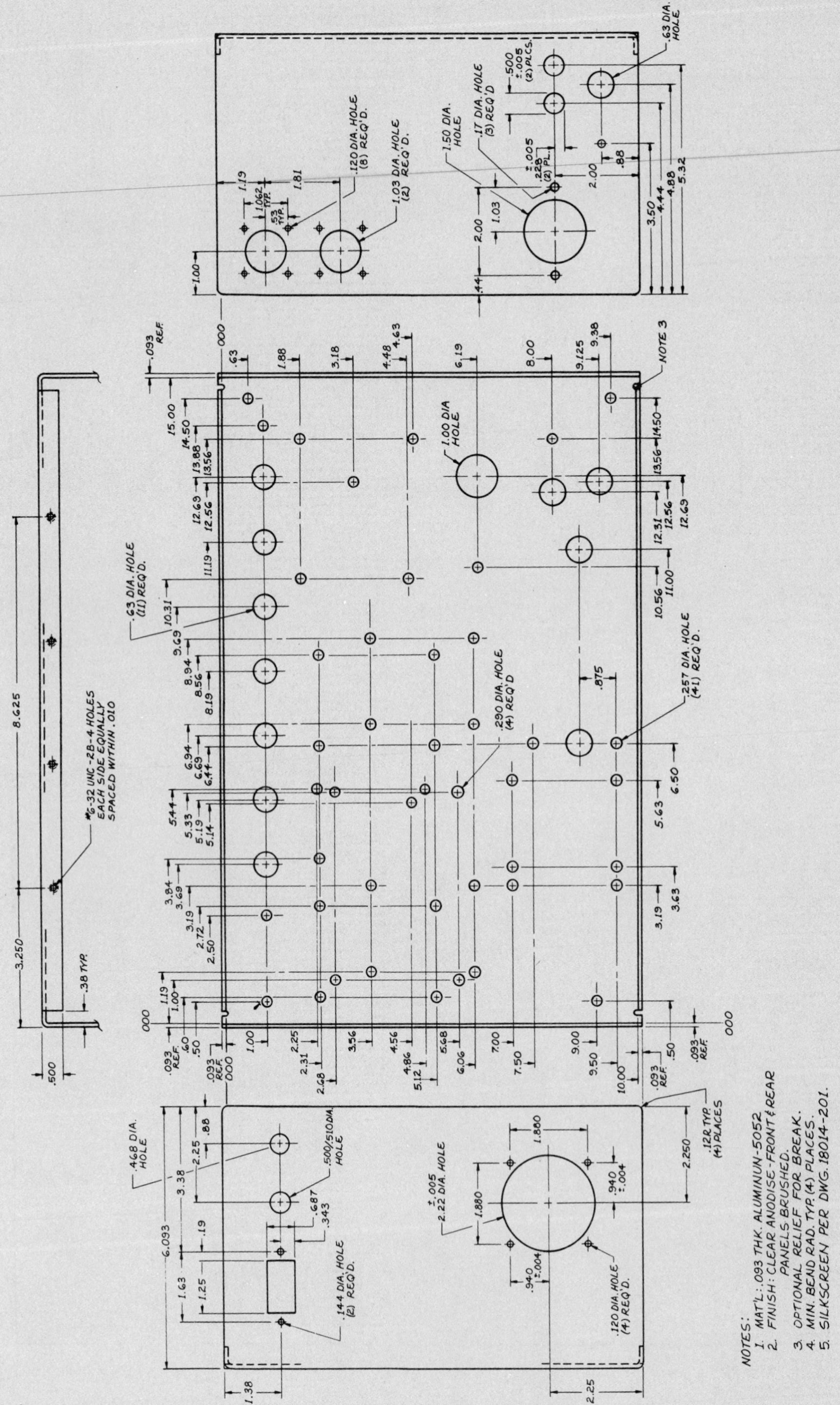

Problem 16.20 Redraw the chassis.

Problem 16.21 Draw and detail the cleaning kit package and dimension completely.

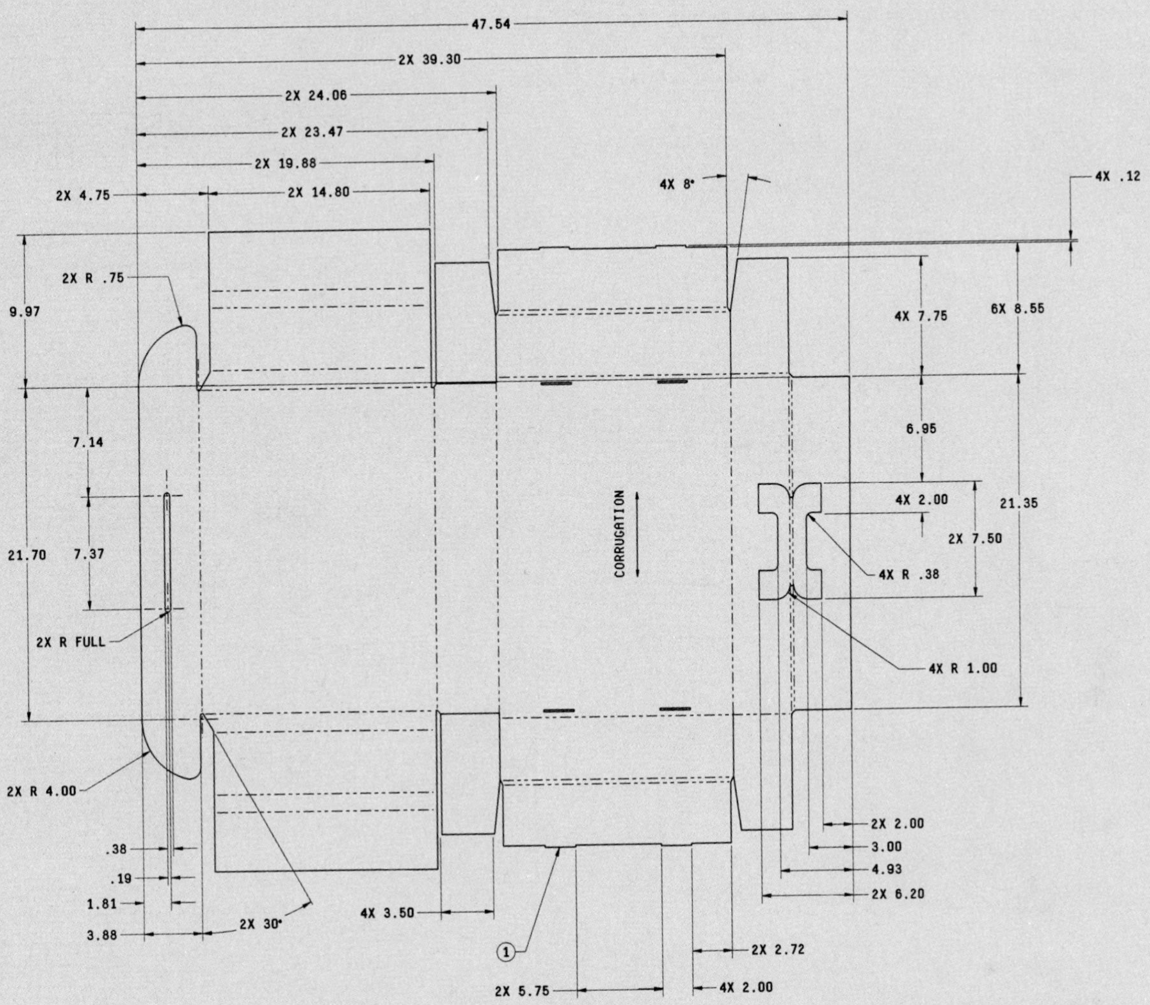

Assembly Drawing Projects

For assembly projects, prepare a layout assembly of the parts by blocking them in for each view required. Be sure to provide sufficient space on the sheet for the assembly and the parts list. The parts list can be generated on a word processor and printed on a separate sheet.

Problem 16.22 Do an assembly and details for the hydraulic valve assembly.

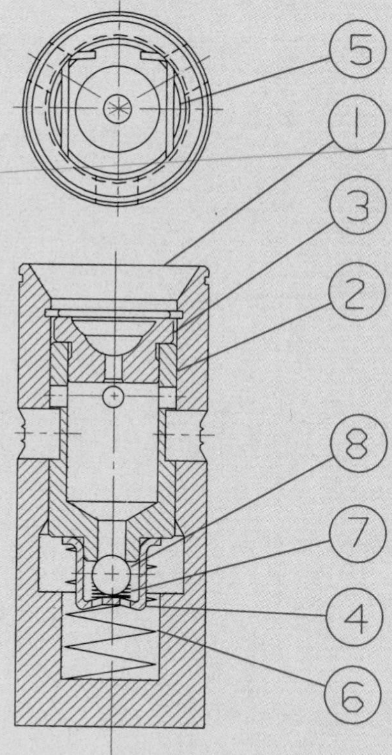

8	BALL BEARING Ø 3/16	1	STEEL
7	BALL CHECK VALVE SPRING	1	SPG STL
6	PLUNGER SPRING	1	SPG STL
5	RETAINER RING	1	SPG
4	BALL RETAINER	1	STL
3	PUSH ROD SEAT	1	STL
2	PLUNGER	1	STL
1	LIFTER BODY	1	STL
NO.	PART NAME	REQD	MATL

Problem 16.22(A)

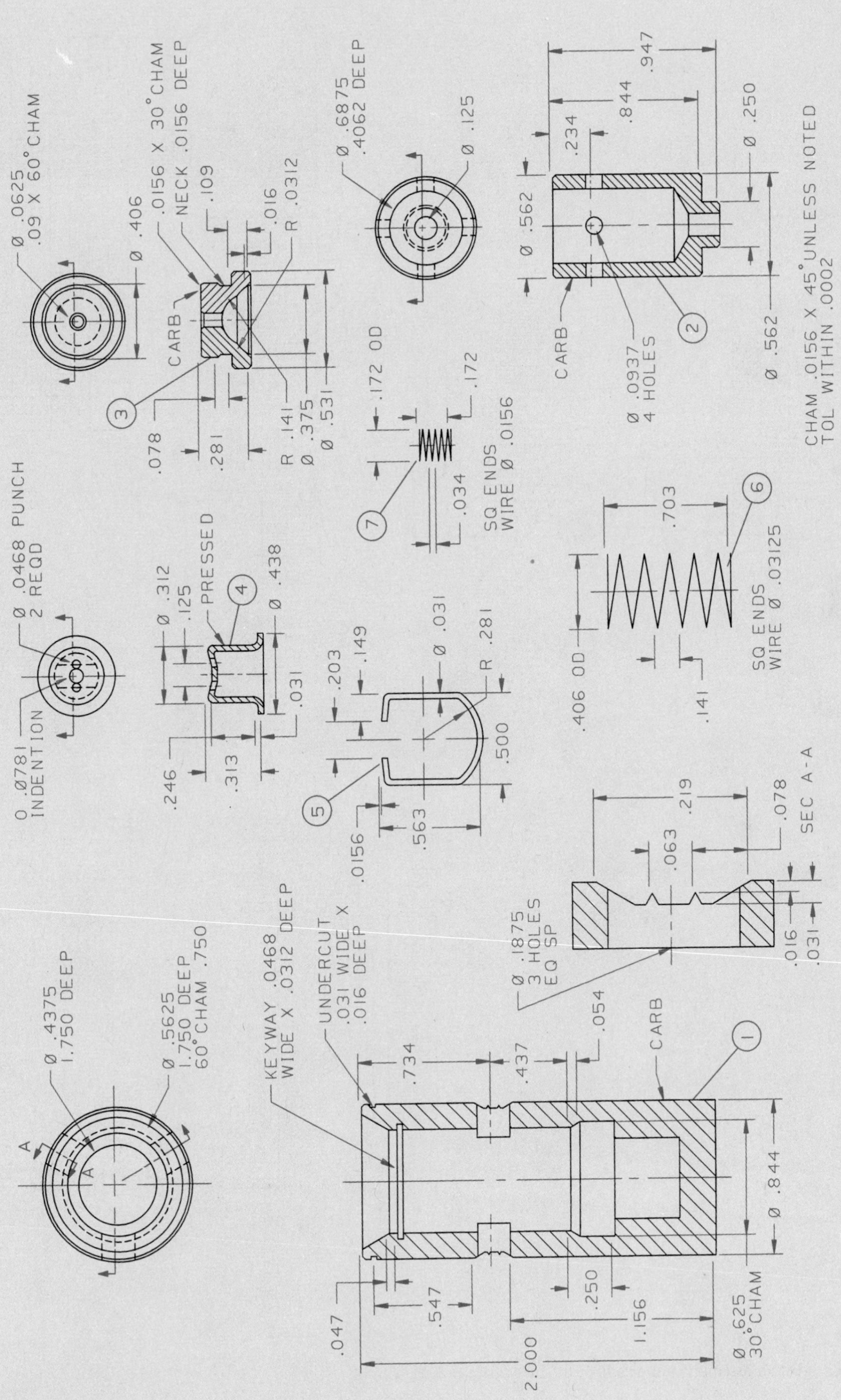

Problem 16.22(B)

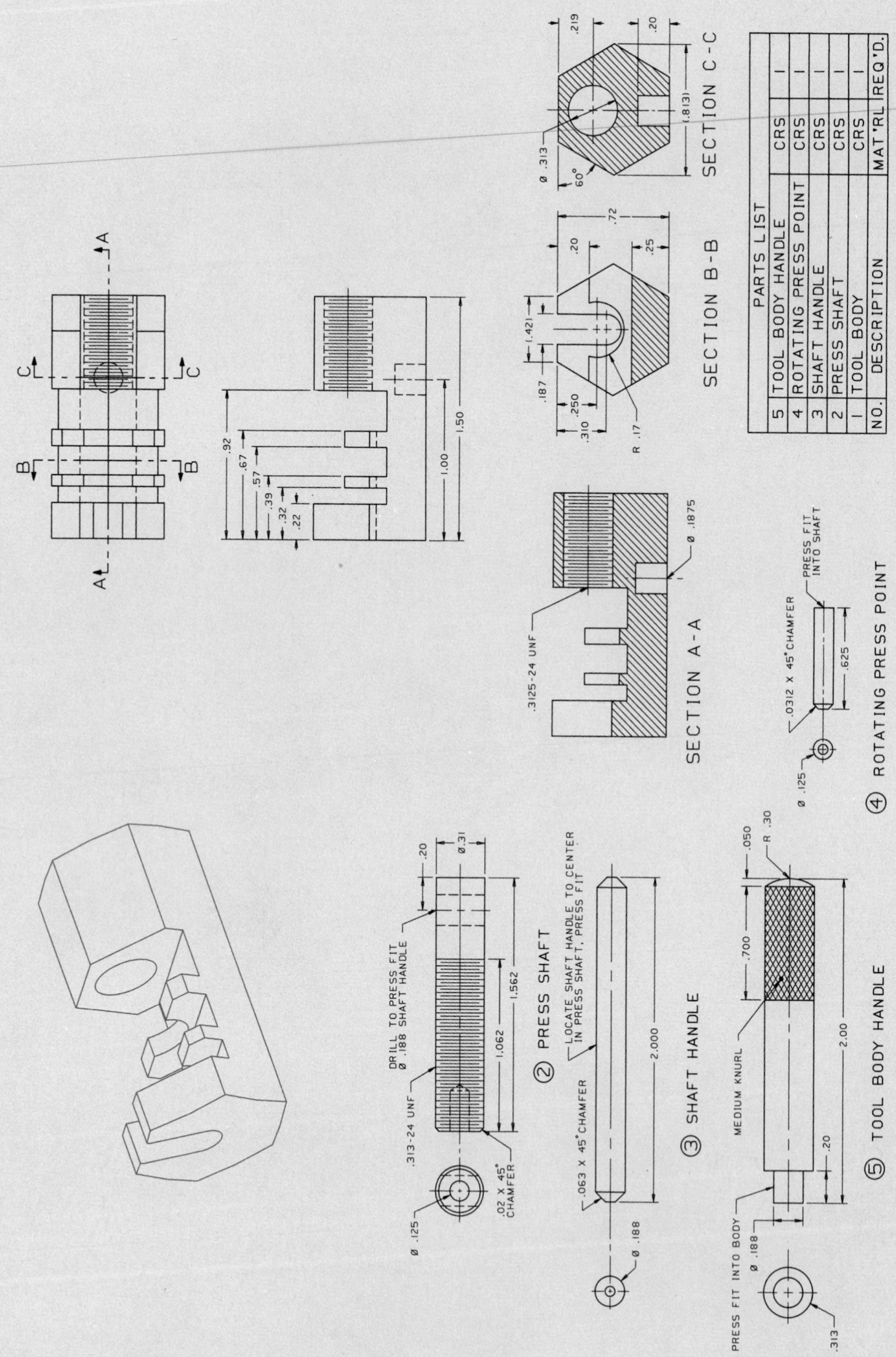

NO.	DESCRIPTION	MAT'RL	REQ'D.
5	TOOL BODY HANDLE	CRS	1
4	ROTATING PRESS POINT	CRS	1
3	SHAFT HANDLE	CRS	1
2	PRESS SHAFT	CRS	1
1	TOOL BODY	CRS	1

Problem 16.23 Do an assembly and details of the bike chain puller assembly.

1
2
3
4
5
6
7
8

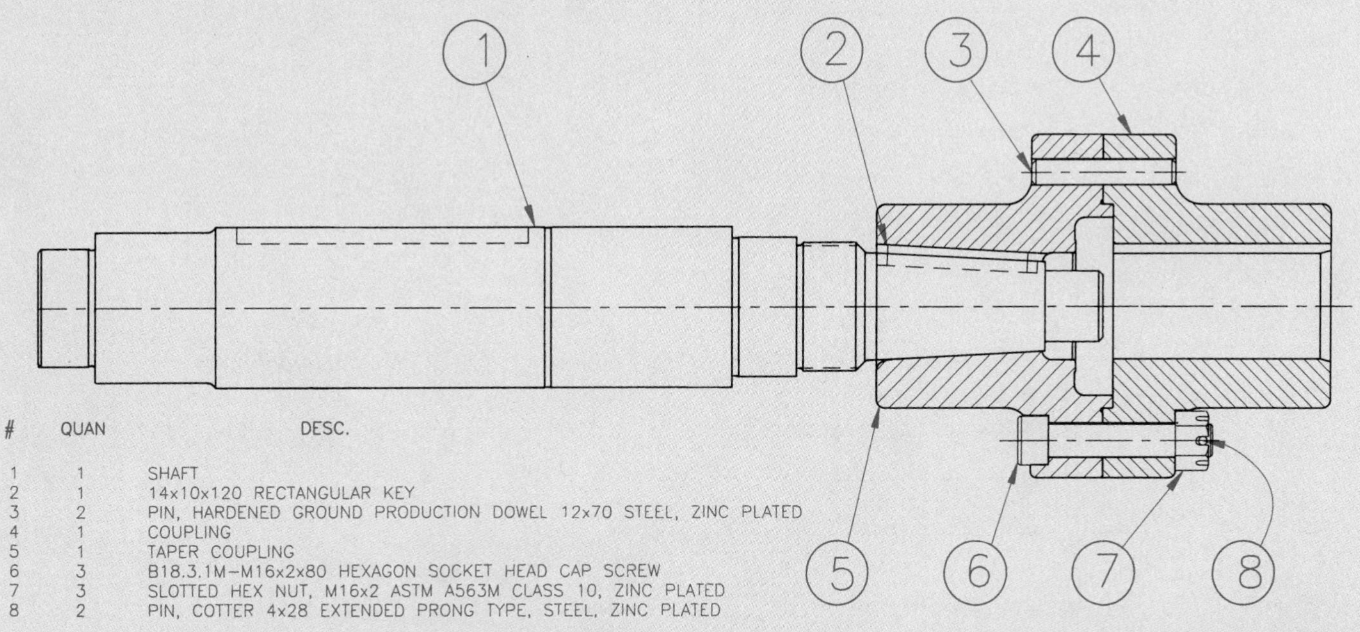

#	QUAN	DESC.
1	1	SHAFT
2	1	14x10x120 RECTANGULAR KEY
3	2	PIN, HARDENED GROUND PRODUCTION DOWEL 12x70 STEEL, ZINC PLATED
4	1	COUPLING
5	1	TAPER COUPLING
6	3	B18.3.1M–M16x2x80 HEXAGON SOCKET HEAD CAP SCREW
7	3	SLOTTED HEX NUT, M16x2 ASTM A563M CLASS 10, ZINC PLATED
8	2	PIN, COTTER 4x28 EXTENDED PRONG TYPE, STEEL, ZINC PLATED

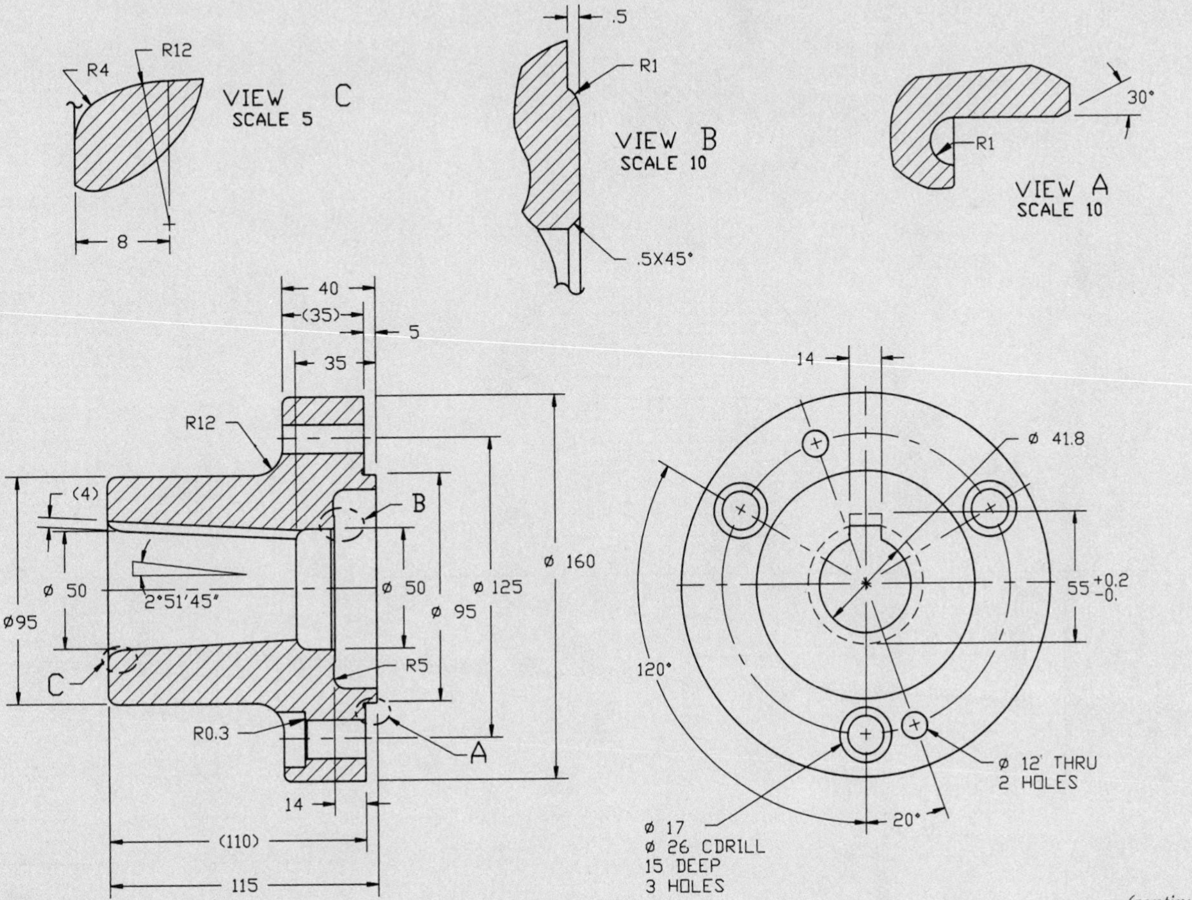

R12
R4
VIEW C
SCALE 5
8
.5
R1
VIEW B
SCALE 10
.5X45°
30°
R1
VIEW A
SCALE 10
40
(35)
5
35
R12
(4)
ø 50
2°51'45"
ø95
C
B
ø 50
ø 95
ø 125
ø 160
R5
R0.3
A
14
(110)
115
14
ø 41.8
55 +0.2 -0.
120°
ø 12' THRU
2 HOLES
20°
ø 17
ø 26 CDRILL
15 DEEP
3 HOLES

(continued)

(B)

Problems 16.24(A) through (B) Draw and detail the assembly and details for the coupling.

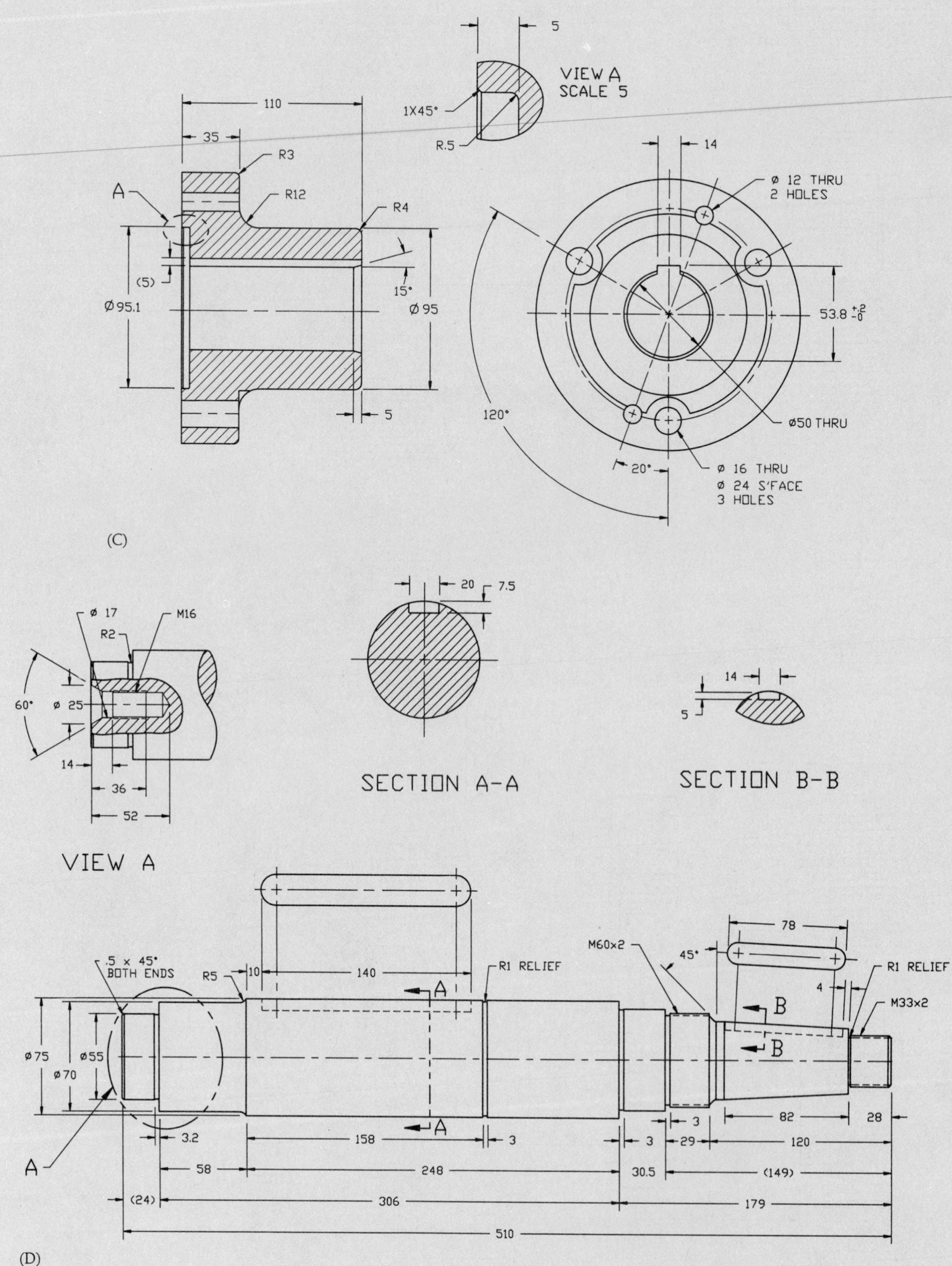

Problems 16.24(C) and (D) Draw and detail the assembly and details for the coupling.

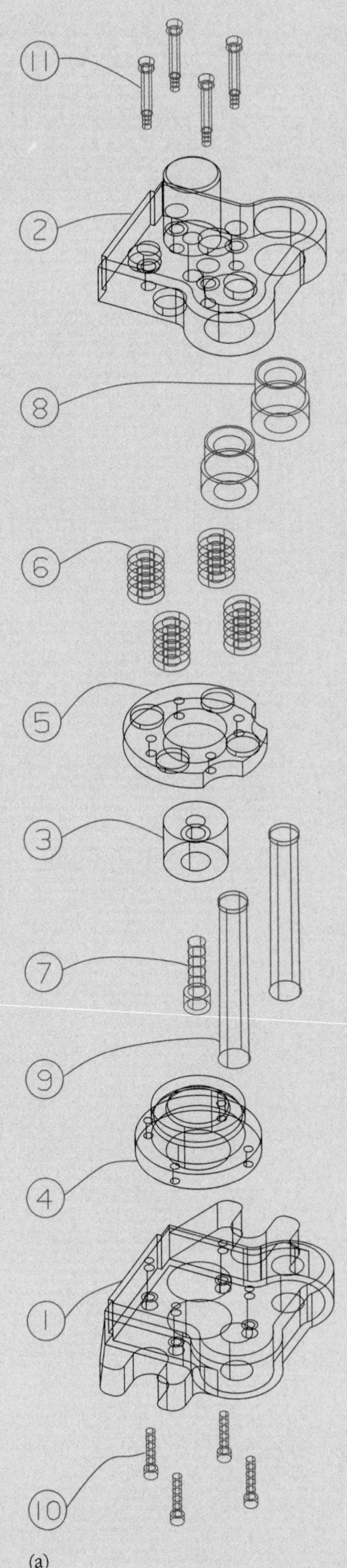

(a)

ITEM	QTY	PART NO.	DESCRIPTION
1	1		DIE BASE, CAST STEEL
2	1		UPPER DIESHOE, CAST STEEL
3	1		DIE PUNCH, STEEL
4	1		WASHER DIE, STEEL
5	1		DIE PLATE, STEEL
6	2	9-1606-21	SPRING, DANLY MEDIUM-HIGH PRESSURE, STEEL RETANGULAR WIRE, HOLE Ø 1.00, ROD Ø .500, 1.50 FREE LENGTH (REF .980 OD .150 ID)
7	1		SCREW, SOCKET HEAD CAP .500-20 X 2.00
8	2	6-07-61	BUSHING, DANLY PRECISION PRESS FIT SHORT SHOULDER, STEEL, .875 ID, 1.375 OD, 1 9/16 SHOULDER ID, LENGTH 1.75, 13/16 LONG SHOULDER
9	2	5-0720-1	GUIDE POST, DANLY MICROME PRECISION STEEL, Ø .875 X 5.00
10	4		SCREW, SOCKET HEAD CAP SHOULDER Ø .3125 X 1.50
11	1		SCREW, SOCKET HEAD CAP Ø .250-20 X 1.25

SH | REV. | DWG. NO.

DRAWN	8-2-91	SIZE A	FSCM NO.	DIE SET	REV.
ISSUED	8-15-91	SCALE 1:1		JAIME GUERRERO	

(b)

(c)

(continued)

Problem 16.25 Draw and detail the die set. Do a complete assembly and set of details.

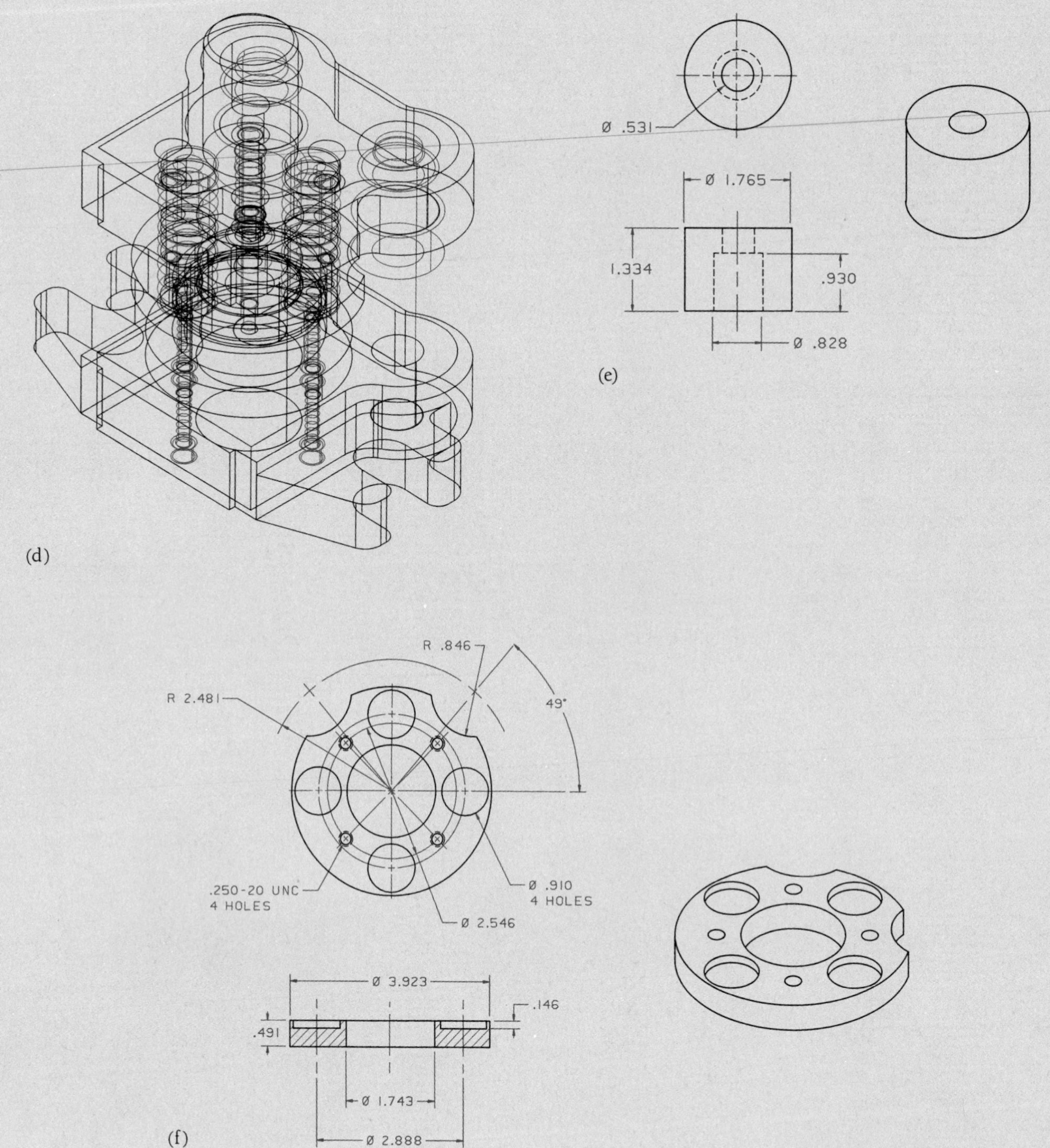

(d)

(e)

(f)

Problem 16.25—*Continued*

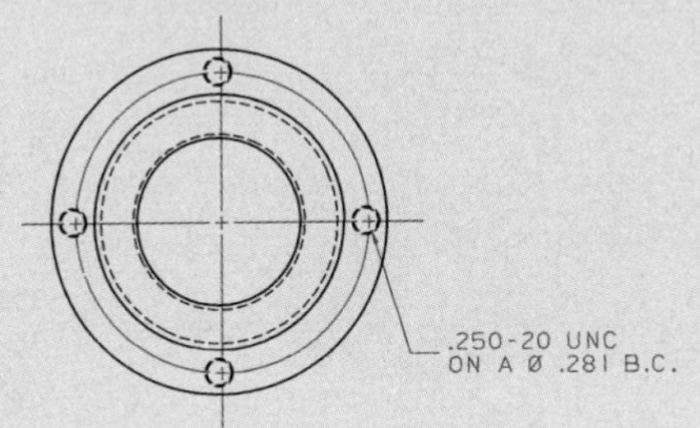
.250-20 UNC
ON A Ø .281 B.C.

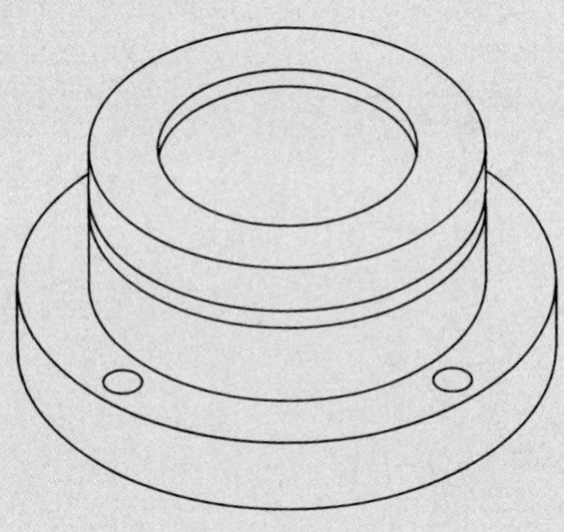

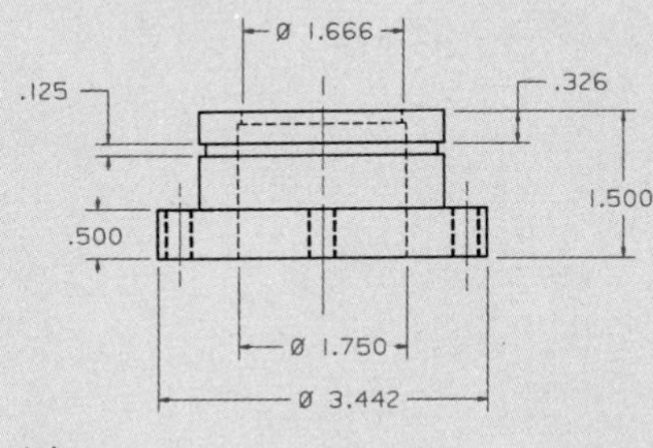
Ø 1.666
.125
.326
1.500
.500
Ø 1.750
Ø 3.442

(g)

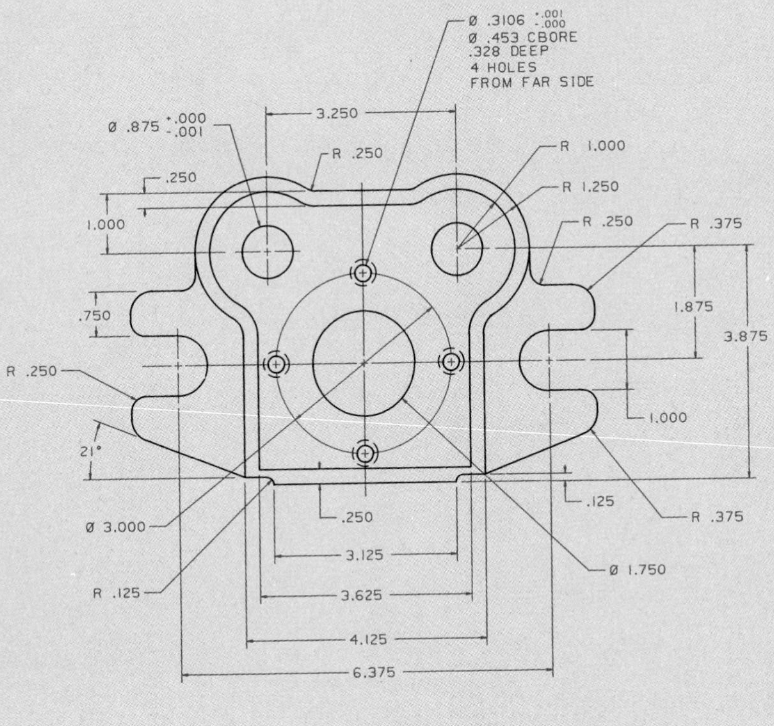
Ø .3106 +.001 -.000
Ø .453 CBORE
.328 DEEP
4 HOLES
FROM FAR SIDE
Ø .875 +.000 -.001
3.250
R .250
R 1.000
R 1.250
.250
1.000
R .250
R .375
1.875
3.875
.750
R .250
1.000
21°
.125
R .375
Ø 3.000
.250
3.125
Ø 1.750
R .125
3.625
4.125
6.375

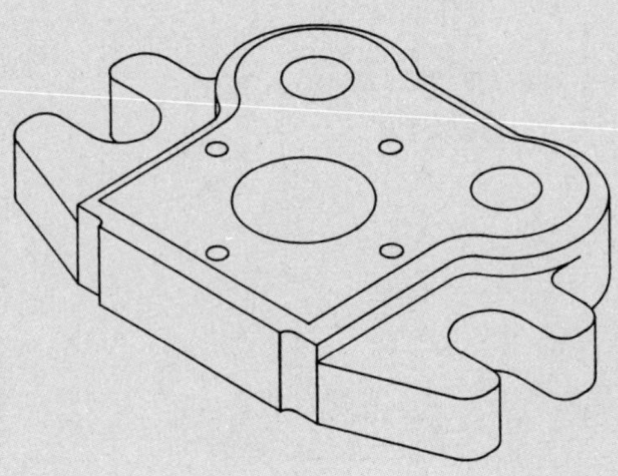

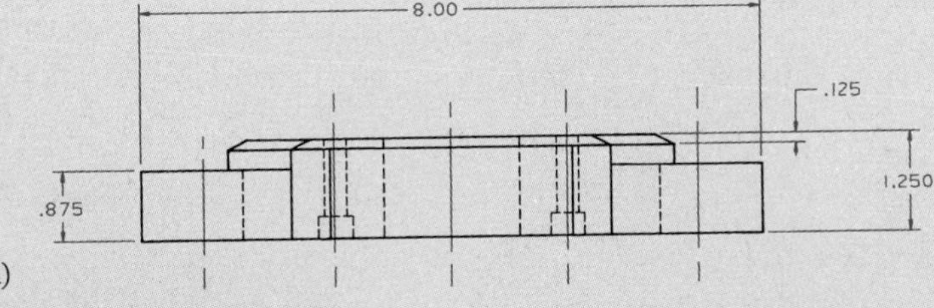
8.00
.125
1.250
.875

(h)

Problem 16.25—*Continued*

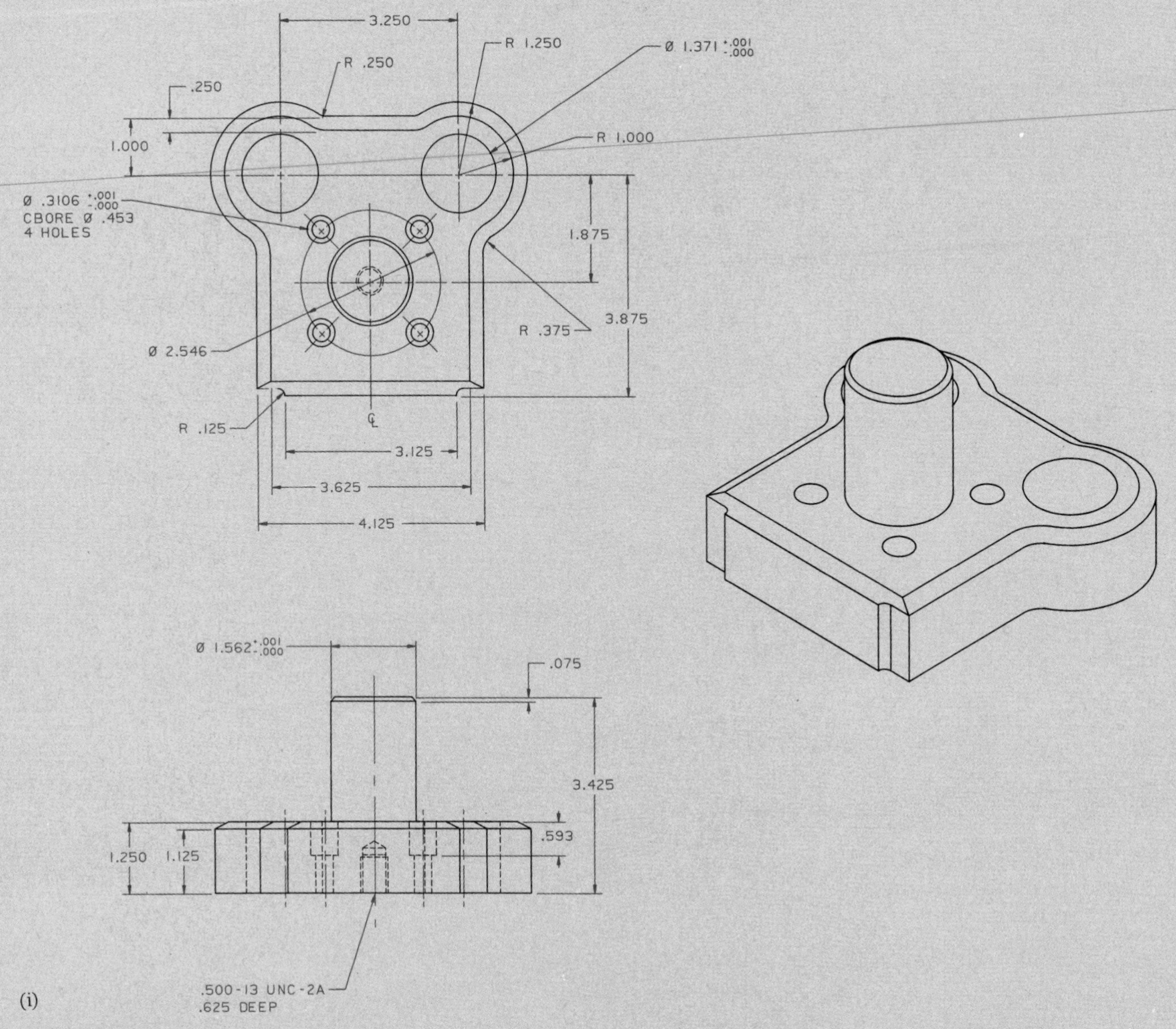

(i)

Problem 16.25—*Continued*

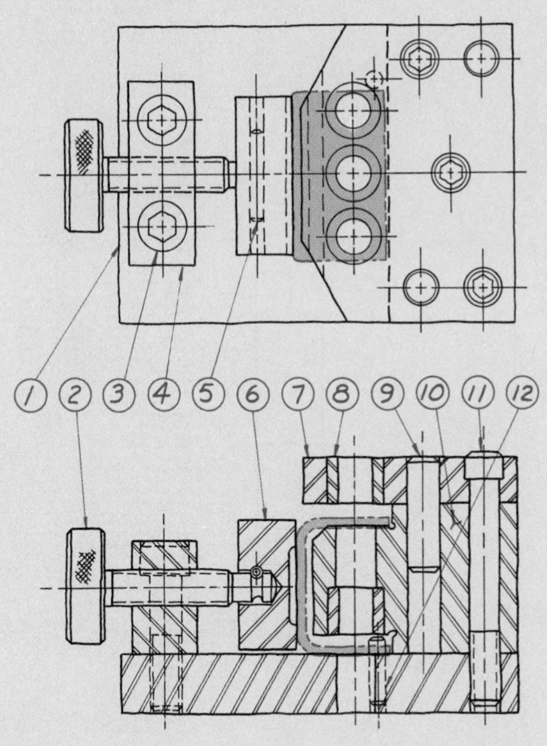

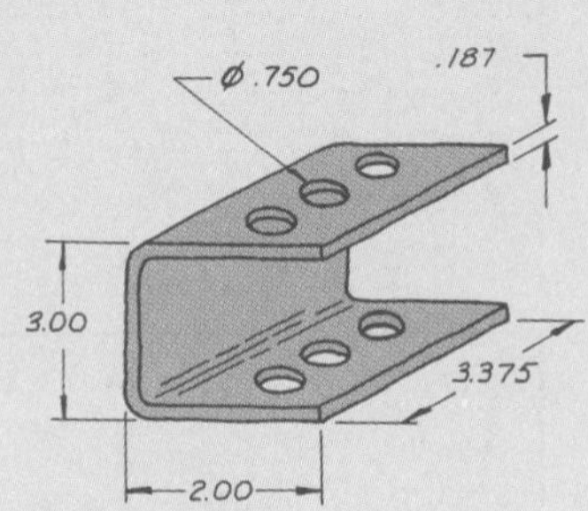

(A)

ITEM	QTY	PART NO.	DESCRIPTION
1	1		BASE PLATE, 1.375 X 6.50 X 9.00, STEEL
2	1		SCREW, KNURLED HEAD, STEEL
3	2		SCREW, HEX SOC HD CAP, .750-10 X 3.00
4	1		CLAMP BLOCK, 1.50 X 2.50 X 4.00, STEEL
5	1		PIN, MACHINE DOWEL, Ø .250 X 2.00
6	1		CLAMP, 1.375 X 3.00 X 3.50, STEEL
7	1		BUSHING PLATE, 1.00 X 4.75 X 6.50, STEEL
8	6		BUSHING, TYPE P HEADLESS PRESS FIT MODIFIED ID = .750, 1.25 OD X 1.00 LG
9	2		PIN, MACHINE DOWEL, Ø .750 X 2.50
10	1		LOCATOR BLOCK, 3.25 X 4.50 X 6.50
11	3		SCREW, HEX SOC HD CAP, .625-11 X 5.00
12	1		PIN, MACHINE DOWEL, Ø .375 X 1.50

DRAWN 3-4-91	SIZE A	FSCM NO.	PLATE JIG, DRILL	REV.
ISSUED 7-30-91	SCALE 1:1			

(B)

Problem 16.26 Draw and detail the jig and fixture assembly.

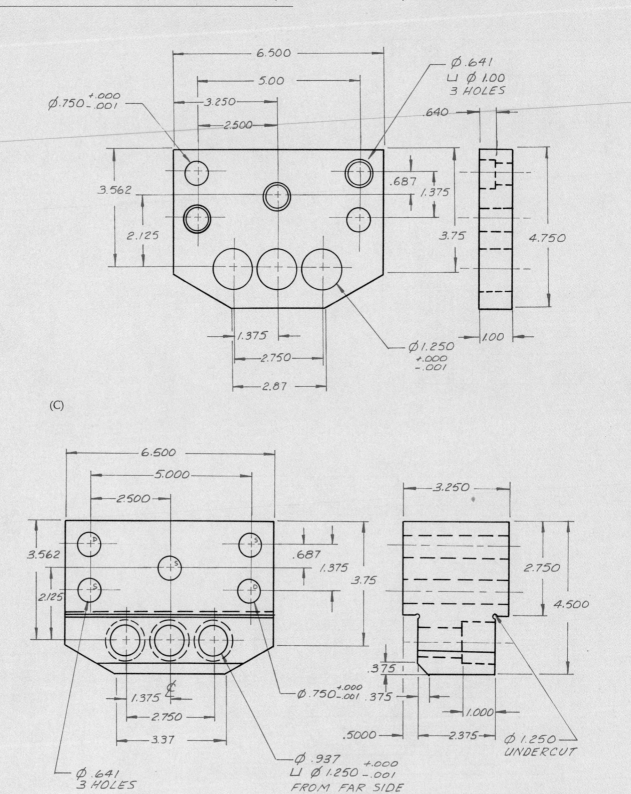

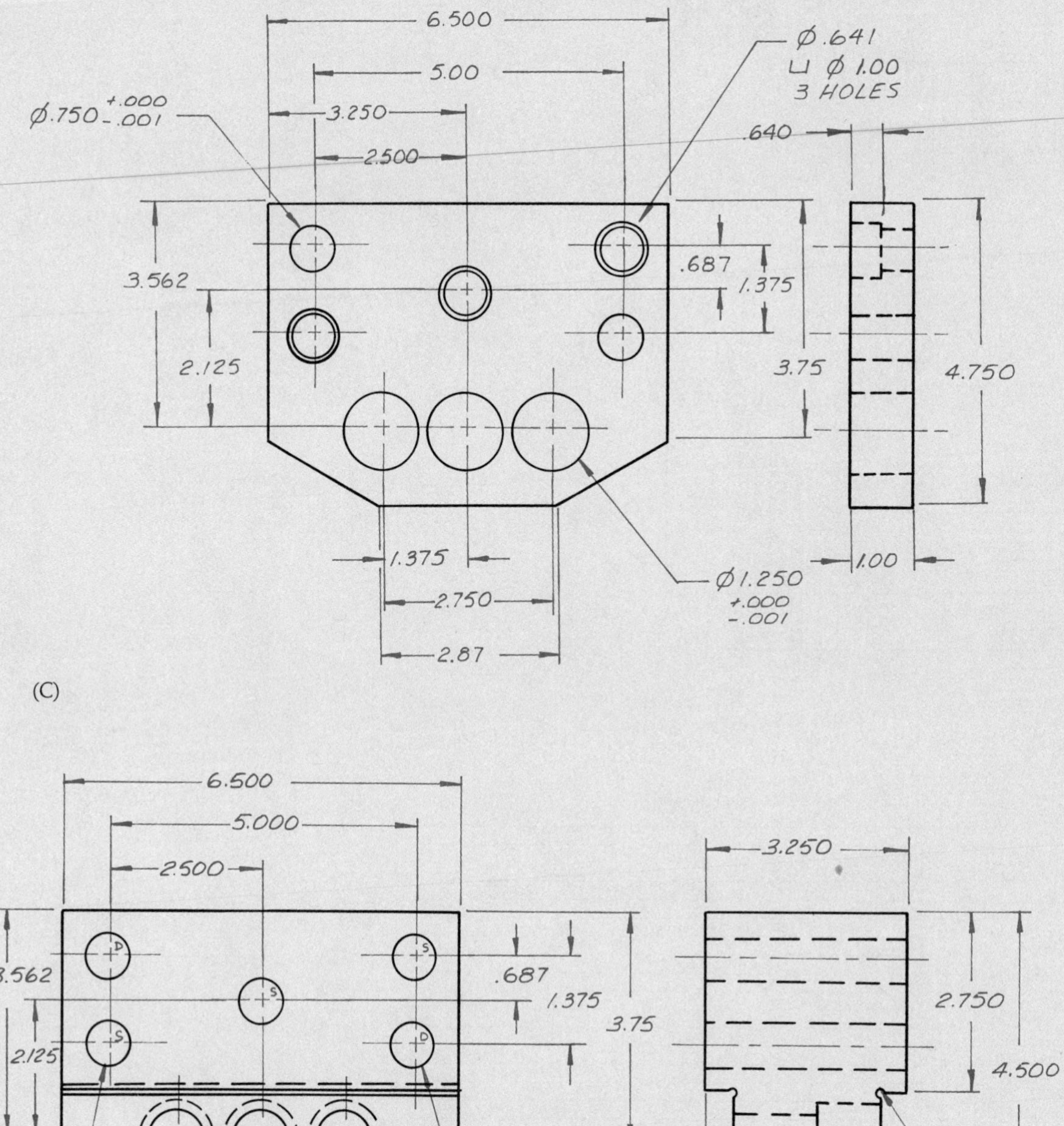

Problem 16.26 –*Continued*

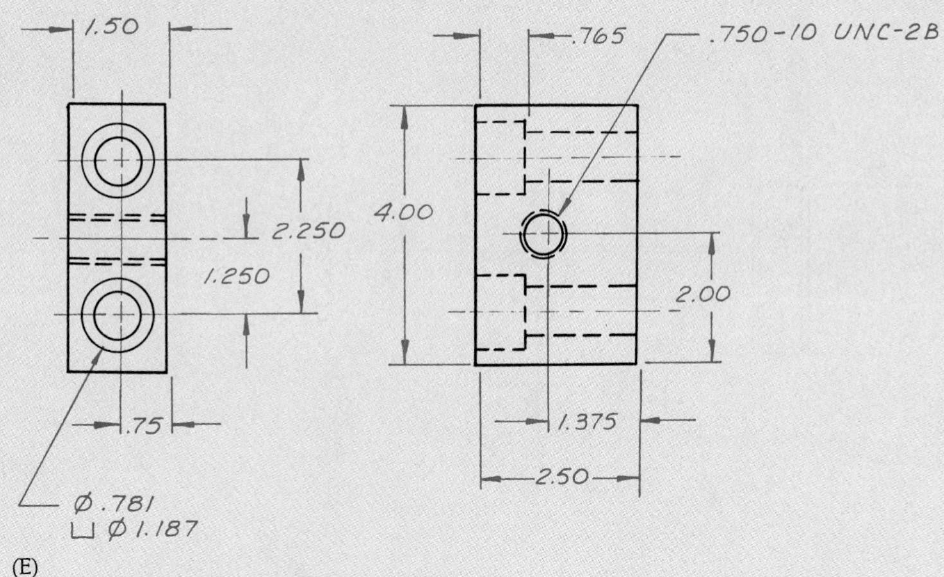

(E)

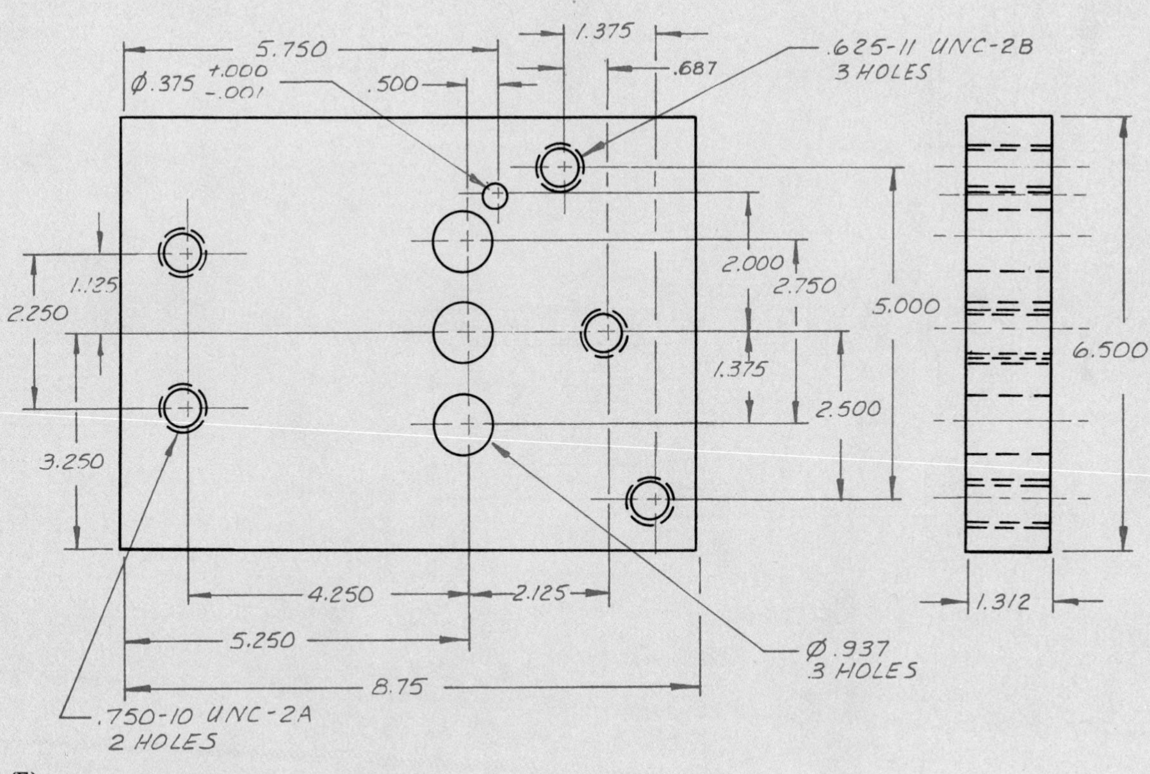

(F)

Problem 16.26—*Continued*

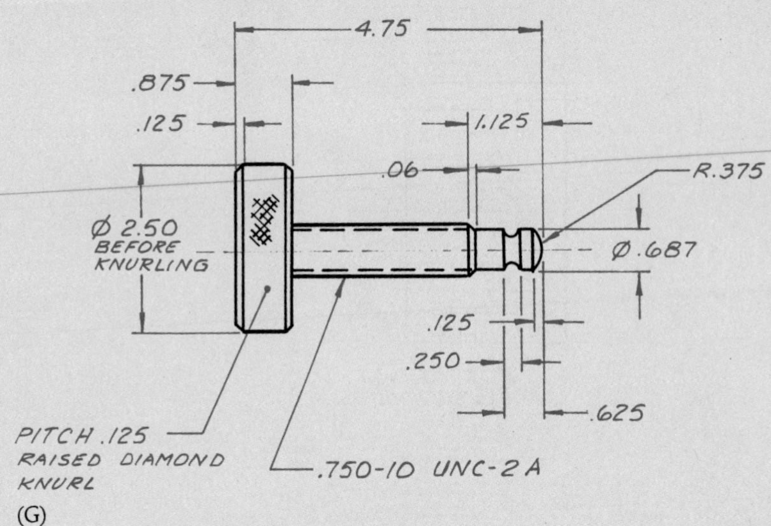

(G)

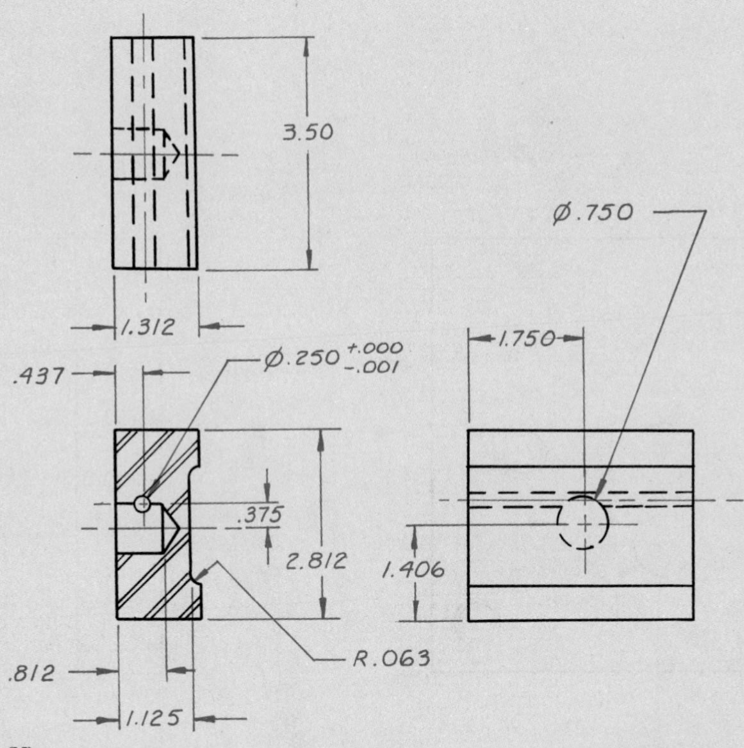

(H)

Problem 16.26—*Continued*

Assembly Grid Problems

The following five assembly projects are not dimensioned. Instead, a scale (inch and metric) is provided on the drawing. Use dividers to transfer the parts sizes, and or use the .25 in. grid to determine dimensions.

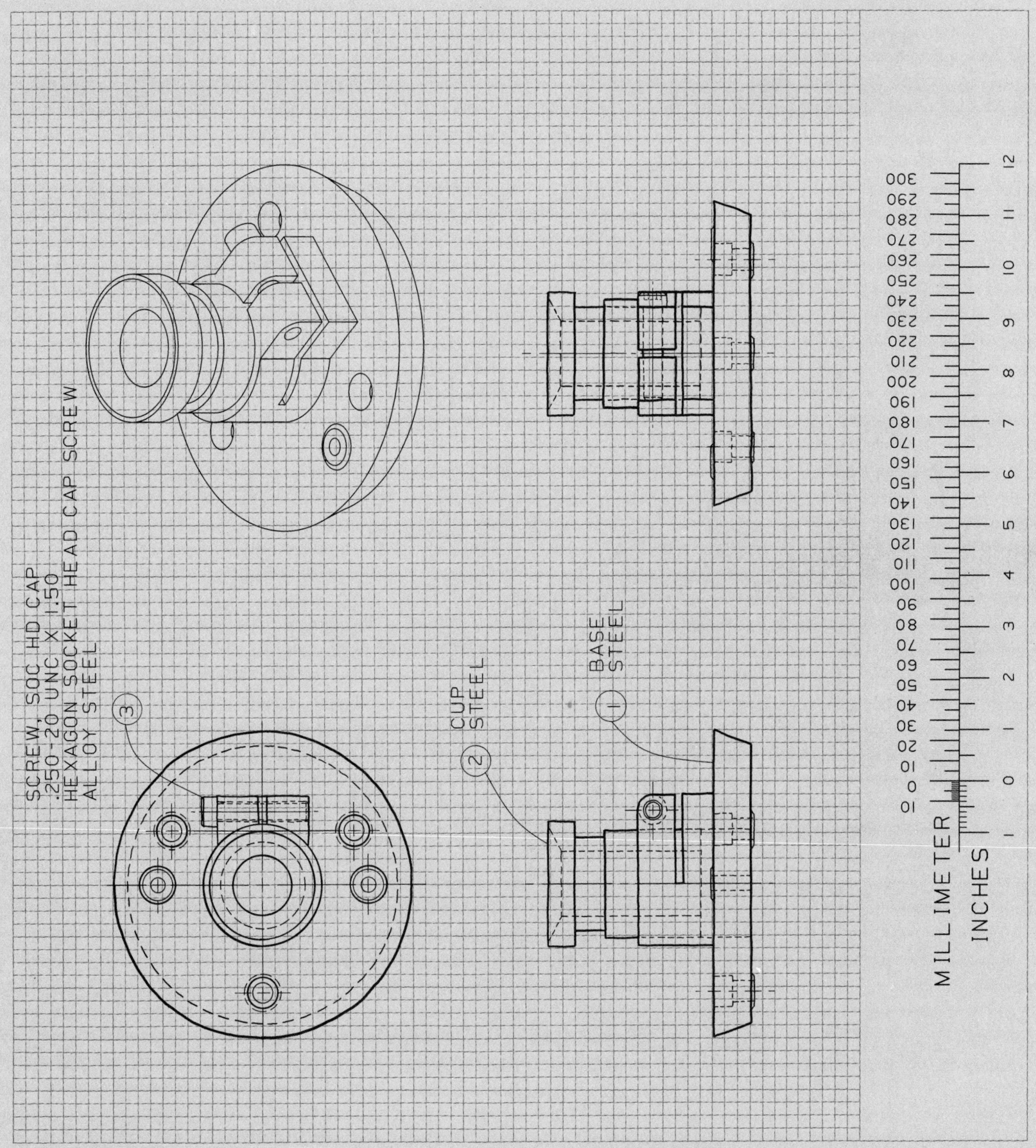

Problem 16.27 Draw and detail the optical cup mount assembly.

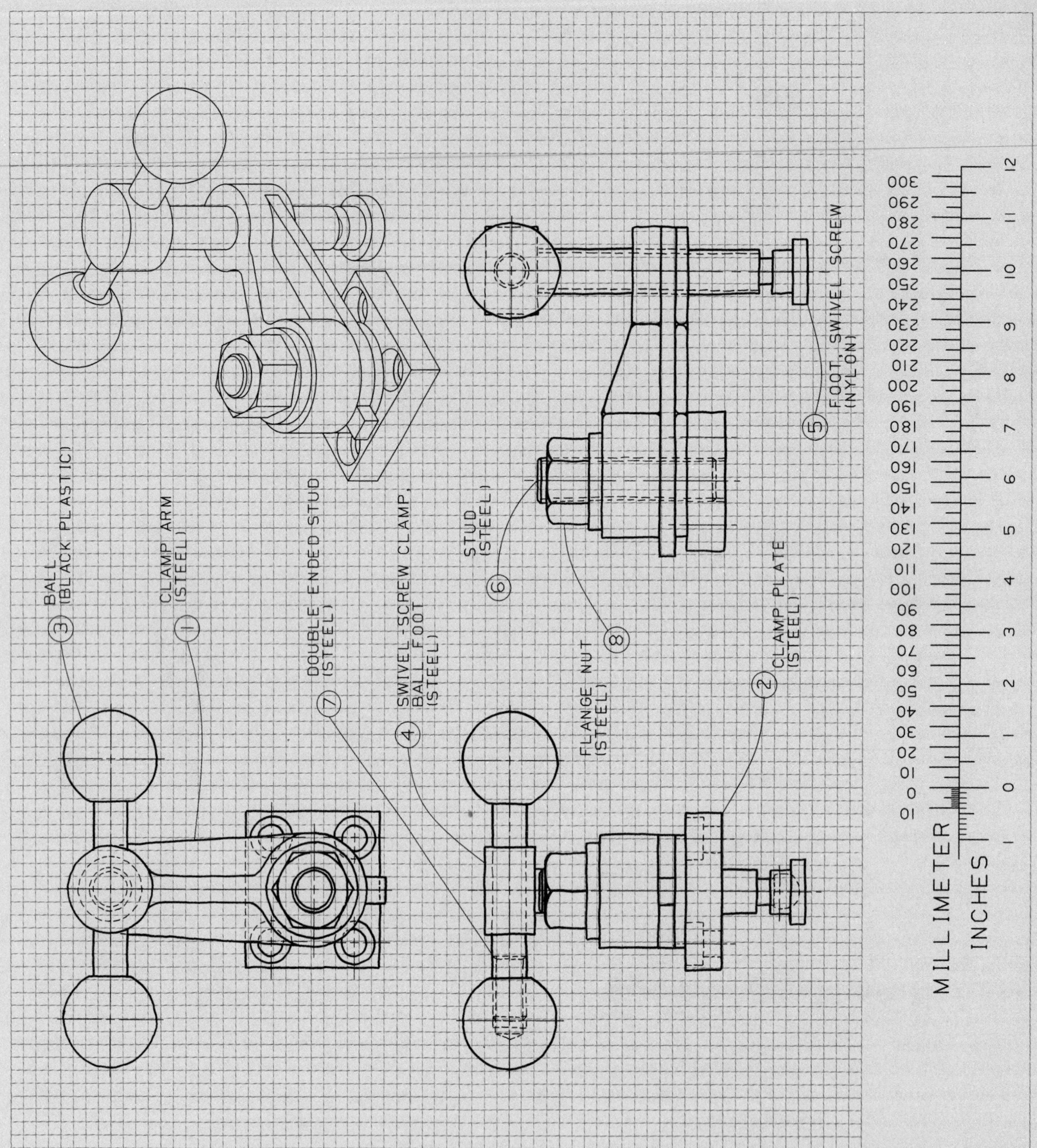

Problem 16.28 Draw and detail the swing clamp flange base assembly.

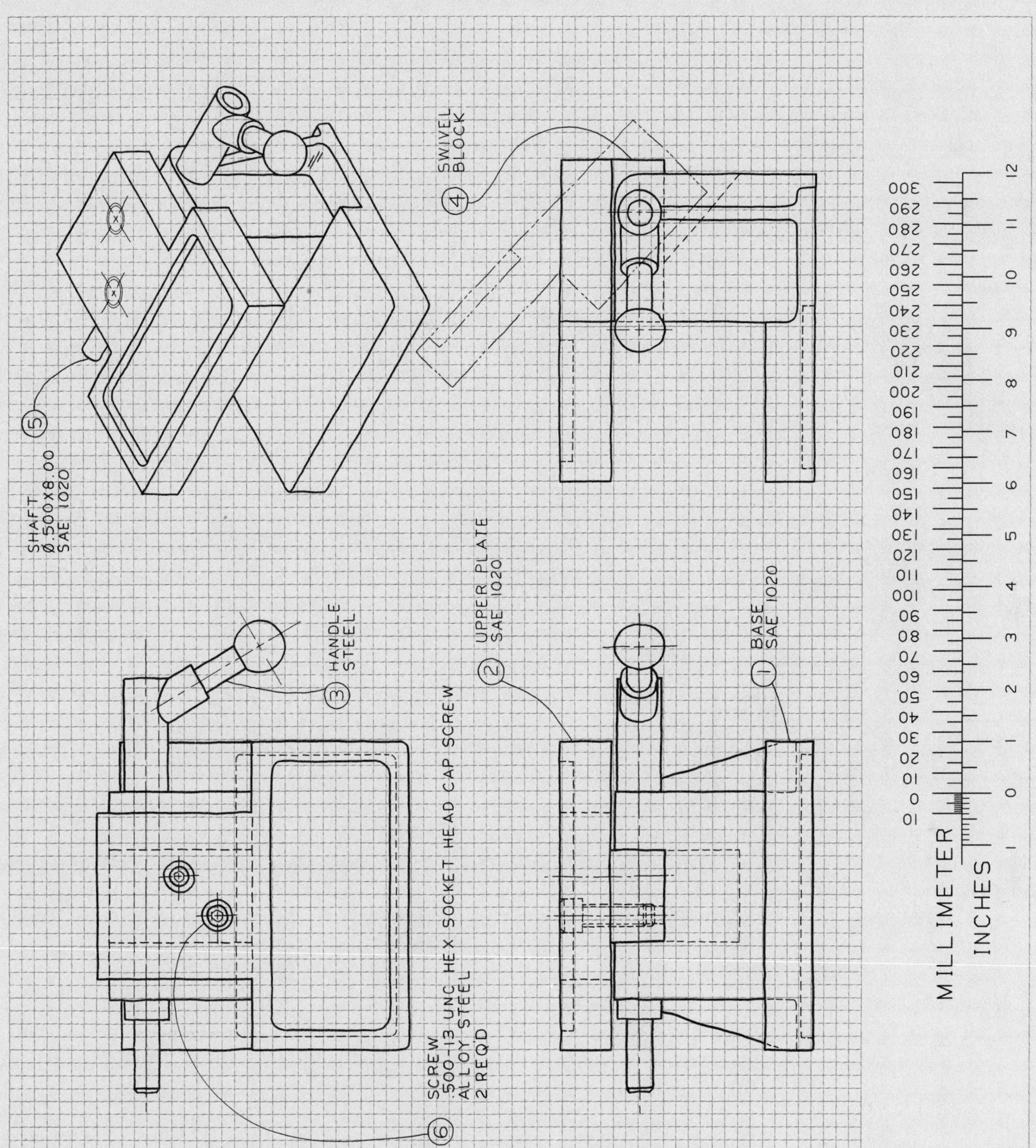

Problem 16.29 Draw and detail the swivel block fixture assembly.

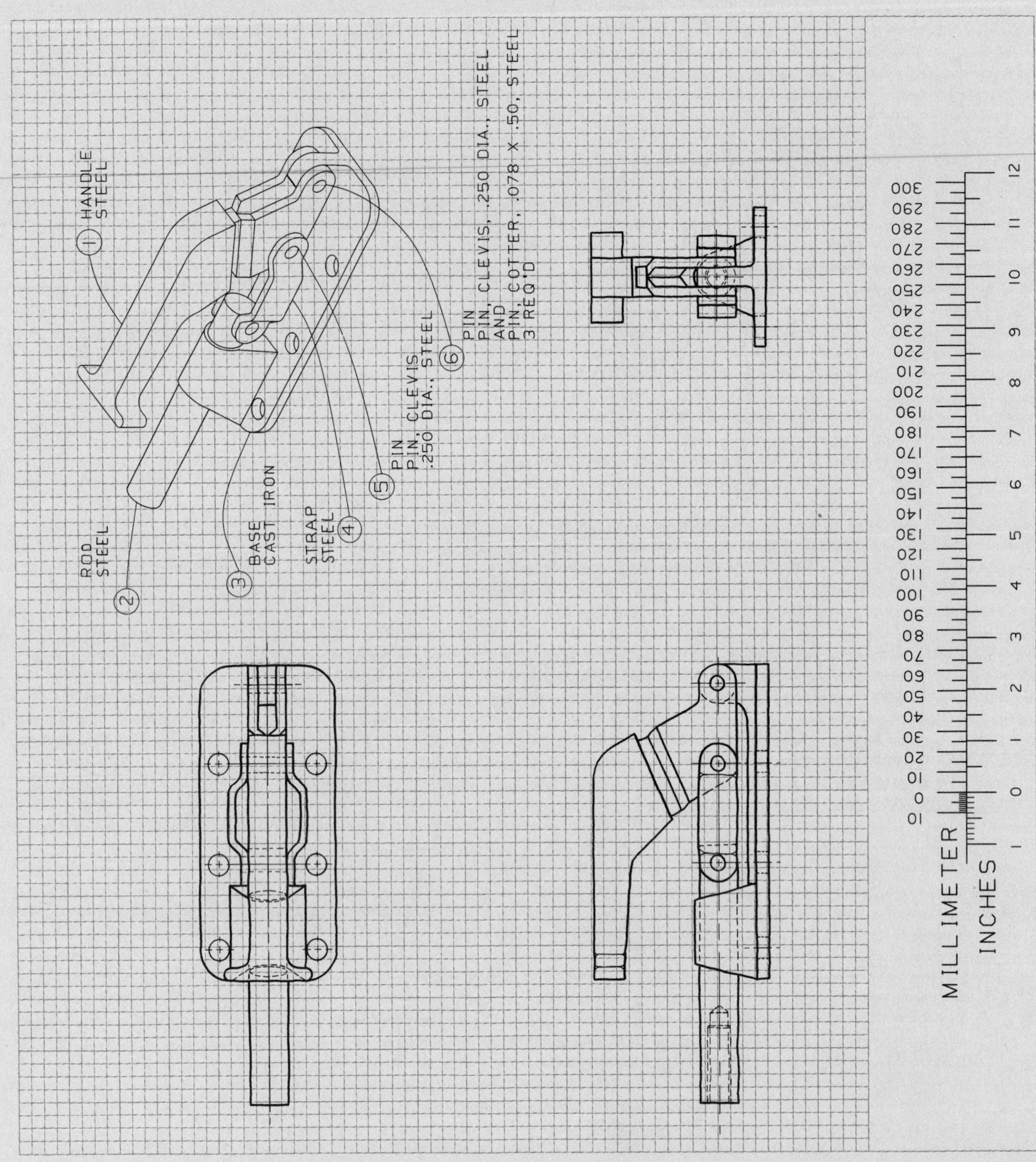

Problem 16.30 Draw and detail the connecting rod assembly.

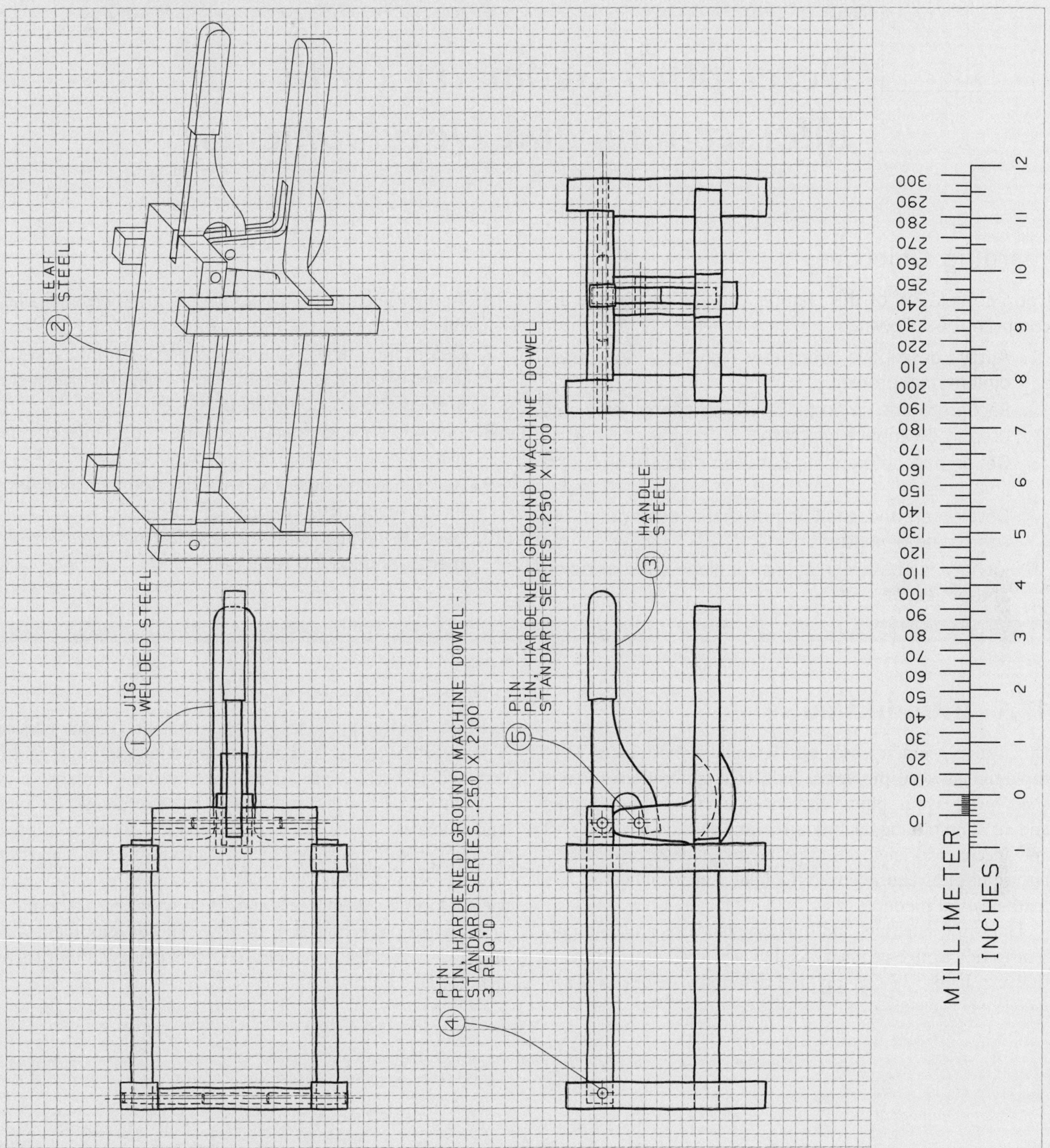

Problem 16.31 Draw and detail the leaf jig assembly.

Problems from Figures in the Chapter

Problem 16.32 Redraw the welded assembly shown in Figure 16.6.

Problem 16.33 Redraw the acoustic cup in Figure 16.8.

Problem 16.34 Model the SmartLevel rail [Fig. 16.25(a)], the wood insert [Fig. 16.25(c)], and the end cap [Fig. 16.25(d)]. Put all parts into an assembly, balloon, and complete a parts list.

Problem 16.35 Draw and detail the housing module shown in Figures 16.25(i) and (j).

Problem 16.36 Draw and detail the front panel shown in Figure 16.25(k).

Problem 16.37 Model and detail the keytop in Figure 16.25(l).

CHAPTER 17

Descriptive Geometry (with Intersections and Developments)

Learning Objectives

Upon completion of this chapter you will be able to accomplish the following:

1. Apply descriptive geometry solutions to three-dimensional problems.
2. Recognize the importance of notational elements used in descriptive geometry.
3. Define and differentiate between principal lines and line types.
4. Develop an understanding of spatial description and coordinate dimensions.
5. Identify the basic conditions for plane representation and projection.
6. Apply the concepts of parallelism and perpendicularity.
7. Determine the line of intersection or common line of joined shapes so that they may be graphically described and economically produced.
8. Develop an understanding of the importance of auxiliary views in solving for intersections.
9. Recognize the significance of development drawings in the manufacturing process.
10. Become familiar with models, flat pattern developments, and joining techniques.

17.1 Introduction

Descriptive geometry uses orthographic projection to solve three-dimensional problems. Industrial applications include sheet metal layout, piping clearances, intersections of heating and air conditioning ducting, structural steel design, and civil and mechanical engineering problems. A descriptive geometry solution is equivalent to the final numerical answer in a mathematical model.

Linework, lettering, and drawing standards are as important to descriptive geometry as they are to all other forms of drafting. Figure 17.1 is a typical descriptive geometry drawing that uses the special language and notation of this subject. The format, symbols, and notation used in descriptive geometry should become part of your technical vocabulary.

Figure 17.2 is a line and symbol key that defines the type and thickness of the lines and symbols used in descriptive geometry. Two new types of lines are the fold line and the development element. The **fold line** is used to divide each view and to establish a reference from which to take dimensions when projecting view to view. The **development element** is used when developing curved surfaces and for triangulation of surfaces.

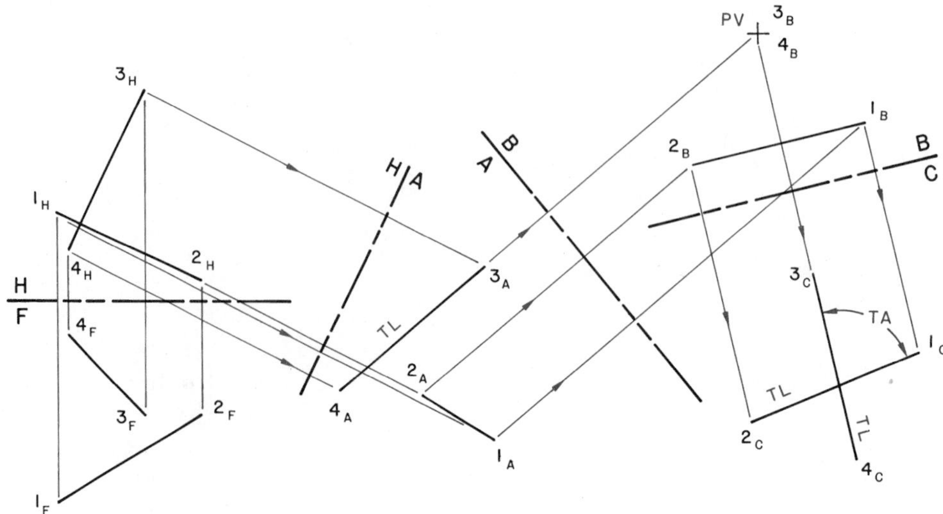

FIGURE 17.1 Descriptive Geometry Problem Setup and Notation

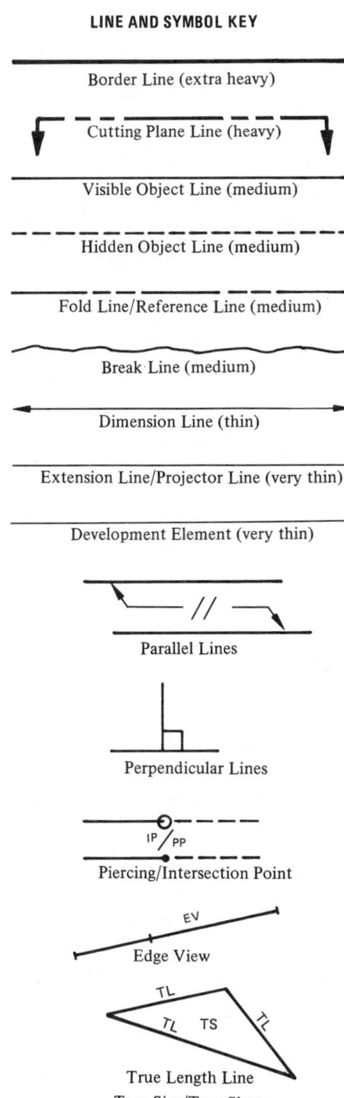

FIGURE 17.2 Descriptive Geometry Line and Symbol Key

17.2 Notation

The notation key gives the abbreviations and notations used in descriptive geometry problems. **EV** is the edge view of a plane. **IP** is the intersection of a line and a surface. **PP** is the piercing point of a line (a part of a plane) and another surface. Theoretically, IP and PP are the same. **PV** is the point view of a line. True shape and true size are essentially the same term and are abbreviated **TS**. **TL** is the true length of a line. **D** is used to note a dimension.

H, **F**, and **P** identify the three primary views: horizontal (top), frontal (front), and profile (side). "**A**" will always be the first auxiliary view, with "**B**", "**C**", "**D**", etc., auxiliary views following.

Whole numbers, 1, 2, 3, 4, 5, 6, etc., establish points in space. They are individual points or determine the extent of lines, planes, or solids. Capital letters may be used as points to add clarity.

Subscripts establish the view that a point is in, such as 2_H. This means that point 2 is in the H horizontal view. Superscripts are used where an aspect of a point appears more than one place in a view. For example, the line of a prism is called $3\text{-}3^1$. Also, the piercing point (for clarity) of a line is noted as an aspect of the original point, 2^1.

Notation Key

EV = Edge view
IP = Intersection point
PP = Piercing point
PV = Point view
TL = True length
TS = True shape
TS = True size
D = Dimension
H = Horizontal view
F = Frontal view
P = Profile view
A, B, C, etc. = Auxiliary views
1, 2, 3, 4, 5, 6, etc. = Points
H, F, P, A, B, C, etc. = View identifications
1_H, 2_F, 3_P, 4_A, etc. = View subscript
1^1, 2^1, 3^2, 4^2, etc. = Superscript
1R, 2R, 3R, 4R, etc. = Revolved points

17.3 Points

Geometric shapes must be reduced to points and the connectors of points, lines. **Points** are the most important geometric element because they are the building blocks for any graphical projection. All projections of lines, planes, and solids can be located and manipulated by identifying a series of points. A point is located in space by establishing it in two or more adjacent views. Two points that are connected are called a **line.**

17.3.1 Views of Points

A point is located by measurements from an established reference line (Fig. 17.3). This figure shows the projection of point 1 in the three principal planes, frontal (1_F), horizontal (1_H), and profile (1_P). *The intersection line of two successive (perpendicular) image planes is called a fold line or reference line.* All measurements are taken from fold lines to locate a point (line, plane, or solid) in space. A fold line/reference line can be visualized as the edge view of a reference plane.

In Figure 17.3, point 1 is below the horizontal plane (D1), to the left of the profile plane (D2), and behind the frontal plane (D3). D1 establishes the elevation or height of the point in the front and side view, D2 the width in the front and top view, and D3 the distance behind (depth) the frontal plane in the top and side view.

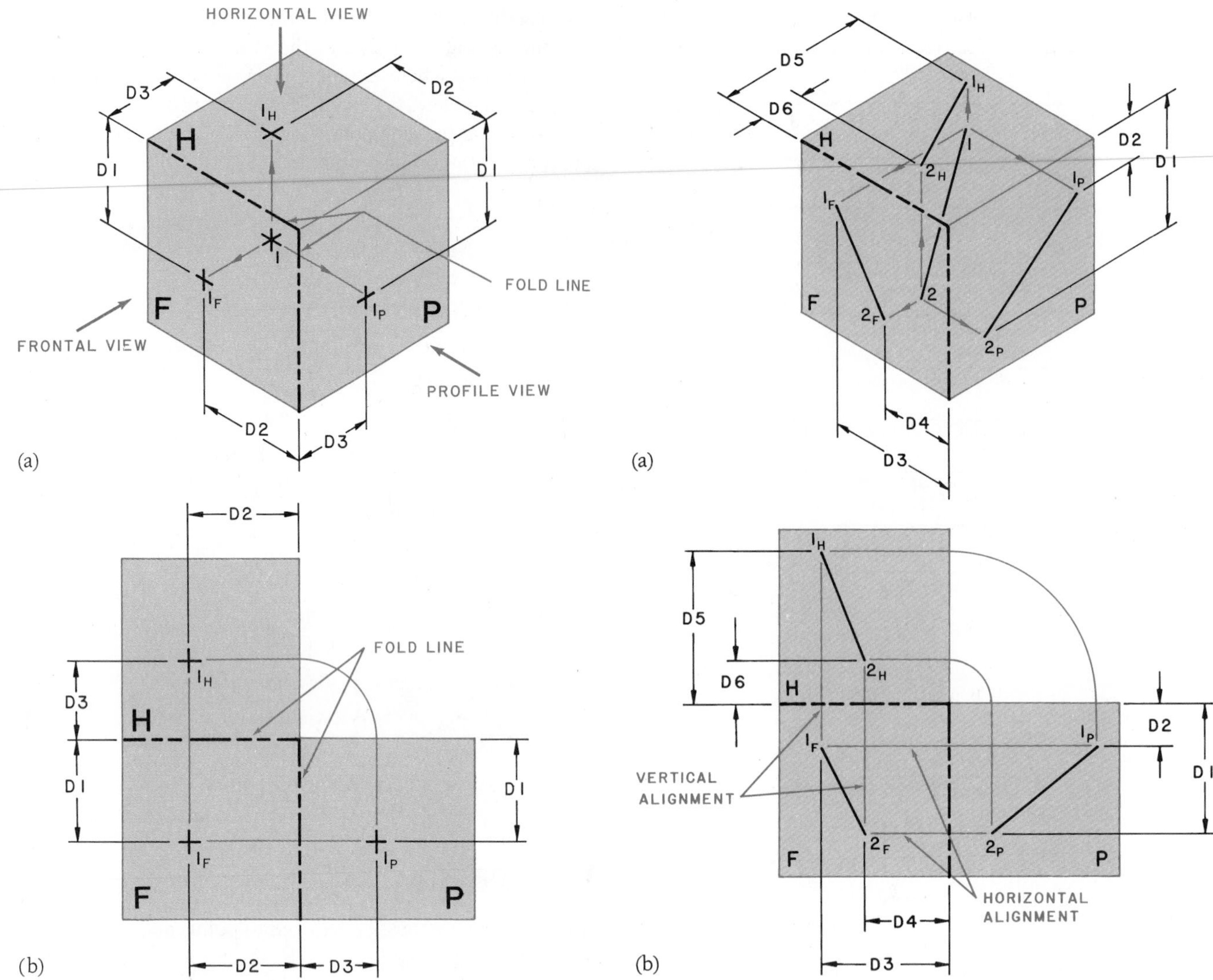

FIGURE 17.3 **Three Views of a Point in Space** Shown here as an isometric pictorial (a) and in Orthographic Projection (b).

FIGURE 17.4 **Three Views of a Line in Space**

17.4 Lines

Lines are a series of points in space that have magnitude (length), but not width. All lines can be extended to solve a problem. Therefore, a theoretical definition of a line is that *they are straight elements that have no width but are infinite in length (magnitude) and can be located by two points that are not at the same location.* If two lines lie in the same plane, they are either parallel or they intersect.

Throughout the text, numbers are used to designate end points of a line. The view of a line and its locating points are labeled with a subscript corresponding to the plane of projection, as in Figure 17.4, where the end points of line 1-2 are notated 1_H and 2_H in the horizontal view, 1_F and 2_F in the frontal view, and 1_P and 2_P in the profile view. For many figures in this chapter, subscripts are eliminated where the view is obvious or only one point is labeled per view.

17.4.1 Multiview Projection of a Line

Lines are classified according to their orientation to the three principal planes of projection or how they appear in a projection plane. They can also be described by their relationship to other lines in the same view. As with points, lines are located from fold lines/reference lines.

In Figure 17.4, line 1-2 is projected onto each principal projection plane and located by dimensions from fold lines. The end points of line 1-2 are located from two fold lines in each view, using dimensions or projection lines that originate in an adjacent view. Dimensions D1 and D2 establish the elevation of the end points in the profile and frontal view, since these points are horizontally in line in these two views. D3 and D4 locate the end points in relation to the F/P fold line (to the left of the profile plane), in both the frontal and horizontal views, since these points are aligned vertically. D5 and D6 locate each end point in relation to the H/F and the F/P fold line since these dimensions are the distance behind the frontal plane and show in both the horizontal and profile views.

17.4.2 Principal Lines

A line that is parallel to a principal plane is called a **principal line** and is true length in the principal plane to which it is parallel. Since there are three principal planes of projection, there are three principal lines: horizontal, frontal, and profile (Fig. 17.5):

1. A **horizontal line** is parallel to the horizontal plane and true length in the horizontal view.
2. A **frontal line** is parallel to the frontal plane and true length in the frontal view.

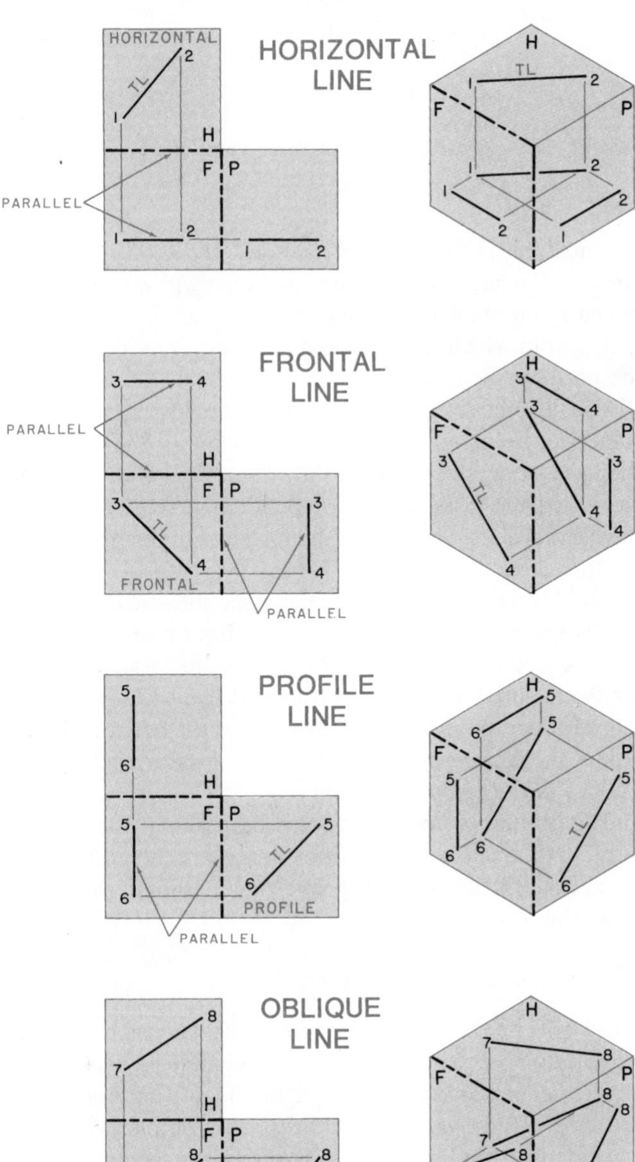

FIGURE 17.5 Types of Lines

3. A **profile line** is parallel to the profile plane and true length in the profile view.
4. An **oblique line** is at an angle to the frontal, horizontal, and profile planes. It does not show true length in the horizontal, frontal, or profile views.

17.4.3 Line Types and Descriptions

The following terms describe lines:

Vertical line Vertical lines are perpendicular to the horizontal plane and appear true length in the frontal and profile views (frontal and profile principal lines). Vertical lines appear as a point (point view) in the horizontal view and show true length in all elevation views.

Level line Any line that is parallel to the horizontal plane is a level line. Level lines are horizontal lines.

Inclined lines Inclined lines are parallel to the frontal or profile planes (profile or frontal principal line) and at an angle to the horizontal plane. An inclined line is always at an angle to the horizontal.

Oblique lines Oblique lines are inclined to all three principal planes and will not be true length in any principal view (Fig. 17.5).

Foreshortened Lines that are not true length in a specific view appear shorter (foreshortened) than their true length.

Point view Where a view is projected perpendicular to a true length line, that line appears as a point view; the end points are coincident. A point view is a view of a line in which the line of sight is parallel to the line.

True length A view in which a line can be measured true distance between its end points shows the line as true length. A line appears true length in any view where it is parallel to the plane of projection.

17.4.4 True Length of a Line

A true length view of an oblique line can be projected from any existing view by establishing a line of sight perpendicular to a view of the line and drawing a fold line parallel to the line. Fold lines are always drawn perpendicular to the line of sight. The following steps describe the procedure for drawing a true length projection of an oblique line from the frontal view (Fig. 17.6):

1. Establish a line of sight perpendicular to oblique line 1-2 in the frontal view.
2. Draw fold line F/A perpendicular to the line of sight and parallel to the oblique line 1-2.
3. Extend projection lines from point 1 and 2 perpendicular to the fold line (parallel to the line of sight). The distance from line 1_F-2_F is random.
4. Transfer the end points of the line from the horizontal view to locate point 1_A and 2_A along the projection lines in auxiliary view A.
5. Connect points 1_A and 2_A. This is the true length of projection of line 1-2.

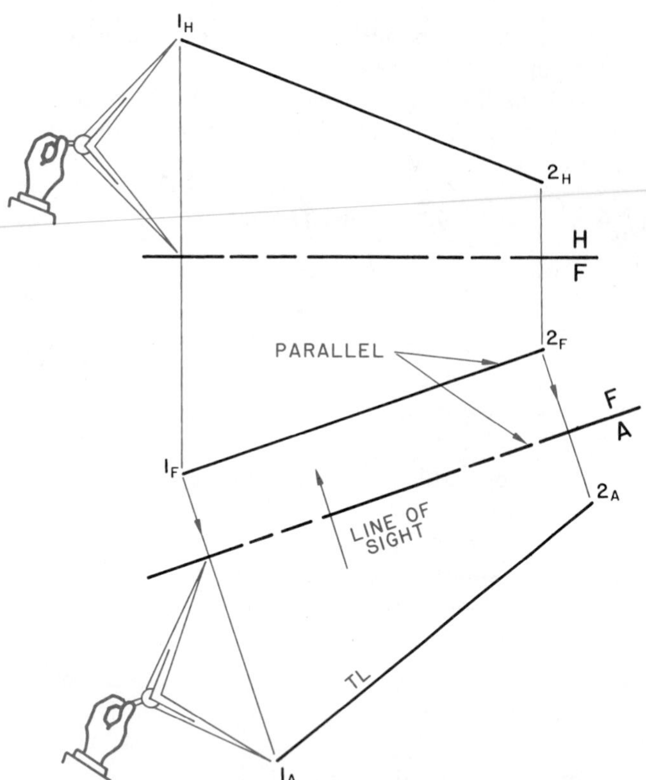

FIGURE 17.6 **Oblique Line Shown in Horizontal Frontal and Auxiliary Views** Note that auxiliary view A shows the lines and a true length projection.

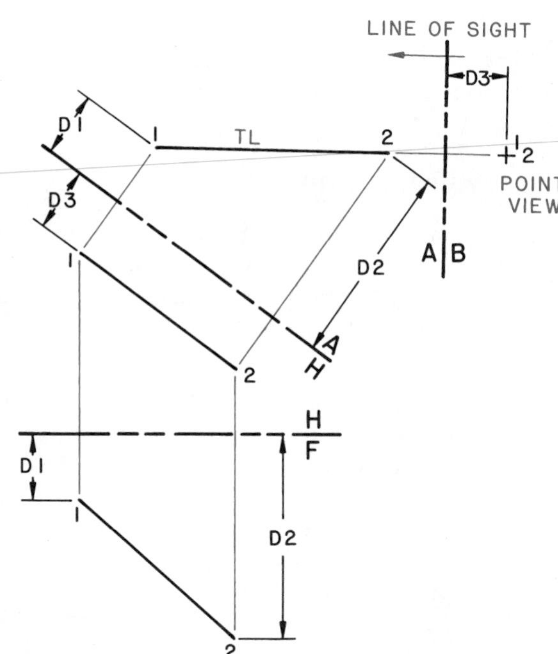

FIGURE 17.7 **True Length and Point View of a Line**

17.4.5 Point View of a Line

A line will project as a *point view* when the line of sight is parallel to a true length view of the line, that is, the point view is projected on a projection plane that is perpendicular to the true length.

The point view of an oblique line is drawn after the line is projected as true length in an auxiliary view. In Figure 17.7, line 1-2 is projected as true length in auxiliary view A. To establish the point view, a secondary auxiliary view (B) is projected perpendicular to the true length line. The following steps describe this process:

1. Project a TL view of line 1-2.
2. Establish a line of sight parallel to the true length line 1-2.
3. Draw the fold line perpendicular to the line of sight (A/B). The fold line is perpendicular to the true length line.
4. Transfer dimension D3 from the horizontal view to locate both points along the projection line in auxiliary view B.

17.4.6 Bearing of a Line

The angle that a line makes with a north-south line in the horizontal view is the **bearing** of that line. *The bearing can be measured only in the horizontal (top) view. The bearing is the map direction of a line and is measured in degrees with a protractor or compass from the north or the south.* The bearing indicates the quadrant in which the line lies and is always measured from the north or the south.

Usually, the originating point is the lowest numerical value. For example, line 1-2 starts at point 1. The *low end* is the lowest point on a line as shown in a frontal or elevation view. In some cases, as for a sloping cross-country pipeline, the bearing is measured from the high end of the line toward the low end.

The bearing is always an acute angle measured from the north or south. In Figure 17.8, line 1-2 has a bearing of N 73° W measured from the north, 73° toward the west. The bearing is measured from the north toward the west, from point 1 toward point 2. Figure 17.8 also shows the horizontal view of line 1-2 located in relation to the compass meridian. Line 1-2 lies in the second quadrant.

The bearing of line 3-4 is S 45° E, meaning that line 3-4 forms a 45° angle with the north-south meridian and is measured from the south toward east. Line 3-4 lies in the fourth quadrant.

17.4.7 Visibility of Lines

When two lines cross in space, one may be visible and the other hidden at the crossing point. A **visibility check** determines the proper relationship of the lines. The visibility of two lines can change in every view. For example, when two pipes or structural members cross in a construction project, one will be above or in front of the other.

In Figure 17.9(a) lines 1-2 and 3-4 cross. A visibility check determines which line lies in front of the other in the frontal

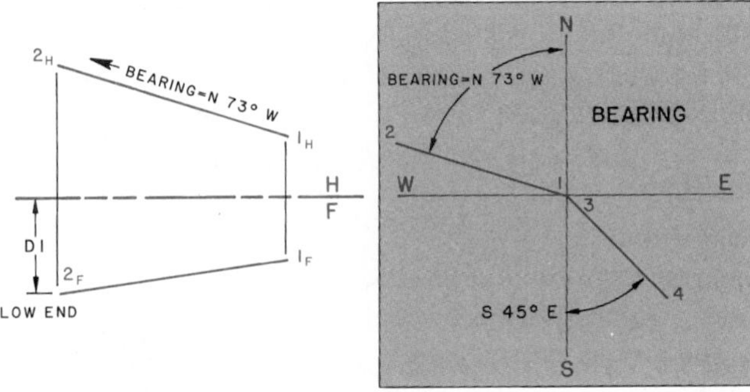

FIGURE 17.8 **Bearing of a Line**

view and which line is on top in the horizontal view. The following steps describe this process:

1. Where line 1_H-2_H crosses line 3_H-4_H in the horizontal view, extend a **sight line** perpendicular to H/F until it meets one of the lines in the frontal view [Fig. 17.9(b)]. Here, line 1_F-2_F is the first line to be encountered, therefore line 1_H-2_H is the visible line in the horizontal view.
2. Extend a sight line from the crossing point of line 1_F-2_F and 3_F-4_F in the frontal view until it meets the first line in its path in the horizontal view. Since line 3_H-4_H was encountered first, it is the visible line in the frontal view.
3. Complete the visibility of lines by showing the proper solid (visible) and dashed (hidden) lines in both views. The visible line has been shaded for clarity in the figure. However, this is not standard practice. Line 1-2 is visible in the horizontal view (is above line 3-4), and line 3-4 is visible in the frontal view (is in front of line 1-2).

17.4.8 Parallelism of Lines

Two lines in space will **intersect, be skew, parallel,** or **perpendicular.** Parallel lines project parallel in all views (Fig. 17.10). Here, lines 1-2 and 3-4 are parallel. Parallel lines may

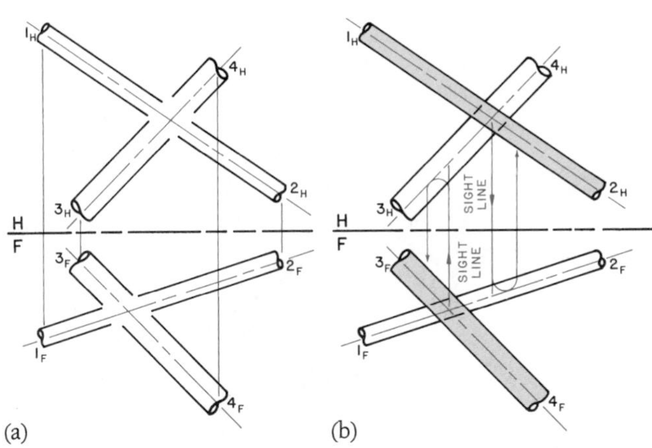

FIGURE 17.9 **Visibility of Lines**

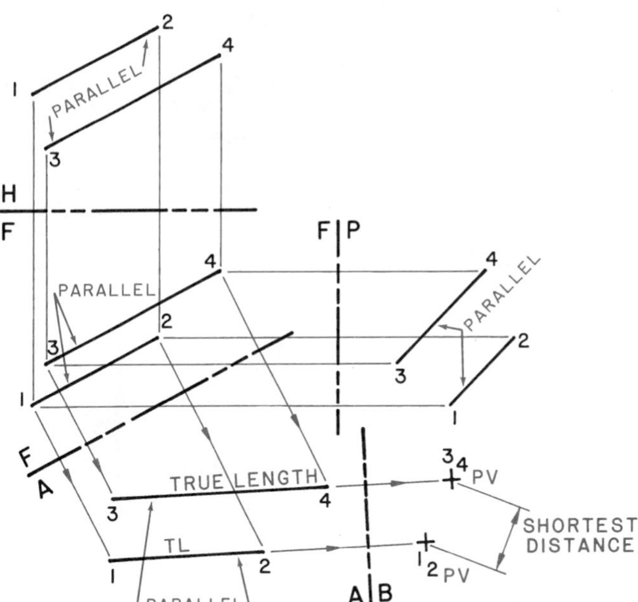

FIGURE 17.10 **Solving for the Shortest Distance Between Two Lines** In this figure the lines happen to be parallel.

appear as points (in the same view) or their projections may coincide.

Two oblique lines that project parallel or coincide in all views will always be parallel. Two lines that are parallel or perpendicular to a principal plane and appear parallel to each other may not be parallel lines. A third view is needed to establish the relationship.

The true distance between two parallel lines is shown in a view where the lines appear as points. In Figure 17.10 oblique lines 1-2 and 3-4 are parallel. Auxiliary view A is projected parallel to both oblique lines from the frontal view (fold line F/A is drawn parallel to 1-2 and 3-4). View A shows both lines as true length. Auxiliary view B is then projected perpendicular to the true length lines (fold line A/B is drawn perpendicular to true length lines 1-2 and 3-4). In auxiliary view B both lines appear as point views, therefore, true distance between the lines can be measured.

You May Complete Exercises 17.1 Through 17.4 at This Time

17.4.9 Perpendicularity of Lines

Lines that are perpendicular will show perpendicularity in any view in which one or both of the lines is true length. Because two lines may be oblique in their given views, it is necessary to project a view that shows one or both of the lines as true length. If two lines appear perpendicular in a given view and neither one is true length, then the lines are not perpendicular. Perpendicular lines can be intersecting or nonintersecting lines.

Frontal perpendicular lines appear parallel in the horizontal and profile views and perpendicular in the frontal view. Both lines show true length in the frontal view (Fig. 17.11). In a view where one line is a point view and the other line is true length, the lines are perpendicular.

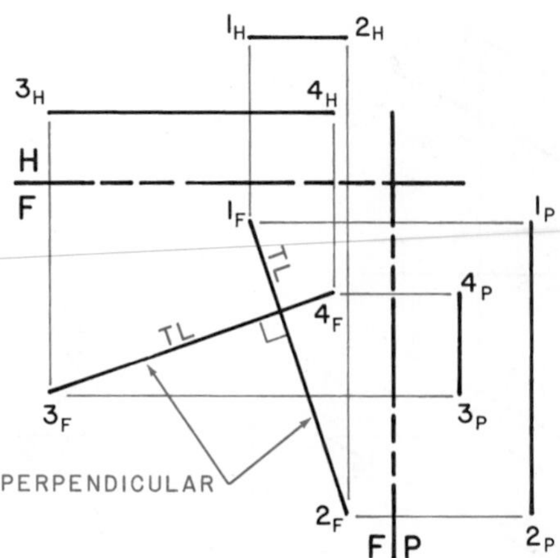

FIGURE 17.11 Nonintersecting Perpendicular Lines

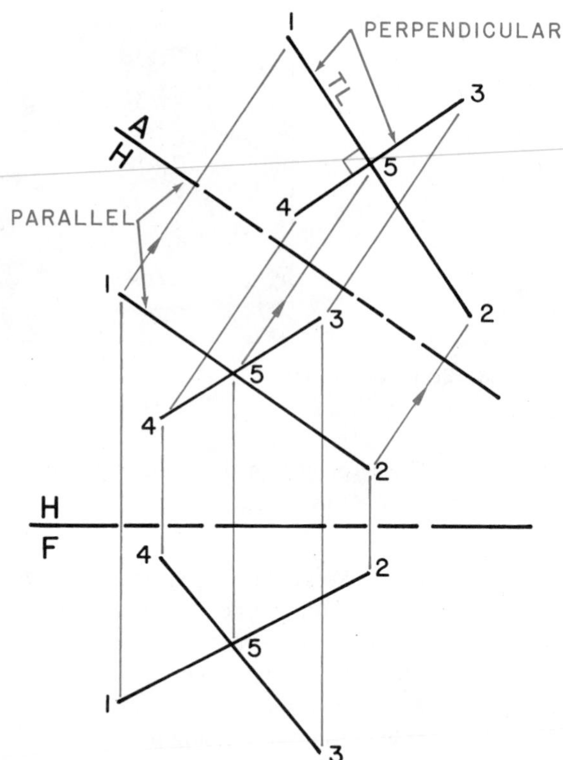

FIGURE 17.12 Intersecting Perpendicular Lines

17.4.10 Intersecting Perpendicular Lines

Intersecting lines have a common point that lies on a single projection line, parallel to all other projection lines between adjacent views. Two lines that intersect at a common point, and form 90° with each other where one or both lines appear as true length, are intersecting perpendicular lines.

When two intersecting lines are oblique in the frontal and horizontal views, project a view where one or both of the lines is true length to check for perpendicularity. Lines 1-2 and 3-4 in Figure 17.12 are intersecting lines since they have a common point that is aligned in adjacent views. Fold line H/A is drawn parallel to oblique line 1-2 (auxiliary view A is parallel to line 1-2). Both lines are then projected into auxiliary view A. Line 1-2 is true length and is perpendicular to line 3-4. Lines 1-2 and 3-4 are perpendicular. Point 5 is the shared point of both lines.

17.4.11 Nonintersecting Perpendicular Lines

Two nonintersecting lines are perpendicular lines if they form right angles in a view where one or both are shown true length. For oblique lines, project an auxiliary view where at least one of the lines is true length and measure the angle between the lines in the new view.

In Figure 17.13, the principal views of the two lines establish that they are nonparallel, nonintersecting, and oblique. Auxiliary view A is projected parallel to line 3-4 by drawing fold line F/A parallel to line 3_F-4_F. Projection lines are then drawn perpendicular to the fold lines from all points in the frontal view. Measurements to locate each point are transferred from the horizontal view to establish the points along the projection lines in auxiliary view A. Line 3_A-4_A is true length and line 1_A-2_A is oblique. The lines are perpendicular since they are at right angles in auxiliary view A.

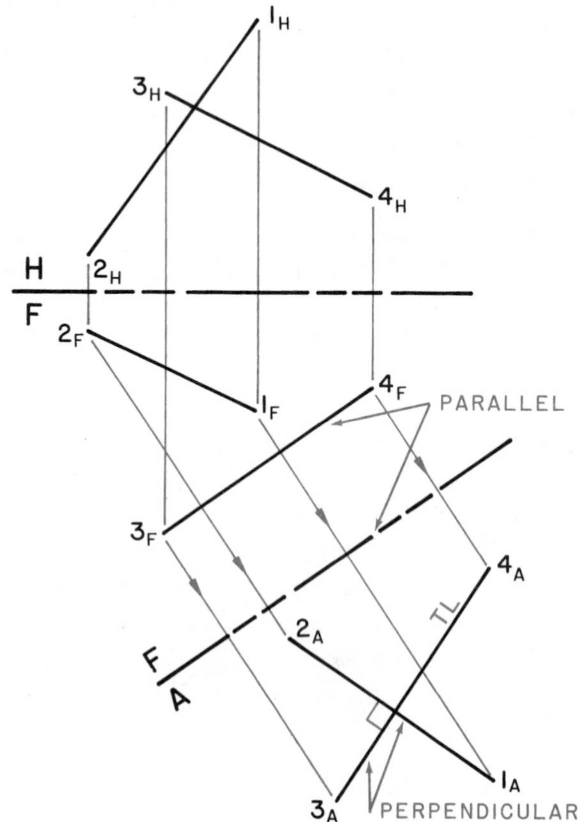

FIGURE 17.13 Perpendicular Lines in Space

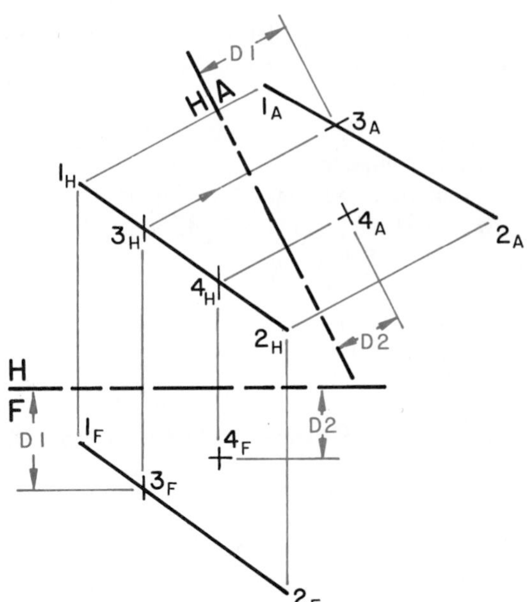

FIGURE 17.14 Points On and Off a Line

17.4.12 Points on Lines

Successive views (principal or auxiliary) of a point on a line may be projected to all adjacent views by extending a projection line from the point, perpendicular to the fold line, until it crosses the line in the next view. In Figure 17.14, point 3 and point 4 appear to be on line 1-2 in the horizontal view, but their frontal and auxiliary views show that only point 3 is on the line; point 4 lies directly above the line (horizontal view). If a point is centered on a line, then it must be centered on the line (true length or oblique) in all views.

17.4.13 Shortest Distance Between a Point and a Line

The **shortest distance** between a point and a line is measured along a perpendicular connector in a view where the line appears as a point view. In Figure 17.15 oblique line 1-2 and point 3 are given. The shortest connector between the line and point is required. This connector must be shown in all views. The following steps describe the procedure:

1. Draw auxiliary view A parallel to oblique line 1-2. Start by drawing fold line FA parallel to the line.
2. Project line 1-2 and point 3 into auxiliary view A. Line 1-2 shows true length.

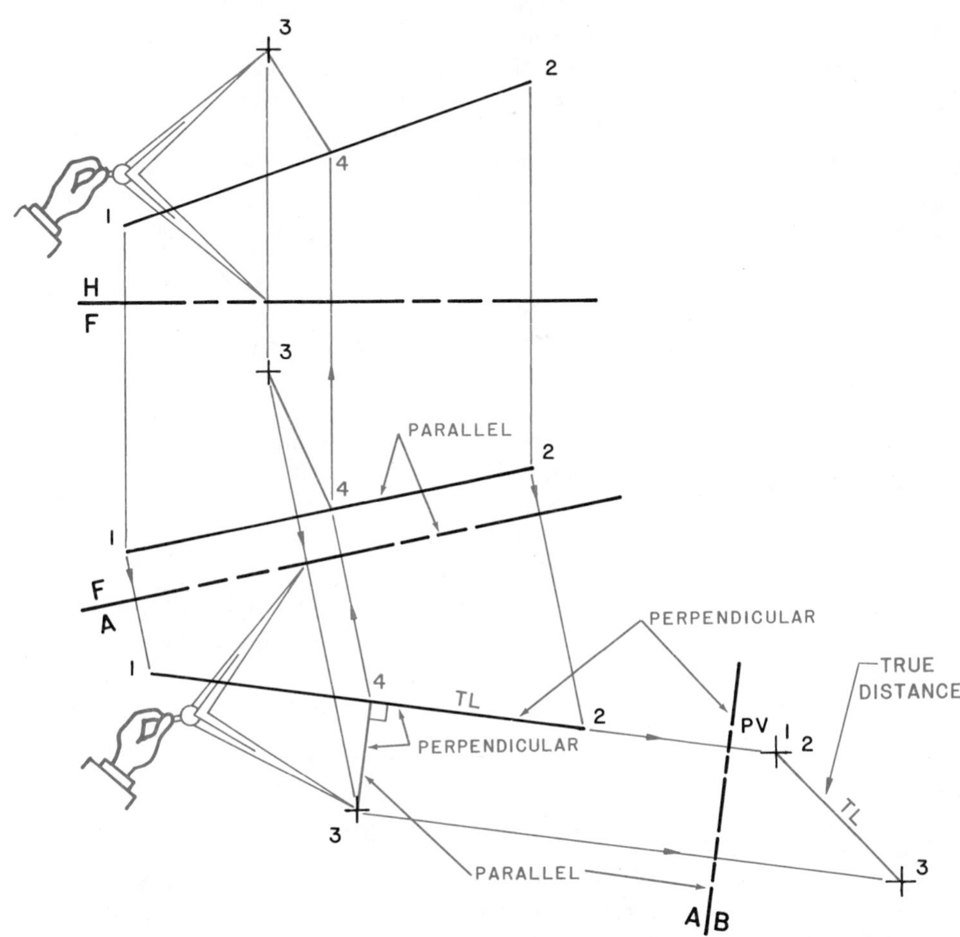

FIGURE 17.15 Shortest Connector (True Distance) Between a Point and a Line

3. Draw a perpendicular connector between point 3 and true length line 1-2. Label this new point 4.
4. Project auxiliary view B parallel to line 3-4 (and perpendicular to true length 1-2). Fold line A/B is parallel to line 3-4 and perpendicular to line 1-2.
5. Auxiliary view B shows line 1-2 as a point view and line 3-4 (the shortest connector) as true length. The true distance between the point and the line is measured here.
6. Project line 3-4 back into each view.

17.4.14 Shortest Distance Between Two Skew Lines

Two nonparallel, nonintersecting lines are called **skew lines.** The shortest distance between two skew lines is a line perpendicular to both lines. Only one solution is possible. This perpendicular is shown as true length in a view where one line appears as a point view and the other oblique or true length. Lines 1-2 and 3-4 are given in the horizontal and frontal view. Use the following steps to solve for the shortest distance (Fig. 17.16):

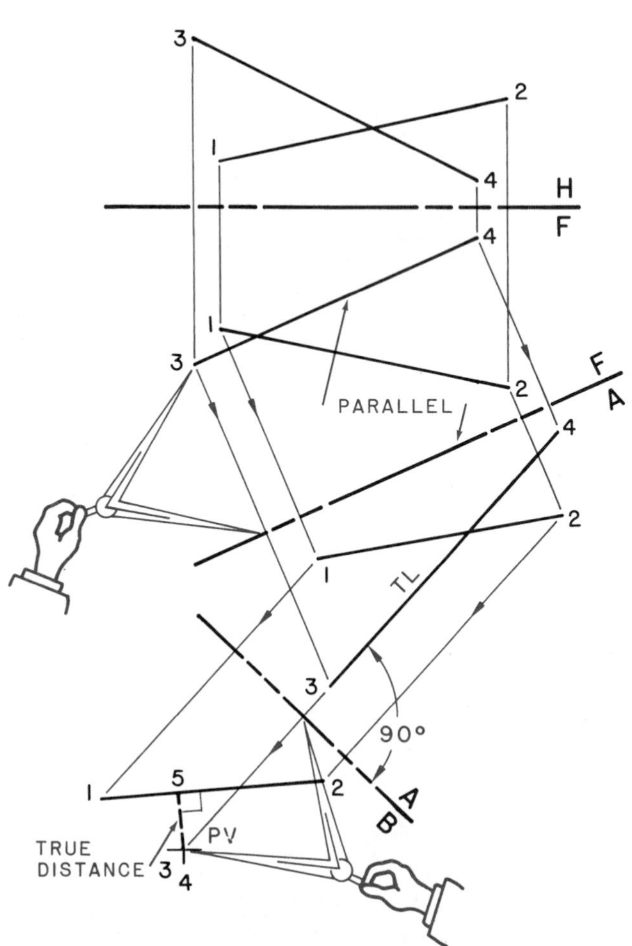

FIGURE 17.16 Shortest Connector Between Two Lines

1. Draw fold line F/A parallel to line 3-4 and project auxiliary view A. Line 3-4 is true length and line 1_A-2_A is oblique.
2. Draw fold line A/B perpendicular to true length line 3_A-4_A and complete auxiliary view B. Line 3-4 projects as a point view and line 1-2 as oblique.
3. Draw a line from point view 3_B-4_B perpendicular to line 1_B-2_B. Point 5B is on line 1_B-2_B. This is the shortest distance between the two skew lines. The distance between PV 3-4 and point 5 is the true distance between lines 1-2 and 3-4.

17.4.15 Angle Between Two Skew Lines

The angle formed by two skew lines is measured in a view where both lines appear as true length. In Figure 17.17, skew lines 1-2 and 3-4 are given in the F and H views; the angle formed by the two lines is required. The following steps are used:

1. Fold line F/A is drawn parallel to line 3_F-4_F.
2. Project primary auxiliary view A. Line 1_A-2_A is oblique and line 3_A-4_A shows as true length.
3. Draw fold line A/B perpendicular to true length line 3_A-4_A.
4. Complete secondary auxiliary view B. Line 1_B-2_B is oblique and line 3_B-4_B appears as a point view.
5. Draw fold line B/C parallel to oblique line 1_B-2_B.
6. Project auxiliary view C parallel to oblique line 1_B-2_B. Line 1_C-2_C shows true length in auxiliary C. Line 1-2 and line 3-4 both show as true length lines in this view.
7. The true angle (acute) formed by the two lines is measured in auxiliary view C since both lines show true length.

17.5 Representation of Planes

A plane can be represented by:

1. Three points not in a straight line
2. A point and a line
3. Two parallel lines
4. Two intersecting lines
5. Three connected lines

In Figure 17.18, a plane is defined in three different ways. In the first method (a) three individual unconnected points define the plane. In the second method (b), two of the points are connected and the plane is defined by a point and a line. In method (c) the same three points are connected to form plane 1-2-3.

17.5.1 Principal Planes

When a plane is parallel to a principal projection plane, it is a **principal plane.** A principal plane can be a horizontal plane, a frontal plane, or a profile plane (Fig.17.19). All lines

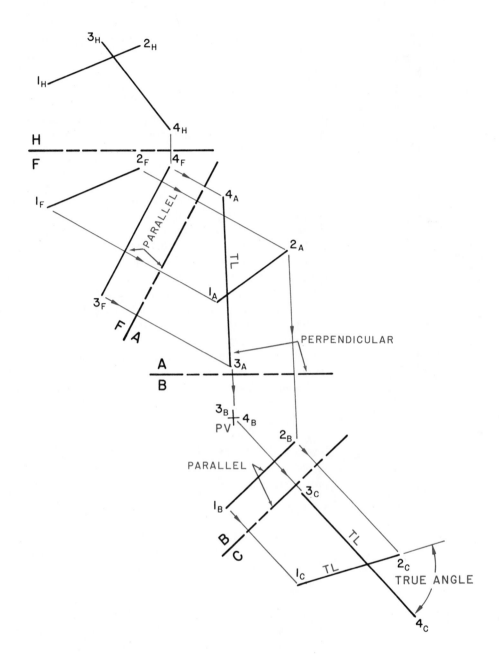

FIGURE 17.17 **Angle Between Two Skew Lines**

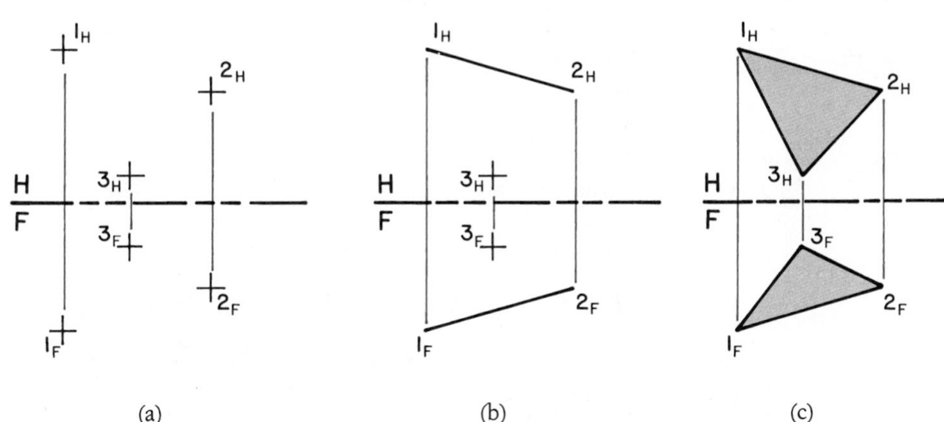

FIGURE 17.18 **Representation of a Plane Using** (a) Three points (b) A line and a point (c) Three lines

(a) HORIZONTAL PLANE

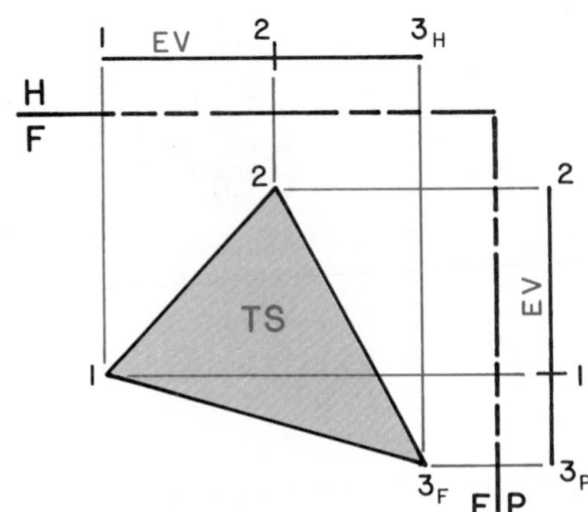

(b) FRONTAL PLANE

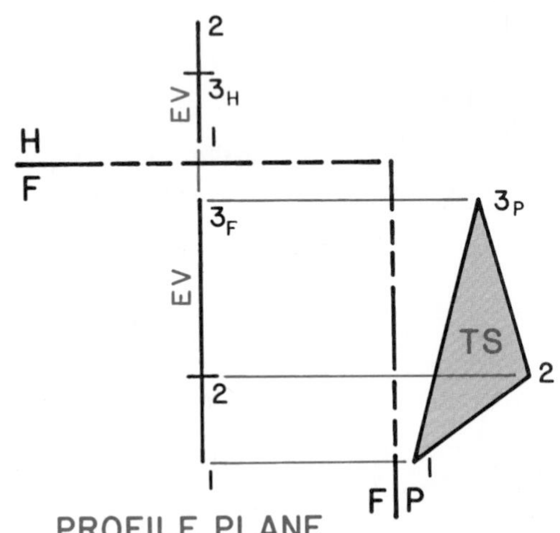

(c) PROFILE PLANE

FIGURE 17.19 Principal Planes (a) Horizontal plane (b) Frontal plane (c) Profile plane

in a horizontal plane, frontal plane, or profile plane are true length lines. Therefore, principal planes are composed of *principal lines.*

To determine if a plane is a principal plane, there must be at least two views, unless the given view shows the plane as parallel to a principal projection plane. In either case, two views are required to fix the position of any plane is space.

A **horizontal plane** [Fig. 17.19(a)] is parallel to the horizontal projection plane. It is true size (true shape) in the horizontal view since all of its lines are principal lines. The frontal and profile view of a horizontal plane always shows the plane as an edge view (EV). It shows as an edge in all elevation projections. Horizontal planes are perpendicular to the frontal and profile projection planes.

A **frontal plane** [Fig. 17.19(b)] lies parallel to the frontal projection plane where it shows as true size. In the horizontal and profile views, the plane appears as an edge view. All lines show true length in the frontal view because they are principal lines (frontal lines). A frontal plane is perpendicular to the horizontal and profile projection planes. Frontal planes are vertical planes because they are always perpendicular to the horizontal projection plane.

A **profile plane** is true size in the profile view and appears as an edge in the frontal and horizontal views [Fig. 17.19(c)]. Every line in the plane is true length in the profile view since they are profile lines. Profile planes are perpendicular to the frontal and horizontal projection planes. Profile planes are vertical planes since they are perpendicular to the horizontal projection plane.

17.5.2 Vertical Planes

Vertical planes are perpendicular to the horizontal proiection plane. The horizontal view of all vertical planes shows the plane as an edge. There are three positions for a vertical plane (Fig. 17.20).

In Figure 17.20(a), the vertical plane appears as an edge in the frontal and horizontal views. Plane 1-2-3 is perpendicular to the frontal and horizontal projection planes. This type of vertical plane is also a profile plane since it shows true shape in the profile view. The frontal and horizontal projections show the edge view of the plane parallel to the profile projection plane.

In Figure 17.20(b), plane 1-2-3 is not parallel to a principal projection plane. It does not show true size in any of the three principal views. The horizontal view of the plane establishes it as a vertical plane since it appears as an edge. The frontal and profile projections are foreshortened.

The example in Figure 17.20(c) is a frontal plane since it is true size in the frontal view. The horizontal and profile views show the plane as an edge and parallel to their adjacent projection planes.

17.5.3 Oblique and Inclined Planes

Planes are classified by their relationship to the three principal projection planes: frontal, horizontal, or profile. An **oblique plane** is inclined to all three principal projection

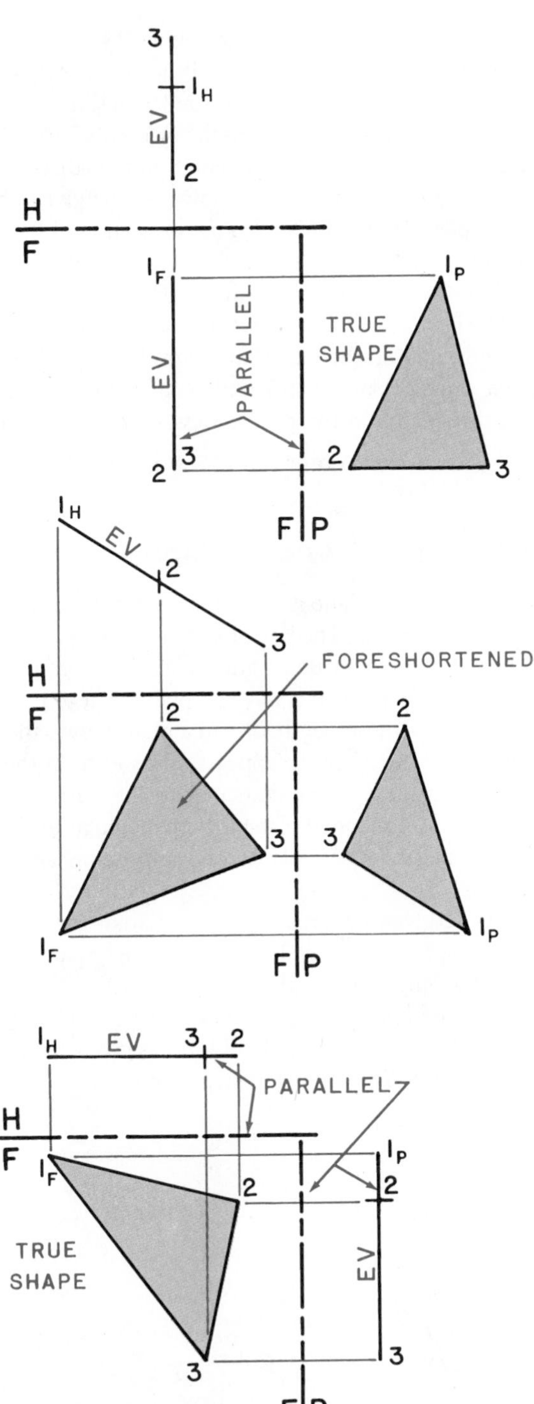

FIGURE 17.20 Vertical Planes

planes; each view is foreshortened. *Oblique planes are not true size in any of the three principal views.*

Oblique planes are not vertical or horizontal and will not be parallel to a principal projection plane. Figure 17.21 shows an example of an oblique plane in isometric and orthographic projection.

17.5.4 True Length Lines on Planes

A true length line is established by drawing a line on the given plane parallel to the fold line. The adjacent projection shows the line as true length and on the given plane. In Figure 17.22, lines have been located in each example so that they are parallel to the fold line in one view and project true length in the adjacent view.

The examples are of oblique planes in the principal projection planes: frontal, horizontal, and profile. These newly introduced lines must be prinicpal lines. The true length of a line can be found in any view using its adjacent projection to

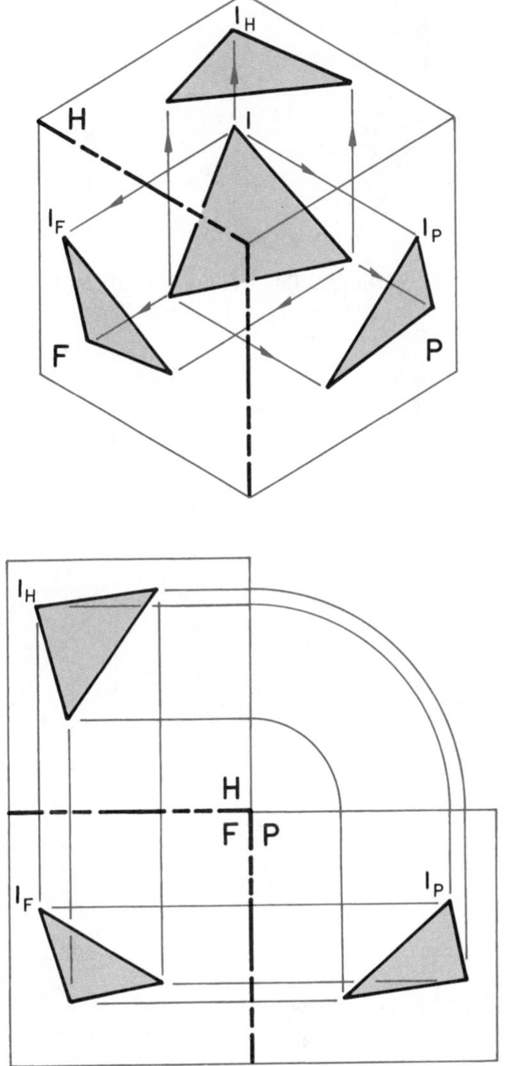

FIGURE 17.21 Three Views of a Plane in Space

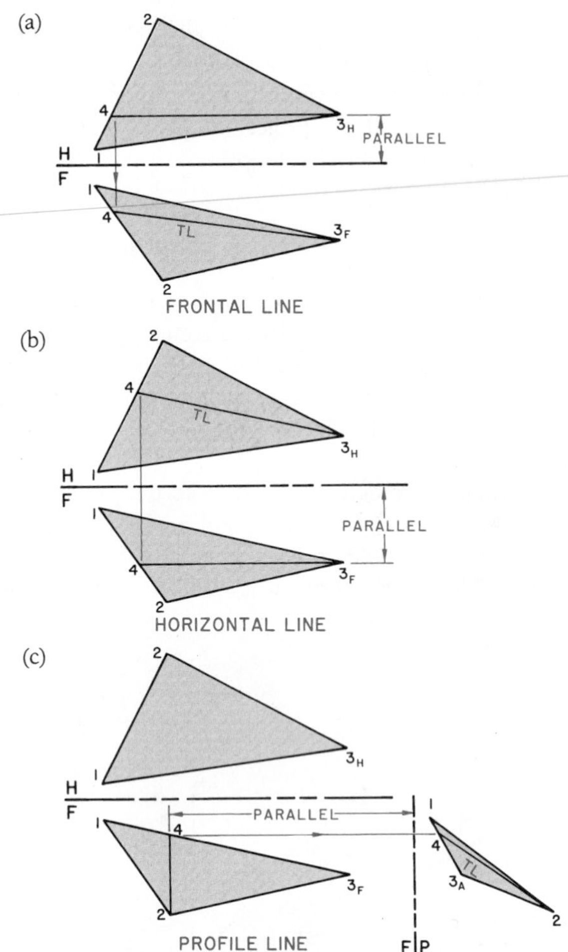

FIGURE 17.22 **Principal Lines (True Length) on Planes** (a) Frontal line (b) Horizontal line (c) Profile line

construct the line parallel to the fold line. When these views are not principal views, the lines will not be principal lines. The only requirement to finding a true length line on a plane is that the line be drawn parallel to the projection plane in one view and, therefore, project true length in the adjacent view.

In Figure 17.22(a) line 3-4 is drawn on the given oblique plane and parallel to H/F. The frontal projection of the line is on the plane and true length (a frontal line). In Figure 17.22(b), line 3-4 is drawn parallel to H/F and on the given plane. The horizontal view shows the line as a horizontal line (true length) and on the plane. In Figure 17.22(c), line 2_F-4_F is drawn on the plane 1-2-3 and parallel to F/P. Line 2-4 appears true length in the profile view; it is a profile line.

You May Complete Exercises 17.5 Through 17.8 at This Time

17.5.5 Edge View of a Plane

The **edge view** of a plane is in a view where the line of sight is parallel to the plane. The line of sight is parallel to the plane when it is parallel to a true length line that lies on the plane. Since a projection plane is always perpendicular to the line of sight, a view drawn perpendicular to a plane shows the plane as an edge. A vertical plane appears as an edge in the horizontal view, since it is perpendicular to the horizontal projection plane. A horizontal plane is perpendicular to the frontal and profile projection planes and appears as an edge in these views.

When the given plane is oblique, an auxiliary projection shows the edge view. To establish a line of sight parallel to the plane, a true length line is drawn on the plane. An auxiliary view where the line appears as a point view shows the plane as an edge. In Figure 17.23, plane 1-2-3 is given and an edge view is required. The following steps are used:

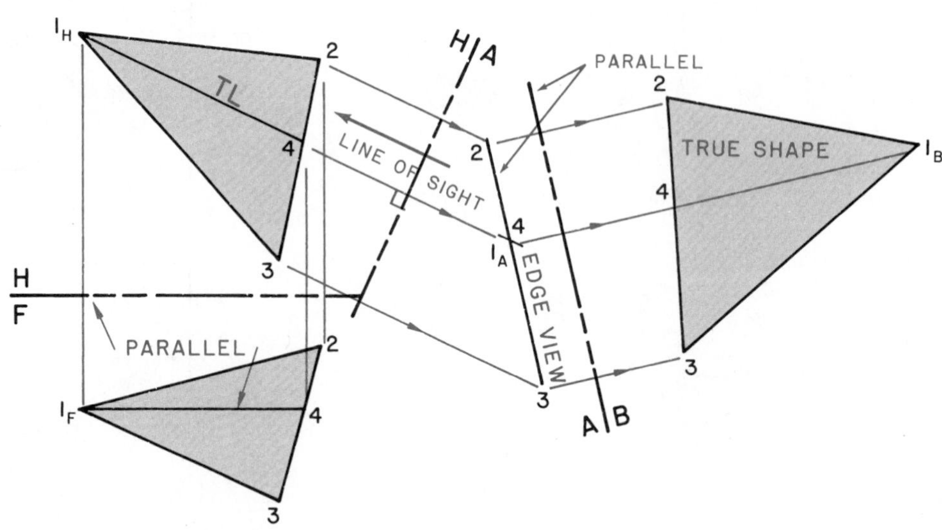

FIGURE 17.23 **Edge View and True Size of an Oblique Plane**

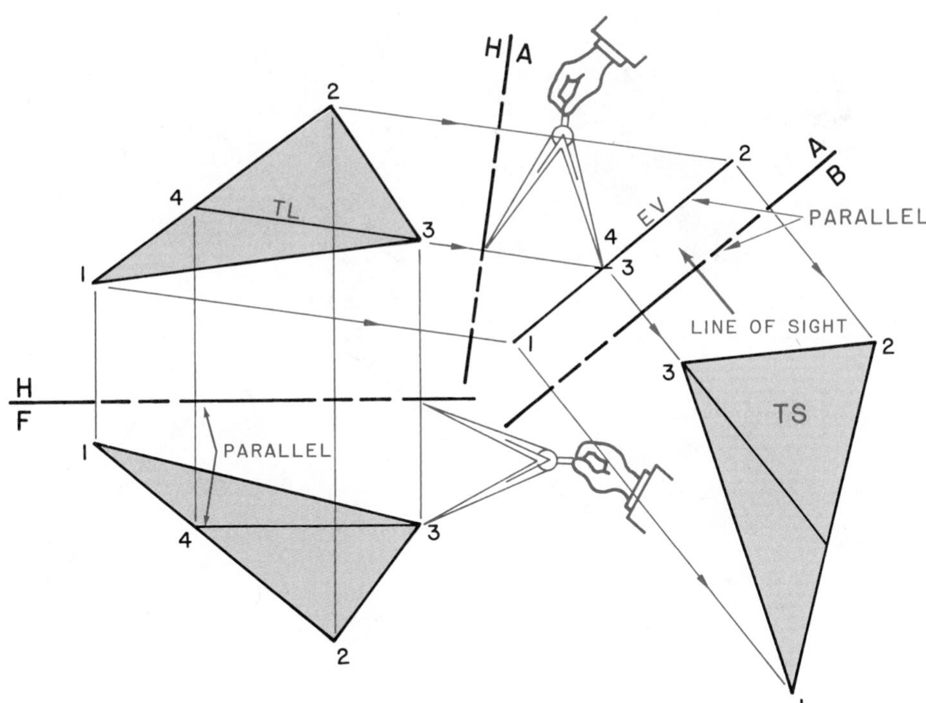

FIGURE 17.24 True Shape View of an Oblique Plane

1. Draw line 1_F-4_F on plane 1-2-3 in the frontal view, parallel to H/F, and complete the horizontal view by projection. Line 3_H-4_H is true length.
2. Project auxiliary view A perpendicular to plane 1-2-3. The line of sight for this projection is parallel to the plane and parallel to true length line 1-4. Draw H/A perpendicular to 1-4 and complete auxiliary view A by projection. Auxiliary view A shows line 1-4 as a point view. Plane 1-2-3 appears as an edge view.
3. A true size view is projected from view A by drawing fold line A/B parallel to the EV of the plane and projecting the points as shown.

17.5.6 True Size (Shape) of an Oblique Plane

When the line of sight is perpendicular to the edge view of a plane, it projects as **true size** (shape). The true size view is projected parallel to the edge view of the plane. The fold line between the views is drawn parallel to the edge view. An oblique plane does not appear as true size in any of the principal projection planes. A primary auxiliary and secondary auxiliary view are needed to solve for the true shape of an oblique plane.

In Figure 17.24, oblique plane 1-2-3 is given and its true shape is required. The following steps were used:

1. Draw horizontal line 3-4 parallel to H/F and project it as true length in the horizontal view.
2. Draw H/A perpendicular to line 3_A-4_A and complete auxiliary view B. Line 3_A-4_A is a point view and plane 1-2-3 is an edge.
3. Project secondary auxiliary view B parallel to the edge view of plane 1-2-3. Draw A/B parallel to the edge view.
4. Complete auxiliary view B; plane 1-2-3 projects true size (shape).

17.5.7 Shortest Distance Between a Point and a Plane

In Figure 17.25, plane 1-2-3 and point 4 are given. The shortest distance between a point and a plane is a perpendicular line drawn between the point and the plane. A line drawn from a point to a plane is its shortest connector if it is drawn perpendicular to an edge view of the plane. Use the following steps:

1. Draw horizontal line 3_F-5_F parallel to H/F and project it as true length in the horizontal view.
2. H/A is drawn perpendicular to horizontal line 3-5. In auxiliary view A, plane 1-2-3 appears as an edge. Draw a line from point 4 perpendicular to the edge view of the plane. Point 6 lies on the plane (at the point where the line pierces the plane). Line 4_A-6_A is the shortest distance between the point and plane.
3. Line 4_A-6_A is true length in auxiliary view A. It projects to the horizontal view as parallel to H/A. Point 6 is fixed by projection from auxiliary view A. Point 6 is located by

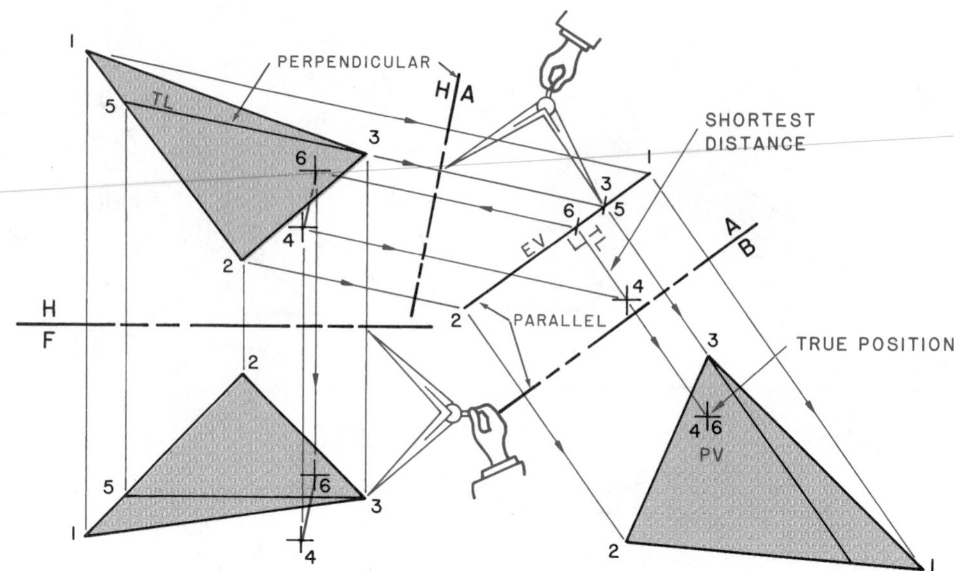

FIGURE 17.25 Shortest Connector (Shortest Distance) Between a Point and a Plane

transferring it from auxiliary view A along its projection line in the frontal view.

4. Project auxiliary view B to establish the true shape of the plane and the true position of the line on the plane.

17.5.8 Parallelism of Planes

Two planes are **parallel** if intersecting lines on one of the planes are parallel to intersecting lines on the other plane. Parallelism is determined by drawing a set of intersecting lines parallel to any two intersecting lines in the other plane. If the two sets of intersecting lines are parallel, then the planes are parallel.

In Figure 17.26, planes 1-2-3 and 4-5-6 are given. Are the two planes parallel? If so, what is the true distance between them? The following steps were used:

1. Draw horizontal line 5_H-7_H on plane 4_H-5_H-6_H, parallel to H/F. It projects as true length in the frontal view.
2. Project auxiliary view A perpendicular to line 5-7 by drawing F/A perpendicular to it.
3. In auxiliary view A, both planes show as edges and also as parallel to one another. The true distance between the planes is measured as the perpendicular distance between the two planes.

17.5.9 Angle Between Two Planes

The angle between two planes is found in a projection where both planes are edge views. *The true angle between two intersecting planes is a dihedral angle.* To solve for the angle between two intersecting planes, a view is necessary where the intersecting line appears as a point view. In this view both planes show as edges and the angle between them can be measured.

The first step is projecting an auxiliary view where the common line is true length. An auxiliary view projected perpendicular to this true length intersection line shows the common line as a point and both planes as edges. The true angle between the planes is measured in this secondary auxiliary view.

In Figure 17.27, the dihedral angle formed by two intersecting oblique planes is required. The following steps were used:

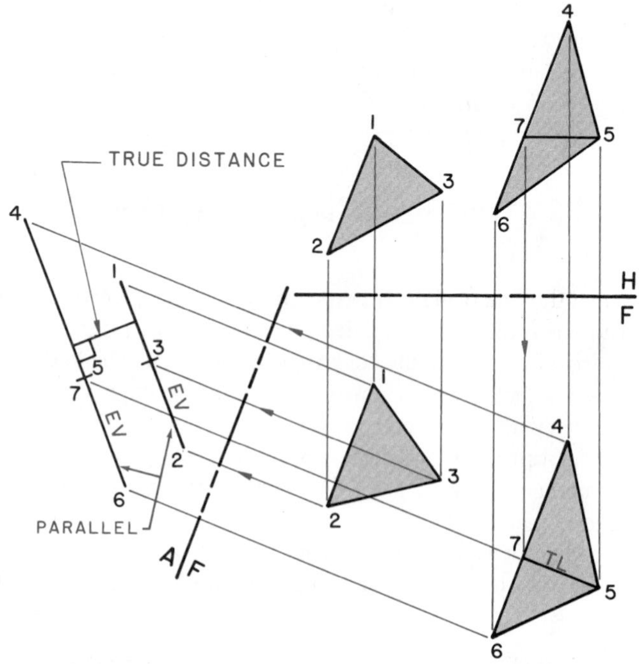

FIGURE 17.26 Shortest Connector (True Distance) Between Two Parallel Planes

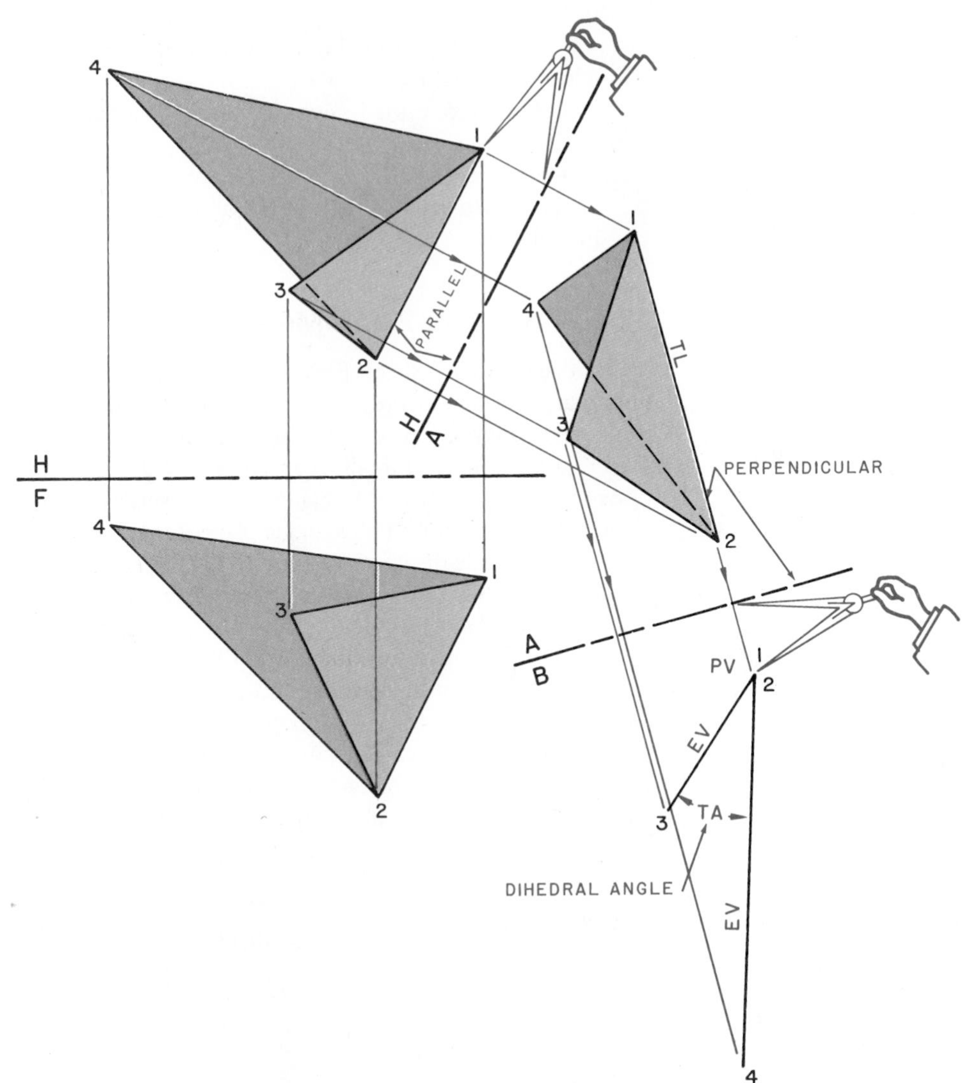

FIGURE 17.27 True Angle Between Connected Planes

1. Draw H/A parallel to line 1-2. Line 1-2 is the intersection line of the two oblique planes.
2. Complete auxiliary view A. Line 1-2 is true length in this projection.
3. Draw A/B perpendicular to true length line 1-2 and complete auxiliary view B by projection and transferring dimensions from the horizontal view.
4. The true angle between the planes is measured in auxiliary view B, since both intersecting planes appear as edges. This is the dihedral angle formed by the two planes.

17.6 Piercing Points

A line and a plane can have three relationships:

A line can lie on a plane.
A line can be parallel to a plane.
A line can intersect (pierce) a plane.

The procedure for finding the intersection of a line and a plane can be applied to intersections in all categories. Both the line and the plane can be extended to solve for theoretical intersections that lie outside the given bounded planes or beyond the given length of the line.

The point at which a line intersects (pierces) a plane is its **piercing point.** This piercing point is obtained by the *edge view (auxiliary view)* method or the cutting plane method.

17.6.1 Edge View Method

In Figure 17.28, plane 1-2-3 and line 4-5 are given. The piercing point and visibility are required. The following steps were used:

1. Draw H/A perpendicular to horizontal line 1_H-2_H and project auxiliary view A. Plane 1-2-3 appears as an edge view and line 4-5 as oblique.

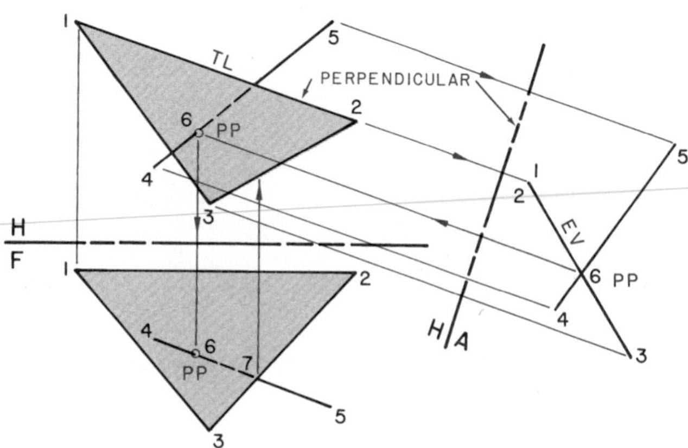

FIGURE 17.28 Piercing Point (Point of Intersection) of a Line and a Plane

2. The piercing point (point 6) is where the line crosses the edge line of the plane.
3. Project point 6 to the horizontal view and frontal views as shown.
4. Visibility is determined by inspection of auxiliary view A or the horizontal view and by the visibility test for the frontal view.

17.6.2 Cutting Plane Method

If two planes intersect, the line of intersection contains all lines that lie on one plane and pierce the other. The cutting plane method forms a new plane that contains the given line. A cutting plane is used to show an edge view in one of the principal projection planes. In Figure 17.29, a vertical cutting plane (VCP) was formed by passing a plane through line 1-2. Line 1-2 represents the edge view of the VCP. Where this VCP "cuts" plane 3-4-5 it forms a line of intersection that is common to both planes, line 6-7. This line of intersection is projected to the adjacent view, where it lies on both planes, line 6-7. The line of intersection between the two planes is parallel to or intersects the given line. If the line of intersection intersects the given line in the adjacent projection, it establishes the piercing point of the line and the plane, PP.

For a vertical cutting plane, the piercing point is established by projecting the line of intersection from the horizon-

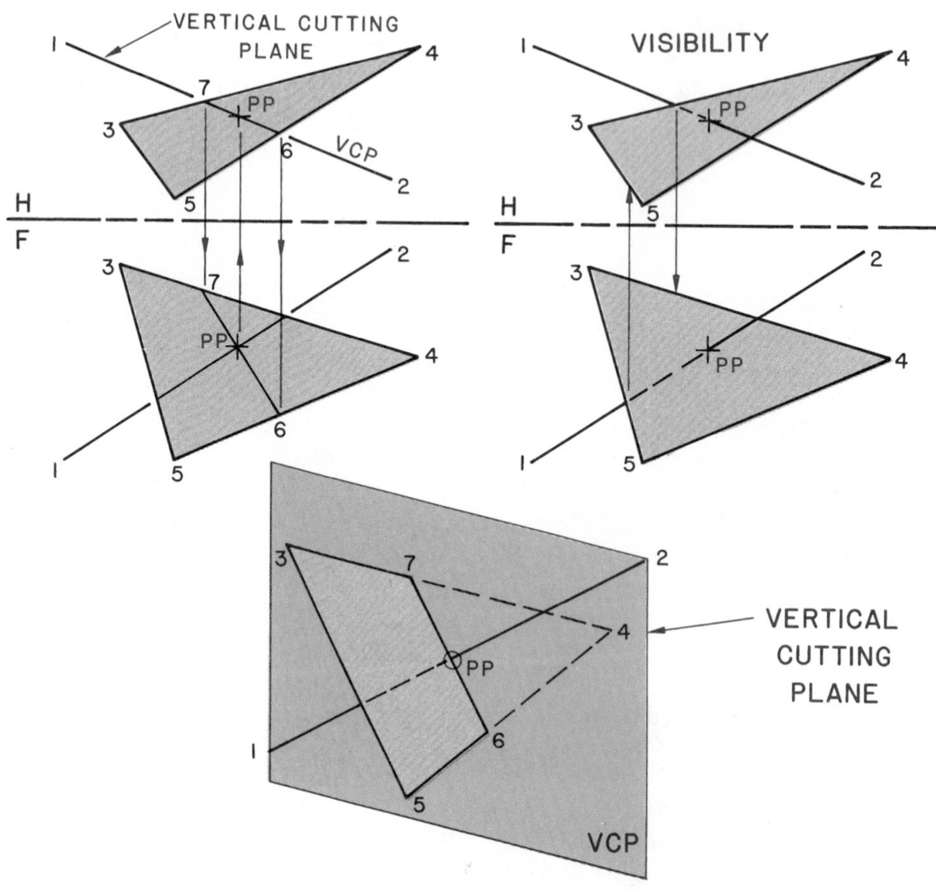

FIGURE 17.29 Intersection of a Line and a Plane Using the Cutting Plane Method

tal view. Points 6 and 7 are projected to the frontal view where they form line 6_F-7_F. Line 6-7 crosses line 1-2 at PP. The horizontal view of PP is located by projection. Use the visibility test to determine visibility.

If the line of intersection does not cross the given line, the line and plane do not intersect. In this case, the given line is parallel to the given plane and lies in front of or behind and above or below the plane.

In Figure 17.29, a pictorial view of this problem is given. The vertical cutting plane contains line 1-2. VCP cuts plane 3-4-5 along a line of intersection, line 6-7. Line 6-7 lies on both planes, as a common line. Where line 6-7 crosses line 1-2, they intersect at PP. PP is the piercing point of line 1-2 and plane 4-5-6.

You May Complete Exercises 17.9 Through 17.12 at This Time

17.7 Revolution/Rotation

Revolution or rotation occurs when the object is revolved or rotated and the observer remains stationary. Fewer views accomplish a specific task. However, the given views may look cluttered or crowded. A revolution will show the rotated position of a mechanical device. The clearance between a fixed point and a rotating form can also be solved by using revolution.

17.7.1 Revolution of a Point

For revolution problems, the observer remains stationary and the object (point) is rotated (revolved) about a straight line axis. Each revolved point moves in a circular path of rotation, perpendicular to the axis line. Since all objects, lines, and planes are composed of points, the principles presented here form the theoretical foundation for all revolution problems:

1. The axis of revolution (rotation) is a straight line and is established before a point can be revolved. The axis is a point view where the path of rotation is a circle, and it appears true length where the path of rotation is an edge.
2. The revolution of a point is always perpendicular to the axis and moves in a circular path around the point view of the axis line. This circular path forms a plane perpendicular to the axis, which appears as a circle (or portion of a circle) when the axis is a point view.
3. The path of rotation is formed by revolving the point and a line connected from it to the axis. This line is the radius of the circle (or arc) formed by revolution. When the axis shows as true length, the path of rotation appears as an angle with a length equal to the diameter of the circle.

In Figure 17.30, point 1 is revolved around vertical axis line 2_H-3_H. Axis 2-3 is true length in the frontal view and is a point view in the horizontal view. Point 1_H is revolved clockwise 135° to position $1R_H$. The path of rotation is an edge view in the frontal view where the axis is true length and is a circular path in the horizontal view.

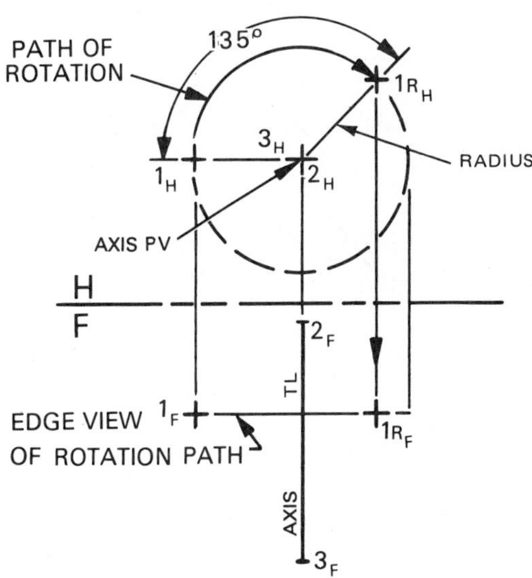

FIGURE 17.30 Rotation of a Point

17.7.2 Revolution of a Point About an Oblique Axis

When a point is revolved about an axis that does not appear as a point in the frontal or horizontal views, an auxiliary projection where the axis appears as a point view is required. In Figure 17.31, point 3 is to be revolved about line 1-2. The path of rotation will appear as an ellipse in the frontal projection. To revolve point 3 about horizontal line 1-2, an auxiliary view is projected perpendicular to the true length of axis 1_H-2_H. Fold line H/A is drawn perpendicular to line 1_H-2_H, and axis 1_A-2_A shows as a point view in this primary auxiliary view. Point 3_A is revolved to position $3R_A$ in this view. The

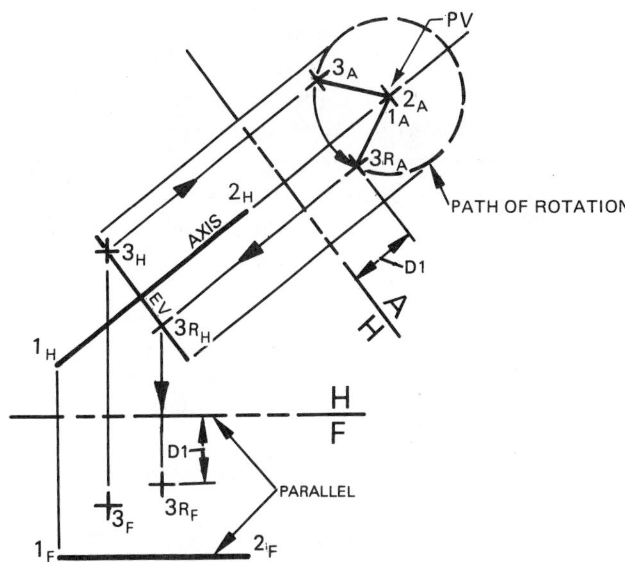

FIGURE 17.31 Rotation of a Point About a Given Line

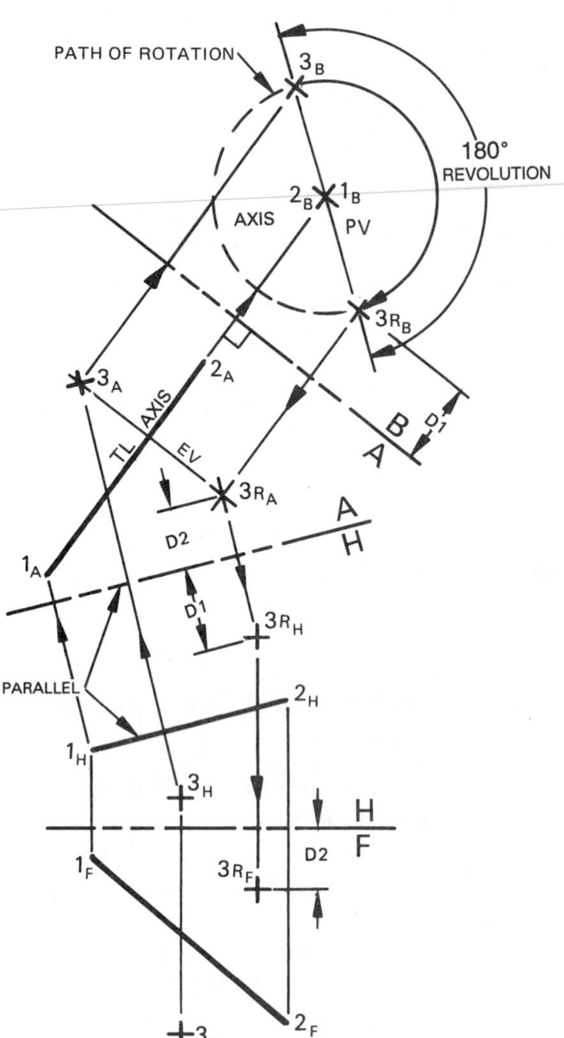

FIGURE 17.32 Rotation of a Point About an Oblique Axis Line

path of rotation generated by moving the point creates a circular plane in this view. Point $3R_A$ is located in the horizontal plane by simple projection since it falls on the edge view of the path of rotation. The frontal position of point 3R is located by transferring D1 from auxiliary view A to the frontal view along its projected line.

In Figure 17.32, point 3 is to be revolved 180° about oblique line 1-2. The following steps describe the procedure:

1. Draw H/A parallel to line 1_H-2_H and project auxiliary view A. Axis line 1_A-2_A is true length.
2. Draw A/B perpendicular to the true length axis line 1_A-2_A and complete auxiliary view B. Axis line 1_B-2_B is a point view.
3. In auxiliary view B, revolve point 3_B 180° about axis 1_B-2_B position $3R_B$.
4. Locate point $3R_A$ in auxiliary view A by projection, where it falls on the edge view of the path of rotation. The horizontal view of point $3R_H$ is found by transferring D1 from auxiliary view B along its projection line. The location of point $3R_F$ is established by drawing its projection line and transferring D2 from auxiliary view A to the frontal view.

17.8 Intersections

Designing products involves lines, planes, and solids that intersect. The power plant model shown in Figure 17.33 is a complex system of pipes and vessels. The power plant has a variety of shapes that were designed and manufactured using intersections. Part of the responsibility of a designer is to establish forms in a way that result in a functional, producible product.

A basic step to find the line of intersection of two geometric shapes is to determine the intersection of a line and a plane (piercing point). The points of intersection are located by the projection of an edge view of the plane and/or the introduction of cutting planes of known orientation. These two methods are used separately or together.

17.8.1 Intersection of Two Planes

The intersection of two planes is determined by finding the edge view of one of the planes. Where any two lines on one plane pierce the edge view of any plane, they determine the end points of the **line of intersection.** Lines and planes are

FIGURE 17.33 Power Plant Model

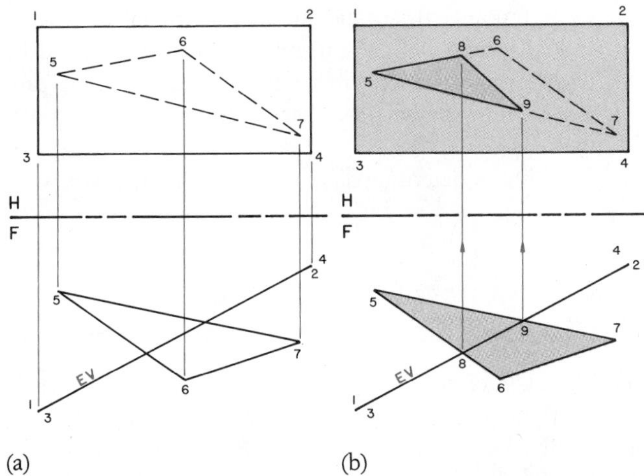

(a) (b)

FIGURE 17.34 Intersection of Two Planes Using the Edge View Method

unlimited in size or length for construction purposes. Their line of intersection can be also be extended as required. The actual intersection of two planes will have a limited line of intersection, which is common to both planes.

It is necessary to find two points common to both planes. These points of intersection form a straight line. In Figure 17.34, the line of intersection and correct visibility are required. The following steps were used to solve the intersection:

1. Inclined plane 1-2-3-4 and oblique plane 5-6-7 are given. Plane 1-2-3-4 appears as an edge in the frontal view.
2. Lines 5-6 and 5-7 pierce the edge view of plane 1-2-3-4 at points 8 and 9, respectively. Project these two piercing points to the horizontal view where they form line 8-9, which is the line of intersection. Visibility is determined by inspection; the portion of the plane formed by points 5-8-9 is above plane 1-2-3-4 in the frontal view. Therefore, it is visible in the horizontal view.

17.8.2 Intersection of Two Oblique Planes (Edge View Method)

To find the intersection of two oblique planes, an auxiliary projection showing one of the planes as an edge is required. In Figure 17.35, oblique planes 1-2-3-4 and 5-6-7 are given. The following steps were used:

1. Lines 1-3 and 2-4 are true length horizontal lines in the horizontal view. Draw H/A perpendicular to line 2-4 and project auxiliary view A. Plane 1-2-3-4 appears as an edge view and plane 5-6-7 is oblique in view A.
2. Line 5-7 pierces the edge view of plane 1-2-3-4 at point 8. Line 5-6 pierces plane 1-2-3-4 at point 9. Project points 8 and 9 to the horizontal view where they form

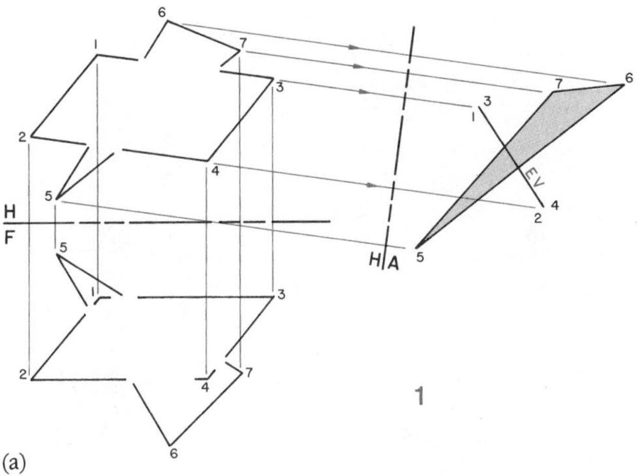

(a)

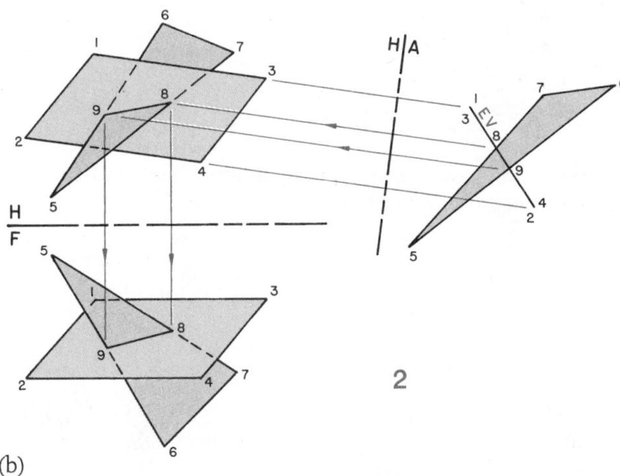

(b)

FIGURE 17.35 Intersection of Two Oblique Planes Using the Edge View Method

the common line of intersection between the two planes, line 8-9. Locate intersection line 8-9 in the frontal view by projection. Visibility is determined by inspection. The portion of the plane formed by line 8-9 and point 5 is above and in front of plane 1-2-3-4. Therefore, it is visible in the frontal and horizontal views.

17.8.3 Intersection of an Oblique Plane and an Oblique Prism (Edge View Method)

An auxiliary view showing the plane as an edge is required if the plane is oblique in the given views. In Figure 17.36, plane 1-2-3-4 and prism 5-6-7 are both oblique in the given frontal and horizontal views. The intersection and correct visibility are required. The following steps were used:

1. Lines 1-2 and 3-4 are horizontal lines (true length in the horizontal view). H/A is drawn perpendicular to line 1-2. Complete auxiliary view A.

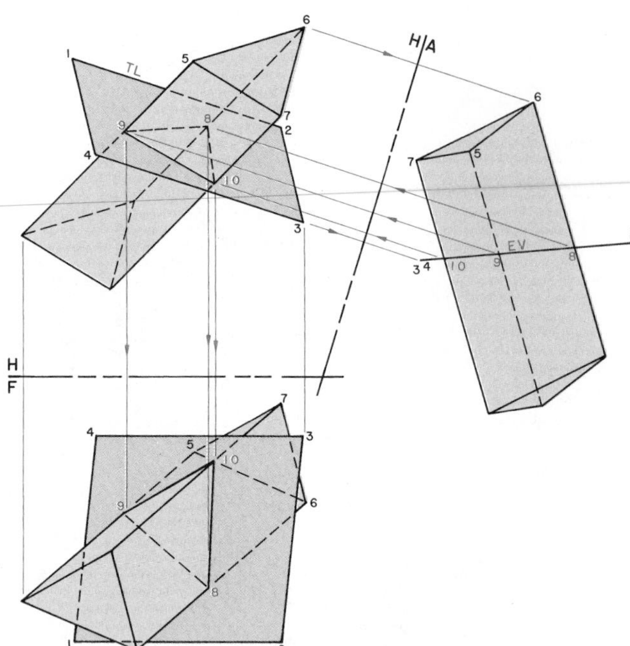

FIGURE 17.36 **Intersection of an Oblique Plane and an Oblique Prism Using the Edge View Method**

2. Plane 1-2-3-4 represents the edge view of a cutting plane in auxiliary view A. Plane 1-2-3-4 intersects the prism at points 8, 9, and 10. The edge lines of the prism pierce the plane at points 8, 9, and 10. Project all three piercing points to the horizontal view. The horizontal view of the piercing points determines the plane section cut from the prism, which is the intersection of the plane and prism.

3. Project points 8, 9, and 10 to the frontal view. The frontal location of each piercing point can be fixed by transferring distances from auxiliary view A along projection lines drawn from each point in the horizontal view.
4. Visibility is determined by inspection of auxiliary view A.

17.8.4 Intersection of a Plane and a Cylinder (Cutting Plane Method)

The line of intersection of a plane and a cylinder is determined by passing a series of cutting planes parallel to the axis of the cylinder. Each cutting plane (CP) cuts elements on the cylinder, which pierce the plane to form an elliptical line of intersection. Accuracy increases with an increasing number of cutting planes.

In Figure 17.37, a series of vertical cutting planes is passed parallel to the axis and through the cylinder. Each cutting plane establishes two elements on the cylinder and a line on the plane. Where these related lines and elements intersect, they establish the required piercing points. The following steps were used:

1. Draw CP1 and CP2 parallel to the axis line (and parallel to the H/F Fold Line). CP1 intersects line 1-3 at point 4 and line 2-3 at point 5. CP2 intersects line 1-2 at point 6 and line 2-3 at point 7. Both CPs establish an element of the cylinder. Project the elements to the frontal view along with lines 4-5 and 6-7. Line 4-5 intersects its element at piercing point A and line 6-7 intersects its corresponding element at point B.
2. Repeat step 1 using CP3, CP4, and CP5. Each of these cutting planes cuts *two* elements on the cylinder. Each locates two piercing points. Connect the piercing points

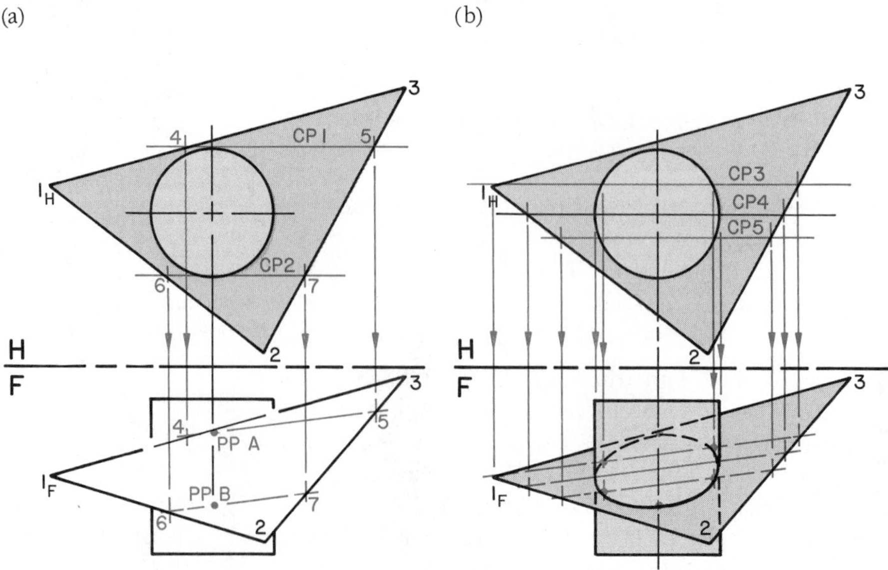

FIGURE 17.37 **Intersection of an Oblique Plane and a Right Cylinder Using the Cutting Plane Method**

in sequence to form a smooth curve. Since point 2 is in the front of the cylinder, lines 1-2 and 2-3 are visible, as is point B.

17.8.5 Intersection of a Plane and a Sphere

The intersection of a plane and a sphere results in a circular line of intersection. If the plane is oblique, the line of intersection appears as an ellipse (Fig. 17.38). The extreme piercing points (the major and minor axes) must be found using the edge view or cutting plane method. The actual ellipse can be constructed with an ellipse template using the major and minor axes, plotting a series of piercing points established by cutting planes in a view showing the plane as an edge, or passing a series of cutting planes through the sphere and the plane where the plane is oblique.

In Figure 17.38, the intersection of the sphere and plane is required. The following steps were used:

1. Pass a series of evenly spaced horizontal cutting planes through the sphere and project them to the horizontal view. Each CP cuts a small circle section.
2. Each CP intersects the edge view of the plane and locates two piercing points, which are projected to the horizontal view to establish the line of intersection.

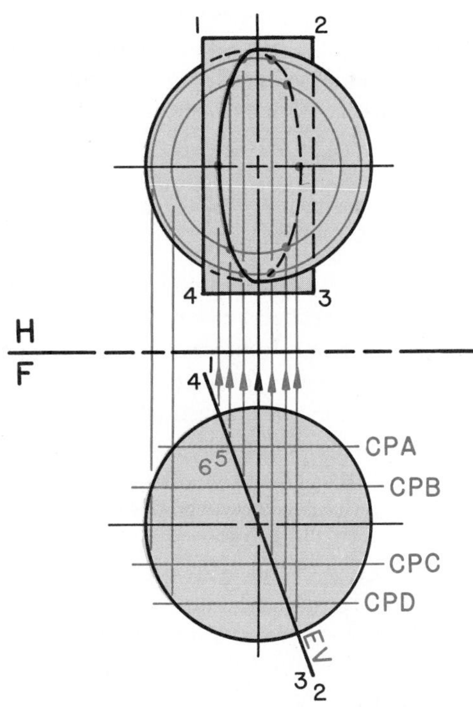

FIGURE 17.38 **Intersection of a Plane and a Sphere Using Cutting Planes**

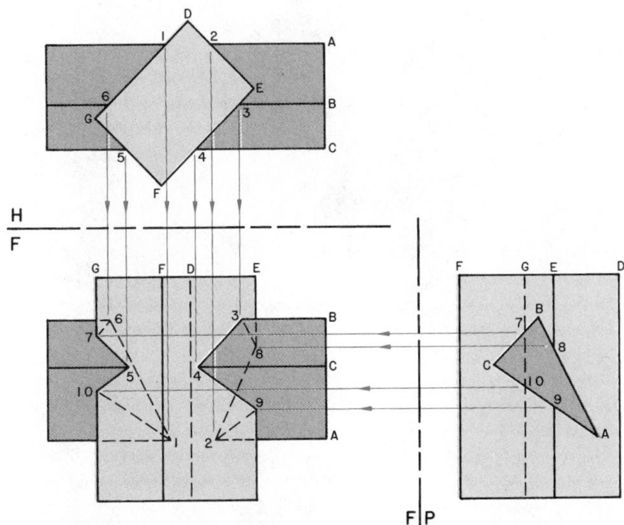

FIGURE 17.39 **Intersection of Two Prisms**

17.8.6 Intersection of Prisms

The intersection of two prisms is found by the edge view method. The piercing point of an edge line of one prism and a surface of the other prism is established in a view where one of the prisms is an edge view. This type of problem can be reduced to finding the piercing point of a line and a plane. Where each edge line of a prism pierces a surface (plane) of the other prism, a point (piercing point) on the line of intersection is established. The line of intersection includes only surface lines of intersection.

In Figure 17.39, two right prisms intersect at right angles. The horizontal view shows the edge view of the rectangular prism and the profile view shows the triangular prism as an edge view. The following steps were used:

1. The edges of the triangular horizontal prism pierce the vertical prism in the horizontal view at points 1 through 6. Edge line A pierces the surface bounded by lines D and G at piercing point 1 and at piercing point 2 on the surface bounded by lines D and E.
2. Project points 1 and 2 to the frontal view until they intersect line A.
3. Repeat this procedure to locate piercing points 3, 4, 5, and 6 in both views.
4. The edges of the vertical rectangular prism pierce the surfaces of the horizontal prism in the profile view at points 7, 8, 9, and 10. Edge line G pierces the surface bounded by lines B and C at piercing point 7.
5. Project point 7 to the frontal view until it intersects line G.
6. Repeat step 5 to locate piercing points 8, 9, and 10.
7. Determine visibility and connect the piercing points to form the line of intersection.

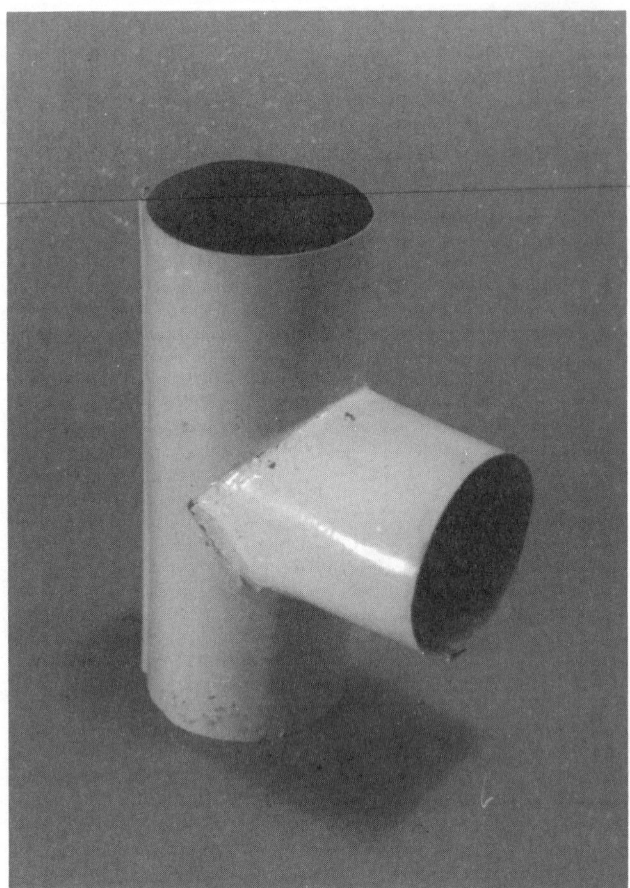

FIGURE 17.40 **Sheet Metal Model of Intersecting Cylinders**

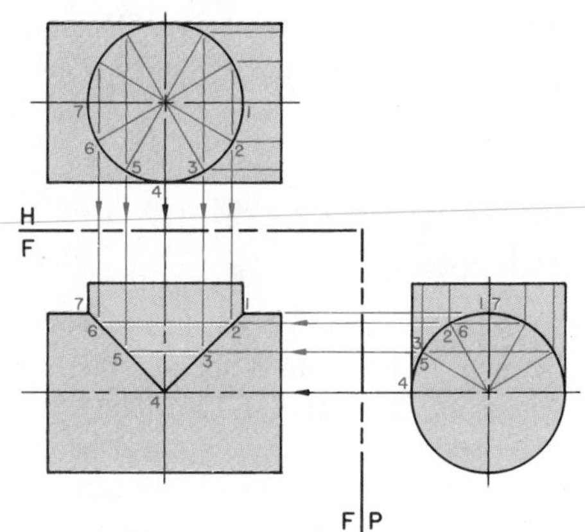

FIGURE 17.41 **Intersection of Two Cylinders at 90°**

17.8.7 Intersection of Cylinders

The intersection of two cylinders (Fig. 17.40) is a common industrial problem in piping and vessel design. Two intersecting perpendicular right cylinders of the same diameter intersect as shown in Figure 17.41. The line of intersection is determined by showing each cylinder as an edge view and passing a series of equally spaced cutting planes through both cylinders. Each cutting plane intersects a cylinder as an element on its surface. The intersection of related elements determines the line of intersection. Each intersection point is actually the piercing point of an element of one cylinder and the surface of the other cylinder.

In Figure 17.41, the resulting curved lines of intersection appear as straight lines in the frontal view. In this case the line of intersection could have been determined by simply drawing the straight lines from point 1 to point 4 to point 7.

To solve for the perpendicular intersection of two cylinders, regardless of their diameters, a series of elements is drawn on the surface of one cylinder by equally dividing the edge view of the vertical cylinder (Fig. 17.41). Each vertical cutting plane passes parallel to the cylinder's axis and cuts a straight-line element on both surfaces. Points 1 through 7 represent the intersection of related elements established by the intersection of a cutting plane and each cylinder. The profile view can also be used to equally divide the horizontal cylinder and establish vertical cutting planes.

17.8.8 Intersection of Two Cylinders (Not at Right Angles)

To find the intersection of two cylinders not at right angles, an edge view of both cylinders is necessary. Project an auxiliary view of the cylinder that does not appear as an edge in a given view. Each cutting plane in the series intersects both cylinders as elements on their surface. Related elements intersect along the line of intersection of the two cylinders. Accuracy increases proportionally to the number of cutting planes. Piercing points are connected by a smooth curve. In Figure 17.42 an industrial drawing of two pipes intersecting at 45° is given. The line of intersection and the development is shown.

The type and number of intersections found in industry is infinite, but the basic procedures and techniques presented here can be applied to all intersecting forms.

17.9 Developments

A variety of industrial structures, products, and manufactured parts are made from flat sheet stock material. Parts produced from flat materials are cut from a pattern called a **development.** The complete layout drawing of a part showing the total surface area in one view is constructed using true length dimensions. This flat plane drawing shows each surface of the part as true shape. All surfaces of the object are connected along adjacent bend lines. Sheet metal parts, cardboard packaging, large-diameter cylindrical vessels, funnels, cans, and ducting are just a few of the many types of parts that use developments. The turbine in Figure 17.43 is example of an industrial product that has a complex sheet metal development in its design. The air intake housing was created from a sheet of metal using a pattern.

A **pattern** is made from the original development drawing and is used in the shop to set up the true shape configuration

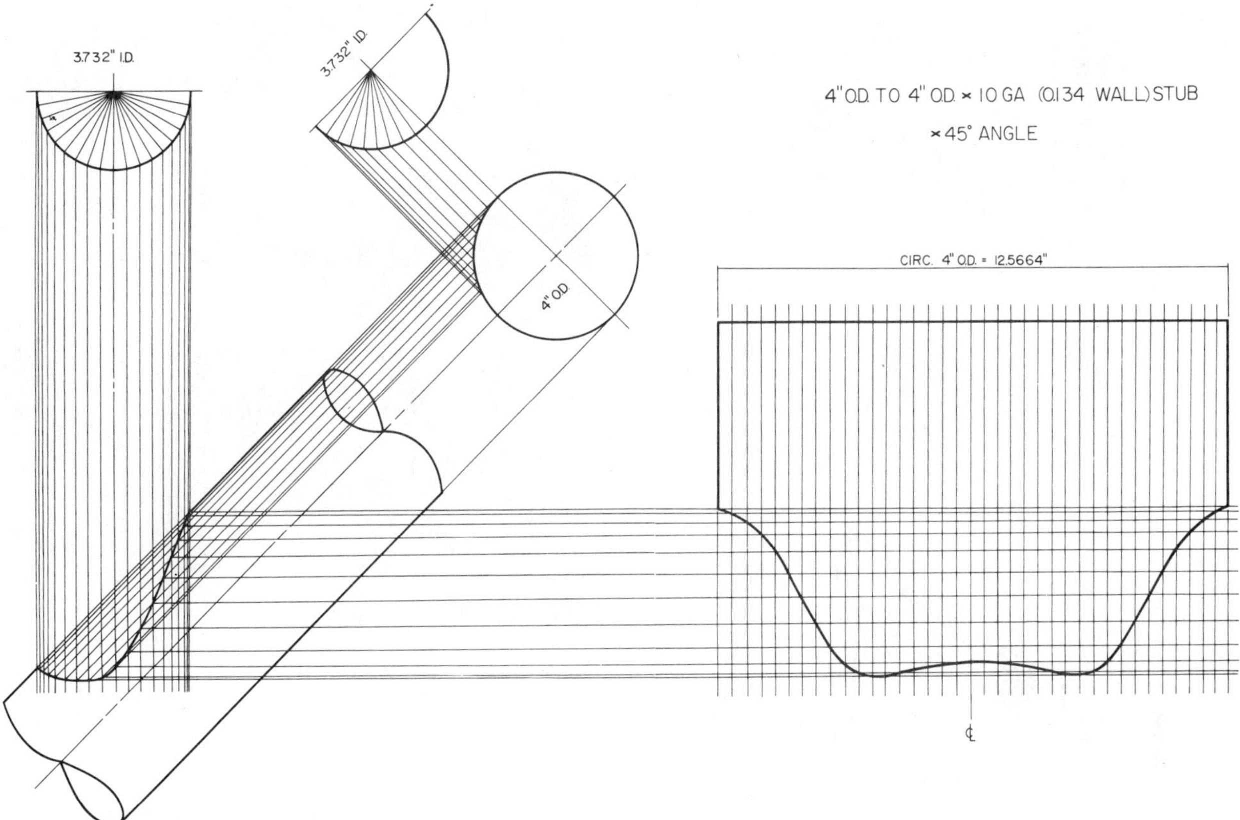

FIGURE 17.42 **4 Inch O.D. to 4 Inch Stub at 45° Is Shown Along with Its Template Development**

FIGURE 17.43 **Turbine**

of a part with tabs. The actual developed flat sheet configuration is then cut according to the pattern. The final steps include bending, folding, or rolling, and stretching the part to its required design. Welding, gluing, soldering, bolting, seaming, or riveting can be used to join the seam edge of the piece.

17.9.1 Basic Developments

Four common shapes that can be developed are the **prism, pyramid, cylinder,** and **cone** (Fig. 17.44). Development is done by unfolding or unrolling surfaces onto the plane of the paper. The actual drawing shows each successive surface as true shape and connected along common edges. One edge line serves as a seam for a shape composed of plane surfaces.

Each part is developed as an *inside-up* pattern drawing. It is unfolded/unrolled so that the inside surface is face-up. Sometimes a pattern may show as an outside-up development. The difference is shown in the representation of the bend lines.

In Figures 17.44(a) and (b), the prism and pyramid have been unfolded inside-up so that each surface is laid flat and connected along common edges. The first and last line of any development represents the same line, because they are joined together along the seam. A right prism unfolds as a rectangle. The length of the rectangle is equal to the perimeter of the base and its width is equal to its altitude.

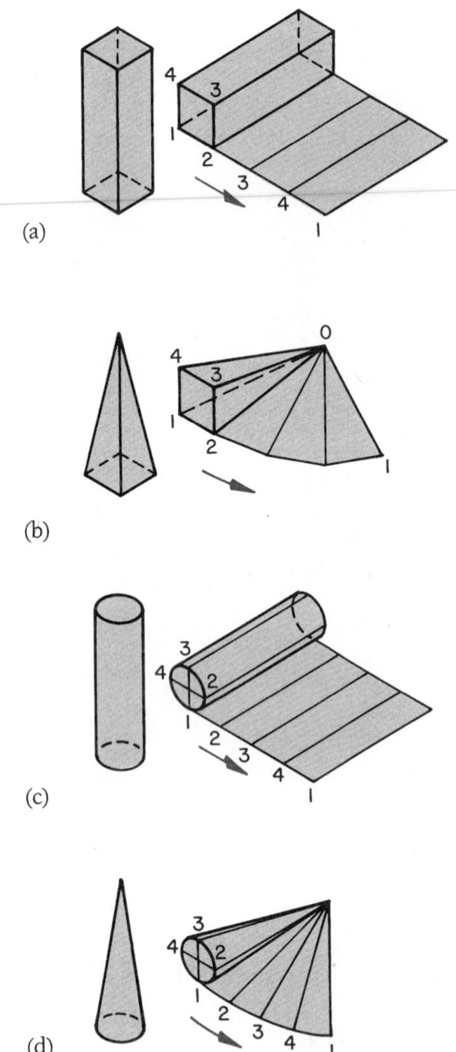

FIGURE 17.44 Basic Developments (a) Prism (b) Pyramid (c) Cylinder (d) Cone

In Figures 17.44(c) and (d), the cylinder and the cone have been unrolled (inside-up). A seam edge for these figures is along a specified line or element on the surface. A cylinder unfolds/unrolls as a rectangle with its length equal to its circumference and the height is equal to the altitude. A cone develops as a portion of a circle (sector).

Double-curved and warped surfaces cannot be developed. Spheres, paraboloids, oblique helicoids, and cylindroids are examples of undevelopable surfaces. However, approximate methods can be used to develop these types of surfaces adequately.

17.9.2 Types of Developments

The division of developments is based on the shape of the surface and/or the method used to construct its development:

1. **Parallel line:** Forms that are composed of parallel straight-line elements or edges, cylinders, prisms. The cylindrical shape of the exhaust ducting for the power plant turbine is a parallel line surface (Fig. 17.45).
2. **Radial line:** Forms whose edges or elements define triangular surface areas, pyramids, cones. Many of the wind tunnel sections shown in Figure 17.46 are radial line surfaces created from frustrums of a cone.
3. **Triangulation:** Forms whose surfaces must be broken into triangular areas. Transition pieces are the most common type of development for this catagory. The bottom portion of the air filtration system shown in Figure 17.47 is a transition piece and must be developed using triangulation.
4. **Approximate:** Forms whose surfaces cannot be truly developed, such as warped and double-curved surfaces, spheres. The water tower shown in Figure 17.48 is an example of a surface that would require an approximate development.

FIGURE 17.45 Power Plant Model

17.9.3 Sheet Metal Developments

Many complex 3D shapes are fabricated from flat sheet materials. The shape to be formed is subdivided into simple elements, which individually have the shapes of prisms, cylinders, cones, pyramids, and spheres. All of these shapes can be formed from a 3D sheet of material by first cutting the proper pattern and then folding or rolling the material into the 3D form.

Patterns are usually made full size and can only be made after the true lengths of all lines that will lie on the pattern have been determined. Therefore, all patterns are true shape/size. Each development must be drawn accurately so that the final product is of the correct shape within given tolerance limits. A bend allowance is usually added to the pattern drawing to accommodate the space taken by the

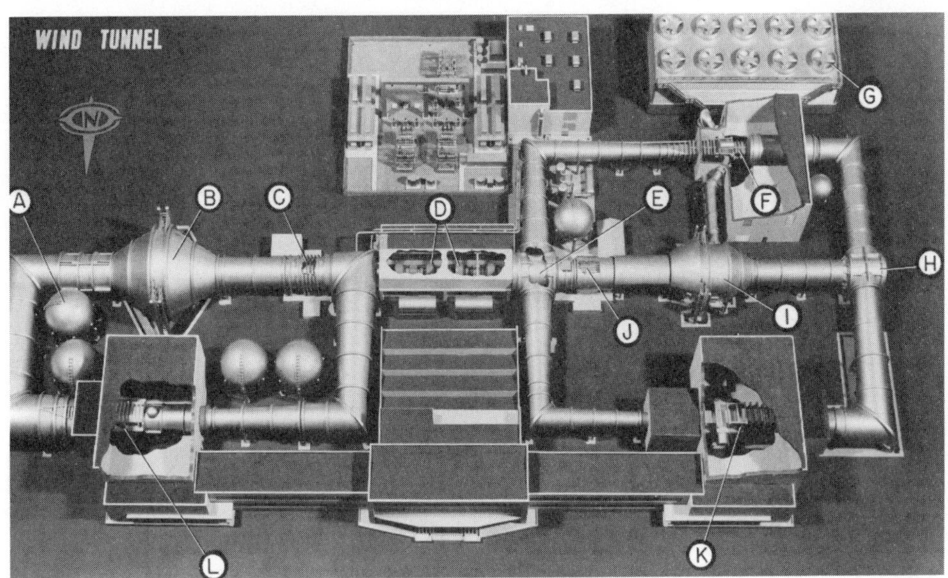

FIGURE 17.46 Scrubber System (a) Dry air storage spheres (b) Aftercooler (c) Three-stage axial flow fan (d) Drive motors (e) Flow diversion valve (f) 8 × 7 foot supersonic test section (g) Cooling tower (h) Flow diversion valve (i) Aftercooler (j) Eleven-stage axial flow compressor (k) 9 × 7 foot supersonic test section (l) 11 × 11 foot transonic test section

bending process. A tab or lap is added to the pattern so that the two adjoining edges that form the seam may be attached. The width of this tab depends on the type of joining process. The length of the lap is established along the shortest edge to limit the length of the seam. In this chapter bend allowance and a lap have been eliminated for the problems and example illustrations. Each development is a true development without bend allowances.

Seams used for sheet metal are shown in Figure 17.49. Seams are either mechanical or welded. Welded and riveted seams are permanent and are used in applications where the pieces to be joined are thicker and heavier. Metal thicknesses are designated by gage numbers from .25 inches. Thickness is designated by inches or metric sizes. See Appendix C for the sheet metal sizes of common gage metals.

Much of electronic packaging uses sheet metal parts for chassis, panels, mounting plates, and a variety of enclosures and envelopes (Fig. 17.50). Sheet metal parts are made from a blank of sheet metal. Panels, mounting plates, and other parts are flat sheets of metal cut to the functional outline with the proper slots, with holes punched or machined as required. The industrial drawing of the chassis enclosure shown in Figure 17.51 has been developed as an inside-up pattern in Figure 17.52. The dashed lines on the pattern development are bend lines.

FIGURE 17.47 Air Filtration System Design Uses a Right Vertical Prism and a Transition Piece

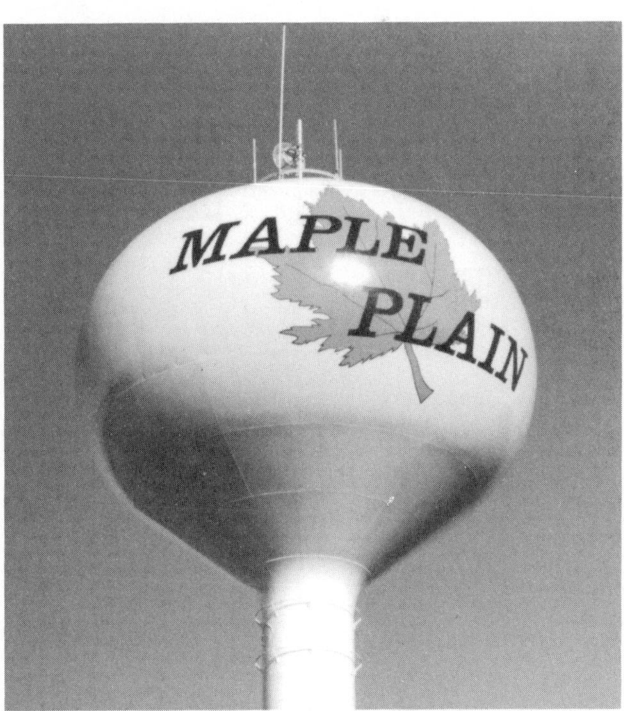

FIGURE 17.48 Ellipsoid-Shaped Water Tower

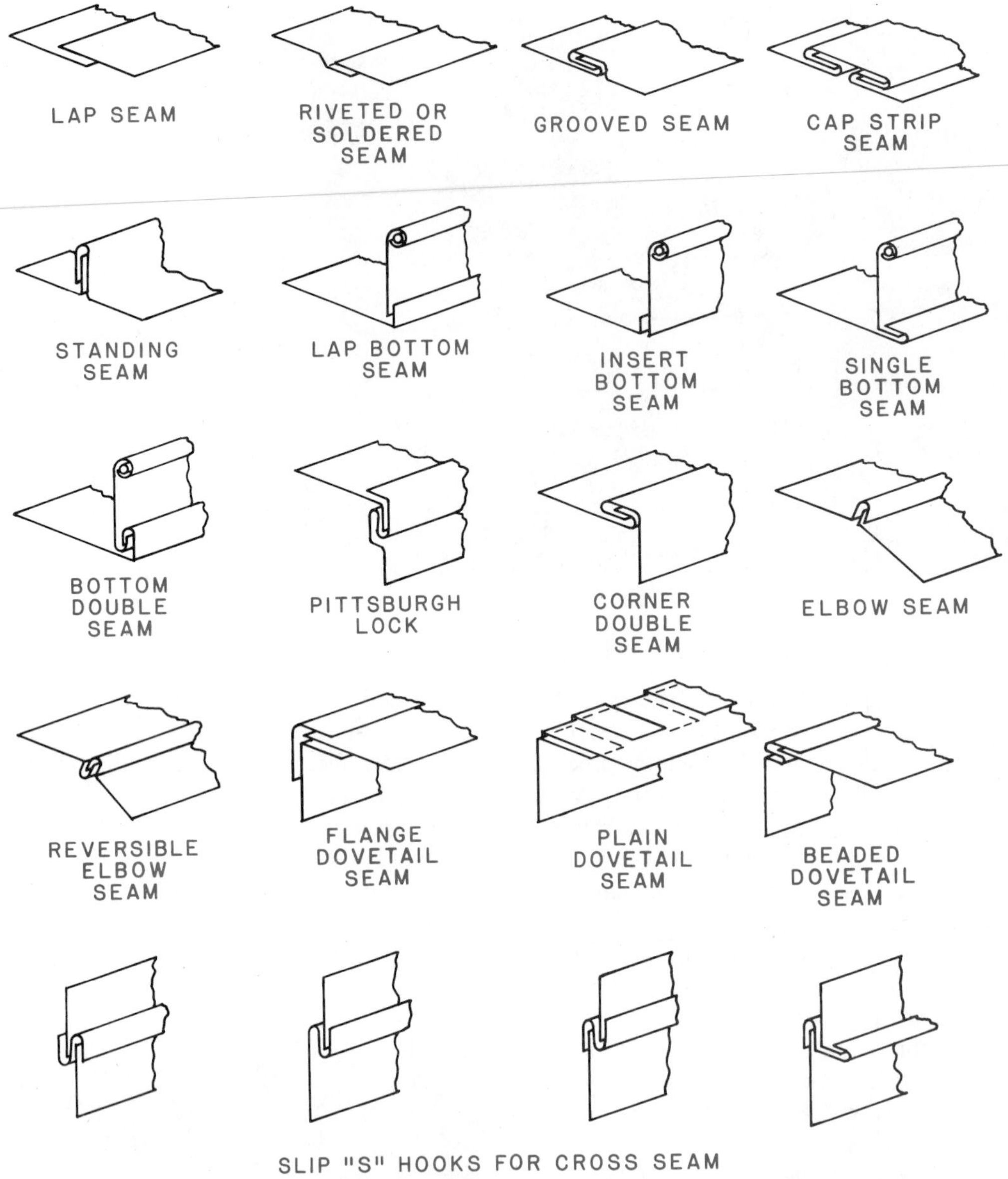

FIGURE 17.49 Standard Types of Seams Used in Sheet Metal Fabrication

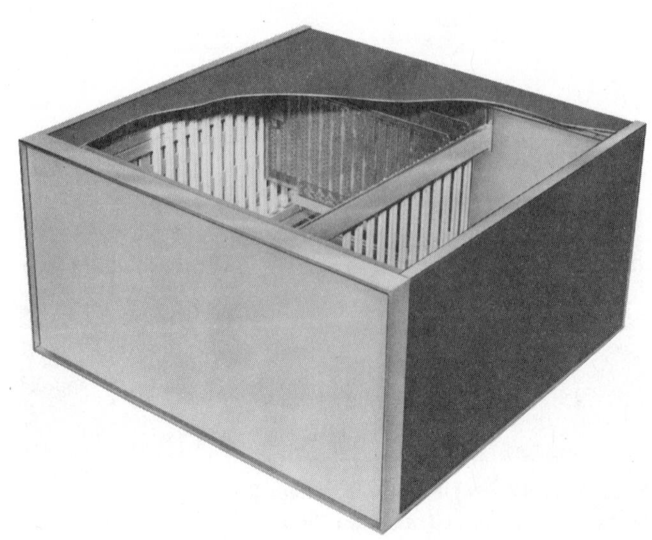

FIGURE 17.50 Sheet Metal Enclosure Used for Electronic Packaging

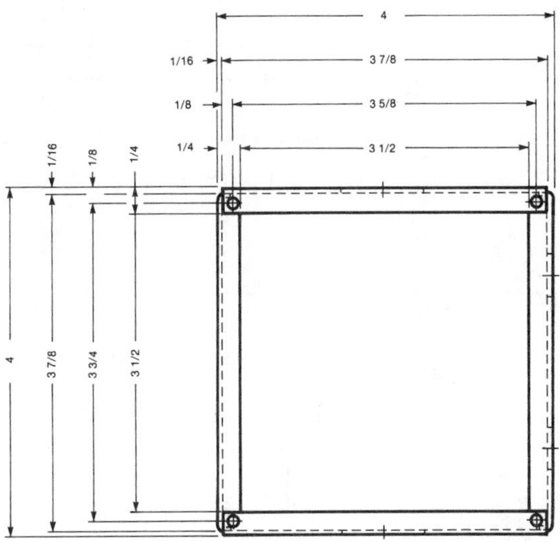

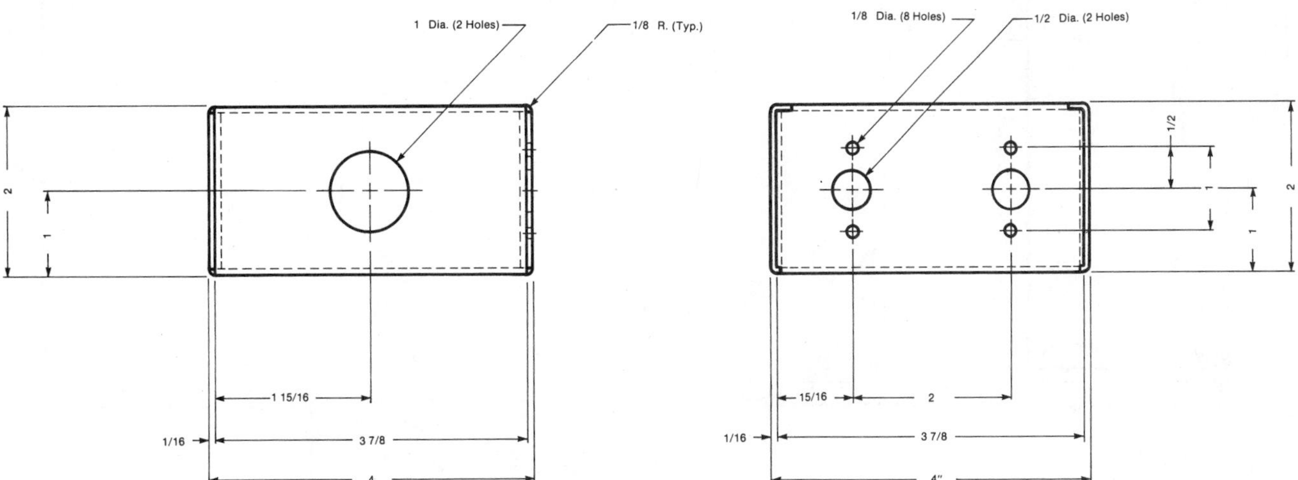

FIGURE 17.51 Sheet Metal Chassis for Electronic Equipment This item was developed from a single sheet of metal shown as a pattern in Figure 17.52.

17.9.4 Development of a Truncated Right Prism

The first step in a parallel line development is to find the true lengths of each edge line and the width of each face plane. A right section view shows the perimeter of the object. The length of the development is equal to the perimeter of the prism as measured in the right section view. A right section view is always taken perpendicular to the true length edge lines of a prism or the axis line of a cylinder (end view). The distance between each edge line/element is measured where they appear as points on the right section. The width of each lateral surface is equal to the distance between points on the right section and is transferred directly to the stretch-out line.

In Figure 17.53, the distance between point 1 and point 2 in the right section view is transferred to the **stretch-out line** to establish the width of the first plane face. A stretch-out line is a construction line along which all perimeter dimensions are laid off. The prism is unfolded clockwise using the shortest edge as the seam when it is required. In Figure 17.53, edge line 1 is used as the seam line. The stretch-out line is drawn perpendicular to the edge lines as shown. The edge lengths are projected from the frontal view. The outline of the development is then completed by connecting the end points of the edge lines. Edge lines in both the front view and the development are true length. Each lateral surface (plane face) is true shape/size. The length of the development is checked by measuring the perimeter of the prism (the distance around the right section view). The development length must equal the perimeter.

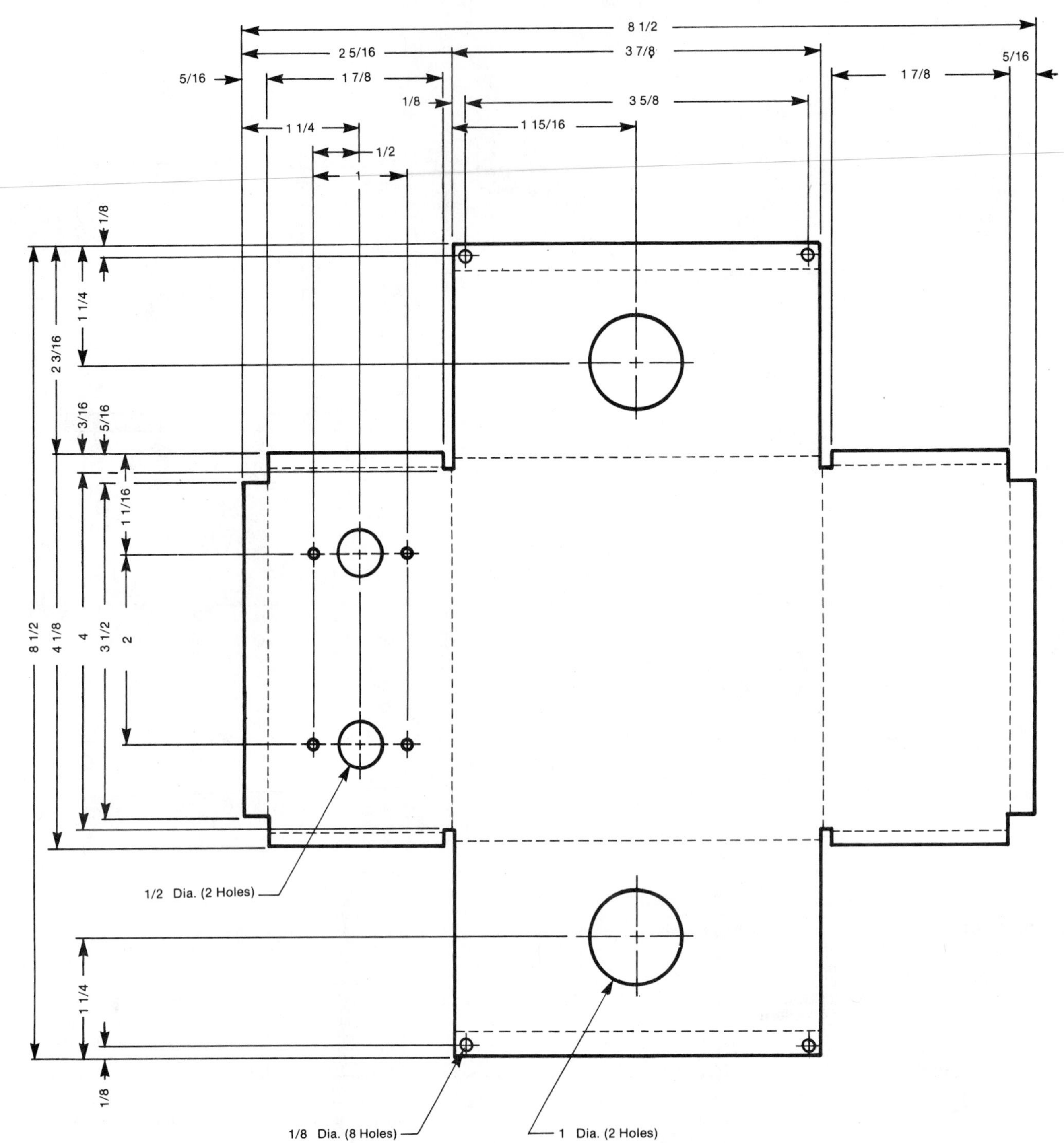

FIGURE 17.52 **Pattern Development Used to Fabricate the Enclosure Shown in Figure 17.51**

17.9.5 Development of a Prism (Top Face and Lower Base Included)

When one end face of a prism is perpendicular to edge lines, a true shape end view is a right section. The stretch-out line is projected parallel to the edge view of an end surface, if that surface is perpendicular to edge lines of the prism. The stretch-out line forms one complete edge of the development outline.

When the lower base and the upper face are required, a view showing these surfaces as true shape must be completed. The true shape of an end surface is established by projecting an auxiliary view perpendicular to the edge view of the base or top face. Each end surface is attached to an appropriate

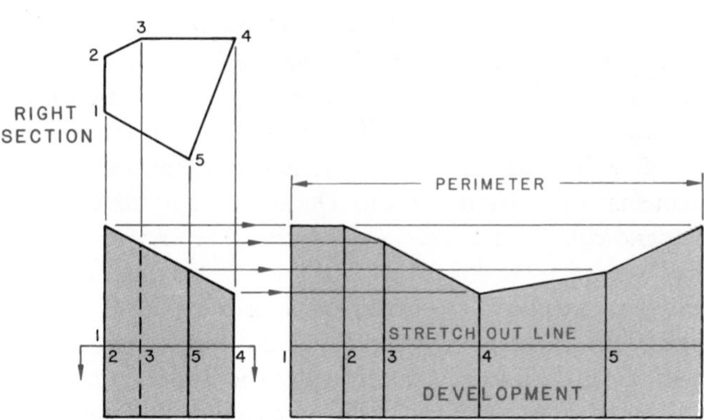

FIGURE 17.53 Development of a Right Truncated Prism

upper or lower border line of the development. A stretch-out line can be established at any convenient location on or off the paper. When this procedure is used, distances above and below the stretch-out line are transferred from the true length view to establish bend line lengths on the development. The face widths are, as before, taken from the right section (or true shape end view).

In Figure 17.54, the development of the prism is required. The bottom surface and the top face are to be included as part of the development. Line 1 is used as the seam. The following steps were used:

1. The edge lines of the prism are frontal lines (true length in the frontal view). The prism is laid on its side. Draw the stretch-out line parallel to the edge view of the top face as shown. The bottom view is given instead of a top view for this example.
2. Project a true shape view of the top face (labeled "Right Section").
3. Transfer the face widths from the true shape/right section view and set off along the stretch-out line.
4. Project the edge line end points to the development and connect to form the outline. The stretch-out line is part of the outline on this development.
5. Attach the top face and the bottom base as shown. The base plane appears as true shape in the bottom view. The upper and lower surfaces can be attached along any related line on the development's outline.

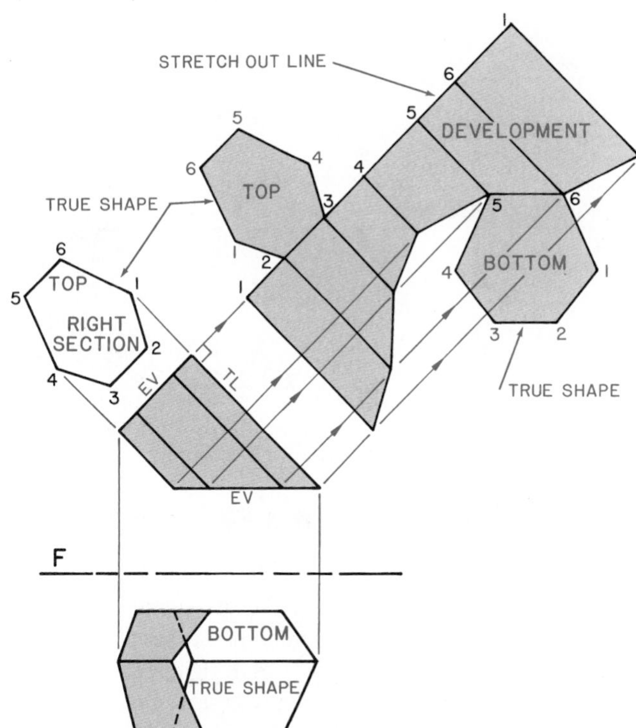

FIGURE 17.54 Prism Development Including End Surfaces

In Figure 17.55, the perimeter of the base is true length in the horizontal (top) view. Revolve an edge line until parallel to the frontal plane to obtain its true length in the frontal view. Use this true length edge line as the true length radius. To start the development, locate vertex point 0 at a convenient location. Swing an arc from point 0 using the TL radius. Starting with point 1, lay off the true length distances transferred from the base edges in the horizontal view. Lines 1-2, 2-3, 3-4, and 4-1 are true length in the top view. Connect each point with vertex point 0 and draw straight-line chords between the points to establish the base perimeter on the development.

17.9.6 Development of a Right Pyramid

Developments of surfaces composed of triangular planes, such as pyramids, or surfaces that can be divided into small triangular areas, cones, are considered radial line developments. Each lateral edge of a pyramid, or element of a cone, radiates from the vertex point.

To develop a pyramid, establish the true length of each of its lateral edges and base lines. The development of a pyramid consists of laying out the true shape of each lateral surface in successive order. If the pyramid is a right pyramid, each of its lateral edges will be of equal length. Therefore, the true length of only one lateral edge is necessary.

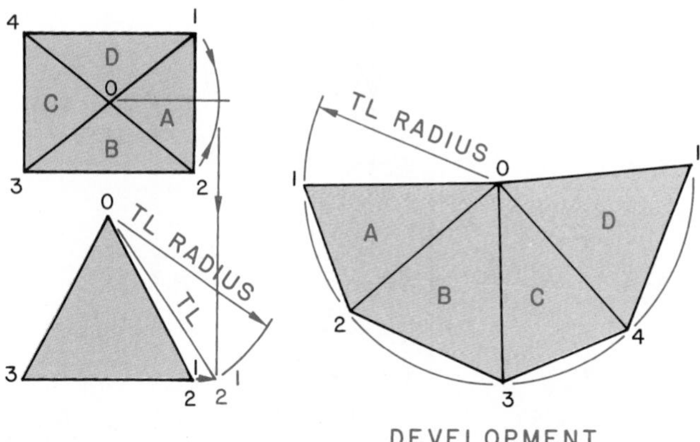

FIGURE 17.55 Development of a Right Pyramid

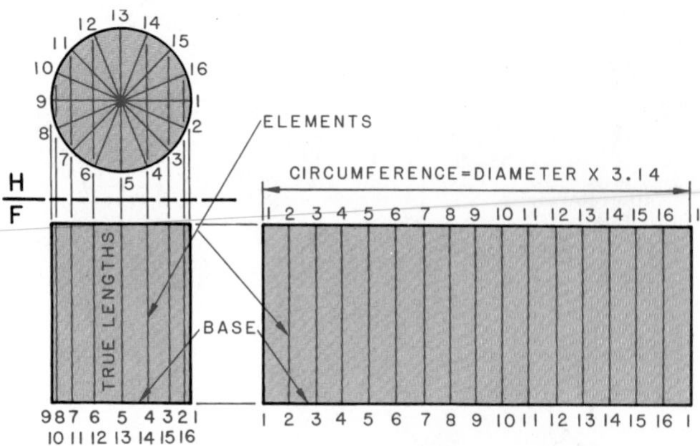

FIGURE 17.56 Development of a Right Cylinder

17.9.7 Development of a Right Circular Cylinder

A cylinder is developed by unrolling its surface, normally inside-up. A right circular cylinder has a stretch-out line equal to its circumference (Fig. 17.56). A right section (axis as a point) and a view showing the axis as true length are necessary to develop a cylinder. The edge view/right section determines the shape of the cylinder and provides a view where elements can be established on its surface. A true length view of the cylinder's axis shows all elements on its surface as true length. A development is made by rolling the lateral surface of the cylinder onto a plane.

In Figure 17.56, the right section of the cylinder is shown in the horizontal view. Elements are established along its surface by dividing the right section view into a number of equal parts. The elements are located by evenly dividing the circumference of the circular section as shown; 12, 16, or 24 radial divisions are commonly used. Each division is projected to the true length view (frontal view) to establish the elements on the lateral surface. The stretch-out line is drawn perpendicular to the true length view. The base perimeter may be used as the stretch-out line if it is perpendicular to the cylinder's axis as in the example. The stretch-out line is divided into the same number of equally spaced parts as the right section and labeled accordingly. The true length of each element is projected to the development, from the true length view, to establish its outline. In Figure 17.56, both bases are perpendicular to the axis. Therefore, all elements are the same length and the development unrolls as a rectangle. Cylinders are a single surface; therefore, the elements are drawn as thin construction lines in all views and on the development.

17.9.8 Development of Intersecting Cylinders

In Figure 17.57, the development of two cylinders intersecting with a 90° **miter bend** is shown. The following steps were used to solve the problem:

1. Draw a half circle and divide it into equal parts as shown. The half section corresponds to the end view (right section) of the cylinder. Label the intersection of the division lines from 1 to 7.

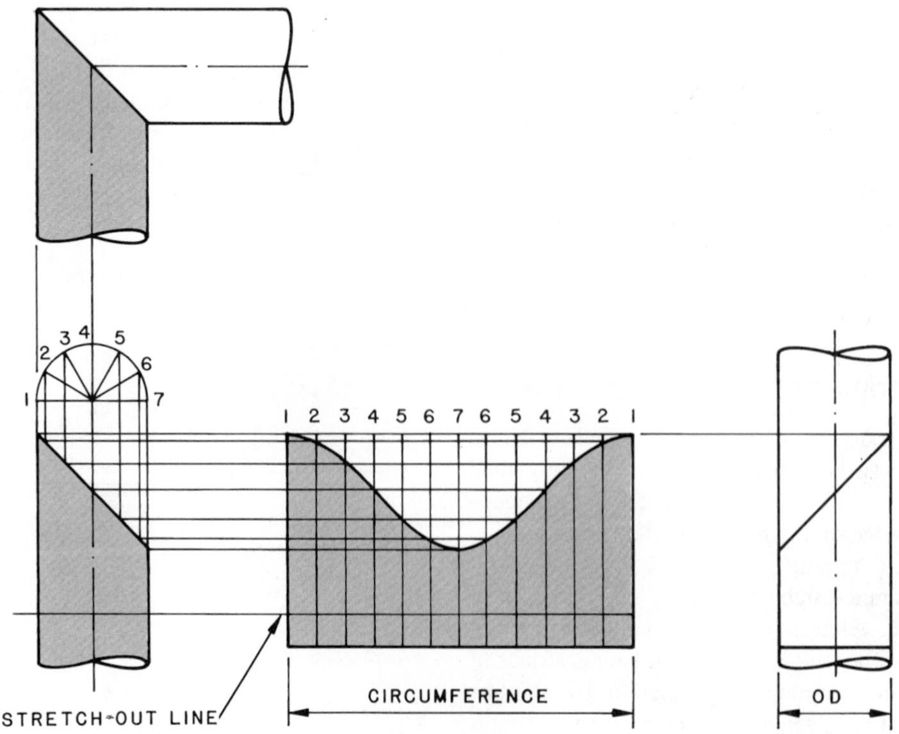

FIGURE 17.57 Development of a 90° Elbow

2. Project points 1 through 7 to the front view where they intersect the miter line.
3. Extend a stretch-out line perpendicular to the front view of the pipe (axis line) and lay off the length of the development using the calculated circumference (or set off the chord distances, 1-2, 2-3, etc.).
4. Divide the circumference into equal parts (12 here) along the stretch-out line and label as shown.
5. Project the height dimension of each element from the front view to the development.
6. Connect points on the development with a smooth curve.
7. The development can now be transferred to a pattern and cutout to use as a wrap-around template.

Figure 17.58 describes how to establish a development of two pipes intersecting at angles other than 90°. Here the pipes intersect at 45° and are the same size diameter. The following steps were used:

1. After drawing the front and side views, construct half-circle end sections and divide into equal parts as shown.
2. Project the end section divisions (points) to the front view to establish the line of intersection. When the pipes are of equal diameters, the distance from the header centerline and the lowest point must be calculated, point 4. Dimension X equals twice the pipe wall thickness of the branch pipe.
3. Draw the stretch-out lines perpendicular to the pipes and calculate their respective circumferences.

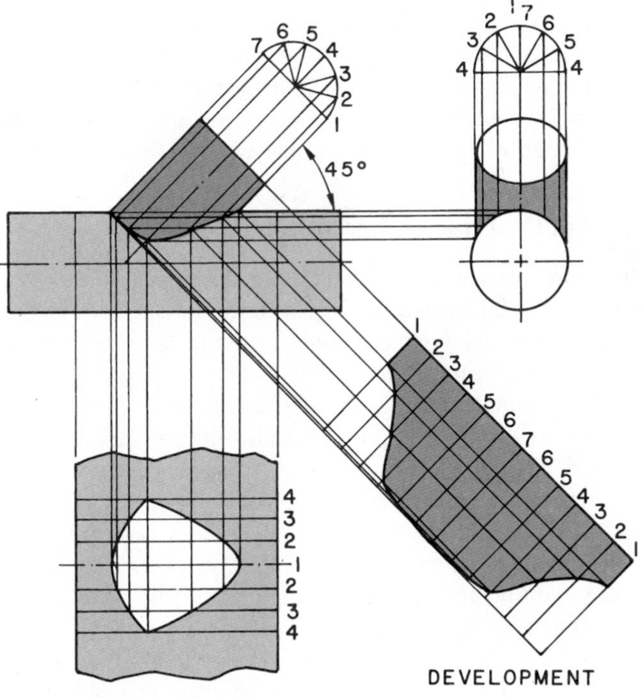

FIGURE 17.58 Pipe Lateral Pattern

4. Divide the cirumference length into 12 equal parts; project the related points to the development and connect the points with a smooth curve.

17.9.9 Development of Cones

Cones are used in the design of industrial products, airplanes, storage tanks, ducting and piping transitions, and numerous structural, architectural, and mechanical designs. *A cone is a single-curved surface generated by the movement of a straight-line generatrix fixed at one end and intersecting a curved directrix.* The fixed point is the vertex and the directrix is a closed curve (usually a circle or ellipse). Each position of the generatrix establishes an element on the surface of the cone. All elements of a cone terminate at the vertex point. The development of a cone is radial line development. The generatrix of a cone is a straight line. The three types of cones are: right circular, oblique, and open. A **right circular cone** is a cone of revolution generated by a directrix and an axis perpendicular to the base plane (directrix plane). An **oblique cone** has an axis that is not perpendicular to its base plane. Its directrix is a closed curve. An **open cone** has an open single-curved or double-curved line as a directrix.

17.9.10 Development of a Right Circular Cone

A right circular cone develops as a sector of a circle, with a radius equal to the slant height of the cone and an arc length equal to the length of the circumference of the cone. The development of a right circular cone uses one of two methods. The graphical method involves dividing the base circle of the cone into equal parts. In Figure 17.59, the base circle is radially divided into16 equal parts. An element on the cone's surface is drawn at each division, elements are of the same length. The true length of an element equals the slant height of the cone. For the development, the slant height is used as the TL radius. The distances between these divisions, chord measurements, are stepped off along the development arc, R1. This method produces a development pattern with an arc length (A) slightly smaller than a true development since the

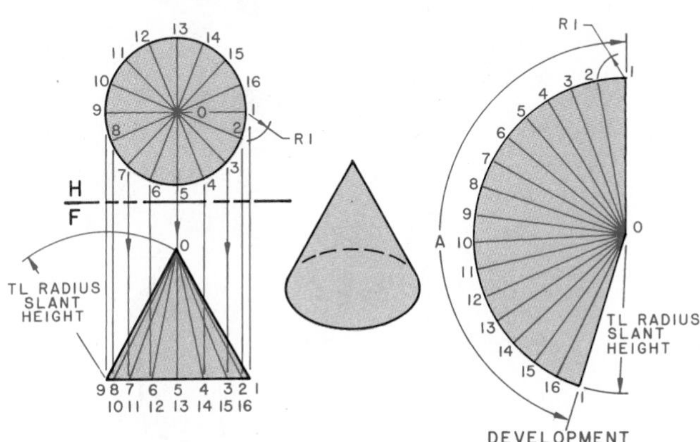

FIGURE 17.59 Development of a Right Circular Cone

chord distance between base divisions is smaller than the arc distance.

When an accurate development is required of a right circular cone, the arc angle (A) is calculated. Angle A is the sector angle of the development. The sector angle (angle A) equals the radius of the base divided by the slant height times 360°. The development is drawn using the computed sector angle to establish the length of the arc of the development.

17.10 Transition Pieces

A transition piece consists of any shape that connects two or more forms of different size. This definition would include types of developments covered under cones and pyramids.

Transition pieces are developed by **triangulation:** dividing the surface of the piece into triangles. Elements are drawn on the surface of the form to be developed and connected by diagonals if adjacent elements do not intersect. The development is laid out as a series of joined triangular areas.

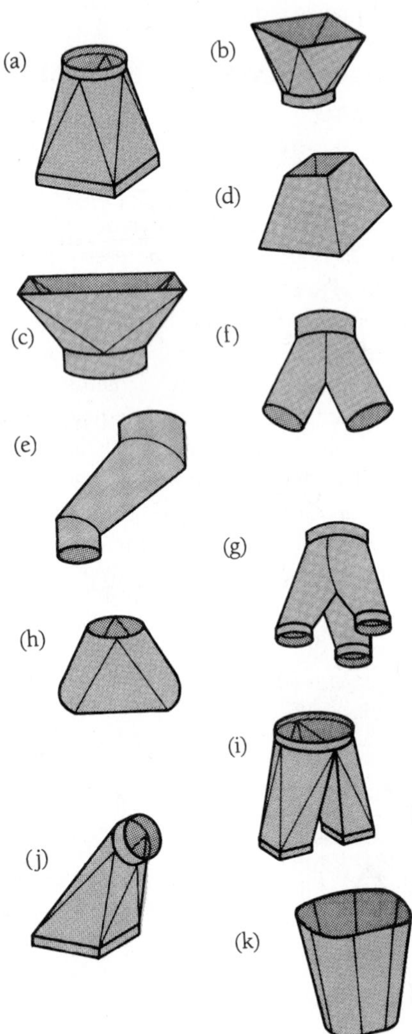

FIGURE 17.60 Examples of Transition Pieces

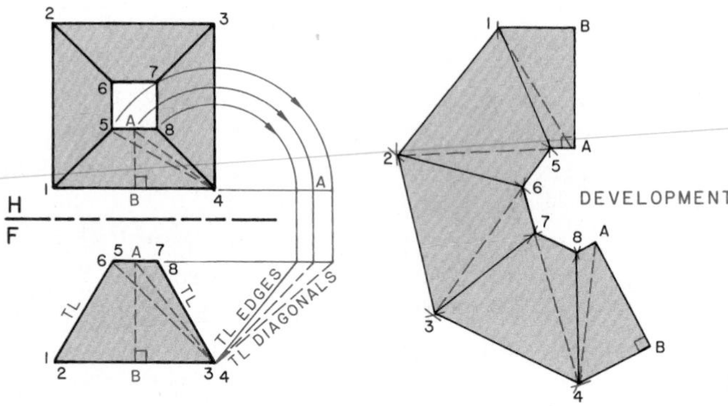

FIGURE 17.61 Triangulation

Each opening of the transition piece will be a different configuration. Transition pieces are usually formed from sheet metal or other materials and connected along a common seam. Hoppers, warped funnels, and vessel bottoms have transition pieces in their design.

In Figure 17.60, eleven possible variations of transition pieces are shown. The possibilities of shapes and sizes are limited only by one's imagination. Types (a) and (b) are symmetrical square to round transitions; (c) is a rectangle to round and is developed with the same general method as (a) and (b); (d) is a square to rectangle transition. It is composed of plane surfaces and can be accurately developed. This type of transition is really a frustrum of a right pyramid. Its given surfaces are developed by triangulation if the vertex is unavailable. The next three examples all involve the connecting of two or more circular or elliptical shapes: (e) is a conical offset connecting two separate pipes of different diameters and axes; (f) is a pipe fitting, connecting two round pipes to one pipe of a larger diameter; (g) is a three-stream transition into a single large-diameter pipe; (h), (i), (j) and (k) are specialized variations of transition pieces: round to oblong (h), two square ducts to one round (i), square to round transition at an angle (j), and a hopper-type (k).

17.10.1 Triangulation

In Figure 17.61, the transition piece is developed by means of triangulation. The square to square form developed in Figure 17.61 has similarly shaped openings, and its edges can be extended to locate a vertex. The form used here is only to provide a simple illustration of the triangulation of a surface. Each surface of the object is identical. Only one surface needs to be divided into a triangular area. A diagonal is drawn so as to divide one of the equal trapezoidal shapes into two triangular planes, 4-5. The true lengths of the edges and diagonals are established by revolution. The true lengths of the upper and lower openings appear in the horizontal view and can be transferred directly to the development.

To establish the shortest seam, divide the front surface in half. Line A-B will be used as the seam edge. This placement of the seam makes the joining method easier, quicker, and along the shortest line. This area must also be divided into

ITEMS OF INTEREST

"Let George Do It"

Have you ever heard the phrase "Let George do it"? If you have, you may have wondered about George. That phrase refers to a bit of advice from World War II that encouraged someone to avoid a burdensome or demanding task by "letting George do it." George is the name given to the automatic pilot on aircraft, which is a device that removes some of the burden from the pilot and keeps the airplane on course during a flight to a particular destination. Automatic navigation is a concept unheard of until recently. For those who once sailed the seas by stars and by plotting complex navigational charts, it must have appeared magical indeed.

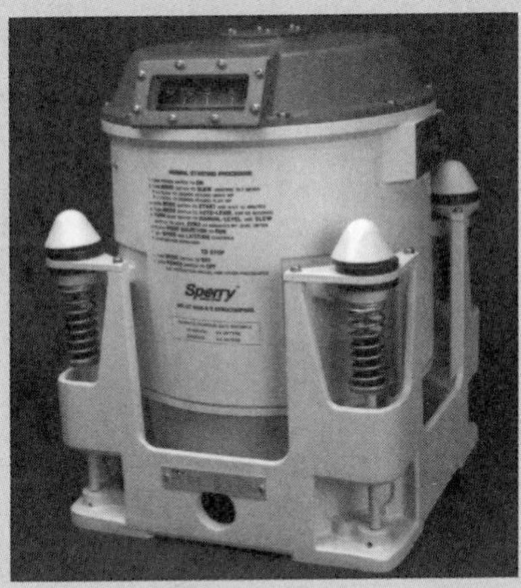

A Gyrocompass

The magic is performed by gyroscopes. Gyroscopes have been around a long time (since the early 1800s), but they weren't applied to useful purposes until the 1900s. At that time, Elmer Sperry was convinced that they had great commercial possibilities. Sperry developed the gyrocompass from the elementary gyroscope. This invention filled an immediate need by the Navy for a compass that would be unaffected by steel ships.

By World War II, 200 of the merchant fleets in the world were guided by the Sperry gyrocompass. Other devices that used the gyroscope were developed, but the most important application was "George," the turn indicator and automatic pilot.

The autopilot relieves the pilot from the physical effort of flying the plane. Today autopilots are found in almost every size aircraft. Electronic navigation and control are used extensively on all modern aircraft. Modern commercial airplanes give George much of the duties involved in navigation and control. If you tell George where you want to go, he can navigate the plane so that it stays on that course.

After you have studied descriptive geometry, you will probably be able to appreciate the relationship between points, lines, and planes in 3D. You may also be able to appreciate navigation at sea during eras before the gyrocompass. Electronic navigation and control play a large part in piloting a modern aircraft. Most pilots will be happy to tell you that life is a lot nicer now that George is around.

triangles. Draw a diagonal from point A to point 4 and establish its true length by revolution.

Start the development by drawing line A-B. Using the true lengths of the edges, diagonals, and upper and lower opening edge lines as arc lengths, complete the development as shown. Swing arcs A-4 and B-4 to locate point 4. Arcs A-8 and 4-8 intersect at point 8.

17.10.2 Development of a Transition Piece: Circular to Rectangular

A transition piece connecting a circular to rectangular geometric form is developed by dividing its surface as in Figure 17.62. The surface of the transition piece is composed of four isosceles triangles and four conical surfaces. The bases of the isosceles triangles form the lower base of the transition piece. The four conical surfaces are portions of an obilque cone. The first step in the development of a circular to rectangular transition is to divide the conical surfaces into triangular areas. In Figure 17.62, the circumference of the circular base is divided into 12 equal parts. Points 1, 4, 7, and 10 already exist as divisions since they correspond to the vertex points of the isosceles triangular areas of the piece's surface. All other points divide the conical surfaces into three separate areas. Since the transition piece is symmetrical, each of the four conical surfaces and their triangular divisions are identical. The true lengths of only one set of elements is established.

A true length diagram is constructed as shown to establish the true lengths of the four elements. The true lengths of the lower rectangular base can be found in the horizontal view as can the chord distances between divisions on the upper circular base. The seam line is established by dividing the frontal triangular surface in half. Line 1-A is used as the seam line.

Start the development by drawing line 1-A. Use the true lengths of the elements, the chord distances and the lower base lengths as arc lengths. Lay out each successive triangular surface to complete the drawing. From point 1 swing arc 1-B. Swing arc A-B from point A. Arc 1-B intersects arc A-B and locates point B. Triangle l-B-2 is laid out next. Each successive triangle is constructed so that the transition piece is unrolled clockwise and inside-up.

The rectangular to circular transition piece shown in Figure 17.63 has an angled base edge. This figure is developed with the same general method as in the previous example. The transition piece is composed of four triangular lateral surfaces whose base lines form the lower base edge of the figure. The corners of the piece are portions of oblique cones. The development is constructed by dividing the surface into triangular areas that approximate the surface of the piece. Each triangle is then laid out in successive order with common elements joined.

The circumference of the upper base circle is divided into equal parts. Elements that define triangular areas on the

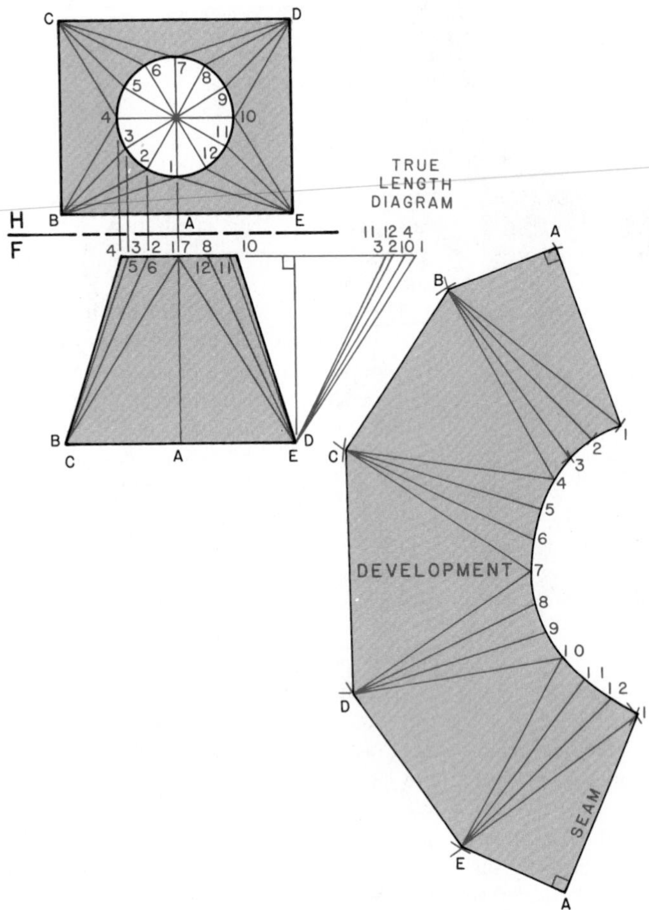

FIGURE 17.62 **Transition Piece Development: Circular to Rectangular**

conical surface are drawn through the division points and connected to one of the lower base corners. The elements correspond to bend lines when the piece is formed by rolling a flat piece of sheet metal that was cut to the outline of the pattern. Since the lower base is at an angle and the circular base is not centered left to right, as was Figure 17.62, eight separate element lengths need to be established before the development can be started. To avoid confusion, two true length diagrams are drawn as shown. Revolution could have been used to determine the true lengths of the elements.

QUIZ

True or False

1. Perpendicular lines show in any view where one or both of the lines is true length.
2. Parallelism of two lines can always be established with only two views.
3. The bearing of a line is measured from the north toward the east or west.

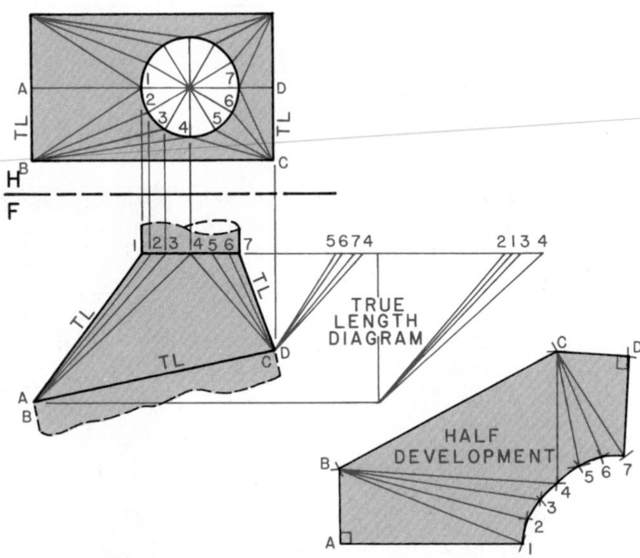

FIGURE 17.63 **Transition Piece Development**

4. Parallel planes can be determined in a view where both planes project as edges.
5. Oblique lines are never true length in a principal view.
6. A cone is generated by a straight-line element passed through the axis and at a specific distance from the vertex.
7. A cutting plane can be introduced at any angle and in any view.
8. A pattern is composed of true length lines.

Fill in the Blanks

9. Two oblique lines that appear ________ in two or more views will always be ________ .
10. The angle between two intersecting lines can be measured in any view where the ________ both appear ________ ________ .
11. Two lines on the same plane must be either ________ or ________ .
12. To establish the angle between a line and a plane the plane must appear as an ________ ________ and the line ________ ________ .
13. To establish the point of intersection point between a line and a plane, project a view where the plane is shown as an ________ ________ .
14. A right cone is generated by revolving a ________ about one of its legs.
15. Most developments should be unfolded with the ________ ________ .
16. A cone is generated by a ________ ________ generatrix.

Answer the Following

17. How many views are necessary to fix the position of a point or line in space?
18. Define vertical line. In what views will it appear vertical?
19. What is the bearing of a line?
20. How do you solve for the edge view of a plane?
21. Explain how to solve for the piercing point of a line and a plane.
22. Explain how to check for perpendicularity of two lines.
23. Describe specific engineering applications for intersecting shapes.
24. What is a transition piece?

EXERCISES

Exercises may be assigned as sketching, instrument, or CAD projects. Transfer the given information to an "A" size sheet of .25 in. grid paper. Complete all views and solve for proper visibility, including centerlines, object lines, and hidden lines. Exercises that are not assigned by the instructor can be sketched in the text to provide practice and understanding of the preceding instructional material.

After Reading the Chapter Through Section 17.4.8 You May Complete the Following Exercises

Exercise 17.1(A) Locate the three points in all views.

Exercise 17.1(B) Locate the following three points in the given views. Point 1 is seven units below point 2. Point 2 is two units behind point 1. Point 3 is three units to the left of point 2. Point 1 is given in the H view, point 2 is given in the F view, and point 3 is given in the P view.

Exercise 17.1(C) Locate points 1 and 2 in all four views. Point 1 is given. Point 2 is .25 in. (6 mm) in front of, .75 in. (20 mm) to the right of, and 1.25 in. (32 mm) below point 1.

Exercise 17.1(D) Locate the following points. Point 1 is four units behind the frontal plane, nine units to the left of the profile plane, and twelve units below the horizontal plane. Point 2 is three units behind the frontal plane, seven units below the horizontal plane, and seven units to the right of point 1. What is the distance between the two points in the front view?

Exercise 17.2(A) Draw the three views of the profile line.

Exercise 17.2(B) Complete the three views of the profile lines.

Exercise 17.2(C) Locate the given line in the required auxiliary views.

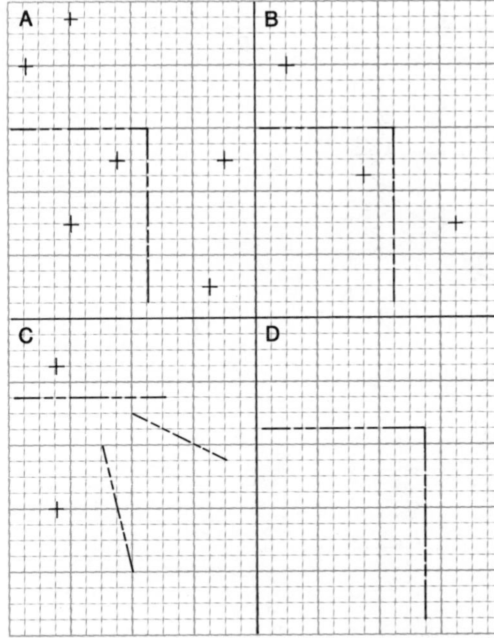

Exercise 17.1

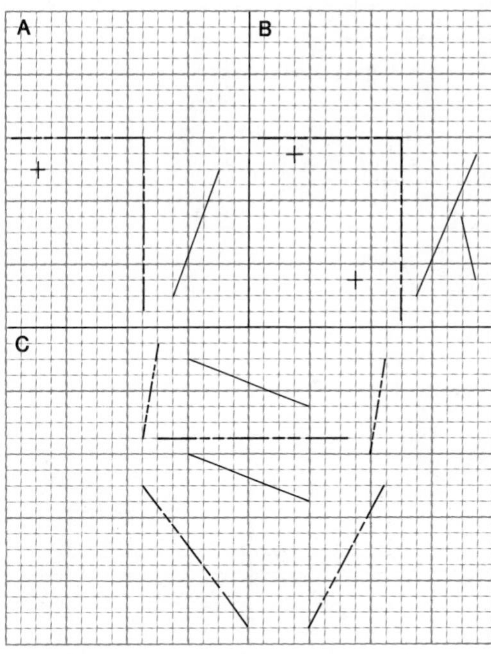

Exercise 17.2

Exercises 17.3(A) and (B) Complete the three views of the given lines. Label lines where they appear as principal lines, and note if a line is oblique, inclined, true length, or parallel with a projection plane. Show all possible solutions.

Exercise 17.3(C) Solve for the true length of the line and the point view. Note the bearing of the line. Point 1 is above point 2.

Exercises 17.4(A) and (B) Complete the views of the pipes and solve for visibility. Shade the pipe that is visible in each view.

Exercise 17.4(C) Complete the required views of the parallel lines.

Exercise 17.4(D) Complete the views of the lines. Are they parallel?

After Reading the Chapter Through Section 17.5.4, You May Complete the Following Exercises

Exercise 17.5(A) Complete the views of the two lines. Line 3-4 shows as a point view in the horizontal view. Are they perpendicular? Note all TL lines.

Exercise 17.5(B) Project the three views of the intersecting perpendicular lines.

Exercise 17.5(C) Construct a line through the point, perpendicular to and on the given line.

Exercise 17.5(D) Draw a line through the point and perpendicular to the line in the horizontal view.

Exercise 17.6(A) Draw the given line in each view and locate the points on the line in each projection.

Exercise 17.6(B) Locate point 3 which is three units to the right of point 1 and lies on the line. Point 4 is eight units below point 1 and on line 1-2. Note point 1 is above point 2.

Exercise 17.6(C) Solve for the true length distance between the line and the point. Show the connector and the point in each view.

Exercise 17.6(D) Find the shortest (perpendicular) distance between the two lines. Project the line back into all views.

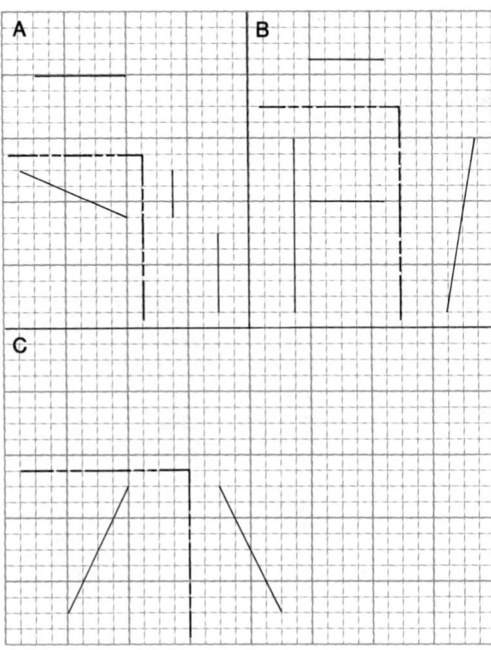

Exercise 17.3

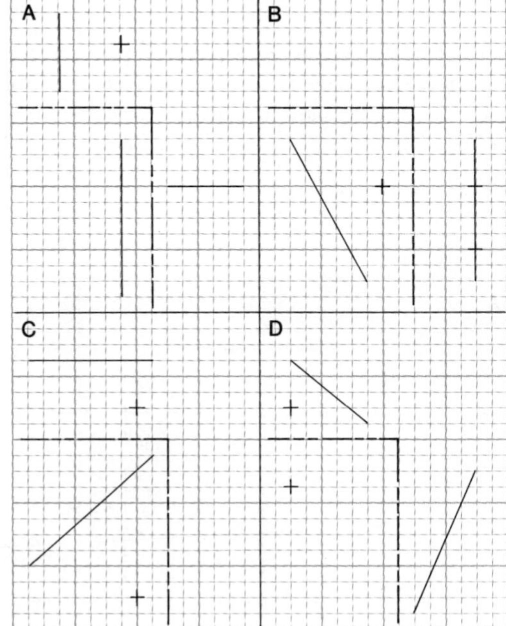

Exercise 17.5

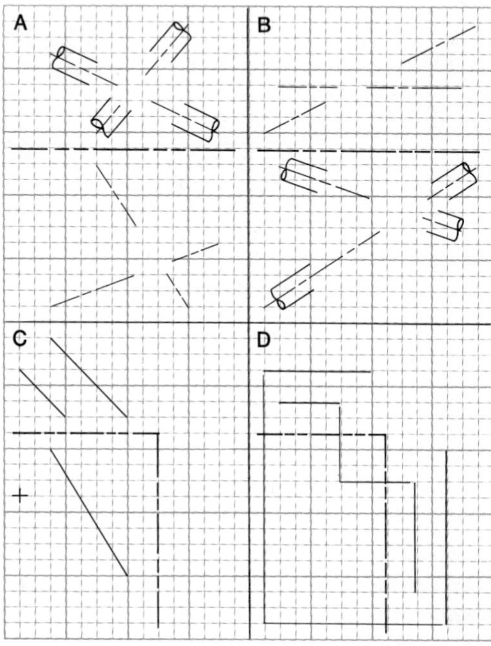

Exercise 17.4

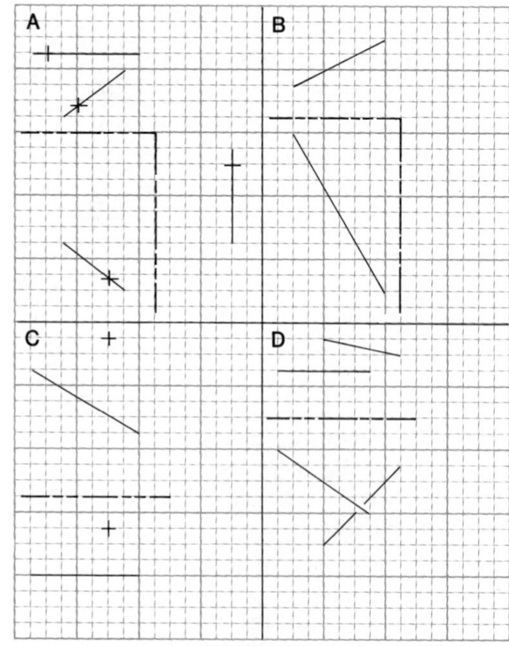

Exercise 17.6

Exercise 17.7 Solve for the angle between the nonintersecting lines. Show the shortest connector in all views.

Exercises 17.8(A) through (D) Complete the views of the planes in each problem. Label for the type of plane: vertical, inclined, oblique, and principal.

After Reading the Chapter Through Section 17.6.2, You May Complete the Following Exercises

Exercise 17.9(A) Project the front view of the plane. Draw three evenly spaced frontal lines on the plane and show in all views.

Exercise 17.9(B) Establish a profile, horizontal, and frontal line on the plane.

Exercise 17.9(C) Show the points and lines on the plane and project in all views.

Exercise 17.9(D) Solve for the EV of the plane.

Exercise 17.10(A) Solve for the largest circle within the plane. Establish an EV view from the horizontal view.

Exercise 17.10(B) Solve for the edge view and true shape of the plane. Project the EV from the frontal view.

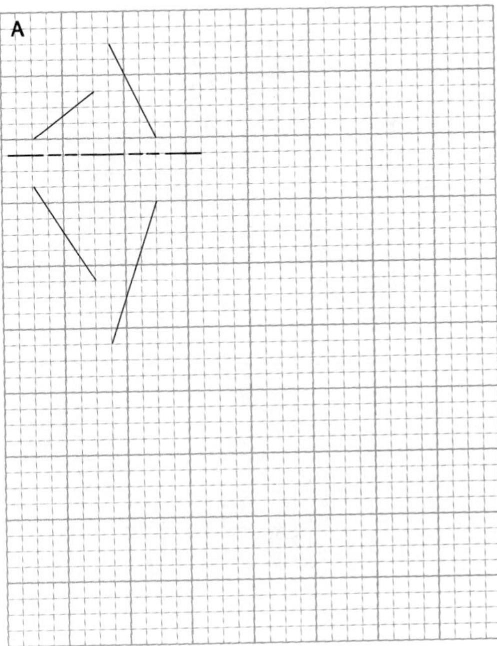

Exercise 17.7

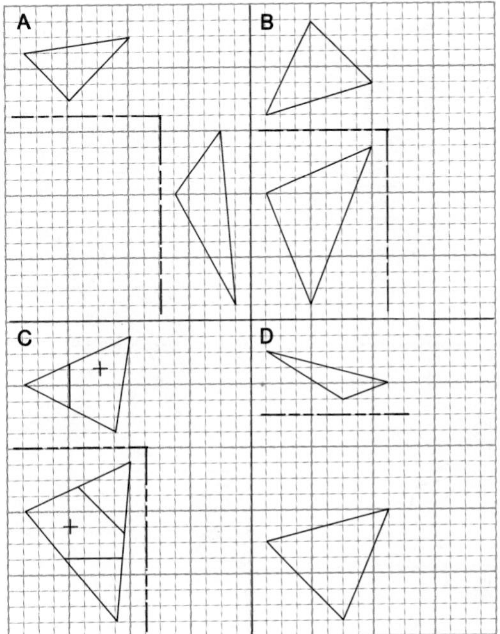

Exercise 17.9

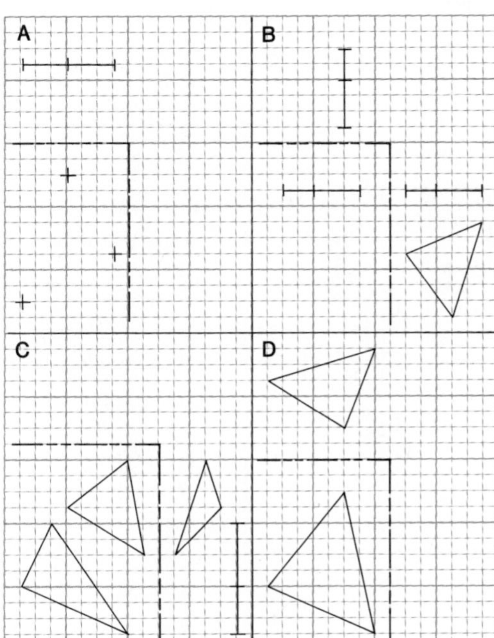

Exercise 17.8

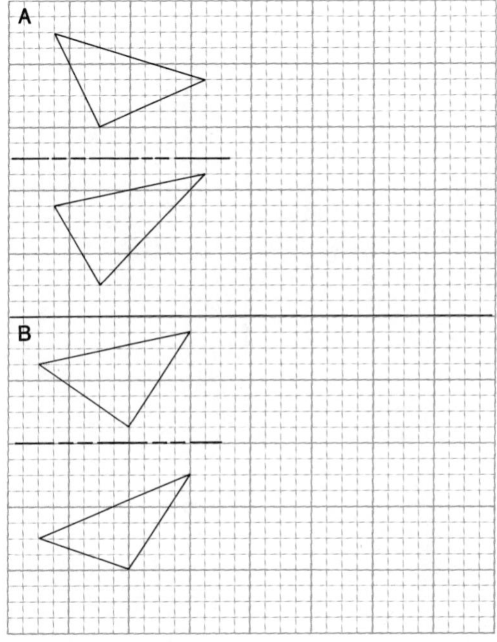

Exercise 17.10

Exercise 17.11(A) Determine the shortest distance between the point and the plane. Show the connecting line in all views. What is the bearing of the line? North is always at the top of the drawing.

Exercise 17.11(B) Construct a plane parallel to the given plane and through the point.

Exercises 17.11(C) and (D) Project and measure the true angle between the two connected planes.

Exercises 17.12(A) and (C) Using the edge view method, determine the piercing point and show in all views along with the proper visibility.

Exercise 17.12(B) Using the cutting plane method, solve for the piercing point of the line and the plane. Complete all views and show the proper visibility.

After Reading the Chapter Through Section 17.10.2, You May Complete the Following Exercises

Exercise 17.13(A) Rotate the point 200° clockwise around the frontal line. Show in all views.

Exercise 17.13(B) Rotate the point around the line 100° counterclockwise.

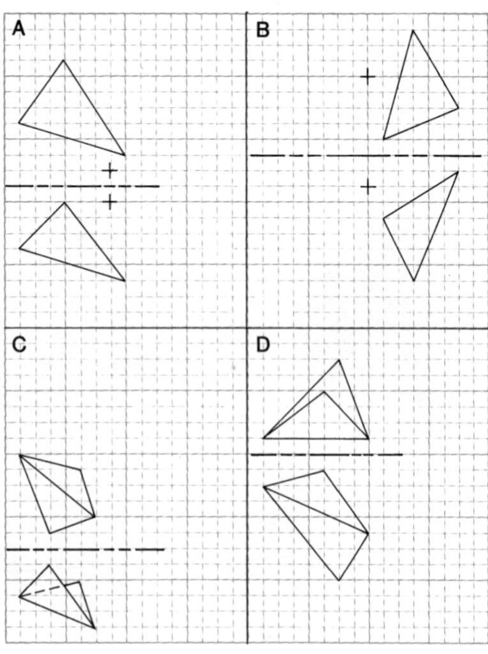

Exercise 17.11

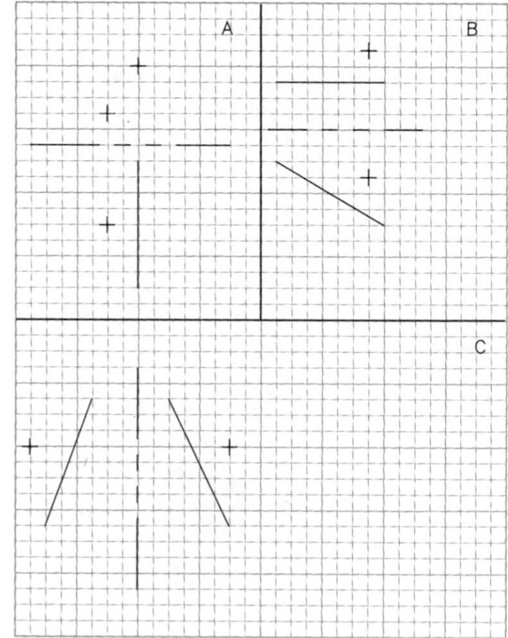

Exercise 17.13

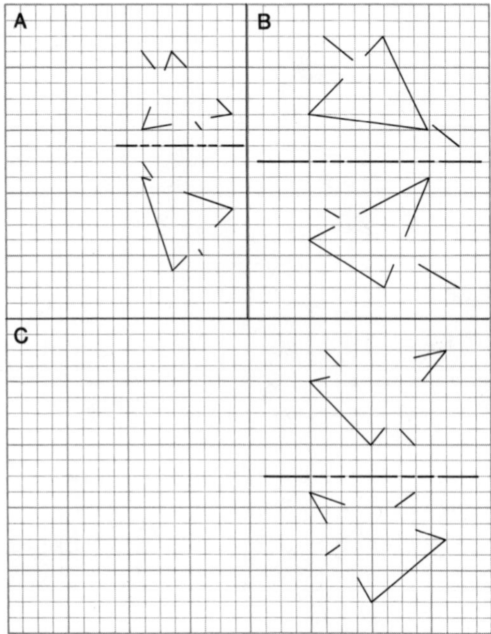

Exercise 17.12

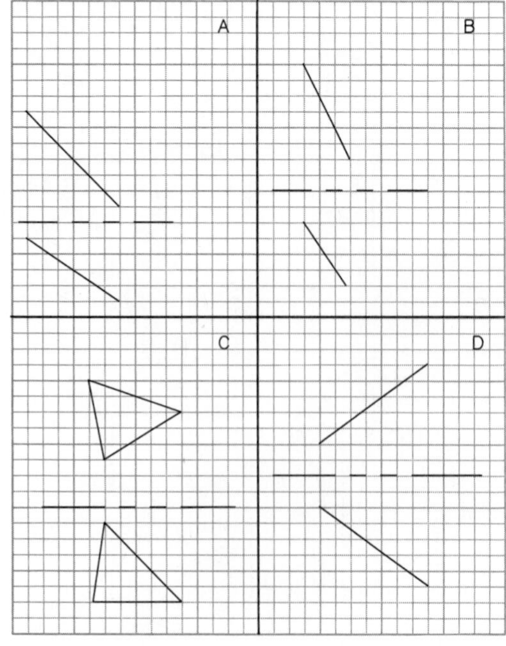

Exercise 17.14

Exercise 17.13(C) Project an auxiliary view to establish a true length of the given line. Then solve for a point view of the line and rotate the given point 180°. Show in all views.

Exercise 17.14(A) Using revolution, determine the true length of the line and the angle the line makes with the horizontal plane. Verify by projection an auxiliary view.

Exercise 17.14(B) Determine the angle between the line and the profile plane.

Exercise 17.14(C) Using revolution, project and measure the true lengths of the sides of the given figure.

Exercise 17.14(D) Solve for the angle that the line makes with the F view and its true length.

Exercises 17.15 (A), (B), and (C) Solve for the intersection of the two planes.

Exercise 17.16(A) Solve for the intersection of the plane and prism.

Exercise 17.16(B) Determine the intersection between the oblique plane and oblique prism.

Exercise 17.17 Complete the two views of the intersection plane and cylinder.

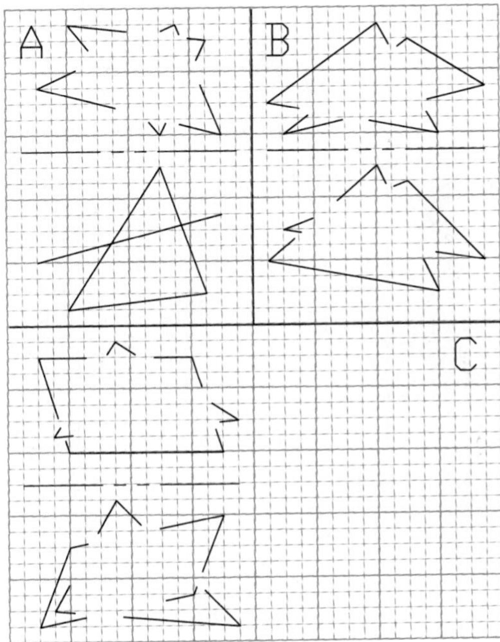

Exercise 17.15

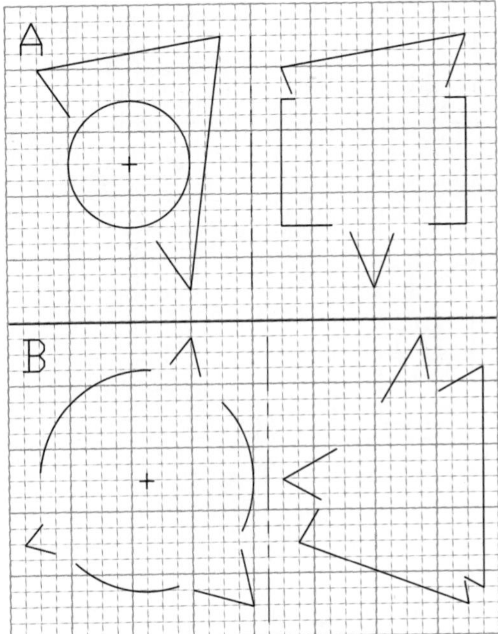

Exercise 17.17

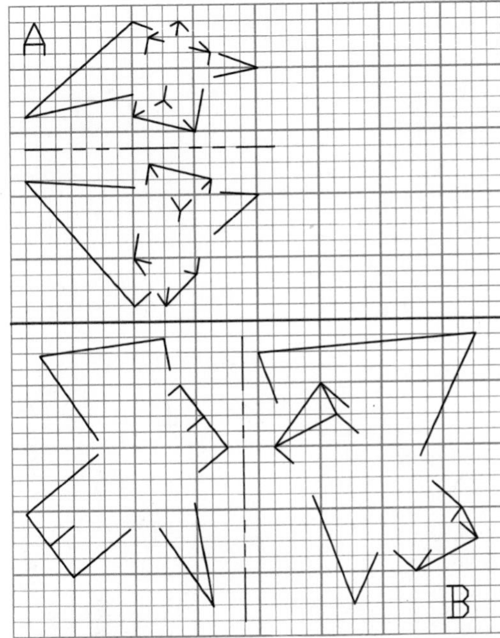

Exercise 17.16

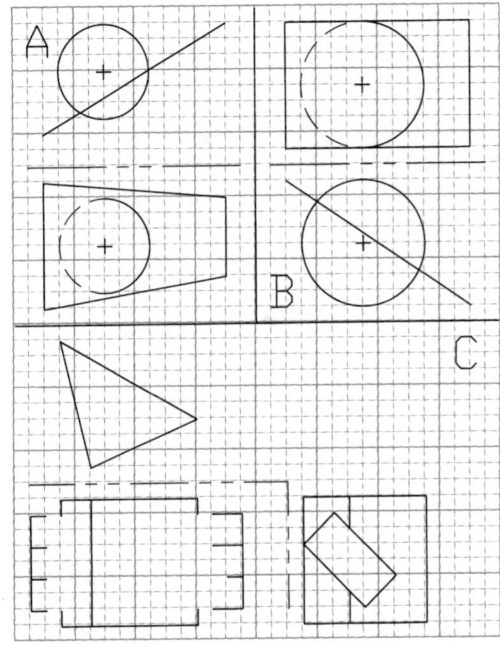

Exercise 17.18

Exercise 17.18 Complete the views and solve for the intersection of the two prisms.

Exercises 17.19 (A) and (B) Develop the inside pattern of each prism. Use the given element to start the roll out.

Exercise 17.20 Develop the inside pattern of each pyramid.

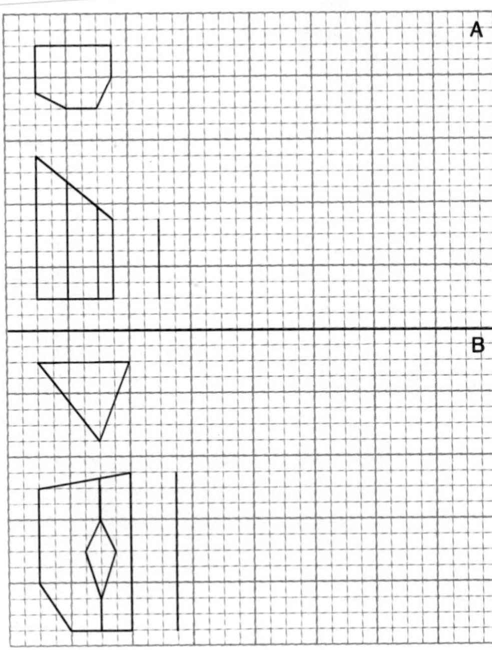

Exercise 17.19

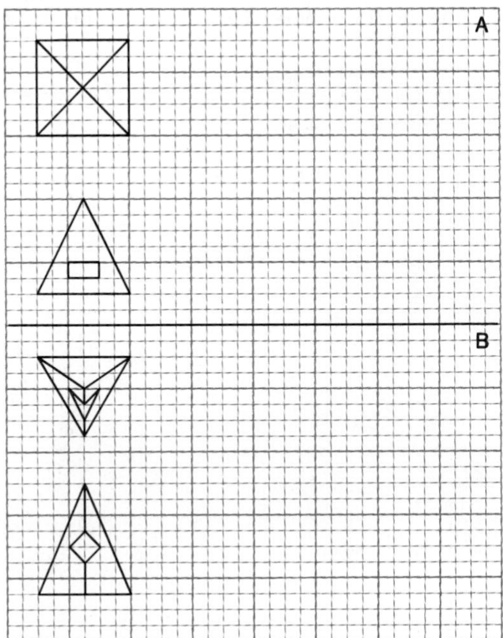

Exercise 17.20

PROBLEMS

Problems may be assigned as sketching, instrument, or CAD projects. For Problems 17.1 and 17.2, transfer the given problem to a separate "A" size sheet. Complete all views and solve for proper visibility, including centerlines, object lines, and hidden lines. Use dividers to transfer the positions of the points and lines. Note that a 1.5 to 1 scale is suggested but a 2 to 1 scale could also be used. Two problems can be put on each sheet. Make a rough trial sketch of the problem and the solution before finalizing its position on drafting paper. This will allow you to avoid placing the problem without enough work space to complete the project.

Problem 17.1(A) Project the profile view of point 1. Locate a point 2 that is .75 in. (1.9 cm) in front of, .50 in. (1.27 cm) to the right of, and 1 in. (2.54 cm) below point 1. Show point 2 in all views.

Problem 17.1(B) Project three views of the line 1-2. Point 2 is .75 in. (1.9 cm) in front of, 1 in. (2.54 cm) to the right of, and 1.25 in. (3.17 cm) below point 1. If there is a true length projection of line 1-2, label it TL.

Problem 17.1(C) Complete the three views of line 2-3 and project an auxiliary view showing the line as true length. Take the auxiliary projection from the frontal view.

Problem 17.1(D) Complete the profile view of line 5-6. Solve for the true length of the line in two separate auxiliary projections and label as TL.

Problem 17.1(E) Solve for the correct visibility of the pipes. Note that the fold line is not shown.

Problem 17.1(F) Project the missing view of the two horizontal lines. Are they parallel? Label any true length projections.

Problem 17.1(G) Construct line 1-2 perpendicular to line 3-4. Point 2 will lie on line 3-4. Project the profile view.

Problem 17.1(H) Construct line 3-4 perpendicular to line 1-2. Point 4 will be at the midpoint of line 1-2.

Problem 17.1(I) Project the shortest connector, line 3-4, between the two skewed lines. Point 4 is to be on line 7-8. Show line 3-4 in the H and F views.

Problem 17.2(A) Project three views of profile line 1-2. Then draw three views of a 1 in. (2.54 cm) line 3-4. The two lines must intersect at their midpoints with line 3-4 appearing as a point in the profile view.

Problem 17.2(B) Project an auxiliary view of line 1-2 so that its projection will be seen in true length. Label TL. Note the bearing of the line.

Problem 17.2(C) Project an auxiliary view of plane 1-2-3 where edge line 1-2 will appear as true length. Label all true length edges TL.

Problem 17.2(D) Determine the bearing and length of the line. The line is assumed to start from point 1. North is at the top of the page. The scale is 1 in. = 50 ft. (or 1 cm = 30 m). List the answers on the plate as below:

a. bearing =

b. length =

Problem 17.2(E) Project the true size of plane 1-2-3 and label it TS.

Problem 17.2(F) Solve for the true shape of the plane. What is the angle between plane 1-2-3 and the F plane?

Problem 17.2(G) Determine the dihedral angle between the two connected planes.

Problem 17.2(H) Solve for the angle between line 1-2 and plane 3-4-5. Does the line pierce the plane? Show the piercing point and the proper visibility in each view.

Problem 17.2(I) Solve for the intersection of the line and the plane.

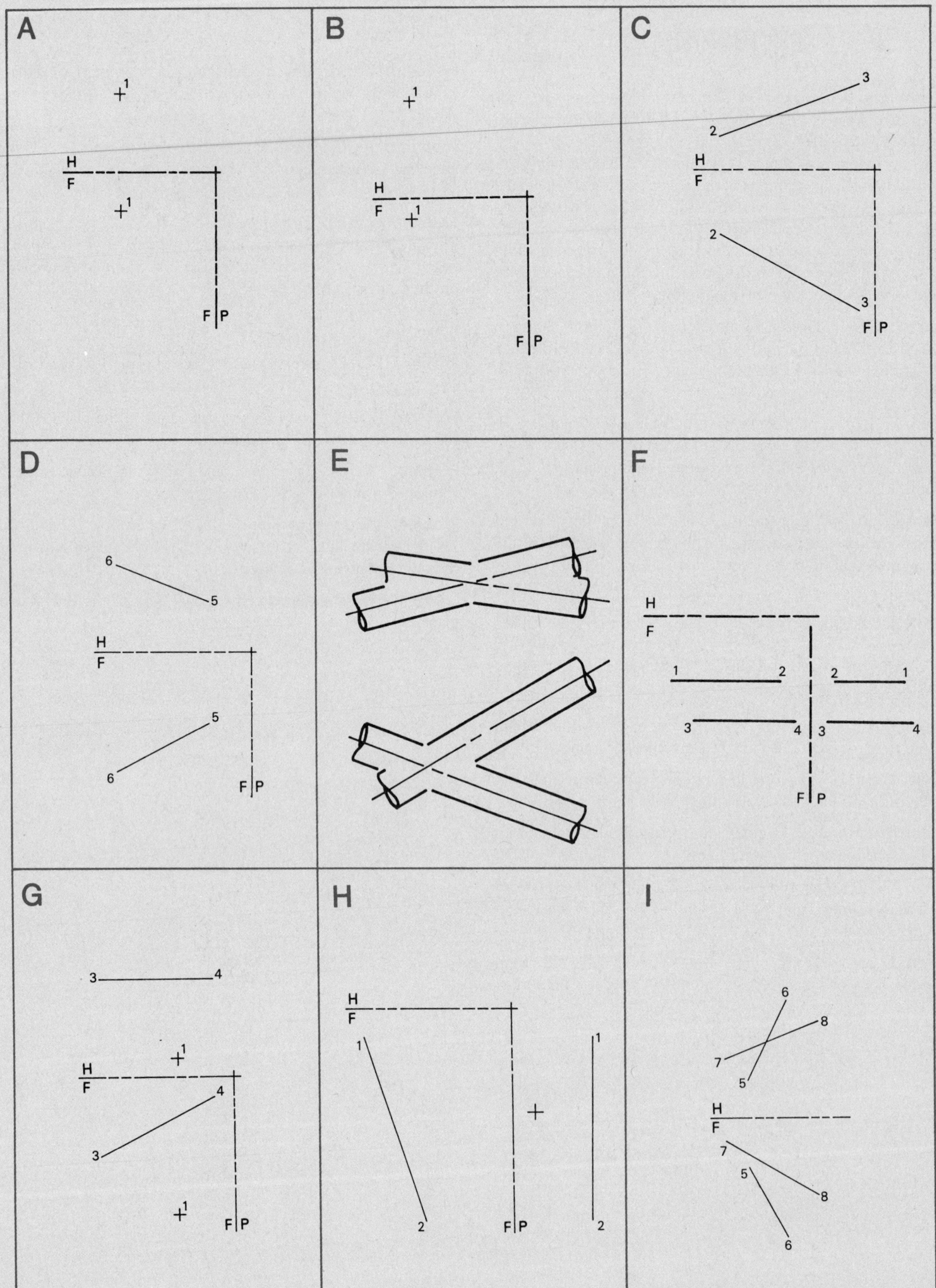
A
1
H
F
1
F P
B
1
H
F
1
F P
C
3
2
H
F
2
3
F P
D
6
5
H
F
5
6
F P
E
F
H
F
1
2
2
1
3
4
3
4
F P
G
3
4
1
H
F
4
3
1
F P
H
H
F
1
1
2
F P
2
I
6
8
7
5
H
F
7
5
8
6

Problem 17.1

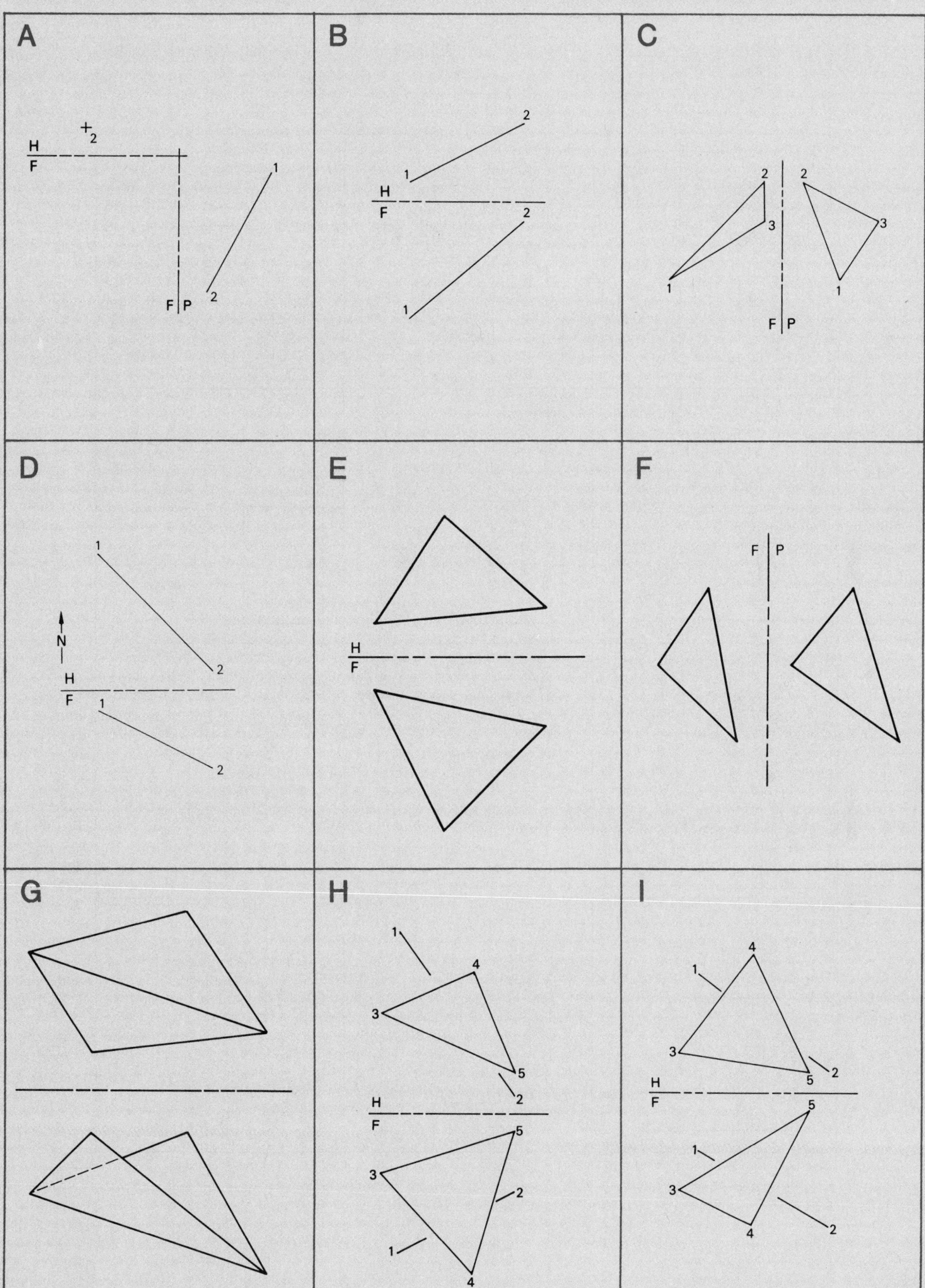
A
H
F
2
1
2
F P
B
2
1
H
F
2
1
C
2
2
3
3
1
1
F P
D
1
N
2
H
F
1
2
E
H
F
F
F P
G
H
1
4
3
5
2
H
F
5
3
2
1
4
I
4
1
3
5
2
H
F
5
1
3
4
2

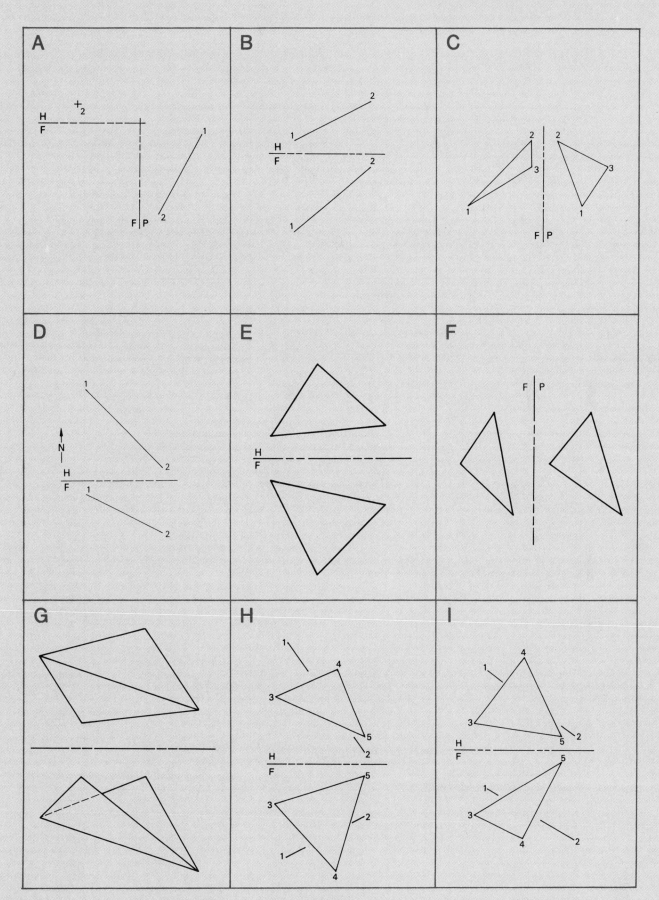

Problem 17.2

Problems 17.3 (A) through (L) Transfer the problem to another sheet. Most problems will fit on an "A" size drawing format. Complete the views of each intersection project. Add any view needed to complete the project. Use the edge view or the cutting plane method. Models of any of the problems can also be assigned.

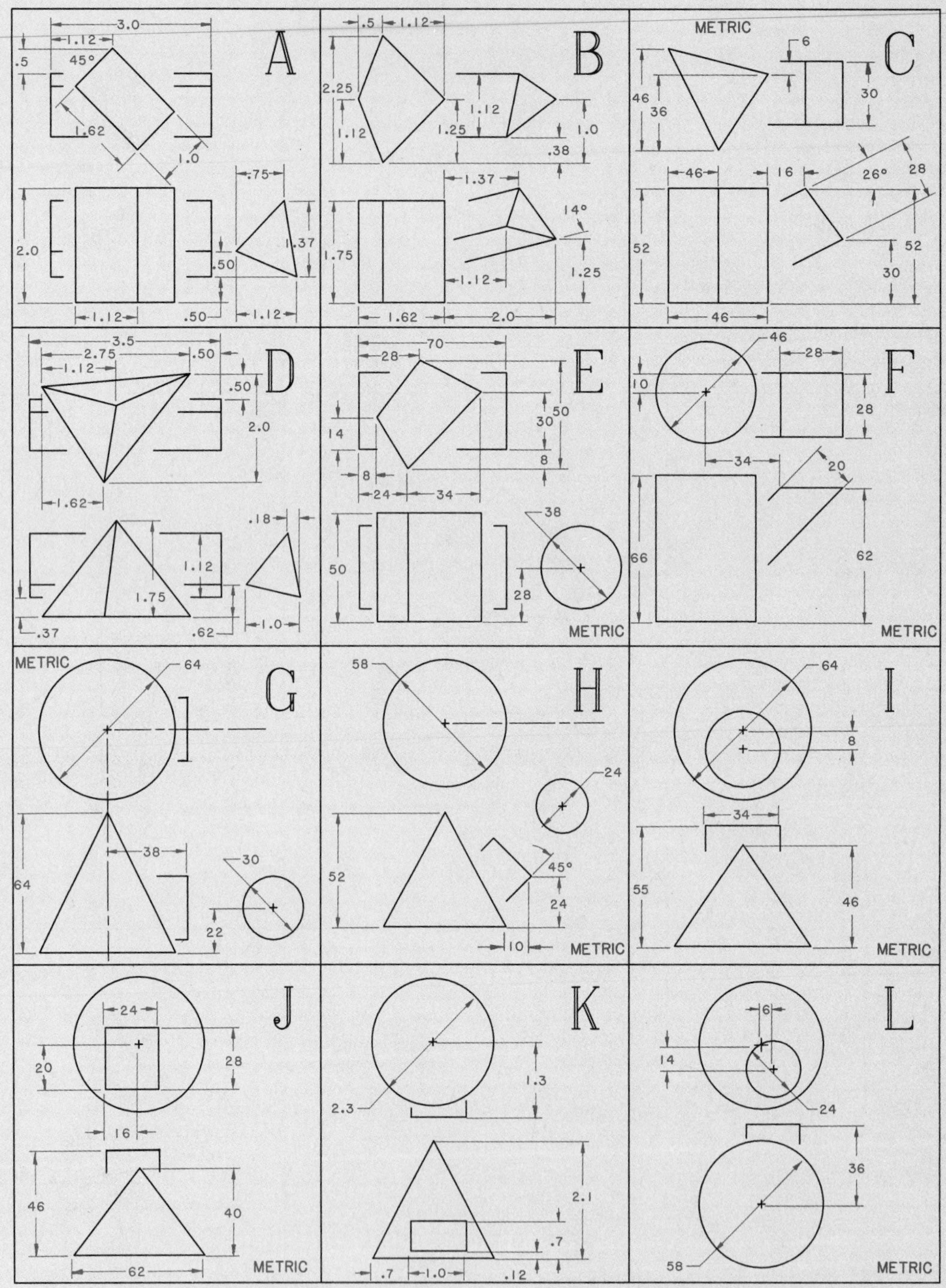

Problem 17.4 Transfer each problem to another sheet. All problems will fit on an "B" size sheet. Develop each part as an inside-up pattern. Instructor may assign some projects as half patterns. Models of any of the problems can also be completed as assigned.

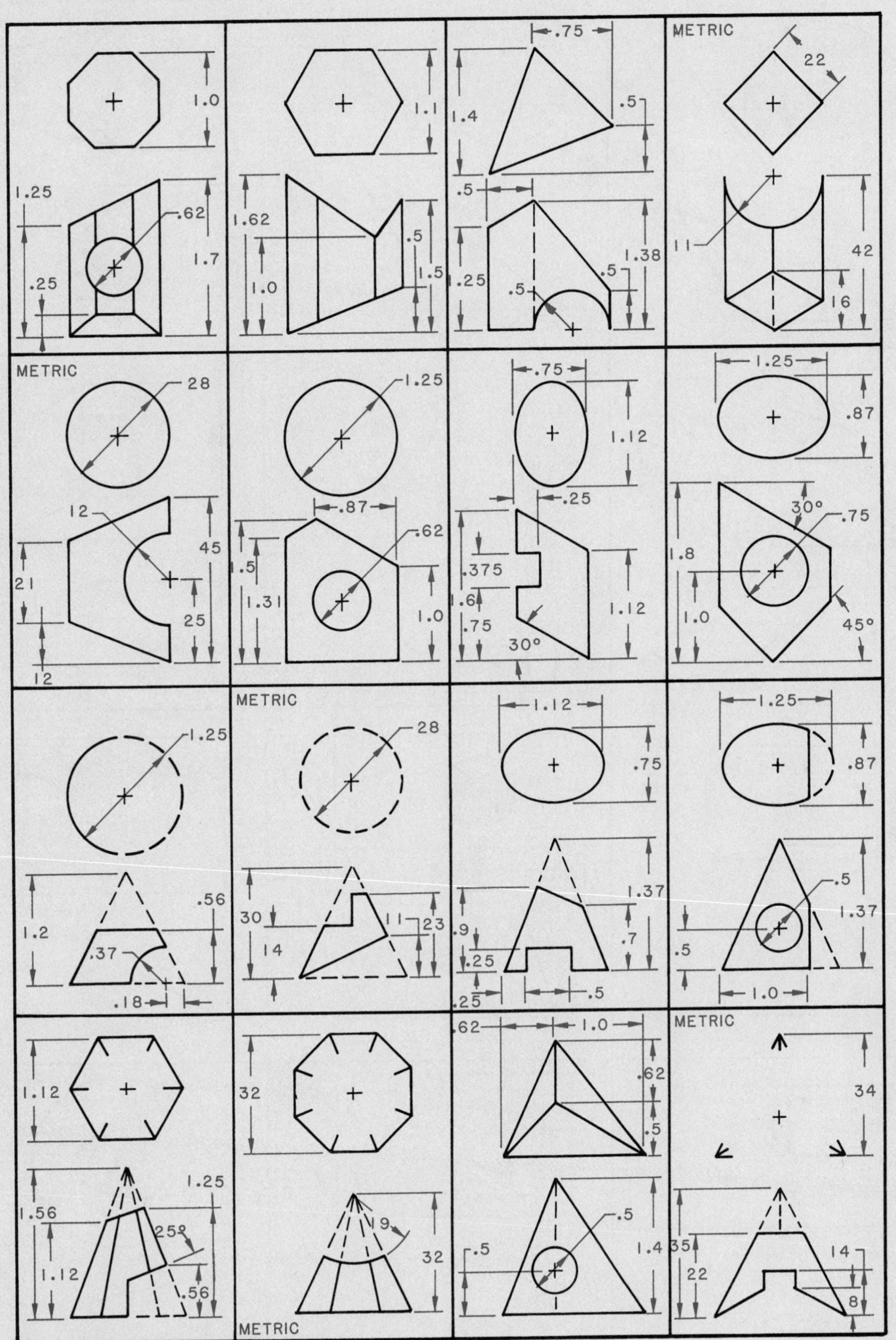

Problem 17.5 Transfer each problem to another sheet. All problems will fit on an "B" size sheet. Develop each part as an inside-up pattern. Instructor may assign some projects as half patterns. Models of any of the problems can also be completed as assigned.

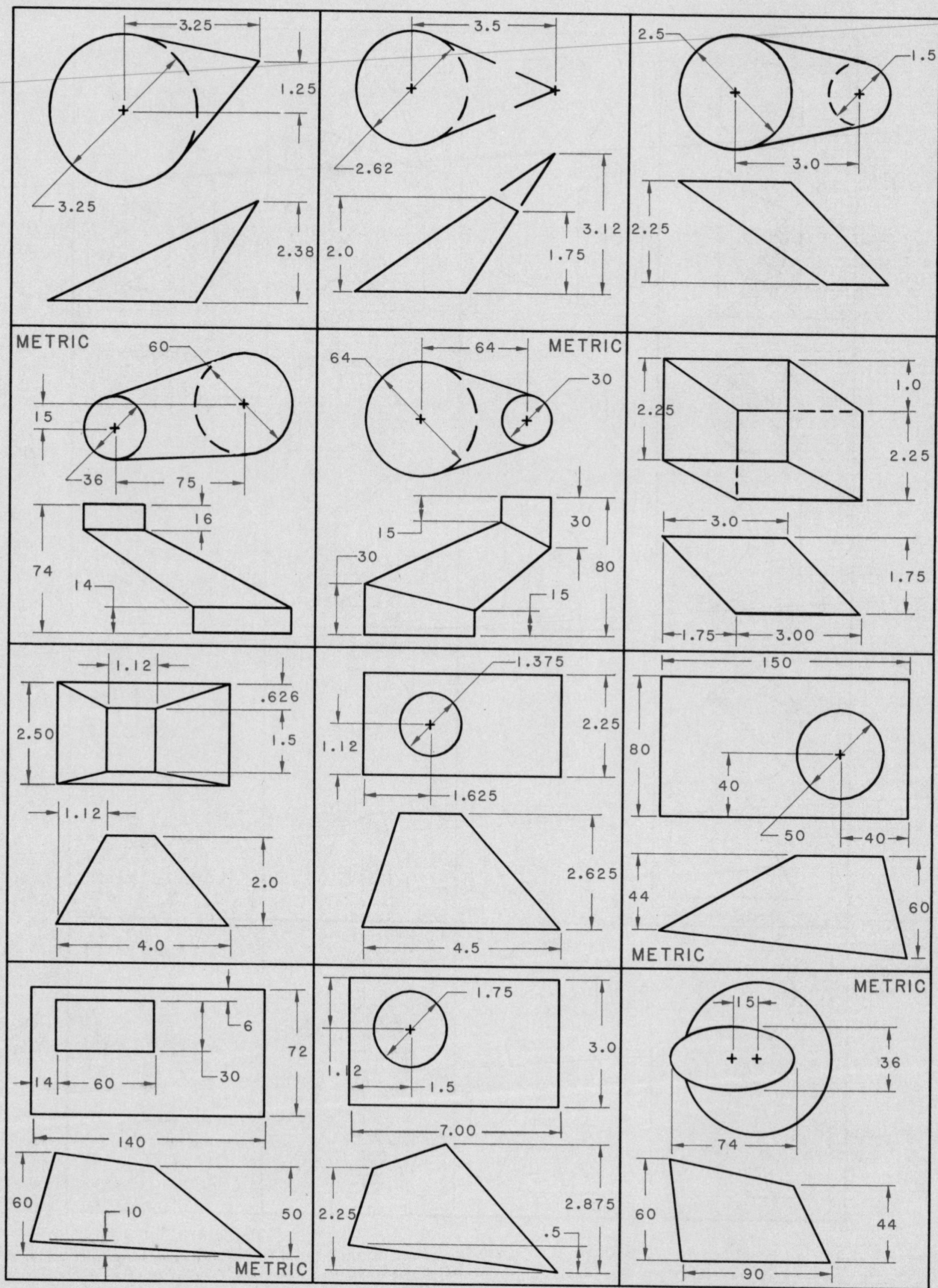

CHAPTER 18

A Guide to Understanding Symbols on Engineering Drawings

Learning Objectives

Upon completion of this chapter, you will be able to accomplish the following:

1. Develop an understanding of welding methods and processes.
2. Develop a basic understanding of the use of welding symbols to represent welds.
3. Develop a basic understanding of piping drafting symbology.
4. Understand basic piping diagrams.
5. Identify types of electronic diagrams and basic electronic symbols.
6. Identify the basic electronic components.
7. Develop a basic understanding of fluid power components.
8. Understand basic fluid power diagrams

18.1 Introduction to Symbols on Drawings

Symbols are used on engineering drawings as a quick way to communicate specific information about the part or process to others involved in the project. For example, a designer can tell the welder in manufacturing exactly how and where to weld two parts together by simply placing the symbol that correctly represents the welding steps in the correct location on the drawing. An entire process may be described graphically by using symbols. The fluid power (hydraulic) system that runs a forklift can be described in an engineering drawing by using fluid power symbols to represent fluid power components and connections. This chapter will help you understand the symbols that are used on engineering drawings in the following areas: welding, piping, electronics, fluid power, and structural steel applications.

18.2 Introduction to Welding

Welding is the process by which two pieces of metal are fused together along a line or a surface between them or at a certain point. Welds are used to fasten an assembly together permanently. The parts to be fastened can be the same type of metal or metals of dissimilar types. Welding is used on assemblies that do not require disassembly for service or maintenance and when only one or a small number of assemblies are required. Since the heat used during the welding process will distort the workpiece, any machining that is required on a welded assembly must be done after the welding process.

18.2.1 Welding Methods

Welding is classified by process or source of energy. With **nonpressure welding** (fusion and brazing), no mechanical pressure is applied. The pieces of metal are welded at the point of contact by the heat created by an electric arc or gas flame. **Pressure welding** (forging) or resistance welding is used to form a joint by the passage of electrical current through the area of the joint as mechanical pressure is applied. It is usually convenient to classify welds into three separate categories: resistance welding, gas welding, and arc welding.

Resistance welding is the process by which heat and pressure are applied at the same time, usually by a machine. Two or more parts can be welded by passing an electric current through the work as pressure is applied. Electronic beam, laser and ultrasonic methods are also used. The main types of resistance welds are spot, seam, projection, flash, and upset.

Heat is created by the combustion of a gas and air or pure oxygen in **fusion welding** (better known as **gas** or **oxyfuel welding**). In oxyacetylene welding, a flame is produced by the combustion of oxygen and acetylene gases. This type of welding is used less frequently today.

Arc welding, the most common method, includes submerged arc welding, shielded arc welding, gas-metal arc welding, and gas-tungsten arc welding. In **submerged arc welding**, coalescence is produced by the heating caused by an electric arc that is generated between the electrode and the work. The work is shielded by a blanket of granular, fusible material called **flux.** The flux protects the weld pool (floats on it). Leftover flux creates slag that must be removed at the end of the process. The filler material is obtained from a supplementary welding rod or from the electrode itself. In this process, loose flux (also called melt or welding composition) is placed over the joint to be welded. After the arc is established, the flux melts to form a shield that coats the molten metal (Fig. 18.1). A bare wire electrode is used in this process instead of a coated electrode, and the flux is supplied separately.

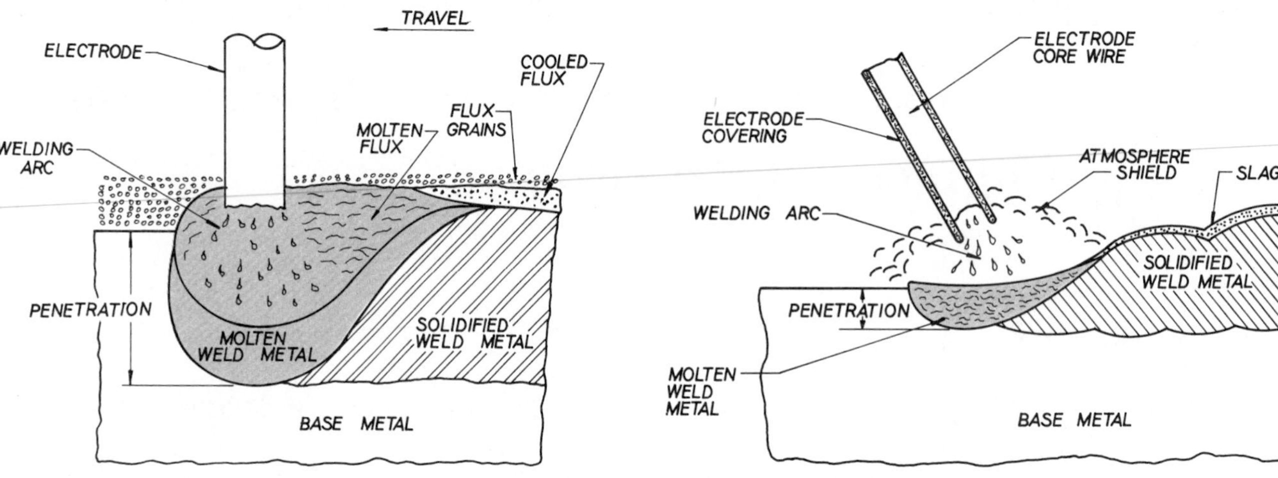

FIGURE 18.1 Submerged Arc Welding

FIGURE 18.2 Manual Shielded Metal Arc Welding

In **shielded arc welding**, the electric arc is produced by the passage of current from a coated metal electrode and the material to be welded. A gap exists between the electrode and the workpiece. Fusion takes place by the intermingling of the molten metal. Figure 18.2 illustrates manual shielded arc welding. Notice how slag is formed on top of the base metal or on top of the solidified weld metal. Slag must be removed after the welding process is completed. The penetration is much deeper than in shielded arc welding. It is also used for **tack welding**, which is holding parts in position prior to final welding.

In **gas-metal arc welding** (GMAW), heat is created electrically, but the shielding is accomplished by a blanket of gas (Fig. 18.3). The term *MIG* is normally used when referring to this process. Pressure may or may not be used, but welding is generally pressureless. The electrode is the filler metal, becoming an integral part of the weld. The filler metal may also be added to the welding zone prior to welding. In this process, inert gases are fed into the welding area to form a blanket. This welding procedure is used for magnesium, aluminum, and carbon steel.

Gas-tungsten arc welding (GTAW) is different than GMAW because the electrode in GTAW is tungsten. The term *TIG* is used when referring to this process. The electrode is used to transmit electric current and is not a filler metal. The TIG process produces root beads of high quality and is seldom used for the entire weld unless there are very high standards to be met. TIG is typically used for aluminum, stainless steel, and exotic materials.

Weldability is the capacity of a metal to be welded in relation to its suitability to the design and service requirements. Carbon steel welding is usually completed with shielded metal arc welding. Rod iron has characteristics similar to those of mild steel, and a similar process is used in its welding. In aluminum and aluminum alloys, most of the commercial welding and brazing processes can be used, although the most common are GTAW and GMAW. Various problems are encountered when using the acetylene process to weld aluminum because of an oxide film that prevents metal flow at welding temperatures. Aluminum is characterized by its low melting point and high thermal conductivity.

18.2.2 Types of Welds

Weld symbols and types of welds are classified by process. Resistance weld symbols are grouped under flash or upset, projection, seam, and spot with supplementary descriptions such as contour weld and field weld.

Arc and gas weld symbols are divided into groove types (bevel, square, J, U, V), bead, fillet, plug, or slot welds. For bead and fillet welds, no special preparations are needed for the metal. The essential difference between the groove welds is the edge preparation of the material to be welded, whether it is to remain square, beveled, or machined in a V, U, or J shape. Although these welds can be combined, an effort should be made to keep welding symbols of similar joints both the same and simple.

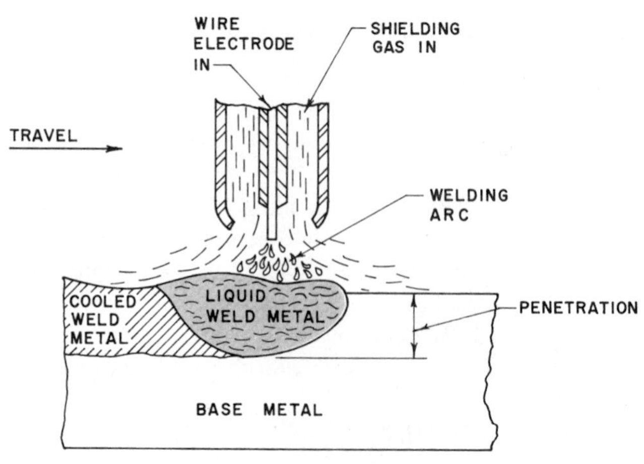

FIGURE 18.3 Gas-Metal Arc Welding

ITEMS OF INTEREST

Using the Laser to Weld

One of the most important and most widely used applications of the laser is in fiber optic communication systems. Laser-based communication systems are prevalent in the United States and Japan and are rapidly spreading throughout Europe. Laser beams that are transmitted by glass fibers carry thousands of times more information than copper cables. Even though lasers are a relatively new technology (1960), they have become one of our most useful tools.

Laser applications in industry, ranging from manufacturing to the space program, have become quite popular. Laser-based tools are used for heat treating, cutting, drilling, and welding. Even though laser cutting and drilling is used, laser welding is by far the most often used process.

There are two different laser welding processes. One is conduction, which occurs at the surface of the material; the other is deep penetration, where heat is moved below the surface of the material.

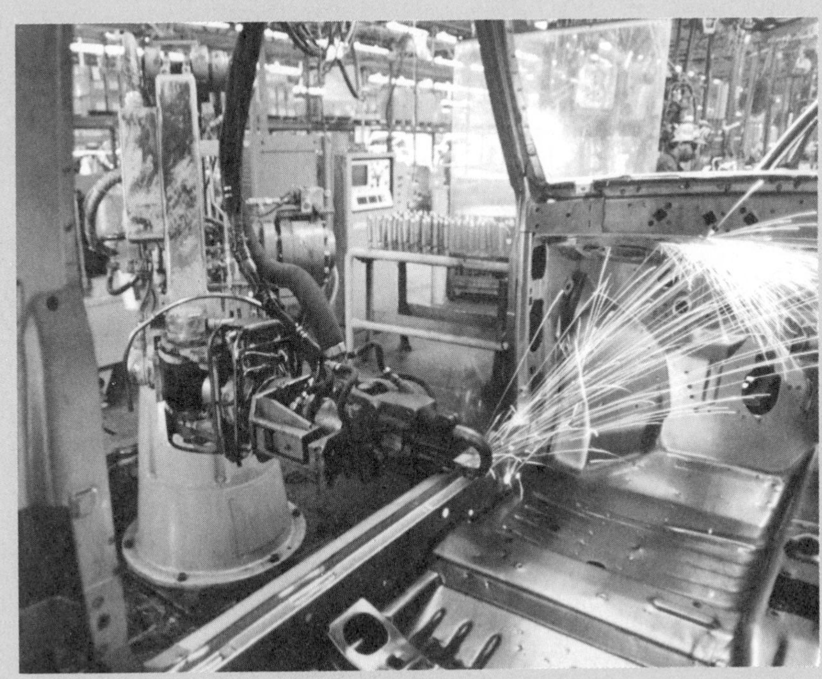
Arc Welding with Robots

Laser Welding Machine

The conduction process is used to join thin sheets. The deep penetration process creates a more efficient weld with high tensile strength and hardness. Laser welding has been used with great success in shipbuilding, pipeline fabrication for the Arctic, nickel steels, and low alloy steels. The National Aeronautics and Space Agency developed a way to weld aluminum effectively using a laser. Aluminum is difficult to weld because it has a low melting temperature. Because of the developed process, aluminum vessels can now contain a high-pressure gas. Other precision aluminum pieces can be fabricated with the same process.

In laser welding, the welding rod is eliminated. The welding is accomplished without the excess heat that distorts and even destroys some materials in conventional welding. Even two dissimilar materials can be joined with laser welding. As larger and more powerful lasers are built, laser welding applications will grow in size and in number. Welding with lasers has made fabrications possible today that were impossible only forty years ago.

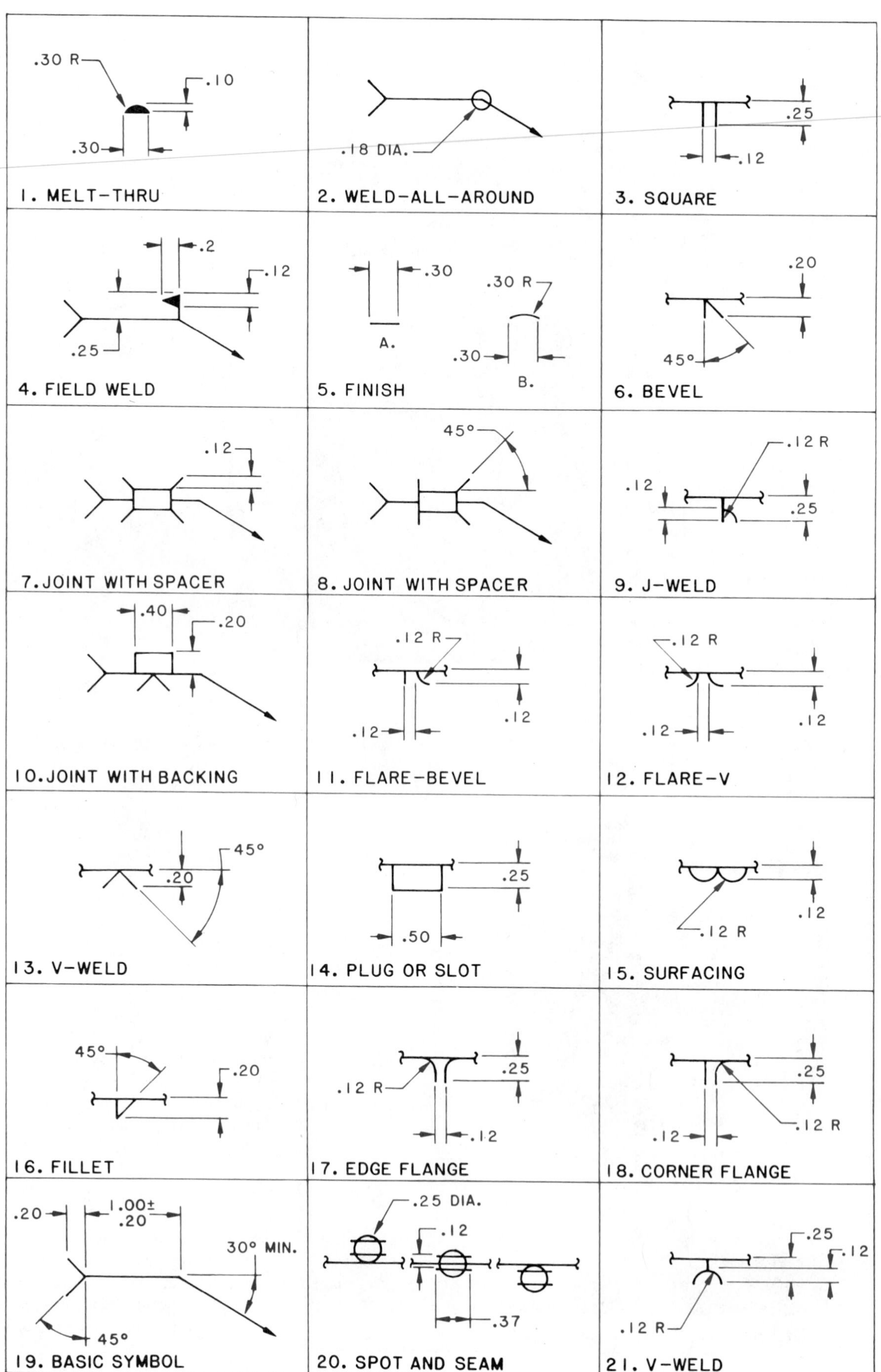

FIGURE 18.4 Weld Symbols

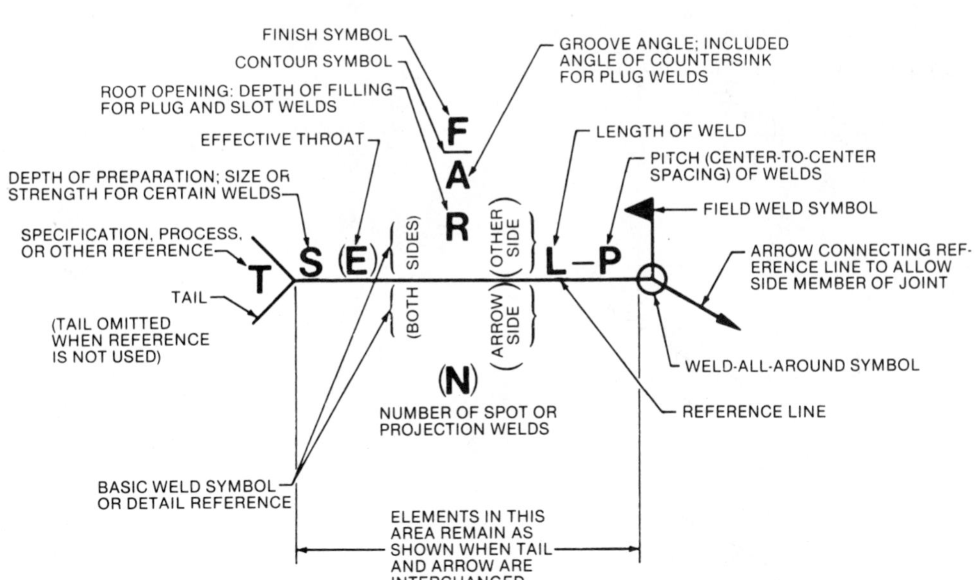

FIGURE 18.5 Standard Location of Elements of a Welding Symbol

18.2.3 Welding Symbols

Welding symbols communicate the weld type, size, and location to the fabricator. All welds can be identified by their profile or cross-sectional view. The welding drawing shows the parts or units that are to be made by welding. Symbols define and locate the specific welds to be used. Each joint in the welding process must be fully described. The weld symbol (Fig. 18.4) denotes the desired type of weld: fillet, square, bevel, J, U, V, flare V, back, weld, arc seam, spot, plug, and slot. The complete welding symbol describes all welding information that might be needed: weld type, size, length, location, place of construction (field or shop). Figure 18.5 shows the standard welding symbol and the location of its elements. This type of symbol is standard throughout industry. Welding symbols are composed of three basic parts:

1. An *arrow* that points to the joint
2. A *reference line* on which all the dimensions and other data are placed
3. The *weld symbol,* which indicates the type of weld required.

The assembled welding symbol consists of eight elements or as many of these elements as are necessary (Fig. 18.5):

Reference Line

- Arrow
- Basic weld symbols
- Dimensions and other data
- Supplementary symbols
- Finish symbols

Tail

- Specifications, process, or other reference

Figure 18.6 shows five different joints that may be encountered in the construction of welded assemblies. Figure 18.7 offers a sample of various welds as applied to the particular joint.

Usually the same welding process is used throughout the part. If this is not the case, such as when the drawing contains submerged arc welding by machine and manual welding, each process must be noted on the symbol when pointing to the joint to be completed. The welding process should be placed at the tail of the welding symbol. The tail is omitted when references are not needed to supplement the symbol.

The welding symbol should be of an adequate size to be readily visible to the fabricator. At the junction of the arrow and the reference line within the welding symbol a flag or a solid field weld designation may be placed. Work that is to be performed on the job site, instead of shop fabricated, will have a **field weld** designation. The edge preparation for such welds is completed at the fabricating plant (seldom at the job site itself), so shop drawings contain all the edge preparation designations. Only field drawings contain field weld symbols. For more information, you should reference ANSI Y32.3.

The arrow in a welding symbol connects the reference line to the joint (see Fig.18.8 on p. 553). The side that the arrow is on is called the **arrow side** of the joint [Fig. 18.8(a)]. The side opposite the arrow is the **opposite** [other side, Fig.18.8(d)] of the joint, except for plug, slot, seam, and other projection welds where the arrow connects the symbol to the surface to be welded. The arrow side of the joint is always considered the near side. Welds on the arrow side of the joint are shown by placing the weld symbol on the side nearest to the reader. To show welds on both sides, the weld symbol is placed on both sides of the reference line [Fig. 18.8(e)].

In many situations one welding symbol is shown on the drawing and a note is included that specifies the type of weld to be used on the entire drawing. For example: All welds to be ¼ inch unless otherwise noted. On welds that are to be on all sides of a particular joint, the **weld-all-around symbol** is used at the junction of the arrow line and the reference line, which may also contain a field weld symbol at the same joint (see Fig. 18.9 on p. 553). When welds must be finished or contoured, this requirement must be shown on the symbol

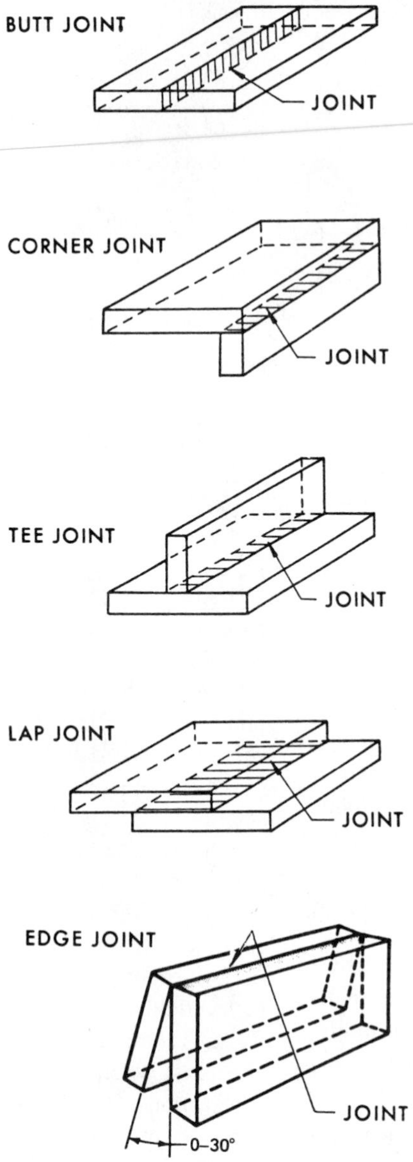

	APPLICABLE WELDS	
BUTT JOINT	SQUARE GROOVE V-GROOVE BEVEL-GROOVE U-GROOVE J-GROOVE	FLARE-V-GROOVE FLARE-BEVEL-GROOVE EDGE-FLANGE BRAZE
CORNER JOINT	FILLET SQUARE-GROOVE V-GROOVE BEVEL-GROOVE U-GROOVE J-GROOVE	FLARE-V-GROOVE CORNER FLANGE EDGE-FLANGE SPOT PROJECTION SEAM FLARE-BEVEL-GROOVE BRAZE
TEE JOINT	FILLET PLUG SLOT BEVEL-GROOVE SQUARE-GROOVE J-GROOVE	FLARE-BEVEL-GROOVE SPOT PROJECTION SEAM BRAZE
LAP JOINT	FILLET PLUG SLOT SPOT SEAM	BEVEL-GROOVE J-GROOVE FLARE-BEVEL-GROOVE PROJECTION BRAZE
EDGE JOINT	SQUARE-GROOVE V-GROOVE BEVEL-GROOVE U-GROOVE	J-GROOVE EDGE-FLANGE CORNER-FLANGE SEAM EDGE

JOINT

0–30°

FIGURE 18.6 Basic Types of Joints

(Fig. 18.9). It is also possible to combine different weld symbols in one welding symbol. Remember, the weld symbol is always placed according to the side of the joint on which the weld is to be made.

18.2.4 Welding Symbol Specifications

Weld symbols are to be drawn as in Figure 18.4. The sizes shown in this figure are used as a guide for symbol construction with their minimum values given. The actual size of the weld symbol and the welding symbol will vary according to the drawing size.

When drawing welding symbols, no distinction is made between arc and gas welding. Weld symbols are shown only as part of the welding symbol. Symbols are drawn "on" the reference line. Fillet, bevel groove, J groove, flare bevel groove, and corner flange weld symbols are shown with the *perpendicular leg always to the left.* Symbols are drawn a uniform size throughout the drawing. If the arrow is directed to the outer surface of one of the members of the joint (plug, slot, seam, and projection welds) at the centerline of the desired weld, the member to which the arrow points is considered the arrow-side member.

18.2.5 Supplementary Welding Symbols

Supplementary welding symbols are shown in Figure 18.9 and are used to define specific welding requirements. The *weld-all-around* symbol is used to indicate welds extending completely around a joint (Fig. 18.10). Welds completely around a joint in which the metal intersections at the points of welding in more than one plane are also indicated by the

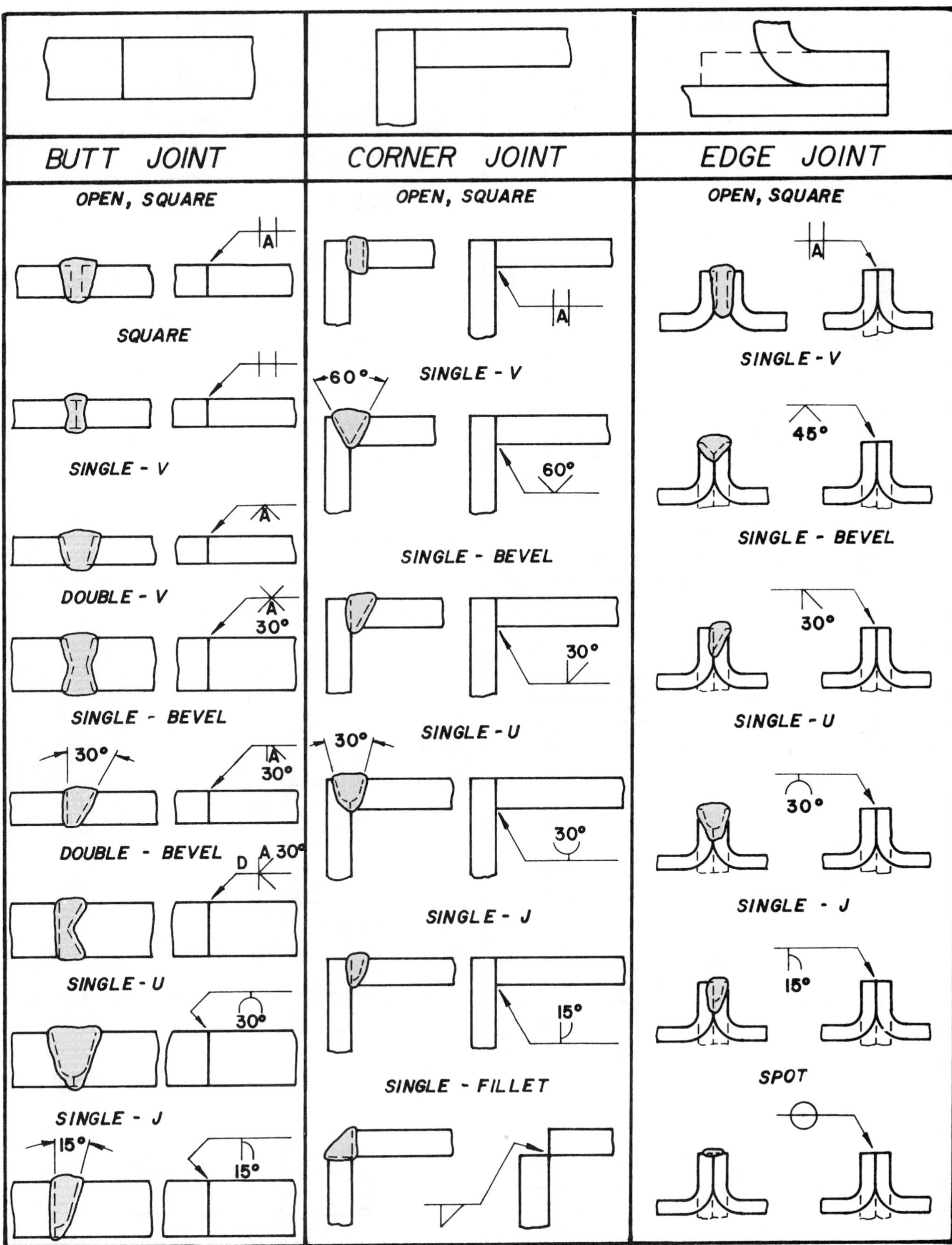

(a)

FIGURE 18.7 Joints and Appropriate Welds

Continued

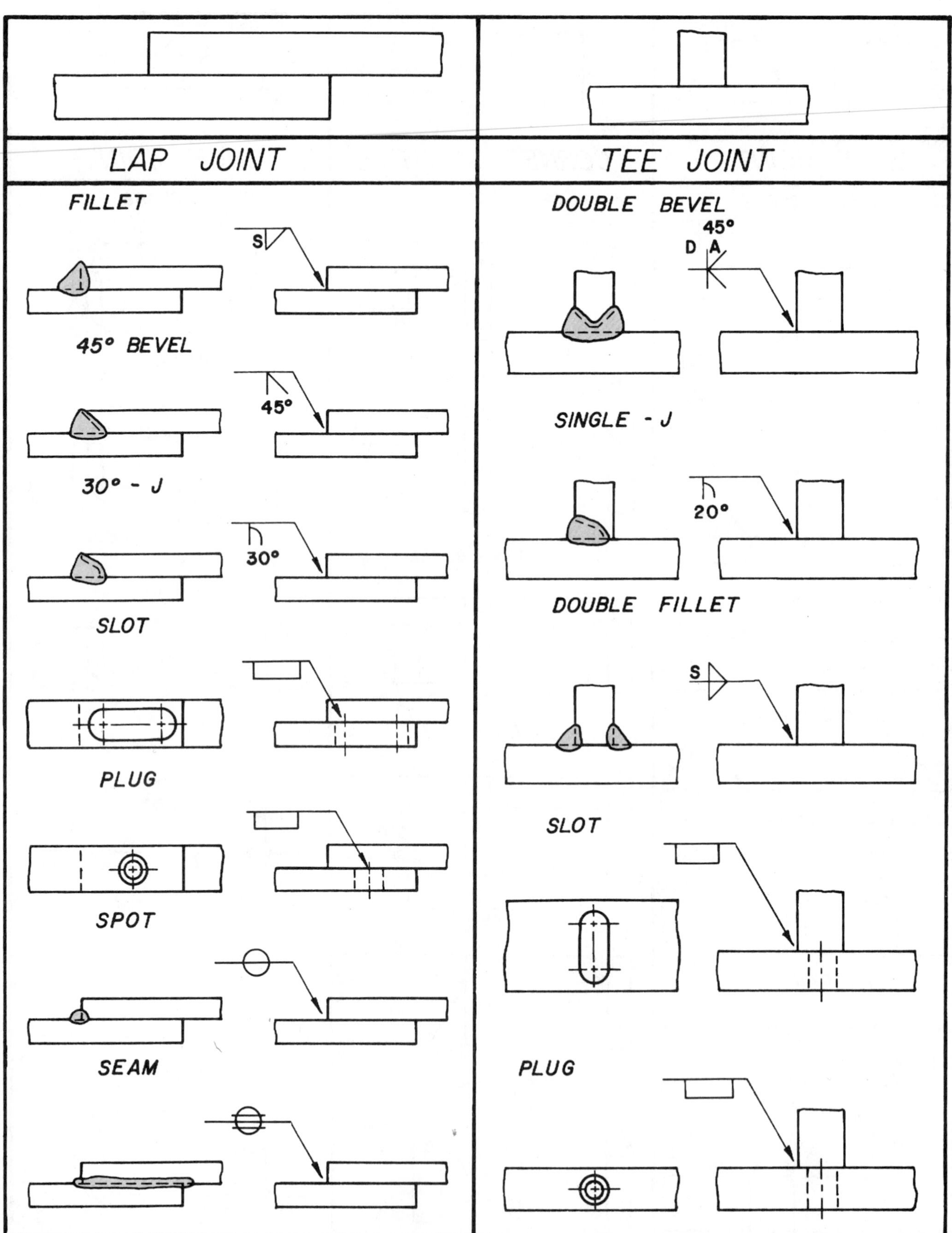

(b)

FIGURE 18.7—*Continued*

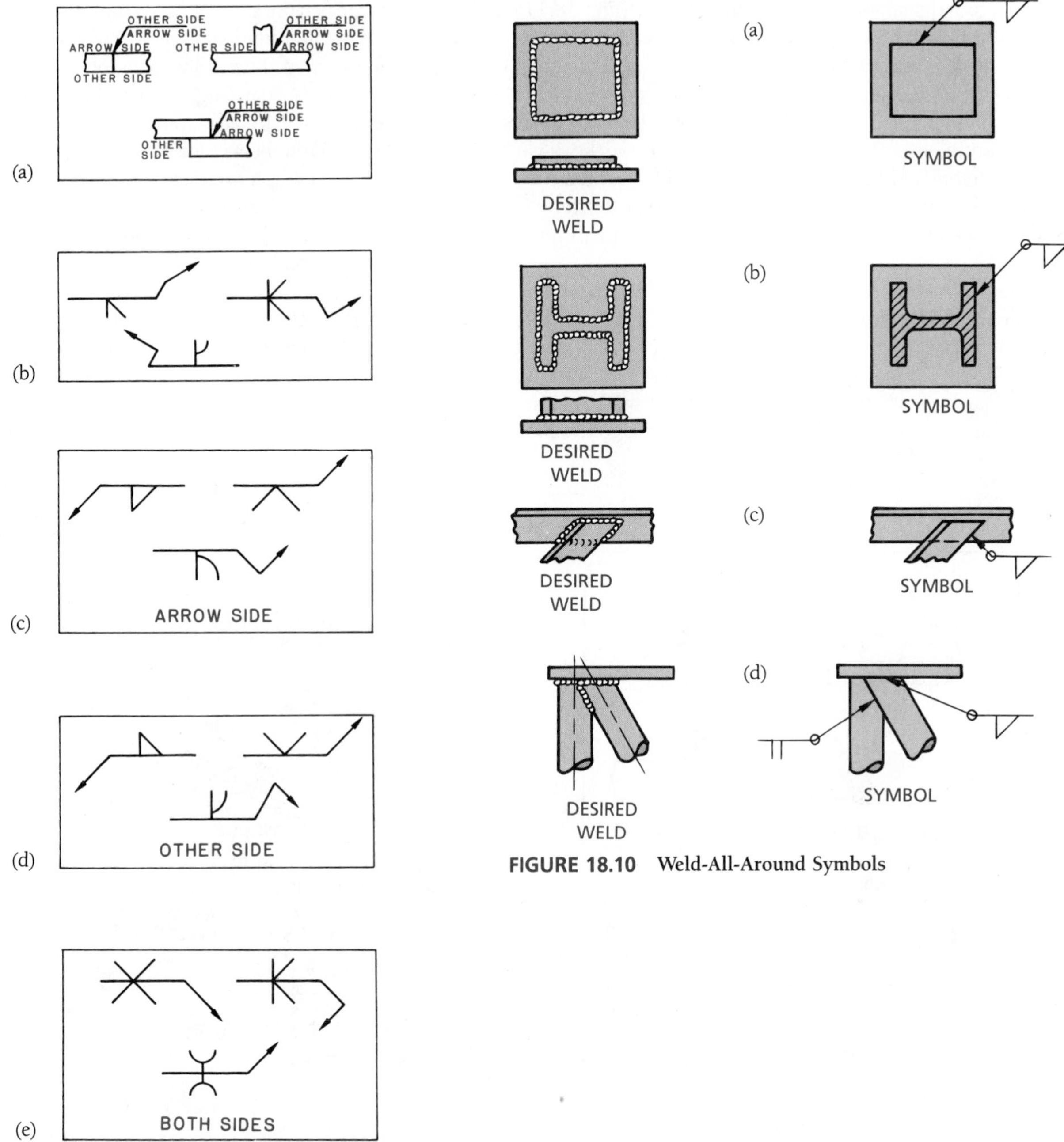

FIGURE 18.8 Example of Welding Symbol Element Locations

FIGURE 18.10 Weld-All-Around Symbols

WELD ALL AROUND	FIELD WELD	BACKING OR SPACER MATERIAL	MELT-THRU	CONTOUR		
				FLUSH	CONVEX	CONCAVE
			✱			

FIGURE 18.9 Supplementary Symbols

weld-all-around symbol. The **melt-thru** symbol (Fig. 18.11) is used only where 100% joint or member penetration plus reinforcement is required in welds made from one side only. Reinforcement (melt-thru) height may be shown on the welding symbol [Fig. 18.11(a)]. Melt-thru that is to be made flush by mechanical means is shown by adding both the flush contour symbol and finish symbol [Fig. 18.11(b)]. Melt-thru that is to be mechanically finished to a convex contour is shown by adding both the convex *contour* symbol and *finish* symbol.

Contour symbols are used, as applicable, to indicate the appropriate weld contour (flax, convex, or concave) desired, either with (in conjunction with a finish symbol) or without mechanical finishing. Finishing of welds, other than cleaning, is indicated by a suitable contour and finish symbol [Figs. 18.11(b) and (c)]. Welds indicated by symbols are continuous between abrupt changes in the direction of the joint except when the weld-all-around symbol is used, or as specified by length dimension on the welding symbol or dimension lines on the view. Welds extending beyond abrupt changes in direction are indicated by additional arrows pointing to each section of the joint to be welded. A symbol is shown for each weld on joints having more than one weld. When the basic weld symbols are inadequate to indicate the desired weld, the weld is shown by a cross section, detail, or other data, with a reference on the welding symbol.

The **pitch** (center-to-center spacing) of intermittent welds is shown to the right of the length dimension and separated from it by a hyphen [Fig. 18.12(a)]. The pitch indicates the distance between centers of the welds on one side of the joint. Chain and staggered intermittent weld dimensions are shown on both sides of the reference line [Figs. 18.12(b) and (c)]. When intermittent welding is called out by itself, the symbol indicates that welds are located at the ends of the joint. When intermittent welding is called out between continuous weld-

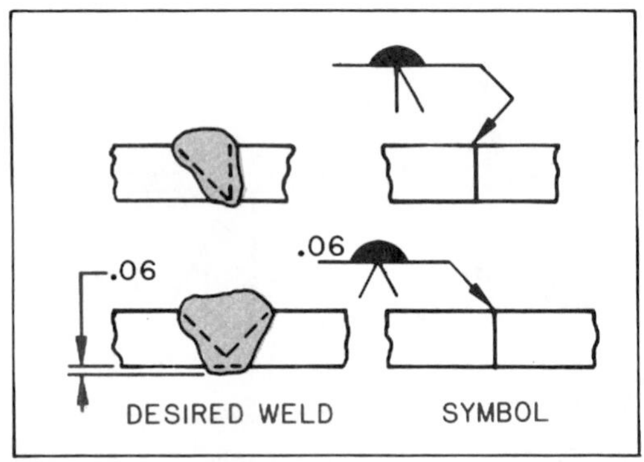

(a) Use of melt-thru symbol

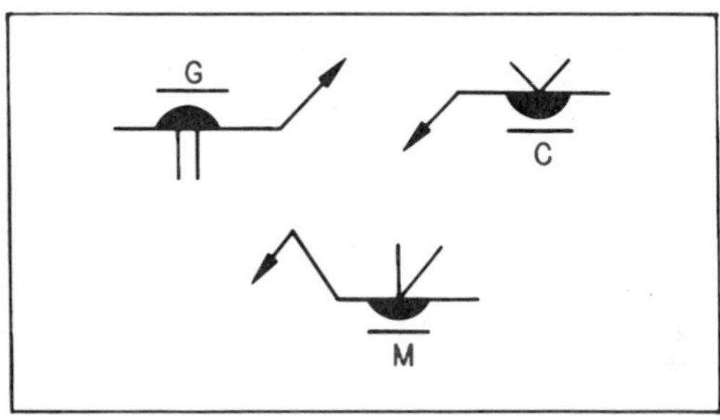

(b) Melt-thru finished flush

METHOD OF FINISH				
CHIP	GRIND	MACHINE	ROLL	HAMMER
C	G	M	R	H

(c) Weld finish symbols

FIGURE 18.11 Melt-Thru and Finish Symbols

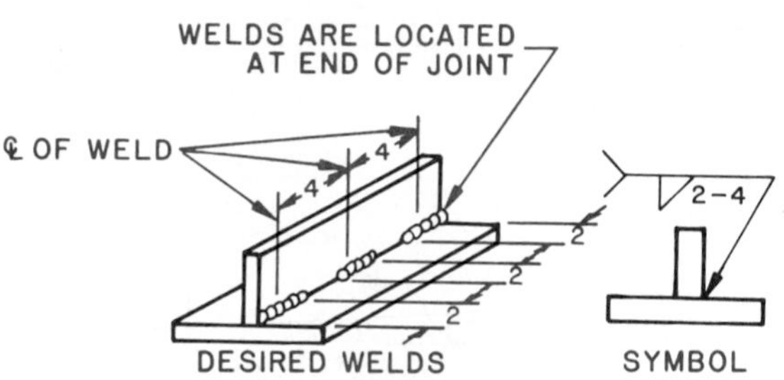

(a) Length and pitch of increments of intermittent welding

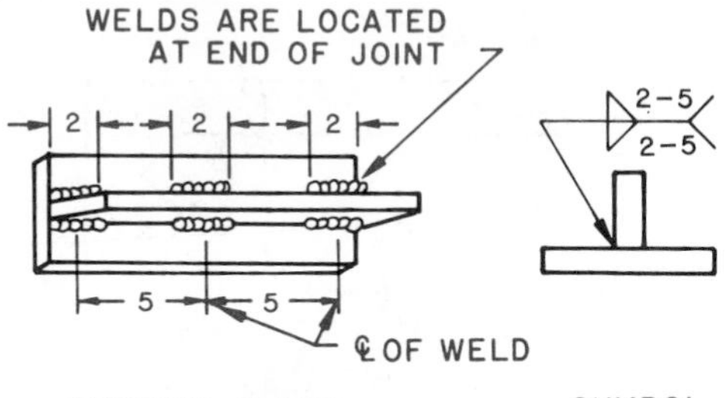

(b) Length and pitch of increments of chain intermittent welding

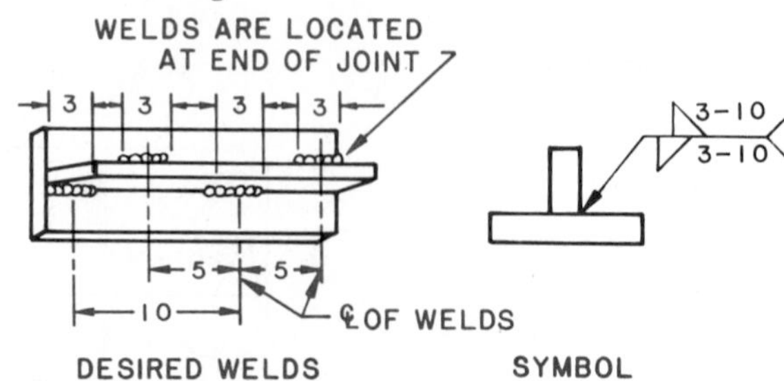

(c) Length and pitch of increments of staggered intermittent welding

FIGURE 18.12 Application of Dimensions to Intermittent Fillet Welding Symbols

ing, the symbol indicates that spaces equal to the pitch minus the length of one increment is left between the end of the continuous weld and the intermittent weld [Fig. 18.12(c)]. Unless otherwise specified, staggered intermittent welds on both sides are symmetrically spaced as shown in Figure 18.12(c). Separate welding symbols are used for intermittent and continuous welding when the two are used in combination along one side of the joint.

Two or more reference lines may be used to indicate a sequence of operations. The first operation is shown on the reference line nearest the arrow. Subsequent operations are shown sequentially on other reference lines. Additional reference lines may be used to show data supplementary to welding symbol information included on the reference line nearest the arrow. Test information may be shown on a second or third reference line away from the arrow. When required, the weld-all-around symbol is placed at the junction of the arrow line and reference line for each operation to which it is applicable. The field weld symbol may also be applied in the same manner. The letters **CP** in the tail of the arrow indicate a *complete penetration* weld regardless of the type of weld or joint preparation.

18.2.6 Welds

Fillet welds have a triangular cross section and join two or more surfaces at right angles—such as lap, tee, and corner joints (Fig. 18.13). They are often used in combination with groove welds for corner joints. The dimensions of fillet welds are placed on the welding symbol. The weld size goes to the left of the fillet weld symbol; the length of the weld is placed to the right of the basic weld symbol when required.

There are two basic types of fillet welds: those with *equal legs* and those with *unequal legs*. The size of fillet welds is shown on the same side of the reference line as the weld symbol and to the left of the weld symbol (Fig. 18.13). When welds on both sides of the joint have the same dimensions, both are dimensioned. The size of a weld with unequal legs is shown in parentheses to the left of the weld symbol.

The rectangular basic weld symbol (Fig. 18.14) is used for designating **plug** and **slot welds.** Plug and slot welds are often used in butt joint and lap joints for reinforcement. Plug welding holes in the arrow-side member of a joint are indicated by placing the weld symbol below the reference line. Holes in the other-side member are indicated by placing the weld symbol above the reference line (Fig. 18.14). Plug weld dimensions are shown on the same side of the reference line as the weld symbol. The diameter of the base of the hole is shown to the left of the weld symbol. The hole is cylindrical unless the included angle of countersink (taper) is shown above (other side) or below (arrow side) the weld symbol. Length, width, spacing, included angle of countersink (taper), orientation, and location of slot welds cannot be shown on the welding symbol. These data are shown on the drawing with a detail referenced on the welding symbol.

When **projection welding** is required, the spot weld symbol is used with the projection welding process reference in the tail of the welding symbol (Fig. 18.15). The spot weld symbol is placed above and below (never on) the reference line to indicate in which member the *embossment* is placed. Dimensions are shown on the same side of the reference line as the weld symbol, or on either side when the symbol is astride the reference line and has no arrow- or other-side significance.

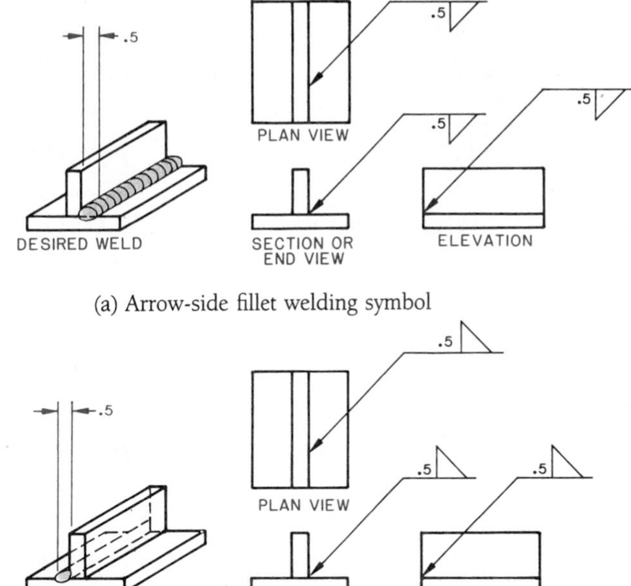

FIGURE 18.13 Application of Fillet Welding Symbols

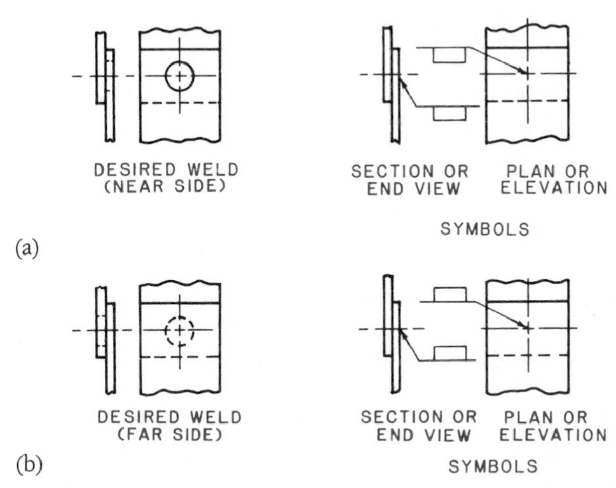

FIGURE 18.14 Plug Welds

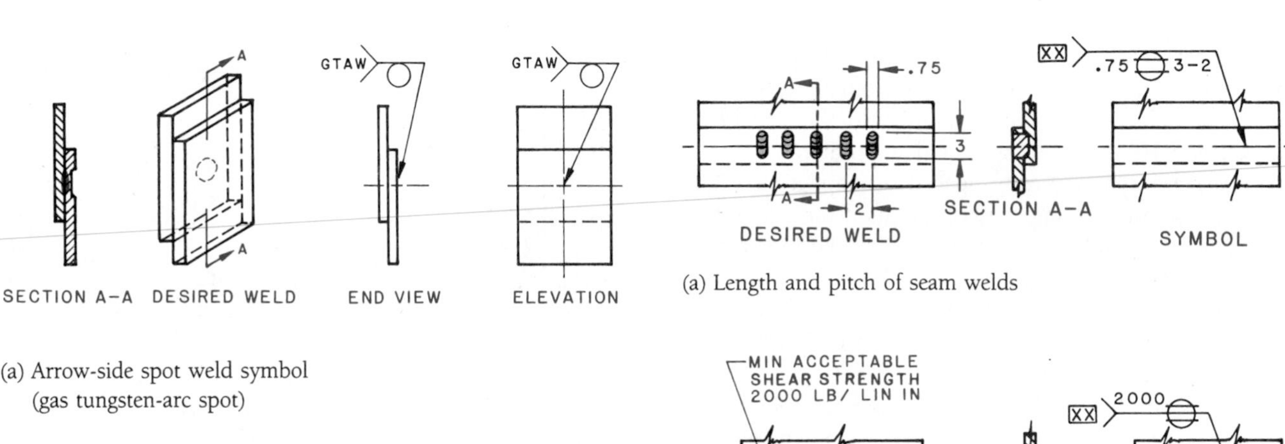

(a) Arrow-side spot weld symbol
(gas tungsten-arc spot)

(a) Length and pitch of seam welds

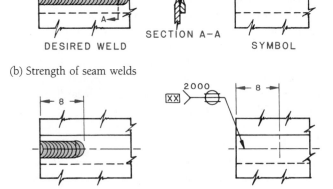

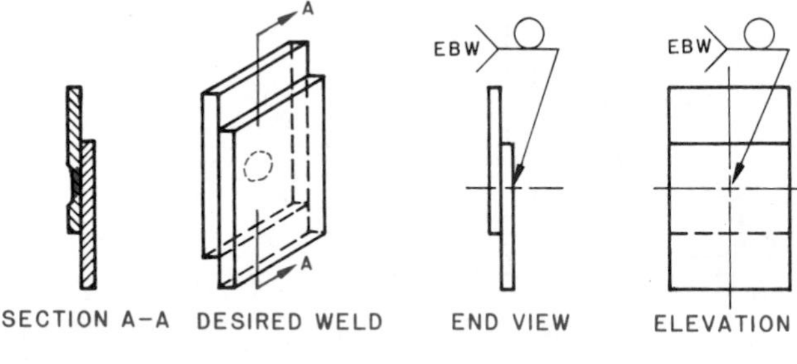

(b) Other-side spot weld symbol
(electron beam spot)

FIGURE 18.15 Spot Welding Symbol Applications

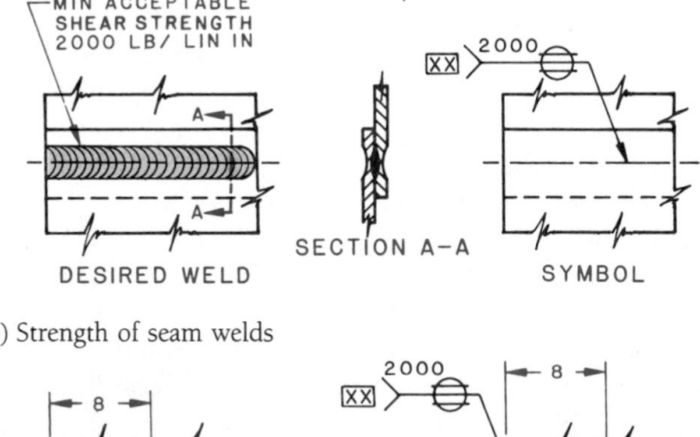

(b) Strength of seam welds

DESIRED WELD

SYMBOL

(c) Extent of seam welds

FIGURE 18.16 Application of Dimension to Seam Welding Symbols

One symbol is used for all **seam welds** regardless of the welding process. The process reference is shown in the tail of the welding symbol. The weld symbol may or may not have location significance, depending on the welding process. Dimensions are shown on the same side of the reference line as the weld symbol, or on either side when the symbol is astride the reference line and has no arrow-side or other-side significance (Fig. 18.16).

There are five basic **groove welds** (beveled, square, J, U, V), which also have various combinations: single V, single bevel, single J, single U, double V, double bevel, double J, and double U. The edges in a groove weld are usually prepared by a flame cutting torch. Whether single or double, these are probably the easiest, most economical welded joints used to join two ends. A general rule to follow when material is thicker than $\frac{1}{8}$ in. (3 mm) is that the edges must be prepared for a groove channel and during the welding process a filler material must be added.

Dimensions of all types of groove welds (Fig. 18.17) are shown on the same side of the reference line as the weld symbol. If double-groove welds have the same dimensions, both are dimensioned. The depth of groove preparation and effective throat of a groove weld are shown to the left of the weld symbol with the effective throat in parentheses. The *effective throat* is the perpendicular depth of the groove cut. The total effective throat never exceeds the thickness of the thinner member of a joint. When no depth of groove preparation or effective throat is shown on the welding symbol for single-groove or symmetrical double-groove welds, complete penetration is required (Fig. 18.17). Unless specified in a general note, the groove angle or groove welds are shown outside the weld symbol. Unless specified in a general note, the root opening of groove welds is shown inside the weld symbol. Groove radii of U groove and J groove welds is specified in a general note, or by a detail view on the drawing, referenced on the welding symbol.

The flush and convex supplementary weld symbols are also applied to groove welds, such as when the outer contour of the weld must be altered by grinding or machining.

Back or **backing welds** of single-groove welds are shown by placing a back or backing weld symbol on the side of the reference line opposite the groove weld symbol (Fig. 18.18). The welding symbol does not indicate the welding sequence (groove weld made before or after backing weld) or backing weld passes (single or multiple). The height of the weld bead is shown to the left of the backing weld symbol when required. No other backing weld dimensions are shown on the welding symbol. Other dimensions may be shown pictorially on a drawing detail.

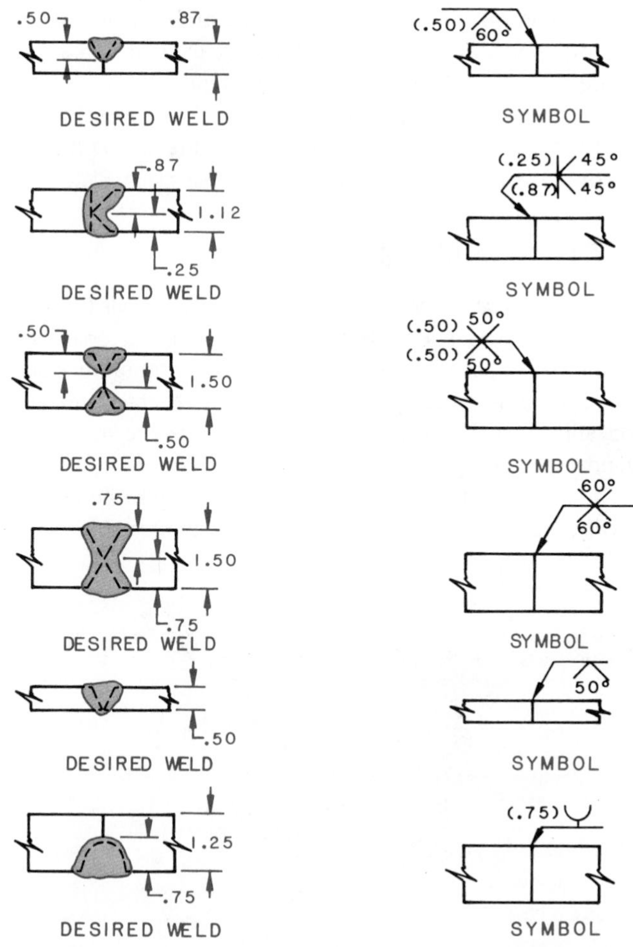

FIGURE 18.17 Designation of Size of Groove Welds with No Specified Root Penetration

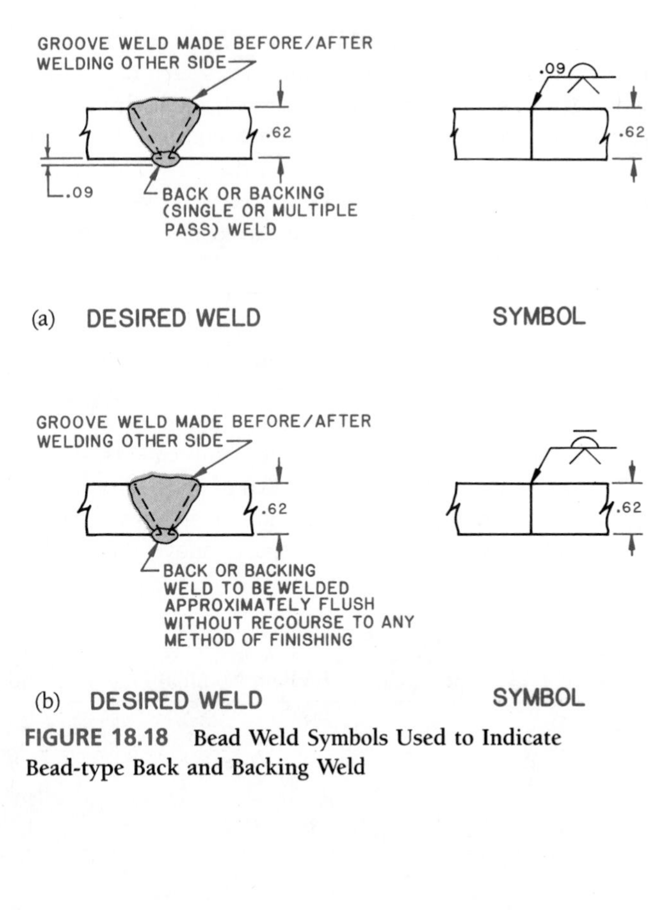

FIGURE 18.18 Bead Weld Symbols Used to Indicate Bead-type Back and Backing Weld

Surface welds are used to reclaim worn part surfaces or add alloying elements to the base metal for added protection. Often, surfaced parts outlast plating parts. The surfacing weld symbol does not indicate the welding of a joint, therefore it does not have arrow-side or other-side significance. The symbol is placed below the reference line and the arrow points clearly to the surface on which the weld is to be deposited.

The minimum thickness of the weld buildup is the only dimension shown on the welding symbol and is placed to its left. When no specific thickness of weld is required, no size dimension is given. When only a portion of the area of a plane or curved surface is to be built up by welding, the extent, location, and orientation of the area to be built up is dimensioned on the drawing (Fig. 18.19).

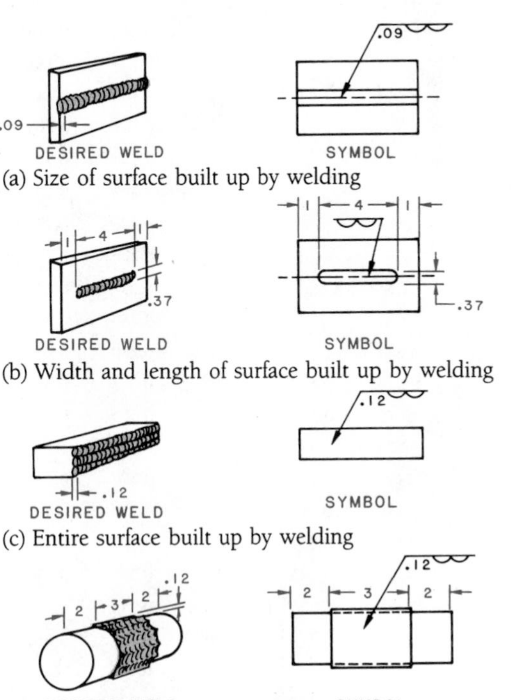

FIGURE 18.19 Dual Bead Weld Symbol to Indicate Surfaces Built Up by Welding

18.3 Introduction to Piping

Piping systems are used to transfer and process fluids. In general, piping systems can be divided between the *low-end* residential and public utility systems and the *high-end* process piping and power piping in industrial piping. Process piping includes oil processing, chemicals, and food and beverage production. Power generation piping covers such diverse areas as nuclear, solar, fossil fuel, geothermal, and hydroelectric plants. The major division between low-end and high-end systems is temperature, pressure, and the size of the pipe. Most low-end systems use small-diameter screwed piping, whereas welded and flanged large-diameter piping predominates in the industrial piping area.

The basic function of all piping is to move fluid from one place to another for storage, processing, or use of the contents. The actual consistency of the line contents varies from steam or air to thick slurries of water and pulverized coal. Design of a piping system must take into account the function of piping and the temperature-pressure requirements.

Piping systems are not just composed of pipes. Piping ties together a vast array of mechanical equipment and vessels. A typical industrial system uses pumps to move the fluids, valves to control the contents movement, and instruments to monitor, control, and record the fluid's state. Tanks and vessels are required to hold, store, or process the medium. Structural elements and pipe supports are used to suspend and hold the system in place. Fittings establish direction and connect different parts of the configuration. (*See Color Plate 47.*)

18.3.1 Pipe Materials

The choice of pipe material is determined by the intended service of the piping system. Pipe is available in many materials including steel, cast iron, wrought iron, lead, aluminum, copper, brass, glass, wood, plastic, and clay. In general, steel and steel alloys, cast iron, and wrought iron are the most frequently specified piping materials.

The majority of pipes are manufactured from carbon steel. Usually, Grade A and Grade B steel pipe is used, although

TABLE 18.1 Pipe Schedule Number, Nominal Pipe Size, and Wall Thickness

		Nominal Wall Thickness for											
Nominal Pipe Size	O.D.	Sched. 20	Sched. 30	SW	Sched. 40	Sched. 60	XS	Sched. 80	Sched. 100	Sched. 120	Sched. 140	Sched. 160	XXS
⅛	0.405	.....	.....	0.068	0.068	.....	0.095	0.095	.....	.....	.....	.....	.....
¼	0.540	.....	.....	0.088	0.088	.....	0.119	0.119	.....	.....	.....	.....	.....
⅜	0.675	.....	.....	0.091	0.091	.....	0.126	0.126	.....	.....	.....	.....	
½	0.840	.....	.....	0.109	0.109	.....	0.147	0.147	.....	.....	.....	0.187	0.294
¾	1.050	.....	.....	0.113	0.113	.....	0.154	0.154	.....	.....	.....	0.218	0.308
1	1.315	.....	.....	0.133	0.133	.....	0.179	0.179	.....	.....	.....	0.250	0.358
1¼	1.660	.....	.....	0.140	0.140	.....	0.191	0.191	.....	.....	.....	0.250	0.382
1½	1.900	.....	.....	0.145	0.145	.....	0.200	0.200	.....	.....	.....	0.281	0.400
2	2.375	.....	.....	0.154	0.154	.....	0.218	0.218	.....	.....	.....	0.343	0.436
2½	2.875	.....	.....	0.203	0.203	.....	0.276	0.276	.....	.....	.....	0.375	0.552
3	3.5	.....	.....	0.216	0.216	.....	0.300	0.300	.....	.....	.....	0.438	0.600
3½	4.0	.....	.....	0.226	0.226	.....	0.318	0.318	.....	.....	.....	.....	.....
4	4.5	.....	.....	0.237	0.237	.....	0.337	0.337	.....	0.438	.....	0.531	0.674
5	5.563	.....	.....	0.258	0.258	.....	0.375	0.375	.....	0.500	.....	0.625	0.750
6	6.625	.....	.....	0.280	0.280	.....	0.432	0.432	.....	0.562	.....	0.718	0.864
8	8.625	0.250	0.277	0.322	0.322	0.406	0.500	0.500	0.593	0.718	0.812	0.906	0.875
10	10.75	0.250	0.307	0.365	0.365	0.500	0.500	0.593	0.718	0.843	1.000	1.125	.....
12	12.75	0.250	0.330	0.375	0.406	0.562	0.500	0.687	0.843	1.000	1.125	1.312	.....
14 O.D.	14.0	0.312	0.375	0.375	0.438	0.593	0.500	0.750	0.937	1.093	1.250	1.406	.....
16 O.D.	16.0	0.312	0.375	0.375	0.500	0.656	0.500	0.843	1.031	1.218	1.438	1.593	.....
18 O.D.	18.0	0.312	0.438	0.375	0.562	0.750	0.500	0.937	1.156	1.375	1.562	1.781	.....
20 O.D.	20.0	0.375	0.500	0.375	0.593	0.812	0.500	1.031	1.281	1.500	1.750	1.968	.....
22 O.D.	22.0	.....	.....	0.375	.....	.....	0.500	.....	.....	.....	.....	.....	.....
24 O.D.	24.0	0.375	0.562	0.375	0.687	0.968	0.500	1.218	1.531	1.812	2.062	2.343	.....
26 O.D.	26.0	.....	.....	0.375	.....	.....	0.500	.....	.....	.....	.....	.....	.....
30 O.D.	30.0	0.500	0.625	0.375	.....	.....	0.500	.....	.....	.....	.....	.....	.....
34 O.D.	34.0	.....	.....	0.375	.....	.....	0.500	.....	.....	.....	.....	.....	.....
36 O.D.	36.0	.....	.....	0.375	.....	.....	0.500	.....	.....	.....	.....	.....	.....
42 O.D.	42.0	.....	.....	0.375	.....	.....	0.500	.....	.....	.....	.....	.....	.....

Grade C is used in a few cases. Other compositions include chrome, moly steel, wrought iron, nickel steel, and stainless steel along with a number of nonferrous metals. Steel pipe is obtained in either black pipe form or galvanized type. Both varieties are available in smaller sizes in lengths up to 40 feet. Available lengths decrease with increasing wall thickness and diameter. A majority of pipe comes in random (±20 ft) or double random (±40 ft) lengths. Steel pipe that has been treated with molten zinc to prevent rust is called **galvanized** pipe. Steel pipe is used throughout industry because it resists high temperatures and pressures. Fittings are also available in steel. Standard steel pipe is specified by nominal diameter. Nominal diameter is less than the actual outside diameter of the pipe from $\frac{1}{8}$ to 12 in. sizes and equals the O.D. for 14 through 42 in. pipe as shown in Table 18.1. To use common fittings, the outer diameter of the different weights of pipe remains constant, while the inside diameter varies to provide for various wall thicknesses.

Originally, steel pipe was available in three traditional designations to distinguish the different weights of pipe: standard wall (SW), extra strong wall (XS), and double extra strong wall (XXS). These designations are still in use (Table 18.1). ANSI has established a range of wall thicknesses corresponding to designations called **Schedules** (SCH). Ten Schedules are available: 10, 20, 30 ,40, 60, 80, 100, 120, 140, and 160.

Schedule numbers are indicative of approximate values of 1,000 times the pressure/stress ratio and can be calculated by using the formula $1{,}000 \times p/s$, where p is equal to the internal pressure of the pipe and s is the allowable fiber stress, both in pounds per square inch. As pressure increases so does the required wall thickness (Schedule number).

Cast iron pipe is used to transfer water and natural gas and in some cases it is used for soil pipe (sewage pipe). The nominal size of cast iron pipe, unlike steel pipe, always indicates the inside diameter (I.D.) regardless of size. When designating cast iron pipe, specify the O.D., regular or heavy, wall thickness, and the nominal diameter. Cast iron pipe is available in a variety of standard sizes and weights. Most cast iron pipe is manufactured with push-on bell and spigot joints, although flanged and screwed joints are available in the smaller sizes.

Aluminum, because it has a weight one-third that of steel, is used throughout the piping industry. Lead pipe or lead-lined pipe resists the chemical activity of acids; therefore, lead pipe is found in chemical work and in systems that transport acids. Brass and copper pipes are excellent for handling liquids containing salts. Brass and copper pipes are manufactured in two weights: regular and extra strong. They are available in sizes of $\frac{1}{8}$ to 12 in. The outside diameters of brass and copper pipe are the same as the corresponding nominal sizes of steel pipe.

Glass pipe is limited to temperature services of 400° and below and it is also vulnerable to vibration and high pressure. Glass pipe is excellent for resisting acids and chemicals and is used throughout the food, beverage, and chemical industries where corrosive contamination of the line contents is undesirable. Plastic piping is resistant to most corrosive chemicals including many acids, alkalis, and organic compounds.

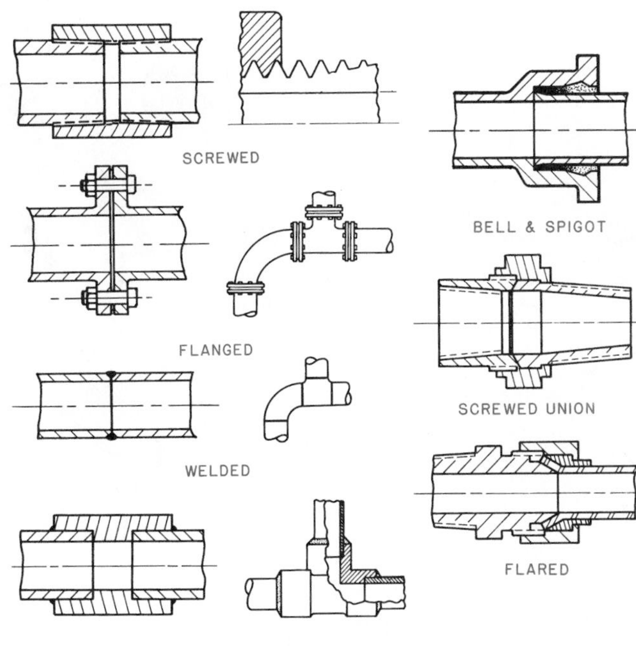

FIGURE 18.20 **Pipe Joint Variations**

18.3.2 Pipe Fittings

Pipe fittings are used to connect pipe, valves, and equipment. Screwed, flanged, and welded fittings are the most common types of connections. Standard fittings come in the same range of materials as pipe and are specified by fitting type, nominal pipe size (NPS), and material. Fittings are joined to the system by welding, bolting, or screwing in most instances. Bell and spigot, flared, and soldered fittings are also used for special services.

The seven major joint variations used on piping systems are shown in Figure 18.20. Welded systems have many advantages over other types. They require less maintenance and they provide a permanent leakproof bond; in effect, the system becomes a closed container. Butt welding fittings are used for larger diameter pipes, while socket welding is confined to smaller pipe sizes. Flanged fittings are bolted together with screwed or welded flanges. This type of joint is found on welded piping systems at points where the system must be dismantled and for valves. Bell and spigot fittings are used for low- temperature/pressure large-diameter piping on sewage, gas, and municipal water lines for underground services. Flared fittings are used on copper and brass tubes.

Threaded or screwed fittings are normally limited to smaller sized piping. They are machined and threaded with standard pipe threads. In Figure 18.21, socket and butt welding fittings are shown. In most cases, these types are available

FIGURE 18.21 Welding Fittings

in all joint variations. Figure 18.22 shows the 20 most common fittings. Bushings are confined to screwed systems and are used to reduce the size of an opening in a fitting and are sometimes used in place of a reducer. Caps close the end of a pipe and come in welded and screwed types. Couplings enable two pieces of pipe to be joined. Crosses are available in all joint types and can be either straight or reducing. Straight crosses have four equal-diameter branch outlets.

The primary purpose for most fittings is to change the direction of the pipe. Elbows change direction at 45 or 90°. Laterals come in straight size and reducing types. Because they lack structural strength, they are confined to low-pressure applications. Plugs are used on screwed piping systems to close off the end of a line or fitting. Reducers are used to decrease the diameter of a pipe run. The 180° bend is used to change the direction of the pipe line and is found on heat exchangers and coils. Stub ends are used on butt-welded systems for lap flanges. A tee is a reinforced fitting that is branched to permit flow at 90° to the main run. A union is a screwed fitting that enables two pieces of threaded pipe to be joined and dismantled easily. True Y fittings allow the pipe to be split into two streams.

18.3.3 Piping Drafting

Piping drafting and design uses a specialized language to transmit information for fabrication and construction. *Pipeline, fitting, valve, and instrument symbols, along with special notations,* are used on drawings to convey design and fabrication information unique to the piping field.

Because piping projects represent "systems," actual detail shown on the drawing must be kept to a minimum. Templates speed the drawing process for manual drawing (Fig. 18.23). Piping components, fittings, valves, instruments, etc., are almost always standard catalog items. Piping diagrams and pictorial spools are normally *not drawn to scale*. However, dimensions for components are still necessary when providing measurements for dimensioning fabrication drawings.

There are six basic types of drawings for most piping projects:

1. Flow diagram
2. Plot plan
3. Plan, elevation, and section
4. Fabrication
5. Pictorial spools
6. Pipe support

The scale of most piping drawings is normally $\frac{3}{8}'' = 1'$, although $\frac{1}{2}'' = 1'$ is often used in the power industry so as to provide more detail and may also be used for modeling power plants.

Two methods are used to draw piping projects: **double line** and **single line** (Fig. 18.24). Both double- and single-line drawings of 12 in. pipe and larger are represented by a centerline and two thicker outside lines corresponding to the O.D. Single-line drawings show only the centerline (as a solid thick line) and short marks representing the O.D. at the ends of the pipe for $1\frac{5}{8}$ in. pipe to 10 in. pipe. Single-line drawings of $1\frac{1}{2}$ in. pipe and smaller only show the pipe's centerline as a thick solid line. Double-line drawings show all pipes with a thin centerline and solid object lines for the pipe's O.D. Figure 18.25 shows three different ways of representing the pipe line and equipment.

18.3.4 Piping Symbols

Piping drafting uses symbols to represent the pipeline, fittings, valves, instruments, vessels, and mechanical equipment such as compressors and pumps. Figure 18.26 shows the types of lines that are found on piping drawings. Pipes are

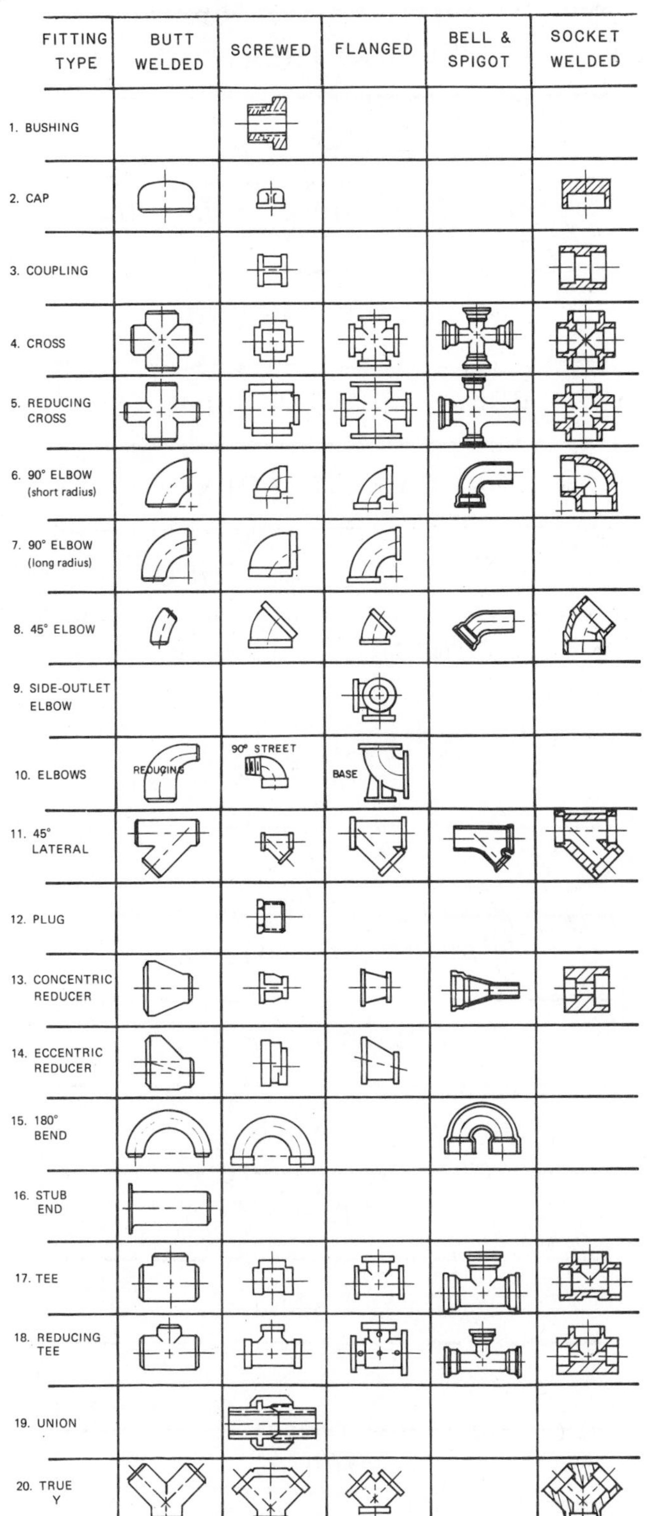

FIGURE 18.22 Common Types of Pipe Fittings and Styles

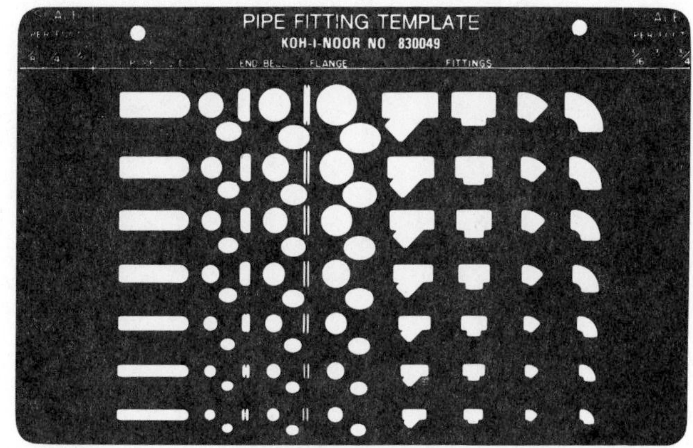

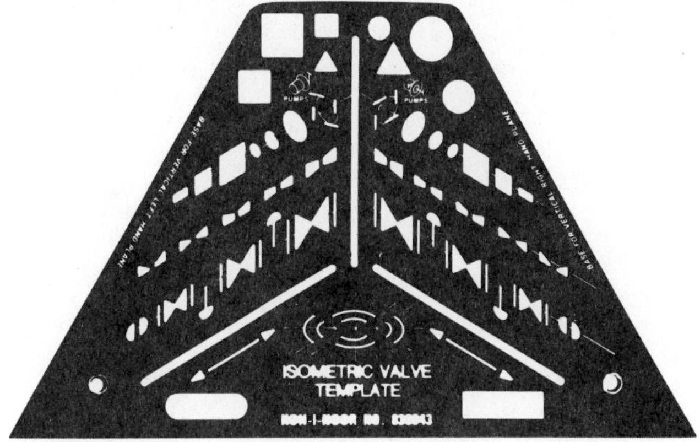

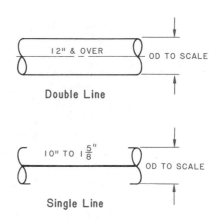

FIGURE 18.23 Pipe Drafting Templates

12" & OVER

OD TO SCALE

Double Line

10" TO $1\frac{5}{8}$"

OD TO SCALE

Single Line

$1\frac{1}{2}$" & BELOW

Single Line

FIGURE 18.24 Pipe Line Representation

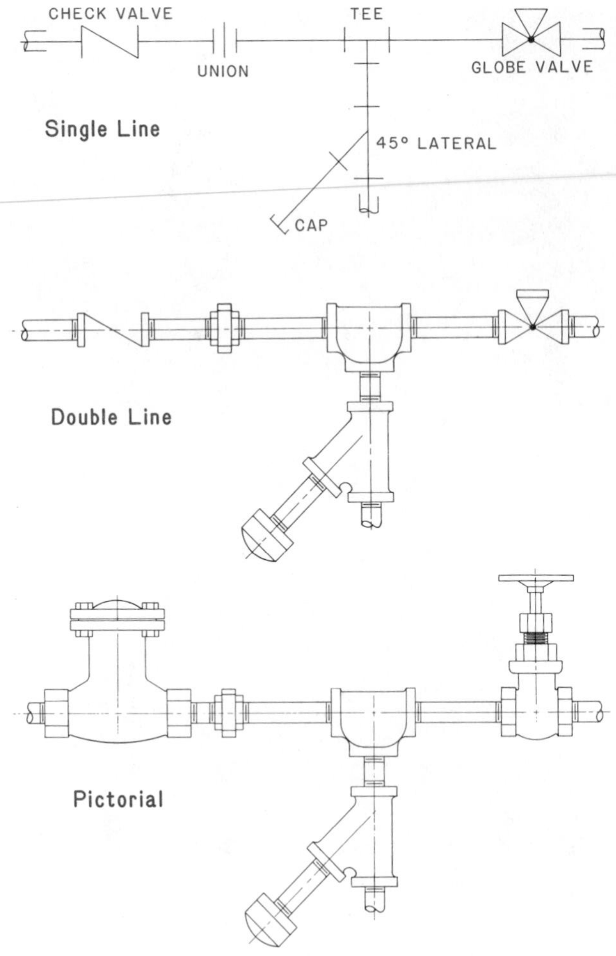

FIGURE 18.25 **The Three Types of Drafting Methods Used for Piping Drawings: Single Line, Double Line, and Pictorial**

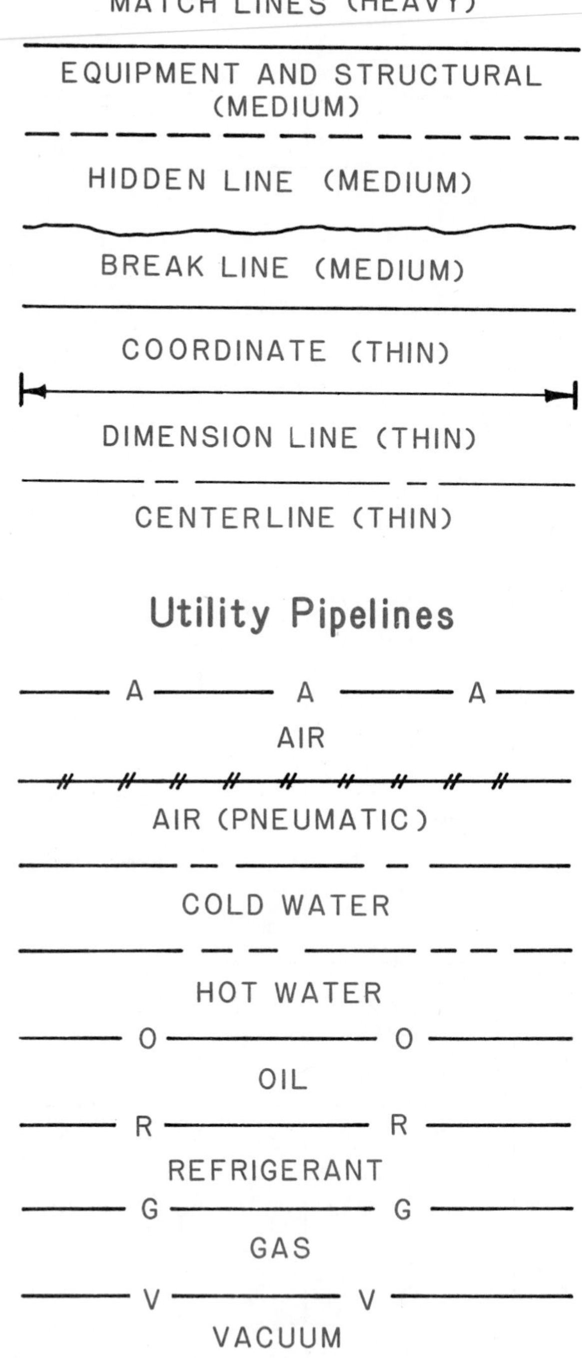

FIGURE 18.26 **Piping Drafting Line Symbols**

drawn with a thick, dense 0.7 mm line weight. Equipment and structural lines are medium thickness (0.45 to 0.5 mm). Centerlines and dimension lines are drawn as in mechanical drafting; thin sharp lines approximately 0.3 to 0.35 mm in thickness. **Utility lines** are used primarily on flow diagrams and are normally represented symbolically.

Symbols are used to represent fittings and valves. Figure 18.27 shows the most common types of fittings. A photograph of each fitting is provided with the fitting as viewed in frontal and end view elevations (when drawn symbolically). An isometric view is also given. Only screwed fittings are shown in this figure. Figure 18.27 shows the fitting as it looks from an end view. Notice that a large dot is shown when the fitting is turned *toward* you. Another method of representing this "end view" is shown in Figure 18.28. Here, the four most typical types of valves are shown. The "turned toward" end view shows a symbolic shaded method of describing the position and direction of the line. Either method is acceptable.

In Figure 18.28, the top (plan), front elevation, and end elevation views of each valve are shown. An isometric view is also given. Flanged valves are shown in this figure. Figures 18.27 and 18.28 show the accepted size of fitting and valve symbols for all types of piping drawings. Symbols for other types of fittings and valves are drawn by using these two figures as examples and consulting Appendixes C and D for the dimensions and shape. In other words, the end-to-end and the height measurements shown in these figures can be applied to all symbols. Note that isometric symbols of elbows use straight lines for the corners.

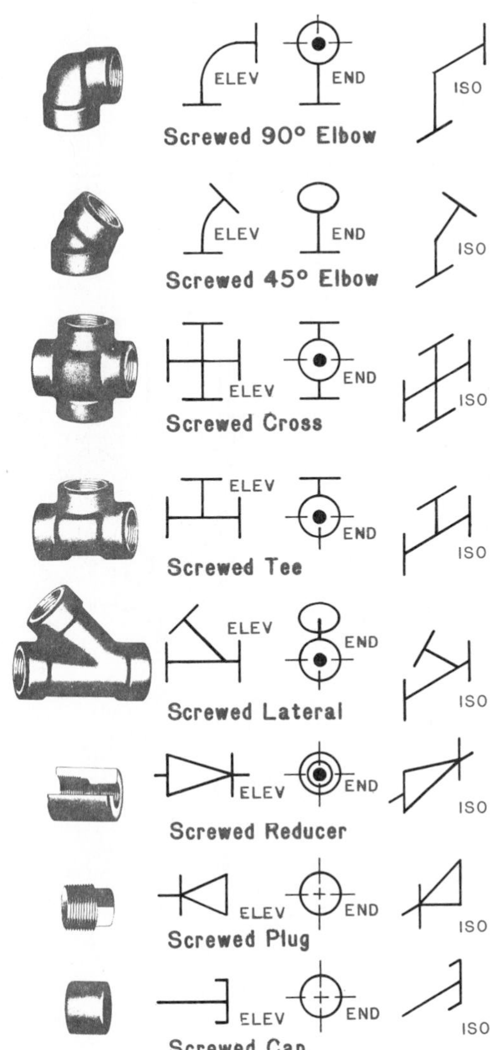

FIGURE 18.27 Pipe Fitting Symbols in Multiview and Isometric/Pictorial Projection

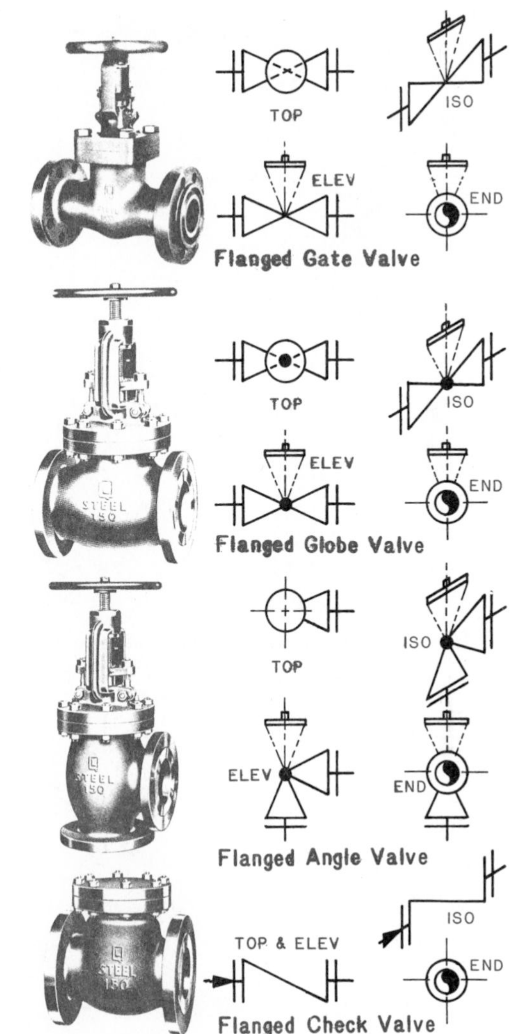

FIGURE 18.28 Valve Symbols in Multiview and Isometric/Pictorial Projection

Handwheels of the valves in Figure 18.28 are symbolically represented. In many piping drawings the handwheel is not shown, as on flow diagrams. Flow diagram symbols are similar but slightly more simplified than symbols used on other piping drawings.

18.3.5 Piping Diagrams and Drawings

Multiview drawings for piping systems are called plan, elevation, and section drawings. In Figure 18.29, three elevation views of a piping and vessel installation are shown. This drawing is a portion of a plan, elevation, and section (the plan view is not shown). Plan, elevation, and section drawings are not drawn to scale.

A north arrow orientation is provided on piping drawings, except on flow and fabrication drawings. Figure 18.30 shows the standard direction of the north arrow. Besides specifying coordinate location dimensions (X and Y or X, Y, and Z), piping drawings have elevation designations. Elevations are supplied for all horizontal runs of a pipe and may also be given to locate points along a sloping pipe. Elevations are established for many different parts of a piping drawing depending on project needs: centerline of pipe, bottom of pipe (BOP), top of concrete (TOC), and top of steel (TOS).

Fabrication drawings are used to detail a section of a pipeline, including flanges, valves, and fittings that are to be produced as one unit. Fabrication drawings are drawn using either pictorial/isometric or multiview drawing techniques, although multiview drawings are usually preferred by the fabricator. In either case, the fabrication is termed a **spool.** Figure 18.31 is an example of a pipe fabrication. Fabrication drawings are either double-line or single-line drawings.

Flow diagrams for piping projects graphically describe the system with simplified symbols used for vessels, equipment, valves, and instrumentation. The flow diagram is not drawn to any scale and it does not show the actual configuration of the piping. Figure 18.32 shows a typical mechanical [piping and instrumentation drawing (P&ID)] flow diagram of a petrochemical project.

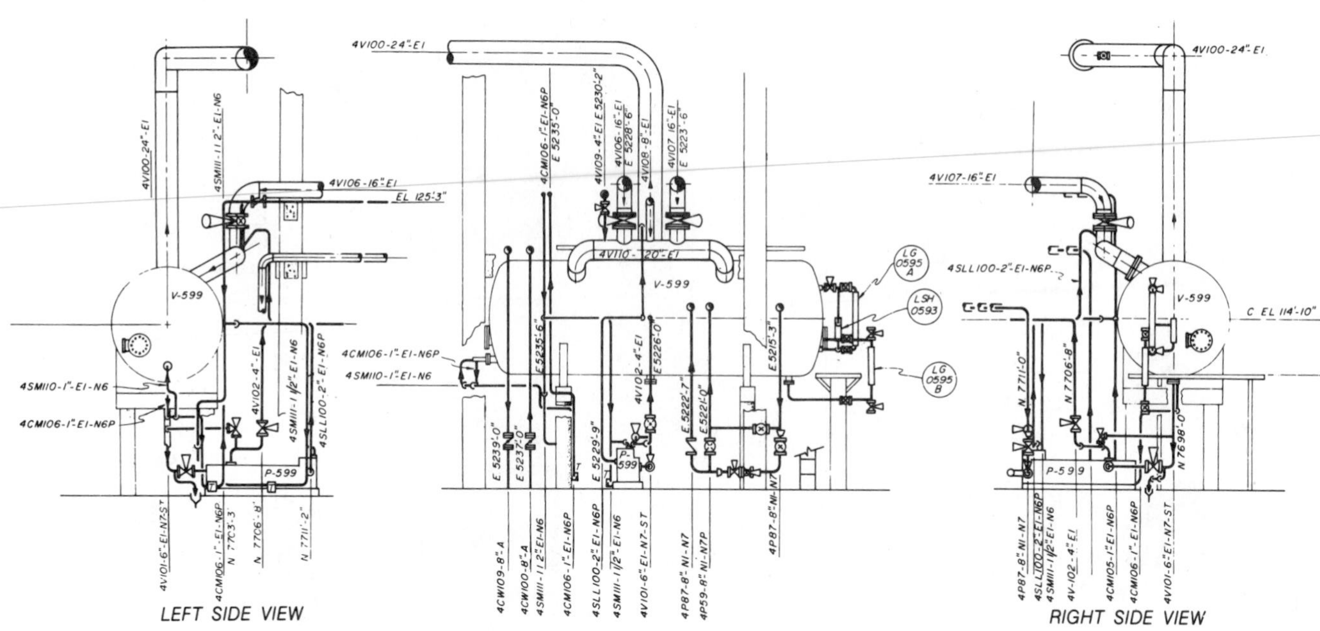

FIGURE 18.29 Elevation Views of Vessel and Associated Piping

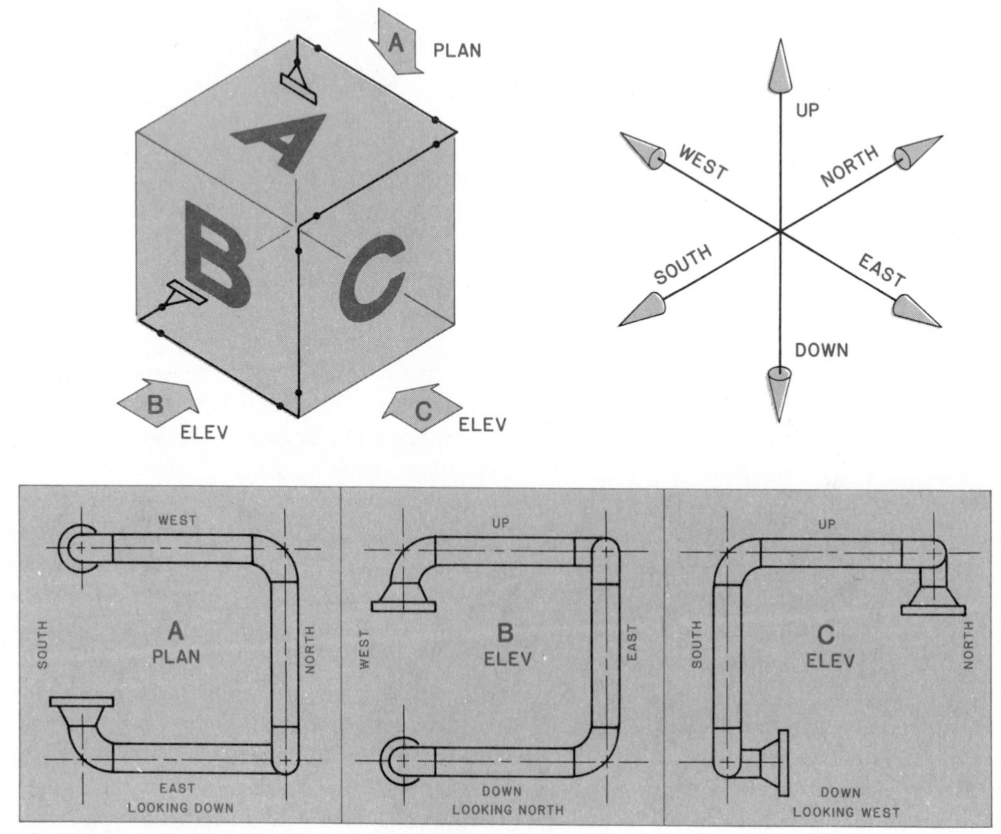

FIGURE 18.30 North Arrow Orientation and Views of a Piping System

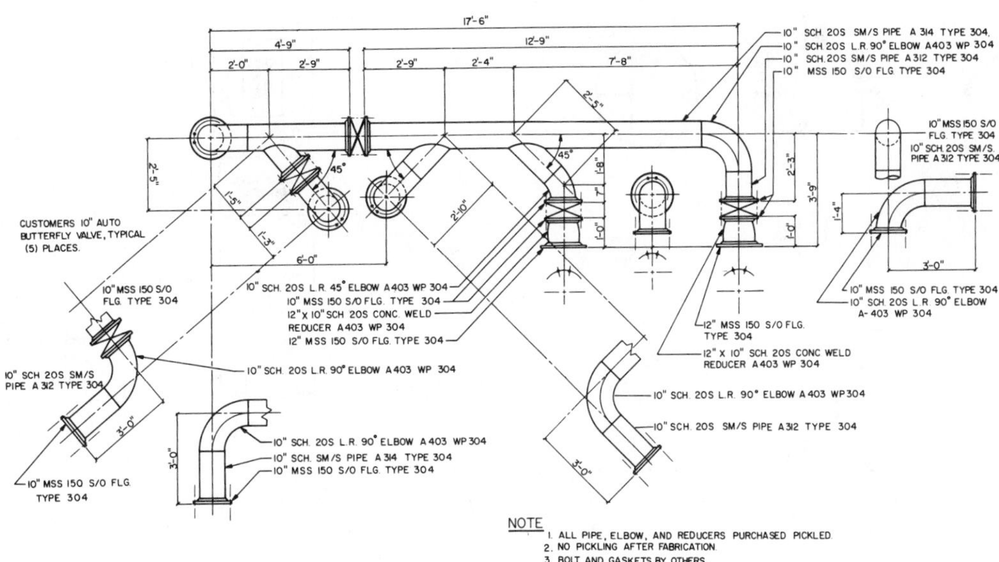

FIGURE 18.31 Fabrication Drawing of Header Assembly

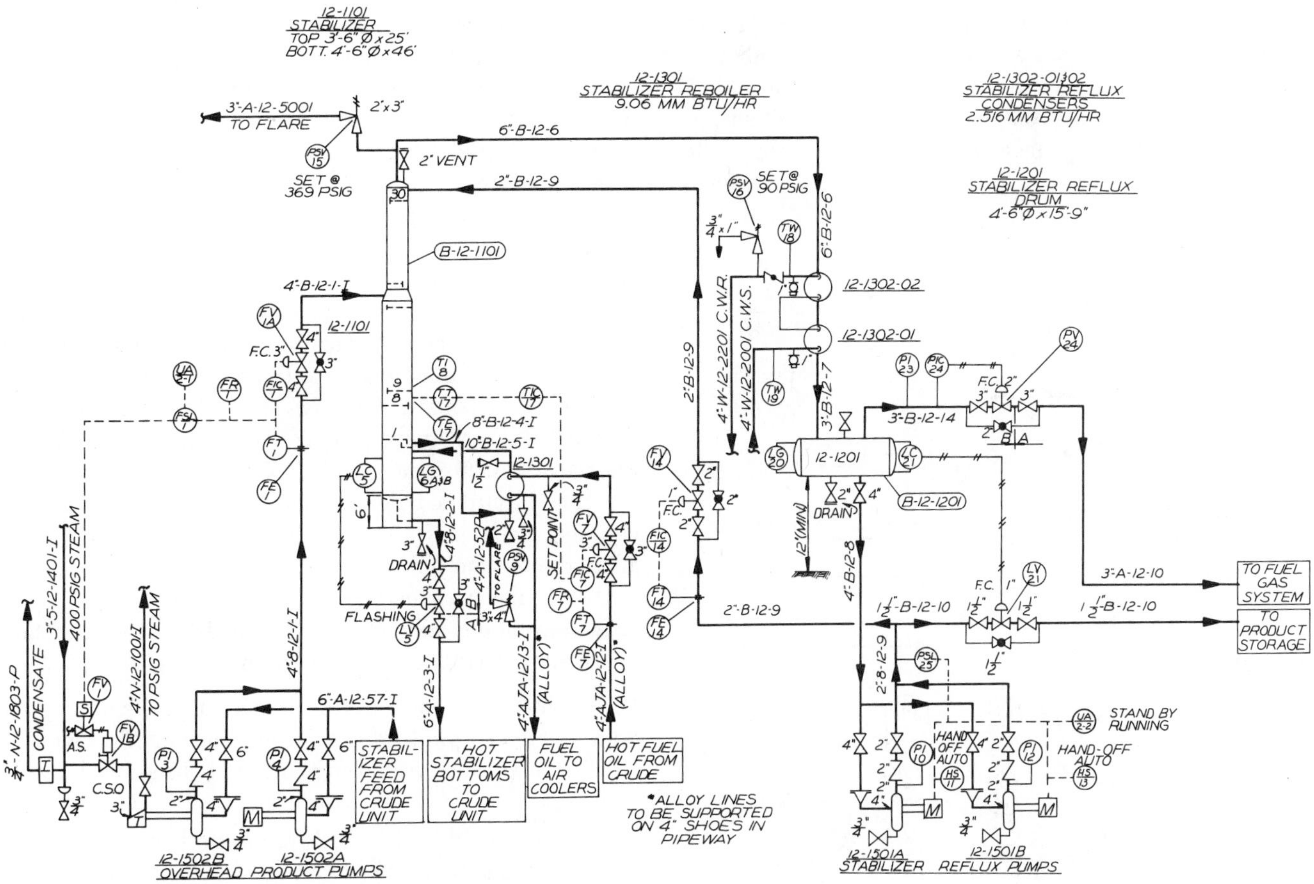

12-1101
STABILIZER
TOP 3'-6"Ø x 25'
BOTT. 4'-6"Ø x 46'

12-1301
STABILIZER REBOILER
9.06 MM BTU/HR

12-1302-01&02
STABILIZER REFLUX CONDENSERS
2.516 MM BTU/HR

12-1201
STABILIZER REFLUX DRUM
4'-6"Ø x 15'-9"

3"-A-12-5001
TO FLARE
SET @ 369 PSIG
SET @ 90 PSIG
2" VENT
TO FUEL GAS SYSTEM
TO PRODUCT STORAGE
STABILIZER FEED FROM CRUDE UNIT
HOT STABILIZER BOTTOMS TO CRUDE UNIT
FUEL OIL TO AIR COOLERS
HOT FUEL OIL FROM CRUDE
*ALLOY LINES TO BE SUPPORTED ON 4" SHOES IN PIPEWAY
STAND BY RUNNING
HAND-OFF AUTO
12-1502B 12-1502A
OVERHEAD PRODUCT PUMPS
12-1501A 12-1501B
STABILIZER REFLUX PUMPS

FIGURE 18.32 Mechanical Flow Diagram (P&ID)

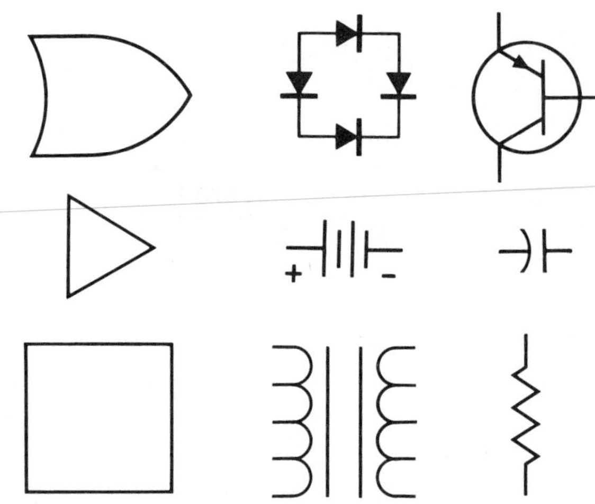

FIGURE 18.33 **Precut Stick-on Transfer Symbols**

18.4 Introduction to Electronics Drafting

Electronic drafting and design involves the creation of drawings associated with a wide range of electronic devices and systems. Drafting tools used in electronics drafting and design are similar to those in other fields of engineering and design work. When prepared manually, electronics drafting makes extensive use of transfer printed letters, numbers, electronic symbols, lines, and standard component shapes (Fig. 18.33). Printed component shapes are available in many configurations and are used for layout of printed circuit boards (Fig. 18.34). Electronic strips, component shapes, and slit shapes are used for lines and symbols in many instances, especially for printed circuit board artwork designs. *(See Color Plates 48-50.)*

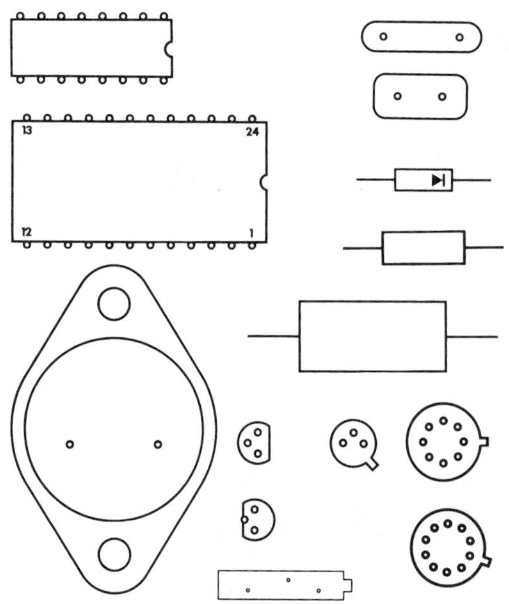

FIGURE 18.34 **Component Assembly Outlines**

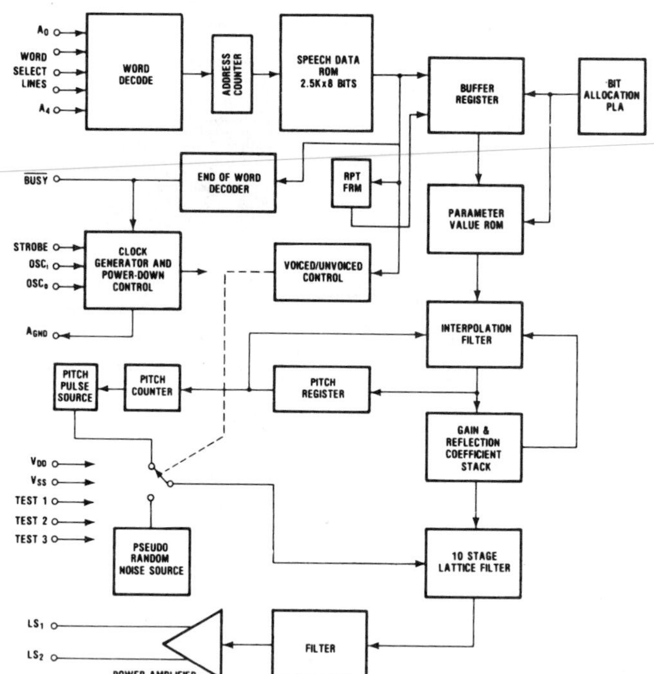

FIGURE 18.35 **Block Diagram**

18.4.1 Lines on Electronic Drawings

The knowledge of lines and their symbolic meanings and usage are fundamental to understanding electronic drawings. In electronics, most drawings are made up of single-line diagrams primarily composed of straight vertical and horizontal lines. Related drawings used for manufacturing electronics equipment include multiview dimensioned drawings, equipment drawings, sheet metal enclosure layouts and developments, and component drawings. Graphs, charts, and tables are frequently used to represent tabulated and related data on electronic drawings.

Understanding electronic drawings requires mastery of the symbol meanings, as well as mastery of the actual symbol and line usage. **Block diagrams** (Fig. 18.35), **logic diagrams** (Fig. 18.36), and **schematic diagrams** (Fig. 18.37) are single-line diagrams. Printed circuit boards (PCBs) have lines of differing thicknesses. ANSI Standard Y32.2, Graphic Symbols for Electrical and Electronic Diagrams, allows a degree of flexibility in choosing lineweights. The width of a drawn line does not affect the meaning of a line or a symbol. All electrical and electronic line weights should conform with ANSI Y14.2, Line Conventions and Lettering. Figure 18.38 illustrates line weights for electronic drawings and diagrams:

Dimension line Same as for mechanical convention; used on dimensioned production drawings, pictorial drawing (1), component multiview drawings (6), and sheet metal details and layouts.

Object line/visible line Same as for mechanical drawings; found on the same type of drawings as dimension lines.

Centerline Same as for mechanical drawings.

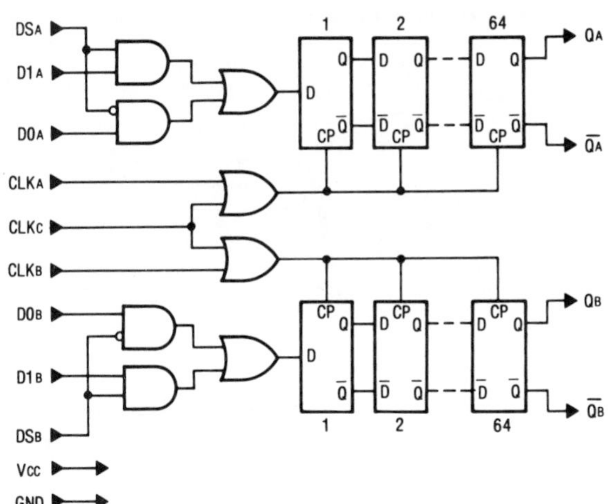

FIGURE 18.36 **Logic Diagram**

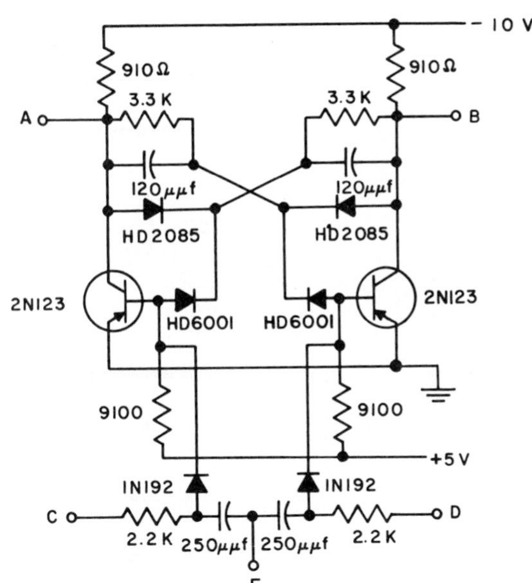

FIGURE 18.37 **Schematic Diagram**

Component outline Can be drawn as thick as an object line, although a medium to thin line is preferred depending on the reduction amount; used on PCB board layouts (7) and a variety of other electronic drawings.

Phantom line Used to represent future or alternative arrangements, such as on the wiring diagram (2).

Diagram line The majority of all lines used in electronic drawing are this type. Diagram lines are not standard on block diagrams (3), logic diagrams (8), schematic diagrams (4, 9) and wiring diagrams (2).

Emphasis line A thicker line used when a particular flow line (or symbol) on a diagram needs to be emphasized; also found on graphs (5).

Dashed line Used on electronic diagrams (10); also used to indicate mechanical linkage or connection between components (10, foot control switch).

18.4.2 Designations, Standards, and Abbreviations in Electronics Drafting

The American National Standards Institute (ANSI) and Department of Defense (DOD) set the standards for component symbols, reference designations, and abbreviations. Reference designations, standards, and abbreviations are used to define components, parts, and functions in electronics. They are used in the preparation of all logic and schematic diagrams to communicate desired designs.

Abbreviations are used to identify quantities or units on electronic drawings. As an example, "microfarads" is not spelled out on a drawing, instead, the abbreviation μF is used. Reference designations are letter symbols or the name of the component and are not to be confused with abbreviations on a drawing. Components can be discussed, drawn, and identified by using an alphanumeric code. Designations are composed of two aspects: a component class letter and a sequence number. As an example, the letter "C" is used for capacitors, the number "1" would be the first capacitor on the schematic or logic diagram and the second would be written C2, and so on until all of the capacitors on the diagram are identified.

18.4.3 Components and Symbols in Electronics Drafting

Because components are unique in size, function, and characteristics, they must have a special symbol to be quickly recognized. **Symbols** are used to represent the physical component when used on a diagram. Color coding and tolerance are used to identify specific components and to give value variation. The color code numbers range from 0 to 9; each number has a color. Components that use color coding also have a multiplier and a tolerance color. Table 18.2 gives the

TABLE 18.2 **Standard Color Code Used for Resistors**

Color	Number	Multiplier	Tolerance
Black	0	1	
Brown	1	10	
Red	2	100	
Orange	3	1000	
Yellow	4	10,000	
Green	5	100,000	
Blue	6	1,000,000	
Violet	7	10,000,000	
Gray	8	100,000,000	
White	9	— — —	
Gold	—	.1	5%
Silver	—	.01	10%
body	—	— — —	20%

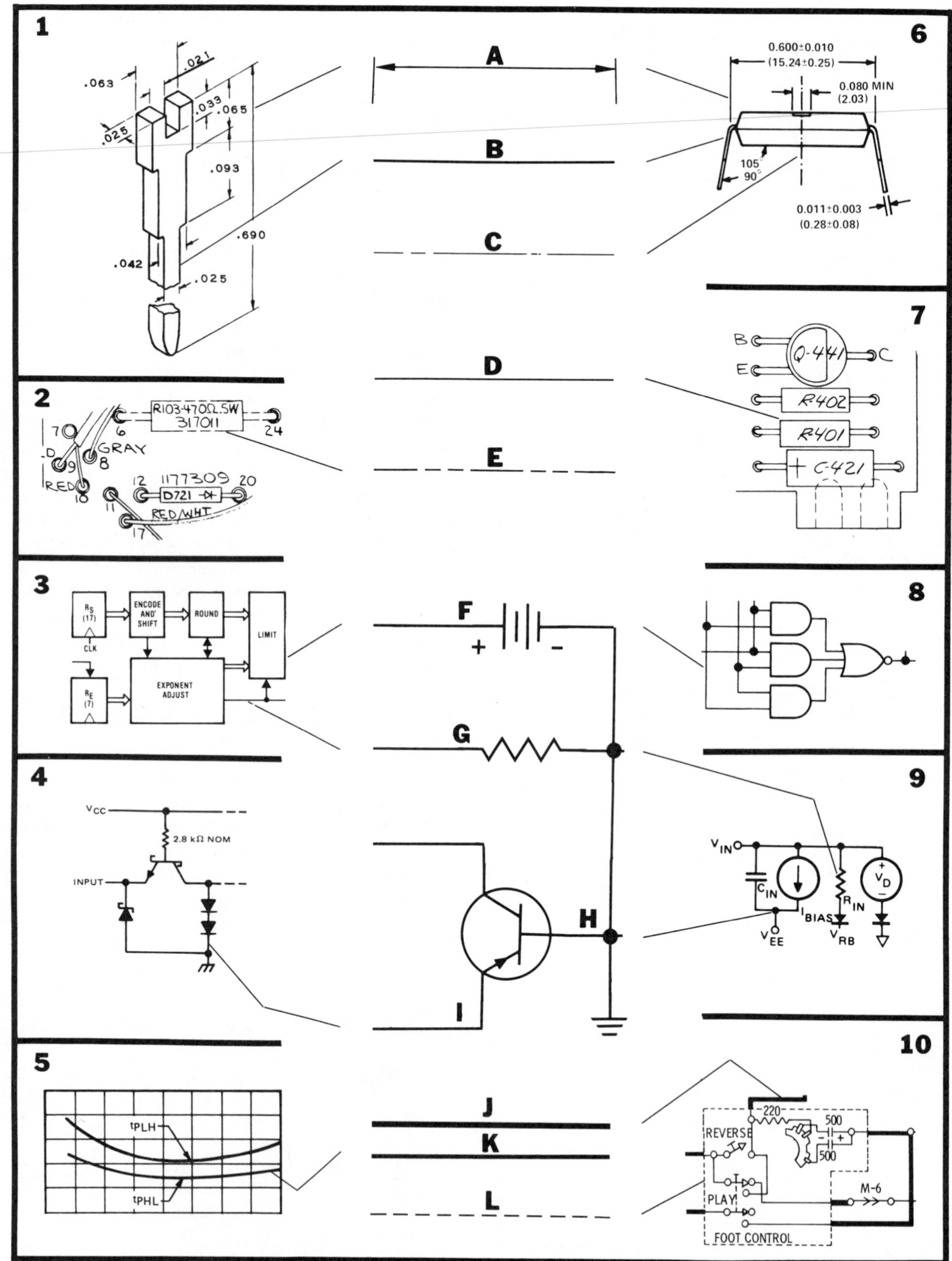

FIGURE 18.38 Line Key for Electronic Drawings

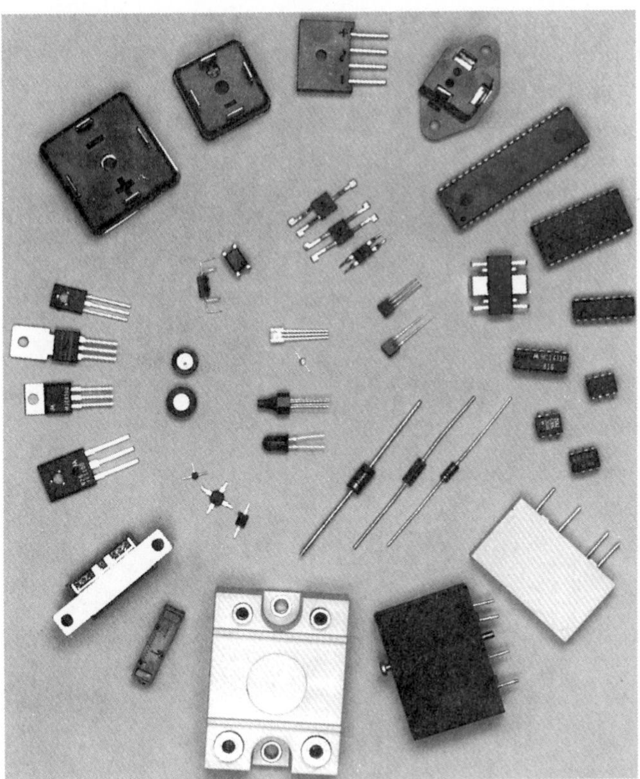

FIGURE 18.39 **Typical Discrete and Integrated Component Packages**

standard color code for resistors. Figure 18.39 shows a few of the many different kinds of components used in electronics. In Figure 18.40 a few electronic component symbols are provided. Refer to Appendix D for a complete list of component symbols.

The most commonly used component is a **resistor** (R). The three basic types of resistors are fixed, variable, and rheostat. The characteristics normally shown on a drawing are resistance value expressed in ohms (Ω) or kilo-ohms (KΩ), or mega ohms (MΩ), the resistance tolerance expressed in percent (%) of component value and the power rating expressed in wattage (W).

Capacitors (C) are components that also have three basic types: fixed, variable, and electrolytic (polarized). The functions of capacitors are to store energy or to block the flow of dc current and to permit the flow of ac current. Capacitor characteristics are expressed on the drawing by capacitance value, farads (F), microfarads (μF), or picofarads (pF), and a tolerance of capacitance expressed as a percent, either ±10% or ±20%. Voltage rates may be called out on the drawing (F/D) or in notes, usually 10 to 200 VDC.

Diodes and **transistors** are both semiconductors. There are many types of diodes, but the three most common are *signal, zener,* and *tunnel.* Transistors (Q) are the *active* or *amplifying devices.* Four of the most commonly used transistors are the bipolar (NPN or PNP), the power transistors, unijunction, and the field-effect (FET). Transistors are used as amplifiers, switches, or detectors.

Transformers (T) are a device consisting of two or more coils coupled together by magnetic induction. Transformers are used to transfer electric energy from one or more other circuits without change in frequency, but usually with changed values of voltage and current.

Another component is a **coil** or **inductor** (L). Coils consist of a number of turns of wire used to introduce inductance into an electric circuit to produce magnetic flux or to react electrically to a changing magnetic flux. The electrical size of a coil is called *inductance* and is expressed in henrys. The opposition that a coil offers to alternating current is called *impedance* and is expressed in ohms.

Integrated circuits (ICs) are an interconnected array of active and passive elements integrated with a single semiconductor substrate or deposited on the substrate by a continuous series of compatible processes. ICs come in many physical configurations from flat paks to TO-5 cans to dual-in-line and from 4-pin to 100-pin arrangements (Fig. 18.41). ICs have microcircuits consisting of diodes, transistors, and resistors, and are made up of silicon, aluminum, isolating barriers, and metal oxide.

18.4.4 Electronic Diagrams

A **block diagram** is a drawing in which the principal divisions of an electronic system, program, or process are indicated by rectangles or other geometric shapes, with signal paths represented by lines (Fig. 18.42). Block diagrams are the easiest form of diagram to draw and read, which makes them ideal for conveying information to people with limited technical knowledge.

Rectangles, diamonds, circles, and ellipses are used on block diagrams (Fig. 18.43). Electronic symbols are also used on some block diagrams. They are usually functional units in their own right and are normally depicted with aspects of their real-life configurations such as antennas, earphones, speakers, and switches (Fig. 18.44).

Blocks and symbols are connected by **flow lines** according to the functional sequence of the system. In Figure 18.45, the signal depicted by the flow line goes from the power supply into each of the three units. The standard left-to-right flow sequence connects the antenna to the RF amplifier to the detector to the audio power amplifier. Flow lines are drawn as solid, medium-weight lines.

Crossovers and connections of flow lines must be clearly presented if the drawing is to be properly understood (see Fig. 18.46 on p. 572). *Connecting lines are drawn with a minimum number of changes in direction.* When laying out the diagram, limit the amount of corners and crossovers to the absolute minimum. In Figure 18.46(a) a crossover is shown. In Figure 18.46(b) an acceptable method of drawing a multiple connection is provided, but Figures 18.46(c) and (d) are the preferred method for showing this form of connection or junction. When a line terminates at another line it is unnecessary to use the dot [Fig. 18.46(e)]. When an array or bank of parallel and perpendicular lines requires connection

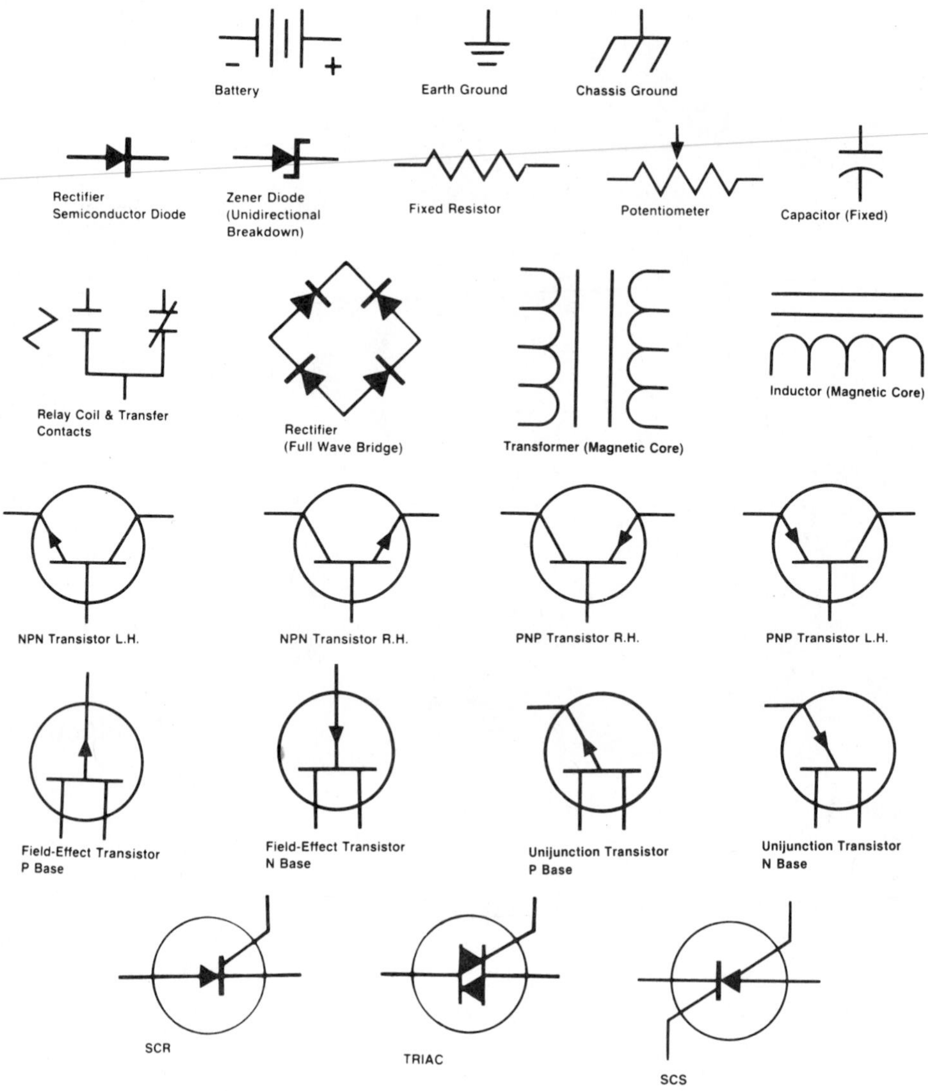

Battery
Earth Ground
Chassis Ground
Rectifier Semiconductor Diode
Zener Diode (Unidirectional Breakdown)
Fixed Resistor
Potentiometer
Capacitor (Fixed)
Relay Coil & Transfer Contacts
Rectifier (Full Wave Bridge)
Transformer (Magnetic Core)
Inductor (Magnetic Core)
NPN Transistor L.H.
NPN Transistor R.H.
PNP Transistor R.H.
PNP Transistor L.H.
Field-Effect Transistor P Base
Field-Effect Transistor N Base
Unijunction Transistor P Base
Unijunction Transistor N Base
SCR
TRIAC
SCS

FIGURE 18.40 Dry Transfer Stick-ons of Electronic Symbols

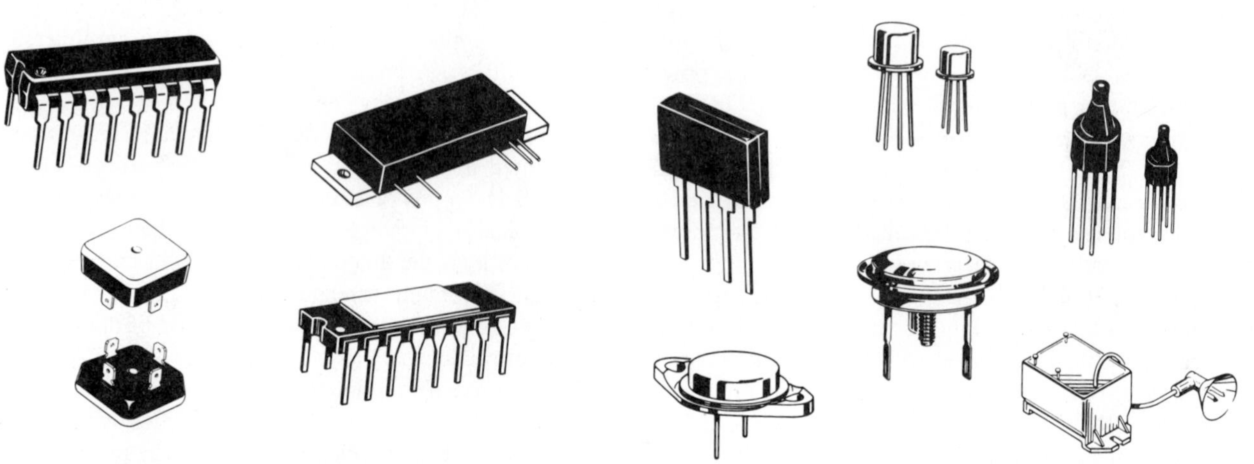

FIGURE 18.41 Electronic Component Cases

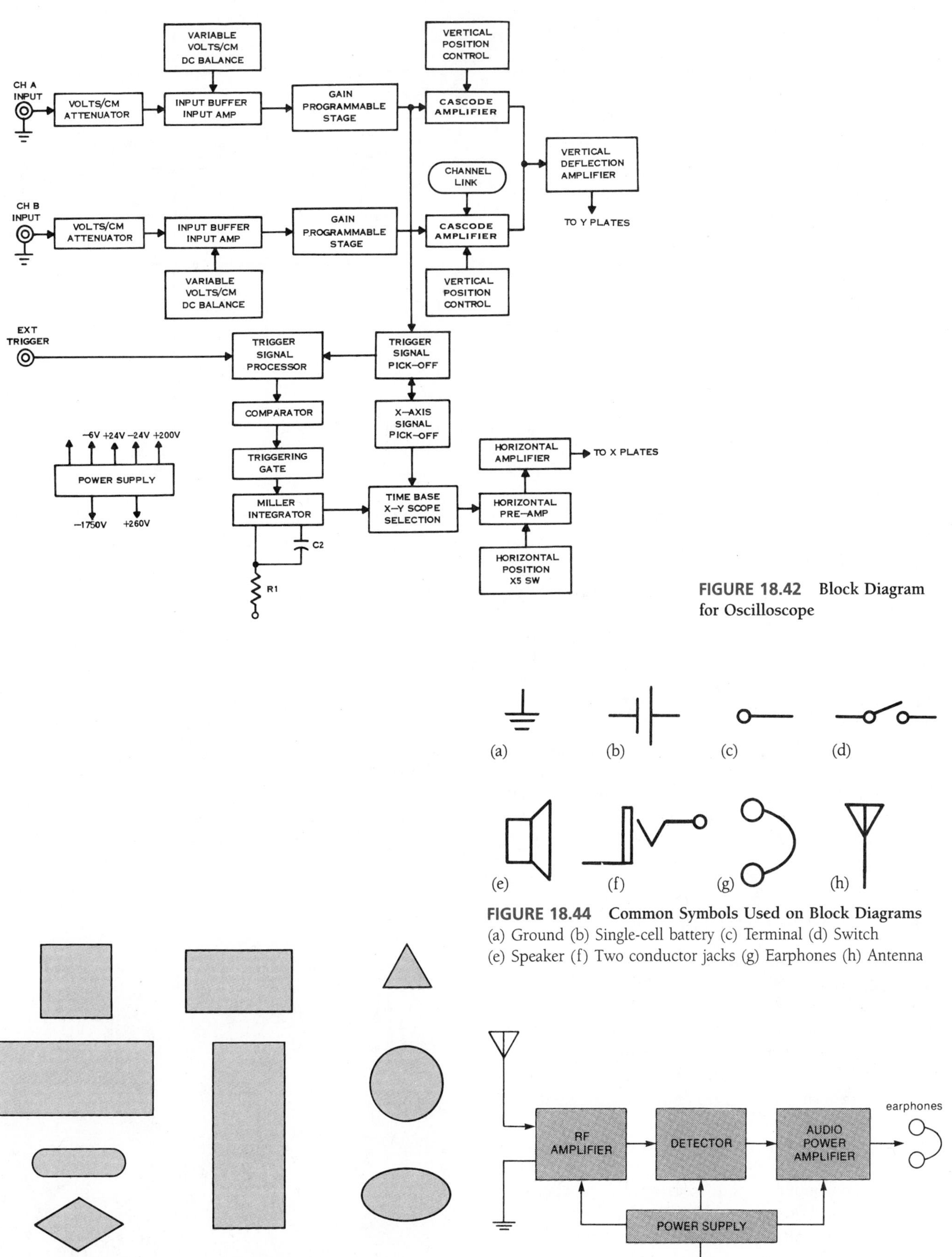

FIGURE 18.42 Block Diagram for Oscilloscope

FIGURE 18.44 Common Symbols Used on Block Diagrams
(a) Ground (b) Single-cell battery (c) Terminal (d) Switch (e) Speaker (f) Two conductor jacks (g) Earphones (h) Antenna

FIGURE 18.43 Common Flow Diagram and Block Diagram Geometric Shapes

FIGURE 18.45 Tuned Frequency Receiver Block Diagram

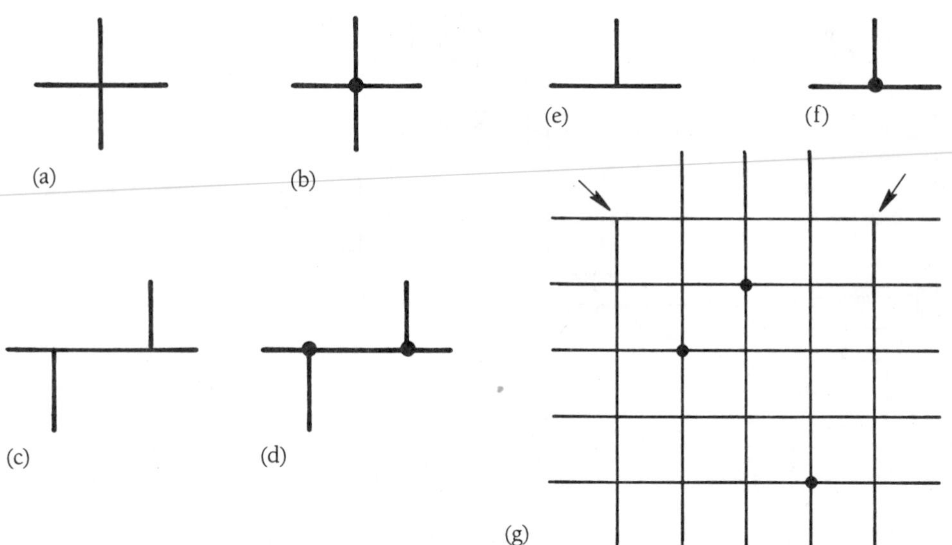

FIGURE 18.46 Crossovers and Dashed Lines (a) Crossover (b) Multiple connection (dot method) (c) Multiple connection (no dot) (d) Preferred multiple connection (dot method) (e) Single connection (no dot) (f) Single connection (dot method) (g) Using dots for array of lines (note that single connections do not require dots)

points, the dot method is used for multiple junctions [Fig. 18.46(f)]. When a line terminates at a perpendicular line in a single connection, the dot may be eliminated [Fig. 18.46(g)].

The schematic diagram is a drawing showing, by means of graphic symbols, the electrical connections and functions of a circuit arrangement. A schematic does not show the size, shape, or location of the component devices or parts. The logic diagram depicts logic symbols and supplementary notations, the details of signal flow, and control but not necessarily the point-to-point wiring existing in a system.

Logic symbols are drawn according to ANSI Y32.14 with the size determined by nomenclature. When drawn manually, the preferred size on gates is half size using a Rapidesign template #R542. For electrical and electronics symbols use ANSI Y32.2 with the .250 grid size preferred using Rapidesign template #315.

18.4.5 Printed Circuit Boards

PCB design is one of the primary functions of electronic drafting. A **printed circuit board** is a baseboard made of a laminated insulating material that contains ICs or discrete components, along with the connections required to implement one or more electronic functions (Fig. 18.47). The connections of a PCB are a thin layer of conductor material. The conductor pattern is established on the board by one of two main processes. The *additive process* involves depositing the conductor material on the base in its required conductor pattern. The *subtractive process* uses an etching solution to eat away the nonconductor path areas on the board after the conductor path pattern is coated or printed with an etch-resisting paint or ink. A PCB may have all the components on one side and the circuitry (conductor pattern) on the other (single-sided boards), or circuitry layered on a number of bonded boards (multilayered boards).

A PCB can be designed with manual or automated processes. Manual methods are frequently used for prototype boards or where the board's tolerance and accuracy specifications are less stringent. Single-sided and double-sided boards are more apt to be manually prepared than multilayered boards with their high-density and critical alignment of layers.

Regardless of whether a PCB is to be a one-of-a-kind prototype or a high-volume production article, it will require graphic documentation. A PCB drawing package may include the drawings shown in Figure 18.48. Exactly how much documentation and how it should be prepared vary with company practice, application, and method of generating the documentation. Adequate documentation conveys to the user the basic electromechanical design concept, the type and quantity of parts and materials required, special manufacturing instructions, and up-to-date revisions. Regardless of the

FIGURE 18.47 A Microcomputer Printed Circuit Board

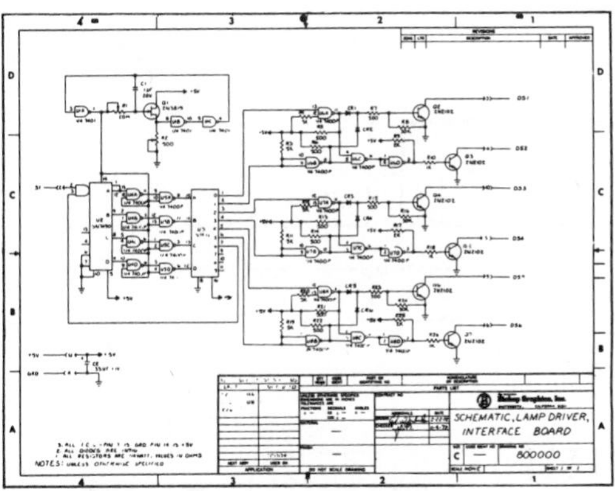

SCHEMATIC DIAGRAM: Consists of graphic symbols indicating the interconnections and functions of an electronic circuit. It is the basis for a printed circuit design. It is also used to test, evaluate and troubleshoot the completed circuit board.

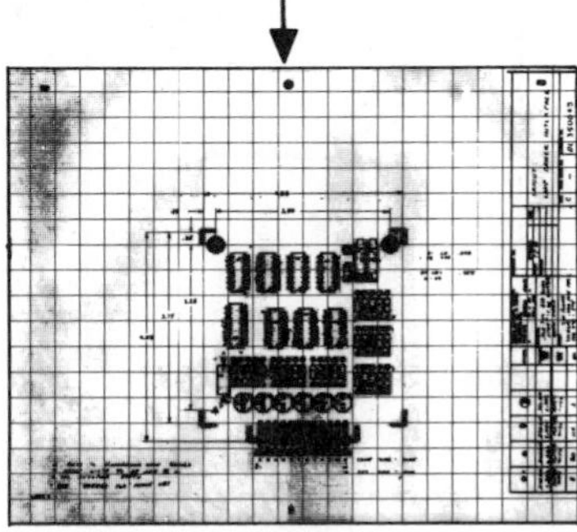

LAYOUT: The conceptual intermediate link between the schematic diagram and the master pattern.

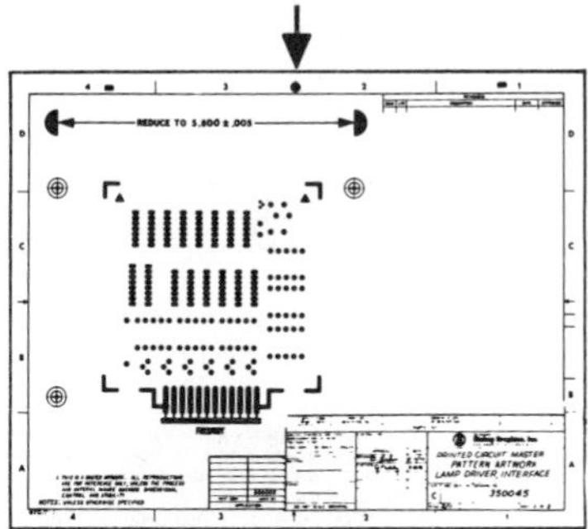

ARTWORK: Accurately scaled configuration of the printed circuit pattern from which the master pattern is photographically produced.

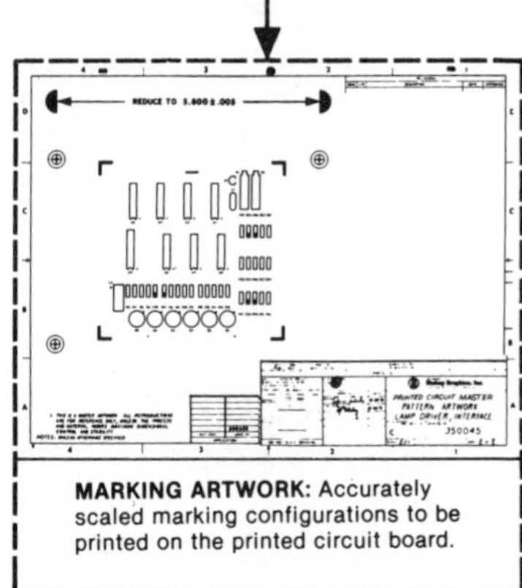

MARKING ARTWORK: Accurately scaled marking configurations to be printed on the printed circuit board.

FIGURE 18.48 Printed Circuit Board Design and Fabrication Sequence

number of design steps, method of design, or the types of documentation, the end result is the assembled functional PC board.

A **terminal area** (pad) is a portion of a printed circuit used for making electrical connections between a component or wire and part of the conductive circuit pattern. The pad size is based on the hole diameter and the hole diameter is determined from the lead diameter.

Component mounting is one of the most important considerations in the design of a PCB. Component type, mounting style, and orientation on the board are all factors in the placement and layout of the board.

One of the primary considerations in PCB design is the selection of the board design and basic outline. In general, a simple rectangular board with a connector strip is the easiest and most simple board geometry design. Boards designed with cutouts, curves, and irregular angles are more expensive to design, fabricate, and manufacture.

Accurately *registered* PCB artwork is produced from pin-registered multisheet artwork techniques when done manually. With this method, a pad master base sheet is created on transparent, dimensionally stable drafting film. An accurate pattern of precisely sized locating holes punched into the sheet is used for pinning successive overlays in precise registration (Fig. 18.49). A PCB designer uses a light table and pin registration for manually laying out printed circuit artwork.

A PCB fabrication drawing presents a complete engineering description of all design features. Information recurring from board to board is placed in the same location on the drawing to facilitate communications between departments and between the company and its outside vendors. The master fabrication drawing is normally drawn at the same scale as the film photomaster. It is viewed from the noncomponent side of the board if only one side is to be shown (Fig. 18.50).

The PCB assembly drawing is a complete engineering description of a printed circuit board assembly (Fig. 18.51) including all components and parts mounted by fastening, soldering, or bolting. A parts list is included on the drawing or on a separate sheet. All parts, other than components, are ballooned on the assembly drawing and listed on the parts list. The assembly shows the board configuration, components, reference designation markings, mechanical parts (clamps, fasteners, and brackets), and dimensions required for accurate assembly and component orientation. The assembly drawing is drawn as viewed from the component side of the board and usually at the same scale as the artwork drawing. In general, the PCB assembly drawing is the same as any engineering assembly drawing. Parts, materials, and assembly instructions should appear on the drawing.

18.4.6 Cables and Harnesses

A **cable** is a group of two or more insulated conductors enclosed in the same outer protection. Generally, the conductors at either end terminate in the same approximate location. Frequently, a cable assembly terminates in a multiple connector.

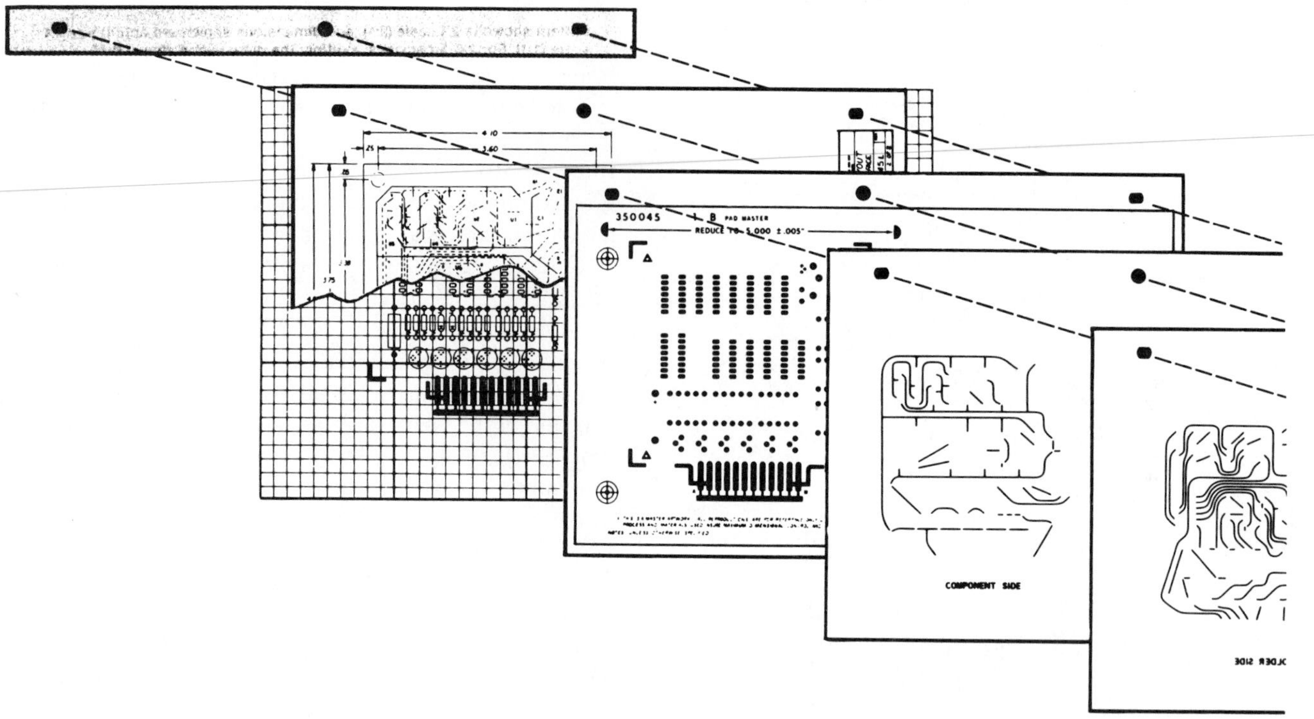

FIGURE 18.49 Pin Registration

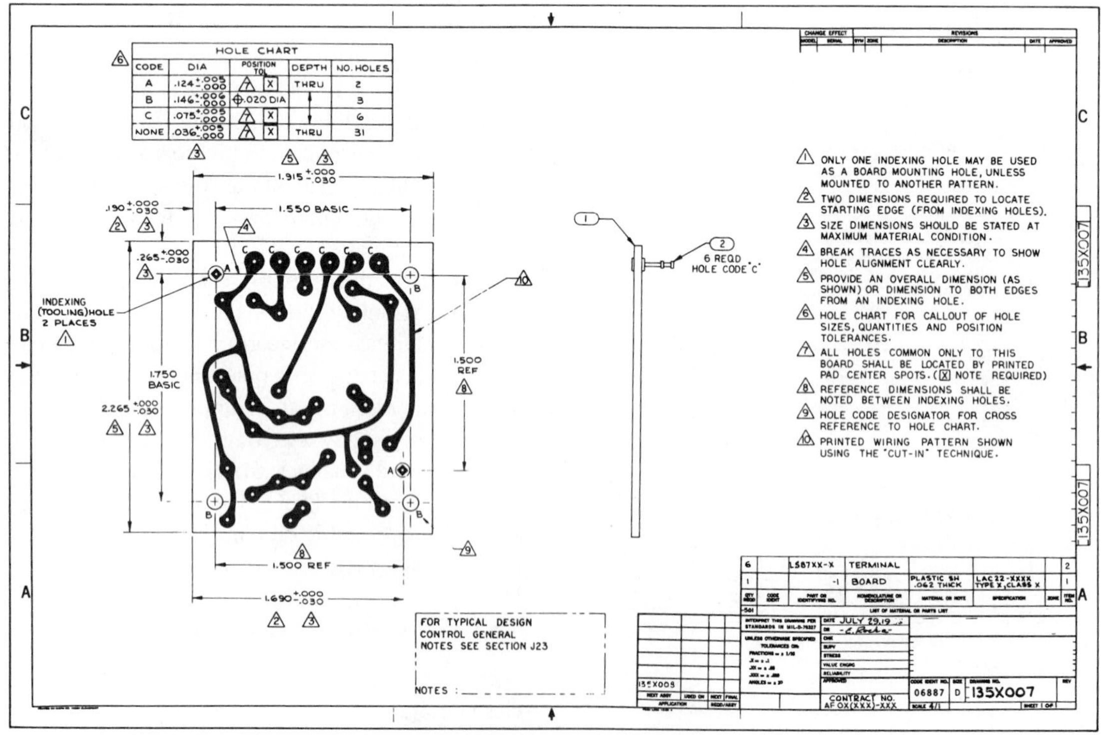

FIGURE 18.50 Printed Circuit Board Detail Drawing

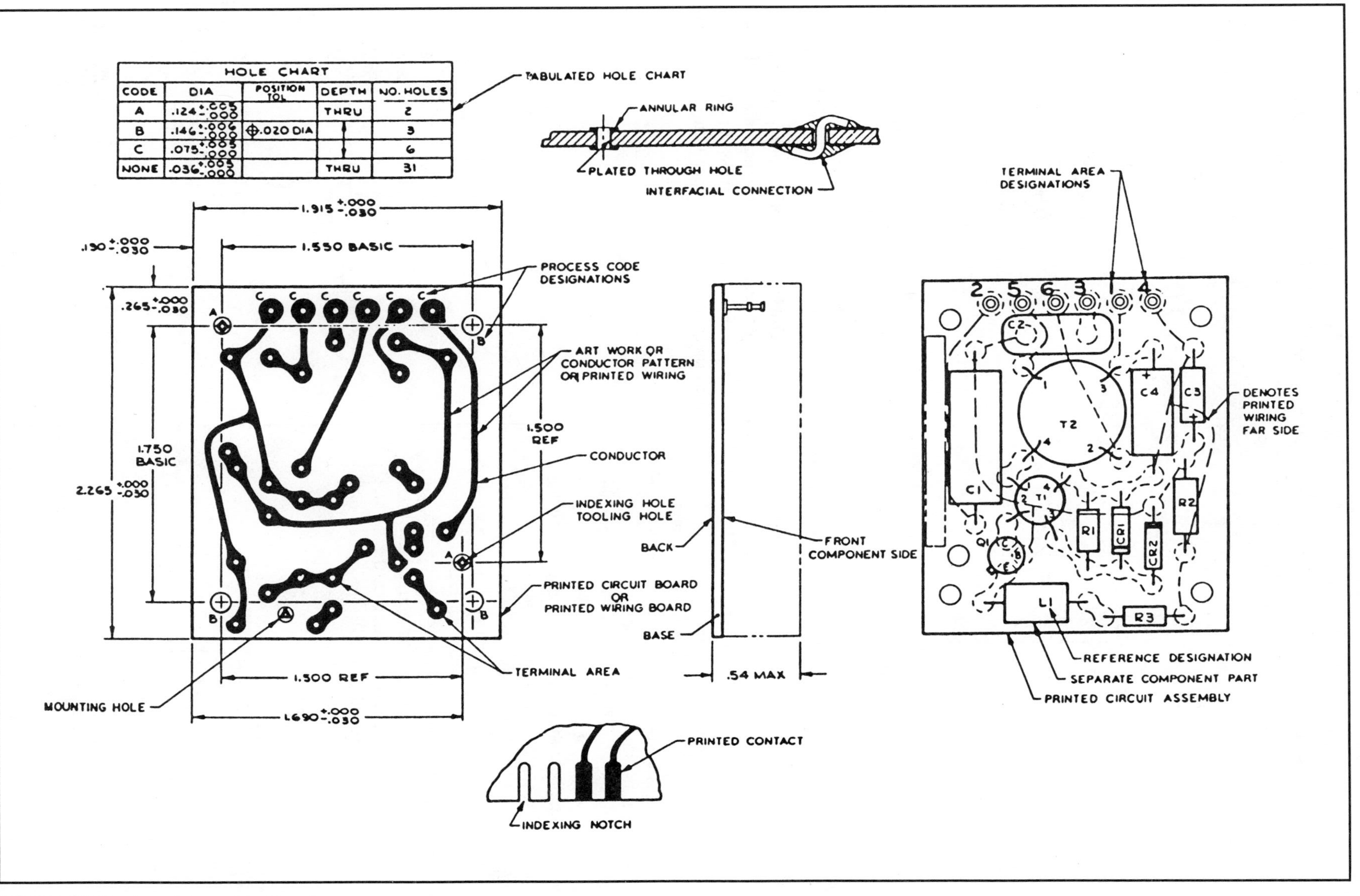

HOLE CHART				
CODE	DIA	POSITION TOL	DEPTH	NO. HOLES
A	$.124^{+.005}_{-.000}$		THRU	2
B	$.146^{+.006}_{-.000}$	⌖ .020 DIA		3
C	$.075^{+.005}_{-.000}$			6
NONE	$.036^{+.005}_{-.000}$		THRU	31

FIGURE 18.51 Board Assembly Drawing

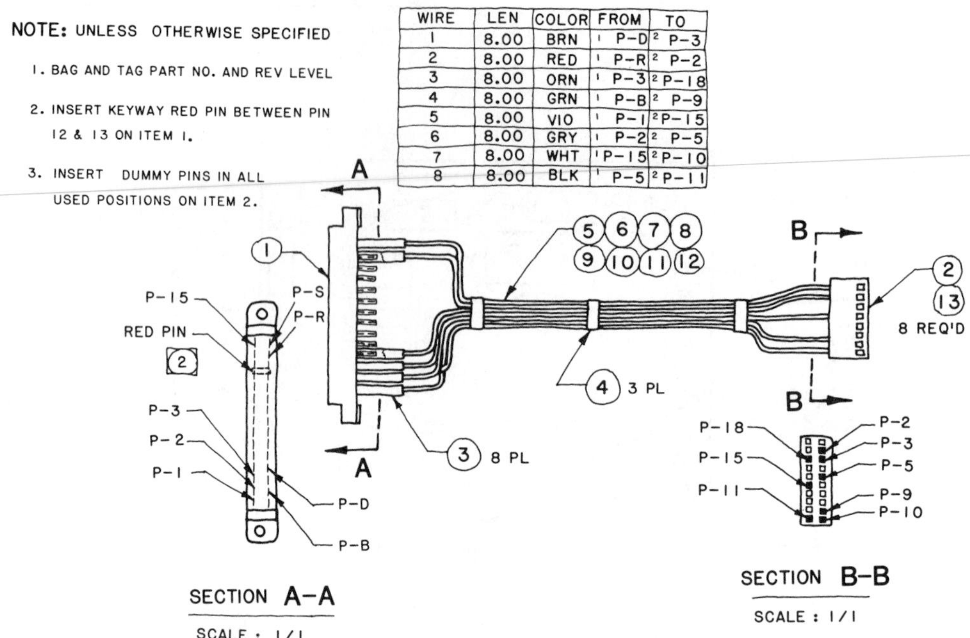

WIRE	LEN	COLOR	FROM	TO
1	8.00	BRN	1 P-D	2 P-3
2	8.00	RED	1 P-R	2 P-2
3	8.00	ORN	1 P-3	2 P-18
4	8.00	GRN	1 P-B	2 P-9
5	8.00	VIO	1 P-1	2 P-15
6	8.00	GRY	1 P-2	2 P-5
7	8.00	WHT	1 P-15	2 P-10
8	8.00	BLK	1 P-5	2 P-11

FIGURE 18.52 Typical Harness Assembly

tor. Typically, **cable assemblies** are used to connect one subassembly to another with mating connectors (Fig. 18.52). One common type of cable is the flat ribbon cable. The conductors on a flat cable are side by side and the assembly looks like a flat ribbon. Although this type of cable is wide, it takes up very little routing space when used to interconnect subassemblies because it fits through small slots.

A wiring harness is shown in Figure 18.53. Unlike the cable, the wiring harness does not require protective covering or coating over the bundled wires, although it may have some type of clear coating to protect it against environmental factors. The primary difference between a cable and a harness is that the cable includes the protective outer covering and a harness is simply a group of wires that for convenience have been bundled together. Also, harness wires can have many different lengths and can branch off at different locations (Fig. 18.53).

After the layout of the harness is complete, the harness diagram assembly is prepared (Fig. 18.54). The overall outline of the harness is used rather than a drawing of individual conductors. The conductors are individually identified where they branch out from the harness. The primary use of the harness diagram is in the preparation of the harness itself,

FIGURE 18.53 Wire Harness Assembly Showing Location of Cable Ties

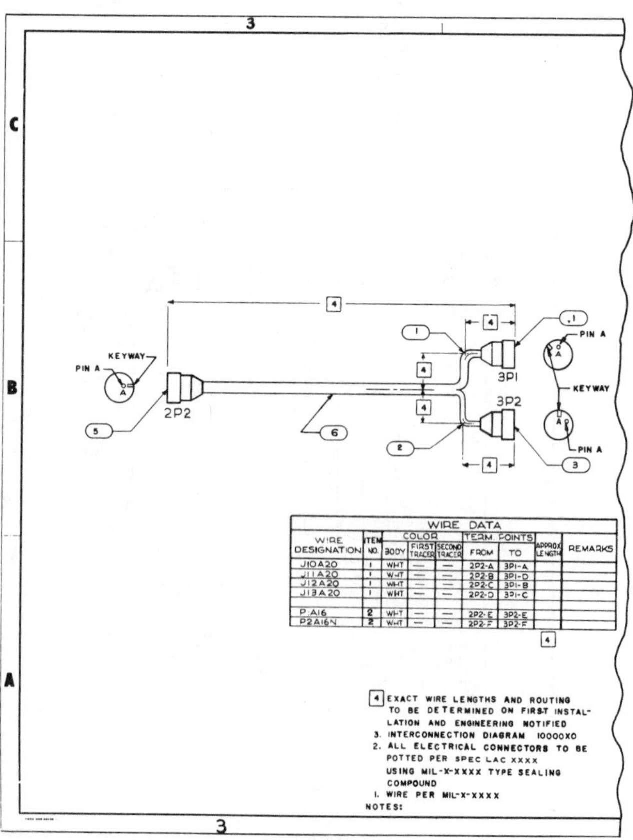

WIRE DESIGNATION	ITEM NO.	COLOR: BODY	COLOR: FIRST TRACER	COLOR: SECOND TRACER	TERM. POINTS: FROM	TERM. POINTS: TO	APPROX LENGTH	REMARKS
J10A20	1	WHT	—	—	2P2-A	3P1-A		
J11A20	1	WHT	—	—	2P2-B	3P1-D		
J12A20	1	WHT	—	—	2P2-C	3P1-B		
J13A20	1	WHT	—	—	2P2-D	3P1-C		
P1A16	2	WHT	—	—	2P2-E	3P2-E		
P2A16N	2	WHT	—	—	2P2-F	3P2-F		

FIGURE 18.54 Harness Assembly with Wire Data and Notes

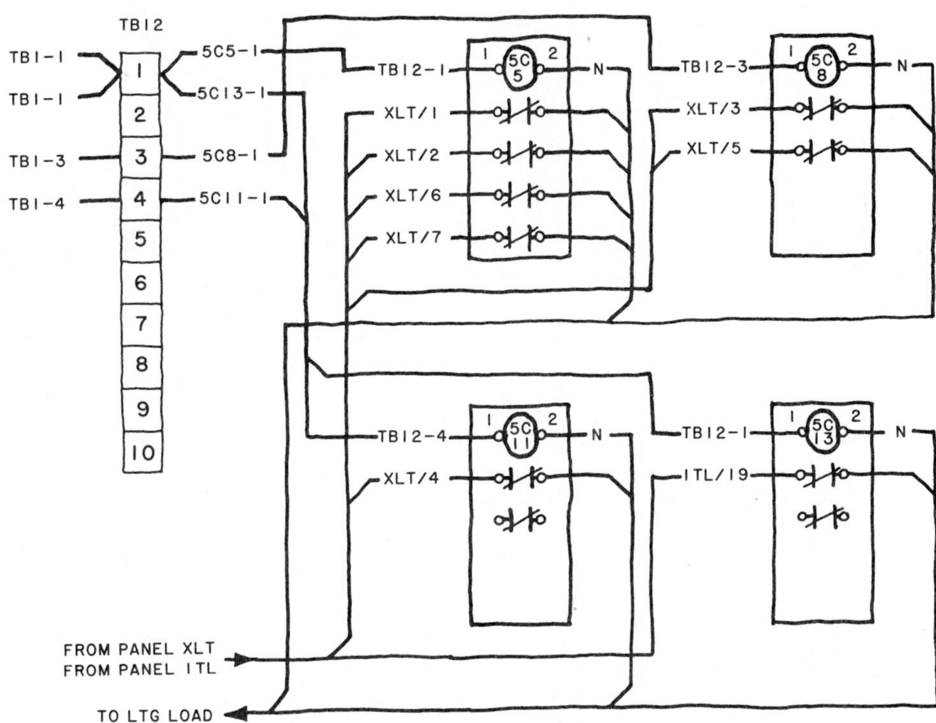

FIGURE 18.55 Wiring Diagram

therefore, it is drawn full size. Details for conductors are included in a table or bill of material along with specifications for the conductors in the harness: color, termination, origination, destination, AWG number, and type. Included on the diagram are the drawing numbers of the schematic diagram, wiring diagram, and any other associated diagrams.

18.4.7 Wiring Diagrams

Unlike schematics, for which connections or junctions are made at convenient locations, on **wiring diagrams** the connecting lines must go from one terminal to another. Junctions, joints, or splices are very seldom made in electronics assemblies. In general, all line weights are the same on wiring diagrams (black, medium-weight lines). If necessary, use thin lines for component outlines or symbols to avoid confusion. In addition to including some type of identification or reference designation for every component, each wire includes identification of all specifications. The lettering can be inclined or vertical and should be aligned to be read from no more than two sides. To avoid confusion, the wire identification lettering may be done with a break in the line representing the wire. The lettering is inserted in the break, as in the sketch of the control panel wiring diagram shown in Figure 18.55. This lettering eliminates any doubt as to which line the identifying code accompanies.

18.5 Introduction to Fluid Power

Fluid power systems use pressurized fluids such as air, water, and oil to control, transmit, or generate power. You are already familiar with many fluid power systems: power steering and power braking systems in automobiles, control systems and landing gears in airplanes, control systems in heavy machinery (forklifts, paving machines, earth movers, etc.), pneumatic tools, fuel pumps and carburation in automobiles, control of spacecraft, automation equipment, such as robots and automated part loaders, hydroelectric power generation systems, and heating, ventilation, and air conditioning (HVAC) systems. Just about everywhere you look today, you can find some kind of fluid power system.

The fluid in a fluid power system can either be a liquid or a gas. **Pneumatic systems** use air because it is convenient to use, inexpensive, and can be vented to the atmosphere. Air, however, is compressible, making pneumatic systems less rigid than hydraulic systems. Water was first used in hydraulic systems because it is so abundant, but it rusts steel and iron components, is not a good lubricant, and freezes easily. Therefore, most modern **hydraulic systems** use hydraulic oils or synthetic oils.

Symbols on engineering drawings of fluid power systems are used to represent the different components of the system. ANSI has defined standard symbols for fluid power components. This section of the chapter describes simple fluid power systems, how they operate, and how they are represented by symbols on engineering drawings.

18.5.1 Fluid Power Systems

Fluid power systems are designed to do useful work, sometimes applying thousands of pounds of force. For example, construction equipment such as excavators, graders, scrapers,

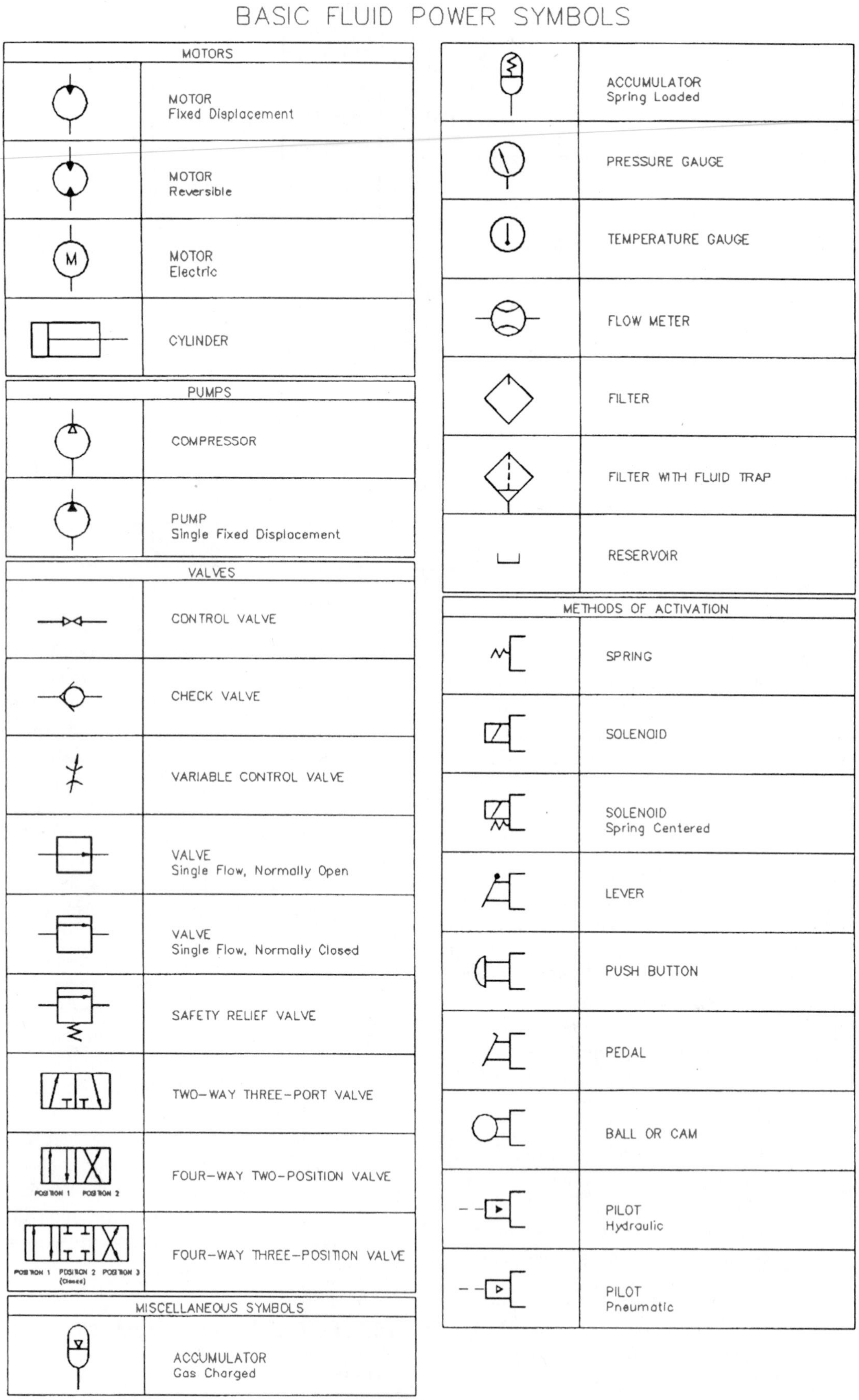

BASIC FLUID POWER SYMBOLS

MOTORS	
	MOTOR Fixed Displacement
	MOTOR Reversible
M	MOTOR Electric
	CYLINDER

PUMPS	
	COMPRESSOR
	PUMP Single Fixed Displacement

VALVES	
	CONTROL VALVE
	CHECK VALVE
	VARIABLE CONTROL VALVE
	VALVE Single Flow, Normally Open
	VALVE Single Flow, Normally Closed
	SAFETY RELIEF VALVE
	TWO-WAY THREE-PORT VALVE
POSITION 1 POSITION 2	FOUR-WAY TWO-POSITION VALVE
POSITION 1 POSITION 2 (Closed) POSITION 3	FOUR-WAY THREE-POSITION VALVE

MISCELLANEOUS SYMBOLS	
	ACCUMULATOR Gas Charged
	ACCUMULATOR Spring Loaded
	PRESSURE GAUGE
	TEMPERATURE GAUGE
	FLOW METER
	FILTER
	FILTER WITH FLUID TRAP
	RESERVOIR

METHODS OF ACTIVATION	
	SPRING
	SOLENOID
	SOLENOID Spring Centered
	LEVER
	PUSH BUTTON
	PEDAL
	BALL OR CAM
	PILOT Hydraulic
	PILOT Pneumatic

FIGURE 18.56 Basic Fluid Power Symbols

and loaders all have fluid power-based systems that apply a great deal of force. Regardless of the complexity of the fluid power and control system, all pneumatic or hydraulic circuits build on the same basic components.

A **basic hydraulic circuit** contains the following components:

1. A *reservoir (tank)* to hold the hydraulic fluid
2. Some *power source,* such as an electric motor, to drive the pump
3. A *pump* to move the fluid through the circuit
4. *Valves* to control the direction, pressure, and flow rate of the fluid
5. An *actuator* that converts the moving fluid into some sort of useful work
   (Hydraulic fluid can turn motors or extend hydraulic cylinders to convert the fluid energy into useful work.)
6. *Piping* to move the fluid from one location to another.

A **basic pneumatic circuit** contains the following components:

1. An *air compressor* to compress air
2. An *air tank* to hold a supply of compressed air
3. An *electric motor* or other device to power the air compressor
4. *Valves* to control the direction, pressure, and flow rate of the air
5. *Actuators* that convert the moving fluid into some sort of useful work (Moving pressurized air could be used to turn motors or extend pneumatic cylinders to convert the fluid energy into useful work.)
6. *Piping* or *tubing* to carry the pressurized air around the circuit.

The basic components of a pneumatic or hydraulic system can be assembled to apply either rotary motion (by using a motor) or linear motion (by using a cylinder) to the workpiece. Control of the fluid power valves can range from simple manual methods to complex microprocessor feedback control systems. In the hydraulic braking system for your car, you manually apply the brakes at the appropriate time. Whether or not the system is power assisted, you choose the braking time. However, on new antilock braking systems (ABS brakes), a microprocessor program determines the optimum way to apply the brakes so that they do not "lock up" while the car is coming to a stop.

Conveying the information about the design of the fluid power system through an engineering drawing is a matter of using the appropriate symbol for each fluid power component and line. ANSI has established standard fluid power symbols for the basic fluid power components. Figure 18.56 shows the standard ANSI symbols for several of the basic components of fluid power systems. Notice that there are symbols for the listed basic components: motors and cylinders, pumps, valves, reservoirs, and instrumentation. Valves can be activated in several ways: springs, solenoids, levers, push buttons, pedal, ball, cam, etc. A **solenoid** is a way to activate a valve that is electrical and not mechanical in nature. For example, depending of the exact design of the valve, a solenoid uses an electrical signal to move the spool in the valve to change the direction of flow in the valve. Since it is electrically operated, solenoid actuation is used extensively in computer- or microprocessor-controlled fluid power circuits.

Figure 18.57 is a drawing of a basic pneumatic fluid power circuit that controls a spring return air cylinder. At first glance the circuit might seem complicated, but there are only four parts to the diagram :

1. **"From air supply"** means that this circuit is hooked to a large air compressor and tank that can supply an unlimited amount of flow to the circuit.
2. **"Filter, regulator and lubricator (frl) unit"** filters and regulates the air in the circuit to a certain pressure value. A lubricant is added to the air to lubricate the metal components. A safety relief valve is in the unit in case the regulator fails to keep the air pressure to a safe value.
3. **"A solenoid (electrically operated) two-position three-port two-way valve"** controls the direction of air flow to the pneumatic cylinder. Pressurized air will be used to extend the cylinder.
4. **"A spring return air cylinder"** is the actuator in the circuit. The pressurized air in the system is used to extend the air cylinder. The air cylinder applies linear motion to accomplish some function, such as moving parts from one position to another. A spring is used to return the air cylinder.

If you wanted to apply a simple rotary motion with the same circuit, you would simply replace the linear pneumatic cylinder with an air motor. Complex fluid power systems are actually just basic systems that are linked together to accomplish some predefined function.

18.5.2 Fluid Power Components

Several basic components are common to many fluid power systems. The design of the device is different if the device is an air (pneumatic) or hydraulic device, but the primary function is unchanged regardless of the fluid involved. Remember that the fluid may be a gas or a liquid. Liquid systems are more rigid than gas systems because liquids are more difficult to compress. Most pneumatic systems are "spongy" because of the compressibility of air. Therefore, pneumatic systems are not used when you need to apply a large amount of force.

A **reservoir** is used in fluid power circuits to hold and supply an appropriate volume of fluid for the circuit. An *air tank* manufactured to exact specifications is the reservoir for pneumatic circuits. The reservoir for hydraulic circuits is more complicated. It may contain a strainer, a filter, an oil level gage, an air breather, and baffles. As hydraulic fluid moves through the circuit, friction losses in the pipes, valves, and joints heat the hydraulic fluid. The *baffles* in the reservoir are designed to remove as much heat as necessary from the circuit fluid.

Pumps are used in fluid power circuits to convert mechanical energy into fluid power energy. Pumps are powered

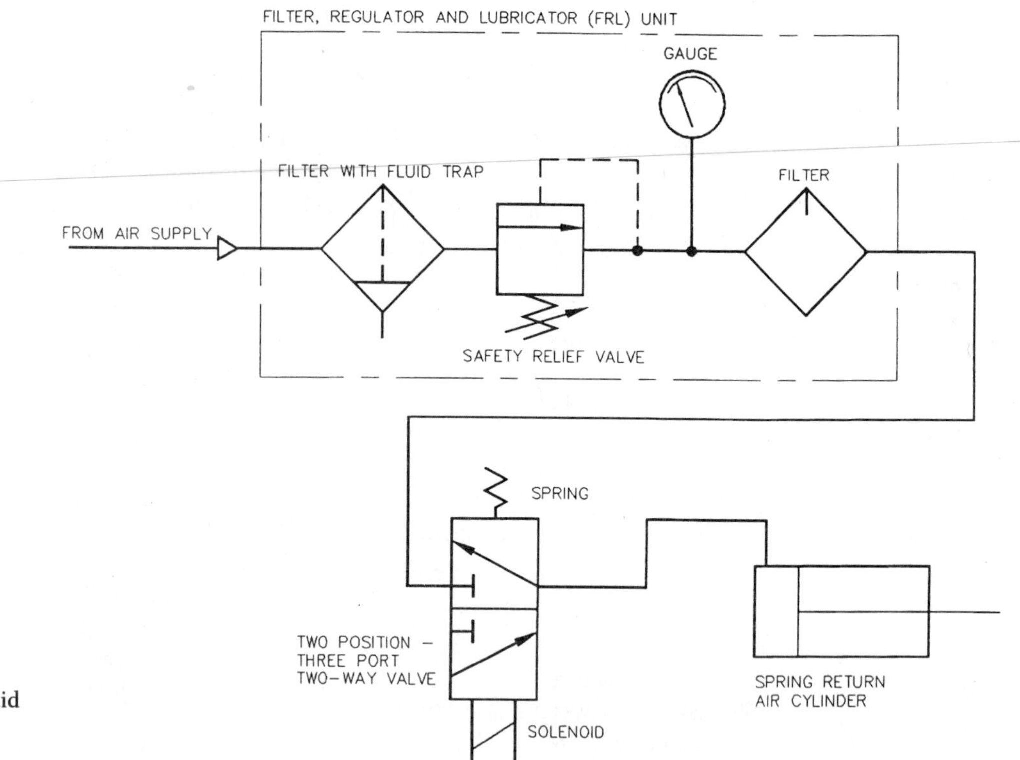

FIGURE 18.57 **Pneumatic Fluid Power Circuit that Controls a Spring Returned Air Cylinder**

by prime movers, such as electrical motors, and convert that energy into fluid moving throughout the circuit. A **positive displacement pump** delivers a fixed quantity of fluid with each revolution of the pump. Most industrial hydraulic power systems use positive displacement pumps to pump the fluid in the system. A **nonpositive displacement pump** or a **kinetic pump** adds energy to the fluid by accelerating it through a rotating impeller. These pumps are used for low-pressure, high-volume applications that move fluids from one location to another.

A positive displacement pump produces a pulsating flow, while a kinetic pump produces a continuous flow. Positive displacement pumps are not affected by variations in system pressure. Types of positive displacement pumps are *gear, vane, screw, cam or lobe, piston, plunger, and diaphragm.* Types of kinetic pumps are *radial flow (centrifugal), axial flow (propeller), and mixed flow.*

Many factors must be considered when selecting a pump for a particular application: type of fluid to be pumped, the volume of fluid to be pumped, the total energy that must be delivered to the system by the pump, the power source for the pump (electric, steam turbine, diesel, etc.), configuration limitations (size), cost of the pump (installation and operating), and type of connections at the inlet and outlet.

Air compressors provide the energy to pneumatic circuits. Air compressors compress air (or another gas) from atmospheric pressure to a higher pressure for the fluid power circuit by reducing the volume of the gas. Air compressors are usually positive displacement machines. They are reciprocating piston, rotary screw, or rotary vane types. As air compresses, heat is generated. Portable and small industrial compressors are air cooled, but large industrial compressors must be water cooled. A single piston air compressor can provide about 150 psi of pressure. Large, multistage compressors can provide 5000 psi.

Valves are important components in fluid power circuits because they control pressure, flow rate, and direction of fluid flow in circuits. Some flow control valve symbols are shown in Figure 18.58. *Flow control valves* control the volume flow rate of fluid through the circuit. The speed of hydraulic or pneumatic cylinders and motors is determined by its own displacement and the amount of fluid available to it. The slower the volume flow rate of fluid, the slower the volume displacement fills and the slower the motor turns or the cyl-

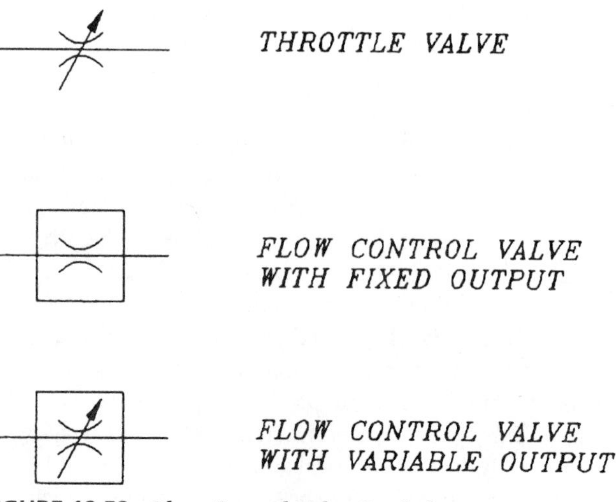

FIGURE 18.58 **Flow Control Valve Symbols**

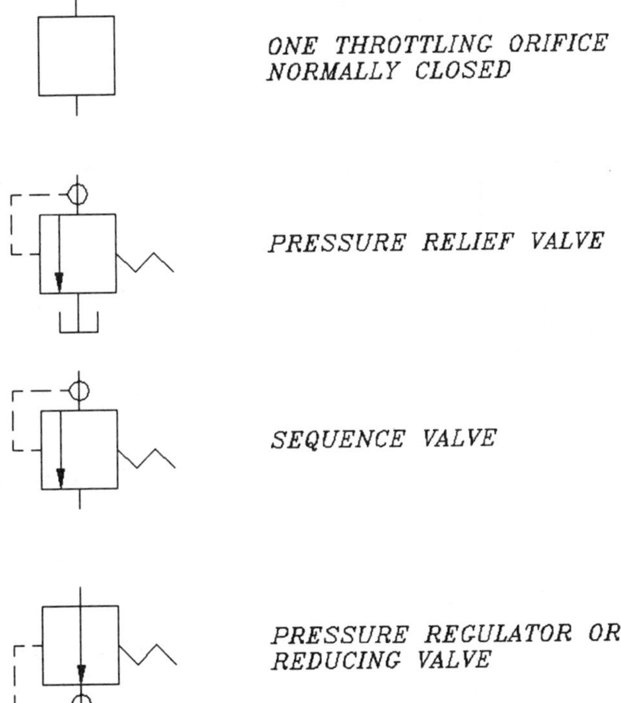

FIGURE 18.59 **Pressure Control Valve Symbols**

inder extends. Therefore, flow control valves in fluid power circuits determine the speed of operation of the various actuators in the circuit.

Pressure control valves (Fig. 18.59) control the pressure for some purpose in fluid power circuits. The most common type of pressure control valve is the pressure relief valve. Pressure relief valves are found in most fluid power circuits because they provide overload (overpressure) protection for the circuit. Since both hydraulic and pneumatic components are designed to operate under specific pressures, overloading the circuit with too much pressure is extremely dangerous. Safety pressure relief valves divert the flow back to the reservoir at a set pressure and, thus, prevent overpressure from developing within the circuit.

Pressure regulating or *reducing valves* maintain specific (reduced) pressures at different locations in the circuit. You have probably seen air pressure regulating valves on small air compressors. Hydraulic pressure reducing valves maintain reduced pressures at different locations in the circuit.

Sequence valves are designed to operate in machines that must operate in proper sequence. Sequence valves operate when pressure has reached a certain level. Although they operate much like a relief valve, a sequence valve diverts fluid flow to another part of the system to do useful work and not just back to the reservoir.

Directional control valves control the direction of fluid flow in a fluid power circuit. A simple directional control valve is a check valve. A check valve (Fig. 18.56) assures fluid flows in one direction in the circuit. Other types of directional control valves are two-way, three-way and four-way directional control valves. Most directional control valves change the path of the fluid by moving a sliding spool inside the valve. The spool is often actuated manually, but circuits that are controlled by computers use an electric solenoid (coil) to move the spool. There are many types of each directional control valve because of the many ways to actuate them and because of the many internal configurations of the spool inside the valve itself. Several of the internal configurations (flow paths) of directional spool valves are shown in Figure 18.60. The first section, marked FLOW PATHS, shows that the fluid may take several paths through the circuit. The fluid flows differently throughout the circuit depending on whether the port is open or closed (blocked). Whatever configuration is required for your circuit, a directional control valve for it is available.

A *two-way valve* may be used to operate a pneumatic cylinder because air can be vented to the atmosphere. In other words, you can use air to extend or retract the cylinder, but you can vent the exhaust gas from the other side of the cylinder to the air. However, you must use a *four-way valve* to control a hydraulic cylinder because you must provide a return path for the hydraulic fluid when you retract an extended cylinder. Hydraulic fluid must have a return path and cannot be vented to the environment. There are also different neutral positions for the valve too (Fig.18.60). Sometimes a cylinder will be locked in place in neutral. Other designs will call for a hydraulic motor to continue operating in neutral.

An **actuator** in a fluid power system is a device that converts the fluid power energy to mechanical energy for the purpose of doing useful work. *Linear actuators* are called pneumatic or hydraulic cylinders or rams. *Rotary actuators* are called pneumatic or hydraulic motors. Thus, the motion that is produced in fluid power systems is either linear or rotary.

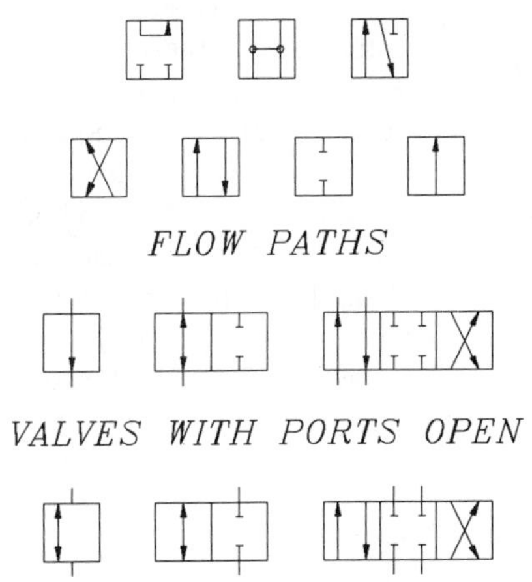

FIGURE 18.60 **Valve Flow Paths and Ports Open and Ports Closed**

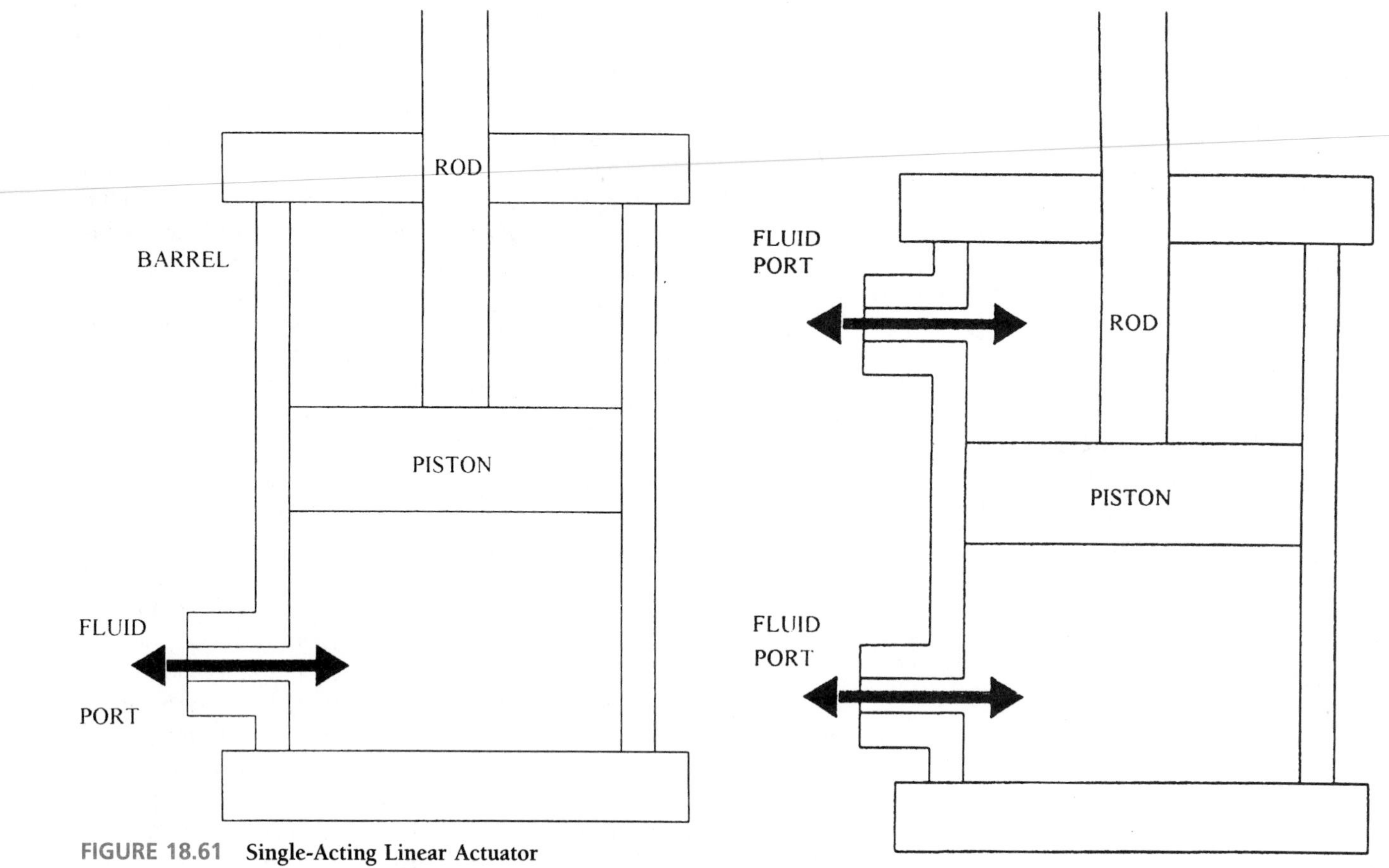

FIGURE 18.61 **Single-Acting Linear Actuator**

FIGURE 18.62 **Double-Acting Differential Linear Actuator**

A hydraulic motor is much like an electric motor except that moving fluid causes the shaft to rotate. Cylinders extend and retract to perform one cycle of linear motion.

The simplest type of **cylinder** is the *single-acting cylinder* (Fig. 18.61). A piston inside the housing (the barrel) is attached to a rod that extends outside the cylinder housing. Fluid force is applied on only one side of the piston. The cylinder retracts by gravity or by some external force only. A *double-acting differential cylinder* is shown in Figure 18.62. Fluid may flow on either side of the piston. Because of the volume taken up by the rod, the rod's extend and retract speeds will not be the same. The extract and retract force will also be different for the same reason. There are also double-acting nondifferential cylinders with rods on both sides of the piston that do extend and retract at the same rate.

There are several different types of *rotary actuators* (motors). Hydraulic pumps can usually be used as hydraulic motors with little or no modification. The function of a motor and pump are opposite. There are gear motors, vane motors, and piston motors.

18.5.3 Fluid Power Diagrams

The four types of fluid power circuit diagrams are *pictorial, cutaway, graphical,* and *combination.* Pictorial and cutaway diagrams are like other pictorial diagrams in engineering graphics—they are used for visualization. *Fluid power pictorials* show the actual layout of the pipe and the symbols used on those drawings are outline symbols that actually show the shape of the components. *Cutaway drawings* show the internal features and are commonly used for understanding circuit operation and for construction. These two types of diagrams contain symbols that are not really standardized and are not covered here.

Graphical diagrams use standard symbols to represent components; the piping is represented by single lines. A diagram showing the operation of a simple reciprocating circuit is shown in Figure 18.63. A diagram showing the operation of a pneumatic power door opener is shown in Figure 18.64.

Interpreting a fluid power diagram is a simple matter of understanding the symbols, how they are connected, and how the valves operate in the circuit. In Figure 18.63, a pump is used to power the circuit and a safety relief valve monitors the pressure in the system for any overloading (overpressure). A gas-charged accumulator keeps the flow and pressure in the system constant. A variable control valve controls the speed of the reversible motor. The directional control valve is used to control the direction of the motor.

Figure 18.64 is a graphical fluid power diagram that describes the operation of a pneumatic powered automatic door opener. A pneumatic cylinder attached to a bracket on the door actually opens and closes the door. Air is supplied from

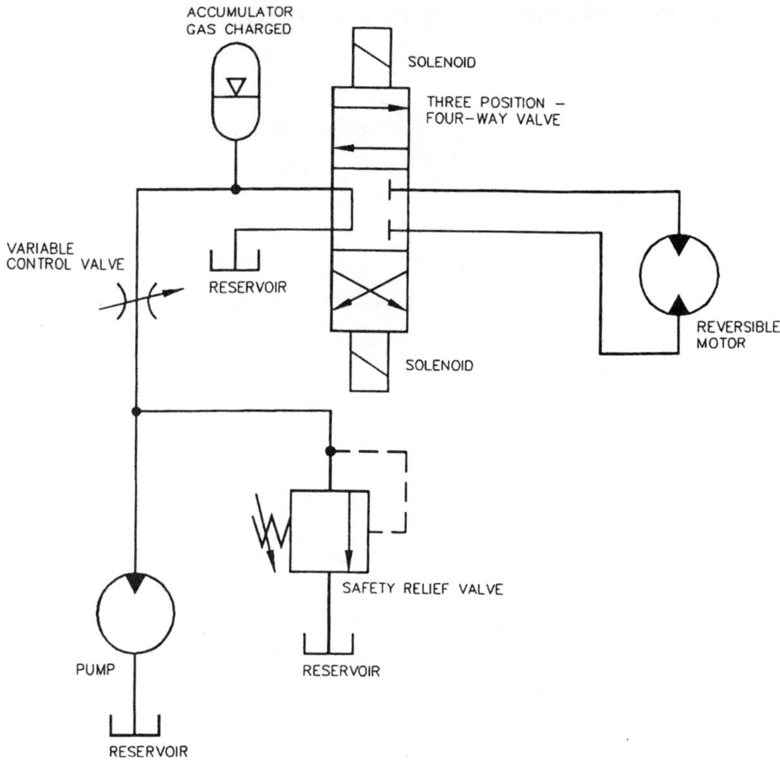

FIGURE 18.63 Simple Reciprocating Hydraulic Circuit

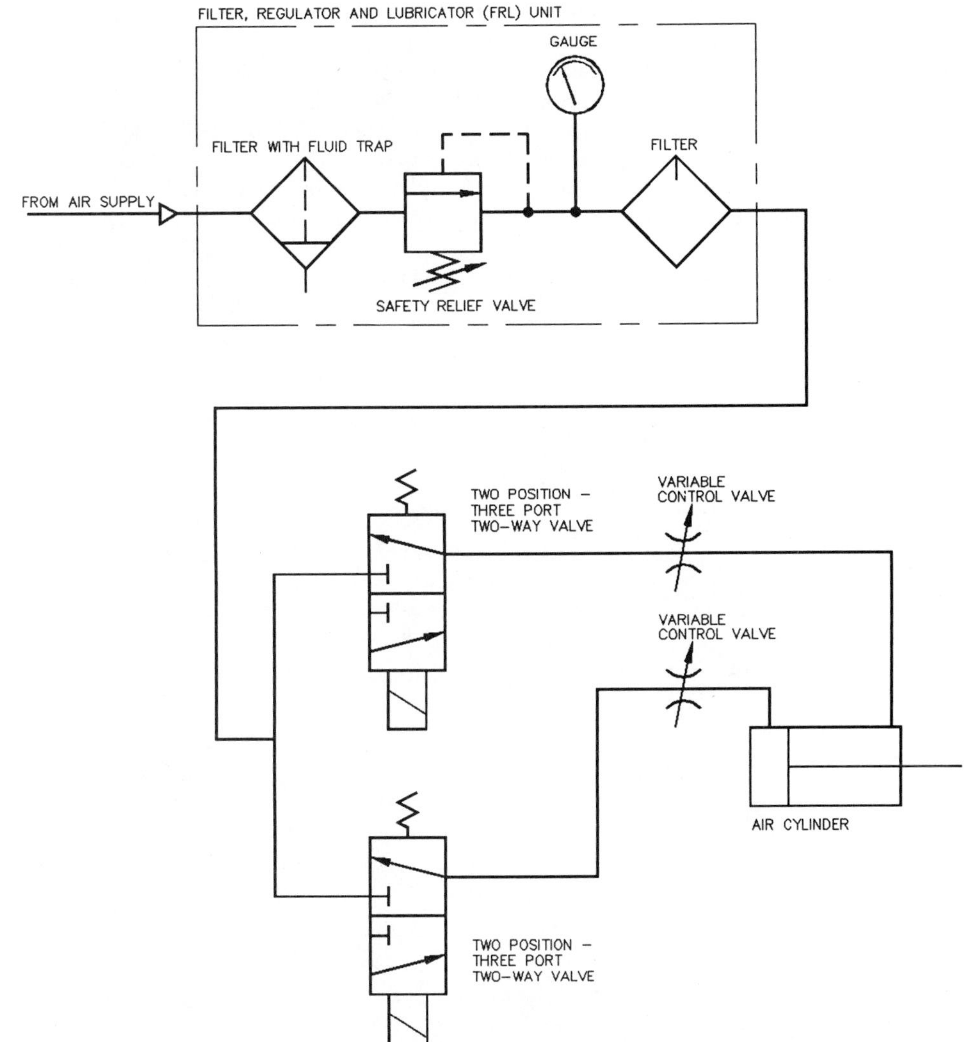

FIGURE 18.64 Circuit for Pneumatic Powered Door Opener

a large compressor and tank that is capable of providing all the needed volume flow rate of fluid. A filter, regulator, and lubricator (FRL) unit regulates the air pressure to an appropriate value (around 100 psi). The air is also filtered and lubricated. A safety relief valve prevents overpressure. Two solenoid-operated two-position three-port two-way valves are used to extend and retract the cylinder (open and close the door). Two variable flow control valves control the speed at which the door opens and closes. Notice that the extend and retract steps are activated independently. A quadriplegic often has trouble going through an open door in the measured amount of time if the circuit is simply timed. Here, there is a way to actuate independently the open and close mechanism. Therefore, the door opens on command and closes only after a different control command.

QUIZ

True or False

1. Welds are classified as resistance, gas, and nonpressure.
2. Normally, different types of welding processes can be encountered on the same drawing.
3. The actual outside diameter of pipe in nominal sizes of .125 in. to 12 in. is larger than the nominal size.
4. Fabrication drawings are usually drawn with the double-line method.
5. Logic diagrams are the first document used to design and lay out a printed circuit board.
6. A symbol on electronics drawings is a graphical representation of standard parts.
7. Safety relief valves are not really necessary components of fluid power circuits.
8. A linear actuator (cylinder) always extends and retract at the same speed.

Fill in the Blanks

9. The assembled welding symbol consists of the following eight parts: a reference line, __________, __________, __________, __________, __________, __________ and specifications.
10. The five basic groove welds are beveled, __________, __________, __________ and __________.
11. Pipe is designed at O.D. pipe for nominal sizes of __________ and above.
12. Flow diagrams do not show __________ or actual pipe __________.
13. Diagram lines represent __________ on an electronics diagram.
14. In electronics, drawings are made up of __________.
15. __________, __________, and __________ are all types of valves used in fluid power circuits.
16. __________, __________, __________, and __________ are types of hydraulic pumps.

Answer the Following

17. Name the elements of a complete weld symbol.
18. What is meant by "arrow side" and "opposite side" and what is the tail of the symbol used for?
19. What is the I.D. of a SCH 40 10 in. pipe?
20. What is the O.D. of a 14 in. pipe?
21. What is a reference designator in electronics drawings? Give an example of one.
22. What is the primary use of a harness diagram?
23. Name the major components of a basic hydraulic fluid power system.
24. Other than holding fluid, what are the features of a hydraulic reservoir?

APPENDIX A

Glossaries

APPENDIX A.1

Mechanical Glossary

A

Accurate Manufactured within the specified tolerances.

Acme thread A screw thread similar to the square thread. The acme has in most cases replaced the square thread because it is stronger and easier to manufacture. It is widely used as a feed screw.

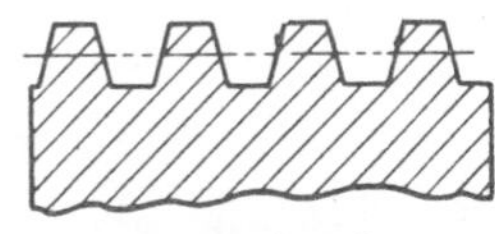

Acme thread

Addendum The radial distance between the top of the tooth and the pitch circle of a gear.

Allen screw Special set screw or cap screw with hexagon socket in head.

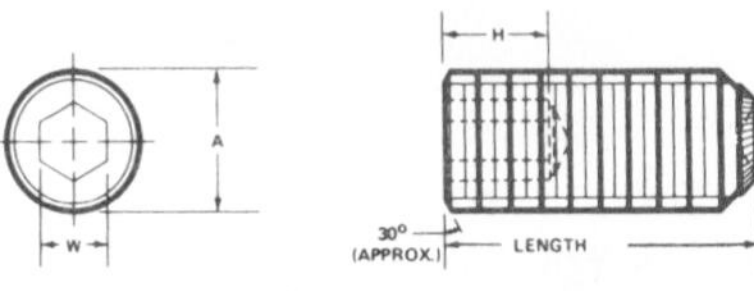

Allen screw

Allowance The intentional difference between the MMC limits of size of mating parts; the minimum clearance (positive allowance) or maximum interference (negative allowance) between such parts.

Alloy A mixture of two or more metals to obtain characteristics similar to the individual metals or different from any displayed by the individual components.

Aluminum A lightweight but strong metal. Principle commercial source is bauxite ore.

Annealing A process of heating steel above the critical range, holding it at that temperature until it is uniformly heated and the grain is refined, and then cooling it very slowly.

ANSI American National Standards Institute. A nongovernmental organization that proposes, modifies, approves, and publishes drafting and manufacturing standards for voluntary use in the United States.

Arc A continuous portion (as of a circle or ellipse) of a curved line.

Arc

Arc welding A process of joining two or more metal parts together by fusing them with an electric arc. The arc melts the welding rod and fuses the parts together.

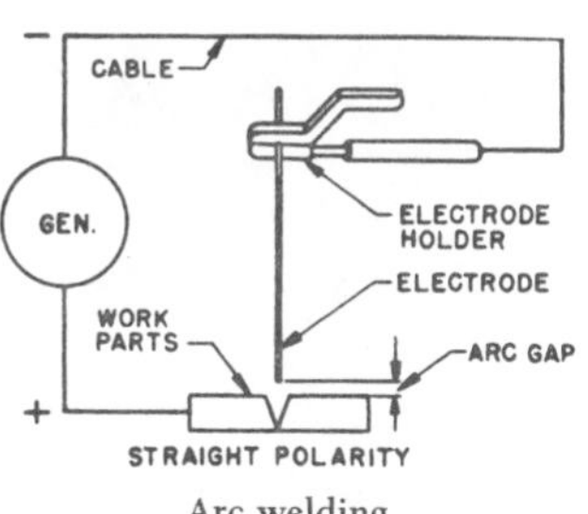

Arc welding

Assembly drawing A drawing representing a group of parts constituting a major subdivision of the final product.

Any view that lies in a projection plane other than the horizontal (top), frontal (front), or profile (side) plane.

Axis A straight line (centerline) about which a feature of revolution revolves; or about which opposite-hand features are symmetrical.

Axonometric One of several forms of single-plane projections giving the pictorial effect of perspective with the possibility of measuring the principal planes directly.

B

Basic dimension A numerical value used to describe the theoretically exact size, profile, orientation, or location of a feature or datum target. It is the basis from which permissible variations are established by tolerances on other dimensions, in notes, or in feature control frames.

Bearing A machine part in which another part turns or slides.

Bend allowance The amount of sheet metal required to make a bend of a specific radius.

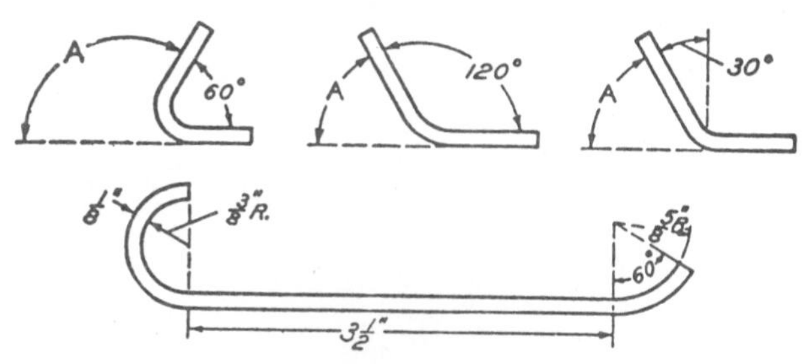

Bend allowance

Bevel The angle that one surface or line makes with another when they are not at right angles.

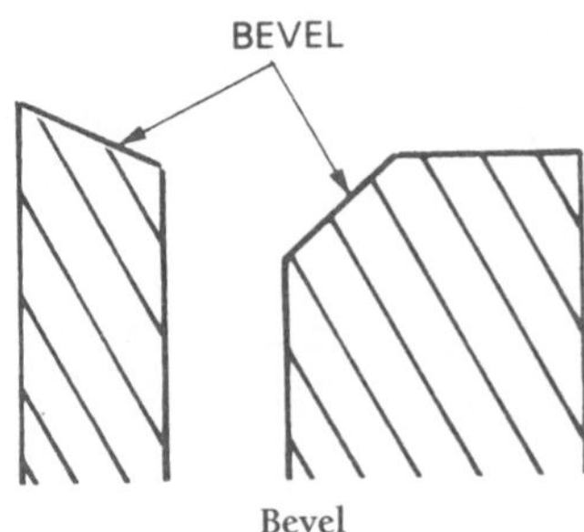

Bevel

Bilateral tolerance A tolerance in which variation is permitted in both directions from the specified dimension.

Blanking A punch press operation that consists of shearing from sheet metal stock a part having a definite contour determined by the punch and die.

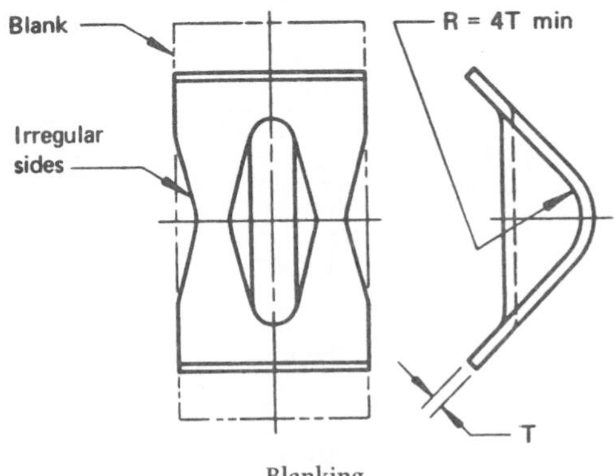

Blanking

Bolt A headed and externally threaded mechanical device designed for insertion through holes in assembled parts. It mates with a nut and is normally intended to be tightened or released by turning that nut.

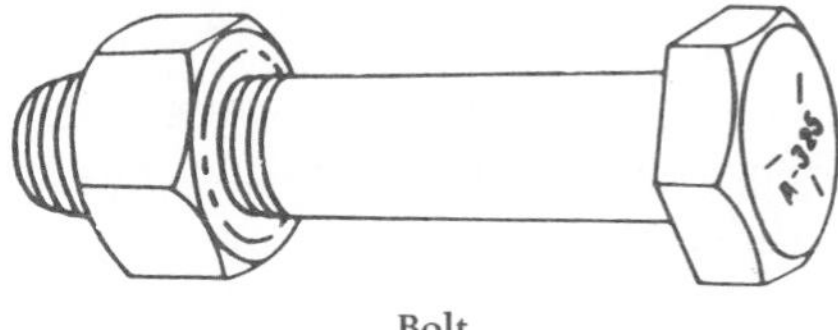
Bolt

Bolt circle A circular centerline on a drawing; contains the centers of holes about a common center.

Bore To enlarge a hole to a specified size by employing a point tool held in a boring bar operated by a lathe or a boring mill.

Boss A circular raised portion of material added around holes in castings or forgings to give strength or bearing surface to the part.

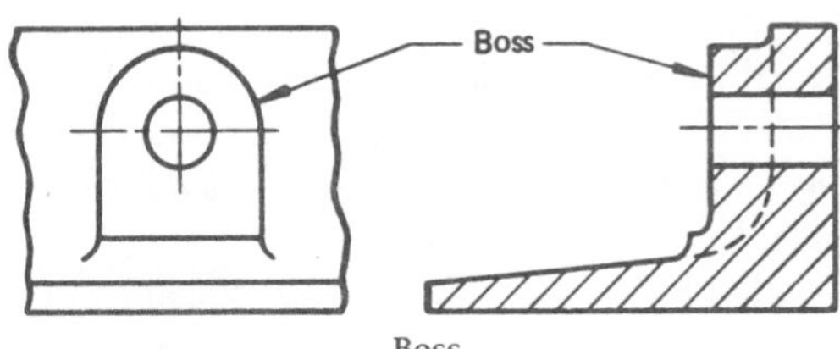

Boss

Brass An alloy of copper and zinc.

Brazing A group of welding processes wherein coalescence is produced by heating to a temperature above 800°F and by using a nonferrous filler metal (solder) having a melting point below that of the base metals (zinc or brass).

Brinell A testing method to determine hardness of metal.

Broaching To machine a hole to a desired shape by planing it with a transverse cutting tool moved in a straight line so that each tooth removes a definite amount of stock.

Broach

Bronze An alloy of nine parts copper and one part tin.

Buffing A polishing operation using a fine abrasive wheel made of discs of cotton or wool impregnated with very fine abrasive, bonded with wax or grease.

Burnishing Smoothly finishing surfaces by compressing with the use of highly polished rolls or by the use of steel balls in rolling contact with the surface.

Burr A rough edge raised on metal along the path of a cutting tool.

Bushing A removable cylindrical sleeve used as a lining of low-friction material that provides a bearing surface. It is either press fitted or removable and provides accurate and quick location of the drilling operation when used in a drill jig.

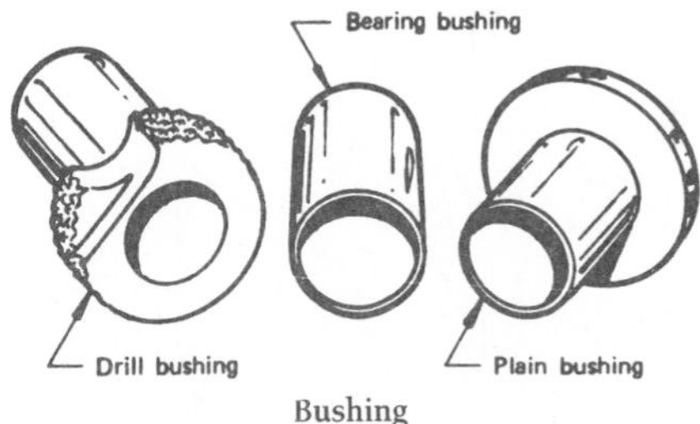

Bushing

C

Caliper A measuring tool used for checking outside or inside measurements.

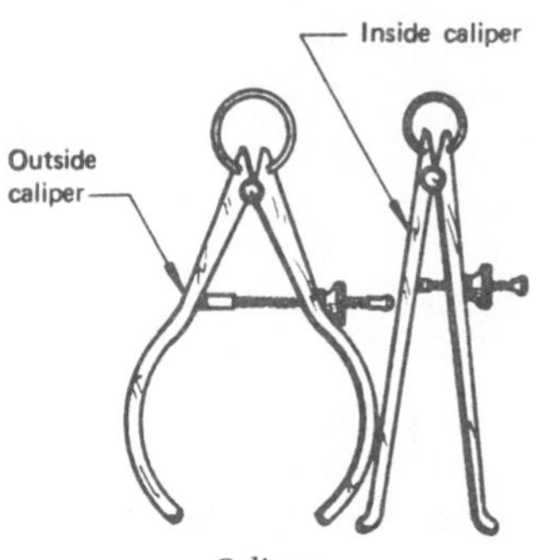

Caliper

Callout A note on a drawing giving a dimension, specification, or manufacturing process.

Cam A mechanical device used on a rotating shaft to transform rotary motion into lateral motion. A face cam is designed so that the follower travels in a groove cut in the face of the cam. A barrel cam is designed with a groove cut in the outer surface of a cylinder to give motion in a direction parallel to the axis of rotation. A disc cam is designed so that the follower travels along its periphery.

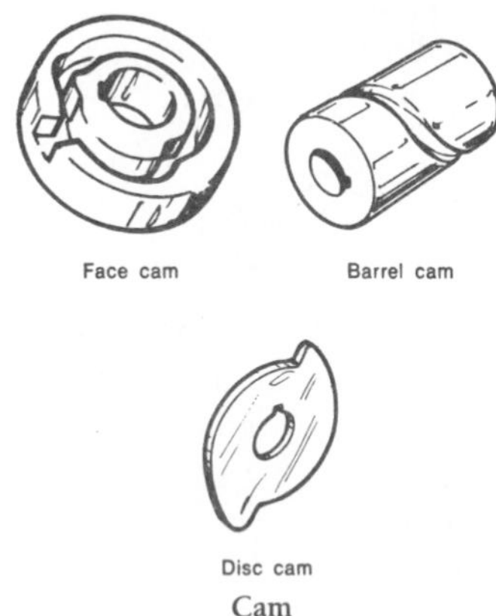

Cam

Carburize To harden the surface of iron-based alloys by heating the metal below its melting point in contact with solids, liquids, or gases that have a high carbon content. Carburizing is best performed on steels containing less than 0.25% carbon content.

Case-harden To harden a surface either by carburizing or through the use of potassium cyanide.

Cast To produce parts by pouring molten material into a mold.

Cast

Center drill A drill to produce bearing holes in the ends of a workpiece. Also called a countersink.

Chamfer A small angular surface on an external edge or corner for purposes of easy assembly or to remove sharp edges. The most frequent application is on shafts and cylinders.

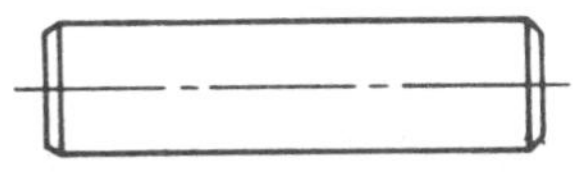

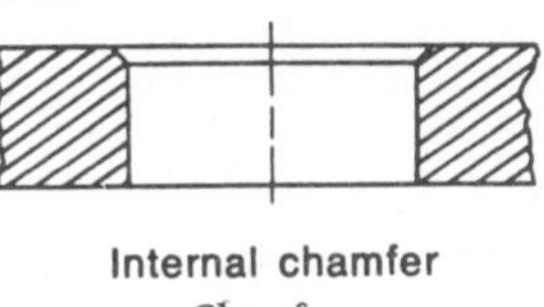

Chamfer

Chill To harden the outer surface of a casting by quick cooling.

Chuck A mechanism for holding a rotating tool or workpiece on a lathe.

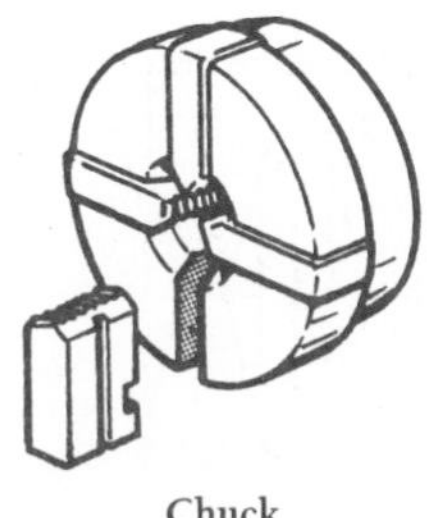
Chuck

Code identification number A five-digit number, assigned to each design activity, used in conjunction with a part or identity number in a parts list [also referred to as a Federal Supply Code for Manufacturers (FSCM)].

Coining A method of cold forging parts to desired size and shape by compressing them under heavy pressure between coining dies.

Cold-rolled steel (CRS) Open hearth or Bessemer steel containing 0.12% to 0.20% carbon that has been rolled while cold to produce a smooth, accurate stock.

Collar A cylindrical ring or round flange fitted on a shaft to prevent a sliding movement.

Collar

Commercial fastener A fastener manufactured to the requirements of published standards or documents and stocked by manufacturers or distributors.

Commercial item A supply or service that is (a) regularly used for other than government purposes and (b) sold or traded in the course of conducting normal business operations.

Core A solid form made of sand that is shaped in a core box and baked and used to shape the interior of a hollow casting.

Core print A projection on a pattern that forms an opening in the sand to hold the end of a core.

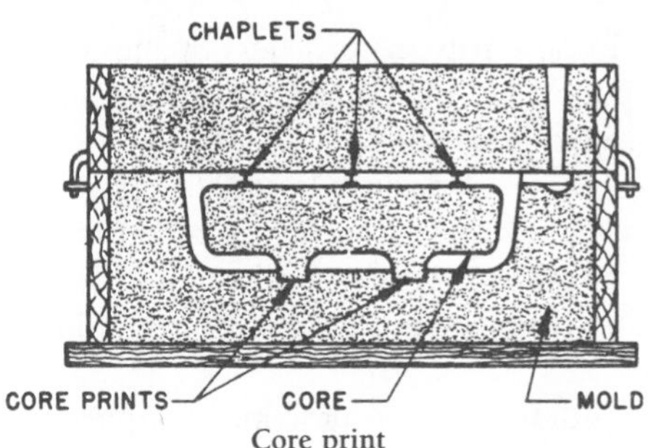

Core print

Cotter pin A half-round stock that is bent so as to have an eye at one end and forms a round split pin when compressed together (used to lock parts of an assembly together).

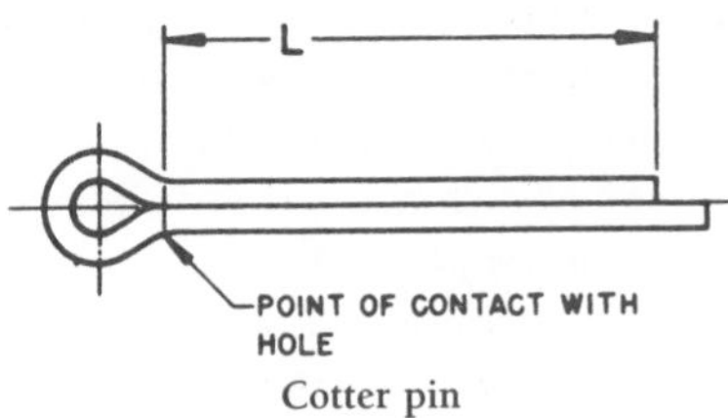

Cotter pin

Counterbore To enlarge the end of a cylindrical hole to a given depth with a flat shoulder; the name of the tool used to produce such a hole.

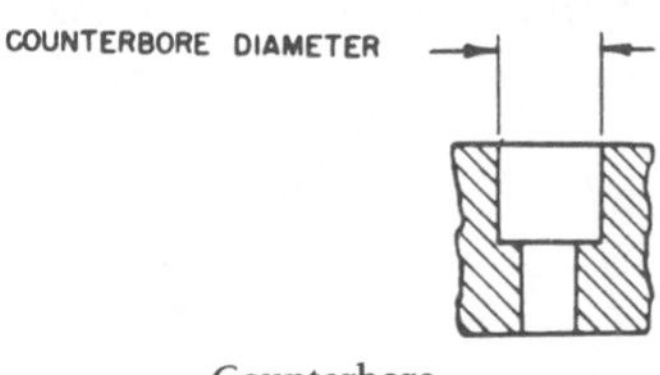

Counterbore

Counterdrill To form a conical shoulder in a drilled hole by enlarging it with a larger drill.

Countersink To recess a hole with a cone-shaped tool to provide a seat for a flathead screw or rivet; also the tool used to make such a hole.

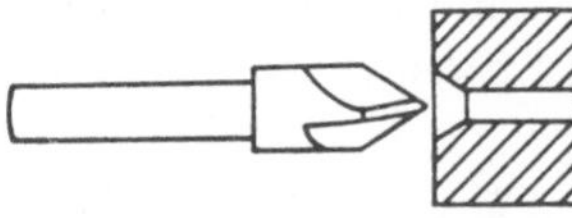
Countersink

Crosshatching Filling in an outline with a series of symbols to highlight part of a design.
Cyaniding Surface hardening of a ferrous alloy by heating at a suitable temperature in contact with a cyanide salt, followed by quenching.

D

Datum A theoretically exact point, axis, or plane derived from the true geometric counterpart of a specified datum feature. A datum is the origin from which the location or geometric characteristics of features of a part are established.
Dedendum Distance from pitch circle to bottom of tooth space on a gear.
Design activity An activity having responsibility for the design of an item; may be a government activity or a contractor, vendor, or others.
Detail drawing A drawing of a single part that provides all the information necessary in the production of that part.
Diameter The length of a straight line running through the center of a circle.

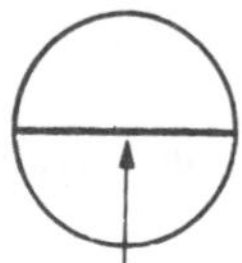
Diameter

Diametral pitch Number of gear teeth per inch of pitch diameter.
Die Hardened metal piece shaped to cut or form a required shape in a sheet of metal by pressing it against a mating die.
Die casting Part produced by forcing a molten alloy into a metal mold composed of two or more parts.
Die stamping A part that has been cut or formed from sheet metal by the use of dies.
Dimension, basic A numerical value used to describe the theoretically exact size, shape, or location of a feature or datum target. It is the basis from which permissible variations are established by tolerances on other dimensions, in notes, or by feature control symbols.
Dimension, coordinate Rectangular coordinate dimensioning is where all dimensions are measured from two or three mutually perpendicular datum planes.
Dog A small auxiliary clamp for preventing work from rotating in relation to the face plate of a lathe.

Dowel A pin used to prevent sliding (and for location) between two contacting flat surfaces.

Draft The taper used on the sides of a pattern so that it can be easily removed from the sand mold; the taper on the sides of a forging die that permit the forging to be removed easily.

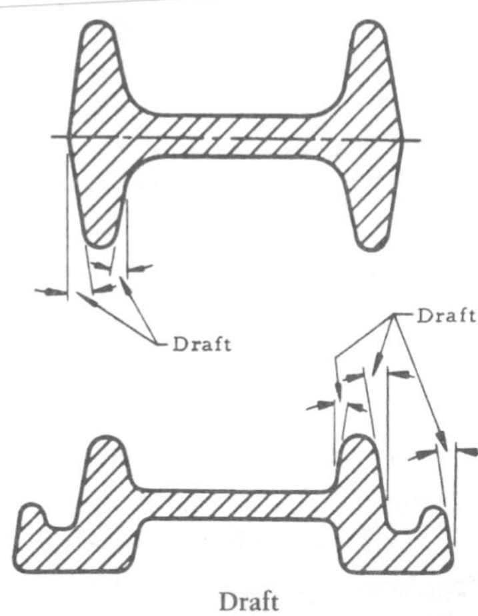

Draft

Draw To form a metal, which may be either hot or cold, by distorting or stretching; to temper steel by gradual or intermittent quenching.

Drawing format The standardized form, usually preprinted, on which various constant information (design activity identification, standard tolerance block, etc.) is provided together with spaces for variable information (drawing number, title, etc.).

Drawing number Consists of letters, numbers, or combination of letters and numbers, which may or may not be separated by dashes. The number is assigned to a particular drawing for identification and file retrieval.

Drawing type Name applied to a drawing, descriptive of its design and end use.

Drill To form a cylindrical hole with a drill; one of a variety of revolving cutting tools designed for cutting at the point.

Drill press A machine used for hole-forming operations.

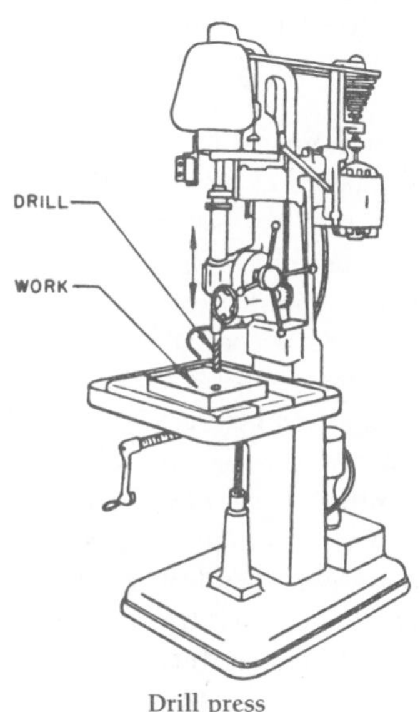

Drill press

Drop forge To form a piece while hot by placing it between dies in a drop hammer.

E

Emboss To raise patterns or letters by impressing with matching punch and die; to form projections in sheet metal prior to projection welding.

Engineering data Drawings, associated lists, accompanying documents, manufacturer specifications, and standards, or other information relating to the design, manufacture, procurement, testing, or inspection of items or services.

Engineering definition A description expressed in engineering terms in sufficient detail to enable meeting the requirements of design, development, engineering, production, procurement, or logistic support.

Engineering document release The process of transferring custody of an engineering document, or change thereto, from the preparing activity to a control activity, which is responsible for its reproduction, distribution, storage, and the maintenance of history records.

Engineering drawing An engineering document that discloses, by means of pictorial and/or textual presentations, the form and function of a part.

Extruding To form a continuous cross section by forcing material through openings designed to a desired shape.

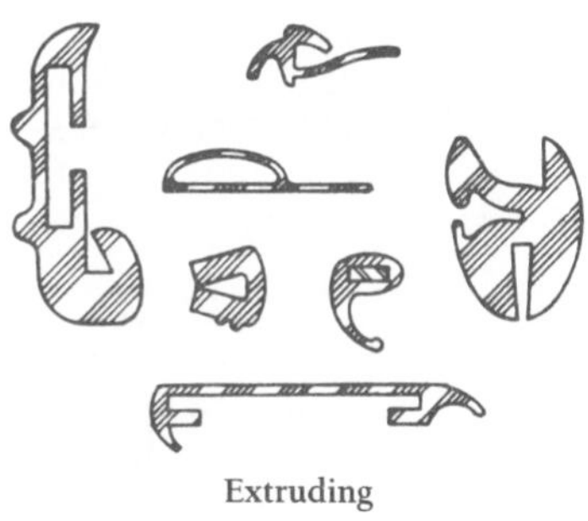
Extruding

F

Face To machine a flat surface on a part using a lathe by turning the surface perpendicular to the axis of rotation.

Fastener A mechanical device designed specifically to hold, join, couple, assemble, or maintain equilibrium of single or multiple components.

Feather A rectangular sliding key that permits a pulley to move along the shaft parallel to its axis.

Feather edge Sharp point on pressed metal stamping.

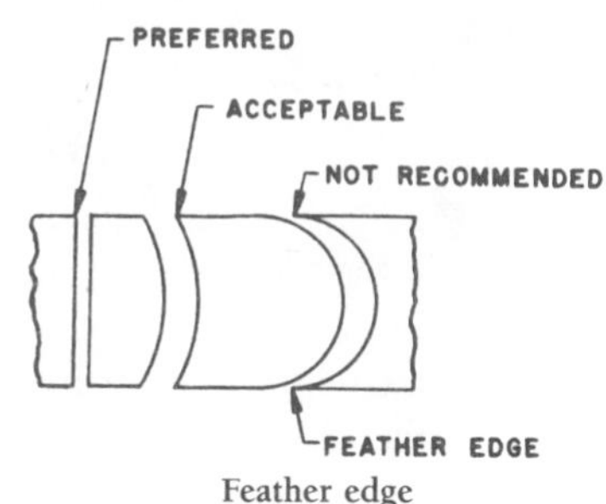

Feather edge

Feather key A flat key, which is partly sunk in a shaft and partly in a hub, permitting the hub to slide lengthwise on the shaft.

Federal supply code for manufacturers (FSCM) Five codes applicable to all activities that have produced or are producing items used by the federal government; also applies to government activities that control design or are responsible for the development of certain specifications, drawings, or standards that control the design of items.

File To shape, finish, or trim with a fine-toothed metal cutting tool that is used in either a rotating arbor or done by hand.

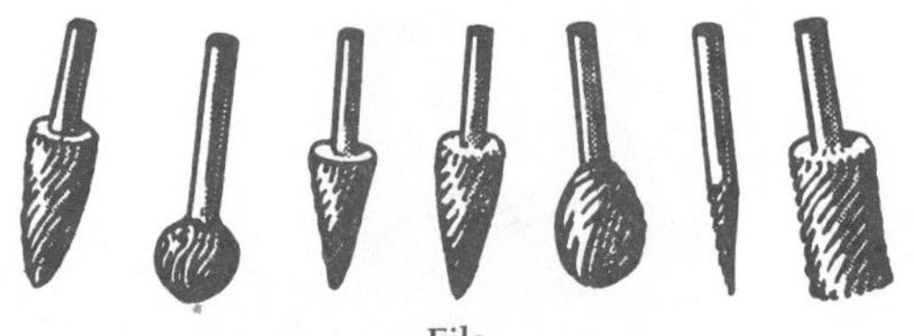

File

Fillet A curved inside corner that increases the strength at the junction of two intersecting surfaces of a part.

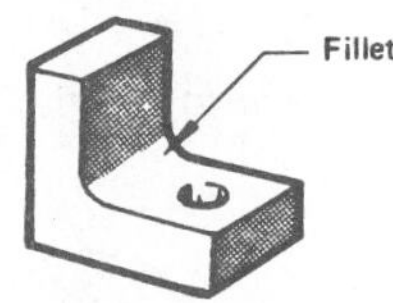

Fillet

Fin A thin extrusion of metal at the intersection of dies or sand molds.

Finish The degree of smoothness or roughness of a surface; the covering applied to a surface such as plating or painting.

Fit Degree of tightness or looseness between two mating parts.

Fixture A tool used for holding a part on which machining operations are being performed.

Flange A rim extending from the main section of a part, such as the top and bottom members of a beam or a projecting rim added at the end of a pipe or fitting for making a connection.

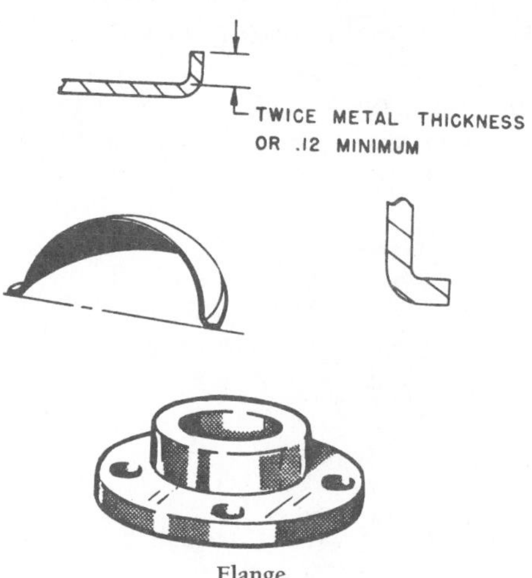

Flange

Flask The container in which sand molds are made; consists of two sections—the cope (upper section) and the drag (lower section). Any midsection is called a cheek.

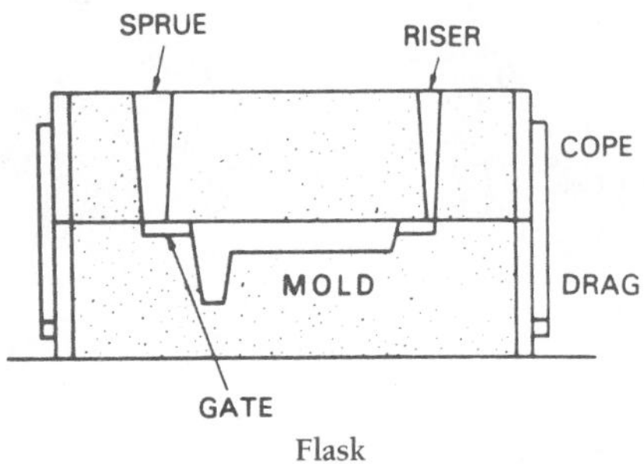

Flask

Flat pattern A layout (development) showing true dimensions of a part before bending.

Flute Groove, as on twist drills, reamers, and taps.

Forge To force metal while it is hot to take on a desired shape by hammering or pressing.

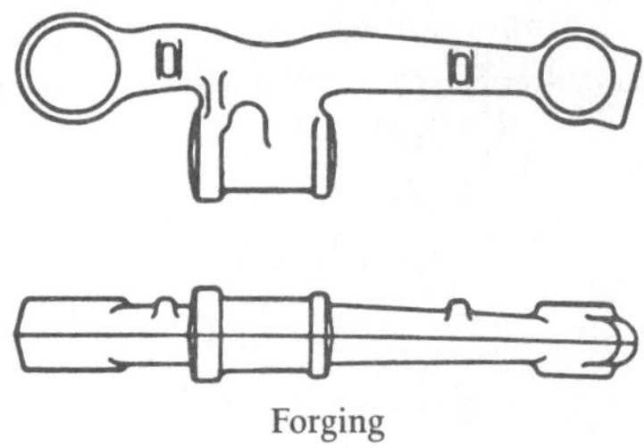

Forging

G

Gage An instrument used for determining correctness of size or strength of manufactured parts within specified limits such as depth gage, dial gage, plug gage, ring gage, snap gage, surface gage, thread gage, and wire gage.

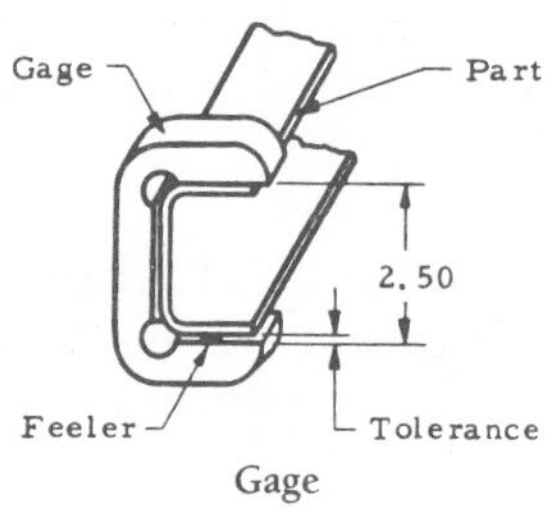

Gage

Galvanize To coat metal parts by immersing in a zinc bath.

Gasket A thin piece of metal, rubber, or other material placed between surfaces to make a tight joint.

Gate The opening in a sand mold at the bottom of the sprue through which the molten metal passes to enter the cavity or mold.

Gears Cylindrical or conical shaped parts having teeth and used in gear trains that transmit power between shafts such as spur gear, helical gear, herringbone gear, bevel gear, gear and rack, and internal spur gear.

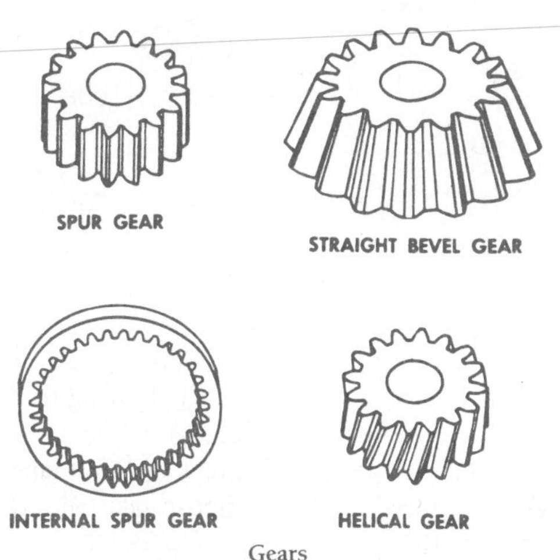

Gears

Grinding Finishing a surface using a revolving abrasive wheel. Abrasive wheels are available in various grades from fine to coarse.

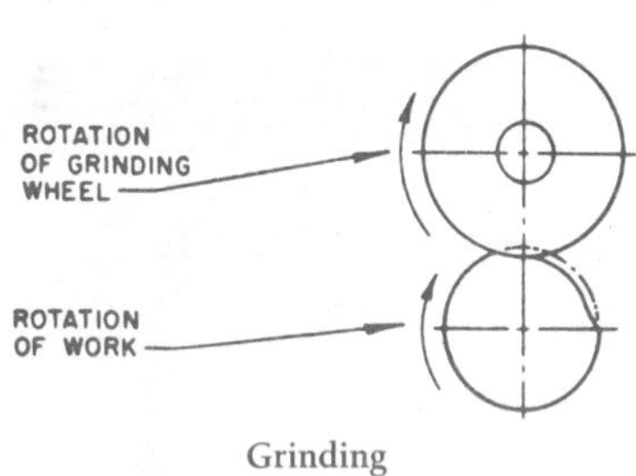

Grinding

Gusset A small plate used in reinforcing assemblies.

H

Hardening To heat steel or aluminum above a critical temperature and then quench in water or oil.
Heat treatment A series of operations that improves the physical properties of a material.

I

Interchangeability A number of similar parts manufactured so that any one part can be used in place of another in an assembly and still function properly.

J

Jig A special type of fixture used to hold and accurately locate, as well as guide, the tools used in manufacture such as that used in drilling operations.

K

Key A part used between a shaft and a hub to prevent movement of one relative to the other.

Keyseat or keyway A groove cut parallel to the axis of a shaft or hub to receive the key. A key rests in a keyseat and slides in a keyway.

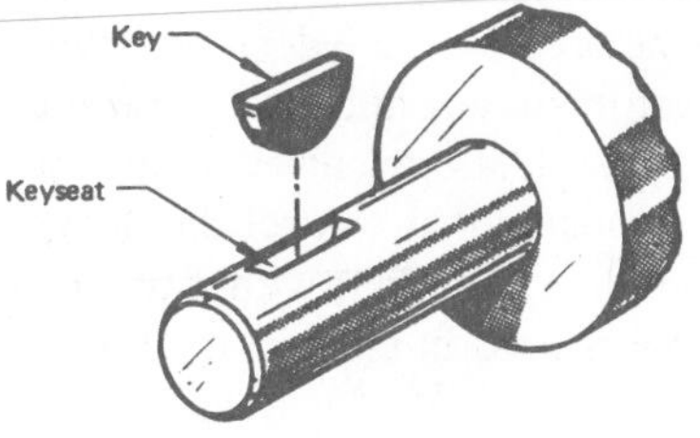

Key and keyseat

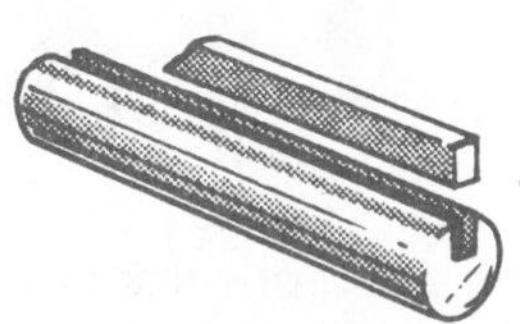
Keyseat or keyway

Knurling The forming of a series of fine ridges to roughen a cylindrical surface to provide a firmer grip for the fingers.

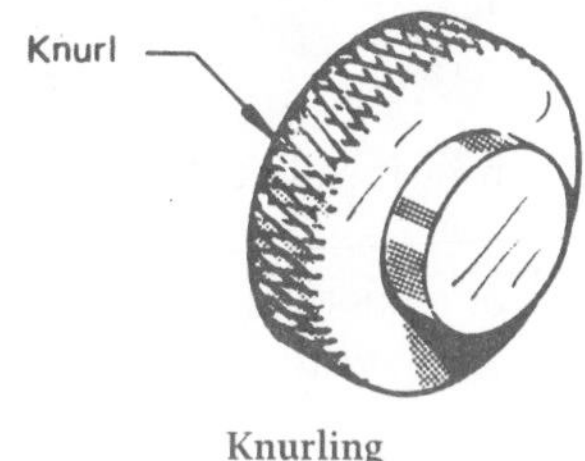

Knurling

L

Lapping To finish or polish a surface with a piece of soft metal, wood, or leather impregnated with abrasive compound.
Lead The axial distance a point will travel on a screw thread when turned one complete revolution of the thread.
Limited production Manufactured under model-shop conditions, as opposed to mass production under factory production line conditions.
Lug A projection or ear that is cast or forged as a portion of a part to provide support or attachment facility with another part.

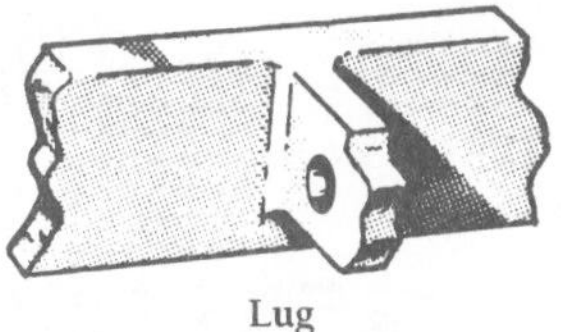
Lug

M

Malleable casting A casting that has been annealed to provide extra strength.

Manufacturer A person or firm who owns, operates, or maintains a factory or establishment that produces on the premises the materials, supplies, articles, or equipment required under the contract or of the general character described by specifications, standards, and publications.

Matched parts Those parts, such as special application parts, which are machine matched, or otherwise mated, and for which replacement as a matched set or pair is essential.

Material allowance Extra material provided for machining to achieve close accuracy and smooth surfaces.

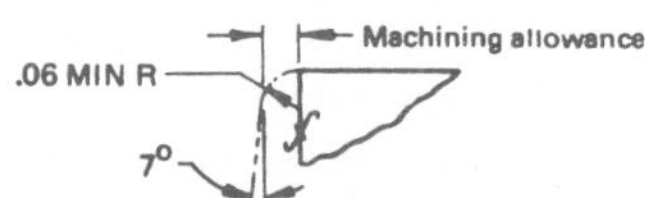

Material allowance

Micrometer caliper A caliper with a micrometer screw attached, used for making accurate measurements.

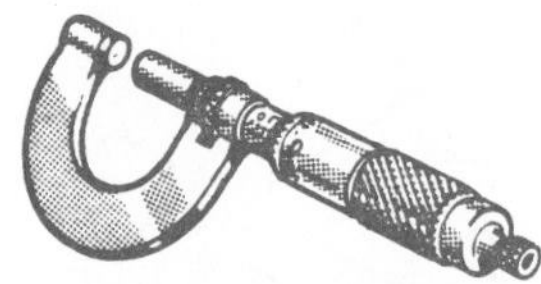

Micrometer caliper

Milling Removing material from a part by means of a revolving cutter. Various cutters are available: end mill, form cutter, straddle milling, and hollow cutter.

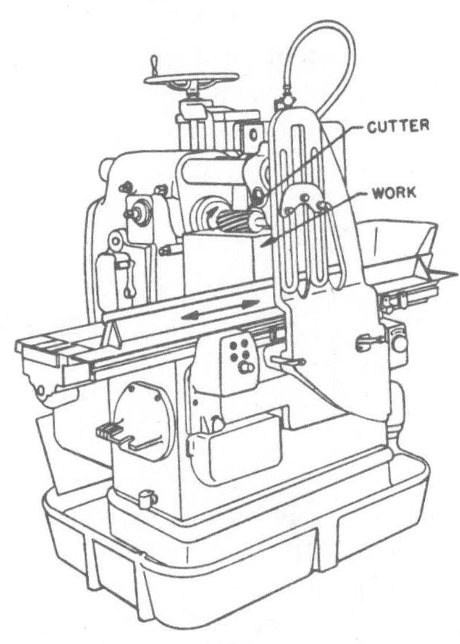

Milling

Mold The form provided for, or the act of forming by pouring molten metal into a hollow during a casting operation to give the part a desired shape when the material solidifies. A mold can be made of sand, plaster, or metal as long as the mold will withstand the temperature required for the material of the part.

N

Neck To cut a circular groove around a shaft to provide firm fitting between the shaft and its mating part in assembly.

Nesting The arrangement of sheet-metal parts on strip stock to provide the least scrap per blanks.

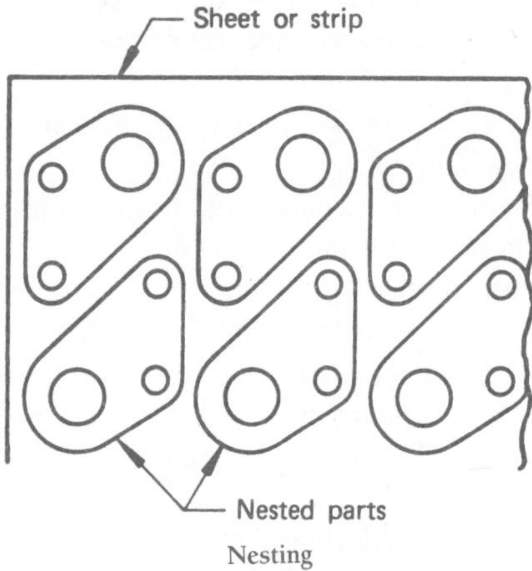

Nesting

Normalize To heat steel above its critical temperature and then cool it in air.

Nut A perforated block (usually of metal) possessing an internal, or female, screw thread, intended for use on an external, or male, screw thread such as a bolt for the purpose of tightening, adjusting, or holding two or more parts in definite relative positions.

P

Pack-harden To carburize, then to case-harden.

Pad A low projection surface, usually rectangular as contrasted with a boss.

Part drawing An engineering drawing that defines an item and assigns a part or control number to identify its configuration.

Part number A number (or combination of numbers and letters) assigned to uniquely identify a specific part. The part number includes the design activity drawing number.

Parting line The line along which the pattern is divided for molding; along the line where the sections of a mold or die separate.

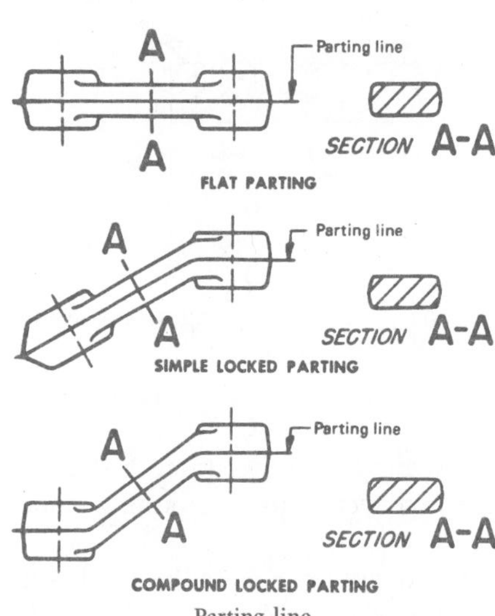

Parting line

Pattern The form used to make the cavity in the mold and which duplicates the shape of the part to be cast, except that it is made proportionately larger to compensate for shrinkage due to the contraction of the metal when cooled.
Peening To stretch metal by battering with the peen end (ball end) of a hammer.
Pickling To remove scale and rust from a casting or forging by immersing it in an acid bath.
Plane To finish a flat surface on a planer machine with a fixed cutter and reciprocating bed on which the part is securely attached; a geometric description of a flat surface.
Polishing The finishing of a surface to a smooth and lustrous condition by the use of a fine abrasive as a basis for plating, etc.
Profiling Using a pattern as a guide to make a similar part in a vertical milling machine operation in which the tool spindle is guided by the master plate made to the required shape of the part.

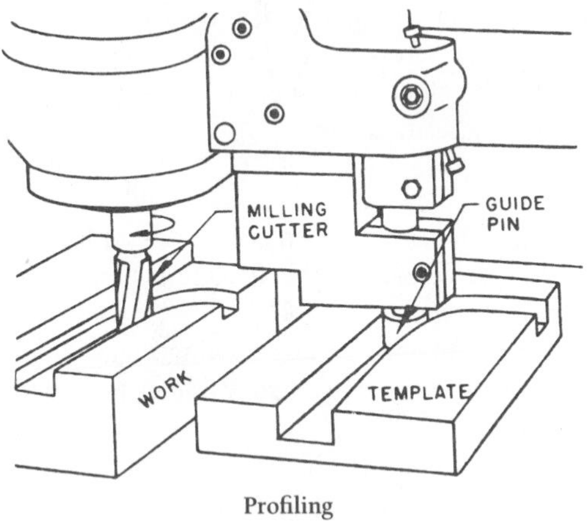

Profiling

Punch That part of a tool that pierces holes in stock or shapes the inside contour of a part in a forming die.

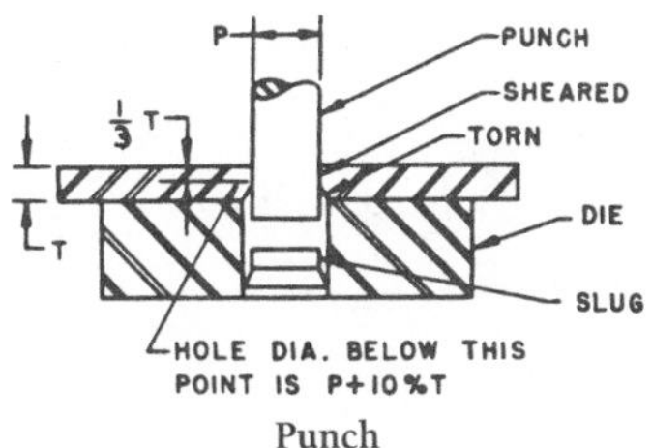

Punch

Q

Quenching The process of cooling a part rapidly by immersing it in liquids, gases, or oil.

R

Rack A bar having gear teeth cut on the face so that rotary motion is converted to reciprocating motion or vice versa.
Reamer A fluted cutting tool used to finish a hole to a desired size within specified limits.

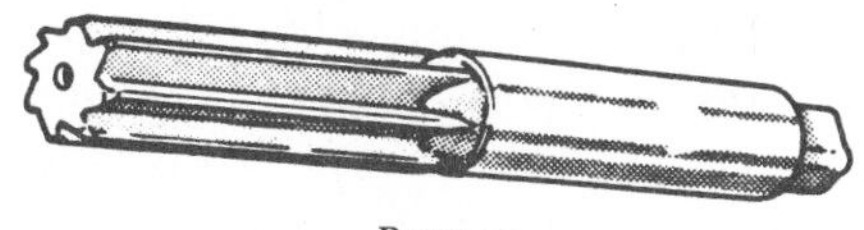
Reamer

Relief A groove on a part such as a cut next to a shoulder.
Revision Any change to an original drawing after that drawing has been released for use.
Revision authorization A document such as a "Notice of Revision," "Engineering Change Notice," or "Revision Directive" that describes the revision in detail and is issued by the source having the authority to revise the drawing.
Revision symbol A letter (which may be accompanied by a suffix number), used to identify particular revisions on the face of the drawing or in a revision description block.
Rib A ridge cast into thin sections of a part to make it stronger.

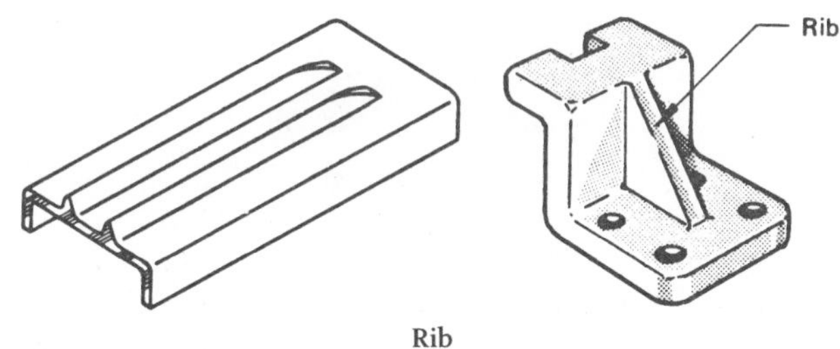

Rib

Rivet A headed and unthreaded mechanical device used to assemble two or more components by an applied force, which deforms the plain rivet end to develop a completed mechanical joint.

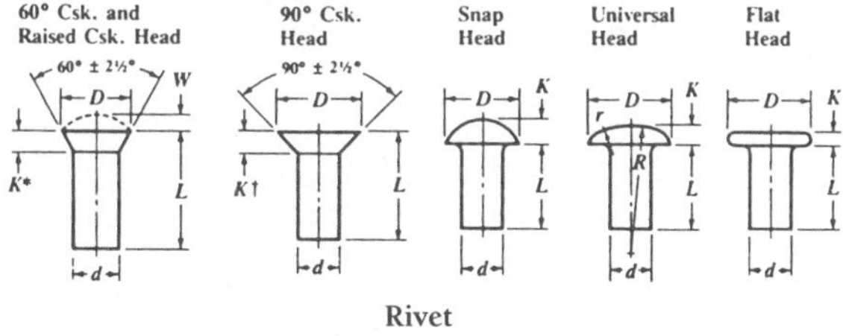

Rivet

Riveting A hammering process by which a rivet is used to fasten two or more parts by passing the shank through mating holes and then peening or pressing down the plain end to form a second head.

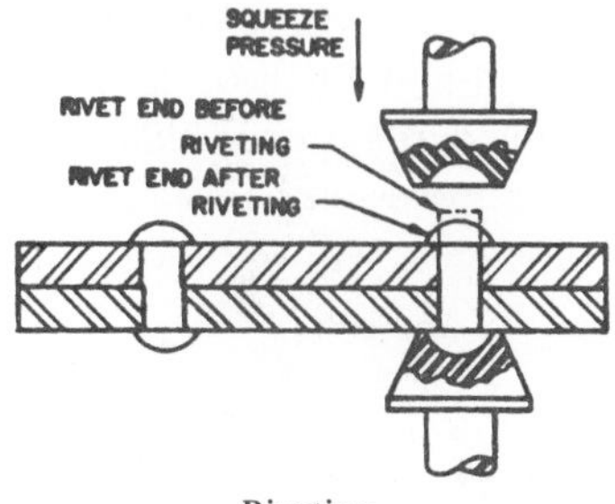

Riveting

Round A rounded external corner on a casting or forging.

S

Sandblasting The cleaning of the surface of a part by means of sand forced from a nozzle at high velocity.

Screw A headed and externally threaded mechanical device possessing capabilities that permit it to be inserted into holes in assembled parts; it is meant to be mated with a preformed internal thread or form its own thread.

Set screw A screw used in a hub which bottoms against a shaft to prevent relative motion between two parts. They are made either headless or with different types of heads as well as points.

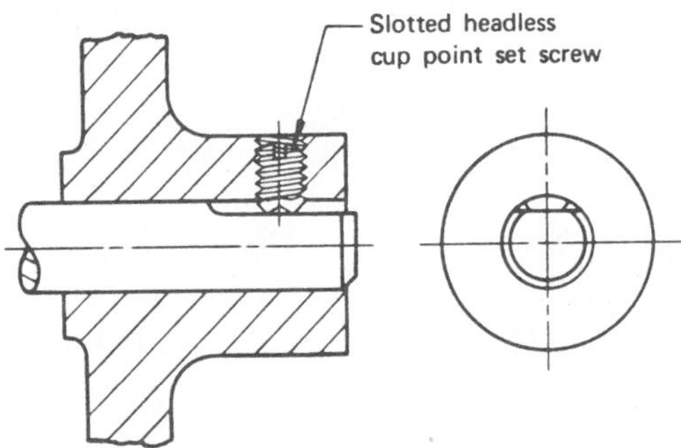

Set screw

Shaper A machine tool with a sliding ram used to finish parts and flat surfaces. The workpiece is clamped in a stationary vise during the cutting stroke.

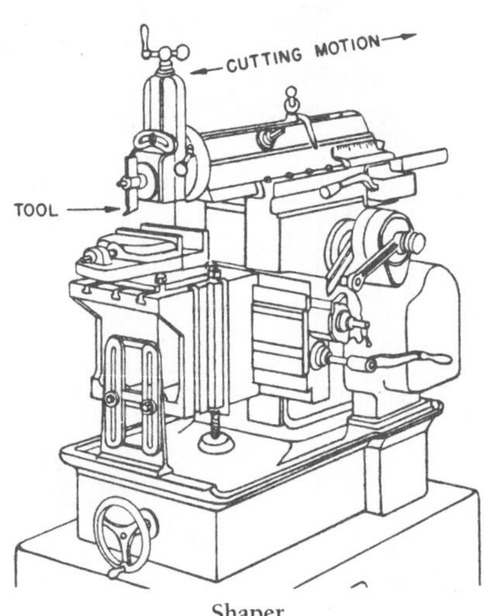

Shaper

Shear To cut off sheet or bar metal with the shearing action of two blades.

Shim A thin metal strip that is inserted between two parts for the purpose of adjustment.

Spline A key for inserting in a slot in a shaft, or a rib that has been machined on the shaft and fits another part having a mating slot.

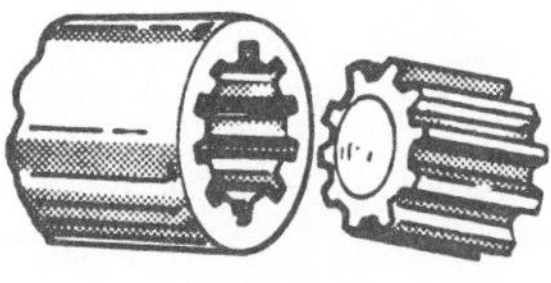
Spline

Spot weld To weld two overlapping sheet-metal parts in spots by means of the heat generated by resistance to an electric current between a pair of electrodes.

Spotfacing To finish the rough surface around a round hole using a counterbore tool to smooth and square the surface to allow a bolt or screw head to seat properly; a shallow counterbore.

Sprocket A gear made for chain-driven rotating mechanisms.

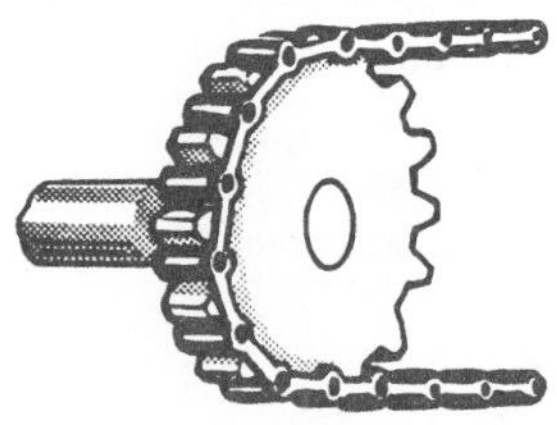
Sprocket

Stamping Any part made by pressed metal operations.

Staple To assemble by use of a U-shaped fastener.

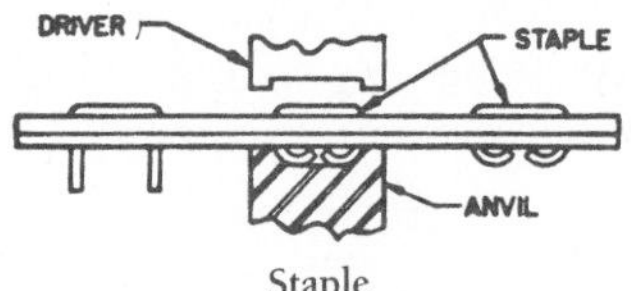

Staple

Stripper The plate used in a die that strips the part from the die.

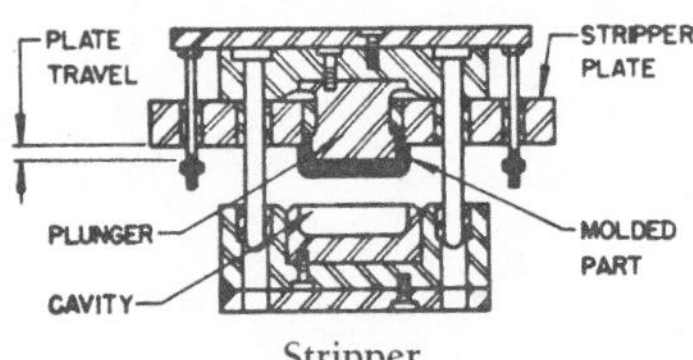

Stripper

Stud A stationary shaft, one end of which is fastened to the body of the part and receives a nut for fastening on the other.

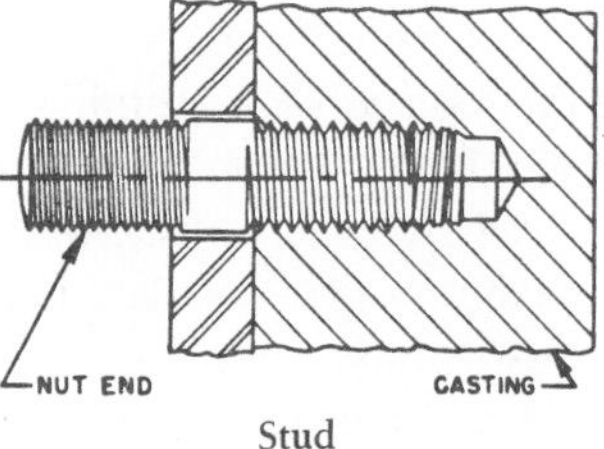

Stud

Surface gage A flat block of steel carrying an adjustable, upright spindle with which a scriber is mounted for layout work or for use as an indicator for transfer readings.

Surface gage

Surface plate A plate with a flat surface used to check parts for flatness.

Surface plate

Swage To form metal while cold by drawing or squeezing or by submitting to a number of blows sufficient to shape to desired form.

Sweat To solder together by clamping the parts in contact with soft solder between them and heating.

T

Tack weld The welding of short intermittent sections.

Tap To cut an internal thread by screwing a fluted tapered cutting tool into the hole.

Taper A gradual and uniform increase or decrease in size.

Taper pin A pin requiring a taper reamed hole at assembly and depending only on a taper lock, which can totally disengage when minor displacement occurs.

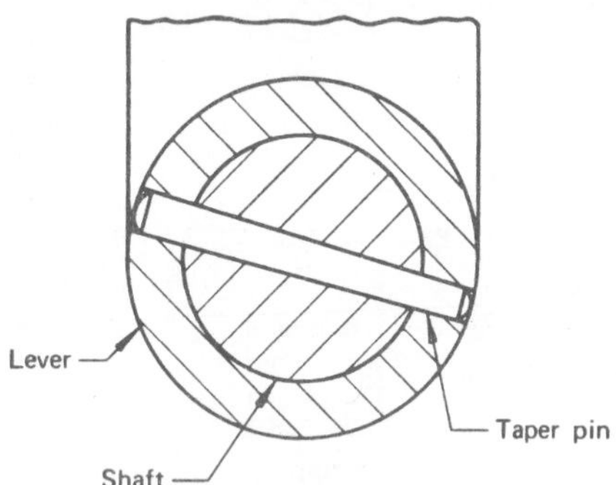

Taper pin

Temper To reheat hardened steel to some temperature below critical temperature, followed by the desired rate of cooling; also called "drawing."

Template A pattern or guide used for laying out duplicate parts or guiding the tools while machining.

Tensile strength (U.T.S.) The maximum tensile load per square inch that a material can withstand. It is computed by dividing the maximum load obtained in a tensile tester by the original cross-sectional area of the test specimen.

Thread pitch The distance between corresponding points on adjacent threads measured parallel to the axis.

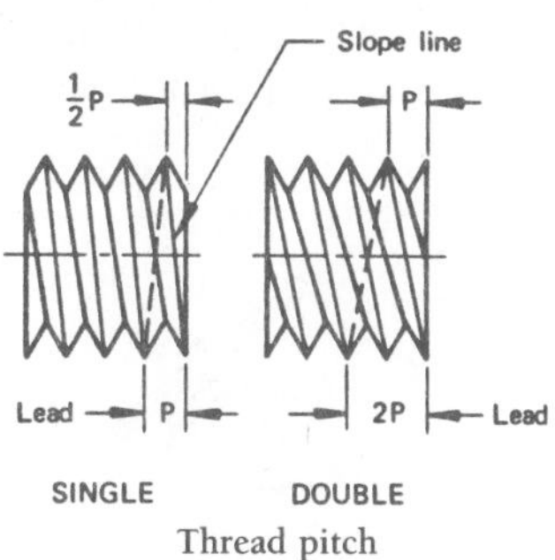

Thread pitch

Tolerance The permissible variation from the basic dimension of a part, expressed either unilaterally (one side given) or bilaterally (a summation at both ends).

Truncate To cut off a geometric solid at an angle to its base.

T-slot A slot machined in a part having a sectional shape resembling the capital letter "T."

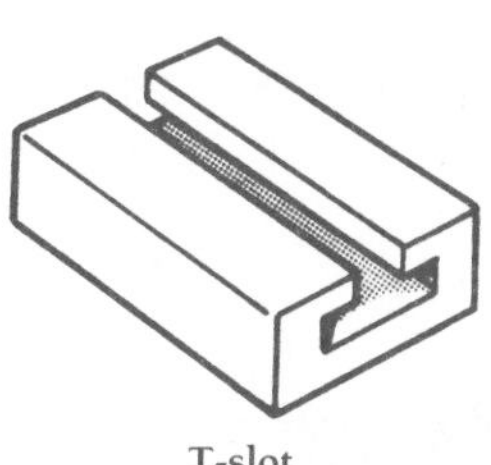

T-slot

Turning A process used for removing material to produce relatively smooth and dimensionally accurate external and internal surfaces by turning the workpiece against the cutting tool as in a lathe.

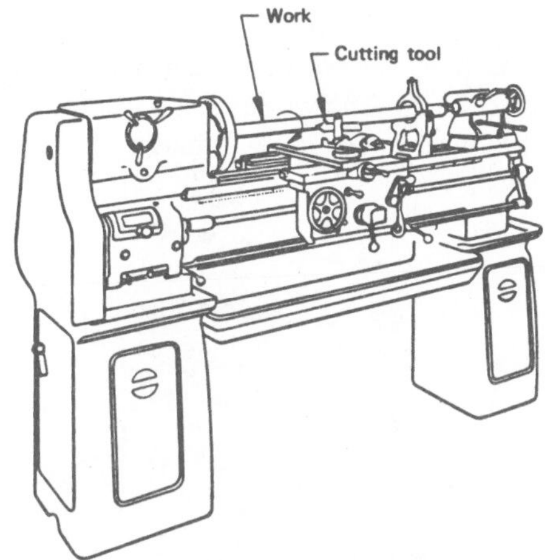

Lathe used for turning

U

Undercut A recessed cut that permits a firm fitting of one part with another.

Union Tube or pipe fitting used to connect lines that carry either gases or liquids at relatively high pressures and that are not subjected to excessive vibration or movement.

Upset To increase the sectional area of a part to form a shoulder during forging.

V

Vernier caliper A small movable scale used on measuring instruments for determining a fractional part of one of the equal divisions of a graduated scale. The vernier caliper yields high accuracy because it is, in effect, one scale on another scale.

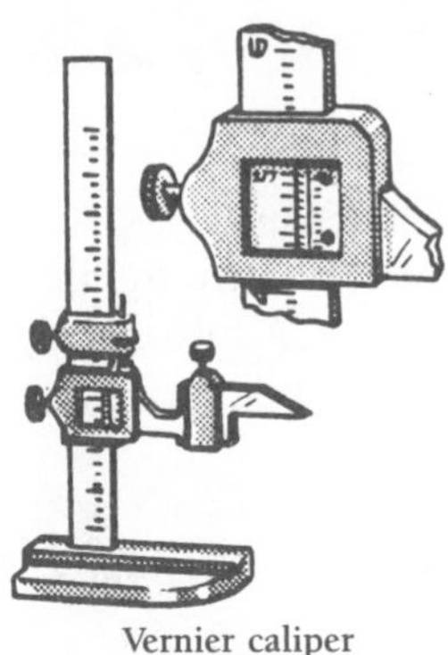

Vernier caliper

W

Washer A round piece of metal (usually steel), either flat or spring type, with a hole in the center and placed on the bolt or screw ahead of the nut to provide positive clamping, bearing surface, or to lock the nut in position.

Washer

Web A thin section of a casting between the ribs, bosses, and flanges to provide additional strength.

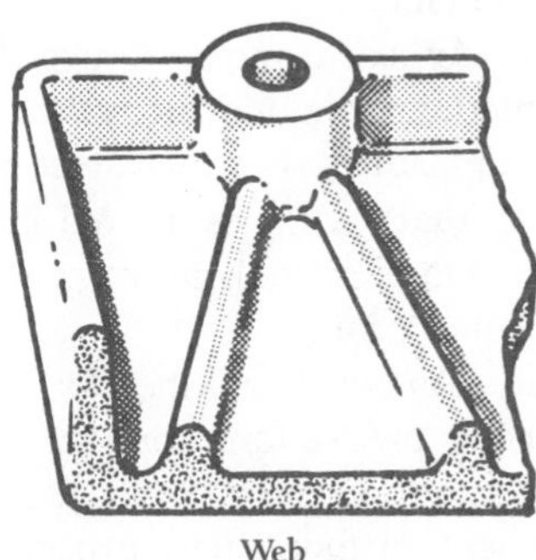

Web

Welding Any one of several methods of permanent assembly whereby parts are joined together by heating the joint to material fusion temperature, such as by arc welding, gas welding, projection welding, and spot welding.

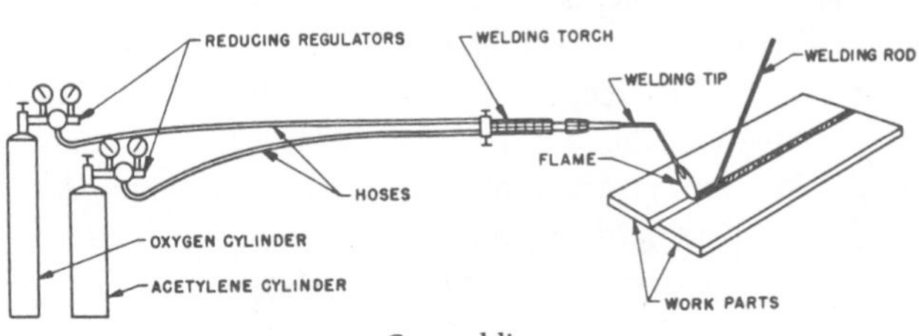

Gas welding

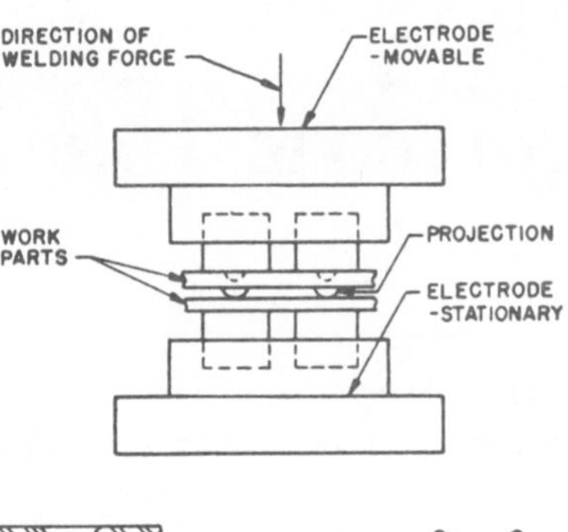

Projection welding

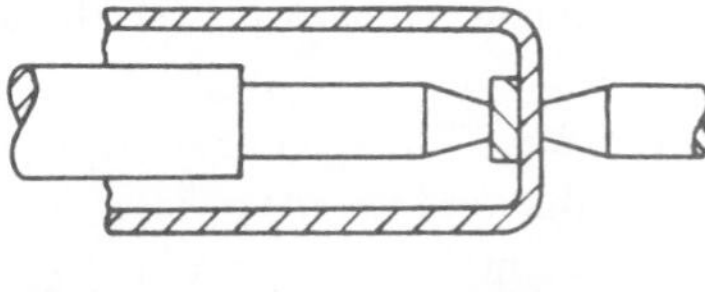

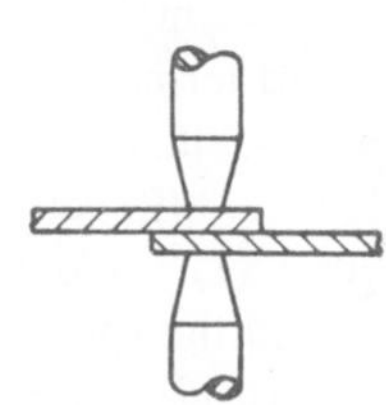

Spot welding

Wing nut Screw fastener especially designed with extensions on the internal threaded part for the convenience of hand tightening.

Wing nut

Worm A gear in the form of a screw used to transmit motion parallel to the axis of the shaft and that provides large reduction of velocity.

Worm wheel The gear that meshes with the worm in an installation, and may be combined with the worm; provides 90° transmission of revolving forces and increases torque and smoothness of action.

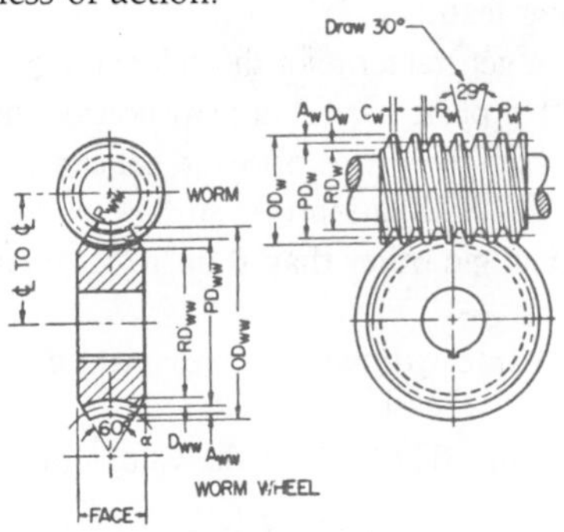

Worm and Worm wheel

APPENDIX A.2

Geometric Tolerancing Glossary

A

Acceptable quality level A quality level which in a sampling plan corresponds to a specific high probability of acceptance.
Acceptance inspection (1) Inspection to determine whether an item delivered or offered for delivery is acceptable. (2) The consent to accept or take a feature, part, assembly, or lot as offered.
Accuracy The verification of a measurement when compared to an ultimate standard or true theoretical value. Also see precision.
Accuracy of the mean The closeness of agreement between the true value and the mean result obtained by applying the experimental procedure numerous times.
Acronym A word formed by the first (or first few) letters of a series of words, such as MMC from maximum material condition, RFS, LSC, MRS, etc.
Actual deviation (1) The difference between the actual (measured size) and the specified (nominal or mean) size (*or for limits and fits*) (2) The algebraical difference between the maximum limit of size and the corresponding basic size.
Actual fit The actual total clearance or interference between two parts at assembly.
Actual local size The actual (measured) individual sizes (distances) at any cross section of a part.
Actual size (1) The size determined from a number of actual local sizes of a dimension of an individual feature. It may be the smallest or largest, for an external or internal shape, respectively, perfect form that inscribes or envelopes the actual local sizes. The aforementioned perfect form refers to that which is specified or depicted. When the size dimension for this form is specified as being oriented, the actual size measurement must also be so oriented. (2) The measured size.
Alignment The condition where features are coplanar, collinear, or coaxial.
Allowance The intentional difference between the MMC limits of mating features.
Angle plate A general term for the different right-angle irons used to establish a plane parallel or perpendicular to a surface plate, against which the part is placed in contact or clamped.
Angularity The inclination of a surface or axis that is at a specified basic angle other than 0 or 90° from a datum plane or line.
Arc length A linear dimension measured on (along) a curved outline.
Arithmetic mean The sum of the values divided by their number.
Assembly A number of separable parts or subassemblies or any combination thereof fitted together to perform one or more specific functions. A subassembly is sometimes referred to as a "section" to differentiate it from the main assembly for clarity.
Attitude tolerance Tolerances of perpendicularity, parallelism, and angularity. Better expressed as (and see *orientation tolerance*.)
Auxiliary datum A datum which has a tolerance relative to the part datum reference frame. A single datum, not related, which is generally erected by contact with a surface.
Auxiliary dimension A reference dimension. A dimension in which size is given solely for information (see *reference dimension*).
Average diameter (1) The diameter of an out-of-round feature in the free state found by averaging the diameter at various azimuths or more precisely by doubling the average radius from the least-squares center or by measuring the perimeter. If forces were applied to this feature constraining it round then, in that state, the average diameter (with its associated tolerance) should be the same as the rounded diameter on a constrained (round) part. (2) The diameter a free state part would assume if it were round; the perimeter (circumference) of the shape divided by π (Pi) or as measured with a Pi tape.
Axis—Common An axis common to two or more coaxial features. The centerline on a drawing is the theoretical common axis of all features shown centered about it.
Axis—Derived (actual) An axis derived from the center points of the actual local sections of the considered feature (such as least-squares axis).
Axis—Simulated An axis established by (or from) a surface of adequately precise form (or a mathematical representation of same). In the RFS case the gage (mathematical model) or tool surface expands (to contact an internal feature) or contracts (to contact an external feature). In the MMC case the gage or mathematical representation is the MMC size of the considered feature thus allowing the tool or gage (datum) axis to "float" when the part feature departs from MMC. The philosophy is that the (MMC) axis must be positioned to the *ONE* best position for part acceptance.
Axis—Theoretical (1) A straight line, real or imaginary, passing through the center of a geometrically perfect form, or for simulation, the axis of a tool or gage. (2) A straight line passing through the center of two defined points. A straight line passing through the center of one defined point and perpendicular to a plane.

Axis—True The axis of revolution. (According to the classic definition in mathematical engineering—see Marks handbook or other source.)

B

Baseline dimensioning Dimensioning from a common datum (specified or implied).
Basic angle The basic dimension of an angle (shown enclosed in a rectangle).
Basic dimension A dimension enclosed in a rectangle that is a numerical value used to describe the theoretically exact size, profile, orientation, or location of a feature or datum target. It is the basis from which permissible variations are established by tolerances in a feature control frame, by other dimensions, or by note.
Basic hole In the ISO System of Limits and Fits wherein the hole lower deviation is always zero. In this case the symbol used is H (as in ∅ 40H11) and the minimum diameter of the hole is the basic specified size.
Basic note "UNTOLERANCED DIMENSIONS LOCATING TRUE POSITION ARE BASIC," delineates basic dimensions by note rather than by symbol—nonpreferred.
Basic pattern The theoretically exact interrelations of the considered features within the group.
Basic shaft In the ISO System of Limits and Fits wherein the shaft upper deviation is always zero. In this case the symbol used is h (as in ∅ 040h8) and the maximum diameter of the shaft is the basic specified size.
Basic size The basic size is that size from which the limits of size are derived by the application of allowances and tolerances. The common value of two mating features from which all limits of size are established.
Bilateral tolerance A tolerance in which variation is permitted in both directions from the specified dimension (e.g., 1.500 ±.005).
Boundary of perfect form An envelope of perfect form established by size limits, such as the boundary of perfect form at MMC (the true geometric form represented on the drawing), requirement for individual features.

C

Cartesian coordinate system Three mutually perpendicular planes (**X**, **Y**, and **Z**) from which points are located in space. Used, especially, for robotic installation of components and CAM (computer-aided manufacturing) or on parts wherein relationships of features (originating from these three planes) are perfect at the maximum material conditions.
Centerlines Theoretical lines, shown on a drawing, depicting an axis of revolution or representing two centerline/planes (at a true 90°) from which rectangular coordinate dimensions are given. Measurements cannot be taken from centerlines alone because they are not definitive—datums are specified for this purpose.
Center plane The median plane of a feature such as a slot.
Center point The center point of a circle, sphere, or other regular form.
Centers—machining Formed shapes on an item (generally in the ends) dimensioned (ANSI B94.11M) to accommodate holding the part during manufacturing, such as on a lathe.
Chain dimensioning Dimensioning in series from feature to feature as opposed to baseline or positional dimensioning. Generally allows a tolerance accumulation equal to the sum of the variances in each direction.
Circular runout The independent measurement at a cross section or at a point on a surface oriented to the axis of circular elements on a solid of revolution when the part is rotated 360°. This composite control verifies out-of-roundness and position on cylindrical items and perpendicularity errors on faces.
Circularity The condition wherein a circular element line is the same radial distance at all azimuths from a center point, independently taken in a plane normal to the axis of the considered feature.
Clearance The algebraic difference between the sizes of the internal feature (hole) and the mating external feature, when the difference is positive.
Clearance fit A fit having limits of size specified such that a clearance always results when mating parts are assembled.
Coaxiality Collinear or coincident axis. Control of coaxiality may be specified as position, runout, or concentricity, as appropriate.
Collinear The condition wherein the center lines or axes of features are in line.
Common axis Axis of two or more coaxial features, which have a common tolerance or together shall act as a single datum.
Complex feature A shape (generally an outline) composed of two or more simple individual features such as a square, rectangle, hexagon, etc.
Concentricity The condition where the centers of all cross-sectional elements (such as the least-squares axis) of a surface of revolution (one or more feature axes) are collinear (or coaxial) with a datum axis.
Conditional tolerance A tolerance or datum reference in a control frame with a modifier other than RFS (e.g., MMC, LMC).
Considered feature The feature under consideration.
Constant A term used with a dimension to indicate that a characteristic (e.g., section thickness, profile tolerance, etc.) is to have the same dimension and tolerance as depicted all over (or around) regardless of the shape of the considered feature.
Contour The outline of a figure, generally a complex shape or portion of an arc. (See *profile of a line or surface.*)
Control drawing A drawing that discloses configurations and configuration limitations; performance and test requirements; mass and space limitations; access clearances; pipe and cable attachments, to the extent necessary that an

item can be developed and procured on the commercial market to meet standard requirements; or for the installation and functioning of an item to be installed with related items.
Coplanarity The condition where axes, center planes, or all elements of a surface are in one plane. (Profile of a surface tolerance is used to control coplanarity.)
Cylindricity The condition of a surface of revolution in which all points on the surface are equidistant from a common axis.

D

Datum Theoretical geometrical reference (generally point, line, or plane) simulated by very precise manufacturing, tooling, or inspection equipment, from which the form, orientation, or position, of other features is established. (Common: the origin of measurement.)
Datum axis The theoretically exact centerline of a true cylinder that would contact the extremities of a cylindrical feature. An axis or straight line established by two defined points such as the intersection of a diameter and a plane at each end of an object. When so specified, the centerline of an MMC cylindrical shape free to float when the feature is not at MMC. A mathematically derived axis such as the least-squares axis.
Datum—Compound A datum established by more than one datum feature, the features being of equal importance/precedence.
Datum equalization Orienting a datum feature relative to a manufacturing or inspection setup with tooling, inspection equipment, or by least-squares analysis.
Datum (identification) feature symbol A rectangle containing the datum letter with a short dash on each side.
Datum letter Any capital letter (preferably other than I, O, M, or Q) used to identify or refer to a datum feature.
Datum plane A theoretically exact plane established by contacting the extremities of an actual feature surface with a simulator (such as a surface plate). In some cases a datum plane is established by the least-squares method.
Datum point Has position in space on a plane but has no size or shape. In roundness measurements, for example, this point is normally the center from which radial measurements are taken. Mathematically the least-squares center is used as a datum point.
Datum reference frame Three mutually perpendicular datum planes relative to which the part is oriented and measured. The datums involved are theoretical and are therefore simulated in manufacturing and inspection.
Datum, simulated The plane, axis, or point established by, or from, a surface of adequately precise form contacting the datum feature. Simulated datums are used as practical embodiments of specified datums.
Datum, single A datum established from one feature.
Datum, specified An explicit datum that has been indicated in a note, datum identification, or feature control symbol.
Datum surface (or feature) The actual (real) surface of the feature (part) from which the datum is established, by contact, simulation, or that which is used to establish the location or orientation of a datum.
Datum system A system consisting of a group of features which are chosen to erect datums (mating of cofunctioning features) on each of two mating or cofunctioning components. A missile datum system may consist of a missile longitudinal axis, missile base, and azimuth datum planes. Generally a three-plane system.
Datum target A specified point, line, or limited areas used to establish datums, especially useful for repeatability of orientation and location of as-forged or as-cast parts.
Derived axis An axis established by a true form contacting the extremities of a feature, or by least-squares analysis.
Design size Design size of a dimension is the size in relation to which the limits for that dimension are assigned. It is therefore usually the size specified on the drawing (especially for limits and fits), and is identical to the basic size when there is no allowance.
Detail assembly drawing A drawing depicting an assembly on which one or more parts are detailed in the assembly view or on separate detail views.
Detail drawing A drawing depicting end product requirements for the parts delineated on the drawing. Mono detail drawings delineate a single part or item.
Deviation Algebraical difference between a size (actual, maximum, etc.) and the corresponding basic size.
Diagrammatic drawing (or diagram) A drawing delineating by means of symbols and lines, characteristics and relationships of items forming an assembly or system.
Dimension A numerical value expressed in appropriate units of measurement. A complete specification for a dimension must include (directly or by association) an origin, direction, magnitude, and allowable variation. Used with lines, symbols, and notes, dimensions are used to describe the size or geometrical characteristics of an object.
Direct dimensioning Dimensions placed directly between functional features to avoid tolerance accumulation.
Drawing A document presenting information pictorially and/or by textual matter. Note: A drawing is normally identified by a drawing number and title.
Dual dimensioning Specifies dimensions in two sets of units (e.g., metric and customary). The primary set of units is specified in the conventional manner and the other is placed in square brackets (after the primary units in feature control frames and below the primary units in other areas). Where no equivalent exists (as for a thread) only the primary units are specified.

E

Eccentricity Radial deviation from a centerline or axis. One half of concentricity.
Effective condition (or size) The actual size to which a feature has been produced plus or minus (external or

internal, respectively) the actual geometrical error [e.g., a shaft at .250 (measured) diameter with a measured straightness deviation on the axis of .007 has an effective size of .257]. Note: The effective condition is differentiated from the virtual condition in that the effective condition deals with actual (or measured) values.

End of full thread The point at which the thread profile ceases to be fully formed.

End product (or item) An item, either an individual part, assembly, structure, or product, in the condition in which it is to be used. It defines a piece ready for assembly or service or it may be the product of a foundry or forge supplied for further processing.

Enveloping boundary (elements) Profile tolerance boundary lines which are separated everywhere by a theoretical circle of the diameter of the specified tolerance centered on the basic (or mean) profile in one plane.

Enveloping boundary (surface) Profile tolerance boundaries which are separated everywhere by a theoretical sphere of the diameter of the specified tolerance centered on the basic (or mean) profile in all directions.

F

Feature A feature is a general term applied to a specific characteristic portion of a part, such as a surface, hole, slot, screw thread, radius, etc.

Feature control (symbol) frame A rectangular frame containing the geometric characteristic symbol, the applicable tolerance, required datums, and modifiers. This symbol describes allowable variations, datum relativity, zone shape information, and conditions under which the tolerances apply.

Feature of size Portion of an item such as (a) a single plane surface or (b) a single cylindrical surface or (c) a single spherical surface or (d) two parallel plane surfaces; each is associated with a size dimension.

Fit The general term used to signify the range of tightness or looseness that results from the application of a specific combination of allowances of two parts to be assembled. Examples of types of fits are: RC: running or sliding clearance; LC: locational clearance fit; LT: transition clearance or interference fit; LN: locational interference fit; and FN: force or shrink fit.

Fit system System of fits comprising shafts and holes belonging to a limit system.

Fixed fastener system The condition where the fastener is positively located (fixed) in one of the mating ports and passes through a clearance hole in the other (e.g., a clearance hole and tapped hole combination, or a press fit dowel hole and clearance hole combination).

Flatness The condition where element lines in all directions are in one plane.

Floating (free) fastener system Condition where a fastener is to be assembled with two or more parts, all of which contain clearance holes or features that do not positively locate (fix) the fastener.

Form tolerance Indicates how far individual surfaces (or elements of a single feature) can vary from the desired form or shape indicated on the drawing. Tolerances of form are straightness, flatness, roundness, cylindricity, and taper (flat or conical) and are not related to datums.

Free state variation (1) The condition where a part of the feature is distorted due to its own weight and geometrical configuration within the elastic limit [i.e., a large unsupported cylindrical shape of small cross section (thin) that is elliptical or oval until restrained round]. (2) The amount of distortion of a feature that occurs when manufacturing forces are removed.

Full indicator movement (FIM) The total movement of the indicator when transversing a surface. The absolute sum of the largest positive and largest negative indicator hand movement. FIM also refers to the total observed deviation with a dial indicator in contact with the part while the part is revolved about the datum axis.

Functional dimension A dimension that is essential to the function of a product.

Functional feature A feature that plays an essential part in the performance of serviceability of the piece to which it belongs. It may be a location feature, i.e., a spigot which serves to locate a component in an assembly, or a working surface, i.e., the bore of bearing.

Functional gage A gage with fixed elements (rather than readout). Provides a go/no-go type of verification. Uses some (generally 10%) percent of the part (feature) tolerance.

Fundamental deviation Defines the position of the tolerance zone relative to the zero line (basic size). This value when added to the basic size gives one limit of size. The fundamental deviation in the ISO system is designated by a letter, uppercase for internal features (holes) and lowercase for external features (shafts), e.g., ⌀ 40H11.

G

Gage (a) *Direct measurement*—The gage is applied (probe, electronic beam, pneumatically, etc.) to the surface and the actual dimension is read on a graduated scale, digitally or from a chart or computer readout. (b) *Fixed limit*—Functional go/no-go type of gages that do not require the operator to make a decision. (c) *Paper or computer*—Actual dimensions from inspection machines or devices are plotted on paper or entered into a computer for mathematical simulation and/or analysis. Echelons of accuracy relative to gages are: International defining standard, National standard, National reference standard, Laboratory standard, Inspection gage, and Production or floor gage.

Gage dimension Theoretically exact (similar to basic dimension), not used on drawings.

Gage size A code number or designation used with cables, rods, sheets, and other materials, the actual dimensions of which are shown on the drawing. Gage or code numbers may be referenced (in parentheses) after the dimension.

General arrangement drawing An arrangement drawing where the item depicted is the end product.

Geometric characteristic symbol The symbols used to describe the type of tolerance (e.g., flat, parallel, etc.) in the tolerance frame.

Geometrical tolerance The general term applied to the category of tolerances used to control maximum permissible variation of form, orientation, location, or runout specified on the drawing.

Go limit The designation applied to the limit of size that corresponds to the maximum material condition, i.e., the upper limit of a shaft (ring gage is GO GAGE), and the lower limit of a hole (plug gage is GO GAGE).

Grade The number symbol designating the tolerance as a function of the basic size (e.g., H8) in the ISO system of limits and fits.

Grade of tolerance In a standardized system of limits and fits, a group of tolerances considered as corresponding to the same level of accuracy for all basic sizes.

H

Hole basis system System of fits in which the different clearances and interferences are obtained by associating various shafts with a single hole (or, possibly, with holes of different grades, but always having the same fundamental deviation).

Hole depth On a blind hole the depth specified is the depth of full-form diameter from the corresponding surface. Where the hole is countersunk, counterbored, or counterdrilled, the depth also applies from the outer surface.

Hole pattern Any group of holes that function as a unit or group, regardless of differences in characteristics of the holes.

I

Implied datum An unspecified (implicit) datum from which dimensions originate, the implied datums are used in any sequence (order of precedence). Implied datums are not recommended except where tolerances are extremely generous.

Interchangeability Where two or more items possess such physical and functional characteristics as to be equivalent in performance and durability and capable of being exchanged one for the other without alteration of the items themselves or of adjoining items except for adjustment and without selection for fit or performance, the items are interchangeable.

Interference fit Limits of size so prescribed such that an interference always results when mating parts are assembled.

International system of units SI System International—An international system of units required to be used by the EEC (European Economic Community) since December 1978. The standardized metric system used worldwide.

ISO International Standards Organization.

L

Least material condition The condition of a feature wherein it contains the least (minimum) amount of material (largest tolerance limit on an internal feature, such as a hole; smallest tolerance limit on an external feature, such as a shaft) within the specified limits of size.

Least material size For an external feature, the minimum limit of size specified on the drawing. For an internal feature, the maximum limit of size specified on the drawing.

Limit dimension A type of dimension wherein the maximum limit of size is expressed above or to the right of the minimum limit of size.

Limit system System of standardized tolerances and deviations.

Limits and fits Standardized fits for holes and shafts—ANSI B4.2 and ISO 286.

Limits of size The two extreme allowable sizes (limits) within which the actual size should lie. For individual features of size the feature surface must lie within an MMC envelope of perfect form and the LMC size is verified by cross-sectional measurement.

Line fit A fit having limits of size (or virtual conditions) so prescribed such that one limit on mating features is the same (line to line or surface to surface) and the other limit results in clearance.

Linear units The units used on the drawing, i.e., metric or customary (inch).

Local size Any individual measurement at a cross section of the feature of a part.

Location tolerance The tolerances of concentricity and position.

Lower deviation The algebraic difference between the minimum limit of size and the corresponding basic size.

M

Mating size (a) *Holes*—The diameter of the largest perfect imaginary cylinder which can be inscribed within the hole so that it just contacts the extremities of the surface.
(b) *Shafts*—The diameter of the smallest perfect imaginary cylinder which can be circumscribed about the shaft so that it just contacts the extremities of the surface. (It will be noted that this definition is unrelated to the position or attitude of the feature.)

Maximum clearance The difference between the maximum specified size of an internal feature and the minimum specified size of the mating external feature.

Maximum dimension The maximum dimension represents the largest acceptable value or the upper limit.

Maximum limit of size The greater of the two limits of size.
Maximum material condition (MMC) (1) The condition of a part or feature wherein it contains the maximum amount of material (within the limits of size). (2) The fit condition producing the smallest amount of clearance (i.e., maximum shaft size, minimum hole size).
Mean dimension One-half of the sum of the limits of a dimension, not necessarily the nominal.
Median plane The center plane of a square or rectangular feature. Median to two parallel planes that contact the extremities of the feature.
Minimum clearance The difference between the minimum specified size of an internal feature and the maximum specified size of the mating external feature. Where geometric tolerances are involved, it is the difference between mating features at their virtual condition.
Minimum dimension The smallest acceptable value or the lower limit.
Minimum limit of size The smaller of the two limits of size.
Modifier A symbol used in a feature control frame to modify the condition under which the tolerance applies. The modifiers are Ⓜ for maximum material condition, Ⓢ for regardless of feature size, Ⓛ for least material condition.
Multiple datum system A system wherein more than one datum feature is used.
Murphy's law A satirical law used in engineering; (does not recognize probability) assumes the worst possible situation will occur regardless of the nature of the data involved.

N

Nominal axis See *derived axis.*
Nominal diameter step (limits and fits) The principal range of limiting sizes, a set increment; the two (extreme) sizes in a given step (range).
Nominal dimension The "ideal" dimension (used in "plus or minus" tolerancing).
Nominal size The size by which an item is designated as a matter of convenience. Examples: M20 screw thread; 5-t crane; 10-kW motor.
Nonmandatory dimensions Processing dimensions identified as "NONMANDATORY (MFG DATA)" such as shrink or finish allowance, etc.
Nonrigid part A part that distorts (within the elastic range) in the free (nonrestrained) state due to its own weight or the release of internal stress buildup during fabrication.
Normal Perpendicular, square.
Not go limit Designation applied to that of the two limits of size which corresponds to the least material condition, i.e., the lower limit of a shaft or the upper limit of a hole.

O

Oblique projection The projection of an object in which the lines of sight are parallel to each other but inclined to the plane of projection and where the object is oriented with the principal face parallel to the plane of projection, thus making this face and parallel faces show in true shape.
Orientation of a line The orientation of a pair of parallel straight lines with the least separation which completely envelops an element or centerline (often found by the least-squares line method).
Orientation of a plane The orientation of a pair of parallel planes with the least separation which completely envelops the actual surface or center plane (often found by the least-squares plane method).
Orientation tolerances Tolerances of parallelism, perpendicularity, and angularity.
Origin of measurement A dimension origin symbol (circle centered on an extension line as one end of a dimension line, the other end terminates with an arrowhead) indicates that the attendant tolerance applies to the feature on the arrow end.
Orthogonal projection The projections of an object in which the lines of sight are perpendicular planes of projection and where the object is oriented so that its three principal axes are parallel to the planes of projection.
Overall dimension The sum of the associated intermediate dimensions (often reference).

P

Parallelepiped A rectangular prism-shaped tolerance zone. Used where a bidirectional tolerance is preferable for a feature of length as in the case of bidirectional position.
Parallelism The condition where lines or planes extend in the same direction; in principle, the considered surface or axis is the same distance from a datum plane or axis at every point.
Part One piece or member or two or more pieces or members joined together that cannot normally be separated without destruction or impairment of designed use. Note: A part is sometimes described as a component.
Pattern Two or more features that are interdimensioned and located (relative to implied or specified datums) as a group.
Perpendicularity The condition where primary surfaces, median planes, or axes are at exactly right angles to a datum.
Perspective projection The projection of an object in which the lines of sight converge to a point of sight located at a finite distance in front of the plane of projection.
Polar dimensioning Dimensioning by indicating the radius and azimuth (angle) with associated tolerances.
Position tolerance (1) The total amount of variation permitted for the location of a feature in the group of which it is a member. (2) A numerical value describing the total zone (cylinder, two parallel planes, etc.) within which

the (axis, median plane, surface, etc.) feature may be positioned (located). This zone is situated relative to a theoretically exact (basic) location or "true position."
Precision Precise duplication traceable to an as-produced article or tooling rather than to a standard or theoretical model.
Primary datum The first datum of a three-plane system; first in order of precedence.
Processing dimensions See *nonmandatory dimensions.*
Profile of a line (element) A uniform two-dimensional boundary along the true profile within which a surface element of an irregular shape, portion of an arc, or other established profile may lie.
Profile of a surface A uniform three-dimensional boundary along the true profile within which the surface of an irregular shape, portion of an arc, or other established contour may lie.
Projected tolerance zone A tolerance zone that extends beyond the fixing feature so that the projection of a fastener (or other feature) located from the fixing feature (thread, press fit hole, etc.) on installation will be controlled over its entire length. Projected zones control the angularity/perpendicularity of fixed fasteners to prevent interference.

R

Rectangular coordinate dimensioning Dimensions giving location of features by abscissa and ordinate (*x, y* coordinates) from datums generally yielding a square or rectangular zone.
Reference dimension A dimension usually without tolerance enclosed in parentheses (for information only) and not to be used to govern production or inspection. Generally used to orient a detail relative to a surface (repeating a dimension) or a value derived or calculated from other dimensions given on the drawing or related drawings.
Regardless of feature size The condition wherein geometric tolerance must be met regardless of the size to which the considered feature, datum, or related features is produced.
Replaceability An item that is a replacement for another item but is not interchangeable, in that the original item may not be interchanged for it.
Rounding Linear units are rounded in accordance with ANSI Z210.1.
Roundness The condition wherein a circular element line is the same radial distance at all azimuths from a center point at all cross sections independently taken normal to a derived axis (or the common center on a sphere) of the considered feature. (Circularity is the preferred term.)
Runout Composite tolerance used to control the functional relationship of one or more features of a part to a datum axis.
Runout, circular Composite two-dimensional tolerance departure of a circular element of the considered feature with respect to a fixed point during one complete revolution about the datum axis, without axial movement of the part or measuring instrument.
Runout, total Composite three-dimensional tolerance departure of a feature from a true geometric form, attitude, or position (as a composite), relative to a datum axis. The tolerance is applied to a zone that is verified over the entire considered feature as the part is rotated.

S

Section The projection of the cut in an object made by a cutting plane.
Sectional view A section of an object including features beyond the cutting plane.
Shaft basis system A system wherein different clearances and interferences are obtained by associating various holes with a single shaft or with shafts of different grades having the same fundamental deviation. The basic shaft in this system generally has an upper deviation of zero.
Size A generic term denoting magnitude of any kind.
Size tolerance A "plus or minus" (unilateral or bilateral) or limit dimension, which designates magnitude and the allowable limits within which the size of the feature must lie.
Slope The inclination of a surface expressed as the ratio of the difference in the heights (at each end) to the distance between the heights.
Specified datum An explicit datum specified in a note, datum identification, or feature control frame.
Standard tolerance Eighteen grades of tolerance (in the ISO system) called IT01, IT0, IT1 to IT16, the numerical values of which are given in tables or can be calculated for each nominal diameter step. For information on the U.S. system refer to ANSI B4.2.
Standard tolerance unit In the ISO or U.S. System of Limits and Fits, a factor expressed only in terms of the basic size and used as a basis for the determination of the standard tolerances of the system. (Each tolerance is equal to the product of the value of the standard tolerance unit, for the considered basic size and a coefficient corresponding to each grade of tolerance.)
Stock Items produced to established government or industry standards such as bars, sheets, tubing, etc., which prescribe dimensional and/or geometric tolerances. When "Stock" is specified on the drawing, the item remains in the as-furnished condition in the end item.
Straightness Is that condition where an element of a surface or an axis is a straight line.
Subassembly Two or more parts that form a portion of an assembly or a unit replaceable as a whole, but having a part or parts which are individually replaceable.
Surface texture The finish on a surface generally specified in microinches, in standardized steps (e.g., 32, 63, 125). Applicable national standards are ANSI Y14.36 and ANSI B46.1.

Symbol A mark, character, letter, or combination thereof that is nationally or internationally accepted for indicating an object, idea, or process. Geometric characteristic symbols represent a type of tolerance in lieu of words.
Symmetry (position) Is that condition whereby a feature can be divided into two identical parts by an indicated line or plane.
System A combination of parts and assemblies fitted together to perform a specific operational function or functions.

T

Tabulated assembly drawing An assembly drawing showing similar configurations with the variations in assembly characteristics given in tabular form.
Tabulated drawing A detail drawing showing similar configurations with variations in dimensions and features given in tabular form. Note: Sometimes referred to as a schedule.
Tangent plane, cylinder, or circle A precise surface or mathematical model of ideal geometry that contacts the extremities (high points) of a surface or surface element.
Taper, conical or flat The ratio of the difference in the height or diameter of the ends of a feature and the distance between the ends.
Tertiary datum The third datum in the order of precedence in a three-plane system of three mutually perpendicular planes (datum reference frame). The datum feature from which the tertiary datum is located.
TIR See *full indicator movement.*
Tolerance The amount by which a specific dimension may vary. The difference between the maximum and the minimum limits. The allowable geometric variation of form, profile, orientation, position, or runout in the case of a geometric tolerance.
Tolerance diagram The geometric reference frame with the tolerance zones superimposed on it.
Tolerance frame A rectangular outline within which the geometric symbol, tolerances, datums, and other data are placed. This frame is used to specify geometric control to a feature.
Tolerance zone The allowable boundaries within which the feature, surface element, center point, axis, or center plane must lie.
Total runout Composite three-dimensional tolerance departure of an entire feature from a true geometrical form, attitude, or position (as a composite) as the part is rotated, relative to a datum axis, without axial movement of a part.
Transition fit Limits of size so prescribed that either a clearance, line, or interference fit may result when mating parts are assembled.
True position The theoretically exact (basic) location of a point, line, plane, or surface.
True position tolerance The former name for positional tolerance.

U

Unilateral tolerance A tolerance that has an allowable variation in only one direction from the nominal, (e.g., 28.00 + .02, 0; 28.00 0, − .02).
Unspecified angles Features shown at 90° (implied); these are controlled by the tolerance block (or general note) for allowable angular variation.
Upper deviation The algebraic difference (in the limit and fit system) between the maximum limit of size and the basic size of a considered feature.

V

Variation of fit Arithmetical sum of the tolerances [and intentional slop (clearance) if designed in] of the two mating parts of a fit.
Verification equipment The technical devices necessary for verification using a specific method.
Verification method The verification methods are the practical applications of the verification principle through use of different equipment and operations.
Verification principle The verification principle is a fundamental geometric basis for the verification of the considered geometrical characteristics.
Virtual condition Of a feature—Limiting functional boundary permitted by the drawing data, which is generated by the collective effect of the maximum material size of the considered feature and the concerned form, orientation, runout, and/or location tolerance.
Of a group of features—Assembly of the virtual condition of all the features comprising the group in perfect geometric relationship as defined by the drawing data.
Virtual size Dimension defining the actual virtual condition of a feature based on measured values.

Z

Zero line In a graphical representation of limits and fits, straight line to which the deviations are referred, the zero line is the line of zero deviation and represents the basic size. By convention, when the zero line is drawn horizontally, positive deviations are shown above the line and negative deviations below it.
Zero tolerance at MMC A tolerance specified as zero at MMC is dependent on the size of the feature involved. When the feature is at MMC the considered geometric tolerance is zero and may be increased an amount equal to the considered feature size departure from MMC.

APPENDIX B

Abbreviations, Formulas, and Standards

APPENDIX B.1

General Abbreviations

Examples of Terms and Corresponding Abbreviations or Symbols

Term	Abbreviation or Symbol
And	&
Across Flats	A/F
American National Standards Institute	ANSI
Angular	ANG
Approximate	APPROX
Assembly	ASSY
Basic	BSC
Bill of Material	B/M
Bolt Circle	BC
Brass	BR
Brown and Sharpe Gage	B & S GA
Bushing	BUSH
Canada Standards Institute	CSI
Casting	CSTG
Cast Iron	CI
Centimeter	cm
Center Line	℄
Center to Center	C to C
Chamfered	CHAM
Circularity	CIR
Cold-Rolled Steel	CRS
Concentric	CONC
Counterbore	⌴ or CBORE
Countersink	⌵ or CSK
Cubic Centimeter	cm^3
Cubic Meter	m^3
Datum	DATUM { ISO: [A] ANSI: [-A-] }
Deep	↧
Degree (Angle)	° or DEG
Diameter	Ø or DIA
Diametral Pitch	DP
Dimension	DIM
Drawing	DWG
Eccentric	ECC
Figure	FIG
Finish All Over	FAO
Gage	GA
Heat Treat	HT TR
Head	HD
Heavy	HVY
Hexagon	HEX
Hydraulic	HYD
Inside Diameter	ID
International Organization for Standardization	ISO
Iron Pipe Size	IPS
Kilogram	kg
Kilometer	km
Large End	LE
Least Material Condition	Ⓛ or LMC
Left Hand	LH
Liter	L
Machined	√ or ✓
Machine Steel	MS or MACH ST
Material	MATL
Maximum	MAX
Maximum Material Condition	Ⓜ or MMC
Meter	m
Metric Thread	M
Micrometer	μm
Millimeter	mm
Minimum	MIN
Minute (Angle)	MIN
Newton	N
Nominal	NOM
Not to Scale	____ or NTS
Number	NO
On Center	OC
Outside Diameter	OD
Parallel	PAR
Pascal	Pa
Perpendicular	PERP
Pitch	P
Pitch Circle Diameter	PCD
Pitch Diameter	PD
Plate	PL
Radian	rad
Radius	R
Reference or Reference Dimension	( ) or REF
Regardless of Feature Size	Ⓢ or RFS
Revolutions per Minute	rev/min
Right Hand	RH
Second (Arc)	(″)
Second (Time)	SEC
Section	SECT
Slotted	SLOT
Socket	SOCK
Spherical	SPHER
Spotface	⌴ or SFACE
Square	□ or SQ
Square Centimeter	cm^2
Square Meter	m^2
Steel	STL
Straight	STR
Symmetrical	‡ or SYM
Thread	THD
Through	THRU
Tolerance	TOL
True Profile	TP
Undercut	UCUT
U.S. Sheet-Metal Gage	USS GA
Watt	W
Wrought Iron	WI

APPENDIX B.2

ANSI and Canadian Standards

ANSI Standards

Column A	Column B	Column A	Column B
Abbreviations	Y1.1-1972	Mechanical and Acoustical Element as Used in Schematic Diagrams	Y32.18-1972(R1978)
American National Standard Drafting Practices		Pipe Fittings, Valves, and Piping	Z32.2.3-1949(R1953)
Size and Format	Y14.1-1980	Heating, Ventilating, and Air Conditioning	Z32.2.4-1949(R1953)
Line Conventions and Lettering	Y14.2M-1979	Heat Power Apparatus	Z32.2.6-1950(R1956)
Multi and Sectional View Drawings	Y14.3-1975(R1980)	Letter Symbols for:	
Pictorial Drawing	Y14.4-1957	Glossary of Terms Concerning Letter Symbols	Y10.1-1972
Dimensioning and Tolerancing	Y14.5M-1982	Hydraulics	Y10.2-1958
Screw Threads	Y14.6-1978	Quantities Used in Mechanics for Solid Bodies	Y10.3-1968
Screws Threads (Metric Supplement)	Y14.6aM-1981	Heat and Thermodynamics	Y10.4-1982
Gears and Splines		Quantities Used in Electrical Science and Electrical Engineering	Y10.5-1968
Spur, Helical, and Racks	Y14.7.1-1971	Aeronautical Sciences	Y10.7-1954
Bevel and Hyphoid	Y14.7.2-1978	Structural Analysis	Y10.8-1962
Forgings	Y14.9-1958	Meteorology	Y10.10-1953(R1973)
Springs	Y14.13M-1981	Acoustics	Y10.11-1953(R1959)
Electrical and Electronic Diagram	Y14.15-1966(R1973)	Chemical Engineering	Y10.12-1955(R1973)
Interconnection Diagrams	Y14.15a-1971	Rocket Propulsion	Y10.14-1959
Information Sheet	Y14.15b-1973	Petroleum Reservoir Engineering and Electric Logging	Y10.15-1958(R1973)
Fluid Power Diagrams	Y14.17-1966(R1980)	Shell Theory	Y10.16-1964(R1973)
Digital Representation for Communication of Product Definition Data	Y14.26M-1981	Guide for Selecting Greek Letters Used as Symbols for Engineering Mathematics	Y10.17-1961(R1973)
Computer-Aided Preparation of Product Definition Data Dictionary of Terms	Y14.26.3-1975	Illuminating Engineering	Y10.18-1967(R1977)
Digital Representation of Physical Object Shapes	Y14 Report	Mathematical Signs and Symbols for Use in Physical Sciences and Technology	Y10.20-1975
Guideline—User Instructions	Y14 Report No. 2	Unified Screw Threads	ANSI B1.1
Guideline—Design Requirements	Y14 Report No. 3	Square and Hex Bolts and Screws	ANSI B18.2.1
Ground Vehicle Drawing Practices	In Preparation	Square and Hex Nuts	ANSI B18.2.2
Chassis Frames	Y14.32.1-1974	Socket Cap, Shoulder, and Setscrews	ANSI B18.3
Parts Lists, Data Lists, and Index Lists	Y14.34M-1982	Slotted-Head Cap Screws, Square-Head Setscrews, Slotted-Headless Setscrews	ANSI B18.6.2
Surface Texture Symbols	Y14.36-1978	Machine Screws and Machine Screw Nuts	ANSI B18.6.3
Illustrations for Publication and Projection	Y15.1M-1979	Woodruff Key and Keyslot Dimensions	ANSI B17.2
Time Series Charts	Y15.2M-1979	Keys and Keyseats	ANSI B17.1
Process Charts	Y15.3M-1979	Lock Washers	ANSI B18.21.1
Graphic Symbols for:		Plain Washers	ANSI B27.2
Electrical and Electronics Diagrams	Y32.2-1975	Surface Texture	ANSI B46.1
Plumbing	Y32.4-1977		
Use on Railroad Maps and Profiles	Y32.7-1972(R1979)		
Fluid Power Diagrams	Y32.10-1967(R1974)		
Process Flow Diagrams in Petroleum and Chemical Industries	Y32.11-1961		

AMERICAN NATIONAL STANDARDS INSITUTE, INC.
1430 BROADWAY,
NEW YORK, N.Y. 10018
THE AMERICAN SOCIETY OF MECHANICAL ENGINEERS
UNITED ENGINEERING CENTER
345 EAST 47TH STREET,
NEW YORK, N.Y. 10017

CSA—Canadian Standards

Column A	Column B	Column A	Column B
Unified and American Screw Threads	CSA B1.1	Surface Texture	CSA B95
Plain Washers	CSA B19.1	Limits and Fits for Engineering and Manufacturing	CSA B97.1
Square and Hexagon Bolts and Nuts, Studs and Wrench Openings	CSA B33.1	Abbreviations for Scientific and Engineering Terms	CSA Z85
Machine Screws, Stove Bolts and Associated Nuts	CSA B35.1	Architectural Drawing Practices (National Research Council, Ottawa, Canada)	33-GP-7
Drawing Standard—General Principles	CSA B78.1		
Drawing Standard—Dimensioning and Tolerancing	CSA B78.2		

CANADIAN STANDARDS ASSOCIATION
178 REXDALE BOULEVARD
REXDALE, ONTARIO, CANADA, M9W 1R3

APPENDIX B.3

Conversions

Conversion Factors

Multiply	by	To Obtain
Absolute viscosity (poise)	1	Gram/second centimeter
Absolute viscosity (centipoise)	0.01	Poise
Acceleration due to gravity (g)	32.174	Feet/second^2
	980.6	Centimeters/second^2
Acres	0.4047	Hectares
	10	Square Chains
	43,560	Square Feet
	4047	Square Meters
	0.001562	Square Miles
	4840	Square Yards
	160	Square Rods
Acre-feet	43,560	Cubic Feet
	325,851	Gallons (US)
	1233.49	Cubic Meters
	1,233,490	Liters
Acre-feet/hour	726	Cubic feet/Minute
	5430.86	Gallons/Minute
Angstroms	10^{-10}	Meters
Ares	0.01	Hectares
	1076.39	Square Feet
	0.02471	Acres
Atmospheres	76.0	Cms of Hg at 32° F
	29.921	Inches of Hg at 32° F
	33.94	Feet of Water at 62° F
	10,333	Kgs/Square meter
	14.6963	Pounds/Square inch
	1.058	Tons/Square foot
	1013.15	Millibars
	235.1408	Ounces/Square inch
Bags of cement	94	Pounds of cement
Barrels of oil	42	Gallons of oil (US)
Barrels of cement	376	Pounds of cement
Barrels (not legal)	31	Gallons (US)
or	31.5	Gallons (US)
Board feet	144 × 1 in*	Cubic inches
Boiler horse power	33,479	BTU/hour
	9.803	Kilowatts
	34.5	Pounds of water evaporated/ hour at 212° F
BTU	252.016	Calories (gm)
	0.252	Calories (Kg)
	777.54	Foot pounds
	0.0003927	Horse power hours
	1054.2	Joules
	107.5	Kilogram meters
	0.0002928	Kilowatt hours
BTU/Cu foot	8.89	Calories (Kg)/Cu meter at 32° F
BTU/Hr/ft^2/°F/ft	0.00413	Cal (gm)/Sec/cm^2/°C/cm
	1.49	Cal (Kg)/Hr/M^2/°C/ Meter
BTU/minute	12.96	Foot pounds/second
	0.02356	Horse power
	0.01757	Kilowatts
	17.57	Watts
BTU/pound	0.556	Calories (Kg)/Kilogram
Bushels	2150.4	Cubic inches
	35.24	Liters
	4	Pecks
	32	Quarts (dry)
Cables	120	Fathoms
Calories (gm)	0.003968	BTU
	0.001	Calories (Kg)
	3.088	Foot pounds
	1.558×10^{-6}	Horse power hours
	4.185	Joules
	0.4265	Kilogram meters
	1.1628×10^{-6}	Kilowatt hours
	0.0011628	Watt hours
Cal (gm)/sec/cm^2/°C/ cm	242.13	BTU/Hr/ft^2/°F/ft
Calories (Kg)	3.968	BTU
	1000	Calories (gm)
	3088	Foot pounds
	0.001558	Horse power hours
	4185	Joules
	426.5	Kilogram meters
	0.0011628	Kilowatt hours
	1.1628	Watt hours
Calories (Kg)/Cu meter	0.1124	BTU/Cu foot at 0° C
Cal (Kg)/Hr/M^2/°C/M	0.671	BTU/Hr/ft^2/°F/foot
Calories (Kg)/Kg	1.8	BTU/pound
Calories (Kg)/minute	51.43	Foot pounds/second
	0.09351	Horse power
	0.06972	Kilowatts
Carats (diamond)	200	Milligram
Centares (Centiares)	1	Square meters
Centigram	0.01	Grams
Centiliters	0.01	Liters
Centimeters	0.3937	Inches
	0.032808	Feet
	0.01	Meters
	10	Millimeters

*For thickness less than 1 in. use actual thickness in decimals of an inch.
Courtesy ITT Grinnell Corporation

Continued

Conversion Factors—*Continued*

Multiply	by	To Obtain
Centimeters of Hg at 32° F	0.01316	Atmospheres
	0.4461	Feet of water at 62°F
	136	Kgs/Square meter
	27.85	Pounds/Square foot
	0.1934	Pounds/Square inch
Centimeters/second	1.969	Feet/minute
	0.03281	Feet/second
	0.036	Kilometers/hour
	0.6	Meters/minute
	0.02237	Miles/hour
	0.0003728	Miles/minute
Centimeters/second^2	0.03281	Feet/second^2
Centipoise	0.000672	Pounds/sec foot
	2.42	Pounds/hour foot
	0.01	Poise
Chains (Gunter's)	4	Rods
	66	Feet
	100	Links
Cheval-vapeur	1	Metric horsepower
	75	Kilogram meters/second
	0.98632	Horse power
Circular inches	10^6	Circular mils
	0.7854	Square inches
	785,400	Square mils
Circular mils	0.7854	Square mils
	10^{-6}	Circular inches
	7.854×10^{-5}	Square inches
Cubic centimeters	3.531×10^{-5}	Cubic feet
	0.06102	Cubic inches
	10^{-6}	Cubic meters
	1.308×10^{-6}	Cubic yards
	0.0002642	Gallons (US)
	0.001	Liters
	0.002113	Pints (liq. US)
	0.001057	Quarts (liq. US)
	0.0391	Ounces (fluid)
Cubic feet	28,320	Cubic centimeters
	1728	Cubic inches
	0.02832	Cubic meters
	0.03704	Cubic yards
	7.48052	Gallons (US)
	28.32	Liters
	59.84	Pints (liq. US)
	29.92	Quarts (liq. US)
	2.296×10^{-5}	Acre feet
	0.803564	Bushels
Cubic feet of water	62.4266	Pounds at 39.2° F
	62.3554	Pounds at 62° F
Cubic feet/minute	472	Cubic centimeters/sec
	0.1247	Gallons (US)/second
	0.472	Liters/second
	62.36	Pounds water/min at 62°F
	7.4805	Gallons (US)/minute
	10,772	Gallons/24 hours
	0.033058	Acre feet/24 hours
Cubic feet/second	646,317	Gallons (US)/24 hours
	448.831	Gallons/minute
	1.98347	Acre feet/24 hours
Cubic inches	16.387	Cubic centimeters
	0.0005787	Cubic feet
	1.639×10^{-5}	Cubic meters
	2.143×10^{-5}	Cubic yards
	0.004329	Gallons (US)
	0.01639	Liters
	0.03463	Pints (liq. US)
	0.01732	Quarts (liq. US)
Cubic meters	10^6	Cubic centimeters
	35.31	Cubic feet
	61,023	Cubic inches
	1.308	Cubic yards
	264.2	Gallons (US)
	1000	Liters
	2113	Pints (liq. US)
	1057	Quarts (liq. US)
Cubic yards	764,600	Cubic centimeters
	27	Cubic feet
	46,656	Cubic inches
	0.7646	Cubic meters
	202	Gallons (US)
	764.6	Liters
	1616	Pints (liq. US)
	807.9	Quarts (liq. US)
Cubic yards/minute	0.45	Cubic feet/second
	3.367	Gallons (US)/second
	12.74	Liters/second
Cubit	18	Inches
Days (mean)	1440	Minutes
	24	Hours
	86,400	Seconds
Days (sidereal)	86,164.1	Solar seconds
Decigrams	0.1	Grams
Deciliters	0.1	Liters
Decimeters	0.1	Meters
Degrees (angle)	60	Minutes
	0.01745	Radians
	3600	Seconds
Degrees F [less 32]	0.5556	Degrees C
Degrees F	1 [plus 460]	Degrees F above absolute 0
Degrees C	1.8 [plus 32]	Degrees F
	1 [plus 273]	Degrees C above absolute 0
Degrees/second	0.01745	Radians/second
	0.1667	Revolutions/minute
	0.002778	Revolutions/second
Dekagrams	10	Grams
Dekaliters	10	Liters
Dekameters	10	Meters
Diameter (circle)	3.14159265359	Circumference
(approx)	3.1416	
(approx)	3.14	
(approx)	$\frac{22}{7}$	
Diameter (circle)	0.88623	Side of equal square
	0.7071	Side of inscribed square

Continued

Conversion Factors—*Continued*

Multiply	by	To Obtain
Diameter^3 (sphere)	0.5236	Volume (sphere)
Diam (major) × diam (minor)	0.7854	Area of ellipse
Diameter^2 (circle)	0.7854	Area (circle)
Diameter^2 (sphere)	3.1416	Surface (sphere)
Diam (inches) × RPM	0.262	Belt speed ft/minute
Digits	0.75	Inches
Drams (avoirdupois)	27.34375	Grains
	0.0625	Ounces (avoir.)
	1.771845	Grams
Fathoms		Feet
Feet	30.48	Centimeters
	12	Inches
	0.3048	Meters
	$\frac{1}{3}$	Yards
	0.06061	Rods
Feet of water at 62	0.029465	Atmospheres
	0.88162	Inches of Hg at 32° F
	62.3554	Pounds/square foot
	0.43302	Pounds/square inch
	304.44	Kilogram/sq meter
Feet/minute	0.5080	Centimeters/second
	0.01667	Feet/second
	0.01829	Kilometers/hour
	0.3048	Meters/minute
	0.01136	Miles/hour
Feet/second	30.48	Centimeters/second
	1.097	Kilometers/ hour
	0.5921	Knots
	18.29	Meters/minute
	0.6818	Miles/hour
	0.01136	Miles/minute
Feet/second^2	30.48	Centimeters/second^2
	0.3048	Meters/second^2
Flat of a hexagon	1.155	Distance across corners
Flat of a square	1.414	Distance across corners
Foot pounds	0.0012861	BTU
	0.32412	Calories (gm)
	0.0003241	Calories (Kg)
	5.05×10^{-7}	Horse power hours
	1.3558	Joules
	0.13826	Kilogram meters
	3.766×10^{-7}	Kilowatt hours
	0.0003766	Watt hours
Foot pounds/minute	0.001286	BTU/minute
	0.01667	Foot pounds/second
	3.03×10^{-5}	Horse power
	0.0003241	Calories (Kg)/minute
	2.26×10^{-5}	Kilowatts
Foot pounds/second	0.07717	BTU/minute
	0.001818	Horse power
	0.01945	Calories (Kg)/minute
	0.001356	Kilowatts
Furlong	40	Rods
	220	Yards
	660	Feet
	0.125	Miles
	0.2042	Kilometers
Gallons (Imperial)	277.42	Cubic inches
	4.543	Liters
	1.20095	Gallons (US)
Gallons (US)	3785	Cubic centimeters
	0.13368	Cubic feet
	231	Cubic inches
	0.003785	Cubic meters
	0.004951	Cubic yards
	3.785	Liters
	8	Pints (liq. US)
	4	Quarts (liq. US)
	0.83267	Gallons (Imperial)
	3.069×10^{-6}	Acre feet
Gallons (US) of water at 62°F	8.3357	Pounds of water
Gallons (US) of water/ minute	6.0086	Tons of water/24 hours
Gallons (US)/minute	0.002228	Cubic feet/second
	0.13368	Cubic feet/minute
	8.0208	Cubic feet/hour
	0.06309	Liters/second
	3.78533	Liters/minute
	0.0044192	Acre feet/24 hours
Grains	1	Grains (avoirdupois)
	1	Grains (apothecary)
	1	Grains (troy)
	0.0648	Grams
	0.0020833	Ounces (troy)
	0.0022857	Ounces (avoir.)
Grains/gallon (US)	17.118	Parts/million
	142.86	Pounds/million gallons (US)
Grams	980.7	Dynes
	15.43	Grains
	0.001	Kilograms
	1000	Milligrams
	0.03527	Ounces (avoir.)
	0.03215	Ounces (troy)
	0.002205	Pounds
Grams/centimeter	0.0056	Pounds/inch
Grams/cubic centimeter	62.43	Pounds/cubic foot
	0.03613	Pounds/cubic inch
	4.37	Grains/100 cubic ft
Grams/liter	58.417	Grains/gallon (US)
	8.345	Pounds/100 gallons (US)
	0.062427	Pounds/cubic foot
	1000	Parts/million
Gravity (g)	32.174	Feet/second^2
	980.6	Centimeters/second^2
Hand	4	Inches
	10.16	Centimeters
Hectares	2.471	Acres
	107,639	Square feet
	100	Ares
Hectograms	100	Grams
Hectoliters	100	Liters
Hectometers	100	Meters
Hectowatts	100	Wattsi

Continued

Conversion Factors—*Continued*

Multiply	by	To Obtain
Hogshead	63	Gallons (US)
	238.4759	Liters
Horse power	42.44	BTU/minute
	33,000	Foot pounds/minute
	550	Foot pounds/second
	1.014	Metric horse power (Cheval vapeur)
	10.7	Calories (Kg)/min
	0.7457	Kilowatts
	745.7	Watts
Horse power (boiler)	33,479	BTU/hour
	9.803	Kilowatts
	34.5	Pounds of water evaporated/ hour at 212° F
Horse power hours	2546.5	BTU
	641,700	Calories (gm)
	641.7	Calories (Kg)
	1,980,000	Foot pounds
	2,684,500	Joules
	273,740	Kilogram meters
	0.7455	Kilowatt hours
	745.5	Watt hours
Inches	2.54	Centimeters
	0.08333	Feet
	1000	Mils
	12	Lines
	72	Points
Inches of Hg at 32° F	0.03342	Atmospheres
	345.3	Kilograms/square meter
	70.73	Pounds/square foot
	0.49117	Pounds/square inch
	1.1343	Feet of water at 62° F
	13.6114	Inches of water at 62° F
	7.85872	Ounces/square inch
Inches of water at 62° F	0.002455	Atmospheres
	25.37	Kilograms/square meter
	0.5771	Ounces/square inch
	5.1963	Pounds/square foot
	0.03609	Pounds/square inch
	0.07347	Inches of Hg at 32° F
Joules	0.00094869	BTU
	0.239	Calories (gm)
	0.000239	Calories (Kg)
	0.73756	Foot pounds
	3.72×10^{-7}	Horse power hours
	0.10197	Kilogram meters
	2.778×10^{-7}	Kilowatt hours
	0.0002778	Watt hours
	1	Watt second
Kilograms	980,665	Dynes
	2.205	Pounds
	0.001102	Tons (short)
	1000	Grams
	35.274	Ounces (avoir.)
	32.1507	Ounces (troy)
Kilogram meters	0.009302	BTU
	2.344	Calories (gm)
	0.002344	Calories (Kg)
	7.233	Foot pounds
	3.653×10^{-6}	Horse power hours
	9.806	Joules
	2.724×10^{-6}	Kilowatt hours
	0.002724	Watt hours
Kilograms/cubic meter	0.06243	Pounds/cubic foot
Kilograms/meter	0.6720	Pounds/foot
Kilograms/sq centimeter	14.223	Pounds/sq inch
	1	Metric atmosphere
Kilogram/sq meter	9.678×10^{-5}	Atmospheres
	0.003285	Feet of water at 62° F
	0.002896	Inches of Hg at 32° F
	0.2048	Pounds/square foot
	0.001422	Pounds/square inch
	0.007356	Centimeters of Hg at 32° F
Kiloliters	1000	Liters
Kilometers	100,000	Centimeters
	1000	Meters
	3281	Feet
	0.6214	Miles
	1094	Yards
Kilometers/hour	27.78	Centimeters/second
	54.68	Feet/minute
	0.9113	Feet/second
	16.67	Meters/minute
	0.6214	Miles/hour
	0.5396	Knots
Kilometers/hr/sec	27.78	Centimeters/sec/sec
	0.9113	Feet/sec/sec
	0.2778	Meters/sec/sec
Kilowatts	56.92	BTU/minute
	44,250	Foot pounds/minute
	737.6	Foot pounds/second
	1.341	Horse power
	14.34	Calories (Kg)/min
	1000	Watts
Kilowatt hours	3413	BTU
	860,500	Calories (gm)
	860.5	Calories (Kg)
	2,655,200	Foot pounds
	1.341	Horse power hours
	3,600,000	Joules
	367,100	Kilogram meters
	1000	Watt hours
Knots	1	Nautical miles/hour
	1.1516	Miles/hour
	1.8532	Kilometers/hour
Leagues	3	Miles
Lines	0.08333	Inches
Links	7.92	Inches
Liters	1000	Cubic centimeters
	0.03531	Cubic feet
	61.02	Cubic inches
	0.001	Cubic meters
	0.001308	Cubic yards
	0.2642	Gallons (US)
	0.22	Gallons (Imp)

Continued

Conversion Factors—*Continued*

Multiply	by	To Obtain
Liters	2.113	Pints (liq. US)
	1.057	Quarts (liq. US)
	8.107×10^{-7}	Acre Feet
	2.2018	Pounds of water at 62° F
Liters/minute	0.0005886	Cubic feet/second
	0.004403	Gallons (US)/second
	0.26418	Gallons (US)/minute
Meters	100	Centimeters
	3.281	Feet
	39.37	Inches
	1.094	Yards
	0.001	Kilometers
	1000	Millimeters
Meters/minute	1.667	Centimeters/second
	3.281	Feet/minute
	0.05468	Feet/second
	0.06	Kilometers/hour
	0.03728	Miles/hour
Meters/second	196.8	Feet/minute
	3.281	Feet/second
	3.6	Kilometers/hour
	0.06	Kilometers/minute
	2.237	Miles/hour
	0.03728	Miles/minute
Microns	10^{-6}	Meters
	0.001	Millimeters
	0.03937	Mils
Mils	0.001	Inches
	0.0254	Millimeters
	25.4	Microns
Miles	160,934	Centimeters
	5280	Feet
	63,360	Inches
	1.609	Kilometers
	1760	Yards
	80	Chains
	320	Rods
	0.8684	Nautical miles
Miles/hour	44.70	Centimeters/second
	88	Feet/minute
	1.467	Feet/second
	1.609	Kilometers/hour
	0.8684	Knots
	26.82	Meters/minute
Miles/minute	2682	Centimeters/second
	88	Feet/second
	1.609	Kilometers/minute
	60	Miles/hour
Millibars	0.000987	Atmosphere
Milliers	1000	Kilograms
Milligrams	0.001	Grams
	0.01543	Grains
Milligrams/liter	1	Parts/million
Milliliters	0.001	Liters
Million gals/24 hours	1.54723	Cubic feet/second
Millimeters	0.1	Centimeters
	0.03937	Inches
	39.37	Mils
	1000	Microns

Multiply	by	To Obtain
Miner's inches	1.5	Cubic feet/minute
Minutes (angle)	0.0002909	Radians
Nautical miles	6080.2	Feet
	1.1516	Miles
Ounces (avoirdupois)	16	Drams (avoir.)
	437.5	Grains
	0.0625	Pounds (avoir.)
	28.349527	Grams
	0.9115	Ounces (troy)
Ounces (fluid)	1.805	Cubic inches
	0.02957	Liters
	29.57	Cubic centimeters
	0.25	Gills
Ounces (troy)	480	Grains
	20	Pennyweights (troy)
	0.08333	Pounds (troy)
	31.103481	Grams
	1.09714	Ounces (avoir.)
Ounces/square inch	0.0625	Pounds/square inch
	1.732	Inches of water at 62° F
	4.39	Centimeters of water at 62° F
	0.12725	Inches of Hg at 32° F
	0.004253	Atmospheres
Palms	3	Inches
Parts/million	0.0584	Grains/gallon (US)
	0.07016	Grains/gallon (Imp)
	8.345	Pounds/million gal (US)
Pennyweights (troy)	24	Grains
	1.55517	Grams
	0.05	Ounces (troy)
	0.0041667	Pounds (troy)
Pints (liq. US)	4	Gills
	16	Ounces (fluid)
	0.5	Quarts (liq. US)
	28.875	Cubic inches
	473.1	Cubic centimeters
Pipe	126	Gallons (US)
Points	0.01389	Inches
Poise	0.0672	Pounds/sec foot
	242	Pounds/hour foot
	100	Centipoise
Poncelots	100	Kilogram meters/second
	1.315	Horse power
Pounds (avoirdupois)	16	Ounces (avoir.)
	256	Drams (avoir.)
	7000	Grains
	0.0005	Tons (short)
	453.5924	Grams
	1.21528	Pounds (troy)
	14.5833	Ounces (troy)
Pounds (troy)	5760	Grains
	240	Pennyweights (troy)
	12	Ounces (troy)
	373.24177	Grams
	0.822857	Pounds (avoir.)
	13.1657	Ounces (avoir.)
	0.00036735	Tons (long)

Continued

Conversion Factors—*Continued*

Multiply	by	To Obtain	Multiply	by	To Obtain
Pounds (troy)	0.00041143	Tons (short)	Revolutions	360	Degrees
	0.00037324	Tons (metric)		4	Quadrants
Pounds of water at 62° F	0.01604	Cubic feet		6.283	Radians
	27.72	Cubic inches	Revolutions/minute	6	Degrees/second
	0.120	Gallons (US)		0.1047	Radians/second
				0.01667	Revolutions/second
Pounds of water/min at 62° F	0.0002673	Cubic feet/second	Revolutions/minute^2	0.001745	Radians/second^2
Pounds/cubic foot	0.01602	Grams/cubic centimeter		0.0002778	Revolutions/second^2
	16.02	Kilograms/cubic meter	Revolutions/second	360	Degrees/second
	0.0005787	Pounds/cubic inch		6.283	Radians/second
Pounds/cubic inch	27.68	Grams/cubic centimeter		60	Revolutions/minute
	27,680	Kilograms/cubic meter	Revolutions/second^2	6.283	Radians/second^2
	1728	Pounds/cubic foot		3600	Revolutions/minute^2
Pounds/foot	1.488	Kilograms/meter	Rods	16.5	Feet
Pounds/inch	178.6	Grams/centimeter		5.5	Yards
Pounds/hour foot	0.4132	Centipoise	Seconds (angle)	4.848×10^{-6}	Radians
	0.004132	Poise grams/sec cm	Sections	1	Square miles
Pounds/sec foot	14.881	Poise grams/sec cm	Side of a square	1.4142	Diameter of inscribed circle
	1488.1	Centipoise		1.1284	Diameter of circle with equal area
Pounds/square foot	0.016037	Feet of water at 62° F			
	4.882	Kilograms/square meter	Span	9	Inches
	0.006944	Pounds/square inch	Square centimeters	0.001076	Square feet
	0.014139	Inches of Hg at 32° F		0.1550	Square inches
	0.0004725	Atmospheres		0.0001	Square meters
Pounds/square inch	0.068044	Atmospheres		100	Square millimeters
	2.30934	Feet of water at 62° F	Square feet	2.296×10^{-5}	Acres
	2.0360	Inches of Hg at 32° F		929.0	Square centimeters
	703.067	Kilograms/square meter		144	Square inches
	27.912	Inches of water at 62° F		0.0929	Square meters
Quadrants (angular)	90	Degrees		3.587×10^{-8}	Square miles
	5400	Minutes		0.1111	Square yards
	324,000	Seconds	Square inches	6.452	Square centimeters
	1.751	Radians		0.006944	Square feet
Quarts (dry)	67.20	Cubic inches		645.2	Square millimeters
Quarts (liq. US)	2	Pints (liq. US)		1.27324	Circular inches
	0.9463	Liters		1,273,239	Circular mils
	32	Ounces (fluid)		1,000,000	Square mils
	57.75	Cubic inches	Square kilometers	247.1	Acres
	946.3	Cubic centimeters		10,760,000	Square feet
Quintal, Argentine	101.28	Pounds		1,000,000	Square meters
Brazil	129.54	Pounds		0.3861	Square miles
Castile, Peru	101.43	Pounds		1,196,000	Square yards
Chile	101.41	Pounds	Square meters	0.0002471	Acres
Metric	220.46	Pounds		10.764	Square feet
Mexico	101.47	Pounds		1.196	Square yards
Quires	25	Sheets		1	Centares
Radians	57.30	Degrees	Square miles	640	Acres
	3438	Minutes		27,878,400	Square feet
	206,625	Seconds		2.590	Square kilometers
	0.637	Quadrants		259	Hectares
				3,097,600	Square yards
Radians/second	57.30	Degrees/second		102,400	Square rods
	0.1592	Revolutions/second		1	Sections
	9.549	Revolutions/minute	Square millimeters	0.01	Square centimeters
Radians/second^2	573.0	Revolutions/minute^2		0.00155	Square inches
	0.1592	Revolutions/second^2		1550	Square mils
				1973	Circular mils
Reams	500	Sheets			

Continued

Conversion Factors—*Continued*

Multiply	by	To Obtain	Multiply	by	To Obtain
Square mils	1.27324	Circular mils	Watts	0.05692	BTU/minute
	0.0006452	Square millimeters		44.26	Foot pounds/minute
	10^{-6}	Square inches		0.7376	Foot pounds/second
Square yards	0.0002066	Acres		0.001341	Horsepower
	9	Square feet		0.01434	Calories (Kg)/minute
	0.8361	Square meters		0.001	Kilowatts
	3.228×10^{-7}	Square miles		1	Joule/second
Stere	1	Cubic meters	Watt hours	3.413	BTU
Stone	14	Pounds		860.5	Calories (gm)
	6.35029	Kilograms		0.8605	Calories (Kg)
Tons (long)	1016	Kilograms		2655	Foot pounds
	2240	Pounds		0.001341	Horsepower hours
	1.12	Tons (short)		3600	Joules
Tons (metric)	1000	Kilograms		367.1	Kilogram meters
	2205	Pounds		0.001	Kilowatt hours
	1.1023	Tons (short)	Watts/square inch	8.2	BTU/square foot/ minute
Tons (short)	2000	Pounds		6373	Foot pounds/sq ft/ minute
	32,000	Ounces		0.1931	Horsepower/square foot
	907.185	Kilograms	Yards	91.44	Centimeters
	0.90718	Tons (metric)		3	Feet
	0.89286	Tons (long)		36	Inches
Tons of refrigeration	12,000	BTU/hour		0.9144	Meters
	288,000	BTU/24 hours		0.1818	Rods
Tons of water/24 hours at 62° F	83.33	Pounds of water/hour	Year (365 days)	8760	Hours
	0.16510	Gallons (US)/minute			
	1.3263	Cubic feet/hour			

Reference Tables

Geometric Formulas

A

Area of a circle = half diameter × half circumference
Area of a circle = square of diameter × 0.7854
Area of a circle = square of circumference × 0.07958
Area of a sector of circle = length of arc × one-half radius
Area of a segment of circle = area of sector of equal radius minus area of triangle, when the segment is less, and plus area of triangle, when segment is greater than the semicircle
Area of ellipse = product of the two diameters × 0.7854
Area of a parabola = base × two-thirds of the altitude
Area of parallelogram = base × altitude
Area of a regular polygon = sum of its sides × perpendicular from its center to one of its sides divided by 2
Area of a rectangle = length × breadth or height
Area of circular ring = sum of the diameter of the two circles × difference of the diameter of the two circles and that product × 0.7854
Area of a square = length × breadth or height
Area of trapezium = divide into two triangles, total their areas
Area of trapezoid = altitude × one-half of the sum of parallel sides
Area of a triangle = base × one-half of the altitude

C

Circumference of circle = diameter × 3.1416
Circumference of circle = radius × 6.283185
Circumference of sphere = square root of surface × 1.772454
Circumference of sphere = cube root of solidity × 3.8978
Contents of pyramid or cone = area of base × one-third of the altitude
Contents of frustum of pyramid or cone = sum of circumference at both ends × one-half of the slant height plus area of both ends
Contents of frustum of pyramid or cone = multiply areas of two ends together and extract square root; add to this root the two areas and × one-third of the altitude
Contents of a sphere = diameter × 0.5236
Contents of segment of sphere = (height squared plus three times the square of radius of base) × (height × 0.5236)
Contents of a wedge = area of base × one-half of the altitude

D

Diameter of circle = circumference × 0.3183
Diameter of circle = square root of area × 1.12838
Diameter of circle that shall contain area of a given square = side of square × 1.1284
Diameter of sphere = cube root of solidity × 1.2407
Diameter of sphere = square root of surface × 0.56419

R

Radius of a circle = circumference × 0.0159155

S

Side of inscribed cube of sphere = radius × 1.1547
Side of inscribed cube of sphere = square root of diameter
Side of inscribed square = diameter × 0.7071
Side of inscribed square = circumference × 0.225
Side of square that shall equal area of circle = diameter × 0.8862
Side of square that shall equal area of circle = circumference × 0.2821
Side of inscribed equilateral triangle = diameter × 0.86
Surface of cylinder or prism = area of both ends plus length and × circumference
Surface of pyramid or cone = circumference of base × one-half of the slant height plus area of base
Surface of sphere = diameter × circumference

V

Volume of sphere = surface × one-sixth of the diameter
Volume of sphere = cube of diameter × 0.5236
Volume of sphere = cube of radius × 4.1888
Volume of sphere = cube of circumference × 0.016887

US Weights and Measures

Weights	
Apothecaries' 20 grains (gr)	1 scruple (s ap. or ℈)
3 scruples	1 dram (dr ap. or ʒ)
8 drams	1 ounce (oz ap. or ℥)
12 ounces	1 pound (lb ap. or ℔)
Avoirdupois 27–11/32 grains (gr)	1 dram (dr)
16 drams	1 ounce (oz)
16 ounces	1 pound (lb)
25 pounds	1 quarter
4 quarters	1 hundredweight (cwt)
20 hundredweights or 2,000 pounds	1 ton (tn or t) or short ton (s.t.)
2,240 pounds	1 long ton (l.t.)
Troy 24 grains (gr)	1 pennyweight (dwt)
20 pennyweights	1 ounce (oz t.)
12 ounces	1 pound (lb t.)

Measures	
Circular 60 seconds (″)	1 minute (′)
60 minutes	1 degree (°)
30 degrees	1 sign
3 signs	1 quadrant or 90 degrees
4 quadrants	1 circle or 1 circumference or 360 degrees
Cubic 1,728 cubic inches (cu in.)	1 cubic foot (cu ft)
27 cubic feet	1 cubic yard (cu yd)
128 cubic feet	1 cord (cd)
Dry 2 pints (pt)	1 quart (qt)
8 quarts	1 peck (pk)
4 pecks	1 bushel (bu) or 2,150.42 cubic inches (cu in.)
Linear or Long 12 inches (in.)	1 foot (ft)
3 feet	1 yard (yd)
5 1/2 yards	1 rod (rd) or pole (p) or perch (p)
40 rods	1 furlong (fur.)
8 furlongs or 1,760 yards or 5,280 feet	1 mile (mi)
3 miles	1 league
Liquid 8 fluid drams (f ʒ)	1 fluid ounce (f ℥)
4 fluid ounces	1 gill (gi)
4 gills	1 pint (pt)
2 pints	1 quart (qt)
4 quarts	1 gallon (gal) or 231 cubic inches (cu in.)
31 ½ gallons	1 barrel (bbl)
Mariners' or Nautical 6 feet (ft)	1 fathom (f or fm)
100 fathoms	1 cable's length (ordinary)
10 cables' lengths	1 nautical mile or 6,080.20 feet
1 nautical mile	1.1516 statute miles
1 knot	a speed of 1 nautical mile, or 1.1516 statute miles per hour
Paper 24 sheets (sh)	1 quire (qr)
20 quires	1 ream (rm)
2 reams	1 bundle (bdl)
5 bundles	1 bale (B/-)
Square 144 square inches (sq in.)	1 square foot (sq ft)
9 square feet	1 square yard (sq yd)
30¼ square yards	1 square rod (sq rd) or square pole (sq p) or square perch (sq p)
160 square rods or 4,840 square yards	1 acre (A)
640 acres	1 square mile (sq mi)
36 square miles	1 township (tp)

APPENDIX C

Standard Catalog Parts and Reference Material

C.1 Threads

C.2 Metric Twist Drills

C.3 Bolts, Nuts, and Screws

C.4 Washers

C.5 Rivets, Retaining Rings

C.6 Pins

C.7 Bushings

C.8 Woodruff Keys

C.9 Fits and Tolerances

APPENDIX C.1

Threads

TABLE C.1.1 Standard Unified Thread Series[a]

Present Unified Thread Nominal Size—diameter			Coarse (NC) (UNC)		Fine (NF) (UNF)		Extra-fine (NEF) (UNEF)	
Inch	*Inch*	*Metric Equiv.*	*Threads per inch*	*Tap drill*[b]	*Threads per inch*	*Tap drill*[b]	*Threads per inch*	*Tap drill*[b]
.060	0	1.52	—	—	80	3/64	—	—
.073	1	1.85	64	No. 53	72	No. 53	—	—
.086	2	2.18	56	No. 50	64	No. 50	—	—
.099	3	2.51	48	No. 47	56	No. 45	—	—
.112	4	2.84	40	No. 43	48	No. 42	—	—
.125	5	3.17	40	No. 38	44	No. 37	—	—
.138	6	3.50	32	No. 36	40	No. 33	—	—
.164	8	4.16	32	No. 29	36	No. 29	—	—
.190	10	4.83	24	No. 25	32	No. 21	—	—
.216	12	5.49	24	No. 16	28	No. 14	32	No. 13
.250	1/4	6.35	20	No. 7	28	No. 3	32	No. 2
.3125	5/16	7.94	18	*F*	24	*I*	32	*K*
.375	3/8	9.52	16	5/16	24	*O*	32	*S*
.4375	7/16	11.11	14	*U*	20	25/64	28	*Y*
.500	1/2	12.70	13	27/64	20	29/64	28	15/32
.5625	9/16	14.29	12	31/64	18	33/64	24	17/32
.625	5/8	15.87	11	17/32	18	37/64	24	19/32
.6875	11/16	17.46	—	—	—	—	24	41/64
.750	3/4	19.05	10	21/32	16	11/16	20	45/64
.8125	13/16	20.64	—	—	—	—	20	49/64
.875	7/8	22.22	9	49/64	14	13/16	20	53/64
.9375	15/16	23.81	—	—	—	—	20	57/64
1.000	1	25.40	8	7/8	12	59/64	20	61/64
1.0625	1 1/16	26.99	—	—	—	—	18	1
1.125	1 1/8	28.57	7	63/64	12	1 3/64	18	1 5/64
1.1875	1 3/16	30.16	—	—	—	—	18	1 9/64
1.250	1 1/4	31.75	7	1 7/64	12	1 11/64	18	1 13/64
1.3125	1 5/16	33.34	—	—	—	—	18	1 17/64
1.375	1 3/8	34.92	6	1 13/64	12	1 19/64	18	1 5/16
1.4375	1 7/16	36.51	—	—	—	—	18	1 3/8
1.500	1 1/2	38.10	6	1 21/64	12	1 27/64	18	1 29/64
1.5625	1 9/16	39.69	—	—	—	—	18	1 1/2
1.625	1 5/8	41.27	—	—	—	—	18	1 9/16
1.6875	1 11/16	42.86	—	—	—	—	18	1 5/8
1.750	1 3/4	44.45	5	1 35/64	—	—	16	1 11/16
2.000	2	50.80	4 1/2	1 25/32	—	—	16	1 15/16
2.250	2 1/4	57.15	4 1/2	2 1/32	—	—	—	—
2.500	2 1/2	63.50	4	2 1/4	—	—	—	—
2.750	2 3/4	69.85	4	2 1/2	—	—	—	—
3.000	3	76.20	4	2 3/4	—	—	—	—
3.250	3 1/4	82.55	4	3	—	—	—	—
3.500	3 1/2	88.90	4	3 1/4	—	—	—	—
3.750	3 3/4	95.25	4	3 1/2	—	—	—	—
4.000	4	101.60	4	3 3/4	—	—	—	—

[a]Adapted from ANSI B1.1-1960.
Bold type indicates Unified threads. To be designated UNC or UNF.
Unified Standard—Classes 1A, 2A, 3A, 1B, 2B, and 3B.
For recommended hole-size limits before threading, see Tables 38 and 39, ANSI B1.1-1960.
[b]Tap drill for a 75% thread (not Unified—American Standard).
Bold-type sizes smaller than 1/4 in. are accepted for limited applications by the British, but the symbols NC or NF, as applicable, are retained.

TABLE C.1.2 Thread Sizes and Dimensions: Fraction/Decimal/Metric

Nominal Size		Diameter (Major)		Diameter (Minor)		Tap Drill (For 75% Th'd.)			Threads Per Inch		Pitch (mm)		T.P.I. (Approx.)	
Inch	*mm*	*Inch*	*mm*	*Inch*	*mm*	*Drill*	*Inch*	*mm*	*UNC*	*UNF*	*Coarse*	*Fine*	*Coarse*	*Fine*
—	M1.4	.055	1.397	—	—	—	—	—	—	—	.3	.2	85	127
0	—	.060	1.524	.0438	1.092	3/64	.0469	1.168	—	80	—	—	—	—
—	M1.6	.063	1.600	—	—	—	—	—	—	—	.35	.2	74	127
1	—	.073	1.854	.0527	1.320	53	.0595	1.499	64	—	—	—	—	—
1	—	.073	1.854	.0550	1.397	53	.0595	1.499	—	72	—	—	—	—
—	M.2	.079	2.006	—	—	—	—	—	—	—	.4	.25	64	101
2	—	.086	2.184	.0628	1.587	50	.0700	1.778	56	—	—	—	—	—
2	—	.086	2.184	.0657	1.651	50	.0700	1.778	—	64	—	—	—	—
—	M2.5	.098	2.489	—	—	—	—	—	—	—	.45	.35	56	74
3	—	.099	2.515	.0719	1.828	47	.0785	1.981	48	—	—	—	—	—
3	—	.099	2.515	.0758	1.905	46	.0810	2.057	—	58	—	—	—	—
4	—	.112	2.845	.0795	2.006	43	.0890	2.261	40	—	—	—	—	—
4	—	.112	2.845	.0849	2.134	42	.0935	2.380	—	48	—	—	—	—
—	M3	.118	2.997	—	—	—	—	—	—	—	.5	.35	51	74
5	—	.125	3.175	.0925	2.336	38	.1015	2.565	40	—	—	—	—	—
5	—	.125	3.175	.0955	2.413	37	.1040	2.641	—	44	—	—	—	—
6	—	.138	3.505	.0975	2.464	36	.1065	2.692	32	—	—	—	—	—
6	—	.138	3.505	.1055	2.667	33	.1130	2.870	—	40	—	—	—	—
—	M4	.157	3.988	—	—	—	—	—	—	—	.7	.35	36	51
8	—	.164	4.166	.1234	3.124	29	.1360	3.454	32	—	—	—	—	—
8	—	.164	4.166	.1279	3.225	29	.1360	3.454	—	36	—	—	—	—
10	—	.190	4.826	.1359	3.429	26	.1470	3.733	24	—	—	—	—	—
10	—	.190	4.826	.1494	3.785	21	.1590	4.038	—	32	—	—	—	—
—	M5	.196	4.978	—	—	—	—	—	—	—	.8	.5	32	51
12	—	.216	5.486	.1619	4.089	16	.1770	4.496	24	—	—	—	—	—
12	—	.216	5.486	.1696	4.293	15	.1800	4.572	—	28	—	—	—	—
—	M6	.236	5.994	—	—	—	—	—	—	—	1.0	.75	25	34
1/4	—	.250	6.350	.1850	4.699	7	.2010	5.105	20	—	—	—	—	—
1/4	—	.250	6.350	.2036	5.156	3	.2130	5.410	—	28	—	—	—	—
5/16	—	.312	7.938	.2403	6.096	F	.2570	6.527	18	—	—	—	—	—
5/16	—	.312	7.938	.2584	6.553	I	.2720	6.908	—	24	—	—	—	—
—	M8	.315	8.001	—	—	—	—	—	—	—	1.25	1.0	20	25
3/8	—	.375	9.525	.2938	7.442	5/16	.3125	7.937	16	—	—	—	—	—
3/8	—	.375	9.525	.3209	8.153	Q	.3320	8.432	—	24	—	—	—	—
—	M10	.393	9.982	—	—	—	—	—	—	—	1.5	1.25	17	20
7/16	—	.437	11.113	.3447	8.738	U	.3680	9.347	14	—	—	—	—	—
7/16	—	.437	11.113	.3726	9.448	25/64	.3906	9.921	—	20	—	—	—	—
—	M12	.471	11.963	—	—	—	—	—	—	—	1.75	1.25	14.5	20
1/2	—	.500	12.700	.4001	10.162	27/64	.4219	10.715	13	—	—	—	—	—
1/2	—	.500	12.700	.4351	11.049	29/64	.4531	11.509	—	20	—	—	—	—
—	M14	.551	13.995	—	—	—	—	—	—	—	2	1.5	12.5	17
9/16	—	.562	14.288	.4542	11.531	31/64	.4844	12.3031	12	—	—	—	—	—
9/16	—	.562	14.288	.4903	12.446	33/64	.5156	13.096	—	18	—	—	—	—
5/8	—	.625	15.875	.5069	12.852	17/32	.5312	13.493	11	—	—	—	—	—
5/8	—	.625	15.875	.5528	14.020	37/64	.5781	14.684	—	18	—	—	—	—
—	M16	.630	16.002	—	—	—	—	—	—	—	2	1.5	12.5	17
—	M18	.709	18.008	—	—	—	—	—	—	—	2.5	1.5	10	17
3/4	—	.750	19.050	.6201	15.748	21/32	.6562	16.668	10	—	—	—	—	—
3/4	—	.750	19.050	.6688	16.967	11/16	.6875	17.462	—	16	—	—	—	—
—	M20	.787	19.990	—	—	—	—	—	—	—	2.5	1.5	10	17
—	M22	.866	21.996	—	—	—	—	—	—	—	2.5	1.5	10	17
7/8	—	.875	22.225	.7307	18.542	49/64	.7656	19.446	9	—	—	—	—	—
.7/8	—	.875	22.225	.7822	19.863	13/16	.8125	20.637	—	14	—	—	—	—
—	M24	.945	24.003	—	—	—	—	—	—	—	3	2	8.5	12.5
1	—	1.000	25.400	.8376	21.2598	7/8	.8750	22.225	8	—	—	—	—	—
1	—	1.000	25.400	.8917	22.632	59/64	.9219	23.415	—	12	—	—	—	—
—	M27	1.063	27.000	—	—	—	—	—	—	—	3	2	8.5	12.5

TABLE C.1.3 Unified Screw Thread Standard Series

Nominal Size		Basic Major Diameter	Graded Pitch Series*			Constant Pitch Series*								Nominal Size
			Coarse UNC	*Fine UNF*	*Extra Fine UNEF*	*4 UN*	*6 UN*	*8 UN*	*12 UN*	*16 UN*	*20 UN*	*28 UN*	*32 UN*	
Primary	*Secondary*		*Threads Per Inch*											
0		0.0600	—	80	—	—	—	—	—	—	—	—	—	0
	1	0.0730	64	72	—	—	—	—	—	—	—	—	—	1
2		0.0860	56	64	—	—	—	—	—	—	—	—	—	2
	3	0.0990	48	56	—	—	—	—	—	—	—	—	—	3
4		0.1120	40	48	—	—	—	—	—	—	—	—	—	4
5		0.1250	40	44	—	—	—	—	—	—	—	—	—	5
6		0.1380	32	40	—	—	—	—	—	—	—	—	UNC	6
8		0.1640	32	36	—	—	—	—	—	—	—	—	UNC	8
10		0.1900	24	32	—	—	—	—	—	—	—	—	UNC	10
	12	0.2160	24	28	32	—	—	—	—	—	—	UNF	UNEF	12
1/4		0.2500	20	28	32	—	—	—	—	—	UNC	UNF	UNEF	1/4
5/16		0.3125	18	24	32	—	—	—	—	—	20	28	UNEF	5/16
3/8		0.3750	16	24	32	—	—	—	—	UNC	20	28	UNEF	3/8
7/16		0.4375	14	20	28	—	—	—	—	16	UNF	UNEF	32	7/16
1/2		0.5000	13	20	28	—	—	—	—	16	UNF	UNEF	32	1/2
9/16		0.5625	12	18	24	—	—	—	UNC	16	20	28	32	9/16
5/8		0.6250	11	18	24	—	—	—	12	16	20	28	32	5/8
	11/16	0.6875	—	—	24	—	—	—	12	16	20	28	32	11/16
3/4		0.7500	10	16	20	—	—	—	12	UNF	UNEF	28	32	3/4
	13/16	0.8125	—	—	20	—	—	—	12	16	UNEF	28	32	13/16
7/8		0.8750	9	14	20	—	—	—	12	16	UNEF	28	32	7/8
	15/16	0.9375	—	—	20	—	—	—	12	16	UNEF	28	32	15/16
1		1.0000	8	12	20	—	—	UNC	UNF	16	UNEF	28	32	1
	1-1/16	1.0625	—	—	18	—	—	8	12	16	20	28	—	1-1/16
1-1/8		1.1250	7	12	18	—	—	8	UNF	16	20	28	—	1-1/8
	1-3/16	1.1875	—	—	18	—	—	8	12	16	20	28	—	1-3/16
1-1/4		1.2500	7	12	18	—	—	8	UNF	16	20	28	—	1-1/4
	1-5/16	1.3125	—	—	18	—	—	8	12	16	20	28	—	1-5/16
1-3/8		1.3750	6	12	18	—	UNC	8	UNF	16	20	28	—	1-3/8
	1-7/16	1.4375	—	—	18	—	6	8	12	16	20	28	—	1-7/16
1-1/2		1.5000	6	12	18	—	UNC	8	UNF	16	20	28	—	1-1/2
	1-9/16	1.5625	—	—	18	—	6	8	12	16	20	—	—	1-9/16
1-5/8		1.6250	—	—	18	—	6	8	12	16	20	—	—	1-5/8
	1-11/16	1.6875	—	—	18	—	6	8	12	16	20	—	—	1-11/16
1-3/4		1.7500	5	—	—	—	6	8	12	16	20	—	—	1-3/4
	1-13/16	1.8125	—	—	—	—	6	8	12	16	20	—	—	1-13/16
1-7/8		1.8750	—	—	—	—	6	8	12	16	20	—	—	1-7/8
	1-15/16	1.9375	—	—	—	—	6	8	12	16	20	—	—	1-15/16
2		2.0000	$4\frac{1}{2}$	—	—	—	6	8	12	16	20	—	—	2
	2-1/8	2.1250	—	—	—	—	6	8	12	16	20	—	—	2-1/8
2-1/4		2.2500	$4\frac{1}{2}$	—	—	—	6	8	12	16	20	—	—	2-1/4
	2-3/8	2.3750	—	—	—	—	6	8	12	16	20	—	—	2-3/8
2-1/2		2.5000	4	—	—	UNC	6	8	12	16	20	—	—	2-1/2
	2-5/8	2.6250	—	—	—	4	6	8	12	16	20	—	—	2-5/8
2-3/4		2.7500	4	—	—	UNC	6	8	12	16	20	—	—	2-3/4
	2-7/8	2.8750	—	—	—	4	6	8	12	16	20	—	—	2-7/8
3		3.0000	4	—	—	UNC	6	8	12	16	20	—	—	3
	3-1/8	3.1250	—	—	—	4	6	8	12	16	—	—	—	3-1/8
3-1/4		3.2500	4	—	—	UNC	6	8	12	16	—	—	—	3-1/4
	3-3/8	3.3750	—	—	—	4	6	8	12	16	—	—	—	3-3/8
3-1/2		3.5000	4	—	—	UNC	6	8	12	16	—	—	—	3-1/2
	3-5/8	3.6250	—	—	—	4	6	8	12	16	—	—	—	3-5/8
3-3/4		3.7500	4	—	—	UNC	6	8	12	16	—	—	—	3-3/4
	3-7/8	3.8750	—	—	—	4	6	8	12	16	—	—	—	3-7/8

Continued

TABLE C.1.3 Unified Screw Thread Standard Series—*Continued*

Nominal Size		Basic Major Diameter	Graded Pitch Series*			Constant Pitch Series*								Nominal Size
			Coarse UNC	*Fine UNF*	*Extra Fine UNEF*	*4 UN*	*6 UN*	*8 UN*	*12 UN*	*16 UN*	*20 UN*	*28 UN*	*32 UN*	
Primary	*Secondary*		*Threads Per Inch*											
4		4.0000	4	—	—	4	6	8	12	16	—	—	—	4
	4-1/8	4.1250	—	—	—	4	6	8	12	16	—	—	—	4-1/8
4-1/4		4.2500	—	—	—	4	6	8	12	16	—	—	—	4-1/4
	4-3/8	4.3750	—	—	—	4	6	8	12	16	—	—	—	4-3/8
4-1/2		4.5000	—	—	—	4	6	8	12	16	—	—	—	4-1/2
	4-5/8	4.6250	—	—	—	4	6	8	12	16	—	—	—	4-5/8
4-3/4		4.7500	—	—	—	4	6	8	12	16	—	—	—	4-3/4
	4-7/8	4.8750	—	—	—	4	6	8	12	16	—	—	—	4-7/8
5		5.0000	—	—	—	4	6	8	12	16	—	—	—	5
	5-1/8	5.1250	—	—	—	4	6	8	12	16	—	—	—	5-1/8
5-1/4		5.2500	—	—	—	4	6	8	12	16	—	—	—	5-1/4
	5-3/8	5.3750	—	—	—	4	6	8	12	16	—	—	—	5-3/8
5-1/2		5.5000	—	—	—	4	6	8	12	16	—	—	—	5-1/2
	5-5/8	5.6250	—	—	—	4	6	8	12	16	—	—	—	5-5/8
5-3/4		5.7500	—	—	—	4	6	8	12	16	—	—	—	5-3/4
	5-7/8	5.8750	—	—	—	4	6	8	12	16	—	—	—	5-7/8
6		6.0000	—	—	—	4	6	8	12	16	—	—	—	6

*For series symbols applying to a particular thread, see dimensional tables for Unified Screw Threads.
Courtesy of American National Standards.

TABLE C.1.4 Drill and Counterbore Sizes for Socket Head Cap Screws (1960 Series)

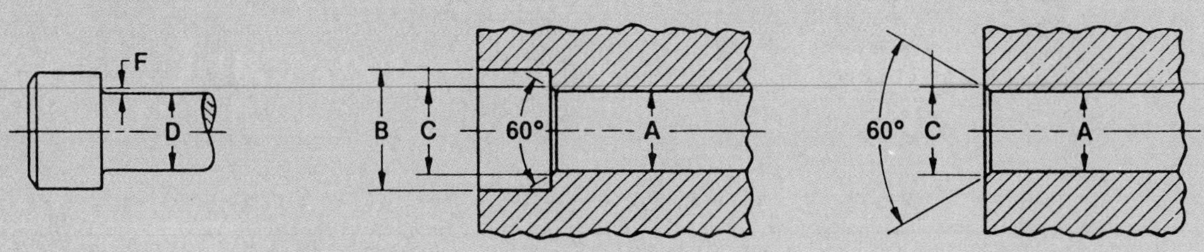

Nominal Size or Basic Screw Diameter		A				B	C
		Nominal Drill Size				Counterbore Diameter	Countersink Diameter *D* (Max) + 2*F* (Max)
		Close Fit		Normal Fit			
		Number or Fractional Size	**Decimal Size**	**Number or Fractional Size**	**Decimal Size**		
0	0.0600	51	0.067	49	0.073	1/8	0.074
1	0.0730	46	0.081	43	0.089	5/32	0.087
2	0.0860	3/32	0.094	36	0.106	3/16	0.102
3	0.0990	36	0.106	31	0.120	7/32	0.115
4	0.1120	1/8	0.125	29	0.136	7/32	0.130
5	0.1250	9/64	0.141	23	0.154	1/4	0.145
6	0.1380	23	0.154	18	0.170	9/32	0.158
8	0.1640	15	0.180	10	0.194	5/16	0.188
10	0.1900	5	0.206	2	0.221	3/8	0.218
1/4	0.2500	17/64	0.266	9/32	0.281	7/16	0.278
5/16	0.3125	21/64	0.328	11/32	0.344	17/32	0.346
3/8	0.3750	25/64	0.391	13/32	0.406	5/8	0.415
7/16	0.4375	29/64	0.453	15/32	0.469	23/32	0.483
1/2	0.5000	33/64	0.516	17/32	0.531	13/16	0.552
5/8	0.6250	41/64	0.641	21/32	0.656	1	0.689
3/4	0.7500	49/64	0.766	25/32	0.781	1 3/16	0.828
7/8	0.8750	57/64	0.891	29/32	0.906	1 3/8	0.963
1	1.0000	1 1/64	1.016	1 1/32	1.031	1 5/8	1.100
1 1/4	1.2500	1 9/32	1.281	1 5/16	1.312	2	1.370
1 1/2	1.5000	1 17/32	1.531	1 9/16	1.562	2 3/8	1.640
1 3/4	1.7500	1 25/32	1.781	1 13/16	1.812	2 3/4	1.910
2	2.0000	2 1/32	2.031	2 1/16	2.062	3 1/8	2.180

APPENDIX C.2

Metric Twist Drills

TABLE C.2.1 American National Standard Combined Drills and Countersinks—Plain and Bell Types (ANSI B94.11M-1979)

PLAIN TYPE

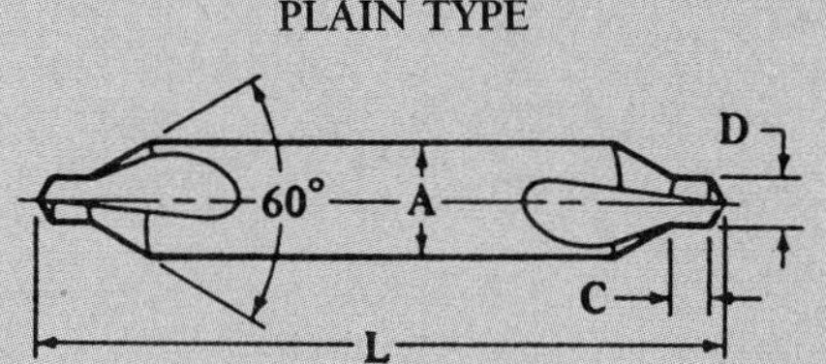

BELL TYPE

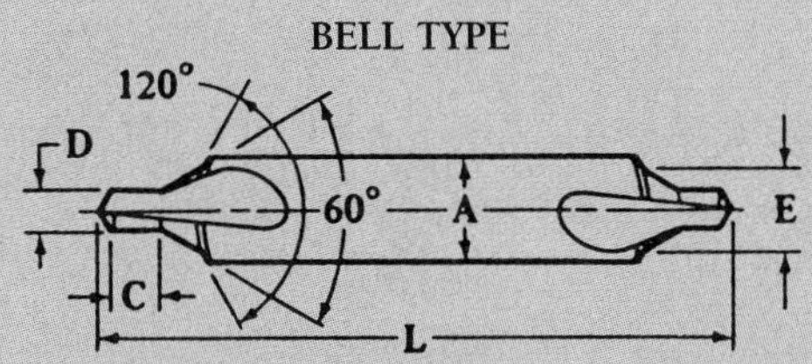

Plain Type								
	Body Diameter		Drill Diameter		Drill Length		Overall Length	
	A		D		C		L	
Size Designation	Inches	Millimeters	Inches	Millimeters	Inches	Millimeters	Inches	Millimeters
00	1/8	3.18	.025	0.64	.030	0.76	1 1/8	29
0	1/8	3.18	1/32	0.79	.038	0.97	1 1/8	29
1	1/8	3.18	3/64	1.19	3/64	1.19	1 1/4	32
2	3/16	4.76	5/64	1.98	5/64	1.98	1 7/8	48
3	1/4	6.35	7/64	2.78	7/64	2.78	2	51
4	5/16	7.94	1/8	3.18	1/8	3.18	2 1/8	54
5	7/16	11.11	3/16	4.76	3/16	4.76	2 3/4	70
6	1/2	12.70	7/32	5.56	7/32	5.56	3	76
7	5/8	15.88	1/4	6.35	1/4	6.35	3 1/4	83
8	3/4	19.05	5/16	7.94	5/16	7.94	3 1/2	89

TABLE C.2.2 Twist Drill Sizes: Decimal/Metric

Number Sizes								Letter sizes			
No. Size	*Decimal Equivalent*	*Metric Equiavalent*	*Closest Metric Drill (mm)*	*No. Size*	*Decimal Equivalent*	*Metric Equivalent*	*Closest Metric Drill (mm)*	*Size Letter*	*Decimal Equivalent*	*Metric Equivalent*	*Closest Metric Drill (mm)*
1	.2280	5.791	5.80	41	.0960	2.438	2.45	A	.234	5.944	5.90
2	.2210	5.613	5.60	42	.0935	2.362	2.35	B	.238	6.045	6.00
3	.2130	5.410	5.40	43	.0890	2.261	2.25	C	.242	6.147	6.10
4	.2090	5.309	5.30	44	.0860	2.184	2.20	D	.246	6.248	6.25
5	.2055	5.220	5.20	45	.0820	2.083	2.10	E	.250	6.350	6.40
6	.2040	5.182	5.20	46	.0810	2.057	2.05	F	.257	6.528	6.50
7	.2010	5.105	5.10	47	.0785	1.994	2.00	G	.261	6.629	6.60
8	.1990	5.055	5.10	48	.0760	1.930	1.95	H	.266	6.756	6.75
9	.1960	4.978	5.00	49	.0730	1.854	1.85	I	.272	6.909	6.90
10	.1935	4.915	4.90	50	.0700	1.778	1.80	J	.277	7.036	7.00
11	.1910	4.851	4.90	51	.0670	1.702	1.70	K	.281	7.137	7.10
12	.1890	4.801	4.80	52	.0635	1.613	1.60	L	.290	7.366	7.40
13	.1850	4.699	4.70	53	.0595	1.511	1.50	M	.295	7.493	7.50
14	.1820	4.623	4.60	54	.0550	1.397	1.40	N	.302	7.671	7.70
15	.1800	4.572	4.60	55	.0520	1.321	1.30	O	.316	8.026	8.00
16	.1770	4.496	4.50	56	.0465	1.181	1.20	P	.323	8.204	8.20
17	.1730	4.394	4.40	57	.0430	1.092	1.10	Q	.332	8.433	8.40
18	.1695	4.305	4.30	58	.0420	1.067	1.05	R	.339	8.611	8.60
19	.1660	4.216	4.20	59	.0410	1.041	1.05	S	.348	8.839	8.80
19	.1610	4.089	4.10	60	.0400	1.016	1.00	T	.358	9.093	9.10
21	.1590	4.039	4.00	61	.0390	0.991	1.00	U	.368	9.347	9.30
22	.1570	3.988	4.00	62	.0380	0.965	0.95	V	.377	9.576	9.60
23	.1540	3.912	3.90	63	.0370	0.940	0.95	W	.386	9.804	9.80
24	.1520	3.861	3.90	64	.0360	0.914	0.90	X	.397	10.084	10.00
25	.1495	3.797	3.80	65	.0350	0.889	0.90	Y	.404	10.262	10.50
26	.1470	3.734	3.75	66	.0330	0.838	0.85	Z	.413	10.491	10.50
27	.1440	3.658	3.70	67	.0320	0.813	0.80				
28	.1405	3.569	3.60	68	.0310	0.787	0.80				
29	.1360	3.454	3.50	69	.0292	0.742	0.75				
30	.1285	3.264	3.25	70	.0280	0.711	0.70				
31	.1200	3.048	3.00	71	.0260	0.660	0.65				
32	.1160	2.946	2.90	72	.0250	0.635	0.65				
33	.1130	2.870	2.90	73	.0240	0.610	0.60				
34	.1110	2.819	2.80	74	.0225	0.572	0.55				
35	.1100	2.794	2.80	75	.0210	0.533	0.55				
36	.1065	2.705	2.70	76	.0200	0.508	0.50				
37	.1040	2.642	2.60	77	.0180	0.457	0.45				
38	.1015	2.578	2.60	78	.0160	0.406	0.40				
39	.0995	2.527	2.50	79	.0145	0.368	0.35				
40	.0980	2.489	2.50	80	.0135	0.343	0.35				

*Fraction-size drills range in size from one-sixteenth—4 in. and over in diameter—by sixty-fourths.

APPENDIX C.3

Bolts, Nuts, and Screws

Socket Flat Countersunk Head Cap Screws (ANSI/ASME B18.3, 1986)

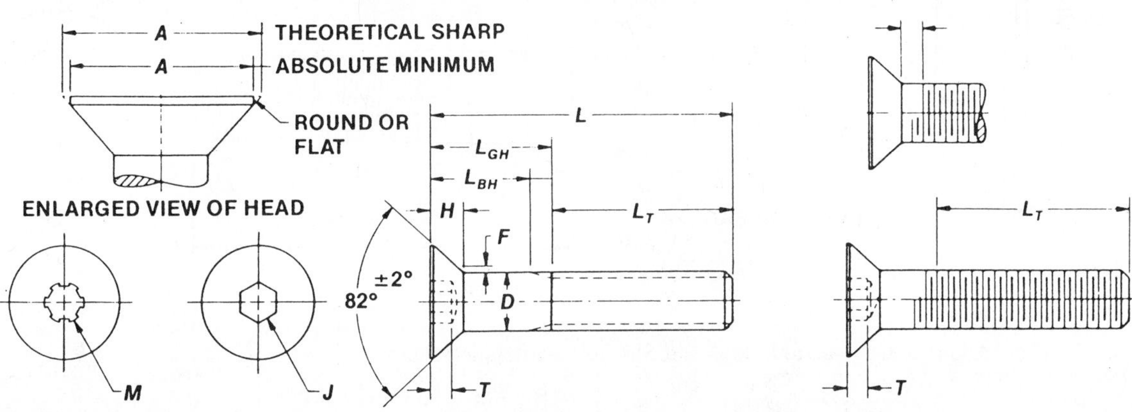

TABLE C.3.1 Dimensions of Hexagon and Spline Socket Flat Countersunk Head Cap Screws

Nominal Size or Basic Screw Diameter		D		A		H		M	J		T	F
		Body Diameter		Head Diameter		Head Height		Spline Socket Size	Hexagon Socket Size		Key Engagement	Fillet Extension Above D Max
		Max	Min	Theoretical Sharp Max	Abs Min	Reference	Flushness Tolerance		Nom		Min	Max
0	0.0600	0.0600	0.0568	0.138	0.117	0.044	0.006	0.048		0.035	0.025	0.006
1	0.0730	0.0730	0.0695	0.168	0.143	0.054	0.007	0.060		0.050	0.031	0.008
2	0.0860	0.0860	0.0822	0.197	0.168	0.064	0.008	0.060		0.050	0.038	0.010
3	0.0990	0.0990	0.0949	0.226	0.193	0.073	0.010	0.072	1/16	0.062	0.044	0.010
4	0.1120	0.1120	0.1075	0.255	0.218	0.083	0.011	0.072	1/16	0.062	0.055	0.012
5	0.1250	0.1250	0.1202	0.281	0.240	0.090	0.012	0.096	5/64	0.078	0.061	0.014
6	0.1380	0.1380	0.1329	0.307	0.263	0.097	0.013	0.096	5/64	0.078	0.066	0.015
8	0.1640	0.1640	0.1585	0.359	0.311	0.112	0.014	0.111	3/32	0.094	0.076	0.015
10	0.1900	0.1900	0.1840	0.411	0.359	0.127	0.015	0.145	1/8	0.125	0.087	0.015
1/4	0.2500	0.2500	0.2435	0.531	0.480	0.161	0.016	0.183	5/32	0.156	0.111	0.015
5/16	0.3125	0.3125	0.3053	0.656	0.600	0.198	0.017	0.216	3/16	0.188	0.135	0.015
3/8	0.3750	0.3750	0.3678	0.781	0.720	0.234	0.018	0.251	7/32	0.219	0.159	0.015
7/16	0.4375	0.4375	0.4294	0.844	0.781	0.234	0.018	0.291	1/4	0.250	0.159	1.015
1/2	0.5000	0.5000	0.4919	0.938	0.872	0.251	0.018	0.372	5/16	0.312	0.172	0.015
5/8	0.6250	0.6250	0.6163	1.188	1.112	0.324	0.022	0.454	3/8	0.375	0.220	0.015
3/4	0.7500	0.7500	0.7406	1.438	1.355	0.396	0.024	0.454	1/2	0.500	0.220	0.015
7/8	0.8750	0.8750	0.8647	1.688	1.604	0.468	0.025	...	9/16	0.562	0.248	0.015
1	1.0000	1.0000	0.9886	1.938	1.841	0.540	0.028	...	5/8	0.625	0.297	0.015
1 1/8	1.1250	1.1250	1.1086	2.188	2.079	0.611	0.031	...	3/4	0.750	0.325	0.031
1 1/4	1.2500	1.2500	1.2336	2.438	2.316	0.683	0.035	...	7/8	0.875	0.358	0.031
1 3/8	1.3750	1.3750	1.3568	2.688	2.553	0.755	0.038	...	7/8	0.875	0.402	0.031
1 1/2	1.5000	1.5000	1.4818	2.938	2.791	0.827	0.042	...	1	1.0000	0.435	0.031

Countersunk Bolts and Slotted Countersunk Bolts (ANSI/ASME B18.5, 1978)

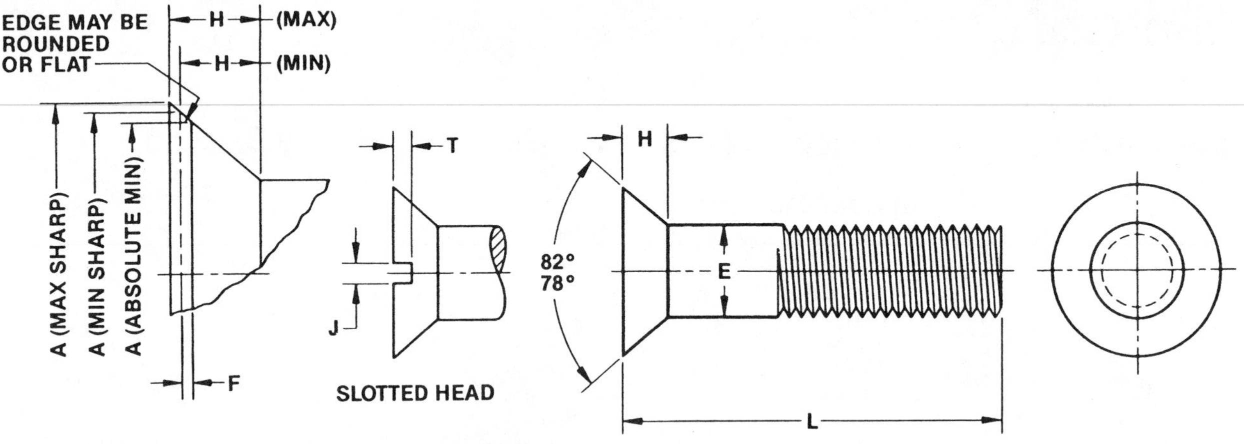

TABLE C.3.2 Dimensions of Countersunk Bolts and Slotted Countersunk Bolts

Nominal Size or Basic Bolt Diameter		*E* Body Diameter		*A* Head Diameter			*F* Flat on Min Dia Head	*H* Head Height		*J* Slot Width		*T* Slot Depth	
		Max	Min	Max Edge Sharp	Min Edge Sharp	Absolute Min Edge Rounded or Flat	Max	Max	Min	Max	Min	Max	Min
1/4	0.2500	0.260	0.237	0.493	0.477	0.445	0.018	0.150	0.131	0.075	0.064	0.068	0.045
5/16	0.3125	0.324	0.298	0.618	0.598	0.558	0.023	0.189	0.164	0.084	0.072	0.086	0.057
3/8	0.3750	0.388	0.360	0.740	0.715	0.668	0.027	0.225	0.196	0.094	0.081	0.103	0.068
7/16	0.4375	0.452	0.421	0.803	0.778	0.726	0.030	0.226	0.196	0.094	0.081	0.103	0.068
1/2	0.5000	0.515	0.483	1.935	1.905	0.845	0.035	0.269	0.233	0.106	0.091	0.103	0.068
5/8	0.6250	0.642	0.605	1.169	1.132	1.066	0.038	0.336	0.292	0.133	0.116	0.137	0.091
3/4	0.7500	0.768	0.729	1.402	1.357	1.285	0.041	0.403	0.349	0.149	0.131	0.171	0.115
7/8	0.8750	0.895	0.852	1.637	1.584	1.511	0.042	0.470	0.408	0.167	0.147	0.206	0.138
1	1.0000	1.022	0.976	1.869	1.810	1.735	0.043	0.537	0.466	0.188	0.166	0.240	0.162
1 1/8	1.1250	1.149	1.098	2.104	2.037	1.962	0.043	0.604	0.525	0.196	0.178	0.257	0.173
1 1/4	1.2500	1.277	1.223	2.337	2.262	2.187	0.043	0.671	0.582	0.211	0.193	0.291	0.197
1 3/8	1.3750	1.404	1.345	2.571	2.489	2.414	0.043	0.738	0.641	0.226	0.208	0.326	0.220
1 1/2	1.5000	1.531	1.470	2.804	2.715	2.640	0.043	0.805	0.698	0.258	0.240	0.360	0.244

Hex Cap Screws (Finished Hex Bolts) (ANSI/ASME B18.2.1, 1981)

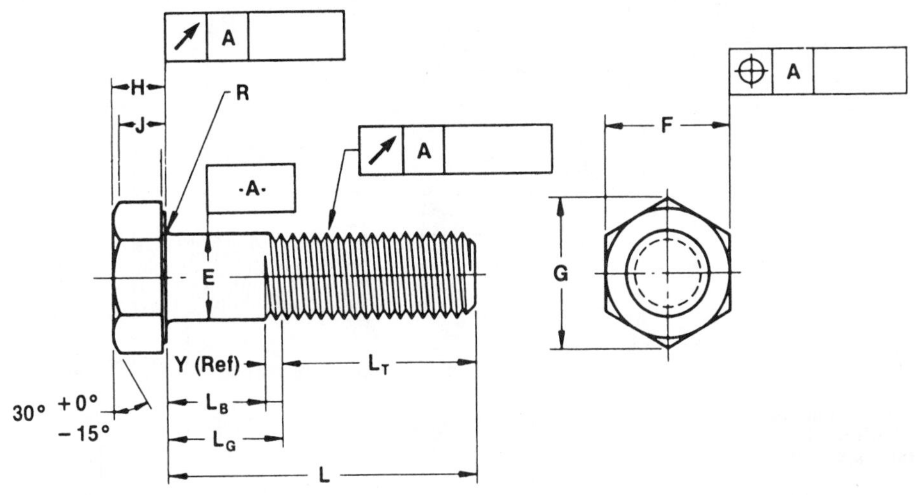

TABLE C.3.3 Dimensions of Hex Cap Screws

Nominal Size or Basic Product Dia		E Body Diameter		F Width Across Flats			G Width Across Corners		H Height			J Wrenching Height	L_T Thread Length For Screw Lengths		Y Transition Thread Length	Runout of Bearing Surface FIM
		Max	Min	Basic	Max	Min	Max	Min	Basic	Max	Min	Min	6 in. and Shorter Basic	Over 6 in. Basic	Max	Max
1/4	0.2500	0.2500	0.2450	7/16	0.438	0.428	0.505	0.488	5/32	0.163	0.150	0.106	0.750	1.000	0.250	0.010
5/16	0.3125	0.3125	0.3065	1/2	0.500	0.489	0.577	0.557	13/64	0.211	0.195	0.140	0.875	1.125	0.278	0.011
3/8	0.3750	0.3750	0.3690	9/16	0.562	0.551	0.650	0.628	15/64	0.243	0.226	0.160	1.000	1.250	0.312	0.012
7/16	0.4375	0.4375	0.4305	5/8	0.625	0.612	0.722	0.698	9/32	0.291	0.272	0.195	1.125	1.375	0.357	0.013
1/2	0.5000	0.5000	0.4930	3/4	0.750	0.736	0.866	0.840	5/16	0.323	0.302	0.215	1.250	1.500	0.385	0.014
9/16	0.5625	0.5625	0.5545	13/16	0.812	0.798	0.938	0.910	23/64	0.371	0.348	0.250	1.375	1.625	0.417	0.015
5/8	0.6250	0.6250	0.6170	15/16	0.938	0.922	1.083	1.051	25/64	0.403	0.378	0.269	1.500	1.750	0.455	0.017
3/4	0.7500	0.7500	0.7410	1-1/8	1.125	1.100	1.299	1.254	15/32	0.483	0.455	0.324	1.750	2.000	0.500	0.020
7/8	0.8750	0.8750	0.8660	1-5/16	1.312	1.285	1.516	1.465	35/64	0.563	0.531	0.378	2.000	2.250	0.556	0.023
1	1.0000	1.0000	0.9900	1-1/2	1.500	1.469	1.732	1.675	39/64	0.627	0.591	0.416	2.250	2.500	0.625	0.026
1-1/8	1.1250	1.1250	1.1140	1-11/16	1.688	1.631	1.949	1.859	11/16	0.718	0.658	0.461	2.500	2.750	0.714	0.029
1-1/4	1.2500	1.2500	1.2390	1-7/8	1.875	1.812	2.165	2.066	25/32	0.813	0.749	0.530	2.750	3.000	0.714	0.033
1-3/8	1.3750	1.3750	1.3630	2-1/16	2.062	1.994	2.382	2.273	27/32	0.878	0.810	0.569	3.000	3.250	0.833	0.036
1-1/2	1.5000	1.5000	1.4880	2-1/4	2.250	2.175	2.598	2.480	1-5/16	0.974	0.902	0.640	3.250	3.500	0.833	0.039
1-3/4	1.7500	1.7500	1.7380	2-5/8	2.625	2.538	3.031	2.893	1-3/32	1.134	1.054	0.748	3.750	4.000	1.000	0.046
2	2.0000	2.0000	1.9880	3	3.000	2.900	3.464	3.306	1-7/32	1.263	1.175	0.825	4.250	4.500	1.111	0.052
2-1/4	2.2500	2.2500	2.2380	3-3/8	3.375	3.262	3.897	3.719	1-3/8	1.423	1.327	0.933	4.750	5.000	1.111	0.059
2-1/2	2.5000	2.5000	2.4880	3-3/4	3.750	3.625	4.330	4.133	1-17/32	1.583	1.479	1.042	5.250	5.500	1.250	0.065
2-3/4	2.7500	2.7500	2.7380	4-1/8	4.125	3.988	4.763	4.546	1-11/16	1.744	1.632	1.151	5.750	6.000	1.250	0.072
3	3.0000	3.0000	2.9880	4-1/2	4.500	4.350	5.196	4.959	1-7/8	1.935	1.815	1.290	6.250	6.500	1.250	0.079

Metric Hex Cap Screws (ANSI B18.2.3.1M, 1979)

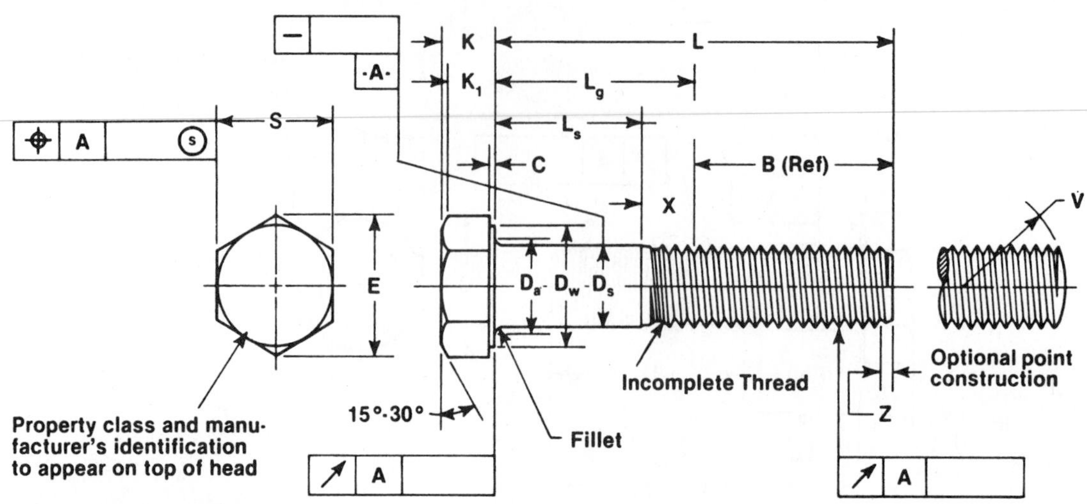

TABLE C.3.4 Dimensions of Hex Cap Screws

D	D_S		*S*		*E*		*K*		K_1	*C*		D_W	Runout of Bearing Surface FIM
Nom Screw Dia and Thread Pitch	Body Diameter		Width Across Flats		Width Across Corners		Head Height		Wrenching Height	Washer Face Thickness		Washer Face Dia	
	Max	Min	Max	Min	Max	Min	Max	Min	Min	Max	Min	Min	Max
M5 × 0.8	5.00	4.82	8.00	7.78	9.24	8.79	3.65	3.35	2.4	0.5	0.2	6.9	0.22
M6 × 1	6.00	5.82	10.00	9.78	11.55	11.05	4.15	3.85	2.8	0.5	0.2	8.9	0.25
M8 × 1.25	8.00	7.78	13.00	12.73	15.01	14.38	5.50	5.10	3.7	0.6	0.3	11.6	0.28
M10 × 1.5	10.00	9.78	16.00	15.73	18.48	17.77	6.63	6.17	4.5	0.6	0.3	14.6	0.32
M12 × 1.75	12.00	11.73	18.00	17.73	20.78	20.03	7.76	7.24	5.2	0.6	0.3	16.6	0.35
M14 × 2	14.00	13.73	21.00	20.67	24.25	23.35	9.09	8.51	6.2	0.6	0.3	19.6	0.39
M16 × 2	16.00	15.73	24.00	23.67	27.71	26.75	10.32	9.68	7.0	0.8	0.4	22.5	0.43
M20 × 2.5	20.00	19.67	30.00	29.16	34.64	32.95	12.88	12.12	8.8	0.8	0.4	27.7	0.53
M24 × 3	24.00	23.67	36.00	35.00	41.57	39.55	15.44	14.56	10.5	0.8	0.4	33.2	0.63
M30 × 3.5	30.00	29.67	46.00	45.00	53.12	50.85	19.48	17.92	13.1	0.8	0.4	42.7	0.78
M36 × 4	36.00	35.61	55.00	53.80	63.51	60.79	23.38	21.62	15.8	0.8	0.4	51.1	0.93
M42 × 4.5	42.00	41.38	65.00	62.90	75.06	71.71	26.97	25.03	18.2	1.0	0.5	59.8	1.09
M48 × 5	48.00	47.38	75.00	72.60	86.60	82.76	31.07	28.93	21.0	1.0	0.5	69.0	1.25
M56 × 5.5	56.00	55.26	85.00	82.20	98.15	93.71	36.20	33.80	24.5	1.0	0.5	78.1	1.47
M64 × 6	64.00	63.26	95.00	91.80	109.70	104.65	41.32	36.68	28.0	1.0	0.5	87.2	1.69
M72 × 6	72.00	71.26	105.00	101.40	121.24	115.60	46.45	43.55	31.5	1.2	0.6	96.3	1.91
M80 × 6	80.00	79.26	115.00	111.00	132.72	126.54	51.58	48.42	35.0	1.2	0.6	105.4	2.13
M90 × 6	90.00	89.13	130.00	125.50	150.11	143.07	57.74	54.26	39.2	1.2	0.6	119.2	2.41
M100 × 6	100.00	99.13	145.00	140.00	167.43	159.60	63.90	60.10	43.4	1.2	0.6	133.0	2.69
*M10 × 1.5	10.00	9.78	15.00	14.73	17.32	16.64	6.63	6.17	4.5	0.6	0.3	13.6	0.31

Socket Head Cap Screws (1960 Series) (ANSI/ASME B 18.3, 1986)

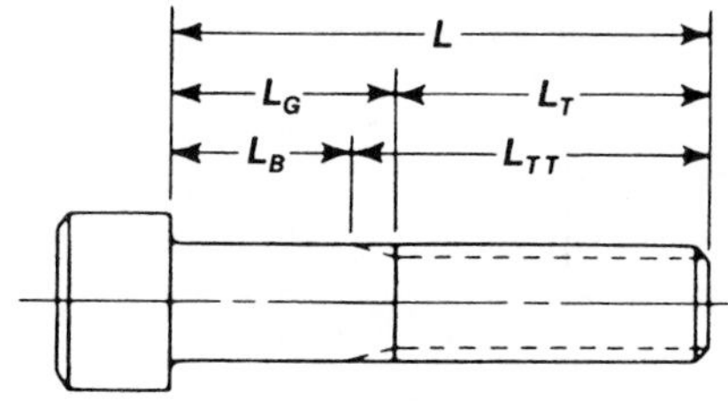

TABLE C.3.5 Screws Beyond Sizes in Table 1C

Nom Size or Basic Screw Dia		L_T Thread Length	L_{TT} Total Thread Length	Nom Size or Basic Screw Dia		L_T Thread Length	L_{TT} Total Thread Length
		Min	Max			Min	Max
0	0.0600	0.50	0.62	7/8	0.8750	2.25	3.69
1	0.0730	0.62	0.77	1	1.0000	2.50	4.12
2	0.0860	0.62	0.80	1 1/8	1.1250	2.81	4.65
3	0.0990	0.62	0.83	1 1/4	1.2500	3.12	5.09
4	0.1120	0.75	0.99	1 3/8	1.3750	3.44	5.65
5	0.1250	0.75	1.00	1 1/2	1.5000	3.75	6.08
6	0.1380	0.75	1.05	1 3/4	1.7500	4.38	7.13
8	0.1640	0.88	1.19	2	2.0000	5.00	8.11
10	0.1900	0.88	1.27	2 1/4	2.2500	5.62	8.99
1/4	0.2500	1.00	1.50	2 1/2	2.5000	6.25	10.00
5/16	0.3125	1.12	1.71	2 3/4	2.7500	6.88	10.87
3/8	0.3750	1.25	1.94	3	3.0000	7.50	11.75
7/16	0.4375	1.38	2.17	3 1/4	3.2500	8.12	12.63
1/2	0.5000	1.50	2.38	3 1/2	3.5000	8.75	13.50
5/8	0.6250	1.75	2.82	3 3/4	3.7500	9.38	14.37
3/4	0.7500	2.00	3.25	4	4.0000	10.00	15.25

Metric Socket Head Cap Screws (ANSI/ASME B18.3.1M, 1982)

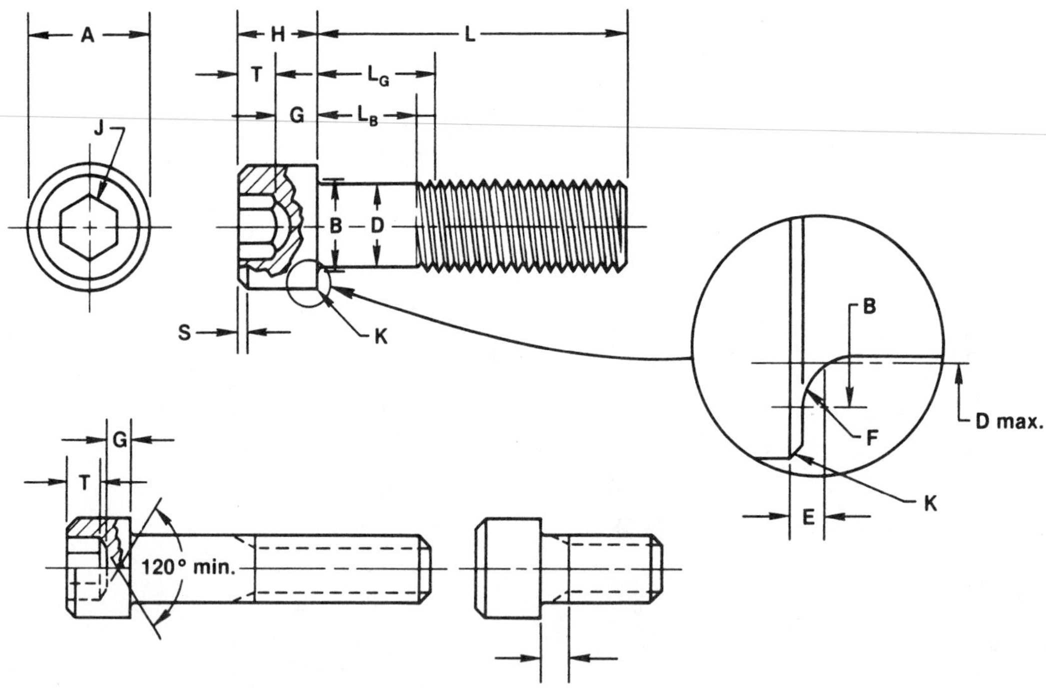

TABLE C.3.6 Dimensions of Metric Socket Head Cap Screws

Nom Screw Dia and Thread Pitch	*D* Body Diameter		*A* Head Diameter		*H* Head Height		*S* Chamfer or Radius	*J* Hexagon Socket Size	*T* Key Engagement	*G* Wall Thickness	*B* Under Head Fillet: Transition Diameter		*E* Under Head Fillet: Transition Length	*F* Under Head Fillet: Juncture Radius	*K* Chamfer or Radius
	Max	Min	Max	Min	Max	Min	Max	Nom	Min	Min	Max	Min	Max	Min	Max
M1.6 × 0.35	1.60	1.46	3.00	2.87	1.60	1.52	0.16	1.5	0.80	0.54	2.0	1.8	0.34	0.10	0.08
M2 × 0.4	2.00	1.86	3.80	3.65	2.00	1.91	0.20	1.5	1.00	0.68	2.6	2.2	0.51	0.10	0.08
M2.5 × 0.45	2.50	2.36	4.50	4.33	2.50	2.40	0.25	2.0	1.25	0.85	3.1	2.7	0.51	0.10	0.08
M3 × 0.5	3.00	2.86	5.50	5.32	3.00	2.89	0.30	2.5	1.50	1.02	3.6	3.2	0.51	0.10	0.13
M4 × 0.7	4.00	3.82	7.00	6.80	4.00	3.88	0.40	3.0	2.00	1.52	4.7	4.4	0.60	0.20	0.13
M5 × 0.8	5.00	4.82	8.50	8.27	5.00	4.86	0.50	4.0	2.50	1.90	5.7	5.4	0.60	0.20	0.13
M6 × 1	6.00	5.82	10.00	9.74	6.00	5.85	0.60	5.0	3.00	2.28	6.8	6.5	0.68	0.25	0.20
M8 × 1.25	8.00	7.78	13.00	12.70	8.00	7.83	0.80	6.0	4.00	3.20	9.2	8.8	1.02	0.40	0.20
M10 × 1.5	10.00	9.78	16.00	15.67	10.00	9.81	1.00	8.0	5.00	4.00	11.2	10.8	1.02	0.40	0.20
M12 × 1.75	12.00	11.73	18.00	17.63	12.00	11.79	1.20	10.0	6.00	4.80	14.2	13.2	1.87	0.60	0.25
(1)M14 × 2	14.00	13.73	21.00	20.60	14.00	13.77	1.40	12.0	7.00	5.60	16.2	15.2	1.87	0.60	0.25
M16 × 2	16.00	15.73	24.00	23.58	16.00	15.76	1.60	14.0	8.00	6.40	18.2	17.2	1.87	0.60	0.25
M20 × 2.5	20.00	19.67	30.00	29.53	20.00	19.73	2.00	17.0	10.00	8.00	22.4	21.6	2.04	0.80	0.40
M24 × 3	24.00	23.67	36.00	35.48	24.00	23.70	2.40	19.0	12.00	9.60	26.4	25.6	2.04	0.80	0.40
M30 × 3.5	30.00	29.67	45.00	44.42	30.00	29.67	3.00	22.0	15.00	12.00	33.4	32.0	2.89	1.00	0.40
M36 × 4	36.00	35.61	54.00	53.37	36.00	35.64	3.60	27.0	18.00	14.40	39.4	38.0	2.89	1.00	0.40
M42 × 4.5	42.00	41.61	63.00	62.31	42.00	41.61	4.20	32.0	21.00	16.80	45.6	44.4	3.06	1.20	0.40
M48 × 5	48.00	47.61	72.00	72.27	48.00	47.58	4.80	36.0	24.00	19.20	52.6	51.2	3.91	1.60	0.40

Socket Head Shoulder Screws (ANSI/ASME B18.3, 1986)

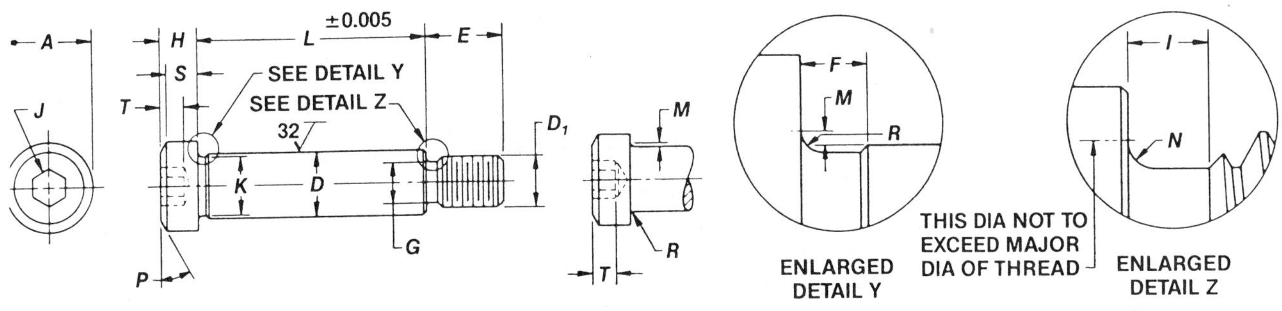

TABLE C.3.7 Dimensions of Hexagon Socket Head Shoulder Screws

Nominal Size or Basic Shoulder Diameter		D Shoulder Diameter		A Head Diameter		H Head Height		S Head Side Height	J Hexagon Socket Size		T Key Engage-ment	M Head Fillet Extension Above D	R Head Fillet Radius
		Max	Min	Max	Min	Max	Min	Min	Nom		Min	Max	Min
1/4	0.250	0.2480	0.2460	0.375	0.357	0.188	0.177	0.157	1/8	0.125	0.094	0.014	0.009
5/16	0.312	0.3105	0.3085	0.438	0.419	0.219	0.209	0.183	5/32	0.156	0.117	0.017	0.012
3/8	0.375	0.3730	0.3710	0.562	0.543	0.250	0.240	0.209	3/16	0.188	0.141	0.020	0.015
1/2	0.500	0.4980	0.4960	0.750	0.729	0.312	0.302	0.262	1/4	0.250	0.188	0.026	0.020
5/8	0.625	0.6230	0.6210	0.875	0.853	0.375	0.365	0.315	5/16	0.312	0.234	0.032	0.024
3/4	0.750	0.7480	0.7460	1.000	0.977	0.500	0.490	0.421	3/8	0.375	0.281	0.039	0.030
1	1.000	0.9980	0.9960	1.312	1.287	0.625	0.610	0.527	1/2	0.500	0.375	0.050	0.040
1 1/4	1.250	1.2480	1.2460	1.750	1.723	0.750	0.735	0.633	5/8	0.625	0.469	0.060	0.050
1 1/2	1.500	1.4980	1.4960	2.125	2.095	1.000	0.980	0.842	7/8	0.875	0.656	0.070	0.060
1 3/4	1.750	1.7480	1.7460	2.375	2.345	1.125	1.105	0.948	1	1.000	0.750	0.080	0.070
2	2.00	1.9980	1.9960	2.750	2.720	1.250	1.230	1.054	1 1/4	1.250	0.937	0.090	0.080

Nominal Size or Basic Shoulder Diameter		K Shoulder Neck Diameter	F Shoulder Neck Width	D_1 Nominal Thread Size or Basic Thread Diameter		Threads per in.	G Thread Neck Diameter		I Thread Neck Width	N Thread Neck Fillet		E Thread Length
		Min	Max				Max	Min	Max	Max	Min	Basic
1/4	0.250	0.227	0.093	10	0.1900	24	0.142	0.133	0.083	0.023	0.017	0.375
5/16	0.312	0.289	0.093	1/4	0.2500	20	0.193	0.182	0.100	0.028	0.022	0.438
3/8	0.375	0.352	0.093	5/16	0.3125	18	0.249	0.237	0.111	0.031	0.025	0.500
1/2	0.500	0.477	0.093	3/8	0.3750	16	0.304	0.291	0.125	0.035	0.029	0.625
5/8	0.625	0.602	0.093	1/2	0.5000	13	0.414	0.397	0.154	0.042	0.036	0.750
3/4	0.750	0.727	0.093	5/8	0.6250	11	0.521	0.502	0.182	0.051	0.045	0.875
1	1.000	0.977	0.125	3/4	0.7500	10	0.638	0.616	0.200	0.055	0.049	1.000
1 1/4	1.250	1.227	0.125	7/8	0.8750	9	0.750	0.726	0.222	0.062	0.056	1.125
1 1/2	1.500	1.478	0.125	1 1/8	1.1250	7	0.964	0.934	0.286	0.072	0.066	1.500
1 3/4	1.750	1.728	0.125	1 1/4	1.2500	7	1.089	1.059	0.286	0.072	0.066	1.750
2	2.000	1.978	0.125	1 1/2	1.5000	6	1.307	1.277	0.333	0.102	0.096	2.000

Square Bolts (ANSI/ASME B18.2.1, 1981)

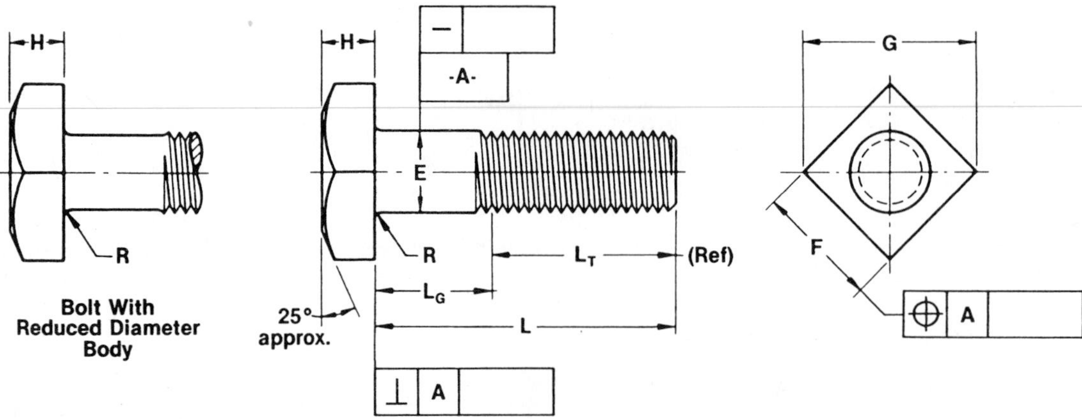

TABLE C.3.8 Dimensions of Square Bolts

Nominal Size or Basic Product Dia		*E* Body Dia	*F* Width Across Flats			*G* Width Across Corners		*H* Height			*R* Radius of Fillet		L_T Thread Length For Bolt Lengths 6 in. and shorter	L_T Thread Length For Bolt Lengths over 6 in.
		Max	Basic	Max	Min	Max	Min	Basic	Max	Min	Max	Min	Basic	Basic
1/4	0.2500	0.260	3/8	0.375	0.362	0.530	0.498	11/64	0.188	0.156	0.03	0.01	0.750	1.000
5/16	0.3125	0.324	1/2	0.500	0.484	0.707	0.665	13/64	0.220	0.186	0.03	0.01	0.875	1.125
3/8	0.3750	0.388	9/16	0.562	0.544	0.795	0.747	1/4	0.268	0.232	0.03	0.01	1.000	1.250
7/16	0.4375	0.452	5/8	0.625	0.603	0.884	0.828	19/64	0.316	0.278	0.03	0.01	1.125	1.375
1/2	0.5000	0.515	3/4	0.750	0.725	1.061	0.995	21/64	0.348	0.308	0.03	0.01	1.250	1.500
5/8	0.6250	0.642	15/16	0.938	0.906	1.326	1.244	27/64	0.444	0.400	0.06	0.02	1.500	1.750
3/4	0.7500	0.768	1-1/8	1.125	1.088	1.591	1.494	1/2	0.524	0.476	0.06	0.02	1.750	2.000
7/8	0.8750	0.895	1-5/16	1.312	1.269	1.856	1.742	19/32	0.620	0.568	0.06	0.02	2.000	2.250
1	1.0000	1.022	1-1/2	1.500	1.450	2.121	1.991	21/32	0.684	0.628	0.09	0.03	2.250	2.500
1-1/8	1.1250	1.149	1-11/16	1.688	1.631	2.386	2.239	3/4	0.780	0.720	0.09	0.03	2.500	2.750
1-1/4	1.2500	1.277	1-7/8	1.875	1.812	2.652	2.489	27/32	0.876	0.812	0.09	0.03	2.750	3.000
1-3/8	1.3750	1.404	2-1/16	2.062	1.994	2.917	2.738	29/32	0.940	0.872	0.09	0.03	3.000	3.250
1-1/2	1.5000	1.531	2-1/4	2.250	2.175	3.182	2.986	1	1.036	0.964	0.09	0.03	3.250	3.500

Socket Button Head Cap Screws (ANSI/ASME B18.3, 1986)

General Note: This product is designed and recommended for light fastening applications such as guards, hinges, etc. It is not suggested for use in critical high strength applications where socket head cap screws should normally be used.

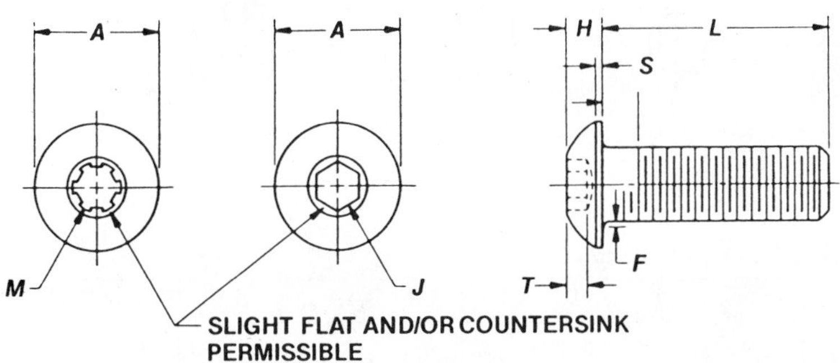

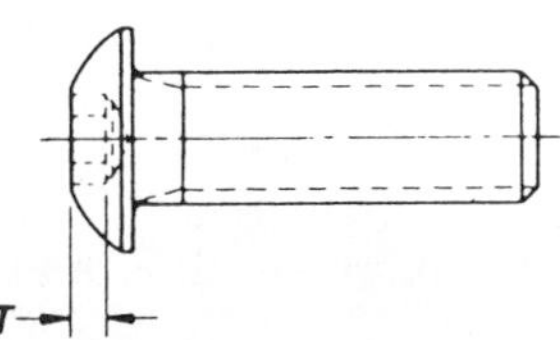

TABLE C.3.9 Dimensions of Hexagon and Spline Socket Button Head Cap Screws

Nominal Size or Basic Screw Diameter		*A* Head Diameter		*H* Head Height		*S* Head Side Height	*M* Spline Socket Size	*J* Hexagon Socket Size		*T* Key Engagement	*F* Fillet Extension		*L* Max Standard Length
		Max	Min	Max	Min	Ref	Nom	Nom		Min	Max	Min	Nom
0	0.0600	0.114	0.104	0.032	0.026	0.010	0.048		0.035	0.020	0.010	0.005	0.50
1	0.0730	0.139	0.129	0.039	0.033	0.010	0.060		0.050	0.028	0.010	0.005	0.50
2	0.0860	0.164	0.154	0.046	0.038	0.010	0.060		0.050	0.028	0.010	0.005	0.50
3	0.0990	0.188	0.176	0.052	0.044	0.010	0.072	1/16	0.062	0.035	0.010	0.005	0.50
4	0.1120	0.213	0.201	0.059	0.051	0.015	0.072	1/16	0.062	0.035	0.010	0.005	0.50
5	0.1250	0.238	0.226	0.066	0.058	0.015	0.096	5/64	0.078	0.044	0.010	0.005	0.50
6	0.1380	0.262	0.250	0.073	0.063	0.015	0.096	5/64	0.078	0.044	0.010	0.005	0.63
8	0.1640	0.312	0.298	0.087	0.077	0.015	0.111	3/32	0.094	0.052	0.015	0.010	0.75
10	0.1900	0.361	0.347	0.101	0.091	0.020	0.145	1/8	0.125	0.070	0.015	0.010	1.00
1/4	0.2500	0.437	0.419	0.132	0.122	0.031	0.183	5/32	0.156	0.087	0.020	0.015	1.00
5/16	0.3125	0.547	0.527	0.166	0.152	0.031	0.216	3/16	0.188	0.105	0.020	0.015	1.00
3/8	0.3750	0.656	0.636	0.199	0.185	0.031	0.251	7/32	0.219	0.122	0.020	0.015	1.25
1/2	0.5000	0.875	0.851	0.265	0.245	0.046	0.372	5/16	0.312	0.175	0.030	0.020	2.00
5/8	0.6250	1.000	0.970	0.331	0.311	0.062	0.454	3/8	0.375	0.210	0.030	0.020	2.00

Socket Set Screws (ANSI/ASME B18.3, 1986)

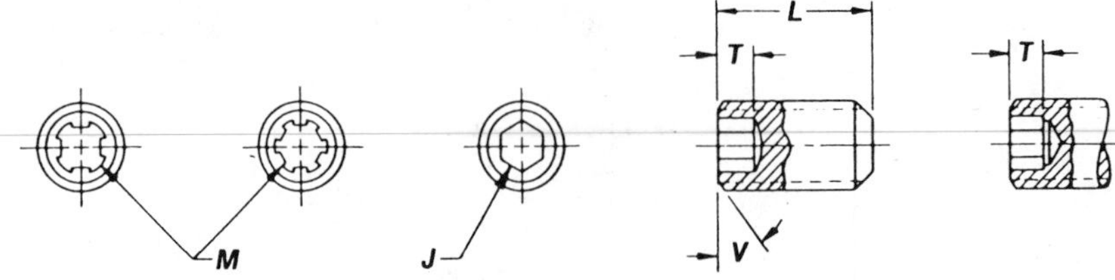

TABLE C.3.10 Dimensions of Hexagon and Spline Socket Set Screws

Nominal Size or Basic Screw Diameter		*J* Hexagon Socket Size		*M* Spline Socket Size	*T* Min Key Engagement to Develop Functional Capability of Key		*C* Cup and Flat Point Diameters		*R* Oval Point Radius	*Y* Cone Point Angle 90 deg ±2 deg for These Nominal Lengths or Longer; 118 deg ±2 deg for Shorter Nominal Lengths
		Nom		Nom	Hex Socket T_H Min	Spline Socket T_S Min	Max	Min	Basic	
0	0.0600		0.028	0.033	0.050	0.026	0.033	0.027	0.045	0.09
1	0.0730		0.028	0.033	0.060	0.035	0.040	0.033	0.055	0.09
2	0.0860		0.035	0.048	0.060	0.040	0.047	0.039	0.064	0.13
3	0.0990		0.050	0.048	0.070	0.040	0.054	0.045	0.074	0.13
4	0.1120		0.050	0.060	0.070	0.045	0.061	0.051	0.084	0.19
5	0.1250	1/16	0.062	0.072	0.080	0.055	0.067	0.057	0.094	0.19
6	0.1380	1/16	0.062	0.072	0.080	0.055	0.074	0.064	0.104	0.19
8	0.1640	5/64	0.078	0.096	0.090	0.080	0.087	0.076	0.123	0.25
10	0.1900	3/32	0.094	0.111	0.100	0.080	0.102	0.088	0.142	0.25
1/4	0.2500	1/8	0.125	0.145	0.125	0.125	0.132	0.118	0.188	0.31
5/16	0.3125	5/32	0.156	0.183	0.156	0.156	0.172	0.156	0.234	0.38
3/8	0.3750	3/16	0.188	0.216	0.188	0.188	0.212	0.194	0.281	0.44
7/16	0.4375	7/32	0.219	0.251	0.219	0.219	0.252	0.232	0.328	0.50
1/2	0.5000	1/4	0.250	0.291	0.250	0.250	0.291	0.270	0.375	0.57
5/8	0.6250	5/16	0.312	0.372	0.312	0.312	0.371	0.347	0.469	0.75
3/4	0.7500	3/8	0.375	0.454	0.375	0.375	0.450	0.425	0.562	0.88
7/8	0.8750	1/2	0.500	0.595	0.500	0.500	0.530	0.502	0.656	1.00
1	1.0000	9/16	0.562	...	0.562	...	0.609	0.579	0.750	1.13
1 1/8	1.1250	9/16	0.562	...	0.562	...	0.689	0.655	0.844	1.25
1 1/4	1.2500	5/8	0.625	...	0.625	...	0.767	0.733	0.938	1.50
1 3/8	1.3750	5/8	0.625	...	0.625	...	0.848	0.808	1.031	1.63
1 1/2	1.5000	3/4	0.750	...	0.750	...	0.926	0.886	1.125	1.75
1 3/4	1.7500	1	1.000	...	1.000	...	1.086	1.039	1.312	2.00
2	2.0000	1	1.000	...	1.000	...	1.244	1.193	1.500	2.25

Metric Socket Set Screws (ANSI B18.3.6M, 1979)

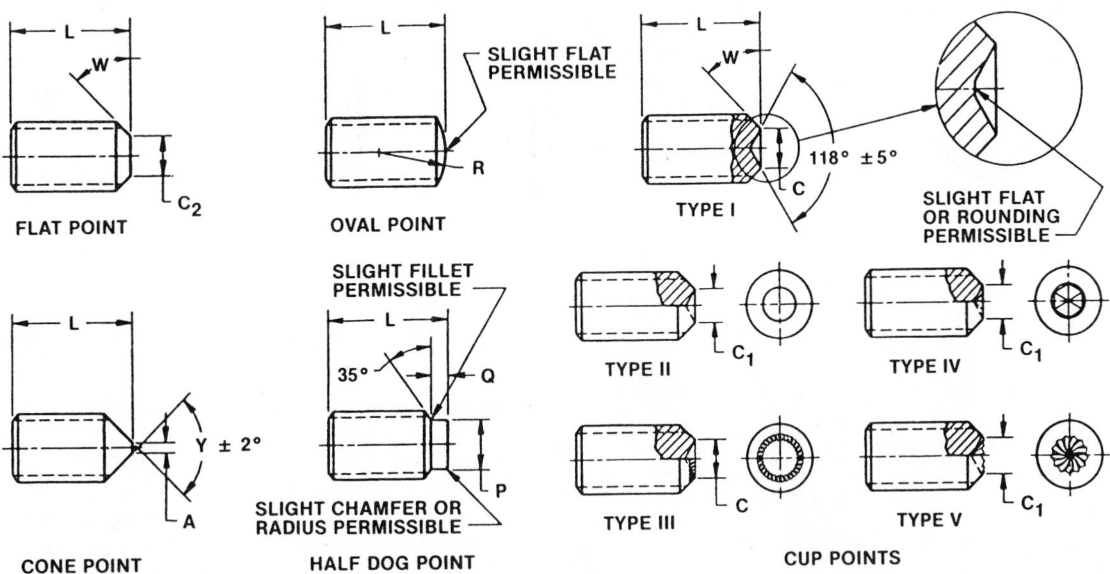

TABLE C.3.11 Dimensions of Points for Metric Socket Set Screws

D	*C*		C_1		C_2		*R*		*Y*	*A*		*P*		*Q*	
Nominal Size of Basic Screw Diameter	Cup Point Diameter For Types I and III		Cup Point Diameter For Types II, IV and V		Flat Point Diameter		Oval Point Radius		*Cone Point Angle 90° For These Lengths And Over; 118° For Shorter Lengths*	Flat of Truncation on Cone Point		Half Dog Point			
												Diameter		Length	
	Max	Min	Max	Min	Max	Min	Max	Min		Max	Min	Max	Min	Max	Min
1.6	0.80	0.55	0.80	0.64	0.80	0.55	1.60	1.20	3	1.16	0	0.80	0.55	0.53	0.40
2	1.00	0.75	1.00	0.82	1.00	0.75	1.90	1.50	3	0.2	0	1.00	0.75	0.64	0.50
2.5	1.20	0.95	1.25	1.05	1.50	1.25	2.28	1.88	4	0.25	0	1.50	1.25	0.78	0.63
3	1.40	1.15	1.50	1.28	2.00	1.75	2.65	2.25	4	0.3	0	2.00	1.75	0.92	0.75
4	2.00	1.75	2.00	1.75	2.50	2.25	3.80	3.00	5	0.4	0	2.50	2.25	1.20	1.00
5	2.50	2.25	2.50	2.22	3.50	3.20	4.55	3.75	6	0.5	0	3.50	3.20	1.37	1.25
6	3.00	2.75	3.00	2.69	4.00	3.70	5.30	4.50	8	1.5	1.2	4.00	3.70	1.74	1.50
8	5.00	4.70	4.00	3.65	5.50	5.20	6.80	6.00	10	2.0	1.6	5.50	5.20	2.28	2.00
10	6.00	5.70	5.00	4.60	7.00	6.64	8.30	7.50	12	2.5	2.0	7.00	6.64	2.82	2.50
12	8.00	7.64	6.00	5.57	8.50	8.14	9.80	9.00	16	3.0	2.4	8.50	8.14	3.35	3.00
16	10.00	9.64	8.00	7.50	12.00	11.57	12.80	12.00	20	4.0	3.2	12.00	11.57	4.40	4.00
20	14.00	13.57	10.00	9.44	15.00	14.57	15.80	15.00	25	5.0	4.0	15.00	14.57	5.45	5.00
24	16.00	15.57	12.00	11.39	18.00	17.57	18.80	18.00	30	6.0	4.8	18.00	17.57	6.49	6.00

Hex Nuts and Hex Jam Nuts (ANSI/ASME B18.2.2, 1986)

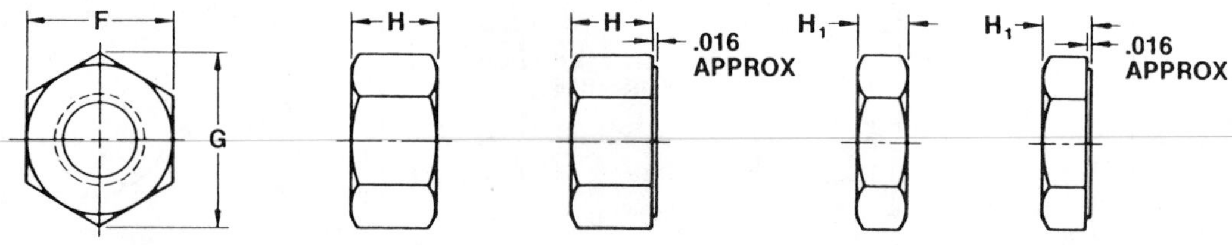

TABLE C.3.12 Dimensions of Hex Nuts and Hex Jam Nuts

Nominal Size or Basic Major Dia of Thread		F			G		H			H_1			*Runout of Bearing Face, FIM*		
		Width Across Flats			Width Across Corners		Thickness Hex Nuts			Thickness Hex Jam Nuts			Hex Nuts — Specified Proof Load		Hex Jam Nuts
													Up to 150,000 psi	150,000 psi and Greater	All Strength Levels
		Basic	Max	Min	Max	Min	Basic	Max	Min	Basic	Max	Min	Max		
1/4	0.2500	7/16	0.438	0.428	0.505	0.488	7/32	0.226	0.212	5/32	0.163	0.150	0.015	0.010	0.015
5/16	0.3125	1/2	0.500	0.489	0.577	0.557	17/64	0.273	0.258	3/16	0.195	0.180	0.016	0.011	0.016
3/8	0.3750	9/16	0.562	0.551	0.650	0.628	21/64	0.337	0.320	7/32	0.227	0.210	0.017	0.012	0.017
7/16	0.4375	11/16	0.688	0.675	0.794	0.768	3/8	0.385	0.365	1/4	0.260	0.240	0.018	0.013	0.018
1/2	0.5000	3/4	0.750	0.736	0.866	0.840	7/16	0.448	0.427	5/16	0.323	0.302	0.019	0.014	0.019
9/16	0.5625	7/8	0.875	0.861	1.010	0.982	31/64	0.496	0.473	5/16	0.324	0.301	0.020	0.015	0.020
5/8	0.6250	15/16	0.938	0.922	1.083	1.051	35/64	0.559	0.535	3/8	0.387	0.363	0.021	0.016	0.021
3/4	0.7500	1 1/8	1.125	1.088	1.299	1.240	41/64	0.665	0.617	27/64	0.446	0.398	0.023	0.018	0.023
7/8	0.8750	1 5/16	1.312	1.269	1.516	1.447	3/4	0.776	0.724	31/64	0.510	0.458	0.025	0.020	0.025
1	1.0000	1 1/2	1.500	1.450	1.732	1.653	55/64	0.887	0.831	35/64	0.575	0.519	0.027	0.022	0.027
1 1/8	1.1250	1 11/16	1.688	1.631	1.949	1.859	31/32	0.999	0.939	39/64	0.639	0.579	0.030	0.025	0.030
1 1/4	1.2500	1 7/8	1.875	1.812	2.165	2.066	1 1/16	1.094	1.030	23/32	0.751	0.687	0.033	0.028	0.033
1 3/8	1.3750	2 1/16	2.062	1.994	2.382	2.273	1 11/64	1.206	1.138	25/32	0.815	0.747	0.036	0.031	0.036
1 1/2	1.5000	2 1/4	2.250	2.175	2.598	2.480	1 9/32	1.317	1.245	27/32	0.880	0.808	0.039	0.034	0.039

Square Nuts (ANSI/ASME B18.2.2, 1986)

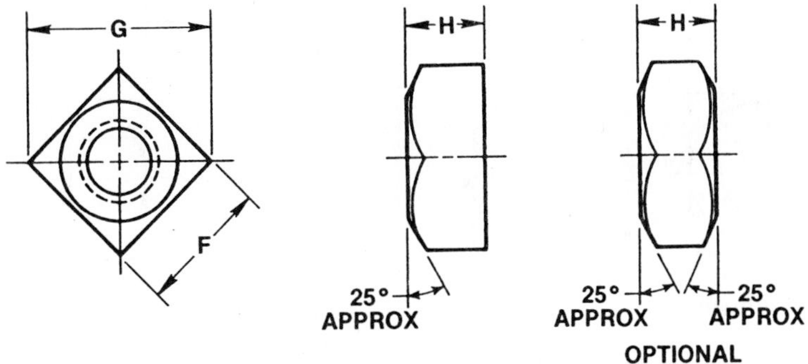

TABLE C.3.13 Dimensions of Square Nuts

Nominal Size or Basic Major Dia of Thread		*F* Width Across Flats			*G* Width Across Corners		*H* Thickness		
		Basic	Max	Min	Max	Min	Basic	Max	Min
1/4	0.2500	7/16	0.438	0.425	0.619	0.554	7/32	0.235	0.203
5/16	0.3125	9/16	0.562	0.547	0.795	0.721	17/64	0.283	0.249
3/8	0.3750	5/8	0.625	0.606	0.884	0.802	21/64	0.346	0.310
7/16	0.4375	3/4	0.750	0.728	1.061	0.970	3/8	0.394	0.356
1/2	0.5000	13/16	0.812	0.788	1.149	1.052	7/16	0.458	0.418
5/8	0.6250	1	1.000	0.969	1.414	1.300	35/64	0.569	0.525
3/4	0.7500	1 1/8	1.125	1.088	1.591	1.464	21/32	0.680	0.632
7/8	0.8750	1 5/16	1.312	1.269	1.856	1.712	49/64	0.792	0.740
1	1.0000	1 1/2	1.500	1.450	2.121	1.961	7/8	0.903	0.847
1 1/8	1.1250	1 11/16	1.688	1.631	2.386	2.209	1	1.030	0.970
1 1/4	1.2500	1 7/8	1.875	1.812	2.652	2.458	1 3/32	1.126	1.062
1 3/8	1.3750	2 1/16	2.062	1.994	2.917	2.708	1 13/64	1.237	1.169
1 1/2	1.5000	2 1/4	2.250	2.175	3.182	2.956	1 5/16	1.348	1.276

Metric Hex Nuts, Style 1 (ANSI B18.2.4.1M, 1979)

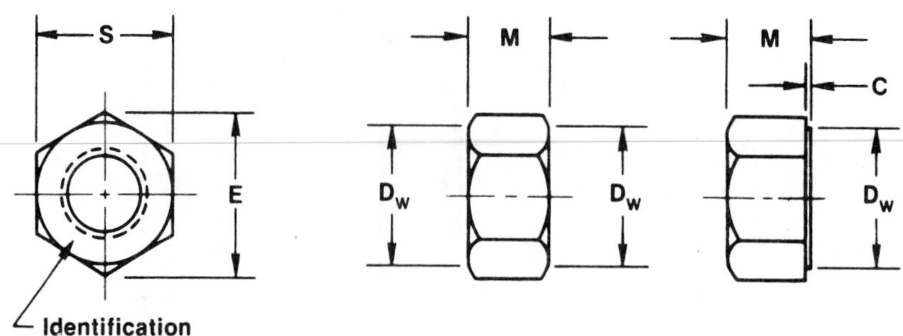

TABLE C.3.14 Dimensions of Hex Nuts, Style 1

Nominal Nut Dia and Thread Pitch	S		E		M		D_W	C		Total Runout of Bearing Surface FIM
	Width Across Flats		Width Across Corners		Thickness		Bearing Face Dia	Washer Face Thickness		
	Max	Min	Max	Min	Max	Min	Min	Max	Min	Max
M1.6 × 0.35	3.20	3.02	3.70	3.41	1.30	1.05	2.4	—	—	—
M2 × 0.4	4.00	3.82	4.62	4.32	1.60	1.35	3.1	—	—	—
M2.5 × 0.45	5.00	4.82	5.77	5.45	2.00	1.75	4.1	—	—	—
M3 × 0.5	5.50	5.32	6.35	6.01	2.40	2.15	4.6	—	—	—
M3.5 × 0.6	6.00	5.82	6.93	6.58	2.80	2.55	5.1	—	—	—
M4 × 0.7	7.00	6.78	8.08	7.66	3.20	2.90	5.9	—	—	—
M5 × 0.8	8.00	7.78	9.24	8.79	4.70	4.40	6.9	—	—	0.30
M6 × 1	10.00	9.78	11.55	11.05	5.20	4.90	8.9	—	—	0.33
M8 × 1.25	13.00	12.73	15.01	14.38	6.80	6.44	11.6	—	—	0.36
M10 × 1.5	16.00	15.73	18.48	17.77	8.40	8.04	14.6	—	—	0.39
M12 × 1.75	18.00	17.73	20.78	20.03	10.80	10.37	16.6	—	—	0.42
M14 × 2	21.00	20.67	24.25	23.35	12.80	12.10	19.6	—	—	0.45
M16 × 2	24.00	23.67	27.71	26.75	14.80	14.10	22.5	—	—	0.48
M20 × 2.5	30.00	29.16	34.64	32.95	18.00	16.90	27.7	0.8	0.4	0.56
M24 × 3	36.00	35.00	41.57	39.55	21.50	20.20	33.2	0.8	0.4	0.64
M30 × 3.5	46.00	45.00	53.12	50.85	25.60	24.30	42.7	0.8	0.4	0.76
M36 × 4	55.00	53.80	63.51	60.79	31.00	29.40	51.1	0.8	0.4	0.89
*M10 × 1.5	15.00	14.73	17.32	16.64	9.1	8.7	13.6	—	—	0.39

APPENDIX C.4

Washers

Plain Washers (ANSI/ASME B18.22.1 1965, (1981)

TABLE C.4.1 Dimensions of Preferred Sizes of Type A Plain Washers

Nominal Washer Size			A Inside Diameter			B Outside Diameter			C Thickness		
			Basic	Tolerance Plus	Tolerance Minus	Basic	Tolerance Plus	Tolerance Minus	Basic	Max	Min
—	—		0.078	0.000	0.005	0.188	0.000	0.005	0.020	0.025	0.016
—	—		0.094	0.000	0.005	0.250	0.000	0.005	0.020	0.025	0.016
—	—		0.125	0.008	0.005	0.312	0.008	0.005	0.032	0.040	0.025
No. 6	0.138		0.156	0.008	0.005	0.375	0.015	0.005	0.049	0.065	0.036
8	0.164		0.188	0.008	0.005	0.438	0.015	0.005	0.049	0.065	0.036
10	0.190		0.219	0.008	0.005	0.500	0.015	0.005	0.049	0.065	0.036
3/16	0.188		0.250	0.015	0.005	0.562	0.015	0.005	0.049	0.065	0.036
12	0.216		0.250	0.015	0.005	0.562	0.015	0.005	0.065	0.080	0.051
1/4	0.250	N	0.281	0.015	0.005	0.625	0.015	0.005	0.065	0.080	0.051
1/4	0.250	W	0.312	0.015	0.005	0.734	0.015	0.007	0.065	0.080	0.051
5/16	0.312	N	0.344	0.015	0.005	0.688	0.015	0.007	0.065	0.080	0.051
5/16	0.312	W	0.375	0.015	0.005	0.875	0.030	0.007	0.083	0.104	0.064
3/8	0.375	N	0.406	0.015	0.005	0.812	0.015	0.007	0.065	0.080	0.051
3/8	0.375	W	0.438	0.015	0.005	1.000	0.030	0.007	0.083	0.104	0.064
7/16	0.438	N	0.469	0.015	0.005	0.922	0.015	0.007	0.065	0.080	0.051
7/16	0.438	W	0.500	0.015	0.005	1.250	0.030	0.007	0.083	0.104	0.064
1/2	0.500	N	0.531	0.015	0.005	1.062	0.030	0.007	0.095	0.121	0.074
1/2	0.500	W	0.562	0.015	0.005	1.375	0.030	0.007	0.109	0.132	0.086
9/16	0.562	N	0.594	0.015	0.005	1.156	0.030	0.007	0.095	0.121	0.074
9/16	0.562	W	0.625	0.015	0.005	1.469	0.030	0.007	0.109	0.132	0.086
5/8	0.625	N	0.656	0.030	0.007	1.312	0.030	0.007	0.095	0.121	0.074
5/8	0.625	W	0.688	0.030	0.007	1.750	0.030	0.007	0.134	0.160	0.108
3/4	0.750	N	0.812	0.030	0.007	1.469	0.030	0.007	0.134	0.160	0.108
3/4	0.750	W	0.812	0.030	0.007	2.000	0.030	0.007	0.148	0.177	0.122
7/8	0.875	N	0.938	0.030	0.007	1.750	0.030	0.007	0.134	0.160	0.108
7/8	0.875	W	0.938	0.030	0.007	2.250	0.030	0.007	0.165	0.192	0.136
1	1.000	N	1.062	0.030	0.007	2.000	0.030	0.007	0.134	0.160	0.108
1	1.000	W	1.062	0.030	0.007	2.500	0.030	0.007	0.165	0.192	0.136
1-1/8	1.125	N	1.250	0.030	0.007	2.250	0.030	0.007	0.134	0.160	0.108
1-1/8	1.125	W	1.250	0.030	0.007	2.750	0.030	0.007	0.165	0.192	0.136
1-1/4	1.250	N	1.375	0.030	0.007	2.500	0.030	0.007	0.165	0.192	0.136
1-1/4	1.250	W	1.375	0.030	0.007	3.000	0.030	0.007	0.165	0.192	0.136
1-3/8	1.375	N	1.500	0.030	0.007	2.750	0.030	0.007	0.165	0.192	0.136
1-3/8	1.375	W	1.500	0.045	0.010	3.250	0.045	0.010	0.180	0.213	0.153
1-1/2	1.500	N	1.625	0.030	0.007	3.000	0.030	0.007	0.165	0.192	0.136
1-1/2	1.500	W	1.625	0.045	0.010	3.500	0.045	0.010	0.180	0.213	0.153

Continued

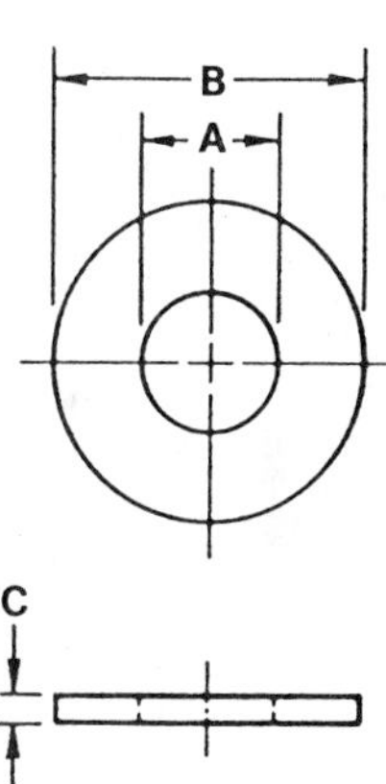

TABLE C.4.1 Dimensions of Preferred Sizes of Type A Plain Washers—*Continued*

Nominal Washer Size		A Inside Diameter			B Outside Diameter			C Thickness		
		Basic	Tolerance Plus	Tolerance Minus	Basic	Tolerance Plus	Tolerance Minus	Basic	Max	Min
1-5/8	1.625	1.750	0.045	0.010	3.750	0.045	0.010	0.180	0.213	0.153
1-3/4	1.750	1.875	0.045	0.010	4.000	0.045	0.010	0.180	0.213	0.153
1-7/8	1.875	2.000	0.045	0.010	4.250	0.045	0.010	0.180	0.213	0.153
2	2.000	2.125	0.045	0.010	4.500	0.045	0.010	0.180	0.213	0.153
2-1/4	2.250	2.375	0.045	0.010	4.750	0.045	0.010	0.220	0.248	0.193
2-1/2	2.500	2.625	0.045	0.010	5.000	0.045	0.010	0.238	0.280	0.210
2-3/4	2.750	2.875	0.065	0.010	5.250	0.065	0.010	0.259	0.310	0.228
3	3.000	3.125	0.065	0.010	5.500	0.065	0.010	0.284	0.327	0.249

Metric Plain Washers (ANSI B18.22M, 1981)

TABLE C.4.2 Dimensions of Metric Plain Washers (General Purpose)

Nom Washer Size	Washer Series	A Inside Dia Max	A Inside Dia Min	B Outside Dia Max	B Outside Dia Min	C Thickness Max	C Thickness Min
1.6	Narrow	2.09	1.95	4.00	3.70	0.70	0.50
	Regular	2.09	1.95	5.00	4.70	0.70	0.50
	Wide	2.09	1.95	6.00	5.70	0.90	0.60
2	Narrow	2.64	2.50	5.00	4.70	0.90	0.60
	Regular	2.64	2.50	6.00	5.70	0.90	0.60
	Wide	2.64	2.50	8.00	7.64	0.90	0.60
2.5	Narrow	3.14	3.00	6.00	5.70	0.90	0.60
	Regular	3.14	3.00	8.00	7.64	0.90	0.60
	Wide	3.14	3.00	10.00	9.64	1.20	0.80
3	Narrow	3.68	3.50	7.00	6.64	0.90	0.60
	Regular	3.68	3.50	10.00	9.64	1.20	0.80
	Wide	3.68	3.50	12.00	11.57	1.40	1.00
3.5	Narrow	4.18	4.00	9.00	8.64	1.20	0.80
	Regular	4.18	4.00	10.00	9.64	1.40	1.00
	Wide	4.18	4.00	15.00	14.57	1.75	1.20
4	Narrow	4.88	4.70	10.00	9.64	1.20	0.80
	Regular	4.88	4.70	12.00	11.57	1.40	1.00
	Wide	4.88	4.70	16.00	15.57	2.30	1.60
5	Narrow	5.78	5.50	11.00	10.57	1.40	1.00
	Regular	5.78	5.50	15.00	14.57	1.75	1.20
	Wide	5.78	5.50	20.00	19.48	2.30	1.60
6	Narrow	6.87	6.65	13.00	12.57	1.75	1.20
	Regular	6.87	6.65	18.80	18.37	1.75	1.20
	Wide	6.87	6.65	25.40	24.88	2.30	1.60
8	Narrow	9.12	8.90	18.80	18.37	2.30	1.60
	Regular	9.12	8.90	25.40	24.48	2.30	1.60
	Wide	9.12	8.90	32.00	31.38	2.80	2.00

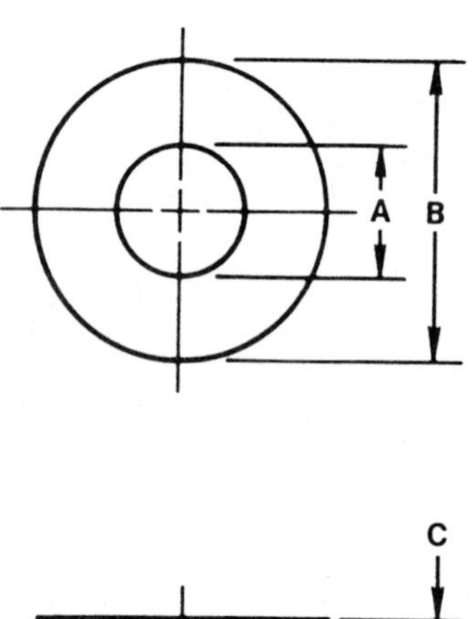

Continued

TABLE C.4.2 Dimensions of Metric Plain Washers (General Purpose)—*Continued*

Nom Washer Size	*Washer Series*	*A* Inside Dia		*B* Outside Dia		*C* Thickness	
		Max	**Min**	**Max**	**Min**	**Max**	**Min**
	Narrow	11.12	10.85	20.00	19.48	2.30	1.60
10	Regular	11.12	10.85	28.00	27.48	2.80	2.00
	Wide	11.12	10.85	39.00	38.38	3.50	2.50
	Narrow	13.57	13.30	25.40	24.88	2.80	2.00
12	Regular	13.57	13.30	34.00	33.38	3.50	2.50
	Wide	13.57	13.30	44.00	43.38	3.50	2.50
	Narrow	15.52	15.25	28.00	27.48	2.80	2.00
14	Regular	15.52	15.25	39.00	38.38	3.50	2.50
	Wide	15.52	15.25	50.00	49.38	4.00	3.00
	Narrow	17.52	17.25	32.00	31.38	3.50	2.50
16	Regular	17.52	17.25	44.00	43.38	4.00	3.00
	Wide	17.52	17.25	56.00	54.80	4.60	3.50
	Narrow	22.32	21.80	39.00	38.38	4.00	3.00
20	Regular	22.32	21.80	50.00	49.38	4.60	3.50
	Wide	22.32	21.80	66.00	64.80	5.10	4.00
	Narrow	26.12	25.60	44.00	43.38	4.60	3.50
24	Regular	26.12	25.60	56.00	54.80	5.10	4.00
	Wide	26.12	25.60	72.00	70.80	5.60	4.50
	Narrow	33.02	32.40	56.00	54.80	5.10	4.00
30	Regular	33.02	32.40	72.00	70.80	5.60	4.50
	Wide	33.02	32.40	90.00	88.60	6.40	5.00
	Narrow	38.92	38.30	66.00	64.80	5.60	4.50
36	Regular	38.92	38.30	90.00	88.60	6.40	5.00
	Wide	38.92	38.30	110.00	108.60	8.50	7.00

NOTES:

1. Nominal washer sizes are intended for use with comparable nominal screw or bolt sizes.
2. See 4.3 for maximum permissible I.D. at the punch exit side.
3. See 4.2 for closeness of fit with coated or plated products.
4. The 18.80/18.37 and 25.40/24.88 mm outside diameters avoid washers which could be used in coin operated devices.

Lock Washers (ANSI/ASME B18.21.1, 1972, R1983)

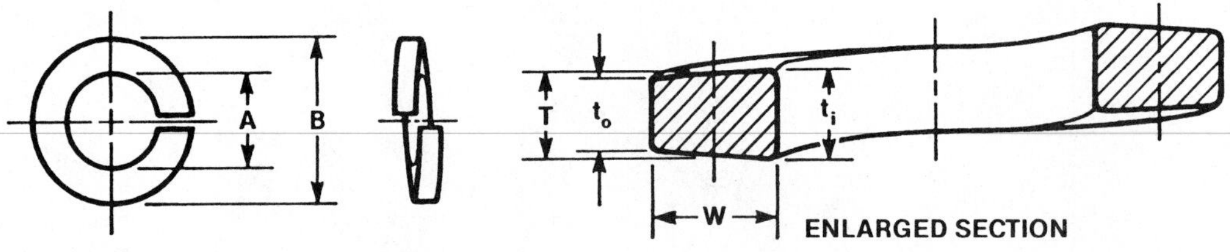

TABLE C.4.3 Dimensions of Regular Helical Spring Lock Washers

Nominal Washer Size		A Inside Diameter Max	A Inside Diameter Min	B Outside Diameter Max[2]	T Mean Section Thickness $\left(\frac{t_i + t_o}{2}\right)$ Min	W Section Width Min
No. 2	0.086	0.094	0.088	0.172	0.020	0.035
No. 3	0.099	0.107	0.101	0.195	0.025	0.040
No. 4	0.112	0.120	0.114	0.209	0.025	0.040
No. 5	0.125	0.133	0.127	0.236	0.031	0.047
No. 6	0.138	0.148	0.141	0.250	0.031	0.047
No. 8	0.164	0.174	0.167	0.293	0.040	0.055
No. 10	0.190	0.200	0.193	0.334	0.047	0.062
No. 12	0.216	0.227	0.220	0.377	0.056	0.070
1/4	0.250	0.262	0.254	0.489	0.062	0.109
5/16	0.312	0.326	0.317	0.586	0.078	0.125
3/8	0.375	0.390	0.380	0.683	0.094	0.141
7/16	0.438	0.455	0.443	0.779	0.109	0.156
1/2	0.500	0.518	0.506	0.873	0.125	0.171
9/16	0.562	0.582	0.570	0.971	0.141	0.188
5/8	0.625	0.650	0.635	1.079	0.156	0.203
11/16	0.688	0.713	0.698	1.176	0.172	0.219
3/4	0.750	0.775	0.760	1.271	0.188	0.234
13/16	0.812	0.843	0.824	1.367	0.203	0.250
7/8	0.875	0.905	0.887	1.464	0.219	0.266
15/16	0.938	0.970	0.950	1.560	0.234	0.281
1	1.000	1.042	1.017	1.661	0.250	0.297
1-1/16	1.062	1.107	1.080	1.756	0.266	0.312
1-1/8	1.125	1.172	1.144	1.853	0.281	0.328
1-3/16	1.188	1.237	1.208	1.950	0.297	0.344
1-1/4	1.250	1.302	1.271	2.045	0.312	0.359
1-5/16	1.312	1.366	1.334	2.141	0.328	0.375
1-3/8	1.375	1.432	1.398	2.239	0.344	0.391
1-7/16	1.438	1.497	1.462	2.334	0.359	0.406
1-1/2	1.500	1.561	1.525	2.430	0.375	0.422

Lock Washers (ANSI/ASME B18.21.1 1972, R1983)

TABLE C.4.4 Dimensions of Internal Tooth Lock Washers

Nominal Washer Size		A		B		C	
		Inside Diameter		Outside Diameter		Thickness	
		Max	Min	Max	Min	Max	Min
No. 2	0.086	0.095	0.089	0.200	0.175	0.015	0.010
No. 3	0.099	0.109	0.102	0.232	0.215	0.019	0.012
No. 4	0.112	0.123	0.115	0.270	0.255	0.019	0.015
No. 5	0.125	0.136	0.129	0.280	0.245	0.021	0.017
No. 6	0.138	0.150	0.141	0.295	0.275	0.021	0.017
No. 8	0.164	0.176	0.168	0.340	0.325	0.023	0.018
No. 10	0.190	0.204	0.195	0.381	0.365	0.025	0.020
No. 12	0.216	0.231	0.221	0.410	0.394	0.025	0.020
1/4	0.250	0.267	0.256	0.478	0.460	0.028	0.023
5/16	0.312	0.332	0.320	0.610	0.594	0.034	0.028
3/8	0.375	0.398	0.384	0.692	0.670	0.040	0.032
7/16	0.438	0.464	0.448	0.789	0.740	0.040	0.032
1/2	0.500	0.530	0.512	0.900	0.867	0.045	0.037
9/16	0.562	0.596	0.576	0.985	0.957	0.045	0.037
5/8	0.625	0.663	0.640	1.071	1.045	0.050	0.042
11/16	0.688	0.728	0.704	1.166	1.130	0.050	0.042
3/4	0.750	0.795	0.769	1.245	1.220	0.055	0.047
13/16	0.812	0.861	0.832	1.315	1.290	0.055	0.047
7/8	0.875	0.927	0.894	1.410	1.364	0.060	0.052
1	1.000	1.060	1.019	1.637	1.590	0.067	0.059
1-1/8	1.125	1.192	1.144	1.830	1.799	0.067	0.059
1-1/4	1.250	1.325	1.275	1.975	1.921	0.067	0.059

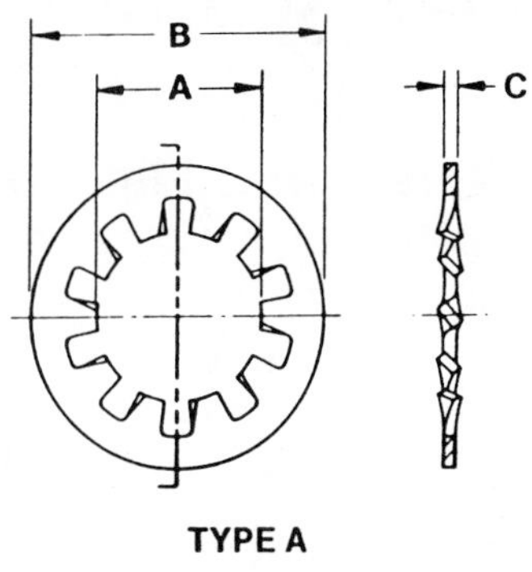

TYPE A

TABLE C.4.5 Dimensions of Heavy Internal Tooth Lock Washers

Nominal Washer Size		A		B		C	
		Inside Diameter		Outside Diameter		Thickness	
		Max	Min	Max	Min	Max	Min
1/4	0.250	0.267	0.256	0.536	0.500	0.045	0.035
5/16	0.312	0.332	0.320	0.607	0.590	0.050	0.040
3/8	0.375	0.398	0.384	0.748	0.700	0.050	0.042
7/16	0.438	0.464	0.448	0.858	0.800	0.067	0.050
1/2	0.500	0.530	0.512	0.924	0.880	0.067	0.055
9/16	0.562	0.596	0.576	1.034	0.990	0.067	0.055
5/8	0.625	0.663	0.640	1.135	1.100	0.067	0.059
3/4	0.750	0.795	0.768	1.265	1.240	0.084	0.070
7/8	0.875	0.927	0.894	1.447	1.400	0.084	0.075

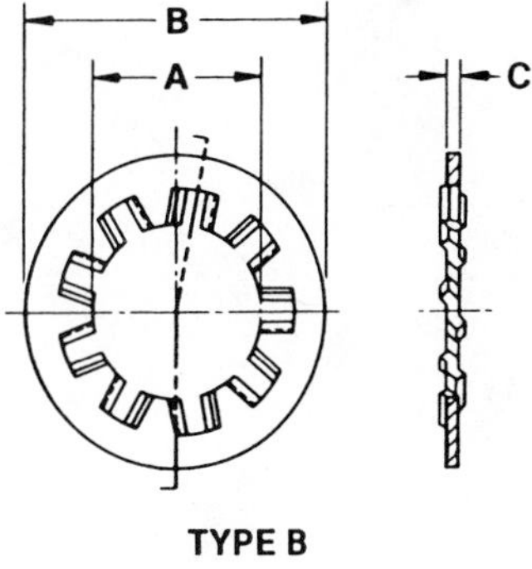

TYPE B

APPENDIX C.5

Rivets, Retaining Rings

Flat Head Rivets and Flat Countersunk Head Rivets (ANSI/ASME B18.1.1 1972, R1981)

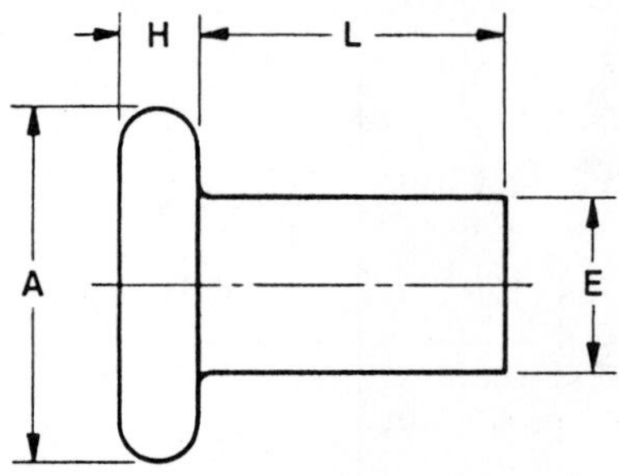

TABLE C.5.1 Dimensions of Flat Head Rivets

Nominal Size or Basic Shank Diameter		*E* Shank Diameter		*A* Head Diameter		*H* Head Diameter	
		Max	Min	Max	Min	Max	Min
1/16	0.062	0.064	0.059	0.140	0.120	0.027	0.017
3/32	0.094	0.096	0.090	0.200	0.180	0.038	0.026
1/8	0.125	0.127	0.121	0.260	0.240	0.048	0.036
5/32	0.156	0.158	0.152	0.323	0.301	0.059	0.045
3/16	0.188	0.191	0.182	0.387	0.361	0.069	0.055
7/32	0.219	0.222	0.213	0.453	0.427	0.080	0.065
1/4	0.250	0.253	0.244	0.515	0.485	0.091	0.075
9/32	0.281	0.285	0.273	0.579	0.545	0.103	0.085
5/16	0.312	0.316	0.304	0.641	0.607	0.113	0.095
11/32	0.344	0.348	0.336	0.705	0.667	0.124	0.104
3/8	0.375	0.380	0.365	0.769	0.731	0.135	0.115
13/32	0.406	0.411	0.396	0.834	0.790	0.146	0.124
7/16	0.438	0.443	0.428	0.896	0.852	0.157	0.135

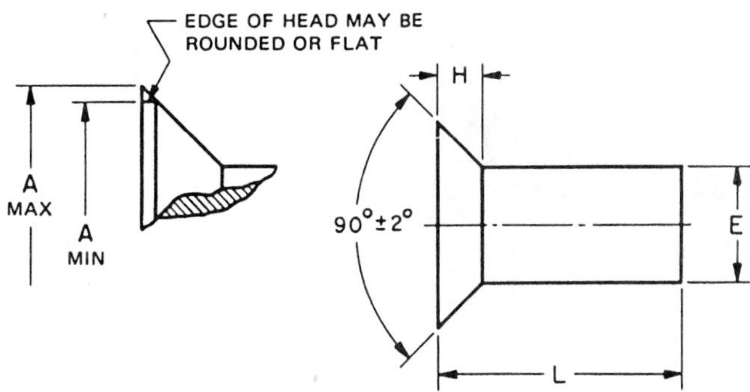

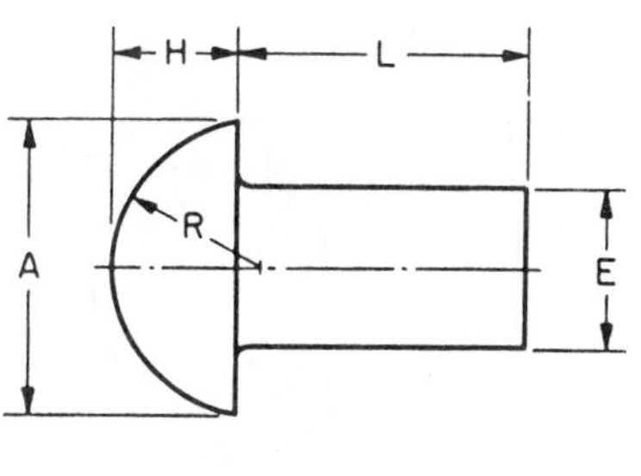

TABLE C.5.2 Dimensions of Flat Countersunk Head Rivets

Nominal Size[1] or Basic Shank Diameter		*E* Shank Diameter		*A* Head Diameter		*H* Head Height
		Max	Min	Max[2]	Min[3]	Ref[4]
1/16	0.062	0.064	0.059	0.118	0.110	0.027
3/32	0.094	0.096	0.090	0.176	0.163	0.040
1/8	0.125	0.127	0.121	0.235	0.217	0.053
5/32	0.156	0.158	0.152	0.293	0.272	0.066
3/16	0.188	0.191	0.182	0.351	0.326	0.079
7/32	0.219	0.222	0.213	0.413	0.384	0.094
1/4	0.250	0.253	0.244	0.469	0.437	0.106
9/32	0.281	0.285	0.273	0.528	0.491	0.119
5/16	0.312	0.316	0.304	0.588	0.547	0.133
11/32	0.344	0.348	0.336	0.646	0.602	0.146
3/8	0.375	0.380	0.365	0.704	0.656	0.159
13/32	0.406	0.411	0.396	0.763	0.710	0.172
7/16	0.438	0.443	0.428	0.823	0.765	0.186

TABLE C.5.3 Dimensions of Button Head Rivets

Nominal Size[1] or Basic Shank Diameter		*E* Shank Diameter		*A* Head Diameter		*H* Head Height		*R* Head Radius
		Max	Min	Max	Min	Max	Min	Approx
1/16	0.062	0.064	0.059	0.122	0.102	0.052	0.042	0.055
3/32	0.094	0.096	0.090	0.182	0.162	0.077	0.065	0.084
1/8	0.125	0.127	0.121	0.235	0.215	0.100	0.088	0.111
5/32	0.156	0.158	0.152	0.290	0.268	0.124	0.110	0.138
3/16	0.188	0.191	0.182	0.348	0.322	0.147	0.133	0.166
7/32	0.219	0.222	0.213	0.405	0.379	0.172	0.158	0.195
1/4	0.250	0.253	0.244	0.460	0.430	0.196	0.180	0.221
9/32	0.281	0.285	0.273	0.518	0.484	0.220	0.202	0.249
5/16	0.312	0.316	0.304	0.572	0.538	0.243	0.225	0.276
11/32	0.344	0.348	0.336	0.630	0.592	0.267	0.247	0.304
3/8	0.375	0.380	0.365	0.684	0.646	0.291	0.271	0.332
13/32	0.406	0.411	0.396	0.743	0.699	0.316	0.294	0.358
7/16	0.438	0.443	0.428	0.798	0.754	0.339	0.317	0.387

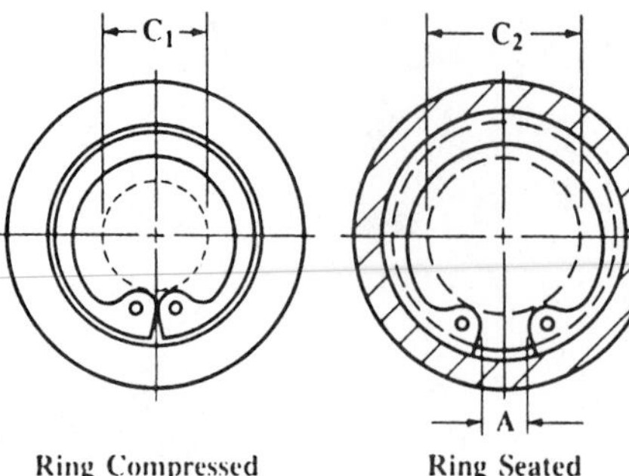

Ring Compressed in Bore

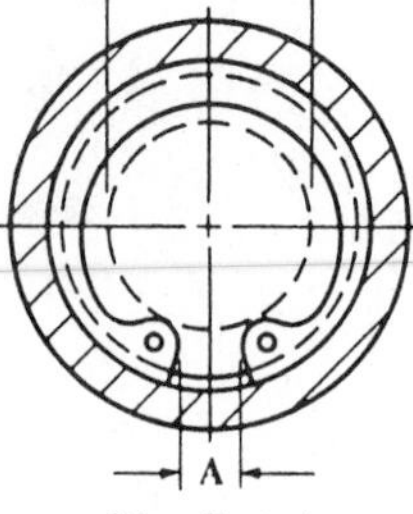

Ring Seated in Groove

Max. Allowable Radius of Retained Part

Chmax.

Max. Allowable Chamfer of Retained Part

Ring Series and Size No.	*Clearance Diam.* Ring in Bore	Ring in Groove	*Gaging Diameter**	*Allowable Thrust Loads Sharp Corner Abutment*		*Maximum Allowable Corner Radii and Chamfers*	
$_3$BMI	C_1	C_2	*A* min	P_r†	P_g‡	*R* max	*Ch* max
No.	mm	mm	mm	kN	kN	mm	mm
−8	4.4	4.8	1.40	2.4	1.0	0.4	0.3
−9	4.6	5.0	1.50	4.4	1.2	0.5	0.35
−10	5.5	6.0	1.85	4.9	1.5	0.5	0.35
−11	5.7	6.3	1.95	5.4	2.0	0.6	0.4
−12	6.7	7.3	2.25	5.8	2.4	0.6	0.4
−13	6.8	7.5	2.35	8.9	2.6	0.7	0.5
−14	6.9	7.7	2.65	9.7	3.2	0.7	0.5
−15	7.9	8.7	2.80	10.4	3.7	0.7	0.5
−16	8.8	9.7	2.80	11.0	4.2	0.7	0.5
−17	9.8	10.8	3.35	11.7	4.9	0.75	0.6
−18	10.3	11.3	3.40	12.3	5.5	0.75	0.6
−19	11.4	12.5	3.40	13.1	6.0	0.8	0.65
−20	11.6	12.7	3.8	13.7	6.6	0.9	0.7
−21	12.6	13.8	4.2	14.5	7.3	0.9	0.7
−22	13.5	14.8	4.3	22.5	8.3	0.9	0.7
−23	14.5	15.9	4.9	23.5	8.9	1.0	0.8
−24	15.5	16.9	5.2	24.8	9.7	1.0	0.8
−25	16.5	18.1	6.0	25.7	11.6	1.0	0.8
−26	17.5	19.2	5.7	26.8	12.7	1.2	1.0
−27	17.4	19.2	5.9	33	14.0	1.2	1.0
−28	18.2	20.0	6.0	34	14.6	1.2	1.0
−30	20.0	21.9	6.0	37	16.5	1.2	1.0
−32	22.0	23.9	7.3	39	17.6	1.2	1.0
−34	24.0	26.1	7.6	42	20.6	1.2	1.0
−35	25.0	27.2	8.0	43	22.3	1.2	1.0
−36	26.0	28.3	8.3	44	23.9	1.2	1.0
−37	27.0	29.3	8.4	45	24.6	1.2	1.0
−38	28.0	30.4	8.6	46	26.4	1.2	1.0
−40	29.2	31.6	9.7	62	27.7	1.7	1.3
−42	29.7	32.2	9.0	65	30.2	1.7	1.3
−45	32.3	34.9	9.6	69	33.8	1.7	1.3
−46	33.3	36.0	9.7	71	36	1.7	1.3
−47	34.3	37.1	10.0	72	38	1.7	1.3
−48	35.0	37.9	10.5	74	40	1.7	1.3
−50	36.9	40.0	12.1	77	45	1.7	1.3
−52	38.6	41.9	11.7	99	50	2.0	1.6
−55	40.8	44.2	11.9	105	54	2.0	1.6
−57	42.2	45.7	12.5	109	58	2.0	1.6

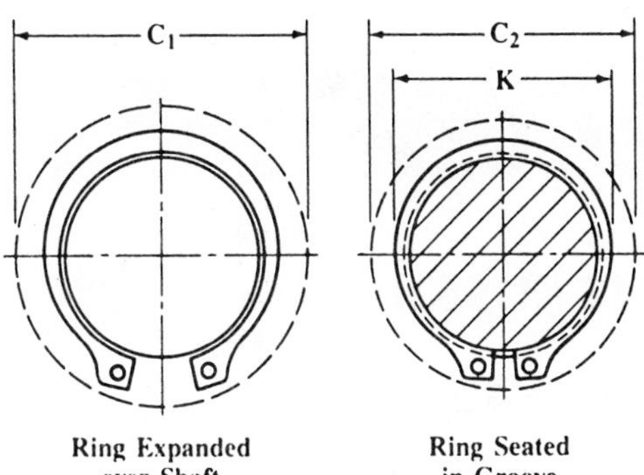

Ring Expanded over Shaft

Ring Seated in Groove

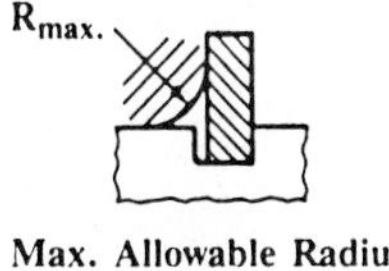

Max. Allowable Radius of Retained Part

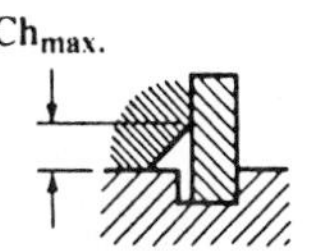

Max. Allowable Chamfer of Retained Pa:

Ring Series and Size No.	*Clearance Diam.* Ring Over Shaft	Ring in Groove	*Gaging Diameter**	*Allowable Thrust Loads Sharp Corner Abutment*		*Maximum Allowable Corner Radii and Chamfers*		*Allowable Assembly Speed§*
3^3 H_3 OAMI	C_1	C_2	*K* max	P_r†	P_g‡	*R* max	*Ch* max	. . .
No.	mm	mm	mm	kN	kN	mm	mm	rpm
−4*	7.0	6.8	4.90	0.6	0.2	0.35	0.25	70 000
−5*	8.2	7.9	5.85	1.1	0.3	0.35	0.25	70 000
−6*	9.1	8.8	6.95	1.4	0.4	0.35	0.25	70 000
−7	12.3	11.8	8.05	2.6	0.7	0.45	0.3	60 000
−8	13.6	13.0	9.15	3.1	1.0	0.5	0.35	55 000
−9	14.5	13.8	10.35	3.5	1.2	0.6	0.35	48 000
−10	15.5	14.7	11.50	3.9	1.5	0.7	0.4	42 000
−11	16.4	15.6	12.60	4.3	1.8	0.75	0.45	38 000
−12	17.4	16.6	13.80	4.7	2.0	0.8	0.45	34 000
−13	19.7	18.8	15.05	7.5	2.2	0.8	0.5	31 000
−14	20.7	19.7	15.60	8.1	2.6	0.9	0.5	28 000
−15	21.7	20.6	17.20	8.7	3.2	1.0	0.6	27 000
−16	22.7	21.6	18.35	9.3	3.5	1.1	0.6	25 000
−17	23.7	22.6	19.35	9.9	4.0	1.1	0.6	24 000

APPENDIX C.6

Pins

Clevis Pins ANSI/ASME B18.8.1 1972 (R1983)

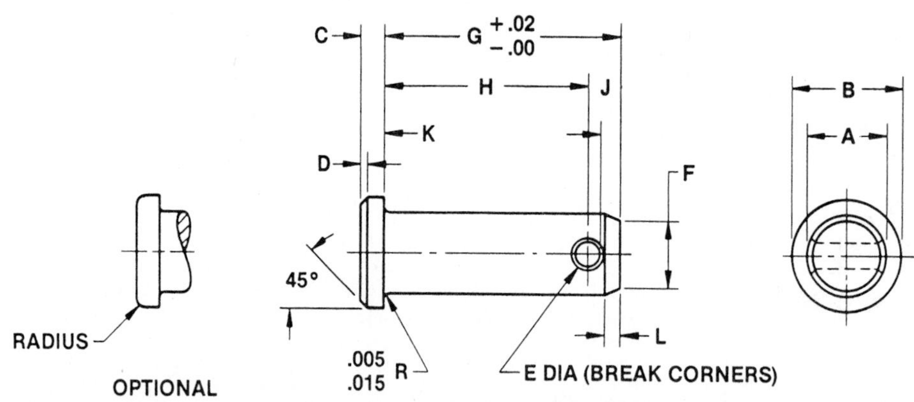

TABLE C.6.1 Dimensions of Clevis Pins

Nominal Size or Basic Pin Diameter	*A* Shank Diameter		*B* Head Diameter		*C* Head Height		*D* Head Chamfer	*E* Hole Diameter		*F* Point Diameter		*G* Pin Length	*H* Head to Center of Hole		*J* End to Center Ref	*K* Head to Edge of Hole Ref		*L* Point Length		*Recommended Cotter Pin Nominal Size*
	Max	Min	Max	Min	Max	Min	±0.01	Max	Min	Max	Min	Basic	Max	Min	Basic	Max	Min	Max	Min	
3/16 0.188	0.186	0.181	0.32	0.30	0.07	0.05	0.02	0.088	0.073	0.15	0.14	0.58	0.504	0.484	0.09	0.548	0.520	0.055	0.035	1/16 0.062
1/4 0.250	0.248	0.243	0.38	0.36	0.10	0.08	0.03	0.088	0.073	0.21	0.20	0.77	0.692	0.672	0.09	0.736	0.708	0.055	0.035	1/16 0.062
5/16 0.312	0.311	0.306	0.44	0.42	0.10	0.08	0.03	0.119	0.104	0.26	0.25	0.94	0.832	0.812	0.12	0.892	0.864	0.071	0.049	3/32 0.093
3/8 0.375	0.373	0.368	0.51	0.49	0.13	0.11	0.03	0.119	0.104	0.33	0.32	1.06	0.958	0.938	0.12	1.018	0.990	0.071	0.049	3/32 0.093
7/16 0.438	0.436	0.431	0.57	0.55	0.16	0.14	0.04	0.119	0.104	0.39	0.38	1.19	1.082	1.062	0.12	1.142	1.114	0.071	0.049	3/32 0.093
1/2 0.500	0.496	0.491	0.63	0.61	0.16	0.14	0.04	0.151	0.136	0.44	0.43	1.36	1.223	1.203	0.15	1.298	1.271	0.089	0.063	1/8 0.125
5/8 0.625	0.621	0.616	0.82	0.80	0.21	0.19	0.06	0.151	0.136	0.56	0.55	1.61	1.473	1.453	0.15	1.548	1.521	0.089	0.063	1/8 0.125
3/4 0.750	0.746	0.741	0.94	0.92	0.26	0.24	0.07	0.182	0.167	0.68	0.67	1.91	1.739	1.719	0.18	1.830	1.802	0.110	0.076	5/32 0.156
7/8 0.875	0.871	0.866	1.04	1.02	0.32	0.30	0.09	0.182	0.167	0.80	0.79	2.16	1.989	1.969	0.18	2.080	2.052	0.110	0.076	5/32 0.156
1 1.000	0.996	0.991	1.19	1.17	0.35	0.33	0.10	0.182	0.167	0.93	0.92	2.41	2.239	2.219	0.18	2.330	2.302	0.110	0.076	5/32 0.156

Cotter Pins ANSI/ASME B18.8.1 1972 (R1983)

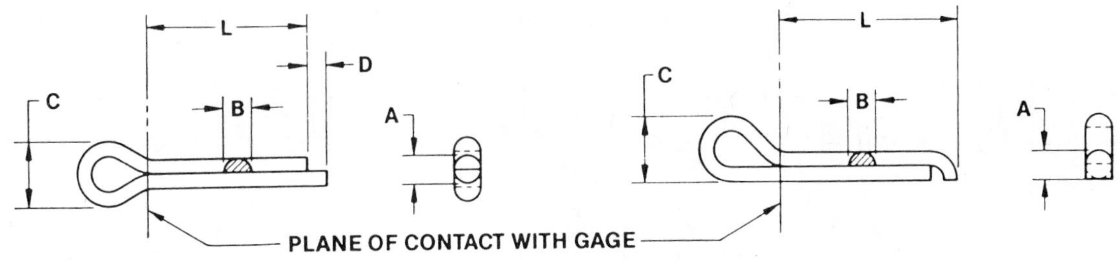

TABLE C.6.2 Dimensions of Cotter Pins

Nominal Size[1] *or Basic Pin Diameter*		A Total Shank Diameter		B Wire Width		C Head Diameter	D Extended Prong Length	*Recommended Hole Size*
		Max	Min	Max	Min	Min	Min	
1/32	0.031	0.032	0.028	0.032	0.022	0.06	0.01	0.047
3/64	0.047	0.048	0.044	0.048	0.035	0.09	0.02	0.062
1/16	0.062	0.060	0.056	0.060	0.044	0.12	0.03	0.078
5/64	0.078	0.076	0.072	0.076	0.057	0.16	0.04	0.094
3/32	0.094	0.090	0.086	0.090	0.069	0.19	0.04	0.109
7/64	0.109	0.104	0.100	0.104	0.080	0.22	0.05	0.125
1/8	0.125	0.120	0.116	0.120	0.093	0.25	0.06	0.141
9/64	0.141	0.134	0.130	0.134	0.104	0.28	0.06	0.156
5/32	0.156	0.150	0.146	0.150	0.116	0.31	0.07	0.172
3/16	0.188	0.176	0.172	0.176	0.137	0.38	0.09	0.203
7/32	0.219	0.207	0.202	0.207	0.161	0.44	0.10	0.234
1/4	0.250	0.225	0.220	0.225	0.176	0.50	0.11	0.266
5/16	0.312	0.280	0.275	0.280	0.220	0.62	0.14	0.312
3/8	0.375	0.335	0.329	0.335	0.263	0.75	0.16	0.375
7/16	0.438	0.406	0.400	0.406	0.320	0.88	0.20	0.438
1/2	0.500	0.473	0.467	0.473	0.373	1.00	0.23	0.500
5/8	0.625	0.598	0.590	0.598	0.472	1.25	0.30	0.625
3/4	0.750	0.723	0.715	0.723	0.572	1.50	0.36	0.750

Spring Pins ANSI/ASME B18.8.2 1978

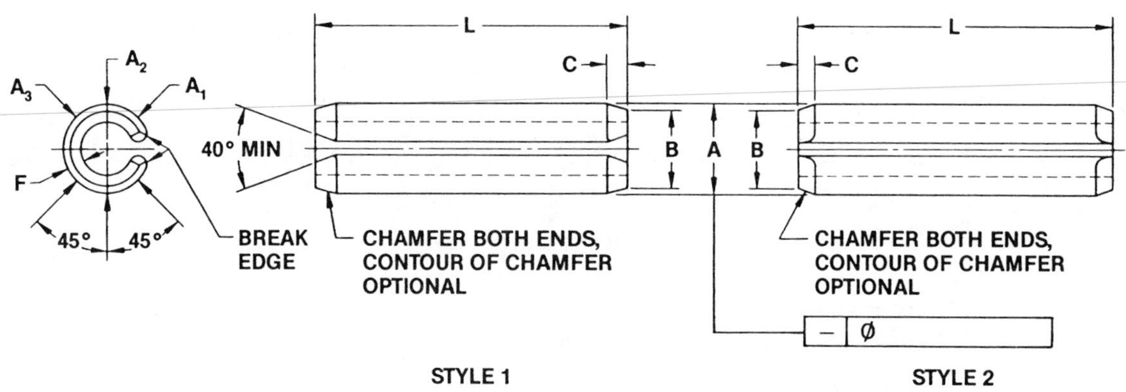

TABLE C.6.3 Dimensions of Slotted Type Spring Pins

Nominal Size or Basic Pin Diameter		*A* Pin Diameter		*B* Chamfer Diameter	*C* Chamfer Length		*F* Stock Thickness	Recommended Hole Size		Double Shear Load, Min, lb Material		
		Max	Min	Max	Max	Min	Basic	Max	Min	AISI 1070-1095 and AISI 420	AISI 302	Beryllium Copper
1/16	0.062	0.069	0.066	0.059	0.028	0.007	0.012	0.065	0.062	425	350	270
5/64	0.078	0.086	0.083	0.075	0.032	0.008	0.018	0.081	0.078	650	550	400
3/32	0.094	0.103	0.099	0.091	0.038	0.008	0.022	0.097	0.094	1,000	800	660
1/8	0.125	0.135	0.131	0.122	0.044	0.008	0.028	0.129	0.125	2,100	1,500	1,200
9/64	0.141	0.149	0.145	0.137	0.044	0.008	0.028	0.144	0.140	2,200	1,600	1,400
5/32	0.156	0.167	0.162	0.151	0.048	0.010	0.032	0.160	0.156	3,000	2,000	1,800
3/16	0.188	0.199	0.194	0.182	0.055	0.011	0.040	0.192	0.187	4,400	2,800	2,600
7/32	0.219	0.232	0.226	0.214	0.065	0.011	0.048	0.224	0.219	5,700	3,550	3,700
1/4	0.250	0.264	0.258	0.245	0.065	0.012	0.048	0.256	0.250	7,700	4,600	4,500
5/16	0.312	0.328	0.321	0.306	0.080	0.014	0.062	0.318	0.312	11,500	7,095	6,800
3/8	0.375	0.392	0.385	0.368	0.095	0.016	0.077	0.382	0.375	17,600	10,000	10,100
7/16	0.438	0.456	0.448	0.430	0.095	0.017	0.077	0.445	0.437	20,000	12,000	12,200
1/2	0.500	0.521	0.513	0.485	0.110	0.025	0.094	0.510	0.500	25,800	15,500	16,800
5/8	0.625	0.650	0.640	0.608	0.125	0.030	0.125	0.636	0.625	46,000[4]	18,800	...
3/4	0.750	0.780	0.769	0.730	0.150	0.030	0.150	0.764	0.750	66,000[4]	23,200	...

Metric Spring Pins IFI 512-S 1982

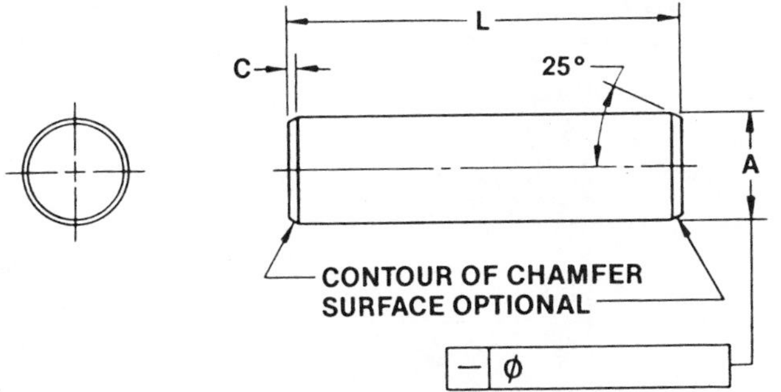

TABLE C.6.4 Slotted Spring Pin Dimensions

Nom Pin Size	*D* Dia		*B* Chamfer Dia	*C* Chamfer Length		*S* Stock Thickness	Recommended Hole Size	
	Max	Min	Max	Max	Min	Nom	Max	Min
1.5	1.68	1.60	1.4	0.7	0.15	0.3	1.60	1.50
2	2.20	2.12	1.9	0.8	0.2	0.4	2.10	2.00
2.5	2.72	2.63	2.4	0.9	0.2	0.5	2.60	2.50
3	3.25	3.15	2.9	1.0	0.2	0.6	3.10	3.00
4	4.28	4.15	3.9	1.2	0.3	0.8	4.12	4.00
5	5.33	5.17	4.8	1.4	0.3	1.0	5.12	5.00
6	6.36	6.20	5.8	1.6	0.4	1.2	6.12	6.00
8	8.40	8.22	7.8	2.0	0.4	1.6	8.15	8.00
10	10.43	10.25	9.7	2.4	0.5	2.0	10.15	10.00
12	12.48	12.28	11.7	2.8	0.6	2.5	12.18	12.00

Dowel Pins ANSI/ASME B18.8.2 1978

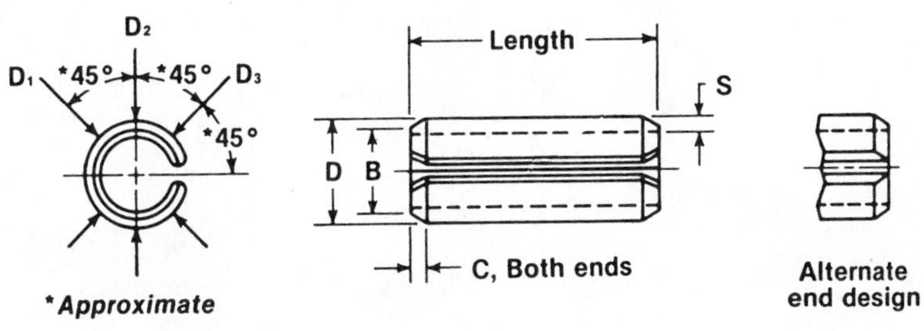

TABLE C.6.5 Dimensions of Unhardened Ground Dowel Pins

Nominal Size or Nominal Pin Diameter		*A* Pin Diameter		*C* Chamfer Length		Double Shear Load Min, lb — Material	
		Max	Min	Max	Min	Carbon Steel	Brass
1/16	0.0625	0.0600	0.0595	0.025	0.005	350	220
3/32	0.0938	0.0912	0.0907	0.025	0.005	820	510
*7/64	0.1094	0.1068	0.1063	0.025	0.005	1,130	710
1/8	0.1250	0.1223	0.1218	0.025	0.005	1,490	930
5/32	0.1562	0.1535	0.1530	0.025	0.005	2,350	1,470
3/16	0.1875	0.1847	0.1842	0.025	0.005	3,410	2,130
7/32	0.2188	0.2159	0.2154	0.025	0.005	4,660	2,910
1/4	0.2500	0.2470	0.2465	0.025	0.005	6,120	3,810
5/16	0.3125	0.3094	0.3089	0.040	0.020	9,590	5,990
3/8	0.3750	0.3717	0.3712	0.040	0.020	13,850	8,650
7/16	0.4375	0.4341	0.4336	0.040	0.020	18,900	11,810
1/2	0.5000	0.4964	0.4959	0.040	0.020	24,720	15,450
5/8	0.6250	0.6211	0.6206	0.055	0.035	38,710	24,190
3/4	0.7500	0.7548	0.7453	0.055	0.035	55,840	34,900
7/8	0.8750	0.8705	0.8700	0.070	0.050	76,090	47,550
1	1.0000	0.9952	0.9947	0.070	0.050	99,460	62,160

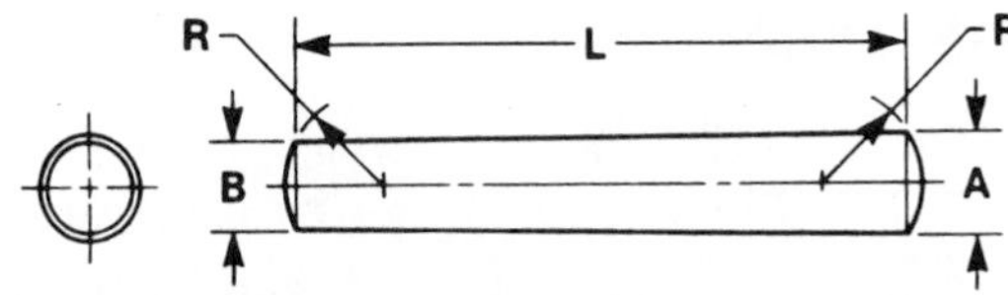

TABLE C.6.6 Dimensions of Taper Pins

Pin Size Number and Basic Pin Diameter		*A* Major Diameter (Large End)				*R* End Crown Radius	
		Commercial Class		Precision Class			
		Max	Min	Max	Min	Max	Min
7/0	0.0625	0.0638	0.0618	0.0635	0.0625	0.072	0.052
6/0	0.0780	0.0793	0.0773	0.0790	0.0780	0.088	0.068
5/0	0.0940	0.0953	0.0933	0.0950	0.0940	0.104	0.084
4/0	0.1090	0.1103	0.1083	0.1100	0.1090	0.119	0.099
3/0	0.1250	0.1263	0.1243	0.1260	0.1250	0.135	0.115
2/0	0.1410	0.1423	0.1403	0.1420	0.1410	0.151	0.131
0	0.1560	0.1573	0.1553	0.1570	0.1560	0.166	0.146
1	0.1720	0.1733	0.1713	0.1730	0.1720	0.182	0.162
2	0.1930	0.1943	0.1923	0.1940	0.1930	0.203	0.183
3	0.2190	0.2203	0.2183	0.2200	0.2190	0.229	0.209
4	0.2500	0.2513	0.2493	0.2510	0.2500	0.260	0.240
5	0.2890	0.2903	0.2883	0.2900	0.2890	0.299	0.279
6	0.3410	0.3423	0.3403	0.3420	0.3410	0.351	0.331
7	0.4090	0.4103	0.4083	0.4100	0.4090	0.419	0.399
8	0.4920	0.4933	0.4913	0.4930	0.4920	0.502	0.482
9	0.5910	0.5923	0.5903	0.5920	0.5910	0.601	0.581
10	0.7060	0.7073	0.7053	0.7070	0.7060	0.716	0.696
11	0.8600	0.8613	0.8593	*	*	0.870	0.850
12	1.0320	1.0333	1.0313	*	*	1.042	1.022
13	1.2410	1.2423	1.2403	*	*	1.251	1.231
14	1.5210	1.5223	1.5203	*	*	1.531	1.511

APPENDIX C.7

Bushings

TABLE C.7.1 Jig Bushings

Range of Hole Sizes in Renewable Bushings	Inside Diameter A			Body Diameter B					Over-all Length C	Radius D	Head Diam. E	Head Thick. F Max	Number
					Unfinished		Finished						
	Nom	Max	Min	Nom	Max	Min	Max	Min					
0.0135 to 0.1562	0.312	0.3129	0.3126	0.500	0.520	0.515	0.5017	0.5014	0.312 0.500 0.750 1.000	0.047	0.625	0.094	HL-32-5 HL-32-8 HL-32-12 HL-32-16
0.1570 to 0.3125	0.500	0.5005	0.5002	0.750	0.770	0.765	0.7518	0.7515	0.312 0.500 0.750 1.000 1.375 1.750	0.062	0.875	0.094	HL-48-5 HL-48-8 HL-48-12 HL-48-16 HL-48-22 HL-48-28
0.3160 to 0.5000	0.750	0.7506	0.7503	1.000	1.020	1.015	1.0018	1.0015	0.500 0.750 1.000 1.375 1.750 2.125	0.062	1.125	0.125	HL-64-8 HL-64-12 HL-64-16 HL-64-22 HL-64-28 HL-64-34
0.5156 to 0.7500	1.000	1.0007	1.0004	1.375	1.395	1.390	1.3772	1.3768	0.500 0.750 1.000 1.375 1.750 2.125 2.500	0.094	1.500	0.125	HL-88-8 HL-88-12 HL-88-16 HL-88-22 HL-88-28 HL-88-34 HL-88-40
0.7656 to 1.0000	1.375	1.3760	1.3756	1.750	1.770	1.765	1.7523	1.7519	0.750 1.000 1.375 1.750 2.125 2.500	0.094	1.875	0.188	HL-112-12 HL-112-16 HL-112-22 HL-112-28 HL-112-34 HL-112-40
1.0156 to 1.3750	1.750	1.7512	1.7508	2.250	2.270	2.265	2.2525	2.2521	1.000 1.375 1.750 2.125 2.500 3.000	0.094	2.375	0.188	HL-144-16 HL-144-22 HL-144-28 HL-144-34 HL-144-40 HL-144-48
1.3906 to 1.7500	2.250	2.2515	2.2510	2.750	2.770	2.765	2.7526	2.7522	1.000 1.375 1.750 2.125 2.500 3.000	0.125	2.875	0.188	HL-176-16 HL-176-22 HL-176-28 HL-176-34 HL-176-40 HL-176-48

All dimensions are in inches.

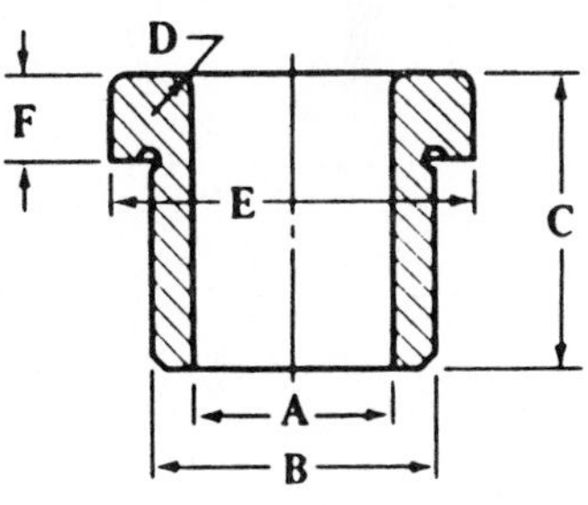

TABLE C.7.2 Headless Type Press Fit Wearing Bushings Type-P

Range of Hole Sizes A	Body Diameter B					*Body Length C*	*Radius D*	*Number*
	Nom	Unfinished		Finished				
		Max	Min	Max	Min			
0.0135 Up To And Including 0.0625	0.156	0.166	0.161	0.1578	0.1575	0.250 0.312 0.375 0.500	0.016	P-10-4 P-10-5 P-10-6 P-10-8
0.0630 To 0.0995	0.203	0.213	0.208	0.2046	0.2043	0.250 0.312 0.375 0.500 0.750	0.016	P-13-4 P-13-5 P-13-6 P-13-8 P-13-12
0.1015 To 0.1405	0.250	0.260	0.255	0.2516	0.2513	0.250 0.312 0.375 0.500 0.750	0.016	P-16-4 P-16-5 P-16-6 P-16-8 P-16-12
0.1406 To 0.1875	0.312	0.327	0.322	0.3141	0.3138	0.250 0.312 0.375 0.500 0.750 1.000	0.031	P-20-4 P-20-5 P-20-6 P-20-8 P-20-12 P-20-16
0.1890 To 0.2500	0.406	0.421	0.416	0.4078	0.4075	0.250 0.312 0.375 0.500 0.750 1.000 1.375 1.750	0.031	P-26-4 P-26-5 P-26-6 P-26-8 P-26-12 P-26-16 P-26-22 P-26-28
0.2570 To 0.3125	0.500	0.520	0.515	0.5017	0.5014	0.312 0.375 0.500 0.750 1.000 1.375 1.750	0.047	P-32-5 P-32-6 P-32-8 P-32-12 P-32-16 P-32-22 P-32-28

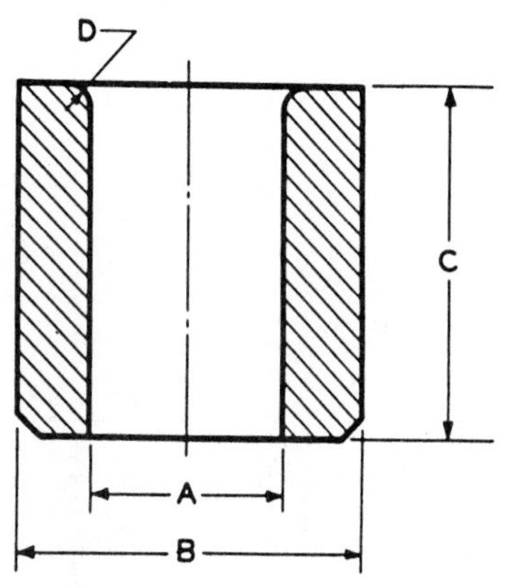

APPENDIX C.8

Woodruff Keys

Keys and Keyseats

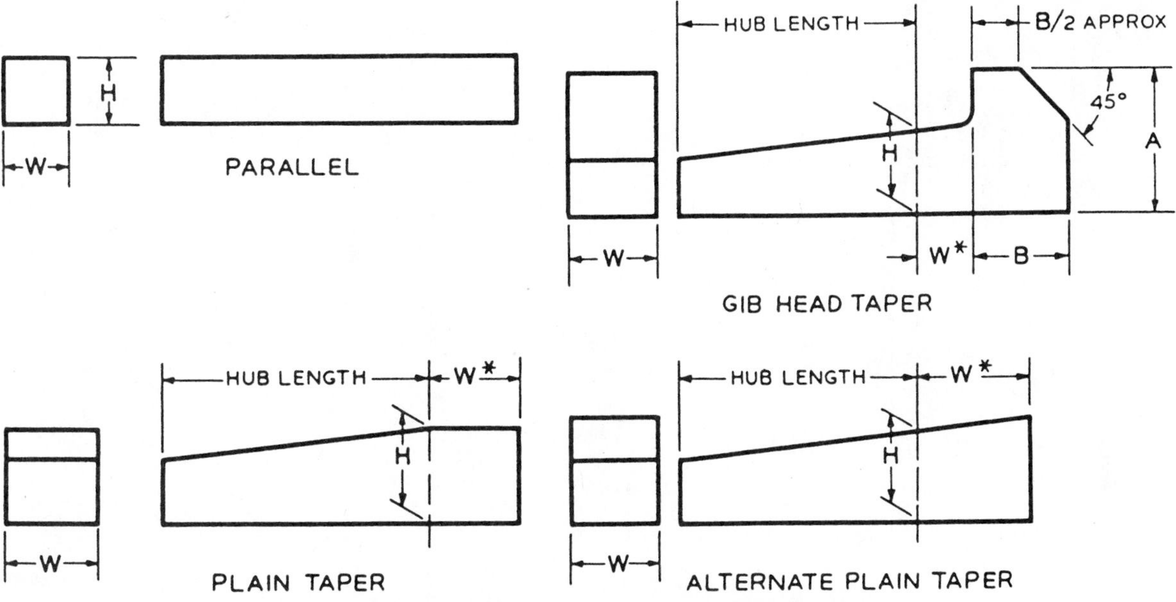

Plain and Gib Head Taper Keys Have a 1/8" Taper in 12"

TABLE C.8.1 Key Dimensions and Tolerances

KEY			NOMINAL KEY SIZE Width, W		TOLERANCE	
			Over	To (Incl)	Width, W	Height, H
Parallel	Square	Bar Stock	—	3/4	+0.000 −0.002	+0.000 −0.002
			3/4	1-1/2	+0.000 −0.003	+0.000 −0.003
			1-1/2	2-1/2	+0.000 −0.004	+0.000 −0.004
			2-1/2	3-1/2	+0.000 −0.006	+0.000 −0.006
		Keystock	—	1-1/4	+0.001 −0.000	+0.001 −0.000
			1-1/4	3	+0.002 −0.000	+0.002 −0.000
			3	3-1/2	+0.003 −0.000	+0.003 −0.000
	Rectangular	Bar Stock	—	3/4	+0.000 −0.003	+0.000 −0.003
			3/4	1-1/2	+0.000 −0.004	+0.000 −0.004
			1-1/2	3	+0.000 −0.005	+0.000 −0.005
			3	4	+0.000 −0.006	+0.000 −0.006
			4	6	+0.000 −0.008	+0.000 −0.008
			6	7	+0.000 −0.013	+0.000 −0.013
		Keystock	—	1-1/4	+0.001 −0.000	+0.005 −0.005
			1-1/4	3	+0.002 −0.000	+0.005 −0.005
			3	7	+0.003 −0.000	+0.005 −0.005
Taper	Plain or Gib Head Square or Rectangular		—	1-1/4	+0.001 −0.000	+0.005 −0.000
			1-1/4	3	+0.002 −0.000	+0.005 −0.000
			3	7	+0.003 −0.000	+0.005 −0.000

*For locating position of dimension *H*. Tolerance does not apply.
All dimensions given in inches.

TABLE C.8.2 Depth Control Values Table 3 Values for *S* and *T*

Nominal Shaft Diameter	Parallel and Taper		Parallel		Taper	
	Square	Rectangular	Square	Rectangular	Square	Rectangular
	S	S	T	T	T	T
1/2	0.430	0.445	0.560	0.544	0.535	0.519
9/16	0.493	0.509	0.623	0.607	0.598	0.582
5/8	0.517	0.548	0.709	0.678	0.684	0.653
11/16	0.581	0.612	0.773	0.742	0.748	0.717
3/4	0.644	0.676	0.837	0.806	0.812	0.781
13/16	0.708	0.739	0.900	0.869	0.875	0.844
7/8	0.771	0.802	0.964	0.932	0.939	0.907
15/16	0.796	0.827	1.051	1.019	1.026	0.994
1	0.859	0.890	1.114	1.083	1.089	1.058
1-1/16	0.923	0.954	1.178	1.146	1.153	1.121
1-1/8	0.986	1.017	1.241	1.210	1.216	1.185
1-3/16	1.049	1.080	1.304	1.273	1.279	1.248
1-1/4	1.112	1.144	1.367	1.336	1.342	1.311
1-5/16	1.137	1.169	1.455	1.424	1.430	1.399
1-3/8	1.201	1.232	1.518	1.487	1.493	1.462
1-7/16	1.225	1.288	1.605	1.543	1.580	1.518
1-1/2	1.289	1.351	1.669	1.606	1.644	1.581
1-9/16	1.352	1.415	1.732	1.670	1.707	1.645
1-5/8	1.416	1.478	1.796	1.733	1.771	1.708
1-11/16	1.479	1.541	1.859	1.796	1.834	1.771
1-3/4	1.542	1.605	1.922	1.860	1.897	1.835
1-13/16	1.527	1.590	2.032	1.970	2.007	1.945
1-7/8	1.591	1.654	2.096	2.034	2.071	2.009
1-15/16	1.655	1.717	2.160	2.097	2.135	2.072
2	1.718	1.781	2.223	2.161	2.198	2.136
2-1/16	1.782	1.844	2.287	2.224	2.262	2.199
2-1/8	1.845	1.908	2.350	2.288	2.325	2.263
2-3/16	1.909	1.971	2.414	2.351	2.389	2.326
2-1/4	1.972	2.034	2.477	2.414	2.452	2.389
2-5/16	1.957	2.051	2.587	2.493	2.562	2.468
2-3/8	2.021	2.114	2.651	2.557	2.626	2.532
2-7/16	2.084	2.178	2.714	2.621	2.689	2.596
2-1/2	2.148	2.242	2.778	2.684	2.753	2.659
2-9/16	2.211	2.305	2.841	2.748	2.816	2.723
2-5/8	2.275	2.369	2.905	2.811	2.880	2.786
2-11/16	2.338	2.432	2.968	2.874	2.943	2.849
2-3/4	2.402	2.495	3.032	2.938	3.007	2.913
2-13/16	2.387	2.512	3.142	3.017	3.117	2.992
2-7/8	2.450	2.575	3.205	3.080	3.180	3.055
2-15/16	2.514	2.639	3.269	3.144	3.244	3.119
3	2.577	2.702	3.332	3.207	3.307	3.182
3-1/16	2.641	2.766	3.396	3.271	3.371	3.246
3-1/8	2.704	2.829	3.459	3.334	3.434	3.309

All dimensions given in inches.

Woodruff Keys and Keyseats

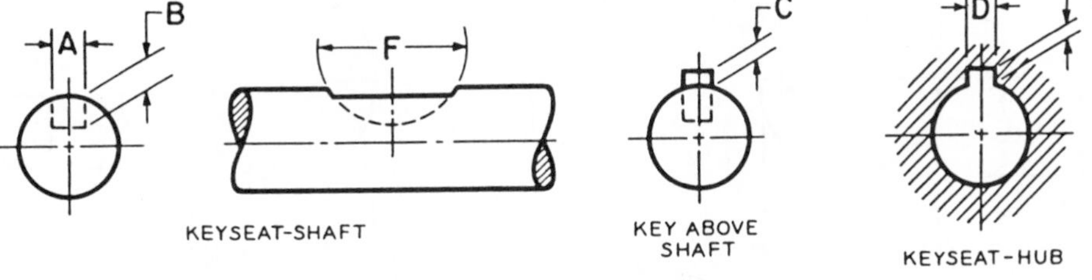

TABLE C.8.3 Keyseat Dimensions

Key Number	Nominal Size Key	Keyseat—Shaft					Key Above Shaft	Keyseat—Hub	
		Width A*		Depth B	Diameter F		Height C	Width D	Depth E
		Min	Max	+0.005 −0.000	Min	Max	+0.005 −0.005	+0.002 −0.000	+0.005 −0.000
202	1/16 × 1/4	0.0615	0.0630	0.0728	0.250	0.268	0.0312	0.0635	0.0372
202.5	1/16 × 5/16	0.0615	0.0630	0.1038	0.312	0.330	0.0312	0.0635	0.0372
302.5	3/32 × 5/16	0.0928	0.0943	0.0882	0.312	0.330	0.0469	0.0948	0.0529
203	1/16 × 3/8	0.0615	0.0630	0.1358	0.375	0.393	0.0312	0.0635	0.0372
303	3/32 × 3/8	0.0928	0.0943	0.1202	0.375	0.393	0.0469	0.0948	0.0529
403	1/8 × 3/8	0.1240	0.1255	0.1045	0.375	0.393	0.0625	0.1260	0.0685
204	1/16 × 1/2	0.0615	0.0630	0.1668	0.500	0.518	0.0312	0.0635	0.0372
304	3/32 × 1/2	0.0928	0.0943	0.1511	0.500	0.518	0.0469	0.0948	0.0529
404	1/8 × 1/2	0.1240	0.1255	0.1355	0.500	0.518	0.0625	0.1260	0.0685
305	3/32 × 5/8	0.0928	0.0943	0.1981	0.625	0.643	0.0469	0.0948	0.0529
405	1/8 × 5/8	0.1240	0.1255	0.1825	0.625	0.643	0.0625	0.1260	0.0685
505	5/32 × 5/8	0.1553	0.1568	0.1669	0.625	0.643	0.0781	0.1573	0.0841
605	3/16 × 5/8	0.1863	0.1880	0.1513	0.625	0.643	0.0937	0.1885	0.0997
406	1/8 × 3/4	0.1240	0.1255	0.2455	0.750	0.768	0.0625	0.1260	0.0685
506	5/32 × 3/4	0.1553	0.1568	0.2299	0.750	0.768	0.0781	0.1573	0.0841
606	3/16 × 3/4	0.1863	0.1880	0.2143	0.750	0.768	0.0937	0.1885	0.0997
806	1/4 × 3/4	0.2487	0.2505	0.1830	0.750	0.768	0.1250	0.2510	0.1310
507	5/32 × 7/8	0.1553	0.1568	0.2919	0.875	0.895	0.0781	0.1573	0.0841
607	3/16 × 7/8	0.1863	0.1880	0.2763	0.875	0.895	0.0937	0.1885	0.0997
707	7/32 × 7/8	0.2175	0.2193	0.2607	0.875	0.895	0.1093	0.2198	0.1153
807	1/4 × 7/8	0.2487	0.2505	0.2450	0.875	0.895	0.1250	0.2510	0.1310
608	3/16 × 1	0.1863	0.1880	0.3393	1.000	1.020	0.0937	0.1885	0.0997
708	7/32 × 1	0.2175	0.2193	0.3237	1.000	1.020	0.1093	0.2198	0.1153
808	1/4 × 1	0.2487	0.2505	0.3080	1.000	1.020	0.1250	0.2510	0.1310
1008	5/16 × 1	0.3111	0.3130	0.2768	1.000	1.020	0.1562	0.3135	0.1622
1208	3/8 × 1	0.3735	0.3755	0.2455	1.000	1.020	0.1875	0.3760	0.1935
609	3/16 × 1 1/8	0.1863	0.1880	0.3853	1.125	1.145	0.0937	0.1885	0.0997
709	7/32 × 1 1/8	0.2175	0.2193	0.3697	1.125	1.145	0.1093	0.2198	0.1153
809	1/4 × 1 1/8	0.2487	0.2505	0.3540	1.125	1.145	0.1250	0.2510	0.1310
1009	5/16 × 1 1/8	0.3111	0.3130	0.3228	1.125	1.145	0.1562	0.3135	0.1622

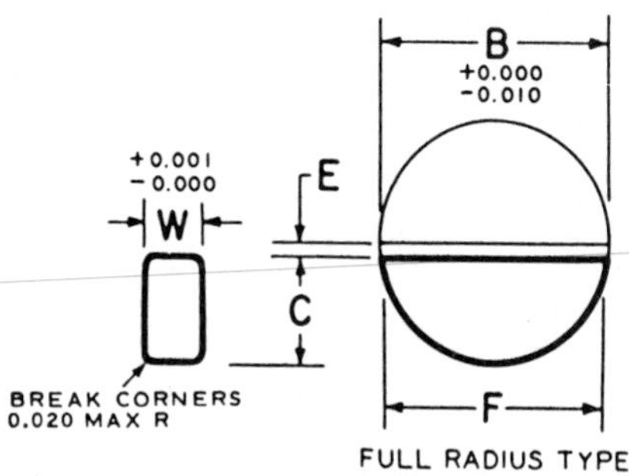

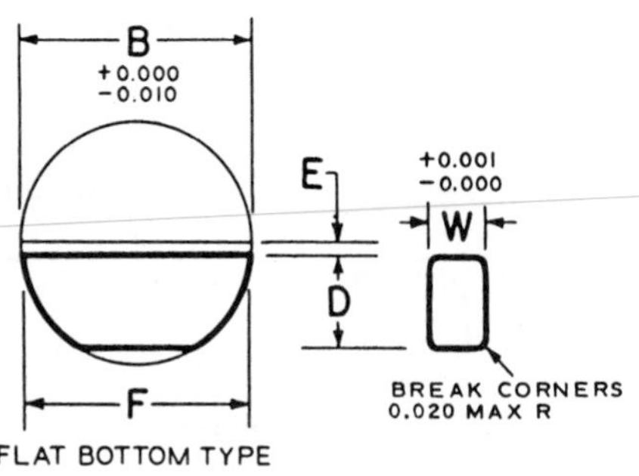

TABLE C.8.4 Woodruff Keys

Key No.	*Nominal Key Size W × B*	*Actual Length F +0.000-0.010*	Height of Key C Max	C Min	D Max	D Min	*Distance Below Center E*
202	1/16 × 1/4	0.248	0.109	0.104	0.109	0.104	1/64
202.5	1/16 × 5/16	0.311	0.140	0.135	0.140	0.135	1/64
302.5	3/32 × 5/16	0.311	0.140	0.135	0.140	0.135	1/64
203	1/16 × 3/8	0.374	0.172	0.167	0.172	0.167	1/64
303	3/32 × 3/8	0.374	0.172	0.167	0.172	0.167	1/64
403	1/8 × 3/8	0.374	0.172	0.167	0.172	0.167	1/64
204	1/16 × 1/2	0.491	0.203	0.198	0.194	0.188	3/64
304	3/32 × 1/2	0.491	0.203	0.198	0.194	0.188	3/64
404	1/8 × 1/2	0.491	0.203	0.198	0.194	0.188	3/64
305	3/32 × 5/8	0.612	0.250	0.245	0.240	0.234	1/16
405	1/8 × 5/8	0.612	0.250	0.245	0.240	0.234	1/16
505	5/32 × 5/8	0.612	0.250	0.245	0.240	0.234	1/16
605	3/16 × 5/8	0.612	0.250	0.245	0.240	0.234	1/16
406	1/8 × 3/4	0.740	0.313	0.308	0.303	0.297	1/16
506	5/32 × 3/4	0.740	0.313	0.308	0.303	0.297	1/16
606	3/16 × 3/4	0.740	0.313	0.308	0.303	0.297	1/16
806	1/4 × 3/4	0.740	0.313	0.308	0.303	0.297	1/16
507	5/32 × 7/8	0.866	0.375	0.370	0.365	0.359	1/16
607	3/16 × 7/8	0.866	0.375	0.370	0.365	0.359	1/16
707	7/32 × 7/8	0.866	0.375	0.370	0.365	0.359	1/16
807	1/4 × 7/8	0.866	0.375	0.370	0.365	0.359	1/16
608	3/16 × 1	0.992	0.438	0.433	0.428	0.422	1/16
708	7/32 × 1	0.992	0.438	0.433	0.428	0.422	1/16
808	1/4 × 1	0.992	0.438	0.433	0.428	0.422	1/16
1008	5/16 × 1	0.992	0.438	0.433	0.428	0.422	1/16
1208	3/8 × 1	0.992	0.438	0.433	0.428	0.422	1/16
609	3/16 × 1 1/8	1.114	0.484	0.479	0.475	0.469	5/64
709	7/32 × 1 1/8	1.114	0.484	0.479	0.475	0.469	5/64
809	1/4 × 1 1/8	1.114	0.484	0.479	0.475	0.469	5/64
1009	5/16 × 1 1/8	1.114	0.484	0.479	0.475	0.469	5/64

APPENDIX C.9

Fits and Tolerances

These fits may be described briefly as follows:

RC 1 *Close sliding fits* are intended for the accurate location of parts which must assemble without perceptible play.

RC 2 *Sliding fits* are intended for accurate location but with greater maximum clearance than class RC1. Parts made to this fit move and turn easily but are not intended to run freely, and in the larger sizes may seize with small temperature changes.

RC 3 *Precision running fits* are about the closest fits which can be expected to run freely, and are intended for precision work at slow speeds and light journal pressures, but are not suitable where appreciable temperature differences are likely to be encountered.

RC 4 *Close running fits* are intended chiefly for running fits on accurate machinery with moderate surface speeds and journal pressures, where accurate location and minimum play is desired.

RC 5, RC 6 *Medium running fits* are intended for higher running speeds, or heavy journal pressures, or both.

RC 7 *Free running fits* are intended for use where accuracy is not essential, or where large temperature variations are likely to be encountered, or under both of these conditions.

RC 8, RC 9 *Loose running fits* are intended for use where wide commercial tolerances may be necessary, together with an allowance, on the external member.

TABLE C.9.1 Running and Sliding Fits

Limits are in thousandths of an inch. Limits for hole and shaft are applied algebraically to the basic size to obtain the limits of size for the parts. Date in bold face are in accordance with ABC agreements. Symbols H5, g5, etc., are Hole and Shaft designations used in ABC System.

Nominal Size Range Inches	*Class RC 1*			*Class RC 2*			*Class RC 3*			*Class RC 4*		
	Limits of Clearance	Standard Limits		Limits of Clearance	Standard Limits		Limits of Clearance	Standard Limits		Limits of Clearance	Standard Limits	
Over To		Hole H5	Shaft g4		Hole H6	Shaft g5		Hole H7	Shaft f6		Hole H8	Shaft f7
0 – 0.12	0.1	+0.2	– 0.1	0.1	+ 0.25	– 0.1	0.3	+ 0.4	– 0.3	0.3	+ 0.6	– 0.3
	0.45	0	– 0.25	0.55	0	– 0.3	0.95	0	– 0.55	1.3	0	– 0.7
0.12 – 0.24	0.15	+0.2	– 0.15	0.15	+ 0.3	– 0.15	0.4	+ 0.5	– 0.4	0.4	+ 0.7	– 0.4
	0.5	0	– 0.3	0.65	0	– 0.35	1.12	0	– 0.7	1.6	0	– 0.9
0.24 – 0.40	0.2	0.25	– 0.2	0.2	+ 0.4	– 0.2	0.5	+ 0.6	– 0.5	0.5	+ 0.9	– 0.5
	0.6	0	– 0.35	0.85	0	– 0.45	1.5	0	– 0.9	2.0	0	– 1.1
0.40 – 0.71	0.25	+0.3	– 0.25	0.25	+ 0.4	– 0.25	0.6	+ 0.7	– 0.6	0.6	+ 1.0	– 0.6
	0.75	0	– 0.45	0.95	0	– 0.55	1.7	0	– 1.0	2.3	0	– 1.3
0.71 – 1.19	0.3	+0.4	– 0.3	0.3	+ 0.5	– 0.3	0.8	+ 0.8	– 0.8	0.8	+ 1.2	– 0.8
	0.95	0	– 0.55	1.2	0	– 0.7	2.1	0	– 1.3	2.8	0	– 1.6
1.19 – 1.97	0.4	+0.4	– 0.4	0.4	+ 0.6	– 0.4	1.0	+ 1.0	– 1.0	1.0	+ 1.6	– 1.0
	1.1	0	– 0.7	1.4	0	– 0.8	2.6	0	– 1.6	3.6	0	– 2.0
1.97 – 3.15	0.4	+0.5	– 0.4	0.4	+ 0.7	– 0.4	1.2	+ 1.2	– 1.2	1.2	+ 1.8	– 1.2
	1.2	0	– 0.7	1.6	0	– 0.9	3.1	0	– 1.9	4.2	0	– 2.4
3.15 – 4.73	0.5	+0.6	– 0.5	0.5	+ 0.9	– 0.5	1.4	+ 1.4	– 1.4	1.4	+ 2.2	– 1.4
	1.5	0	– 0.9	2.0	0	– 1.1	3.7	0	– 2.3	5.0	0	– 2.8
4.73 – 7.09	0.6	+0.7	– 0.6	0.6	+ 1.0	– 0.6	1.6	+ 1.6	– 1.6	1.6	+ 2.5	– 1.6
	1.8	0	– 1.1	2.3	0	– 1.3	4.2	0	– 2.6	5.7	0	– 3.2
7.09 – 9.85	0.6	+0.8	– 0.6	0.6	+ 1.2	– 0.6	2.0	+ 1.8	– 2.0	2.0	+ 2.8	– 2.0
	2.0	0	– 1.2	2.6	0	– 1.4	5.0	0	– 3.2	6.6	0	– 3.8
9.85 – 12.41	0.8	+0.9	– 0.8	0.8	+ 1.2	– 0.8	2.5	+ 2.0	– 2.5	2.5	+ 3.0	– 2.5
	2.3	0	– 1.4	2.9	0	– 1.7	5.7	0	– 3.7	7.5	0	– 4.5
12.41 – 15.75	1.0	+1.0	– 1.0	1.0	+ 1.4	– 1.0	3.0	+	– 3.0	3.0	+ 3.5	– 3.0
	2.7	0	– 1.7	3.4	0	– 2.0	6.6	0	– 4.4	8.7	0	– 5.2
15.75 – 19.69	1.2	+1.0	– 1.2	1.2	+ 1.6	– 1.2	4.0	+ 1.6	– 4.0	4.0	+ 4.0	– 4.0
	3.0	0	– 2.0	3.8	0	– 2.2	8.1	0	– 5.6	10.5	0	– 6.5
19.69 – 30.09	1.6	+1.2	– 1.6	1.6	+ 2.0	– 1.6	5.0	+ 3.0	– 5.0	5.0	+ 5.0	– 5.0
	3.7	0	– 2.5	4.8	0	– 2.8	10.0	0	– 7.0	13.0	0	– 8.0
30.09 – 41.49	2.0	+1.6	– 2.0	2.0	+ 2.5	– 2.0	6.0	+ 4.0	– 6.0	6.0	+ 6.0	– 6.0
	4.6	0	– 3.0	6.1	0	– 3.6	12.5	0	– 8.5	16.0	0	–10.0
41.49 – 56.19	2.5	+2.0	– 2.5	2.5	+ 3.0	– 2.5	8.0	+ 5.0	– 8.0	8.0	+ 8.0	– 8.0
	5.7	0	– 3.7	7.5	0	– 4.5	16.0	0	–11.0	21.0	0	–13.0
56.19 – 76.39	3.0	+2.5	– 3.0	3.0	+ 4.0	– 3.0	10.0	+ 6.0	–10.0	10.0	+10.0	–10.0
	7.1	0	– 4.6	9.5	0	– 5.5	20.0	0	–14.0	26.0	0	–16.0
76.39 – 100.9	4.0	+3.0	– 4.0	4.0	+ 5.0	– 4.0	12.0	+ 8.0	–12.0	12.0	+12.0	–12.0
	9.0	0	– 6.0	12.0	0	– 7.0	25.0	0	–17.0	32.0	0	–20.0
100.9 – 131.9	5.0	+4.0	– 5.0	5.0	+ 6.0	– 5.0	16.0	+10.0	–16.0	16.0	+16.0	–16.0
	11.5	0	– 7.5	15.0	0	– 9.0	32.0	0	–22.0	36.0	0	–26.0
131.9 – 171.9	6.0	+5.0	– 6.0	6.0	+ 8.0	– 6.0	18.0	+ 8.0	–18.0	18.0	+20.0	–18.0
	14.0	0	– 9.0	19.0	0	–11.0	38.0	0	–26.0	50.0	0	–30.0
171.9 – 200	8.0	+6.0	– 8.0	8.0	+10.0	– 8.0	22.0	+16.0	–22.0	22.0	+25.0	–22.0
	18.0	0	–12.0	22.0	0	–12.0	48.0	0	–32.0	63.0	0	–38.0

Continued

TABLE C.9.1 Running and Sliding Fits—*Continued*

Class RC 5			*Class RC 6*			*Class RC 7*			*Class RC 8*			*Class RC 9*			*Nominal Size Range Inches*	
Limits of Clearance	Standard Limits		Limits of Clearance	Standard Limits		Limits of Clearance	Standard Limits		Limits of Clearance	Standards Limits		Limits of Clearance	Standard Limits			
	Hole H8	Shaft e7		Hole H9	Shaft e8		Hole H9	Shaft d8		Hole H10	Shaft c9		Hole H11	Shaft	Over	To
0.6	+ 0.6	− 0.6	0.6	+ 1.0	− 0.6	1.0	+ 1.0	− 1.0	2.5	+ 1.6	− 2.5	4.0	+ 2.5	− 4.0	0 −	0.12
1.6	− 0	− 1.0	2.2	− 0	− 1.2	2.6	0	− 1.6	5.1	0	− 3.5	8.1	0	− 5.6		
0.8	+ 0.7	− 0.8	0.8	+ 1.2	− 0.8	1.2	+ 1.2	− 1.2	2.8	+ 1.8	− 2.8	4.5	+ 3.0	− 4.5	0.12 −	0.24
2.0	− 0	− 1.3	2.7	− 0	− 1.5	3.1	0	− 1.9	5.8	0	− 4.0	9.0	0	− 6.0		
1.0	+ 0.9	− 1.0	1.0	+ 1.4	− 1.0	1.6	+ 1.4	− 1.6	3.0	+ 2.2	− 3.0	5.0	+ 3.5	− 5.0	0.24 −	0.40
2.5	− 0	− 1.6	3.3	− 0	− 1.9	3.9	0	− 2.5	6.6	0	− 4.4	10.7	0	− 7.2		
1.2	+ 1.0	− 1.2	1.2	+ 1.6	− 1.2	2.0	+ 1.6	− 2.0	3.5	+ 2.8	− 3.5	6.0	+ 4.0	− 6.0	0.40 −	0.71
2.9	− 0	− 1.9	3.8	− 0	− 2.2	4.6	0	− 3.0	7.9	0	− 5.1	12.8	− 0	− 8.8		
1.6	+ 1.2	− 1.6	1.6	+ 2.0	− 1.6	2.5	+ 2.0	− 2.5	4.5	+ 3.5	− 4.5	7.0	+ 5.0	− 7.0	0.71 −	1.19
3.6	− 0	− 2.4	4.8	− 0	− 2.8	5.7	0	− 3.7	10.0	0	− 6.5	15.5	0	− 10.5		
2.0	+ 1.6	− 2.0	2.0	+ 2.5	− 2.0	3.0	+ 2.5	− 3.0	5.0	+ 4.0	− 5.0	8.0	+ 6.0	− 8.0	1.19 −	1.97
4.6	− 0	− 3.0	6.1	− 0	− 3.6	7.1	0	− 4.6	11.5	0	− 7.5	18.0	0	− 12.0		
2.5	+ 1.8	− 2.5	2.5	+ 3.0	− 2.5	4.0	+ 3.0	− 4.0	6.0	+ 4.5	− 6.0	9.0	+ 7.0	− 9.0	1.97 −	3.15
5.5	− 0	− 3.7	7.3	− 0	− 4.3	8.8	0	− 5.8	13.5	0	− 9.0	20.5	0	− 13.5		
3.0	+ 2.2	− 3.0	3.0	+ 3.5	− 3.0	5.0	+ 3.5	− 5.0	7.0	+ 5.0	− 7.0	10.0	+ 9.0	− 10.0	3.15 −	4.73
6.6	− 0	− 4.4	8.7	− 0	− 5.2	10.7	0	− 7.2	15.5	0	− 10.5	24.0	0	− 15.0		
3.5	+ 2.5	− 3.5	3.5	+ 4.0	− 3.5	6.0	+ 4.0	− 6.0	8.0	+ 6.0	− 8.0	12.0	+ 10.0	− 12.0	4.73 −	7.09
7.6	− 0	− 5.1	10.0	− 0	− 6.0	12.5	0	− 8.5	18.0	0	− 12.0	28.0	0	− 18.0		
4.0	+ 2.8	− 4.0	4.0	+ 4.5	− 4.0	7.0	+ 4.5	− 7.0	10.0	+ 7.0	− 10.0	15.0	+ 12.0	− 15.0	7.09 −	9.85
8.6	−0	− 5.8	11.3	0	− 6.8	14.3	0	− 9.8	21.5	0	− 14.5	34.0	0	− 22.0		
5.0	+ 3.0	− 5.0	5.0	+ 5.0	− 5.0	8.0	+ 5.0	− 8.0	12.0	+ 8.0	− 12.0	18.0	+ 12.0	− 18.0	9.85 −	12.41
10.0	0	− 7.0	13.0	0	− 8.0	16.0	0	− 11.0	25.0	0	− 17.0	38.0	0	− 26.0		
6.0	+ 3.5	− 6.0	6.0	+ 6.0	− 6.0	10.0	+ 6.0	− 10.0	14.0	+ 9.0	− 14.0	22.0	+ 14.0	− 22.0	12.41 −	15.75
11.7	0	− 8.2	15.5	0	− 9.5	19.5	0	13.5	29.0	0	− 20.0	45.0	0	− 31.0		
8.0	+ 4.0	− 8.0	8.0	+ 6.0	− 8.0	12.0	+ 6.0	− 12.0	16.0	+ 10.0	− 16.0	25.0	+ 16.0	− 25.0	15.75 −	19.69
14.5	0	− 10.5	18.0	0	− 12.0	22.0	0	− 16.0	32.0	0	− 22.0	51.0	0	− 35.0		
10.0	+ 5.0	− 10.0	10.0	+ 8.0	− 10.0	16.0	+ 8.0	− 16.0	20.0	+ 12.0	− 20.0	30.0	+ 20.0	− 30.0	19.69 −	30.09
18.0	0	− 13.0	23.0	0	− 15.0	29.0	0	− 21.0	40.0	0	− 28.0	62.0	0	− 42.0		
12.0	+ 6.0	− 12.0	12.0	+ 10.0	− 12.0	20.0	+ 10.0	− 20.0	25.0	+ 16.0	− 25.0	40.0	+ 25.0	− 40.0	30.09 −	41.49
22.0	0	− 16.0	28.0	0	− 18.0	36.0	0	− 26.0	51.0	0	− 35.0	81.0	0	− 56.0		
16.0	+ 8.0	− 16.0	16.0	+ 12.0	− 16.0	25.0	+ 12.0	− 25.0	30.0	+ 20.0	− 30.0	50.0	+ 30.0	− 50.0	41.49 −	56.19
29.0	0	− 21.0	36.0	0	− 24.0	45.0	0	− 33.0	62.0	0	− 42.0	100	0	− 70.0		
20.0	+ 10.0	− 20.0	20.0	+ 16.0	− 20.0	30.0	+ 16.0	− 30.0	40.0	+ 25.0	− 40.0	60.0	+ 40.0	− 60.0	56.19 −	76.39
36.0	0	− 26.0	46.0	0	− 30.0	56.0	0	− 40.0	81.0	0	− 56.0	125	0	− 85.0		
25.0	+ 12.0	− 25.0	25.0	+ 20.0	− 25.0	40.0	+ 20.0	− 40.0	50.0	+ 30.0	− 50.0	80.0	+ 50.0	− 80.0	76.39 −	100.9
45.0	0	− 33.0	57.0	0	− 37.0	72.0	0	− 52.0	100	0	− 70.0	160	0	− 110		
30.0	+ 16.0	− 30.0	30.0	+ 25.0	− 30.0	50.0	+ 25.0	− 50.0	60.0	+ 40.0	− 60.0	100	+ 60.0	− 100	100.9 −	131.9
56.0	0	− 40.0	71.0	0	− 46.0	91.0	0	− 66.0	125	0	− 85.0	200	0	− 140		
35.0	+ 20.0	− 35.0	35.0	+ 30.0	− 35.0	60.0	+ 30.0	− 60.0	80.0	+ 50.0	− 80.0	130	+ 80.0	− 130	131.9 −	171.9
67.0	0	− 47.0	85.0	0	− 55.0	110.0	0	− 80.0	160	0	− 110	260	0	− 180		
45.0	+ 25.0	− 45.0	45.0	+ 40.0	− 45.0	80.0	+ 40.0	− 80.0	100	+ 60.0	− 100	150	+ 100	− 150	171.9 −	200
86.0	0	− 61.0	110.0	0	− 70.0	145.0	0	− 105.0	200	0	− 140	310	0	− 210		

TABLE C.9.2 Locational Clearance Fits

Limits are in thousandths of an inch. Limits for hole and shaft are applied algrebraically to the basic size to obtain the limits of size for the parts. Data in bold face are in accordance with ABC agreements. Symbols H6, h5, etc., are Hole and Shaft designations used in ABC System.

Nominal Size Range Inches	*Class LC 1*			*Class LC 2*			*Class LC 3*			*Class LC 4*			*Class LC 5*		
Over To	Limits of Clearance	Standard Limits Hole H6	Standard Limits Shaft h5	Limits of Clearance	Standard Limits Hole H7	Standard Limits Shaft h6	Limits of Clearance	Standard Limits Hole H8	Standard Limits Shaft h7	Limits of Clearance	Standard Limits Hole H10	Standard Limits Shaft h9	Limits of Clearance	Standard Limits Hole H7	Standard Limits Shaft g6
0 – 0.12	0 0.45	+ 0.25 − 0	+0 −0.2	0 0.65	+ 0.4 − 0	+ 0 − 0.25	0 1	+ 0.6 − 0	+ 0 − 0.4	0 2.6	+ 1.6 − 0	+ 0 − 1.0	0.1 0.75	+ 0.4 − 0	− 0.1 − 0.35
0.12 – 0.24	0 0.5	+ 0.3 − 0	+0 −0.2	0 0.8	+ 0.5 − 0	+ 0 − 0.3	0 1.2	+ 0.7 − 0	+ 0 − 0.5	0 3.0	+ 1.8 − 0	+ 0 − 1.2	0.15 0.95	+ 0.5 − 0	− 0.15 − 0.45
0.24 – 0.40	0 0.65	+ 0.4 − 0	+0 −0.25	0 1.0	+ 0.6 − 0	+ 0 − 0.4	0 1.5	+ 0.9 − 0	+ 0 − 0.6	0 3.6	+ 2.2 − 0	+ 0 − 1.4	0.2 1.2	+ 0.6 − 0	− 0.2 − 0.6
0.40 – 0.71	0 0.7	+ 0.4 − 0	+0 −0.3	0 1.1	+ 0.7 − 0	+ 0 − 0.4	0 1.7	+ 1.0 − 0	+ 0 − 0.7	0 4.4	+ 2.8 − 0	+ 0 − 1.6	0.25 1.35	+ 0.7 − 0	− 0.25 − 0.65
0.71 – 1.19	0 0.9	+ 0.5 − 0	+0 −0.4	0 1.3	+ 0.8 − 0	+ 0 − 0.5	0 2	+ 1.2 − 0	+ 0 − 0.8	0 5.5	+ 3.5 − 0	+ 0 − 2.0	0.3 1.6	+ 0.8 − 0	− 0.3 − 0.8
1.19 – 1.97	0 1.0	+ 0.6 − 0	+0 −0.4	0 1.6	+ 1.0 − 0	+ 0 − 0.6	0 2.6	+ 1.6 − 0	+ 0 − 1	0 6.5	+ 4.0 − 0	+ 0 − 2.5	0.4 2.0	+ 1.0 − 0	− 0.4 − 1.0
1.97 – 3.15	0 1.2	+ 0.7 − 0	+0 −0.5	0 1.9	+ 1.2 − 0	+ 0 − 0.7	0 3	+ 1.8 − 0	+ 0 − 1.2	0 7.5	+ 4.5 − 0	+ 0 − 3	0.4 2.3	+ 1.2 − 0	− 0.4 − 1.1
3.15 – 4.73	0 1.5	+ 0.9 − 0	+0 −0.6	0 2.3	+ 1.4 − 0	+ 0 − 0.9	0 3.6	+ 2.2 − 0	+ 0 − 1.4	0 8.5	+ 5.0 − 0	+ 0 − 3.5	0.5 2.8	+ 1.4 − 0	− 0.5 − 1.4
4.73 – 7.09	0 1.7	+ 1.0 − 0	+0 −0.7	0 2.6	+ 1.6 − 0	+ 0 − 1.0	0 4.1	+ 2.5 − 0	+ 0 − 1.6	0 10	+ 6.0 − 0	+ 0 − 4	0.6 3.2	+ 1.6 − 0	− 0.6 − 1.6
7.09 – 9.85	0 2.0	+ 1.2 − 0	+0 −0.8	0 3.0	+ 1.8 − 0	+ 0 − 1.2	0 4.6	+ 2.8 − 0	+ 0 − 1.8	0 11.5	+ 7.0 − 0	+ 0 − 4.5	0.6 3.6	+ 1.8 − 0	− 0.6 − 1.8
9.85 – 12.41	0 2.1	+ 1.2 − 0	+0 −0.9	0 3.2	+ 2.0 − 0	+ 0 − 1.2	0 5	+ 3.0 − 0	+ 0 − 2.0	0 13	+ 8.0 − 0	+ 0 − 5	0.7 3.9	+ 2.0 − 0	− 0.7 − 1.9
12.41 – 15.75	0 2.4	+ 1.4 − 0	+0 −1.0	0 3.6	+ 2.2 − 0	+ 0 − 1.4	0 5.7	+ 3.5 − 0	+ 0 − 2.2	0 15	+ 9.0 − 0	+ 0 − 6	0.7 4.3	+ 2.2 − 0	− 0.7 − 2.1
15.75 – 19.69	0 2.6	+ 1.6 − 0	+0 −1.0	0 4.1	+ 2.5 − 0	+ 0 − 1.6	0 6.5	+ 4 − 0	+ 0 − 2.5	0 16	+10.0 − 0	+ 0 − 6	0.8 4.9	+ 2.5 − 0	− 0.8 − 2.4
19.69 – 30.09	0 3.2	+ 2.0 − 0	+0 −1.2	0 5.0	+ 3 − 0	+ 0 − 2	0 8	+ 5 − 0	+ 0 − 3	0 20	+12.0 − 0	+ 0 − 8	0.9 5.9	+ 3.0 − 0	− 0.9 − 2.9
30.09 – 41.49	0 4.1	+ 2.5 − 0	+0 −1.6	0 6.5	+ 4 − 0	+ 0 − 2.5	0 10	+ 6 − 0	+ 0 − 4	0 26	+16.0 − 0	+ 0 −10	1.0 7.5	+ 4.0 − 0	− 1.0 − 3.5
41.49 – 56.19	0 5.0	+ 3.0 − 0	+0 −2.0	0 8.0	+ 5 − 0	+ 0 − 3	0 13	+ 8 − 0	+ 0 − 5	0 32	+20.0 − 0	+ 0 −12	1.2 9.2	+ 5.0 − 0	− 1.2 − 4.2
56.19 – 76.39	0 6.5	+ 4.0 − 0	+0 −2.5	0 10	+ 6 − 0	+ 0 − 4	0 16	+10 − 0	+ 0 − 6	0 41	+25.0 − 0	+ 0 −16	1.2 11.2	+ 6.0 − 0	− 1.2 − 5.2
76.39 – 100.9	0 8.0	+ 5.0 − 0	+0 −3.0	0 13	+ 8 − 0	+ 0 − 5	0 20	+12 − 0	+ 0 − 8	0 50	+30.0 − 0	+ 0 −20	1.4 14.4	+ 8.0 − 0	− 1.4 − 6.4
100.9 – 131.9	0 10.0	+ 6.0 − 0	+0 −4.0	0 16	+10 − 0	+ 0 − 6	0 26	+16 − 0	+ 0 −10	0 65	+40.0 − 0	+ 0 −25	1.6 17.6	+10.0 − 0	− 1.6 − 7.6
131.9 – 171.9	0 13.0	+ 8.0 − 0	+0 −5.0	0 20	+12 − 0	+ 0 − 8	0 32	+20 − 0	+ 0 −12	0 8	+50.0 − 0	+ 0 −30	1.8 21.8	+12.0 − 0	− 1.8 − 9.8
171.9 – 200	0 16.0	+10.0 − 0	+0 −6.0	0 26	+16 − 0	+ 0 −10	0 41	+25 − 0	+ 0 −16	0 100	+60.0 − 0	+ 0 −40	1.8 27.8	+16.0 − 0	− 1.8 −11.8

Continued

TABLE C.9.2 Locational Clearance Fits—*Continued*

Class LC 6			*Class LC 7*			*Class LC 8*			*Class LC 9*			*Class LC 10*			*Class LC 11*			*Nominal Size Range Inches*	
Limits of Clearance	Standard Limits Hole H9	Standard Limits Shaft f8	Limits of Clearance	Standard Limits Hole H10	Standard Limits Shaft e9	Limits of Clearance	Standard Limits Hole H10	Standard Limits Shaft d9	Limits of Clearance	Standard Limits Hole H11	Standard Limits Shaft c10	Limits of Clearance	Standard Limits Hole H12	Standard Limits Shaft	Limits of Clearance	Standard Limits Hole H13	Standard Limits Shaft	Over	To
0.3 1.9	+1.0 0	−0.3 −0.9	0.6 3.2	+1.6 0	−0.6 −1.6	1.0 3.6	+1.6 −0	−1.0 −2.0	2.5 6.6	+2.5 −0	−2.5 −4.1	4 12	+4 −0	−4 −8	5 17	+6 −0	−5 −11	0	0.12
0.4 2.3	+1.2 0	−0.4 −1.1	0.8 3.8	+1.8 0	−0.8 −2.0	1.2 4.2	+1.8 −0	−1.2 −2.4	2.8 7.6	+3.0 −0	−2.8 −4.6	4.5 14.5	+5 −0	−4.5 −9.5	6 20	+7 −0	−6 −13	0.12	0.24
0.5 2.8	+1.4 0	−0.5 −1.4	1.0 4.6	+2.2 0	−1.0 −2.4	1.6 5.2	+2.2 −0	−1.6 −3.0	3.0 8.7	+3.5 −0	−3.0 −5.2	5 17	+6 −0	−5 −11	7 25	+9 −0	−7 −16	0.24	0.40
0.6 3.2	+1.6 0	−0.6 −1.6	1.2 5.6	+2.8 0	−1.2 −2.8	2.0 6.4	+2.8 −0	−2.0 −3.6	3.5 10.3	+4.0 −0	−3.5 −6.3	6 20	+7 −0	−6 −13	8 28	+10 −0	−8 −18	0.40	0.71
0.8 4.0	+2.0 0	−0.8 −2.0	1.6 7.1	+3.5 0	−1.6 −3.6	2.5 8.0	+3.5 −0	−2.5 −4.5	4.5 13.0	+5.0 −0	−4.5 −8.0	7 23	+8 −0	−7 −15	10 34	+12 −0	−10 −22	0.71	1.19
1.0 5.1	+2.5 0	−1.0 −2.6	2.0 8.5	+4.0 0	−2.0 −4.5	3.0 9.5	+4.0 −0	−3.0 −5.5	5 15	+6 −0	−5 −9	8 28	+10 −0	−8 −18	12 44	+16 −0	−12 −28	1.19	1.97
1.2 6.0	+3.0 0	−1.2 −3.0	2.5 10.0	+4.5 0	−2.5 −5.5	4.0 11.5	+4.5 −0	−4.0 −7.0	6 17.5	+7 −0	−6 −10.5	10 34	+12 −0	−10 −22	14 50	+18 −0	−14 −32	1.97	3.15
1.4 7.1	+3.5 0	−1.4 −3.6	3.0 11.5	+5.0 0	−3.0 −6.5	5.0 13.5	+5.0 −0	−5.0 −8.5	7 21	+9 −0	−7 −12	11 39	+14 −0	−11 −25	16 60	+22 −0	−16 −38	3.15	4.73
1.6 8.1	+4.0 0	−1.6 −4.1	3.5 13.5	+6.0 0	−3.5 −7.5	6 16	+6 −0	−6 −10	8 24	+10 −0	−8 −14	12 44	+16 −0	−12 −28	18 68	+25 −0	−18 −43	4.73	7.09
2.0 9.3	+4.5 0	−2.0 −4.8	4.0 15.5	+7.0 0	−4.0 −8.5	7 18.5	+7 −0	−7 −11.5	10 29	+12 −0	−10 −17	16 52	+18 −0	−16 −34	22 78	+28 −0	−22 −50	7.09	9.85
2.2 10.2	+5.0 0	−2.2 −5.2	4.5 17.5	+8.0 0	−4.5 −9.5	7 20	+8 −0	−7 −12	12 32	+12 −0	−12 −20	20 60	+20 −0	−20 −40	28 88	+30 −0	−28 −58	9.85	12.41
2.5 12.0	+6.0 0	−2.5 −6.0	5.0 20.0	+9.0 0	−5 −11	8 23	+9 −0	−8 −14	14 37	+14 −0	−14 −23	22 66	+22 −0	−22 −44	30 100	+35 −0	−30 −65	12.41	15.75
2.8 12.8	+6.0 0	−2.8 −6.8	5.0 21.0	+10.0 0	−5 −11	9 25	+10 −0	−9 −15	16 42	+16 −0	−16 −26	25 75	+25 −0	−25 −50	35 115	+40 −0	−35 −75	15.75	19.69
3.0 16.0	+8.0 0	−3.0 −8.0	6.0 26.0	+12.0 −0	−6 −14	10 30	+12 −0	−10 −18	18 50	+20 −0	−18 −30	28 88	+30 −0	−28 −58	40 140	+50 −0	−40 −90	19.69	30.09
3.5 19.5	+10.0 0	−3.5 −9.5	7.0 33.0	+16.0 −0	−7 −17	12 38	+16 −0	−12 −22	20 61	+25 −0	−20 −36	30 110	+40 −0	−30 −70	45 165	+60 −0	−45 −105	30.09	41.49
4.0 24.0	+12.0 0	−4.0 −12.0	8.0 40.0	+20.0 −0	−8 −20	14 46	+20 −0	−14 −26	25 75	+30 −0	−25 −45	40 140	+50 −0	−40 −90	60 220	+80 −0	−60 −140	41.49	56.19
4.5 30.5	+16.0 0	−4.5 −14.5	9.0 50.0	+25.0 −0	−9 −25	16 57	+25 −0	−16 −32	30 95	+40 −0	−30 −55	50 170	+60 −0	−50 110	70 270	+100 −0	−70 −170	56.19	76.39
5.0 37.0	+20.0 0	−5 −17	10.0 60.0	+30.0 −0	−10 −30	18 68	+30 −0	−18 −38	35 115	+50 −0	−35 −65	50 210	+80 −0	−50 −130	80 330	+125 −0	−80 −205	76.39	100.9
6.0 47.0	+25.0 0	−6 −22	12.0 67.0	+40.0 −0	−12 −27	20 85	+40 −0	−20 −45	40 140	+60 −0	−40 −80	60 260	+100 −0	−60 −160	90 410	+160 −0	−90 −250	100.9	131.9
7.0 57.0	+30.0 0	−7 −27	14.0 94.0	+50.0 −0	−14 −44	25 105	+50 −0	−25 −55	50 180	+80 −0	−50 −100	80 330	+125 −0	−80 −205	100 500	+200 −0	−100 −300	131.9	171.9
7.0 72.0	+40.0 0	−7 −32	14.0 114.0	+60.0 −0	−14 −54	25 125	+60 −0	−25 −65	50 210	+100 −0	−50 −110	90 410	+160 −0	−90 −250	125 625	+250 −0	−125 −375	171.9	200

TABLE C.9.3 Locational Transition Fits

Limits are in thousandths of an inch. Limits for hole and shaft are applied algebraically to the basic size to obtain the limits of size for the mating parts. Data in bold face are in accordance with ABC agreements. "Fit" represents the maximum interference (minus values) and the maximum clearance (plus values). Symbols H7, js6, etc., are Hole and Shaft designations used in ABC System.

Nominal Size Range Inches		*Class LT 1*			*Class LT 2*			*Class LT 3*			*Class LT 4*			*Class LT 5*			*Class LT 6*		
		Fit	Standard Limits		Fit	Standard Limits		Fit	Standard Limits		Fit	Standard Limits		Fit	Standard Limits		Fit	Standard Limits	
Over	To		Hole H7	Shaft js6		Hole H8	Shaft js7		Hole H7	Shaft k6		Hole H8	Shaft k7		Hole H7	Shaft n6		Hole H7	Shaft n7
0	0.12	−0.10 +0.50	+0.4 −0	+0.10 −0.10	−0.2 +0.8	+0.6 −0	+0.2 −0.2							−0.5 +0.15	+0.4 −0	+0.5 +0.25	−0.65 +0.15	+0.4 −0	−0.65 +0.25
0.12	0.24	−0.15 +0.65	+0.5 −0	+0.15 −0.15	−0.25 +0.95	+0.7 −0	+0.25 −0.25							−0.6 +0.2	+0.5 −0	+0.6 +0.3	−0.8 +0.2	+0.5 −0	+0.8 +0.3
0.24	0.40	−0.2 +0.8	+0.6 −0	+0.2 −0.2	−0.3 +1.2	+0.9 −0	+0.3 −0.3	−0.5 +0.5	+0.6 −0	+0.5 +0.1	−0.7 +0.8	+0.9 −0	+0.7 +0.1	−0.8 +0.2	+0.6 −0	+0.8 +0.4	−1.0 +0.2	+0.6 −0	+1.0 +0.4
0.40	0.71	−0.2 +0.9	+0.7 −0	+0.2 −0.2	−0.35 +1.35	+1.0 −0	+0.35 −0.35	−0.5 +0.6	+0.7 −0	+0.5 +0.1	−0.8 +0.9	+1.0 −0	+0.8 +0.1	−0.9 +0.2	+0.7 −0	+0.9 +0.5	−1.2 +0.2	+0.7 −0	+1.2 +0.5
0.71	1.19	−0.25 +1.05	+0.8 −0	+0.25 −0.25	−0.4 +1.6	+1.2 −0	+0.4 −0.4	−0.6 +0.7	+0.8 −0	+0.6 +0.1	−0.9 +1.1	+1.2 −0	+0.9 +0.1	−1.1 +0.2	+0.8 −0	+1.1 +0.6	−1.4 +0.2	+0.8 −0	+1.4 +0.6
1.19	1.97	−0.3 +1.3	+1.0 −0	+0.3 −0.3	−0.5 +2.1	+1.6 −0	+0.5 −0.5	−0.7 +0.9	+1.0 −0	+0.7 +0.1	−1.1 +1.5	+1.6 −0	+1.1 +0.1	−1.3 +0.3	+1.0 −0	+1.3 +0.7	−1.7 +0.3	+1.0 −0	+1.7 +0.7
1.97	3.15	−0.3 +1.5	+1.2 −0	+0.3 −0.3	−0.6 +2.4	+1.8 −0	+0.6 −0.6	−0.8 +1.1	+1.2 −0	+0.8 +0.1	−1.3 +1.7	+1.8 −0	+1.3 +0.1	−1.5 +0.4	+1.2 −0	+1.5 +0.8	−2.0 +0.4	+1.2 −0	+2.0 +0.8
3.15	4.73	−0.4 +1.8	+1.4 −0	+0.4 −0.4	−0.7 +2.9	+2.2 −0	+0.7 −0.7	−1.0 +1.3	+1.4 −0	+1.0 +0.1	−1.5 +2.1	+2.2 −0	+1.5 +0.1	−1.9 +0.4	+1.4 −0	+1.9 +1.0	−2.4 +0.4	+1.4 −0	+2.4 +1.0
4.73	7.09	−0.5 +2.1	+1.6 −0	+0.5 −0.5	−0.8 +3.3	+2.5 −0	+0.8 −0.8	−1.1 +1.5	+1.6 −0	+1.1 +0.1	−1.7 +2.4	+2.5 −0	+1.7 +0.1	−2.2 +0.4	+1.6 −0	+2.2 +1.2	−2.8 +0.4	+1.6 −0	+2.8 +1.2
7.09	9.85	−0.6 +2.4	+1.8 −0	+0.6 −0.6	−0.9 +3.7	+2.8 −0	+0.9 −0.9	−1.4 +1.6	+1.8 −0	+1.4 +0.2	−2.0 +2.6	+2.8 −0	+2.0 +0.2	−2.6 +0.4	+1.8 −0	+2.6 +1.4	−3.2 +0.4	+1.8 −0	+3.2 +1.4
9.85	12.41	−0.6 +2.6	+2.0 −0	+0.6 −0.6	−1.0 +4.0	+3.0 −0	+1.0 −1.0	−1.4 +1.8	+2.0 −0	+1.4 +0.2	−2.2 +2.8	+3.0 −0	+2.2 +0.2	−2.6 +0.6	+2.0 −0	+2.6 +1.4	−3.4 +0.6	+2.0 −0	+3.4 +1.4
12.41	15.75	−0.7 +2.9	+2.2 −0	+0.7 −0.7	−1.0 +4.5	+3.5 −0	+1.0 −1.0	−1.6 +2.0	+2.2 −0	+1.6 +0.2	−2.4 +3.3	+3.5 −0	+2.4 +0.2	−3.0 +0.6	+2.2 −0	+3.0 +1.6	−3.8 +0.6	+2.2 −0	+3.8 +1.6
15.75	19.69	−0.8 +3.3	+2.5 −0	+0.8 −0.8	−1.2 +5.2	+4.0 −0	+1.2 −1.2	−1.8 +2.3	+2.5 −0	+1.8 +0.2	−2.7 +3.8	+4.0 −0	+2.7 +0.2	−3.4 +0.7	+2.5 −0	+3.4 +1.8	−4.3 +0.7	+2.5 −0	+4.3 +1.8

TABLE C.9.4 Locational Interference Fits

Limits are in thousandths of an inch. Limits for hole and shaft are applied algebraically to the basic size to obtain the limits of size for the parts. Data in bold face are in accordance with ABC agreements, Symbols H7, p 6, etc., are Hole and Shaft designations used in ABC System.

Nominal Size Range Inches	Class LN 1			Class LN 2			Class LN 3		
	Limits of Intolerance	Standard Limits		Limits of Tolerance	Standard Limits		Limits of Intolerance	Standard Limits	
Over To		Hole H6	Shaft n5		Hole H7	Shaft p6		Hole H7	Shaft r6
0 – 0.12	0 0.45	+0.25 −0	+0.45 +0.25	0 0.65	+ 0.4 − 0	+ 0.65 + 0.4	0.1 0.75	+ 0.4 − 0	+ 0.75 + 0.5
0.12– 0.24	0 0.5	+0.3 −0	+0.5 +0.3	0 0.8	+ 0.5 − 0	+ 0.8 + 0.5	0.1 0.9	+ 0.5 0	+ 0.9 + 0.6
0.24– 0.40	0 0.65	+0.4 −0	+0.65 +0.4	0 1.0	+ 0.6 − 0	+ 1.0 + 0.6	0.2 1.2	+ 0.6 − 0	+ 1.2 + 0.8
0.40– 0.71	0 0.8	+0.4 −0	+0.8 +0.4	0 1.1	+ 0.7 − 0	+ 1.1 + 0.7	0.3 1.4	+ 0.7 − 0	+ 1.4 + 1.0
0.71– 1.19	0 1.0	+0.5 −0	+1.0 +0.5	0 1.3	+ 0.8 − 0	+ 1.3 + 0.8	0.4 1.7	+ 0.8 − 0	+ 1.7 + 1.2
1.19– 1.97	0 1.1	+0.6 −0	+1.1 +0.6	0 1.6	+ 1.0 − 0	+ 1.6 + 1.0	0.4 2.0	+ 1.0 − 0	+ 2.0 + 1.4
1.97– 3.15	0.1 1.3	+0.7 −0	+1.3 +0.7	0.2 2.1	+ 1.2 − 0	+ 2.1 + 1.4	0.4 2.3	+ 1.2 − 0	+ 2.3 + 1.6
3.15– 4.73	0.1 1.6	+0.9 −0	+1.6 +1.0	0.2 2.5	+ 1.4 − 0	+ 2.5 + 1.6	0.6 2.9	+ 1.4 − 0	+ 2.9 + 2.0
4.73– 7.09	0.2 1.9	+1.0 −0	+1.9 +1.2	0.2 2.8	+ 1.6 − 0	+ 2.8 + 1.8	0.9 3.5	+ 1.6 − 0	+ 3.5 + 2.5
7.09– 9.85	0.2 2.2	+1.2 −0	+2.2 +1.4	0.2 3.2	+ 1.8 − 0	+ 3.2 + 2.0	1.2 4.2	+ 1.8 − 0	+ 4.2 + 3.0
9.85– 12.41	0.2 2.3	+1.2 −0	+2.3 +1.4	0.2 3.4	2.0 − 0	+ 3.4 + 2.2	1.5 4.7	+ 2.0 − 0	+ 4.7 + 3.5
12.41– 15.75	0.2 2.6	+1.4 −0	+2.6 +1.6	0.3 3.9	+ 2.2 − 0	+ 3.9 + 2.5	2.3 5.9	+ 2.2 − 0	+ 5.9 + 4.5
15.75– 19.69	0.2 2.8	+1.6 −0	+2.8 +1.8	0.3 4.4	+ 2.5 − 0	+ 4.4 + 2.8	2.5 6.6	+ 2.5 − 0	+ 6.6 + 5.0
19.69– 30.09		+2.0 −0		0.5 5.5	+ 3 − 0	+ 5.5 + 3.5	4 9	+ 3 − 0	+ 9 + 7
30.09– 41.49		+2.5 −0		0.5 7.0	+ 4 − 0	+ 7.0 + 4.5	5 11.5	+ 4 − 0	+11.5 + 9
41.49– 56.19		+3.0 −0		1 9	+ 5 − 0	+ 9 + 6	7 15	+ 5 − 0	+15 +12
56.19– 76.39		+4.0 −0		1 11	+6 − 0	+11 + 7	10 20	+ 6 − 0	+20 +16
76.39–100.9		+5.0 −0		1 14	+8 − 0	+14 + 9	12 25	+ 8 − 0	+25 +20
100.9 –131.9		+6.0 −0		2 18	+10 − 0	+18 +12	15 31	+10 − 0	+31 +25
131.9 –171.9		+8.0 −0		4 24	+12 − 0	+24 +16	18 38	+12 − 0	+38 +30
171.9 –200		+10.0 −0		4 30	+16 − 0	+30 +20	24 50	+16 − 0	+50 +40

TABLE C.9.5 Force and Shrink Fits

Limits are in thousandths of an inch. Limits for hole and shaft are applied algebraically to the basic size to obtain the limits of size for the parts. Data in bold face are in accordance with ABC agreements. Symbols H7, s6, etc., are Hole and Shaft designations used in ABC System.

Nominal Size Range Inches		*Class FN 1*			*Class FN 2*			*Class FN 3*			*Class FN 4*			*Class FN 5*		
		Limits of Interference	Standard Limits		Limits of Interference	Standard Limits		Limits of Interference	Standard Limits		Limits of Interference	Standard Limits		Limits of Interference	Standard Limits	
Over	To		Hole H6	Shaft		Hole H7	Shaft s6		Hole H7	Shaft t6		Hole H7	Shaft u6		Hole H8	Shaft x7
0	0.12	0.05	+0.25	+0.5	0.2	+0.4	+0.85				0.3	+0.4	+ 0.95	0.3	+0.6	+ 1.3
		0.5	−0	+0.3	0.85	−0	+0.6				0.95	−0	+ 0.7	1.3	−0	+ 0.9
0.12	0.24	0.1	+0.3	+0.6	0.2	+0.5	+1.0				0.4	+0.5	+ 1.2	0.5	+0.7	+ 1.7
		0.6	−0	+0.4	1.0	−0	+0.7				1.2	−0	+ 0.9	1.7	−0	+ 1.2
0.24	0.40	0.1	+0.4	+0.75	0.4	+0.6	+1.4				0.6	+0.6	+ 1.6	0.5	+0.9	+ 2.0
		0.75	−0	+0.5	1.4	−0	+1.0				1.6	−0	+ 1.2	2.0	−0	+ 1.4
0.40	0.56	0.1	−0.4	+0.8	0.5	+0.7	+1.6				0.7	+0.7	+ 1.8	0.6	+1.0	+ 2.3
		0.8	−0	+0.5	1.6	−0	+1.2				1.8	−0	+ 1.4	2.3	−0	+ 1.6
0.56	0.71	0.2	+0.4	+0.9	0.5	+0.7	+1.6				0.7	+0.7	+ 1.8	0.8	+1.0	+ 2.5
		0.9	−0	+0.6	1.6	−0	+1.2				1.8	−0	+ 1.4	2.5	−0	+ 1.8
0.71	0.95	0.2	+0.5	+1.1	0.6	+0.8	+1.9				0.8	+0.8	+ 2.1	1.0	+1.2	+ 3.0
		1.1	−0	+0.7	1.9	−0	+1.4				2.1	−0	+ 1.6	3.0	−0	+ 2.2
0.95	1.19	0.3	+0.5	+1.2	0.6	+0.8	+1.9	0.8	+0.8	+ 2.1	1.0	+0.8	+ 2.3	1.3	+1.2	+ 3.3
		1.2	−0	+0.8	1.9	−0	+1.4	2.1	−0	+ 1.6	2.3	−0	+ 1.8	3.3	−0	+ 2.5
1.19	1.58	0.3	+0.6	+1.3	0.8	+1.0	+2.4	1.0	+1.0	+ 2.6	1.5	+1.0	+ 3.1	1.4	+1.6	+ 4.0
		1.3	−0	+0.9	2.4	−0	+1.8	2.6	−0	+ 2.0	3.1	−0	+ 2.5	4.0	−0	+ 3.0
1.58	1.97	0.4	+0.6	+1.4	0.8	+1.0	+2.4	1.2	+1.0	+ 2.8	1.8	+1.0	+ 3.4	2.4	+1.6	+ 5.0
		1.4	−0	+1.0	2.4	−0	+1.8	2.8	−0	+ 2.2	3.4	−0	+ 2.8	5.0	−0	+ 4.0
1.97	2.56	0.6	+0.7	+1.8	0.8	+1.2	+2.7	1.3	+1.2	+ 3.2	2.3	+1.2	+ 4.2	3.2	+1.8	+ 6.2
		1.8	−0	+1.3	2.7	−0	+2.0	3.2	−0	+ 2.5	4.2	−0	+ 3.5	6.2	−0	+ 5.0
2.56	3.15	0.7	+0.7	+1.9	1.0	+1.2	+2.9	1.8	+1.2	+ 3.7	2.8	+1.2	+ 4.7	4.2	+1.8	+ 7.2
		1.9	−0	+1.4	2.9	−0	+2.2	3.7	−0	+ 3.0	4.7	−0	+ 4.0	7.2	−0	+ 6.0
3.15	3.94	0.9	+0.9	+2.4	1.4	+1.4	+3.7	2.1	+1.4	+ 4.4	3.6	+1.4	+ 5.9	4.8	+2.2	+ 8.4
		2.4	−0	+1.8	3.7	−0	+2.8	4.4	−0	+ 3.5	5.9	−0	+ 5.0	8.4	−0	+ 7.0
3.94	4.73	1.1	+0.9	+2.6	1.6	+1.4	+3.9	2.6	+1.4	+ 4.9	4.6	+1.4	+ 6.9	5.8	+2.2	+ 9.4
		2.6	−0	+2.0	3.9	−0	+3.0	4.9	−0	+ 4.0	6.9	−0	+ 6.0	9.4	−0	+ 8.0
4.73	5.52	1.2	+1.0	+2.9	1.9	+1.6	+4.5	3.4	+1.6	+ 6.0	5.4	+1.6	+ 8.0	7.5	+2.5	+11.6
		2.9	−0	+2.2	4.5	−0	+3.5	6.0	−0	+ 5.0	8.0	−0	+ 7.0	11.6	−0	+10.0
5.52	6.30	1.5	+1.0	+3.2	2.4	+1.6	+5.0	3.4	+1.6	+ 6.0	5.4	+1.6	+ 8.0	9.5	+2.5	+13.6
		3.2	−0	+2.5	5.0	−0	+4.0	6.0	−0	+ 5.0	8.0	−0	+ 7.0	13.6	−0	+12.0
6.30	7.09	1.8	+1.0	+3.5	2.9	+1.6	+5.5	4.4	+1.6	+ 7.0	6.4	+1.6	+ 9.0	9.5	+2.5	+13.6
		3.5	−0	+2.8	5.5	−0	+4.5	7.0	−0	+ 6.0	9.0	−0	+ 8.0	13.6	−0	+12.0
7.09	7.88	1.8	+1.2	+3.8	3.2	+1.8	+6.2	5.2	+1.8	+ 8.2	7.2	+1.8	+10.2	11.2	+2.8	+15.8
		3.8	−0	+3.0	6.2	−0	+5.0	8.2	−0	+ 7.0	10.2	−0	+ 9.0	15.8	−0	+14.0
7.88	8.86	2.3	+1.2	+4.3	3.2	+1.8	+6.2	5.2	+1.8	+ 8.2	8.2	+1.8	+11.2	13.2	+2.8	+17.8
		4.3	−0	+3.5	6.2	−0	+5.0	8.2	−0	+ 7.0	11.2	−0	+10.0	17.8	−0	+16.0
8.86	9.85	2.3	+1.2	+4.3	4.2	+1.8	+7.2	6.2	+1.8	+ 9.2	10.2	+1.8	+13.2	13.2	+2.8	+17.8
		4.3	−0	+3.5	7.2	−0	+6.0	9.2	−0	+ 8.0	13.2	−0	+12.0	17.8	−0	+16.0
9.85	11.03	2.8	+1.2	+4.9	4.0	+2.0	+7.2	7.0	+2.0	+10.2	10.0	+2.0	+13.2	15.0	+3.0	+20.0
		4.9	−0	+4.0	7.2	−0	+6.0	10.2	−0	+ 9.0	13.2	−0	+12.0	20.0	−0	+18.0

Continued

TABLE C.9.5 Force and Shrink Fits—*Continued*

Nominal Size Range Inches	*Class FN 1*			*Class FN 2*			*Class FN 3*			*Class FN 4*			*Class FN 5*		
	Limits of Interference	Standard Limits		Limits of Interference	Standard Limits		Limits of Interference	Standard Limits		Limits of Interference	Standard Limits		Limits of Interference	Standard Limits	
Over To		Hole H6	Shaft		Hole 17	Shaft s6		Hole H7	Shaft t6		Hole H7	Shaft u6		Hole H8	Shaft x7
11.03 – 12.41	2.8 4.9	+ 1.2 – 0	+ 4.9 + 4.0	5.0 8.2	+ 2.0 – 0	+ 8.2 + 7.0	7.0 10.2	+ 2.0 – 0	+ 10.2 + 9.0	12.0 15.2	+ 2.0 – 0	+ 15.2 + 14.0	17.0 22.0	+ 3.0 – 0	+ 22.0 + 20.0
12.41 – 13.98	3.1 5.5	+ 1.4 – 0	+ 5.5 + 4.5	5.8 9.4	+ 2.2 – 0	+ 9.4 + 8.0	7.8 11.4	+ 2.2 – 0	+ 11.4 + 10.0	13.8 17.4	+ 2.2 – 0	+ 17.4 + 16.0	18.5 24.2	+ 3.5 + 0	+ 24.2 + 22.0
13.98 – 15.75	3.6 6.1	+ 1.4 – 0	+ 6.1 + 5.0	5.8 9.4	+ 2.2 – 0	+ 9.4 + 8.0	9.8 13.4	+ 2.2 – 0	+ 13.4 + 12.0	15.8 19.4	+ 2.2 – 0	+ 19.4 + 18.0	21.5 27.2	+ 3.5 – 0	+ 27.2 + 25.0
15.75 – 17.72	4.4 7.0	+ 1.6 – 0	+ 7.0 + 6.0	6.5 10.6	+ 2.5 – 0	+ 10.6 + 9.0	9.5 13.6	+ 2.5 – 0	+ 13.6 + 12.0	17.5 21.6	+ 2.5 – 0	+ 21.6 + 20.0	24.0 30.5	+ 4.0 – 0	+ 30.5 + 28.0
17.72 – 19.69	4.4 7.0	+ 1.6 – 0	+ 7.0 + 6.0	7.5 11.6	+ 2.5 – 0	+ 11.6 + 10.0	11.5 15.6	+ 2.5 – 0	+ 15.6 + 14.0	19.5 23.6	+ 2.5 – 0	+ 73.6 + 22.0	26.0 32.5	+ 4.0 – 0	+ 32.5 + 30.0
19.69 – 24.34	6.0 9.2	+ 2.0 – 0	+ 9.2 + 8.0	9.0 14.0	+ 3.0 – 0	+ 14.0 + 12.0	15.0 20.0	+ 3.0 – 0	+ 20.0 + 18.0	22.0 27.0	+ 3.0 – 0	+ 27.0 + 25.0	30.0 38.0	+ 5.0 – 0	+ 38.0 + 35.0
24.34 – 30.09	7.0 10.2	+ 2.0 – 0	+ 10.2 + 9.0	11.0 16.0	+ 3.0 – 0	+ 16.0 + 14.0	17.0 22.0	+ 3.0 – 0	+ 22.0 + 20.0	27.0 32.0	+ 3.0 – 0	+ 32.0 + 30.0	35.0 43.0	+ 5.0 – 0	+ 43.0 + 40.0
30.09 – 35.47	7.5 11.6	+ 2.5 – 0	+ 11.6 + 10.0	14.0 20.5	+ 4.0 – 0	+ 20.5 + 18.0	21.0 27.5	+ 4.0 – 0	+ 27.5 + 25.0	31.0 37.5	+ 4.0 – 0	+ 37.5 + 35.0	44.0 54.0	+ 6.0 – 0	+ 54.0 + 50.0
35.47 – 41.49	9.5 13.6	+ 2.5 – 0	+ 13.6 + 12.0	16.0 22.5	+ 4.0 – 0	+ 22.5 + 20.0	24.0 30.5	+ 4.0 – 0	+ 30.5 + 28.0	36.0 43.5	+ 4.0 – 0	+ 43.5 + 40.0	54.0 64.0	+ 6.0 – 0	+ 64.0 + 60.0
41.49 – 48.28	11.0 16.0	+ 3.0 – 0	+ 16.0 + 14.0	17.0 25.0	+ 5.0 – 0	+ 25.0 + 22.0	30.0 38.0	+ 5.0 – 0	+ 38.0 + 35.0	45.0 53.0	+ 5.0 – 0	+ 53.0 + 50.0	62.0 75.0	+ 8.0 – 0	+ 75.0 + 70.0
48.28 – 56.19	13.0 18.0	+ 3.0 – 0	+ 18.0 + 16.0	20.0 28.0	+ 5.0 – 0	+ 28.0 + 25.0	35.0 43.0	+ 5.0 – 0	+ 43.0 + 40.0	55.0 63.0	+ 5.0 – 0	+ 63.0 + 60.0	72.0 85.0	+ 8.0 – 0	+ 85.0 + 80.0
56.19 – 65.54	14.0 20.5	+ 4.0 – 0	+ 20.5 + 18.0	24.0 34.0	+ 6.0 – 0	+ 34.0 + 30.0	39.0 49.0	+ 6.0 – 0	+ 49.0 + 45.0	64.0 74.0	+ 6.0 – 0	+ 74.0 + 70.0	90.0 106	+ 10.0 – 0	+ 106 + 100
65.54 – 76.39	18.0 24.5	+ 4.0 – 0	+ 24.5 + 22.0	29.0 39.0	+ 6.0 – 0	+ 39.0 35.0	44.0 54.0	+ 6.0 – 0	+ 54.0 + 50.0	74.0 84.0	+ 6.0 – 0	+ 84.0 + 80.0	110 126	+ 10.0 – 0	+ 126 + 120
76.39 – 87.79	20.0 28.0	+ 5.0 – 0	+ 28.0 + 25.0	32.0 45.0	+ 8.0 – 0	+ 45.0 + 40.0	52.0 65.0	+ 8.0 – 0	+ 65.0 + 60.0	82.0 95.0	+ 8.0 – 0	+ 95.0 + 90.0	128 148	+ 12.0 – 0	+ 148 + 140
87.79 – 100.9	23.0 31.0	+ 5.0 – 0	+ 31.0 + 28.0	37.0 50.0	+ 8.0 – 0	+ 50.0 + 45.0	62.0 75.0	+ 8.0 – 0	+ 75.0 + 70.0	92.0 105	+ 8.0 – 0	+ 105 + 100	148 168	+ 12.0 – 0	+ 168 + 160
100.9 – 115.3	24.0 34.0	+ 6.0 – 0	+ 34.0 + 30.0	40.0 56.0	+ 10.0 – 0	+ 56.0 + 50.0	70.0 86.0	+ 10.0 – 0	+ 86.0 + 80.0	110 126	+ 10.0 – 0	+ 126 + 120	164 190	+ 16.0 – 0	+ 190 + 180
115.3 – 131.9	29.0 39.0	+ 6.0 – 0	+ 39.0 + 35.0	50.0 66.0	+ 10.0 – 0	+ 66.0 + 60.0	80.0 96.0	+ 10.0 – 0	+ 96.0 + 90.0	130 146	+ 10.0 – 0	+ 146 + 140	184 210	+ 16.0 – 0	+ 210 + 200
131.9 – 152.2	37.0 50.0	+ 8.0 – 0	+ 50.0 + 45.0	58.0 78.0	+ 12.0 – 0	+ 78.0 + 70.0	88.0 108	+ 12.0 – 0	+ 108 + 100	148 168	+ 12.0 – 0	+ 168 + 160	200 232	+ 20.0 – 0	+ 232 + 220
152.2 – 171.9	42.0 55.0	+ 8.0 – 0	+ 55.0 + 50.0	68.0 88.0	+ 12.0 – 0	+ 88.0 + 80.0	108 128	+ 12.0 – 0	+ 128 + 120	168 188	+ 12.0 – 0	+ 188 + 170	230 262	+ 20.0 – 0	+ 262 + 250
171.9 – 200	50.0 66.0	+ 10.0 – 0	+ 66.0 + 60.0	74.0 100	+ 16.0 – 0	+ 100 + 90	124 150	+ 16.0 – 0	+ 150 + 140	184 210	+ 16.0 – 0	+ 210 + 200	275 316	+ 2.5 – 0	+ 316 + 300

TABLE C.9.6 Preferred Metric Hole Basis Clearance Fits

BASIC SIZE		*LOOSE RUNNING*			*FREE RUNNING*			*CLOSE RUNNING*			*SLIDING*			*LOCATIONAL CLEARANCE*		
		Hole H11	**Shaft c11**	**Fit**	**Hole H9**	**Shaft d9**	**Fit**	**Hole H8**	**Shaft f7**	**Fit**	**Hole H7**	**Shaft g6**	**Fit**	**Hole H7**	**Shaft h6**	**Fit**
1	MAX	1.060	0.940	0.180	1.025	0.980	0.070	1.014	0.994	0.030	1.010	0.998	0.018	1.010	1.000	0.016
	MIN	1.000	0.880	0.060	1.000	0.955	0.020	1.000	0.984	0.006	1.000	0.992	0.002	1.000	0.994	0.000
1.2	MAX	1.260	1.140	0.180	1.225	1.180	0.070	1.214	1.194	0.030	1.210	1.198	0.018	1.210	1.200	0.016
	MIN	1.200	1.080	0.060	1.200	1.155	0.020	1.200	1.184	0.006	1.200	1.192	0.002	1.200	1.194	0.000
1.6	MAX	1.660	1.540	0.180	1.625	1.580	0.070	1.614	1.594	0.030	1.610	1.598	0.018	1.610	1.600	0.016
	MIN	1.600	1.480	0.060	1.600	1.555	0.020	1.600	1.584	0.006	1.600	1.592	0.002	1.600	1.594	0.000
2	MAX	2.060	1.940	0.180	2.025	1.980	0.070	2.014	1.994	0.030	2.010	1.998	0.018	2.010	2.000	0.016
	MIN	2.000	1.880	0.060	2.000	1.955	0.020	2.000	1.984	0.006	2.000	1.992	0.002	2.000	1.994	0.000
2.5	MAX	2.560	2.440	0.180	2.525	2.480	0.070	2.514	2.494	0.030	2.510	2.498	0.018	2.510	2.500	0.016
	MIN	2.500	2.380	0.060	2.500	2.455	0.020	2.500	2.484	0.006	2.500	2.492	0.002	2.500	2.494	0.000
3	MAX	3.060	2.940	0.180	3.025	2.980	0.070	3.014	2.994	0.030	3.010	2.998	0.018	3.010	3.000	0.016
	MIN	3.000	2.880	0.060	3.000	2.955	0.020	3.000	2.984	0.006	3.000	2.992	0.002	3.000	2.994	0.000
4	MAX	4.075	3.930	0.220	4.030	3.970	0.090	4.018	3.990	0.040	4.012	3.996	0.024	4.012	4.000	0.020
	MIN	4.000	3.855	0.070	4.000	3.940	0.030	4.000	3.978	0.010	4.000	3.988	0.004	4.000	3.992	0.000
5	MAX	5.075	4.930	0.220	5.030	4.970	0.090	5.018	4.990	0.040	5.012	4.996	0.024	5.012	5.000	0.020
	MIN	5.000	4.855	0.070	5.000	4.940	0.030	5.000	4.978	0.010	5.000	4.988	0.004	5.000	4.992	0.000
6	MAX	6.075	5.930	0.220	6.030	5.970	0.090	6.018	5.990	0.040	6.012	5.996	0.024	6.012	6.000	0.020
	MIN	6.000	5.855	0.070	6.000	5.940	0.030	6.000	5.978	0.010	6.000	5.988	0.004	6.000	5.992	0.000
8	MAX	8.090	7.920	0.260	8.036	7.960	0.112	8.022	7.987	0.050	8.015	7.995	0.029	8.015	8.000	0.024
	MIN	8.000	7.830	0.080	8.000	7.924	0.040	8.000	7.972	0.013	8.000	7.986	0.005	8.000	7.991	0.000
10	MAX	10.090	9.920	0.260	10.036	9.960	0.112	10.022	9.987	0.050	10.015	9.995	0.029	10.015	10.000	0.024
	MIN	10.000	9.830	0.080	10.000	9.924	0.040	10.000	9.972	0.013	10.000	9.986	0.005	10.000	9.991	0.000
12	MAX	12.110	11.905	0.315	12.043	11.950	0.136	12.027	11.984	0.061	12.018	11.994	0.035	12.018	12.000	0.029
	MIN	12.000	11.795	0.095	12.000	11.907	0.050	12.000	11.966	0.016	12.000	11.983	0.006	12.000	11.989	0.000
16	MAX	16.110	15.905	0.315	16.043	15.950	0.136	16.027	15.984	0.061	16.018	15.994	0.035	16.018	16.000	0.029
	MIN	16.000	15.795	0.095	16.000	15.907	0.050	16.000	15.966	0.016	16.000	15.983	0.006	16.000	15.989	0.000
20	MAX	20.130	19.890	0.370	20.052	19.935	0.169	20.033	19.980	0.074	20.021	19.993	0.041	20.021	20.000	0.034
	MIN	20.000	19.760	0.110	20.000	19.883	0.065	20.000	19.959	0.020	20.000	19.980	0.007	20.000	19.987	0.000

Continued

TABLE C.9.6 Preferred Metric Hole Basis Clearance Fits—*Continued*

BASIC SIZE		LOOSE RUNNING Hole H11	Shaft c11	Fit	FREE RUNNING Hole H9	Shaft d9	Fit	CLOSE RUNNING Hole H8	Shaft f7	Fit	SLIDING Hole H7	Shaft g6	Fit	LOCATIONAL CLEARANCE Hole H7	Shaft h6	Fit
25	MAX	25.130	24.890	0.370	25.052	24.935	0.169	25.033	24.980	0.074	25.021	24.993	0.041	25.021	25.000	0.034
	MIN	25.000	24.760	0.110	25.000	24.883	0.065	25.000	24.959	0.020	25.000	24.980	0.007	25.000	24.987	0.000
30	MAX	30.130	29.890	0.370	30.052	29.935	0.169	30.033	29.980	0.074	30.021	29.993	0.041	30.021	30.000	0.034
	MIN	30.000	29.760	0.110	30.000	29.883	0.065	30.000	29.959	0.020	30.000	29.980	0.007	30.000	29.987	0.000
40	MAX	40.160	39.880	0.440	40.062	39.920	0.204	40.039	39.975	0.089	40.025	39.991	0.050	40.025	40.000	0.041
	MIN	40.000	39.720	0.120	40.000	39.858	0.080	40.000	39.950	0.025	40.000	39.975	0.009	40.000	39.984	0.000
50	MAX	50.160	49.870	0.450	50.062	49.920	0.204	50.039	49.975	0.089	50.025	49.991	0.050	50.025	50.000	0.041
	MIN	50.000	49.710	0.130	50.000	49.858	0.080	50.000	49.950	0.025	50.000	49.975	0.009	50.000	49.984	0.000
60	MAX	60.190	59.860	0.520	60.074	59.900	0.248	60.046	59.970	0.106	60.030	59.990	0.059	60.030	60.000	0.049
	MIN	60.000	59.670	0.140	60.000	59.826	0.100	60.000	59.940	0.030	60.000	59.971	0.010	60.000	59.981	0.000
80	MAX	80.190	79.850	0.530	80.074	79.900	0.248	80.046	79.970	0.106	80.030	79.990	0.059	80.030	80.000	0.049
	MIN	80.000	79.660	0.150	80.000	79.826	0.100	80.000	79.940	0.030	80.000	79.971	0.010	80.000	79.981	0.000
100	MAX	100.220	99.830	0.610	100.087	99.880	0.294	100.054	99.964	0.125	100.035	99.988	0.069	100.035	100.000	0.057
	MIN	100.000	99.610	0.170	100.000	99.793	0.120	100.000	99.929	0.036	100.000	99.966	0.012	100.000	99.978	0.000
120	MAX	120.220	119.820	0.620	120.087	119.880	0.294	120.054	119.964	0.125	120.035	119.988	0.069	120.035	120.000	0.057
	MIN	120.000	119.600	0.180	120.000	119.793	0.120	120.000	119.929	0.036	120.000	119.966	0.012	120.000	119.978	0.000
160	MAX	160.250	159.790	0.710	160.100	159.855	0.345	160.063	159.957	0.146	160.040	159.986	0.079	160.040	160.000	0.065
	MIN	160.000	159.540	0.210	160.000	159.755	0.145	160.000	159.917	0.043	160.000	159.961	0.014	160.000	159.975	0.000
200	MAX	200.290	199.760	0.820	200.115	199.830	0.400	200.072	199.950	0.168	200.046	199.985	0.090	200.046	200.000	0.075
	MIN	200.000	199.470	0.240	200.000	199.715	0.170	200.000	199.904	0.050	200.000	199.956	0.015	200.000	199.971	0.000
250	MAX	250.290	249.720	0.860	250.115	249.830	0.400	250.072	249.950	0.168	250.046	249.985	0.090	250.046	250.000	0.075
	MIN	250.000	249.430	0.280	250.000	249.715	0.170	250.000	249.904	0.050	250.000	249.956	0.015	250.000	249.971	0.000
300	MAX	300.320	299.670	0.970	300.130	299.810	0.450	300.081	299.944	0.189	300.052	299.983	0.101	300.052	300.000	0.084
	MIN	300.000	299.350	0.330	300.000	299.680	0.190	300.000	299.892	0.056	300.000	299.951	0.017	300.000	299.968	0.000
400	MAX	400.360	399.600	1.120	400.140	399.790	0.490	400.089	399.938	0.208	400.057	399.982	0.111	400.057	400.000	0.093
	MIN	400.000	399.240	0.400	400.000	399.650	0.210	400.000	399.881	0.062	400.000	399.946	0.018	400.000	399.964	0.000
500	MAX	500.400	499.520	1.280	500.155	499.770	0.540	500.097	499.932	0.228	500.063	499.980	0.123	500.063	500.000	0.103
	MIN	500.000	499.120	0.480	500.000	499.615	0.230	500.000	499.869	0.068	500.000	499.940	0.020	500.000	499.960	0.000

TABLE C.9.7 Preferred Metric Hole Basis Transition and Interference Fits

	LOCATIONAL TRANSN.			*LOCATIONAL TRANSN.*			*LOCATIONAL INTERF.*			*MEDIUM DRIVE*			*FORCE*		
BASIC SIZE	Hole H7	Shaft k6	Fit	Hole H7	Shaft n6	Fit	Hole H7	Shaft p6	Fit	Hole H7	Shaft s6	Fit	Hole H7	Shaft u6	Fit
1 MAX	1.010	1.006	0.010	1.010	1.010	0.006	1.010	1.012	0.004	1.010	1.020	−0.004	1.010	1.024	−0.008
MIN	1.000	1.000	−0.006	1.000	1.004	−0.010	1.000	1.006	−0.012	1.000	1.014	−0.020	1.000	1.018	−0.024
1.2 MAX	1.210	1.206	0.010	1.210	1.210	0.006	1.210	1.212	0.004	1.210	1.220	−0.004	1.210	1.224	−0.008
MIN	1.200	1.200	−0.006	1.200	1.204	−0.010	1.200	1.206	−0.012	1.200	1.214	−0.020	1.200	1.218	−0.024
1.6 MAX	1.610	1.606	0.010	1.610	1.610	0.006	1.610	1.612	0.004	1.610	1.620	−0.004	1.610	1.624	−0.008
MIN	1.600	1.600	−0.006	1.600	1.604	−0.010	1.600	1.606	−0.012	1.600	1.614	−0.020	1.600	1.618	−0.024
2 MAX	2.010	2.006	0.010	2.010	2.010	0.006	2.010	2.012	0.004	2.010	2.020	−0.004	2.010	2.024	−0.008
MIN	2.000	2.000	−0.006	2.000	2.004	−0.010	2.000	2.006	−0.012	2.000	2.014	−0.020	2.000	2.018	−0.024
2.5 MAX	2.510	2.506	0.010	2.510	2.510	0.006	2.510	2.512	0.004	2.510	2.520	−0.004	2.510	2.524	−0.008
MIN	2.500	2.500	−0.006	2.500	2.504	−0.010	2.500	2.506	−0.012	2.500	2.514	−0.020	2.500	2.518	−0.024
3 MAX	3.010	3.006	0.010	3.010	3.010	0.006	3.010	3.012	0.004	3.010	3.020	−0.004	3.010	3.024	−0.008
MIN	3.000	3.000	−0.006	3.000	3.004	−0.010	3.000	3.006	−0.012	3.000	3.014	−0.020	3.000	3.018	−0.024
4 MAX	4.012	4.009	0.011	4.012	4.016	0.004	4.012	4.020	0.000	4.012	4.027	−0.007	4.012	4.031	−0.011
MIN	4.000	4.001	−0.009	4.000	4.008	−0.016	4.000	4.012	−0.020	4.000	4.019	−0.027	4.000	4.023	−0.031
5 MAX	5.012	5.009	0.011	5.012	5.016	0.004	5.012	5.020	0.000	5.012	5.027	−0.007	5.012	5.031	−0.011
MIN	5.000	5.001	−0.009	5.000	5.008	−0.016	5.000	5.012	−0.020	5.000	5.019	−0.027	5.000	5.023	−0.031
6 MAX	6.012	6.009	0.011	6.012	6.016	0.004	6.012	6.020	0.000	6.012	6.027	−0.007	6.012	6.031	−0.011
MIN	6.000	6.001	−0.009	6.000	6.008	−0.016	6.000	6.012	−0.020	6.000	6.019	−0.027	6.000	6.023	−0.031
8 MAX	8.015	8.010	0.014	8.015	8.019	0.005	8.015	8.024	0.000	8.015	8.032	−0.008	8.015	8.037	−0.013
MIN	8.000	8.001	−0.010	8.000	8.010	−0.019	8.000	8.015	−0.024	8.000	8.023	−0.032	8.000	8.028	−0.037
10 MAX	10.015	10.010	0.014	10.015	10.019	0.005	10.015	10.024	0.000	10.015	10.032	−0.008	10.015	10.037	−0.013
MIN	10.000	10.001	−0.010	10.000	10.010	−0.019	10.000	10.015	−0.024	10.000	10.023	−0.032	10.000	10.028	−0.037
12 MAX	12.018	12.012	0.017	12.018	12.023	0.006	12.018	12.029	0.000	12.018	12.039	−0.010	12.018	12.044	−0.015
MIN	12.000	12.001	−0.012	12.000	12.012	−0.023	12.000	12.018	−0.029	12.000	12.028	−0.039	12.000	12.033	−0.044
16 MAX	16.018	16.012	0.017	16.018	16.023	0.006	16.018	16.029	0.000	16.018	16.039	−0.010	16.018	16.044	−0.015
MIN	16.000	16.001	−0.012	16.000	16.012	−0.023	16.000	16.018	−0.029	16.000	16.028	−0.039	16.000	16.033	−0.044
20 MAX	20.021	20.015	0.019	20.021	20.028	0.006	20.021	20.035	−0.001	20.021	20.048	−0.014	20.021	20.054	−0.020
MIN	20.000	20.002	−0.015	20.000	20.015	−0.028	20.000	20.022	−0.035	20.000	20.035	−0.048	20.000	20.041	−0.054

Continued

TABLE C.9.7 Preferred Metric Hole Basis Transition and Interference Fits—*Continued*

	LOCATIONAL TRANSN.			*LOCATIONAL TRANSN.*			*LOCATIONAL INTERF.*			*MEDIUM DRIVE*			*FORCE*		
BASIC SIZE	Hole H7	Shaft k6	Fit	Hole H7	Shaft n6	Fit	Hole H7	Shaft p6	Fit	Hole H7	Shaft s6	Fit	Hole H7	Shaft u6	Fit
25 MAX	25.021	25.015	0.019	25.021	25.028	0.006	25.021	25.035	−0.001	25.021	25.048	−0.014	25.021	25.061	−0.027
MIN	25.000	25.002	−0.015	25.000	25.015	−0.028	25.000	25.022	−0.035	25.000	25.035	−0.048	25.000	25.048	−0.061
30 MAX	30.021	30.015	0.019	30.021	30.028	0.006	30.021	30.035	−0.001	30.021	30.048	−0.014	30.021	30.061	−0.027
MIN	30.000	30.002	−0.015	30.000	30.015	−0.028	30.000	30.022	−0.035	30.000	30.035	−0.048	30.000	30.048	−0.061
40 MAX	40.025	40.018	0.023	40.025	40.033	0.008	40.025	40.042	−0.001	40.025	40.059	−0.018	40.025	40.076	−0.035
MIN	40.000	40.002	−0.018	40.000	40.017	−0.033	40.000	40.026	−0.042	40.000	40.043	−0.059	40.000	40.060	−0.076
50 MAX	50.025	50.018	0.023	50.025	50.033	0.008	50.025	50.042	−0.001	50.025	50.059	−0.018	50.025	50.086	−0.045
MIN	50.000	50.002	−0.018	50.000	50.017	−0.033	50.000	50.026	−0.042	50.000	50.043	−0.059	50.000	50.070	−0.086
60 MAX	60.030	60.021	0.028	60.030	60.039	0.010	60.030	60.051	−0.002	60.030	60.072	−0.023	60.030	60.106	−0.057
MIN	60.000	60.002	−0.021	60.000	60.020	−0.039	60.000	60.032	−0.051	60.000	60.053	−0.072	60.000	60.087	−0.106
80 MAX	80.030	80.021	0.028	80.030	80.039	0.010	80.030	80.051	−0.002	80.030	80.078	−0.029	80.030	80.121	−0.072
MIN	80.000	80.002	−0.021	80.000	80.020	−0.039	80.000	80.032	−0.051	80.000	80.059	−0.078	80.000	80.102	−0.121
100 MAX	100.035	100.025	0.032	100.035	100.045	0.012	100.035	100.059	−0.002	100.035	100.093	−0.036	100.035	100.146	−0.089
MIN	100.000	100.003	−0.025	100.000	100.023	−0.045	100.000	100.037	−0.059	100.000	100.071	−0.093	100.000	100.124	−0.146
120 MAX	120.035	120.025	0.032	120.035	120.045	0.012	120.035	120.059	−0.002	120.035	120.101	−0.044	120.035	120.166	−0.109
MIN	120.000	120.003	−0.025	120.000	120.023	−0.045	120.000	120.037	−0.059	120.000	120.079	−0.101	120.000	120.144	−0.166
160 MAX	160.040	160.028	0.037	160.040	160.052	0.013	160.040	160.068	−0.003	160.040	160.125	−0.060	160.040	160.215	−0.150
MIN	160.000	160.003	−0.028	160.000	160.027	−0.052	160.000	160.043	−0.068	160.000	160.100	−0.125	160.000	160.190	−0.215
200 MAX	200.046	200.033	0.042	200.046	200.060	0.015	200.046	200.079	−0.004	200.046	200.151	−0.076	200.046	200.265	−0.190
MIN	200.000	200.004	−0.033	200.000	200.031	−0.060	200.000	200.050	−0.079	200.000	200.122	−0.151	200.000	200.236	−0.265
250 MAX	250.046	250.033	0.042	250.046	250.060	0.015	250.046	250.079	−0.004	250.046	250.169	−0.094	250.046	250.313	−0.238
MIN	250.000	250.004	−0.033	250.000	250.031	−0.060	250.000	250.050	−0.079	250.000	250.140	−0.169	250.000	250.284	−0.313
300 MAX	300.052	300.036	0.048	300.052	300.066	0.018	300.052	300.088	−0.004	300.052	300.202	−0.118	300.052	300.382	−0.298
MIN	300.000	300.004	−0.036	300.000	300.034	−0.066	300.000	300.056	−0.088	300.000	300.170	−0.202	300.000	300.350	−0.382
400 MAX	400.057	400.040	0.053	400.057	400.073	0.020	400.057	400.098	−0.005	400.057	400.244	−0.151	400.057	400.471	−0.378
MIN	400.000	400.004	−0.040	400.000	400.037	−0.073	400.000	400.062	−0.098	400.000	400.208	−0.244	400.000	400.435	−0.471
500 MAX	500.063	500.045	0.058	500.063	500.080	0.023	500.063	500.108	−0.005	500.063	500.292	−0.189	500.063	500.580	−0.477
MIN	500.000	500.005	−0.045	500.000	500.040	−0.080	500.000	500.068	−0.108	500.000	500.252	−0.292	500.000	500.540	−0.580

TABLE C.9.8 Preferred Metric Shaft Basis Clearance Fits

BASIC SIZE		LOOSE RUNNING			FREE RUNNING			CLOSE RUNNING			SLIDING			LOCATIONAL CLEARANCE		
		Hole C11	Shaft h11	Fit	Hole D9	Shaft h9	Fit	Hole F8	Shaft h7	Fit	Hole G7	Shaft h6	Fit	Hole H7	Shaft h6	Fit
1	MAX	1.120	1.000	0.180	1.045	1.000	0.070	1.020	1.000	0.030	1.012	1.000	0.018	1.010	1.000	0.016
	MIN	1.060	0.940	0.060	1.020	0.975	0.020	1.006	0.990	0.006	1.002	0.994	0.002	1.000	0.994	0.000
1.2	MAX	1.320	1.200	0.180	1.245	1.200	0.070	1.220	1.200	0.030	1.212	1.200	0.018	1.210	1.200	0.016
	MIN	1.260	1.140	0.060	1.220	1.175	0.020	1.206	1.190	0.006	1.202	1.194	0.002	1.200	1.194	0.000
1.6	MAX	1.720	1.600	0.180	1.645	1.600	0.070	1.620	1.600	0.030	1.612	1.600	0.018	1.610	1.600	0.016
	MIN	1.660	1.540	0.060	1.620	1.575	0.020	1.606	1.590	0.006	1.602	1.594	0.002	1.600	1.594	0.000
2	MAX	2.120	2.000	0.180	2.045	2.000	0.070	2.020	2.000	0.030	2.012	2.000	0.018	2.010	2.000	0.016
	MIN	2.060	1.940	0.060	2.020	1.975	0.020	2.006	1.990	0.006	2.002	1.994	0.002	2.000	1.994	0.000
2.5	MAX	2.620	2.500	0.180	2.545	2.500	0.070	2.520	2.500	0.030	2.512	2.500	0.018	2.510	2.500	0.016
	MIN	2.560	2.440	0.060	2.520	2.475	0.020	2.506	2.490	0.006	2.502	2.494	0.002	2.500	2.494	0.000
3	MAX	3.120	3.000	0.180	3.045	3.000	0.070	3.020	3.000	0.030	3.012	3.000	0.018	3.010	3.000	0.016
	MIN	3.060	2.940	0.060	3.020	2.975	0.020	3.006	2.990	0.006	3.002	2.994	0.002	3.000	2.994	0.000
4	MAX	4.145	4.000	0.220	4.060	4.000	0.090	4.028	4.000	0.040	4.016	4.000	0.024	4.012	4.000	0.020
	MIN	4.070	3.925	0.070	4.030	3.970	0.030	4.010	3.988	0.010	4.004	3.992	0.004	4.000	3.992	0.000
5	MAX	5.145	5.000	0.220	5.060	5.000	0.090	5.028	5.000	0.040	5.016	5.000	0.024	5.012	5.000	0.020
	MIN	5.070	4.925	0.070	5.030	4.970	0.030	5.010	4.988	0.010	5.004	4.992	0.004	5.000	4.992	0.000
6	MAX	6.145	6.000	0.220	6.060	6.000	0.090	6.028	6.000	0.040	6.016	6.000	0.024	6.012	6.000	0.020
	MIN	6.070	5.925	0.070	6.030	5.970	0.030	6.010	5.988	0.010	6.004	5.992	0.004	6.000	5.992	0.000
8	MAX	8.170	8.000	0.260	8.076	8.000	0.112	8.035	8.000	0.050	8.020	8.000	0.029	8.015	8.000	0.024
	MIN	8.080	7.910	0.080	8.040	7.964	0.040	8.013	7.985	0.013	8.005	7.991	0.005	8.000	7.991	0.000
10	MAX	10.170	10.000	0.260	10.076	10.000	0.112	10.035	10.000	0.050	10.020	10.000	0.029	10.015	10.000	0.024
	MIN	10.080	9.910	0.080	10.040	9.964	0.040	10.013	9.985	0.013	10.005	9.991	0.005	10.000	9.991	0.000
12	MAX	12.205	12.000	0.315	12.093	12.000	0.136	12.043	12.000	0.061	12.024	12.000	0.035	12.018	12.000	0.029
	MIN	12.095	11.890	0.095	12.050	11.957	0.050	12.016	11.982	0.016	12.006	11.989	0.006	12.000	11.989	0.000
16	MAX	16.205	16.000	0.315	16.093	16.000	0.136	16.043	16.000	0.061	16.024	16.000	0.035	16.018	16.000	0.029
	MIN	16.095	15.890	0.095	16.050	15.957	0.050	16.016	15.982	0.016	16.006	15.989	0.006	16.000	15.989	0.000
20	MAX	20.240	20.000	0.370	20.117	20.000	0.169	20.053	20.000	0.074	20.028	20.000	0.041	20.021	20.000	0.034
	MIN	20.110	19.870	0.110	20.065	19.948	0.065	20.020	19.979	0.020	20.007	19.987	0.007	20.000	19.987	0.000

Continued

TABLE C.9.8 Preferred Metric Shaft Basis Clearance Fits—*Continued*

		LOOSE RUNNING			*FREE RUNNING*			*CLOSE RUNNING*			*SLIDING*			*LOCATIONAL CLEARANCE*		
BASIC SIZE		Hole C11	Shaft h11	Fit	Hole D9	Shaft h9	Fit	Hole F8	Shaft h7	Fit	Hole G7	Shaft h6	Fit	Hole H7	Shaft h6	Fit
25	MAX	25.240	25.000	0.370	25.117	25.000	0.169	25.053	25.000	0.074	25.028	25.000	0.041	25.021	25.000	0.034
	MIN	25.110	24.870	0.110	25.065	24.948	0.065	25.020	24.979	0.020	25.007	24.987	0.007	25.000	24.987	0.000
30	MAX	30.240	30.000	0.370	30.117	30.000	0.169	30.053	30.000	0.074	30.028	30.000	0.041	30.021	30.000	0.034
	MIN	30.110	29.870	0.110	30.065	29.948	0.065	30.020	29.979	0.020	30.007	29.987	0.007	30.000	29.987	0.000
40	MAX	40.280	40.000	0.440	40.142	40.000	0.204	40.064	40.000	0.089	40.034	40.000	0.050	40.025	40.000	0.041
	MIN	40.120	39.840	0.120	40.080	39.938	0.080	40.025	39.975	0.025	40.009	39.984	0.009	40.000	39.984	0.000
50	MAX	50.290	50.000	0.450	50.142	50.000	0.204	50.064	50.000	0.089	50.034	50.000	0.050	50.025	50.000	0.041
	MIN	50.130	49.840	0.130	50.080	49.938	0.080	50.025	49.975	0.025	50.009	49.984	0.009	50.000	49.984	0.000
60	MAX	60.330	60.000	0.520	60.174	60.000	0.248	60.076	60.000	0.106	60.040	60.000	0.059	60.030	60.000	0.049
	MIN	60.140	59.810	0.140	60.100	59.926	0.100	60.030	59.970	0.030	60.010	59.981	0.010	60.000	59.981	0.000
80	MAX	80.340	80.000	0.530	80.174	80.000	0.248	80.076	80.000	0.106	80.040	80.000	0.059	80.030	80.000	0.049
	MIN	80.150	79.810	0.150	80.100	79.926	0.100	80.030	79.970	0.030	80.010	79.981	0.010	80.000	79.981	0.000
100	MAX	100.390	100.000	0.610	100.207	100.000	0.294	100.090	100.000	0.125	100.047	100.000	0.069	100.035	100.000	0.057
	MIN	100.170	99.780	0.170	100.120	99.913	0.120	100.036	99.965	0.036	100.012	99.978	0.012	100.000	99.978	0.000
120	MAX	120.400	120.000	0.620	120.207	120.000	0.294	120.090	120.000	0.125	120.047	120.000	0.069	120.035	120.000	0.057
	MIN	120.180	119.780	0.180	120.120	119.913	0.120	120.036	119.965	0.036	120.012	119.978	0.012	120.000	119.978	0.000
160	MAX	160.460	160.000	0.710	160.245	160.000	0.345	160.106	160.000	0.146	160.054	160.000	0.079	160.040	160.000	0.065
	MIN	160.210	159.750	0.210	160.145	159.900	0.145	160.043	159.960	0.043	160.014	159.975	0.014	160.000	159.975	0.000
200	MAX	200.530	200.000	0.820	200.285	200.000	0.400	200.122	200.000	0.168	200.061	200.000	0.090	200.046	200.000	0.075
	MIN	200.240	199.710	0.240	200.170	199.885	0.170	200.050	199.954	0.050	200.015	199.971	0.015	200.000	199.971	0.000
250	MAX	250.570	250.000	0.860	250.285	250.000	0.400	250.122	250.000	0.168	250.061	250.000	0.090	250.046	250.000	0.075
	MIN	250.280	249.710	0.280	250.170	249.885	0.170	250.050	249.954	0.050	250.015	249.971	0.015	250.000	249.971	0.000
300	MAX	300.650	300.000	0.970	300.320	300.000	0.450	300.137	300.000	0.189	300.069	300.000	0.101	300.052	300.000	0.084
	MIN	300.330	299.680	0.330	300.190	299.870	0.190	300.056	299.948	0.056	300.017	299.968	0.017	300.000	299.968	0.000
400	MAX	400.760	400.000	1.120	400.350	400.000	0.490	400.151	400.000	0.208	400.075	400.000	0.111	400.057	400.000	0.093
	MIN	400.400	399.640	0.400	400.210	399.860	0.210	400.062	399.943	0.062	400.018	399.964	0.018	400.000	399.964	0.000
500	MAX	500.880	500.000	1.280	500.385	500.000	0.540	500.165	500.000	0.228	500.083	500.000	0.123	500.063	500.000	0.103
	MIN	500.480	499.600	0.480	500.230	499.845	0.230	500.068	499.937	0.068	500.020	499.960	0.020	500.000	499.960	0.000

TABLE C.9.9 Preferred Metric Shaft Basis Transition and Interference Fits

		LOCATIONAL TRANSN.			*LOCATIONAL TRANSN.*			*LOCATIONAL INTERF.*			*MEDIUM DRIVE*			*FORCE*		
BASIC SIZE		Hole K7	Shaft h6	Fit	Hole N7	Shaft h6	Fit	Hole P7	Shaft h6	Fit	Hole S7	Shaft h6	Fit	Hole U7	Shaft h6	Fit
1	MAX	1.000	1.000	0.006	0.996	1.000	0.002	0.994	1.000	0.000	0.986	1.000	−0.008	0.982	1.000	−0.012
	MIN	0.990	0.994	−0.010	0.986	0.994	−0.014	0.984	0.994	−0.016	0.976	0.994	−0.024	0.972	0.994	−0.028
1.2	MAX	1.200	1.200	0.006	1.196	1.200	0.002	1.194	1.200	0.000	1.186	1.200	−0.008	1.182	1.200	−0.012
	MIN	1.190	1.194	−0.010	1.186	1.194	−0.014	1.184	1.194	−0.016	1.176	1.194	−0.024	1.172	1.194	−0.028
1.6	MAX	1.600	1.600	0.006	1.596	1.600	0.002	1.594	1.600	0.000	1.586	1.600	−0.008	1.582	1.600	−0.012
	MIN	1.590	1.594	−0.010	1.586	1.594	−0.014	1.584	1.594	−0.016	1.576	1.594	−0.024	1.572	1.594	−0.028
2	MAX	2.000	2.000	0.006	1.996	2.000	0.002	1.994	2.000	0.000	1.986	2.000	−0.008	1.982	2.000	−0.012
	MIN	1.990	1.994	−0.010	1.986	1.994	−0.014	1.984	1.994	−0.016	1.976	1.994	−0.024	1.972	1.994	−0.028
2.5	MAX	2.500	2.500	0.006	2.496	2.500	0.002	2.494	2.500	0.000	2.486	2.500	−0.008	2.482	2.500	−0.012
	MIN	2.490	2.494	−0.010	2.486	2.494	−0.014	2.484	2.494	−0.016	2.476	2.494	−0.024	2.472	2.494	−0.028
3	MAX	3.000	3.000	0.006	2.996	3.000	0.002	2.994	3.000	0.000	2.986	3.000	−0.008	2.982	3.000	−0.012
	MIN	2.990	2.994	−0.010	2.986	2.994	−0.014	2.984	2.994	−0.016	2.976	2.994	−0.024	2.972	2.994	−0.028
4	MAX	4.003	4.000	0.011	3.996	4.000	0.004	3.992	4.000	0.000	3.985	4.000	−0.007	3.981	4.000	−0.011
	MIN	3.991	3.992	−0.009	3.984	3.992	−0.016	3.980	3.992	−0.020	3.973	3.992	−0.027	3.969	3.992	−0.031
5	MAX	5.003	5.000	0.011	4.996	5.000	0.004	4.992	5.000	0.000	4.985	5.000	−0.007	4.981	5.000	−0.011
	MIN	4.991	4.992	−0.009	4.984	4.992	−0.016	4.980	4.992	−0.020	4.973	4.992	−0.027	4.969	4.992	−0.031
6	MAX	6.003	6.000	0.011	5.996	6.000	0.004	5.992	6.000	0.000	5.985	6.000	−0.007	5.981	6.000	−0.011
	MIN	5.991	5.992	−0.009	5.984	5.992	−0.016	5.980	5.992	−0.020	5.973	5.992	−0.027	5.969	5.992	−0.031
8	MAX	8.005	8.000	0.014	7.996	8.000	0.005	7.991	8.000	0.000	7.983	8.000	−0.008	7.978	8.000	−0.013
	MIN	7.990	7.991	−0.010	7.981	7.991	−0.019	7.976	7.991	−0.024	7.968	7.991	−0.032	7.963	7.991	−0.037
10	MAX	10.005	10.000	0.014	9.996	10.000	0.005	9.991	10.000	0.000	9.983	10.000	−0.008	9.978	10.000	−0.013
	MIN	9.990	9.991	−0.010	9.981	9.991	−0.019	9.976	9.991	−0.024	9.968	9.991	−0.032	9.963	9.991	−0.037
12	MAX	12.006	12.000	0.017	11.995	12.000	0.006	11.989	12.000	0.000	11.979	12.000	−0.010	11.974	12.000	−0.015
	MIN	11.988	11.989	−0.012	11.977	11.989	−0.023	11.971	11.989	−0.029	11.961	11.989	−0.039	11.956	11.989	−0.044
16	MAX	16.006	16.000	0.017	15.995	16.000	0.006	15.989	16.000	0.000	15.979	16.000	−0.010	15.974	16.000	−0.015
	MIN	15.988	15.989	−0.012	15.977	15.989	−0.023	15.971	15.989	−0.029	15.961	15.989	−0.039	15.956	15.989	−0.044
20	MAX	20.006	20.000	0.019	19.993	20.000	0.006	19.986	20.000	−0.001	19.973	20.000	−0.014	19.967	20.000	−0.020
	MIN	19.985	19.987	−0.015	19.972	19.987	−0.028	19.965	19.987	−0.035	19.952	19.987	−0.048	19.946	19.987	−0.054

Continued

TABLE C.9.9 Preferred Metric Shaft Basis Transition and Interference Fits—*Continued*

	LOCATIONAL TRANS.			*LOCATIONAL TRANSN.*			*LOCATIONAL INTERF.*			*MEDIUM DRIVE*			*FORCE*		
BASIC SIZE	**Hole K7**	**Shaft h6**	**Fit**	**Hole N7**	**Shaft h6**	**Fit**	**Hole P7**	**Shaft h6**	**Fit**	**Hole S7**	**Shaft h6**	**Fit**	**Hole U7**	**Shaft h6**	**Fit**
25 MAX	25.006	25.000	0.019	24.993	25.000	0.006	24.986	25.000	−0.001	24.973	25.000	−0.014	24.960	25.000	−0.027
MIN	24.985	24.987	−0.015	24.972	24.987	−0.028	24.965	24.987	−0.035	24.952	24.987	−0.048	24.939	24.987	−0.061
30 MAX	30.006	30.000	0.019	29.993	30.000	0.006	29.986	30.000	−0.001	29.973	30.000	−0.014	29.960	30.000	−0.027
MIN	29.985	29.987	−0.015	29.972	29.987	−0.028	29.965	29.987	−0.035	29.952	29.987	−0.048	29.939	29.987	−0.061
40 MAX	40.007	40.000	0.023	39.992	40.000	0.008	39.983	40.000	−0.001	39.966	40.000	−0.018	39.949	40.000	−0.035
MIN	39.982	39.984	−0.018	39.967	39.984	−0.033	39.958	39.984	−0.042	39.941	39.984	−0.059	39.924	39.984	−0.076
50 MAX	50.007	50.000	0.023	49.992	50.000	0.008	49.983	50.000	−0.001	49.966	50.000	−0.018	49.939	50.000	−0.045
MIN	49.982	49.984	−0.018	49.967	49.984	−0.033	49.958	49.984	−0.042	49.941	49.984	−0.059	49.914	49.984	−0.086
60 MAX	60.009	60.000	0.028	59.991	60.000	0.010	59.979	60.000	−0.002	59.958	60.000	−0.023	59.924	60.000	−0.057
MIN	59.979	59.981	−0.021	59.961	59.981	−0.039	59.949	59.981	−0.051	59.928	59.981	−0.072	59.894	59.981	−0.106
80 MAX	80.009	80.000	0.028	79.991	80.000	0.010	79.979	80.000	−0.002	79.952	80.000	−0.029	79.909	80.000	−0.072
MIN	79.979	79.981	−0.021	79.961	79.981	−0.039	79.949	79.981	−0.051	79.922	79.981	−0.078	79.879	79.981	−0.121
100 MAX	100.010	100.000	0.032	99.990	100.000	0.012	99.976	100.000	−0.002	99.942	100.000	−0.036	99.889	100.000	−0.089
MIN	99.975	99.978	−0.025	99.955	99.978	−0.045	99.941	99.978	−0.059	99.907	99.978	−0.093	99.854	99.978	−0.146
120 MAX	120.010	120.000	0.032	119.990	120.000	0.012	119.976	120.000	−0.002	119.934	120.000	−0.044	119.869	120.000	−0.109
MIN	119.975	119.978	−0.025	119.955	119.978	−0.045	119.941	119.978	−0.059	119.899	119.978	−0.101	119.834	119.978	−0.166
160 MAX	160.012	160.000	0.037	159.988	160.000	0.013	159.972	160.000	−0.003	159.915	160.000	−0.060	159.825	160.000	−0.150
MIN	159.972	159.975	−0.028	159.948	159.975	−0.052	159.932	159.975	−0.068	159.875	159.975	−0.125	159.785	159.975	−0.215
200 MAX	200.013	200.000	0.042	199.986	200.000	0.015	199.967	200.000	−0.004	199.895	200.000	−0.076	199.781	200.000	−0.190
MIN	199.967	199.971	−0.033	199.940	199.971	−0.060	199.921	199.971	−0.079	199.849	199.971	−0.151	199.735	199.971	−0.265
250 MAX	250.013	250.000	0.042	249.986	250.000	0.015	249.967	250.000	−0.004	249.877	250.000	−0.094	249.733	250.000	−0.238
MIN	249.967	249.971	−0.033	249.940	249.971	−0.060	249.921	249.971	−0.079	249.831	249.971	−0.169	249.687	249.971	−0.313
300 MAX	300.016	300.000	0.048	299.986	300.000	0.018	299.964	300.000	−0.004	299.850	300.000	−0.118	299.670	300.000	−0.298
MIN	299.964	299.968	−0.036	299.934	299.968	−0.066	299.912	299.968	−0.088	299.798	299.968	−0.202	299.618	299.968	−0.382
400 MAX	400.017	400.000	0.053	399.984	400.000	0.020	399.959	400.000	−0.005	399.813	400.000	−0.151	399.586	400.000	−0.378
MIN	399.960	399.964	−0.040	399.927	399.964	−0.073	399.902	399.964	−0.098	399.756	399.964	−0.244	399.529	399.964	−0.471
500 MAX	500.018	500.000	0.058	499.983	500.000	0.023	499.955	500.000	−0.005	499.771	500.000	−0.189	499.483	500.000	−0.477
MIN	499.955	499.960	−0.045	499.920	499.960	−0.080	499.892	499.960	−0.108	499.708	499.960	−0.292	499.420	499.960	−0.580

APPENDIX D

Symbols

APPENDIX D.1

Welding Symbols

Welding Symbols and Processes—American Welding Society Standard (ANSI/AWS A2.4-79.)

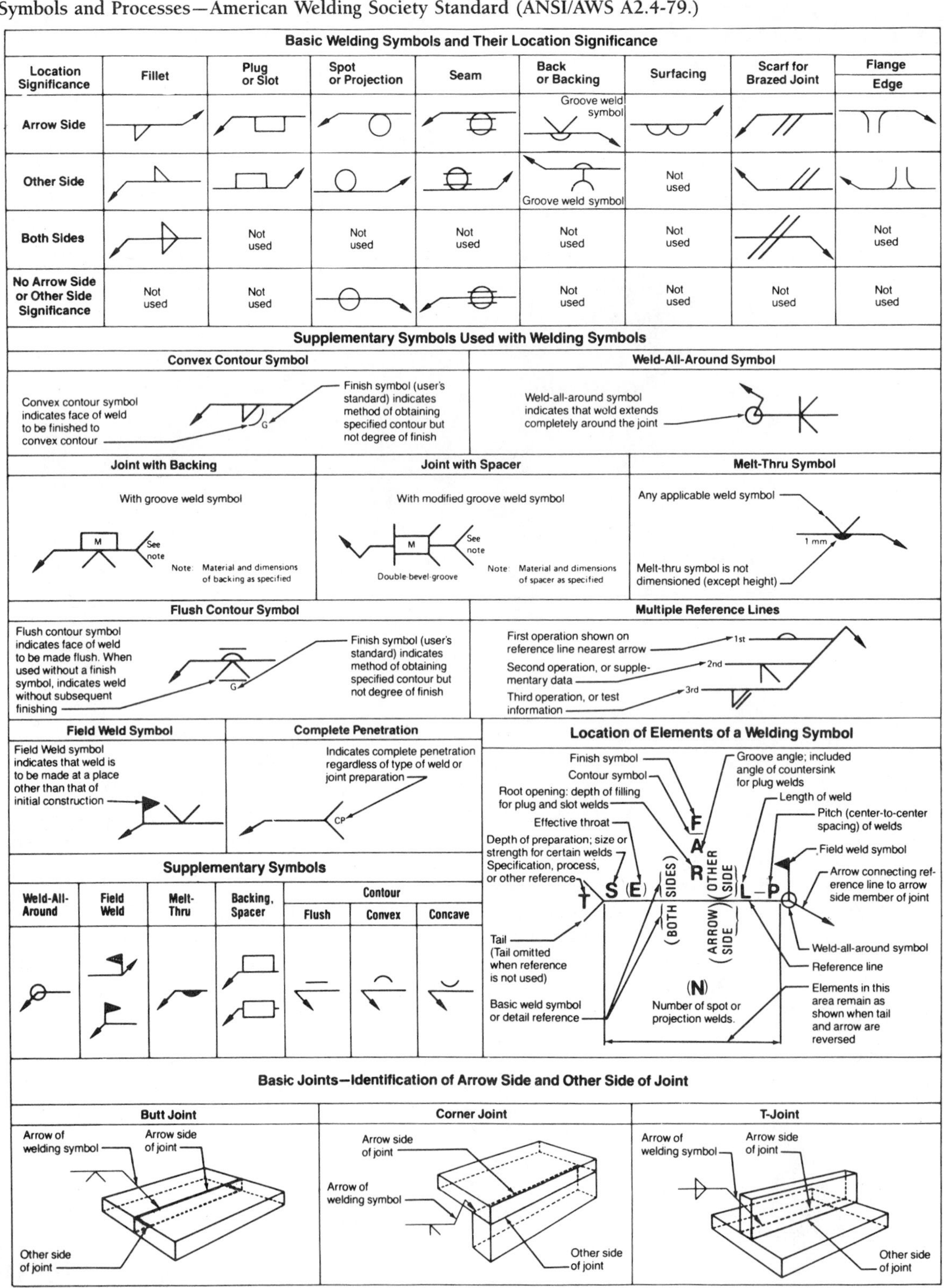

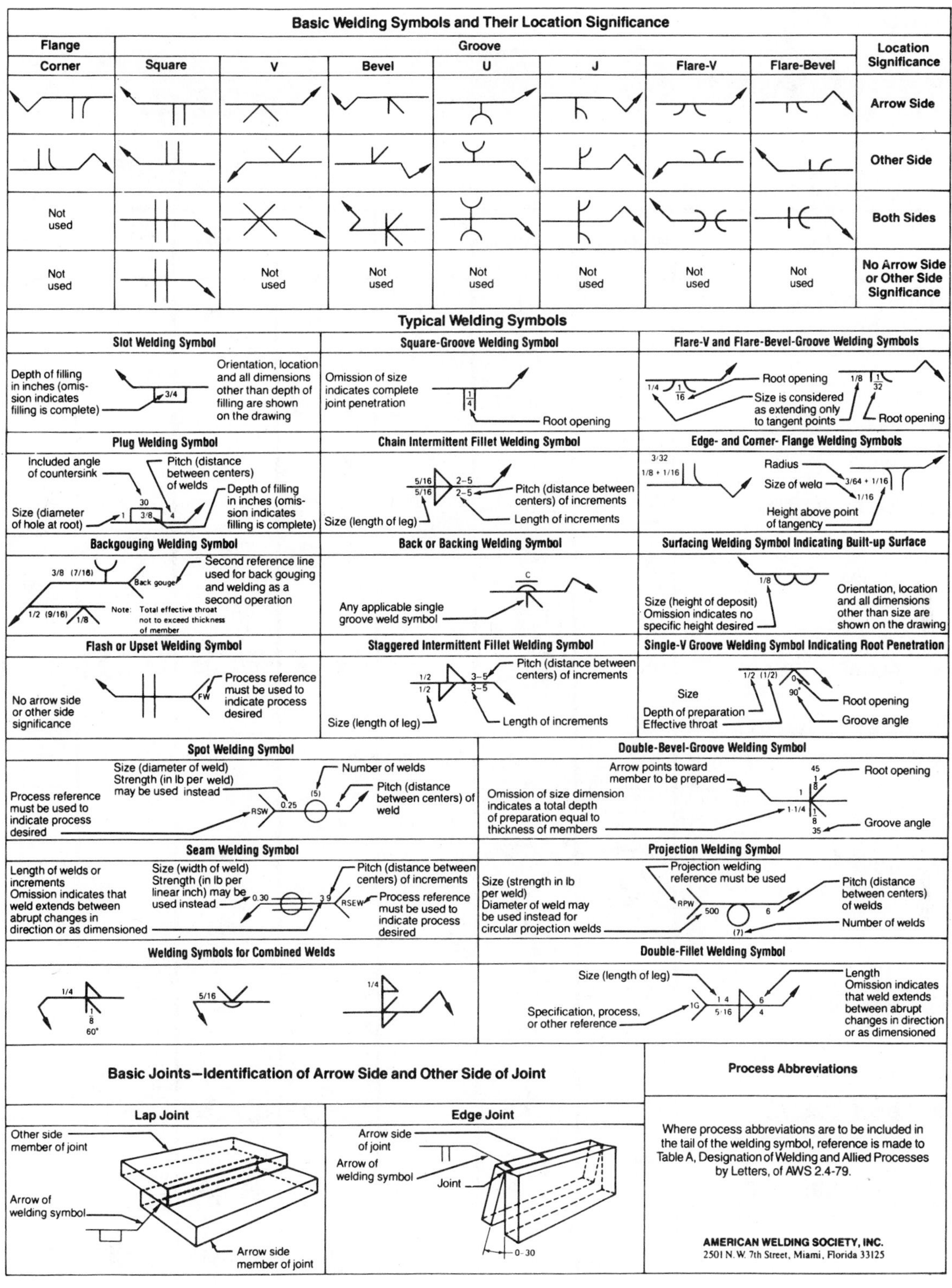
Basic Welding Symbols and Their Location Significance
Flange
Corner
Groove
Square
V
Bevel
U
J
Flare-V
Flare-Bevel
Location Significance
Arrow Side
Other Side
Both Sides
No Arrow Side or Other Side Significance
Not used
Typical Welding Symbols
Slot Welding Symbol
Depth of filling in inches (omission indicates filling is complete)
Orientation, location and all dimensions other than depth of filling are shown on the drawing
Square-Groove Welding Symbol
Omission of size indicates complete joint penetration
Root opening
Flare-V and Flare-Bevel-Groove Welding Symbols
Root opening
Size is considered as extending only to tangent points
Plug Welding Symbol
Included angle of countersink
Pitch (distance between centers) of welds
Depth of filling in inches (omission indicates filling is complete)
Size (diameter of hole at root)
Chain Intermittent Fillet Welding Symbol
Pitch (distance between centers) of increments
Size (length of leg)
Length of increments
Edge- and Corner- Flange Welding Symbols
Radius
Size of weld
Height above point of tangency
Backgouging Welding Symbol
Second reference line used for back gouging and welding as a second operation
Back gouge
Note: Total effective throat not to exceed thickness of member
Back or Backing Welding Symbol
Any applicable single groove weld symbol
Surfacing Welding Symbol Indicating Built-up Surface
Size (height of deposit) Omission indicates no specific height desired
Orientation, location and all dimensions other than size are shown on the drawing
Flash or Upset Welding Symbol
No arrow side or other side significance
Process reference must be used to indicate process desired
Staggered Intermittent Fillet Welding Symbol
Pitch (distance between centers) of increments
Size (length of leg)
Length of increments
Single-V Groove Welding Symbol Indicating Root Penetration
Size
Depth of preparation
Effective throat
Root opening
Groove angle
Spot Welding Symbol
Size (diameter of weld) Strength (in lb per weld) may be used instead
Number of welds
Pitch (distance between centers) of weld
Process reference must be used to indicate process desired
Double-Bevel-Groove Welding Symbol
Arrow points toward member to be prepared
Root opening
Omission of size dimension indicates a total depth of preparation equal to thickness of members
Groove angle
Seam Welding Symbol
Length of welds or increments Omission indicates that weld extends between abrupt changes in direction or as dimensioned
Size (width of weld) Strength (in lb per linear inch) may be used instead
Pitch (distance between centers) of increments
Process reference must be used to indicate process desired
Projection Welding Symbol
Projection welding reference must be used
Size (strength in lb per weld) Diameter of weld may be used instead for circular projection welds
Pitch (distance between centers) of welds
Number of welds
Welding Symbols for Combined Welds
Double-Fillet Welding Symbol
Size (length of leg)
Specification, process, or other reference
Length Omission indicates that weld extends between abrupt changes in direction or as dimensioned
Basic Joints—Identification of Arrow Side and Other Side of Joint
Lap Joint
Other side member of joint
Arrow of welding symbol
Arrow side member of joint
Edge Joint
Arrow side of joint
Arrow of welding symbol
Joint
Process Abbreviations
Where process abbreviations are to be included in the tail of the welding symbol, reference is made to Table A, Designation of Welding and Allied Processes by Letters, of AWS 2.4-79.
AMERICAN WELDING SOCIETY, INC.
2501 N.W. 7th Street, Miami, Florida 33125

APPENDIX D.2

Piping Symbols

Orthographic Piping Symbols

TYPE	FLANGED	SCREWED	BELL AND SPIGOT	WELDED X OR ●	SOLDERED	DOUBLE LINE	PICTORIAL
ANGLE VALVES							
1. CHECK							
2. GATE (ELEVATION)							
3. GATE (PLAN)							
4. GLOBE (ELEVATION)							
5. GLOBE (PLAN)							
AUTOMATIC VALVES							
6. BY-PASS							
7. GOVERNORED OPERATED							
8. REDUCING							
9. BALL VALVE							
10. BUSHING							
11. BUTTERFLY VALVE							
CHECK VALVES							
12. STRAIGHTWAY							
13. COCK OR PLUG VALVE							
14. CAP							

TYPE	FLANGED	SCREWED	BELL AND SPIGOT	WELDED X OR ●	SOLDERED	DOUBLE LINE	PICTORIAL
15. COUPLING							
16. CROSS, STRAIGHT							
17. CROSS, REDUCING							
18. CROSS							
19. DIAPHRAGM VALVE							
ELBOWS 20. 45°							
21. 90°							
22. TURNED DOWN							
23. TURNED UP							
24. BASE							
25. DOUBLE BRANCH							
26. LONG RADIUS							
27. REDUCING							
28. SIDE OUTLET (TURNED DOWN)							
29. SIDE OUTLET (TURNED UP)							
30. ELBOWLET							
FLANGES 31. BLIND							

TYPE	FLANGED	SCREWED	BELL AND SPIGOT	WELDED X OR ●	SOLDERED	DOUBLE LINE	PICTORIAL
32. ORIFICE							
33. REDUCING							
34. SOCKET WELD							
35. WELD NECK							
36. FLOAT VALVE							
37. GATE VALVE							
38. MOTOR OPERATED GATE VALVE	M	M		M			
39. GLOBE VALVE							
40. MOTOR OPERATED GLOBE VALVE	M	M		M			
HOSE VALVE							
41. ANGLE							
42. GATE							
43. GLOBE							
JOINTS							
44. CONNECTING PIPE							
45. EXPANSION							
46. LATERAL							
47. LOCKSHIELD VALVE							
48. MOTOR CONTROL VALVE							
PLUGS							
49. BULL							
50. PIPE							

TYPE	FLANGED	SCREWED	BELL AND SPIGOT	WELDED X OR ●	SOLDERED	DOUBLE LINE	PICTORIAL
51. QUICK OPENING							
REDUCERS							
52. CONCENTRIC							
53. ECCENTRIC							
54. SOLENOID VALVE							
55. RELIEF VALVE							
56. SAFETY VALVE							
57. SLEEVE							
58. STRAINER							
TEES							
59. STRAIGHT SIZE							
60. OUTLET UP							
61. OUTLET DOWN							
62. DOUBLE SWEEP							
63. REDUCING							
64. SINGLE SWEEP							
65. SIDE OUTLET DOWN							
66. SIDE OUTLET UP							
67. UNION							
68. Y-VALVE							

APPENDIX D.3

Electronic Symbols

Typical Graphic Symbols for Electronic Diagrams with Basic Device Designations

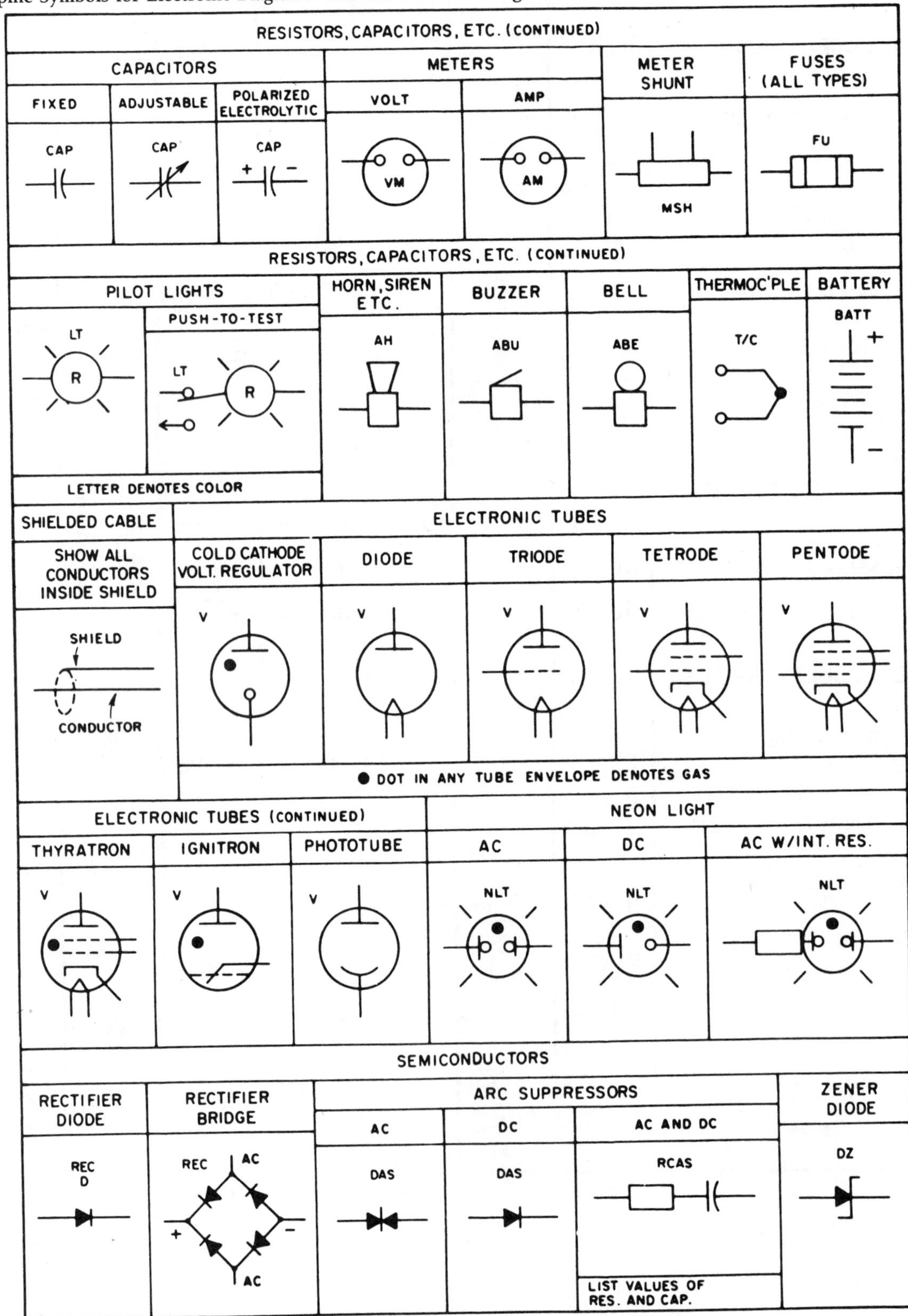

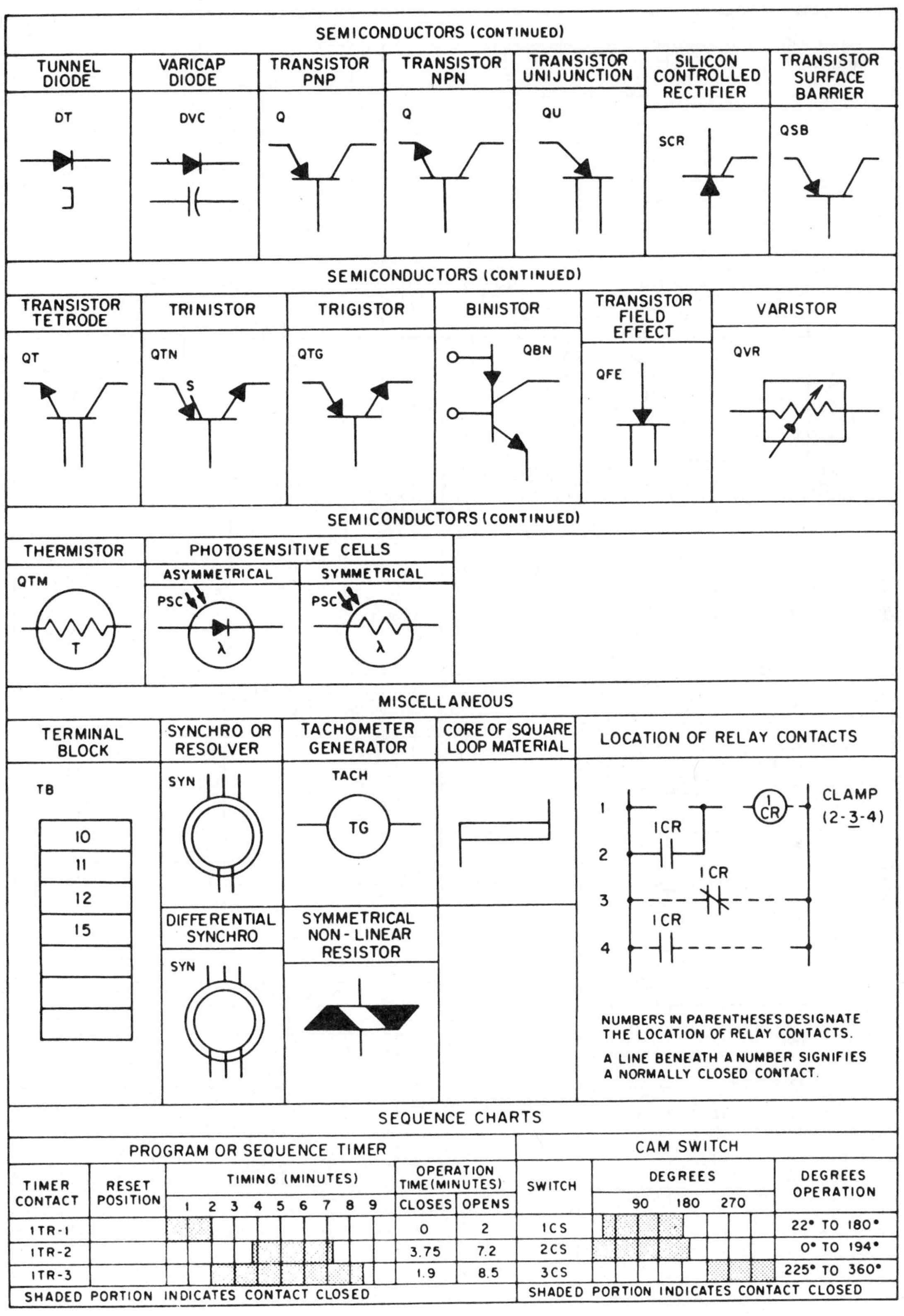

SEQUENCE CHARTS

PROGRAM OR SEQUENCE TIMER

TIMER CONTACT	RESET POSITION	TIMING (MINUTES) 1 2 3 4 5 6 7 8 9	OPERATION TIME (MINUTES) CLOSES	OPENS
1TR-1			0	2
1TR-2			3.75	7.2
1TR-3			1.9	8.5

SHADED PORTION INDICATES CONTACT CLOSED

CAM SWITCH

SWITCH	DEGREES 90 180 270	DEGREES OPERATION
1CS		22° TO 180°
2CS		0° TO 194°
3CS		225° TO 360°

SHADED PORTION INDICATES CONTACT CLOSED

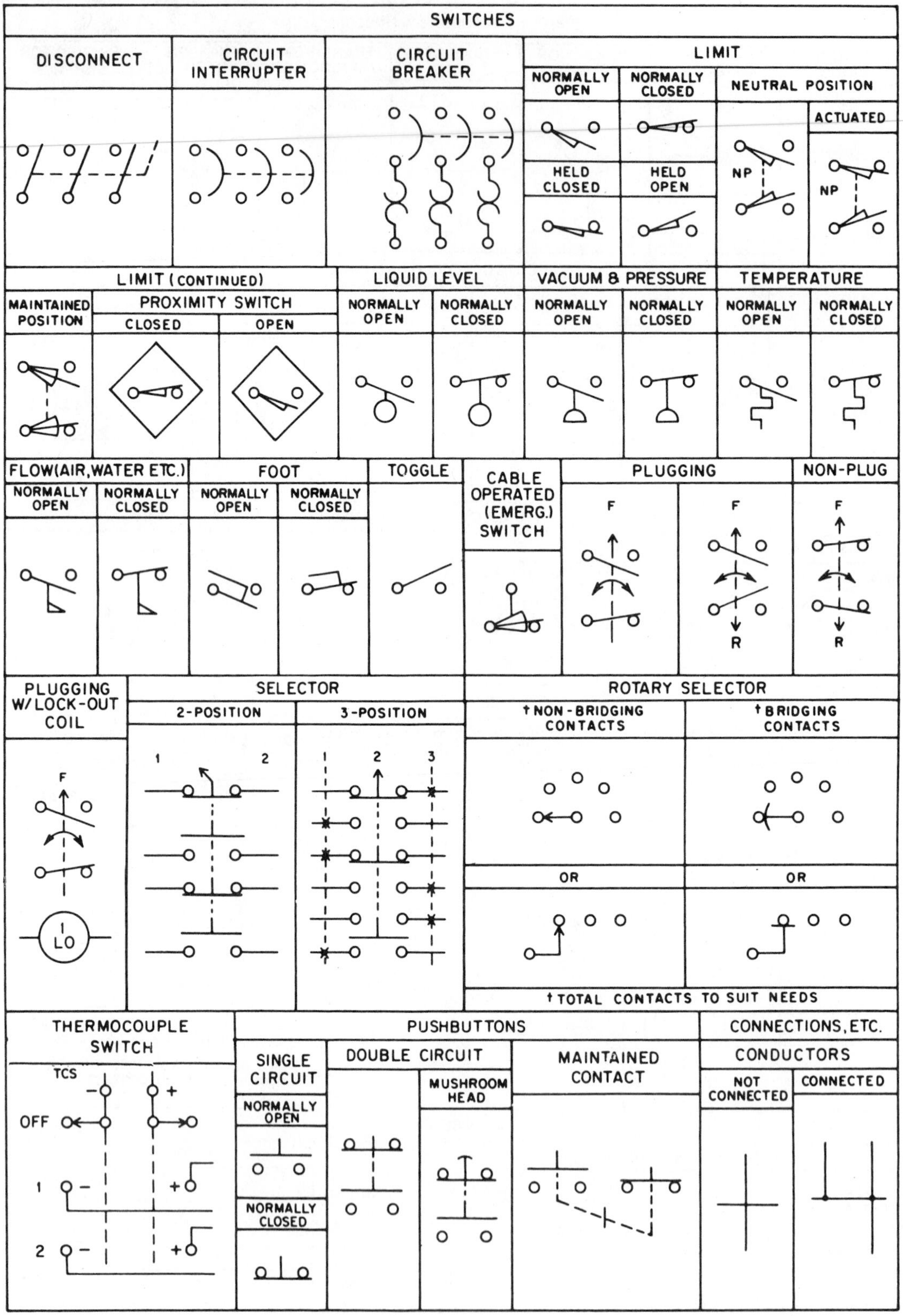

SWITCHES

DISCONNECT

CIRCUIT INTERRUPTER

CIRCUIT BREAKER

LIMIT

NORMALLY OPEN

NORMALLY CLOSED

HELD CLOSED

HELD OPEN

NEUTRAL POSITION

NP

ACTUATED

NP

LIMIT (CONTINUED)

MAINTAINED POSITION

PROXIMITY SWITCH

CLOSED

OPEN

LIQUID LEVEL

NORMALLY OPEN

NORMALLY CLOSED

VACUUM & PRESSURE

NORMALLY OPEN

NORMALLY CLOSED

TEMPERATURE

NORMALLY OPEN

NORMALLY CLOSED

FLOW(AIR, WATER ETC.)

NORMALLY OPEN

NORMALLY CLOSED

FOOT

NORMALLY OPEN

NORMALLY CLOSED

TOGGLE

CABLE OPERATED (EMERG.) SWITCH

PLUGGING

F

R

F

R

NON-PLUG

F

R

PLUGGING W/LOCK-OUT COIL

F

LO

SELECTOR

2-POSITION

1 2

3-POSITION

1 2 3

ROTARY SELECTOR

† NON-BRIDGING CONTACTS

OR

† BRIDGING CONTACTS

OR

† TOTAL CONTACTS TO SUIT NEEDS

THERMOCOUPLE SWITCH

TCS

OFF

1

2

PUSHBUTTONS

SINGLE CIRCUIT

NORMALLY OPEN

NORMALLY CLOSED

DOUBLE CIRCUIT

MUSHROOM HEAD

MAINTAINED CONTACT

CONNECTIONS, ETC.

CONDUCTORS

NOT CONNECTED

CONNECTED

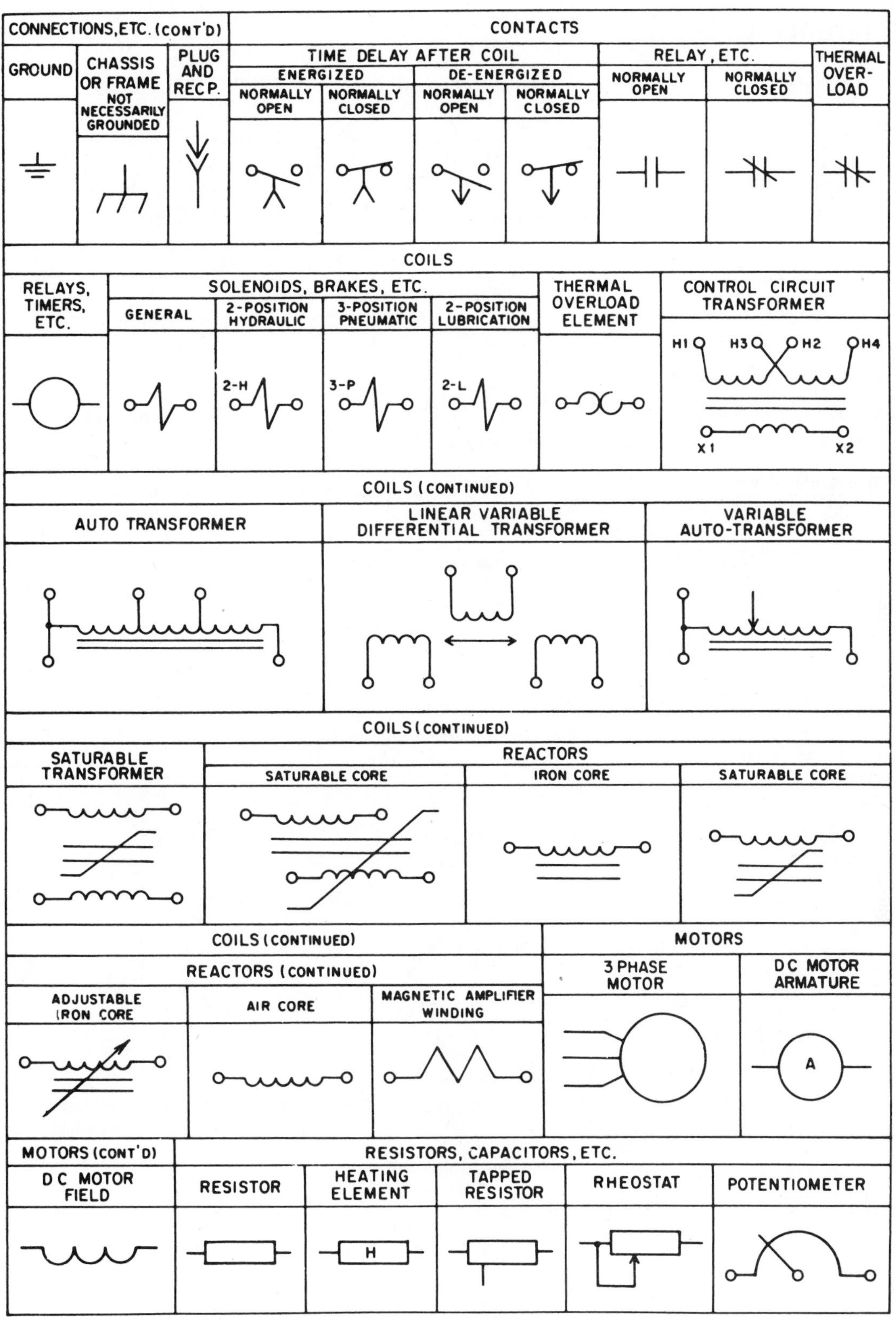
CONNECTIONS, ETC. (CONT'D)
CONTACTS
GROUND
CHASSIS OR FRAME NOT NECESSARILY GROUNDED
PLUG AND RECP.
TIME DELAY AFTER COIL
ENERGIZED
DE-ENERGIZED
NORMALLY OPEN
NORMALLY CLOSED
NORMALLY OPEN
NORMALLY CLOSED
RELAY, ETC.
NORMALLY OPEN
NORMALLY CLOSED
THERMAL OVER-LOAD
COILS
RELAYS, TIMERS, ETC.
SOLENOIDS, BRAKES, ETC.
GENERAL
2-POSITION HYDRAULIC
3-POSITION PNEUMATIC
2-POSITION LUBRICATION
2-H
3-P
2-L
THERMAL OVERLOAD ELEMENT
CONTROL CIRCUIT TRANSFORMER
H1
H3
H2
H4
X1
X2
COILS (CONTINUED)
AUTO TRANSFORMER
LINEAR VARIABLE DIFFERENTIAL TRANSFORMER
VARIABLE AUTO-TRANSFORMER
COILS (CONTINUED)
SATURABLE TRANSFORMER
REACTORS
SATURABLE CORE
IRON CORE
SATURABLE CORE
COILS (CONTINUED)
REACTORS (CONTINUED)
ADJUSTABLE IRON CORE
AIR CORE
MAGNETIC AMPLIFIER WINDING
MOTORS
3 PHASE MOTOR
DC MOTOR ARMATURE
A
MOTORS (CONT'D)
DC MOTOR FIELD
RESISTORS, CAPACITORS, ETC.
RESISTOR
HEATING ELEMENT
H
TAPPED RESISTOR
RHEOSTAT
POTENTIOMETER

Quick Reference to Symbols

1. Qualifying Symbols

1.1 Adjustability
Variability

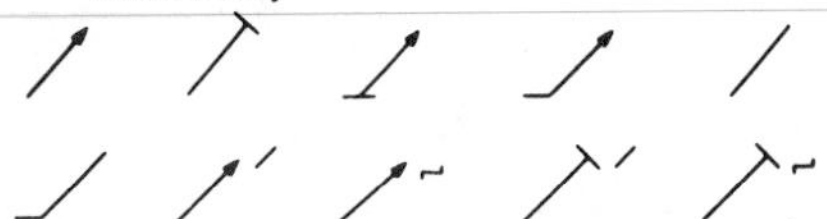

1.2 Special-Property Indicators

1.3 Radiation Indicators

1.4 Physical State Recognition Symbols

1.5 Test-Point Recognition Symbol

1.6 Polarity Markings

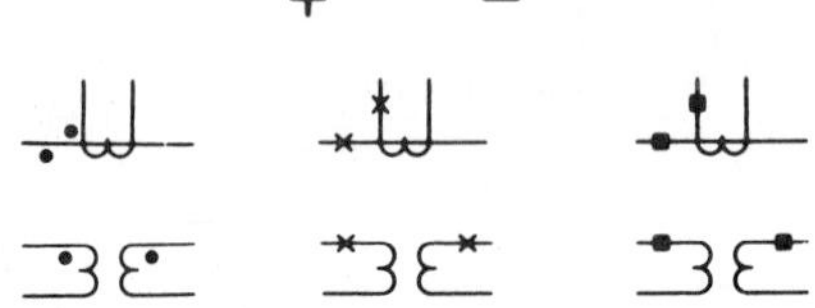

1.7 Direction of Flow of Power, Signal, or Information

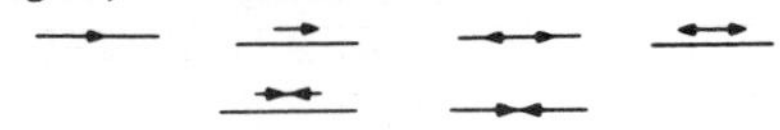

1.8 Kind of Current

1.9 Connection Symbols

1.10 Envelope
Enclosure

1.11 Shield
Shielding

1.12 Special Connector or Cable Indicator

1.13 Electret

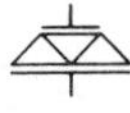

2. Fundamental Items

2.1 Resistor

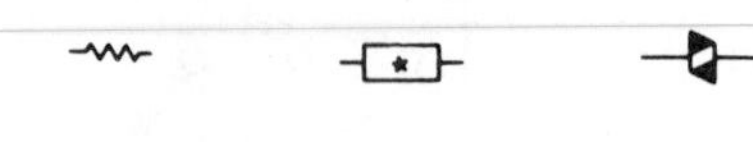

2.2 Capacitor

2.3 Antenna

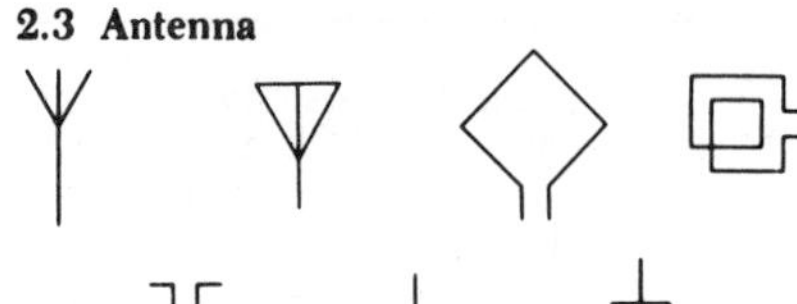

2.4 Attenuator

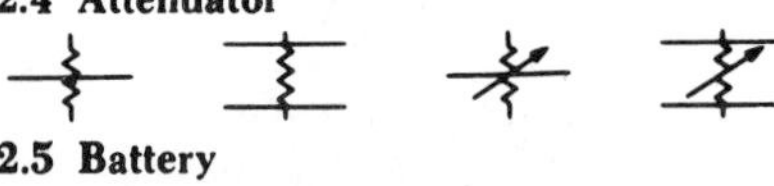

2.5 Battery

2.6 Delay Function
Delay Line
Slow-Wave Structure

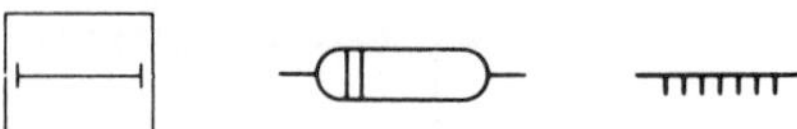

2.7 Oscillator
Generalized Alternating-Current Source

2.8 Permanent Magnet

2.9 Pickup
Head

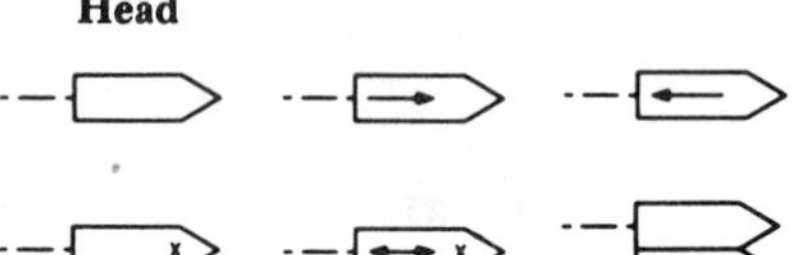

2.10 Piezoelectric Crystal Unit

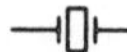

2.11 Primary Detector
Measuring Transducer

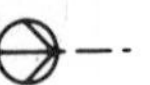

2.12 Squib, Electrical

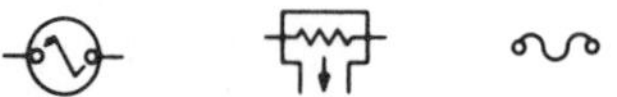

2.13 Thermocouple

2.14 Thermal Element
Thermomechanical Transducer

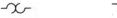

2.15 Spark gap
Igniter gap

2.16 Continuous Loop Fire Detector (temperature sensor)

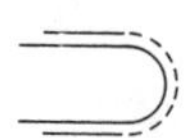

2.17 Ignitor Plug

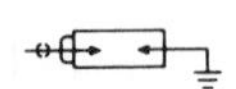

3. Transmission Path

3.1 Transmission Path
Conductor
Cable
Wiring

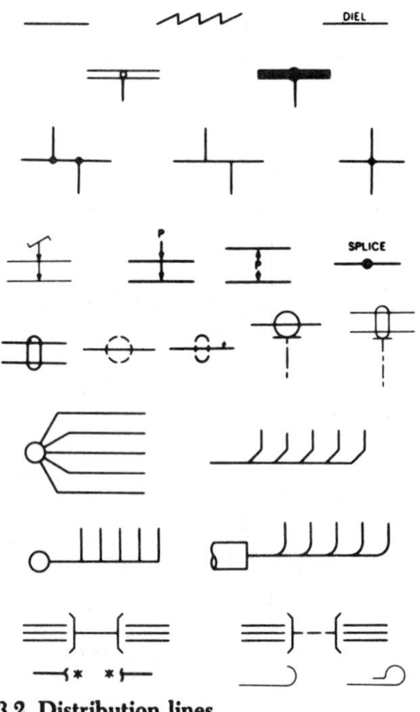

3.2 Distribution lines
Transmission lines

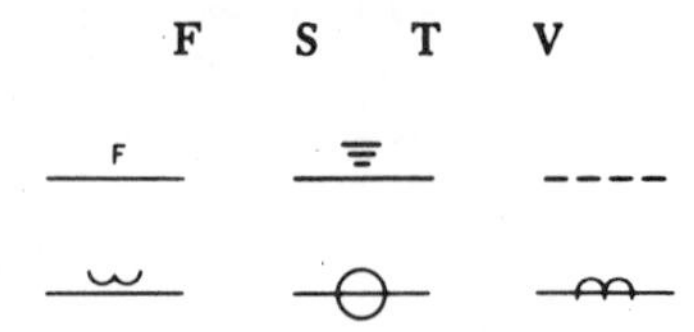

3.3 Alternative or Conditioned Wiring

3.4 Associated or Future

3.5 Intentional Isolation of Direct-Current Path in Coaxial or Waveguide Applications

3.6 Waveguide

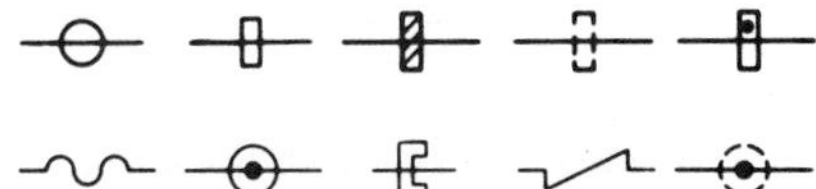

3.7 Strip-Type Transmission Line

3.8 Termination

3.9 Circuit Return

3.10 Pressure-Tight Bulkhead Cable Gland
Cable Sealing End

4. Contacts, Switches, Contactors, and Relays

4.1 Switching Function

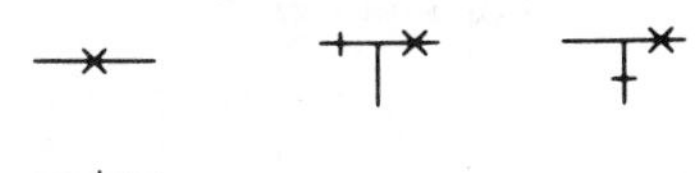

4.2 Electrical Contact

4.3 Basic Contact Assemblies

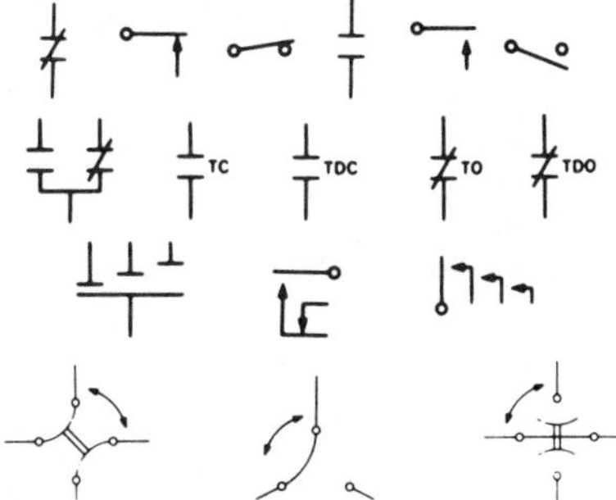

4.4 Magnetic Blowout Coil

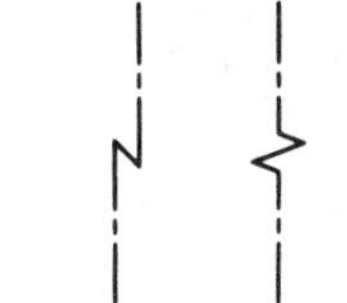

4.5 Operating Coil
Relay Coil

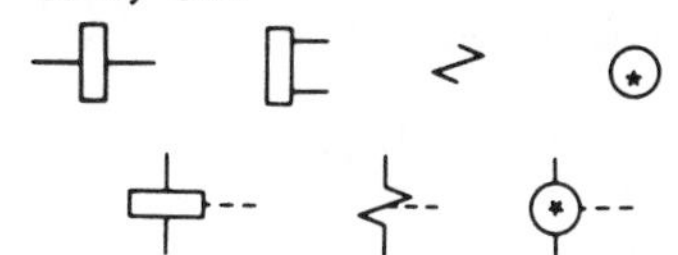

4.6 Switch

4.7 Pushbutton, Momentary or Spring-Return

4.8 Two-Circuit, Maintained or Not Spring-Return

4.9 Nonlocking Switch, Momentary or Spring-Return

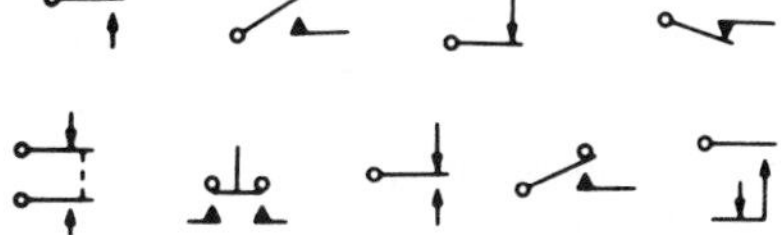

4.10 Locking Switch

4.11 Combination Locking and Non-locking Switch

4.12 Key-Type Switch
Lever Switch

4.13 Selector or Multiposition Switch

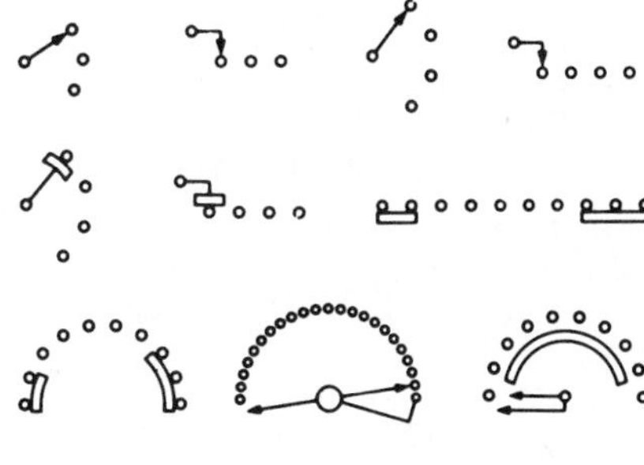

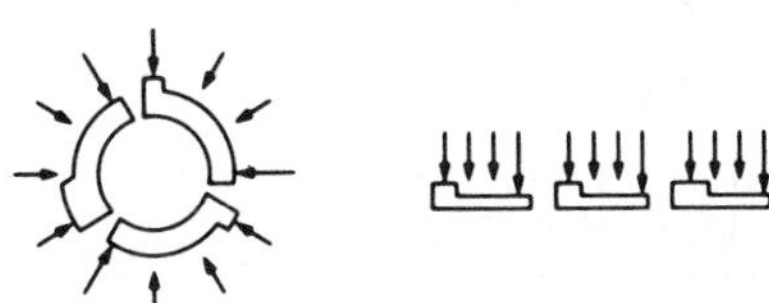

4.14 Limit Switch
Sensitive Switch

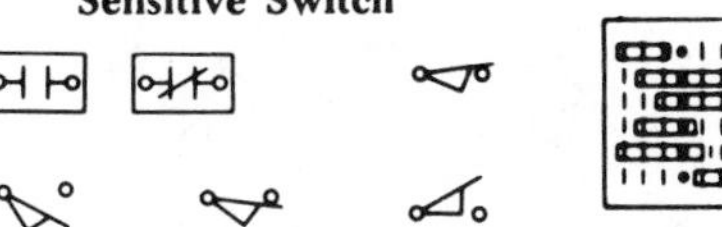

4.15 Safety Interlock

4.16 Switches with Time-Delay Feature

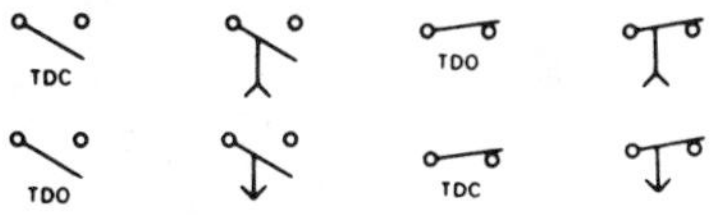

4.17 Flow-Actuated Switch

4.18 Liquid-Level-Actuated Switch

4.19 Pressure- or Vacuum-Actuated Switch

4.20 Temperature-Actuated Switch

4.21 Thermostat

4.22 Flasher
Self-interrupting switch

4.23 Foot-Operated Switch
Foot Switch

4.24 Switch Operated by Shaft Rotation and Responsive to Speed or Direction

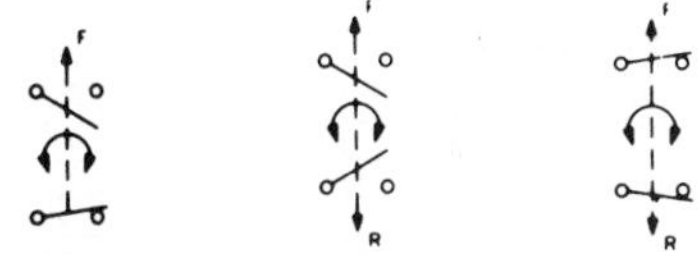

4.25 Switches with Specific Features

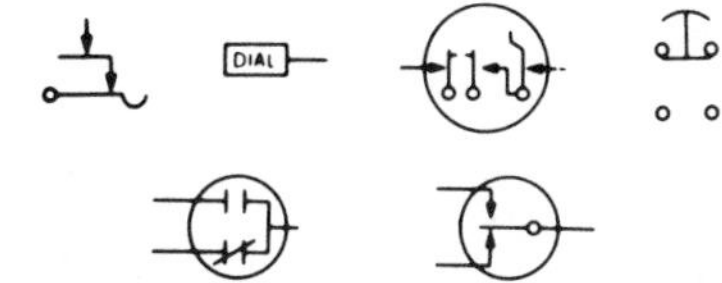

4.26 Telegraph Key

4.27 Governor
Speed Regulator

4.28 Vibrator
Interrupter

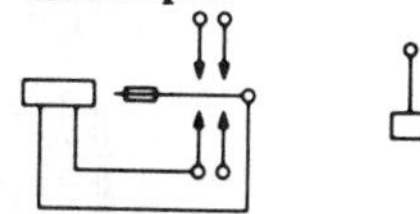

4.29 Contactor

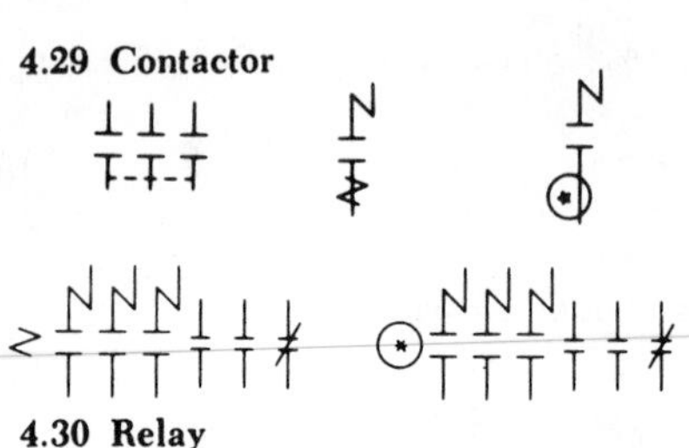

4.30 Relay

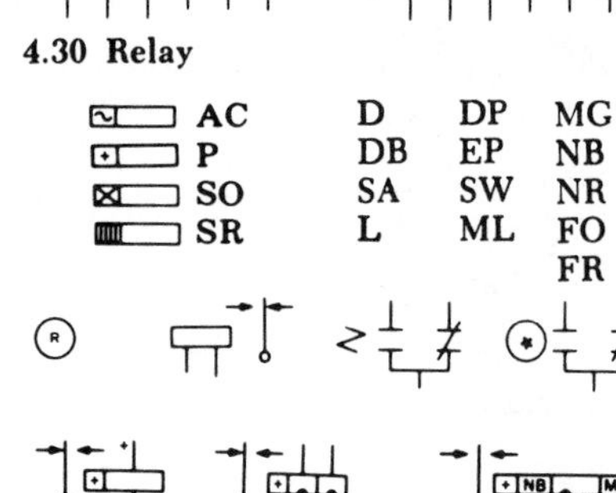

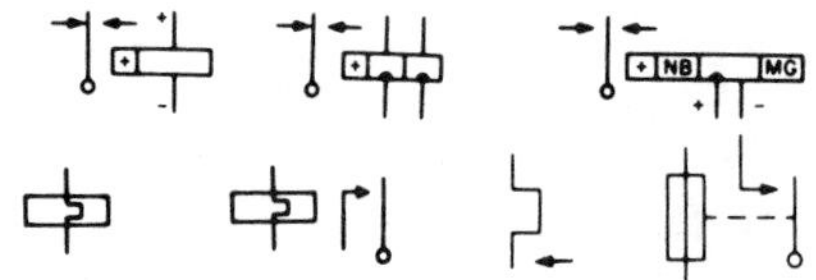

4.31 Inertia Switch

4.32 Mercury Switch

4.33 Aneroid Capsule

5. Terminals and Connectors

5.1 Terminals

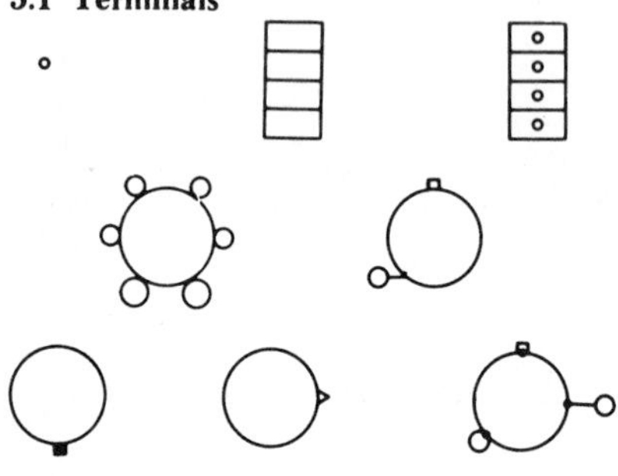

5.2 Cable Termination

5.3 Connector
Disconnecting Device

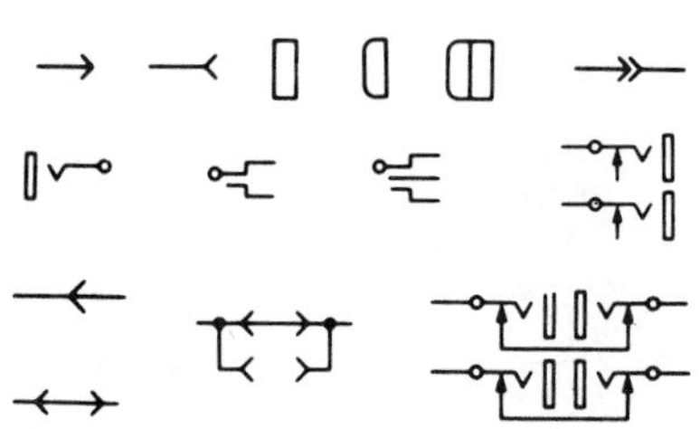

5.4 Connectors of the Type Commonly Used for Power-Supply Purposes

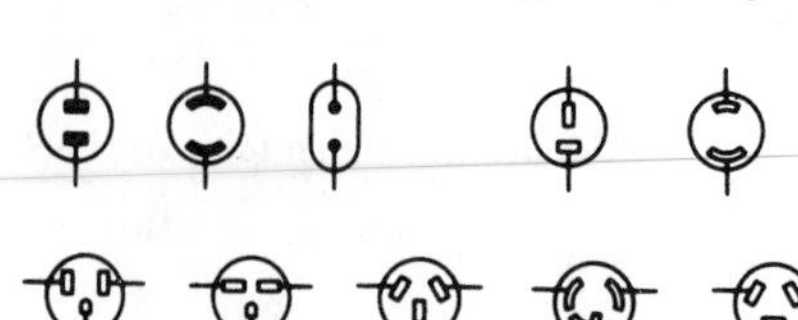

5.5 Test Blocks

5.6 Coaxial Connector

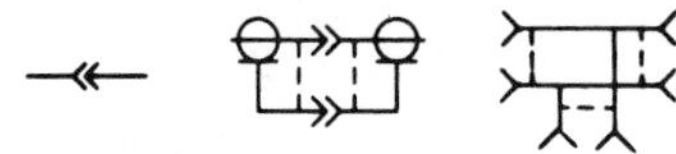

5.7 Waveguide Flanges
Waveguide junction

6. Transformers, Inductors, and Windings

6.1 Core

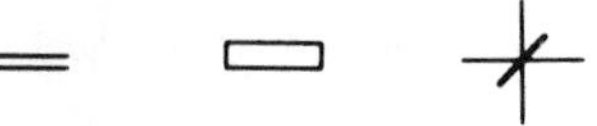

6.2 Inductor
Winding
Reactor
Radio frequency coil
Telephone retardation coil

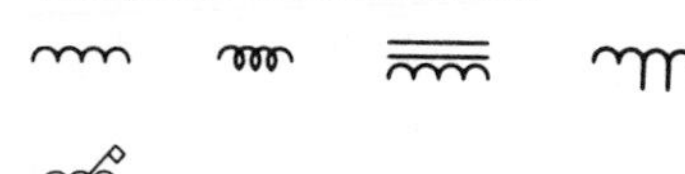

6.3 Transductor

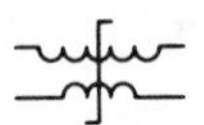

6.4 Transformer
Telephone induction coil
Telephone repeating coil

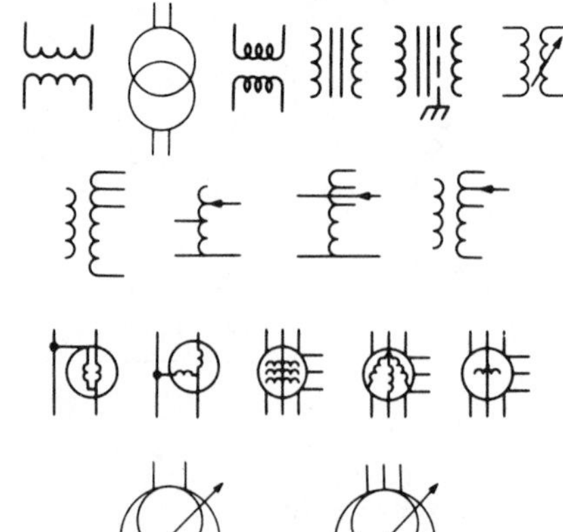

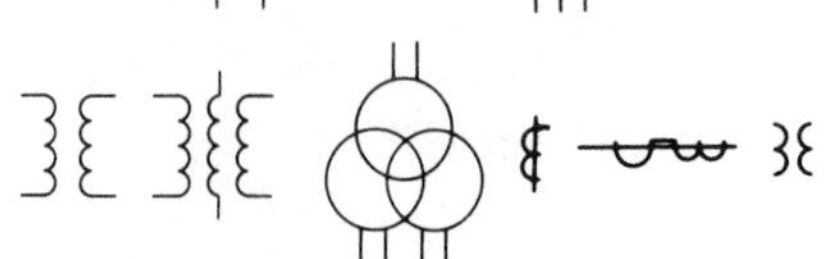

6.5 Linear Coupler

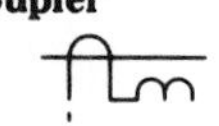

7. Electron Tubes and Related Devices

7.1 Electron Tube

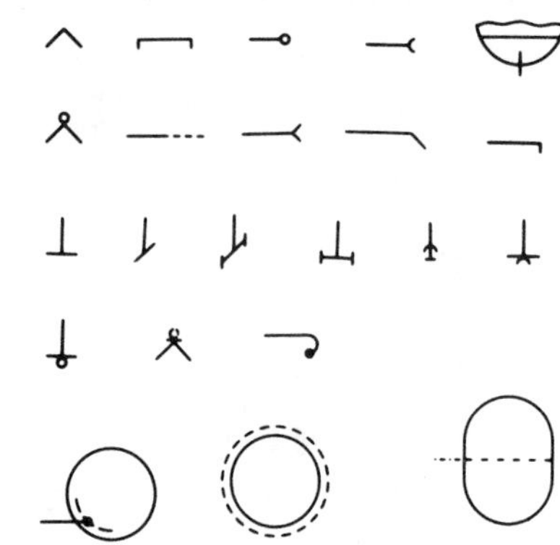

7.2 General Notes

7.3 Typical Applications

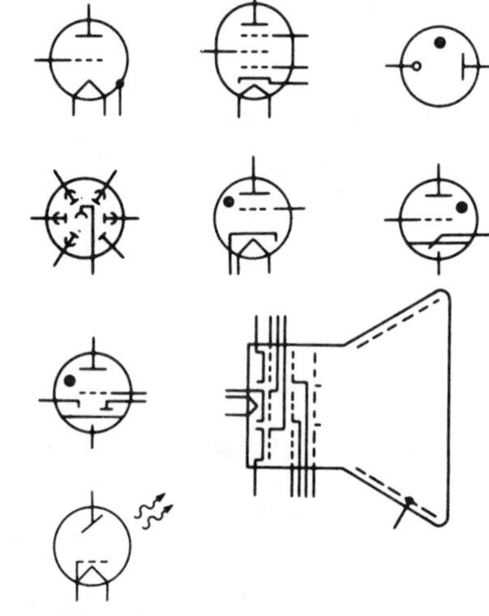

7.4 Solion
Ion-Diffusion Device

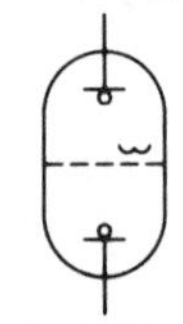

7.5 Coulomb Accumulator
Electrochemical Step-Function Device

7.6 Conductivity cell

7.7 Nuclear-Radiation Detector
Ionization Chamber
Proportional Counter Tube
Geiger-Müller Counter Tube

8. Semiconductor Devices

8.1 Semiconductor Device
Transistor
Diode

8.2 Element Symbols

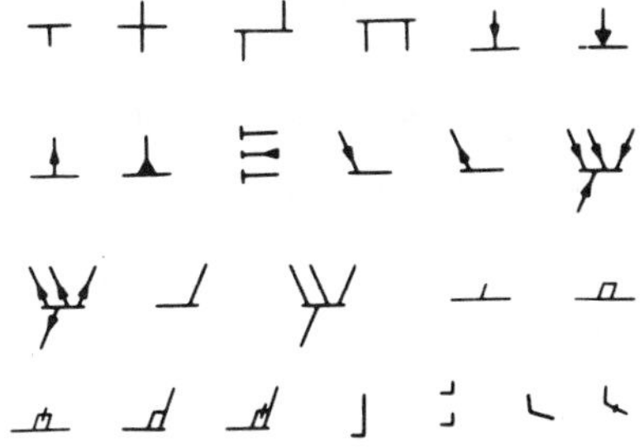

8.3 Special Property Indicators

8.4 Rules for Drawing Style 1 Symbols

8.5 Typical Applications: Two-Terminal Devices

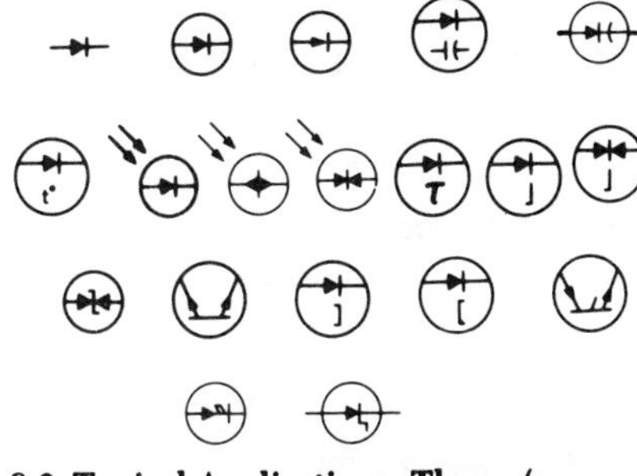

8.6 Typical Applications: Three- (or More) Terminal Devices

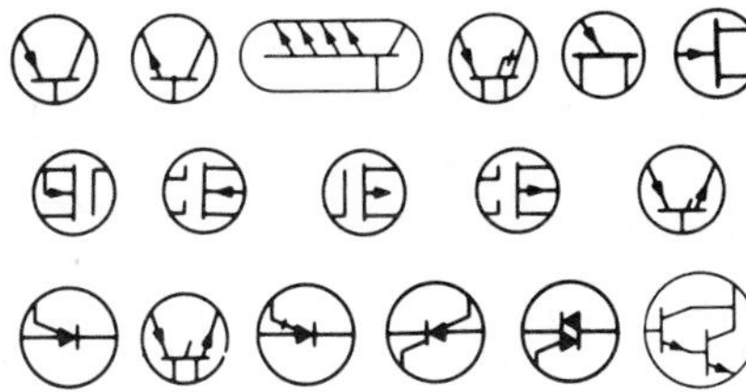

8.7 Photosensitive Cell

8.8 Semiconductor Thermocouple

8.9 Hall Element
Hall Generator

8.10 Photon-coupled isolator

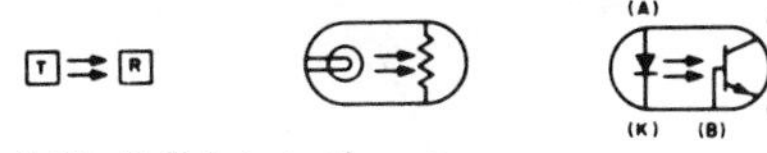

8.11 Solid-state-thyratron

9. Circuit Protectors

9.1 Fuse

9.2 Current Arrester

9.3 Lightning Arrester
Arrester
Gap

9.4 Circuit Breaker

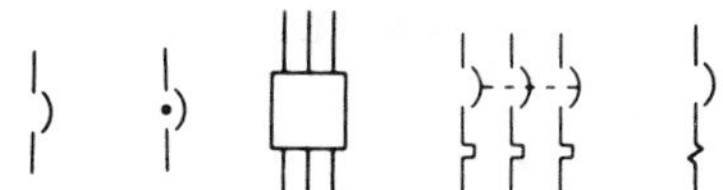

9.5 Protective Relay

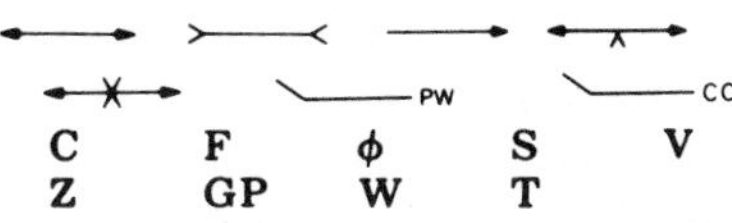

C F ϕ S V
Z GP W T

10. Acoustic Devices

10.1 Audible-Signaling Device

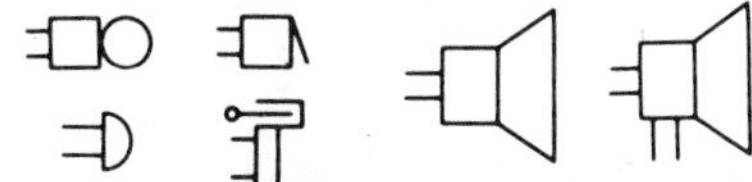

10.2 Microphone

10.3 Handset
Operator's Set

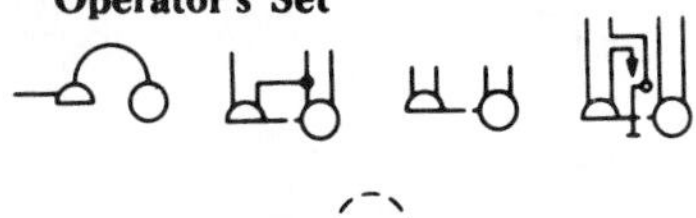

10.4 Telephone Receiver
Earphone
Hearing-Aid Receivers

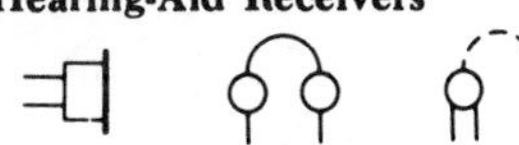

11. Lamps and Visual-Signaling Devices

11.1 Lamp

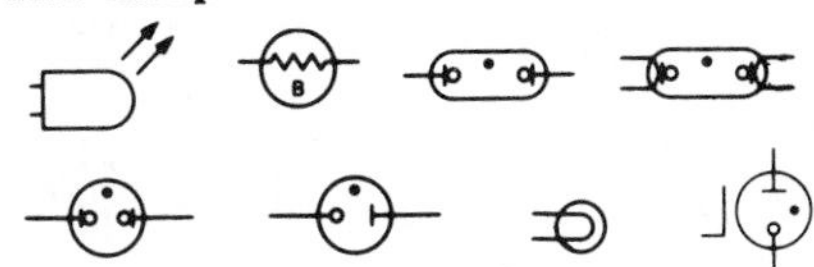

11.2 Visual-Signaling Device

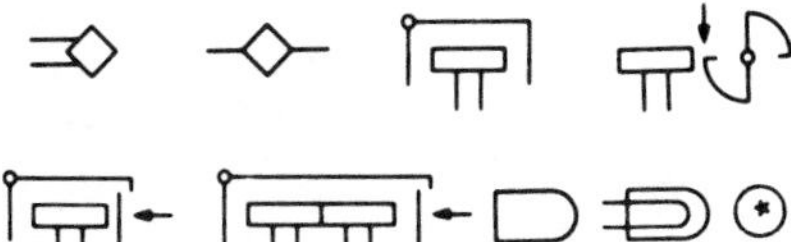

12. Readout Devices

12.1 Meter
Instrument

A	DB	I	OP	RF	VA
AH	DBM	INT	OSCG	SY	VAR
C	DM	μA	PH	TLM	VARH
CMA	DTR	UA	PI	t°	VI
CMC	F	MA	PF	THC	VU
CMV	G	NM	RD	TT	W
CRO	GD	OHM	REC	V	WH

12.2 Electromagnetically Operated Counter
Message Register

13. Rotating Machinery

13.1 Rotating Machine

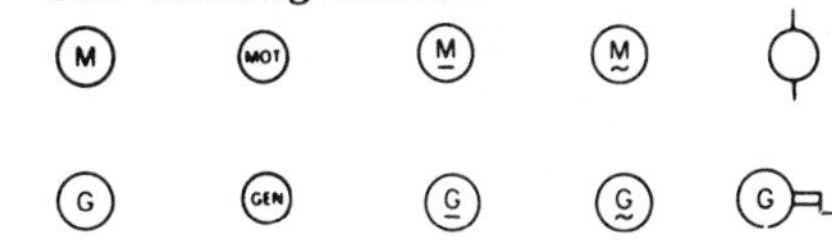

13.2 Field, Generator or Motor

13.3 Winding Connection Symbols

13.4 Applications: Direct-Current Machines

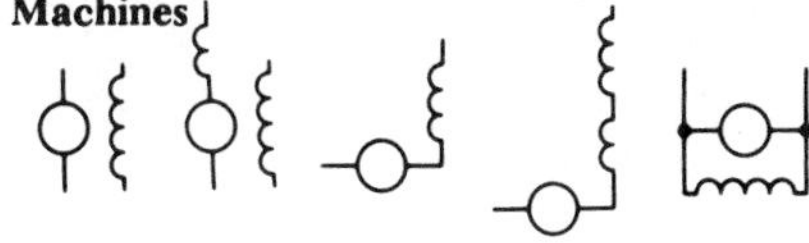

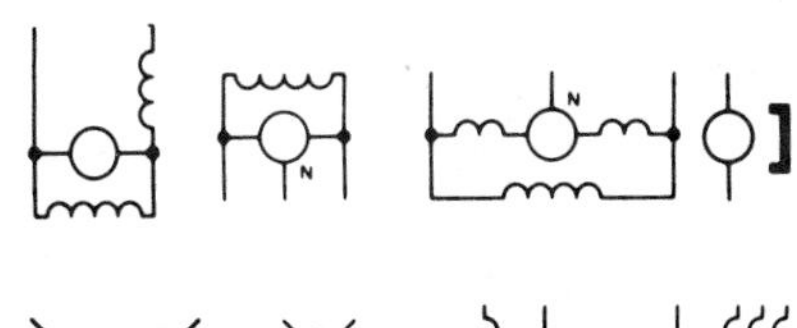

13.5 Applications: Alternating-Current Machines

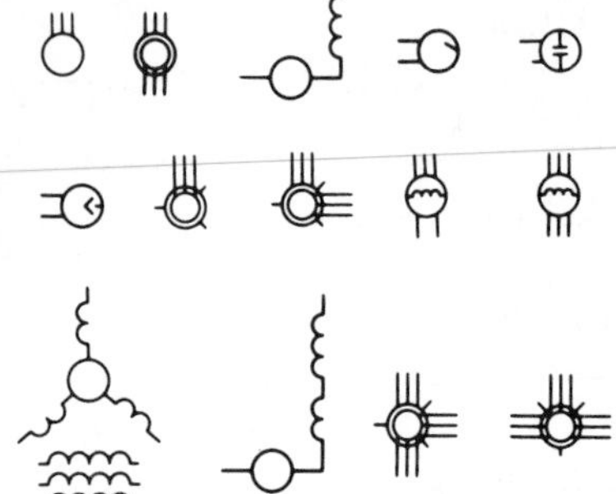

13.6 Applications: Alternating-Current Machines with Direct-Current Field Excitation

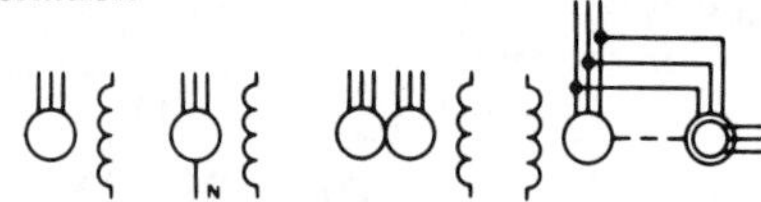

13.7 Applications: Alternating- and Direct-Current Composite

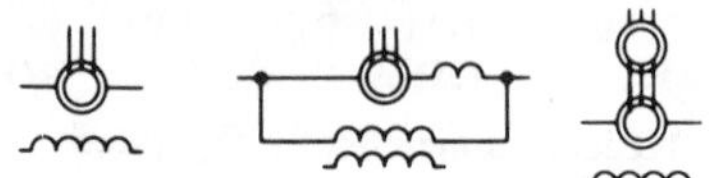

13.8 Synchro

CDX	TDX
CT	TR
CX	TX
TDR	RS

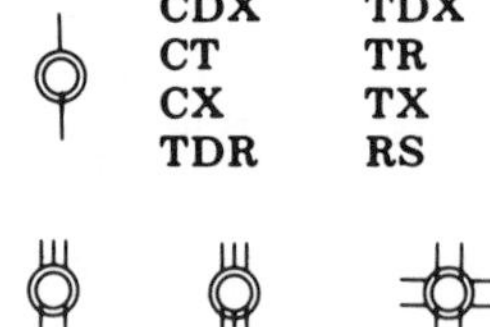

14. Mechanical Functions

14.1 Mechanical Connection
Mechanical Interlock

14.2 Mechanical Motion

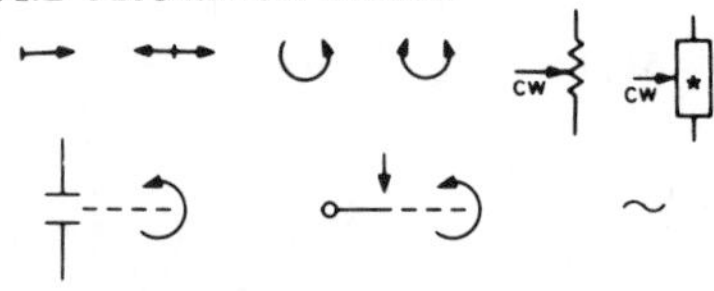

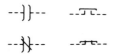

14.3 Clutch
Brake

14.4 Manual Control

15. Commonly Used in Connection with VHF, UHF, SHF Circuits

15.1 Discontinuity

15.2 Coupling

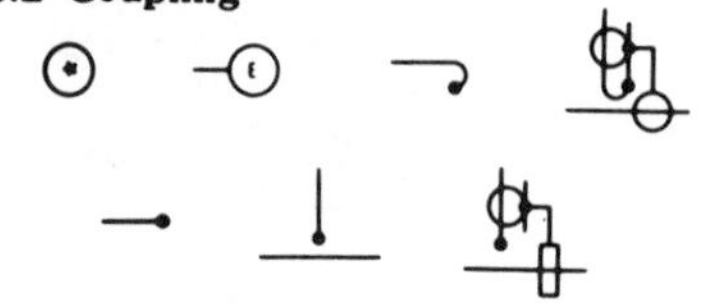

15.3 Directional Coupler

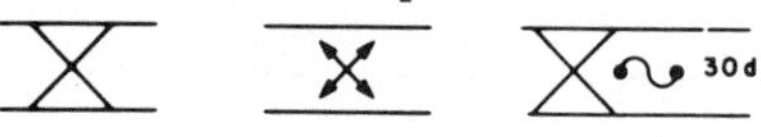

15.4 Hybrid
Directionally Selective
Transmission Devices

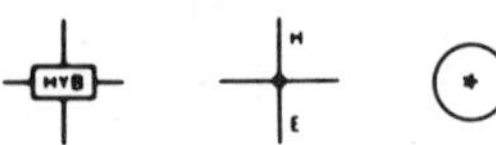

15.5 Mode Transducer

15.6 Mode Suppression

15.7 Rotary Joint

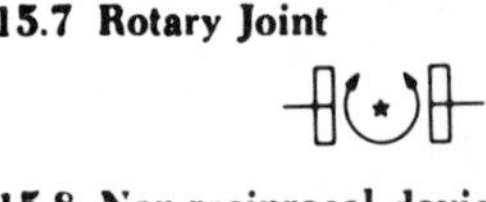

15.8 Non-reciprocal devices

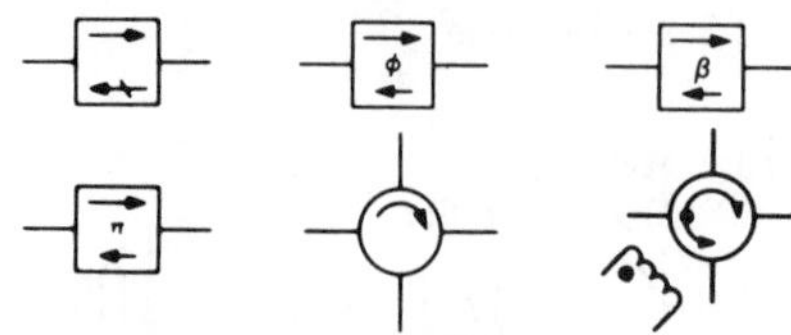

15.9 Resonator
Tuned Cavity

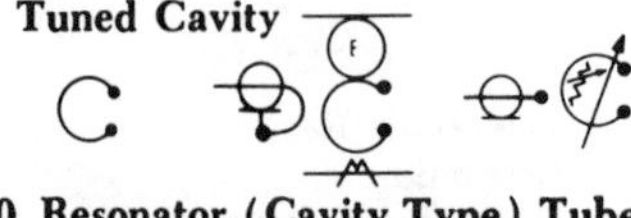

15.10 Resonator (Cavity Type) Tube

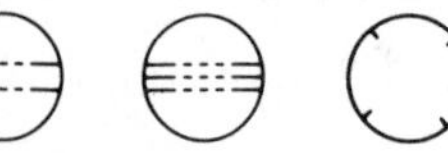

15.11 Magnetron

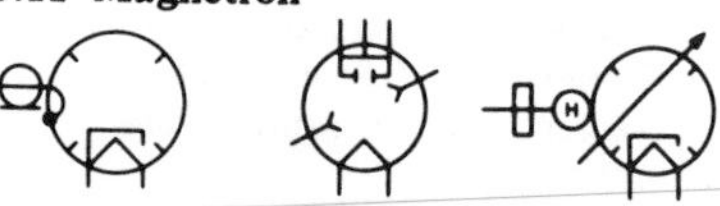

15.12 Velocity-Modulation (Velocity-Variation) Tube

15.13 Transmit-Receive (TR) Tube

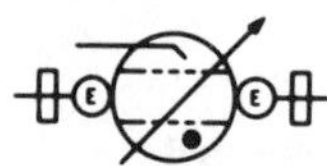

15.14 Traveling-Wave-Tube

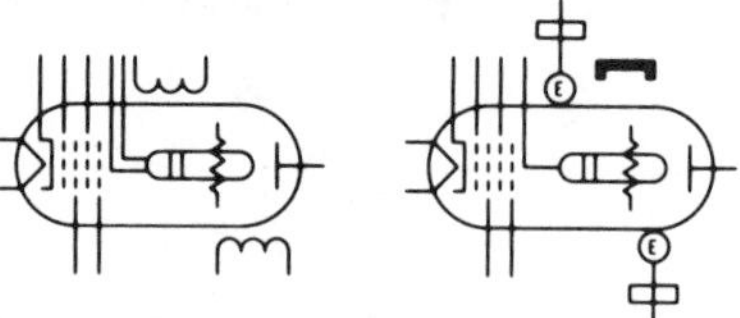

15.15 Balun

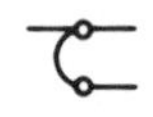

15.16 Filter

15.17 Phase shifter

15.18 Ferrite bead rings

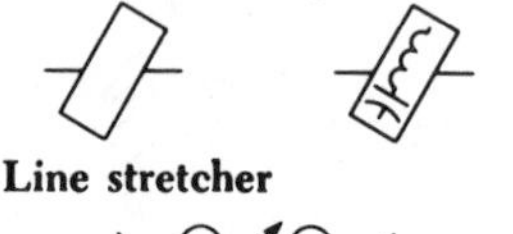

15.19 Line stretcher

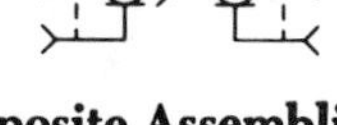

16. Composite Assemblies

16.1 Circuit assembly
Circuit subassembly
Circuit element

EQ	FL-BP	RG	TPR
FAX	FL-HP	RU	TTY
FL	FL-LP	DIAL	CLK
FL-BE	PS	TEL	IND
ST-INV			

16.2 Amplifier

BDG	EXP	PRE
BST	LIM	PWR
CMP	MON	TRQ
DC	PGM	

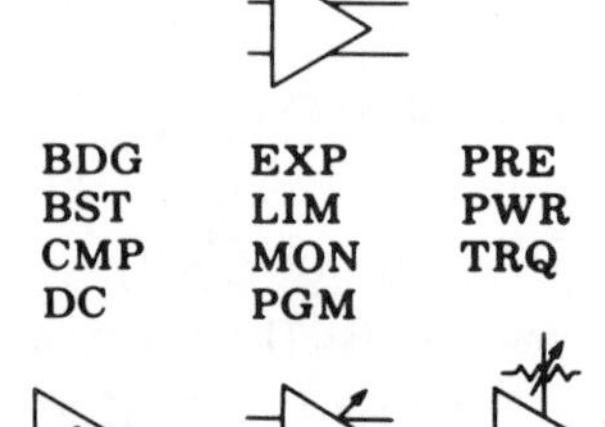

16.3 Rectifier

16.4 Repeater

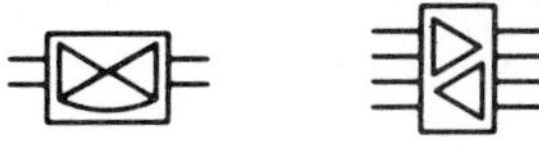

16.5 Network

16.6 Phase Shifter
Phase-Changing Network

16.7 Chopper

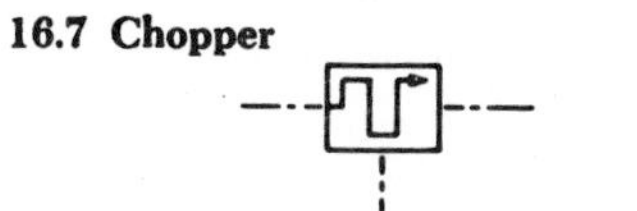

16.8 Diode-type ring demodulator
Diode-type ring modulator

16.9 Gyro
Gyroscope
Gyrocompass

16.10 Position Indicator

16.11 Position Transmitter

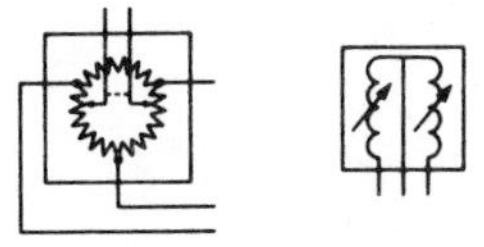

16.12 Fire Extinguisher Actuator Head

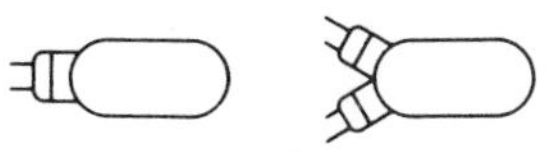

17. Analog Functions

17.1 Operational Amplifier

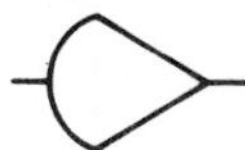

17.2 Summing Amplifier

17.3 Integrator

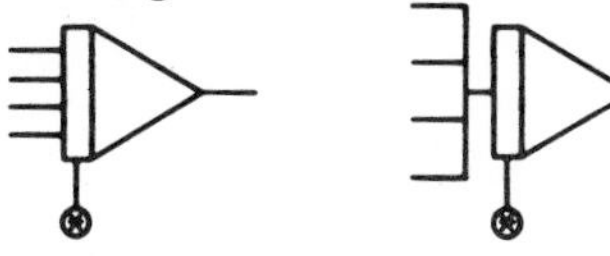

17.4 Electronic Multiplier

17.5 Electronic Divider

17.6 Electronic Function Generator

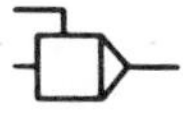

17.7 Generalized Integrator

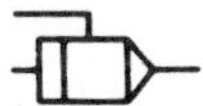

17.8 Positional Servo-mechanism

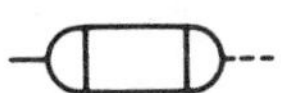

17.9 Function Potentiometer

18. Digital Logic Functions

18.1 Digital Logic Functions
(See cross references)

19. Special Purpose Maintenance Diagrams

19.1 Data flow code signals

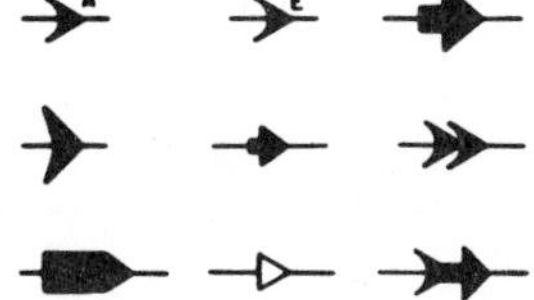

19.2 Functional Circuits

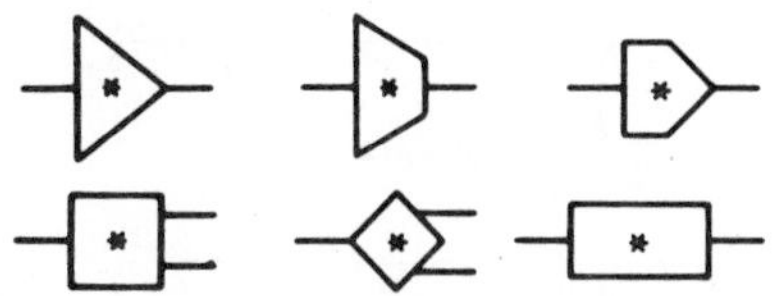

20. System Diagrams, Maps and Charts

20.1 Radio station

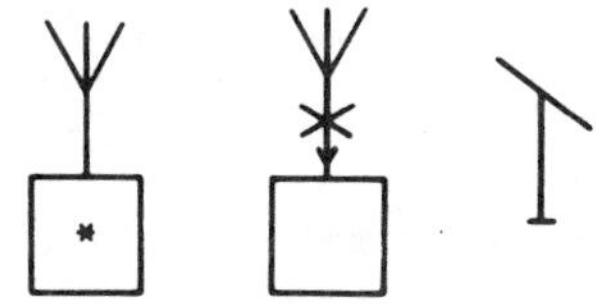

20.2 Space station

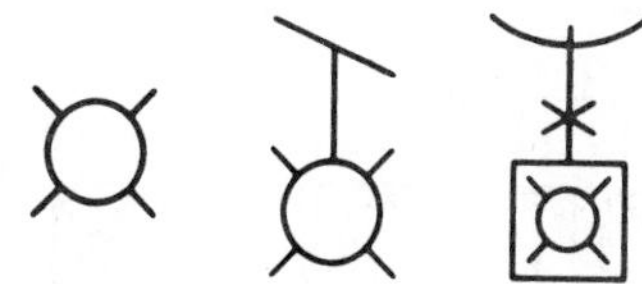

20.3 Exchange equipment

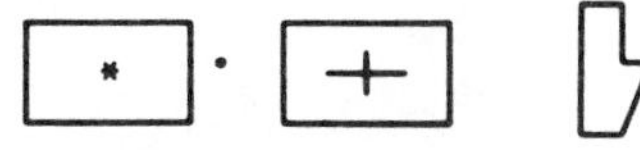

20.4 Telegraph repeater

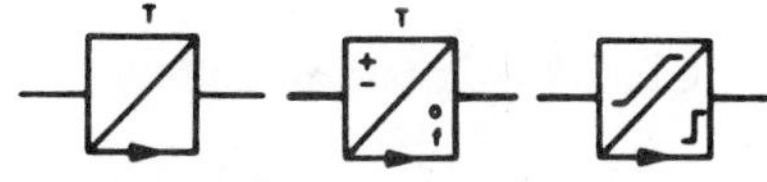

20.5 Telegraph equipment

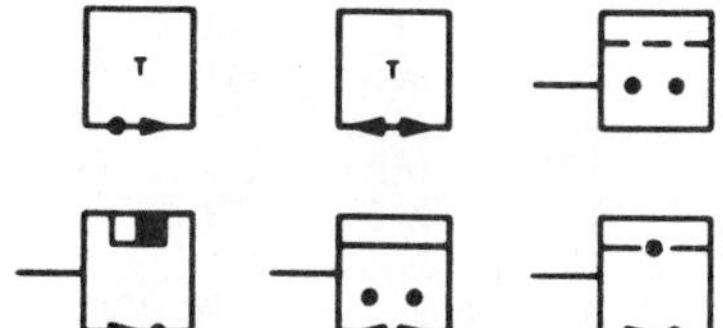

20.6 Telephone set

21. System Diagrams, Maps and Charts

21.1 Generating station

21.2 Hydroelectric generating station

 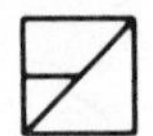 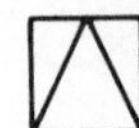

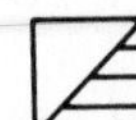  

21.3 Thermoelectric generating station

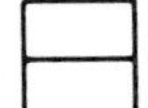  

21.4 Prime mover

21.5 Substation

22. Class Designation Letters

A	DS	J	PU	TP
AR	E	K	Q	TR
AT	EQ	L	R	U
B	F	LS	RE	V
BT	FL	M	RT	VR
C	G	MG	RV	W
CB	H	MK	S	WT
CP	HP	MP	SQ	X
CR	HR	MT	SR	Y
D	HS	N	T	Z
DC	HT	P	TB	
DL	HY	PS	TC	

Index

E

F

M

Credits

Chapter 1
1.1 Lockheed-California Co.; 1.2 Hewlett-Packard Co.; 1.4 Computervision; 1.5 and 1.6 Engineering Model Associates, Inc.; 1.7 NASA; 1.10 Lockheed-California Co.; 1.12 Macola, Inc.; 1.13 J.I. CASE; 1.15 Evolution Engineering; 1.16 Chicago Bridge and Iron Co.; 1.18 ISICAD, Inc.; 1.19 NASA; 1.21 Adage, Inc.; 1.22 AUTODESK; 1.23 Hampton Roads Naval Museum; 1.24 Evolution Engineering; 1.26 AUTODESK.

Chapter 2
2.1 Teledyne Post; 2.2 and 2.3 Hamilton; 2.4 through 2.5 3M; 2.6 Pickett; 2.7 through 2.9 Keuffel & Esser/Kratos; 2.11 Koh-I-Noor; 2.13 and 2.14 Berol RapiDesign; 2.15 Koh-I-Noor; 2.18 Hearlihy & Co.; 2.21 Keuffel & Esser/Kratos; 2.22 Hearlihy & Co.; 2.26 Staedtler; 2.28 Hearlihy & Co.; 2.29 and 2.30 Berol RapiDesign; 2.31 Hearlihy & Co.; 2.32 Keuffel & Esser/Kratos; 2.33 and 2.34 VEMCO; 2.35 Keuffel & Esser/Kratos; 2.38 Koh-I-Noor; 2.40 ANSI; 2.41 ANSI; 2.42 ANSI; 2.44 and 2.45 ANSI; 2.56 Koh-I-Noor.

Chapter 3
3.4 ANSI; 3.7 Koh-I-Noor; 3.19 Hearlihy & Co.; 3.20 Keuffel & Esser/Kratos; 3.21 Lettering Guide; 3.22 Varigraph, Inc.; 3.24 Graphic Products; 3.25 and 3.26 Kroy, Inc.; 3.27 NASA; Items of Interest The Bettmann Archive.

Chapter 5
Items of Interest The Bettmann Archive.

Chapter 6
6.3 Adapted ANSI; 6.8 Adapted ANSI; Items of Interest Outboard Marine Co.

Chapter 7
7.2 through 7.4 Adapted ANSI; 7.20 Rachael Svit; 7.22 ANSI; 7.27 NASA; 7.30 Rachael Svit; 7.31 Adapted ANSI; 7.34 Lunkenheimer Co.; Items of Interest The Bettmann Archive.

Chapter 8
8.1 Lunkenheimer Co.; 8.6 Adapted ANSI; 8.16 ANSI; 8.20 ANSI; 8.22 ANSI; 8.26 ANSI; 8.29 ANSI; 8.32 AUTODESK; Items of Interest Techsonic Industries, Inc. Humminbird's TCR Color 1, 8-color liquid crystal TFT screen with over 114,000 pixels.

CHAPTER 9
9.3 ANSI; Items of Interest Johnson & Johnson.

Chapter 10
10.22 and 10.23 Adapted ANSI; 10.26 Adapted ANSI; 10.37 Control Data Corp.; Items of Interest Ford Motor Co.

Chapter 11
11.42 ANSI; 11.44 Adapted ANSI; 11.69 Adapted ANSI; 11.71 Adapted ANSI; 11.72 Adapted ANSI; 11.73 Adapted ANSI; Items of Interest The Bettmann Archive.

Chapter 12
12.1 Adapted ANSI; 12.3 ANSI; 12.4 ANSI; 12.5 ANSI; 12.7 Adapted ANSI; 12.11 Adapted ANSI; 12.17 Adapted ANSI; 12.18 Adapted ANSI; 12.19 Adapted ANSI; 12.20 Adapted ANSI; 12.23 Adapted ANSI; 12.24 Adapted ANSI; 12.25 Adapted ANSI; 12.26 Adapted ANSI; 12.45 Adapted ANSI; 12.56 Adapted ANSI.

Chapter 13
13.6 Adapted IFI; 13.7 Adapted IFI; 13.14 ANSI; 13.15 ANSI; 13.16 ANSI; 13.17 ANSI; 13.18 ANSI; 13.19 ANSI; 13.32 American National Pipe Thread; 13.33 Adapted ANSI; 13.34 Federal Screw Products, Inc.; 13.38 CARR LANE Manufacturing; 13.38 General Motors Corp.; 13.39 General Motors Corp.; 13.40 IFI; 13.41 General Motors Corp.; 13.42 IFI; 13.43 IFI; 13.44 IFI; 13.45 Great Lakes Screw; 13.46 IFI; 13.47 Holo-Krome; 13.48 IFI; 13.49 IFI; 13.50 IFI; 13.51 IFI; 13.53 IFI; 13.54 IFI; 13.55 IFI; 13.56 IFI; 13.57 IFI; 13.58 IFI; 13.59 IFI; 13.68 IFI; 13.70 IFI; 13.71 IFI; 13.72 IFI; 13.73 IFI; 13.74 IFI; 13.75 IFI; 13.77 ANSI; 13.79 IFI; 13.80 IFI; 13.81 IFI; 13.82 IFI; 13.83 Koh-I-Noor; 13.84 House of Fasteners; 13.86 ANSI; 13.87 ANSI; 13.88 ANSI; 13.89 ANSI; 13.90 ANSI; 13.91 ANSI; 13.92 ANSI.

Chapter 14
14.1 ANSI; 14.2 ANSI; 14.3 ANSI; 14.5 ANSI; 14.6 ANSI; 14.7 ANSI; 14.9 ANSI; 14.10 Associated Spring, Barnes Grouping; 14.11 ANSI; 14.13 ANSI.

Chapter 15
15.3 Logitech, Inc.; 15.6 NASA; 15.7 Heathkit; 15.10 Engineering Model Associates, Inc.; 15.12 NASA; 15.20 and 15.21 Grove Valve & Regulator Co.; 15.23 David Cunningham; 15.26 and 15.27 NASA; 15.28 Engineering Model Associates, Inc.; 15.31 ANSI; 15.32 Strux Corp.; 15.34 NASA; 15.37 and 15.38 NASA; 15.40 NASA; 15.41 The Scale Model Makers; 15.43 Magee-Bralla, Inc.; 15.48 Innovative Design Systems Corp.

Chapter 16
16.1 Hewlett-Packard Co.; 16.5 Grove Valve & Regulator Co.; 16.7 FAG Bearing Corp.; 16.10 Drawn by P. Jenkins; 16.15 Adapted ANSI; 16.20 Teledyne Corp.; 16.22 Gerber Systems Technology; 16.24 and 16.25 Wedge Innovations.

Chapter 17
17.46 NASA; 17.48 Bonestroo, Rosene, Anderlik & Assoc.

Chapter 18
18.4 Adapted ANSI; 18.5 ANSI; 18.6 ANSI; 18.10 Adapted ANSI; 18.11 ANSI; 18.12 Adapted ANSI; 18.13 Adapted ANSI; 18.14 Adapted ANSI; 18.15 Adapted ANSI; 18.16 Adapted ANSI; 18.17 Adapted ANSI; 18.18 Adapted ANSI; 18.19 Adapted ANSI; 18.20 Adapted ANSI; 18.33 Bishop Graphics, Inc.; 18.39 Motorola, Inc., Semiconductor Products Sector; 18.40 Bishop Graphics, Inc.; 18.41 Motorola, Inc., Semiconductor Products Sector; 18.47 Motorola, Inc., Semiconductor Products Sector; 18.48 and 18.49 Bishop Graphics, Inc.; 18.53 Motorola, Inc., Semiconductor Products Sector; Items of Interest Cincinnati Milacron.

Color Plates
1 through 11 IBM 12 through 20 Control Data Corp. 21 Evans & Sutherland Computer Corp. 22 Evans & Sutherland Computer Corp. 23 Lockheed-California Co. 24 Hewlett-Packard Co. 25 through 31 AUTODESK 32 through 36 Calma, a division of Computervision 37 CADKEY, Inc. 38 Megatek Corp. 39 through 41 PATRAN, a division of PDA Engineering 42 Evans & Sutherland Computer Corp. 43 PATRAN, a division of PDA Engineering 44 IBM 45 AUTODESK 46 IBM 47 IBM 48 Calma, a division of Computervision 49 Intergraph Corp. 50 AUTODESK

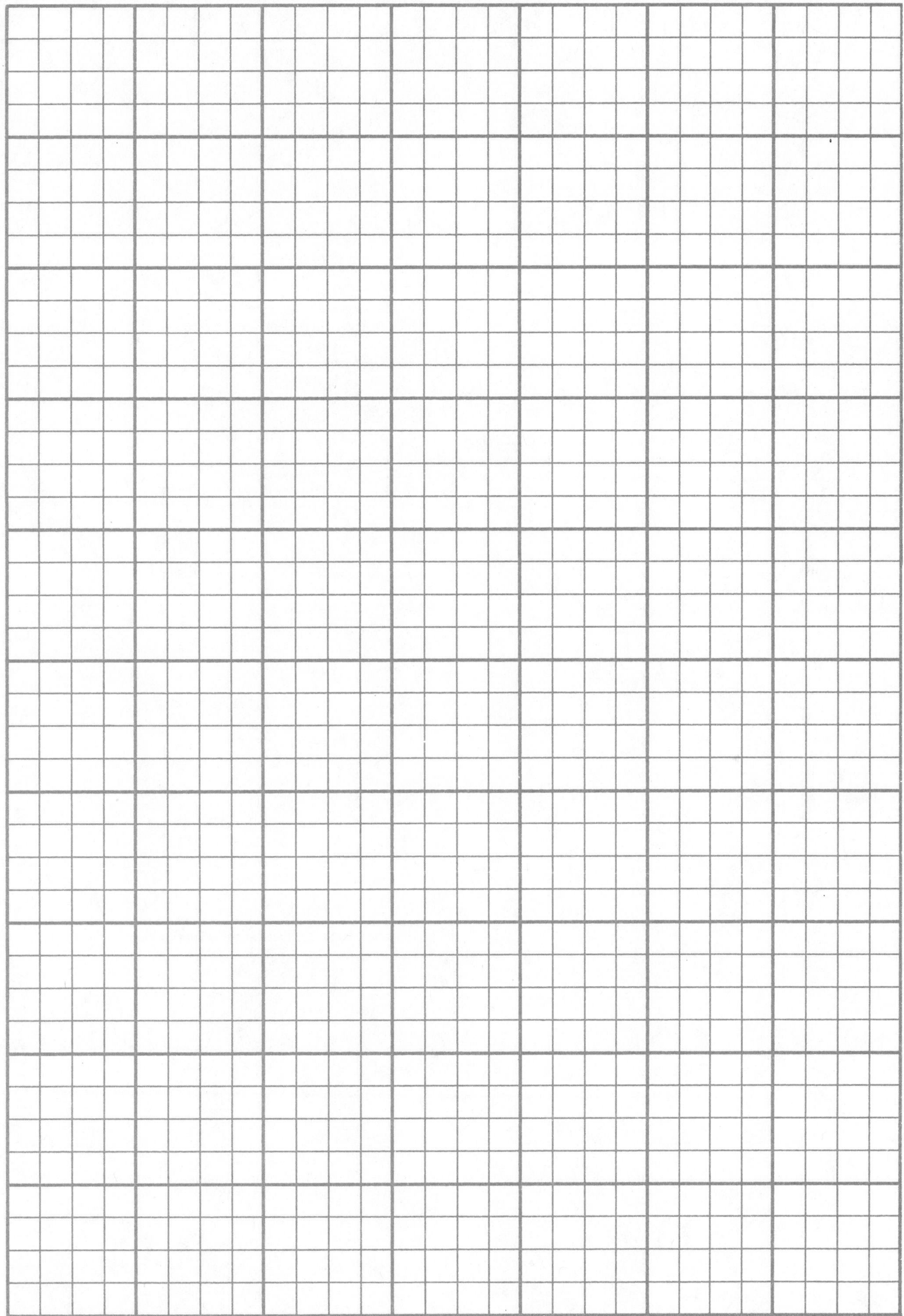

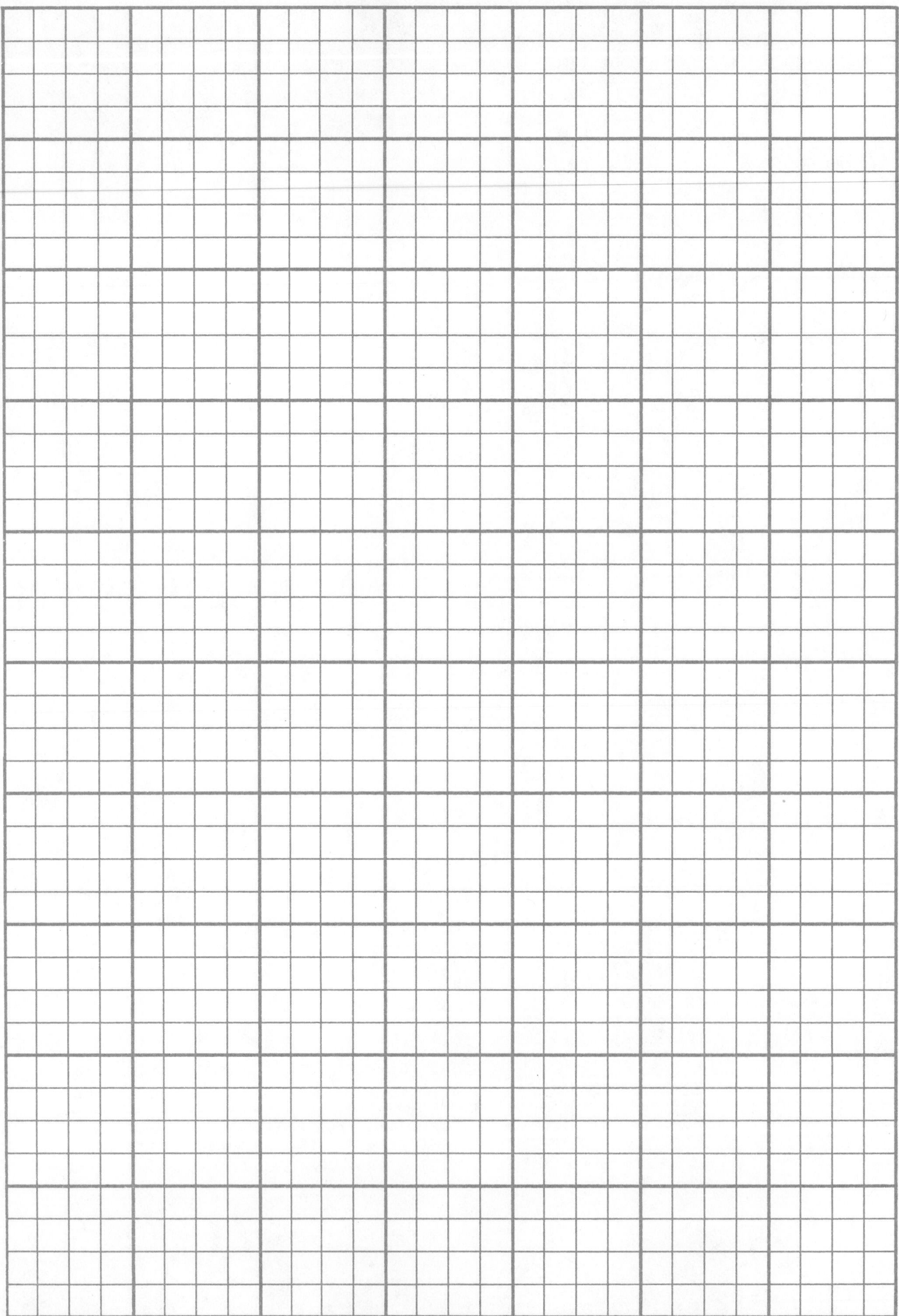

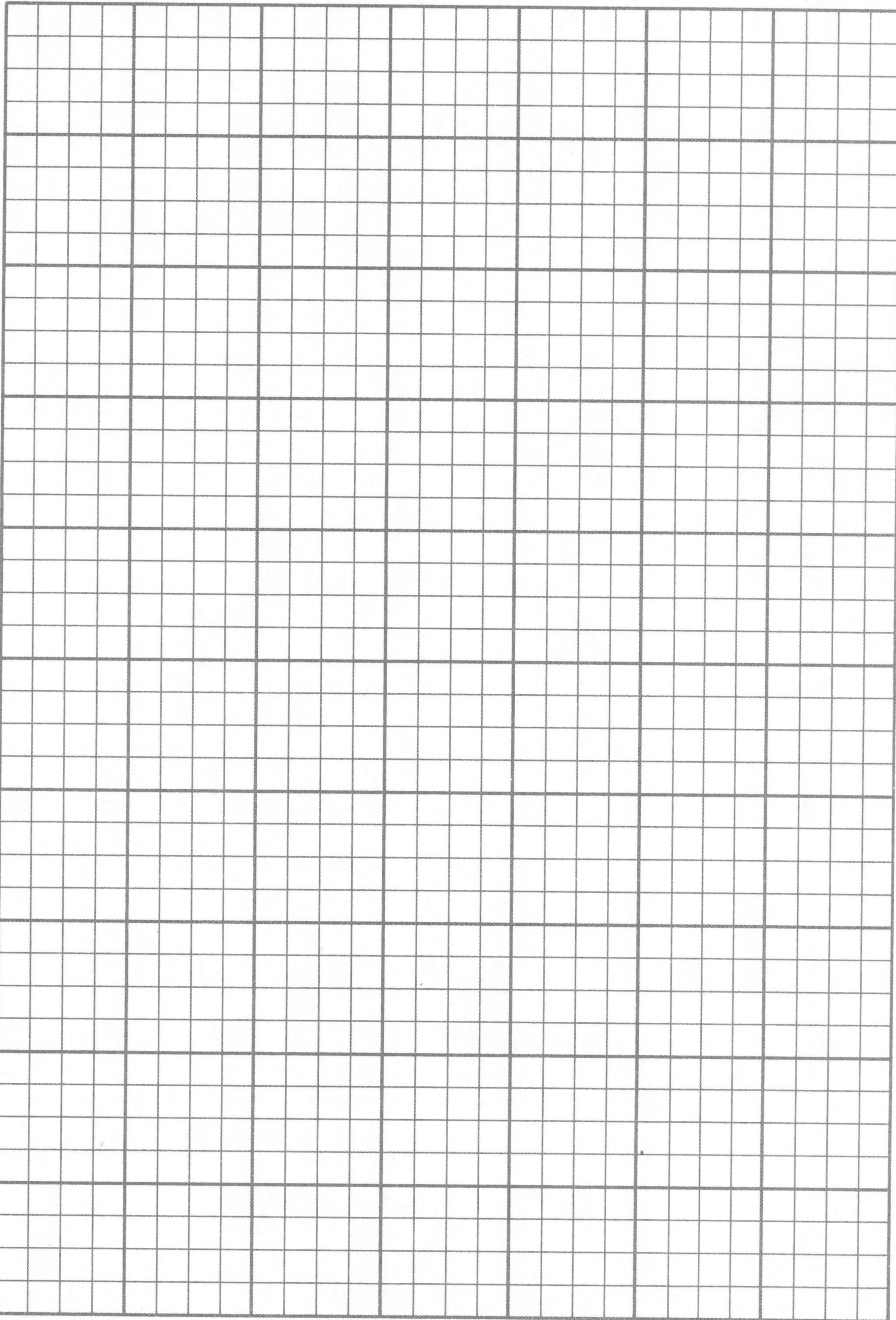

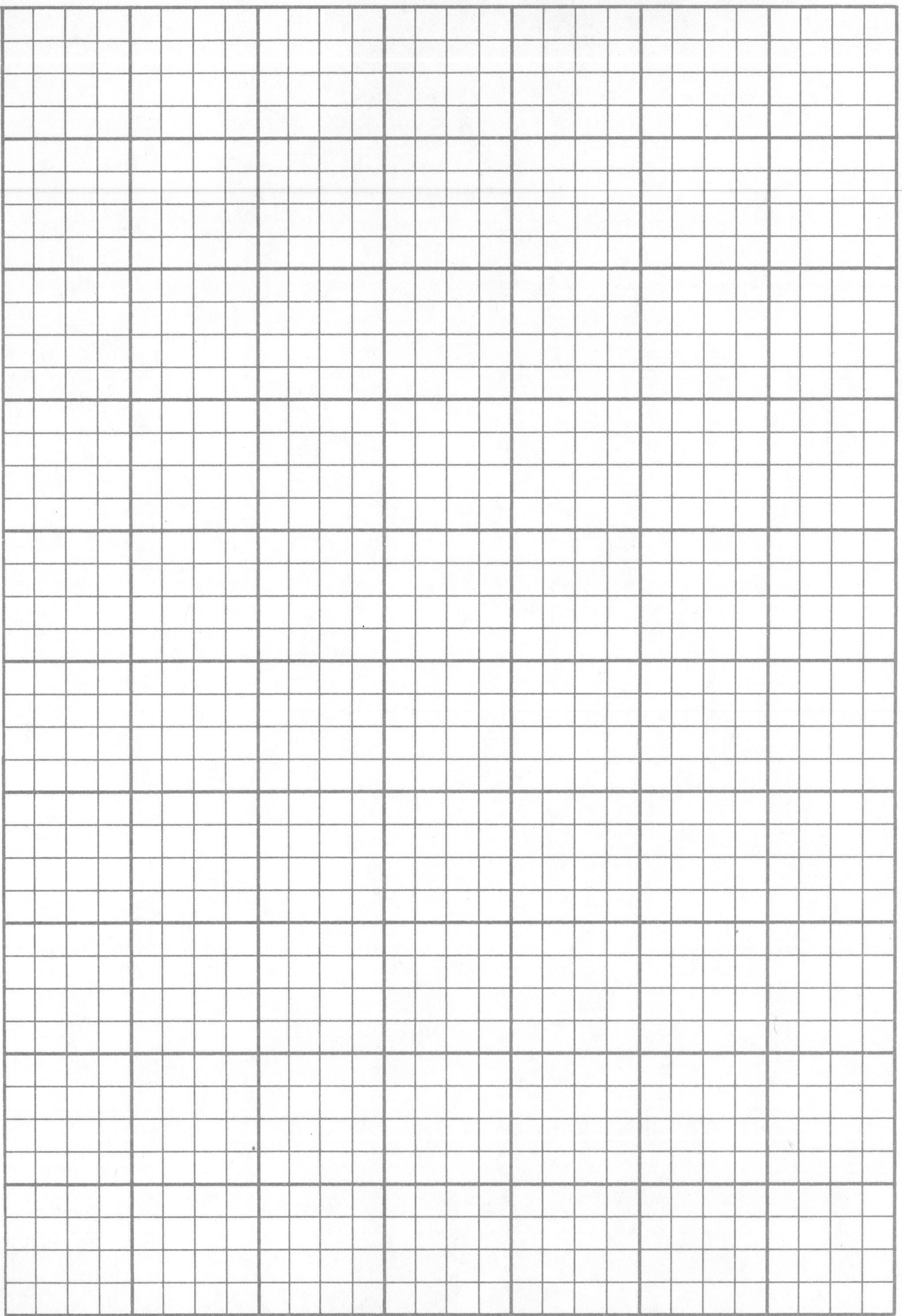

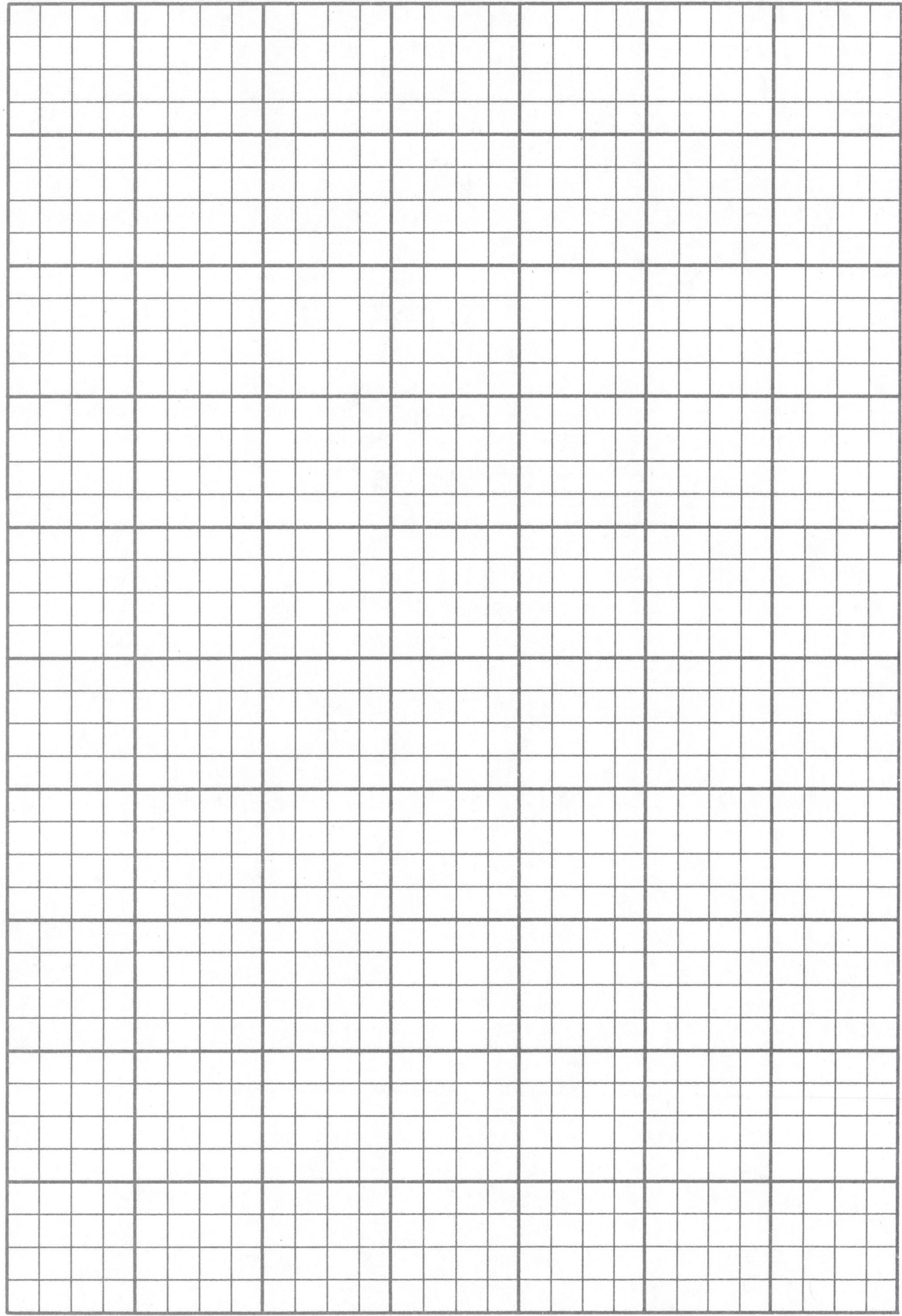

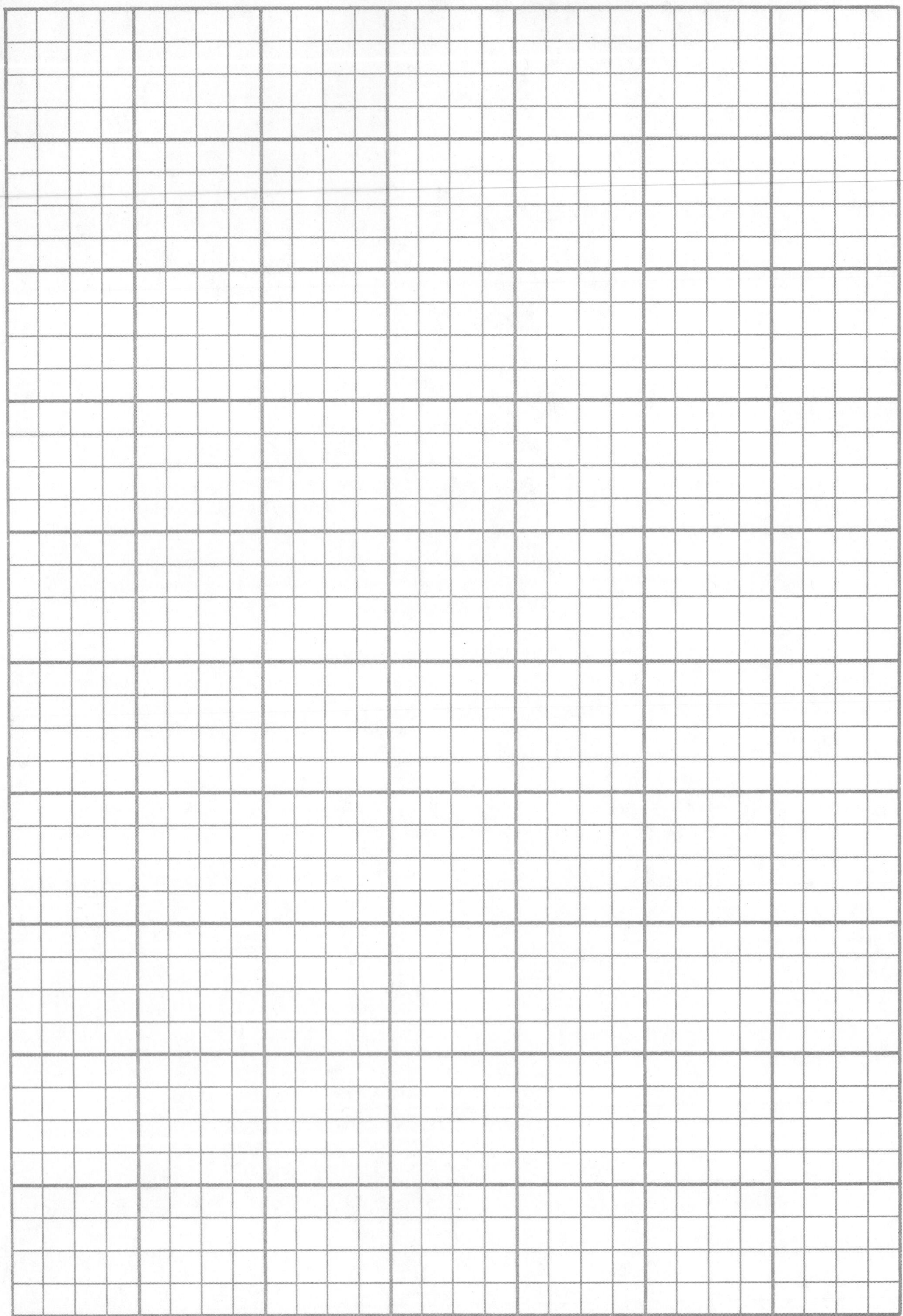

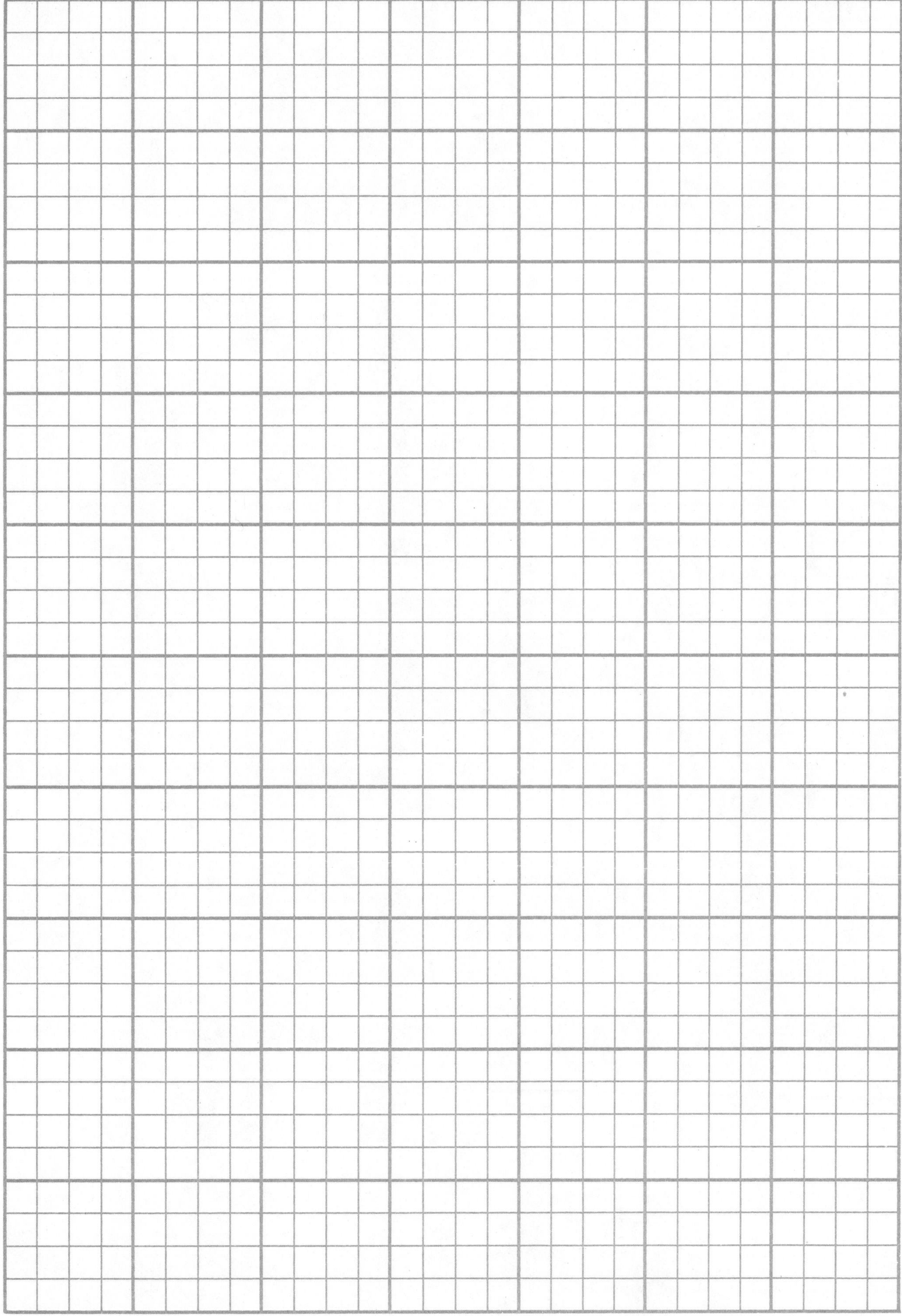

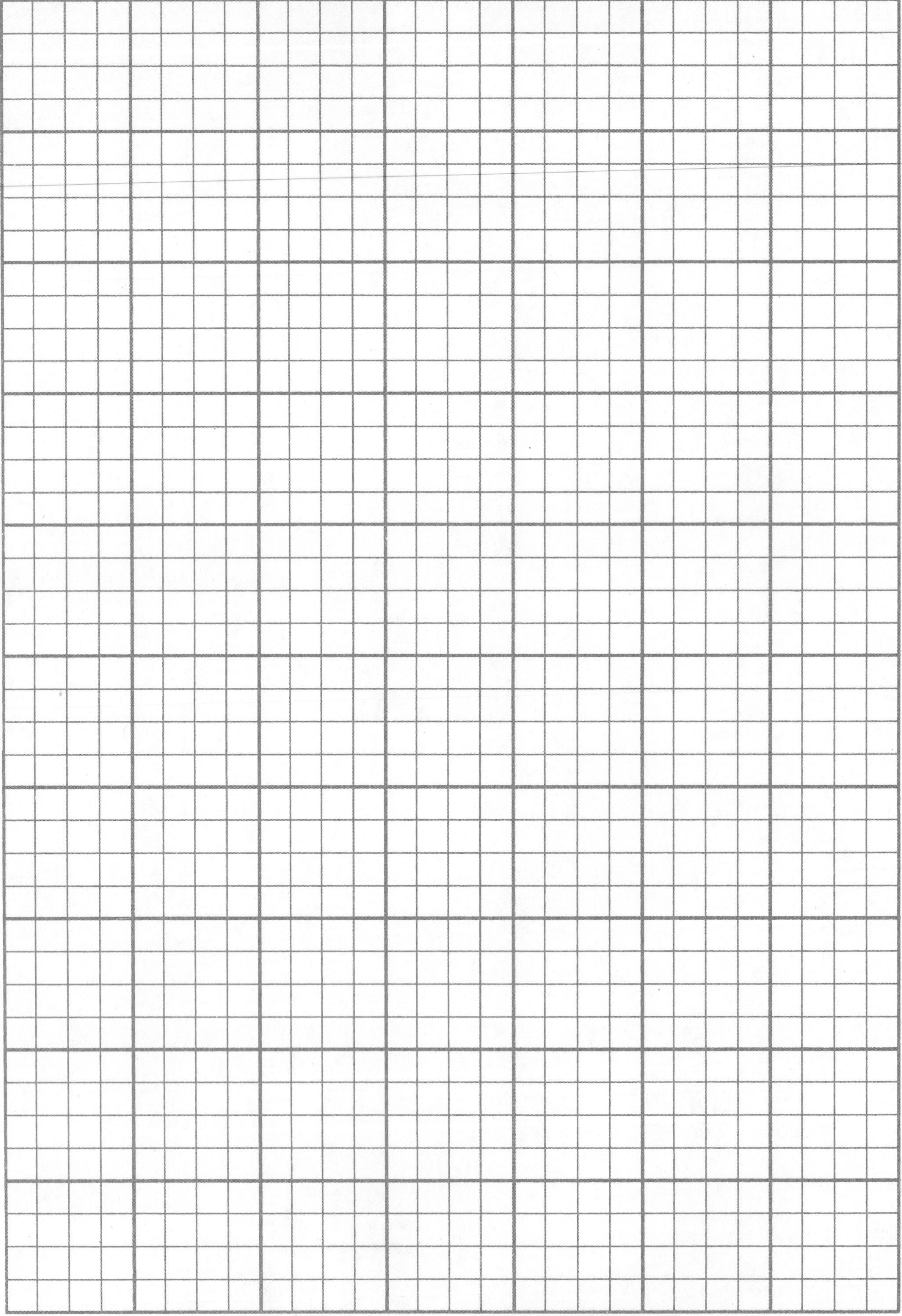

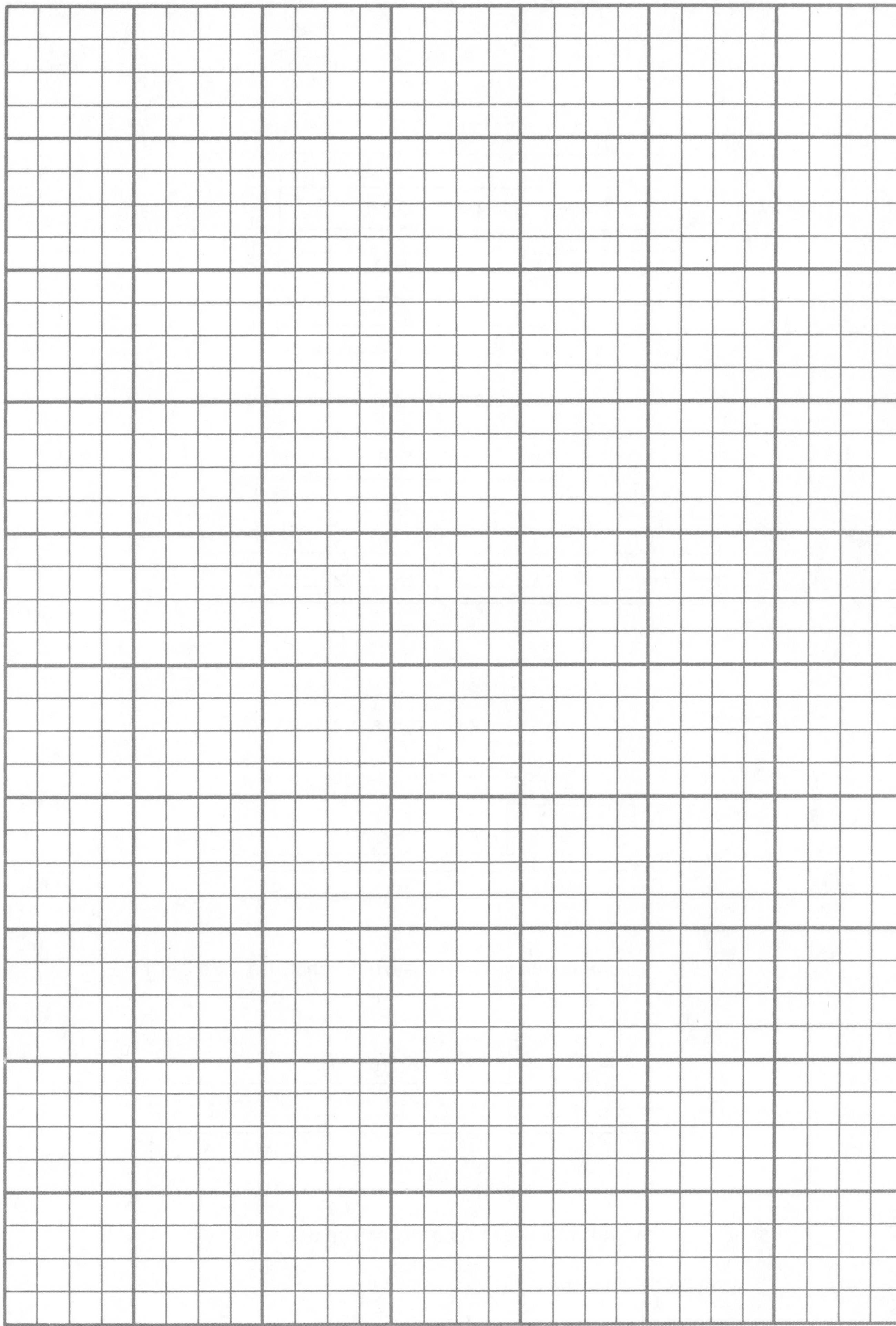

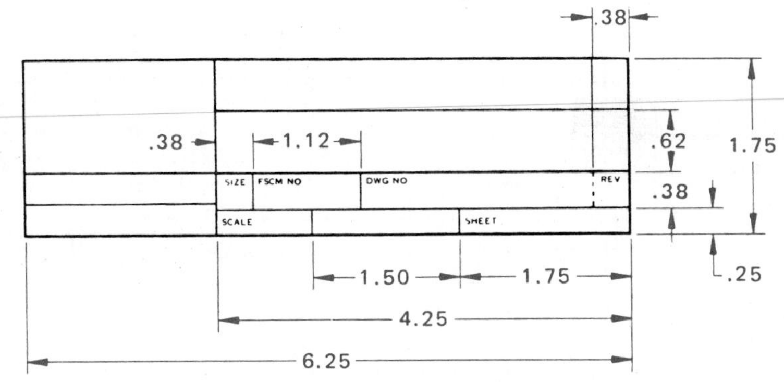

TITLE BLOCK FOR A, B, C, AND G – SIZES

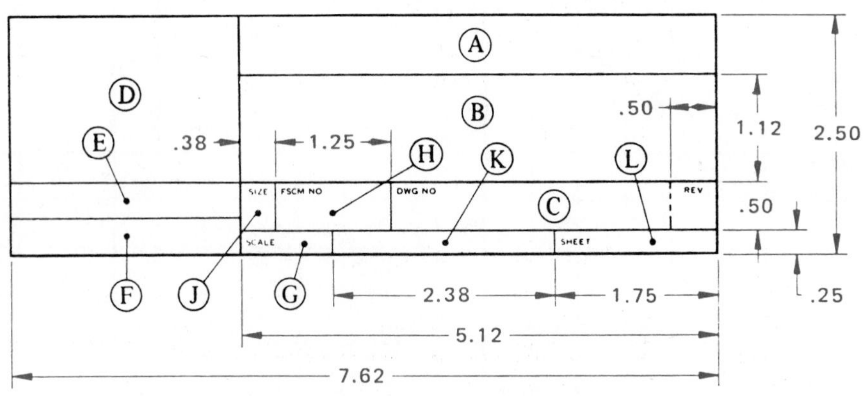

TITLE BLOCK FOR D, E, F, H, J, AND K – SIZES

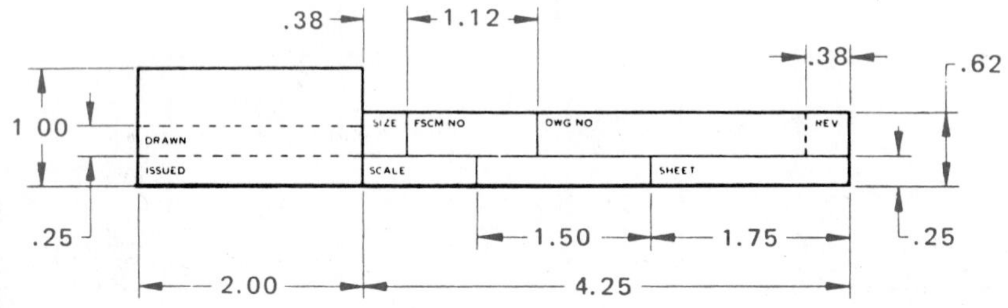

CONTINUATION SHEET TITLE BLOCK FOR A, B, C, and G – SIZES

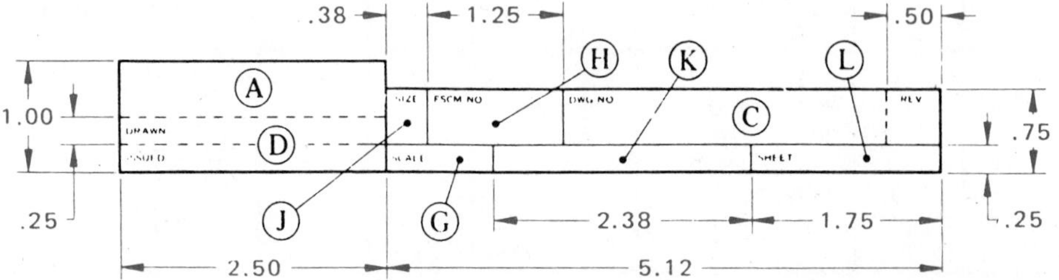

CONTINUATION SHEET TITLE BLOCK FOR D, E, F, H, J, AND K – SIZES

ANSI Title Block Dimensions